# Laboratory Techniques in Electroanalytical Chemistry

# Laboratory Techniques in Electroanalytical Chemistry

Second Edition, Revised and Expanded

edited by

Peter T. Kissinger
*Purdue University and*
*Bioanalytical Systems, Inc.*
*West Lafayette, Indiana*

William R. Heineman
*University of Cincinnati*
*Cincinnati, Ohio*

Marcel Dekker, Inc. New York • Basel • Hong Kong

**Library of Congress Cataloging-in-Publication Data**

Laboratory techniques in electroanalytical chemistry / edited by Peter T. Kissinger, William R. Heineman. — 2nd ed., rev. and expanded.
p. cm.
Includes bibliographical references and index.
ISBN 0-8247-9445-1 (hardcover : alk. paper)
1. Electrochemical analysis—Laboratory manuals. I. Kissinger, Peter T. II. Heineman, William R.
QD115.L23 1996
543'.087—dc20 95-46373
CIP

The publisher offers discounts on this book when ordered in bulk quantities. For more information, write to Special Sales/Professional Marketing at the address below.

This book is printed on acid-free paper.



MARCEL DEKKER, INC.
270 Madison Avenue, New York, New York 10016

Current printing (last digit):
10 9 8 7 6 5 4 3 2 1

**PRINTED IN THE UNITED STATES OF AMERICA**

# Preface

The text that had the most influence on the education of American electrochemists between the late 1950s and the late 1960s was Delahay's *New Instrumental Methods in Electrochemistry* (1954). For well over a decade, this book played a dominant role in placing finite current electroanalytical chemistry on its modern course—beyond the titrations, coulometry, polarography, and electrogravimetry of an earlier age. The membrane-covered oxygen electrode developed by Clark in 1955 rapidly became one of the most important analytical tools. Modern engineering mathematics began to conquer mass transport problems in the 1950s, and toward the end of that decade, operational amplifiers revolutionized experimental electrochemistry.

In the 1960s, modern organic, inorganic, and biological electroanalytical chemistry advanced rapidly behind the advent of cyclic voltammetry, pulse chronoamperometry, rotating ring disk electrodes, and the use of electron spin resonance and optical measurements coupled to electrochemical experiments. After a burst of tremendous activity, it became clear during this period that chronopotentiometry was not competitive with controlled-potential techniques. In the second half of the 1960s, thin-layer electrochemistry had its heyday and electrodes other than mercury began to attract serious attention. It was discovered that platinum was surprisingly well behaved for studies in nonaqueous solvents, and even the likes of glassy carbon and carbon paste, to say nothing of thin films of metals and semiconductors deposited on glass, could actually solve chemical problems. Ralph Adams' book, *Electrochemistry at Solid Electrodes* (1969), surprised more than a few who did not believe it was possible. Anodic stripping voltammetry became a commercial reality and even was able to compete, for some elements, with atomic spectroscopy. Polarography gradually diminished in importance for practical analysis through the early 1960s. A little company in Princeton, New Jersey, brought it out of hibernation by making differential pulse polarography commercially available via the first significant op-amp-based electrochemical instruments.

The 1970s saw electroanalytical chemistry refine its fundamental basis and develop more powerful and more reliable instrumentation. Chemistry became more important than technique. For the first time, electrochemists had a significant clientele of users whose primary interest was not electrochemistry per se. Two important fundamental new directions in the 1970s were toward chemically modified electrodes and photoelectrochemistry. In addition, high-vacuum surface spectroscopies began to play a serious role in electrochemical investigations. A significant new development of practical analytical importance was the use of electrochemical flow cells for liquid chromatography, a development that introduced thousands of pharmacologists, clinical chemists, and toxicologists to electrochemistry for the first time. A small company in West Lafayette, Indiana, made this methodology easily accessible to non-electrochemists.

In the 1980s, it became widely recognized that there are advantages from doing electrochemistry at very small electrodes. In the 1990s, benefits from speed and access to electrochemistry in unusual media make microelectrodes quite routine. It is interesting that the microelectrodes of today are at least a thousand times smaller than the "microelectrodes" of the '30s, '40s, and '50s. What will the microelectrodes of tomorrow look like?

This book was conceived in 1970 by Peter Kissinger to provide a means of enabling a neophyte in electroanalytical chemistry to get started in the lab. During preparation of the first edition, the emphasis was expanded to include a pedagogical component. In the 12 years since the first edition, we've received a number of suggestions based on actual classroom experience. Many of these have been incorporated in this second edition.

These days, research papers are necessarily brief with respect to experimental insights. Review chapters are often replete with unexplained and generally confusing jargon and mathematics. Textbooks on instrumental methods have their good practical moments, but such moments are usually short, incomplete, and often out of date. None of these sources really does justice to the limited selection of material for the "bench chemist" wanting to get things moving in a hurry.

The emphasis of this book is entirely on analytical, mechanistic (homogeneous), kinetic (homogeneous), and synthetic (laboratory-scale) applications. Physical electrochemistry is not a direct concern, and equilibrium methods (potentiometry) are intentionally omitted. There is no attempt to include specific chemical examples except where they are particularly illustrative and have pedagogical value. No extensive review of the original literature is included, but references to key reviews and papers of historical interest are emphasized. Authors have selected experimental approaches that work best and have commented freely on outmoded or underdeveloped methods. The authors and editors have made value judgments that undoubtedly will disappoint some readers.

Electrochemistry is a very broad subject. Those interested in batteries, fuel cells, corrosion, membrane potentials, and so forth will not satisfy their needs here.

The organization of the book flows from principles through methodology to applications. Chapters 2–5 are devoted to what the editors perceive as the most commonly used techniques in electroanalytical chemistry today. These techniques have passed the development stage and can now be considered routine. The approach in these chapters is designed to give the readers an intuitive understanding of each technique without mathematical rigor. This is achieved by considering the excitation signal for each technique and the resulting concentration-distance profiles that determine the consequent response signal. Value judgments are given to permit an educated selection of techniques to use in a given situation. Chapter 2 provides the fundamental concepts that are important throughout the book. It is the editors' opinion that Chapters 2–5 are suitable for use in graduate-level introductory courses in electroanalytical chemistry.

Instrumentation for selected aspects of electroanalytical chemistry is covered in Chapters 6–8. Although computers have made a tremendous impact on electroanalytical instrumentation, many aspects of these chapters are timeless. The basic configurations of a potentiostat have not changed since the early 1960s, although the electronic components themselves are dramatically different! Learn to build your own potentiostat in Chapter 6, then see how to fine-tune it in Chapter 7.

Chapters 9–19 deal with some practical aspects of electroanalytical chemistry. These chapters are aimed at giving the novice some insight into the nuts and bolts of electrochemical cells and solutions. In this second edition, further emphasis has been given to obtaining and maintaining clean solutions, and new chapters have been added on chemically modified electrodes and electrochemical studies at reduced temperature.

In the early 1970s, many electrochemists learned about digital simulation from Steve Feldberg's papers and Joseph Maloy's "underground" chapter, which is now revised as Chapter 20. This field is 30 years old, and only recently has commercial software extended its reach to a wider audience.

Chapters 21–23 are for the neophyte trying to determine which technique can be useful for unraveling a mechanism and/or preparing a strategy for synthesis. The virtues of the key techniques described in Chapter 3 are illustrated here with specific chemical examples.

Chapters 24–29 are a potpourri of electroanalytical techniques and applications. Hybrid techniques in which electroanalytical chemistry is combined with luminescence, spectroscopy, or chromatography are discussed.

It is risky to predict the future, but the 1990s continue the trends of the 1970s: more chemistry; more practical organic, inorganic, and biochemical problem-solving; and more detailed knowledge of the chemical nature of the interface and how it can be manipulated to human advantage. Publication of the first edition in 1984 coincided with the 150th anniversary of Faraday's law; there seems no end of uses for this important principle describing the relationship between chemistry and electricity.

We hope that this book is of some use to you in the course of your research, for which we wish you good luck—it certainly helps.

Many people have assisted with this project. We would like to especially thank Peggy Sue Precup for her expertise in organizing numerous details, helping us overcome our English language deficiencies, and coordinating with the publisher. Rod Yoder and David Michell turned many rough figures into comprehensible art. Drs. Adrian Bott and Jon Howell helped refine several chapters and their suggestions were invaluable.

*Peter T. Kissinger*
*William R. Heineman*

# Contents

# Contributors

**James L. Anderson, Ph.D.** Professor, Department of Chemistry, The University of Georgia, Athens, Georgia

**Andrew B. Bocarsly, Ph.D.** Professor, Department of Chemistry, Princeton University, Princeton, New Jersey

**Marvin A. Brooks, Ph.D.** Senior Director, Pharmaceutical Analysis and Control, Pharmaceutical Research and Development, Merck Research Laboratories, West Point, Pennsylvania

**Kristin K. Cline, Ph.D.** Assistant Professor, Department of Chemistry, Wittenberg University, Springfield, Ohio

**David J. Curran, Ph.D.** Professor, Department of Chemistry, University of Massachusetts at Amherst, Amherst, Massachusetts

**Christie G. Enke, Ph.D.** Professor, Department of Chemistry, The University of New Mexico, Albuquerque, New Mexico

**Dennis H. Evans, Ph.D.** Professor, Department of Chemistry and Biochemistry, University of Delaware, Newark, Delaware

**Larry R. Faulkner, Ph.D.** Provost and Vice Chancellor for Academic Affairs, University of Illinois at Urbana-Champaign, Urbana, Illinois

**Colby A. Foss, Jr., Ph.D.** Assistant Professor, Department of Chemistry, Georgetown University, Washington, D.C.

**Steven N. Frank, Ph.D.** Texas Instruments, Dallas, Texas

**Albert J. Fry, Ph.D.** Professor, Department of Chemistry, Wesleyan University, Middletown, Connecticut

**Zbigniew Galus, Ph.D.** Professor, Department of Chemistry, The University of Warsaw, Warsaw, Poland

**William E. Geiger, Ph.D.** Professor, Department of Chemistry, University of Vermont, Burlington, Vermont

**Ira B. Goldberg, Ph.D.** Technical Staff, Department of Materials Science, Rockwell Science Center, Thousand Oaks, California

**Fred M. Hawkridge, Ph.D.** Professor, Department of Chemistry, Virginia Commonwealth University, Richmond, Virginia

**M. Dale Hawley, Ph.D.** Professor and Department Head, Department of Chemistry, Kansas State University, Manhattan, Kansas

**William R. Heineman, Ph.D.** Distinguished Research Professor, Department of Chemistry, University of Cincinnati, Cincinnati, Ohio

**F. James Holler, Ph.D.** Professor, Department of Chemistry, University of Kentucky, Lexington, Kentucky

**Charles L. Hussey, Ph.D.** Professor, Department of Chemistry, The University of Mississippi, University, Mississippi

**Vladimir Katovic, Ph.D.** Professor, Department of Chemistry, Wright State University, Dayton, Ohio

**Csaba P. Keszthelyi, Ph.D.** Department of Chemistry, Louisiana State University, Baton Rouge, Louisiana

**Peter T. Kissinger, Ph.D.** Professor, Department of Chemistry, Purdue University, and President, Bioanalytical Systems, Inc., West Lafayette, Indiana

**Susan A. Lerke, Ph.D.** Senior Research Scientist, Analytical R & D, DuPont Merck Pharmaceutical Company, Wilmington, Delaware

**Craig E. Lunte, Ph.D.** Associate Professor, Department of Chemistry, The University of Kansas, Lawrence, Kansas

**Susan M. Lunte, Ph.D.** Director, Center for Bioanalytical Research, and Associate Professor, Department of Pharmaceutical Chemistry, The University of Kansas, Lawrence, Kansas

**J. T. Maloy, Ph.D.** Associate Professor, Department of Chemistry, Seton Hall University, South Orange, New Jersey

**Charles R. Martin, Ph.D.** Professor, Department of Chemistry, Colorado State University, Fort Collins, Colorado

**Michael A. May, Ph.D.** Research Associate, Department of Chemistry, Central State University, Wilberforce, Ohio

**Richard L. McCreery, Ph.D.** Professor, Department of Chemistry, The Ohio State University, Columbus, Ohio

**Ted M. McKinney, Ph.D.** Technical Staff, Department of Materials Science, Rockwell Science Center, Thousand Oaks, California

**Adrian C. Michael, Ph.D.** Assistant Professor, Department of Chemistry, University of Pittsburgh, Pittsburgh, Pennsylvania

**Su-Moon Park, Ph.D.** Professor, Department of Chemistry, The University of New Mexico, Albuquerque, New Mexico

**Carl R. Preddy, Ph.D.*** Research Assistant, Department of Chemistry, Purdue University, West Lafayette, Indiana

**Thomas H. Ridgway, Ph.D.** Professor, Department of Chemistry, University of Cincinnati, Cincinnati, Ohio

**David K. Roe, Ph.D.** Professor, Department of Chemistry, Portland State University, Portland, Oregon

**Ronald E. Shoup, Ph.D.** President, BAS Analytics Division, Bioanalytical Systems, Inc., West Lafayette, Indiana

**Eberhard Steckhan, Ph.D.** Professor, Institute for Organic Chemistry and Biochemistry, Rheinische Friedrich-Wilhelms-Universität Bonn, Bonn, Germany

**Hiroyasu Tachikawa, Ph.D.** Professor, Department of Chemistry, Jackson State University, Jackson, Mississippi

**Eric W. Tsai, Ph.D.** Research Fellow, Pharmaceutical Analysis and Control, Pharmaceutical Research and Development, Merck Research Laboratories, West Point, Pennsylvania

**Joseph Wang, Ph.D.** Professor, Department of Chemistry and Biochemistry, New Mexico State University, Las Cruces, New Mexico

**R. Mark Wightman, Ph.D.** Kenan Professor, Department of Chemistry, The University of North Carolina at Chapel Hill, Chapel Hill, North Carolina

**Nicholas Winograd, Ph.D.** Evan Pugh Professor, Department of Chemistry, The Pennsylvania State University, University Park, Pennsylvania

---

**Current affiliation*: Kodak Research Labs, Eastman Kodak Company, Rochester, New York

# Laboratory Techniques in Electroanalytical Chemistry

# 1

# An Overview

**Peter T. Kissinger** *Purdue University and Bioanalytical Systems, Inc., West Lafayette, Indiana*

## I. SOME PHILOSOPHY

Much of the initial progress in a science derives from passive observation of a natural system. This approach is severely limited in that one remains at the mercy of the system for what qualitative and quantitative properties it chooses to reveal. Nevertheless, passive observation is the first step in all areas of research and remains of ultimate importance in some (e.g., astronomy). A more pragmatic approach is one in which the experimenter plays an active role by forcing information out of the object of his or her affection. The experimenter first tickles the system with a forcing function thought to be appropriate by passive observation and intuition. Every so often the experimenter will even be inclined to try anything. This is not as bad as it sounds because the "anything" *selected* will be inseparable from previous experience and will invariably be better than doing nothing. The observed response function to the chosen excitation signal is analyzed and this information is then used to design new active probes. Such a closed-loop modus operandi is shown in Figure 1.1.

Experiments of this kind are most easily interpreted when the system behaves linearly. Linear performance is recognized by the validity of two simple rules:

1. The amplitudes of the response and excitation functions are directly proportional (i.e., an increase in excitation amplitude results in the same *relative* increase in response amplitude).
2. The response to a collection of excitations is given by the sum of responses to the individual excitations (i.e., superposition theory holds).

Problems (nonlinearity) will certainly develop if the system under study significantly influences the excitation signal applied to it. It is often therefore essen-

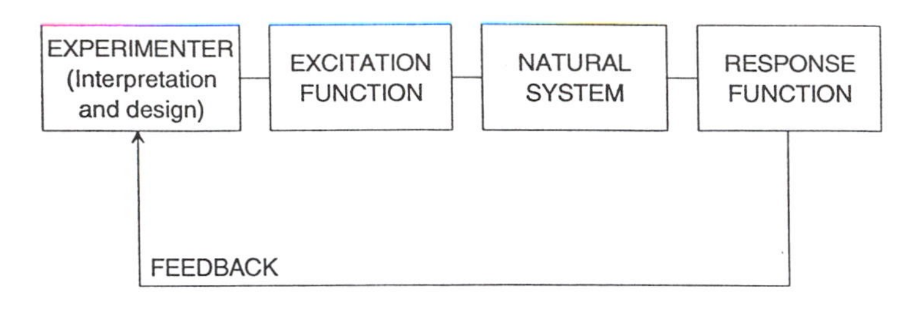


**Figure 1.1** Block diagram illustrating modification of experimental design using feedback from previous experiments.

tial to monitor (via negative feedback) the excitation *at its site of application* to ensure that this is not the case. In some systems, an error in the excitation function may be directly related to the response function. This circumstance permits compensation of the error by a positive feedback link between the two critical functions. These modifications are depicted in Figure 1.2.

The simple concepts we have just discussed can be given a sound mathematical basis which is useful in the design of new experiments and instrumentation systems. The practice of interfacial electrochemistry involves analysis of the response of a phase boundary to various stimuli. In this book we examine the most frequently chosen perturbations and their experimental implementations. Some understanding of the properties of the phase boundary between an electrode and a solution can help tie together the seemingly endless variety of electrochemical techniques.

## II. PROGRESS AND PROGNOSIS

Electrochemical progress has benefited tremendously from the application of feedback principles. It is essential for the modern electrochemist to be aware

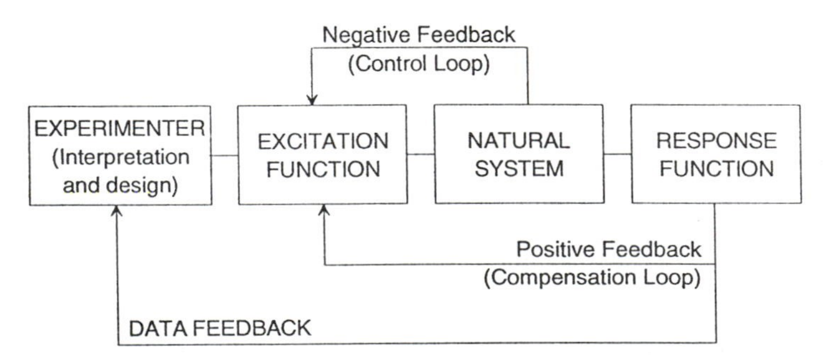


**Figure 1.2** Modern experiments employ several feedback loops.

of them and use the opportunities they provide to design improved experiments. Unfortunately, the technological imbalance in our world has afforded these opportunities to relatively few of us. Even in the more fortunate economies, investigators have been tied down to methods that provide insufficient data for many purposes. Although economic constraints are still extant, the availability of inexpensive integrated circuits and personal computers has largely eliminated any excuse for not applying state-of-the-art instrumentation for electroanalytical techniques.

A deficient knowledge of electronics remains a significant problem. For the electrochemist this is easily overcome, because integrated subsystems (linear and nonlinear, analog and digital) for building highly sophisticated electrochemical apparatus can be effectively utilized with *working* knowledge of only a few basic laws of electricity. In fact, practitioners of our art need know far less physical electronics than did their predecessors several decades ago. Many students find that there is a considerable psychological barrier that must be exceeded before the "fear of electronics" is overcome. Once this is accomplished, the rest is easy.

In the text that follows, we examine the methodology of electrochemistry as it may be usefully applied to a wide variety of chemical problems. In short, when a molecule can be coupled to an electrode reaction, that reaction may be used for analytical, mechanistic, or kinetic studies involving this substance. We make no pretense about the fact that this is not a book for physical electrochemists. Those who wish to direct their efforts toward unraveling the microsteps in heterogeneous electrode processes per se are best served by other monographs. Those focused on generation of electrical power from chemical reactions will also find little here of direct interest.

The great facility we have for dealing with electricity leads us to convert or transduce information in virtually every other data domain (e.g., chemical, optical, magnetic) into an easily processed electrical signal. Similarly, there is much virtue in controlling some event in a nonelectrical domain with an electrical signal. Faraday's discovery of the one-to-one equivalence that exists between chemistry and electricity provides the basis for all electroanalytical techniques. The electrode in solution is in many respects an ideal transducer between the chemical and electrical domains. The role of the electrode may be either to monitor species in solution or to generate new species that will interact with the medium in an interesting or perhaps useful way.

Molecular specificity is afforded by control of the potential energy difference across the electrode-solution interface. The selectivity or resolution of electrochemistry is not nearly as great as we would like, and the inability to distinguish between two or more species can be a serious weakness. Solution electrochemistry suffers by comparison with mass spectrometry in the same sense that optical spectra in solution suffer by comparison with vapor-phase spectra.

Nevertheless, it is an immutable fact of nature that every molecule participating in the electron transfer reaction will be related to an integral number of electrons in the external circuitry. Often this fact can be used to great advantage.

Before considering instrumentation in some detail in later chapters, it will be helpful to outline the kinds of experiments that we wish to implement electronically. It is useful to characterize electroanalytical techniques as either static or dynamic. Static methods are philosophically akin to the passive observation mentioned earlier. They entail measurements of potential difference at zero current such that the system defined by the solid-solution interphase is not disturbed and Nernstian equilibrium is maintained. Although such potentiometric measurements (e.g., pH, pM) are of great practical importance, our focus here will be on the dynamic techniques, in which a system is intentionally disturbed from equilibrium by excitation signals consisting of a wide variety of potential and current programs.

While amperometric titrations and cyclic voltammetry are the only dynamic electroanalytical techniques familiar to most chemists, others are becoming increasingly important. The teaming of many of these methods with spectroscopic techniques (both optical and magnetic) and the developing commercial and academic role of electrosynthesis lend support to this trend. Of particular importance is the realization that great benefits result from studies at stationary (both solid and liquid) electrodes. Consequently, polarography has been relegated to second place in most modern electroanalytical laboratories and in many ways is becoming a thing of the past.

This text places electrochemical techniques in perspective by introducing the most promising methods and evaluating their relative merits for specific applications. Although an almost endless variety of excitation functions have been applied to the electrode-solution interface, most of these have had little following even among electrochemists, and some have been superseded by more modern ideas. Emphasis is given to practical considerations. Although the important phenomenological aspects are covered, a detailed mathematical development has been avoided in deference to other sources.

The 1950s and 1960s were characterized by an enormous exploratory effort in the theory and methodology of electrochemical techniques and their possible application to chemical problems. Chemistry seemed ancillary to much of this work in that the model systems examined were more often than not uninteresting to chemists. This is not to say that no exciting things happened electro*chemically* (e.g., electrochemiluminescence). It has just taken a while for the technologists and chemists to get together. This began in the 1970s and it is now clear that dynamic electrochemistry has a secure place in studies of solution chemistry. It is still necessary to recognize the limitations. The competitive position vis-à-vis optical, magnetic, and other analytical probes must be kept in mind for any specific chemical problem.

## III. SO MUCH NOMENCLATURE, SO MUCH JARGON

### A. Techniques

The nonequilibrium behavior associated with dynamic electrochemical techniques is largely controlled by diffusion, which may or may not be coupled with chemical reactions in solution or on the surface. The voltage, current, and time variables may be adjusted (although not independently) by the experimenter to obtain maximum information from the system at hand. The system response to a given voltage or current excitation program can sometimes be predicted by a linear systems approach utilizing mathematical transforms. Many individual circumstances are not amenable to this approach, and in fact, few of the differential equations describing real chemical situations are analytically soluble. Nevertheless, individual techniques are best recognized by their excitation-response characteristics. As we shall see, the various dynamic electrochemical methods both mathematically and experimentally fall into two convenient categories—large-amplitude and small-amplitude—related to the magnitude of the excitation signal.

Within these broad categories two principal classes may be distinguished, depending on whether current or voltage is the controlled parameter. The individuals in each class are often described by an operational nomenclature consisting of an independent-variable part followed by a dependent-variable part (i.e., volt-ammetry, chrono-potentiometry) with some system-specific modifiers (i.e., rotating disk voltammetry). Unfortunately, the whims of history have left the electrochemical nomenclature in a rather confused state. The operational approach has been only partially adopted but seems to be gaining popularity.

In general, electrochemists apply an electrical *excitation* signal (often the independent variable) to some *system* (particular electrode-solution composition and geometry) and then monitor a *response* signal (dependent variable), which together with the excitation allows some description of the properties of the system at hand. The nomenclature ideally should incorporate enough information about these three items to describe the experiment (technique) to some reasonable degree without being cumbersome. The nomenclature cannot be expected to provide a phenomenological explanation of the response beyond the individual's ability to understand the experiment thereby described. The operational approach implicitly ignores the subtleties of individual techniques, but does describe the experiment.

The term "polarography" provides a particular bone of contention. The continued use of "polarography" as a general appellation for all faradaic electrochemistry is very undesirable. Polarography is actually a relatively old technique from the 1920s (i.e., dc voltammetry at the dropping mercury electrode) for which the Nobel prize was awarded in 1959 to Jaroslav Heyrovsky. The term "polarography" is generally accepted in connection with techniques employing a dropping mercury electrode; however, more general usage is strongly discouraged. Figure 1.3 illustrates a family tree relating the more important methods. The reader unfamiliar with electroanalytical chemistry should be able to see the logic in these associations after reading Chapters 3 to 5. It would be a good self-test to be able to identify the essential characteristics of each technique in Figure 1.3.

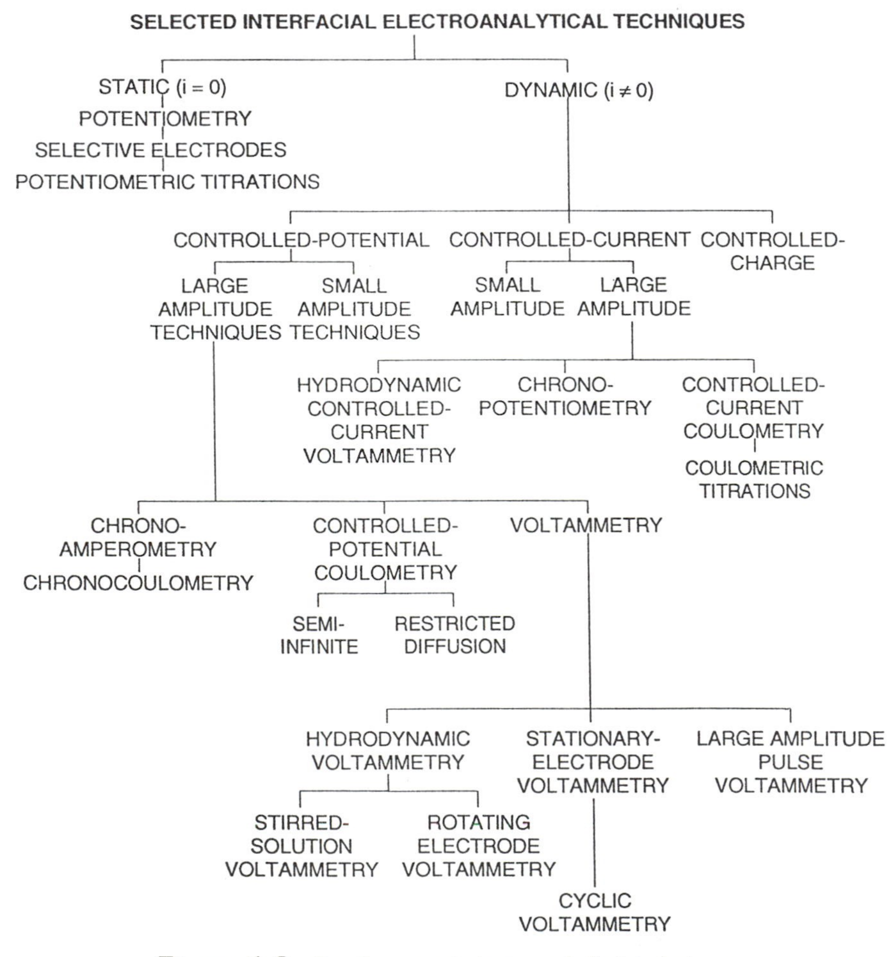


**Figure 1.3** Family tree of electroanalytical techniques.

Figure 1.4 is an illustration of a typical dynamic electrochemical experiment in which the reduced form of a substance (white circles) is initially present. Current or potential is applied to oxidize this substance. The oxidized substance (black circles) can then be reconverted to the starting material. The electrochemical cell can be represented as a circuit element as depicted in the upper left of the figure. The potential of the working electrode is monitored in relation to the reference electrode. The current passes between the auxiliary and working electrodes. How and why this is done is the subject of Chapters 2 to 7. The motion of molecules or ions to and from the electrode surface is critical. The electron transfer occurs at the working electrode and its surface properties are therefore crucial. While students new to chemistry are introduced to redox couples such as Fe(II)/Fe(III) and Ce(III)/Ce(IV), many redox active substances are far more complex and frequently exhibit instability.

## B. Chemical Complications

For introducing techniques and how to instrument them, we will assume a generalized system

$$O + ne \rightleftarrows R$$

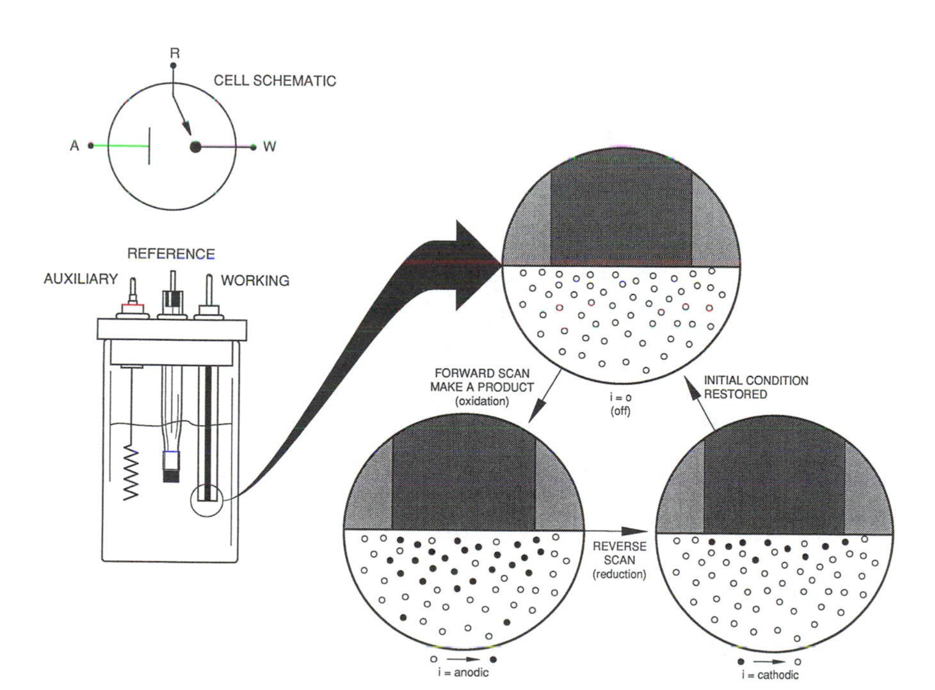


**Figure 1.4** The typical electroanalytical cell consists of three electrodes in a few milliliters of solution. Electron transfer occurs at the surface of the working electrode. Reactant and product are transported to and from this surface by diffusion.

where O and R are the oxidized and reduced forms of the *electroactive species*, in a polar solvent containing a relatively high concentration of inert *supporting electrolyte*, and the heterogeneous electron transfer (ET) rate constant is large (fast, reversible). For dynamic experiments, the sample is often completely in either the O or the R form, whereas for static experiments both forms are usually present.

In the real world, the simple redox couple may be perturbed by finite ET rates, by adsorption of O and/or R on the electrode surface, and by homogeneous (i.e., in solution) chemical kinetics involving O and/or R. Various combinations of heterogeneous ET steps (E) with homogeneous chemical steps (C) are encountered. It should be clear that if one or more species in equilibrium in solution are electroactive, electrochemistry can be used to perturb the equilibrium and study the solution chemistry.

Many other complications occur, including heterogeneous chemical steps and homogeneous ET steps. Our understanding of these complications when built into real chemical systems is at a primitive stage. There is much work to be done before evaluation of such problems becomes a routine matter. The literature often displays more optimism than is warranted by the facts. This has caused many disappointments among practical problem solvers. Some investigators have thus come to view analytical electrochemistry with considerable skepticism. One should recognize that electrochemistry will always be more useful as a quantitative tool than as a diagnostic tool because of its relatively low resolution. There is a tendency in some circles to overinterpret electrochemical results without stating the appropriate reservations. Whenever possible, supporting evidence from other sources (usually spectroscopic or chromatographic techniques) should be brought to bear on any mechanistic conclusions suggested by electrochemical data. A large portion of the early literature has ignored this principle and should therefore be examined with appropriate caution.

The following bibliography lists a few of the more recent texts covering topics of interest to both beginners and experts in electroanalytical chemistry.

## BIBLIOGRAPHY

A. J. Bard and L. R. Faulkner, *Electrochemical Methods*, John Wiley & Sons, New York, 1980.

H. H. Bauer, *Electrodics: Modern Ideas Concerning Electrode Reactions*, Georg Thieme, Stuttgart, 1972.

A. M. Bond, *Modern Polarographic Methods in Analytical Chemistry*, Marcel Dekker, New York, 1980.

C. M. A. Brett and A. M. O. Brett, *Electrochemistry Principles, Methods, and Applications*, Oxford University Press, New York, 1993.

D. R. Crow, *Principles and Applications of Electrochemistry*, Blackie Academic & Professional, New York, 1994.

E. A. M. F. Dahmen, *Electroanalysis: Theory and Applications in Aqueous and Non-Aqueous Media and in Automated Chemical Control*, Elsevier, New York, 1986.

A. J. Fry and W. E. Britton, eds., *Topics in Organic Electrochemistry*, Plenum Press, New York, 1986.

R. J. Gale, *Spectroelectrochemistry Theory and Practice*, Plenum Press, New York, 1988.

Z. Galus, *Fundamentals of Electrochemical Analysis*, 2nd ed., John Wiley & Sons, New York, 1994.

E. Gileadi, *Electrode Kinetics for Chemists, Chemical Engineers, and Materials Scientists*, VCH, New York, 1993.

D. K. Gossar, Jr., *Cyclic Voltammetry: Simulation and Analysis of Reaction Mechanisms*, VCH, New York, 1993.

H. Lund and M. M. Baizer, eds., *Organic Electrochemistry*, 3rd ed., Marcel Dekker, New York, 1991.

R. W. Murray, *Molecular Design of Electrode Surfaces*, John Wiley & Sons, New York, 1992.

K. B. Oldham and J. C. Myland, *Fundamentals of Electrochemical Science*, Academic Press, San Diego, 1994.

J. A. Plambeck, *Electroanalytical Chemistry: Basic Principles and Applications*, John Wiley & Sons, New York, 1982.

P. H. Rieger, *Electrochemistry*, Prentice Hall, Englewood Cliffs, NJ, 1987.

I. Rubinstein, *Physical Electrochemistry: Principles, Methods, and Applications*, Marcel Dekker, New York, 1995.

*Southampton Electrochemistry Group Instrumental Methods in Electrochemistry*, Halsted Press, New York, 1985.

J. T. Stock, and M. V. Orna, *Electrochemistry, Past and Present*, American Chemical Society, Washington, DC, 1989.

H. R. Thirsk, and J. A. Harrison, *A Guide to the Study of Electrode Kinetics*, Academic Press, New York, 1972.

B. H. Vassos and G. W. Ewing, *Electroanalytical Chemistry*, John Wiley & Sons, New York, 1983.

J. Wang, *Analytical Electrochemistry*, VCH Publishers, New York, 1994.

# 2

# Fundamental Concepts of Analytical Electrochemistry

**Peter T. Kissinger** *Purdue University and Bioanalytical Systems, Inc., West Lafayette, Indiana*

**Carl R. Preddy*** *Purdue University, West Lafayette, Indiana*

**Ronald E. Shoup** *Bioanalytical Systems, Inc., West Lafayette, Indiana*

**William R. Heineman** *University of Cincinnati, Cincinnati, Ohio*

## I. INTRODUCTION

The common thread binding all electroanalytical methods is heterogeneity. The very act of placing an electrode in contact with a solution creates a phase boundary that differentiates otherwise identical solute molecules into two types: those at a distance from the electrode, and those close enough to participate in the fascinating mutual interactions known collectively as electrochemistry. This is not a trivial distinction, for often it is the bulk-phase properties alone that are of analytical concern. Unlike most spectroscopic methods, electrochemical measurements are actually made on only a minute fraction of the sample confined to a highly nonhomogeneous environment, the electrode-solution interphase. The coupling and interplay within this region of such phenomena as interfacial charge transfer, diffusional mass transport, adsorption, chemisorption, homogeneous-phase chemical reaction, convection, and dissolution can cloud the interpretation of electrochemical data and discourage the practically minded analyst. On the other hand, electrochemistry offers an invaluable tool for fundamental investigation of these processes, each important in its own right. In either case, the ultimate success of the experimenter will depend on a firm grasp of the underlying physical principles.

**Current affiliation*: Kodak Research Labs, Eastman Kodak Company, Rochester, New York

We have chosen to dwell on certain theoretical aspects only to the extent that they will serve as a foundation for the methodology presented in the ensuing chapters. It is practice rather than theory that is of primary concern here. We acknowledge extensive, intentional bias and abridgment in derivations and discussion. Our intent is to convey the feeling that much may be gained from electrochemical investigation without recourse to an entanglement of mathematical expressions, provided that one develops an insight into the physical realities of electrode processes.

## II. MASS TRANSPORT: LINEAR DIFFUSION

Since the electrochemical reduction or oxidation of a molecule occurs at the electrode-solution interface, molecules dissolved in solution in an electrochemical cell must be transported to the electrode for this process to occur. Consequently, the transport of molecules from the bulk liquid phase of the cell to the electrode surface is a key aspect of electrochemical techniques. This movement of material in an electrochemical cell is called mass transport. Three modes of mass transport are important in electrochemical techniques: hydrodynamics, migration, and diffusion.

Hydrodynamic mass transport is caused by the movement of solution by stirring the solution, rotating the electrode, or flowing the solution through a cell. The moving solution transports reactant to the electrode surface while carrying electrogenerated product away.

Migration is the movement of a charged particle due to its interaction with an electric field, such as that which exists in the vicinity of an electrode. For example, cations are attracted by a negatively charged electrode and repelled by a positively charged electrode. In most analytical techniques, the effect of migration is minimized by the addition of an inert electrolyte, called a supporting electrolyte, which decreases the field strength near the electrode.

In many respects, the simplest and best understood process influencing electrochemistry is diffusion. Diffusion is a factor in virtually every type of electroanalytical measurement, yet it is most often introduced as a set of elemental laws, devoid of physical significance. The governing mathematical relationships are apt to seem abstruse, but are in fact quite elegant and readily grasped. Their derivation follows from simple finite mathematics, using as a model a collection of inert particles in random motion. This microscopic-level approach relies on statistical rather than thermodynamic concepts, and although less rigorous, Fick's well-known laws arise directly with few assumptions and without losing sight of physical reality.

### A. Impulse Relaxation

As with the familiar theoretical treatment of ideal gases, the logical starting point for understanding the process of diffusion is consideration of the motion of a single solute molecule or ion. In fluid media, a molecule undergoes frequent collisions with neighboring molecules. This continuous rebounding imparts randomness to each molecule's trajectory. In a closed, isotropic system (no phase boundaries or thermal, electromagnetic, or concentration gradients), there is no favored direction of motion. An imaginary plane dividing the system will be continuously traversed by many molecules moving from both sides, but in any significant time interval, the number that has passed from one direction will exactly equal the number from the other. Thus there will be no *net* movement of material from one side to the other. The concentration, or more specifically the chemical potential, of the solute is uniform throughout, and the system is therefore at equilibrium.

When regions of dissimilar chemical potential are created, solute molecules will move between them until a homogeneous condition is restored. This relaxation process involves a temporary net *mass transport* across some imaginary plane. The transport of matter from a region of higher chemical potential to one of lower chemical potential is the process of diffusion. The motive force behind this movement is maximization of entropy. The fully relaxed system is in its most random configuration.

The transition between regions of high and low chemical potential in real systems cannot be infinitely sharp. There must be a zone over which the concentration varies with distance. A place in which $dC/dx$ is nonzero is said to exhibit a *concentration gradient*. A plot of concentration as a function of distance is called a *concentration profile*, so the gradient is the first derivative of the profile function. Since diffusional mass transport is always from high to low concentration, the direction of movement will be determined by the sign of the gradient. One imagines that molecules diffuse *down* their respective concentration gradient, meaning that direction in which the sign of $dC/dx$ is negative from the molecules' frame of reference. This is a point worth pondering. It is fundamental to the understanding of unrestricted motion under the influence of any type of potential field—thermal, electrical, or gravitational. The mathematical description of the transfer of heat along temperature gradients is identical to mass diffusion in every respect. Just as objects cool more rapidly at high temperature than near room temperature, so the rate of mass transport is reasonably expected to be related to the magnitude of the concentration gradient. The form of this relationship will be examined presently.

Another useful concept is *flux*. Flux is defined as the number of molecules penetrating a unit area of an imaginary plane in a unit of time. The usual units are mol/($cm^2$ s), and the sign identifies the direction of motion, positive toward and negative away from the plane. The prior assertion that equilibrium demands no net mass transport is equivalent to a requirement that the sum of the fluxes of all components is exactly zero at any test plane within the system. Flux is a measure of the rate of mass transport at a fixed point. Its electrochemical relevance stems from the direct relationship it holds to electrode current.

Consider an infinite volume of pure solvent at equilibrium. If we imagine a portion of this system divided into a number of thin, adjacent slabs of uniform thickness $\Delta x$, the picture resembles the left side of Figure 2.1. Into the central box, arbitrarily labeled 0, we introduce a number $N_0$ of solute molecules at time $t = 0$. At the starting time, the concentration profile across this region exhibits a sharp spike at the center. The right side of Figure 2.1 depicts the profile as a histogram of a quantity of molecules versus distance from the center. Since the concentration profile is a step function at either boundary of box 0, the gradient at these points is some huge negative value. According to our notion of molecular behavior, this impulse of concentration should be short-lived when left to itself. Bombardment by solvent molecules and collision with each other will cause the molecules to wander out randomly from the central box. Isotropy assures that motion to the left is no more probable than motion to the right. The fluxes across either boundary should be equal at all times. After some time $T_1$, half of the original $N_0$ molecules will have left the 0th box. One-fourth will be found in box 1, and an equal number will have made their way into box −1. The situation is now as depicted in the second set of Figure 2.1. The model system now consists of three nonempty boxes with $N_0/2$ molecules in the center and $N_0/4$ in each of the neighboring boxes. Each box is entirely independent of the other in the sense that each will behave exactly as box 0 did in the first step. At some later time $T_2$, half the original molecules in each box will have meandered into the adjacent boxes. The distribution ratio across the boxes has progressed from 1:2:1 after the first time increment, to 1:4:6:4:1 after the second, then to 1:6:15:20:15:6:1 at some later time $T_3$. The fraction of molecules in the ith box after the kth step is given by the familiar binomial distribution function

$$\frac{N_{i,k}}{N_0} = \frac{2k!}{(k+i)!\,(k-i)!} \times 2^{-2k} \qquad k = 0, 1, \ldots; \quad i = 0, \ldots, \pm k \tag{2.1}$$

It should be emphasized that the time steps are *not* evenly spaced.

If the total number of molecules is very large and the distance increments, $\Delta x$, become infinitesimally small ($\Delta x \rightarrow dx$), the distribution will be described by a smooth function (a Gaussian distribution) following the familiar normal curve of error,

$$\frac{dN}{N_0} = \frac{1}{(4\pi Dt)^{1/2}} \exp\left(\frac{-x^2}{4Dt}\right) dx \tag{2.2}$$

where $dN/N_0$ is the fraction of the total population of molecules located between x and $(x + dx)$, and D is a constant known as the diffusion coefficient and has

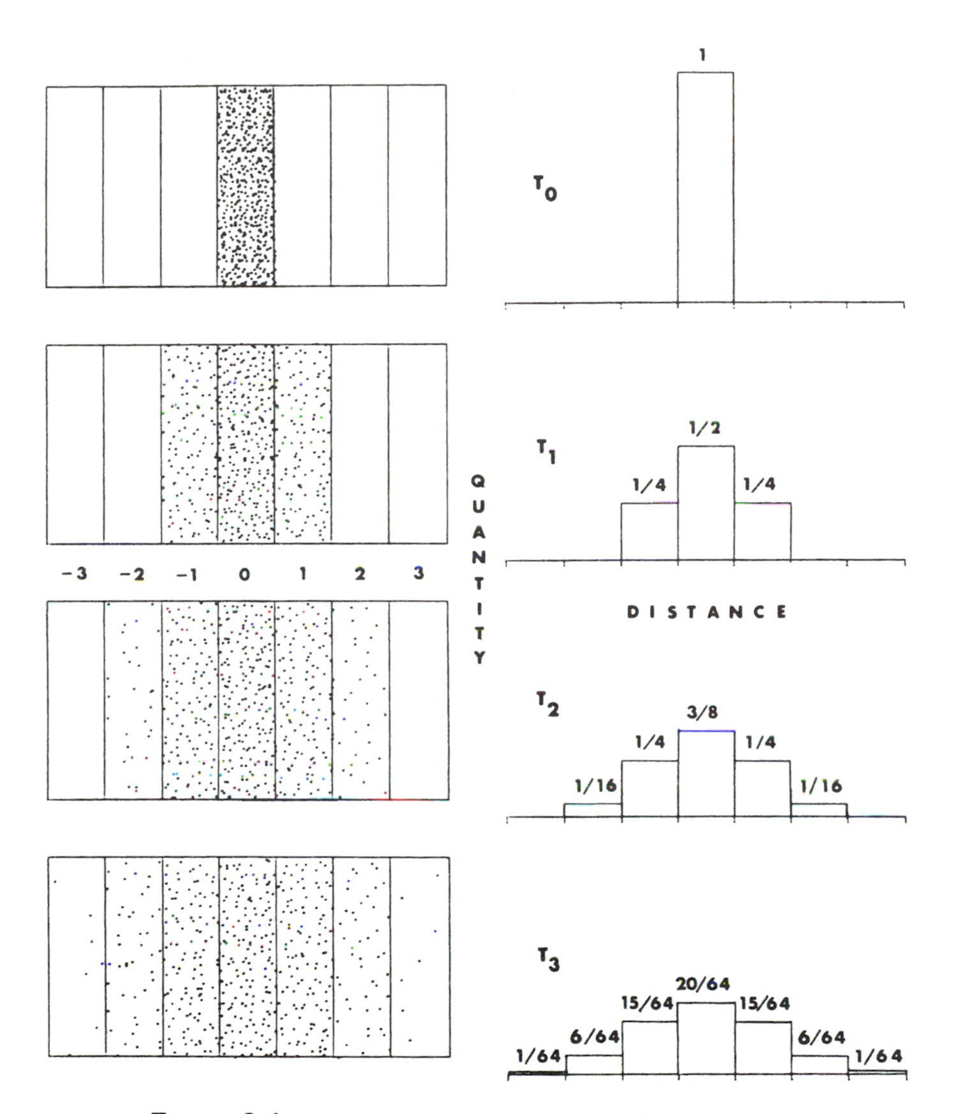


**Figure 2.1** Schematic representation of the diffusion process.

units of $cm^2/s$. The preexponential factor in the normal distribution is usually described in terms of $[1/\sigma\ (2\pi)^{1/2}]$, where $\sigma$, the population standard deviation, is a measure of distribution width. In the present case,

$$\sigma = \sqrt{2Dt} \tag{2.3}$$

If one considers a unit cross-sectional area (i.e., 1 $cm^2$), dN/dx is a measure of concentration in the region of space between x and (x + dx), and

$$C_{x,t} = \frac{N_0}{(4\pi Dt)^{1/2}} \exp\left(\frac{-x^2}{4Dt}\right) \tag{2.4}$$

Figure 2.2 illustrates a plot of concentration versus distance profiles at various times following the creation of an impulse of solute molecules at t = 0. Note

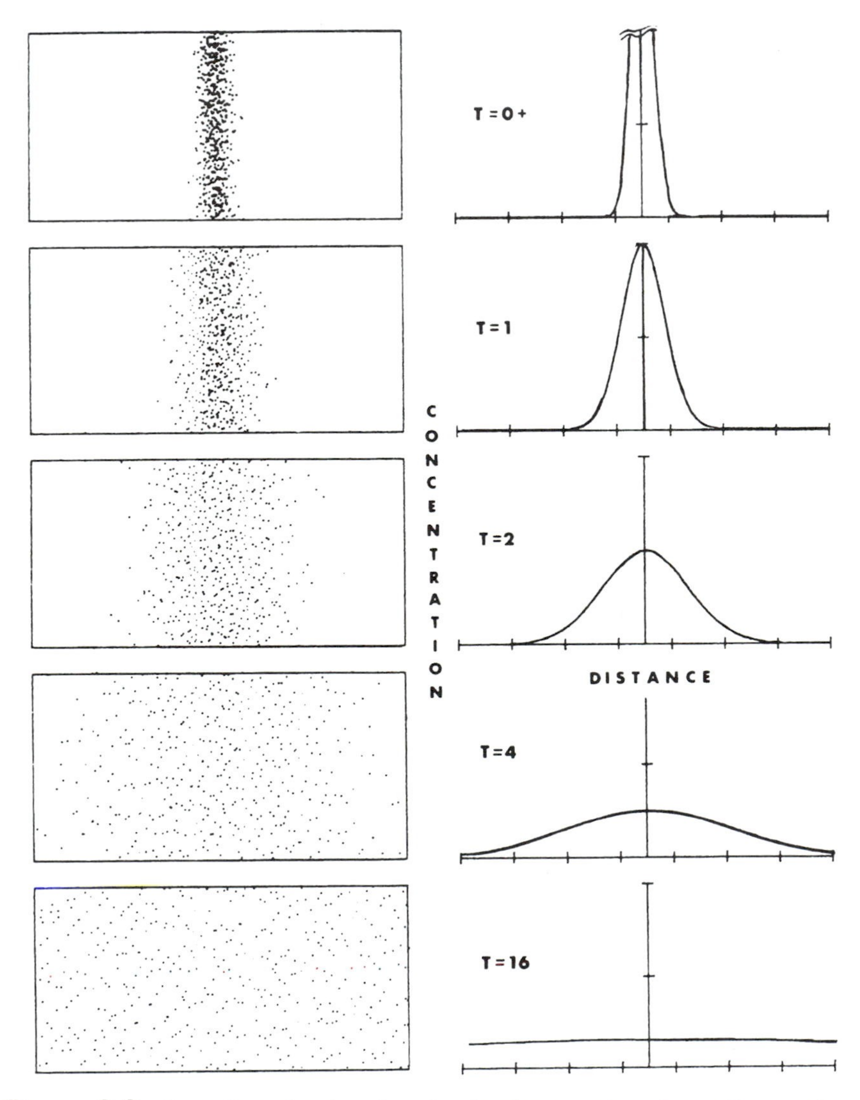


**Figure 2.2** Spreading of an impulse of molecules at various times after creation.

that the concentration gradient $dC_{x,t}/dx$ is zero at $x = 0$ but nonzero in the neighboring solution. The driving force for the diffusion process is nonuniformity of chemical potential, so in the limit as $t \rightarrow \infty$, $dC/dx$ will be zero throughout.

Examining a given concentration profile in greater detail, we may construct the model illustrated in Figure 2.3, which approximates the bell shape by many hypothetical impulse functions of different heights located $\Delta x$ distance increments apart. Simple diffusion is a linear process that fits the two criteria discussed earlier (Chap. 1). According to the superposition principle, the impulse functions may be treated completely independently if there is no solute-solute interaction (e.g., dimerization). The future of the total profile will be described accurately by a summation of the behavior of each independent impulse.

The net movement of molecules per unit area in a diffusion process at point x as a function of time t is described as the flux per unit area, $J_{x,t}$, which has units of mol/(cm$^2$ s). Referring to the inset in Figure 2.3, which shows a magnified view of the volume region from $x_1$ to $(x_1 + \Delta x)$, one can see that in time $\Delta t$, one-half of the molecules in column $x_1$, on the average, will have migrated to position $(x_1 + \Delta x)$, and one-half will have gone in the opposite direction, to $(x_1 - \Delta x)$. Similarly, one-half of the molecules at $(x_1 + \Delta x)$ will migrate to $x_1$, and so on. Therefore, one can define the flux at $(x_1 + \Delta x/2)$ at time t (denoted $J_{x_1+\Delta x/2,t}$) as the net flow from $x_1$ to $(x_1 + \Delta x)$. Thus

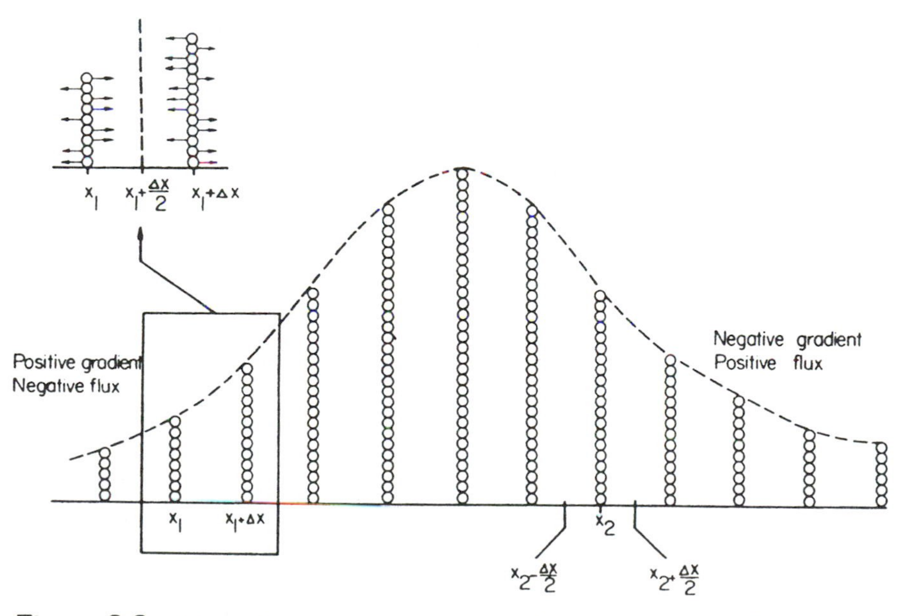


**Figure 2.3** Schematic representation of a Gaussian concentration profile by closely spaced spikes (or impulses) of molecules.

$$J_{x_1+\Delta x/2,t} = \frac{1}{\Delta t}\left(\frac{1}{2}N_{x_1,t} - \frac{1}{2}N_{x_1+\Delta x,t}\right) \tag{2.5}$$

describes the flow of molecules from $x_1$ to $(x_1 + \Delta x)$ minus that from $(x_1 + \Delta x)$ to $x_1$ across the hypothetical plane located at $(x_1 + \Delta x/2)$. This then gives the flux at point $(x_1 + \Delta x/2)$, where $\Delta t$ is the time required to traverse the distance $\Delta x$. In terms of concentration $(N/\Delta x)$,

$$J_{x_1+\Delta x/2,t} = \frac{\Delta x}{2\Delta t}\left(C_{x_1,t} - C_{x_1+\Delta x,t}\right) \tag{2.6}$$

or

$$J_{x_1+\Delta x/2,t} = \frac{(\Delta x)^2}{2\Delta t}\frac{C_{x_1,t} - C_{x_1+\Delta x,t}}{\Delta x} \tag{2.7}$$

Such a partial differential equation, expressed in terms of discrete units of x and t, is called a finite difference equation. It is subject to some "discretization error" depending on the magnitude of the finite differences chosen. In the limit as x and t become infinitesimal, the equation converges to reality (i.e., the discretization error becomes zero) and thus

$$\lim_{\substack{\Delta x\to 0\\ \Delta t\to 0}} J_{x_1+\Delta x/2,t} = \lim_{\substack{\Delta x\to 0\\ \Delta t\to 0}} \frac{(\Delta x)^2}{2\Delta t}\frac{C_{x_1,t} - C_{x_1+\Delta x,t}}{\Delta x} \tag{2.8a}$$

$$J_{x_1,t} = D\left(-\frac{\partial C_{x,t}}{\partial x}\right)_{x=x_1} \tag{2.8b}$$

$$= -D\left(\frac{\partial C_{x,t}}{\partial x}\right)_{x=x_1} \tag{2.8c}$$

Note that D is defined by the Einstein equation using finite differences as $D = (\Delta x)^2/2\,\Delta t$. Equation 2.8c is known as Fick's first law and $\partial C/\partial x$ is, of course, the concentration gradient. It should be clear that the net transport of solute mass per unit time across a plane intersecting a concentration profile will be proportional to the steepness of the profile and in the downhill direction. This fact must be intuitively obvious to anyone who wishes to use electrochemical techniques effectively.

It is often of interest to examine how the concentration at a given point changes as a function of time. Generally, it is not possible to investigate diffusion characteristics of a system under conditions of constant concentration gradient, since this implies the existence of a steady state. It is usually more convenient to evaluate such parameters from a determination of the change of concentration with time, as it is caused by diffusion in the system in question. Although qualitatively useful, Fick's first law of diffusion is of less value to the practical experimenter in that it does not directly relate to the easily measurable quantities of concentration, time, and distance. Fick's second law of diffusion, to be derived presently, partially alleviates these difficulties.

Referring to Figure 2.4, one can see that as long as $J_{x_2-\Delta x/2,t} = J_{x_2+\Delta x/2,t}$, the concentration of material bounded by the walls of unit cross-sectional area at $(x_2 \pm \Delta x/2)$ will be time invariant. On the other hand, if $J_{x_2-\Delta x/2,t} > J_{x_2+\Delta x/2,t}$, a net increase of material into the box would be expected, and concentration would increase with time. Since J is defined as the flux per unit area, dividing J by the distance between the walls yields the time rate of change of concentration within the box. More formally,

$$\left(\frac{\Delta C}{\Delta t}\right)_{x=x_2} = \frac{J_{x_2-\Delta x/2,t} - J_{x_2+\Delta x/2,t}}{\Delta x} \tag{2.9}$$

Substituting for the flux terms using Equation 2.7 gives us

$$\left(\frac{\Delta C}{\Delta t}\right)_{x=x_2} = \frac{(\Delta x)^2}{2\Delta t}\left(\frac{C_{x_2-\Delta x,t} - C_{x_2,t}}{\Delta x} - \frac{C_{x_2,t} - C_{x_2+\Delta x,t}}{\Delta x}\right)\Bigg/\Delta x \tag{2.10a}$$

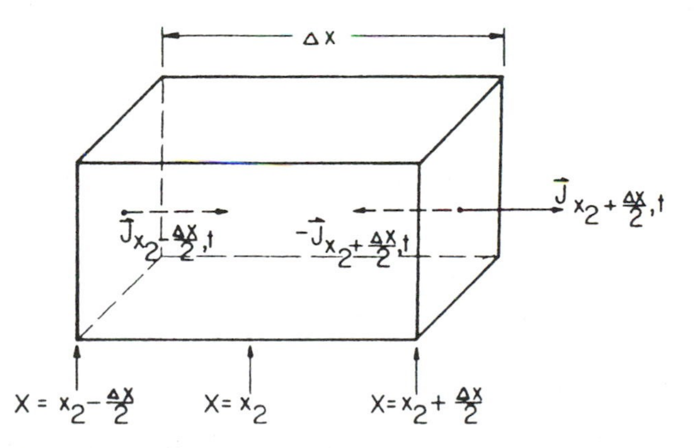


**Figure 2.4** Diagram of portion of solution of length x and unit cross-sectional area, located about $x = x_2$.

or

$$\left(\frac{\Delta C}{\Delta t}\right)_{x=x_2} = D\left[\frac{C_{x_2+\Delta x,t} - 2C_{x_2,t} + C_{x_2-\Delta x,t}}{(\Delta x)^2}\right] \tag{2.10b}$$

In the limit as $\Delta x \to 0$ and $\Delta t \to 0$, the finite difference equation becomes

$$\left(\frac{\partial C}{\partial t}\right)_{x=x_2} = D\left(\frac{\partial^2 C}{\partial x^2}\right)_{x=x_2} \tag{2.11}$$

which is known as Fick's second law of diffusion. As can easily be verified, Equation 2.4 describing $C_{x,t}$ is an analytical solution of Fick's second law. As predicted earlier, at the point of inflection on the instantaneous concentration profile curve, where $J_{x_2-\Delta x/2,t} = J_{x_2+\Delta x/2,t}$, the second partial derivative is zero, so the concentration is invariant. Above the point of inflection, the second derivative is negative; the concentration of material at points in this region is decreasing as time goes on. Similarly, the concentration at points below the inflection point is increasing with time.

If the diffusion process is *coupled* with other influences (chemical reactions, adsorption at an interface, convection in solution, etc.), additional concentration dependences will be added to the right side of Equation 2.11, often making it analytically insoluble. In such cases it is profitable to retreat to the finite difference representation and model the experiment on a digital computer. Modeling of this type, when done properly, is not unlike carrying out the experiment itself (provided that the discretization error is equal to or smaller than the accessible experimental error). The method is known as *digital simulation,* and the result obtained is the finite difference solution. This approach is described in more detail in Chapter 20.

Analytical solutions of Fick's laws are most easily derived using Laplace transforms, a subject described in every undergraduate book on differential equations. The solution of diffusion equations has fascinated academic electroanalytical chemists for years, and they naturally have a tendency to expound on them at the slightest provocation. Fortunately, the chemist using electrode reactions can accomplish a great deal without more than a cursory appreciation of the mathematics. Our intention here is to provide this qualitative appreciation on a level sufficient to understand laboratory techniques.

## B. Step Response

A frequently encountered diffusion experiment is initiated with a step functional change in concentration. In this case we envision an imaginary barrier separating a region of finite solute concentration from a region of zero concentration as shown in Figure 2.5a. When the barrier is hypothetically removed, the sharp boundary is gradually made more diffuse as solute moves to the right. The flux decreases with time as the concentration gradient becomes more gradual, as indicated in Figure 2.5b.

Simple diffusion is a linear process. Therefore, the original step function may be broken into a series of spatially separated impulses (Fig. 2.5c). Each impulse can be treated independently according to the results described above (Fig. 2.5d). The total concentration profile will be described by the summation of response functions to the original excitation pulses. Thus, making use of

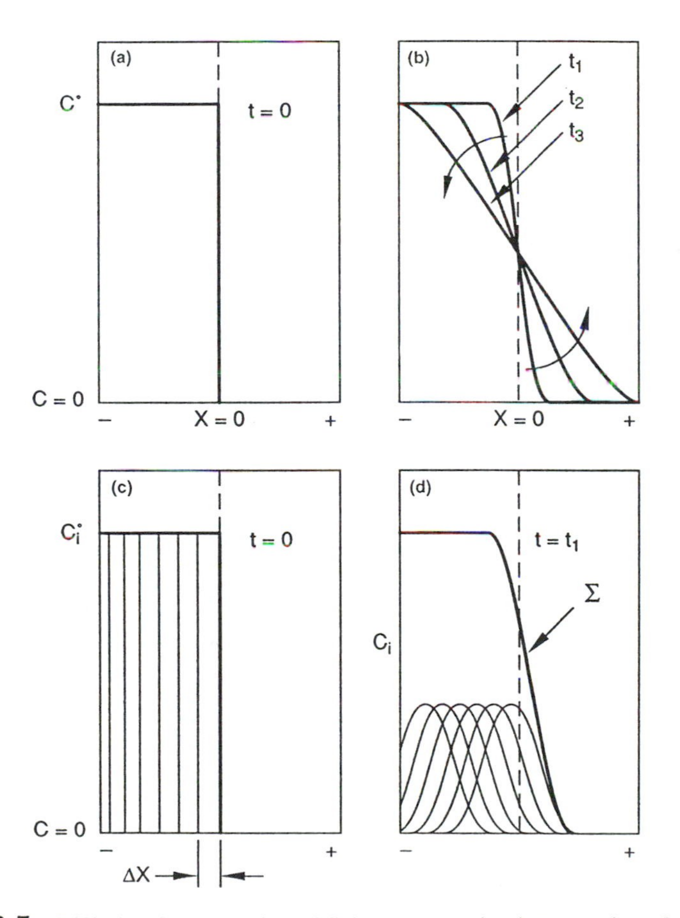


**Figure 2.5** Diffusion from a region of finite concentration into a region of zero concentration. Diffusional response to a step-function excitation under linear conditions.

Equation 2.4 for each impulse, we have

$$C_{x,t} = \frac{C^\circ}{(4\pi Dt)^{1/2}} \sum_{n=0}^{m} \exp\left[\frac{-(x - n\Delta x)^2}{4Dt}\right]\Delta x \tag{2.12}$$

for the total, which in the limit as $\Delta x \rightarrow dx$ and $m \rightarrow \infty$ becomes*

$$C_{x,t} = \frac{C^\circ}{2\pi^{1/2}}\left[\pi^{1/2} - 2\int_0^{(4Dt)^{1/2}} \exp\left(\frac{-x^2}{4Dt}\right) d\left(\frac{x}{4Dt}\right)^{1/2}\right] \tag{2.13}$$

and then differentiation with respect to x yields the concentration gradient

$$\left(\frac{\partial C}{\partial x}\right)_{x,t} = \frac{C^\circ}{(4\pi Dt)^{1/2}} \exp\left(\frac{-x^2}{4Dt}\right) \tag{2.14}$$

This expression has the form of the normal curve of error, and therefore we anticipate the *gradients* in this case to have a Gaussian shape.

It is important to recognize the unique relationship that exists between the responses to an impulse and step change in concentration. The derivative of the step response (Eq. 2.14) is identical to the impulse response (Eq. 2.4), and the integral of the impulse response is identical to the step response. This reciprocity is an important property of linear systems in general. The reader should now appreciate that under linear conditions, the time dependence of *any* concentration profile can be treated by adding the response functions for its component impulses.

### C. Barriers to Free Diffusion

Thus far we have examined diffusion under infinite conditions, where no phase boundaries exist. Some practical situations may be described by the above treatment. More frequently, the diffusion process will be initiated in the neighborhood of one or more phase boundaries as, for example, in chromatography and electrochemistry. The phase boundaries may be either permeable or impermeable to the diffusing solute. In electrochemical techniques, the boundary (e.g., the working electrode) is usually impermeable; however, this is not always so (e.g., some ion-selective electrodes, membranes, liquid-liquid interfaces). In the

---

*Equation 2.13 is commonly expressed in terms of the complementary error function such that

$$C_{x,t} = \frac{C^\circ}{2} \operatorname{erfc}\left(\frac{-x}{\sqrt{4Dt}}\right)$$

most interesting cases, the diffusion barrier is not passive, but somehow interacts with the solute (e.g., an electrochemical reaction occurs).

Let us first consider what would result if the impulse described earlier were created near an impenetrable inert barrier. Such a barrier would behave as a reflector. In Figure 2.6a, an impulse of molecules is suddenly created in solution a short distance away from the interface. The dashed line is the profile that would result in the absence of the barrier. The dotted line results from "folding" this concentration profile about the interfacial plane. Because diffusion is a linear process, the actual concentration profile in the presence of the barrier is the sum of the dotted and dashed lines, as shown in Figure 2.6b.

If the impulse is created *at* the interface, as in Figure 2.6c, then the hypothetical (dashed) concentration profile would result in an actual profile (Fig. 2.6d) with twice the amplitude of what would have been the case had the barrier not been present. The results for later times can be attained by the same logic. Note that at infinite time, the profile will have completely "relaxed" away. Diffusion under these conditions is most commonly encountered in electrochem-

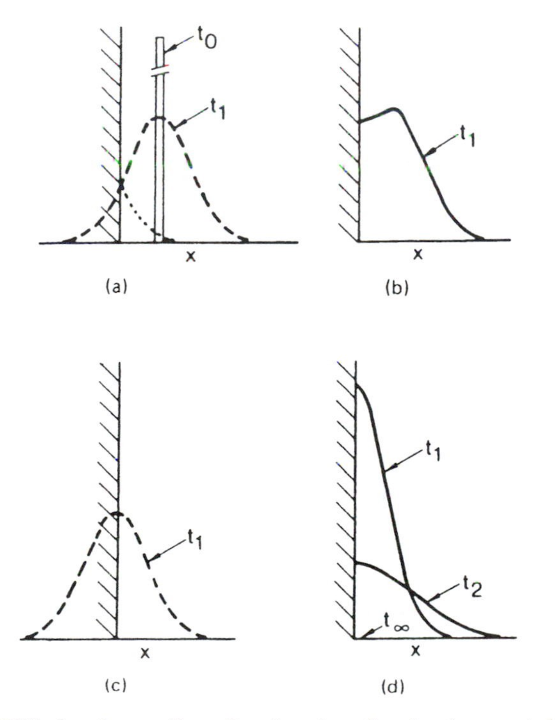


**Figure 2.6** Diffusional spreading of an impulse of molecules created (a,b) near, and (c,d) at an impenetrable barrier.

istry and is described as being semi-infinite. We can, in fact, create a close approximation to an impulse of molecules by generating product molecules in response to a pulse excitation (potential, current, or charge). A good example is when the surface itself is the reactant, such as when metal ions are suddenly released from a metal surface by oxidative dissolution.

Occasionally (e.g., thin-layer electrochemistry, porous-bed electrodes, metal atoms dissolved in a mercury film), diffusion may be further confined by a second barrier. Figure 2.7 illustrates the case of *restricted diffusion* when the solution is confined between two parallel barrier plates. Once again, the folding technique quickly enables a prediction of the actual result. In this case, complete relaxation of the profile results in a uniform finite concentration across the slab of solution, in distinct contrast to the semi-infinite case. When the slab thickness $\ell$ is given, the time for the average molecule to diffuse across the slab is calculable from the Einstein equation such that

$$t = \frac{\ell^2}{2D} \tag{2.15}$$

Thus the relaxation time for a concentration profile in the thin layer will vary with the thickness squared.

Although we will not belabor this point here, the folding technique can easily be adapted to the mathematical description. The reader should recognize that the presence of the barrier does not invalidate either of Fick's laws but does influence the boundary conditions chosen to obtain a specific solution, whether analytically or by the finite difference approach.

It is also important to recognize that the preceding considerations apply if we create a solute *vacancy* (i.e., a region of zero concentration) rather than an impulse. For example, it is possible to create a vacancy in a layer of solution

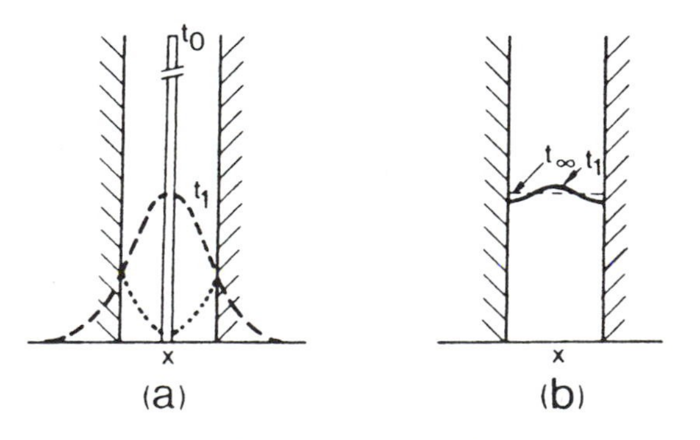


**Figure 2.7** Diffusional spreading of an impulse of molecules created between two barrier plates.

adjacent to an electrode by instantly converting the solute electrochemically. This is illustrated in Figure 2.8 for a semi-infinite situation. The initial condition (Fig. 2.8a) is perturbed (Fig. 2.8b and c) and the resulting vacancy then relaxes as indicated in Figure 2.8d–f until the initial condition is restored at infinite time (Fig. 2.8a).

There is an important electrochemical technique called chronoamperometry (i.e., current measured as a function of time), where a vacancy is maintained at the surface (Chap. 3). The electrode potential is controlled at a value sufficient to immediately react any sample molecule that diffuses to the surface. This results in the sequence shown in Figure 2.9. The top drawings indicate consumption of the reactant at three times following the application of the potential. The corresponding concentration profiles are also shown. The bottom sequence indicates creation of product at the same three times. The sample molecules try their best to fill in the vacancy, but the electrode reaction prevents this. If the

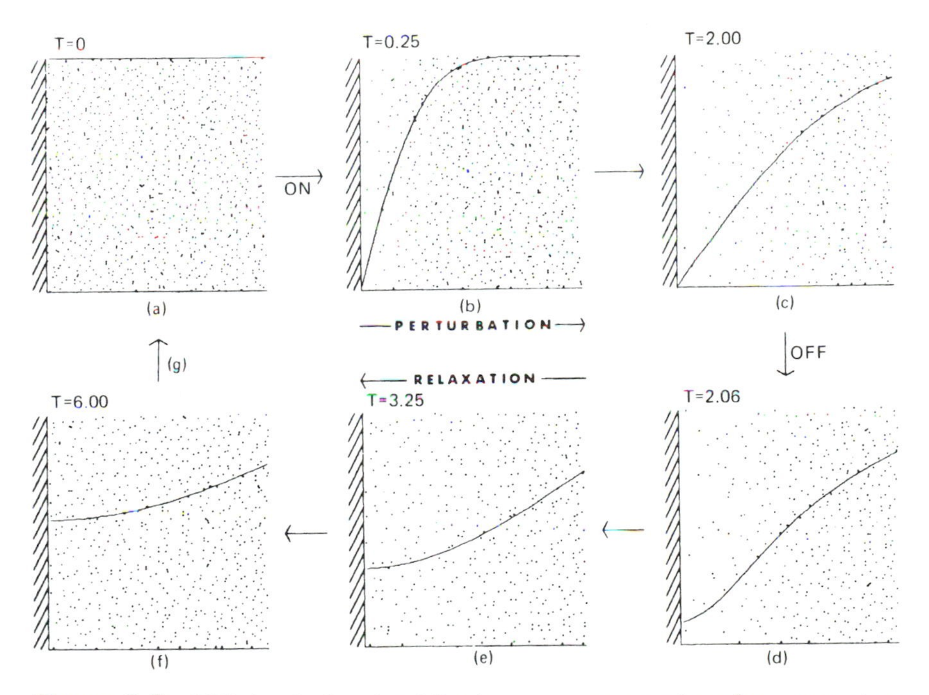


**Figure 2.8** Diffusional relaxation following momentary creation of a vacancy at an inert barrier. Semi-infinite conditions prevail. Density map and concentration-distance profiles are shown. (a) Initial condition; (b) vacancy creation; (c) vacancy extends farther into bulk of medium; (d) relaxation begins; (e) relaxation continues; (f) relaxation continues; (g) initial condition restored.

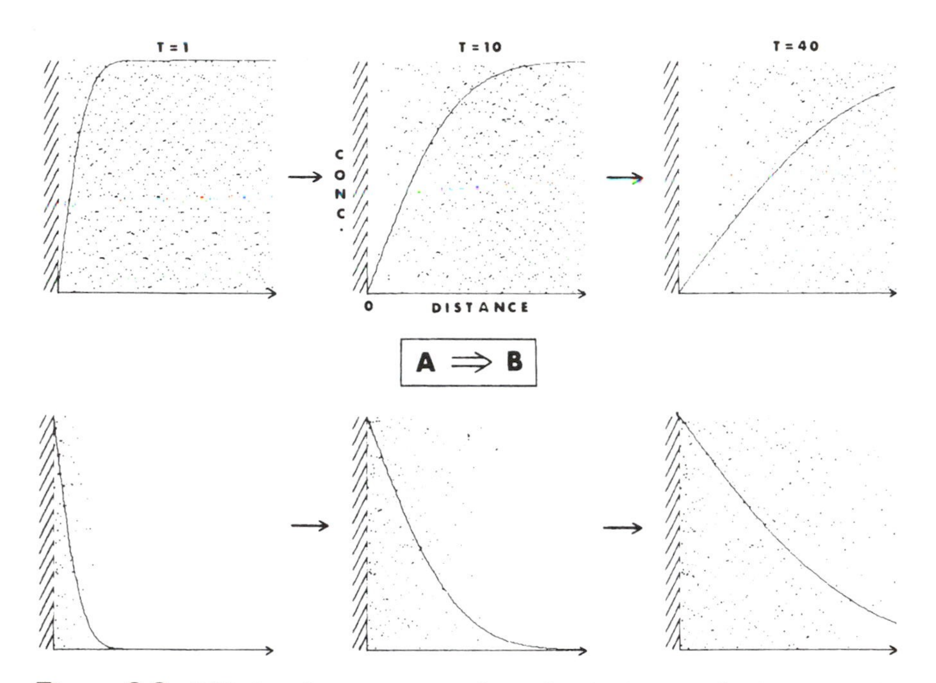


**Figure 2.9** Diffusion of a reactant toward a surface (top) occurs simultaneously with diffusion of product away from the surface (bottom).

electrode is disconnected so no further reaction is possible, the initial condition will be restored.

The diffusion problems encountered in chronoamperometry are described in more detail in Chapter 3. The reader should be able to make mental pictures of the transport of molecules and ions to and from an electrode surface. In our experience, this is one of the most difficult concepts to grasp for beginning students of electrochemistry. Many hours of practice at sketching concentration-distance profiles is highly recommended. As an alternative, digital simulation programs are available which provide a graphic display of concentration profile dynamics on a personal computer screen as the actual experiment is modeled. A review of this section after reading Chapters 3 to 5 would probably prove worthwhile to reinforce diffusion concepts.

## III. THE CHARGED INTERPHASE

Every chemist at one time or another has encountered the Debye-Hückel theory of electrolyte solutions. It is essential to understand the nature of ions in homogeneous solution in order to appreciate the interaction of these ions with a charged surface. If one considers the assumptions of the Debye-Hückel theory for a central ion (i.e., the solvated ion is a hard nonpolarizable sphere, a point charge, in a continuous medium of uniform dielectric constant), it is reasonable to apply both the assumptions and the resulting conclusions to a charged surface. After all, the concept of a planar charged surface is no different from that of an ion of very large radius in contact with much smaller ions and solvent molecules. Theories that deal with either situation on a microscopic level have been grossly approximate. Nevertheless, simple electrostatic models are available in which the attraction between an ion and a surface is countered by thermal agitation.

Let us begin by going back to some very fundamental ideas from physics. First consider two parallel inert metallic conductors separated by an evacuated space, d. If an electrical potential difference is applied across the plates (Fig. 2.10), the difference in potential between the bulk conductor phases implies that an electric field exists between them of magnitude $\Delta\phi_m/d$ (V/cm). Any molecule or ion in free space between the plates may sense the field ("potential gradient") but not the absolute magnitude of the inner potentials, $\phi_m$, for each plate.

If we place a neutral molecule in the field, very little will happen unless the field strength is extremely great (e.g., $10^6$–$10^9$ V/cm). On such occasions, there will be significant orientation of dipoles, polarization of orbitals, and possibly even ionization. In the latter case, an organic molecule loses an electron from its highest filled molecular orbital, forming a cation radical. Field ionization in this manner is extremely difficult to achieve in free space. On the other hand, the ionization process can often be stimulated by radiation (e.g., photoionization) or the nearness of a metal electrode (e.g., field ionization mass spectrometry). In either case, an electron leaves the molecule and is accelerated toward the positive electrode (anode) while the cation radical accelerates toward the negative electrode (cathode). Each particle picks up kinetic energy along the way, depending only on its charge and the potential difference through which it falls. Gas-phase redox reactions are quite familiar in field ionization mass

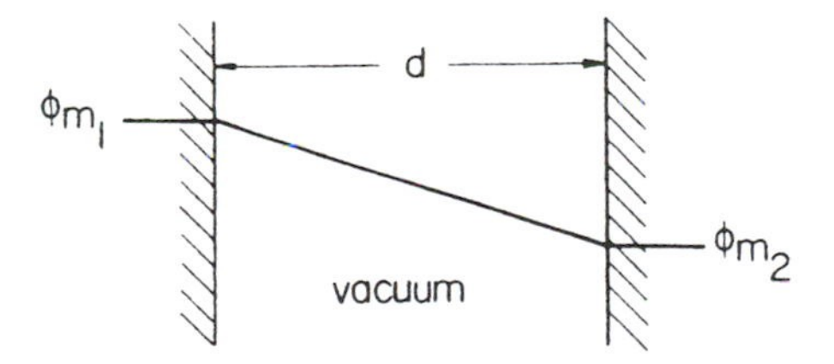


**Figure 2.10** Linear field between two conducting plates with a vacuum between them.

spectrometry. Molecules passing through a strong field near a fine wire or sharp-edged anode are oxidized to cations with little internal energy and therefore little tendency to fragment. The cations are focused and accelerated into the analyzer region of the mass spectrometer and then counted at a detector.

Suppose that we now fill the space between our planar electrodes with a solution. First let us choose a pure solvent of low dielectric constant (e.g., hexane) with no charge carriers present. How does this compare to the previous situation? First, we are limited in the field we can achieve before breakdown in the dielectric occurs. It is virtually impossible to field ionize a molecule in such a medium. On the other hand, photoionization can be accomplished with the field providing an impetus to charge separation. As in a vacuum, the photoionized molecule and the electron are accelerated in opposite directions, but now a terminal velocity is readily achieved depending on the viscous drag of each charged particle. The solvated photoelectron will, of course, move far more rapidly than the ion.

The complexity of the situation can be increased significantly by replacing the hexane with a medium of high dielectric constant (e.g., water) containing many charge carriers. Now the potential difference we can apply at zero current is limited to a few electron volts (at most). A greater difference would result in electrochemical reaction of the solvent and/or the ions present. Why is this so?

If we consider the potential gradient across the present cell we find that it exists only in the narrow *interphase* regions near the two surfaces (Fig. 2.11). Furthermore, the thickness of the potential gradient is related to the ionic strength. At high ionic strengths, the potential will decay more quickly at each interphase. This is expected from the Debye-Hückel theory, since the potential falls away from a central ion faster at higher ionic strength. This should make sense if we consider the competition between ordering by electrostatic attraction and disordering by thermal motion in the medium. At high ionic strength, the large number of ions permits a *time-averaged distribution* closer to the

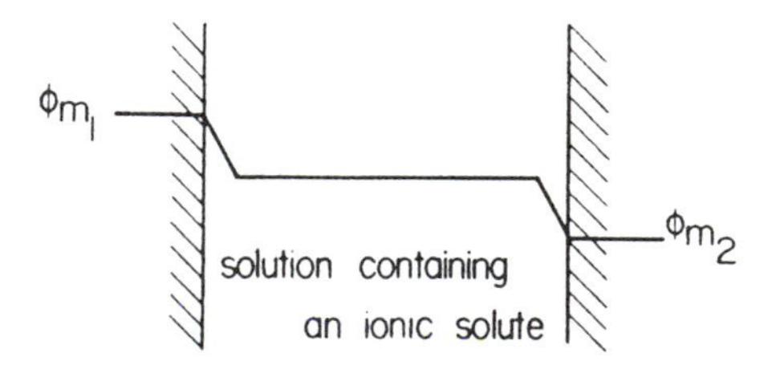


**Figure 2.11** Potential gradients are near the plates when a solution containing ions is placed between them.

surface, whereas at low ionic strength, the thermal agitation competes more effectively and the time-averaged distribution of an ion excess (or ion imbalance) extends farther from the surface. In the bulk of the solution, electroneutrality exists and the potential, $\phi_s$, is independent of distance.

The interphase between an electrolyte solution and an electrode has become known as the electrical double layer. It was recognized early that the interphase behaves like a capacitor in its ability to store charge. Helmholtz therefore proposed a simple electrostatic model of the interphase based on charge separation across a constant distance as illustrated in Figure 2.12. This parallel-plate capacitor model survives principally in the use of the term "double layer" to describe a situation that is quite obviously far more complex. Helmholtz was unable to account for the experimentally observed potential dependence and ionic strength dependence of the capacitance. For an *ideal* capacitor, $Q = CV$, and the capacitance C is not a function of V.

Figure 2.13 illustrates what is currently a widely accepted model of the electrode-solution interphase. This model has evolved from simpler models, which first considered the interphase as a simple capacitor (Helmholtz), then as a Boltzmann distribution of ions (Gouy-Chapman). The electrode is covered by a sheath of oriented solvent molecules (water molecules are illustrated). Adsorbed anions or molecules, A, contact the electrode directly and are not fully solvated. The plane that passes through the center of these molecules is called the inner Helmholtz plane (IHP). Such molecules or ions are said to be *specifically adsorbed* or *contact adsorbed*. The molecules in the next layer carry their primary (hydration) shell and are separated from the electrode by the monolayer of oriented solvent (water) molecules adsorbed on the electrode. The plane passing through the center of these solvated molecules or ions is referred to as the outer Helmholtz plane (OHP). Beyond the compact layer defined by the OHP is a Boltzmann distribution of ions determined by electrostatic interaction between the ions and the potential at the OHP and the random jostling of ions and

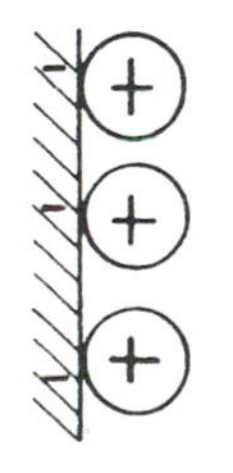


**Figure 2.12** Simple capacitor model of electrode-solution interface as a charged double layer (original Helmholtz model). Negatively charged surface. Positively charged ions are attracted to the surface, forming an electrically neutral interphase.

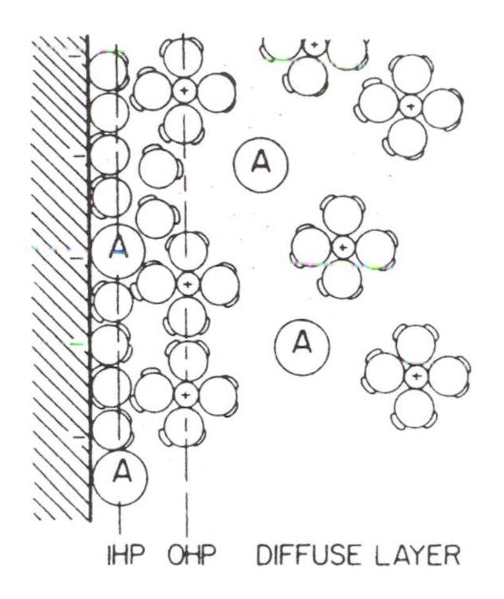


**Figure 2.13** Model of the electrode-solution interphase as described by Bockris, Devanathan, and Muller [J. O'M. Bockris and A. K. N. Reddy, *Modern Electrochemistry*, Vol. 2, Chap. 7, Plenum, New York, 1970.]

solvent molecules. This region is referred to as the *diffuse layer*. In the diffuse layer, random thermal motion tends to distribute the ions evenly throughout the solution, whereas electrostatic forces tend to attract or dispel ions from the surface depending on their charge. These two tendencies counterbalance each other, resulting in a nonuniform distribution of ions near the surface. As one looks beyond the diffuse layer, *ions in the homogeneous bulk solution phase cannot feel the presence of the electrode*. In fact, the potential diminishes exponentially with distance from the surface, and the field strength becomes zero very close (usually considerably less than $10^{-6}$ m) to the electrode.

As a result of the availability of charge carriers, all the potential difference between two electrodes is dropped across the two interphases for an electrolyte solution and not across the bulk solution phase. When a current passes across the solution, there is a possibility that a potential difference will develop due to the finite conductivity of the solution. In most electroanalytical experiments this is very small compared to the interfacial potential difference and always results in a comparatively weak electric field (small potential dropped across a large distance). This matter will be dealt with beginning in Chapter 6.

Knowledge of the structure of the interphase is necessary before a detailed understanding of the electron-transfer process is possible. A great deal of progress has been made in the last 50 years. Nevertheless, many of the details remain beyond the reach of current experimental techniques.

## IV. THE NERNST EQUATION AND ELECTROCHEMICAL REVERSIBILITY

Let us consider the surface electron-transfer reaction occurring at a single electrode maintained at a fixed potential, where both O and R are soluble and the reaction shown is rate-controlling (i.e., no other processes limit this heterogeneous reaction):

$$O + ne \underset{k_{b,h}}{\overset{k_{f,h}}{\rightleftarrows}} R \tag{2.16}$$

The forward and backward heterogeneous rate constants, $k_{f,h}$ and $k_{b,h}$, are formal constants, which implies that activity coefficients are assumed to be unity. In normal homogeneous solution kinetics, for example,

$$O \underset{k_b}{\overset{k_f}{\rightleftarrows}} R \tag{2.17}$$

and the formal rate equation is given by

$$\frac{d[R]_t}{dt} = -\frac{d[O]_t}{dt} = k_f[O]_t - k_b[R]_t \tag{2.18}$$

where the rate constants are expressed in units of $s^{-1}$ and the concentrations in mol/L. On the other hand, in electrode kinetics, the reaction occurs only near the interface between the electrode and the solution, where the electric field strength is great. It is therefore convenient to deal in terms of mol/cm$^2$ ($\underline{N}$) transformed per unit time [i.e., mol/(cm$^2$ s)] such that

$$\frac{d\underline{N}_R}{dt} = -\frac{d\underline{N}_O}{dt} = k_{f,h}C_O - k_{b,h}C_R \tag{2.19}$$

where $C_O$ and $C_R$ are the surface concentrations in mol/cm$^3$ (note the use of bracketed symbols for mol/L, C with subscripts for mol/cm$^3$, and underlined symbols to indicate per unit area). $C_O$ and $C_R$ are first assumed to be the same as the bulk concentrations, and independent of cell current and time. It should be clear that the heterogeneous rate constants will have units of cm/s. The *net* transformation of O to R ($-d\underline{N}_O/dt$) can be expressed as

$$\frac{d\underline{N}_{O,f}}{dt} - \frac{d\underline{N}_{O,b}}{dt} \tag{2.20}$$

or [mol/(cm$^2$ s) of O lost in the forward reaction] – [mol/(cm$^2$ s) of O gained via the reverse reaction].

From Faraday's law, $Q = nFN$, the number of coulombs passed per square centimeter in the forward reaction is $\underline{Q}_f = nF\underline{N}_{O,f}$ and in the reverse reaction $\underline{Q}_b = nF\underline{N}_{O,b}$ where $\underline{N}_{O,f}$ and $\underline{N}_{O,b}$ are the mol/cm$^2$ of O lost and gained. The current density $\underline{i}$ (A/cm$^2$) is given by $d\underline{Q}/dt$, so that the partial current densities for the forward and reverse reactions are

$$\underline{i}_f = \frac{d\underline{Q}_f}{dt} = nF\frac{d\underline{N}_{O,f}}{dt} \tag{2.21}$$

and

$$\underline{i}_b = \frac{d\underline{Q}_b}{dt} = nF\frac{d\underline{N}_{O,b}}{dt} \tag{2.22}$$

By analogy, the net current density (A/cm$^2$) is given by

$$\begin{aligned} \underline{i}_{net} &= -nF\frac{d\underline{N}_O}{dt} = \underline{i}_f - \underline{i}_b = nF\left(\frac{d\underline{N}_{O,f}}{dt} - \frac{d\underline{N}_{O,b}}{dt}\right) \\ &= nF(k_{f,h}C_O - k_{b,h}\,C_R) \end{aligned} \tag{2.23}$$

The actual current (amperes) measured is

$$i_{net} = \underline{i}_{net}A = nFA(k_{f,h}C_O - k_{b,h}C_R) \tag{2.24}$$

The *formal rate constants* $k_{f,h}$ and $k_{b,h}$ are potential dependent. At any given potential, in order for the transition from the oxidized form to occur, it will be necessary to pass over an *activation free-energy barrier*, $\Delta G_f^{\ddagger}$, as illustrated by the Morse curves in Figure 2.14. *The rate of reaction will be proportional to* $\exp(-\Delta G_f^{\ddagger}/RT)$.

At equilibrium, $\Delta G_f^{\ddagger} = \Delta G_b^{\ddagger}$ as shown in Figure 2.15, where the probability of electron transfer is the same in each direction, $-\underline{i}_b = \underline{i}_f = \underline{i}_0$, where $\underline{i}_0$ is the *exchange current density*. Under nonequilibrium conditions favoring the cathodic reaction ($O \rightarrow R$), $\Delta G_f^{\ddagger\prime} < \Delta G_b^{\ddagger\prime}$ by an amount of free energy given by $-nF(E - E^{\circ\prime})$, where $E^{\circ\prime}$ is the formal potential of the couple O/R.

Now let us idealize the intersection region of the overlapping Morse potential energy curves as shown in Figure 2.15. In the nonequilibrium situation, we arbitrarily define a and b so that the change in barrier height for the reverse reaction is

$$b = \Delta G_b^{\ddagger\prime} - \Delta G_0^{\ddagger} \tag{2.25}$$

and

$$a + b = \Delta G_b^{\ddagger\prime} - \Delta G_f^{\ddagger\prime} \tag{2.26}$$

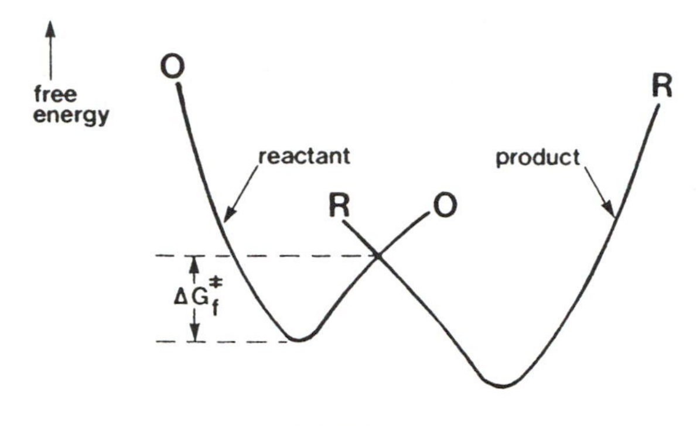


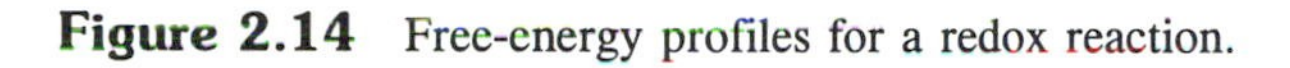

**Figure 2.14** Free-energy profiles for a redox reaction.

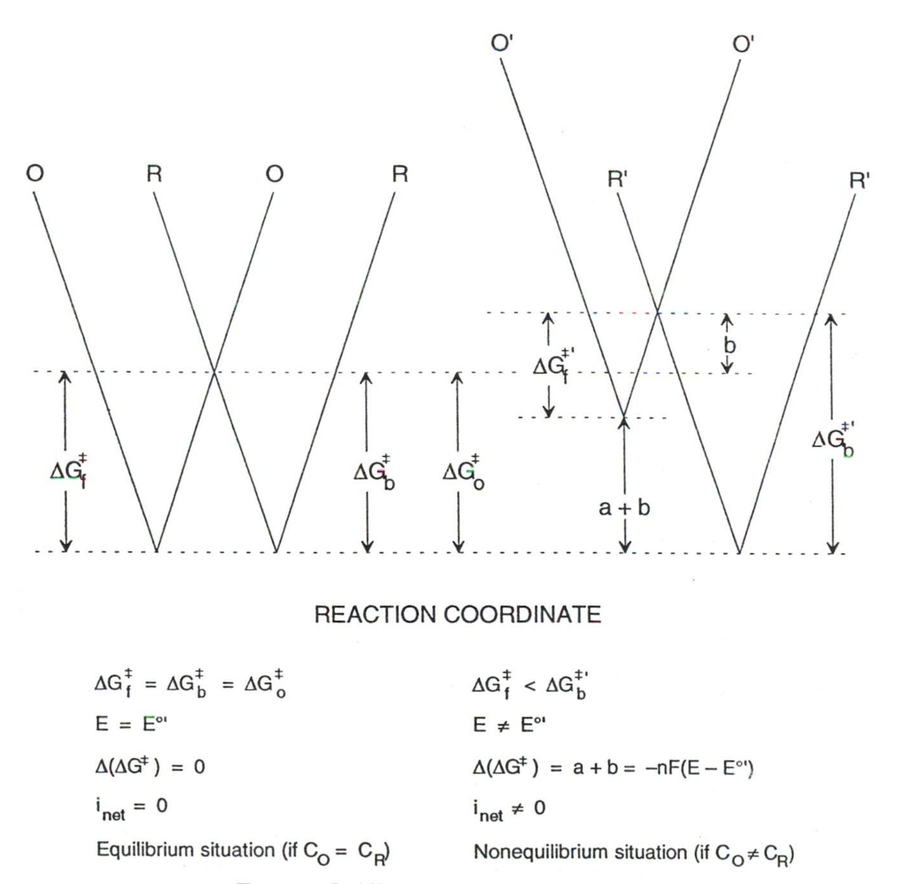


**Figure 2.15** Linearized Morse curves.

In order to move from the state of equilibrium on the left of Figure 2.15 to the nonequilibrium situation on the right, we apply a new potential E, which causes more O to be transformed into R per unit time than R is transformed into O (i.e., a net reduction). This shifts the potential curves to a new relative position such that

$$\Delta G_b^{\ddagger'} - \Delta G_f^{\ddagger'} = -nF(E - E^{\circ\prime}) \tag{2.27}$$

$$\Delta G_f^{\ddagger'} - nF(E - E^{\circ\prime}) = \Delta G_b^{\ddagger'} = \Delta G_0^{\ddagger'} + b \tag{2.28}$$

and

$$-nF(E - E^{\circ\prime}) = a + b \tag{2.29}$$

If we define the *transfer coefficient* $\alpha$ as $a/(a + b)$, then

$$a = -\alpha nF(E - E^{\circ\prime}) \tag{2.30}$$

and

$$b = -(1 - \alpha)nF(E - E^{\circ\prime}) \tag{2.31}$$

The transfer coefficient depends on the symmetry of the two potential curves, and for the idealized curves shown, $\alpha = 0.5$. Now since

$$\begin{aligned} \Delta G_f^{\ddagger'} &= \Delta G_0^{\ddagger} + b + nF(E - E^{\circ\prime}) \\ &= \Delta G_0^{\ddagger} + b - (a + b) \\ &= \Delta G_0^{\ddagger} - a \end{aligned} \tag{2.32}$$

then by substitution

$$\Delta G_f^{\ddagger'} = \Delta G_0^{\ddagger} + \alpha nF(E - E^{\circ\prime}) \tag{2.33}$$

and

$$\Delta G_b^{\ddagger'} = \Delta G_0^{\ddagger} + b = \Delta G_0^{\ddagger} - (1 - \alpha)\, nF(E - E^{\circ\prime}) \tag{2.34}$$

Using the Arrhenius equation,

$$k = A\, e^{-\Delta G^{\ddagger}/RT} \tag{2.35}$$

where A is a constant pre-exponential factor, we have

$$k_{f,h} = Ae^{-\Delta G_f^{\ddagger'}/RT} = Ae^{-\Delta G_0^{\ddagger}/RT}\, e^{-\alpha nF(E-E^{\circ\prime})/RT} \tag{2.36}$$

and

$$k_{b,h} = Ae^{-\Delta G_b^{\ddagger'}/RT} = Ae^{-\Delta G_0^{\ddagger}/RT}\, e^{(1-\alpha)nF(E-E^{\circ\prime})/RT} \tag{2.37}$$

At the standard potential, $k_{f,h} = k_{b,h} = k_{s,h}$, for all reactants in the standard state, where $k_{s,h}$ is the standard heterogeneous rate constant, defined according to the Arrhenius equation as

$$k_{s,h} = Ae^{-\Delta G_0^{\ddagger}/RT} \tag{2.38}$$

Substituting Equation 2.38 into Equations 2.36 and 2.37, we have

$$k_{f,h} = k_{s,h}\, e^{-\alpha nF(E-E^{\circ\prime})/RT} \tag{2.39}$$

and

$$k_{b,h} = k_{s,h}\, e^{(1-\alpha)nF(E-E^{\circ\prime})/RT} \tag{2.40}$$

Thus we have succeeded in expressing the individual heterogeneous rate constants in terms of a standard rate constant and a potential-dependent exponential term. Since $\underline{i}_f = nFC_Ok_{f,h}$ and $\underline{i}_b = nFC_Rk_{b,h}$, the individual current densities can now be written as follows:

$$\underline{i}_f = nFC_Ok_{s,h}\, e^{-\alpha nF(E-E^{\circ\prime})/RT} \tag{2.41}$$

and

$$\underline{i}_b = nFC_Rk_{s,h}\, e^{(1-\alpha)nF(E-E^{\circ\prime})/RT} \tag{2.42}$$

The net current was shown earlier to be given by $A(\underline{i}_f - \underline{i}_b)$. Now we can explicitly include the potential dependence such that

$$i_{net} = nFAk_{s,h}\,[C_O\, e^{-\alpha nF(E-E^{\circ\prime})/RT} - C_R\, e^{(1-\alpha)nF(E-E^{\circ\prime})/RT}] \tag{2.43}$$

This important expression is known as the Eyring equation. Note that $i_{net}$ depends on both the potential and the surface concentration of each form of the couple. For example, a high concentration of R and a very positive potential combine to result in a large anodic current.

At equilibrium, $i_{net} = 0$, so

$$C_Oe^{-\alpha nF(E_{eq}-E^{\circ\prime})/RT} = C_Re^{(1-\alpha)nF(E_{eq}-E^{\circ\prime})/RT} \tag{2.44}$$

or

$$\frac{C_O}{C_R} = e^{nF(E_{eq}-E^{\circ\prime})/RT} \tag{2.45}$$

and

$$2.303 \log \frac{C_O}{C_R} = \frac{nF(E_{eq} - E^{\circ\prime})}{RT} \tag{2.46}$$

Happily then, in agreement with Nernst,

$$E_{eq} = E^{\circ\prime} + 2.303\left(\frac{RT}{nF}\right)\log\frac{C_O}{C_R} \tag{2.47}$$

It is important to recognize that the Nernst equation is valid only at the equilibrium condition, determined by specifying $i_{net} = 0$. Most electrochemical techniques (chronoamperometry, chronocoulometry, voltammetry, etc.) involve nonequilibrium conditions and therefore cannot be expected to exhibit a Nernstian response unless the rates are very fast and equilibrium is quickly reestablished at the surface.

To derive the equation for $\underline{i}_0$ (and obtain the Butler-Volmer equation), substitute $E^{\circ\prime} = E_{eq} - (RT/nF) \ln (C_O/C_R)$ from the Nernst equation into equations 2.41 and 2.42.

$$\underline{i}_f = nFk_{s,h} C_O \exp\left(\frac{-\alpha nF}{RT}E + \frac{\alpha nF}{RT}E_{eq} - \frac{\alpha nF}{RT}\frac{RT}{nF}\ln\frac{C_O}{C_R}\right) \tag{2.48}$$

$$= nFk_{s,h}C_O \exp\left[\frac{-\alpha nF}{RT}(E - E_{eq}) + \ln\left(\frac{C_O}{C_R}\right)^{-\alpha}\right] \tag{2.49}$$

$$= nFk_{s,h} C_O^{(1-\alpha)} C_R^{\alpha} e^{-\alpha nF\eta/RT} \tag{2.50}$$

where the overpotential $\eta = E - E_{eq}$. Similarly,

$$\underline{i}_b = nFk_{s,h} C_O^{(1-\alpha)} C_R^{\alpha} e^{(1-\alpha)nF\eta/RT} \tag{2.51}$$

At $E = E_{eq}$, $\eta = 0$,

$$\underline{i}_{f(eq)} = \underline{i}_{b(eq)} = \underline{i}_0 = nFk_{s,h} C_O^{(1-\alpha)} C_R^{\alpha} \tag{2.52}$$

At $E \neq E_{eq}$, $i_{net} = A(\underline{i}_f - \underline{i}_b)$, so

$$i_{net} = nFAk_{s,h} C_O^{(1-\alpha)} C_R^{\alpha} [e^{-\alpha nF\eta/RT} - e^{(1-\alpha)nF\eta/RT}] \tag{2.53}$$

$$= \underline{i}_0 A [e^{-\alpha nF\eta/RT} - e^{(1-\alpha)nF\eta/RT}] \tag{2.54}$$

Two limiting forms of the Butler-Volmer equation of experimental interest are concerned with the current response of the system at both small and large overpotentials. For small overpotentials ($\eta < 8$ mV/n), the exponential terms may be linearized (remember that $e^{-x} = 1 - x$ for small x), so that

$$i_{net} = i_0 [(1 - \alpha nf\eta) - (1 - nf\eta - \alpha nf\eta)] \tag{2.55}$$

or

$$i_{net} = i_0 nf\eta \tag{2.56}$$

Under this small-amplitude condition, current is directly proportional to overpotential, as shown in Figure 2.16. By analogy with Ohm's law ($i = E/R = EY$), $i_0nf$ defines a transfer admittance and $1/i_0nf$ an impedance. The latter is often called the *faradaic resistance*, meaning that near the equilibrium potential, an electrochemical reaction gives a linear current versus potential curve as shown later, with zero current intercept and slope equal to the transfer admittance. Systems with higher admittances demonstrate faster heterogeneous electron transfer than those with lower values for $i_0nf$. The other limiting form of the Butler-Volmer equation deals with large overpotentials ($\eta > 120$ mV/n), where it may be assumed that the rate of one of the reactions becomes negligible. For example,

$$i_{net} = i_0 \, e^{-\alpha nf\eta} \tag{2.57}$$

or

$$\ln i_{net} = \ln i_0 - \alpha nf\eta \tag{2.58}$$

which is known as the Tafel equation. Differentiation gives

$$\frac{d \ln i_{net}}{d\eta} = -\alpha f n \tag{2.59}$$

Thus for large amplitudes, the current is logarithmically related to overpotential as shown in Figure 2.17. Tafel plots (Fig. 2.17) are frequently employed by physical electrochemists to determine exchange currents and transfer coefficients. There are many other ways to obtain these parameters experimentally, but such numbers are rarely of interest to the analytical chemist. As we will see later, the rate of the heterogeneous electron transfer *relative* to other controlling factors (e.g., diffusion and coupled chemical reactions) is of critical importance to most experiments.

The complete Eyring equation (good for all values of $\eta$) is plotted in Figure 2.18, illustrating that the net current is the sum of two opposing exponen-

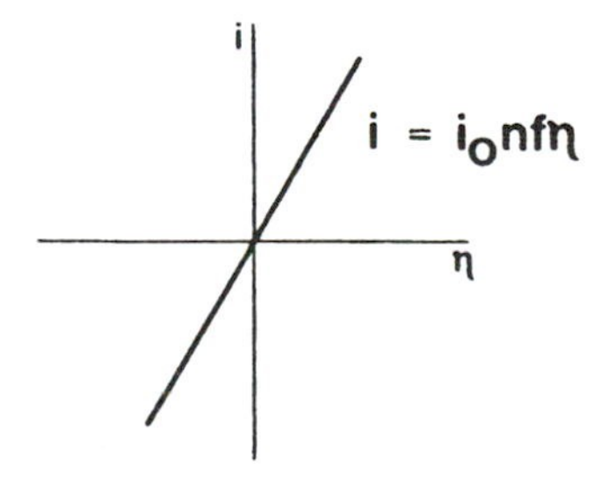


**Figure 2.16** Ohmic behavior of heterogeneous electron transfer at low overpotentials.

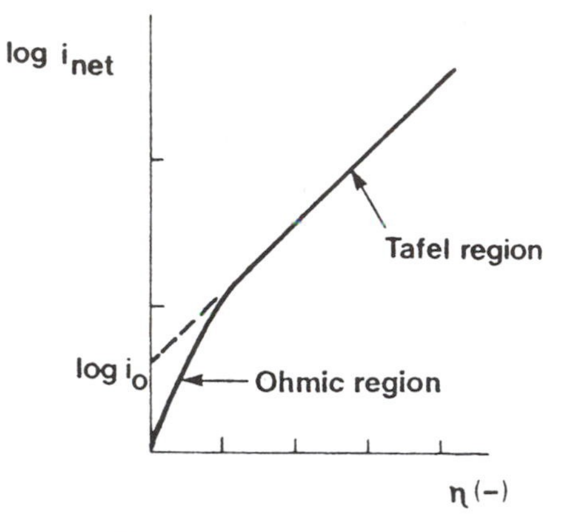


**Figure 2.17** Tafel plot.

tial branches, one for the cathodic (reduction) reaction and one for the anodic (oxidation) reaction. Notice that the $i_{net}$ curve passes through the origin where $\eta = 0$, since the partial currents are equal and opposite at this point. This is where the electric field at the interface is such that $\Delta G^{\ddagger}$ is the same for both the forward and reverse reactions. This does not imply that there is no electric field!

Once again, it is important to recognize that these current-voltage curves pertain only to the situation in which the activation parameters of the electron transfer itself are controlling. This circumstance is unusual for most analytical electrochemical techniques, where it is not desirable to be rate-limited by the actual exchange of electrons at the surface.

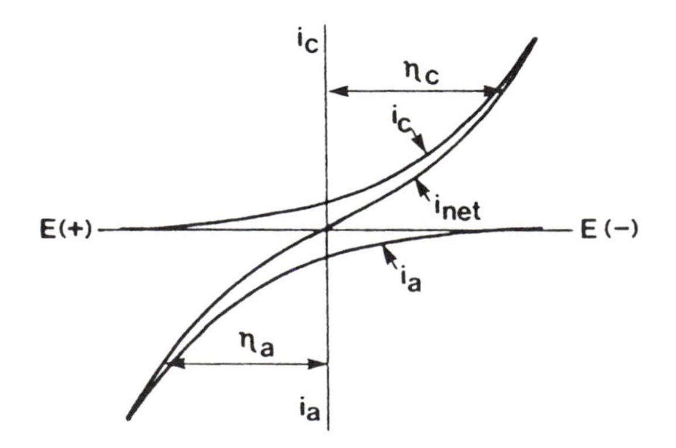


**Figure 2.18** Example of current-potential curve assuming that electron transfer is rate-limiting and concentrations of O and R are equal.

At the same time, it is significant that the rate constants for electrochemical reactions are an exponential function of the potential. Thus the electrochemist has a great deal of control over kinetics. The exponential potential dependence also means that there is not a sharp threshold for electron transfer, so that the resolution between different chemical species is not nearly as good as one would hope. The sharpness of transitions in atomic (gas-phase) spectroscopy or m/e discrimination in mass spectrometry is so much greater than in electrochemistry that the latter must be considered a low-resolution technique by comparison. A great many compounds will not exchange electrons with a metal in ionic solution due to the limited accessible range of electron free energy. Reaction of the medium and/or electrode material itself limits the available range to ca. 2–5 eV, depending on the specific medium and electrode chosen. Many practical examples of this limitation will be described in later chapters.

## V. COUPLED CHEMICAL REACTIONS AND CHEMICAL REVERSIBILITY

One of the most intriguing aspects of electrochemistry involves the homogeneous chemical reactions that often accompany heterogeneous electron-transfer processes occurring at the electrode-solution interface. The addition or removal of an electron from a molecule generates a new redox state, which can be chemically reactive. A variety of mechanisms, some of which involve complicated sequences of electrode and chemical reactions, have been characterized. Several of the more common mechanisms with examples of applicable chemical systems are described next. More examples are given in Chaps. 21 and 23.

One of the simplest electrode reactions is the *EC mechanism* (also called a *following chemical reaction*) in which the electrogenerated species (R) rearranges or reacts with some other solution component (Z) at a rate characterized by the rate constant k. The EC mechanism is summarized by the following reaction sequence, in which the labels E and C identify the heterogeneous electron-transfer reaction (electrode reaction) and the subsequent homogeneous solution reaction (chemical reaction), respectively:

$$\text{Electrode reaction, E: } \mathrm{O} + \mathrm{e} \rightleftharpoons \mathrm{R} \tag{2.60a}$$

$$\text{Chemical reaction, C: } \mathrm{R} + \mathrm{Z} \xrightarrow{k} \text{product(s)} \tag{2.60b}$$

An example of the EC mechanism is the oxidation of ascorbic acid [1]:

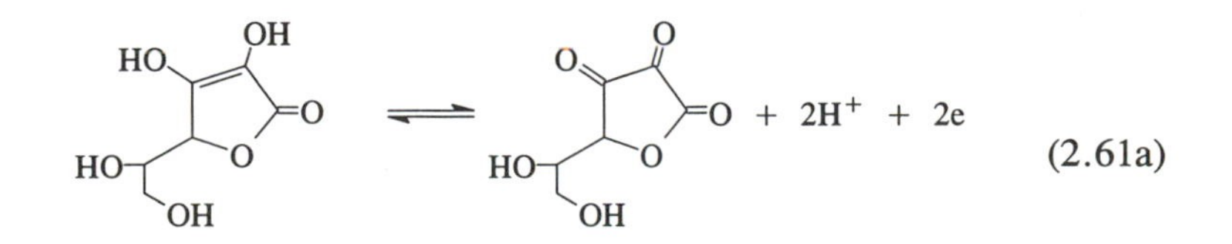


(2.61a)

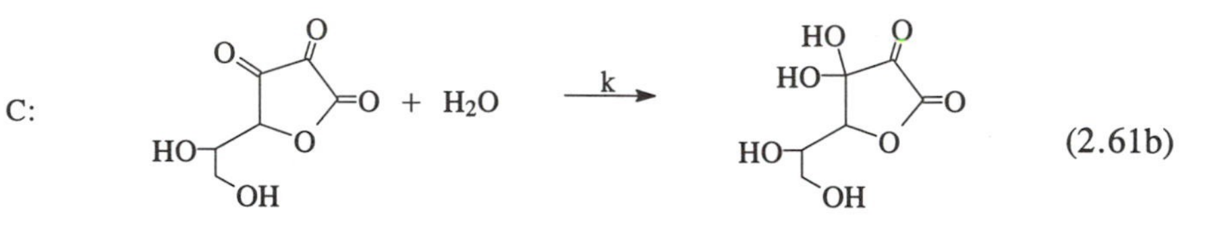


Other examples of the EC mechanism include nucleophilic addition of Z (such as $H_2O$ or $CN^-$) to electrogenerated organic radical cations, ligand exchange reactions in the case of coordination compounds, and redox reactions between R and Z (or O and Z in the case of a reduction).

The *catalytic regeneration* mechanism is a variation of the EC mechanism in which the initial electroactive species is regenerated by the homogeneous chemical reaction, as follows:

An interesting aspect of this mechanism is the implicit unreactivity of Z at the electrode, although it reacts with electrogenerated O. This situation has been frequently encountered in the reaction of certain types of biological materials such as heme proteins. An example is the reduction of cytochrome c by electrogenerated methyl viologen cation radical, $MV^{\dot{+}}$ [2]:

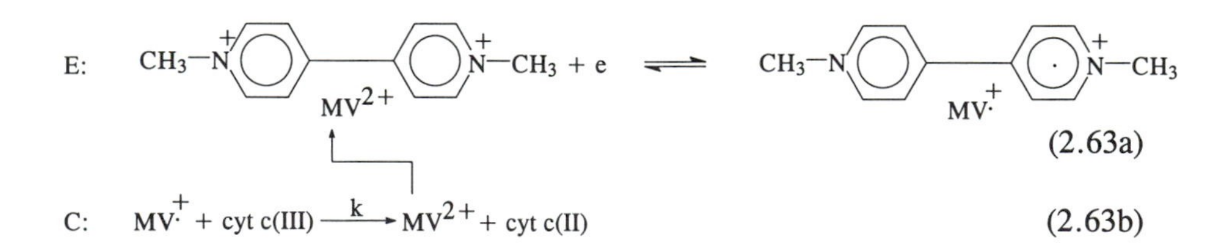


Cytochrome c contains a heme iron, which is reduced by $MV^{\dot{+}}$ from Fe(III) to Fe(II). The unreactivity of cytochrome c at the electrode is possible due to physical isolation of the iron redox center from the electrode by the surrounding protein structure.

A more complicated mechanism involving regeneration of starting material is the *ECC* or *half-regeneration* mechanism:

$$\text{E:} \quad R \rightleftharpoons O + e \tag{2.64a}$$

$$\text{C:} \quad O + Z \longrightarrow OZ \tag{2.64b}$$

$$\text{C:} \quad OZ + O \longrightarrow OZ^{+} + R \tag{2.64c}$$

The oxidation of 9,10-diphenylanthracene (DPA) in acetonitrile in the presence of nucleophiles such as water and pyridine (Py) has been shown to follow this mechanism [3]:

$$\text{E:} \quad DPA \rightleftharpoons DPA^{\dot{+}} + e \tag{2.65a}$$

$$\text{C:} \quad DPA^{\dot{+}} + Py \xrightarrow{k_1} DPA(Py)^{\dot{+}} \tag{2.65b}$$

$$\text{C:} \quad DPA(Py)^{\dot{+}} + DPA^{\dot{+}} \xrightarrow{k_2} DPA(Py)^{2+} + DPA \tag{2.65c}$$

$DPA(Py)^{2+}$ → further reaction

The *ECE mechanism* involves electrochemical generation of a species that then reacts with some other component in solution. The product of this reaction is more easily oxidized/reduced than the starting material and so is immediately electrolyzed. The sequence of reactions can be summarized as follows:

$$\text{E:} \quad R \rightleftharpoons O + e \tag{2.66a}$$

$$\text{C:} \quad O + Z \xrightarrow{k} OZ \tag{2.66b}$$

$$\text{E:} \quad OZ \longrightarrow OZ^{+} + e \tag{2.66c}$$

This mechanism is encountered in the oxidation of organic compounds in the presence of a nucleophile. A typical example is the oxidation of a hydroquinone in the presence of a thiol [4]:

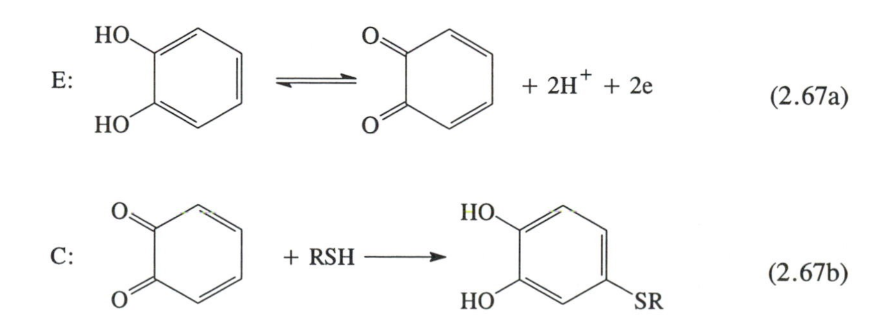

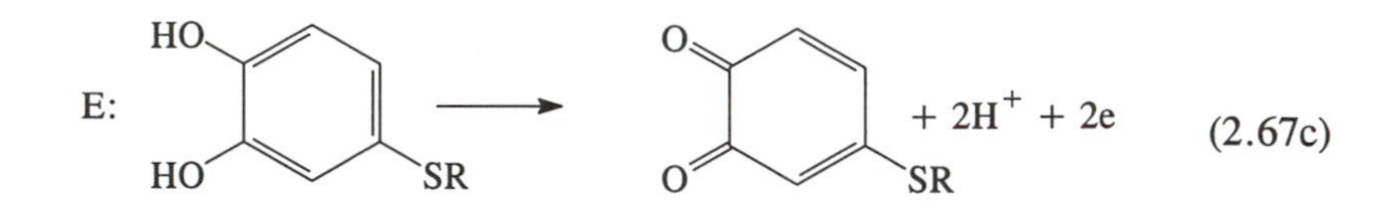


Another variation of an ECE mechanism involves a reproportionation redox reaction:

E: $R \rightleftharpoons O^{2+} + 2e$ (2.68a)

C: $R + O^{2+} \rightleftharpoons 2O^{+}$ (2.68b)

E: $O^{+} \rightleftharpoons O^{2+} + e$ (2.68c)

This mechanism can be found in the electrochemistry of compounds with two or more oxidation states. Oxidation of the tantalum cluster compound $Ta_6Br_{12}{}^{2+}$ is a typical case [5]:

E: $Ta_6Br_{12}^{2+} \rightleftharpoons Ta_6Br_{12}^{4+} + 2e$ (2.69a)

C: $Ta_6Br_{12}^{2+} + Ta_6Br_{12}^{4+} \rightleftharpoons 2Ta_6Br_{12}^{3+}$ (2.69b)

E: $Ta_6Br_{12}^{3+} \rightleftharpoons Ta_6Br_{12}^{4+} + e$ (2.69c)

In the *CE mechanism* (or preceding chemical reaction) the electroactive species is generated from an electroinactive species by a chemical reaction:

C: $A \xrightarrow{k} O$ (2.70a)

E: $O + e \rightleftharpoons R$ (2.70b)

Thus the rate of electrolysis of O is determined by the magnitude of k in concert with the diffusion process. Metal complexes such as cadmium nitrilotriacetic acid $Cd(NTA)^-$, in which the metal complex first dissociates to form a hydrated metal ion that then undergoes reduction, fall into this category [6]:

C: $Cd(NTA)^- \rightleftharpoons Cd^{2+} + NTA^{3-}$ (2.71a)

E: $Cd^{2+} + 2e \rightleftharpoons Cd$ (2.71b)

The interest in characterization of electrode mechanisms has motivated the development of a multitude of electrochemical techniques. These techniques enable the sequence of reactions to be determined and the rate constant(s) of the homogeneous chemical reactions to be measured. Such information is deduced from the effect of the coupled chemical reactions on the response signal to a particular excitation signal that is impressed on the electrode. (This is a subject of considerable discussion in subsequent chapters.) Spectroscopic techniques (UV-visible-IR, NMR, ESR, Raman) have been effectively coupled with electrochemistry to enable monitoring of homogeneous chemical reactions. Intermediates and products can sometimes be identified from their spectra.

Electrochemistry has proven to be a valuable technique for generating reactive oxidation states and studying the attendant solution chemistry of such electrogenerated species. The ease of oxygen removal from electrochemical cells greatly facilitates the study of oxygen-sensitive species that are electrogenerated. Small quantities of valuable material can be studied with ease. Very rapid reactions can be monitored, since some electrochemical techniques are capable of measuring reactions up to the limit of diffusion-controlled rates. Electrochemistry often has an additional advantage over traditional chemical approaches in that the solution is not complicated by a reagent added to generate the redox state of interest.

## VI. LIQUID-SOLID ADSORPTION

### A. Types of Adsorption

Many species dissolved in solution exhibit a tendency to adsorb on the electrode surface, a phenomenon that can markedly affect the results of electrochemical experiments. For example, the course of an electrode reaction can be altered, or the rate of electron exchange enhanced or virtually stopped. Adsorption is responsible for much "unusual" electrochemical behavior and is frequently blamed for unexplained results. Thus it is important for the chemist using electrochemical techniques to recognize phenomena that are attributable to adsorption and to realize which techniques are useful for studying adsorbed species.

Adsorption can be considered in terms of the equilibrium (after Anson [7]) in Equation 2.72,

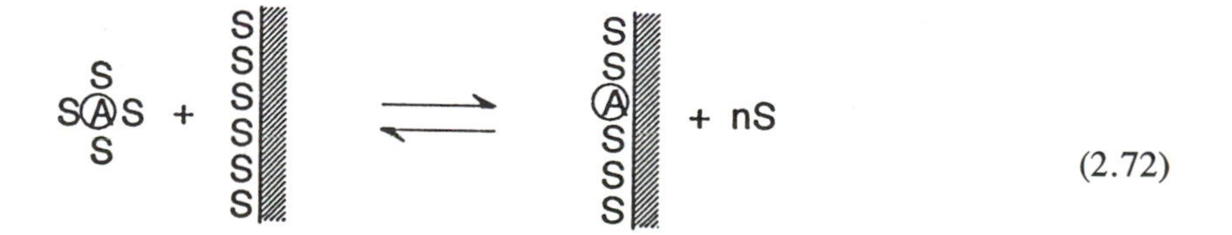


(2.72)

where the adsorbing molecule or ion, A, loses some or all of its solvate molecules, S, and displaces oriented solvent molecules from the electrode surface so that contact between electrode and adsorbate is made. The driving forces for Reaction 2.72 are varied: the hydrophobicity of the adsorbate, the electrostatic attraction between an ionic adsorbate and a charged electrode, and the adsorbate-electrode bonding. Anson has classified adsorbing species into five categories [7].

*Class 1A* consists of simple inorganic anions and cations that are so weakly solvated that electrostatic attraction between ionic charge and an oppositely charged electrode surface pulls Reaction 2.72 to the right. Examples of class 1A ions are $ClO_4^-$, $NO_3^-$, $H_2PO_4^-$, $PF_6^-$, $Cs^+$, and $R_4N^+$. The hydrophobic character of some class 1A anions is attributed to a tendency to disrupt the local solvent structure when they are dissolved in water.

*Class 1B* ions adsorb to a much greater extent than do class 1A ions because the attraction between adsorbate and electrode extends beyond simple coulombic forces to covalent bonding with the electrode via donation of adsorbate electrons to orbitals on the electrode surface. Examples of class 1B ions, in order of increasing strength of interaction with the electrode, are the anions $Cl^- < Br^- < I^-$ ; $SO_4^{2-} << S_2O_3^{2-}$ ; $NCO^- << NCS^-$.

Most cations are too strongly solvated in water to follow Reaction 2.72 and contact the electrode directly, and therefore are not specifically adsorbed.

*Class II* consists of neutral organic molecules that adsorb on the electrode surface primarily because of hydrophobicity in aqueous solutions. Generally, the less soluble a molecule is, the more strongly it adsorbs. Bonding between the electrode surface and $\pi$ electrons and nitrogen lone-pair electrons of the molecule can also enhance adsorption. Class II molecules can be found in virtually every category of organic compound. Typical examples are camphor, *n*-butanol, and quinoline.

Figure 2.19 shows the dependence of adsorption of class 1A, 1B, and II species on the electronic charge density of the electrode. Adsorption of anions and cations is enhanced by electrostatic attraction with an oppositely charged electrode. The greater adsorption of 1B anions compared to 1A anions is evident. Neutral organic molecules adsorb to the greatest extent on an uncharged electrode. As the electrode becomes charged (positively or negatively), the neutral molecules are displaced by oriented water molecules whose dipole charges become attracted to the electrode, or by specifically adsorbed anions (for a positively charged electrode). Some organic molecules have been shown to orient in a specific geometry on the electrode surface, and some exhibit more than one orientation depending on the electrode potential and the solution concentration [8].

Class 1 and II species constitute the classical examples of adsorption that are typically encountered by the electrochemist. The following classes have been recognized more recently and are quite different from the classical cases.

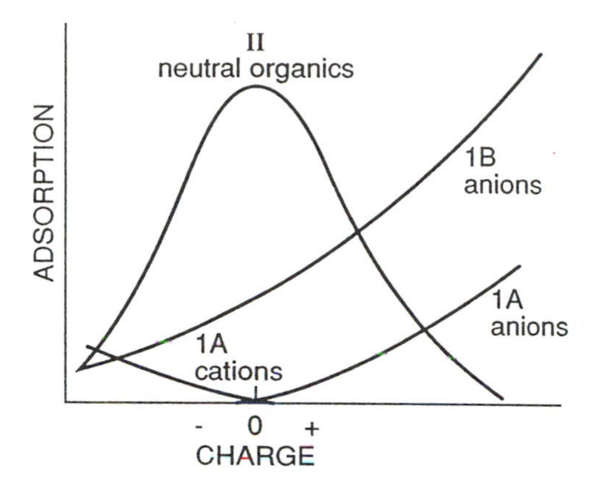


**Figure 2.19** Dependence of adsorption of class 1A, 1B, and II adsorbates on electrode charge. [From Ref. 8, adapted with permission.]

*Class III* adsorbates are cations such as Zn(II), Cd(II), Tl(I), and Pb(II), which do not adsorb when in solution with nonadsorbing anions but which exhibit an "induced adsorption" by strongly adsorbing class IB anions such as $I^-$. Such induced adsorption involves the formation of surface complexes or crystalline bilayers with the adsorbed anion.

*Class IV* adsorbates are transition metal cations whose adsorption is induced by complexation with $NCS^-$ and $N_3^-$ anions, both of which adsorb strongly on mercury.

*Class V* adsorbates exhibit metal-metal bonding between an adsorbed complex such as $Ru(en)_2^+$ and a mercury electrode.

Electrodes can become coated with interfering films via two processes that are quite different from the classes of adsorption just described.

1. Rapid polymerization of electrochemically generated species such as organic radicals can cover an electrode with a polymeric film. Such films are sometimes impenetrable and difficult to remove, which results in passivation of the electrode surface. A typical case is the oxidation of 1,2-diaminobenzene [9]. The modification of electrode properties by coating with thin polymer films is currently an area of active investigation [10–12].
2. The oxidation of an electrode such as mercury in the presence of an anion such as bromide forms a film of insoluble $Hg_2Br_2$ on the electrode surface. The formation of analogous films with sulfur-containing com-

pounds causes substantial difficulties when mercury electrodes are used in biological media, since the films tend to passivate the electrode surface.

Adsorption has been used as a means of purposely modifying the electrode surface to form a "chemically modified electrode" [10]. This topic has received such attention since the first edition of this book that a separate chapter on modified electrodes is included in this edition (Chap. 13). An example is the self-assembly of a mixed thiol monolayer on a gold electrode such as the one illustrated in Figure 2.20 [11].

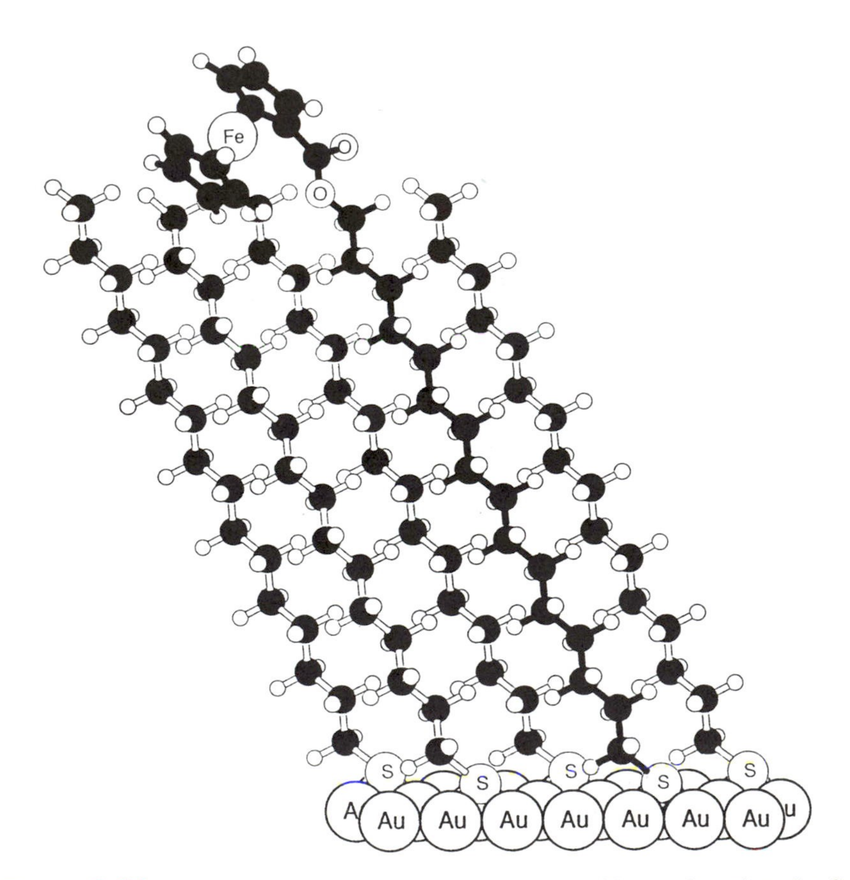


**Figure 2.20** Inferred structure of the monolayer formed by coadsorption of a ferrocene-terminated alkane thiol (highlighted with black bonds) and an unsubstituted alkane thiol on Au(111). [From Ref. 11, with permission. Copyright 1991 American Association for the Advancement of Science.]

## B. Effects of Adsorption

The effects of adsorption on an electrochemical response signal are varied, ranging from complete passivation of the electrode, to catalytic enhancement of a redox process, to the appearance of new electrochemical response signals. Commonly encountered situations are dealt with briefly below. The reader is referred to the works by Bauer [13] and Reilley and Stumm [14] for a more detailed discussion.

*Inhibition of Electrode Processes*

A common phenomenon of adsorption is physical blocking of the electrode surface by the adsorbate, forcing a redox reactant to penetrate an adsorbed film before exchanging electrons with the electrode. This is particularly characteristic of strongly adsorbed organic molecules. The effect on a redox process can vary from mild distortion of the electrochemical signal to complete inhibition of the electrode process. A typical example is the effect of adsorbed camphor on the reduction of the Cu(II)-EDTA complex. Figure 2.21A shows an electrocapillary curve [a plot of the drop time of a dropping mercury electrode (Chap. 14) as a function of potential] of supporting electrolyte with and without camphor. The depression of drop time indicates adsorption of camphor in the potential range −0.2 to −1.1 V versus SCE. Figure 2.21B shows strong inhibition of the polarographic reduction (see Chap. 3, Sec. III.D, for a discussion of polarography) of copper within the potential range of camphor adsorption. The chapter by Reilley and Stumm provides more detailed descriptions of penetration of adsorbed films [14].

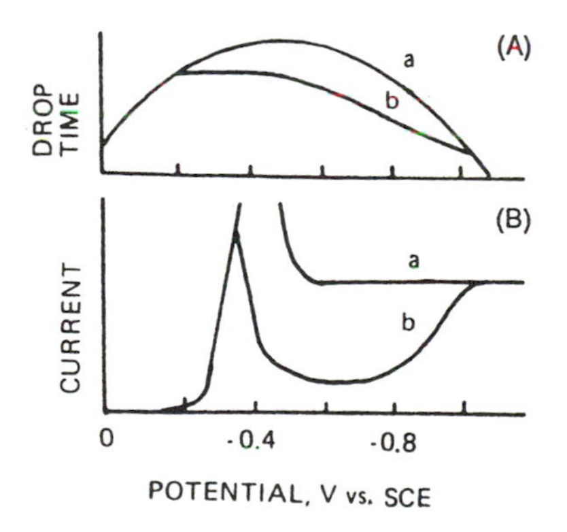


**Figure 2.21** Effect of adsorbed camphor on the electrocapillary curve of mercury and on the polarographic reduction of Cu(II)-EDTA. (A) Electrocapillary curve of camphor in acetate buffer: a, without; b, with $3 \times 10^{-4}$ M camphor. (B) Polarograms of a solution containing 0.009 M EDTA, 0.001 M Cu(II), 0.1 M acetate buffer, pH 4.5: a, without; b, with $3 \times 10^{-4}$ M camphor. [Adapted from Ref. 14, p. 91.]

*Adsorption Waves*

Electrochemical response signals attributable to adsorbed species are observable with most electroanalytical techniques. An example of an adsorption postwave is shown in Figure 2.22 for the voltammogram (see Chap. 3, Sec. III.A and III.B, for a discussion of this type of voltammetry) of $Pb^{2+}$ in the presence of a large excess of $I^-$ [15]. The first, rounded peak is due to the reduction of solution-soluble lead diffusing to the electrode; the second, sharp peak is attributed to reduction of a lead-iodide complex that is adsorbed on the electrode surface. Such adsorption waves are frequently very sharp "spikes" compared to the diffusion-controlled waves. This is due both to the immediate availability of material for reaction at the surface (i.e., mass transport is not needed) and to the limited amount of reactant present. Since the adsorbed species is stabilized by its interaction with the electrode, it is more difficult to reduce than the solution species. Hence the adsorption wave is a postwave, occurring at a more negative potential than the diffusion wave. Adsorption of the *product* of the electrode reaction typically results in a prewave, since electron transfer is then energetically easier. Such pre- and postwaves are often observable under the condition of large surface coverage by a strongly adsorbed electroactive species.

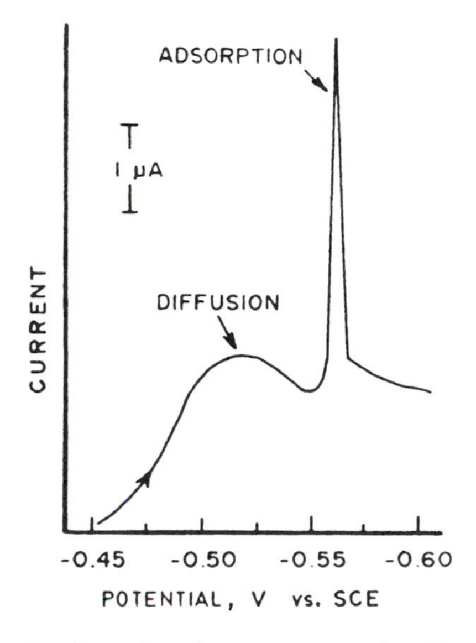


**Figure 2.22** Stationary-electrode voltammogram at a hanging mercury drop electrode for 0.495 mM $Pb^{2+}$ in 1.0 M NaI, 0.01 M HClO4, scan rate = 0.005 V/s. [From Ref. 15, reprinted with permission.]

Current spikes that are attributable to rapid adsorption or desorption of an adsorbate are sometimes observable for strongly adsorbing but electroinactive species such as camphor at a mercury electrode. The spike is a nonfaradaic current caused by the change in capacitance resulting from the sudden alteration in double-layer structure when the molecule adsorbs or desorbs.

*Potential of the OHP*

The presence of adsorbed ions can alter the potential of the OHP and thereby influence the rate of charge transfer at the electrode. For example, a specifically adsorbed anion causes the potential at the OHP to be more negative than in the absence of adsorption.

### C. Techniques

Adsorption phenomena have been studied by means of virtually every electrochemical technique, including recently developed spectroelectrochemical methods. Electrocapillary methods and measurements of double-layer capacitance have played a central role in the understanding of adsorption. AC studies have also been very useful and are very sensitive to adsorption effects. More recently, chronocoulometry (Chap. 3, Sec. II.C) has been applied effectively to the measurement of quantities of adsorbed electroactive species. The interested reader is referred to the sections that deal with these techniques for more detailed information.

## VII. CONCLUSION

In this chapter we have given an overview of several of the more important fundamental concepts that will often reappear in the following chapters. The reader wishing to have a better feeling for how these concepts influence electroanalytical experiments would do well to return to this chapter after gaining an appreciation of how the experiments are actually carried out in the laboratory.

## REFERENCES

1. S. P. Perone and W. J. Kretlow, *Anal. Chem. 38*:1760 (1966).
2. E. Steckhan and T. Kuwana, *Ber. Bunsenges. Phys. Chem. 78*:253 (1974).
3. J. F. Evans and H. N. Blount, *J. Am. Chem. Soc. 100*:4191 (1978).
4. K. T. Findley, in *The Chemistry of the Quinoid Compounds*, Part 2 (S. Patai, ed.), Wiley, New York, 1974, pp. 877–1144.
5. N. Winograd and T. Kuwana, *J. Am. Chem. Soc. 92*:224 (1970); *93*:4343 (1971).
6. J. Koryta, *Coll. Czech. Chem. Commun. 24*:3057 (1959).
7. F. C. Anson, *Acc. Chem. Res. 8*:400 (1975).

8. G. N. Salaita and A. T. Hubbard in *Molecular Design of Electrode Surfaces* (R. W. Murray, ed.), Wiley-Interscience, New York, 1992, Chapter 11.
9. A. M. Yacynych and H. B. Mark, Jr., *J. Electrochem. Soc. 123*:1346 (1976).
10. R. W. Murray, ed., *Molecular Design of Electrode Surfaces*, Wiley-Interscience, New York, 1992.
11. C. E. D. Chidsey, *Science 251*:919 (1991).
12. R. W. Murray, in *Electroanalytical Chemistry*, Vol. 13 (A. J. Bard, ed.), Marcel Dekker, New York, 1984, pp. 191–368.
13. H. H. Bauer, *Electrodics*, Georg Thieme, Stuttgart, West Germany, 1972, p. 77.
14. C. N. Reilley and W. Stumm, in *Progress in Polarography*, Vol. 1 (P. Zuman and I.M. Kolthoff, eds.), Wiley-Interscience, New York, 1962.
15. R. W. Murray and D. J. Gross, *Anal. Chem. 38*:392 (1966).

# 3

# Large-Amplitude Controlled-Potential Techniques

**William R. Heineman** *University of Cincinnati, Cincinnati, Ohio*

**Peter T. Kissinger** *Purdue University and Bioanalytical Systems, Inc., West Lafayette, Indiana*

## I. INTRODUCTION

### A. Classification of Dynamic Techniques via Large- and Small-Amplitude Excitations

In Chapter 2, the concept of faradaic impedance was introduced, whereby a heterogeneous electron transfer reaction can be modeled as an equivalent electrical circuit. In the simplest case, *at small overpotentials*, we saw how it was possible to represent the electrochemical reaction kinetics by a single resistor, since the small net current is linearly related to $\eta$ and is therefore said to be *ohmic*. Electrochemical techniques in which the current and potential fall within this linear region are referred to as *small-amplitude techniques* and will be discussed briefly in Chapter 5. Mathematical and experimental advantages are sometimes associated with such techniques, particularly in the study of the heterogeneous electron transfer process per se. In the present and following chapter we will consider *large-amplitude* controlled-potential and controlled-current techniques. These techniques operate in the nonlinear region, where current is exponentially related to overpotential. The large-amplitude techniques prove to be useful for certain analytical applications and for studies of coupled chemical reactions.

## II. POTENTIAL-STEP TECHNIQUES IN STATIONARY SOLUTION

### A. Potential-Step Excitation

The electroanalytical techniques *chronoamperometry*, *chronocoulometry*, and *chronoabsorptometry* are all based on the same excitation function of one or more potential steps that are applied to an electrode immersed in a nonstirred solution. The system response is thus identical for all three techniques. They differ only in the data domain of the monitored response. Consequently, the common excitation aspect is dealt with here, whereas each monitored response is considered individually in subsequent sections.

The important concept in these dynamic electrochemical methods is *diffusion-controlled* oxidation or reduction. Consider a planar electrode that is immersed in a quiescent solution containing O as the only electroactive species. This situation is illustrated in Figure 3.1A, where the vertical axis represents concentration and the horizontal axis represents distance from the electrode-solution interface. This interface or boundary between electrode and solution is indicated by the vertical line. The dashed line is the initial concentration of O, which is homogeneous in the solution; the initial concentration of R is zero. The excitation function that is impressed across the electrode-solution interface consists of a potential step from an initial value $E_i$, at which there is no current due to a redox process, to a second potential $E_s$, as shown in Figure 3.2. The value of this second potential is such that essentially all of O at the electrode surface is instantly reduced to R as in the generalized system of Reaction 3.1:

$$O + ne \rightarrow R \tag{3.1}$$

The electrochemical depletion of O in the solution immediately adjacent to the electrode prompts a net diffusion of O from the bulk solution into the depleted region. However, as rapidly as O diffuses to the electrode, it is converted to R. Thus the rate of loss of O and the consequent rate of formation of R are both controlled only by the rate of diffusion of O to the electrode surface under the influence of a concentration gradient. This depletion of O and accumulation of R as a function of time is graphically represented in Figure 3.1B by the concentration-distance profiles for the two species at different times following application of the potential step. It is important that the reader feel comfortable with this figure. Note that a net mass transport always occurs from a region of high concentration to a region of low concentration. In other words, the species diffuse down their respective concentration profiles. The profiles shown in the figure are for the case in which the diffusion coefficients of O and R are equal. It is also assumed that sufficient supporting electrolyte is present in solution to eliminate migration.

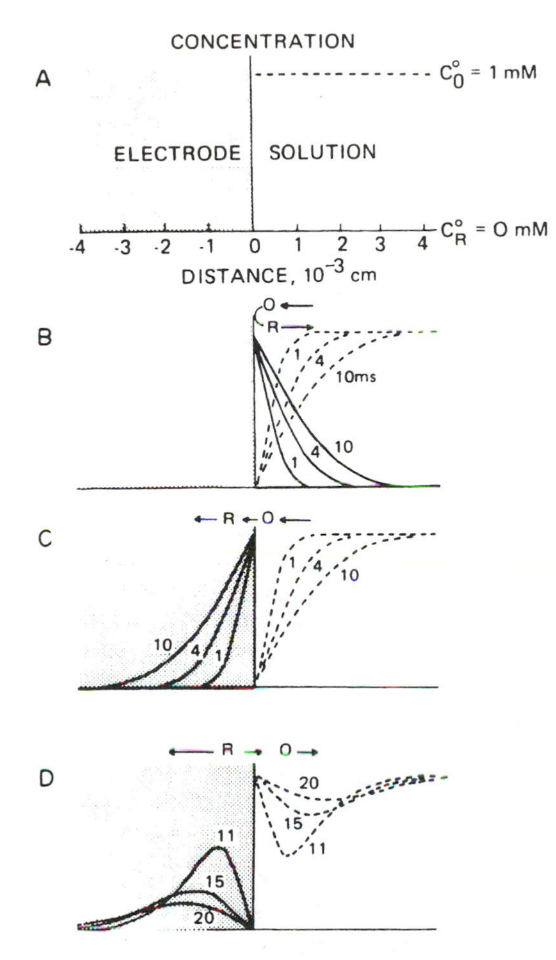


**Figure 3.1** Concentration-distance profiles during diffusion-controlled reduction of O to R at a planar electrode. $D_O = D_R$. (A) Initial conditions prior to potential step. $C^\circ_O = 1$, $C^\circ_R = 0$ mM. (B) Profiles for O (dashed line) and R (solid line) at 1, 4, and 10 ms after potential step to $E_s$. R is soluble in solution. (C) Profiles for O and R at 1, 4, and 10 ms after potential step to $E_s$. R is soluble in the electrode. (D) Potential stepped from $E_s$ to $E_f$ at $\tau$ = 10 ms so that oxidation of R to O is now diffusion-controlled. Profiles are for 11, 15, and 20 ms after step to $E_s$. R is soluble in the electrode.

Figure 3.1C illustrates what happens when the species R is soluble in the electrode material rather than in the solution. Such is the situation during the reduction of certain metal ions (e.g., $Pb^{2+}$, $Cd^{2+}$) to neutral atoms at a mercury electrode; the metal atoms dissolve readily in the mercury. Consequently,

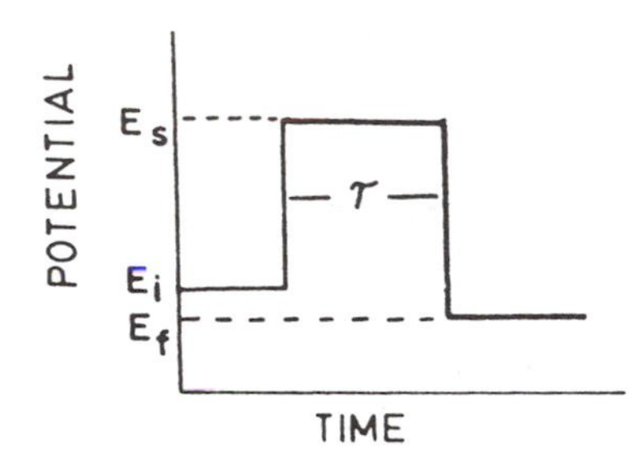


**Figure 3.2** Generalized excitation signal for potential-step techniques. Step from initial potential ($E_i$) to step potential ($E_s$) to final potential ($E_f$). $\tau$ is the duration of the potential step at $E_s$.

the concentration-distance profiles for R extend *into the electrode* rather than into the solution. The area under a particular R profile is proportional to the amount of R generated up to that specific time; the area depleted above a particular O profile is proportional to the amount of O consumed by electrolysis.

The duration of the potential step (the time interval $\tau$ of $E_s$) is usually determined by the type of information that the experimenter desires of the particular system. The time may vary from as little as 10 $\mu$s to as long as several seconds. The minimum time is limited by the ability of the potentiostat to charge the electrode, whereas the maximum time is determined by the onset of convection from vibrations or density gradients. Special shielded electrodes enable times on the order of 100 s to be reached with no apparent disturbance of the diffusional process.

The potential step $E_s$ is generally terminated by switching the potential to some final value $E_f$ at which R is now oxidized back to O. If this final potential is sufficiently positive, the concentration of R at the electrode surface is made to be essentially zero. Consequently, accumulated R now diffuses to the electrode, where it is consumed by oxidation back to O, the original species in solution. This is illustrated in Figure 3.1D. Since the concentration of R is zero both at the surface and in the bulk of the electrode, R diffuses both toward and away from the interface under the influence of both downhill sides of its concentration-distance profile. For this reason, all of the R originally generated is not oxidized back to O unless the potential is maintained for a considerable length of time.

The exact ratio of O and R concentrations at the *electrode surface* is related to the imposed potential by the Nernst equation for a reversible couple:

$$E = E^{\circ\prime} - \frac{0.059}{n} \log \frac{C_R^s}{C_O^s} \tag{3.2}$$

To impose the diffusion-controlled conversion of O to R as described earlier, the potential E impressed across the electrode-solution interface must be a value such that the ratio $C_R^s/C_O^s$ is large. Table 3.1 shows the potentials that must be applied to the electrode to achieve various ratios of $C_R^s/C_O^s$ for the case in which $E_{O,R}^{\circ\prime} = 0$. For *practical* purposes, $C_R^s/C_O^s = 1000$ is equivalent to reducing the concentration of O to zero at the electrode surface. According to Table 3.1, an applied potential of −177 mV (vs. E°′) for n = 1 (or −88.5 mV for n = 2) will achieve this ratio. Similar arguments apply to the selection of the final potential. On the reverse step, a small $C_R^s/C_O^s$ is desired to cause diffusion-controlled oxidation of R. Impressed potentials of +177 mV beyond the E°′ for n = 1 (and +88.5 mV for n = 2) correspond to $C_R^s/C_O^s = 10^{-3}$. These calculations are valid only for reversible systems. Larger potential excursions from E°′ are necessary for irreversible systems. Also, the effects of iR drop in both the electrode and solution must be considered and compensated for as described in Chapter 6.

## B. Chronoamperometry

The excitation signal in chronoamperometry is a square-wave voltage signal, as shown in Figure 3.3A, which steps the potential of the working electrode from a value at which no faradaic current occurs, $E_i$, to a potential, $E_s$, at which the surface concentration of the electroactive species is effectively zero. The potential can either be maintained at $E_s$ until the end of the experiment or be stepped to a final potential $E_f$ after some interval of time $\tau$ has passed. The latter experiment is termed double-potential-step chronoamperometry. The reader is referred to Section II.A for a detailed description of the resulting physical phenomena that occur in the vicinity of the electrode.

**Table 3.1** Relationship of $C_R^s/C_O^s$ to E for $E_{O,R}^{\circ} = 0$ for a Reversible System

| $C_R^s/C_O^s$ | n = 1 | n = 2 |
|---|---|---|
| 1/10,000 | 236 | 118 |
| 1/1000 | 177 | 88.5 |
| 1/100 | 118 | 59 |
| 1/10 | 59 | 29.5 |
| 1 | 0 | 0 |
| 10/1 | -59 | -29.5 |
| 100/1 | -118 | -59 |
| 1000/1 | -177 | -88.5 |
| 10,000/1 | -236 | -118 |

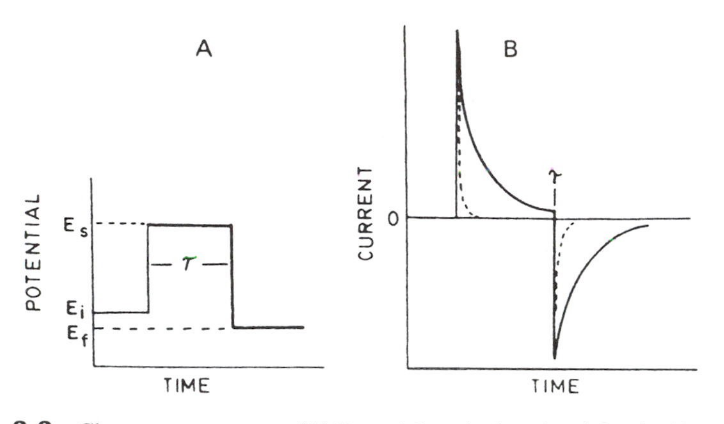


**Figure 3.3** Chronoamperometry. (A) Potential excitation signal for double potential step. (B) Current-time response signal (chronoamperogram).

Current as a function of time is the system response as well as the monitored response in chronoamperometry. A typical double-potential-step chronoamperogram is shown by the solid line in Figure 3.3B. (The dashed line shows the background response to the excitation signal for a solution containing supporting electrolyte only. This current decays rapidly when the electrode has been charged to the applied potential.) The potential step initiates an instantaneous current as a result of the reduction of O to R. The current then drops as the electrolysis proceeds.

Understanding the shape of the chronoamperogram requires consideration of concentration-distance profiles for a potential-step excitation in conjunction with Faraday's law. Faraday's law is so fundamental to dynamic electrochemical experiments that it cannot be emphasized too much. It is important to keep in mind that the charge Q passed across the interface is related to the *amount* of material that has been converted, and the current i is related to the instantaneous *rate* at which this conversion occurs. Current is physically defined as the rate of charge flow; therefore,

$$Q = nFN \tag{3.3}$$

where N is the number of moles converted, and the instantaneous current at time t is

$$i_t = \left(\frac{dQ}{dt}\right)_t = nF\left(\frac{dN}{dt}\right)_t \tag{3.4}$$

The rate of conversion, dN/dt, is directly proportional to the electrode area and to the flux of material to the electrode as described by the following equation (which is derived from Fick's first law):

$$i_t = nFAD_O\left(\frac{\partial C_O}{\partial x}\right)_{x=0,t} \qquad (3.5)$$

where

$i_t$ = current at time t, A
n = number of electrons, eq/mol
F = Faraday's constant, 96,485 C/eq
A = electrode area, $cm^2$
$D_O$ = diffusion coefficient of O, $cm^2/s$
$C_O$ = concentration of O, $mol/cm^3$ (not mol/L!)
x = distance from the electrode, cm
t = time, s

The product $D_O\,(\partial C_O/\partial x)_{x=0,t}$ is the flux or the number of moles of O diffusing per unit time to unit area of the electrode in units of $mol/(cm^2\,s)$. (The reader should perform a dimensional analysis on the equations to justify the units used.) Since $(\partial C_O/\partial x)_{x=0,t}$ is the slope of the concentration-distance profile for species O at the electrode surface at time t, the expected behavior of the current during the chronoamperometry experiment can be determined from the behavior of the slope of the profiles shown in Figure 3.1B. Examination of the profiles for O at x = 0 reveals a decrease in the slope with time, which means a decrease in current. In fact, the current decays smoothly from an expected value of ∞ at t = 0 and approaches zero with increasing time as described by the Cottrell equation for a planar electrode,

$$i_t = \frac{nFAC_O^\circ D_O^{1/2}}{\pi^{1/2}t^{1/2}} = Kt^{-1/2} \qquad (3.6)$$

where the units are identical to those in Equation 3.5. The current is inversely proportional to the square root of time.

The Cottrell equation states that the product $it^{1/2}$ should be a constant K for a diffusion-controlled reaction at a planar electrode. Deviation from this constancy can be caused by a number of situations, including nonplanar diffusion, convection in the cell, slow charging of the electrode during the potential step, and coupled chemical reactions. For each of these cases, the variation of $it^{1/2}$ when plotted against t is somewhat characteristic.

Nonplanar electrodes such as spheres and cylinders exhibit an increase in $it^{1/2}$ with increasing t. However, planar diffusion can be closely approximated at sufficiently short times. A common example is the use of the hanging mercury drop electrode. At times less than about 1 s, this electrode follows the Cottrell equation for planar diffusion with negligible variation of $it^{1/2}$ with t. At such short times, the curvature of the electrode surface is negligible relative to the depth of the diffusion layer. When time becomes sufficiently long that the effects of spherical diffusion become significant, $it^{1/2}$ begins to increase. The time restriction for the approximation of planar diffusion is determined for electrodes of different geometries by the time at which departure from $it^{1/2}$ constancy occurs. (An interesting application of this concept involves the dual-area nature of a minigrid electrode [1].) In order to retain the relative simplicity of planar diffusion mathematics, it is often desirable to operate within such a time restriction when nonplanar electrodes are used. The Cottrell equation has been appropriately modified for electrodes that are spherical, cylindrical, and so on [2].

Positive deviations of $it^{1/2}$ with increasing time can also be evidence for convection within an electrochemical cell. Convection can be caused by external vibrations or by density gradients created by the local concentration differences resulting from the electrochemical perturbation. While the influence of external vibrations can be largely eliminated by isolation of the cell with a damped table, the natural convection due to unequal densities of O and R is an unavoidable consequence of the experiment, the importance of which depends on the particular species involved. The effect of natural convection at planar electrodes is most serious when the surface is mounted vertically. It is therefore desirable to carry out electrochemical experiments at surfaces facing up or down whenever possible.

Variations in $it^{1/2}$ can signify slow attainment of the imposed potential by the electrode. Although the input voltage to a potentiostat is a square wave for a potential-step experiment, the actual charging of the electrode to the new potential may lag significantly because of iR drop within the cell or limitations in the potentiostat itself. (Details of these important problems are considered in Chapter 7.) The result of slow charging is diminished faradaic current until the correct potential (i.e., the state of diffusion control) is achieved. Consequently, $it^{1/2}$ is correspondingly diminished during this time. Thus the onset of negative $it^{1/2}$ deviation at shorter values of t is roughly characteristic of the time required to achieve diffusion control at the electrode. As a point of reference, charging times on the order of microseconds have been reported for optimized systems at conventional electrodes [3] and even shorter times at ultramicroelectrodes (Chap. 12).

In all the preceding evaluations, it is prudent to first employ a redox system that is known to be reversible and free from the complications of coupled chemical reactions and adsorption. Deficiencies in the electronics and/or cell design will thus be revealed before the critical experiments are begun. Typical examples of reversible systems are the reduction of $Tl^+$ at a mercury electrode [4] and the ferri/ferrocyanide system in 1–3 M KCl at solid electrodes [5].

Chronoamperometry has proven useful for the measurement of diffusion coefficients of electroactive species. An average value of $it^{1/2}$ over a range of time is determined at an electrode, the area of which is accurately known, and with a solution of known concentration. The diffusion coefficient can then be calculated from $it^{1/2}$ by the Cottrell equation. Although the electrode area can be physically measured, a common practice is to measure it electrochemically by performing the chronoamperometric experiment on a redox species whose diffusion coefficient is known [6]. The value of A is then calculated from $it^{1/2}$. Such an electrochemically measured surface area takes into account any unusual surface geometry that may be difficult to measure geometrically.

If the *heterogeneous* electron transfer of the redox species with the electrode itself is slow, the current after the potential step is necessarily less than in a system in which the electron transfer is rapid. This aspect of chronoamperometry has been used for the measurement of heterogeneous rate constants [7].

The behavior of $it^{1/2}$ as a function of time can be influenced substantially by the presence of chemical reactions that are coupled to the electrode process (see Chap. 2). Consequently, characteristic variations of $it^{1/2}$ versus t have been effectively utilized for the quantitative study of such homogeneous chemical reactions. The ECE reaction in which a chemical step exists between two electron transfer steps is one mechanism that has been investigated by means of chronoamperometry:

$$\text{E:}\quad \text{O} + \text{e} \rightarrow \text{R} \tag{3.7a}$$

$$\text{C:}\quad \text{R} \xrightarrow{k} \text{X} \tag{3.7b}$$

$$\text{E:}\quad \text{X} + \text{e} \rightarrow \text{P} \tag{3.7c}$$

As shown in the preceding reaction sequence, a rate-determining chemical step is interposed between the two electrode reactions. (See Chap. 2 for an explanation and an example of this mechanism.) The two dashed lines in Figure 3.4A show hypothetical chronoamperograms for the 1e reduction of O to R and for the direct 2e reduction of O to P with no kinetic complications. The solid line shows a typical chronoamperogram for an ECE mechanism. The current is intermediate between the 1e and 2e reductions, since the reduction of X to P is controlled by the rate of the chemical reaction of R to generate X. The exact position of the solid line is determined by the value of the rate constant k.

Figure 3.4B displays the data in terms of normalized i versus $t^{-1/2}$. For the 1e and 2e cases, the two straight lines are simple plots of the Cottrell equation (Eq. 3.6) for which the slope is determined by the respective n values. The shape

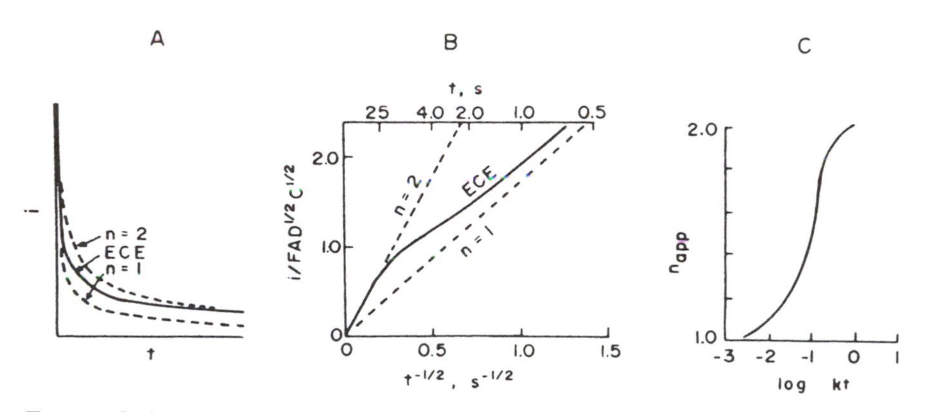


**Figure 3.4** Chronoamperometry for ECE mechanism. (A) Current-time response. (B) Data plotted as i versus $t^{-1/2}$. (C) Calculated working curve for obtaining rate constant k from experimental value for $n_{app}$. [From Ref. 8, adapted with permission.]

of the ECE curve can be understood in terms of the amount of time allowed for the chemical reaction of R. At short times (large $t^{-1/2}$) the ECE curve approaches the 1e Cottrell plot since insufficient time is allowed for R to generate a measurable amount of X. Hence on this time scale the mechanism is essentially the 1e reduction of O to R. At long times (small $t^{-1/2}$), ample time is allowed for R to generate X, which is then reduced to P. Consequently, on this time scale the mechanism is indistinguishable from a simple 2e reduction of O directly to P. The transition between these two extremes occurs during the chronoamperometry experiment when the time is comparable to the half-life of the chemical reaction. It is in this time frame that the effect of the ECE mechanism is evident.

Figure 3.4C is a calculated working curve that relates the apparent value for n ($n_{app}$) calculated from $it^{1/2}$ using the Cottrell equation to the rate constant k and the time t during the chronoamperogram. Such a curve is used to calculate the value of k. For example, the value of $n_{app}$ for a particular time t is determined and the corresponding value of kt is obtained from the working curve. Knowing t, one can then calculate k. Chronoamperometry has been applied to the study of several electrode mechanisms [8,9]. Such studies should be undertaken with an awareness that different mechanisms may exhibit similar responses [9,10]. It is always best to obtain mechanistic information from several independent sources. The reader is referred to Chapter 21 for further examples of the use of chronoamperometry in studying mechanisms of electrode processes.

Double-potential-step chronoamperometry is particularly suited for studying systems that follow EC [11] or dimerization [12] mechanisms.

## C. Chronocoulometry

Chronocoulometry is simply chronoamperometry in which the current is integrated so that the monitored response is now charge, Q [13–15]. As such, the excitation is a potential step, as shown in Figure 3.5A, the details of which are dealt with in Section II.A. It should be emphasized that the system response to this excitation is still current as explained in Section II.B. However, by integrating the current and presenting the charge as a function of time (Fig. 3.5B), some of the information obtained in the current response becomes more easily extractable. Although the integration of the current could be achieved by manually integrating a chronoamperometric i-t curve, the task is usually accomplished by analog integration with an operational amplifier or by digital integration following analog-to-digital conversion of the data.

The measured charge represents contributions from three possible *sources*:

1. Electrolysis of electroactive species in solution at a rate that is controlled by diffusion to the electrode
2. Electrolysis of electroactive species that is adsorbed on the electrode surface
3. Charging of the electrode-electrolyte double-layer capacitance to the new potential

In a simplified mathematical treatment, the three sinks of charge are considered separately as additive functions,

$$\text{Total charge} = \text{diffusing component} + \text{adsorbed component} + \text{double-layer charging} \tag{3.8a}$$

$$Q_{total} = Q_{diff} + Q_{ads} + Q_{dl} \tag{3.8b}$$

$$Q_{total} = \int_0^t i\,dt = \frac{2nFA\,C_O^\circ D_O^{1/2} t^{1/2}}{\pi^{1/2}} + nFA\Gamma_O + Q_{dl} \tag{3.8c}$$

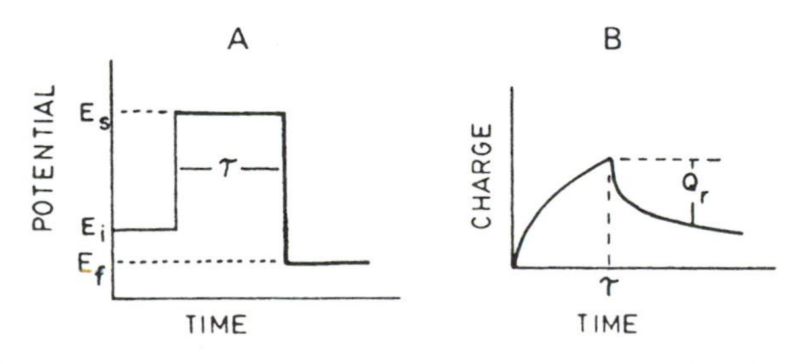


**Figure 3.5** (A) Potential excitation for double potential step. (B) Charge-time monitored response (chronocoulogram).

where the total charge $Q_{total}$ and the double-layer charge $Q_{dl}$ are in coulombs, $\Gamma_O$ is the amount of adsorbed reactant in mol/cm$^2$, and the other terms are identified with Equation 3.5. (The reader is referred to Chapter 2 for a pertinent discussion of adsorption and double-layer charging.) The expression for the diffusing component of charge is simply the Cottrell equation integrated with respect to time.

A typical chronocoulometric Q-t curve is illustrated in Figure 3.6A for a system in which the electroactive species O adsorbs on the electrode surface. The contributions of the three individual components of Equation 3.8b are also shown. It should be apparent that the total charge required for double-layer charging and for reducing the adsorbed O is "instantly" consumed whereas the charge required for reducing the O in solution exhibits a diffusion-controlled time dependency. This is also expressed mathematically in Equation 3.8c, which shows that a plot of Q versus $t^{1/2}$ should be a straight line with an intercept equal to ($nFA\Gamma_O + Q_{dl}$). Such a plot of Q versus $t^{1/2}$ is shown in Figure 3.6B.

That the adsorption and double-layer components can be so easily separated from the diffusional component by plotting Q versus $t^{1/2}$ is the forte of chronocoulometry. Since $Q_{dl}$ can usually be measured in a separate experiment on the supporting electrolyte alone, the contribution of the adsorbed species can be

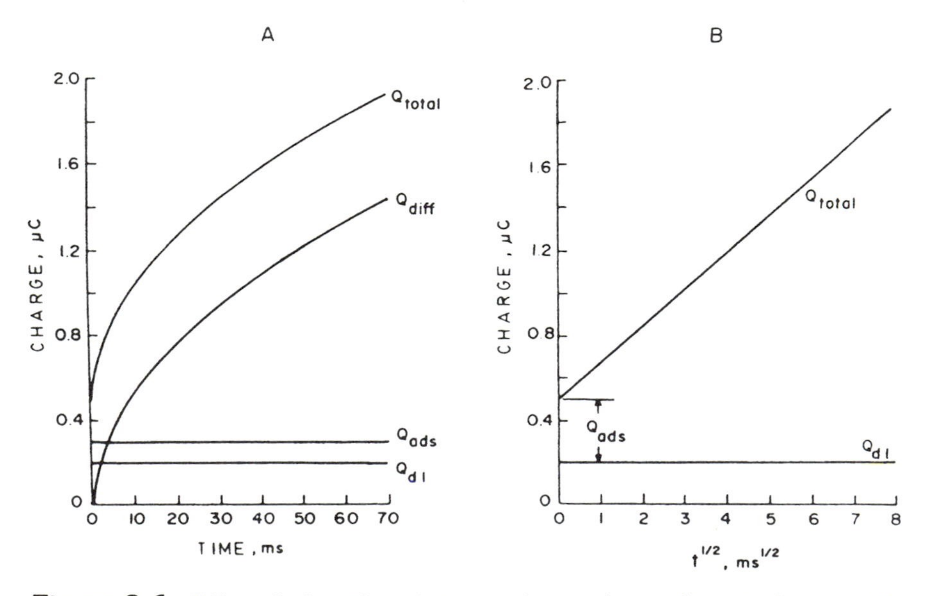


**Figure 3.6** Effect of adsorption of electroactive species on chronocoulometry. (A) Charge-time curves for individual components contributing to total charge ($Q_{total}$) as in Equation 3.8b. (B) Plots of charge versus $t^{1/2}$ for $Q_{total}$ and $Q_{dl}$.

determined and $\Gamma_O$ calculated. This is illustrated in Figure 3.6B, in which the difference between the intercepts of $Q_{total}$ and $Q_{dl}$ is the amount of charge required to reduce the adsorbed species, $Q_{ads}$. An implied assumption is that $Q_{dl}$ is the same in the presence and absence of the adsorbed species. Thus chronocoulometry is a most useful technique for determining the surface excess $\Gamma$ of an *electroactive* species. The information is also contained in a chronoamperometry experiment, but in a much less accessible form. The adsorption of many electroactive inorganic ions, metal complexes, and organic compounds at electrode surfaces has been investigated by means of chronocoulometry [15–17].

Chronocoulometry as just described involves measurement of charge after what is frequently referred to as the forward potential step. A logical extension of the method is to step the potential to its original value or some other value $E_f$ at which the redox process is reversed, and to monitor the resulting charge behavior (see Fig. 3.5) [13].

Chronocoulometry can also be used for the study of homogeneous chemical reactions that are coupled to the electrode reaction. The technique has not yet seen extensive application to this end, although the theory is quite highly developed [18].

### D. Chronoabsorptometry

The development of electrodes that exhibit optical transparency has enabled spectral observations to be made directly through the electrode simultaneously with electrochemical perturbations [19–21]. These electrodes typically consist of a very thin film of conductive material such as Pt, Au, carbon, or a semiconductor such as doped tin oxide that is deposited on a glass or quartz substrate. Miniature metal screens, minigrid electrodes in which the presence of very small holes (6–40 $\mu$m) lends transparency, have also been used. Optically transparent electrodes (OTE) and the cells that incorporate them are discussed in Chapters 9 and 11.

A beam of light can be directed perpendicularly through the transparent electrode and the solution as shown in Figure 3.7A. The electrochemical cell is now also a spectral cell (one optical face being the OTE) that is positioned in the light beam of a spectrometer. This electrode transparency enables changes in absorbance that occur in the solution adjacent to the electrode to be monitored during an electrochemical perturbation. If such optical monitoring is performed in conjunction with a potential step as the electrochemical excitation signal (Fig. 3.8A), the technique is termed *potential-step chronoabsorptometry*. The resulting absorbance-time response is illustrated by the following example. (The reader is referred to Section II.A for a discussion of the potential-step perturbation, which is most useful at this point.) Let us again assume that spe-

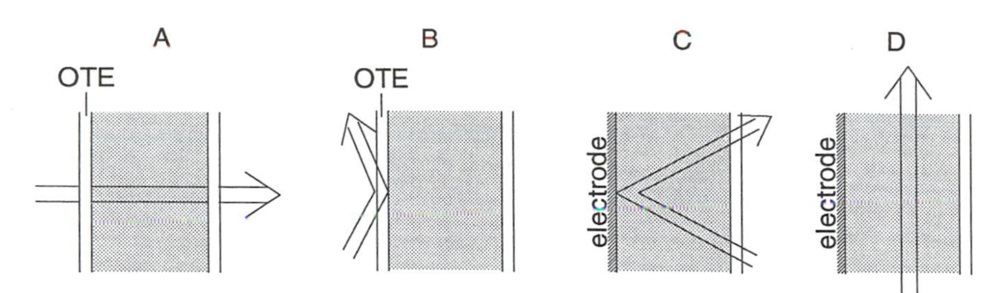


**Figure 3.7** Spectroelectrochemical techniques: (A) transmission, (B) internal reflectance, (C) specular reflectance, (D) parallel.

cies O is present in solution and is reducible to R at the electrode surface according to Equation 3.1. In this case we shall also assume that R absorbs light at the visible wavelength of 600 nm. Consequently, any R that is formed at the electrode will be detected by a change in absorbance of a 600-nm light beam passing through the electrode and the solution as in Figure 3.7A. When the potential of the transparent electrode is stepped so that the formation of R is diffusion-controlled, the resulting increase in absorbance is proportional to the accumulation of R. The amount of R that has accumulated at any time t is proportional to the area beneath the concentration-distance profile for R at that particular time, as shown in Figure 3.1B.

The actual change in absorbance is related to concentration and optical path length by Beer's law,

$$A = \varepsilon bc \tag{3.9}$$

For chronoabsorptometry this law must be couched in a form that takes into account the continuous increase in the thickness of the diffusion layer, which is the optical cell, and the fact that the concentration of the absorbing species

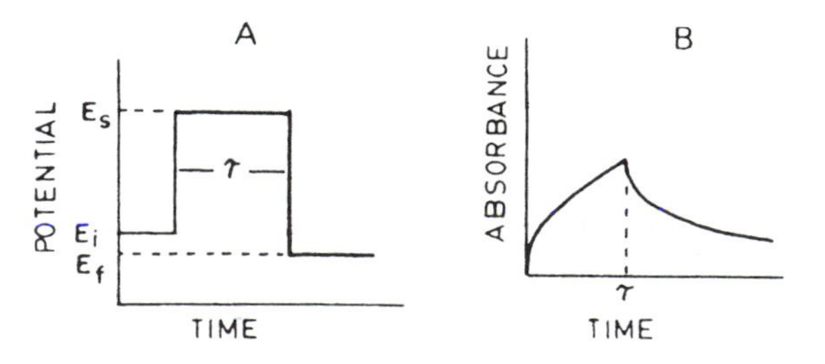


**Figure 3.8** Chronoabsorptometry. (A) Potential excitation signal for double potential step. (B) Absorbance-time monitored response (chronoabsorptogram).

is not homogeneous within this layer. The optical integration of the concentration can be expressed mathematically by Equation 3.10.

$$A_t = \varepsilon_R \int_0^{\infty} C_{R,x,t}\, dx \tag{3.10}$$

where $\varepsilon_R$ is the molar absorptivity ($M^{-1}$ $cm^{-1}$) of R at the wavelength being monitored, and $C_{R,x,t}$ is the concentration-distance profile of R, which is changing during the electrogeneration. This equation is a more general form of Beer's law. Substitution of the diffusion equations that describe the behavior of $C_{R,x,t}$ gives the following absorbance-time relationship:

$$A_t = \frac{2}{\pi^{1/2}} \varepsilon_R C_O^{\circ} D_O^{1/2} t^{1/2} \tag{3.11}$$

where $C_O^{\circ}$ and $D_O$ are the molar concentration and diffusion coefficient of O, and t is time in seconds. The reader should note that the concentration and the diffusion coefficient are for the oxidized species O, even though it is the reduced species R that is being monitored optically; the formation of R is controlled by the diffusion of O to the electrode. A typical absorbance-time curve is illustrated in Figure 3.8B. As described by Equation 3.11, A increases at a $t^{1/2}$ rate.

It is apparent from Equation 3.11 that a plot of A versus $t^{1/2}$ is a straight line. The value for $\varepsilon_R$ can be obtained from the slope of this line since $C_O^{\circ}$ is usually known and $D_O$ can be measured by means of chronoamperometry. Thus this method is useful for obtaining molar absorptivities of electrogenerated species.

The analogy between chronoabsorptometry and chronocoulometry is evident from the $t^{1/2}$ response of both A and Q and the similarity of Equations 3.11 and 3.8c. Whereas chronoabsorptometry optically integrates the *amount of R present* at any time in accordance with Beer's law, chronocoulometry integrates the *amount of R that has been generated* at any time by integrating the current required to form R.

Chronoabsorptometry is useful for the study of homogeneous chemical reactions that involve an electrogenerated species. For example, consider the situation in which an additional species, Z, is also present in solution and is capable of a rapid homogeneous chemical reaction with R at a rate characterized by the second-order rate constant k to form products P (see Chap. 2 for a discussion of the EC mechanism).

$$\text{E: } O + e \rightleftharpoons R \tag{3.12}$$

$$\text{C: } R + Z \xrightarrow{k} P \tag{3.13}$$

In this case, the electrogenerated R will encounter Z as it diffuses away from the electrode and will react to form P. This *diffusional mixing* of two reactants (R and Z in this example) is the basis for chronoabsorptometry (and other electrochemical techniques) as a kinetic method. The homogeneous chemical reaction of R will perturb its accumulation by a magnitude that is proportional both to k and to the concentration of Z. The influence of this perturbation is apparent from the concentration-distance profiles for R, which are illustrated for a rate constant of $10^7$ L/(mol s) in Figure 3.9.

Comparison of the areas under the profiles for R in Figures 3.1B and 3.9 shows a significant decrease in the amount of R accumulating in solution in Figure 3.9 because of its reaction with Z. Note that in both cases R is being generated at the same rate. Since the absorbance response of the monitored species R is proportional to the area beneath the profile $C_{R,x,t}$ for R, this kinetic perturbation is also manifested in the absorbance-time response when R is monitored optically. The A-t curves for different rate constants are shown in Figure 3.10. It is apparent that the absorbance response is quite sensitive to the magnitude of the rate constant, k. Such a rate constant can be determined by fitting an experimental curve to calculated curves such as those in Figure 3.10 or by fitting to a dimensionless working curve [20,21]. This has been demonstrated for several types of homogeneous reactions [19–21].

The transparency of the electrode also enables spectra to be recorded of electrogenerated species as well as of any species produced as a result of a homogeneous chemical reaction. Such spectra have been recorded with rapid scanning spectrometers that are capable of recording as many as 100 or more spectra per second in the UV-visible range [22]. Spectra can be useful for structural identification of intermediate components in the reaction sequence and for

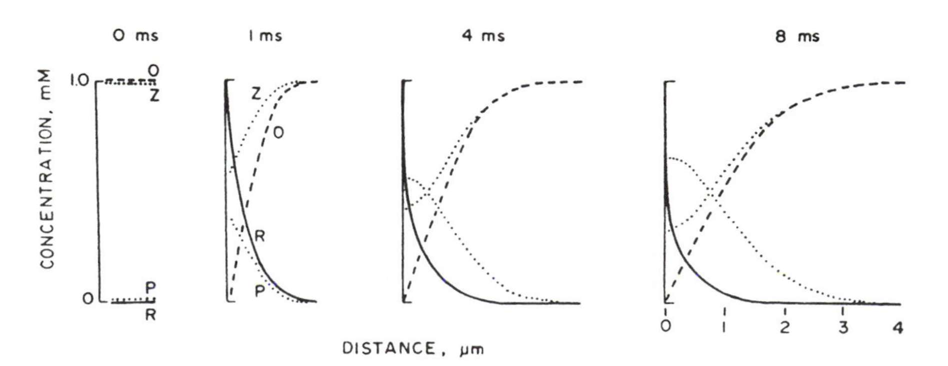


**Figure 3.9** Concentration-distance profiles for EC mechanism with k = $10^7$ L/(mol s). See Figure 3.1 legend for conditions. [Adapted from R. F. Broman, W. R. Heineman, and T. Kuwana, *Faraday Discuss. Chem. Soc. 56*:16 (1974).]

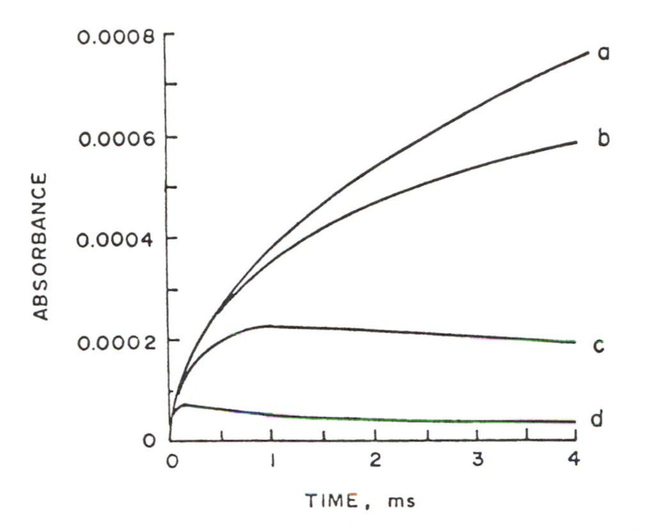


**Figure 3.10** Absorbance-time curves for transmission spectroelectrochemistry. EC mechanism for rate constants k of (a) 0, (b) $10^5$, (c) $10^6$, and (d) $10^7$ L/(mol s). R monitored optically, $D_O = 10^{-6}$ cm²/s, $\varepsilon_R = 10^4$ L/(mol cm), concentrations of A and Z = 1.0 mM. [Adapted from R. F. Broman, W. R. Heineman, and T. Kuwana, *Faraday Discuss. Chem. Soc. 56*:16 (1974).]

the determination of the wavelengths of maximum absorbance for such components so that set wavelength experiments can be performed at maximum sensitivity to measure kinetic parameters as described earlier.

Optical measurements through the electrode can also be made by *internal reflectance spectroscopy* (IRS). In this technique the light beam is introduced to the electrode at an angle of incidence $\theta_i$ as illustrated in Figure 3.7B. When this angle is greater than the critical angle $\theta_c$, the beam will be totally reflected at the electrode-solution interface as shown. However, a small standing wave component will exist within the solution itself. The penetration depth of this component into the solution is typically 2000–4000 Å for visible light. Light-absorbing species in solution interact with this evanescent wave so that spectra can be recorded by varying the wavelength of the incident beam and analyzing the reflected beam. Consequently, spectroelectrochemical experiments analogous to the types described earlier for transmission chronoabsorptometry can also be performed with IRS chronoabsorptometry. Because of the limited penetration depth of the IRS wave, the *optical* cell is a very thin layer of solution adjacent to the electrode. Thus the optical advantages of an optically transparent thin-layer cell (see Sec. II.E) are realized, but the disadvantages (high solution resistance, poor current distribution) are not. The IRS mode offers an enhanced sensitivity in absorbance change compared to transmission, and is therefore useful for very fast reactions in which the absorbance change is quite small. Signal-averaging techniques are used for these fast reactions in order to detect the small absorbance changes. Rate constants for chemical reactions that approach diffusion control have been measured by this method [19–21].

Spectroelectrochemical measurements can be made at conventional nontransparent electrodes by specular reflectance [23–25]. The optical beam is passed through the electrolyte and reflected from the electrode surface as shown in Figure 3.7C. This technique has been used effectively to study electrode mechanisms and to observe changes in the electrode surface itself.

A significant disadvantage in using strictly electrochemical techniques for studying an electrode process is the general lack of structural information about the molecules involved that can be obtained from electrochemistry. The combination of spectroscopic techniques with electrochemistry is an attractive path for overcoming this difficulty. Since the previous edition of this book, significant advances have been made in combining x-ray techniques, surface-enhanced Raman scattering, infrared and UV-visible reflectance spectroscopy, nonlinear optical methods, and Mossbauer spectroscopy with electrochemistry [25–27]. However, with the possible exception of thin-layer spectroelectrochemistry (Sec. II.E), few of the many spectroelectrochemical methods have seen general applicability. They remain within the province of the experts.

## E. Potential-Step Excitation in Thin Layers of Solution

An operational definition of thin-layer electrochemistry is that area of electrochemical endeavor in which special advantage is taken of restricting the diffusional field of electroactive species and products. Typically, the solution under study is confined to a well-defined layer, less than 0.2 mm thick, trapped between an electrode and an inert barrier, between two electrodes, or between two inert barriers with an electrode between. Diffusion under this restricted condition has been described in Chapter 2 (Sec. II.C). Solution trapped in a porous-bed electrode will have qualitatively similar electrochemical properties; however, geometric complexities make this configuration less useful for analytical purposes. The variety of electrical excitation signals applicable to thin-layer electrochemical work is large. Three reviews of the subject have appeared [28–30].

All of the principles of semi-infinite potential-step experiments discussed so far apply to thin-layer work. Some modification in the quantitative response is necessary to account for the presence of a diffusion barrier. Figure 3.11 illustrates the diffusion phenomena occurring during a chronoamperometric experiment in a thin-layer cell of typical dimensions. It will be useful to compare this figure with the semi-infinite situation depicted in Figure 3.1. Notice that the supply of reactant in the bulk solution phase is effectively infinite in Fig-

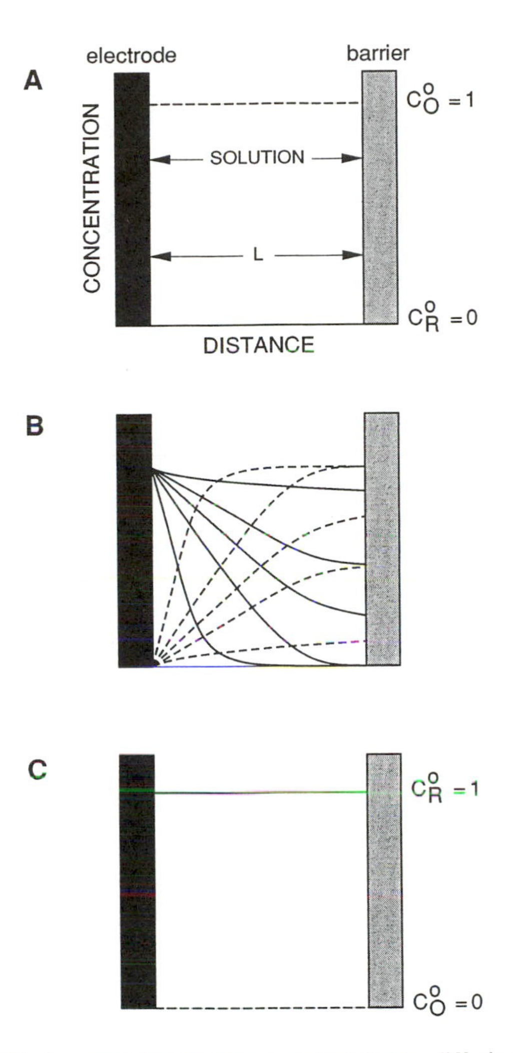


**Figure 3.11** Thin-layer system response to potential-step, diffusion-controlled reduction of O to R with $D_O = D_R$. (A) Initial condition. (B) Profiles for O (dashed lines) and R (solid line) after potential step. (C) Final condition.

ure 3.1, whereas in Figure 3.11 there is no bulk solution and all of the available reactant is readily consumed. The two experiments are, of course, identical until the concentration profiles strike the barrier. If at any time the diffusion is stepped back to a value where R $\rightarrow$ O occurs at a diffusion-controlled rate, the condition shown in Figure 3.11A would be restored. For stable species the complete interconversion of O to R and R to O could be repeated indefinitely. The reader should attempt to sketch the concentration profiles extant when the barrier plate in Figure 3.11 is replaced by another electrode identical to the first, a situation sometimes used in practice.

The measured i-t and Q-t responses to a potential step are shown in Figure 3.12 for an electrode both with and without a diffusion barrier. The mathematical description is somewhat more complex for the restricted diffusion case.

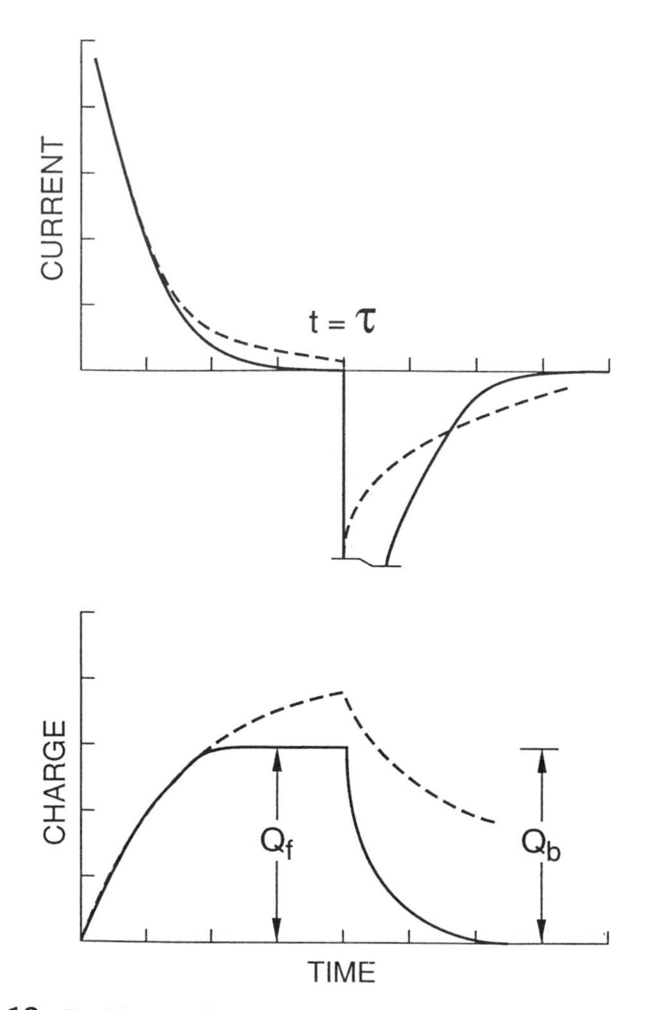


**Figure 3.12** Double-potential-step thin-layer electrochemistry. (A) Chronoamperometry. (B) Chronocoulometry. Dashed line, same experiment for semi-infinite situation.

The simple boundary value problem describing this experiment is commonly solved in elementary texts on differential equations under the guise of "heat conduction in a slab of finite width" [30,31]. Differentiation of the concentration profile, $C_{O,x,t}$, at the electrode surface immediately affords the chronoamperometric response

$$i_t = nFAD_O\left(\frac{\partial C_{O,x,t}}{\partial x}\right)_{x=0,t} = \frac{2nFAC_O^\circ D_O}{L}\sum_{k=0}^{\infty}\exp(P) \tag{3.14}$$

where $P = -D_O\,(2k + 1)^2\pi^2 t/4L^2$. Integration of this series readily provides an expression for the coulometric response

$$Q_t = \int_0^t i\,dt = nFALC_O^\circ\left[1 - \frac{8}{\pi^2}\sum_{k=0}^{\infty}\frac{1}{(2k+1)^2}\exp(P)\right] \tag{3.15}$$

In the limit as $t \to \infty$, Equation 3.15 converges to Faraday's law,

$$Q_{inf} = nFALC_O^\circ = nFVC_O^\circ = nFN \tag{3.16}$$

where V is the volume of the thin-layer cavity.

While Equations 3.14 and 3.15 provide an understanding of the time response, to date they have had no practical significance. There is no advantage to be gained from data on the rising part of a thin-layer chronocoulogram, since semi-infinite chronocoulometry can provide the same information more efficaciously. The important datum from the present technique is $Q_{inf}$, which can be used to calculate n, V, or $C_O^\circ$ when the other parameters are known. The number of electrons involved in a process is essential for elucidating mechanisms. Thin-layer coulometry is eminently suitable for this purpose. Considering that V is typically less than 10 $\mu$L, such experiments require very little material and very little time (typically a few seconds). A characteristic time constant for the theoretical response is $L^2/D$ seconds, the time at which the electrolysis will be 93% complete. This is one-half the mean time for a molecule to diffuse across L as given by the Einstein equation (Chap. 2, Sec. II.C). The challenging details of constructing useful thin-layer cells are described in Chapter 9.

A typical example of the use of thin-layer coulometry to determine an n value is for the drug chlorpromazine (CPZ). A thin-layer coulogram for the oxidation of CPZ by a potential step to 575 mV vs. SCE is shown in Figure 3.13. A coulogram is also shown for the same potential step repeated in supporting electrolyte.

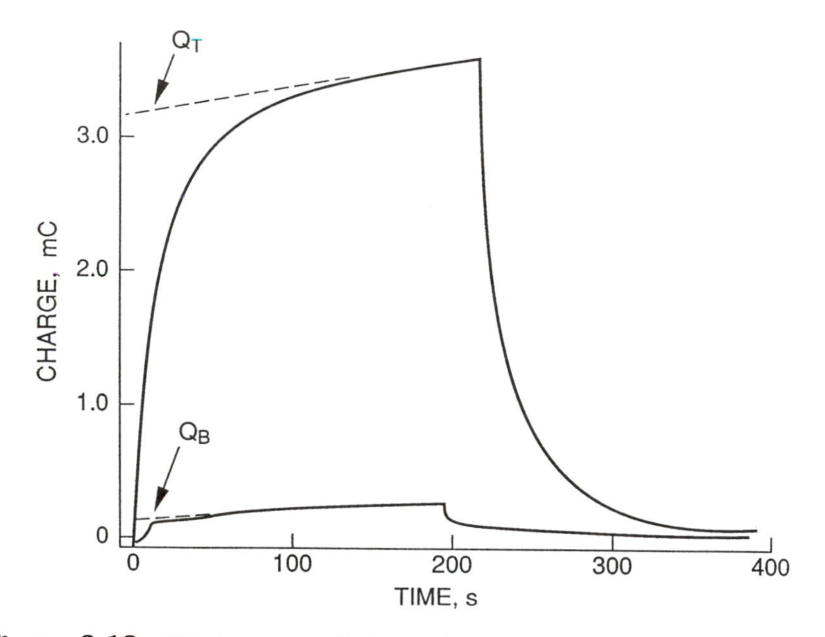


**Figure 3.13** Thin-layer controlled-potential coulometry of chlorpromazine (CPZ) oxidation in an optically transparent thin-layer cell. $E_i$ = 250 mV; $E_s$ = 575 mV; $E_f$ = 250 mV vs. SCE. $4.8 \times 10^{-4}$ M CPZ, 3 M $H_2SO_4$, and 3 M $H_2SO_4$ alone. [From T. B. Jarbawi, Ph.D. dissertation, University of Cincinnati, 1981.]

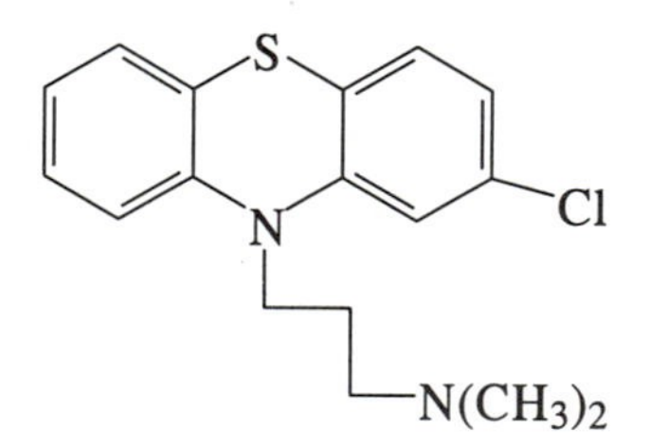


Chlorpromazine (CPZ)

The total charge $Q_T$ and the background charge $Q_B$ are measured by extrapolating the linear portion of the Q-t curve back to t = 0. The difference in $Q_T$ and $Q_B$ is the faradaic charge from which a value for n is calculated from Equation 3.16. (The thin-layer cell is precalibrated with respect to its volume V by performing a similar experiment on a species of known n and C° and calculating V.) In the case of CPZ oxidation, a calculated value of 1e suggests oxidation to $CPZ^{\dot{+}}$ [32].

Under ideal conditions, charge consumed by the double-layer capacitance and adsorbed reactants will follow the same time course as discussed earlier for ordinary chronocoulometry. Since the ratio of electrode area to solution volume is larger for thin-layer experiments, charge thus accounted for may represent a much greater proportion of the total. This fact points to an advantage of restricted diffusion experiments for studying some surface phenomena.

One area in which the thin-layer approach has been popular is spectroelectrochemistry [20]. Transparent or reflecting electrodes permit spectral observation of reaction product(s) without interference from starting material [33]. The stability of product species may be assessed either spectrally or by comparing the ratio of $Q_b$ to $Q_f$ (Fig. 3.12B) at various switching times $\tau$. In semi-infinite spectroelectrochemical experiments, the light beam must pass through a large volume of solution before reaching the diffusion layer unless, of course, IRS techniques are used (see Sec. II.D). This means that it will be very difficult to monitor *reactant,* since the concentration (and absorbance) change will be very small relative to the bulk solution phase. This is not the case in thin-layer spectroelectrochemistry, in which either reactant or product can be monitored with equal facility. Because of the difficulty in designing thin-layer electrodes that behave ideally at very short times, the semi-infinite experiment is preferred for following rapid kinetics. Thin-layer cells, however, are most effective for measuring the spectra of products and of reasonably stable intermediates using conventional or rapid scanning spectrophotometers.

An important characteristic of thin-layer cells is rapid diffusional mixing. For example, consider Figure 3.11B. If the experiment is interrupted at this point by disconnection of the working electrode, the concentrations of O and R will rapidly become homogeneous throughout the cell as illustrated in Figure 3.14. If either species is subject to a chemical reaction that is slow in relation to this concentration profile relaxation, the kinetics may be treated as for an initially homogeneous concentration of reactant. This greatly simplifies the mathematics when contrasted to the usual electrochemical experiment, where the reacting species are not uniformly distributed. Naturally, rate processes examined in this way will have to be much slower than those accessible by semi-infinite experiments.

The oxidation of 5,6-diaminouracil provides a good example of the use of thin-layer spectroelectrochemistry to determine the rate constant of an electrogenerated intermediate [34]. The reaction sequence is as follows:

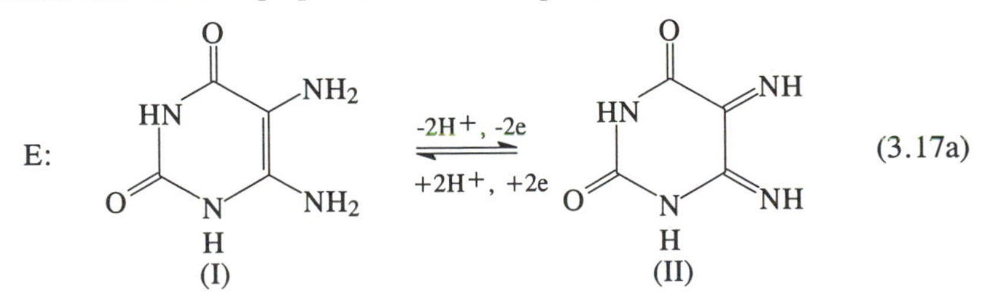

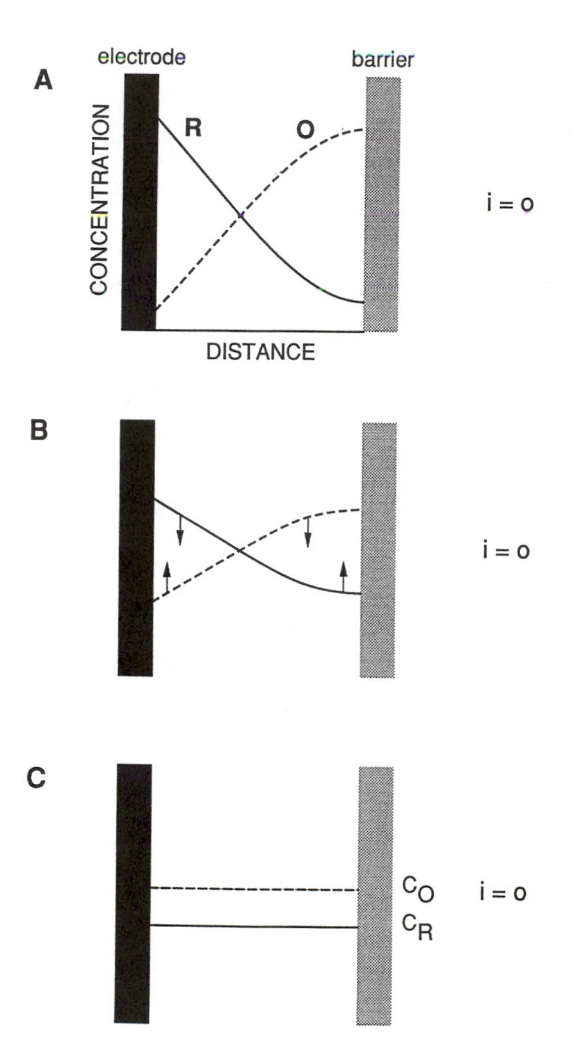


**Figure 3.14** Concentration profile relaxation in thin-layer cell. (A) Initial condition (solid line, product; dashed line, reactant). (B) Relaxation in progress. (C) Final condition.

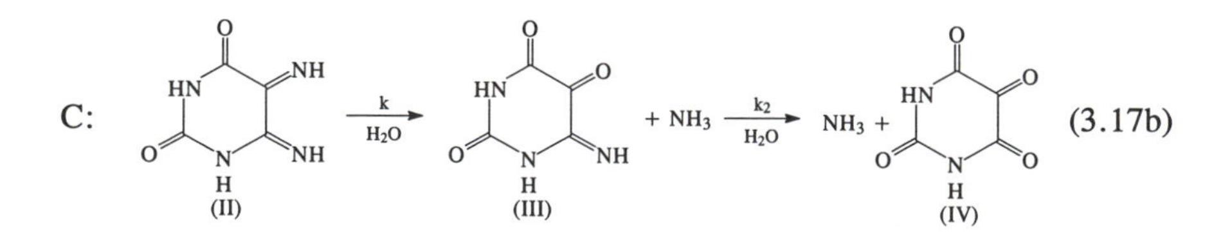

A thin-layer chronoabsorptometry experiment in which the diimine intermediate (species II) is monitored at 330 nm is shown in Figure 3.15A. The absorbance increases when the potential is stepped to 0.35 V to electrogenerate the imine and then decreases due to the hydrolysis reaction. A first-order kinetic plot of the absorbance decrease (Fig. 3.15B) gives a pseudo-first-order value for k of $7.6 \times 10^{-3}$ $s^{-1}$. This technique gains sensitivity when performed with a long optical pathlength thin-layer spectroelectrochemical cell in which the optical beam is passed parallel to the working electrode, as in Figure 3.7D [35].

Since the redox-active species in a thin-layer cell rapidly equilibrate with the potential applied to the electrode, precise values of $E^{\circ\prime}$ and n can be obtained from a spectropotentiostatic technique [36]. In this method, a series of potentials is applied to the cell and a spectrum is recorded at each potential after equilibration has been achieved. A typical example is the study of *o*-tolidine in aqueous acidic solution [36].

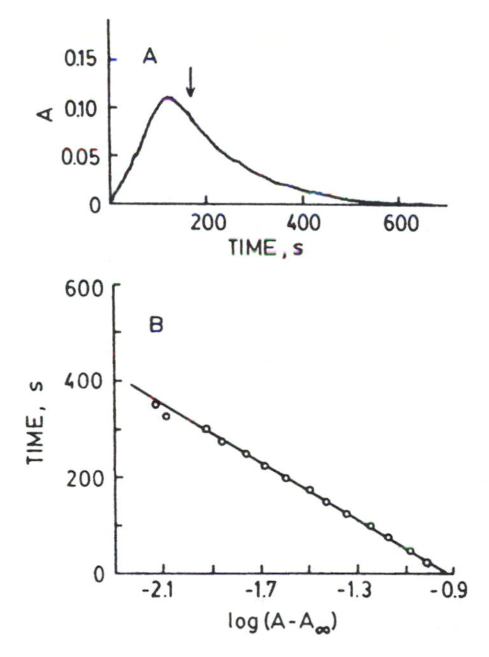


**Figure 3.15** Thin-layer chronoabsorptometry of electrogenerated reactant in an optically transparent thin-layer electrode. (A) Variation of the absorbance of the intermediate species ($A_{320\ nm}$) during the electrooxidation of 10 mM 5,6-diaminouracil in McIlvaine buffer pH 5 at 0.35 V. Arrow indicates time at which all 5,6-diaminouracil has been electrolyzed. (B) Kinetic plot of time versus the absorbance at 330 nm of intermediate species. Absorbance data were taken from curve A. Setting time = 0 at the point where the concentration of 5,6-diaminouracil was zero. [From Ref. 34.]

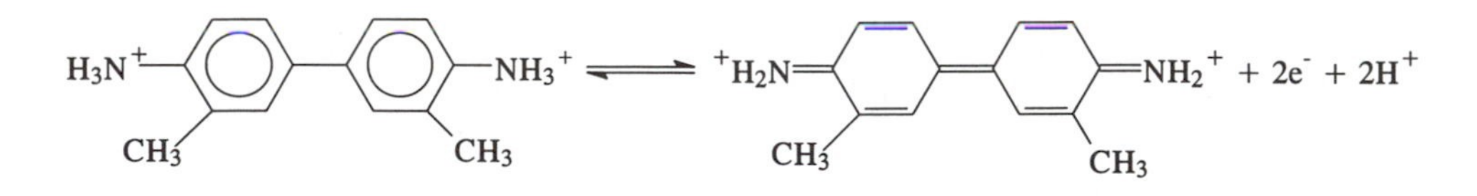


Figure 3.16A shows spectra of *o*-tolidine in an optically transparent thin-layer electrode (OTTLE) for a series of applied potentials. Curve a was recorded after application of +0.800 V, which caused complete oxidation of *o*-tolidine ([O]/[R] > 1000). Curve g was recorded after application of +0.400 V, causing complete reduction ([O]/[R] < 0.001). The intermediate spectra correspond to intermediate values of $E_{applied}$. Since the absorbance at 438 nm reflects the amount of *o*-tolidine in the oxidized form via Beer's law, the ratio [O]/[R] that corresponds to each value of $E_{applied}$ can be calculated from the spectra by Equation 3.18.

$$\frac{[O]}{[R]} = \frac{\dfrac{A_2 - A_1}{\varepsilon b}}{\dfrac{A_3 - A_2}{\varepsilon b}} = \frac{A_2 - A_1}{A_3 - A_2} \tag{3.18}$$

As shown for curve d in Figure 3.16A, $A_3$ is the absorbance when *o*-tolidine is completely oxidized; $A_1$, the absorbance when entirely reduced; and $A_2$, the absorbance for the mixture of oxidized and reduced forms.

Figure 3.16B shows a plot of $E_{applied}$ versus log([O]/[R]) for the data in Figure 3.16A. The plot is linear as predicted by the Nernst equation. The slope of the plot is 30.8 mV, which corresponds to an n value of 1.92, and the intercept is 0.612 V vs. SCE, which corresponds to the $E^{\circ\prime}$ of *o*-tolidine vs. SCE.

## III. POTENTIAL-SCAN TECHNIQUES IN STATIONARY SOLUTION

Some of the most useful electroanalytical techniques are based on the concept of continuously varying the potential that is applied across the electrode-solution interface and measuring the resulting current. This section describes large-amplitude techniques in this category for a working electrode immersed in a stationary solution.

Two voltammetric techniques, stationary-electrode voltammetry (SEV) and cyclic voltammetry (CV), are among the most effective electroanalytical methods available for the *mechanistic* probing of redox systems. In part, the basis for their effectiveness is the capability for rapidly observing redox behavior over the entire potential range available. Since CV is an extension of SEV, many points pertinent to CV are discussed in the SEV section.

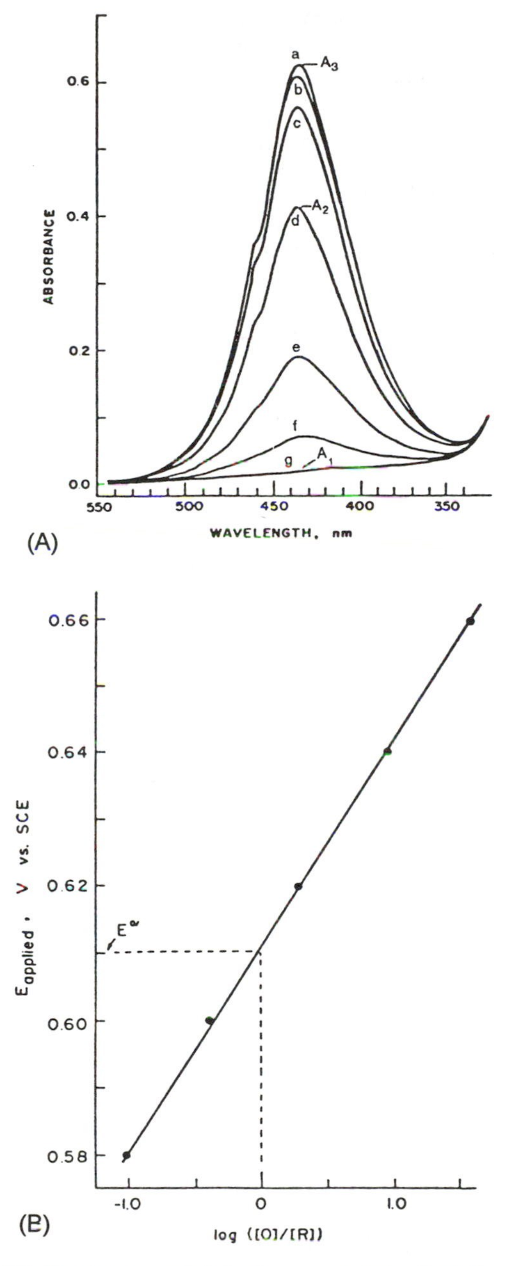


**Figure 3.16** (A) Spectra recorded during spectropotentiostatic experiment in optically transparent thin-layer electrode on 0.97 mM *o*-tolidine, 0.5 M acetic acid, 1.0 M $HClO_4$. Applied potentials: A, 800; B, 660; C, 640; D, 620; E, 600; F, 580; G, 400 mV vs. SCE. (B) Nernst plot at 438 nm. [From Ref. 36.]

The voltammetric techniques, which are also discussed in this section, are important in that they are among the more sensitive, reproducible, and easily used electroanalytical methods for *analysis*.

## A. Stationary-Electrode Voltammetry

In stationary-electrode voltammetry (SEV), the potential applied across the electrode-solution interface is scanned linearly from an initial value $E_i$ to a final value $E_f$ at a constant rate $\nu$ (V/s) [6,7,37]. This potential excitation is illustrated by the E-time profiles in Figure 3.17 for several rates of scan and several values

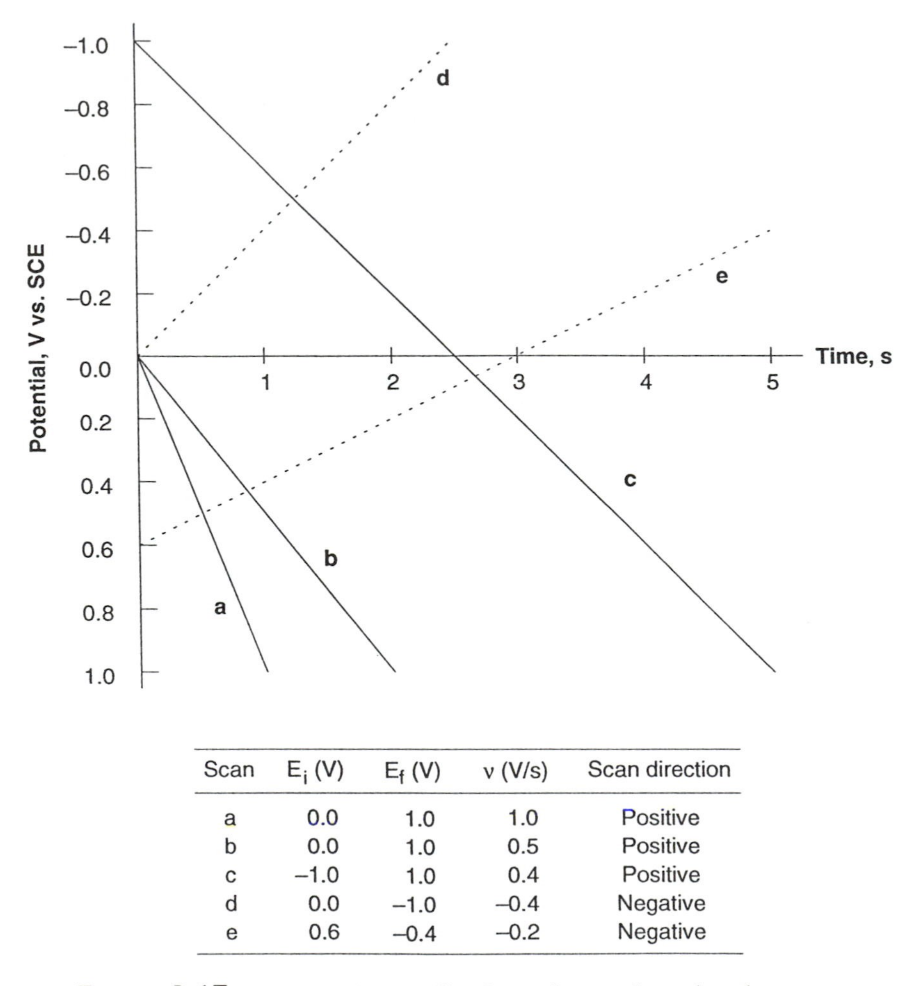


| Scan | $E_i$ (V) | $E_f$ (V) | $\nu$ (V/s) | Scan direction |
|---|---|---|---|---|
| a | 0.0 | 1.0 | 1.0 | Positive |
| b | 0.0 | 1.0 | 0.5 | Positive |
| c | −1.0 | 1.0 | 0.4 | Positive |
| d | 0.0 | −1.0 | −0.4 | Negative |
| e | 0.6 | −0.4 | −0.2 | Negative |

**Figure 3.17** Potential-time profiles for stationary electrode voltammetry.

of $E_i$ and $E_f$. An excursion in which the potential becomes increasingly positive (i.e., is heading in the positive direction) is termed a positive scan (*not* an *anodic* scan). Since the time required for scan a is 1.0 s, the scan rate is +1.0 V/1.0 s, or +1.0 V/s. Scan e is termed a negative scan since the potential is becoming less positive (i.e., is heading in the negative direction, even though in this example the actual potential is positive part of the time).

To obtain a stationary-electrode voltammogram, the current is measured during the potential scan in an unstirred solution. A voltammogram that was obtained for a solution of O with supporting electrolyte is shown in Figure 3.18. The potential scan that was applied to the working electrode to obtain this voltammogram is identical to excitation signal e in Figure 3.17. The potential axis in Figure 3.17 is the horizontal axis of Figure 3.18. Note that the experiment is completed in only 5 s.

In SEV one generally wants to initiate the potential scan at an $E_i$ at which no electrolysis occurs and then scan through the standard electrode potential(s) of the species in solution. Thus the composition of the reaction layer will not be altered in an unknown manner at the beginning of the experiment by appli-

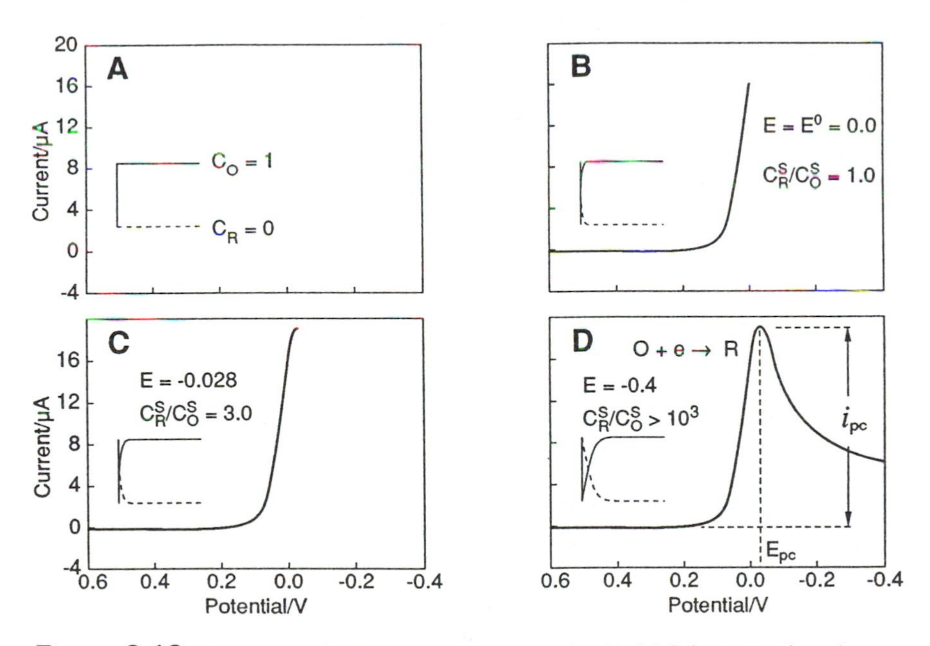


**Figure 3.18** Stationary-electrode voltammogram for 1 mM O in supporting electrolyte. Electrode reaction is $O + e \rightleftharpoons R$; $E^{\circ\prime}_{O,R} = 0$ V vs. SCE. Dashed line, SEV for supporting electrolyte without O. Scan rate = –0.2 V/s. Concentration-distance profiles during the potential scan are shown to the left in each block. (A) Start, (B) 3.0 s, (C) 3.15 s, (D) 5.0 s.

cation of $E_i$. The formal electrode potential of the O/R couple, $E^{\circ\prime}_{O,R}$, is taken as 0 V vs. reference electrode in this example. Since the solution to be examined contained the oxidized form of the couple, O, an $E_i$ that was *positive* of $E^{\circ\prime}_{O,R}$ was selected. This selection enabled the $E_i$ to be applied across the electrode-solution interface without substantially altering the redox state of O at the electrode surface. In order to do this, the $E_i$ should be positive of the standard electrode potential by at least 177 mV for an $n = 1$ reversible system. (The reader is referred to Table 3.1 and the accompanying text for correlation of ratios of redox states at the electrode surface with applied potential via the Nernst equation.) Note that selection of an $E_i$ negative of the standard electrode potential (e.g., −0.4 V) would cause immediate conversion of O to R at the electrode surface (chronoamperometry) before the potential scan was initiated. Also note that in the case in which the redox couple is initially present in solution as R, −0.4 V would then be a good selection for $E_i$.

In a practical vein, one does not always know either the standard redox potential or the redox state for the species in solution. In this case a potential can usually be found (by trial and error or by measuring the open-circuit potential) that when applied does not cause faradaic current to occur. Such a potential is generally suitable for $E_i$ since the absence of current (other than charging current) can be taken as evidence that no significant electrolytic perturbation of the reaction layer is occurring. One can then "find" the redox couple by scanning in either direction from this $E_i$.

In the Figure 3.18 example, after imposing $E_i$ across the electrode-solution interface, the potential is scanned negatively toward the standard redox potential of the O/R couple. The ratios of O and R that must exist at the electrode surface ($C^s_R/C^s_O$) at several potentials during the scan are given in each figure. These values are dictated by the Nernst equation for a reversible system (see Table 3.1). Since the solution initially contained "only" O, the R required to satisfy the Nernst equation is obtained from O by reduction, causing cathodic current.

The physical situation in the solution adjacent to the electrode during the potential scan is illustrated by the concentration-distance profiles included in Figure 3.18 for selected potentials [38]. The C-x profiles in Figure 3.18A are for O and R when $E_i$ is imposed. Note that the application of $E_i$ does not measurably alter the concentration of O at the electrode surface as compared to the solution bulk. As the potential is scanned positively, the concentration of O at the electrode surface decreases in order to establish a $C^s_R/C^s_O$ ratio that satisfies the Nernst equation for the applied potential at any particular instant. This is illustrated by profiles in B–D. Note that the profile in B (for which the concentration of O at the electrode surface equals the concentration of R) corresponds to an $E_{applied}$ that is the formal electrode potential (vs. SCE) of the couple.

The profiles in D correspond to a potential that is sufficiently negative of the formal electrode potential that the concentration of O is *effectively* zero at the electrode surface. The conditions for these profiles are analogous to those for chronoamperometry (Sec. II.A, Fig. 3.1). Once the potential has reached a value sufficient for a zero reactant-surface concentration, the potential and its rate of change become immaterial to the diffusion-controlled current. In other words, should the scan be stopped at the potential for D, the current will follow the same time course as if the scan had been continued.

The behavior of the current during the potential scan can be understood by carefully examining the C-x profiles for O in Figure 3.18. Recalling that Equation 3.5 shows the current to be determined by the slope of the C-x profile for the electroactive species at the electrode surface, one can correlate the magnitude of the slopes of the profiles in Figure 3.18 with current on the voltammogram at those potentials. The slope of the profile in A is zero and negligible current exists at that potential. As the potential is then scanned negatively, $(\partial C/\partial x)_{x=0}$ increases for the profiles in B and C and the current in Figure 3.18 increases correspondingly. When the profile in C is reached, $(\partial C/\partial x)_{x=0}$ then begins to decrease, as shown by the profile in D, because of depletion of O near the electrode. At the profile in D the chronoamperometric condition of diffusion-controlled conversion of O has been achieved (i.e., $C_O^s \approx 0$). Correspondingly, the current is dropping with a $t^{-1/2}$ dependence as in chronoamperometry. Thus the observed current behavior for the voltammogram is an increase to a peak current with a subsequent decay due to depletion of electroactive species near the electrode.

Two important parameters in SEV are the peak potential $E_p$ and the peak current $i_p$, which are the potential and current at the characteristic voltammogram peak as shown in Figure 3.18. For a reversible system the peak current is defined by the Randles-Sevcik equation,

$$i_p = (2.69 \times 10^5) n^{3/2} A D^{1/2} C^\circ v^{1/2} \tag{3.19}$$

at 25°C where $i_p$ is the peak current (A), n is the number of electrons transferred, A is the electrode area ($cm^2$), D is the diffusion coefficient of the species being oxidized/reduced ($cm^2/s$), $C^\circ$ is the concentration of this same species in the bulk solution ($mol/cm^3$), and $v$ is the scan rate (V/s). The peak potential is defined (for a reduction) by

$$E_p = E^{\circ\prime}_{O,R} - \frac{0.029}{n} \tag{3.20}$$

at 25°C where $E_p$ and $E^{\circ\prime}_{O,R}$ are expressed in volts, $D_O = D_R$, and $E^{\circ\prime}_{O,R}$ is the formal electrode potential corrected to the reference electrode being used.

According to Equation 3.19, $i_p$ is proportional to $\nu^{1/2}$ and a plot of $i_p$ versus $\nu^{1/2}$ should be a straight line for a reversible system. Although the peak current increases with scan rate, the potential at which the peak occurs is invariant with scan rate as indicated by Equation 3.20. The reason for the increase in current with faster scan rate is illustrated in Figure 3.19. Figure 3.19 shows voltammograms and the C-x profiles for SEV at two scan rates. Comparison of the profiles for the faster scan rate a with the profiles for scan b reveals a difference in the slope of the profiles at x = 0. For any given potential, $(\partial C/\partial x)_{x=0}$ is larger for the faster scan. Consequently, the current is larger. In both cases, the value of the concentration at the electrode surface (x = 0) is determined by the Nernst equation and is therefore independent of the scan rate. However, in the faster scan, less time is available for consequent depletion of the adjacent solution, resulting in steeper profiles. [If the current for both scans is integrated where $Q = \int_0^t i\,dt$, for which scan rate will Q (the total charge passed) be larger?]

In the case of an irreversible system, the equation for $i_p$ is

$$i_p = (2.99 \times 10^5)n\,(\alpha n_a)^{1/2}\,AD^{1/2}C^{\circ}\,\nu^{1/2} \tag{3.21}$$

where $\alpha$ is the transfer coefficient and $n_a$ is the number of electrons in the rate-determining step of the electrode process. The peak potential $E_p$ is no longer independent of scan rate, as shown by Equation 3.22,

$$E_p = E^{\circ\prime} + \frac{RT}{\alpha n_a F}\left[-0.78 + \ln\frac{k_1}{D_O^{1/2}} - \tfrac{1}{2}\ln\left(\frac{\alpha n_a F}{RT}\nu\right)\right] \tag{3.22}$$

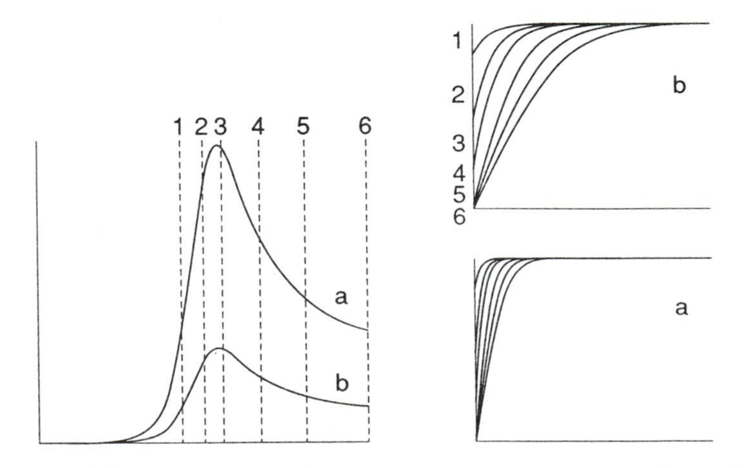


**Figure 3.19** Voltammograms and concentration-distance profiles for (a) fast, and (b) slow scan rate. Simulation by DigiSim.®

where $k_1$ is related to the heterogeneous rate constant of the electron transfer reaction. In general, irreversible behavior gives voltammograms that are more spread out and flatter.

An important aspect of voltammetry is the background current, which is the current obtained on a solution containing all ingredients except the electroactive species of interest. It is composed of (1) residual current, oxidation or reduction of adventitious electroactive material (e.g., dissolved oxygen, the solvent, or the electrode itself), and (2) charging current, nonfaradaic current required to charge the electrode to a given potential (behavior analogous to a capacitor). The peak current should be measured in a manner that does not include the background current. A common technique is to extrapolate the baseline current that precedes the faradaic current and measure $i_p$ by difference, as shown in Figure 3.18D. Alternatively, a separate scan (at the *same* scan rate) can be made on the supporting electrolyte solution. Here $i_p$ is taken as the difference between the peak current (measured from zero current) and the background current at the same potential.

The current due to charging the electrode to a given potential is directly proportional to the scan rate as given by Equation 3.23:

$$i_c = C \frac{dE}{dt} = AC_{dl}\nu \qquad (3.23)$$

where $i_c$ is the charging current, A is the electrode area ($cm^2$), and $C_{dl}$ is the capacitance of the double layer per square centimeter. Consequently, the correction for this component of the background current becomes increasingly significant as the scan rate increases. It is important to recognize that $C_{dl}$ is a function of potential and therefore $i_c$ is a function of E as well as of $\nu$. Also, $C_{dl}$ is not necessarily a linear function of E, so that the linear extrapolation of the baseline over a large distance (potential) can lead to significant error in the measurement of $i_p$.

SEV has a reasonable detection limit, being useful at $10^{-5}$ M in some cases. Sensitivity can be improved by increasing the scan rate, which increases the magnitude of $i_p$ as in Equation 3.19. However, charging current becomes a problem at higher scan rates, since $i_c$ increases more rapidly with $\nu$ than does $i_p$ ($i_c \propto \nu$, $i_p \propto \nu^{1/2}$). This significant charging current can be compensated for by a differential technique in which simultaneous scans are performed on two matched electrodes, one immersed in sample and one in supporting electrolyte. The background current from the supporting electrolyte scan is electronically subtracted from the voltammogram of the sample. The detection limit of the rapid scan and differential SEV techniques is better than that of conventional polarography but not as good as that of the current sampled pulse techniques that are discussed later in this chapter and in Chapter 5.

The detection limit of SEV for analysis is greatly extended by its use in conjunction with an electrochemical preconcentration step. This technique, termed dc stripping voltammetry, is discussed in Chapter 24.

SEV is an effective means of probing homogeneous chemical reactions that are coupled to electrode reactions, especially when it is extended to cyclic voltammetry as described in the next section. Considerable information can be obtained from the dependence of $i_p$ and $E_p$ on the rate of potential scan. Figure 3.20 illustrates the behavior of $i_p$ and $E_p$ with variation in scan rate for a reversible heterogeneous electron transfer reaction that is coupled to various types of homogeneous chemical reactions. The current function $\psi_p$ is proportional to $i_p$ according to the equation

$$\psi_p = \frac{i_p}{nFAD^{1/2}C^{\circ}\left(nF\nu/RT\right)^{1/2}} \tag{3.24}$$

Different types of coupled chemical reactions (or mechanisms) can be identified by their characteristic currents and potential behaviors with varying scan rate as shown in Figure 3.20. Theory and diagnostic criteria have been developed for a variety of kinetic cases in which the homogeneous chemical reactions are coupled to either reversible or irreversible heterogeneous electron exchange [39]. Adsorption of the electroactive species has also been considered [40]. In general, conclusions based on current measurements are more reliable diagnostics of electrode mechanism than those based on potential measurements because experimental factors (iR drop, junction potentials, state of electrode surface) can shift the potential axis significantly.

In view of the fact that SEV is a transient technique and that the potential axis may also be thought of as a time axis, it has often been called chronoamperometry with potential sweep. The entire sweep may be applied during the lifetime of a single mercury drop in a polarographic experiment. Historically, this hybrid technique was termed single-sweep oscillographic polarography or cathode-ray polarography. The IUPAC suggests single-sweep polarography.

## B. Cyclic Stationary-Electrode Voltammetry

Cyclic stationary-electrode voltammetry, usually called cyclic voltammetry (CV), is perhaps the most effective and versatile electroanalytical technique available for the mechanistic study of redox systems [37,39,41–44]. It enables the electrode potential to be scanned rapidly in search of redox couples. Once located, a couple can then be characterized from the potentials of peaks on the cyclic voltammogram and from changes caused by variation of the scan rate. CV is often the first experiment performed in an electrochemical study. Since cyclic voltammetry is a logical extension of stationary-electrode voltammetry (SEV), some important aspects of CV were treated in the preceding section.

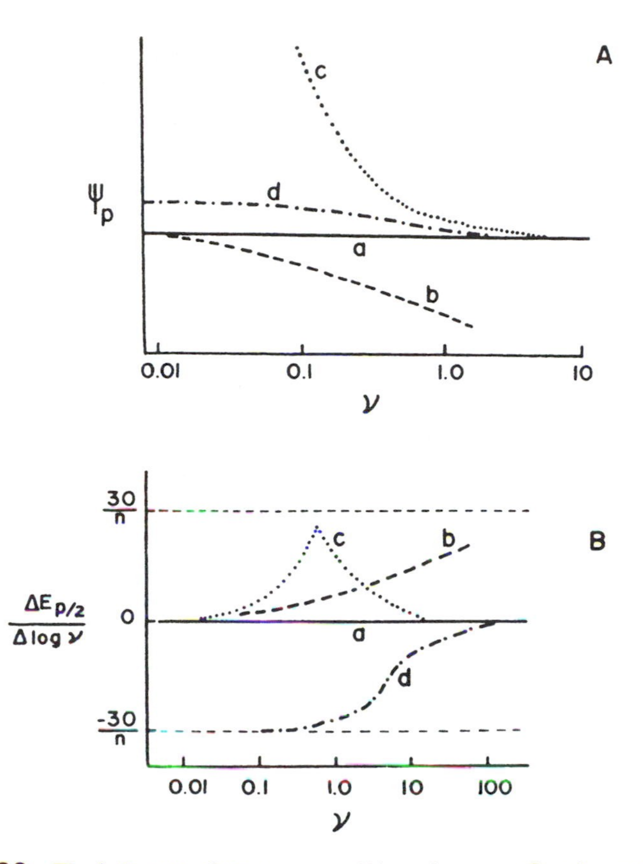


**Figure 3.20** The influence of scan rate on (A) peak current function and (B) half-peak potential for several electrode processes with reversible electron transfer.

a. Reversible electron transfer — E: $O + e \rightleftharpoons R$

b. Preceding chemical reaction — C: $Z \rightleftharpoons O$; E: $O + e \rightleftharpoons R$

e. Catalytic regeneration — E: $O + e \rightleftharpoons R$; C: $R + Z \longrightarrow O$ (O regenerated, returning to the E step)

c. EC mechanism — E: $O + e \rightleftharpoons R$; C: $R \longrightarrow Z$

The repetitive triangular potential excitation signal for CV causes the potential of the working electrode to sweep back and forth between two designated values (the switching potentials). The potential excitation that is applied across the electrode-solution interface in order to obtain a cyclic voltammogram is illustrated by the potential-time profiles in Figure 3.21. The labeled segments in Figure 3.21 can be interpreted in the following way:

a. Negative potential scan from +0.60 to -0.40 V
b. Direction of scan reversed at switching potential -0.40 V
c. Positive potential scan from -0.40 to +0.60 V
d. Termination of the first cycle

Note that up to point b the excitation signal is that of SEV, as described in the preceding section (see Fig. 3.17, scan e).

Although the potential scan is frequently terminated at the end of the first cycle (point d), it can be continued for any number of cycles, hence the terminology *cyclic* voltammetry. The dotted line in Figure 3.21 denotes a second cycle. A scan in which the potential is becoming increasingly positive (i.e., heading in a positive direction) is termed a positive scan. A scan in which the potential is becoming increasingly negative is a negative scan (even though the

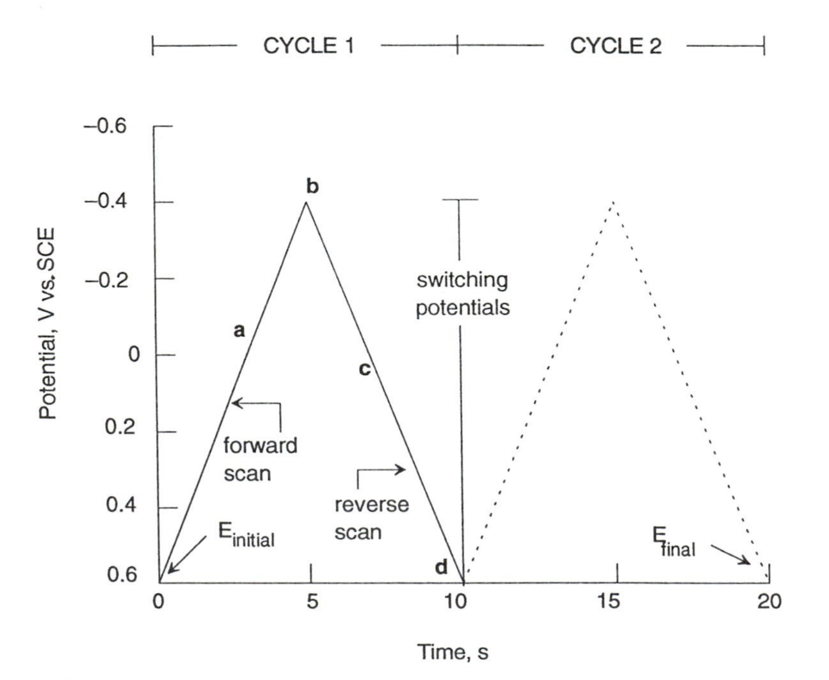


**Figure 3.21** Typical potential-time excitation signal for cyclic voltammetry.

potential may actually be positive for all or part of the scan, as is the case in this example). Since the time required for the first cycle is 10 s, the scan rate is 2 V/10 s or 0.2 V/s. The scan profile used in a particular experiment is generally determined by the location of the redox couple of interest.

To obtain a cyclic voltammogram, the current at the working electrode in an unstirred solution is measured during the potential scan as shown in Figure 3.22. The potential scan that was applied across the electrode-solution interface to obtain this simulated cyclic voltammogram is identical to the profile depicted in Figure 3.21. Correlate the potential of Figure 3.21 with the potential (horizontal) axis of Figure 3.22. Note that the negative scan from +0.60 to -0.40 V in Figure 3.22 corresponds to segment a in Figure 3.21 and that the reverse (positive) scan corresponds to segment c.

During the scan from +0.60 to -0.40 V, the applied potential becomes sufficiently "negative" at +0.1 V to cause reduction of O to occur at the electrode surface. This reduction is accompanied by cathodic current, which increases rapidly until the surface concentration of O approaches zero as signaled by peaking of the current Figure 3.22C. The current then decays (after C) as the solution surrounding the electrode is depleted of O due to its conversion to R. This depletion of O and accumulation of R near the electrode is depicted by the inset concentration-distance profiles for O and R in A–D in Figure 3.22. The magnitude of the cathodic current during the negative scan is related to the slope of the C-x profile for O as described by Equation 3.5. The reader is referred to Section III.A for further explanation of the characteristic current behavior during this negative scan as well as a discussion concerning the selection of the initial potential, $E_i$.

During the negative scan in which O is reduced to R, the depletion of O in the vicinity of the electrode is accompanied by an accumulation of R. This can be seen by the concentration-distance profiles at various potentials in Figure 3.22. After the direction of potential scan is switched at -0.40 V to a positive scan, reduction continues (as is evident by the cathodic current and the C-x profiles in E) until the applied potential becomes sufficiently positive to cause oxidation of the accumulated R. Oxidation of R is signaled by the appearance of anodic current. Once again, the current increases as the potential moves increasingly positive, until the oxidation of R sufficiently depletes the region near the electrode of R to cause the current to peak and then to decrease. Thus the physical phenomena that caused a current peak during the oxidation cycle also cause a current peak during the reduction cycle. This can be seen by comparing the concentration-distance profiles for the two scans.

Simply stated, in the forward scan, R is electrochemically generated as indicated by the cathodic current. In the reverse scan, this R is oxidized back to O as indicated by the anodic current. Thus CV is capable of rapidly generating a new species during the forward scan and then probing its fate on the

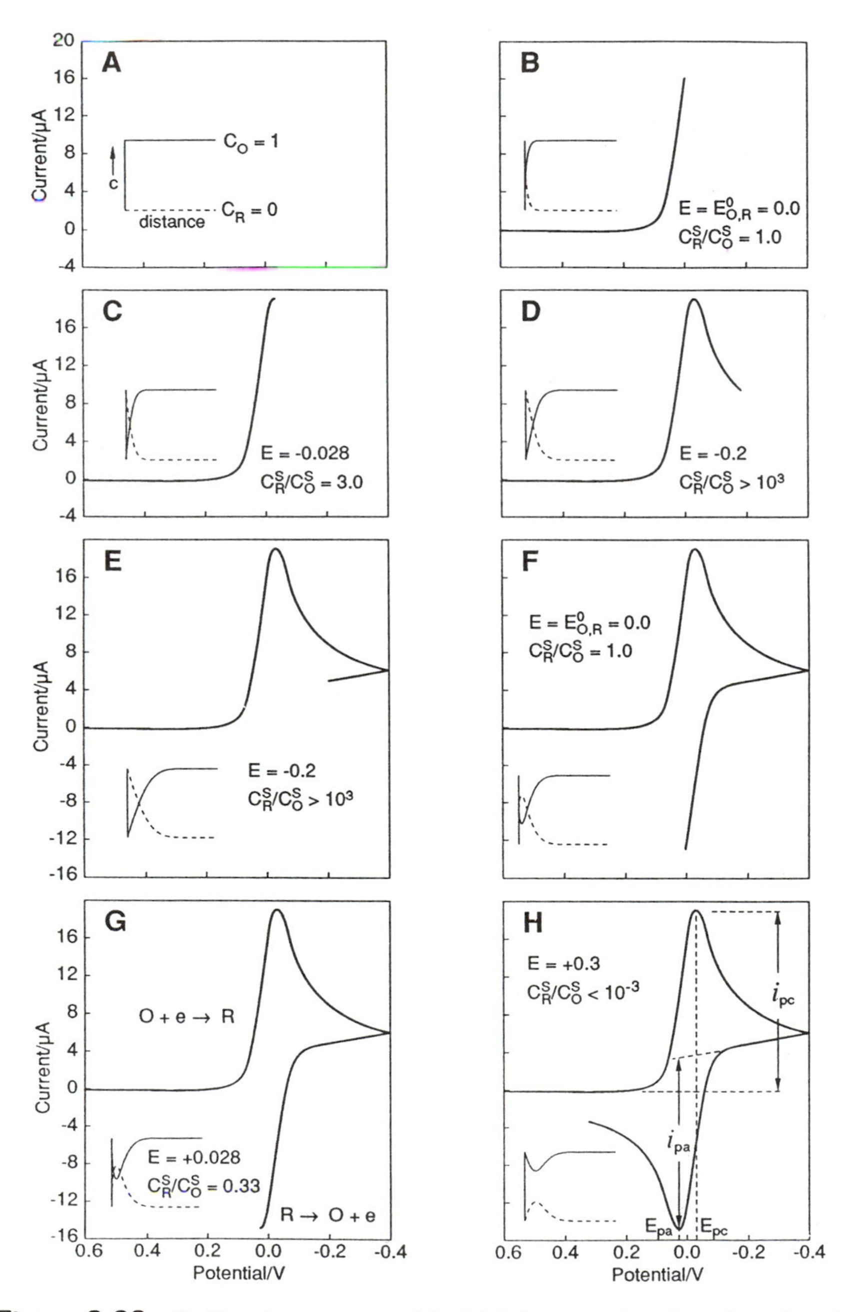


**Figure 3.22** Cyclic voltammogram of 1 mM O in supporting electrolyte. Scan initiated at 0.6 V vs. SCE in negative direction at 200 mV $s^{-1}$. Concentration-distance profiles a–h keyed to voltammogram. $E^\circ_{O,R} = 0$ V vs. SCE. Simulation by DigiSim.®

reverse scan. This is a very important aspect of the technique. (Would an anodic current be expected if the same experiment were performed in a stirred solution?)

The important parameters of a cyclic voltammogram are the magnitudes of the peak currents, $i_{pa}$ and $i_{pc}$, and the potentials at which the peaks occur, $E_{pa}$ and $E_{pc}$. These parameters are labeled in Figure 3.22H. One method for measuring $i_p$ involves extrapolation of a baseline current. The establishment of a correct baseline is essential for the accurate measurement of peak currents. This is not always easy, particularly for more complicated systems. The second sweep generally poses the main problem since the baseline is not the same as the residual current obtained by an identical experiment in supporting electrolyte. Difficulty in obtaining accurate peak currents is perhaps the biggest liability of CV. The reader is referred to Adams' monograph for a more detailed discussion of this sticky situation [6, pp. 152–158].

A redox couple in which both species rapidly exchange electrons with the working electrode is termed an *electrochemically* reversible couple. Such a couple can be identified from a cyclic voltammogram by measurement of the potential difference between the two peak potentials. Equation 3.25 applies to a system that is both electrochemically and chemically reversible:

$$\Delta E_p = E_{pa} - E_{pc} \approx \frac{0.058}{n} \tag{3.25}$$

where n is the number of electrons transferred and $E_{pa}$ and $E_{pc}$ are the anodic and cathodic peak potentials, respectively, in volts. This 0.058/n separation of peak potentials is independent of scan rate for a reversible couple, but is slightly dependent on switching potential and cycle number (see Ref. 7, pp. 229–230). The potential midway between the two peak potentials is the formal reduction potential (corrected for the reference electrode being used) of the couple:

$$E^{\circ\prime} = \frac{E_{pa} + E_{pc}}{2} \tag{3.26}$$

When the scan rate $v$ is increased, $i_{pa}$ and $i_{pc}$ both increase in proportion to $v^{1/2}$. The same reasoning for this current increase in SEV applies here. Plots of $i_{pa}$ and $i_{pc}$ versus $v^{1/2}$ should be linear with intercepts at the origin for a reversible couple.

The values of $i_{pa}$ and $i_{pc}$ are similar in magnitude for a reversible couple with no kinetic complications. That is,

$$\frac{i_{pa}}{i_{pc}} \approx 1 \tag{3.27}$$

(This equation is strictly applicable after several cycles. The first voltammogram is not quite the same as the reproducible curve obtained after several cycles.) However, the ratio of peak currents can be significantly influenced by chemical reactions coupled to the electrode process.

Figure 3.23 illustrates variations in $i_{pa}/i_{pc}$ as a function of scan rate for several electrode mechanisms [39]. Such characteristic behavior in peak currents makes cyclic voltammetry a powerful technique for studying mechanisms of electrode reactions. The effect of a coupled chemical reaction on a cyclic voltammogram is illustrated in Figure 3.24 for the EC mechanism (Chap. 2, Sec. V). Voltammograms are shown for a series of rate constants with the current normalized for scan rate. A very slow rate constant essentially freezes out the homogeneous chemical reaction, and a cyclic voltammogram with equal heights of the two peaks (Eq. 3.27) is obtained. However, as the rate constant increases,

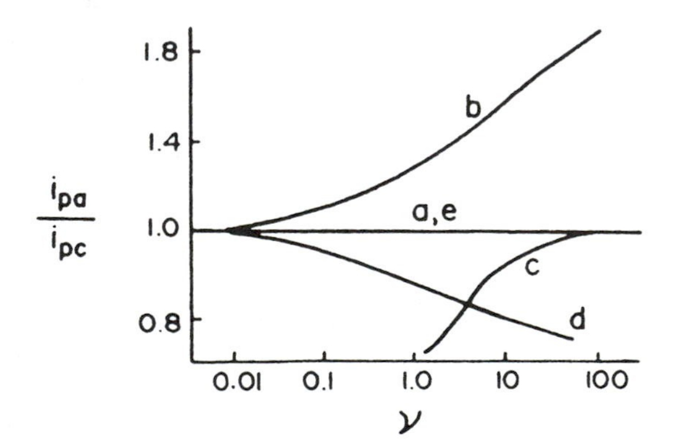


**Figure 3.23** Variation in the ratio of anodic to cathodic peak currents as a function of scan rate for several electrode processes with reversible electron transfer. [From Ref. 39, reprinted with permission.]

a. Reversible electron transfer — E: $O + e \rightleftharpoons R$

b. Preceding chemical reaction — C: $Z \rightleftharpoons O$; E: $O + e \rightleftharpoons R$

c. EC mechanism — E: $O + e \rightleftharpoons R$; C: $R \longrightarrow Z$

d. EC mechanism — E: $O + e \rightleftharpoons R$; C: $R \rightleftharpoons Z$

e. Catalytic regeneration — E: $O + e \rightleftharpoons R$; C: $R + Z \longrightarrow O$ (O regenerated)

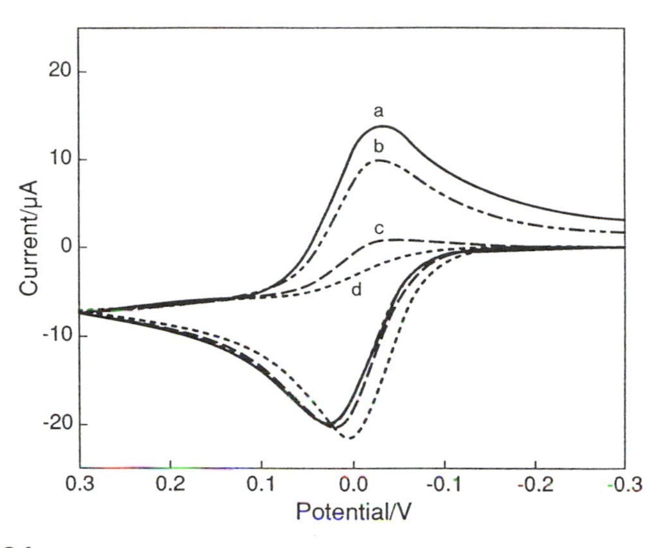


**Figure 3.24** Cyclic voltammograms for the EC mechanism for various rate constants. Current normalized to $v^{1/2}$. (a) $k_c = 0.01$, (b) $k_c = 0.1$, (c) $k_c = 1$, (d) $k_c = 10$. Scan rate = 0.1 V $s^{-1}$. $R \rightleftharpoons O + e^-$; $O \xrightarrow{k_c} P$. Simulation by DigiSim.®

the anodic peak diminishes relative to the cathodic peak due to consumption of electrogenerated R by chemical conversion to Z. This is shown by curve c in Figure 3.23. A sufficiently fast rate constant results in disappearance of the anodic peak, since ample time is allowed for the coupled chemical reaction to essentially deplete the solution of R in the vicinity of the electrode. Computer programs for the simulation of various mechanisms are now commercially available (BAS DigiSim,® Gosser). Examples are given in Chapters 21 and 23.

Electrochemical irreversibility is caused by slow electron exchange of the redox species with the working electrode. (The reader is referred to Chap. 2 for a discussion of irreversibility.) In this case Equation 3.25 is not applicable. Electrochemical irreversibility is characterized by a separation of peak potentials that is greater than 0.059/n V and that is *dependent* on the scan rate (Examples are given in Chapter 23.) Figure 3.25A illustrates voltammograms for reversible and irreversible systems.

The dependence of $\Delta E_p$ on the heterogeneous rate constant $k_s$ and the scan rate v is shown in Figure 3.25B. It is apparent that at a given scan rate $\Delta E_p$ increases as the heterogeneous rate constant $k_s$ of the system decreases. Alternatively, for a system with a given rate constant, the peak separation increases with increasing scan rate. The magnitude of $k_s$ can be determined from a plot

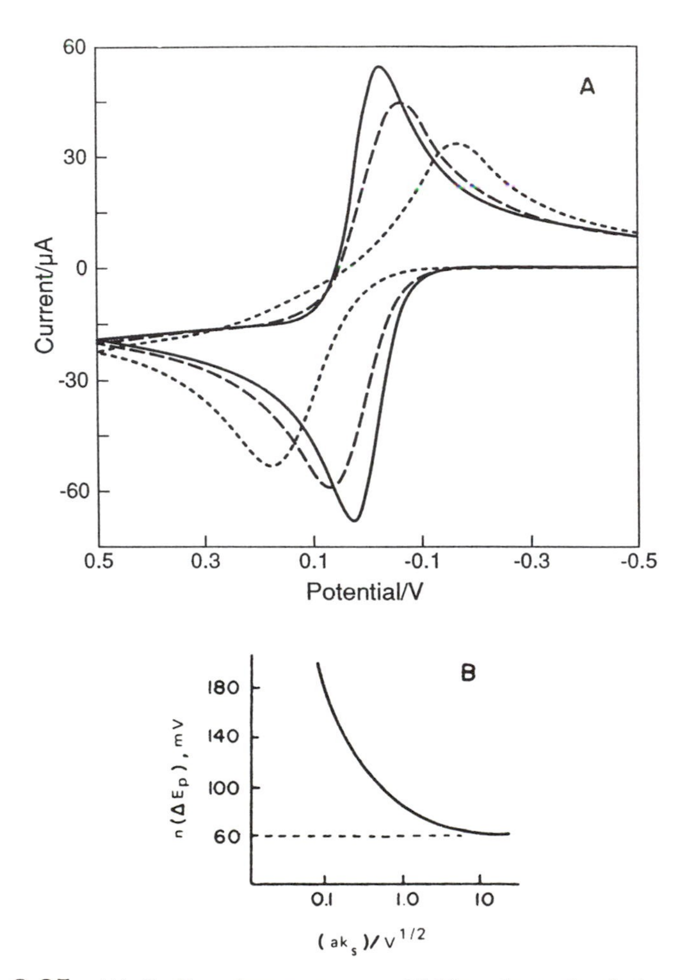


**Figure 3.25** (A) Cyclic voltammograms exhibiting electrochemical reversibility: effect of variation of $k_s$: $k_s$ = 1 (solid lines), 0.01 (dashed lines), 0.001 (dotted lines). Scan rate = 1 V $s^{-1}$. [Simulation by DigiSim®.] (B) Dependence of $\Delta E_p$ on $k_s$ and $\nu^{1/2}$. [From Ref. 45, reprinted with permission.]

such as Figure 3.25B if $\Delta E_p$ is known for a given scan rate. (Examples are given in Chap. 23.) Figure 3.25B also points out that a system that appears reversible at one scan rate can "become irreversible" at faster scan rates. This illustrates the somewhat arbitrary nature of reversibility. Uncompensated iR drop also causes peak separation to increase with ν. Elimination of iR drop by one of the methods described in Chapters 6 and 7 is crucial to the measurement of "correct" $\Delta E_p$ values. It is important to recognize that electrochemical irreversibility also influences the peak current ratio. The more irreversible a couple becomes, the smaller will be the $i_p$ on the reverse scan. This is often due to both the smaller $k_s$ and the fact that significant product has diffused away from the surface by the time the reverse $E_p$ is reached. Because of these complications, it can be dangerous to infer quantitative information from the $i_p$ ratio for extremely irreversible couples.

In most CV experiments, there is little advantage to be gained by carrying on the potential scan for more than two to three cycles. One exception is the use of repetitive cycling to monitor the accumulation of electroactive species in films of chemically modified electrodes (Chap. 13). Data are typically obtained via XY recorder at slow scans (i.e., less than 500 mV/s) and storage oscilloscope or computer at faster rates. Scan rates up to 1,000,000 V/s have been used; however, rates faster than 100 V/s are rarely practicable because of iR drop and charging current. Very small electrodes now make it easier to implement high-speed experiments (see Chap. 12).

With the advent of digital implementation of electroanalytical experiments came the technique called "staircase cyclic voltammetry," wherein the triangular wave is approximated by a series of small potential steps. The reason for such an approximation is partly due to the impossibility of digitally generating a "pure" ramp and more importantly to the realization that substantial improvements accrue from sampling the current at the end of each step, where double-layer charging has decayed away. If the steps are small, the data will fit theory based on pure ramps quite well.

CV has become a standard technique in all fields of chemistry as a means of studying redox states. The method enables a wide potential range to be rapidly scanned for reducible or oxidizable species. This capability, together with its variable time scale and good sensitivity, makes CV the most versatile electroanalytical technique thus far developed. It must, however, be emphasized that its merits are largely in the realm of qualitative or "diagnostic" experiments. Quantitative measurements (of rates or concentrations) are *best* obtained via other means (e.g., step, pulse, or hydrodynamic techniques). Because of the kinetic control of many CV experiments, some caution is advisable when evaluating the results in terms of thermodynamic parameters (e.g., measurement of $E^{\circ\prime}$ for irreversible couples).

Perhaps the most useful aspect of CV is its application to the qualitative diagnosis of electrode reactions that are coupled to homogeneous chemical reactions [37,41–44]. The forte of CV is its ability to generate a species during one scan and then probe its fate with subsequent scans. Examples of CV usage in unraveling electrode mechanisms are presented in Chapters 21 and 23.

A chemical system should be examined using several carefully chosen electrochemical techniques in combination with spectroscopic information and good chemical intuition. It is dangerous to draw too many conclusions from voltammetric information alone. Chapters 21 and 23 are devoted to selected examples from the literature that illustrate this theme.

## C. Derivative Stationary-Electrode Voltammetry

The excitation signal for derivative stationary-electrode voltammetry (derivative SEV) is identical to that for SEV (i.e., a potential scan). The system response signal, current, is also the same. However, in derivative SEV this response signal is electronically differentiated, and the resulting derivative, di/dt, is recorded as a function of applied potential. Second and third derivatives can also be displayed.

Figure 3.26 illustrates voltammograms obtained for SEV and derivative SEV for reduction of $Cd^{2+}$ at a hanging mercury drop electrode (HMDE) [46].

The first derivative peak height is directly proportional to the concentration of the electroactive species. Several advantages accrue from monitoring the derivative response, including increased sensitivity, resolution, and ease of obtaining rate parameters of coupled chemical reactions [47,48]. An enhanced signal-to-background ratio derives from the relatively small variation in double-layer charging current with potential, making its contribution to the derivative small. This advantage becomes important at high scan rates where charging current grows in relation to the faradaic response.

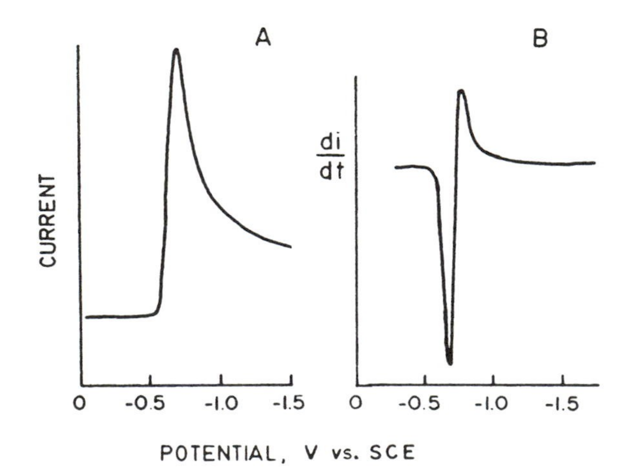


**Figure 3.26** (A) Stationary electrode voltammogram of 2.00 mM $Cd^{2+}$ in 0.2 M KCl, $\nu = 0.0327$ V/s, HMDE. (B) Derivative stationary electrode voltammogram, identical conditions. [From Ref. 46, reprinted with permission.]

Although derivative SEV does exhibit these apparent advantages over SEV, they are not really significant and the technique has been used very little.

## D. DC Polarography

Polarography is the classical electrochemical technique involving the dropping mercury electrode (DME). The excitation signal is a slow potential scan, and the resulting current is displayed as a function of this applied potential. Although the term "polarography" has been used over the years in connection with a variety of electrochemical techniques, "polarography" is now reserved as a name for voltammetric experiments in which the DME is employed (see Chap. 14). Many texts have been written on the subject [49–52].

A typical polarogram is illustrated in Figure 3.27. In this case the solution contains the electroactive species $Pb^{2+}$ and $Cd^{2+}$ as well as 0.1 M KCl supporting electrolyte.

As the applied potential is scanned negatively from an initial value of 0.0 V vs. SCE, the current changes only slightly until a sufficiently negative potential is reached to cause reduction of $Pb^{2+}$ to $Pb^{\circ}$. As the potential becomes more negative, the current for this reduction increases. The increase is caused by the greater slope for the concentration-distance profile for $Pb^{2+}$ at the electrode surface [i.e., $(\partial C/\partial x)_{x=0}$ becomes greater]. This is shown by C-x profiles a–e for various potentials. The current reaches a plateau when the potential becomes sufficiently negative that the concentration of $Pb^{2+}$ is effectively zero at the electrode surface. Now the rate of $Pb^{2+}$ reduction (which determines the current) is governed (limited) by the rate of diffusion of $Pb^{2+}$ to the electrode surface. Consequently, the current plateau is referred to as the diffusion current ($i_d$) or the limiting current ($i_l$).

The insert in Figure 3.27 correlates the repetitive current fluctuation that is characteristic of the polarogram with the "life" of a mercury drop at the DME. The current increases as the drop grows, declining abruptly the moment the drop falls. This current-time behavior is determined by (1) the changing surface area of the drop, and (2) the diffusion of $Pb^{2+}$ to the electrode surface. The electrode area increases at a $t^{2/3}$ rate, which would cause current (which is proportional to electrode area) to also increase at $t^{2/3}$. Counterbalancing this increase is the $t^{-1/2}$ decrease in current caused by depletion of $Pb^{2+}$ in the adjacent solution under diffusion-controlled conditions. (The diffusional current of polarography can be considered in terms of chronoamperometry at an expanding mercury sphere since the potential does not vary significantly during the life of a single drop. See Eq. 3.6 for the $t^{-1/2}$ relationship.) Combining these two phenomena gives a current that increases at a $t^{1/6}$ rate during the lifetime of a drop. When the drop falls, the electrode area instantly decreases, causing the current to drop abruptly. Each falling drop stirs the solution sufficiently to erase almost all concentration depletion caused by electrolysis at the drop.

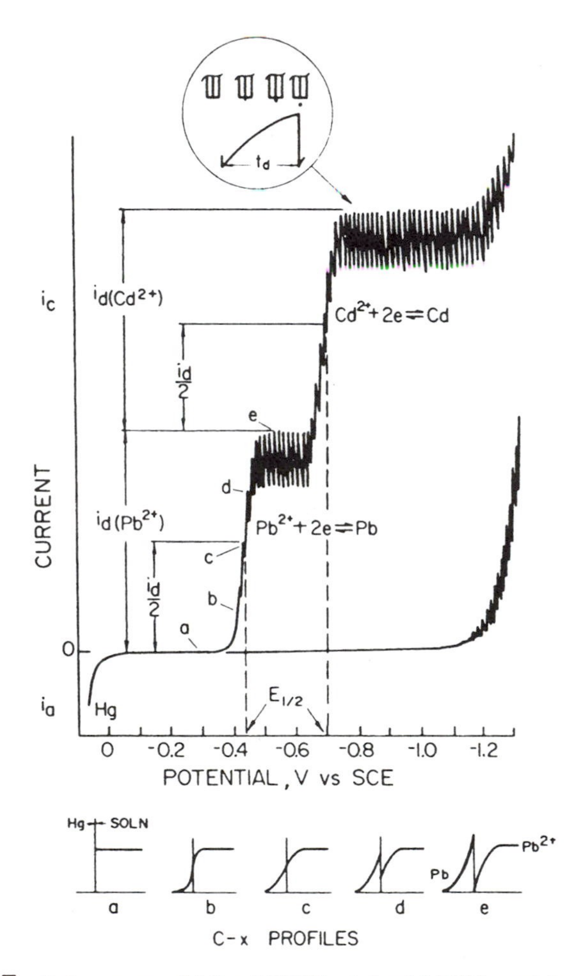


**Figure 3.27** Polarogram of 1.0 mM $Pb^{2+}$, 1.0 mM $Cd^{2+}$, and 0.1 M KCl. $O_2$ removed. Concentration-distance profiles during potential scan shown below polarogram.

The diffusion current (over the lifetime of a drop) is expressed by the Ilkovic equation,

$$i_d(\max) = 708\,n m^{2/3} D^{1/2} t_d^{1/6} C^{\circ} \tag{3.28}$$

where

$i_d$(max) = diffusion current, $\mu$A
n = number of electrons transferred per molecule or ion
m = flow rate of mercury, mg/s

$D$ = diffusion coefficient of electroactive species, $cm^2/s$
$t_d$ = drop time, s
$C°$ = concentration of electroactive species, mM

The applied potential is related to the polarographic current for a reversible system by the Nernst equation as expressed in the form

$$E = E_{1/2} - \frac{0.059}{n} \log \frac{i}{i_d - i} \tag{3.29}$$

where i is the current at a particular applied potential E. The half-wave potential $E_{1/2}$ is defined as the potential at which the current is equal to $i_d/2$. It is also related to the formal reduction potential, $E°'$, of the electroactive species according to

$$E_{1/2} = E°' - \frac{0.059}{n} \log \frac{D_O^{1/2}}{D_R^{1/2}} \tag{3.30}$$

where $D_O$ and $D_R$ are the diffusion coefficients of the oxidized and reduced species, respectively. Because $E_{1/2}$ is directly related to the formal reduction potential of the redox couple, it is useful for qualitative analysis of electroactive species. Tables of $E_{1/2}$ values of redox couples in a variety of media have been compiled for the purpose of identification.

As Equation 3.28 shows, the diffusion current $i_d$ is directly proportional to the concentration of the electroactive species. Consequently, polarography has been used extensively for the analysis of solutions containing electroactive materials. The optimum concentration range for determinations by conventional polarography as described here is $10^{-2}$–$10^{-4}$ M. Reproducibility is generally ±3%.

The diffusion current is typically measured from a baseline that is obtained by extrapolating the residual current prior to the wave. Alternatively, the baseline can be recorded in a separate experiment on a solution of deoxygenated supporting electrolyte. The residual current arises from capacitive current required to maintain the expanding drop at the applied potential and from reduction of trace electroactive impurities.

The diffusion currents of electroactive species are additive, as shown for $Pb^{2+}$ and $Cd^{2+}$ in Figure 3.27. When the applied potential becomes sufficiently negative, reduction of $Pb^{2+}$ and $Cd^{2+}$ occurs simultaneously and the resulting current reflects the sum of these two processes. The baseline for measuring the $i_d$ of the second process is obtained by extrapolation of the $i_d$ of the first process. Thus the determination of mixtures of electroactive species is possible.

The "potential window" of the DME is limited by hydrogen evolution or reduction of supporting electrolyte cation on the negative extreme, and by oxidation of mercury on the positive extreme. In certain cases the oxidation of mercury can be used effectively for the determination of species that form complexes (e.g., $CN^-$, EDTA) or insoluble salts (e.g., $Cl^-$, $Br^-$) with mercurous or mercuric ion. In this case the potential is scanned in the positive direction. The resulting anodic waves are a result of the oxidation of the mercury electrode (not of the ligand or *anion*) at a rate controlled by the diffusion of the anion or ligand (A) to the electrode surface. The general electrode reaction can be expressed as

$$Hg + A \rightleftharpoons HgA^{n+} + ne \tag{3.31}$$

In this case, $i_d$ is proportional to the concentration of A in solution. The variation in half-wave potentials corresponds to the variation in stabilities of the $HgA^{n+}$ compounds (i.e., $K_{sp}$ or $K_f$).

Many organic compounds undergo reduction or oxidation at a DME. Consequently, polarographic techniques have been used extensively for determinations of organic compounds and for studying the mechanisms of their electrode reactions. In aqueous solution, the reduction of organic compounds is frequently a 2e process accompanied by protonation as in Equation 3.32:

$$O + 2e + 2H^+ \longrightarrow RH_2 \tag{3.32}$$

Polarograms for the quinone-hydroquinone couple are shown in Figure 3.28. The electrode mechanism for quinone reduction is

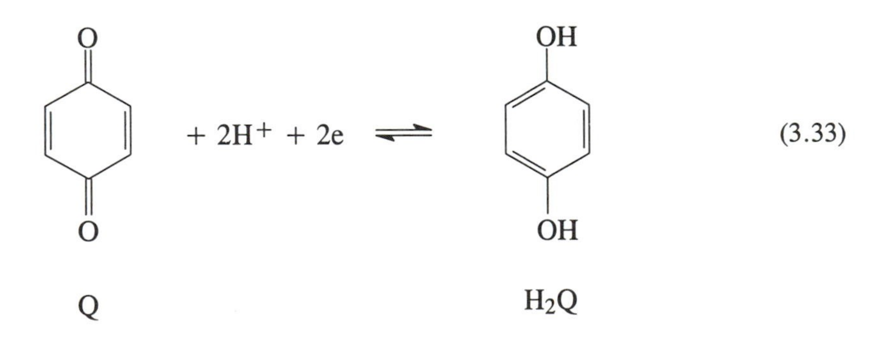


The polarograms were obtained on solutions of Q, $H_2Q$, and an equimolar mixture of Q, $H_2Q$. In the latter case, both reduction and oxidation waves were obtained.

The $E_{1/2}$ for an electrode reaction of this type is pH dependent. As the pH increases, $E_{1/2}$ shifts negatively due to decreasing availability of protons. A typical $E_{1/2}$-pH plot is shown in Figure 3.29. The magnitude of negative shift per pH unit is related to the acid dissociation constants, $K_{a_1}$ and $K_{a_2}$, for the

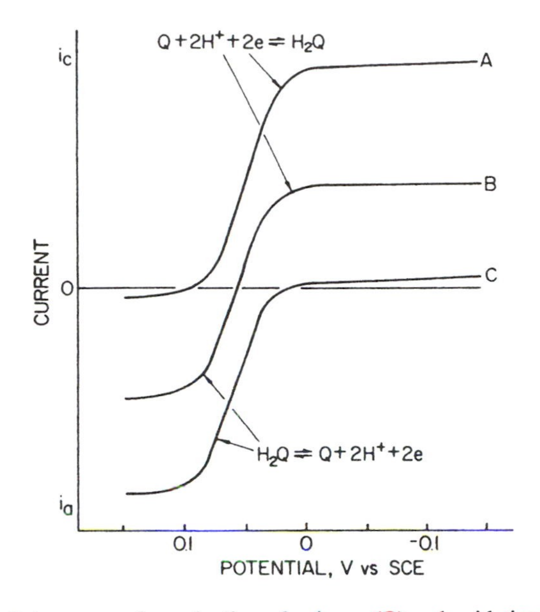


**Figure 3.28** Polarograms for reduction of quinone (Q) and oxidation of hydroquinone ($H_2Q$), in buffered solution: A, quinone; B, quinone-hydroquinone equimolar mixture; C, hydroquinone.

species $RH_2$ for a mechanism as shown by the equations in the caption of Figure 3.29.

In nonaqueous solvents, nonprotonated species can be generated. Consequently, many electrochemical studies of organic compounds employ nonaqueous solvents such as acetonitrile, dimethylformamide, and dimethyl sulfoxide [53]. Electrochemistry in nonaqueous solvents is addressed in Chapters 15–18.

In addition to its extensive use for analytical purposes, polarography has been widely employed for the measurement of formation constants of complexes by the effect of complexation on $E_{1/2}$, the kinetics of homogeneous chemical reactions that are coupled to the electrode process, the kinetics of heterogeneous electron transfer between a mercury electrode and an electroactive species, and the effects of adsorption of both electroactive and nonelectroactive species on the electrode surface. Polarography is also useful for detecting endpoints in titrations that involve electroactive species (amperometric titrations). Many of these measurements can now be made much more effectively by the more recently developed potential-step and potential-sweep methods. Polarography derives great advantage from the clean and reproducible electrode surface of pure

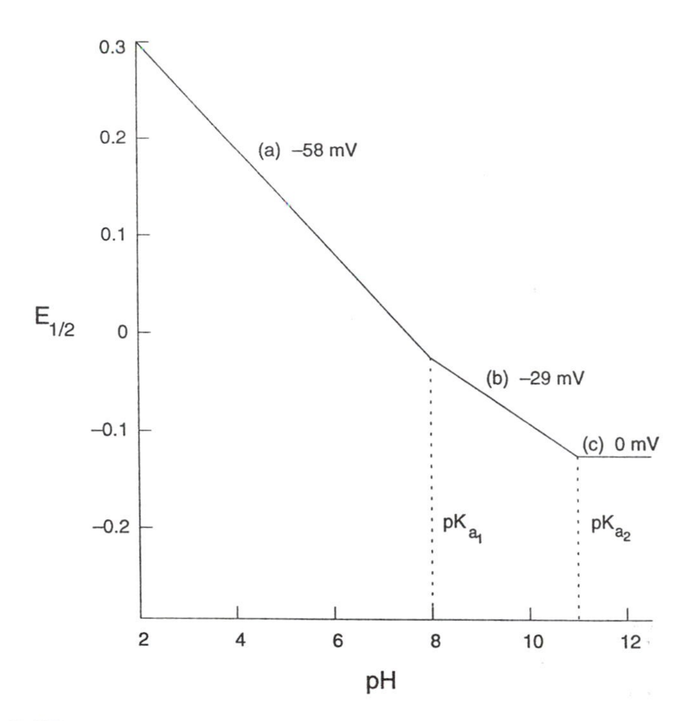


**Figure 3.29** Dependence of $E_{1/2}$ on pH for typical 2e reduction of an organic compound. (a) $O + 2e + 2H^+ \rightarrow RH_2$; (b) $O + 2e + H^+ \rightarrow RH^-$; (c) $O + 2e \rightarrow R^{2-}$.

mercury (and its excellent hydrogen overpotential) obtainable from the DME, thus eliminating many of the difficulties attendant on stationary electrodes. This is almost its only advantage; however, it remains of such significance that the chance of polarography completely disappearing is scant indeed. It has nevertheless lost its position as the paramount electroanalytical technique.

## E. Current-Sampled Polarography

Current-sampled polarography (sometimes called Tast polarography) is a variation of dc polarography in which the response current is only registered during a small interval of time (typically 16.7 ms) during the latter part of drop life for each drop. This enables the contribution of the charging current to the overall current to be minimized, as shown in Figure 3.30A. As we have already noted, the drop surface area increases with $t^{2/3}$. Because capacitance is directly proportional to area and the charging current at a fixed voltage is related to the rate of change of capacitance (since $Q = CV$, $i_{dl} = V\ dC/dt$, and $i_{dl} \propto dA/dt$),

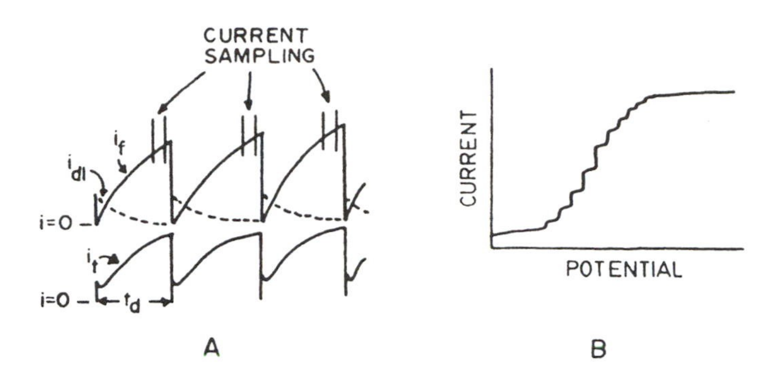


**Figure 3.30** (A) Current-time curves illustrating the sampling sequence for current-sampled polarography: $i_f$, diffusion-controlled faradaic current; $i_{dl}$, double-layer charging current; $i_t = i_f + i_{dl}$. (B) Current-sampled polarogram.

then $i_{dl} \propto t^{-1/3}$. The double-layer charging current will therefore be large early in drop life and fall off as the drop grows.

A sample-and-hold circuit maintains the current value of the previous sampling interval and integrates out noise and interference. This gives the polarogram a staircase appearance as shown in Figure 3.30B. The slower the scan rate and the faster the drop rate, the more samples will be taken to represent a given polarographic wave and the smoother the wave will appear. The improved detection limit realized by minimization of charging current is slight; however, many workers prefer the sampled data presentation because it eliminates the annoying current fluctuations associated with drop growth. This saves ink as well as wear and tear on laboratory recorders.

It is important to recognize that interesting adsorption phenomena can have a profound influence on events early in drop life [54]. Data of this sort would be missed in a sampled experiment. Therefore, it is best to look first at the response without sampling. If no unusual behavior is observed, further experiments may be performed in the sampling mode.

In this technique the excitation and fundamental system response is identical to that for dc polarography. The sample-and-hold measuring scheme is like that used to great advantage in LAPP (see Sec. III.F).

### F. Large-Amplitude Pulse Voltammetry

In large-amplitude pulse voltammetry (LAPV, often called "normal" pulse voltammetry for historical reasons) the excitation waveform consists of successive pulses of gradually changing amplitude between which a constant "initial" potential is applied. The initial potential is usually chosen in a region where none of the sample components is electrochemically reactive. The current response may be measured during the forward step, after return to the initial condition, or both. Usually, the current is sampled only at the end of the forward step, and the remainder of the response waveform is ignored. The sampled current is then plotted against the pulse height (thus "voltammetry"). Sometimes the difference between successive sampled currents is plotted as a function of potential. As we shall see, this results in a response having a derivative shape. It is important to recognize that all variations of LAPV are in essence a series of potential pulse chronoamperometry experiments.

Figure 3.31 illustrates the simplest example of LAPV at a stationary electrode. In this case we assume that the time delay $t_d$ between pulses is such that the initial condition is restored. This implies that all concentration profiles are completely relaxed before the next pulse is applied and every pulse initiates a new chronoamperometric experiment. The length of $t_d$ required to accomplish this will depend on the chemical reversibility of the system, being shorter for reversible reactions and longer for irreversible reactions.

For convenience, let us assume a fully (chemically and electrochemically) reversible system, which gives a response to the ramp excitation illustrated in Figure 3.31. Although many pulses will be applied, we have selected three for closer examination.

*Response to pulse A*: Here the pulse amplitude is insufficient to initiate significant faradaic reaction and the sample concentration profile is not disturbed. The current response is then primarily due to charging the double-layer capacitance to the new potential ($E_i + \Delta E$). The current sampled just before $\tau$ is negligible, for by that time the double layer is fully charged.

*Response to pulse B*: Now the pulse amplitude extends to the half-wave potential for the reactive species, concentration gradients will form, and significant faradaic current will be sampled. Note that the double-layer charging current will have decayed away when the sample is taken. This time-domain discrimination against $i_{dl}$ is a principal advantage of pulse voltammetry. The faradaic current will follow a modified Cottrell equation:

$$i = nFAC^{\circ}\left(\frac{D}{\pi t}\right)^{1/2}\left[1 + \left(\frac{C_O}{C_R}\right)_{x=0}\right]^{-1} \tag{3.34}$$

where for a reversible system

$$\frac{C_O^s}{C_R^s} = \exp\left(\frac{nF}{RT}\right)\left[(E_i + \Delta E) - E_{1/2}\right] \tag{3.35}$$

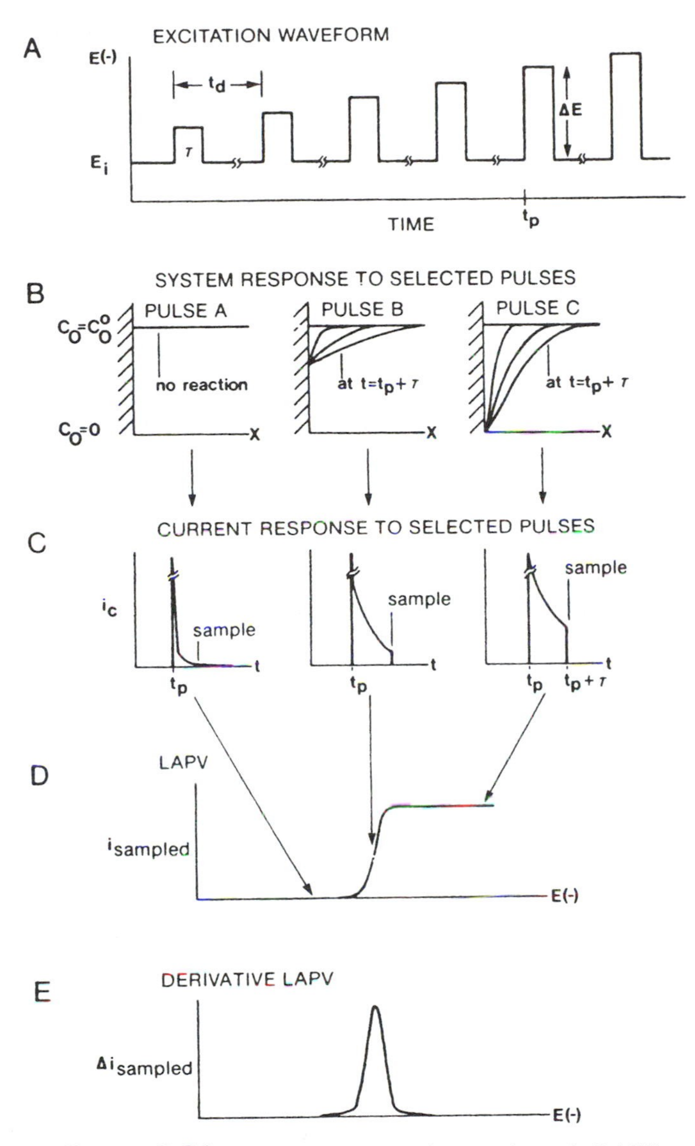


**Figure 3.31** Introduction to stationary electrode LAPV.

In the case of pulse B, $(E_i + \Delta E) = E_{1/2}$, so $(C_O^s/C_R^s) = 1$ and $(1 + C_O^s/C_R^s)^{-1} = 0.5$.

*Response to pulse C*: Now we have the *limiting* case of chronoamperometry. The surface concentration has become effectively zero and the response will follow the Cottrell equation.

Plotting the sampled current versus the applied potential results in the complete pulse voltammogram shown in Figure 3.31D. The reader should determine what effect taking current samples earlier or later would have on this response. Note that there is a trade-off between sensitivity and discrimination against the charging current! If the difference between successive samples is plotted (Fig. 3.31E), a peaked response is obtained.

When LAPV is carried out with a dropping mercury electrode, the technique is large-amplitude pulse polarography, or LAPP (usually referred to as "normal" pulse polarography). One pulse is applied per drop at a fixed time late in the life of the drop. The current is sampled just before the drop is dislodged by mechanical means, a time at which the rate of change of area is smallest. Because the pulse widths are usually in the 50-ms region, the electrode is effectively stationary and of fixed area during each pulse. Since each drop is a new electrode, the time between pulses $t_d$ is made to be the drop time. The timing circuitry that is used to generate the pulses also functions to trigger an electromechanical "drop knocker" or a static mercury drop electrode (see Chap. 14).

Sometimes it is useful to apply identical pulses to several drops before increasing the amplitude. This allows for time averaging and some improvement in the signal-to-noise ratio. LAPP is analytically useful, being sensitive to $10^{-6}$ M in some circumstances.

In an interesting variation of LAPP (originally ascribed to Kalousek), the potential applied to a DME is alternately switched from an initial potential (which is sufficiently negative of the $E_{1/2}$ for a reducible species to cause a limiting cathodic current) to a series of potential pulses changing in height from values more positive than $E_{1/2}$ toward more negative values. The current is sampled late in the life of each pulse, as indicated in Figure 3.32A. At first, the reduction product formed at the initial potential is oxidized during each pulse (unless the couple is totally irreversible) as shown by the anodic wave in Figure 3.32B. After the pulse heights pass through $E_{1/2}$ and reach potentials for reduction, a cathodic current is recorded. The relative values of $i_a$ and $i_c$ will depend on the reversibility of the reaction and the specific pulse widths and sampling times employed.

In yet another variation the current is recorded at the initial potential following a series of pulses increasing in amplitude from the initial potential toward more negative values (i.e., the same excitation shown in Fig. 3.31A). When the pulses reach sufficiently negative potentials to cause reduction of the species, the product of the reduction is oxidized at $E_i$, and the sampled anodic current is recorded unless the couple is totally irreversible. This is equivalent to a series of potential pulse chronoamperometry experiments with samples taken on the reverse step. Techniques of this sort have been used occasionally in reversibility studies and for identification of unstable products of electrode pro-

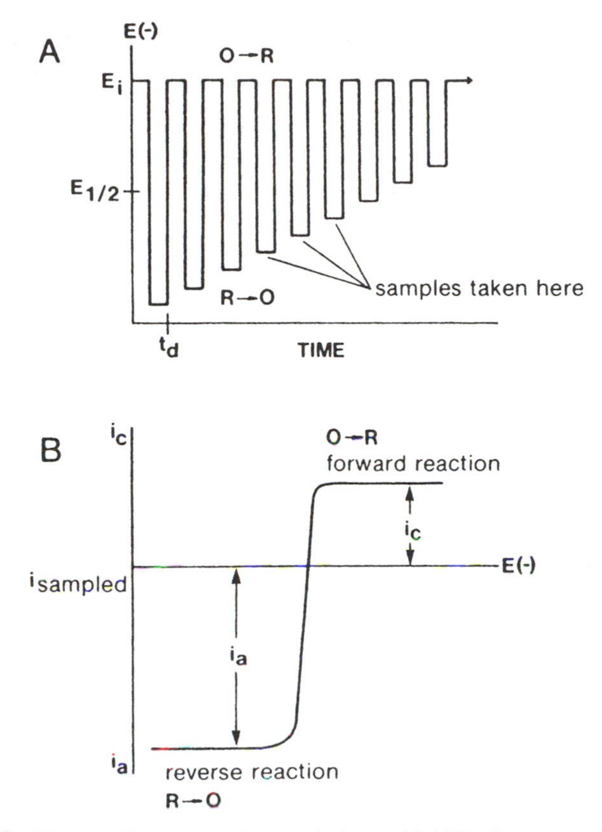


**Figure 3.32** In one of many possible variations of LAPP, the current is sampled on a reverse step just prior to drop fall: (A) excitation, (B) response.

cesses. Although potentially useful, such methods were largely ignored until a few years ago, when a number of such "reverse pulse" experiments were reported [55]. We owe a great deal to the imaginative ideas conceived by Kalousek some 60 years ago. In many respects there is a close relationship between large- and small-amplitude pulse methods. The latter, generally the more popular, are described in Chapter 5.

## G. Thin-Layer Voltammetry

Thin-layer voltammetry (TLV) is a very slow technique, and this fact is used to great advantage in quantitative studies of electrode reactions with small heterogeneous rate constants. This area has been pioneered and reviewed by Hubbard [29,30]. TLV is also applicable to examination of metal deposition and adsorption. The principal work here was carried out by Schmidt and co-workers [56]. The technique has thus far not been particularly useful for analytical determinations; however, in combination with optical monitoring of products, it has proved helpful in mechanism studies of redox-controlled solution chemistry.

The number of coulombs required to oxidize a substance R confined within a thin-layer cell is given by Faraday's law,

$$Q = nFVC_R^\circ \tag{3.36}$$

where $C_R^\circ$ is the initial ($t = 0$) concentration of R, and V is the volume of the thin-layer cavity. If the experiment is carried out slowly enough, $C_{R,t}$ will be uniform across the cell due to rapid diffusional mixing in thin solution films (see Sec. II.E). The current at any time will be given simply by

$$i = \frac{dQ}{dt} = nFV\frac{dC_{R,t}}{dt} \tag{3.37}$$

where $nV(dC_{R,t}/dt)$ is the number of equivalents consumed per second. We know that as R is depleted, the product, O, is formed. Since the amount of material in the cavity is constant,

$$C_{O,t} = C_R^\circ - C_{R,t} \tag{3.38}$$

For a reversible system, the potential will be defined by the Nernst equation, such that

$$E_t = E^{\circ\prime} + \frac{RT}{nF}\ln\frac{C_{O,t}}{C_{R,t}} \tag{3.39}$$

Substituting for $C_{O,t}$, we have

$$E_t = E^{\circ\prime} + \frac{RT}{nF}\ln\frac{C_R^\circ - C_{R,t}}{C_{R,t}} \tag{3.40}$$

It is important to recognize that this equation will hold true regardless of whether R is converted to O in response to a small current, a slowly varying potential, or even via a photooxidation reaction.

For a slow potential ramp excitation

$$\frac{dE_t}{dt} = \nu \tag{3.41}$$

where $\nu$ is the scan rate in V/s. If we differentiate Equation 3.40 with respect to time, substitute $\nu$ for $dE/dt$ and $i/nFV$ for $dC_R/dt$, and combine the result with the exponential form of Equation 3.40, we have

$$i = -\frac{n^2F^2VC_R^\circ \nu \exp[(nF/RT)(E - E^{\circ\prime})]}{RT\{1 + \exp[(nF/RT)(E - E^{\circ\prime})]\}^2} \qquad (3.42)$$

The current is maximum at E = $E^{\circ\prime}$, so

$$i_p = \frac{n^2F^2VC_R^\circ \nu}{4RT} \qquad (3.43)$$

A plot of i versus E (Eq. 3.42) defines the ideal thin-layer voltammogram shown in Figure 3.33. The contrast with voltammetry under quasi-infinite conditions should be evident. Note that the peak is symmetrical about $E^{\circ\prime}$, and that the current drops to zero after the potential passes through $E_p$, and that $i_p$ is *directly* proportional to scan rate. The number of coulombs passed in fully traversing the peak is given by Equation 3.36 since the conversion of R to O is complete.

If we reverse the scan direction, a reduction wave is obtained for the reverse reaction (dotted line in Fig. 3.33), which is an exact mirror image of the oxidation wave. This result is obtained only for a fully reversible electrode

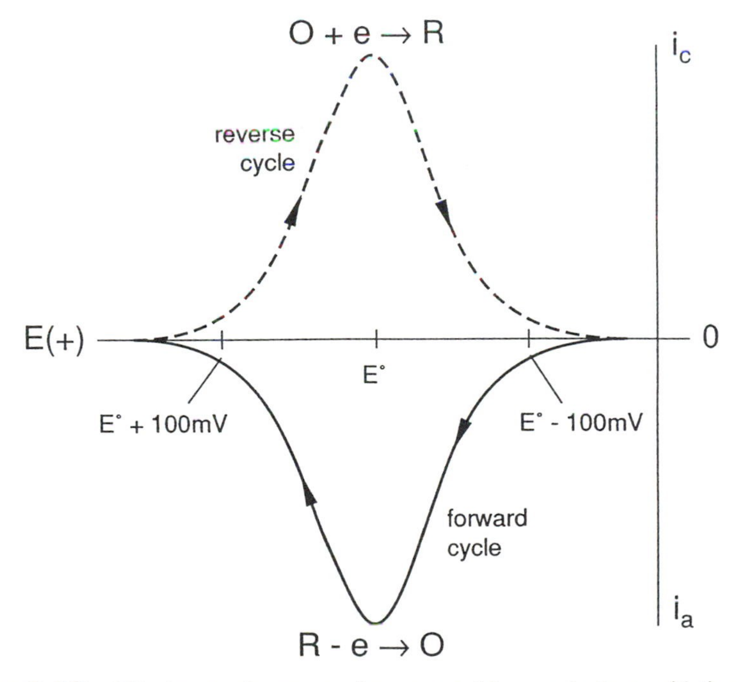


**Figure 3.33** Thin-layer voltammetry for a reversible one-electron oxidation.

reaction. When the heterogeneous rate constant is very low, the forward and reverse waves will lose their symmetry and the peak potentials will separate.

Thin-layer CV is frequently used in conjunction with an optically transparent thin-layer electrode to obtain spectra, $E^{\circ\prime}$, and n for redox couples by the "spectropotentiostatic" technique and thin-layer coulometry, as described earlier in this chapter. CV is used initially to locate the redox couple and give an estimate of $E^{\circ\prime}$. Once the CV is obtained, appropriate potentials can be selected for the spectropotentiostatic experiment and potential-step coulometry. A typical thin-layer cyclic voltammogram for the Schiff base complex Co($sal_2en$) in nonaqueous solvent is shown in Figure 3.34.

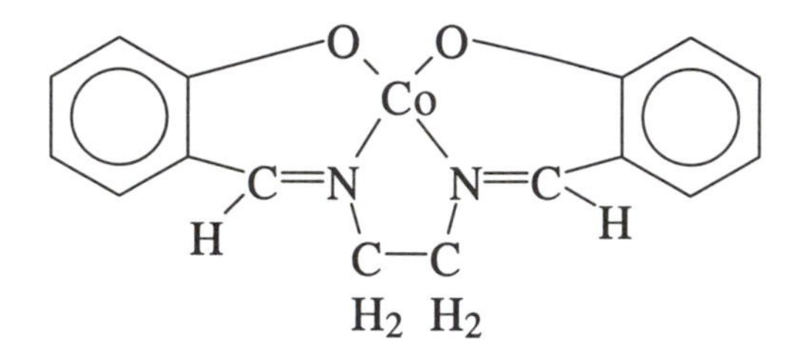


Co($sal_2en$)

The waves for the Co(III)/Co(II) and Co(II)/Co(I) couples are well formed. The peak potentials, which should be identical for the oxidation and reduction halves of a couple, are split due to iR drop in the cell.

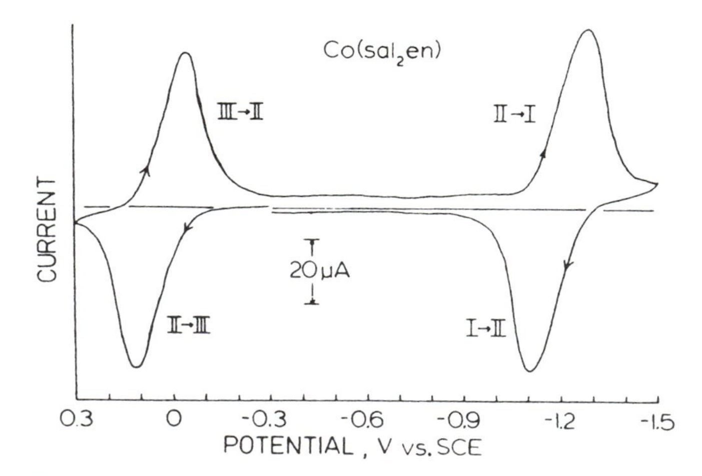


**Figure 3.34** Thin-layer cyclic voltammogram of 1.3 mM Co($sal_2en$) in 0.5 M TEAP/DMF. Scan rate 2.0 mV/s, initial potential –0.3 V vs. SCE. [From Ref. 57.]

A unique form of thin-layer voltammetry was devised by Anderson and Reilley [58]. In this case, two opposing working electrodes are used and the potential difference across each interface is controlled independently. The electronics necessary to accomplish this are described in Chapter 6. It is thus possible to achieve a cyclic steady-state condition whereby one electrode serves as cathode and the other as anode. Both forms of the redox couple are continuously interconverted. Figure 3.35 illustrates how this condition can be achieved. The cell initially contains only O and the potential of both electrodes is such that only O is stable. As $E_1$ is made more negative, conversion of O to R begins at $W_1$. $E_2$ is kept sufficiently positive to reconvert R back to O at $W_2$. Thus a steady-state current, $i_{ss} \neq f(t)$, is achieved. The flux of O toward $W_1$ will equal the flux of R toward $W_2$, with each molecule moving along its respective concentration profile. The profiles will have different slopes if the diffusion coefficients are different, since the flux is equal to D(dC/dx) for each species. Under limiting conditions, both surface concentrations become zero and the current, $i_{ss}^{lim}$, is independent of potential.

Suppose that one species (e.g., R) of a redox couple is subjected to a chemical reaction leading to nonelectroactive products. If this reaction occurs during a limiting steady-state condition (Fig. 3.36), $i_{ss}^{lim}$ will decay until both R and O are completely exhausted. The rate constant for the coupled reaction can be calculated from the decay curve [59].

The twin-electrode technique has numerous potential applications both to analysis and to mechanism studies [28]; however, because of the difficulty of manufacturing cells, experiments fall well behind the theoretical developments. One application uses the twin electrodes to selectively deposit Zn on one elec-

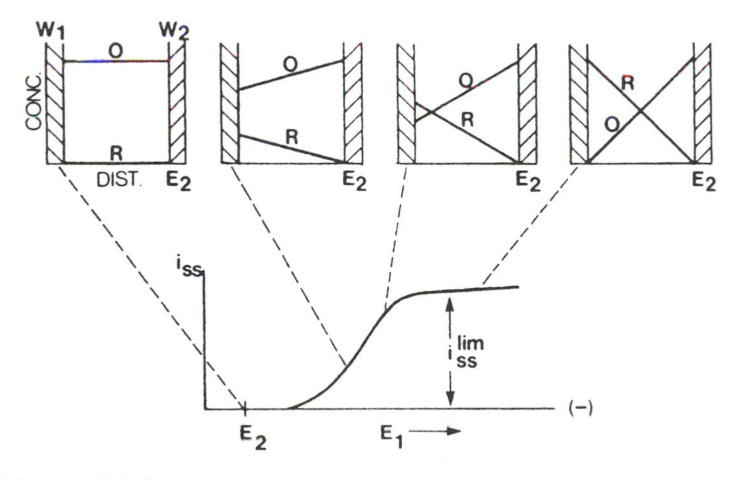


**Figure 3.35** Thin-layer twin-electrode voltammetry. [From Ref. 58.]

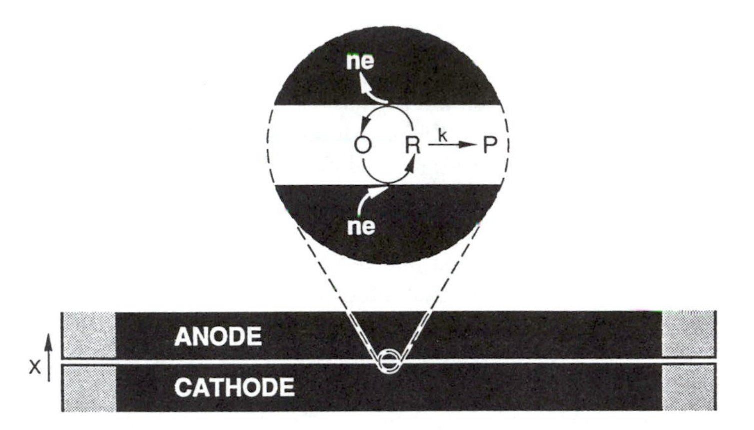


**Figure 3.36** Destruction of a steady-state redox cycle by a coupled chemical reaction. [From Ref. 59, reprinted with permission.]

trode and Cu on the other electrode to avoid the formation of an interfering intermetallic compound of Cu-Zn [60]. In general, though, there is little activity with twin-electrode TLV at this time.

The detection limit for TLV has been improved substantially by using differential pulse and square-wave voltammetry (Chap. 5). For example, detection limits in the $10^{-8}$ M range and below have been demonstrated in thin-layer cells requiring less than 100 $\mu$L of sample [61,62]. One practical application of twin-electrode thin-layer cells is in the automatic electrochromic rearview mirror for automobiles. A cell with optically transparent electrodes is placed in front of a mirrored surface. At night, electrolysis in the cell to generate colored material can rapidly reduce glare from following vehicles.

## IV. CONTROLLED-POTENTIAL TECHNIQUES IN FLOWING SOLUTION

### A. Introduction to Hydrodynamic Voltammetry

Several electroanalytical techniques involve the hydrodynamic flow of electrolyte over the electrode surface. These methods take advantage of enhanced sensitivity resulting from the enhanced mass transport of electroactive substance to the electrode that occurs under hydrodynamic conditions.

Some understanding of liquid flow along a solid-solution interface is useful in understanding these techniques. Consider a stationary platinum electrode immersed in a stirred solution. Three regions of solution flow can be identified.

(1) *Turbulent flow* comprises the solution bulk. (2) As the electrode surface is approached, a transition to *laminar flow* occurs. This is a nonturbulent flow in which adjacent layers slide by each other parallel to the electrode surface. (3) The rate of this laminar flow decreases near the electrode due to frictional forces until a thin layer of *stagnant solution* is present immediately adjacent to the electrode surface. It is convenient, although not entirely correct, to consider this thin layer of stagnant solution as having a discrete thickness $\delta$, called the Nernst diffusion layer.

Figure 3.37 illustrates the Nernst diffusion layer in terms of concentration-distance profiles for a solution containing species O. As pointed out previously, the concentration of redox species in equilibrium at the electrode-solution interface is determined by the Nernst equation. Figure 3.37A illustrates the concentration-distance profile for O under the condition that its surface concentration has not been perturbed. Either the cell is at open circuit, or a potential has been applied that is sufficiently positive of $E^{\circ\prime}_{O,R}$ not to alter measurably the surface concentrations of the O,R couple.

The profiles in Figure 3.37B represent the situation in which a potential is applied that requires equal concentrations of O and R at the electrode surface to satisfy the Nernst equation (i.e., $E = E^{\circ\prime}_{O,R}$). To fulfill this requirement, the electrode electrolyzes O to R at the rate required to maintain equal concentrations of O and R at the surface. If this potential is maintained, a continuous electrolysis of O to R is necessary to maintain surface concentrations because R diffuses away from the interface across the stagnant layer and is then swept away by the laminar flow.

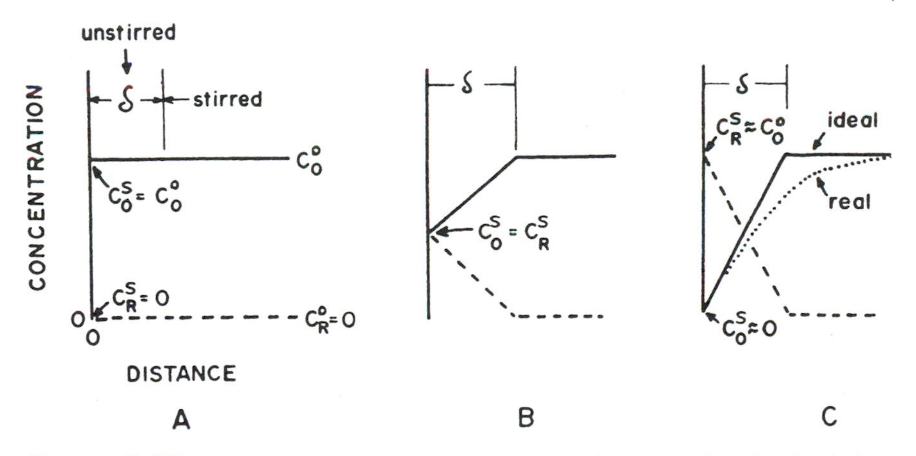


**Figure 3.37** Concentration-distance profiles for voltammetry in stirred solution. $D_O = D_R$.

Two regions of this profile can be considered.

1. At distances greater than $\delta$, concentrations are maintained homogeneous by the stirring action. As long as the electrode area is small (microelectrode) relative to the solution volume and the experiment is not prolonged, the bulk concentrations will not be altered appreciably by the electrolytic conversion of O to R at the surface.
2. The removal of O at the electrode surface sets up a concentration gradient across the stagnant solution layer. Species O diffuses across this layer to the electrode surface where it is electrolyzed to R, which then diffuses back across the stagnant layer to the bulk solution.

Thus the electrolysis process is controlled by a combination of (1) mass transport of O to the edge of the stagnant layer by the laminar flow, and (2) subsequent diffusion of O across the stagnant layer under the influence of the concentration gradient (caused by electrolysis of O to R at the electrode surface) to satisfy the Nernst equation.

Figure 3.37C shows the profiles that result when the applied potential is sufficiently negative that the concentration of O at the electrode surface is effectively zero. In this case, essentially all of O at the electrode surface must be electrolyzed to R in order to satisfy the Nernst equation. Consequently, O is converted to R as rapidly as it can diffuse to the electrode surface. Since this is the limiting condition, application of even more negative potentials causes no measurable change in the profiles.

Although the transition between stagnant and flowing solution is considered to be abrupt in this example, the transition is in reality gradual. Consequently, the profiles will be rounded as shown by the dotted line in Figure 3.37C. However, the hypothetical situation of an abrupt transition is a useful approximation in mathematical treatments as shown below. (See Ref. 6, Chap. 4, for a critique of the Nernst diffusion layer.)

This discussion has focused on a stationary electrode immersed in a stirred or flowing solution. Similar arguments apply to the case of a rotating electrode.

## B. Stirred-Solution Voltammetry

Stirred-solution voltammetry utilizes current-voltage relationships that are obtained at a stationary electrode immersed in a stirred solution. In order to understand this aspect of electrochemistry, it is extremely useful to consider a typical current-voltage curve (voltammogram) in terms of the concept of concentration-distance profiles presented in the preceding section. The discussion will consider the potential, rather than the current, as the controlled variable.

The current resulting from an applied potential is determined by the slope of the concentration-distance profile of the reactant at the electrode surface as described by Equation 3.5. This equation can be expressed in terms of the Nernst diffusion-layer concept by simply approximating $\partial x$ by $\delta$ and $\partial C$ by $C_O^\circ - C_O^s$ for the case in Figure 3.37.

$$i = nFAD_O \frac{C_O^\circ - C_O^s}{\delta} \tag{3.44}$$

Examination of the profiles in Figure 3.37 shows that this is a valid substitution for the slope of the profile at the electrode surface for the ideal case of an abrupt transition at $x = \delta$.

Figure 3.38 illustrates a typical current-voltage curve that would be obtained for a solution containing equal concentrations of O and R. The shape of this curve can be understood by considering the slopes of the concentration-distance profiles that are depicted for several representative potentials. During oxidation the current is determined by the slope of the profile for R, whereas the profile

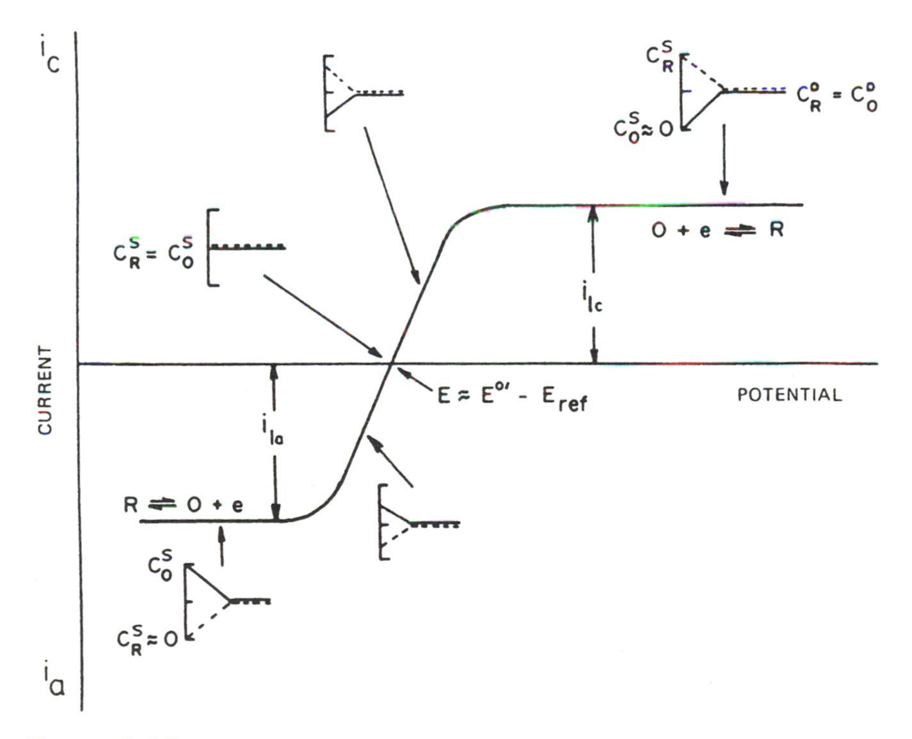


**Figure 3.38** Voltammogram with representative concentration-distance profiles for a solution containing equal concentrations of O (solid line) and R (dotted line). $D_O = D_R$.

for O is the determining factor during reduction. A limiting current ($i_l$) is reached when the surface concentration $C^s$ of the controlling species becomes effectively zero. Substitution of $C_O^s = 0$ into Equation 3.44 gives the equation for the limiting current.

$$i_l = \frac{nFAD_O C_O^\circ}{\delta} \tag{3.45}$$

The equation for the sigmoidal part of the voltammogram is analogous to that described for polarography (Sec. III.D).

$$E = E_{1/2} - \frac{RT}{nF} \ln \frac{i - i_{la}}{i_{lc} - i} \tag{3.46}$$

$$E_{1/2} = E^{\circ\prime}_{O,R} - \frac{RT}{nF} \ln \frac{D_O}{D_R} \tag{3.47}$$

It is important to recognize that all forms of hydrodynamic voltammetry are *steady-state methods* in that the current at a given potential is independent of both scan direction and time. In fact, identical voltammograms can be obtained by controlling the current and monitoring the potential response. This is, however, rarely done in practice.

## C. Rotating-Disk Voltammetry

Stirred-solution voltammetry per se has been largely superseded by rotating-disk voltammetry, in which the hydrodynamics are more easily controlled [63–67]. A typical rotating disk electrode is shown in Figure 3.39. The hydrodynamic flow pattern resulting from rapid rotation of the disk moves liquid horizontally out and away from the center of the disk with a consequent upward axial flow to replenish liquid at the surface as shown by the arrows in Figure 3.39. A completely rigorous hydrodynamic treatment has been given by Levich [63,64]. The equation for the limiting current that is derived from this treatment is

$$i_l = 0.620 nFAC^\circ D^{2/3} \nu^{-1/6} \omega^{1/2} \tag{3.48}$$

where

$C^\circ$ = solution concentration, $mol/cm^3$
$i_l$ = limiting current, A
$\nu$ = kinematic viscosity of the fluid, $cm^2/s$
$\omega$ = angular velocity of the disk ($\omega = 2\pi N$, N = rps)

Good electrode design is necessary for strict adherence to this equation.

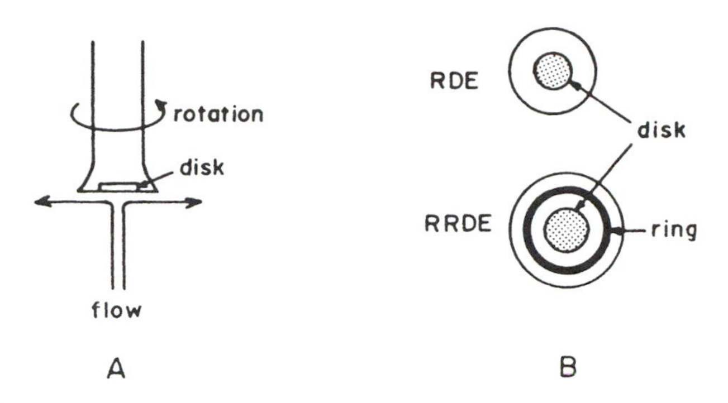


**Figure 3.39** (A) Rotating-disk electrode with hydrodynamic flow pattern. (B) Bottom view of rotating-disk electrode (RDE) and rotating ring-disk electrode (RRDE).

The shapes of rotating-disk voltammograms are identical to the stirred-solution voltammograms. Thus a rotating disk voltammogram for a solution containing equal concentrations of O and R would be identical in shape to the voltammogram in Figure 3.38.

Considerable information can be obtained about a redox system by rotating-disk voltammetry. An initial voltammogram will locate redox couples in solution, giving information about redox potentials and reversibility. Electrode mechanisms that involve coupled homogeneous chemical reactions are sometimes amenable to study by the rotating disk. Generally, mechanisms and rate constants are deduced from variations in the limiting current as a function of the rotation speed of the disk. A classical example of such a study is the ECE mechanism. The "apparent n" values obtained from the limiting current via Equation 3.48 for a series of angular velocities of the disk are shown in Figure 3.40. At low angular velocities, the entire reaction sequence has time to occur as reactant is swept past the disk by laminar flow and the apparent n value is 2. However, as the angular velocity is increased, species B has less time to react before it is swept past the electrode surface. At sufficiently fast rotational rates, all of species B is swept past the surface before any significant reaction has occurred and the apparent n is 1. The magnitude of the rate constant k can be determined from the angular velocity required for the transitional region from $n = 1$ to $n = 2$. A variety of other mechanisms are also amenable to study by rotating-disk electrodes [65,66].

Since the limiting current is proportional to concentration, a rotating-disk electrode can be used for analytical purposes (but almost never is in practice). In summary, the rotating disk electrode is one of the valuable techniques available for the fundamental study of electrode reactions as evidenced by its substantial popularity.

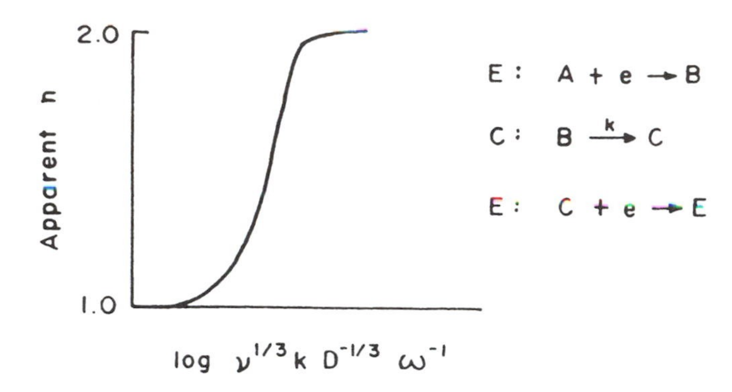


**Figure 3.40** Variation in "apparent n" as a function of angular velocity of an RDE for the ECE mechanism.

## D. Rotating Ring-Disk Voltammetry

The rotating ring-disk electrode is an extension of the previously discussed disk electrode in that the disk is encircled by a ring, which functions as a second working electrode. The configuration is shown in Figure 3.39B. The idea here is to electrochemically generate a reactive species at the disk and then to monitor the species electrochemically as it is swept past the ring by laminar flow. This technique has proven exceedingly useful in the study of reaction mechanisms since electroactive intermediates of coupled homogeneous chemical reactions can be monitored at the ring electrode.

The reduction of oxygen illustrates the use of rotating ring-disk voltammetry [6,68]. Figure 3.41A shows the voltammogram obtained at the disk for the reduction of oxygen. Figure 3.41B shows the resulting ring "voltammogram" when the potential of the ring is maintained at a value for oxidation of $H_2O_2$ and the resulting current plotted as a function of the potential applied to the *disk*. In the potential region in which the disk is generating $H_2O_2$, anodic current is recorded at the ring due to its oxidation of the $H_2O_2$ that is being swept past by laminar flow. When the potential applied to the disk becomes sufficiently negative that $O_2$ is being reduced to $OH^-$ at the disk, the ring current diminishes, since $OH^-$ rather than $H_2O_2$ is being swept past. Such experiments can confirm the existence of a particular intermediate, such as $H_2O_2$ in this case, in an electrode reaction.

The rotating ring-disk electrode has been applied to a variety of electrode mechanisms [66,67]. It is one of the most useful techniques for studying electrode reactions.

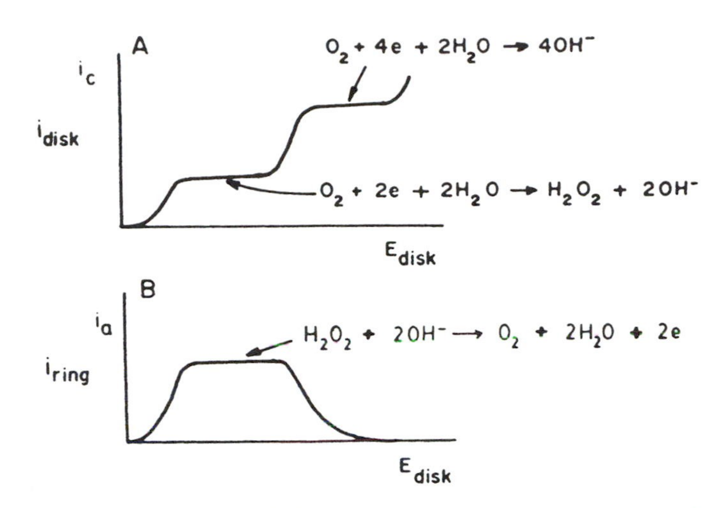


**Figure 3.41** (A) Disk and (B) ring currents for reduction of oxygen at the rotating ring-disk electrode. [Adapted from Ref. 6.]

## E. Voltammetry in a Channel

Voltammetry experiments are occasionally undertaken in the form of a tubular or rectangular channel through which the electrolyte solution is pumped at a more or less constant velocity. The electrode may form the channel itself or be embedded in the wall of an inert material, which defines the flow pattern. Sometimes the channel is packed with small particles of electrode material in contact with each other. The latter situation is designed to improve the conversion efficiency of the cell. When all the electroactive molecules are converted during passage through such a porous bed, the efficiency is 100% and the cell is said to be operating coulometrically (see Sec. IV.F).

It should be clear from Faraday's law that the limiting current monitored for a coulometric flow cell will be given by

$$i_l = nFC_iu \tag{3.49}$$

where $C_i$ is the concentration of reactant and u is the volume flow rate in $cm^3/s$. Note that the current is independent of the diffusion coefficient for the reactant. In most flow cells, the efficiency is far below 100%, and the operation is said to be amperometric. In such cases the current will be given by

$$i_l = nF(C_i - C_f)u \tag{3.50}$$

where $C_f$ is the reactant concentration at the outlet side. The problem is not simple, however, because the conversion efficiency $[(C_f/C_i) \times 100\%]$ is a complicated function of u and the cell geometry. Nevertheless, understanding the essential characteristics of voltammetry under these conditions is perfectly analogous with our earlier discussion for stirred solutions and rotating electrodes. The mass transport-controlled steady-state current is the same function of potential, and the voltammetric curves have the same characteristic shape.

Voltammetry experiments are not often performed in flow cells for analytical purposes. One reason for this is the special problem of ohmic potential losses (iR drops) at an electrode in a confined stream. Another reason is the problem of precisely pumping solution at a carefully controlled velocity. In general, rotating electrodes are more easily controlled and do not involve serious plumbing problems. On the other hand, flow cells operated at a fixed potential (i.e., at one point along the steady-state voltammetric curve) are eminently useful for electrosynthesis, chromatographic detection, and automated analysis systems. These features will be described in later chapters.

## F. Controlled-Potential Coulometry

Coulometric techniques involve the determination of the quantity of material electrolyzed from the amount of charge passed through an electrochemical cell during electrolysis. Faraday's law relates the measured charge to the amount of material electrolyzed,

$$N = \frac{Q}{nF} \tag{3.51}$$

where N is the number of moles of substance being electrolyzed, Q is the total charge passed (coulombs), n is the number of electrons transferred per molecule, and F is Faraday's constant (96,485 C/eq).

In controlled-potential coulometry, the potential of a working electrode is applied at a value such that complete electrolysis of the desired species occurs. The total charge passed during the exhaustive electrolysis is obtained by integrating (usually electronically) the current. The completion of electrolysis is indicated when the current becomes indistinguishable from the residual current. The charge due to the residual current must be subtracted from the total measured charge in order to obtain the faradaic charge for the electrode reaction of interest. This residual charge is estimated by repeating the experiment on supporting electrolyte and integrating the current over the same time interval. There is no guarantee, however, that the processes responsible for the residual current are completely independent of the presence of sample and/or product.

Selection of the correct potential to apply can be made by examining a voltammogram of the species being electrolyzed. The applied potential should be well on the plateau of limiting current $i_l$ for the redox process in order to effect rapid and complete electrolysis.

Maximizing the magnitude of the current during electrolysis minimizes the time required for the complete electrolysis. Equation 3.45 relates the concentration of the electroactive species to the magnitude of the limiting current obtained during a controlled-potential electrolysis at an electrode immersed in a stirred solution. The current is usually increased by increasing the surface area of the electrode, A, and by minimizing the diffusion layer δ with efficient stirring. Of course, as the electrolysis proceeds, C° decreases, causing $i_l$ to drop. The current at any time t during electrolysis is given by

$$i = i_0 e^{-(AD/V\delta)t} \tag{3.52}$$

where $i_0$ is the limiting current at the onset of electrolysis, V is the solution volume, and the other terms were identified with Equation 3.45. It is important to consider these parameters to cause the electrolysis to proceed at a reasonable rate.

Controlled-potential coulometry has been applied extensively to precision determinations [69] and to the establishment of n values for electrode reactions. The technique has also been used for the elucidation of electrode reactions [70,71]. It is a very useful technique when used in conjunction with a thin-layer cell where complete electrolysis is rapid (see Sec. II.E).

A variation of controlled-potential coulometry is the indirect coulometric titration with simultaneous spectral monitoring (i.e., spectrocoulometry) [72]. This approach has been applied to the spectrocoulometric titration of biological redox components such as cytochrome c, cytochrome c oxidase, and ferredoxin [73–75]. Many of these biological species undergo direct electrolysis at an electrode at a negligible or very slow rate. Thus coulometric titration by means of an EC catalytic regeneration mechanism (see Chap. 2, Sec. V) is a particularly useful technique for these systems. An example is the titration of cytochrome c oxidase with electrogenerated methyl viologen radical cation ($MV^{\dot{+}}$).

E: $CH_3-^{+}N(C_5H_4)-(C_5H_4)N^{+}-CH_3 + e \rightleftharpoons MV^{\dot{+}}$ (3.53a)

($MV^{2+}$)

C: $4MV^{\dot{+}} + cyt_{ox} \rightleftharpoons 4MV^{2+} + cyt_{red}$ (3.53b)

The titrant $MV^{\dot{+}}$ is generated by controlled-potential coulometry. Since $MV^{\dot{+}}$ reacts with the cytochrome, the amount of charge passing through the cell can be related to reduction of cytochrome c oxidase. Since cytochrome c oxidase contains two units of heme a and two atoms of copper, four electrons are required to reduce the oxidized form completely (two electrons to reduce the two atoms of $Fe^{3+}$ in the hemes and two electrons to reduce the two atoms of $Cu^{2+}$ to $Cu^{+}$). Figure 3.42A shows spectra recorded during the titration at an optically transparent electrode. The reduction of the two hemes causes the absorbance increase at 605 nm. The equivalence point (e.p.) of the titration can be determined by monitoring the absorbance changes at 605 nm during the titration as shown in Figure 3.42B. Titrations of this sort have been used mainly to measure n values of biological redox systems, rather than concentrations.

## G. Preconcentration by Electrodeposition

Preconcentration by electrodeposition is one of the more valuable techniques for electroanalytical trace analysis. In this method the electroactive species to be determined is concentrated by electrodeposition into or onto an electrode placed

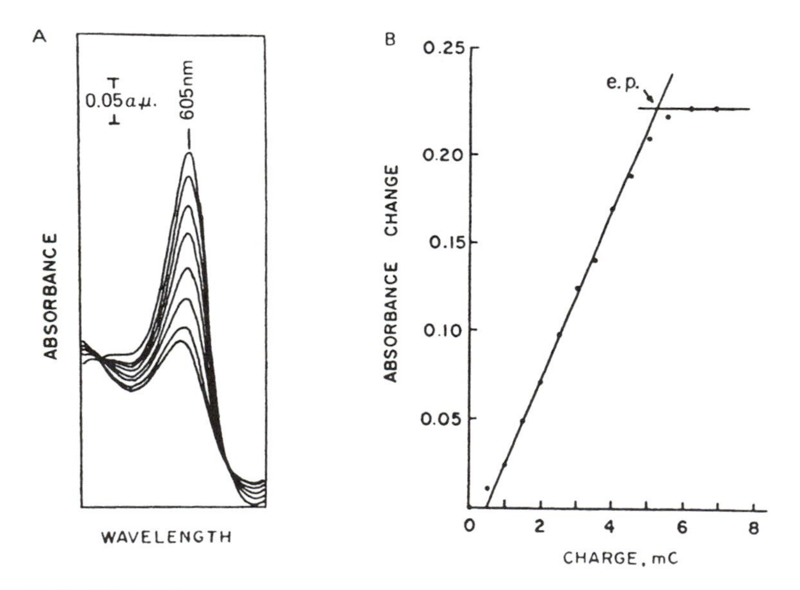


**Figure 3.42** Indirect coulometric titration of cytochrome c oxidase by incremental generation of $MV^{\dot{+}}$ at $SnO_2$ optically transparent electrode. (A) Spectra recorded during titration. (B) Titration curve, $MV^{\dot{+}}$ added in increments of 0.25 mC. [Adapted from W. R. Heineman, T. Kuwana, and C.R. Hartzell, *Biochem. Biophys. Res. Commun. 49*:1 (1972).]

in a stirred (usually) sample solution. For example, $Pb^{2+}$ can be concentrated by reducing it to $Pb^{\circ}$ at a hanging mercury drop electrode or at a thin mercury film electrode. This is an electrochemical extraction, the $Pb^{2+}$ being extracted from an aqueous solution into the liquid mercury phase as $Pb^{\circ}$. Since the volume of the mercury electrode is usually substantially less than that of the dilute aqueous solution being analyzed, the concentration of $Pb^{\circ}$ in the mercury becomes much greater than the concentration of $Pb^{2+}$ in the water as the electrodeposition proceeds. A concentration enhancement of as great as $10^6$ can take place. After a sufficiently high concentration level is reached in the drop, the stirring is stopped, and the electrodeposited $Pb^{\circ}$ is electrochemically "stripped" out of the mercury and back into the aqueous solution as $Pb^{2+}$ by a positive potential scan from the deposition potential. The resulting current signal is proportional to the concentration of electrodeposited material, which in turn is proportional to the concentration of electroactive species in solution.

The combination of preconcentration by electrodeposition with stripping by voltammetry is probably the most sensitive electroanalytical method in common use today. Consequently, this popular technique is discussed in more detail in Chapter 24. Preconcentration of material by an electrode reaction has been used in sample preparation for atomic absorption, neutron activation, x-ray fluorescence, microprobe, and several other spectroscopic techniques.

## H. Amperometric Titrations

A titration in which the endpoint is determined by the current resulting from a potential applied across two electrodes is an amperometric titration [76].

*Working Electrode-Reference Electrode*

In the case in which the potential of a working electrode is controlled relative to a reference electrode, the potential is applied so that a limiting current that is proportional to the concentration of one or more of the reactants or products of the titration is measured. A titration curve is obtained by plotting the limiting current as a function of volume of titrant added. The shape of the titration curve can be predicted from hydrodynamic voltammograms of the solution obtained at various stages of the titration. Figure 3.43A shows voltammograms recorded during the titration of $Cl^-$ with $Ag^+$. The titration reaction is

$$Cl^- \text{ (sample)} + Ag^+ \text{ (titrant)} \rightleftharpoons AgCl(s) \tag{3.54}$$

The working electrode is silver metal. At 0% titration, the anodic limiting current is controlled by the mass transport of $Cl^-$ to the electrode for the oxidation of Ag to form AgCl. As the titration proceeds, the decrease in this limiting current reflects the decrease in $Cl^-$ concentration due to addition of $Ag^+$ titrant until the current becomes zero at the equivalence point (100% titration).

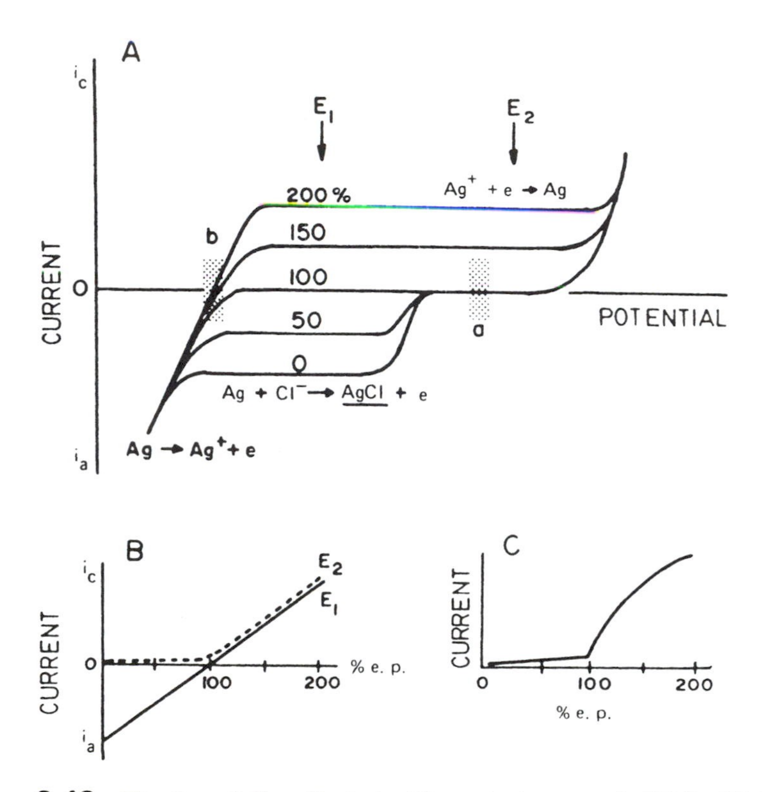


**Figure 3.43** Titration of $Cl^-$ with $Ag^+$. ($Cl^- + Ag^+ \rightleftharpoons AgCl(s)$). (A) Voltammograms at Ag electrode for 0, 50, 100, 150, and 200% of the equivalence point during the titration. (B) Amperometric titration curve for applied potential at $E_1$ and $E_2$. (C) Amperometric titration curve for two Ag electrodes. $\Delta E$ indicated by the shaded areas in A.

The appearance of a cathodic limiting current after the equivalence point reflects the reduction of excess $Ag^+$ titrant. Figure 3.43B shows resulting amperometric titration curves for two values of applied potential. Their shapes are determined by the behavior of the limiting current of the voltammogram at the particular potential during the titration.

*Two Working Electrodes*

A useful variation of the amperometric titration involves measuring the current resulting from a small fixed potential applied across two working electrodes. One electrode functions as an anode and the other as a cathode. Once again, the expected current behavior during a titration can be explained by means of hydrodynamic voltammograms. Position a in Figure 3.43A shows the small potential, $\Delta E$, applied across two Ag electrodes for the example of $Cl^-$ titrated with $Ag^+$. Before the equivalence point is reached, the measured current is only the small residual current for reduction at the more negative electrode and oxidation at the more positive electrode. (Note that for a significantly larger current, the value for $\Delta E$ would have to be increased to a value where $H_2$ would evolve at the cathode and AgCl would deposit on the anode.) However, after the equivalence point is reached, the potential of the two electrodes shifts positively to position b, where the value of $\Delta E$ is sufficiently great to cause reduction of $Ag^+$ at the cathode and oxidation of Ag at the anode. The current is limited by the mass transport of $Ag^+$ to the cathode surface, which increases as additional excess $Ag^+$ is added. The resulting amperometric titration curve is shown in Figure 3.43C.

It is important to realize that the actual potential of the two electrodes is not fixed with respect to any reference electrode. Consequently, in Figure 3.43A the $\Delta E$ increment is free to move on the potential axis. It locates itself at whatever potential is necessary for the current at the anode to balance the current at the cathode. This means that the $\Delta E$ increment will always straddle the potential at which the current voltage curve shows zero current (intercepts the potential axis).

One advantage of the amperometric titration is its ease of automation. A titrator can be signaled to shut off when a specified current level is reached. The advantage of the two-working-electrode variation is the elimination of a reference electrode, which can be troublesome in nonaqueous solvents.

## REFERENCES


1. M. Petek, T. E. Neal, and R. W. Murray, *Anal. Chem. 43*:1069 (1971).
2. P. Delahay, *New Instrumental Methods in Electrochemistry*, Interscience, New York, 1954.
3. J. E. Davis and N. Winograd, *Anal. Chem. 44*:2152 (1972).
4. I. Shain and K. J. Martin, *J. Phys. Chem. 65*:254 (1961).
5. M. von Stackelberg, M. Pilgram, and V. Toome, *Z. Elektrochem. 57*:342 (1953).
6. R. N. Adams, *Electrochemistry at Solid Electrodes*, Marcel Dekker, New York, 1969.
7. A. J. Bard and L. R. Faulkner, *Electrochemical Methods: Fundamentals and Applications*, Wiley, New York, 1980.
8. G. S. Alberts and I. Shain, *Anal. Chem. 35*:1859 (1963).
9. M. D. Hawley and S. W. Feldberg, *J. Phys. Chem. 70*:3459 (1966).
10. S. Feldberg, *J. Phys. Chem. 73*:1238 (1969).
11. W. M. Schwarz and I. Shain, *J. Phys. Chem. 69*:30 (1965).
12. M. L. Olmstead and R. S. Nicholson, *Anal. Chem. 41*:851 (1969).
13. F. C. Anson, *Anal. Chem. 38*:54 (1966).
14. J. H. Christie, R. A. Osteryoung, and F. C. Anson, *J. Electroanal. Chem. 13*:236 (1967).

15. F. C. Anson and R. A. Osteryoung, *J. Chem. Educ. 60*:293 (1983).
16. B. Case and F. C. Anson, *J. Phys. Chem. 71*:402 (1967).
17. G. W. O'Dom and R. W. Murray, *Anal. Chem. 39*:51 (1967).
18. J. H. Christie, *J. Electroanal. Chem. 13*:79 (1967).
19. T. Kuwana and W. R. Heineman, *Acc. Chem. Res. 9*:241 (1976).
20. T. Kuwana and N. Winograd, in *Electroanalytical Chemistry*, Vol. 7 (A. J. Bard, ed.), Marcel Dekker, New York, 1974, Chap. 1.
21. W. R. Heineman, F. M. Hawkridge, and H. N. Blount, in *Electroanalytical Chemistry*, Vol. 13 (A. J. Bard, ed.), Marcel Dekker, New York, 1984, Chap. 1.
22. J. W. Strojek, G. A. Gruver, and T. Kuwana, *Anal. Chem. 41*:481 (1969).
23. A. W. B. Aylmer-Kelly, A. Bewick, P. R. Cantrill, and A. M. Tuxford, *Faraday Discuss. Chem. Soc. 56*:96 (1974).
24. J. D. E. McIntyre, in *Advances in Electrochemistry and Electrochemical Engineering*, Vol. 9 (R. H. Muller, ed.), Wiley-Interscience, New York, 1973.
25. R. J. Gale, ed., *Spectroelectrochemistry: Theory and Practice*, Plenum Press, New York, 1988.
26. H. D. Abruña, ed., *Modern Techniques for In-Situ Interface Characterization*, VCH, New York, 1991.
27. R. G. Compton and A. Hamnett, eds., *New Techniques for the Study of Electrodes and Their Reactions*, Elsevier, New York, 1989.
28. C. N. Reilley, *Rev. Pure Appl. Chem. 18*:137 (1968).
29. A. T. Hubbard and F. C. Anson, in *Electroanalytical Chemistry*, Vol. 4 (A. J. Bard, ed.), Marcel Dekker, New York, 1970, pp. 129–214.
30. A. T. Hubbard, *CRC Crit. Rev. Anal. Chem. 3*:201 (1973).
31. D. M. Oglesby, S. H. Omang, and C. N. Reilley, *Anal. Chem. 37*:1312 (1965).
32. T. B. Jarbawi and W. R. Heineman, *J. Electroanal. Chem. 132*:323 (1982).
33. P. T. Kissinger and C. N. Reilley, *Anal. Chem. 42*:12 (1970).
34. J. L. Owens and G. Dryhurst, *J. Electroanal. Chem. 80*:171 (1977).
35. J. Y. Gin, G. W. Hance, and T. Kuwana, *J. Electroanal. Chem. 309*:73 (1991).
36. T. P. DeAngelis and W. R. Heineman, *J. Chem. Educ. 53*:594 (1976).
37. D. K. Gosser, Jr., *Cyclic Voltammetry: Simulation and Analysis of Reaction Mechanisms*, VCH, New York, 1994.
38. J. T. Maloy, *J. Chem. Educ. 60*:285 (1983).
39. R. S. Nicholson and I. Shain, *Anal. Chem. 36*:706 (1964).
40. M. H. Hulbert and I. Shain, *Anal. Chem. 42*:162 (1970).
41. D. H. Evans, *Acc. Chem. Res. 10*:313 (1977).
42. J. Heinze, *Angew. Chem. 23*:831 (1984).
43. D. H. Evans, K. M. O'Donnell, R. A. Peterson, and M. J. Kelly, *J. Chem. Educ. 60*:290 (1983).
44. P. T. Kissinger and W. R. Heineman, *J. Chem. Educ. 60*:702 (1983).
45. R. S. Nicholson, *Anal. Chem. 37*:1351 (1965).
46. S. P. Perone and T. R. Mueller, *Anal. Chem. 37*:2 (1965).
47. S. P. Perone and C. V. Evins, *Anal. Chem. 37*:1061 (1965).
48. S. P. Perone and C. V. Evins, *Anal. Chem. 37*:1643 (1965).
49. I. M. Kolthoff and J. J. Lingane, *Polarography*, 2nd ed., Wiley-Interscience, New York, 1952.

50. L. Meites, *Polarographic Techniques*, 2nd ed., Wiley-Interscience, New York, 1965.
51. J. Heyrovsky and J. Kuta, *Principles of Polarography*, Academic Press, New York, 1966.
52. A. M. Bond, *Modern Polarographic Methods in Analytical Chemistry*, Marcel Dekker, New York, 1980.
53. C. K. Mann, in *Electroanalytical Chemistry*, Vol. 3 (A. J. Bard, ed.), Marcel Dekker, New York, 1969, pp. 57–134.
54. C. N. Reilley and W. Stumm, in *Progress in Polarography*, Vol. 1 (P. Zuman and I. M. Kolthoff, eds.), Wiley-Interscience, New York, 1962.
55. S. A. Norman, *Anal. Chem. 54*:698A (1982).
56. E. Schmidt and H. R. Gygax, *J. Electroanal. Chem. 12*:300 (1966).
57. D. F. Rohrbach, E. Deutsch, and W. R. Heineman, in *Characterization of Solutes in Nonaqueous Solvents* (G. Mamantov, ed.), Plenum Press, New York, 1978. pp. 177–195.
58. L. B. Anderson and C. N. Reilley, *J. Electroanal. Chem. 10*:295, 538 (1965).
59. B. McDuffie, L. B. Anderson, and C. N. Reilley, *Anal. Chem. 38*:883 (1966).
60. D. A. Roston, E. E. Brooks, and W. R. Heineman, *Anal. Chem. 51*:1728 (1979).
61. T. P. DeAngelis, R. E. Bond, E. E. Brooks, and W. R. Heineman, *Anal. Chem. 49*:1792 (1977).
62. V. Kumar and W. R. Heineman, *Anal. Chem. 59*:842 (1987).
63. V. G. Levich, *Acta Physicochim. URSS 17*:257 (1942); *19*:133 (1944).
64. V. G. Levich, *Zh. Fiz. Khim. 18*:355 (1944); *22*:575 (1948); *22*:711 (1948).
65. F. Opekar and P. Beran, *J. Electroanal. Chem. 69*:1 (1976).
66. S. Bruckenstein and B. Miller, *Acc. Chem. Res. 10*:54 (1977).
67. W. J. Albery and M. L. Hitchman, *Ring-Disc Electrodes*, Oxford University Press, London, 1971.
68. A. Frumkin, L. Nekrasov, B. Levich, and J. Ivanov, *J. Electroanal. Chem. 1*:84 (1959).
69. J. E. Harrar, in *Electroanalytical Chemistry*, Vol. 8 (A. J. Bard, ed.), Marcel Dekker, New York, 1975, pp. 1–167.
70. A. J. Bard and K. S. V. Santhanam, in *Electroanalytical Chemistry*, Vol. 4 (A. J. Bard, ed.), Marcel Dekker, New York, 1970, pp. 215–315.
71. J. Janata and H. B. Mark, Jr., in *Electroanalytical Chemistry*, Vol. 3 (A. J. Bard, ed.), Marcel Dekker, New York, 1969, pp. 1–55.
72. R. Szentrimay, P. Yeh, and T. Kuwana, in *Electrochemical Studies of Biological Systems* (D.T. Sawyer, ed.), ACS Symposium Series No. 38, American Chemical Society, Washington, DC, 1977, Chap. 9.
73. F. M. Hawkridge and T. Kuwana, *Anal. Chem. 45*:1021 (1973).
74. J. L. Anderson, T. Kuwana, and C. R. Hartzell, *Biochemistry 15*:3847 (1976).
75. L. H. Rickard, H. L. Landrum, and F. M. Hawkridge, *Bioelectrochem. Bioenerg. 5*:686 (1978).
76. R. W. Murray and C. N. Reilley, in *Treatise on Analytical Chemistry*, Part I, Vol. 4 (I. M. Kolthoff and P. J. Elving, eds.), Wiley-Interscience, New York, 1963, pp. 2187–2191.

# 4

# Large-Amplitude Controlled-Current Techniques

**William R. Heineman** *University of Cincinnati, Cincinnati, Ohio*

**Peter T. Kissinger** *Purdue University and Bioanalytical Systems, Inc., West Lafayette, Indiana*

## I. CONTROLLED-CURRENT TECHNIQUES IN STATIONARY SOLUTION

Controlled-current techniques in stationary solution saw extensive development and application in the 1960s. However, they were largely supplanted by controlled-potential techniques, especially cyclic voltammetry, in the 1970s. Today, controlled-current techniques in stationary solutions are used occasionally.

### A. Current-Step Excitation

In electrochemical techniques based on current excitation, a current signal is impressed on a quiescent electrochemical cell, and a response signal, usually the potential of the working electrode, is measured as a function of time. The most commonly used excitation signal is the application of a constant-current-step function to the cell as shown in Figure 4.1.

As was the case for techniques based on potential excitation, current-excitation methods are best understood by studying the time-dependent concentration changes in solution caused by the excitation signal applied to the electrode. Concentration-distance profiles for the case of species O being reduced to R by a current-step excitation signal (application of constant current to the cell) are shown in Figure 4.2. Consider first the profiles in Figure 4.2A for the reactant, O. An important concept is the relationship between the applied current and the slope of the profile at the electrode surface as expressed by

**Figure 4.1** General excitation signals for controlled-current techniques. (A) Current step. (B) Double current step.

$$-\frac{i}{nFA} = -D_O\left(\frac{\partial C_O}{\partial x}\right)_{x=0} = J \tag{4.1}$$

which has been described previously (Eq. 3.5). As the equation shows, the slope of the profile at the electrode surface is controlled by the magnitude of the applied current i. In other words, the flux J of reactant to the electrode surface is the exact amount necessary to support the applied current. Note that while $(\partial C_O/\partial x)_{x=0}$ remains constant, the profiles in Figure 4.1A drop as solution near the electrode is depleted of O by the constant-current electrolysis. When the profile intersects the origin, the surface concentration of O is negligible. At this time, referred to as the transition time $\tau$, the rate of reduction of O becomes independent of the magnitude of i and is now controlled by the rate of diffusion of O to the electrode. The slope of the profile at the electrode surface now falls off in a manner analogous to the profiles in Figure 3.1. Since the flux of O to the electrode is no longer sufficiently great to support the applied current, some other reducible species in solution must begin to react so that all the impressed current can pass through the cell. Now the total flux is defined as

$$-\frac{i}{nFA} = -\left[D_O\left(\frac{\partial C_O}{\partial x}\right)_{x=0} + D_{O'}\left(\frac{\partial C_{O'}}{\partial x}\right)_{x=0}\right] = J \tag{4.2}$$

where O′ is a more difficult to reduce species, which only reacts after $\tau$. Since the flux of O is diffusion-controlled after $\tau$ is reached, the value of $(\partial C_O/\partial x)_{x=0}$ will decrease with time as described by the Cottrell equation presented earlier (Eq. 3.6). This means that $(\partial C_{O'}/\partial x)_{x=0}$ must increase by a proportionate amount in order to supply reactant (O and O′) to the electrode surface at the rate demanded by i.

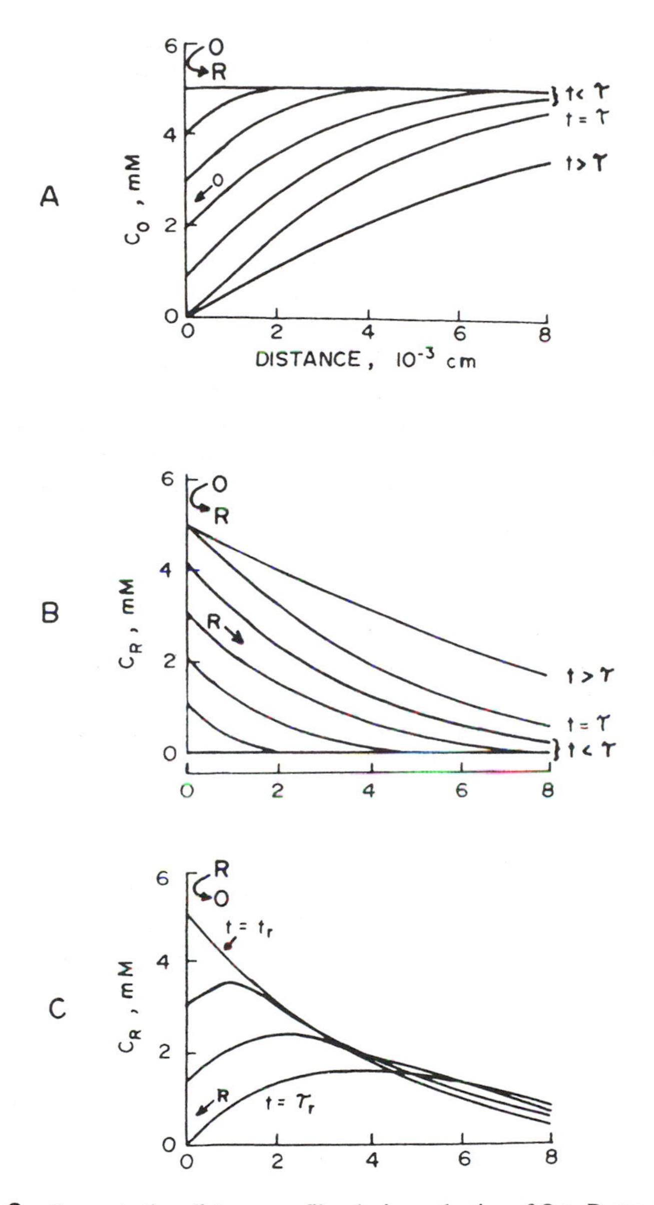


**Figure 4.2** Concentration-distance profiles during reduction of O to R at a planar electrode by current-step excitation. Quiescent solution. $D_O = D_R = 10^{-5}$ cm²/s, $C_O^\circ = 5$ mM, $C_R^\circ = 0$ mM, $i = 10^{-2}$ A/cm². (A) C-x profiles for O. Curves for 0, 0.07, 0.3, 0.7, 1.2, and 1.8 s. (B) C-x profiles for R. Curves for 0, 0.07, 0.3, 0.7, 1.2, and 1.8 s. (C) C-x profiles for R after current reversed at $t = \tau$. Curves for 0, 0.007, 0.3, and 0.6 s after current reversal. [Adapted from Ref. 1.]

The concentration profile of product R is determined by the flux of R away from the electrode surface. The flux of R is controlled by the flux of O to the electrode, so that

$$D_R \left(\frac{\partial C_R}{\partial x}\right)_{x=0} = - D_O \left(\frac{\partial C_O}{\partial x}\right)_{x=0} \qquad (4.3)$$

If $D_O = D_R$, the concentration-distance profiles of O and R will have slopes of the same magnitude but opposite sign as those shown in Figure 4.2B. At $\tau$, all of the reactant O at the electrode surface is converted to product R.

The effect of reversing the polarity of the current excitation signal on the concentration-distance profiles is shown in Figure 4.2C. The original reactant O is regenerated at a rate now determined by the flux of R back to the electrode. When the surface concentration of R becomes zero, a reverse transition time $\tau_r$ occurs. At this time, another species must undergo oxidation to enable the cell to accept the applied current.

## B. Chronopotentiometry

Chronopotentiometry is the electrochemical technique in which a current step is impressed across an electrochemical cell containing unstirred solution [1–5]. The resulting potential response of the working electrode versus a reference electrode is measured as a function of time (thus, chronopotentiometry), giving a chronopotentiogram as shown in Figure 4.3.

The shape of the potential-time response is determined by the concentration changes of O and R at the electrode surface during electrolysis. The potential is related to $C_O^s/C_R^s$ via the Nernst equation for a reversible system. The initial potential before current application is simply the rest potential or open-circuit potential ($E_{oc}$) of the solution, which reflects the initial $C_O^s/C_R^s$ in solution. At the instant of current application, this ratio becomes finite and the potential changes to a value consistent with the Nernst equation. Early in the

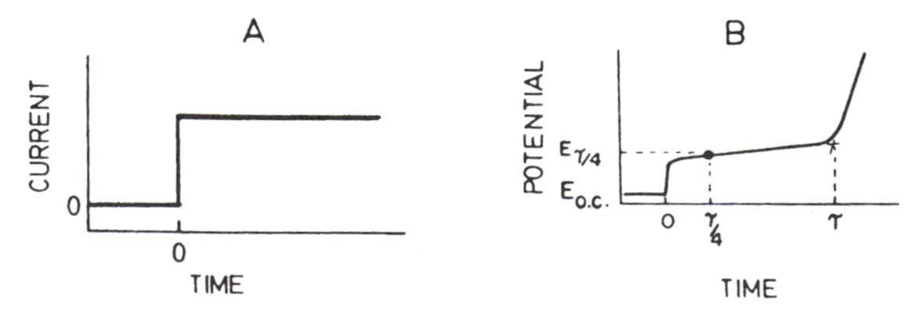


**Figure 4.3** Chronopotentiometry. (A) Current excitation signal. (B) Potential response signal (chronopotentiogram).

experiment, the potential changes dramatically from the rest potential to a value that is in the "range" of $E^{\circ\prime}_{O,R}$ vs. the reference electrode.

The subsequent gradual change of potential during continued electrolysis reflects the conversion of O to R and the consequent decrease of $C^s_O/C^s_R$ at the electrode surface as shown in Figures 4.2A and B. At the transition time $\tau$, the surface concentration of O approaches zero, causing the potential to shift rapidly to a more negative value so that electrolysis of another electroactive component can occur to support the constant current. If a second reducible species is present in solution, a second chronopotentiometric wave will occur. If not, the potential will shift to the reduction of solvent or supporting electrolyte. Oxidation of solvent, supporting electrolyte, or the electrode itself (e.g., Hg) occurs when an anodic current is applied.

For a reversible couple, the surface concentrations of the electroactive species for a simple reduction can be substituted into the Nernst equation to give an expression describing the potential-time response curve. At $x = 0$, $C^s_O/C^s_R$ can be related to time by $(\tau^{1/2} - t^{1/2})/t^{1/2}$, which gives

$$E = E_{\tau/4} - \frac{RT}{nF} \ln \frac{t^{1/2}}{\tau^{1/2} - t^{1/2}} \tag{4.4}$$

where RT/nF has its usual significance, t is time (s), and $\tau$ is the transition time (s). The value $E_{\tau/4}$ is analogous to the voltammetric $E_{1/2}$ and is defined by

$$E_{\tau/4} = E^{\circ\prime} - \frac{RT}{nF} \ln \frac{D_O^{1/2}}{D_R^{1/2}} \tag{4.5}$$

where D is the diffusion coefficient. $E_{\tau/4}$ corresponds to the potential at which the surface concentrations of the redox couple species are equal (i.e., $C^s_O = C^s_R$).

Under conditions of linear diffusion with no kinetic complications, the relationship between the applied current and the transition time is given by the *Sand equation,*

$$i\tau^{1/2} = \frac{nFA\pi^{1/2} D^{1/2} C^{\circ}}{2} \tag{4.6}$$

where i is the applied current (A), $\tau$ is the transition time (s), A is the electrode area ($cm^2$), D is the diffusion coefficient of electroactive species ($cm^2/s$), $C^{\circ}$ is the bulk concentration ($mol/cm^3$), n is the number of electrons transferred per ion/molecule, and F is Faraday's constant (96,485 C/eq).

The product $i\tau^{1/2}$ is a significant diagnostic parameter in chronopotentiometry. It is apparent from the Sand equation that the quantity $i\tau^{1/2}$ is a constant for a given concentration of electroactive species. The application of a larger i results in a smaller $\tau$ due to the more rapid electrolysis. The magnitude of $i\tau^{1/2}$ can be used analytically for concentration measurements down to about $10^{-4}$–$10^{-5}$ M. However, chronopotentiometry is rarely used for analysis; the voltammetric techniques are more popular because of their lower detection limits and the difficulty in measuring $\tau$, *vide infra*.

A very important feature of chronopotentiometry is the characteristic variation in $i\tau^{1/2}$ that occurs for certain electrode reactions when $\tau$ is varied over a wide range by varying i. The behavior of $i\tau^{1/2}$ can be effectively used to diagnose certain mechanisms of electrode reactions. Diagnostic curves for $i\tau^{1/2}$ versus i are shown in Figure 4.4 for several mechanistic situations. A constant value of $i\tau^{1/2}$ over a wide range of $\tau$ (Fig. 4.4A) is characteristic of an uncomplicated, diffusion-controlled electrode reaction with no kinetic or adsorption phenomena at a planar electrode.

An electron-transfer reaction that is coupled to a *preceding chemical reaction* is characterized by a decrease in $i\tau^{1/2}$ as i increases (Fig. 4.4B). At very large i (short $\tau$), $i\tau^{1/2}$ is determined by the equilibrium concentration of O, since $\tau$ is reached before X can generate O by chemical reaction. As $\tau$ is increased by decreasing i, $i\tau^{1/2}$ begins to increase when the rate constant k is sufficiently large to generate additional O by the homogeneous chemical reaction. Further decrease in i causes further increase in $i\tau^{1/2}$ as more time is allowed for X to generate O. Extrapolation of the $i\tau^{1/2}$ increase to zero current gives an $i\tau^{1/2}$ intercept that is related to the sum of the concentrations of X and O, since sufficient time is allowed for all of X to convert to O as i approaches zero ($\tau \rightarrow \infty$). Note that the steepness of the $i\tau^{1/2}$ decrease with increasing i decreases as k increases. The magnitude of k can be calculated from this slope [1]. Curves of this type would be obtained for the reduction of $Cd(NTA)^-$, which is the example of a CE mechanism described in Chapter 2.

In the case of the *catalytic chemical reaction*, $i\tau^{1/2}$ increases to infinity as i decreases, since no transition time is ever reached if R is given enough time to recycle O (Fig. 4.4C). The constant $i\tau^{1/2}$ that is reached with a sufficiently large current is proportional to the equilibrium concentration of O in solution and reflects the fact that the electrolysis time before $\tau$ is too short for significant reaction to occur. An example of this mechanism is the reduction of Ti(IV) in the presence of hydroxylamine [6].

In the case in which the electroactive species O is adsorbed on the electrode, $i\tau^{1/2}$ increases with increasing i (Fig. 4.4D). The Sand equation considers only those molecules of O that have reached the surface by diffusion. As $\tau$ decreases, the charge contribution to the reduction of $O_{ads}$ consumes an increasingly large fraction of the total current, causing $i\tau^{1/2}$ to increase. A typical example of this behavior is the reduction of Alizarin Red S at mercury [7].

Inhibition of electrode reactions caused by material adsorbed on the electrode surface causes $i\tau^{1/2}$ to decrease with increasing i (Fig. 4.4E). As i in-

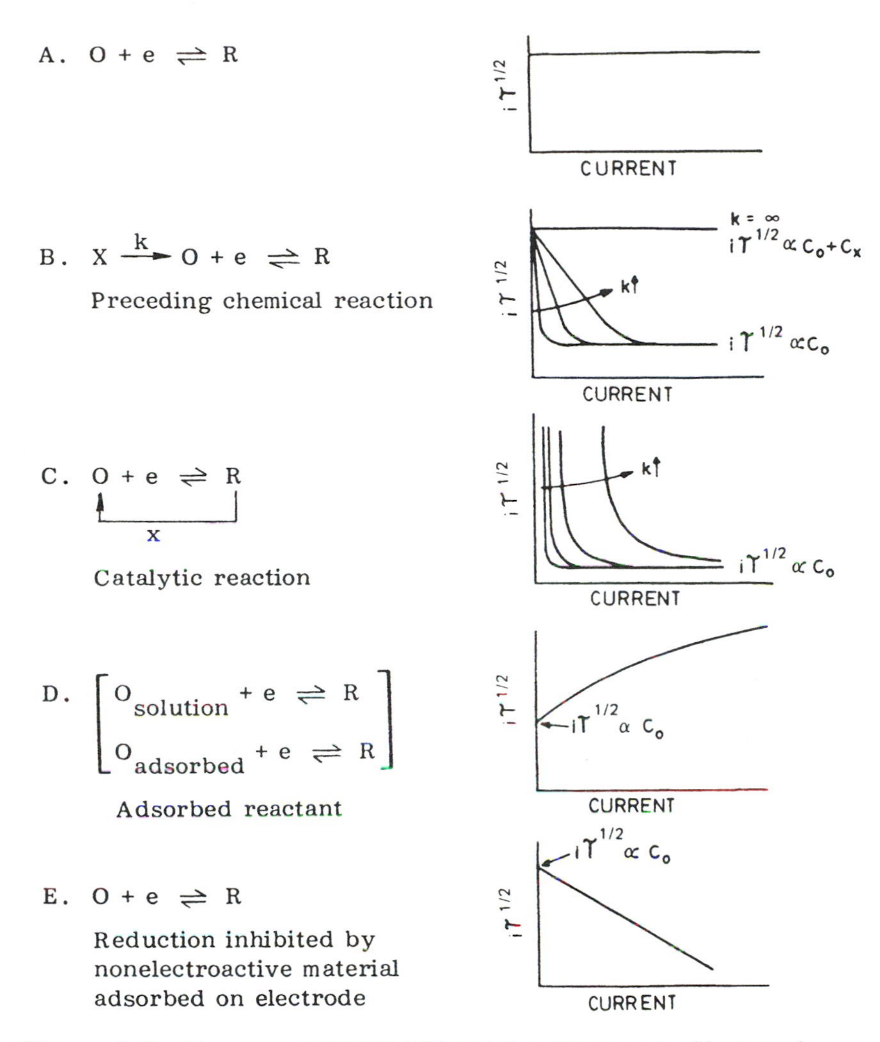


**Figure 4.4** Chronopotentiometric $i\tau^{1/2}$ variations for a range of impressed currents as a diagnostic for electrode mechanisms.

creases, less time is allowed for the electroactive species to penetrate the adsorbed layer. This behavior is analogous to the inhibition of the polarogram shown in Figure 2.21. The reduction of a copper tartrate complex at a mercury electrode exhibits this typical behavior [8].

Chronopotentiometry suffers from difficulty in experimentally determining the transition time and the fact that the charging current is not easily resolved from the faradaic response, since the electrode potential is varying throughout the experiment. Various extrapolation procedures have been devised, but usually without rigorous theoretical justification [1–5].

## C. Current-Reversal Chronopotentiometry

If the polarity of the applied current in an ordinary chronopotentiometry experiment is reversed during the recording of the chronopotentiogram, the product R of the initial electrochemical reaction may now undergo the reverse reaction to give a current-reversal chronopotentiogram, as shown in Figure 4.5 [1–5]. A reverse transition time $\tau_r$ will result when the concentration of R becomes zero at the electrode surface (see Fig. 4.2C). Such reverse potential-time curves can be treated quantitatively for reversible and irreversible couples.

For a system devoid of coupled chemical reactions and adsorption phenomena, the relationship between the forward electrolysis time $t_f$ and the reverse transition time $\tau_r$ is

$$\tau_r = \frac{1}{3} t_f \tag{4.7}$$

where $t_f$ is the time of current reversal. The current must be reversed before or at the forward transition time $\tau$ for this relationship to be valid. Two-thirds of the electrogenerated R is "lost" by diffusion away from the electrode.

Electrode mechanisms with following chemical reactions (EC mechanism; see Chap. 2) are particularly amenable to study by current-reversal chronopotentiometry. A good example is the oxidation of p-aminophenol (PAP), which undergoes the following reaction sequence [9]:

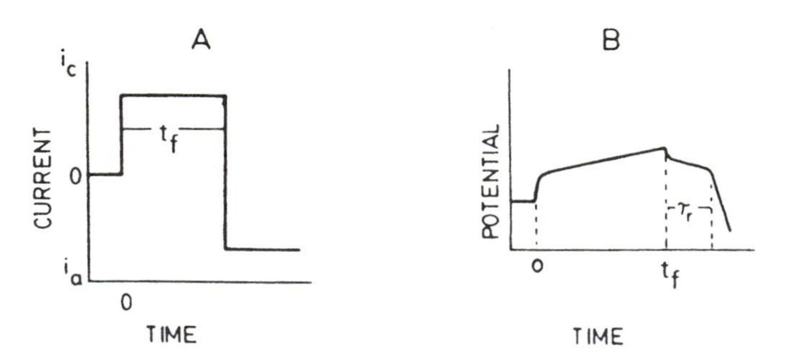


**Figure 4.5** Current-reversal chronopotentiometry. (A) Current excitation signal. (B) Potential response.

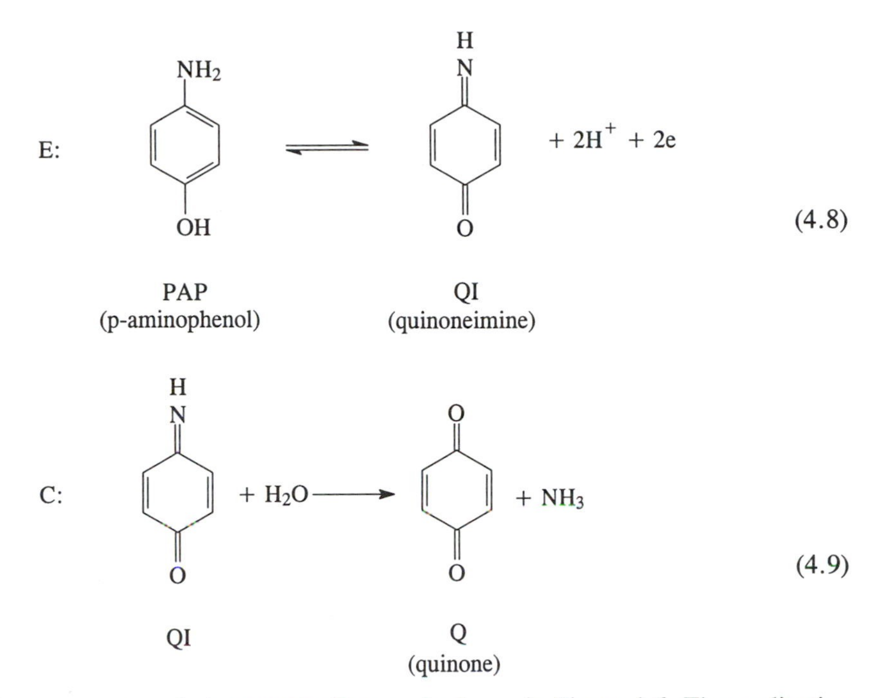


A current-reversal chronopotentiogram is shown in Figure 4.6. The application of an anodic current for the period $t_f$ generates QI at the electrode by oxidation of PAP. When the current is reversed, QI is reduced back to PAP during the reverse transition labeled $\tau_r$. The chemical reaction of QI to form Q causes $\tau_r/t_f$ to be less than 1/3 by an amount that depends on the magnitude of the rate constant k. Figure 4.7 shows a calculated working curve which relates $\tau_r/t_f$ to $kt_f$. The rate constant can be calculated from this curve after the experimental value of $\tau_r$ is determined for a given $t_f$. The second reverse transition time in Figure 4.6 is due to reduction of the quinone, Q, to hydroquinone, $H_2Q$, at the electrode (see Eq. 3.33).

Current-reversal chronopotentiometry is also useful for detecting adsorption of the product generated during the forward electrolysis time. Complete adsorption of the product on the electrode surface causes $\tau_r$ to equal $t_f$ since no product is lost by diffusion from the electrode. This approach has been used to determine the amount of adsorbed material formed during the reduction of riboflavin and the oxidation of iodide [10].

## D. Cyclic Chronopotentiometry

In cyclic chronopotentiometry, the current is continually reversed at potentials corresponding to the forward and reverse transition times as shown in Figure

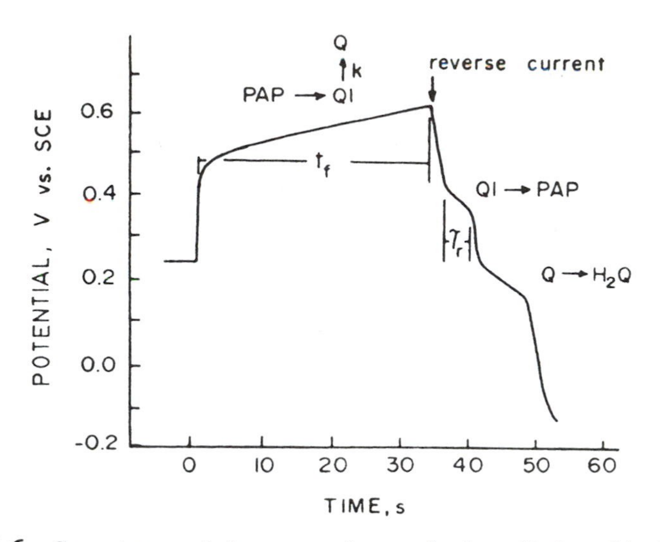


**Figure 4.6** Current-reversal chronopotentiogram for the oxidation of 1 mM PAP in 0.1 M $H_2SO_4$. Pt electrode, i = 100 $\mu A/cm^2$. [From Ref. 9, adapted with permission. Copyright 1960 American Chemical Society.]

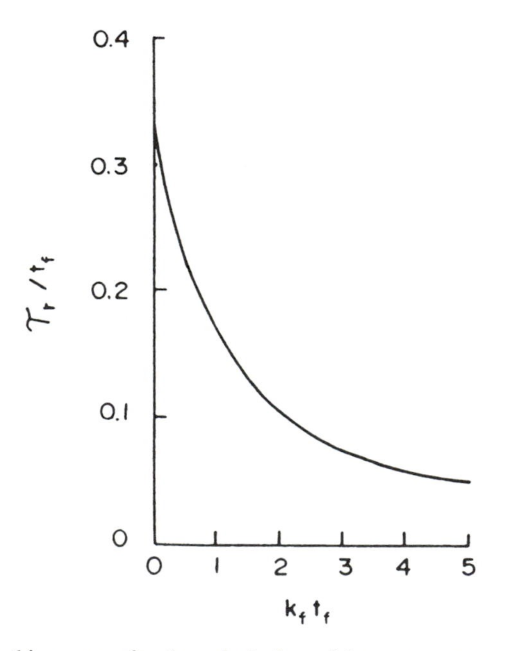


**Figure 4.7** Working curve for the calculation of the rate constant $k_f$ for a following chemical reaction from current-reversal chronopotentiometry. [From Ref. 9, adapted with permission. Copyright 1960 American Chemical Society.]

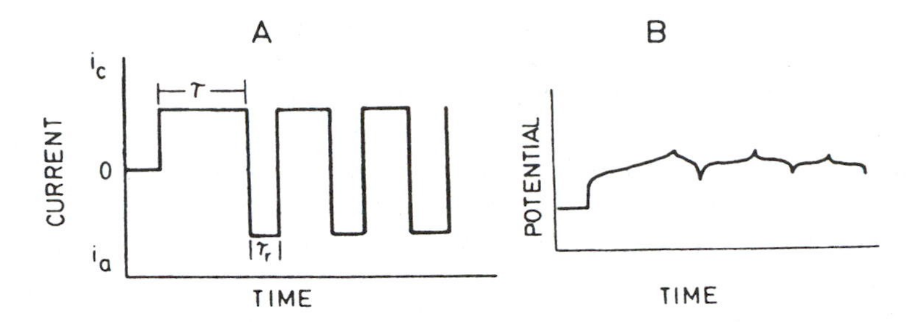


**Figure 4.8** Cyclic chronopotentiometry. (A) Current excitation signal. (B) Potential response signal.

4.8 [11]. For a system with no kinetic or adsorption complications, the forward transition time $\tau$ decreases while $\tau_r$ increases until finally $\tau = \tau_r$ in the limit, at steady state. (Because the convergence rate is slow, equality of $\tau$ and $\tau_r$ is not commonly achieved experimentally before the onset of natural convection and nonplanar diffusion effects.) Quantitative treatments for single component systems, multicomponent systems, stepwise reactions, and systems involving chemical kinetics have been derived. The technique has not been used extensively.

## E. Derivative Chronopotentiometry

In derivative chronopotentiometry, the potential response signal of a normal chronopotentiometry experiment is electronically differentiated, and this rate of change of potential with time, dE/dt, is recorded as a function of time, as shown in Figure 4.9 [12]. The minimum in a derivative chronopotentiogram is quantitatively related to the transition time. Thus for a reversible couple,

$$\left(\frac{dE}{dt}\right)_{min} = -\frac{27RT}{8nF\tau} \qquad (4.10)$$

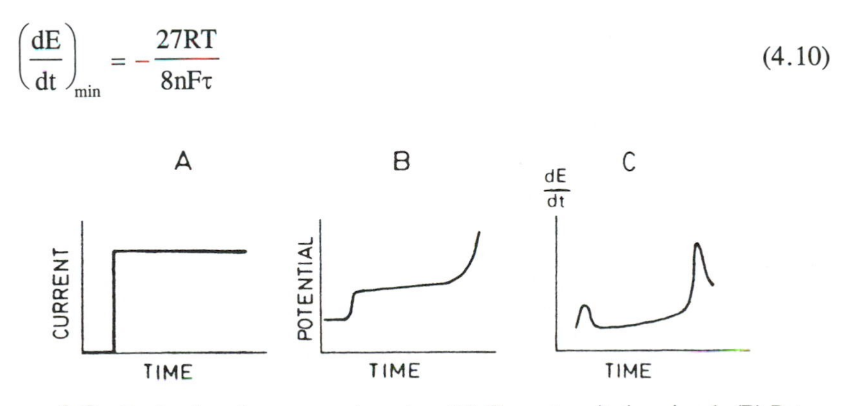


**Figure 4.9** Derivative chronopotentiometry. (A) Current excitation signal. (B) Potential response signal. (C) Monitored response: dE/dt.

Calculating $\tau$ this way is somewhat advantageous since the rate of charging of the double layer is minimal (not zero) at $(dE/dt)_{min}$. Derivative chronopotentiometry has not been widely used, probably because the improvement over the ordinary E versus t response is insufficient to justify the more complex instrumentation required. A more serious criticism is that the minimum dE/dt occurs at a time that is dependent on the kinetics of the heterogeneous reaction. For fast (reversible) reactions, $t_{min} = 4\tau/9$, and for very slow (irreversible) reactions, $t_{min} = \tau/4$. Unfortunately, a great many reactions fall somewhere in between, and the relation of $t_{min}$ to $\tau$ is not likely to be clearly defined.

## F. Programmed Current Chronopotentiometry

In general, chronopotentiometry consists of the application of a programmed current function to an electrode, the potential of which is measured as a function of time [13]. Thus the current program may be any function that is electronically feasible. Examples of potentially useful current functions (in addition to those described in preceding sections) are ramp current chronopotentiometry for the elucidation of adsorption mechanisms, and square-root-of-time current chronopotentiometry for which concentration is directly proportional to the transition time:

$$\tau = \frac{2nFD^{1/2}C^{\circ}}{\beta\pi^{1/2}} \tag{4.11}$$

where $\beta$ is a proportionality factor. In current-cessation chronopotentiometry, a constant current is applied, then abruptly discontinued, and the resulting potential decay is observed.

## G. Controlled-Current Chronoabsorptometry

Controlled-current chronoabsorptometry involves the simultaneous optical monitoring of the product or other redox component in the electrode mechanism during a chronopotentiometry experiment [14]. Although this technique has been demonstrated with $SnO_2$ optically transparent electrodes, it has generally received little use, since the resistance effects in thin-film electrodes can give unequal current densities across the electrode face. This results in distorted potential-time and absorbance-time responses. Consequently, the more prevalent spectroelectrochemical methods utilize potential rather than current as the excitation signal.

## H. Current-Step Excitation in Thin Layers of Solution

Although chronopotentiometric experiments were included among the earliest thin-layer studies, there has been very little work in recent years even though the theory is well developed [15]. Difficulties with iR losses in thin-layer cells compound the problem of defining the transition time and greatly limit the current and thus the time scale that can be studied. At this point in time there is little to recommend in thin-layer chronopotentiometry. Even the instrumental advantage of constant current for stationary-solution coulometric experiments and controlled-potential coulometry can be carried out with ease. The constant-current excitation is most valuable for the hydrodynamic experiments referred to in the following section. Thin-layer flow cells operated at constant current are useful for precise generation of micromolar titrants.

## II. CONTROLLED-CURRENT TECHNIQUES IN FLOWING SOLUTION

Controlled-current electrolysis in flowing solution has been extremely useful for analytical purposes. The prevalent techniques are constant-current coulometry and coulometric titrations, which are discussed in Chapter 25.

## REFERENCES

1. P. Delahay, *New Instrumental Methods in Electrochemistry*, Interscience, New York, 1954.
2. D. G. Davis, in *Electroanalytical Chemistry*, Vol. 1 (A. J. Bard, ed.), Marcel Dekker, New York, 1966, pp. 157–196.
3. M. Paunovic, *J. Electroanal. Chem. 14*:447 (1967).
4. P. Delahay, in *Treatise in Analytical Chemistry*, Part I, Vol. 4 (I. M. Kolthoff and P. J. Elving, eds.), Wiley-Interscience, New York, 1963, p. 2233.
5. R. N. Adams, *Electrochemistry at Solid Electrodes*, Marcel Dekker, New York, 1969, Chap. 6.
6. P. Delahay, C. C. Mattax, and T. Berzins, *J. Am. Chem. Soc. 76*:5319 (1954).
7. H. A. Laitinen and L. M. Chambers, *Anal. Chem. 36*:5 (1964).
8. M. Kodama and R. W. Murray, *Anal. Chem. 37*:1638 (1965).
9. A. C. Testa and W. H. Reinmuth, *Anal. Chem. 32*:1512 (1960).
10. H. B. Herman, S. V. Tatwawadi, and A. J. Bard, *Anal. Chem. 35*:2210 (1963).
11. H. B. Herman and A. J. Bard, *Anal. Chem. 35*:1121 (1963); *36*:510, 971 (1964).
12. D. G. Peters and S. L. Burden, *Anal. Chem. 38*:530 (1966).
13. R. W. Murray and C. N. Reilley, *J. Electroanal. Chem. 3*:64,182 (1962).
14. W. N. Hansen, T. Kuwana, and R. A. Osteryoung, *Anal. Chem. 38*:1810 (1966).
15. A. T. Hubbard and F. C. Anson, in *Electroanalytical Chemistry*, Vol. 4 (A. J. Bard, ed.), Marcel Dekker, New York, 1970, pp. 129–214.

# 5

# Small-Amplitude Controlled-Potential Techniques

**Peter T. Kissinger** *Purdue University and Bioanalytical Systems, Inc., West Lafayette, Indiana*

**Thomas H. Ridgway** *University of Cincinnati, Cincinnati, Ohio*

## I. INTRODUCTION

Small-amplitude techniques were originally developed in an attempt to isolate the faradaic response of an electrochemical cell from the associated effects of solution resistance and capacitance. To understand the need for these techniques, it will be helpful to briefly consider an electrical model of the typical three-electrode electrochemical cell as represented by Figure 5.1a. The solution between electrodes can be well represented by a resistance whose magnitude is directly proportional to distance and inversely proportional to the concentration and mobility of the ionized species. These are normally predominately electrochemically inert electrolytes added to decrease the solution resistance (the supporting electrolyte). This is represented in Figure 5.1b by the resistors, $R_c$, $R_u$, and $R_{ref}$. The electrochemical charge transfer, or faradaic process, takes place only at the interface between the solution and an electrode. The current at an electrode depends upon the concentration of electrochemically active species and the potential of the electrode relative to the E°s of the available electrochemical couples. The electron flux resulting from each couple will range from zero to an amount limited only by the rate of mass transport to the electrode surface. This is represented by the boxes labeled $Z_f$ and $Z'_f$ in Figure 5.1b . The supporting electrolyte is chosen such that its ions are not reducible or oxidizable at potentials expected for the working electrode. When a potential is applied to an electrode, either the anions or cations of the supporting electrolyte will be attracted to the surface electrostatically, but will not be able to undergo a charge

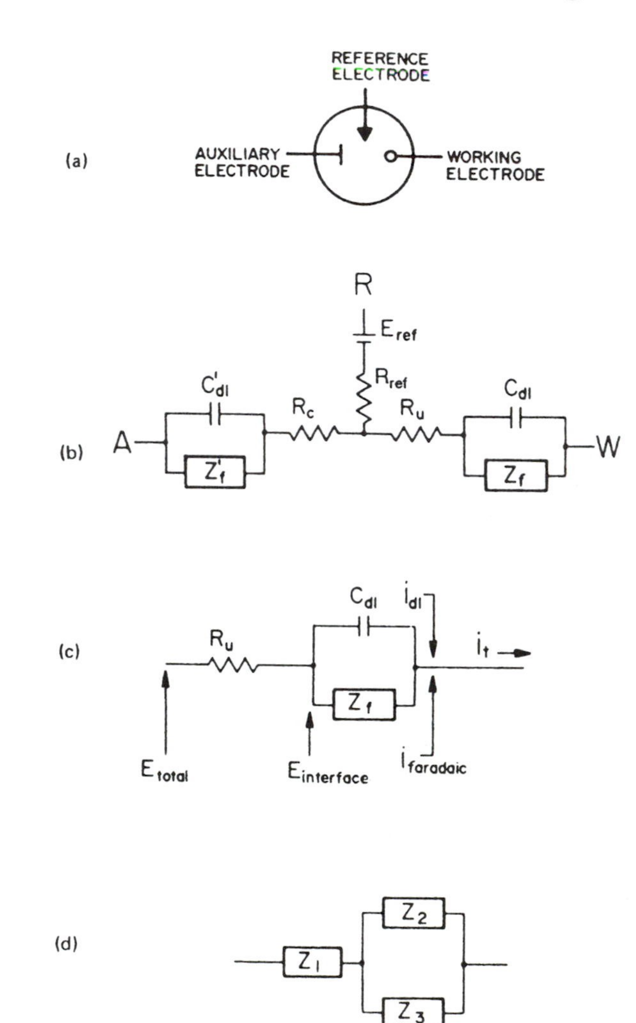


**Figure 5.1** Schematic representation of an electrochemical cell: (a) three electrodes; (b) equivalent circuit for three-electrode cell; (c) equivalent circuit for the working-electrode interphase; (d) a "solution" impedance in series with two parallel "surface" impedances.

transfer reaction because the potential is not sufficiently positive or negative. This leads to a sheet of ions very close (angstroms) to a conductor of opposite charge with no electron transfer between them. This is the classical definition of a capacitor and is termed the double-layer capacitance. This is represented in Figure 5.1b by the capacitors $C'_{dl}$ and $C_{dl}$.

Both the double-layer capacitance and the faradaic charge transfer process occur at the electrode-solution interface and act as parallel paths for electron flow, and they are both in series with the solution resistance. This is shown schematically in Figure 5.1b–c. Both working and auxiliary electrodes have a $Z_f$ and $C_{dl}$ associated with them and they are connected by the resistance of the solution path between them (as well as the resistance of other barriers such as conducting membranes, glass frits, etc.) This resistance is symbolized in Figure 5.1b by the two components $R_c$ and $R_u$. The reference electrode is modeled by a simple battery and a series resistance, which represents the internal resistance of the reference electrode and its junction. The characteristics of all three electrodes are important for purposes of instrumentation and cell design and will be treated in some detail elsewhere, but for our purposes here we can focus only on the working electrode as represented by the three components $R_u$, $C_{dl}$, and $Z_f$ in Figure 5.1c. The consequences of $R_u$ and $C_{dl}$ on the electronics and cell design is considered further in Chapters 6, 7, and 9. It should also be noted that the solution resistance is somewhat more complex than an ideal resistor, and the double-layer capacitance has a number of nonideal characteristics, including a fairly strong potential dependency. Even though this cell model is somewhat flawed, it is still very useful in understanding the behavior of real electrochemical cells.

A combination of correct cell design and control electronics normally allows us to consider only the working electrode portion of the model as shown in Figure 5.1c. The resistive element $R_u$ is nominally the solution resistance between the reference electrode tip and the working electrode. We can only control (or monitor) the potential between the reference electrode tip and the working electrode, which is represented in Figure 5.1c by the right side junction of capacitance and faradaic process. Any passage of electrons through a resistive element results in a voltage drop across this resistance; this voltage drop will be in series with (additive to) the voltage drop across the electrode interface. The result of the presence of this resistance is to distort the potential signal seen by the electrode solution interface for potential excitations and to complicate the potential response for current excitations. One can minimize the magnitude of this resistance in various ways, such as by placing the reference probe tip close to the electrode, or by employing electronic cancellation techniques [1], but one cannot totally eliminate it.

The net effect of the presence of the solution resistance on potential excitation methods is that the potential seen by the electrode solution interface is different from the potential applied by the potentiostat. This difference is current-dependent and the current is itself potential-dependent. The resistance also makes it more difficult to separate current components arising from the double-layer capacitance from the faradaic process. Similar complications arise for current excitations.

### A. What Is a Small-Amplitude Technique?

The objective of most electrochemical experiments is to allow the experimenter to investigate one or more of three types of parameters: (1) the concentration and identity of one or more solution components, (2) the kinetics of chemical, charge transfer, or adsorption processes, and (3) the nature of the double-layer capacitance associated with the electrode-solution interface. Historically, most small-amplitude techniques have been developed in an attempt to allow an easier separation of the contributions of these basic parameters.

In order for a method to be useful for these purposes, it is at least necessary to be able to predict what the current-time-voltage-concentration relationship should be like for a given set of conditions. Only then can one evaluate experimental results and determine concentration or kinetic contributions.

We have seen that the instantaneous faradaic current at an electrode is related to surface concentrations and charge transfer rate constants, and exponentially to the difference of the electrode potential from the E° of the electrochemical couple. This is represented in Figure 5.1c by $Z_f$. With very few exceptions, this leads to intractable nonlinear differential equations. These systems have no closed form solutions and are treatable only by numerical integrations or numerical simulations (e.g., cyclic voltammetry). In addition, the double-layer capacitance itself is also nonlinear with respect to potential.

One way around this problem is to restrict the potential changes to very small excursions. This approach is based on the principle that any curve can be approximated by a series of *small* straight line segments. Applying this principle to the exponential functions that cause all the trouble leads to a far more manageable set of linear differential equations, which can usually be solved, allowing a prediction of the expected response.

This immediately leads to a question: How small must these excursions be in order for the predictions to be valid? Theoretically, the answer is zero millivolts, a clever but uninteresting answer. Practically the answer usually found in the literature is between 8/n and 12/n mV where n is the number of electrons transferred in the electrochemical reaction. These numbers are arrived at by estimating what kind of deviation from theoretical behavior can be detected experimentally. For purposes of this discussion we will use 10 mV. At this point it is useful to remember that the exponential terms are of the form $anF(E - E°)RT$, where T is the absolute temperature and a is either $\alpha$ or $1 - \alpha$. The 10/n mV figure is based on an $\alpha$ of 0.5 at 25 °C. Any change in these parameters from their nominal value would influence this limit (particularly in the case of low-temperature electrochemistry in nonaqueous solvents). This leads to the obvious next question: What happens if you exceed this limit? The answer is that the response begins to deviate noticeably from the ideal, theoretical model. How great the deviation is depends upon how far one exceeds the limit and what experimental properties are being studied. In analytical applications, where concentration determination is the goal, it is not uncommon to see excitations employed that are five or even ten times the theoretical limit. This will tend to produce somewhat distorted shapes, but they are often quite acceptable for concentration determinations.

As in large-amplitude methods, one can control the potential and observe the current changes, or control the current and observe the change in potential. Irrespective of whether current or potential is controlled, the small-amplitude limitation applies to the potential excursion.

When a small-amplitude excitation (current or potential) is applied to an electrode, it causes a change in the concentration profile of the electroactive species at the surface. Since we are considering only small changes in electrode potential, this will result in a small perturbation in the surface concentration from its original value prior to the application of the small-amplitude excitation. This initial unperturbed concentration is often referred to as the dc surface concentration, for historical reasons. The ratio of the concentrations of the species comprising an electrochemical couple is related to the potential of that electrode via the Nernst equation. This implies that we can think of the electrode as having a dc potential (established by the dc surface concentration ratios) regardless of how this concentration ratio is established, even if the surface concentration ratio (and hence the potential) is slowly changing with time.

We can conceptually separate the response of the electrode to the small-amplitude perturbation into two components—a time dependency controlled primarily by the type of excitation and electrode kinetics, and a magnitude component independent of the type of perturbation but dependent on the concentration of the electroactive species and the dc potential (which is related to the concentration ratios). It will be helpful to consider the magnitude component first. Sigmoidal current (y-axis) versus potential (x-axis) plots such as Figure 5.2b are commonly encountered in classical polarography, as well as rotating-disk and other forms of hydrodynamic voltammetry. A current of 1 A represents a flux of $6.2 \times 10^{18}$ electrons per second, so that for a one-electron redox process such as the reduction of Fe(III) to Fe(II), a 1-$\mu$A current represents the conversion of $6.2 \times 10^{12}$ of the Fe(III) species per second. It would be equally valid to replace the y-axis current units with the percent of the bulk species undergoing electrolysis. At the foot of the wave, there is little or no conversion of the bulk species. On the plateau, 100% of the bulk species would be electrolyzed at the surface. The 50% point (halfway to the limiting current) is the $E_r^{1/2}$ (reversible half-wave potential), which differs from the E° for the couple by the log of the ratio of the diffusion coefficients of the two species in the couple. Taking the first derivative of this sigmoidal shape (di/dE or d%C/dE, etc.) produces a maximum at $E_r^{1/2}$, which indicates that the concentration

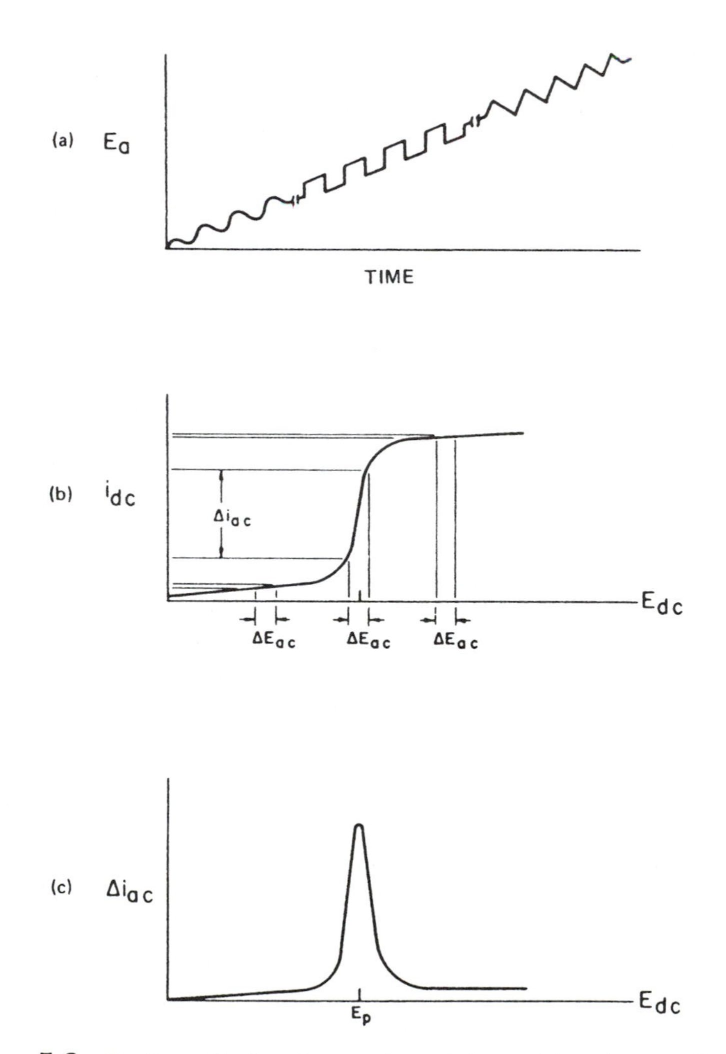


**Figure 5.2** Small-amplitude voltammetric techniques: (a) various small-amplitude waveforms are imposed on a "dc" ramp (normally only one waveform is used in a given experiment); (b) the sigmoidal dc response is typical of dc polarography and hydrodynamic voltammetry. The greatest amplitudes for the small-amplitude current ($\Delta i_{ac}$) are achieved on the rising part of the dc current, where the small-amplitude voltage signal causes the greatest change in the surface concentrations; (c) small-amplitude current response versus applied dc potential.

change at the electrode surface is greatest at that point (see Fig. 5.2c). One can also think of the x axis as a logarithmic representation of the concentration ratio of the two species in the redox couple. In this formalism, the maximum occurs where the ratio of the two species is nearly unity (nominally the E°).

This allows us to make predictions concerning the magnitude component of the response to either current or voltage excitations, irrespective of the method employed to control the potential or concentration ratio. The response to potential excitations will be a maximum at dc potentials near E° (concentration ratios near unity). Conversely, the response to current excitations will be a minimum in this region because the flux of the same number of electrons will produce a smaller change in the potential (a maximum in dI/dE means a minimum in dE/dI).

There are three ways to control the concentration ratios at the electrode surface. One can control the current, the electrode potential, or the bulk concentrations of the oxidized and reduced forms of the main couple. The two forms of small-amplitude control and three methods of controlling the overall surface concentrations produce a total of six overall control methods; examples of all six have appeared in the literature. When the small-amplitude current is controlled, the experimenter has only a very indirect influence on whether the small-amplitude limit is exceeded. In addition, the magnitude of the potential response is reciprocally related to the bulk concentration of the electroactive species. For these reasons, small-amplitude controlled-current methods are encountered very infrequently. Because of the problems associated with small-amplitude current-control methods, we will not consider them further.

Methods based on current control of the overall surface concentration coupled with small-amplitude potential excitations have been reported, but in general appear to provide insufficient advantage to justify their greater instrumental complexity.

There have been very few methods that establish the surface concentrations by directly manipulating the bulk solution concentrations. Essentially, the only technique that still employs this method is the faradaic impedance method, which will be discussed below.

The vast majority of small-amplitude methods are based on small-amplitude potential excitations with potential control of the surface concentrations. In earlier chapters, the relationship between surface concentration and electrode potential was explored and the concept of concentration profiles was presented. Whenever there is a flux of electrons at the electrode surface, the concentration profiles of at least two species will exhibit nonzero slopes at the electrode surface, as the electrochemical conversion of one member of a couple into another takes place and mass transport processes act to reestablish a uniform concentration distribution. These processes occur irrespective of whether the current flux arises from a potential or current excitation of the cell. In either case, they result in a perturbation from the previously existing concentration profile. The initial surface concentrations (which existed prior to the application of the new perturbation) are often termed the "dc surface concentrations." It is useful to note that at any time, the distance integral of the concentration excess or defect is directly proportional to the charge passed due to that perturbation. This implies that if the current or voltage conditions of the electrode are changed such that the resulting net change in the concentration profile is small, the response due to the change can be separated from any response arising from the initial concentration profile. There are two kinds of experimental situations that meet this condition.

For an experimental system in which the initial concentration profiles are flat, the unperturbed current is zero and the electrode potential is constant. Any changes in current or potential are due to the new perturbation. This situation is only encountered when one controls the surface concentrations by adjusting the bulk concentrations of the members of the electrochemical couple. This method is employed in faradaic impedance studies of electrode kinetics (detailed below) but is not generally applicable (particularly for analytical purposes).

The next case is one in which the overall change in concentration profile is small compared to the original slope of the profile; this condition can be achieved in two ways. The most common approach is to restrict the time duration of the change such that the new gradients do not propagate very far out into solution. In this situation, one can essentially assume that the unperturbed profile would have remained constant during the time interval of the small-amplitude excitation. Separation of the response due to the small amplitude is reduced to simple subtraction of the preexcitation current from the total response. The other case is encountered when a symmetric and periodic perturbation is applied. In this circumstance, a steady-state (repeating) response is observed. The separation method employed is to high-pass filter the total signal, which removes the relatively low frequency response components that are not due to the small-amplitude excitation.

The magnitude component of the response is totally independent of the nature of the excitation. The response always has a shape similar to that shown in Figure 5.2c no matter what type of excitation is employed, and is also independent of the method employed to establish the dc potential. The time functionality of the response will depend on the excitation signal, but the magnitude as a function of dc potential is independent of the technique. For multicomponent systems, this presents a significant advantage relative to the shapes normally encountered in polarography and voltammetry. The shape observed for small-amplitude techniques is narrower and it is easier to resolve overlapping processes. The width at half height of a reversible small-amplitude peak is $3.52RT/nF$ or $90/n$ mV at 25°C; $0.154/n$ V on either side of a peak, the signal has dropped to 1% of its original value.

## II. FARADAIC IMPEDANCE

The faradaic impedance method is the grandparent of all other small-amplitude methods. The first known experiments were reported by Warburg in 1899 [2].

A small-amplitude voltage is impressed across a two-electrode cell in a bridge circuit such as the one shown in Figure 5.3. The resistive and capacitive components (R and C) are adjusted until a balance is achieved. The component values required depend on the characteristics of the cell and the frequency employed. When a voltage is applied across a circuit element, a flow of electrons occurs. When the voltage is a time-varying sinusoid and the circuit element is a linear type such as a resistor, capacitor, or inductor, the resulting current is also a sinusoid of the same frequency whose phase differs from that of the voltage by a phase angle $\theta$. For a resistor, $\theta$ is 0°, and for a capacitor, 90°. The current amplitude for a resistive element is independent of the exciting frequency. For a capacitive element, the current amplitude is directly proportional to the frequency. Warburg showed that the faradaic impedance ($Z_f$) portion of the circuit in Figure 5.1b can be considered as just another circuit element. For electrochemical systems that are diffusion-controlled (and diffusion is planar) the characteristic current phase angle $\theta$ is 45° relative to the exciting voltage and the current amplitude is proportional to the square root of the frequency. For more complicated systems where the charge transfer rate constant is finite or chemical kinetic complications occur, the phase angle will differ from 45° and will usually be a function of the excitation frequency. The current amplitude versus frequency dependency also changes. These dependencies can be used to unravel the mechanism and determine rate constants [1].

No simple combination of resistors correctly models the phase angle and current response of $Z_f$ at all frequencies, but it can be modeled at any *single* frequency by a resistor and capacitor arranged either in series or in parallel. This effective impedance, now termed the Warburg impedance ($Z_w$), is a function of frequency and can be extracted from experimental data either numerically or

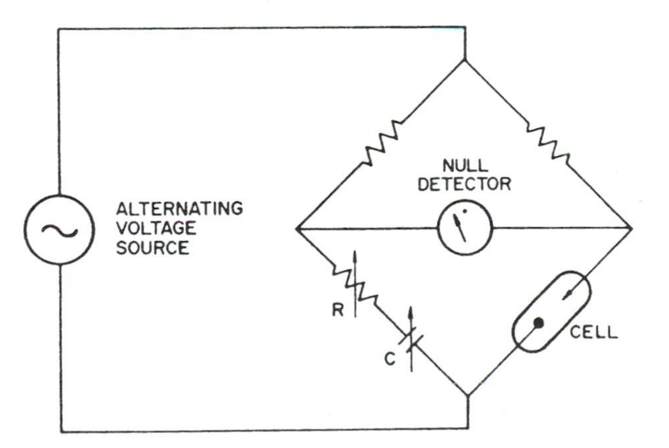


**Figure 5.3** Classical impedance bridge for a two-electrode cell.

graphically. The relative simplicity of the model has allowed the development of mathematical predictions or systems with slow kinetics, chemical complications, adsorption, etc. The steady-state or dc (direct current as opposed to alternating current, ac) electrode potential is controlled by changing the ratio of the solution or electrode concentrations of the species involved. Although the method yields highly precise data, its practical application is laborious and time-consuming. It is still sometimes employed in electrodynamic studies but not in analytical measurements. Its inclusion here is primarily due to its historical importance and because of its effect on the nomenclature employed in other small-amplitude methods.

## III. SINUSOIDAL ALTERNATING-CURRENT VOLTAMMETRY

This technique is a direct descendant of the faradaic impedance method. The small-amplitude sinusoidal voltage (nominally 1 to 10 mV) with a frequency of 10 Hz to 100 kHz rides on a constant or at least slowly changing potential ramp (nominally 5 mV/s). The nonsinusoidal (dc) potential is essentially constant during any single cycle of the sine wave and can be considered time-invariant. In the case of polarography where a dropping mercury electrode is used, the potential is also essentially constant during the lifetime of any individual drop. The resulting current signal is transduced into a voltage and filtered to remove unwanted signal components. These unwanted components are primarily the response due to the slow voltage ramp (the dc signal) and would have made up the desired signal in simple polarography or voltammetry. With a sufficiently slow ramp, the major frequency component of this response is lower than the ac excitation frequency and a simple high-pass filter is sufficient for the separation. Band-pass filtering at the excitation frequency improves the signal-to-noise ratio for low-concentration cases. After the filtering, the signal is full-wave rectified and analyzed as a function of potential. For a simple system with no charge transfer, chemical kinetics, or adsorption, the response looks similar to Figure 5.2c. The differences arise from the presence of $C_{dl}$ and $R_u$ in the cell as shown in Figure 5.1. Applying a sinusoidal voltage across a capacitor results in a sinusoidal current of the same frequency which lags behind the voltage by 90° and whose amplitude is directly proportional to the frequency, capacitance, and voltage amplitude as in Equation 5.1.

$$I(wt) = dEwC_{dl} \tag{5.1}$$

The total current signal monitored is the sum of the current through the double-layer capacitance $C_{dl}$ and the current through the faradaic impedance $Z_f$. Applying a sinusoidal voltage of amplitude dE across an isolated faradaic impedance (if only we could) would result in a sinusoidal current that lags the voltage by 45° and whose magnitude is

$$I(wt) = (nF)^2 AC_{ox}(wD_{ox})^{1/2} F(E_{dc})dE/4RT \quad (5.2)$$

In this expression $F(E_{dc})$ represents the potential-dependency of the current (the shape function that leads to Figure 5.2c).

This signal can be thought of as being composed of two components, one in phase with the excitation voltage ($\theta = 0°$) and one whose phase angle $\theta$ is $-90°$. For a signal of magnitude I with phase angle $\theta$ relative to the excitation, the in-phase component $I_0 = I \sin(\theta)$, and the out-of-phase or quadrature component $I_{90} = I \cos(\theta)$.

The total cell current is the vector sum of the in-phase and quadrature components. The faradaic component, the information we are usually interested in, is combined with and often buried in the double-layer capacitance information. The ratio of the faradaic to double-layer signal is proportional to the reciprocal of the square root of the excitation frequency. Practical considerations limit the low end of the frequency to about 10 Hz, so other means of improving the faradaic to double-layer ratio must be found. A commonly employed solution is found in the phase relationships previously mentioned. An ideal capacitance has all of its current response 90° to the excitation signal, while for an ideal, reversible isolated faradaic impedance, the current signal is composed of two components of equal magnitude at 90° and 0° relative to the excitation. While the signal component at 90° is only 70.7% of the total faradaic in magnitude, it is not competing with a strong signal from the double-layer capacitance. This is termed phase-selective ac voltammetry. This property is shown in Figure 5.4, which also illustrates one of the problems associated with this method. The signal at 0° relative to the excitation still contains some contribution from the double-layer capacitance. The primary reason for this deviation from the predictions of our simple picture is not so much that the double layer is not a simple, ideal capacitor, but rather the problem arises from the presence of the uncompensated solution resistance, shown in Figure 5.1, which couples the faradaic and capacitive components together. The amount of double-layer capacity information "leaking" into the 0° faradaic measurement depends on the magnitude of the uncompensated resistance. To some extent this effect can be compensated for by electronic solution resistance cancellation techniques and by cell design. Even though the separation is not ideal, the use of phase-selective ac voltammetry provides a major improvement in detection limit. There are excellent theoretical [1,2–5] and practical [1,2–6] treatments available for ac voltammetry and polarography. Theoretical predictions are available for a great variety of chemical and kinetic complications. An underexploited variant on this technique was developed by Smith in the 1970s [7,8]. Smith employed a digitally synthesized signal, which represented a set of carefully chosen frequencies. Fourier transformation of the response allowed the simultaneous determination of the cell response (both amplitude and phase) to these different frequencies. Although this

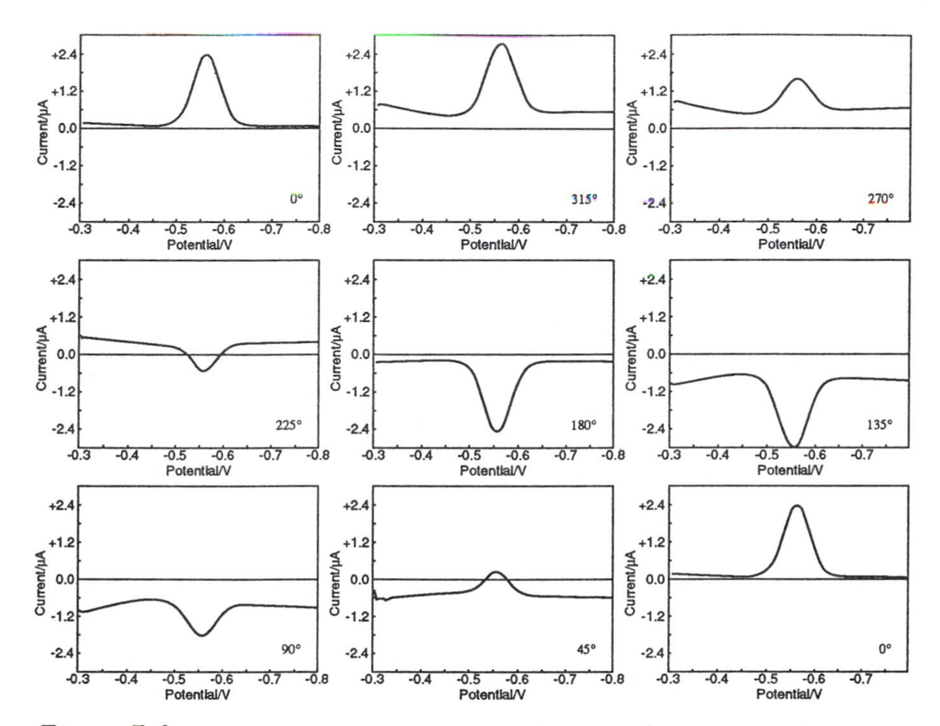


**Figure 5.4** Phase-selective ac polarograms for 20 ppm $Cd^{2+}$ at different phase angles ($\Delta E$ = 25 mV, ac frequency = 50 Hz).

methodology was developed primarily for mechanistic studies, the concept should be applicable to analytical studies where it could allow a better separation of double-layer and faradaic responses.

## IV. CYCLIC ALTERNATING-CURRENT VOLTAMMETRY

This analog to normal cyclic voltammetry is formed by the addition of a small-amplitude sinusoidal voltage to the large-amplitude potential triangular wave of cyclic voltammetry previously discussed in Chapter 3. As long as the ramp potential changes only a fraction of a millivolt in the period of the sinusoidal voltage, it is possible to obtain full information from both techniques simultaneously [6,9,10]. The CV portion of the response (nominally the dc or slowly changing component) provides mechanistic information, while the ac portion is most useful for quantitative evaluations of rate constants.

## V. TENSAMMETRY

One further relative of ac voltammetry should also be mentioned here. The technique, known as tensammetry [11,12], can be used to monitor the presence of many types of surfactants. The experimental procedure is identical to that of ac voltammetry but the response is nonfaradaic, arising instead from the double-layer capacitance. Many surfactants adsorb strongly and in a highly potential-dependent manner to the surface of electrode materials. The double-layer capacitance of the electrode with an adsorbed layer of surfactant is far different than that of the bare material, and in many instances the adsorption-desorption process occurs over a very narrow potential window (see Fig. 5.5). While we normally think of the current through a capacitor as being proportional to the capacitance multiplied by the time derivative of the voltage across it, the more general expression is the derivative of the product of capacitance times the voltage drop (the capacitance is generally a constant). In the case of tensammetry, the capacitance actually changes fairly abruptly as a function of potential (which is changing as a function of time) and sharp current spikes re-

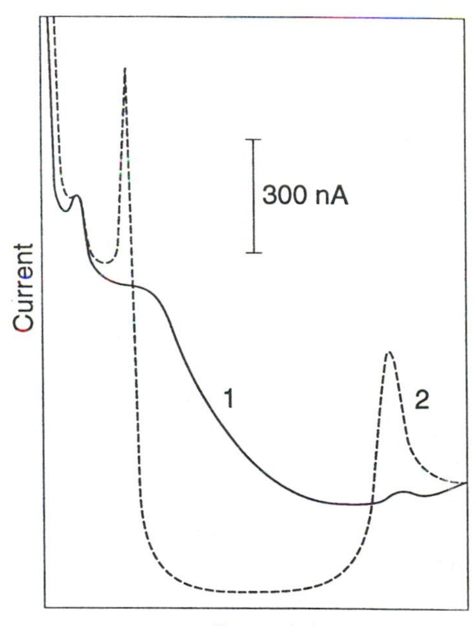


**Figure 5.5** AC polarograms of 1 N $Na_2SO_4$ (curve 1) and 1-octanol in 1 N $Na_2SO_4$ (curve 2).

sult (normally one at the potential where adsorption occurs and at another where desorption occurs). The shape of the response is not nearly as predictable as that observed for the faradaic process and is usually strongly dependent on the amplitude of the excitation voltage, rate of change of the dc potential, and often on the concentration of the surfactant. One can even encounter systems where the tensammetric peak height decreases with increasing excitation amplitude. These and other factors tend to limit the applicability of tensammetry to systems with one or perhaps two surfactants, but in the proper circumstances it can provide valuable information rapidly and fairly inexpensively.

## VI. STEP-BASED METHODS

A variety of electrochemical techniques are based upon the application of combinations of potential steps to the electrode. The majority of these gain their utility due to the fact that the current component arising from charging the double-layer capacitance decays much more rapidly than the faradaic current for a step excitation. For an isolated series resistance-capacitance combination, the signal is an exponential decay whose time constant is the reciprocal of the $C_{dl}$ times $R_u$ in Figure 5.1. The maximum signal is the step size divided by $R_u$. The response for an isolated reversible faradaic process decays as $t^{-1/2}$. Although a real cell does not behave like the simple combination of these two responses, the ratio of faradaic to double-layer current increases dramatically as a function of time after the step. Historically the first step-based technique was developed by Kemula in 1930 [13] in a study of adsorption at a dropping mercury electrode (DME). Almost two decades later, Kalousek [14] employed a commutator to add a 5-Hz square-wave potential to the ramp potential applied to a DME in a study of electrochemical reversibility. Ishibashi and Fujinaga [15,16] developed a low-frequency mechanical square-wave instrument, and G. C. Barker [17] reported a higher frequency electromechanical unit and predicted the response for simple systems. Kambara also developed theoretical predictions for the method [18]. Rather than trace the historical development of these techniques, we will consider them from the fairly arbitrary standpoint of the complexity of the excitation. At this point, it is worth mentioning that the utility of any technique is directly proportional to the availability of instruments that implement it. In the 1960s, practical electroanalytical chemistry had become effectively moribund due to the lack of commercial instrumentation that implemented the new methods being reported, primarily from academic and governmental laboratories. This condition was rectified by Princeton Applied Research, which produced the first "modern" commercial instrument that implemented many of the new techniques. Today a number of companies, including Amel, Bioanalytical Systems, Cypress Systems, Metrohm, Princeton Applied Research, and Tacussel, are active in the field.

## VII. STAIRCASE VOLTAMMETRY

Normal dc polarography, cyclic, and linear sweep voltammetry are based on potential ramp waveforms. In the case of polarography (which implies the use of a DME) the ramp rate is so slow that, at least in theory, the potential is essentially constant over the lifetime of a single drop. The application of a fixed potential to each drop more closely approximates theory than the slow ramp that was originally employed for polarography, but the major advantage of the staircase waveform is best seen in the case of fixed electrodes. Consider Figure 5.6 which shows a typical staircase waveform. Small steps of amplitude dE (normally 10 mV or less) are applied every p seconds and the cell response current is sampled at times S (S is often expressed as a fraction of $t_p$) ($0 < S < 1$). The method was first suggested by Barker [19]. The first staircase instruments were built prior to the availability of inexpensive laboratory computers and employed purely analog methods to develop the staircase waveform [20]. Some of the allure of staircase voltammetry and other step methods is that these are the waveforms that are the easiest to produce via digital computer-based instrumentation [19], but its true value arises from the improved faradaic signal ratio even at high sweep rates [21–23]. The exact shapes of the responses depend not only upon the electrochemistry and sweep rate (controlled by dE and $t_p$) but also upon the value of S employed (when the current is sampled within a step) [24–26]. For a simple electrochemically reversible system, a value of $S = 0.707$ has been shown to yield results identical to those predicted for a ramp excitation [27]. Other values of S are required to match ramp theory for other electrochemical reactions [28]. The overall response pattern will be quite similar, but peak shapes and the fine details of potential shifts and current ratios will differ somewhat from ramp theory for almost any sampling ratio. From the standpoint of analytical utility, the use of $S = 1$ is obviously optimal. The

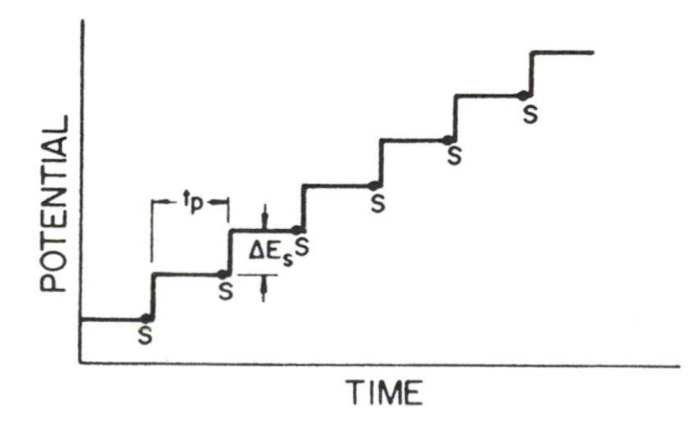


**Figure 5.6** Excitation signal for staircase voltammetry.

smaller the step size employed, the better the match with ramp theory, but contrary to intuitive expectations, a deviation still occurs.

## VIII. DIFFERENTIAL PULSE METHODS

This technique was first proposed by Barker [29] as an outgrowth of earlier square-wave work but to a large extent was popularized by the work of the Osteryoungs [30–33]. The excitation waveform is shown in Figure 5.7. It is basically the staircase waveform with tread height $dE_s$ and width T with an additional pulse of height $dE_p$ and width $t_p$ applied at the end. Current is sampled at times $S_1$ and $S_2$. When the electrode employed is a DME, the technique is termed differential pulse polarography and the drop is terminated just after time $S_2$ so that the next drop grows in at the new constant potential and total drop life is T. When some electrode other than a DME is employed the technique is known as differential pulse voltammetry. The typical range of T values is 0.5 to 5.0 s, and $t_p$ is on the order of 50 ms. In practice, the currents at $S_1$ and $S_2$ ($I_1$ and $I_2$) are averaged over a time interval of one period of the power-line frequency (16.7 ms for a 60-Hz system) to minimize noise. The signal $I_1$ will be either the dc polarogram or staircase cyclic voltammogram of the system. Since $t_p$ is very small compared to T, the signal $I_1$ will be almost identical to $I_2$ with a pulse height $dE_p$ of zero (the signal would change a negligible amount in the time interval $t_p$ in the absence of a pulse). The difference $I_2 - I_1$ represents the current due to the application of the pulse. For pulses whose heights meet the small-amplitude criterion, the difference versus potential plot closely resembles the results for ac polarography.

The correction represented by the sample at $S_1$ is not perfect because the time delay $t_p$ is not infinitely small. In the case of the DME we can improve this by means of the excitation signal shown in Figure 5.8a, which is a modification of the original. Here we use two drops, one in which we apply a pulse, and one where no pulse is employed. The difference in the two signals just prior to drop detachment represents a much better estimate of the current due to the

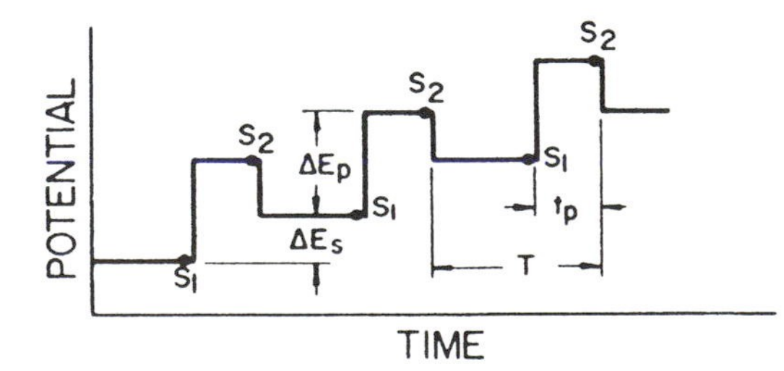


**Figure 5.7** Excitation signal for differential pulse voltammetry.

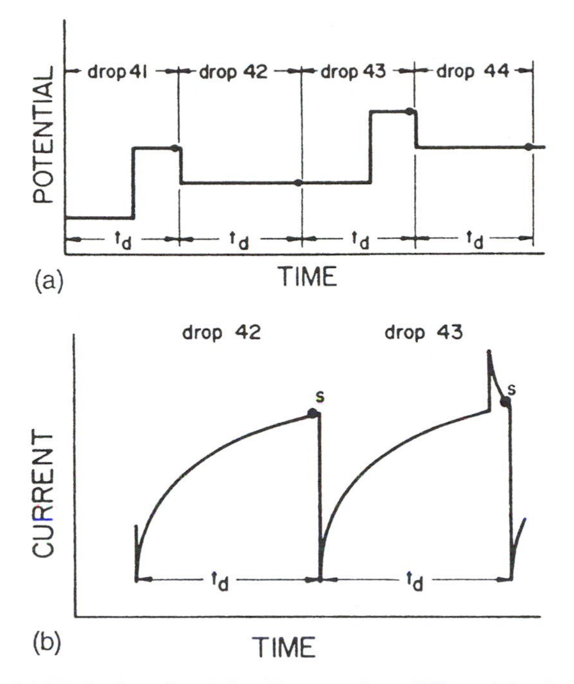


**Figure 5.8** (a) Excitation signal for alternate drop differential pulse polarography. (b) Total current response for alternate drop differential pulse polarography.

pulse. This method, termed alternate drop differential pulse polarography, is attributed to Osteryoung and co-workers [31,32]. This approach doubles the experimental time and is not generally applicable to other electrode types, but it is a clever solution to the problem and provides an optimal response.

The faradaic signal increases linearly with $dE_p$ in the small-amplitude region, as does the double-layer current. Increasing the amplitude of $dE_p$ improves signal strength and increases the signal-to-electronic-noise ratio while the faradaic to double-layer signal ratio stays constant. As the excitation signal $dE_p$ increases, some deviations from theoretical predictions occur, but analytically useful results are often obtained even for values in excess of 0.100 V. The response typically begins to broaden out, becoming increasingly asymmetric, and plots of peak current $i_p$ versus $dE_p$ are not linear at high values of $dE_p$. As the amplitude of $dE_p$ increases, the ratio of faradaic to double-layer current decreases (double-layer current is still linearly proportional to $dE_p$ but the faradaic is not) and most applications stay below 0.100 V. Under favorable conditions, concentrations as low as $10^{-7}$ M can be monitored for reversible systems.

The peak shape and peak current magnitude are strongly influenced by the chemical and electrochemical processes involved. This is true of all step-based techniques since they tend to stress the charge-transfer rate constant. Slow rates lead to lower currents and broader peaks.

One serious limitation common to most small-amplitude techniques is the greatly reduced response for systems with slow charge-transfer kinetics. Due to the high activation energy of slow electron-transfer processes, they are particularly sensitive to the presence of other species in the solution. Real-world environmental samples are notoriously "dirty" and these matrix effects can be difficult to deal with. Applications notes from the instrument manufacturers are frequently an invaluable source of practical information for dealing with these problems for specific elements and matrices.

## IX. SQUARE-WAVE METHODS

These methods are the source of a fair amount of confusion. The problem arises from the number of waveforms employed, which are frequently described as simply square-wave voltammetry. In this discussion we will consider three basic groups: the Barker, Kalousek, and Osteryoung formats. Barker square-wave voltammetry or polarography is the simplest to visualize. The waveform is a direct analog to sinusoidal ac voltammetry with a symmetric square wave of frequency f and amplitude dE riding on either a ramp or slow staircase waveform. Multiple cycles of the square wave are applied for each drop of the DME or step of the staircase. Just prior to drop detachment (or new step in the staircase) the current is measured at the end of each half cycle of the square wave and the differences are displayed as a function of the ramp or staircase potential. The original Barker [17] instrument employed a 225-Hz square wave whose half height was variable between 2 and 17.5 mV and terminated the drop 1.75 seconds after the beginning of growth. This signal possesses good faradaic to double-layer signal characteristics for reasons analogous to those discussed for differential pulse polarography. It is important to note that this signal represents the *steady-state* response of the cell to the square-wave excitation. Kalousek [14] square-wave polarography (termed Kalousek Type III by Heyrovsky [34]) is a lower frequency method (5 Hz), which measures the current only on the reverse half cycle of the square wave. This is in fact not really a small-amplitude method since the total current is displayed. Its inclusion here is due to its historical importance and its similarity to the next method. The most common form of square wave today was first proposed by Ramaley and Krause [35] but is most closely associated with the Osteryoungs because of their many publications in the area [36–40]. The waveform employed is shown in Figure 5.9. It differs from the previous methods in that the base potential (potential of the staircase) increases by dE for each full cycle of the square wave whose half height is $E_{sw}$

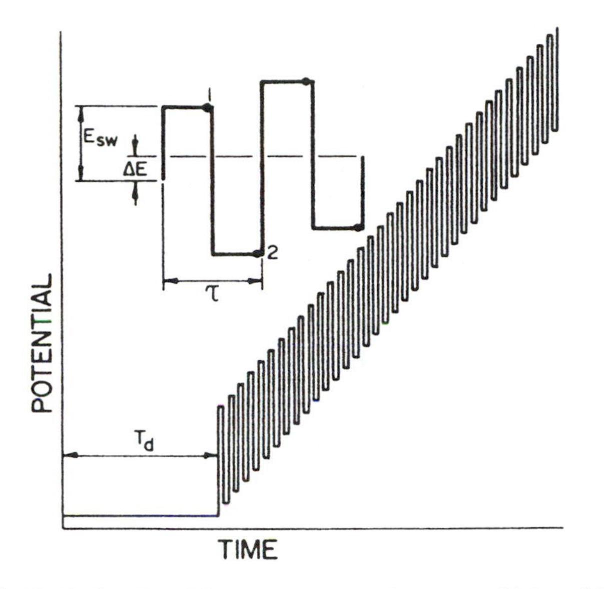


**Figure 5.9** Excitation signal for square-wave voltammetry. [Adapted from Ref. 40.]

and whose period is $\tau$. The current is measured at the end of each half cycle (times 1 and 2). This wave form can be applied to a stationary electrode or, as is implied in Figure 5.9, to a single drop from a DME. In this case the time interval $t_d$ is designed to allow the drop to grow to a predetermined size. The results are shown in Figure 5.10. The current component labeled A is the current for the forward half cycle (time 1), component B is due to the reverse half cycle (time 2), and component C is the difference. Note that component B is essentially the response for Kalousek type III and, like it, is extremely sensitive to the electrochemical reversibility of the couple and the chemical stability of the product of the forward step. The method can be quite rapid and lends itself to the monitoring of rapid processes such as liquid chromatography, as shown in Figure 5.11. J. Osteryoung has presented an excellent description of the method with predictions for the response with many types of complications and a valuable review of pulse methods in general [40].

## X. DIFFERENTIAL "NORMAL PULSE" VOLTAMMETRY

Large-amplitude ("normal") pulse voltammetry techniques were introduced in Chapter 3. The differential "normal pulse" (DNP) method combines several features of both the small- and large-amplitude pulse techniques. This technique is normally performed at a DME and is actually a form of polarography. The

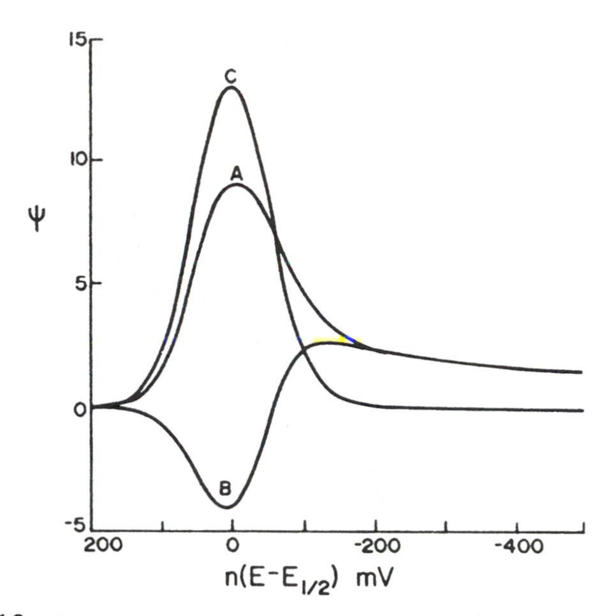


**Figure 5.10** Calculated square-wave voltammograms for reversible electron transfer: (A) forward current; (B) reverse current; (C) net current, dimensionless units. [Adapted from Ref. 40.]

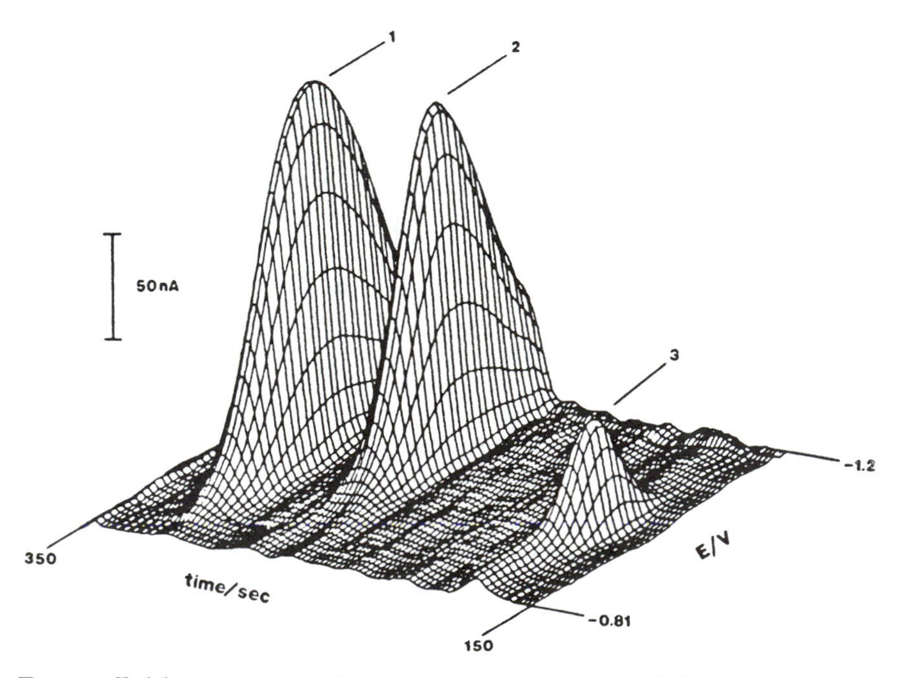


**Figure 5.11** Three-dimensional "chromatovoltammogram" for two nitrosamines (peaks 1 and 2) and an unknown impurity (peak 3). [From Ref. 39, reproduced with permission.]

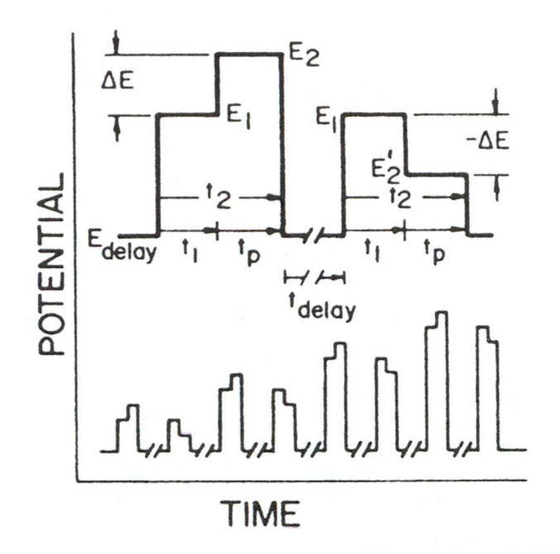


**Figure 5.12** Excitation signal for differential "normal pulse" voltammetry. [Adapted from Ref. 42.]

waveform shown in Figure 5.12 represents the excitation signal employed for DNP in the alternate mode [41–43]. The pulses begin at a potential where no faradaic reaction occurs. The potential is initially changed to a potential $E_1$ and held for $t_1$ seconds and then changed by an amount dE to the new value $E_2$ for a time $t_p$. The current is sampled at times $t_1$ and $t_2$ and the difference $(i_2 - i_1)$ is recorded. A new mercury drop is initiated. The sign of dE alternates between + and – for each pulse. The potential $E_1$ is changed to a new value after every pair of pulses so that there is + and – dE data for every value of $E_1$. The pulse for which the sign of dE has the same sign as the direction of increase for $E_1$ is considered the *forward* pulse and the other the *reverse* pulse. The results for a typical experiment for lead ion in an acetate buffer in are shown in Figure 5.13. The current $i_{for}$ is the current difference for forward pulses and $i_{rev}$ is the difference for the reverse pulses; dI is $(i_{for} - i_{rev})$. The small-amplitude potential changes dE are the analytic component, and the large-amplitude pulses $E_1$ are used to establish the initial concentration profiles. The use of the fairly narrow $(t_1)$ $E_1$ pulse as opposed to a staircase as in the case of DPV makes the technique less sensitive to adsorption processes and follow-up reactions since the drop spends most of its time at a potential where no reaction occurs.

## XI. CONCLUSION

Frequency-domain and time-domain small-amplitude techniques provide unique advantages for both analytical and fundamental studies. They tend to be more sensitive than large-amplitude techniques and are often more precise. These advantages result from the ease with which the signal can be distinguished from

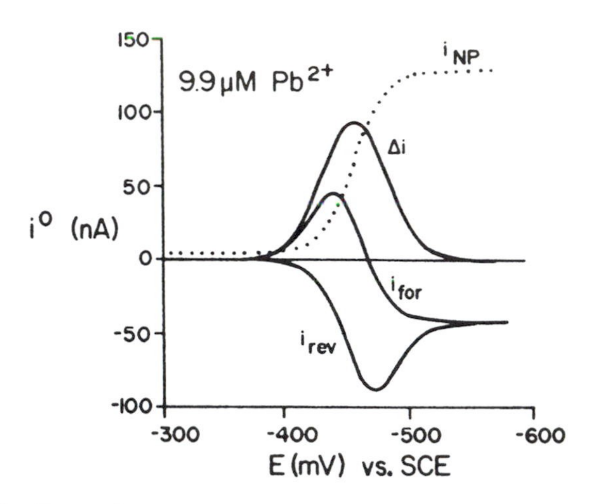


**Figure 5.13** Forward, reverse, and difference current DNP voltammograms of 9.9 μm (2.05 ppm) $Pb^{2+}$ in 0.5 M $CH_3COOH$/0.5 M $CH_3$ COONa. Conventional NP voltammogram is also shown (dotted line): $t_1 = t_p = 50$ ms; $t_s = 16.67$ ms; $t_{delay} = 0.4$ s; $E_{delay} = -300$ mV; $\Delta E = -25$ mV; T = 20°C. Solid lines are experimental curves. Dotted line is theoretical response. [From Ref. 42, reprinted with permission.]

the background, the ease of minimizing charging current influences, and the fact that steady-state measurements can be made. In most cases, small-amplitude techniques are coupled to large-amplitude techniques, with the latter setting the surface concentration as an initial condition for the former. Understanding how to use small-amplitude techniques properly for fundamental purposes (e.g., determination of rates and transfer coefficients) requires a mathematical appreciation beyond the interest of most chemists. There are excellent sources, and Bard and Faulkner is the best place to start [1].

If nothing else has been accomplished, it should be very clear from Chapters 3 to 5 that there is an extraordinary number of finite current electroanalytical techniques. There is no doubt that this can cause considerable confusion for novices. Fortunately, all of these methods are based on relatively few fundamental concepts. It must be understood that (1) electron transfer rates and equilibrium constants vary with potential, (2) mass transport to an electrode surface is precisely defined and reproducible, and (3) the charge required to establish an electrode potential can be temporally distinguished from that utilized by a redox couple. These concepts are addressed in Chapter 2. Now that we have covered the more important electrochemical techniques, it is strongly recommended that Chapter 2 be reviewed with these techniques in mind.

## REFERENCES

1. A. J. Bard and L. R. Faulkner, *Electrochemical Methods*, Wiley, New York, 1980.
2. E. Warburg, *Ann. Physik* 67:493 (1899).
3. D. E. Smith, in *Electroanalytical Chemistry,* Vol. 1 (A. J. Bard, ed.), Marcel Dekker, New York, 1966.
4. D. E. Smith, *Crit. Rev. Anal. Chem.* 2:247 (1971).
5. D. D. Macdonald, *Transient Techniques in Electrochemistry*, Plenum, New York, 1977.
6. A. M. Bond, *Modern Polarographic Methods in Analytical Chemistry*, Marcel Dekker, New York, 1980.
7. S. C. Creson, J. W. Hayes, and D. E. Smith, *J. Electroanal. Chem. 47*:9 (1973).
8. D. E. Smith, *Anal. Chem. 48*:231A, 517A (1976).
9. A. M. Bond, R. J. O'Halloran, I. Ruzic, and D. E. Smith, *Anal. Chem. 48*:872 (1976).
10. A. M. Bond, R. J. O'Halloran, I. Ruzic, and D. E. Smith, *Anal. Chem. 50*:216 (1978).
11. B. Breyer and H. H. Bauer, *Alternating Current Polarography and Tensammetry*, Interscience, Wiley, New York, 1963.
12. P. M. Bersier and J. Bersier, *Analyst 113*:3 (1988).
13. W. Kemula, *Coll. Czech. Chem. Commun.* 2:502 (1930).
14. M. Kalousek, *Coll. Czech. Chem. Commun. 13*:105 (1948).
15. M. Ishibashi and T. Fujinaga, *Bull. Chem. Soc. Jpn. 23*:261 (1950).
16. M. Ishibashi and T. Fujinaga, *Bull. Chem. Soc. Jpn. 25*:68, 233 (1952).
17. G. C. Barker and I. L. Jenkins, *Analyst 77*:685 (1952).
18. T. Kambara, *Bull. Chem. Soc. Jpn. 27*:253, 527, 529 (1954).
19. G. C. Barker, *Adv. Polarogr. 1*:144 (1960).
20. C. K. Mann, *Anal. Chem. 33*:1484 (1961).
21. G. Lauer and R. A. Osteryoung, *Anal. Chem.40*:30A (1968).
22. C. K. Mann, *Anal. Chem. 37*:326 (1965).
23. R. S. Nigmatullin and M. R. Vyaselev, *Zh. Fiz. Khim. 19*:545 (1964).
24. S. Stefani and R. Steeber, *Anal. Chem. 54*:2524 (1982).
25. L. H. Miaw, P. A. Boudreau, M. A. Pichler, and S. P. Perone, *Anal. Chem. 50*:1988 (1978).
26. J. J. Zipper and S. P. Perone, *Anal. Chem. 45*:452 (1973).
27. H. L. Suprenant, T. H. Ridgway, and C. N. Reilley, *J. Electroanal. Chem. 75*:125 (1977).
28. (a) M. M. Murphy, J. J. O'dea, D. Arn, and J. G. Osteryoung, *Anal. Chem. 61*:2249 (1989); (b) M. M. Murphy, J. J. O'dea, D. Arn, and J. G. Osteryoung, *Anal. Chem. 62*:903 (1990).
29. G. C. Barker and A. W. Gardner, *Z. Anal. Chem. 173*:79 (1960).
30. E. P. Parry and R. A. Osteryoung, *Anal. Chem. 37*:1634 (1965).
31. B. H. Vassos and R. A. Osteryoung, *Chem. Instrum. 5*:257(1973).
32. J. H. Christie, L. L. Jackson, and R. A. Osteryoung, *Anal. Chem. 48*:242 (1976).
33. J. Osteryoung and K. Hasebe, *Rev. Polarography (Jpn.) 22*:1 (1976).

34. J. Heyrovsky and J. Kuta, *Principles of Polarography*, Czech Academy of Science, Prague, 1965.
35. L. Ramaley and M. S. Krause, Jr., *Anal. Chem. 41*:1362, 1365 (1969).
36. J. H. Christie, J. A. Turner, and R. A. Osteryoung, *Anal. Chem. 49*:1899 (1977).
37. J. A. Turner, J. H. Christie, M. Vukovic, and R. A. Osteryoung, *Anal. Chem. 49*:1904 (1977).
38. J. J. O'Dea, J. Osteryoung, and R. A. Osteryoung, *Anal. Chem. 53*:695 (1981).
39. R. Samuelsson, J. J. O'Dea, and J. Osteryoung, *Anal. Chem. 52*:2215 (1980).
40. J. Osteryoung and J. J. O'Dea, in *Electroanalytical Chemistry,* Vol. 14 (A.J. Bard, ed.), Marcel Dekker, New York, 1986.
41. K. Aoki, J. Osteryoung, and R. A. Osteryoung, *J. Electroanal. Chem. 110*:1 (1980).
42. T. Brumleve, J. J. O'Dea, R. A. Osteryoung, and J. Osteryoung, *Anal. Chem. 53*:702 (1981).
43. T. Brumleve, R. A. Osteryoung, and J. Osteryoung, *Anal. Chem. 54*:782 (1982).

# 6

# Introduction to Analog Instrumentation

**Peter T. Kissinger** *Purdue University and Bioanalytical Systems, Inc., West Lafayette, Indiana*

## I. CLASSICAL CONTROLLED-POTENTIAL INSTRUMENTATION

Instrumenting the controlled-potential techniques described in Chapter 3 is not difficult within certain limits of time and current. Nevertheless, beginners and old-timers alike often seem to have only the vaguest notion of how their instruments function and the logic implicit in various experimental designs. One reason for this is that there seems to be no good introduction to the subject. The literature is either too superficial or inordinately complex. Our goal here is the intermediate ground. We attempt to provide a phenomenological basis for several fundamental instruments on a level satisfactory for utilizing them intelligently. The steps up the ladder to more complex situations (e.g., very fast experiments, alternating-current experiments, two working electrodes at different potentials in the same solution, iR compensation, digital computer control and analysis, etc.) are relatively trivial ones. In our experience, most students encounter difficulties with electrochemical instrumentation at the outset primarily due to a lack of appreciation of basic electricity. The meaning of such terms as charge, voltage, current, and resistance must be intuitively clear.

Analytical techniques are conveniently discussed in terms of the excitation-system-response parlance described earlier. In most cases the system is some molecular entity in a specific chemical environment in some physical container (the cell). The cell is always an important consideration; however, its role is normally quite passive (e.g., in absorption spectroscopy, fluorescence, nuclear magnetic resonance, electron spin resonance) because the phenomena of interest are homogeneous throughout the medium. Edge or surface effects are most often negligible. On the other hand, interactions between phases are the central issue in chromatography and electrochemistry. In such heterogeneous techniques, the physical characteristics of the sample container become of critical importance. In electrochemical instrumentation, the cell is a circuit element with electrical properties that influence the performance of the overall instrument.

The potential difference that one either measures or controls in an electrochemical experiment is that said to exist between the electrode and the solution. These tasks are not as easy to implement as they might first appear. For one thing, the potential difference (strictly, the inner potential difference) $\Delta\phi$ between the two phases is not measurable. It quickly becomes obvious that a second electrode* in contact with the solution will be necessary and that this electrode will have its own metal-solution potential difference (again not measurable). It is important to recognize that in the media of interest (high dielectric constant, charge carriers present in great numbers) these potential differences exist only across a very small region contiguous with the metal (typically nanometers in width). There is no potential gradient across the bulk of the solution phase under zero-current conditions. This means that an ion (or molecule) in the bulk solution phase will have no way of knowing whether or not electrodes exist. The same ion will recognize that no potential gradient exists across this phase, and will have no way of ascertaining its own potential relative to that of either of the two electrodes.

The region across which the $\Delta\phi$ is impressed is broadly defined as the "electrical double layer," and it is in this region of very high potential gradient (typically $10^6$ V/cm) that all electrochemical reactions take place. The double layer has electrical characteristics similar to that of a conventional capacitor, and one often hears about "double-layer capacitance" and "double-layer charging currents." The microscopic structural and kinetic details of the double layer are the subjects of continuing research. Charged interfaces (metal-solutions, membranes-solutions, solutions-solutions, etc.) are of paramount importance in numerous physical, chemical, and biological phenomena. It is, however, naive to assume that these things are well understood by anyone.

Figure 6.1 illustrates the classical apparatus for applying a potential difference to an electrochemical cell consisting of two electrodes. The electrode where the chemistry of interest occurs is called the working electrode (W) and the other electrode, for want of a better name, is dubbed the counter electrode. The applied potential $E_a$ is measured between the counter and working electrodes, and the resulting current (if any) is measured in either the working-electrode lead (as shown) or the counter electrode lead with equal facility. Notice that the working electrode is tied to ground potential and remains so, irrespective of the value of $E_a$ selected with the potentiometer. It should be clear that to ions in the solution, it makes absolutely no difference whether or not this electrode is grounded.

---

*For our purposes an electrode is any solid conductor or semiconductor immersed in solution.

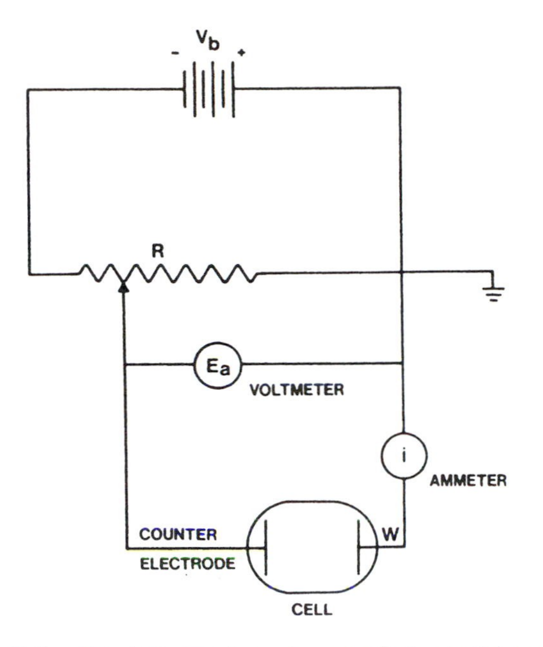


**Figure 6.1** Classical two-electrode controlled-potential apparatus.

In Figure 6.2a the situation in the cell is schematically illustrated in greater detail for the zero-current case. Note that the applied potential is divided (not necessarily equally) between $\Delta\phi_c$ and $\Delta\phi_w$ and that no potential difference exists across the bulk solution. Although the working electrode is at ground as far as the outside world is concerned, it is *positive relative to the bulk solution* inner potential. The potential difference we wish to control is $\Delta\phi_w$ and the desired control point (DCP) is indicated. The actual control point (ACP) as far as the circuit is concerned is on the metal side of the counter electrode interface. Thus $\Delta\phi_w = E_a - \Delta\phi_c$ and one considers the working electrode to have a potential $E_a$ versus $\Delta\phi_c$. In general, the counter electrode is chosen to be one with a very rapid redox couple available, which means that $\Delta\phi_c$ does not change significantly as a function of current density if the latter is kept small (e.g., by using a large surface area). Such familiar electrodes as the calomel electrode or the silver-silver chloride electrode fit this bill quite well. Thus, although we never know $\Delta\phi_w$ in an absolute sense, we are perfectly capable of knowing it in a relative sense.

The two-electrode system is just fine under the zero-current condition, but this does not allow very much chemistry to be accomplished. Figure 6.2b represents the finite-current case, where $\Delta\phi_w$ is the same as in Figure 6.2a. If the

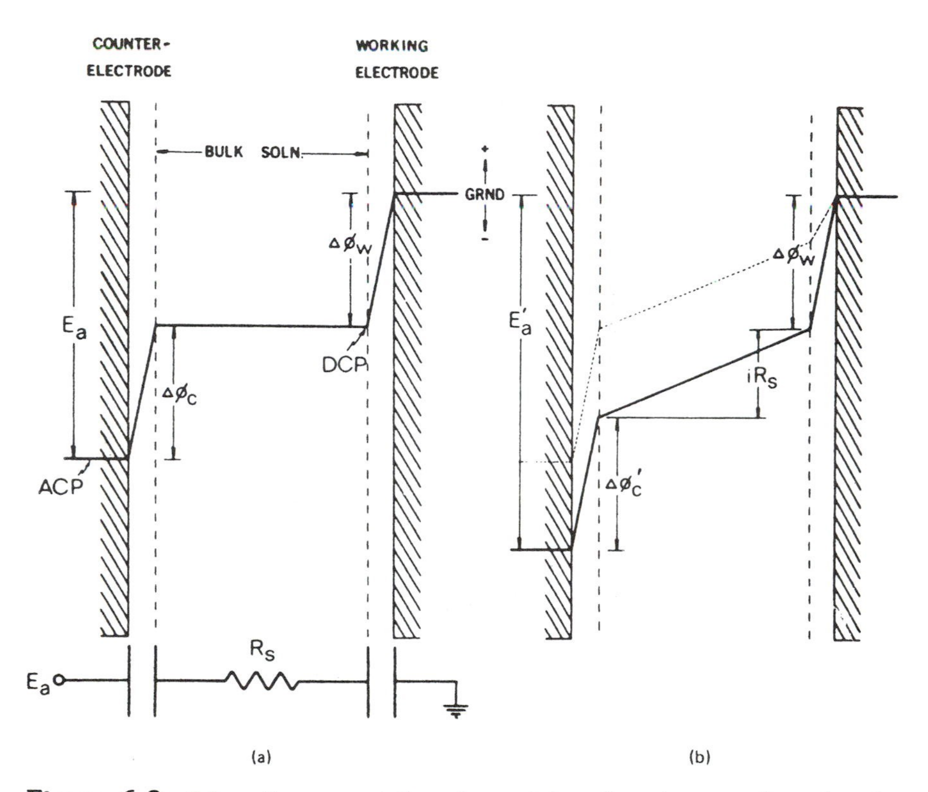


**Figure 6.2** Schematic representation of potential gradients in a two-electrode cell.

current is small enough and the counterelectrode is large enough (i.e., the current density at *this* electrode is kept small), $\Delta\phi_c$ may be assumed to be constant. If the solution resistance and the current are small enough, $E_a$ may be kept the same as in the zero-current case to maintain $\Delta\phi_w$ at the desired value. When the iR drop across the bulk solution phase is significant (as illustrated), $E_a$ must be increased by the value of iR or control will be lost. This undesirable situation derives from the remoteness of the ACP from the DCP. Fortunately, we can do much better. The problem just described is illustrated again in Figure 6.3. The importance of understanding this figure cannot be overestimated.

*The counter electrode in the two-electrode system serves two functions. First, it completes the circuit, allowing charge to flow through the cell; second, it is assumed to maintain a constant interfacial potential difference regardless of the current.* These two functions are mutually exclusive only under very restricted circumstances. Both needs are better served by two separate electrodes:

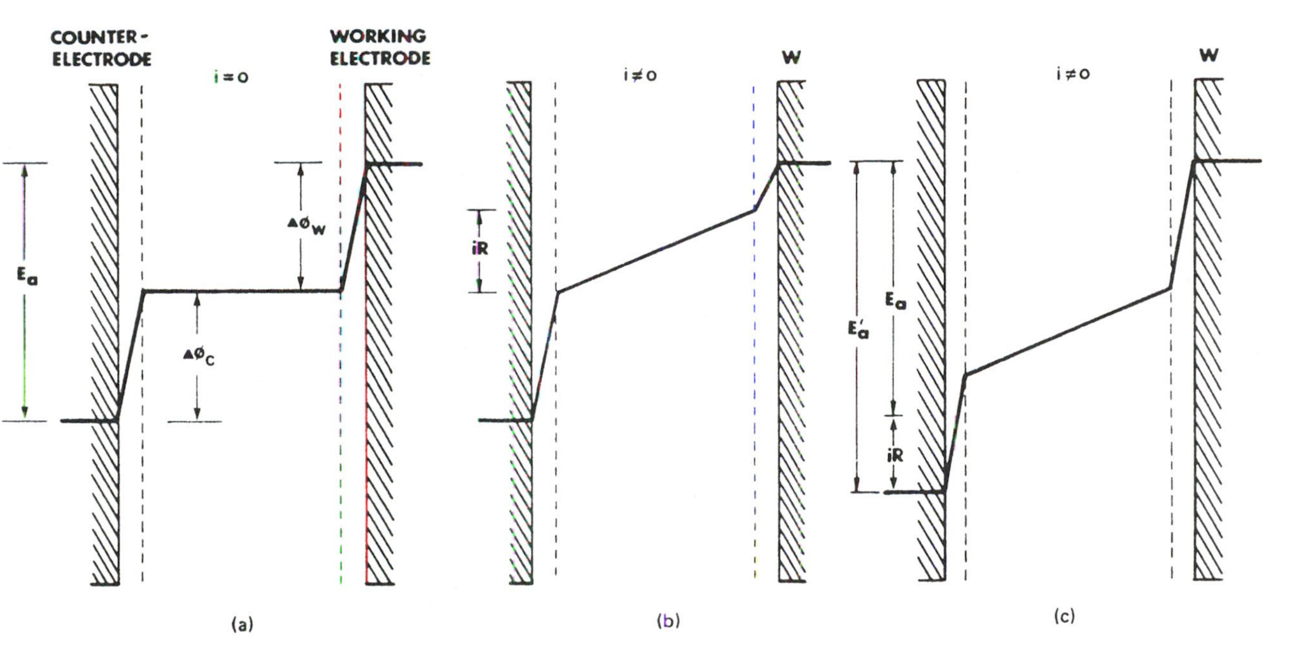


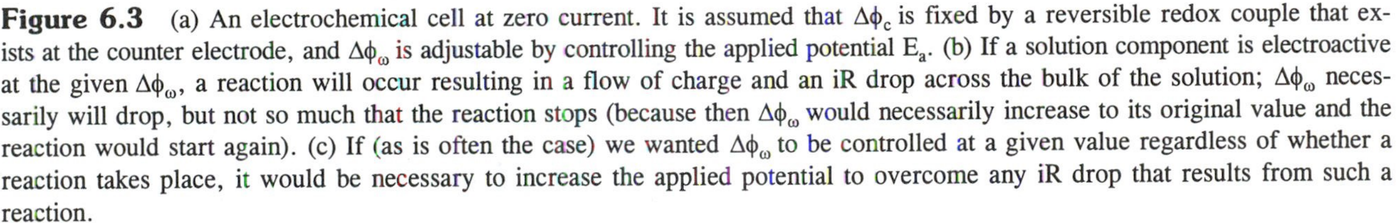

**Figure 6.3** (a) An electrochemical cell at zero current. It is assumed that $\Delta\phi_c$ is fixed by a reversible redox couple that exists at the counter electrode, and $\Delta\phi_\omega$ is adjustable by controlling the applied potential $E_a$. (b) If a solution component is electroactive at the given $\Delta\phi_\omega$, a reaction will occur resulting in a flow of charge and an iR drop across the bulk of the solution; $\Delta\phi_\omega$ necessarily will drop, but not so much that the reaction stops (because then $\Delta\phi_\omega$ would necessarily increase to its original value and the reaction would start again). (c) If (as is often the case) we wanted $\Delta\phi_\omega$ to be controlled at a given value regardless of whether a reaction takes place, it would be necessary to increase the applied potential to overcome any iR drop that results from such a reaction.

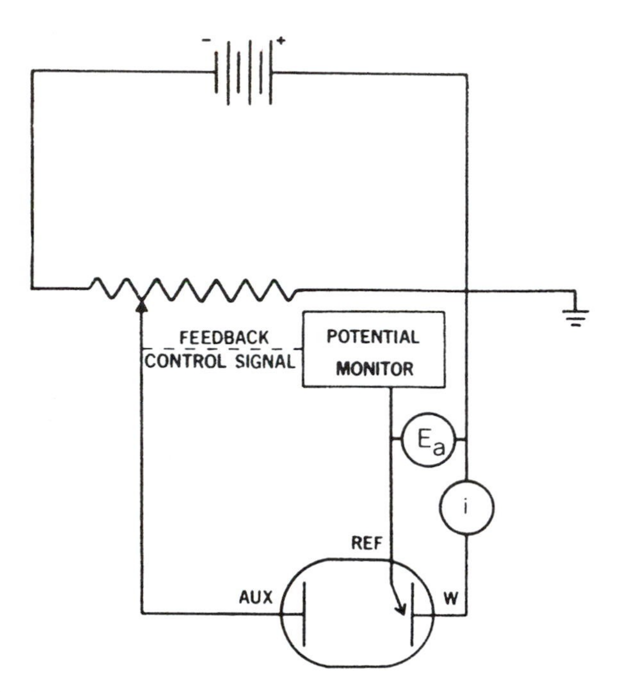


**Figure 6.4** Schematic representation of a primitive three-electrode controlled-potential apparatus.

an auxiliary electrode to assume the first function and a reference electrode to assume the second.* This approach is illustrated schematically in Figures 6.4 and 6.5.

The reference electrode is used as a potentiometric (always zero-current) probe to monitor $\Delta\phi_w$ relative to its own $\Delta\phi_r$. This value is compared with $E_a$ and if a difference (i.e., an error signal) exists, the potential impressed across the cell by the potentiometer is adjusted until balance (i.e., no error signal) is achieved. A device that accomplishes this control function automatically is called a "potentiostat" for obvious reasons. Such behavior can be mimicked by the experimenter. Although this is assuredly almost never done these days, it is useful to think about a manual potentiostat as a pedagogical device.

**Nomenclature note*: "Counterelectrode" is a term which is best applied in connection with two-electrode experiments and "auxiliary electrode" is a term reserved for three-electrode experiments. Although this rule has not always been followed in the literature, there is a healthy trend in this direction.

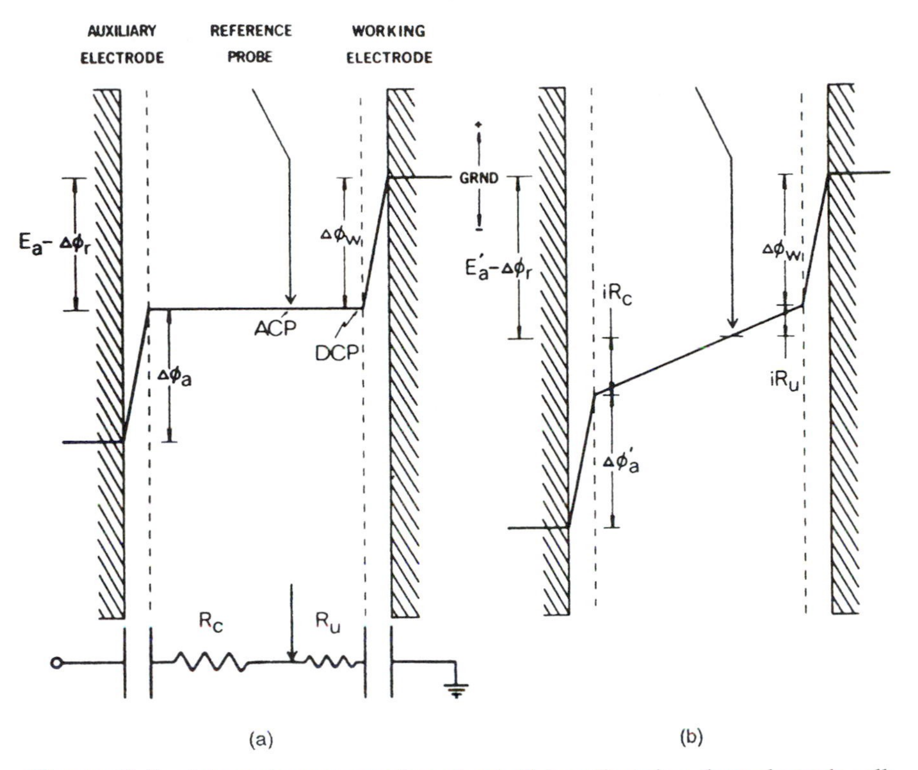


**Figure 6.5** Schematic representation of potential gradients in a three-electrode cell: (a) $i = 0$; (b) $i \neq 0$.

Imagine the situation given in Figure 6.4 where the potential monitor is a high-impedance voltmeter connected to a saturated calomel electrode (SCE) as reference. The potentiometer is now adjusted to a value sufficient to initiate a reaction at the working electrode. The potential at which the reaction ensues ($E_a$ vs. SCE) is monitored with the meter. As the reactive species is consumed at W, the current through (and the iR drop across) the cell will decrease. This causes $E_a$ to *increase*. The experimenter observes this increase on the monitor and judiciously adjusts the potentiometer to *decrease* the voltage applied to the auxiliary electrode until $E_a$ returns to its initial value. Implicit in this process of monitoring and adjusting is *negative feedback*. The result is a constant $E_a$. How constant $E_a$ will be depends on the sensitivity of the meter and the response time of the experimenter.

Notice (Fig. 6.5) that although the DCP and ACP are much closer together in the three-electrode approach, they are still not one and the same. There will be some iR drop error, albeit small. The total cell resistance has been divided into two parts: a compensated resistance $R_c$ and an uncompensated resistance $R_u$. Only the ohmic loss $iR_u$ will be effective in moving the ACP from the DCP. The effectiveness of the potentiostat depends largely on the ratio of $R_c$ to $R_u$. The larger this ratio, the better the result. This suggests that the geometry of the reference electrode and its placement may be of critical importance, depending on the solution resistance. The efficacious design of electrochemical cells and the placement of electrodes is discussed in Chapter 9.

We can summarize the virtues of the three-electrode potentiostat in two statements:

1. The iR drop across a cell is largely compensated for by bringing actual experimental control of potential closer to that desired.
2. Since the monitoring circuit draws negligible current, the reference electrode redox couple need not be perfectly reversible (i.e., need not have a large exchange current density) and need not have a very large surface area.

These advantages are very much independent, and either one may predominate in particular circumstances. The second advantage is quite often forgotten (particularly by those accustomed to two-electrode methodology). Limitations and experimental details are described later. It is also important to understand the unique circumstances when the working electrode is extremely small. This is the focus of Chapter 12.

## II. CONTROLLED-POTENTIAL INSTRUMENTATION BASED ON OPERATIONAL AMPLIFIERS

### A. Introduction

In order to assimilate the following mateial, the reader should be familiar with a few simple concepts in instrumentation elecronics. These topics are more than adequately discussed in numerous texts and so we do not belabor them here. For the convenience of those who need to review, we have prepared a list of subjects *essential* to understanding these introductory sections. This list is all-inclusive, so study of topics not on the list (e.g., ac circuit theory, inductance, transformers, power supplies, digital electronics, transducers, transistors, etc.) will be of no immediate value.

*Essential Topics in Electronics*
Resistors (Ohm's law) and capacitors
Voltage dividers
"Loaded" voltage dividers
Passive differentiators

Passive integrators
Solid-state diode (including zener diodes)
Ideal operational amplifier
Real operational amplifier (compliance limitations, offset current and voltage)
- Voltage follower
- Follower with gain
- Inverter
- Integrator
- Differentiator
- Comparators

Concept of input impedance and output impedance

## B. The Voltage Follower as a Potentiostat

In the next few pages we illustrate the evolution of modern controlled-potential instrumentation. The voltage follower (Fig. 6.6a) is a device that satisfies the need for an "impedance buffer" to isolate high-impedance sources (with no capability of supplying current) from low-impedance sinks (desirous of current). In other words, the follower transforms a *soft* voltage point, $E_i$, which is subject to *loading*, into a *hard* voltage point, $E_o$, which can supply considerable current. In this configuration the operational amplifier (OA) has no closed-loop voltage gain since $E_i = E_o$. The quality of the follower will depend on the impedance seen at the two inputs (the higher the better) and the current capability of the output stage of the OA. Of particular interest is the finite (though

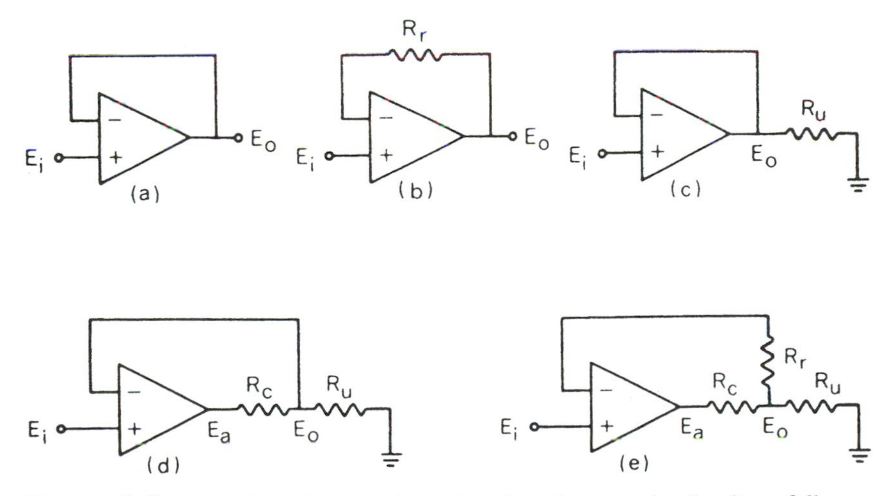


**Figure 6.6** Evolution of a potentiostat based on the operational voltage follower.

very small) "bias current" that will flow across the input terminals and an "offset voltage" that will exist between the input terminals. The latter will, of course, cause a difference to exist between $E_i$ and $E_o$. For most electrochemical applications it is desirable that this difference be less than about 1 mV, a feat easily achieved in practice.

If a resistance is placed in the feedback loop (Fig. 6.6b), the bias current $i_b$ will also create a difference between $E_i$ and $E_o$ by an amount $i_bR_r$. Even very inexpensive (< \$1) OAs can have bias currents of less than $10^{-9}$ A, which means that the value of $R_r$ will have to exceed 1 MΩ to create a 1-mV error. Amplifiers with bias currents of less than 0.1 pA ($10^{-13}$ A) are available. Using the same criterion, $R_r$ may then reach 1000 MΩ, a value well beyond any resistance commonly encountered in dynamic electroanalytical techniques. Such amplifiers are, however, eminently useful for constructing pH meters and pH stats and measuring potentials in electrophysiology, where very small high-impedance electrodes are often used.

Figure 6.6c depicts the voltage follower with a resistive load $R_u$. Limitations on $R_u$ are imposed only by the maximum output current of the OA, $i_o$(max). Quite obviously, $E_o/R_u$ cannot be allowed to exceed $i_o$(max). In other words, $R_u$ must be larger than $R_{min}$, where $R_{min} = E_o/i_o(\max)$.

In Figure 6.6d the change is much more dramatic. Here we have added an additional resistance $R_c$ between the OA output and the control point $E_o$. If readers have done their homework, they will immediately recognize this circuit as the familiar "follower with gain" drawn in an atypical way. The $E_o$ will still equal $E_i$ and it should be clear that $E_a = (1 + R_c/R_u)E_i$. If this expression is not clear, readers are advised to doodle on some scrap paper until it becomes obvious. Otherwise, there is no point in continuing with this material.

Notice that in Figure 6.6d, $i_o$(max) will still place a lower limit on $R_u$ (and only $R_u$), because the current through the two resistors is defined only by $E_o$ and $R_u$. The other resistance, $R_c$, is not controlling. Understanding this fact is also crucial to the reader's progress. There is, however, a limitation on $R_c$ imposed by the maximum *voltage* swing, $E_a$(max), permitted by the output stage of the OA. The limitation is defined by the fact that $i(R_c + R_u)$ cannot exceed $E_a$(max). Figure 6.6e incorporates the three resistors just discussed, and all of the limitations will apply.

The symbology chosen for Figure 6.6 will become clear for the reader who reviews the three-electrode configuration introduced in Figures 6.4 and 6.5. To use the voltage follower as a potentiostat, the auxiliary electrode is placed at the OA output, the reference electrode is connected to the negative input, $R_r$ is the reference electrode impedance, $R_c$ is the compensated resistance, $R_u$ is the uncompensated resistance, and the working electrode is connected to ground.

The use of passive components (resistors and capacitors) to construct an equivalent circuit for the electrochemical cell is carried one step further in Figure

6.7a. In addition to the bulk-phase resistances ($R_c$ and $R_u$), the interphase components $C_d$ (double-layer capacitance) and $Z_f$ (faradaic impedance) have been added. When a faradaic reaction occurs, the electrode interface takes on the character of a leaky capacitor. Phenomenologically, one considers the charge stored in the double layer to leak away through some heterogeneous electron-transfer reaction represented by the impedance, $Z_f$. Naturally, in order to maintain the potential difference $\Delta\phi_w$, additional charge must flow to replenish that being lost through $Z_f$. This is, of course, the function of our potentiostat. Note that the potentiostat in this case consists of a single operational amplifier. At this point in time, amplifiers suitable for many experiments can be purchased for as little as $0.50. A mathematical analysis of the equivalent circuit in Figure 6.7a will not be necessary until instrumentation for fast transient experiments is described in the following chapter.

Controlled-potential experiments (e.g., voltammetry, chronoamperometry) require some way of measuring the current without disturbing the controlled parameter. The most convenient way of accomplishing this is to connect an operational amplifier current-to-voltage converter (sometimes called a "current follower") to the working electrode as shown in Figure 6.7b. The current-to-voltage converter is a simplified version of the familiar operational amplifier inverter. One can think of the input resistor in an inverter as being a current source ($E_i/R_i$) for the feedback resistance, $R_f$. Replacing *this* current source with the working electrode is a trivial matter. The summing point of the OA will

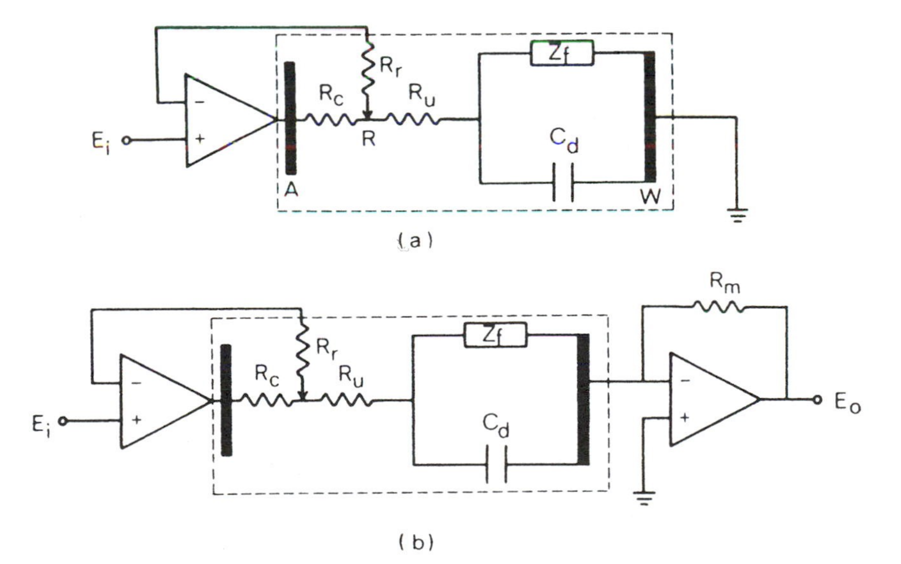


**Figure 6.7** Evolution of a potentiostat (continued). (a) Equivalent circuit of a three-electrode cell. (b) Addition of current-to-voltage converter.

under most conditions remain at *virtual ground,* and therefore connecting the current follower does not perturb the experiment.

The output of the converter, $E_o$, is a function of the current applied to it by the electrode reaction ($E_o = -iR_m$). Normally, current is measured by using a current-measuring resistor ($R_m$) to convert it into voltage. The circuit is most sensitive when $R_m$ is large. For any given current, $R_m$ must not be so large that $iR_m$ exceeds the maximum output voltage of the OA (typically, ca. ±14 V). In addition, i (as determined by the concentration of electroactive species, area of the working electrode, convection of the solution, etc.) may not exceed the current capability of the amplifier (typically, ±10 mA). If the potentiostat (or "control amplifier") and current-to-voltage converter have different current capabilities, i must not exceed the $i_{max}$ of the less powerful OA.

## C. The Adder as a Potentiostat

Figure 6.8 depicts an alternative potentiostat design based on the operational amplifier adder rather than the voltage follower. This two-amplifier potentiostat or DeFord-type potentiostat is widely used. As far as we can determine, the only advantage of this configuration (developed before differential-input OAs were widely available) is that excitation programs can be easily synthesized by adding a number of different waveforms via a sufficient number of input resistances $R_i$. For example, the basic ac voltammetry experiment can be performed by adding a sinusoid to a ramp via two input resistors. The contribution of each $E_i$ to the total excitation signal will depend on the relative magnitudes of the $R_i$ values, although in most applications these will all be the same.

The reader should study the adder-type potentiostat in detail and attempt to analyze it based on qualitative negative feedback principles (monitor a control point, compare with desired control, make adjustments as necessary). In your analysis first assume that $R_i$ and $R_f$ are equal. Why is the voltage follower required in the reference electrode lead? If a species oxidizes at +1.0 V vs. SCE, what should be the sign of $E_i$ to accomplish this chemistry in the two-amplifier potentiostat? In the voltage follower potentiostat?

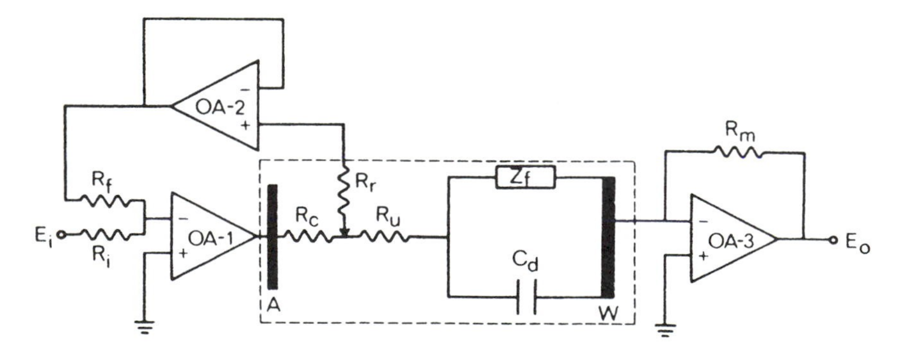


**Figure 6.8** Two-amplifier adder-type potentiostat with current-to-voltage converter.

## D. Current Measurements

Figure 6.9 illustrates four methods of measuring the current which are equally applicable to either potentiostat design. It is *essential* that the reader be able to calculate $E_o$ as a function of the current and the various resistors. If this exercise is difficult, you are in *serious trouble* and will have to go back to the study topics listed earlier. Circuit 6.9a has already been described. The other three can be useful in high-current experiments when only the control amplifier (and not the current-to-voltage converter) is capable of carrying the maximum current anticipated. In circuit 6.9b, we cannot let $R_m$ get too large. Why not? In circuit 6.9c there is a limitation both on the value of $R_m$ and the value of $R_m/R_1$. What is this limitation? In 6.9d there is also a limitation of the magnitude of $R_m$ and the ratio $R_m/R_i$. Can you figure this one out? If these questions are easy, you are doing just fine. There are several ways in which 6.9d may be improved if two additional amplifiers are available. Think about it.

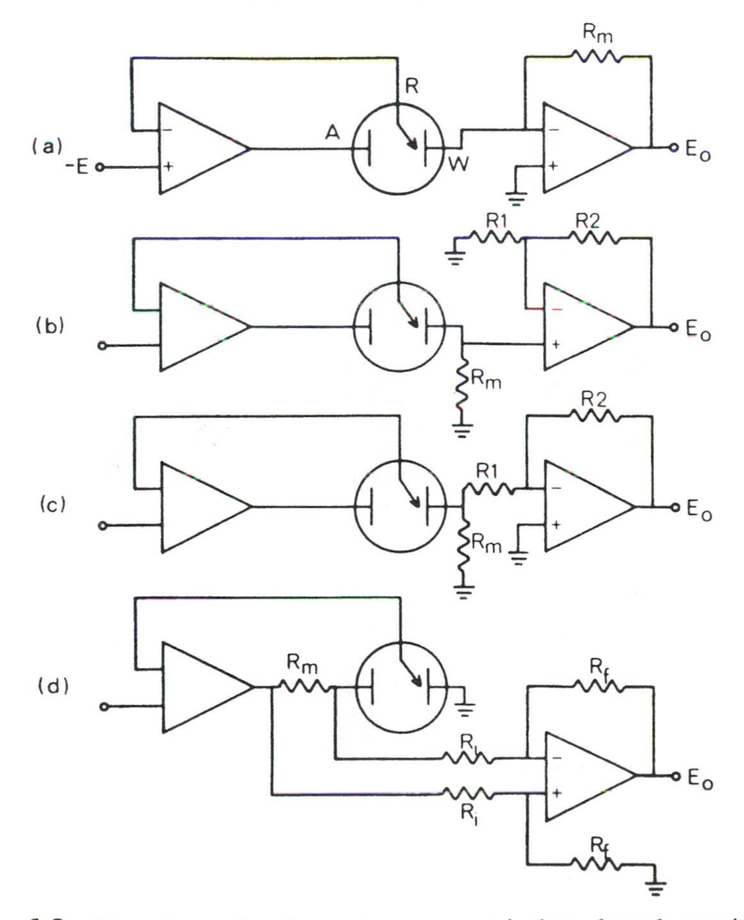


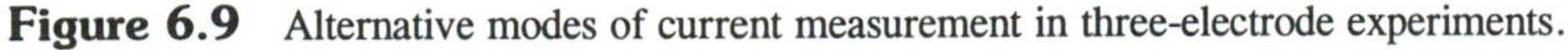

**Figure 6.9** Alternative modes of current measurement in three-electrode experiments.

## E. Cyclic Voltammetry Instrumentation

It is not the purpose of this volume to catalog the wide variety of circuits that have been used to generate excitation signals and measure the corresponding response of an electrode-solution interface. The apparatus required for a given experiment may range from a battery, a resistor, and a microammeter to sophisticated systems operated under computer control. Since electrometric titrations and voltammetry are the most common electroanalytical experiments, we make specific reference to the experimental requirements in several chapters. In the following section the operation of an analog cyclic voltammetry instrument (Bioanalytical Systems Model CV-1B, Fig. 6.10) is described in some detail.

Most cyclic voltammetry instruments can be conveniently thought of as the combination of three circuits: a signal generator, a potentiostat, and a current-measuring amplifier. The latter two subunits consist of an operational amplifier voltage follower and current-to-voltage converter and are used in a conventional manner. On the other hand, the waveform generator is unique and requires careful study. We examine a simple version of the circuit first and add refinements until the full design is explained.

A straightforward approach to triangular-wave generation is to integrate a square wave. *The amplitude of the square wave will determine the slope of the*

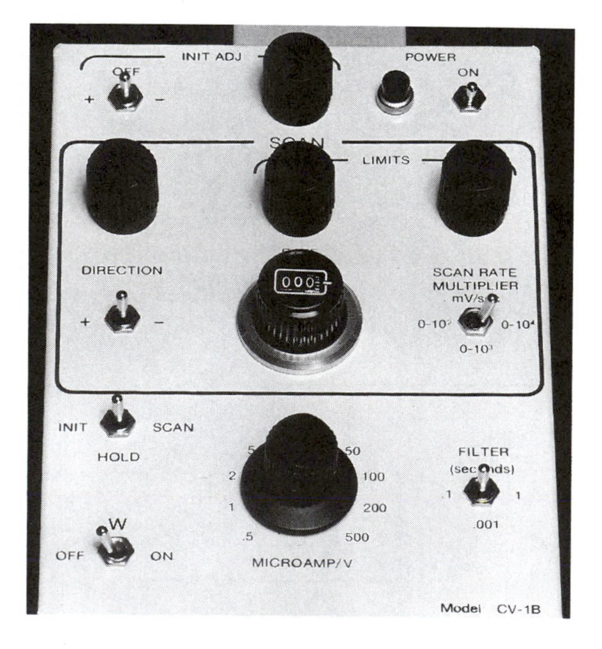


**Figure 6.10** Top view of BAS CV-1B.

*triangular wave and the period of the square wave will determine the amplitude of the triangular wave.*

Figure 6.11 illustrates the basic triangular-wave generator. Assume ideal amplifiers and that R1 = R2. In this configuration, OA-1 functions as a voltage comparator (square-wave generator) and OA-2 functions as the integrator with a time constant of R3 · C seconds. For the moment we will assume the output of OA-1 will be in either of two states (e.g., +10 or −10 V) depending on the polarity of the voltage at its positive input (i.e., a positive voltage at the + input results in a +10 V output). Note that the polarity at the positive input will depend on the voltage impressed across the series combination of R1 and R2.

If the output of OA-1 is +10 V, the output of OA-2 will be a negative-going ramp of slope −10/(R3 · C). As the voltage of the ramp becomes more negative, the + input of OA-1 also moves in a negative direction. When the output of OA-2 becomes more negative than −10 V, the + input of OA-1 will cross zero and become negative. At this point OA-1 must necessarily change state and its output swings to −10 V. The integrator responds by changing the ramp to a positive direction. When the integrator output reaches +10 V, OA-1 changes state again (−10 V to +10 V) and the integrator again changes di-

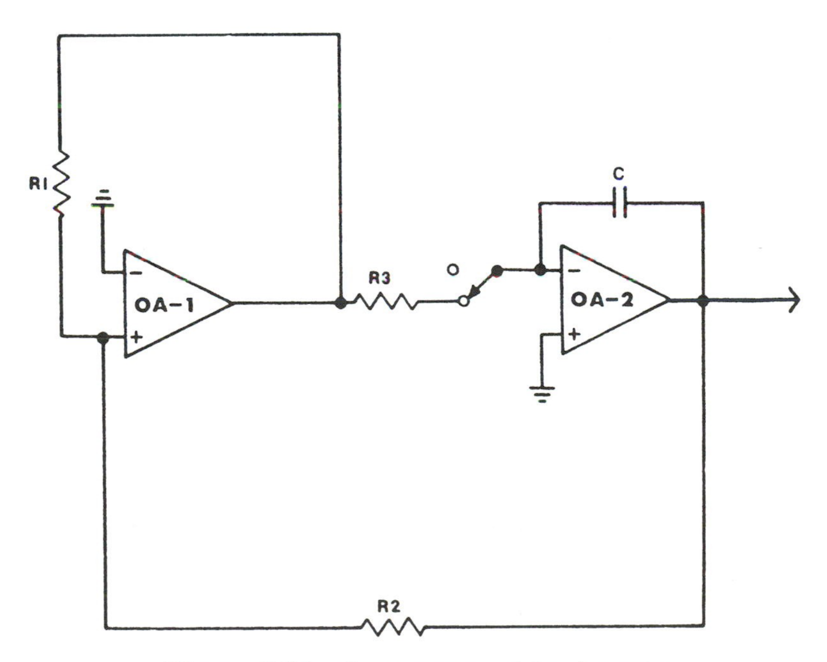


**Figure 6.11** Comparator coupled to integrator.

rection. This process continues indefinitely as long as the amplifiers are powered. The signal generator is said to be *free running*. To understand the events just described, it is useful to study the relation between the square wave (OA-1 output) and triangular wave (OA-2 output) as depicted in Figure 6.12.

The circuit of Figure 6.11 demonstrates the principles, but it is of little practical value for cyclic voltammetry. The embellishments added in Figure 6.13 improve utility by allowing the switching points of the triangular wave to be adjusted independently via two variable resistors. Now the voltage at which the sweep changes direction can be adjusted to values less than 10 V. In effect, only a portion of the OA-1 output is fed back via R1 as a reference to control the zero crossing for the + input of OA-1. The diodes, D1 and D2, permit independent control of the reference voltage for the two output states of OA-1.

The three-position switch at the input to OA-2 enables the ramp voltage to be held at any value by opening the integrator input (center position). The third position leaves the integrator input open, but drains charge from the capacitor via R4, resetting the output of OA-2 to zero. The improvements illustrated in Figure 6.13 are not quite sufficient for a practical CV instrument and several additional features must be added. A completely satisfactory design is presented in Figure 6.14.

Momentary contact switch SW1 enables the state of OA-1 to be changed at will. A control (R25) has been added that permits the triangular wave to be moved (offset) along the voltage axis without influencing its time course. Note

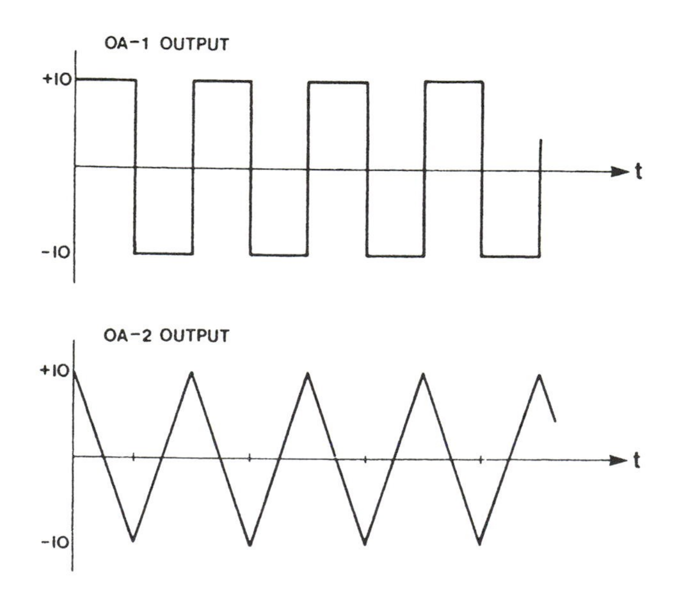


**Figure 6.12** Integration of square wave to generate a triangle wave.

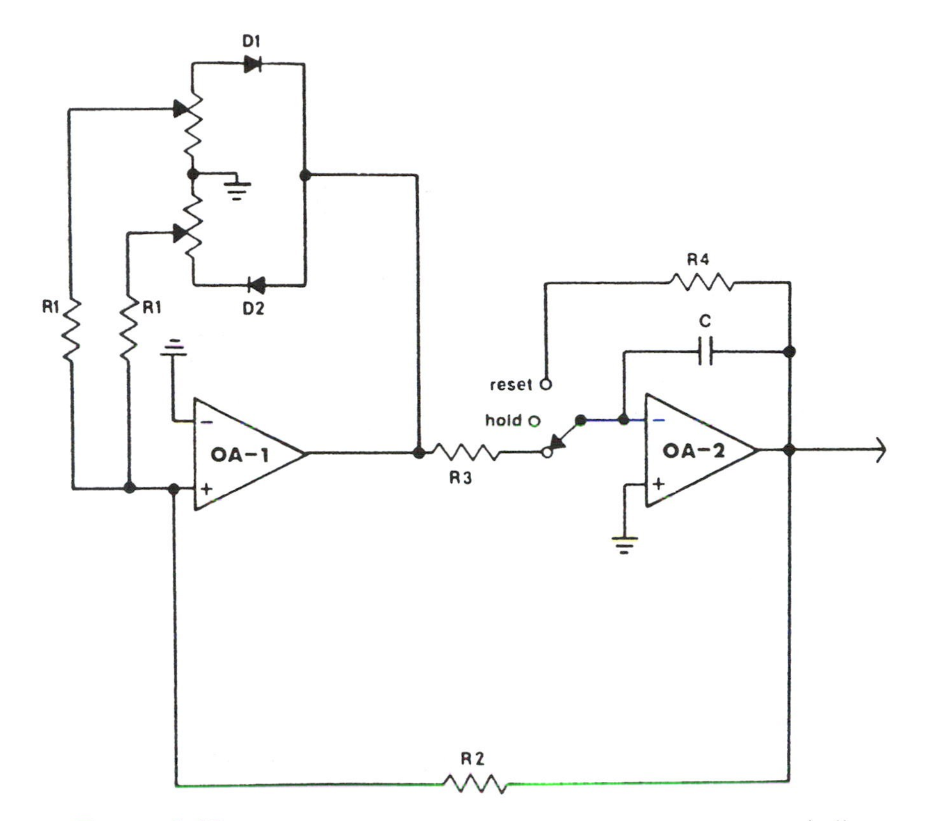


**Figure 6.13** Adding circuitry to change comparator states asymmetrically.

that the voltage set on R25 is not dependent on the square wave. An initial voltage control (R29) has been added to enable a nonzero initial condition at the output of OA-2.

We have assumed ideality for OA-1 and OA-2. As long as the offset current for OA-2 is small (i.e., FET input stage), ideality is not a bad approximation of performance for OA-2 for the usual CV experiment. The nonideality of OA-1 does present problems. For example, its inherent output swing cannot be assumed to be symmetrical. Symmetry in the square wave is crucial because we must ensure that both legs of the triangular wave have the same absolute slope. Thus we must operate on the output of OA-1 to obtain a symmetrical swing around zero. A complementary pair of transistors (Q1 and Q2) is added to the output stage, and a trimmer (R30) is added to the collector Q1. To set a symmetrical $\pm 10$ V square wave at the top of R2, we adjust R30 to achieve symmetry and R31 to achieve accuracy. The other problems with OA-1 are finite

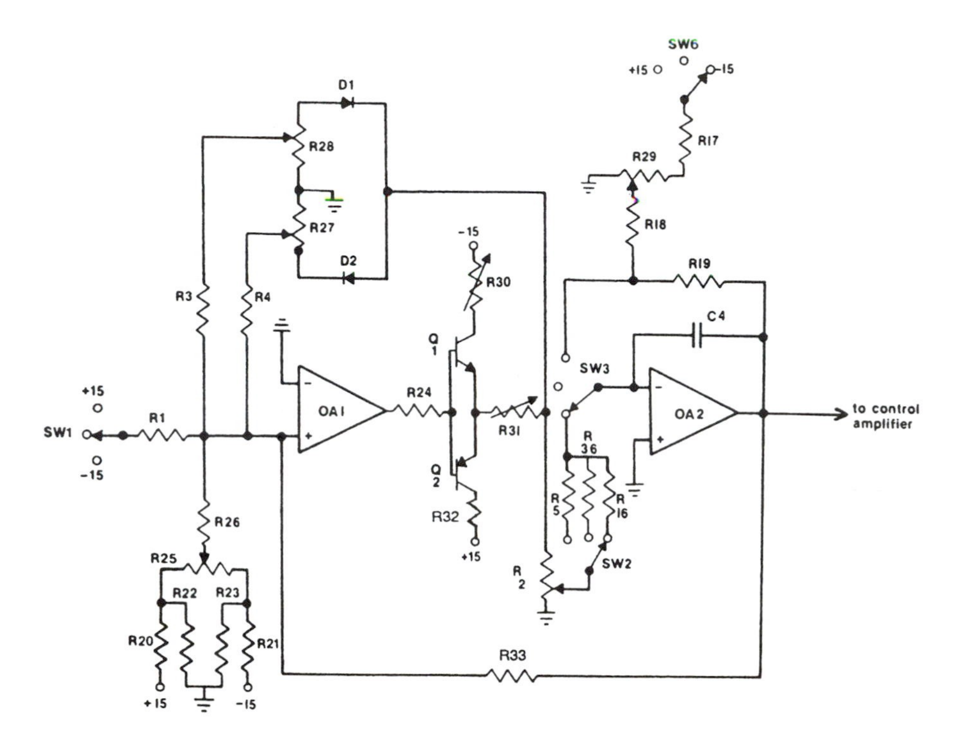


**Figure 6.14** Signal generator for cyclic voltammetry experiments.

slew rate and finite input current, which limit the rate and linearity of the triangular wave and the integrity of the apices.

The precision resistors R5, R16, and R36 provide 10-fold changes in time constant (and therefore scan rate) via SW2. Figure 6.15 shows the control and current conversion circuitry that when combined with Figure 6.14 give a complete cyclic voltammetry instrument.

There are numerous cyclic voltammetry circuits in the literature. In the 1960s and early 1970s, most laboratories built their own units because commercial versions were simply not available.

## F. Test Circuits for Controlled-Potential Instrumentation

Figure 6.16 illustrates two common arrangements whereby simple performance checks may be implemented. Both experienced and novice electrochemists often encounter situations in which it is clear that there is some difficulty with their apparatus. This is certainly one of the reasons electrochemical experiments have not been popular among chemists in general. An understanding of the appara-

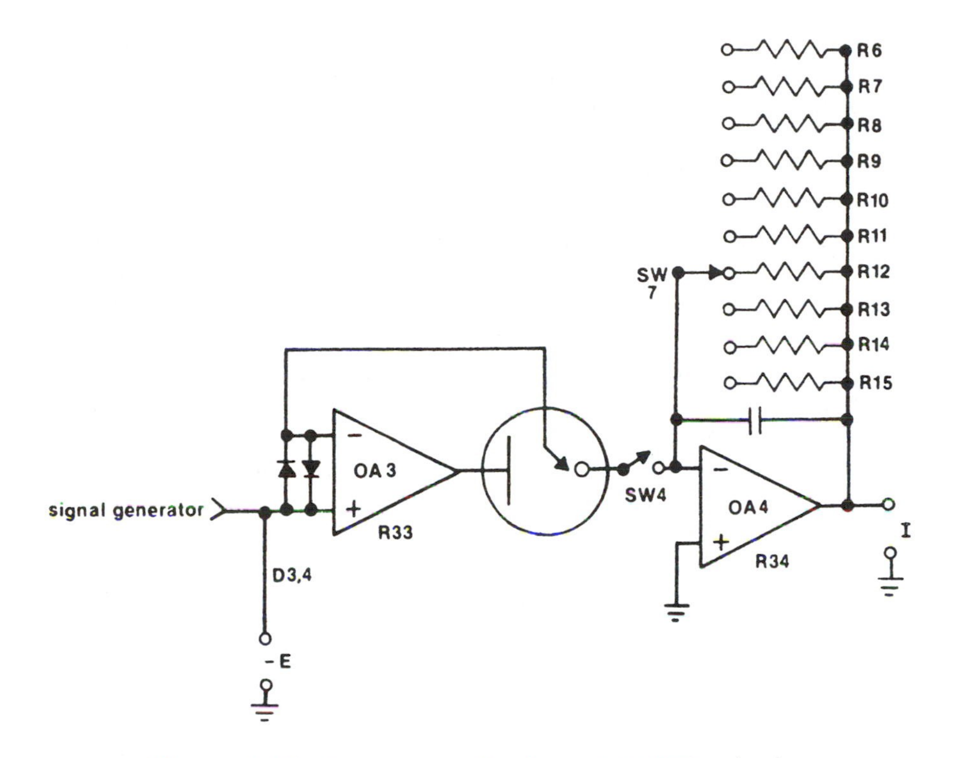


**Figure 6.15** Potential-control and current-measuring circuitry.

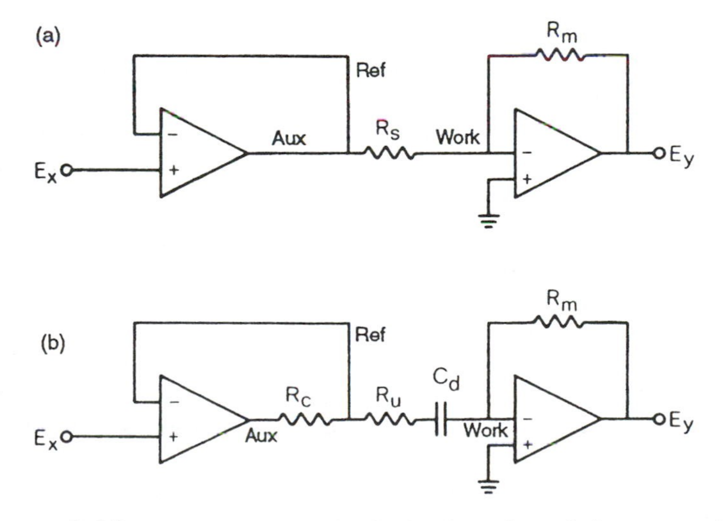


**Figure 6.16** Dummy-cell test circuits for three-electrode instrumentation.

tus, coupled with a little experience, should enable one to become operational without too much swearing and foot stomping. As is usual, the first job is to isolate the source of unhappiness and the second is to apply appropriate medication.

Almost always the problem resides in the vicinity of the cell and involves improper connection of the electrodes (one or more leads not connected or connected to the wrong electrode), shorting of the electrode leads together, an electrode not immersed in the solution, a gas bubble on an electrode or in a salt bridge, a broken lead, a broken electrode, no solution, no supporting electrolyte, no sample, clogged capillary, poor platinum-to-glass seal, and so on. Careful observation of the cell will usually lead to a sound diagnosis and cure.

There is a tendency for beginners to connect their electrodes without regard for the electronics. This can not only ruin the best experimental intentions, but can also be *extremely dangerous* in the case of high-powered potentiostats. It is important to remember that a potentiostat will not function as such unless the feedback loop is connected. This loop will always include the auxiliary electrode, the reference electrode, and the solution between them. It is always best to set up the electronics first, and then connect the electrodes! Many students attempting a cyclic voltammetry experiment, for example, will connect their cell while the triangular-wave generator is sweeping willy-nilly all over the place. This approach can change the state of a working electrode (by exposing it to extremes in potential) and the desired initial condition will be lost. The initial potential and switching potentials should be chosen, then the cell connected, and only then should the sweep be initiated. If an electrode has been exposed to a potential extreme, it may have to be replaced or repolished before a meaningful experiment is achieved.

The undesirable response of the instrument will be a helpful datum. For example, if the reference electrode is not connected, then when the working electrode is turned on (i.e., connected), the current axis will be driven to high values (perhaps the saturation current or voltage of the current follower, whichever comes first), even at applied potentials where no reaction should occur. This results from breaking the control loop and thereby driving the control amplifier (and auxiliary electrode) into voltage saturation. Under such conditions the voltage applied to the control amplifier is meaningless with regard to what is actually going on at the working electrode interface.

The next most frequent problems involve the recording device. The sensitivity may be too high or too low or the leads may not be connected properly. Neglect of ground leads is extremely common among novitiates (particularly with regard to differential-input XY recorders where four of the six input terminals will normally *have to be* grounded by the user). In the 1990s, these recorders are becoming rare as digital plotters and laser printers become output devices of choice. These are never connected to the instrument, but instead to the computer which is itself operating the instrument.

Malfunction of the instrumentation per se rarely occurs. To convince oneself that all is well with the electronics, it is useful to replace the cell with a passive analog frequently referred to as the "dummy cell." The simplest dummy cell consists of a single resistor as illustrated in Figure 6.16a. The reference and auxiliary leads are connected to one side and the working electrode lead to the other. The applied potential $E_x$ will thus be impressed across $R_s$ and the measured current, $-E_y/R_m$, will be equal to $E_x/R_s$. When a potential scan is applied, the response will take the form of an oblique straight line crossing the origin. The line will be independent of sweep rate. Typically, $R_s$ is a precision resistor of 10 kΩ. Although we have a tendency to avoid two-electrode experiments, the same configuration may be used for this purpose where the counter electrode is connected to *both* the auxiliary and reference leads. This further emphasizes the dual role of counter electrodes discussed earlier.

A more advanced dummy cell is employed in Figure 6.16b. In this case a time invariant $E_x$ will obviously result in no current response. A step change in $E_x$ will result in a current transient decaying away to zero. Note that the combination of $R_u$ and $C_d$ forms a passive integrator, through which the current will have a derivative format. If a sweep is applied, the responding current will have a steady nonzero value. If the sweep direction is reversed, a current value of the same magnitude but opposite sign will be recorded. It is important to understand both test circuits because their operation is closely analogous to that with real cells. Circuit 6.16b is discussed at length when fast transient experiments are described in detail. It is, for example, quite useful in determining the capabilities of an instrument to charge the double layer in as short a time as possible. For a step change in $E_x$ it will take a finite time for the solution side of $C_d$ to achieve $E_x$. How would you expect this time to depend on the value of $C_d$, $R_u$, $R_c$, and the ratio of $R_c$ to $R_u$?

## G. Multiple-Electrode Controlled-Potential Instrumentation

To carry out amperometric or voltammetric experiments simultaneously at different electrodes in the same solution is not difficult. In principle, any number of working electrodes could be studied; however, it is unlikely that more than two or three would ever be widely used in practice. The bulk of the solution can have only one *controlled* potential at a time (if there are significant iR drops, there will be severe control problems with multiple-electrode devices). It is necessary to use a single reference electrode to monitor the difference between this "inner" solution potential and the "inner" potential of W1 at the summing point of an operational amplifier current-to-voltage converter (this is the potential of the circuit common; see OA-2 in Fig. 6.17). The potential difference between

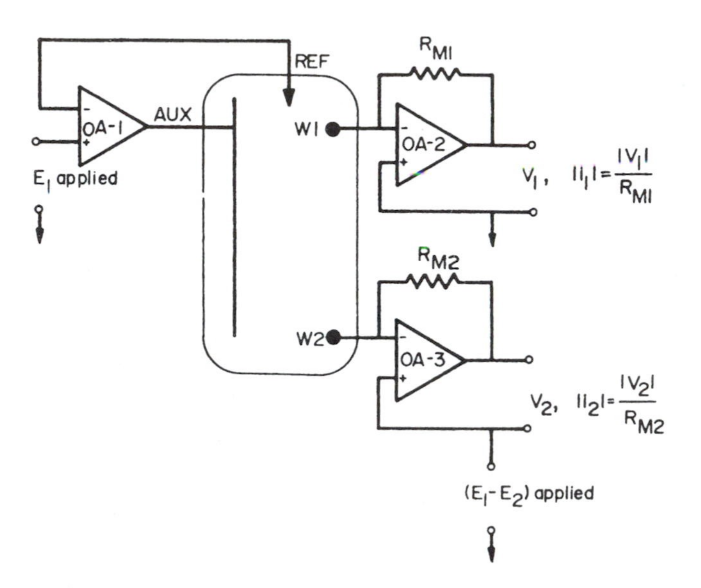


**Figure 6.17** Circuit for dual-electrode controlled-potential experiments.

the inner potential of the solution and the working electrode determines the chemistry that will occur at the W1 interface. Nothing can be done to incorporate an independent *second* working electrode if this action would change the potential of the bulk solution. Otherwise, the interfacial potential of the first working electrode would be changed. The only recourse is to vary the interfacial potential difference of the second electrode by changing its potential *external to the electrochemical cell.*

If the external potential of W2 were at the potential of the circuit common, it would necessarily have the same interfacial potential as W1, since the bulk solution potential would also be the same for both electrodes. The remedy is to float the circuit common for the second (third, etc.) current-to-voltage converter (OA-3) to a potential that is the difference between the desired electrode potentials for W1 and W2 (W1 and W3, etc.).

The circuit illustrated in Figure 6.17 is a simple schematic designed to introduce the principles of the system. Obviously, there is a great deal more involved in producing a practical device. One has to consider such factors as time constants, background current offset circuits, and a convenient means of controlling the two (or more) potentials without having to make inconvenient calculations. There are several approaches to development of a practical circuit. While these details are beyond the scope of the present chapter, suffice it to say that such devices are now available commercially and have been used for many years by electrochemists, usually in connection with ring-disk electrodes.

As was hinted at earlier, if there is a significant potential drop across the bulk of the solution, problems arise. This is especially significant with respect to thin-layer electrochemical transducers for liquid chromatography detection (Chapter 27) because the resistance of the bulk solution can be quite high. In our favor, however, are the very low currents (typically $10^{-10}$–$10^{-9}$ A) usually encountered in such cells. Nevertheless, these iR-drop problems can be very severe with multiple-electrode detectors, causing the events at one electrode to influence the events at the other electrodes. For example, if a reaction at one electrode results in a large current at that electrode, this current might cause an iR drop across the solution, which would change the interfacial potential of all other electrodes.

Remember that the interfacial potential is a *difference* between the inner potential of the electrode material and the potential established in the bulk of the solution. An iR drop can cause the latter to change. One way to minimize this problem is to construct the cell with an auxiliary electrode(s) that is directly across from the working electrodes, resulting in a very short current path and, therefore, a very small iR drop. Multiple-electrode transducers constructed in this way minimize "cross-talk" between different electrodes.

## III. CLASSICAL CONTROLLED-CURRENT INSTRUMENTATION

The reader familiar with controlled-potential methodology will have no trouble understanding a controlled-current apparatus. Figure 6.18 illustrates classical approaches to two- and three-electrode constant-current chronopotentiometric experiments (see Chap. 4). The simplicity of these circuits was for many years an attractive feature of chronopotentiometry. Improvements in potentiostats have been largely responsible for a decline in the popularity of "chronopot" in recent years. Nevertheless, constant-current experiments are even more important with respect to coulometric titrations and stripping potentiometry (Chap. 24).

The necessity for a constant-current source is satisfied (Fig. 6.18) simply by impressing a known voltage across a resistance large in comparison with the equivalent resistance of the cell. If the applied voltage is very much larger than that developed across the cell, the potential-time response of the cell will not significantly influence the applied current. Variation of V and/or R affords convenient adjustment of the excitation current. The interfacial potential of the counter electrode must be independent of the applied current for meaningful cell potentials to be measured using the two-electrode approach.

Figure 6.18b depicts the classical three-electrode constant-current experiment. The counter electrode is now replaced by an auxiliary electrode and a reference electrode. The reference electrode enables one to monitor the working electrode potential as a function of time using a high-impedance (zero-current) measuring device. In many early experiments, carried out on a time scale

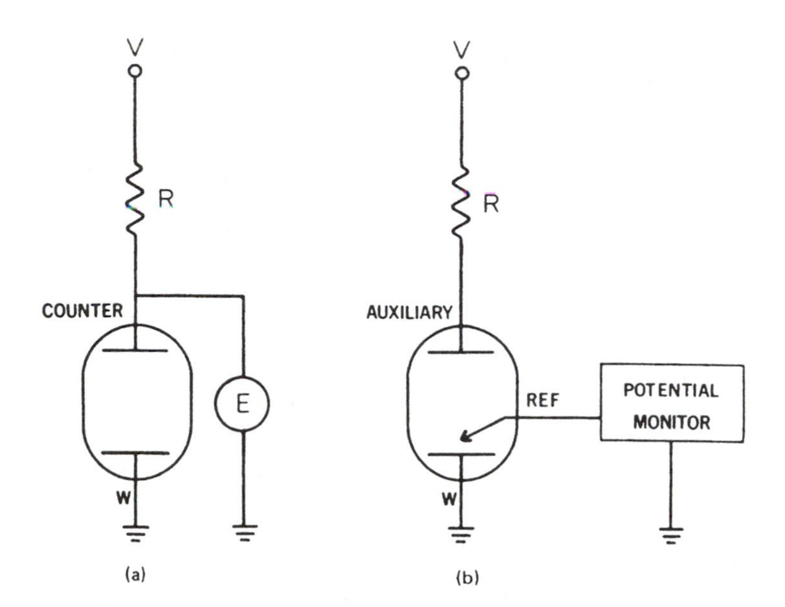


**Figure 6.18** (a) Classical two-electrode controlled-current apparatus. (b) Primitive three-electrode controlled-current apparatus.

of tens of seconds, a conventional pH meter served this purpose nicely. It is important to note that unlike the three-electrode potentiostat, the three-electrode amperostat shown here does not employ negative feedback principles to maintain control over the excitation signal.

## IV. CONTROLLED-CURRENT INSTRUMENTATION BASED ON OPERATIONAL AMPLIFIERS

The evolution of operational amplifier-based amperostats is illustrated in Figure 6.19. The OA inverter is a satisfactory starting point. As mentioned earlier, $E_i$ and $R_i$ function together as a current source for $R_f$. Replacing $R_f$ with the electrochemical cell is all that is necessary. Notice (Figure 6.19b) that the working electrode is placed at the summing point and the auxiliary electrode at the output of the OA. Consider the electron flow in Figure 6.19b. What will the sign of $E_i$ have to be when a cathodic current is applied to W? The addition of a reference electrode and a voltage follower permits one to monitor the working electrode potential as illustrated in Figure 6.19c. The presence of the reference probe has no influence whatsoever on the processes occurring at the working electrode. This is contradistinct to potentiostatic experiments, where removing the reference probe destroys the control loop. There are, in fact, many

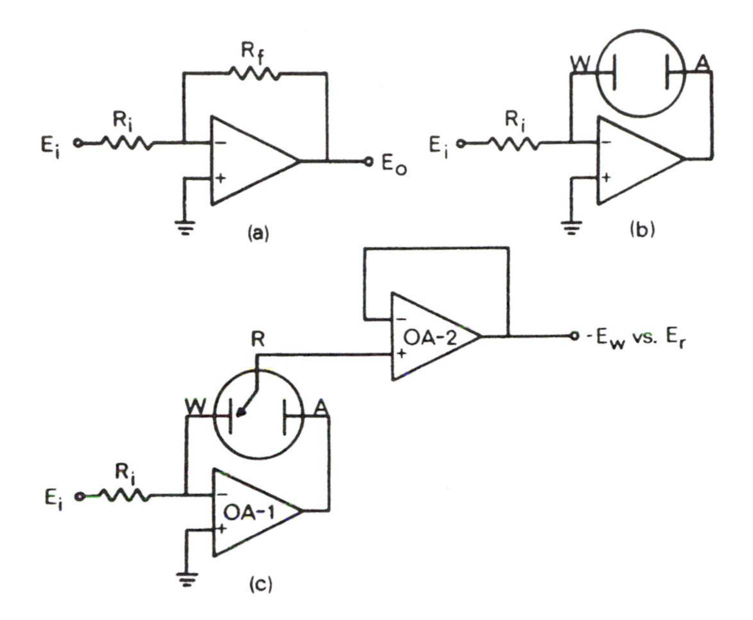


**Figure 6.19** Evolution of an operational amplifier three-electrode amperostat.

constant-current experiments in which a reference electrode is not needed (e.g., coulometric titrations, constant-current chronoabsorptometry) because the dependent variable of interest is not the potential.

Figure 6.20 illustrates a circuit that has been widely used. SW1 affords choice of anodic or cathodic current, SW2 initiates the experiment, and then SW1 may be used for current reversal. OA-3 is available for differentiating E with respect to t. A more advanced circuit (Fig. 6.21) incorporates an additional feedback loop and a comparator to perform cyclic chronopotentiometry with automatic switching. Operation of this circuit is perfectly analogous to the cyclic voltammetry circuit discussed in Section II.E.

## V. MICROPROCESSOR-BASED ELECTROCHEMICAL INSTRUMENTATION

The arrival of large-scale integrated circuits in the last 20 years has revolutionized chemical instrumentation just as it has kitchens, automobiles, and television sets. With respect to electrochemistry, the microprocessor has been incorporated in signal generation and data processing, while the basic instrumentation (e.g., potentiostat and current-to-voltage converter) remains as described in earlier sections of this chapter. Microprocessor instruments provide *flexibility*

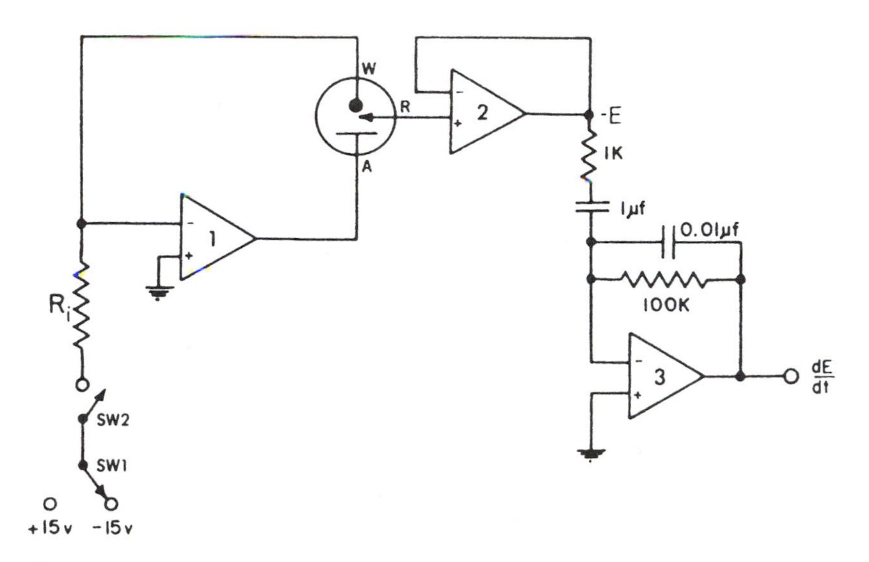


**Figure 6.20** Current-reversal chronopotentiometry with derivative output.

in that different experiments can be implemented via software changes. Such instruments can also be *interactive* in the sense that decisions can be made from one experiment that define the conditions for subsequent experiments (with or without human intervention). Finally, *automated sample processing* and *data storage* can be incorporated more easily than for conventional instruments.

The major difficulty with processor-based instruments is the enormous investment in software development required to make all of these advantages a reality. This can take a great deal of time (person-years in some cases). While such units are flexible, it is necessary to limit flexibility to some degree to provide convenience of operation. This implies that some decisions have to be made early in the development phase that might be difficult to countermand later. In my laboratory, documentation of software by graduate students was such a problem that it was sometimes easier to start from scratch rather than to modify existing programs! Fortunately, there are so many good software packages for statistics, graphics, and instrument control that it is no longer necessary to start coding from scratch and we can concentrate on chemistry.

Commercial processor-based electrochemical instruments are available in two forms. In the first configuration, a general-purpose laboratory or "personal" computer is interfaced to the analog instrumentation. In the other approach, the package is integrated in such a way that the processor is dedicated to the electrochemical experiments. Several dedicated, processor-based pulse polarographs

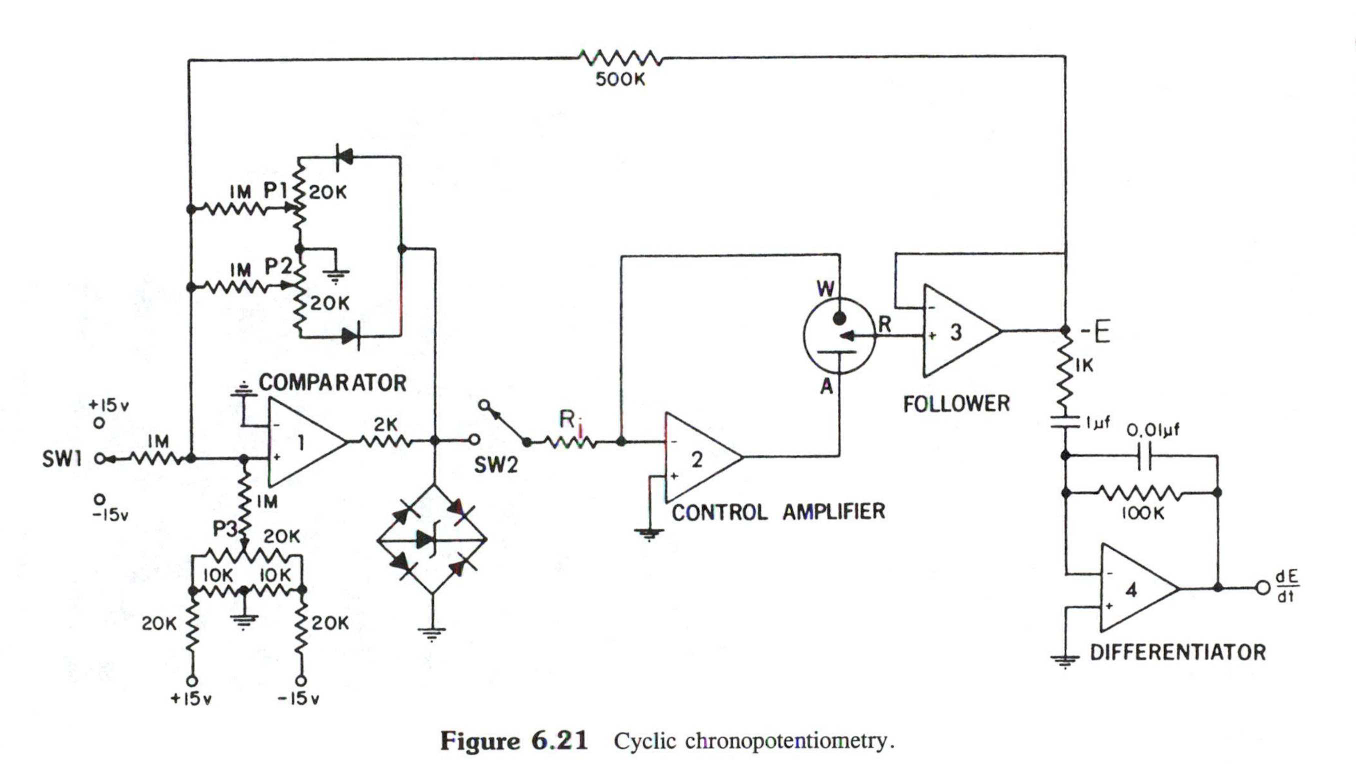


**Figure 6.21** Cyclic chronopotentiometry.

have been available since the mid-1970s. The first units were developed by Princeton Applied Research at a time when microprocessors were just beginning to influence all chemical instrumentation.

In early 1983, Bioanalytical Systems introduced a new class of integrated processor-driven instrumentation based on a concept first developed by Faulkner and his co-workers [1] at the University of Illinois. This unit (Figs. 6.22 and 6.23) has evolved over the years and now includes a repertoire of some 35 electrochemical techniques, including the most popular large-amplitude (Chap. 3) and small-amplitude (Chap. 5) controlled-potential methods. The unit also is capable of determining electrocapillary curves and can automatically measure and compensate for solution resistance ($R_u$ in Fig. 6.5). Thus in a *single instrument* it is possible to utilize virtually all of the diagnostic criteria introduced in Chapters 3 and 5 and also to explore quickly which technique is optimum for

**Figure 6.22** Microprocessor-based electrochemical analyzer coupled to a 486 personal computer. [Courtesy of Bioanalytical Systems, Inc., West Lafayette, IN.]

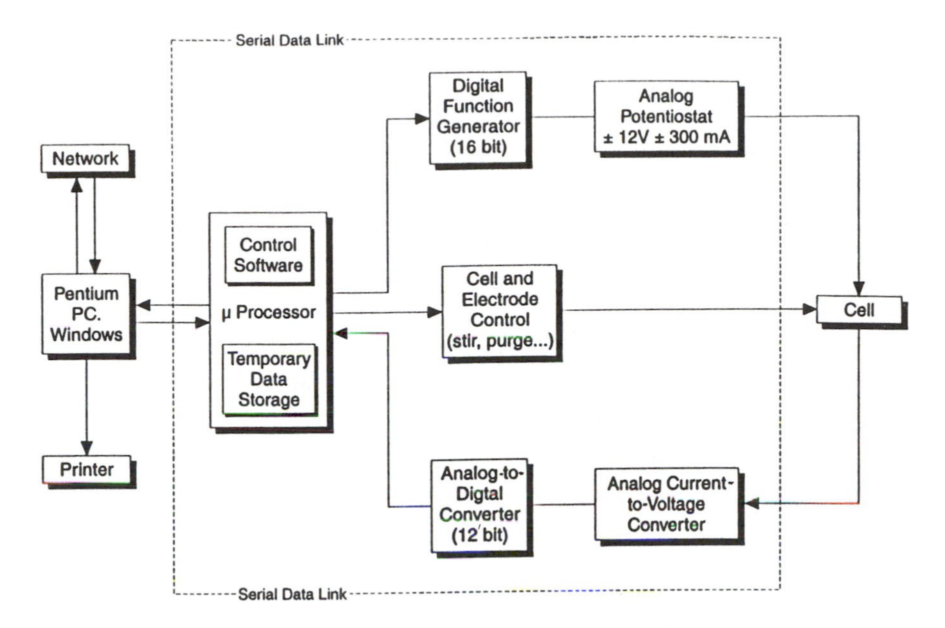


**Figure 6.23** Block diagram for BAS 100B/W electrochemical analyzer illustrating various accessories.

analytical purposes. Data can be digitally smoothed, signal-averaged, and background-subtracted, and such things as peak currents and peak potential are provided directly in numerical form. The excitation signal parameters and the electrochemical response can be displayed on a video terminal, printer, and/or digital plotter. Digital simulation software (DigiSim®) is an option that enables the experimenter to model electrochemical mechanisms with both thermodynamic and kinetic details. The simulated results can be compared with experimental to refine the model (see Chap. 20). This type of instrumentation provides unique opportunities that save a great deal of time in electrochemical investigations. A wide variety of experiments can be accomplished on the same solution without the need to change the electronics.

At this writing it is possible to link digital instrumentation with "notebook" computers and even achieve portability. Developments have been so rapid in this area that it is difficult to imagine what will come next. General-purpose processor-based electroanalytical instruments are now available from Amel, Bioanalytical Systems, Cypress Systems, EcoChemie, Metrohm, Tacussel, and Princeton Applied Research.

## VI. CONCLUSION

The goal of this chapter was to present an overview of electrochemical instrumentation, not to carry the reader to the point of instrument construction. Most universities have one or more courses in electronics at an adequate level to build rudimentary apparatus. The nature of electrochemistry is such that some knowledge of electronics is extremely useful to good science.

## REFERENCE

1. P. He, J. P. Avery, and L. R. Faulkner, *Anal. Chem.* *54*:1313A (1982).

# 7

# Overcoming Solution Resistance with Stability and Grace in Potentiostatic Circuits

**David K. Roe** *Portland State University, Portland, Oregon*

## I. INTRODUCTION

Three advantages accrue from the use of operational amplifiers (OAs) in controlled-potential instrumentation, as described in Chapter 6. These are: (1) many independent input signals can be accurately summed and applied to the cell; (2) potential control is relative to a reference electrode which is subject to negligible current; and (3) accuracy of potential control is essentially independent of the reactions at the auxiliary electrode and also of the solution resistance between the auxiliary and reference electrodes. The improvements over the traditional scheme (low-impedance voltage divider and two-electrode cell) are especially significant with measurement techniques involving moderate to high frequencies or step changes in potential.

However, potential control is not yet perfect since some resistance remains between the reference and working electrodes; the value depends on solution conductivity, the distance between the two electrodes, and electrode geometry. Overcoming this last hurdle with poise is not an easy task, so it is worthwhile to first examine the magnitude of this resistance and its consequences.

In a cell with spherical symmetry, the resistance between the large auxiliary electrode (radius $r_a$) and a small working electrode (radius $r_w$) is given by

$$R_s = \frac{1}{4\pi K}\left(\frac{1}{r_w} - \frac{1}{r_a}\right) \tag{7.1}$$

The specific conductivity of the solution, K $(\Omega\text{-cm})^{-1}$, can be found in reference handbooks or estimated from limiting ionic conductances.

Between these two spheres, the reference electrode is subject to a part of the total resistance. If the distance between the surface of the working electrode and the reference tip is r, this resistance is

$$R_u = \frac{1}{4\pi K}\left(\frac{1}{r_w} - \frac{1}{r_w + r}\right) \tag{7.2}$$

It is $R_u$ that is not compensated by the feedback action of a potentiostat.

Many other electrode geometries in common use approximate parallel, circular planes, and these resistances can be adequately estimated from the following equations (which are approximate in that only straight-line current paths are considered):

$$R_s = \frac{l}{K\pi r_w (r_w + r_a)} \tag{7.3}$$

$$R_u = \frac{r}{K\pi r_w (r_w + r_a)} \tag{7.4}$$

The distance between the planes is l, and the reference electrode tip is located at a distance r from the working electrode of radius $r_w$. Often, the auxiliary electrode is isolated from the cell by a porous plug; in this case, the radius of the plug should be used for $r_a$, instead of the radius of the auxiliary electrode. Additionally, the resistance between the plug and the auxiliary electrode should be included by another calculation based on the geometry of the auxiliary electrode and its compartment. Representative values of cell resistance are listed in Table 7.1 for several common electrolytes.

In addition to solution resistance, electrodes may have appreciable resistance in themselves. The common example is the mercury in a dropping mercury electrode (DME) capillary. Such resistances are summed with solution resistances since they are in series and the treatment that follows is unchanged.

The control amplifier OA1 in Figure 7.1A sums the input voltage and applies an output voltage, $V_o$, such that $V_r$ is equal in magnitude but opposite in sign to the scaled sum of input voltages; that is, $V_r = -\Sigma V_i' R_f/R_i$. Considering only one input signal and letting the symbol $V_1$ include the scale factor $R_f/R_1$, the reference electrode potential $E_r$ is equal to $-V_1$ since OA2 has a gain of exactly unity. The potential across the solution-working electrode interface is given by

$$E_e = -V_1 - iR_u \tag{7.5a}$$

or

$$E_e = E_r - iR_u \tag{7.5b}$$

**Table 7.1** Representative Cell Resistances (Ω)

| Cell geometry | 0.1 M KCl | | 1 M KCl | | 0.1 M HCl | | 1 M HCl | |
|---|---|---|---|---|---|---|---|---|
| | $R_s$ | $R_u$ | $R_s$ | $R_u$ | $R_s$ | $R_u$ | $R_s$ | $R_u$ |
| Spherical | | | | | | | | |
| $r_a$ = 10 cm | | | | | | | | |
| $r_w$ = 0.1 cm | | | | | | | | |
| r = 1 cm | 61 | 56 | 7.9 | 7.2 | 20 | 19 | 3.9 | 3.6 |
| r = 0.1 cm | 61 | 31 | 7.9 | 4.0 | 20 | 10 | 3.9 | 2.0 |
| $r_a$ = 1 cm | | | | | | | | |
| $r_w$ = 0.1 cm | | | | | | | | |
| r = 0.2 cm | 55 | 41 | 7.2 | 5.3 | 18 | 14 | 3.6 | 2.7 |
| r = 0.05 cm | 55 | 20 | 7.2 | 2.7 | 18 | 6.8 | 3.6 | 1.3 |
| Planar | | | | | | | | |
| $r_a$ = 10 cm | | | | | | | | |
| $r_w$ = 0. cm | | | | | | | | |
| l = 10 cm | | | | | | | | |
| r = 0.2 cm | 242 | 4.8 | 32 | 0.64 | 81 | 1.6 | 16 | 0.32 |
| r = 0.05 cm | 242 | 1.2 | 32 | 0.16 | 81 | 0.40 | 16 | 0.08 |
| $r_a$ = 1 cm | | | | | | | | |
| $r_w$ = 0.1 cm | | | | | | | | |
| l = 10 cm | | | | | | | | |
| r = 0.2 cm | 2200 | 44 | 2900 | 58 | 740 | 15 | 145 | 2.9 |
| r = 0.05 cm | 2200 | 11 | 2900 | 14 | 740 | 3.7 | 145 | 0.72 |

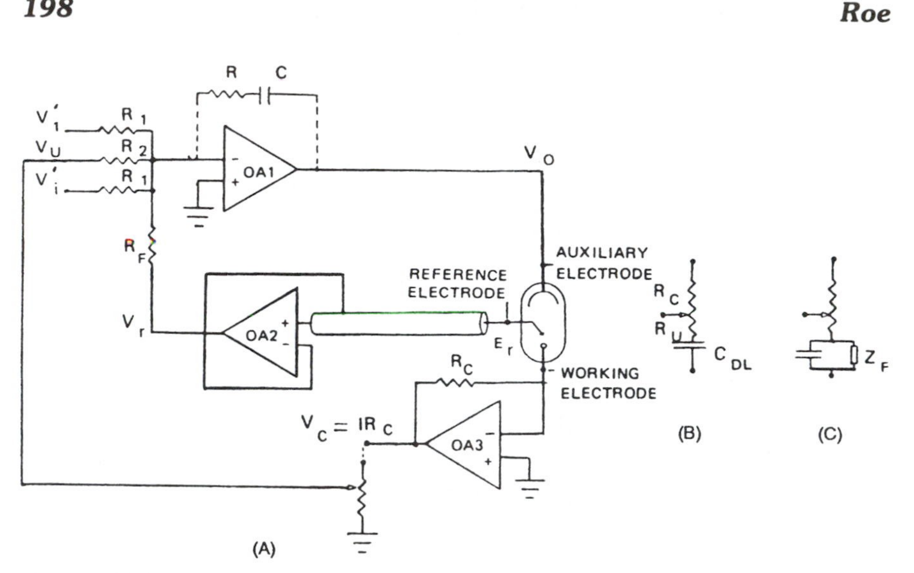


**Figure 7.1** (A) Typical controlled-potential circuit and cell: OA1, the control amplifier; OA2, the voltage follower ($V_r = E_r$); OA3, the current-to-voltage converter. (B) Equivalent circuit of cell: $R_c$, solution resistance between auxiliary and working electrodes; $R_u$, solution resistance between reference and working electrodes, $R_s = R_c + R_u$; and $C_{dl}$, capacitance of interface between solution and working electrode. (C) Equivalent circuit with the addition of faradaic impedance $Z_f$ due to charge transfer. Potentials are relative to circuit common, and working electrode is effectively held at circuit common ($E_w = 0$) by OA3.

The potential of the working electrode relative to the reference electrode, as usually measured, is opposite in sign to $E_r$. Here the current i at the working electrode is given a positive sign if anodic to agree with Ohm's law. The potential given by $iR_u$ is the error in potentiostatic control of the working electrode. Whatever the tolerable error is (e.g., $\leq 1$ mV), one can simply apply Ohm's law and arrive at acceptable trade-offs between i and $R_u$. The range of possible values of i and $R_u$ is so great that any given situation must be examined and decisions reached from experimental values. However, a few generalities can be proffered with qualifications. Conventional polarography and (slow) linear sweep voltammetry involving currents less than about 10 $\mu$A will usually be free of significant potential errors by proper positioning of the reference electrode. Coulometry and electrosynthesis experiments routinely involve much larger currents and the resulting $iR_u$ terms are also larger, but the acceptable error level may be much larger (10–100 mV). Note that for a given current density, the term $iR_u$ is independent of electrode area for the planar arrangement, but for a spherical electrode the magnitude of $iR_u$ depends on $r_w$ and r. If $r >> r_w$, $iR_u$ is proportional to $r_w$, but if $r_w >> r$, this term is proportional to r.

Fast linear scans, step changes of potential, and ac methods may be subject to significant errors because of large faradaic currents and large capacitive currents. Even if the currents are large for a moment, perturbations may be introduced that persist long enough to cause artifacts in the measurements. Some care toward minimization of $R_u$ is usually necessary with this group of techniques.

There are three methods for decreasing the magnitude of the error term $iR_u$: (1) use highly conducting solutions; (2) position the reference tip close to the working electrode; and (3) devise an electronic means to compensate for $iR_u$ by increasing the output of the control amplifier (positive feedback). Although the first method should not be overlooked, it is often impractical with organic solvents. While the second and third methods are feasible means for increasing the accuracy of potential control, they cannot be applied without some attention to stability of the feedback system. In the sections that follow, stability requirements for these two methods are examined from a practical viewpoint and illustrated with well-defined examples. From this information, some general guidelines are developed that allow approximate, but easy and useful, assessments of the most direct path to optimum performance. Detailed derivations of equations are relegated to the appendix.

## II. INPUT-OUTPUT RELATIONS OF CELLS AND POTENTIOSTATS

The equivalent circuit of an electrochemical cell is shown in Figure 7.1B; it is adequate to illustrate the response of a potentiostat system to small ($<100$ mV) changes in potential. A potentiometer equal in value to $R_s$ ( $= R_c + R_u$) allows $R_u$ to be changed without changing the auxiliary-to-working electrode resistance. This corresponds to changing the position of the reference electrode while the two other electrodes in the cell are fixed; thus only $R_u$ is changed and all other parameters of the system remain constant. To avoid nonlinear effects, the demands on the potentiostat must not exceed its capabilities in terms of output voltage, current, and slew rate. Small values of potential changes and some attention to choices of values of $R_s$, $R_u$, and $C_{dl}$ will ensure this condition.

A few comments are in order on the probable validity of conclusions based on this equivalent circuit to real cells. Quite simply stated, real cells that are properly designed will have the same properties as dummy cells of the same values of $R_s$, $R_u$, and $C_{dl}$. Important design features of a cell are (1) equal resistance between all points on the surface of the working electrode and the auxiliary electrode; (2) low-impedance reference electrode; and (3) low stray capacitance between electrodes, between leads, and to shields. Spherical symmetry is a good, but somewhat inconvenient, method of meeting the first requirement; a parallel arrangement also works with planar electrodes. At the very least, two auxiliary electrodes on opposite sides of a spherical or cylindrical working electrode should be used. A low-impedance reference electrode is one that does not introduce an additional time constant into the loop. This means that electrodes designed for use with pH meters should be avoided and the shield of the lead should be connected to the output of the voltage follower, as shown in Figure 7.1. Common sense will help to minimize the third requirement. In particular, avoid long leads between the cell and the potentiostat. Chapter 9 contains many examples and details of cell designs. Further information on the high-frequency characteristics of cells can be found in the bibliography at the end of this chapter.

Perhaps the most obvious difference between the equivalent circuit to be used and a real cell is the lack of a faradaic impedance in Figure 7.1B. The rationale for omitting $Z_f$ is that the most difficult stability problems occur when only a capacitive load is present. Placing $Z_f$ in parallel with $C_{dl}$, as in Figure 7.1C, actually increases the stability of the circuit when the reaction is either rate- or diffusion-controlled. However, if $Z_f$ includes a capacitive component due to adsorption-desorption of a reactant or product, or if $C_{dl}$ changes markedly with potential, a sudden change in circuit stability may occur. As will be seen, it will then be important to assess the range of values of $C_{dl}$, or to allow extra margin for stability. So for the majority of situations, this equivalent circuit represents the worst case for design purposes.

From the viewpoint of the control OA, OA1 in Figure 7.1A, the cell is simply a voltage divider, producing $E_r$ at the reference electrode terminal when $V_o$ is applied to the auxiliary electrode (note that $E_r$ and $V_o$ are small signal *changes*, not the dc signal). In terms of the equivalent cell of Figure 7.1B, the relation (or transfer function) between the two voltages as vectors, when $V_o$ is a sine wave of frequency f (Hz), is

$$\frac{\overline{E_r}}{\overline{V_o}} = \frac{R_u + X_c}{R_s + X_c} \tag{7.6}$$

Making use of the capacitive reactance equation, $X_c = 1/j2\pi fC_{dl}$, and solving for the ratio of the magnitudes of the peak values gives

$$\frac{E_r}{V_o} = \sqrt{\frac{1 + (2\pi fR_uC_{dl})^2}{1 + (2\pi fR_sC_{dl})^2}} \tag{7.7}$$

There is a phase lag between $E_r$ and $V_o$ because of the capacitance; the angle is

$$\phi = \tan^{-1}\left[-\frac{2\pi fC_{dl}(R_s - R_u)}{1 + R_uR_s(2\pi fC_{dl})^2}\right] \tag{7.8}$$

These equations express two facts: $E_r$ is less than $V_o$ and lags behind $V_o$ in time. Depending on the frequency, the effects may be very small or significant. It is quite easy to assess the frequency range over which the phase shift is significant: it occurs between $f_s = 1/2\pi R_s C_{dl}$ and $f_u = 1/2\pi R_u C_{dl}$, and decreases just below and above these points. In Figure 7.2A are plotted the ratio of voltages and the phase lag on linear scales and the frequency, as both f (Hz) and w (rad/s), on a log scale. At these two characteristic frequencies, the phase lag is 45° ($\phi = -45°$). As $R_u$ is decreased, the frequency range between $f_s$ and $f_u$ increases and the phase lag dips toward its limiting value of 90°. For example, if $R_u/R_s = 0.0001$, $\phi = -88.9°$ at the midpoint, regardless of all other parameters. Another characteristic of the graph is that $E_r/V_o = 0.707$ at $f_s$ and has a limiting value of $R_u/R_s$ at $f >> f_u$.

In the case of potential steps, it is more direct to think in terms of characteristic time constants than in terms of impedances. Two time constants can be defined and used to describe the cell transfer function: $\tau_s = R_s C_{dl}$. The units

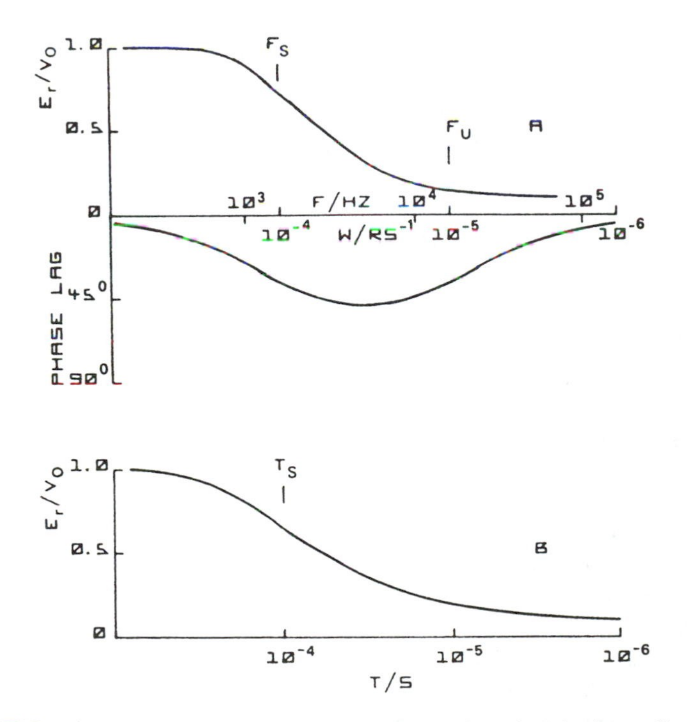


**Figure 7.2** (A) Voltage-transfer function and phase lag of equivalent cell with $R_s C_{dl} = 1 \times 10^{-4}$ s and $R_u C_{dl} = 1 \times 10^{-5}$ s. $V_o$ is amplitude of sine wave of frequency f. (B) Transient response of same equivalent cell. $V_o$ is magnitude of potential step applied at $t = 0$.

will be in seconds if ohms and farads are used. Note that Equations 7.7 and 7.8 could be cast in terms of these time constants. In these terms, the time dependence of $E_r$ of the equivalent cell after a step change $V_o$ is

$$\frac{E_r}{V_o} = 1 - \left(1 - \frac{\tau_u}{\tau_s}\right) \exp\left(\frac{-t}{\tau_s}\right) \tag{7.9}$$

From the plot of this equation in Figure 7.2B, it is seen that immediately after a step change in $V_o$, the ratio $E_r/V_o$ has the value $\tau_u/\tau_s$, which is 0.1 in the example. Following this, the increase with time is exponential. Note that the abscissa is logarithmic, with time increasing to the left, and that the frequency and time scales of Figures 7.2A and B are related by $f = 1/2\pi t$. This allows a direct comparison to be made between the response in the frequency domain (Fig. 7.2A) and the time domain (Fig. 7.2B). Now, the significant point is that over a certain frequency-time region there is a phase shift-time delay. The role of $R_u$, holding $R_s$ constant, in changing the shape of the curves in Figure 7.2 should be evident. The phase or time delay in $E_r$ relative to $V_o$ is an inherent property of electrolytic cells. The consequence of all this is that $E_r$ can fool the control amplifier, which then overcompensates, leading to overshoot or even sustained oscillations. We should remember that $R_u$ is the adjustable parameter of interest and we are attempting to see how its value will influence circuit stability and accuracy.

Amplifier characteristics also enter into the list of factors that influence the accuracy of the circuit and its stability. The two parameters of particular interest to this analysis are the dc (or low-frequency limit) open-loop gain and the frequency at which the open-loop gain has decreased to $1/\sqrt{2}$ of the dc value. These parameters are specified by the manufacturer of all OAs and most potentiostats. They can also be easily measured. A common method of presenting performance characteristics of OAs is by means of a graph of log (gain) versus log (frequency), often referred to as a Bode plot. An example is given in Figure 7.3, curve A, for which the open-loop dc gain, $A_o$, has a value of $1 \times 10^5$ V/V and the frequency at which the gain drops to $7.07 \times 10^4$, $f_a$, is 10 Hz. Alternative terms for this frequency are corner frequency, roll-off frequency, dominant pole frequency, and –3-dB point. Extrapolation of the two straight lines of Figure 7.3, curve A, results in their intersection at this frequency, as shown in dashed lines. If only $A_o$ and the upper bandwidth frequency ($f_o$ in Fig. 7.2A) are available, $f_a$ can be calculated from $f_a = F_o/A_o$; this is a consequence of the slope of gain versus frequency being –1; at least it should be in a traditional feedback amplifier.

Phase shift always accompanies the decreasing gain of an OA; the plot in Figure 7.3 includes this information. In a manner similar to the equivalent cell

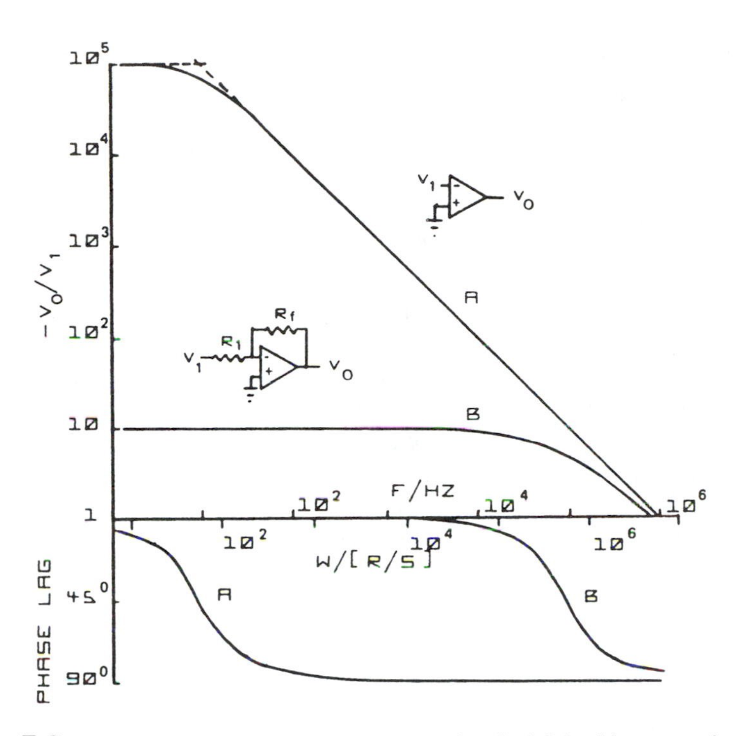


**Figure 7.3** (A) Bode plot of gain and phase lag of typical OA without negative feedback. $V_1$ is amplitude of input sine wave of frequency f. (B) Bode plot of gain and phase lag of typical OA with feedback; $R_f/R_1 = 10$.

circuit, the phase shift is –45° at $f_a$, relative to much lower frequencies, and increases beyond this point to -90° for the remaining, usable frequency region. This phase shift is in addition to the inverting action of the amplifier, so the total phase shift is ($\phi$ – 180°) between the inverting input and the output, but only $\phi$ between the noninverting input (which is not used in this circuit) and output. It is common to omit the –180° in discussing stability, since it is constant and understood to be present.

Such amplifiers are commonly used with a feedback connection; curve B of Figure 7.3 illustrates the relation between input and output for a ratio of feedback resistor $R_f$ to input resistor $R_1$ of Equation 7.10. Under this condition, a gain of –10 is obtained to about f = 10 kHz, beyond which the gain drops and the phase lag commences. The same results would be obtained with the control circuit of Figure 7.1 if $R_s/R_u = 10$ and if $C_{dl}$ is shorted. Curve B can be calculated from the equation

$$\frac{V_o}{V_1} = \frac{R_f/R_1}{1 + (R_f + R_1)/R_1 A} \tag{7.10}$$

in which A is $A_o/(1 + jf/f_a)$, the gain in complex notation. The term $(R_f + R_1)/R_1A$ must be much smaller than unity to ensure accuracy; this implies a high open-loop gain. The reciprocal, $AR_1/(R_f + R_1)$, is the loop gain under feedback and represents that fraction of $V_o$ that is mixed with the input signal at the inverting input.

A time constant $\tau_a$ can also be defined to represent the delay between output and input of an OA without feedback. It can be calculated from $f_a$ or $f_o$ using the relations $\tau_a = 1/2\pi f_a$ or $A_o/2\pi f_o$. A plot of the open-loop response to a step input of an OA having $\tau_a = 1.6 \times 10^{-2}$ s and $A_o = 1 \times 10^5$ V/V is given in Figure 7.4, curve A. This is the same OA as was used for Figure 7.3, as can be verified with the preceding relations. This plot shows the time lag between output and input and the time region over which it occurs. Again, a log scale was used for the time axis, and time increases to the left to emphasize the similarity between frequency and time domains. The equation of the curve is

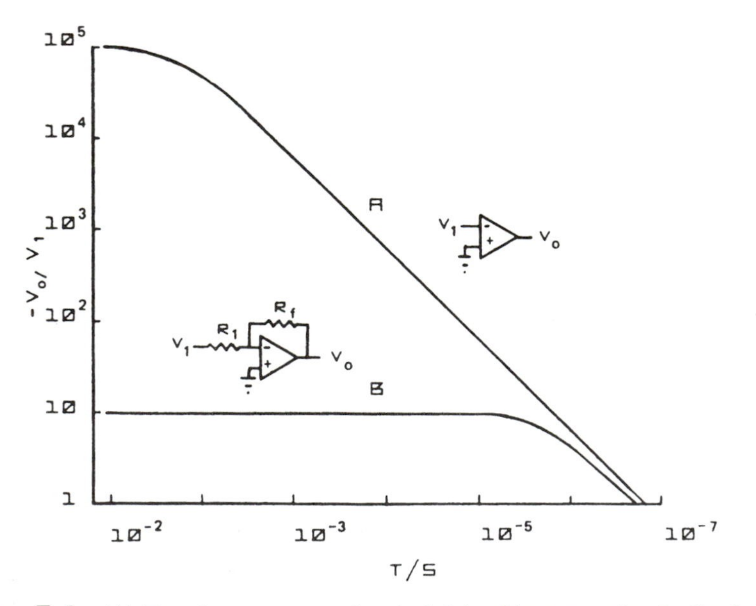


**Figure 7.4** (A) Transient response of typical OA without negative feedback. $V_1$ is magnitude of potential step applied at t = 0. (B) Transient response of typical OA with negative feedback; $R_f/R_1 = 10$.

$$\frac{V_o}{V_{in}} = -A_o\left[1 - \exp\left(\frac{-t}{\tau_a}\right)\right] \tag{7.11}$$

and applies only if $V_o$ does not exceed the ratings of the amplifier, which is quite likely without feedback, since $A_o$ is rather large. Normally, the step response of this type of amplifier is measured with feedback ($R_f$) and input ($R_1$) resistors so that the ratio $V_o/V_{in}$ has a reasonable value. Curve B of Figure 7.4 shows the response for $R_f/R_1 = 10$. Under these conditions, the voltage-transfer ratio to a step input is

$$\frac{V_o}{V_{in}} = -\frac{R_f}{R_1 - (R_1 + R_f)/A_o}\left(1 - \exp\left\{-\frac{t[R_f + R_1(A_o - 1)]}{\tau_a(R_1 + R_f)}\right\}\right) \tag{7.12}$$

or more simply, but accurate to $\leq 0.1\%$ when $R_f/R_1 \leq 100$,

$$\frac{V_o}{V_{in}} = -\frac{R_f}{R_1}\left\{1 - \exp\left[\frac{tA_oR_1}{\tau_a(R_f + R_1)}\right]\right\} \tag{7.13}$$

Now the effective time constant is decreased by a factor of $A_oR_1/(R_f + R_1)$, which is the loop gain in the long-time limit.

## III. STABILITY OF POTENTIOSTAT-CELL CIRCUITS AND THE ROLE OF $R_u$

So far, the individual transfer functions of cells and potentiostats have been described. It is time to put them together and seek answers to the question: How does the value of $R_u$ affect the stability of the circuit? An intuitive answer will be developed before delving into a quantitative relation.

Connecting an electrochemical cell or its equivalent to an OA introduces a phase or time lag between the output and the input (Fig. 7.2A and B). Depending on the amount of phase or time lag relative to the characteristics of the OA, we can anticipate that the output $V_o$ may become excessively large while the corrective feedback signal is delayed. Sinusoid signals in this frequency range will have a larger-than-expected amplitude and increased phase shift, leading to artifacts in ac voltammograms. Step and pulse response will have overshoot and perhaps damped oscillations. In both cases, sustained oscillations may result if the total phase or time lag is extreme.

With sinusoid input signals, the output signal, and therefore $E_r$, will be larger than anticipated where the frequency is in the range where the phase shift from the cell adds to the phase shift of the amplifier. The magnitude of the error depends on the loop gain, and it is at its worst near the upper frequency limit of the amplifier.

Figure 7.5, curve A, represents the combination of the equivalent cell of Figure 7.2A and the amplifier of Figure 7.3. The phase lag caused by $C_{dl}$ produced the rise in the ratio $E_r/V_1$ and also resulted in a rapid change in phase just beyond 10 kHz. The magnitude of both effects is related to $R_u$, as shown by curves B and C. In assessing these changes, it is useful to refer to Figure 7.2A; decreasing $R_u$ allows the phase lag to increase and also makes the loop gain decrease beyond $f_u$.

Before introducing further examples, the significance of phase shifts upon measurement error in ac voltammetry should be explored. Since the basic pur-

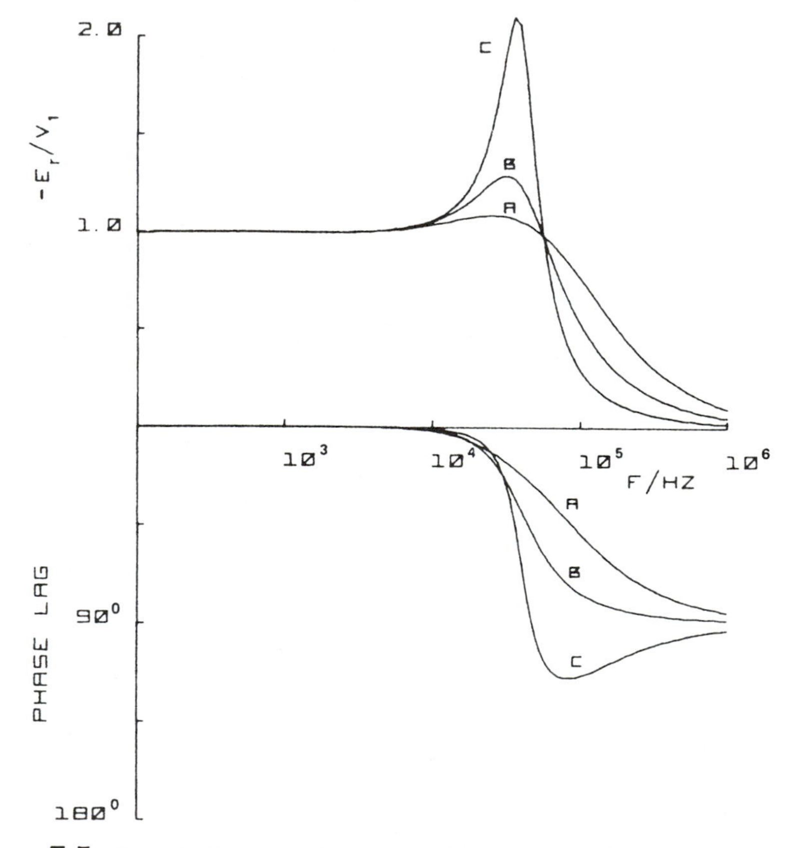


**Figure 7.5** Transfer function and phase lag of potentiostat and equivalent cell at reference electrode. $A_o = 1 \times 10^5$, $f_a = 10$ Hz, $R_sC_{dl} = 1 \times 10^{-4}$ s with $R_uC_{dl} = 1 \times 10^{-5}$ s (A), $5 \times 10^{-6}$ s (B), and $2 \times 10^{-6}$ s (C).

pose is to measure an ac current due to a faradaic process, the current due to $C_{dl}$ must be rejected as completely as possible. This is accomplished by making use of the fact that the capacitive current $i_c$ and the faradaic current $i_f$ are not in phase but differ by 45 to 90°, depending on the reversibility of the charge-transfer reaction. If the phase of the capacitive current relative to $V_1$ (or $E_r$) is precisely known, the rejection of $i_c$ is easily visualized if $i_f$ is measured the instant that $i_c$ passes through zero. Most instruments rely on a technique known variously as synchronous, lock-in, or phase-selective detection. If $E_e$ is the potential across the solution-electrode interface, it can be stated with certainty that $i_c$ leads $E_e$ by 90°. Unfortunately, $E_e$ is not experimentally available, but it is possible to illustrate the phase difference between $E_e$ and $V_1$. Then, using $V_1$ as the reference, the error can be predicted as a function of frequency and circuit parameters. Figure 7.6 gives the ratio $E_e/V_1$ and the phase lag for the same

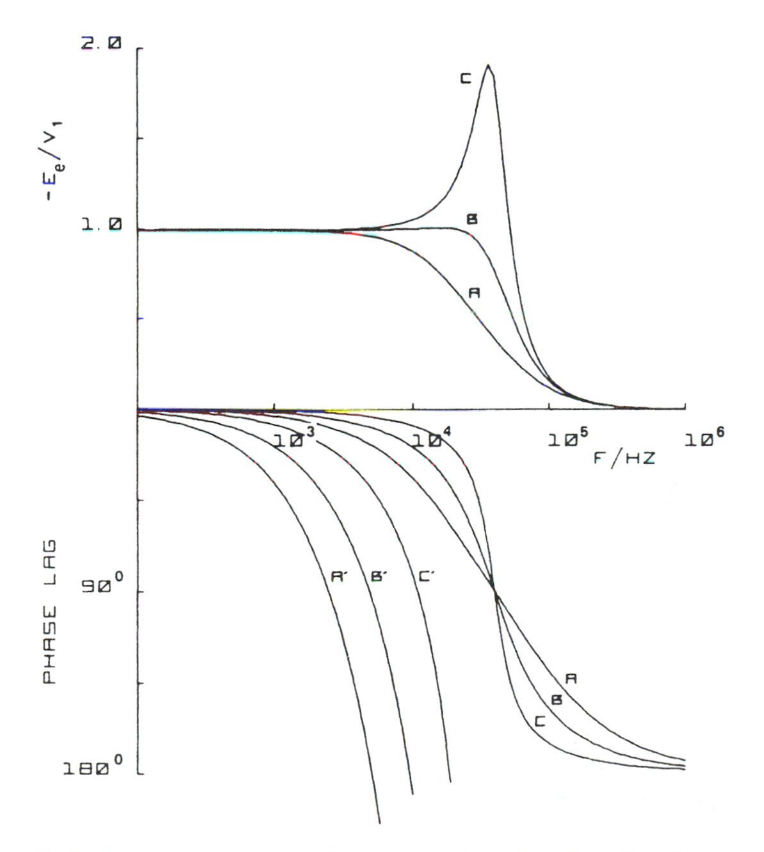


**Figure 7.6** Transfer function and phase lag at solution interface of working electrode; same parameters as Figure 7.5. A′, B′, and C′ are phase lag × 10.

system as in the preceding figure. The main differences are the increase in phase lag and the decrease in the peak of voltage ratios. Both effects are due to the low-pass-filter action of $R_uC_{dl}$.

As long as $i_c$ does not change in phase relative to $V_1$, the measurement will be essentially free of this error after setting the synchronous detector's phase to 90°. Any change in the relative phase of $i_c$ introduces an error proportional to the cosine of the phase angle. For example, on curve A′ of Figure 7.6 at f = $10^3$ Hz, the phase of $E_e$ has changed by about 4° and the same change must occur in the phase of $i_c$. Therefore, a fraction of the peak value of $i_c$ is measured and this fraction is cos 86° or 0.07. Depending on the circumstances, this may be an acceptable error in the baseline of an ac voltammogram. Figure 7.7

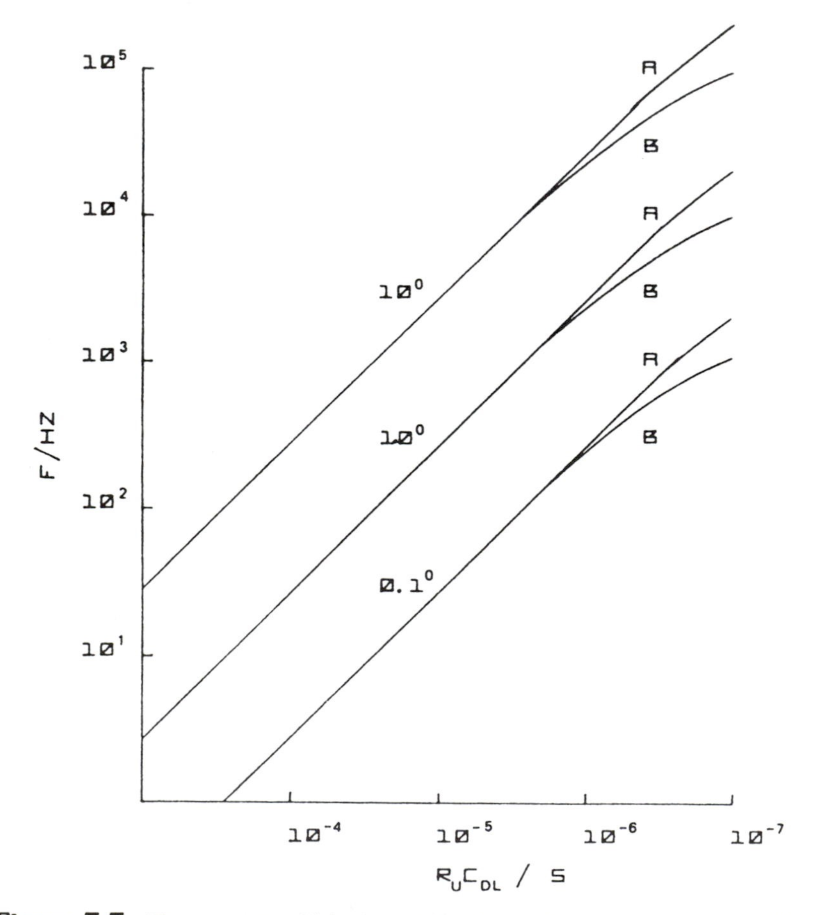


**Figure 7.7** Frequency at which phase shift of capacitive current relative to $V_1$ equals 0.1°, 1.0°, and 10° as a function of $R_uC_{dl}$. Curve A applies if $f_a$ = 100 Hz, and curve B is $f_a$ = 10 Hz. $R_s/R_u \leq 50$ for all curves.

summarizes the frequency at which the phase error equals 0.1°, 1.0°, and 10° as a function of $\tau_u = R_uC_{dl}$. Over the range where the plots are straight lines, the phase error is due solely to $\tau_u$, but as this time constant is decreased, the control amplifier and $\tau_s$ cause the plots to curve.

Several further examples of response to ac signals are given in Figures 7.8 to 7.11 to show the changes caused by increasing amplifier bandwidth and decreasing $\tau_s$. It is evident that $R_u$ plays a pivotal role in all of these examples, in determining both total phase shift (stability) and maximum operating frequency for a given level of measurement error. While the acceptable error can be selected to meet experimental requirements, the total phase shift between $E_r$ and $V_o$ should be less than −135° due to the factors cited. Real systems introduce

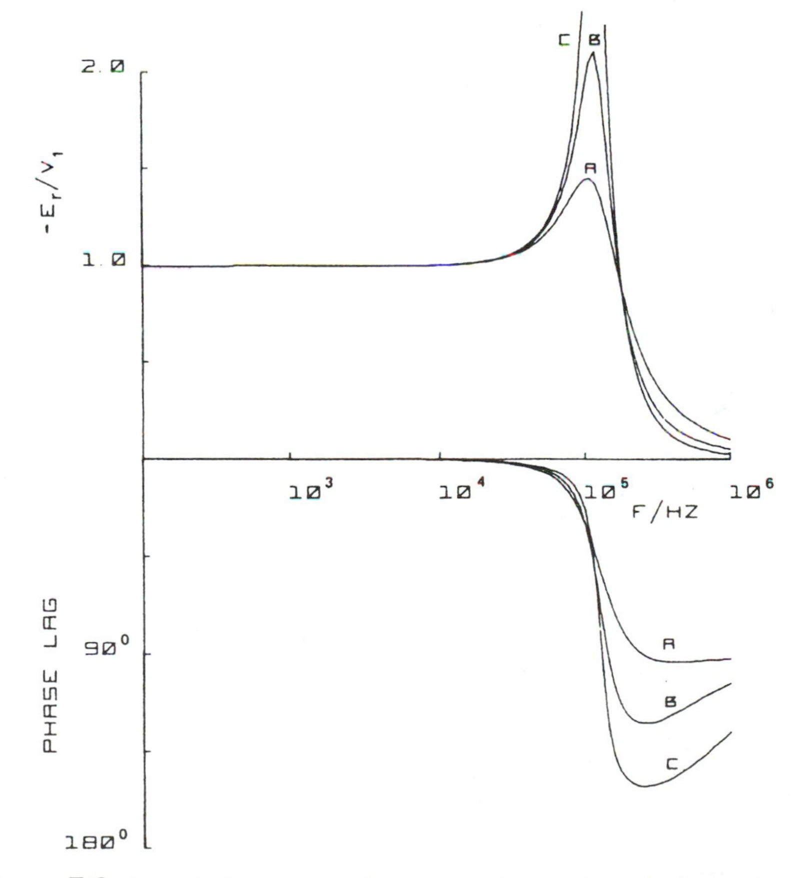


**Figure 7.8** Transfer function and phase lag at reference electrode of potentiostat and equivalent cell; $A_o = 1 \times 10^5$, $f_a = 10$ Hz, $R_sC_{dl} = 1 \times 10^{-5}$ s with $R_uC_{dl} = 1 \times 10^{-7}$ s (B), and $2 \times 10^{-6}$ s (C).

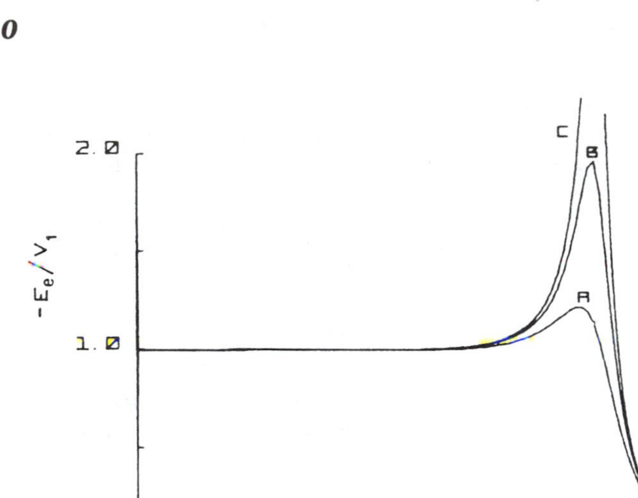


**Figure 7.9** Transfer function and phase lag at solution interface of working electrode; same parameters as Figure 7.8. A′, B′, and C′ are phase lag × 10.

further phase errors (usually in the wrong direction) and some leeway must be retained. When the phase lag due to all causes reaches −180°, the potentiostat becomes an oscillator at the peak frequency.

Time-domain response of feedback amplifiers has been regularly correlated with frequency-domain behavior, and vice versa. Examples have usually been restricted to those situations in which only the amplifier contributes phase shift (single pole) or where a second source of phase was included (two-pole), such as from nonideal amplifier design or from the effects of stray capacitance at the input terminal. The system of interest in electrochemistry is more complicated than a two-pole system because there is also a decreasing phase shift caused by

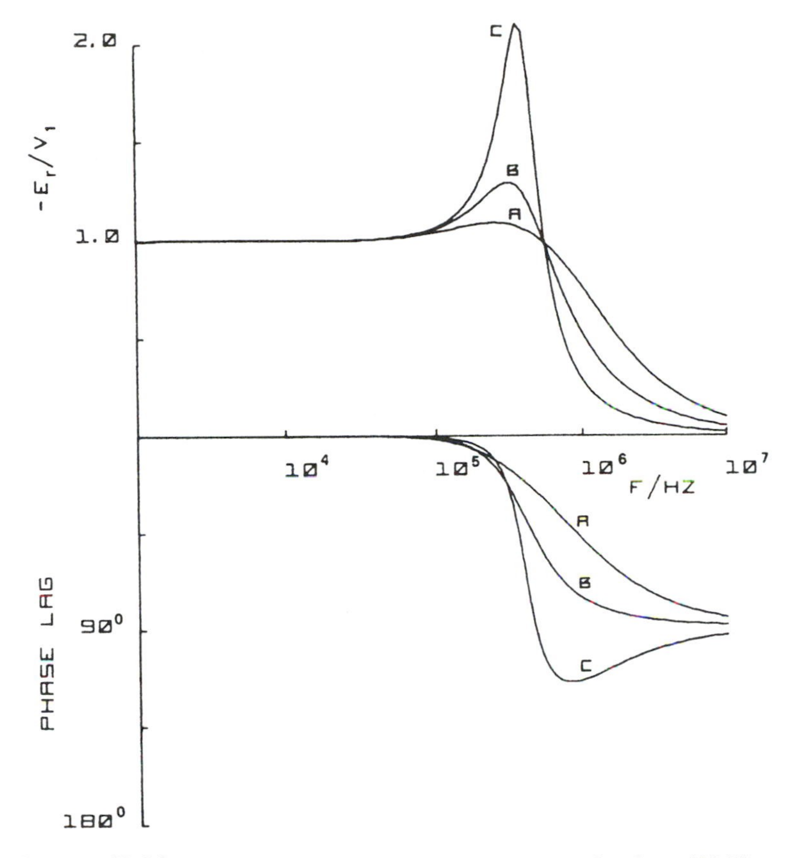


**Figure 7.10** Same parameters as Figure 7.8 except that $f_a$ = 100 Hz.

$R_uC_{dl}$. This is referred to as a "zero." Consequently, predicting rise times and overshoot in the time domain from transfer functions in the frequency domain may be unreliable. So exact solutions have been obtained, and from them the detailed influence of $R_u$ on system stability and accuracy will be examined. From these results, a useful empirical relation was found to summarize the behavior for the control amplifier and equivalent cell of Figure 7.1.

Upon application of a potential step $V_1$ to the control amplifier, the potential $E_r$ rises with time at a rate determined by the combined effects of the amplifier gain $A_o$ and the three time constants $\tau_a$, $\tau_s$, and $\tau_u$. The potential across the solution-working electrode interface $E_e$ is delayed relative to $E_r$ by the time constant $\tau_u$. Calculated curves for a range of parameters are given in Figures 7.12 to 7.14.

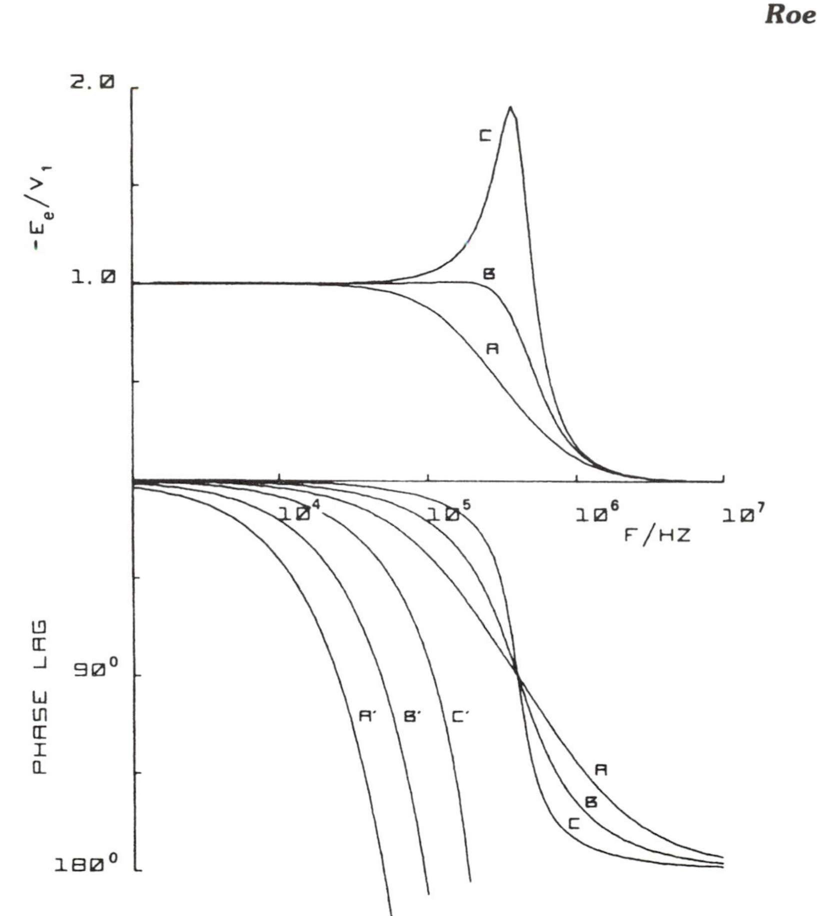


**Figure 7.11** Same parameters as Figure 7.9 except that $f_a$ = 100 Hz.

Three general trends are evident from these graphs.

1. As $\tau_u$ is decreased ($\tau_s$ and $\tau_a$ held constant), the rise time of $E_r$ increases, the rise time of $E_e$ decreases, and overshoot increases.
2. As $\tau_s$ is decreased ($\tau_u/\tau_s$ and $\tau_a$ held constant), the rise times of $E_r$ and $E_e$ decrease and overshoot decreases.
3. As $\tau_a$ is decreased ($\tau_u$ and $\tau_s$ held constant), the rise times of $E_r$ and $E_e$ decrease and overshoot decreases.

A reflection on the curves of Figures 7.2B and 7.4 should reveal the same conclusions. The relative benefits of decreasing $R_u$, and therefore $\tau_u$, in order to decrease the $iR_u$ drop must be carefully compared with the disadvantage of marginal stability. In order to have a reasonable means of predicting instability, as measured by overshoot of $E_r/V_1$, a large number of curves were calcu-

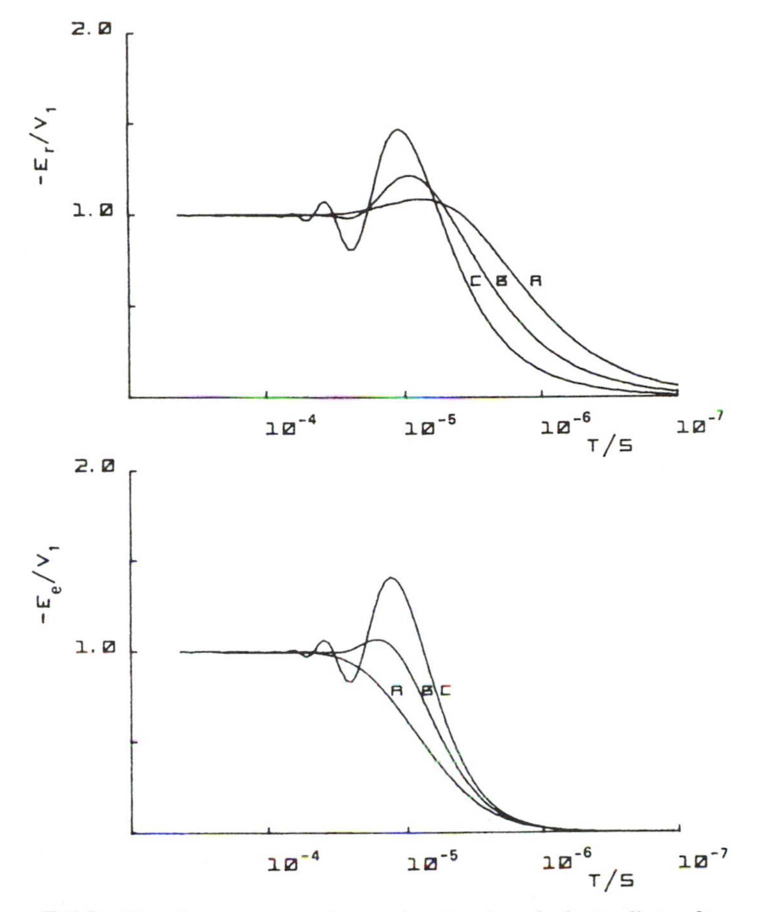


**Figure 7.12** Transient response of potentiostat and equivalent cell at reference electrode (top) and at solution interface of working electrode (bottom); $A_o = 1 \times 10^5$, $f_a = 10$ Hz, $R_sC_{dl} = 1 \times 10^{-4}$ s with $R_uC_{dl} = 1 \times 10^{-5}$ s (A), $5 \times 10^{-6}$ s (B), and $2 \times 10^{-6}$ s (C).

lated and the percent overshoot was plotted versus $(\tau_s\tau_a/A_o)^{1/2}\tau_u$, as shown in Figure 7.15. Three curves are plotted to account for the relative contributions of the time delays of the amplifier and the cell. The curves are valid for any reasonable combination of cell and amplifier parameters of interest in electrochemistry. A real system will no doubt exhibit greater overshoot than predicted by those curves, due to stray capacitance and lead inductance. However, results very close to these are obtainable in practice, and the graph at least provides a good starting point for system optimization.

A small amount of overshoot in $E_r$ is quite acceptable, as a comparison of curves in Figures 7.12 to 7.14 reveals. If $E_r$ has $<10\%$ overshoot, $E_e$ will have a negligible amount for a very wide range of cell parameters. Also, the pres-

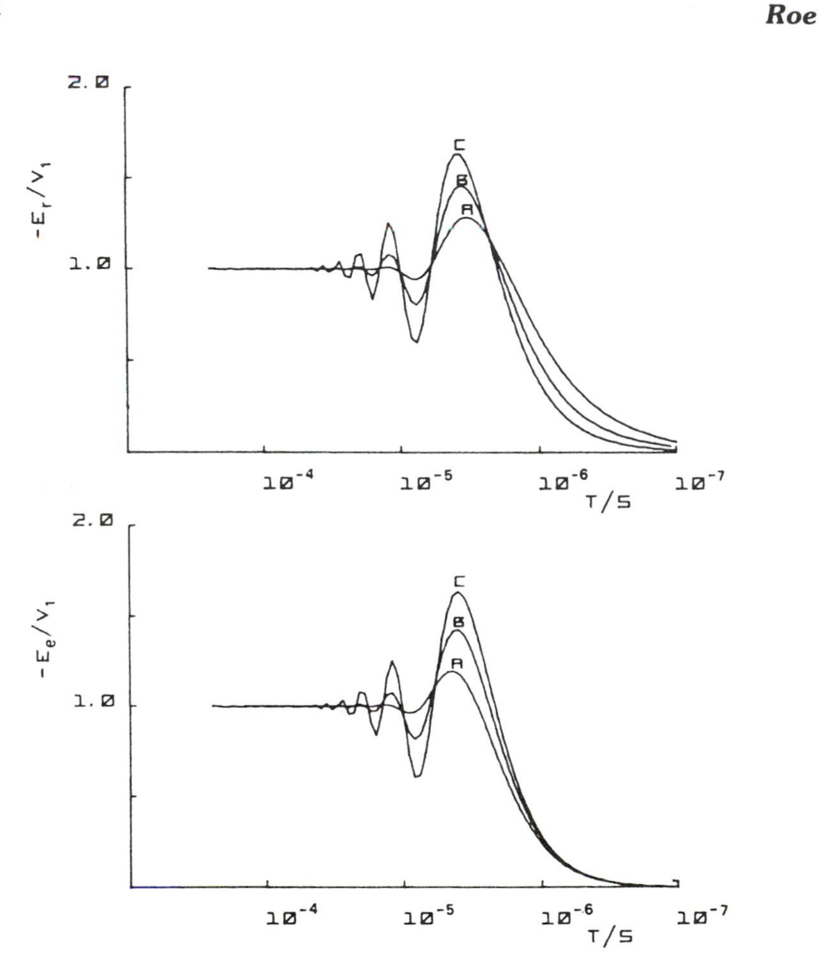


**Figure 7.13** Same as Figure 7.12 except that $R_sC_{dl} = 1 \times 10^{-5}$ s with $R_uC_{dl} = 1 \times 10^{-6}$ s (A), $5 \times 10^{-7}$ s (B), and $2 \times 10^{-7}$ s (C).

ence of a faradaic process that does not alter $C_{dl}$ will tend to decrease overshoot in both $E_r$ and $E_e$ due to a decrease in phase and time lag, but the faradaic current increases the error due to $iR_u$.

In all the numerical examples illustrated thus far, the potential changes across the interface were primarily governed by $\tau_u$ rather than by amplifier characteristics. Under these conditions, it is a simple matter to calculate the current due to double-layer charging if $R_u$ and $C_{dl}$ are known or can be estimated. The peak value of this current is equal to the change in potential divided by $R_u$ and it decays exponentially thereafter. Following a given number of time constants, the contribution of the charging current to the total current measured can be determined and thereby the extent of error in the quantity of interest, the faradaic current, can be estimated.

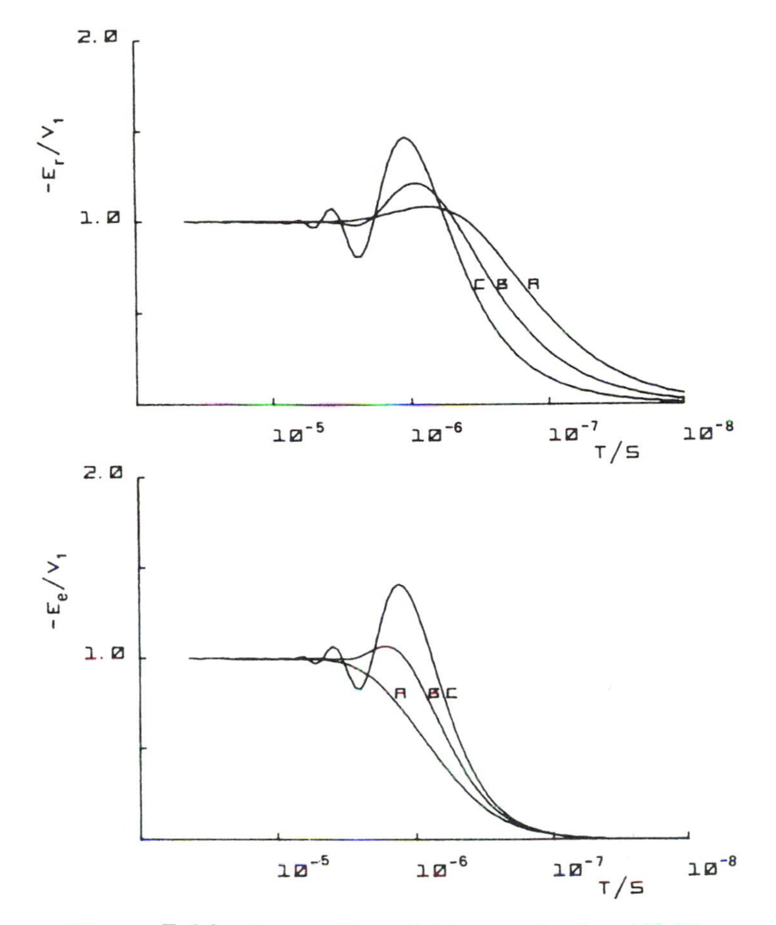


**Figure 7.14** Same as Figure 7.13 except that $f_a$ = 100 Hz.

The preceding suggests that a wide range of values of $\tau_s$ and $\tau_u$ might be worth examining. For example, if $\tau_s$ and $\tau_u$ are both less than $\tau_o$, the phase and time lag due to the cell may become negligible in the control loop and instability is decreased. Figures 7.16 and 7.17 show that this is indeed true; the overshoot becomes negligible as $\tau_s/\tau_o$ is made less than unity. There are two methods for meeting this condition: (1) decrease the cell resistance and/or electrode capacitance (area), and (2) increase $\tau_o$ by decreasing amplifier bandwidth. The first method is somewhat limited in practice, but calls attention to the advantages of low cell resistance and capacitance. The second approach should always work and has been extensively used. It is usually implemented by connecting a capacitor between the inverting input and output of the control amplifier; some potentiostats have this or an equivalent provision built in. There is a second

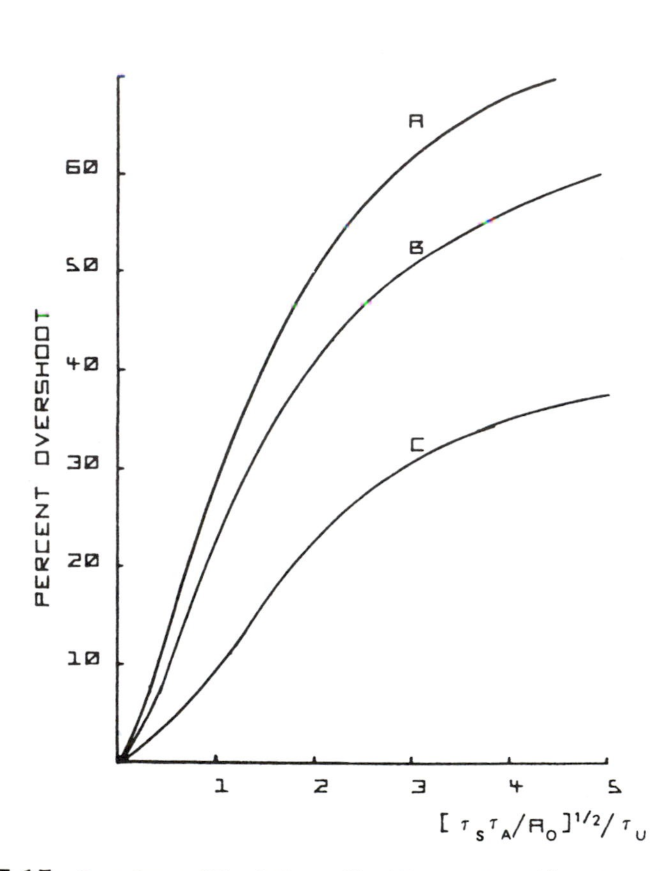


**Figure 7.15** Overshoot of $E_r$ relative to $V_1$; (A) $\tau_s = 1 \times 10^3\ (\tau_a/A_o)$; (B) $\tau_s = 100\ (\tau_a/A_o)$; (C) $\tau_s = 10\ (\tau_a/A_o)$.

benefit from this decrease in bandwidth—the system noise level decreases. Although this follows general rules of noise-bandwidth relations, it is particularly noticeable when changes in bandwidth eliminate a peak in $E_r/V_1$ because noise in the frequency region of the peak receives excess amplification. Further details on bandwidth tailoring are given in a later section.

Finally, it should be noted that the condition $\tau_s \cong \tau_u$ also avoids instability regardless of the values of the parameters; at worst there will be a small amount of overshoot if $\tau_o$ is close to either time constant and the latter are separated by a small amount. When the loop gain is high, system response is smooth, as is illustrated for $\tau_s = 10^{-3}$ s, $\tau_u = 5 \times 10^{-4}$ s, and $\tau_o = 1.6 \times 10^{-7}$ s in Figures 7.18 and 7.19. The trade-off is a loss of potential control; $E_e$ will depend on cell current, and the acceptability of the error must be determined in each case.

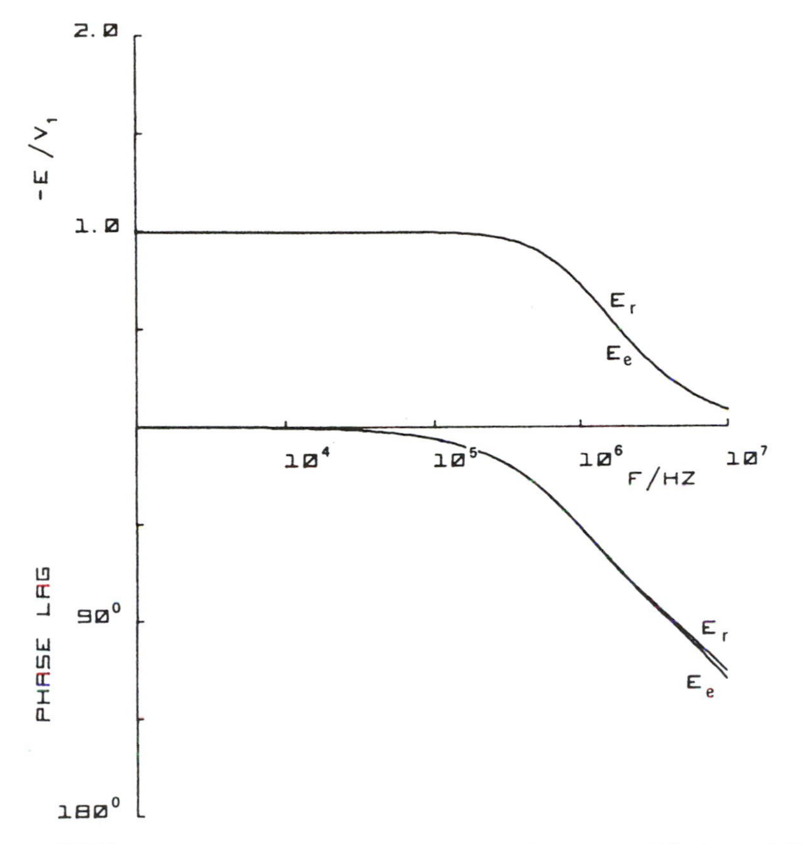


**Figure 7.16** Transfer function and phase lag for $A_o = 1 \times 10^5$, $f_a = 10$ Hz, $R_sC_{dl} = 1 \times 10^{-8}$ s, and $R_uC_{dl} = 1 \times 10^{-9}$ s.

## IV. COMPENSATION FOR $iR_u$ BY POSITIVE FEEDBACK

Whenever cell parameters do not allow a reduction in $R_u$ commensurate with potentiostatic performance requirements, the technique of positive feedback can be implemented to achieve the goal. This method is nearly as old as the analog potentiostat, and it has been applied in a variety of ways with highly variable results. In principle, the error due to $iR_u$ can be reduced to zero (or even made negative) through positive feedback. Fortunately, stability problems are unchanged in principle relative to the examples presented thus far, but they may be somewhat more demanding of careful analysis in practice. Through the illustrations of this section, one point emerges: positive feedback of a signal proportional to cell current causes exactly the same changes in system response as moving the reference electrode closer to the working electrode. In the final

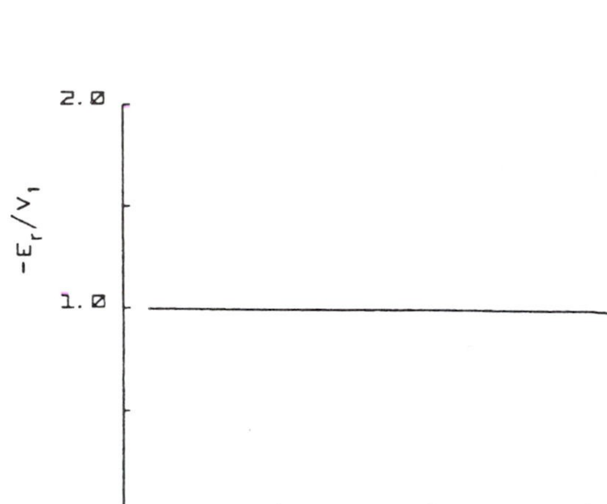


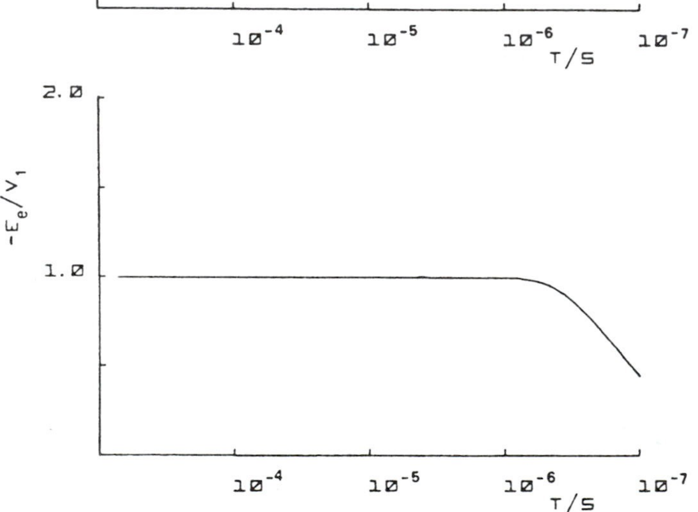


**Figure 7.17** Transient response to step input. Same parameters as Figure 7.16.

section, methods are given whereby graceful control may be achieved with or without positive feedback.

The signal $V_u$, for compensation of $iR_u$, must be proportional to this quantity and identical in phase. A convenient source of $V_u$ is the output of the current follower, OA3, of Figure 7.1A. Adjustment of the magnitude is provided by a potentiometer so that some fraction F of $iR_u$ is compensated. Because of the wide range of resistor values that may be used in the circuit, $V_u$ will not generally be equal to $F(iR_u)$, but will be scaled according to the equation

$$V_u = V_c \frac{FR_uR_2}{R_cR_f} \tag{7.14}$$

For any given set of resistors, the potentiometer may be calibrated in ohms of compensated resistance through the relation

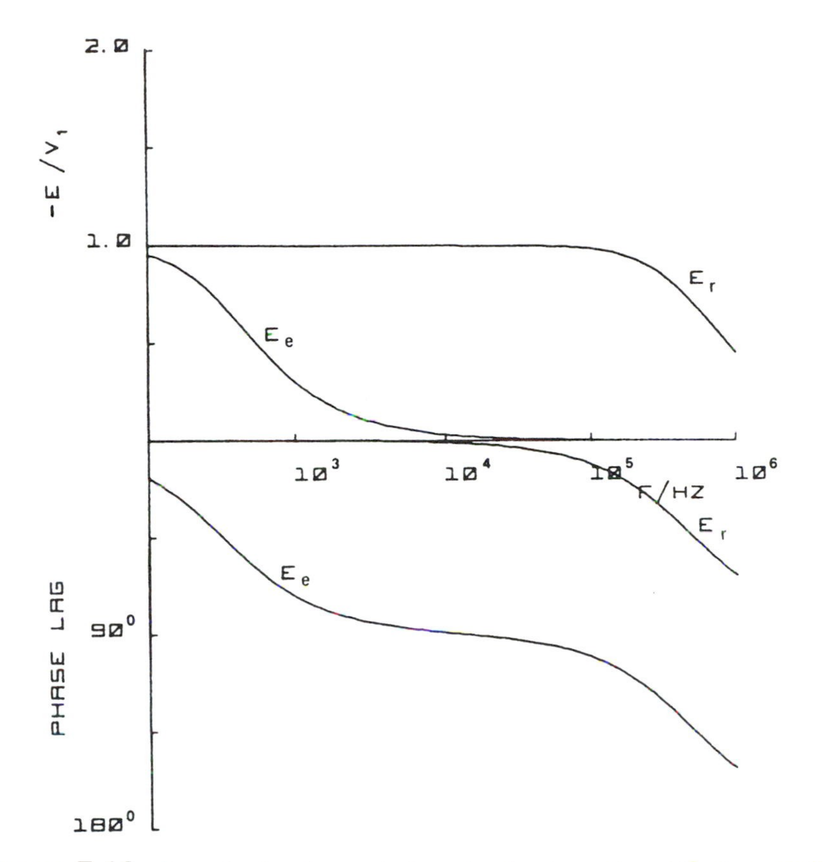


**Figure 7.18** Transfer function and phase lag for $A_o = 1 \times 10^5$, $f_a = 10$ Hz, $R_sC_{dl} = 1 \times 10^{-3}$ s, and $R_uC_{dl} = 5 \times 10^{-4}$ s.

$$R_u = \frac{\beta R_c R_f}{R_2} \tag{7.15}$$

where ß is the fraction of $V_c$ applied to $R_2$. For clarity, the examples refer to the fraction F of $R_u$ that is compensated, not to ß.

Before attempting to complete the positive feedback loop, the signal $V_u$ (and therefore $V_c$) must meet the same requirement that was imposed on $V_f$: no phase-time lag in the frequency region of operation of the control loop around OA1. This is a demanding task and requires that OA3 be a high-performance device in the upper frequency regions. The task is made easier if $R_c$ is as small as practical while allowing β to equal unity for the maximum value of $R_u$ anticipated. An additional OA in a noninverting configuration may be added to amplify $V_c$, allowing $R_c$ to be kept small. Two OAs in series will always outperform a single OA of the same type in terms of gain-bandwidth product and therefore phase/time lag will be reduced.

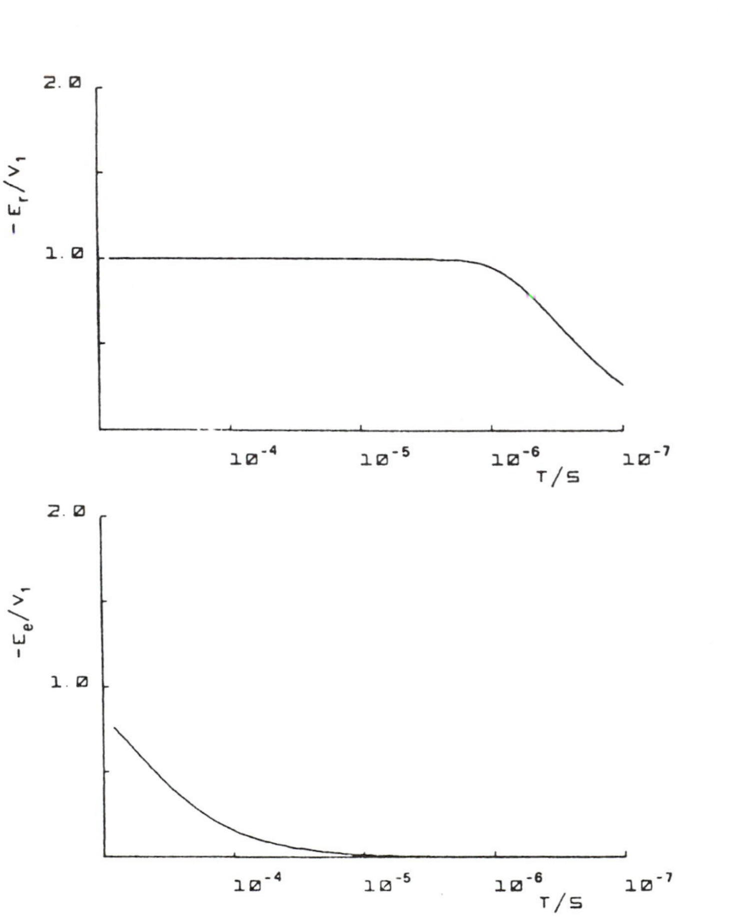


**Figure 7.19** Transient response to step input. Same parameters as Figure 7.18.

With the compensation signal, $V_u$, applied to $R_2$, the frequency-domain responses for several values of F are plotted in Figure 7.20. The control OA and cell parameters are the same as those used for Figures 7.5 and 7.6, but here only $E_e/V_1$ is given. Curves A to C are identical with their counterparts of Figure 7.6, since values of F are 0, 0.5, and 0.8, corresponding to $C_{dl}R_u$ values of $10^{-5}$, $5 \times 10^{-6}$, and $2 \times 10^{-6}$ s. Complete compensation of the effects of $R_u$ results in curve D. Again it should be noted that a real system will undoubtedly have a small amount of additional phase lag in the high-frequency region, which will add to the plotted values and produce a sustained oscillation when $F = 1$. The frequency of the oscillation will be about $5 \times 10^4$ Hz.

Time-domain response to a step change is given in Figure 7.21 for the same system. Again the first three curves are identical with $E_e/V_1$ curves of Figure 7.12 and the fourth results when $F = 1$. That the system is teetering on the edge of instability is very evident from curve D.

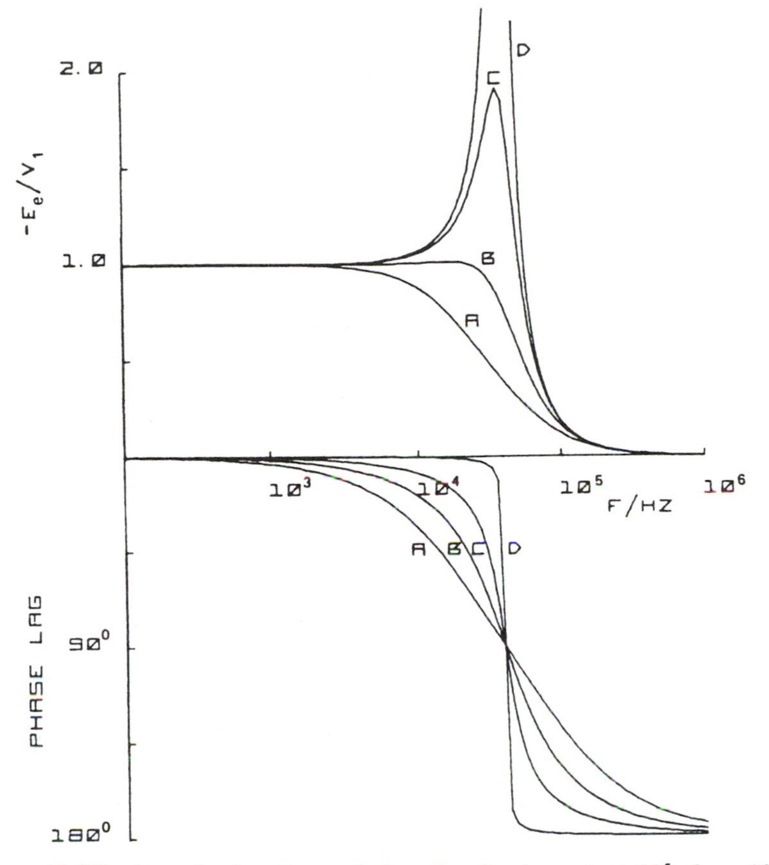


**Figure 7.20** Transfer function and phase lag for $A_o = 1 \times 10^5$, $f_a = 10$ Hz, $R_sC_{dl} = 1 \times 10^{-4}$ s, $R_uC_{dl} = 1 \times 10^{-5}$ s, and fractional compensation of $R_u$ equal to (A) 0, (B) 0.5, (C) 0.8, and (D) 1.0.

Accepting these results for an idealized system confirms the intuitive notion that the use of positive feedback from the current follower is identical to physically moving the reference electrode closer to the working electrode. There are no additional concepts to cope with in relating response to physical parameters. It will become evident that this situation allows a rather simple assessment of loop gain and a clear view of corrective measures that will lead to stability with $R_u$ compensation.

There is a possibility that concern over the new feedback loop and system instability has overshadowed the improvement in potentiostatic control when F = 1. This change is not readily apparent in the graphs, but the low-frequency, long-time value of $E_e/V_1$ is unity in the presence of faradaic current. After the instability is cured, the dependence of potential on current will be more evident than it is now.

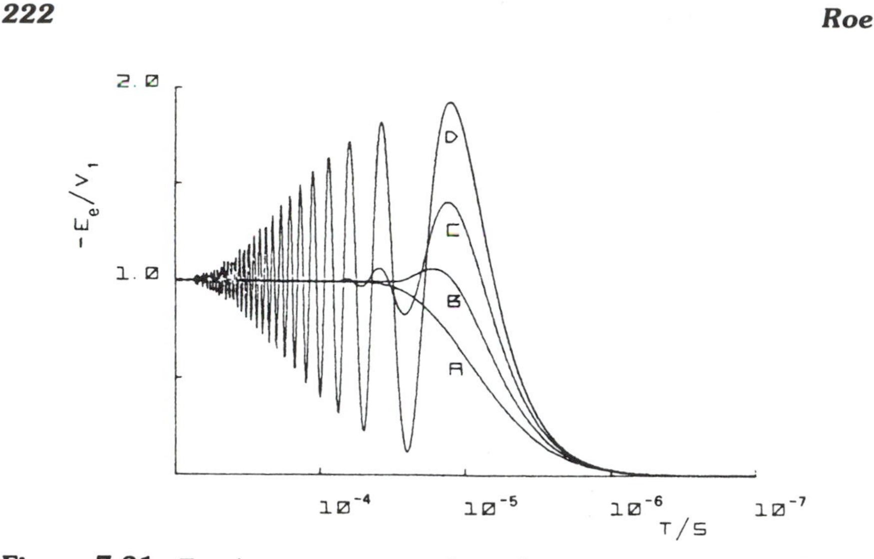


**Figure 7.21** Transient response to step input. Same parameters as Figure 7.12.

## V. ACHIEVING STABILITY THROUGH GAIN-FREQUENCY SHAPING

Reduction of the actual or effective value of $R_u$ is the basic reason for upsetting the feedback loop. The phase-time lag related to $C_{dl}R_s$ becomes extended as $R_u$ is decreased; this causes the feedback signal $V_r$ to be delayed over a wide frequency-time region. The effect is most apparent when the delay occurs over a frequency-time region where the loop gain is small but greater than unity. Attempts to change cell parameters will not lead to a general solution of the problem, but there is the possibility of altering the gain-frequency characteristics of the control OA. As has been usually implemented, the method involves trading gain for stability, but in certain cases it is possible to arrive at stable control by increasing the gain. The principle involved is to decrease the phase-time lag of the control OA in a selected frequency-time region. A graphical method will be used to clarify the method.

Figure 7.22 is a straight-line approximation of the information presented in Figures 7.2 and 7.3. Amplifier gain A (inverting for the circuit of Fig. 7.1) is represented by line segments A and B over the frequency range of interest, and cell attenuation A (unity or less) is represented in the lower part of the graph by the abscissa below $f_s$ and then by lines C and D. These lines intersect at $f_u$, and decreasing $R_u$ ($C_{dl}$ constant) moves the intersection to the lower right, along C. For the present purpose, this can also be assumed to be the case when positive feedback of $V_u$ is used to compensate for $R_u$. Line C has been extended to show cell attenuation when $R_u$ has been effectively compensated; if F = 1, the

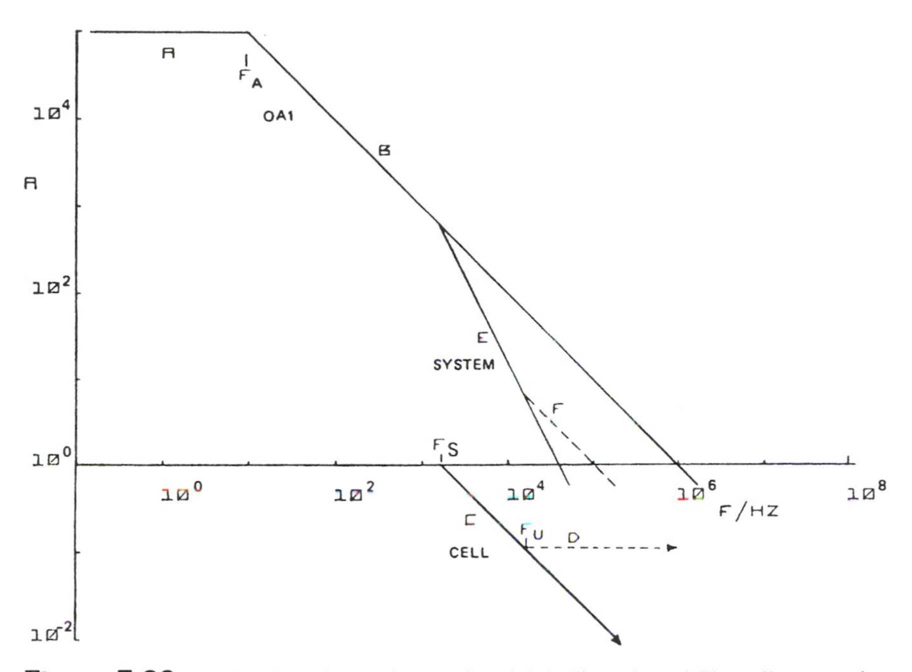


**Figure 7.22** Bode plot of open-loop gain of OA (lines A and B), cell attenuation (lines C and D), and their product (lines E and F). See the text for details.

slope of C is unchanged to infinite frequency. Note that phase lag was not plotted; it can be estimated by reference to previous figures. It is only necessary to remember that decreasing gain is the product of amplifier gain and cell attenuation, and this is obtained by addition of ordinate values on the log scale. Loop gain follows lines A and B, branching to E (and F if $R_u$ is uncompensated). The assumption is made that attenuation due to the resistor network on the input of OA1 is negligible, that is, $R_f \ll R_1 || R_2$. A segment with a slope of –2 signals a phase lag increasing toward 180°, and this condition leads to overshoot and oscillations. The situation becomes worse when this steep segment occurs in the region of low loop gain. All of the examples illustrated by frequency- or time-response curves can be quickly analyzed by constructing this type of straight-line Bode plot from values of $f_a$, $f_s$, $f_u$, and $f_o$ and comparing response curves in a relative fashion.

Graphically speaking, stability or first-order response can be achieved by elimination of the line segment with a slope of –2. One method depends on the introduction of a horizontal section in the amplifier's gain-frequency curve, as shown in Figure 7.23. A series capacitor and resistor between the inverting input and the output of OA1 as shown in Figure 7.1A will tailor the gain in this way.

Segment G is due to the capacitor, and segment H is due to the resistor; the normal behavior then dominates beyond the intersection of H and B at $f_2$. Loop gain now follows the segments A to G at $f_a$ and to E at $f_i$. Now the slope is $-1$ below frequency $f_1$, indicating that the phase lag is decreased in the critical region identified in Figure 7.22. The slope does increase to $-2$ to the right of $f_2$ where $A < 1$, and this change means the total phase lag is 135° at the frequency of unity loop gain.

Curves A of Figures 7.24 and 7.25 show the benefits of this technique of gain-frequency shaping; comparison to curves D of Figures 7.20 and 7.21 immediately reveals a dramatic improvement in stability, with potentiostatic conditions ($F = 1$) prevailing smoothly over the frequency region below about 1 kHz (phase lag $<1°$) or at times longer than $5 \times 10^{-5}$ s ($E_e/V_1 > 0.99$). To the right of these points, the control OA becomes less able to maintain potentiostatic conditions because of limited loop gain. There is about 20% overshoot in $E_e/V_i$ because of the excess phase lag just before the loop gain reaches unity. By moving line G to the left and lowering line H by the same amount, so that their intersection is unchanged on the abscissa ($f_1 = f_s$), the overshoot

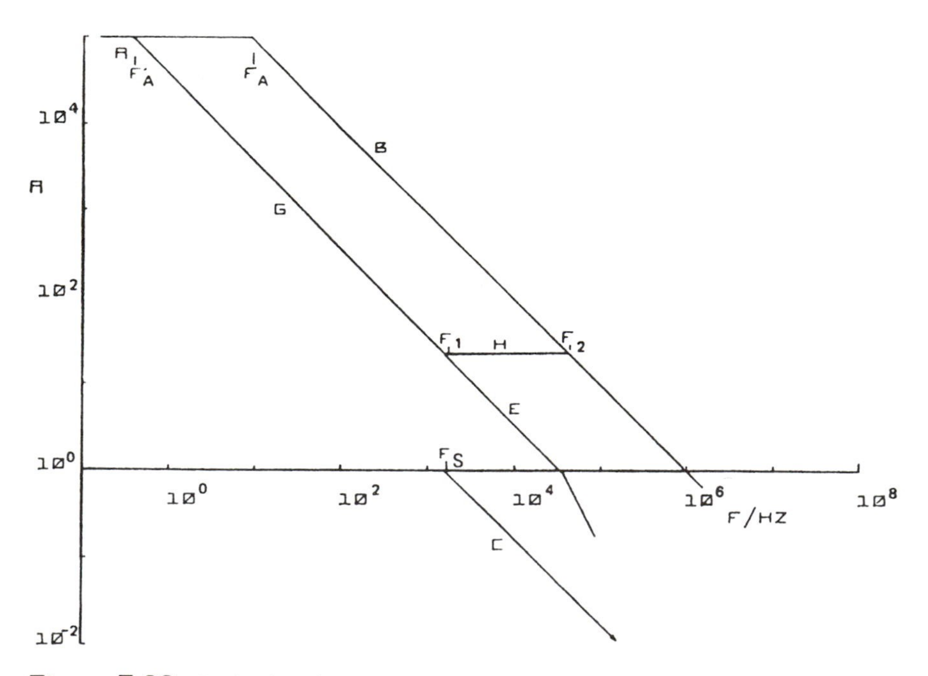


**Figure 7.23** Bode plot of open-loop gain of OA (lines A and B), with gain-frequency shaping (lines G and H), cell attenuation (line C), and the product of H and C (line E). See the text for details.

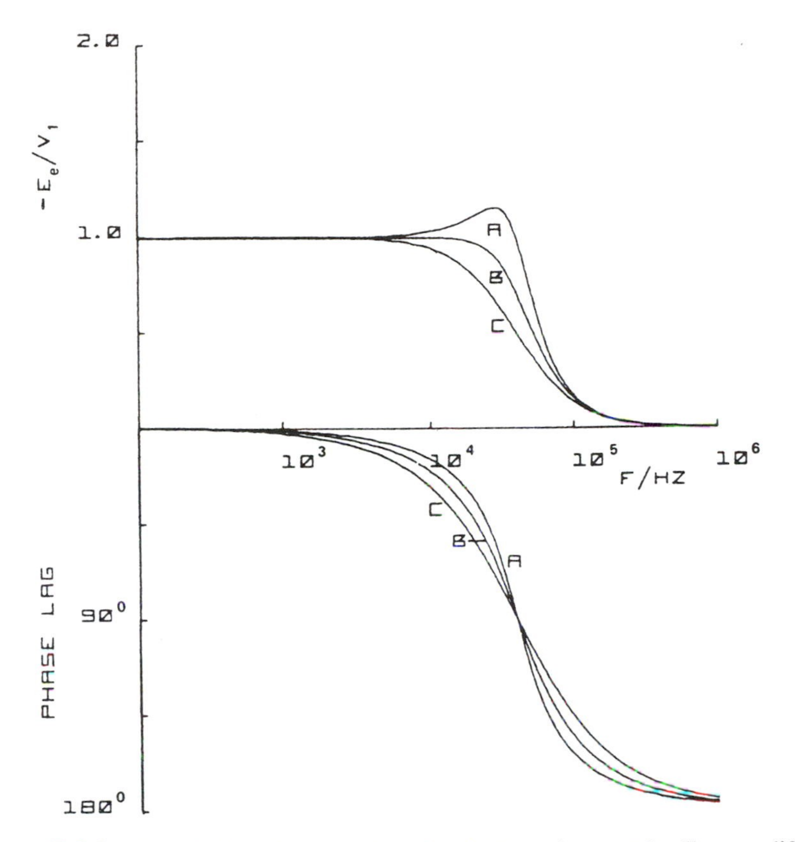


**Figure 7.24** Transfer function and phase lag for potentiostat and cell as modified in Figure 7.23. A, $f_a = 0.4$ Hz, $f_1 = 1.6 \times 10^3$ Hz, $f_2 = 4 \times 10^4$ Hz; B, $f_a = 0.28$ Hz, $f_1 = 1.6 \times 10^3$ Hz, $f_2 = 5.6 \times 10^4$ Hz; C, $f_a = 0.20$ Hz, $f_1 = 1.6 \times 10^3$ Hz, $f_2 = 4.0 \times 10^4$ Hz. Other parameters as in Figure 7.20, curve D.

can be eliminated. Curves B and C of Figures 7.24 and 7.25 were plotted after shifting the line segments by factors of 0.71 and 0.5 in the directions noted. Taking the response represented by curves C as optimum, simple expressions can be obtained to guide the choice of $f_a$, $f_1$, and $f_2$.

$$f'_a = \frac{1}{2}\left(\frac{f_a f_s}{A_o}\right)^{1/2} \tag{7.16}$$

$$f_1 = f_s \tag{7.17}$$

$$f_2 = 2(A_o f_a f_s)^{1/2} \tag{7.18}$$

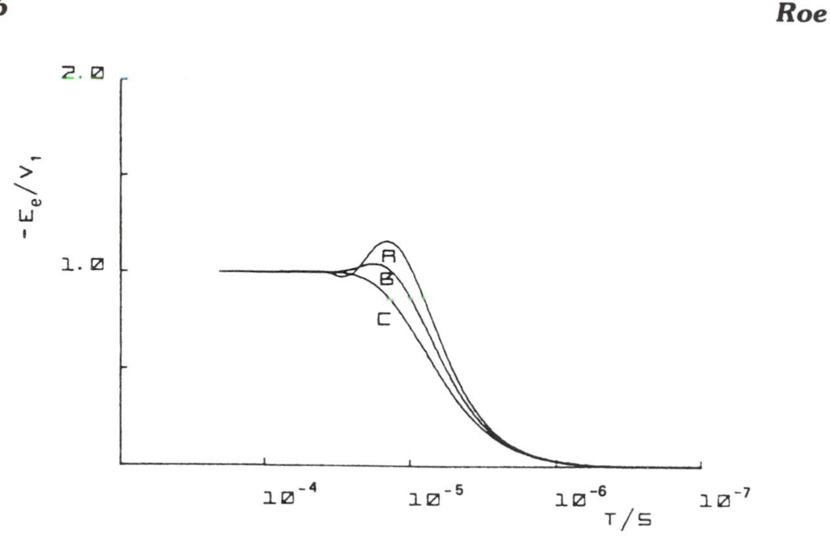


**Figure 7.25** Transient response to step input. Same conditions as in Figure 7.24.

Often it is easier to think in terms of time constants; the corresponding equations are

$$\tau'_a = 2(\tau_a \tau_s A_o)^{1/2} \tag{7.19}$$

$$\tau_1 = \tau_s \tag{7.20}$$

$$\tau_2 = \frac{1}{2}\left(\frac{\tau_a \tau_s}{A_o}\right)^{1/2} \tag{7.21}$$

Elimination of the numerical factors of 1/2 and 2 gives the conditions of curves A in Figures 7.24 and 7.25.

Shaping the gain-frequency curve of the control OA requires knowing the value of feedback resistor $R_f$ (Fig. 7.1) and an estimate of the cell roll-off frequency $f_s$. First-order approximations for the stabilizing elements are

$$R \cong 0.1R_f \tag{7.22}$$

$$C \cong \frac{1}{2\pi f_s} = \frac{\tau_s}{R} \tag{7.23}$$

The optimum value for R depends on the cell attenuation, so it is advantageous to use a potentiometer so that some adjustment is possible.

In practice, the one quantity that will be estimated is $f_s$ (or $\tau_s$), and it is important to have a rough idea of the consequence of uncertainty in this number. Again, using the same cell and control OA parameters as used for Figures 7.24 and 7.25, responses to sine waves and a step input are presented in Figures 7.26 and 7.27 for variations of factors of 0.5 and 2 in $f_1$. That is, $f_s$ is unchanged, but curves were calculated on the assumption that the estimated value was higher and lower than the actual cell break frequency and these values were set equal to $f_1$. Corresponding adjustments of $f_a$ and $f_2$ were calculated to agree with Equations 7.16 and 7.18. One may conclude that an error of about a factor of 2 in the estimate of $f_s$ is just acceptable for this particular situation. However, the amount of overshoot or undershoot and the time constant for the final

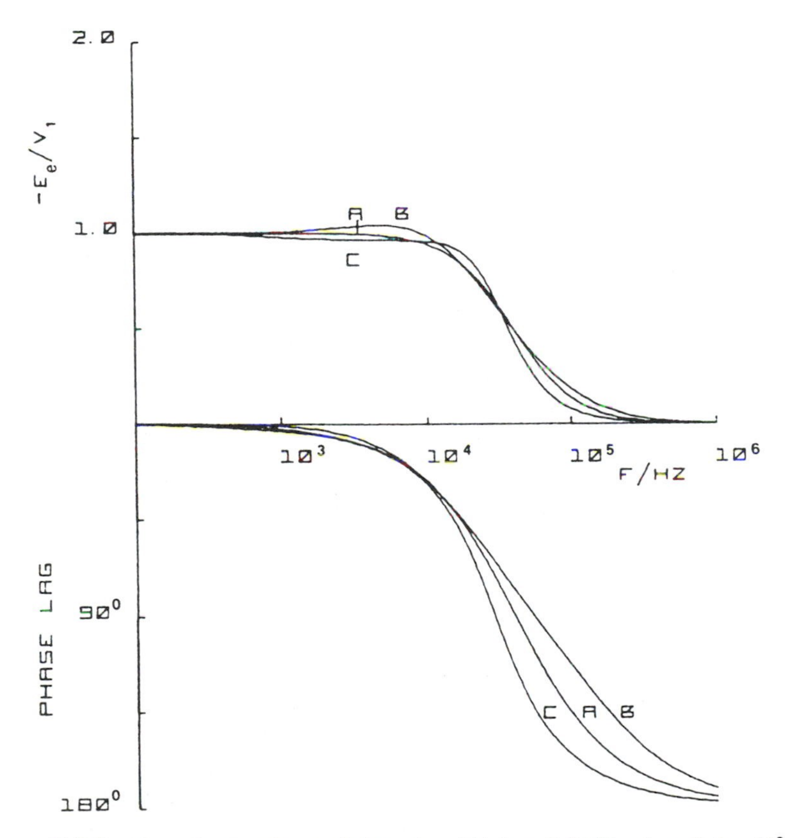


**Figure 7.26** Transfer function and phase lag. (A) $f_a = 0.20$ Hz, $f_1 = 1.6 \times 10^3$ Hz, $f_2 = 8 \times 10^4$ Hz; (B) $f_a = 0.40$ Hz, $f_1 = 3.2 \times 10^3$ Hz, $f_2 = 1.6 \times 10^5$ Hz; (C) $f_a = 0.10$ Hz, $f_1 = 8 \times 10^2$ Hz, $f_2 = 4 \times 10^4$ Hz. Other parameters as in Figure 7.20, curve D.

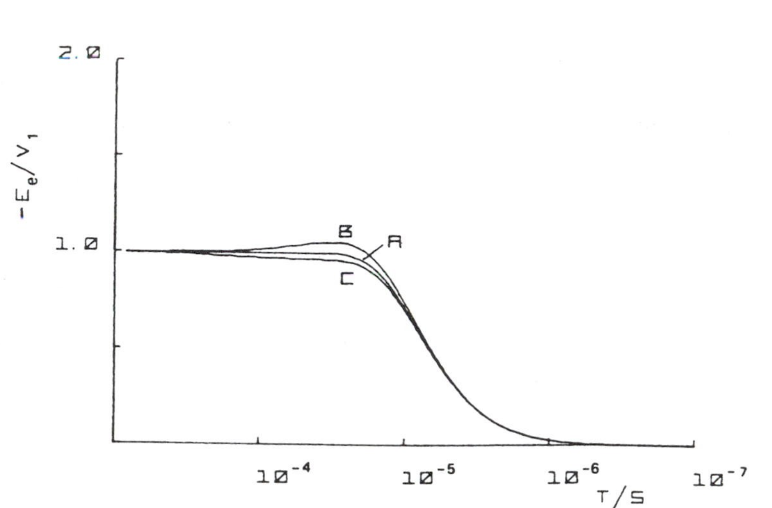


**Figure 7.27** Transient response to step input. Same conditions as in Figure 7.26.

settling do depend on the loop gain in the vicinity of $f_1$. Increased loop gain produces less of a deviation, and such would result from a lower value of $f_s$ than was used in this example.

When experimental requirements are not met in terms of upper frequency limit or rise time by the preceding method, a second approach may be considered for tailoring amplifier response. This method greatly improves the gain-bandwidth limits but requires circuit modification of commercial instrumentation. The principle is obvious when the graphical problem of slopes discussed relative to Figure 7.22 is readdressed but with the view that line segment B is not a fixed boundary to gain-frequency capabilities of an amplifier. If line B can be moved to the right, a horizontal section can be constructed that allows the loop gain line to maintain a slope of −1. This is the same stability objective used in the first method. Figure 7.28 shows the locations of $f_1$ and $f_2$ that meet the basic requirement. The dashed line, E, is an estimated gain-frequency response of an OA without any internal compensation (which is usually provided by a single capacitor). This dashed line is an absolute limit for an OA. If the internal compensation network is modified to shape the gain-frequency curve successfully as shown by line segments A, B to D at $f_1$ and to E at $f_2$, it is to be expected that high-frequency performance and stability will be obtained. Figures 7.29 and 7.30 show the results, which are one to two orders of magnitude faster than those obtained with the same basic system used to calculate responses shown in Figures 7.24 and 7.25. There is another advantage and there is a disadvantage. The advantage is that this method always gives a higher loop gain in the vicinity of $f_1$ so that errors in the estimated value of $f_s$ will produce less

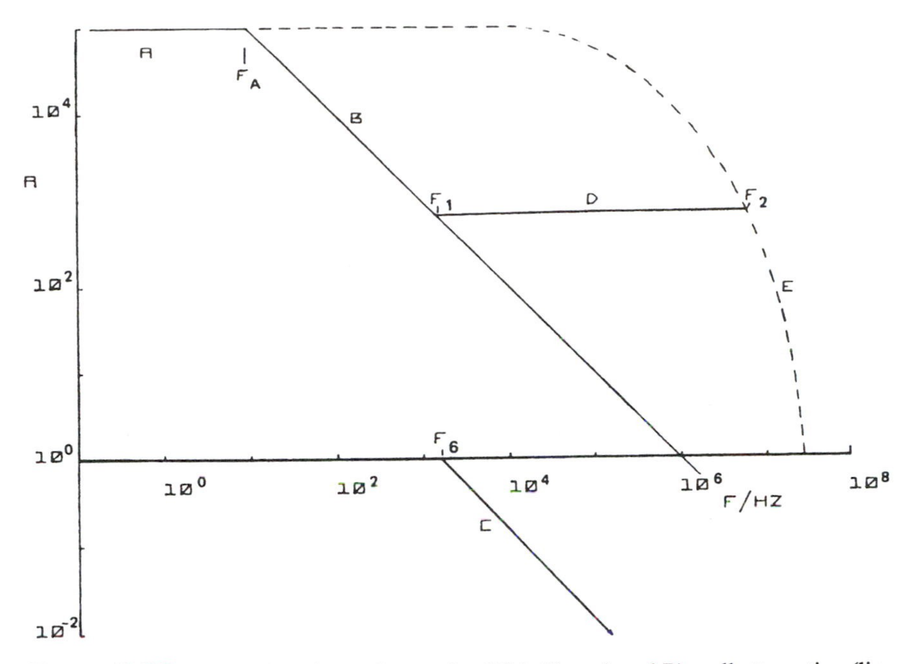


**Figure 7.28** Bode plot of open-loop gain of OA (lines A and B), cell attenuation (line C), modified open-loop gain (curve D), and uncompensated open-loop gain (E). See the text for details.

distortion than with the first method (compare curves B and C of the figures). The disadvantage is that system noise will be increased by the increased bandwidth, so that the signal-to-noise ratio will suffer. In cases in which fast response is needed, this is the best approach. The details of altering internal compensation circuits will be left to the reader (usually, the addition of a potentiometer will suffice), but it is advisable to know curve E and then to adjust $f_s$ so that OA phase lag at $f_2$ is less than 135°. In calculation of the curves it was assumed that phase lag at $f_2$ was only 90°, although the Bode plot shows otherwise.

There are several other methods of achieving stability in potentiostatic circuits. A capacitor may be added between the counter and reference electrodes to reduce phase shift in the critical frequency region. Some caution must be exercised since a low-resistance reference electrode then becomes the counterelectrode at high frequencies. A particularly interesting method is known as input lead-lag compensation; a series RC is connected between the input terminals of the control amplifier, and a second resistor is connected between the noninverting input and common. This form of compensation has minimum effect on the slew rate of the control amplifier. Further details can be found in the book by Stout and Kaufman listed in the bibliography.

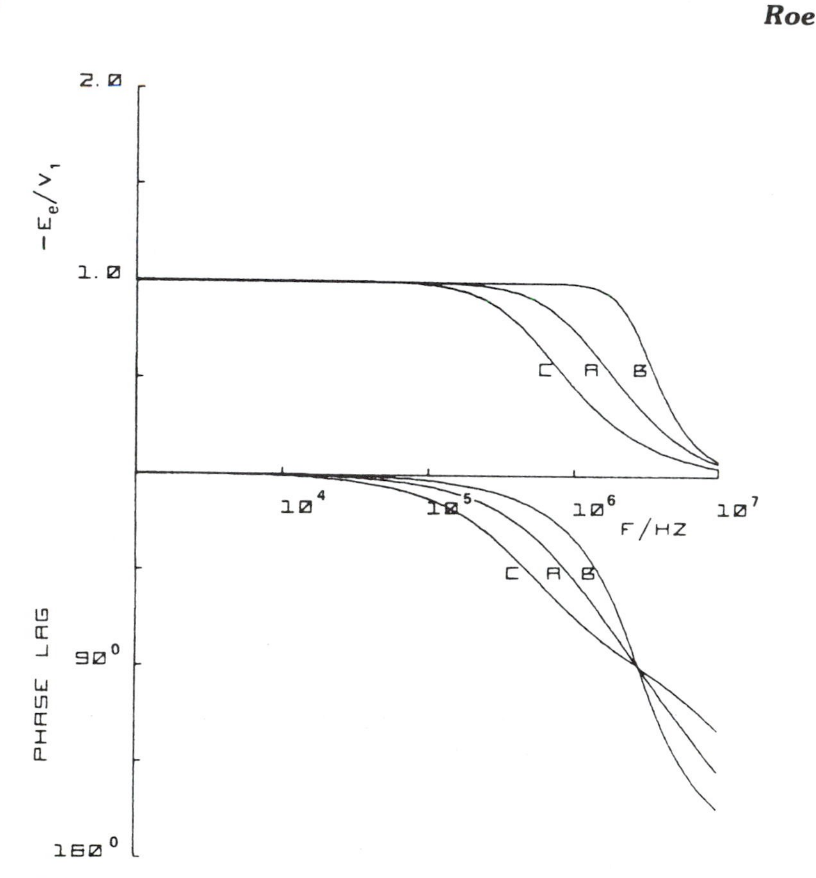


**Figure 7.29** Transfer function and phase lag for potentiostat and cell as modified in Figure 7.28. (A) $f_a = 10$ Hz, $f_1 = 1.6 \times 10^3$ Hz, $f_2 = 8 \times 10^6$ Hz; (B) $f_1 = 8 \times 10^2$ Hz, $f_2 = 4 \times 10^6$ Hz; (C) $f_1 = 3.2 \times 10^3$ Hz, $f_2 = 1.6 \times 10^7$ Hz. Other parameters as in Figure 7.20, curve D.

It is reasonable to ask at this point: Are there other approaches to reach stability with grace in true potentiostatic circuits? The answer is indeed affirmative, but unfortunately with qualifications. One technique is discontinuous control of cell potential. It is not a new approach; it was, in fact, the method used in the very first electronic potentiostat by Hickling in 1942. The principle is quite simple: Current pulses are applied to the counterelectrode so that the desired potential is maintained between reference and working electrode. Since the potential can be measured between pulses, there is no iR drop. Cell potential is not steady; it depends on the sensitivity of the comparator circuit and the rate at which current demand can be met.

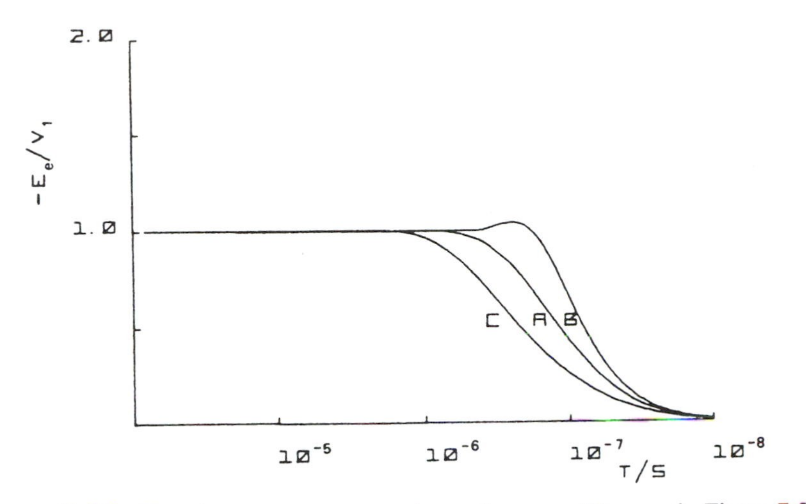


**Figure 7.30** Transient response to step input. Same conditions as in Figure 7.29.

Another method is to apply feedback control as in Figure 7.1, but to periodically interrupt the circuit for a brief moment so that the magnitude of $iR_u$ can be determined and a correction applied. Several variations of such a control system have been published, and one version is available as an accessory for a potentiostat. The outstanding feature of these approaches is that $R_u$ does not have to be known and it can vary during an experiment with only minor decrease in control accuracy. In many situations, the brief interruption of the feedback loop is a small price to pay for such convenience. Microprocessors will foster further developments in this direction; the advent of digital potentiostats is imminent.

## VI. REAL SYSTEMS

It should be anticipated that there will not be a smooth transition from these idealized, simple systems into the real world. Some precautions and pitfalls have been cited, but usually in a parenthetical manner that lacked proper emphasis. The systems selected to illustrate the general principles of potentiostatic control have shown what can be expected under ideal conditions, but real systems have additional parameters that may tilt the balance from graceful control to chaos. Cell design is of paramount importance, and a guide to transfer characteristics of cells is included in the bibliography. To bring the information in this chapter effectively into use, it is necessary to acknowledge the role that cells play and to design them to meet the measurement requirements as part of the system. A classical H cell with 10-ft coaxial cables is not likely to give satisfactory performance with even the best potentiostat available.

If one of the main points of this chapter is to be put in the form of a maxim, it would be: An operational amplifier is *not* a potentiostat; a potentiostat is a system comprising operational amplifiers *and* a cell.

## APPENDIX

All the system response curves in frequency and time domains were calculated numerically from equations that are much too involved to reproduce in detail here. Transfer functions in Laplace transform notation are easily defined for the potentiostat and cell of Figure 7.1. Appropriate combinations of these functions then yield system transfer functions that may be cast into time- or frequency-dependent equations by inverse Laplace transformation or by using complex number manipulation techniques. These methods have become rather common in electrochemical literature and are not described here. The interested reader will find several citations in the bibliography to be helpful in clarifying details.

Transfer functions are as follows:

Control OA1:

$$\frac{V_o(s)}{V_-(s)} = -A(s) = -\frac{A_o(1 + s\tau_1)}{(a + s\tau_a)(1 + s\tau_s)} \tag{A7.1}$$

where $V_o(s)$ is the output, $V_-(s)$ is the voltage at the inverting input, $A_o$ is the dc gain, and the time constants $\tau_a$, $\tau_1$, and $\tau_2$ are defined in the frequency domain by Figures 7.23 and 7.28. When gain-frequency shaping is not used, $\tau_1$ and $\tau_2$ are set equal to zero.

Auxiliary electrode to reference electrode of cell:

$$H(s) = \frac{E_r(s)}{V_o(s)} = \frac{1 + s\tau_u}{1 + s\tau_s} \tag{A7.2}$$

Auxiliary electrode to solution-working electrode interface:

$$K(s) = \frac{E_e(s)}{V_o(s)} = \frac{1}{1 + s\tau_s} \tag{A7.3}$$

where $\tau_u = R_uC_{dl}$ and $\tau_s = R_sC_{dl}$.

Relative value of voltage drop, $R_ui(s)$, to be compensated by positive feedback:

$$L(s) = \frac{V_u(s)}{V_o(s)} = \frac{Fs\tau_u}{1 + s\tau_u} \tag{A7.4}$$

where F is the fraction of $R_u$ compensated.

In order to combine the transfer functions, the relation between $V_-(s)$, the summing resistors, and the other voltages is required. From Kirchhoff's current law, this equation is

$$V_-(s) = \left[\frac{V_1'(s)}{R_1} + \frac{V_u'(s)}{R_2} + \frac{V_r(s)}{R_f}\right]R_p \tag{A7.5}$$

where

$$R_p = \left(\frac{1}{R_1} + \frac{1}{R_2} + \frac{1}{R_f}\right)^{-1} \tag{A7.6}$$

It is advantageous to make $R_1$, $R_2 >> R_f$ to avoid loss of amplifier gain, so $R_p = R_f$, and a simplified form results:

$$V_-(s) = V_1(s) + V_u(s) + V_r(s) \tag{A7.7}$$

where $V_1(s)$ and $V_u(s)$ are the scaled values of the actual signal voltages.

Combining transfer functions A7.1, A7.2, and A7.4 with Equation A7.7 gives the system transfer function between the signal at the reference electrode and the input, provided that $E_r(s) = V_r(s)$:

$$\frac{E_r(s)}{V_1(s)} = -\frac{A(s)H(s)}{1 + A(s)[H(s) - L(s)]} \tag{A7.8}$$

By using Equation A7.3 instead of Equation A7.4, the system transfer function at the solution-working electrode interface relative to the input signal is obtained:

$$\frac{E_e(s)}{V_1(s)} = -\frac{A(s)K(s)}{1 + A(s)[K(s) - L(s)]} \tag{A7.9}$$

Substitutions for the transfer functions in Equations A7.8 and A7.9, noting that $s = j\omega$ for steady-state sinusoid signals, give equations in complex notation. The magnitude ratio and phase relation follow by the usual methods. Transient response results by setting $V_1(s)$ equal to $V_1/s$, where $V_1$ is the magnitude of the voltage step. Inverse Laplace transformation then gives the time dependence of either signal of the cell relative to $V_1$.

The response curves were plotted by numerically solving the final equations at intervals of 0.01 or 0.04 or major divisions on the abscissa scale using a Tektronix 31 calculator and plotter.

## BIBLIOGRAPHY

The following is only a partial listing of significant literature on the design and use of potentiostats. References cited in these publications provide a thorough bibliography of chemical and control theory papers on the subject.


C. Amatore and C. Lefrou, "New Concept for a Potentiostat for On-Line Ohmic Drop Compensation in Cyclic Voltammetry Above 300 kV $s^{-1}$," *J. Electroanal. Chem. 324*:33–58 (1992).

C. Amatore, C. Lefrou, and F. Pflüger, "On-Line Compensation of Ohmic Drop in Submicrosecond Time Resolved Cyclic Voltammetry at Ultramicroelectrodes," *J. Electroanal. Chem. 270*:43–59 (1989).

A. Bewick, "Analysis of the Use of iR Compensation in Potentiostatic Investigations," *Electrochim. Acta 13*:825 (1968).

G. L. Booman and W. B. Holbrook, "Electroanalytical Controlled-Potential Instrumentation," *Anal. Chem. 35*:1793 (1963). (Related publications in same issue.)

G. L. Booman and W. B. Holbrook, "Optimum Stabilization Networks for Potentiostats with Application to a Polarograph Using Transistor Operational Amplifiers," *Anal. Chem. 37*:795 (1965).

D. Britz, "iR Elimination in Electro-chemical Cells," *J. Electroanal. Chem. 88*:309 (1978).

E. R. Brown, H. L. Hung, T.G. McCord, D. E. Smith, and G. L. Booman, "A Study of Operational Amplifier Potentiostats Employing Positive Feedback for iR Compensation: II. Application to AC Polarography," *Anal. Chem. 40*:1411 (1968).

D. Garreau, P. Hapiot, and J.-M. Savéant, "Instrumentation for Fast Voltammetry at Ultramicroelectrodes: Stability and Bandpass Limitations," *J. Electroanal. Chem. 272*:1–16 (1989).

D. Garreau and J.-M. Savéant, "Linear Sweep Voltammetry. Compensation of Cell Resistance and Stability. Determination of Residual Uncompensated Resistance," *J. Electroanal. Chem. 35*:309 (1972).

J. E. Harrar, "Techniques, Apparatus and Analytical Applications of Controlled-Potential Coulometry," *Electroanalytical Chemistry*, Vol. 8 (A. J. Bard, ed.), Marcel Dekker, New York, 1975.

J. E. Harrar and C. L. Pomernacki, "Linear and Nonlinear System Characteristics of Controlled-Potential Electrolysis Cells," *Anal. Chem. 45*:57 (1973).

J. E. Mumby and S. P. Perone, "Potentiostat and Cell Design for the Study of Rapid Electrochemical Systems," *Chem. Instrum. 3*:191 (1971).

A. A. Pilla, "Influence of the Faradaic Process on Nonfaradaic Resistance Compensation in Potentiostatic Techniques," *J. Electrochem. Soc. 118*:702 (1971).

R. R. Schroeder and I. Shain, "Application of Feedback Principles to Instrumentation for Potentiostatic Studies," *Chem. Instrum. 1*:233 (1969).

M. Souto, "Electronic Configurations in Potentiostats for the Correction of Ohmic Losses," *Electroanalysis 6*: 531–542 (1994).

D. Stout and M. Kaufman, *Handbook of Operational Amplifier Circuit Design*, McGraw-Hill, New York, 1976.

D. E. Tallman, G. Shepherd, and W. J. MacKellar, "A Wide Bandwidth Computer Based Potentiostat for Fast Voltammetry at Microelectrodes," *J. Electroanal. Chem. 280*:327–340 (1990).

# 8

# Conductivity and Conductometry

**F. James Holler** *University of Kentucky, Lexington, Kentucky*

**Christie G. Enke** *The University of New Mexico, Albuquerque, New Mexico*

Electrochemical cells of all types are composed of various electrically conductive materials. The nature and degree of conductivity of the cell can substantially affect the current through the cell, the electric field strengths throughout the cell, and the transport of charged particles from one cell region to another. Thus the phenomenon of conductance is highly relevant to all nonequilibrium electrochemical techniques. For studies that emphasize interfacial behavior, it is helpful to understand the potential fields and ionic transport that occur in the bulk materials adjoining the interfaces. For the study of bulk electrical properties, factors that affect the accuracy and precision of the conductivity measurement determine the limits of usefulness of conductometric detection and determination.

As we shall see, the solution conductivity depends on the ion concentration and the characteristic mobility of the ions present. Therefore, conductivity measurements of simple, one-solute solutions can be interpreted to indicate the concentration of ions (as in the determination of solubility or the degree of dissociation) or the mobility of ions (as in the investigations of the degree of solvation, complexation, or association of ions). In multiple-solute solutions, the contribution of a single ionic solute to the total solution conductivity cannot be determined by conductance measurements alone. This lack of specificity or selectivity of the conductance parameter combined with the degree of tedium usually associated with electrolytic conductivity measurements has, in the past, discouraged the development of conductometry as a widespread electroanalytical technique. Today, there is a substantial reawakening of interest in the practical applications of conductometry. Recent electronic developments have resulted in automated precision conductometric instrumentation and applications are being developed in which conductometry is used to follow the course of highly specific chemical reactions or varying solute concentrations for chromatographic detection.

In this chapter we take a careful look at the phenomenon of electrical conductivity of materials, particularly electrolytic solutions. In the first section, the nature of electrical conductivity and its relation to the electrolyte composition and temperature is developed. The first section and the second (which deals with the direct-current contact methods for measuring conductance) introduce the basic considerations and techniques of conductance measurement. This introduction to conductance measurements is useful to the scientist, not only for electrolytic conductance, but also for understanding the applications of common resistive indicator devices such as thermistors for temperature, photoconductors for light, and strain gauges for mechanical distortion. The third section of this chapter describes the special techniques that are used to minimize the effects of electrode phenomena on the measurement of electrolytic conductance. In that section you will encounter the most recent solutions to the problems of conductometric measurements, the solutions that have sparked the resurgent interest in analytical conductometry.

## I. SOME BASIC RELATIONSHIPS

A material exhibits the property of electrical conductivity when it contains charged particles that are free to move through the material. When an electrical potential is applied across such a material, the charged particles will experience a force along the field in a direction opposite to their charge. The resulting net motion of the charged particles is an electric current. The greater the current produced by a given electric field, the greater the conductivity of the material. Since conductivity is related to the rate of flow of charge in response to an electric field, the magnitude of the conductivity is dependent on the concentration, mobility, and charge of the charged particles. In this section, the relationship between the conductivity and the mobility, charge, and concentration of the charged species is developed. Finally, the property of conductivity is related to the measured parameter, conductance.

### A. Conductivity

When a charged particle i (an electron or ion) in a liquid or solid material is subjected to an electric field, it quickly reaches a limiting average velocity of motion in the direction of the field opposite to the sign of its charge. The velocity $v_i$ (cm/s) is given by

$$v_i = u_i \mathcal{E} \tag{8.1}$$

where $u_i$ is the particle's mobility and $\mathcal{E}$ is the electric field strength (V/cm). The resulting electric current density ($J_i$ in C/s-cm$^2$) is given by

$$J_i = v_i N_i q_i \tag{8.2}$$

where $N_i$ is the number of i particles per cubic centimeter and $q_i$ is the coulombic charge on each particle. Note that the sign of $v_i$ will change with the sign of $q_i$ so that J is positive for particles of either sign. Substituting Equation 8.1 in Equation 8.2, we obtain

$$J_i = \mathcal{E} N_i |q_i| u_i \tag{8.3}$$

where the absolute value of $q_i$ is required since the vector quantity $v_i$ has been eliminated.

## B. Ionic Conductivity

The total current density J is the sum of the individual current density charge carriers in the substance.

$$J = \sum_{i=1}^{n} J_i = \mathcal{E} \sum_{i=1}^{n} N_i |q_i| u_i \tag{8.4}$$

The electrical conductivity K has been defined as the current density per unit electric field, that is,

$$K \equiv \frac{J}{\mathcal{E}} \tag{8.5}$$

Combining Equations 8.4 and 8.5, we have

$$K = \sum_{i=1}^{n} N_i |q_i| u_i \tag{8.6}$$

From Equation 8.6 it is clear that any change in the substance that affects the concentration or mobility of any of the charge carriers affects the conductivity of the substance. Ionic concentration can be affected by shifts in chemical equilibrium between charged and uncharged species (such as a weak acid and the weak acid anion) or among variously charged species (such as a metal cation with a varying number of anionic ligands). Shifts in chemical equilibrium can be caused by titration with a reaction, or by changes in ionic strength, solvent, or temperature. These latter factors also affect the ionic mobility in significant and complex ways.

In chemical applications, the concentration units of mol/L are more useful than ions/cm$^3$. From Equation 8.3 and the relationship $N_i = C_iN_{av}/1000$, where $C_i$ is the molar concentration of i and $N_{av}$ is Avogadro's number,

$$J_i = \mathcal{E}\frac{C_iN_{av}}{1000}|q_i|\,u_i \tag{8.7}$$

The charge on an ion $q_i$ can be given by $q_i = Z_iq_e$, where $Z_i$ is the charge number of the ion and $q_e$ is the unit electron charge in coulombs. Also, the quantity $N_{av}q_e$ is defined as the faraday, F. Thus $|q_i| = |Z_i|\ F/N_{av}$. The individual ion current density can now be written

$$J_i = \mathcal{E}\frac{C_iF_{av}}{1000}|Z_i|u_i \tag{8.8}$$

and the conductivity, considering all species, is

$$K = \frac{F}{1000}\sum_{i=1}^{n} C_i|Z_i|u_i \tag{8.9}$$

In the case of a dissolved simple salt, the mobile charges are the cations and anions resulting from the solvation and dissociation of the salt. The conductivity is thus

$$K = \frac{F}{1000}\left(C_+|Z_+|u_+ + C_-|Z_-|u_-\right) \tag{8.10}$$

Electrical neutrality requires that the charge concentration for positive and negative charges be equal, that is,

$$C_+|Z_+| = C_-|Z_-| \tag{8.11}$$

If the salt is *completely* dissociated, the normality of the salt solution, C*, is equal to $C_+\ |Z_+| = C_-\ |Z_-|$. The equivalent conductance $\Lambda$ of a salt solution is defined as

$$\Lambda = \frac{1000K}{C^*} \tag{8.12}$$

whether the salt is completely dissociated or not. In Equation 8.12, C* is the normality of the salt solution, *not* just the concentration of the ions. Clearly, $\Lambda$ depends on the degree of dissociation of the salt, having a lower value for a lower degree of dissociation. The dissociation reaction can be studied by measuring $\Lambda$ as a function of C* [1,2].

The definition in Equation 8.12 does not require complete dissociation, but *if the salt is completely dissociated*, from Equation 8.10,

$$K = \frac{C^*F}{1000}(u_+ + u_-) \quad \text{complete dissociation} \tag{8.13}$$

Combining Equations 8.12 and 8.13 gives us

$$\Lambda = F(u_+ + u_-) \quad \text{complete dissociation} \tag{8.14}$$

Now the ionic equivalent conductance $\lambda$ can be defined for each ion.

$$\lambda_+ \equiv Fu_+ \qquad \lambda_- \equiv Fu_- \qquad \lambda_i \equiv Fu_i \tag{8.15}$$

Substituting the expressions for $\lambda_+$ and $\lambda_-$ into Equations 8.13 and 8.14, we have, for the case of complete dissociation,

$$K = \frac{C^*}{1000}(\lambda_+ + \lambda_-) \quad \text{complete dissociation} \tag{8.16}$$

and

$$\Lambda = \lambda_+ + \lambda_- \quad \text{complete dissociation} \tag{8.17}$$

Since $\lambda$ depends on the mobility of the ion, it is a function of the ion type and of the solution parameters (solvent, solute concentration, temperature, etc.). At very low solute concentrations, $\lambda$ reaches a steady "limiting" value, $\lambda^0$, that is often tabulated for the common ions in water solvent at specific temperatures. A table of $\lambda^0$ values is given in Table 8.1. The $\lambda^0$ values from such a table can be used to estimate the conductivity of a completely dissociated salt solution by application in Equation 8.10. For finite salt concentrations, the calculated values are indicative, but not exact. Onsager, Fuoss, and others [2,3] have developed equations by which quite accurate values of $\lambda$ can be calculated from $\lambda^0$ values under certain circumstances.

In analytical applications of conductometry, the sample often contains many species of ions, each contributing to the total conductivity. The total conductivity is then

$$K = \frac{1}{1000}\sum_{i=1}^{n} C_i^*\lambda_i = \sum_{i=1}^{n} C_i|Z_i|\lambda_i \quad \text{complete dissociation or actual ion concentrations} \tag{8.18}$$

where $C_i^*$ and $C_i$ are the normality and molarity, respectively, of the ionic species i in the solution. Thus, in using Equation 8.18 to calculate (or estimate) the conductivity of a solution, *all* ionic species present must be included. Also, Equation 8.18 shows that the variation in concentration or mobility of any of

**Table 8.1** Limiting Equivalent Conductance of Ions in Water at 25°C

| Cations | $\lambda^0_+$ | Anions | $\lambda^0_-$ |
|---|---|---|---|
| $H^+$ | 249.8 | $OH^-$ | 199.2 |
| $Li^+$ | 38.6 | $F^-$ | 55.4 |
| $Na^+$ | 50.1 | $Cl^-$ | 76.4 |
| $K^+$ | 73.5 | $Br^-$ | 78.1 |
| $Rb^+$ | 77.8 | $I^-$ | 76.8 |
| $Ag^+$ | 61.9 | $NO_3^-$ | 71.5 |
| $NH^{4+}$ | 73.5 | $ClO_3^-$ | 64.6 |
| $(CH_3)_2NH_2^+$ | 51.8 | $ClO_4^-$ | 67.4 |
| $(CH_3)_3NH^+$ | 47.2 | $IO_4^-$ | 54.5 |
| $(CH_3)_4N^+$ | 44.9 | Formate | 54.6 |
| $Mg^{2+}$ | 53.0 | Acetate | 40.9 |
| $Ca^{2+}$ | 59.5 | Benzoate | 32.4 |
| $Ba^{2+}$ | 63.6 | $SO_4^{2-}$ | 80.0 |
| $Cu^{2+}$ | 53.6 | $CO_3^{2-}$ | 69.3 |
| $Zn^{2+}$ | 52.8 | $Fe(CN)_6^{4-}$ | 111.0 |
| $La^{3+}$ | 69.7 | — | — |
| $Ce^{3+}$ | 69.8 | — | — |

*Source*: Reprinted in part from Reference 2, p. 463, with permission. Copyright 1968 Butterworths.

the ionic species will result in a change in K. A change in temperature can affect both ion concentration and mobility by affecting ion dissociation, complexation, and solvation equilibria as well as solvent viscosity. The combined effects are often complex and rarely negligible.

## C. Conductance

When a voltage source of E volts is connected across the contacts of conducting material as shown in Figure 8.1, the mobile charges move in the conducting material in response to the field. The electric field $\mathcal{E}$ applied to the conductor is $E/\ell$. From Equation 8.5 the resulting current density J is

$$J = \mathcal{E}K = \frac{EK}{\ell} \tag{8.19}$$

The total current I is current density times the area a:

$$J = Ja = \frac{EKa}{\ell} \tag{8.20}$$

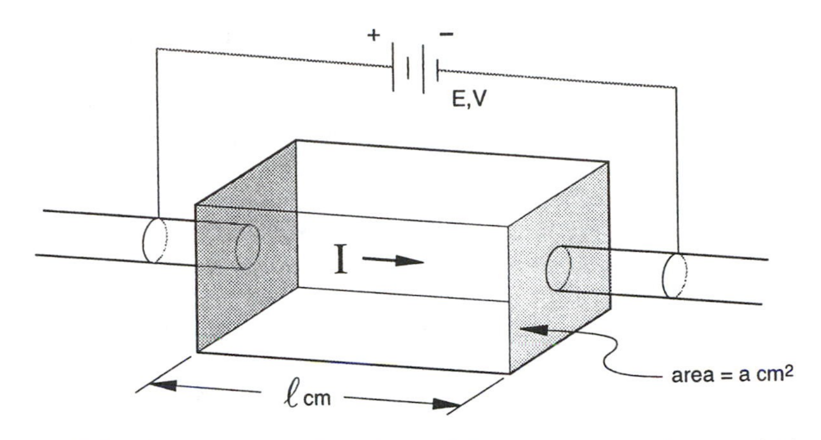


**Figure 8.1** The application of an electric field across a conducting material is accomplished by placing the material between metallic contacts and applying an electrical potential to those contacts.

From Equation 8.20 we see that the current in a conductor is proportional to the voltage across it. The proportionality constant $K a/\ell$ is defined as the conductance G:

$$G \equiv K \frac{a}{\ell} \tag{8.21}$$

The current-voltage relationship is now simply

$$I = EG \tag{8.22}$$

This relationship was first articulated by Ohm, who also defined the resistance R of a conductor as the reciprocal of the conductance,

$$R \equiv \frac{1}{G} \tag{8.23}$$

so that Ohm's law is also often written

$$I = \frac{E}{R} \qquad \text{or} \qquad E = IR \tag{8.24}$$

Another way to state Equation 8.24 is that the voltage across a resistor is proportional to the current through the resistor. These simple relationships are the basis of virtually all techniques of measuring conductance (or resistance); the measurement of the I/E ratio for a conductor yields G directly, while the E/I ratio is equal to R. The unit of R (volts/ampere) is the ohm ($\Omega$) and the unit of G (amperes/volt) is the mho ($\Omega^{-1}$).

In the case of electrolytic solutions, it is G that is measured; that is, the conductance of a cell filled with the sample solution. The conductance of an actual cell is related to the conductivity of the solution and the geometry of the cell as given by Equation 8.21. The conductance of the solution is proportional to the area a of the electrodes and inversely proportional to the distance between them, so that K is calculated from a measurement of G by $K = G\ell/a$, where the factor $\ell/a$ is often called the cell constant. Actually, the exact cell constant is rarely obtained by geometric measurement. It is generally determined empirically by measuring the conductance of a cell filled with a solution for which the conductivity is exactly known.

## II. DC CONTACT MEASUREMENT OF CONDUCTANCE

The conductance of some materials can be measured by the scheme shown in Figure 8.1 and Ohm's law. The current can be measured for a given applied voltage, or the voltage across the contacts can be measured when a known current is passed between them. This approach depends on a free and uniform transport of charge carriers through the interface between the metallic contacts and the test material. Since the charge carriers in the metallic contacts are electrons, the material to be tested also needs to be an electronic conductor. Some nonmetallic electronic conductors (such as semiconductors) may require special surface treatments at the contact interface to provide free electronic transport through it. There are, fortunately, many devices, substances, and measurements for which metallic contacts are sufficiently ideal and for which the simple relationships of Ohm's law can be used as the basis of measurement. The ohmmeters and Wheatstone bridge described in the next two parts of this section can be used where the metallic contact is effective. The consideration of special contact problems will conclude this section.

### A. Practical Ohmmeters

The resistance measurement concept used in a modern digital ohmmeter (or the "ohms" scales of a digital multimeter) is shown in Figure 8.2. The readout device is a digital voltmeter (DVM) composed of a fixed-range analog-to-digital converter and decimal display. The full-scale reading of the digital voltmeter is frequently 200.0 mV. A switchable constant-current source is applied across the parallel combination of the DVM and the unknown resistance. It is assumed that the DVM input resistance is so much larger than that of the unknown resistance that all of the current passes through the test resistance. If the constant current were exactly 1.000 mA, the voltage at the DVM input would be 1.000

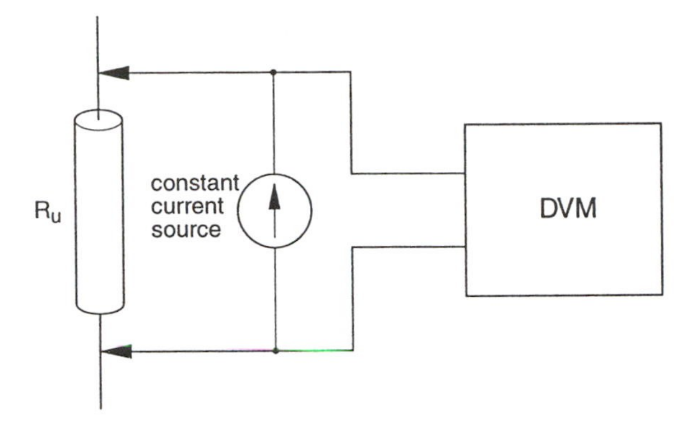


**Figure 8.2** The digital ohmmeter uses a constant-current source to produce a voltage across the unknown resistance proportional to its resistance, and a digital voltmeter to read that voltage.

V for each ohm of resistance in the unknown. If the full-scale reading of the DVM were 200.0 mV, the maximum resistance that could be measured would be 200.0 Ω. To measure higher values of resistance, smaller decade values of current are applied; to measure lower values of resistance without loss of significant figures, higher values of current are applied. The overall useful range of this approach is determined on the low end by the contact resistance of the probes and the maximum constant-current source available, and on the high end by the lowest practical constant-current source and highest practical DVM input resistance.

*Operational Amplifier Resistance-to-Voltage Converters*

For continuous laboratory monitoring of resistance, the operational amplifier current follower circuit offers an easy means of converting either resistance or conductance to a proportional voltage. In Figure 8.3a, $R_u$ is connected to the voltage source E, and the result is a current $i = E/R_u$ that is proportional to the conductance of $R_u$. The operational amplifier maintains an output voltage such that point S is at the common potential. Therefore, the input current, i, produces a voltage $e_o = iR_u$ at the current-follower output. Also, the connection of G-to-I converter circuit to point S is equivalent to a connection to common; that is, essentially no additional voltage or current is added to the input circuit. Since $i = E/R_u$ and $e_o = -iR_u$,

$$e_o = \frac{-ER_f}{R_u} = -ER_fG_u \tag{8.25}$$

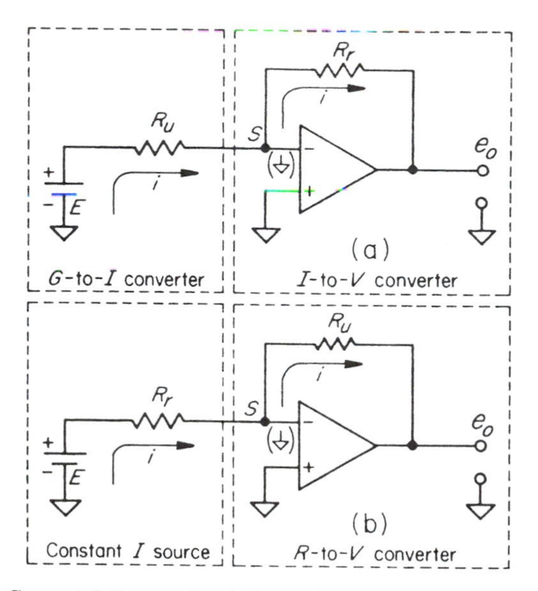


**Figure 8.3** Current-follower circuit for resistance measurement: (a) $e_o = -ER_r/R_u$; (b) $e_o = -ER_u/R_r$.

This results in an output voltage directly proportional to the conductance of $R_u$ within the output voltage and current capabilities of the operational amplifier.

In the circuit of Figure 8.3b, a fixed resistance $R_r$ is used in the input current-generating circuit. Since this same current passes through $R_u$ and the operational amplifier maintains point S at the common potential, the output voltage $e_o$ is equal to $-iR_u$. From

$$i = \frac{E}{R_r} \quad \text{and} \quad e_o = -iR_u \tag{8.26}$$

$$e_o = -\frac{E}{R_r} R_u \tag{8.27}$$

A very useful application of the circuit of Figure 8.3b is produced if we replace the feedback resistor $R_u$ with a resistive transducer such as a coated semiconductor thermistor. The commercial availability of small, rapid-response, chemically inert thermistors that have conveniently measurable resistances at temperatures of 200–600 K makes them an excellent choice as the transducer in chemical applications that require the rapid and accurate measurement of temperature. Unfortunately, a continuous current in $R_u$ may produce undesirable

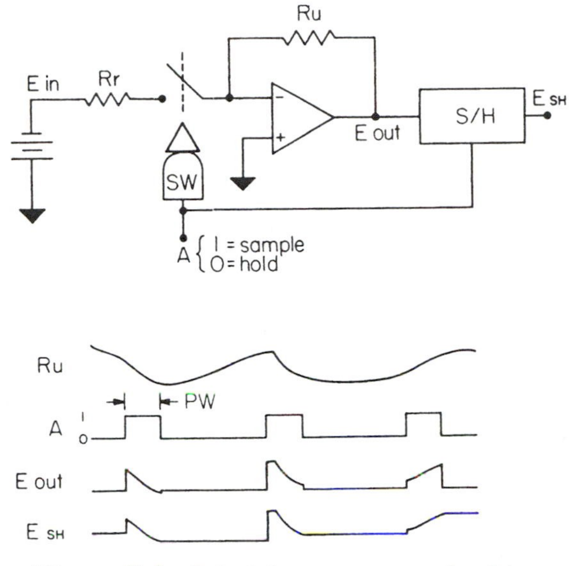


**Figure 8.4** Pulsed dc measurement of resistance.

Joule heating in the thermistor and its surroundings and thus cause an error in the measurement. This error may be reduced by utilizing a pulsed voltage source as illustrated in Figure 8.4. The logic signal A closes the electronic switch SW and produces a current pulse in $R_u$, the duration of which is designated by PW in the figure. The logic signal also switches the sample-and-hold module (S/H) into the sample mode and causes the transient analog signal that is proportional to $R_u$ to be held at the S/H output between pulses. The pulses need only be applied at the frequency of the desired temperature sampling rate and the output of the S/H module may easily be connected to an analog-to-digital converter for computer data acquisition.

## B. Null Comparison Measurement

In a null comparison measurement of resistance, the effect of an unknown resistance must be compared with the effect of a variable standard resistance under conditions as identical as possible. Therefore, the unknown and standard resistances are placed in identical circuits in such a way that the resulting voltage or current in each circuit can be compared. Then the standard is varied until the difference in voltage or current between the two circuits is zero. Several methods for performing this comparison have been devised, of which the Wheatstone bridge is by far the most common. Comparison methods for resistance measurement offer great accuracy, resolution, and relative independence of the electronic sources and detectors used.

*The Wheatstone Bridge*

The Wheatstone bridge shown in Figure 8.5 provides the most direct and best known circuit for comparison of unknown resistances against standard resistances. Resistances $R_A$, $R_B$, and R are standard resistance values that are used in the measurement of the unknown resistance $R_u$. Resistance R is made variable and is adjusted until the null detector indicates that the bridge is balanced.

When the circuit is at balance, there is no current through the null detector and no potential difference between terminals x and y. At balance, four significant conditions exist:

1. The current through $R_A$ and $R_u$ is $I_1$.
2. The current through $R_B$ and R is $I_2$.
3. $I_1R_u = I_2R$.
4. $I_1R_A = I_2R_B$.

Therefore, $R_u/R = R_A/R_B$ and

$$R_u = R\frac{R_A}{R_B} \tag{8.28}$$

It can be seen from Equation 8.28 that the unknown resistance $R_u$ is determined from the values of the three standard resistances R, $R_A$, and $R_B$. It is common practice to make the ratio $R_A/R_B$ some exact decimal fraction or multiple such as 0.01, 0.1, 10, or 100 and to refer to this ratio as the "multiplier."

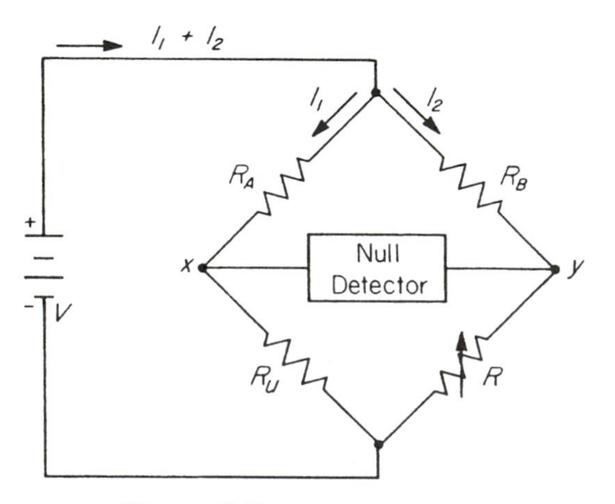


**Figure 8.5** Wheatstone bridge.

The resistance R is variable continuously or in small increments so that the dial reading of R times the multiplier setting equals the unknown resistance $R_u$. The main sources of error in a Wheatstone bridge are the inaccuracies of the three standard resistances R, $R_A$, and $R_B$ (as low as 0.001%), establishing the null point, and thermal electromotive force (EMF) values.

The magnitude of the off-balance indication of the null detector of a Wheatstone bridge is sometimes used as a measure of the change in resistance of the unknown resistor $R_u$. In this application, $R_u$ is frequently a light-, temperature-, or strain-dependent resistor. When matched devices are used for R and $R_u$, or $R_A$ and $R_u$, an off-balance output is obtained that is related to the ratio of the resistances of the two devices. This "comparative" resistance measurement with matched devices can tend to cancel the effects of resistance changes due to environmental factors other than the measured quantity. For example, the resistance of strain-gauge resistors depends on temperature and humidity as well as strain. If twin strain gauges are used in adjacent arms of a Wheatstone bridge with one gauge under strain but both in the same environment, the environmental effects will not affect the resistance ratio and only the strain effect will affect the off-balance output signal.

## C. Contact Considerations

In all the preceding examples of conductance or resistance measurement, it was assumed that the probe leads and the contacts to the measured conductor were ideal. An ideal lead and contact have zero resistance and no thermally generated voltages or uncompensated contact potentials. In a very wide range of measurements of electron-conducting devices, the ideal conditions are met within the desirable or practical error limits. However, the measurement of very low resistances can pose a problem in that the lead and contact resistance must be negligible compared to the resistance measured. Normal lead and contact resistance can be several tenths of an ohm, which limits 1% accuracy measurements to values greater than about 50 Ω.

The best technique for lower resistances is the four-contact method shown in Figure 8.6. The leads and contacts that measure the voltage across the resistor, caused by the current through the resistor, are not the same leads and contacts that supply the current. Since the current does not appear in the voltmeter leads or contacts, no error due to an iR voltage in the voltmeter contacts will occur.

Direct contact techniques do not work with electrolytic solutions. One cannot, for example, measure the resistance of salt water in a cell with an ordinary ohmmeter. This is due to the fact that the mobile charge carriers in an electrolytic solution are ions, not electrons. The conversion of ionic to electronic conduction at the electrode interface can only take place through an electrochemi-

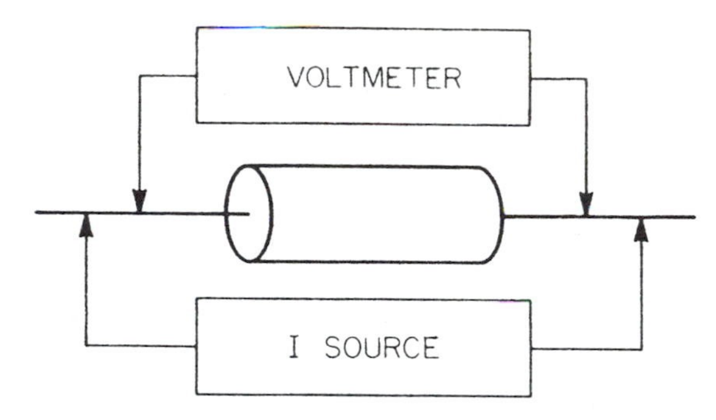


**Figure 8.6** Four-contact technique for low-R measurement.

cal oxidation or reduction. The establishment of such a process involves a substantial potential difference between the electrode and the solution and has finite impedance characteristics as well. If the electrodes used to supply the measurement current are separate from the iR drop measuring electrodes, as in the four-electrode scheme described above, the effects of the interfacial reactions at the current electrodes can be minimized. This approach is shown in Figure 8.7. The technique actually measures the conductance of the part of the solution that is in the capillary tubing between the two vials of the H cell. When a current is applied from the current source, this current must appear in the capillary section of the cell. The iR voltage in the capillary section is very much

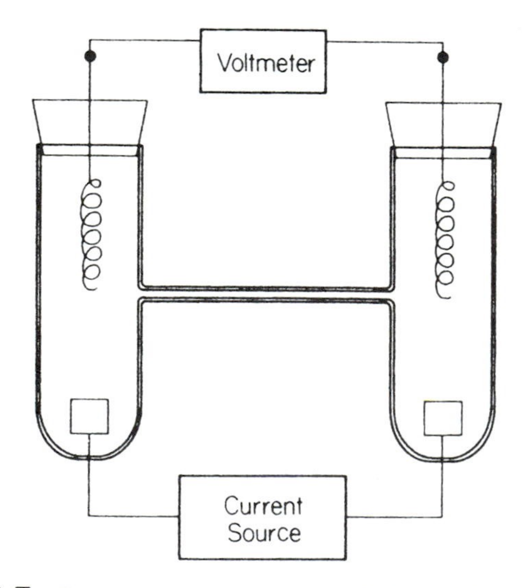


**Figure 8.7** Four-contact measurement of electrolytic conductance.

larger than in the vials because of the much smaller cross-sectional area of the capillary. Thus the voltmeter contacts to the solution in the vials measure the voltage (iR drop) developed in the capillary section. If the voltmeter probe contacts have stable and identical electrode-solution potentials, the interface voltages will cancel in the voltmeter circuit. Reference-type electrodes are often used for the voltmeter probes.

## III. CAPACITIVE CONTACT MEASUREMENT OF CONDUCTANCE

A method of avoiding the effect of potential differences arising at the electrode-solution interface is to take advantage of the capacitive behavior of the double layer at the electrode surface to make ac (alternating current) contact with the solution. To understand how this may be accomplished, it is necessary to consider a basic model of a conductance cell and examine its behavior under the influence of ac excitation. A review of ac circuit principles at a level sufficient for understanding the behavior of conductance cells and the instrumentation for conductance measurement is presented. The reader who desires a more thorough study of this topic is directed to material contained in the references [4–7].

### A. AC Excitation

At relatively low applied frequencies, a conductance cell may be represented as the double-layer capacitance $C_s$ in series with the solution resistance R, as shown in Figure 8.8a. When a sinusoidal voltage $e_s$ is applied to the series RC circuit, the instantaneous current i is the same in every part of the circuit and is given by

$$i = I \sin \omega t \qquad (8.29)$$

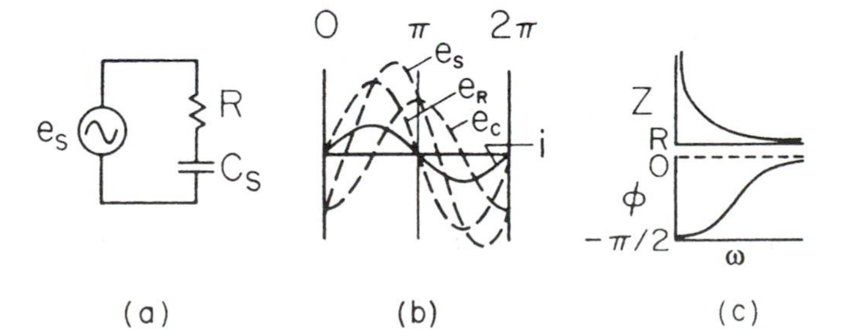


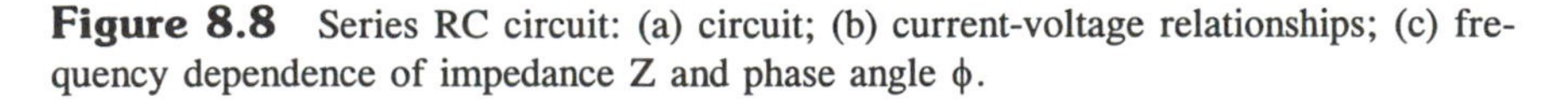
**Figure 8.8** Series RC circuit: (a) circuit; (b) current-voltage relationships; (c) frequency dependence of impedance Z and phase angle $\phi$.

where I is the maximum current, $\omega$ is the angular frequency (rad/s) ($\omega = 2\pi f$, where f is the frequency in Hz), and t is time (s). From Ohm's law the instantaneous voltage $e_R$ across R is then

$$e_R = iR = IR \sin \omega t \tag{8.30}$$

Since the instantaneous current in a capacitor is given by $i = C_s(de_C/dt)$,

$$e_C = \frac{1}{C_s}\int i\,dt = \frac{1}{C_s}\int I \sin \omega t\,dt \tag{8.31}$$

$$= \frac{I}{\omega C_s}(-\cos \omega t) \tag{8.32}$$

It follows that the maximum voltage $E_C$ across $C_s$ occurs when $-\cos \omega t = 1$ and therefore,

$$E_C = \frac{I}{\omega C_s} = I\frac{1}{\omega C_s} = IX_c \tag{8.33}$$

This expression has a form exactly analogous to Ohm's law in dc circuits. The quantity $1/\omega C_s$ is called the capacitive reactance $X_C$. It is a measure of the opposition to the flow of charge in a capacitor, and it is therefore measured in ohms. If we now substitute the definition of $X_C$ into the expression for $e_C$, it follows that

$$e_C = IX_C(-\cos \omega t) \tag{8.34}$$

Finally, applying the appropriate trigonometric identity, we obtain

$$e_C = IX_C \sin\left(\omega t - \frac{\pi}{2}\right) \tag{8.35}$$

A comparison of Equations 8.35 for $e_C$ and 8.30 for $e_R$ reveals two very important points: $e_C$ lags $e_R$ by $\pi/2$ (a quarter cycle or 90° of phase) and $e_C$ is directly proportional to $X_C$, while $e_R$ is proportional to R.

Capacitive reactance is a frequency-dependent quantity, decreasing with increasing frequency. Typically, double-layer capacitances for aqueous solutions are 10–100 $\mu F/cm^2$. Thus the capacitive reactance for a 1-$cm^2$ electrode with a 10-$\mu F$ capacitance at an applied frequency of $10^4$ rad/s (1.6 kHz) is

$$X_C = \frac{1}{\omega C} = \frac{1}{(10^4)(10^{-5})} = 10\,\Omega \tag{8.36}$$

The relationships among $e_R$, $e_C$, $e_s$, and i for the series RC circuit are illustrated in Figure 8.8b. Since the instantaneous voltages across R and C are additive, $e_s$ may be obtained by graphically summing $e_R$ and $e_C$. Note that the current in the circuit is in phase with $e_R$ but is out of phase with $e_C$ by $\pi/2$ (90°), and therefore the signal voltage is always between 0 and 90° out of phase with the current. This phase difference is called the phase angle.

The total opposition to the current in an ac circuit is referred to as its impedance Z, and in a simple RC network is given by the vector sum of R and $X_C$; that is,

$$Z = \sqrt{X_C{}^2 + R^2} = \sqrt{\left(\frac{1}{\omega C_s}\right)^2 + R^2} \tag{8.37}$$

From the preceding expression and the expressions for $e_C$ and $e_R$, we see that as the signal frequency is increased, $X_C$ decreases, Z approaches R, and G approaches 1/Z. The potential across $C_s$ also decreases gradually, and the phase angle $\phi$ between i and $e_s$ approaches zero, as is illustrated in the plot of $\phi$ versus $\omega$ in Figure 8.8c. For an RC circuit, there is a frequency for which $X_C$ = R. The reciprocal of this frequency is called the time constant $\tau$. When R = $1/\omega C$, $1/\omega$ = RC. Thus $\tau$ = RC, in seconds.

It is desirable to measure R (and thus G) at a high frequency in order to reduce $X_C$ to a negligible value compared to R. Unfortunately, other complications arise if the frequency is increased above a few kilohertz, and therefore other means must be devised to decrease $X_C$. A commonly used remedy is to increase the surface area and thus the capacitance of the electrodes as much as possible. A 100-fold area increase is obtained by platinizing the electrodes, that is, electrodepositing a layer of platinum black onto the platinum electrodes, usually from a solution of chloroplatinic acid [8].

A high-frequency limit for the applied potential is encountered above several kilohertz where the impedance of the conductance cell again begins to deviate from the resistance R. Since the solution medium itself is a dielectric situated between two parallel charged surfaces, it can assume the characteristics of a capacitor placed in parallel across the solution resistance as shown in Figure 8.9a. The magnitude of this capacitance is given by

$$C_p = \frac{10^9 D}{4\pi c^2 (l/a)} \tag{8.38}$$

where the quantities D, c, and l/a are the dielectric constant of the medium, the speed of light, and the cell constant, respectively. The dilute aqueous solution dielectric constant D is approximately 80, and if we assume a cell constant of

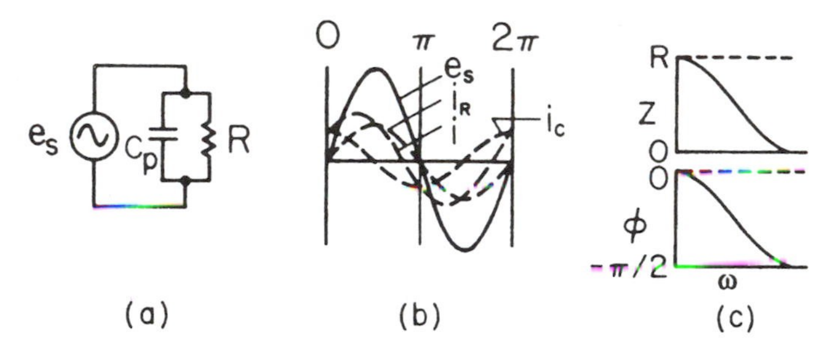


**Figure 8.9** Parallel RC circuit: (a) circuit; (b) current-voltage relationships; (c) frequency dependence of impedance Z and phase angle $\phi$.

1 $cm^{-1}$, the parallel capacitance is approximately 10 pF. At an applied frequency of $10^4$ rad/s (1.6 kHz) this capacitance results in $X_C = 10$ MΩ.

In order to assess the effect of $C_p$ on the conductance measurements, the analysis of the parallel RC network of Figure 8.9a may be carried out in a manner analogous to the series network discussed previously. The situation differs from the series circuit in that in this case, the *voltage* is the same across R and $C_p$, and the currents $i_C$ and $i_R$ in $C_p$ and R are different. The instantaneous currents are given by

$$i_C = \frac{e_s}{X_C} \qquad \text{and} \qquad i_R = \frac{e_s}{R} \tag{8.39}$$

The current in the resistor, $i_R$, is in phase with the applied voltage while the current in the capacitor leads the applied voltage by 90° as illustrated in Figure 8.9b. To obtain an expression for the total current i, we must find the vector sum of $i_C$ and $i_R$ as follows:

$$i = \sqrt{\left(\frac{e_s}{X_C}\right)^2 + \left(\frac{e_s}{R}\right)^2} = e_s\sqrt{\left(\frac{1}{X_C}\right)^2 + \left(\frac{1}{R}\right)^2} \tag{8.40}$$

The cell impedance is then

$$Z = \frac{e_s}{i} = \frac{1}{\sqrt{(1/X_C)^2 + (1/R)^2}} \tag{8.41}$$

Thus if $X_C \approx 10$ MΩ as already shown, the current path through $C_p$ only becomes appreciable (for a 1% measurement) when the solution resistance approaches 100 kΩ. Such situations are usually avoided by a judicious choice of cell constant and concentration of electrolyte in the solution of interest.

Unfortunately, other experimental factors, such as contact capacitance at the junction of the cell leads and the measurement system, lead capacitance, and capacitance due to the dielectric properties of the thermostatting medium, may contribute substantially to the parallel capacitance. These effects may be minimized by proper choice of cell design and use of oil rather than water in the thermostatting bath. The art of making ac conductance measurements has been refined to a high degree of precision and accuracy, and detailed discussions of the rather elaborate procedures that are often necessary are available [9,10].

The simple high- and low-frequency models of a conductance cell may be combined in the network shown in Figure 8.10a. The response of this network

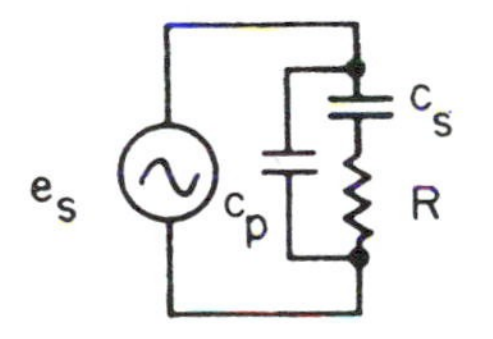

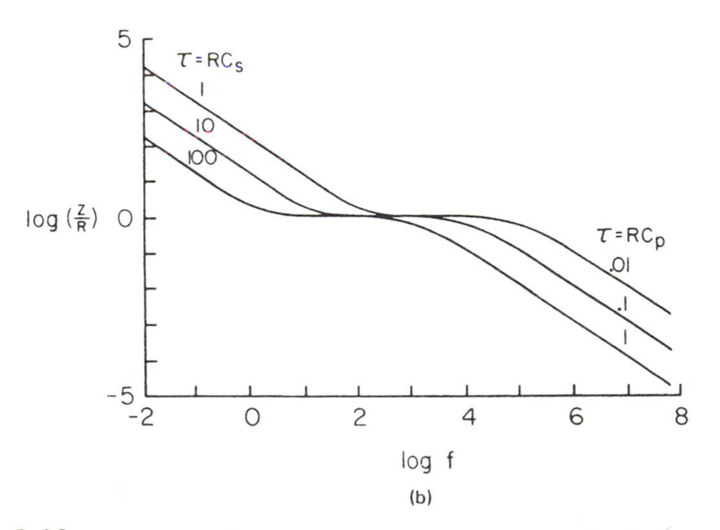


**Figure 8.10** Resistance with series and parallel capacitors: (a) circuit; (b) frequency dependence of impedance.

to ac excitation is illustrated in Figure 8.10b, in which the logarithm of the ratio of the total impedance to the resistance is plotted as a function of the logarithm of the frequency for several values of series and parallel RC time constants. It is obvious from the curves that the solution resistance is best measured over a fairly narrow midfrequency range or plateau. For some solutions this plateau may not exist at all. However, if the plateau can be located, the solution resistance is easily determined by measuring the cell impedance as discussed in the following section.

## B. Impedance Measurements

It is apparent from the previous discussion that in order to measure electrolytic conductance we must devise a means for compensating for the effects of $C_p$ and $C_s$. Figure 8.11 illustrates in block diagram form the elements of the impedance measurement process. Basically, a sinusoid is impressed across the conductance cell and the resulting signal is amplified and demodulated in such a way as to produce a signal proportional to the resistive component of the cell impedance. In the last step of the process, the signal is filtered to provide a dc voltage suitable for driving a convenient readout device.

The most straightforward method for measuring impedance is accomplished by choosing an excitation frequency for which $Z = R$ (i.e., $X_{C_s} >> R >> X_{C_p}$), thus making the demodulation step trivial. A very basic circuit for measuring impedance at such frequencies is illustrated in Figure 8.12 [11]. This circuit functions in much the same way as the current-follower circuit of Figure 8.3 except that diodes $D_1$ and $D_2$ have been added to rectify (demodulate) the signal resulting from the ac excitation of the cell [12]. As the input signal A drops below zero volts, the summing point of the operational amplifier drops a very small amount below ground potential. This causes the output to swing positive, reverse biasing $D_1$ and forward biasing $D_2$, causing the selected feedback resistor, 10R, to be in the feedback loop. At this point, and for the remainder of the negative half-cycle, the gain of the operational amplifier is $10R/Z_{cell}$.

As the input signal swings above 0 V, the output of the operational amplifier swings negative, causing $D_1$ to conduct and $D_2$ to block. This results in a zero-gain situation and the output of the circuit at point B becomes zero. The resulting rectified output is shown in waveform B. This waveform is filtered by the passive RC network at the output to provide a dc signal that may be fed to

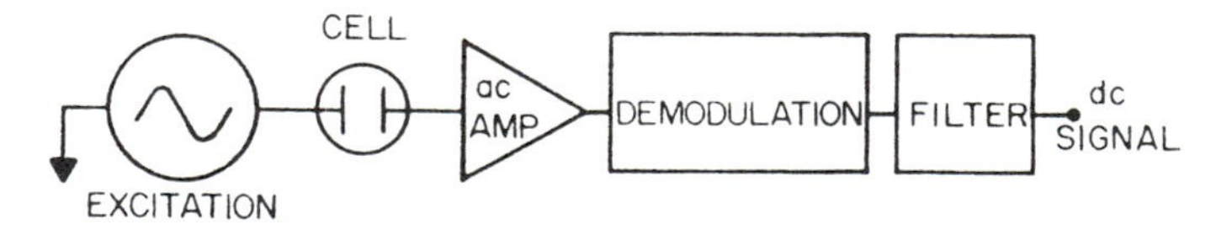


**Figure 8.11** Impedance measurement system.

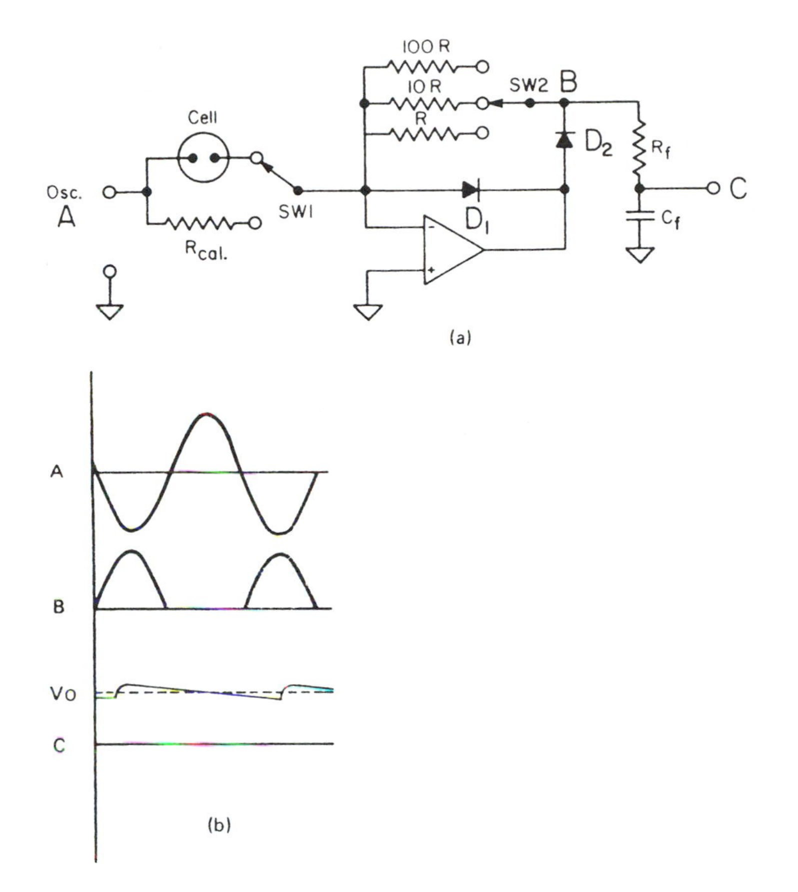


**Figure 8.12** Impedance measurement by phase-selective demodulation: (a) circuit diagram; (b) waveforms. [From Ref. 11.]

a strip-chart recorder, digital voltmeter, or other readout device. The switches SW1 and SW2 facilitate calibration of the output and selection of proper amplification for full-scale deflection of the readout device.

Since the output of the circuit is directly proportional to the conductance of the cell ($1/Z_{cell}$ in this case), the circuit may be used as a direct-reading detector for conductometric titration, reaction kinetic studies of fairly slow reactions, liquid chromatography, or ion-exchange chromatography.

The circuit may be easily constructed for a modest cost using inexpensive operational amplifiers and any of a number of integrated circuit oscillators that are commercially available. This basic piece of instrumentation is suitable for student laboratory experiments, conductometric monitoring of distilled water or other flowing streams, and other routine applications that do not require high precision or absolute accuracy. The instrument is limited to systems in which the capacitive effects are negligible.

Another technique that is useful at frequencies of less than ~10 kHz is phase-selective demodulation. A distinct advantage of this technique is that it enables the separation of the resistive (in-phase or "real") and capacitive (90°-out-of-phase or "quadrature") components of the cell impedance. This is accomplished through the process of cross-correlation [13] (selecting that component of $e_s$ that correlates with the phase of i). The excitation voltage waveform is multiplied by a square wave that is in phase with the cell current waveform.

We can understand how this is carried out by considering the waveforms of Figure 8.13a. At frequencies for which parallel capacitive components of the conductance cell impedance are negligible, sinusoidal excitation of the cell produces the waveforms of A, where $e_s$, $e_R$, $e_C$, and i have the same significance as previously discussed. In order to measure the real component of the impedance, the magnitude of the correlation integral cc must be determined.

$$cc = \int re_s \, dt \tag{8.42}$$

The function r is the bipolar square wave illustrated by waveform B, which has an amplitude of unity and is in phase with $e_R$. We shall demonstrate that the integral of Equation 8.42 is proportional to R over the first half-cycle. As an

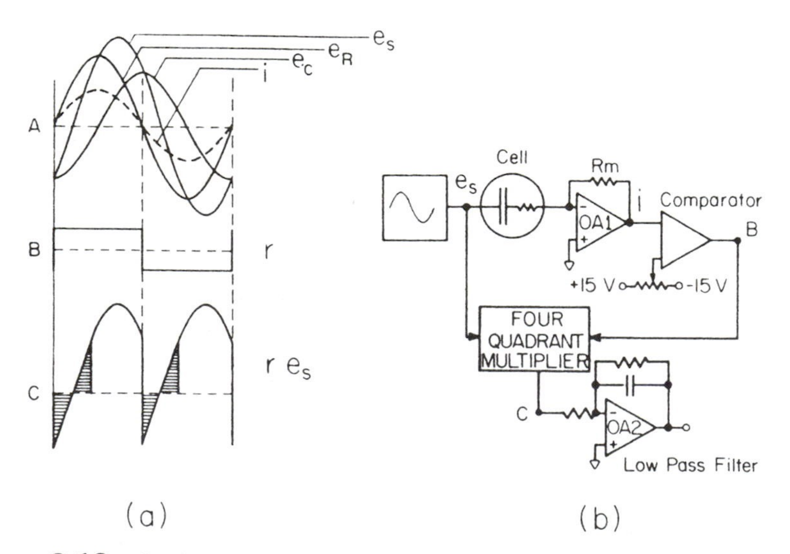


**Figure 8.13** Conductance measurement by phase-selective demodulation: (a) waveforms; (b) circuit schematic.

exercise, the reader may verify that this is true over the second half-cycle and that cross-correlation of $e_s$ with a quadrature square wave results in an integral that is proportional to $X_C$.

For r = 1 over the first half-cycle, the correlation integral is

$$cc = \int_0^{\pi}(+1)e_s\ dt \tag{8.43}$$

Since $e_s$ may be expressed as the sum of $e_R$ and $e_C$, we have the following:

$$cc = \int_0^{\pi}(+1)\left(IR \sin \omega t - \frac{I}{\omega C}\cos \omega t\right) dt \tag{8.44}$$

Separation of the sine and cosine terms gives

$$cc = IR\int_0^{\pi} \sin \omega t - \frac{1}{\omega C}\int_0^{\pi} \cos \omega t \tag{8.45}$$

$$= -\left(\frac{IR}{\omega}\cos \omega t\right)_0^{\pi} - \left(\frac{I}{\omega^2 C}\sin \omega t\right)_0^{\pi} \tag{8.46}$$

$$= \frac{2I}{\omega}R \tag{8.47}$$

Therefore, the magnitude of the correlation integral is proportional to the resistive component of the cell impedance at constant frequency. This is shown qualitatively in waveforms A to C of Figure 8.13a. Graphical multiplication of waveform B and $e_s$ in A gives the correlation waveform C. The shaded areas represent areas that cancel upon integration and the unshaded area represents $\int re_s\ dt$.

Instrumentally, the cross-correlation and integration may be easily carried out by the circuit shown in Figure 8.13b. The cell is arranged so that it is the input impedance to the familiar current follower. If a sinusoidal voltage $e_s$ is impressed across the cell, the output of operational amplifier OA1 is proportional to and in phase with the current i (see waveform A). Since i is in phase with $e_R$, we may use the output of OA1 to generate the bipolar square wave of waveform B. This is accomplished by the comparator, which is adjusted so that its output alternates between its positive and negative limits as the current waveform changes sign. The output of the comparator is connected to the four-quadrant multiplier, which carries out the multiplication of the square wave with $e_s$. The output of the multiplier is then filtered by the active low-pass filter to provide a dc output voltage proportional to R.

Circuitry similar to that presented in Figure 8.13b has been used to analyze cells with impedances ranging from $10^2$ to $10^{11}$ Ω with 1% accuracy and resolution better than 1 part in $10^4$ over a frequency range of 0.005 Hz to 10 kHz [14]. The technique has been especially useful for studies of the reaction kinetics of moderately fast chemical reactions. Kadish et al. [15] used phase-selective techniques to make ac impedance measurements to evaluate reference electrodes for use in nonaqueous solvents. Recent decreases in the cost of integrated function modules such as analog multipliers, oscillators, and phase-locked loops make this type of phase-selective instrumentation more accessible than ever.

Traditionally, the instrument of choice for accurate conductance measurements that are relatively free of capacitance effects has been the ac Wheatstone bridge illustrated in Figure 8.14. The details of operation and the derivation of the balance condition of the ac bridge are presented in considerable detail elsewhere [16,17]. The balance condition is exactly analogous to that of the dc bridge except that impedance vectors must be substituted for resistances in the arms of the bridge when reactive circuit elements are present.

At balance, the points at d and b on the ac bridge must be equal in magnitude as well as in phase. The simplest method for determining when this condition exists is to use an oscilloscope connected as shown in the figure as the null detector. If precautions are taken to ensure that the excitation and the null signal are electronically isolated, the balance condition is easily, albeit slowly, obtained [16].

Assuming that the bridge is operated at frequencies for which the parallel capacitance of the cell has negligible conductance, the bridge will balance when $R_s = R_{cell}$ and $C_s = C_{cell}$. Bridges that have been carefully calibrated against known standard resistors may achieve accuracies of 0.01%. With careful prepa-

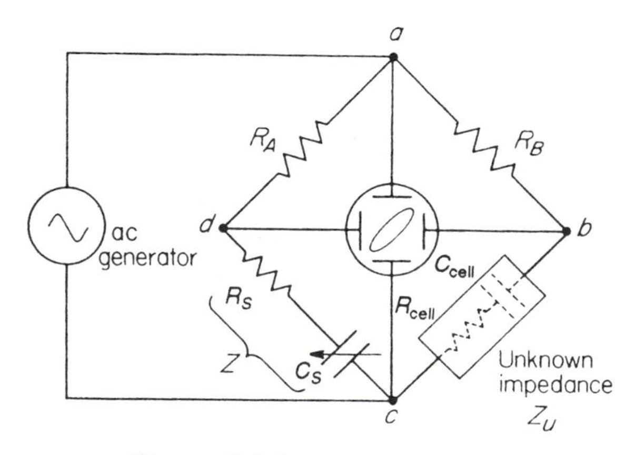


**Figure 8.14** Impedance bridge.

ration of solutions and control of experimental conditions, the measurement of conductances by this method ranks among the most accurate and precise of all electrochemical techniques.

## C. Bipolar Pulse Measurements

The bipolar pulse technique for measuring solution resistance minimizes the effects of both the series and parallel cell capacitances in a unique way. The instrumentation for this technique is illustrated in Figure 8.15. The technique consists of applying two consecutive voltage pulses of equal magnitude and pulse width but of opposite polarity to a cell and then measuring the cell current precisely at the end of the second pulse [18].

The pulses are provided by a precision bipolar voltage source, which is switched into the input of the pulsing amplifier by the switch at point A in the circuit. A very accurate crystal-controlled timing circuit (not shown) drives the switch to ensure that the pulses are symmetrical. The pulsing amplifier inverts the signal as shown by waveform B and supplies current to the cell. The cell current is amplified by the current follower, the output of which is illustrated by waveform C.

At the beginning of the first pulse, any parallel capacitance in the cell is rapidly charged, producing the small spike in waveform C. As long as $t_1 << R_sC_s$, the potential developed across $C_s$ will be small compared to the excitation potential $E_{in}$. During $t_1$, the current will decrease slightly as the poten-

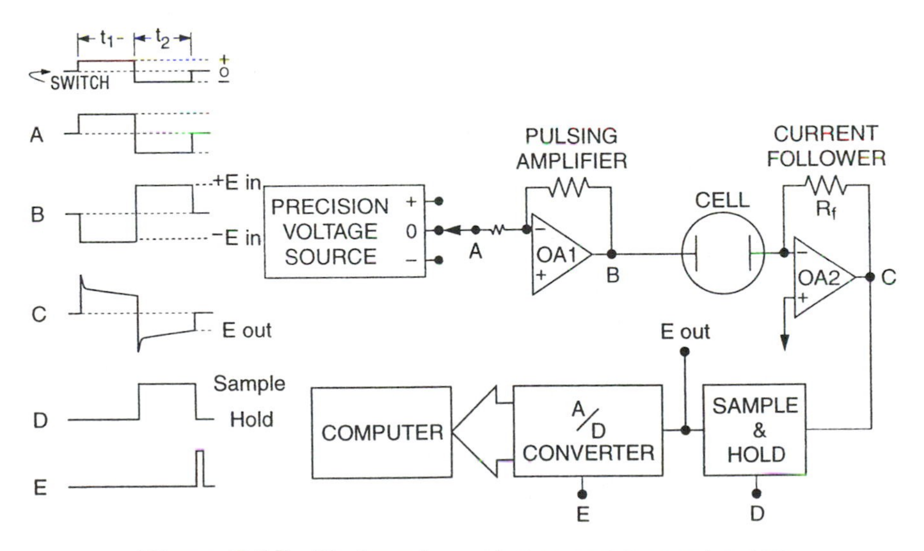


**Figure 8.15** Bipolar pulse conductance measurement system.

tial across $C_s$ increases. At the end of the first pulse, the polarity is quickly reversed, again resulting in the small spike due to charging of the parallel capacitance. Assuming that $t_1 = t_2$, the double layer is exactly discharged during the second half of the cycle, and the cell current at the end of $t_2$ is due to the resistive component of the cell impedance alone.

The sample-and-hold amplifier samples the output of OA2 during $t_2$ as shown by waveform D and holds the signal $E_{out}$ at the exact end of the second pulse. The falling edge of signal D fires a monostable multivibrator generating a trigger pulse (E) for the analog-to-digital conversion of the sample-and-hold output. Between pulses the cell is held at ground potential in order to eliminate any spurious currents.

The resistance of the cell is calculated from the following expressions. Since, at the end of $t_2$,

$$i_{cell} = \frac{E_{out}}{R_f} \qquad \text{and} \qquad R_s = \frac{E_{in}}{i_{cell}} \tag{8.48}$$

we have

$$R_s = \frac{E_{in}}{E_{out}} R_f \tag{8.49}$$

If the pulse widths of the two pulses are equal to within 1% and the double-layer voltage can be kept to less than 1% of the applied voltage by keeping the pulses short (i.e., 10 $\mu$s to 10 ms in duration), the conductance may be determined to within 0.01% [18].

The advantages of the bipolar pulse technique include speed (discrete measurements at a rate as high as 30 kHz), accuracy, and signal-to-noise ratio. The system has been employed as a detector in automated conductometric titrations and in stopped-flow mixing systems with excellent results.

In another variation of the bipolar pulse technique, a bipolar current pulse is applied to a conductance cell, and the voltage is sampled at the end of the second pulse [19]. Analogously, the solution resistance is calculated from $R_s = E_{measured}/i_{applied}$. This technique has also been applied with good success to chemical problems similar to those mentioned above.

Applications of the bipolar pulse technique have demonstrated its utility in a variety of experiments, but it is particularly useful in monitoring reaction kinetics [18]. The technique has been shown to be useful on the stopped-flow time scale by the investigation of the dehydration of carbonic acid [20]. The study of this widely used text reaction demonstrates the accuracy and precision of the method. A sample data set from a single experiment is shown in Figure 8.16, and the excellent precision obtainable in such experiments is evident. The

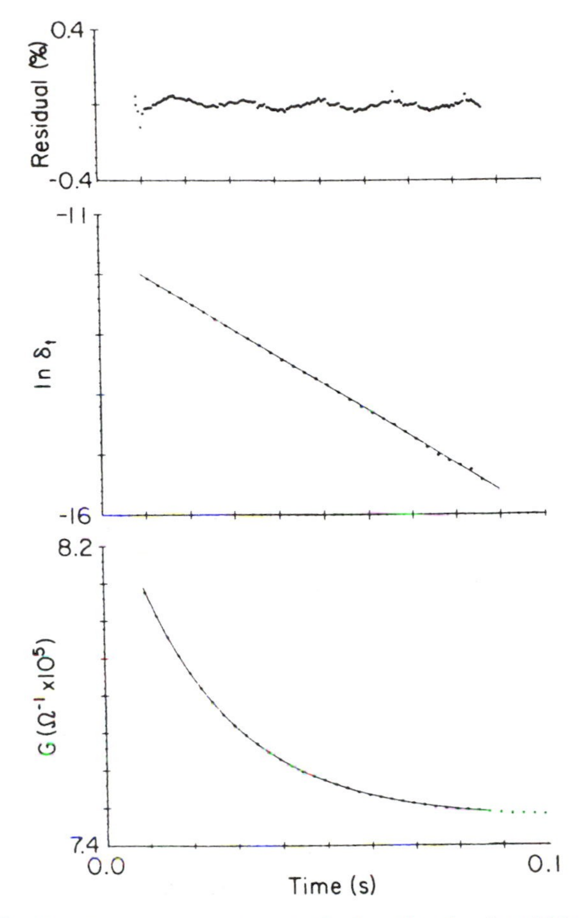


**Figure 8.16** Reaction curve for the dehydration of carbonic acid by conductometric detection. Lower: dots, every tenth experimental point; solid line, fitted exponential curve. Middle: dots, logarithmic plot, every tenth point; solid line, least-squares line. Upper: residuals ($G - G_{calcd}$). [From Ref. 20, reprinted with permission. Copyright 1978 American Chemical Society.]

standard error of the estimate of the fitted first-order logarithmic plot (middle curve) amounts to 0.2% of the conductance change or 0.015% of the average conductance during the course of the reaction. Using a computer-controlled instrument with simultaneous measurement of conductance and temperature, it is also possible to correct conductance measurements to constant temperature.

More recently, the kinetics of the urease-catalyzed decomposition of urea were investigated using a wide-range bipolar pulse instrument capable of both current and voltage pulse modes with either integrated or sampled data acquisition [21]. With this instrument, conductance accuracy of 1% was attained over the range of $5 \times 10^{-9}$ mho to 10 mho. Accuracy of 0.001–0.01% was achieved over the range of $10^{-6}$ mho to 1 mho while measurements were made in as few as 32 μs.

A novel application of the bipolar pulse technique that introduces a degree of selectivity into conductometric methods is found in the measurement of the conductance of ion-selective electrodes as a means of rapidly measuring ion concentrations [22]. For example, working curves of conductance versus log concentration for $Ca^{2+}$ are linear over at least four decades of concentration with a detection limit of $10^{-6}$ M. The response time for measuring the conductance of a calcium ion-selective electrode is about 10 ms. Other applications of the bipolar pulse technique, including the measurement of critical micelle concentrations [23] and high-precision conductometric titrations [24], have appeared in the literature, but none is more significant than its use as a detector in ion chromatography [25]. A large fraction of the commercial conductometric detectors for ion chromatography are based on the bipolar pulse method. The advent of the bipolar pulse conductance technique has generated renewed interest in one of our oldest instrumental methods.

## REFERENCES


1. J. E. Prue, *Ionic Equilibria*, Topic 15, Vol. 3 of *International Encyclopedia of Physical Chemistry and Chemical Physics*, Pergamon Press, New York, 1966.
2. R. A. Robinson and R. H. Stokes, *Electrolyte Solutions*, 2nd ed., revised, Butterworth, Sevenoaks, Kent, England, 1968.
3. R. M. Fuoss and L. Onsager, *J. Phys. Chem. 66*:1722 (1962); *67*:621 (1963); *68*:1 (1964).
4. H. V. Malmstadt, C. G. Enke, and S. R. Crouch, *Electronics and Instrumentation for Scientists*, Benjamin-Cummings, Menlo Park, CA, 1981, pp. 115–121.
5. H. V. Malmstadt, C. G. Enke, and E. C. Toren, Jr., *Electronics for Scientists*, Benjamin-Cummings, Menlo Park, CA, 1962, pp. 545–581.
6. R. D. Sacks and H. B. Mark, Jr., *Simplified Circuit Analysis*, Marcel Dekker, New York, 1972, Chap. 1.
7. A. G. Diefenderfer, *Principles of Electronic Instrumentation*, 2nd ed., W. B. Saunders, Philadelphia, 1979, Chap. 2.
8. D. F. Evans and M. A. Matesich, in *Techniques of Electrochemistry* (E. Yeager and A. J. Salkind, eds.), Wiley-Interscience, New York, 1973, p. 49.
9. *Ibid.*, p. 30.
10. T. Shedlovsky, in *Techniques of Organic Chemistry*, 2nd ed., Vol. I, Part II (A. Weissberger, ed.), Interscience, New York, 1959.
11. C. N. Reilley, *J. Chem. Educ. 39*:A853 (1962).
12. J. I. Smith, *Modern Operational Circuit Design*, Wiley-Interscience, New York, 1971, Chap. II.
13. Ref. 4, p. 420.

14. A.J. Bentz, J.R. Sandifer, and R.P. Buck, *Anal. Chem.* *46*:543 (1974).
15. S. Cai, T. Malinski, X. Lin, J. Ding, and K. M. Kadish, *Anal. Chem.* *55*:161 (1983); K. M. Kadish, S. Cai, T. Malinski, J. Ding, and X. Lin, *Anal. Chem.* *55*:163 (1983).
16. J. Braunstein and G. D. Robbins, *J. Chem. Educ.* *48*:52 (1971).
17. Ref. 1, p. 481.
18. D. E. Johnson and C. G. Enke, *Anal. Chem.* *42*:329 (1970).
19. P. H. Daum and D. F. Nelson, *Anal. Chem.* *45*:463 (1973).
20. K. J. Caserta, F. J. Holler, S. R. Crouch, and C. G. Enke, *Anal. Chem.* *50*:1534 (1978).
21. R. K. Calhoun, F. J. Holler, R. F. Geiger, Jr., T. A. Nieman, and K. J. Caserta, *Anal. Chim. Acta* *252*:29 (1991).
22. C. R. Powley and T. A. Nieman, *Sel. Electrode Rev.* *10*:185 (1988).
23. J. Baxter-Hammond, C. R. Powley, K. D. Cook, and T. A. Nieman, *J. Colloid Interface Sci.* *76*:434 (1980).
24. M. E. Hail and F. J. Holler, *Mikrochim. Acta* *111*:295 (1986).
25. J. M. Keller, *Anal. Chem.* *53*:344 (1981).

# 9

# Electrochemical Cells

**Fred M. Hawkridge** *Virginia Commonwealth University, Richmond, Virginia*

## I. DESIGN CONCEPTS

### A. Introduction

This chapter deals with the design and use of electrochemical cells for cases in which the dimensions that define the electrode area (e.g., diameter, length, width) are large relative to the dimensions of the diffusion layer. The design of electrochemical cells for use with microelectrodes (i.e., electrodes with dimensions on the order of micrometers) may require consideration of different characteristics and are considered in Chapter 12.

During a nonexhaustive "semi-infinite" electroanalytical experiment, the initial concentration of the electroactive material remains unchanged at some distance into the solution. Under these conditions the solution might just as well extend an infinite distance from the working electrode surface, since the chemical changes that occur are confined to a small volume adjacent to the interface. Nevertheless, the electrochemical cell is an important part of the total experimental system (the control instrumentation, cell, and sample), and its characteristics must be understood to ensure that meaningful data are obtained. The physical design of the cell and the materials used in its construction must be chosen with both the nature of the sample and the experimental objectives in mind. A configuration that yields valid results under one set of conditions might cause experimental artifacts when the objectives are changed.

The investigator must define the experimental goals and select the technique(s) most appropriate to satisfy them. Viewing the system as a whole, one can then optimize the characteristics of the electronics, the cell, and the sample. The geometry of the cell body and each electrode, as well as the placement of the electrodes within the cell, are important factors contributing to overall cell performance. The transfer function of the cell must be compatible with both the electronics and the electrochemical technique used.

The controlled-potential three-electrode apparatus can be conveniently used to discuss the matching of the cell, the sample, and the instrument. Controlled-potential experiments are common and the following discussion is relevant to many electroanalytical techniques. The operation of a potentiostat is discussed in Chapters 6 and 7, and the reader should be familiar with the characteristics of potentiostats. In short, feedback of an error signal to the input of the potentiostat maintains control of the potential difference between the working and reference electrodes (Chap. 6).

An equivalent circuit of the three-electrode cell discussed in Chapters 6 and 7 is illustrated in Figure 9.1. In this simple model, $R_r$ is the resistance of the reference electrode (including the resistance of a reference electrode probe, i.e., salt bridge), $R_c$ is the resistance between the reference probe tip and the auxiliary electrode (which is compensated for by the potentiostat), $R_u$ is the uncompensated resistance between the reference probe and the working-electrode interphase ($R_t$ is the total cell resistance between the auxiliary and working electrodes and is equal to the sum of $R_c$ and $R_u$), $C_{dl}$ is the double-layer capacitance of the working-electrode interface, and $Z_f$ is the faradaic impedance of the electrode reaction.

Electrochemical techniques can be broadly divided into two categories based on whether the time is long and the current is low or the time is short and the current is high. It is the latter category that presents major challenges in the design of an electrochemical cell.

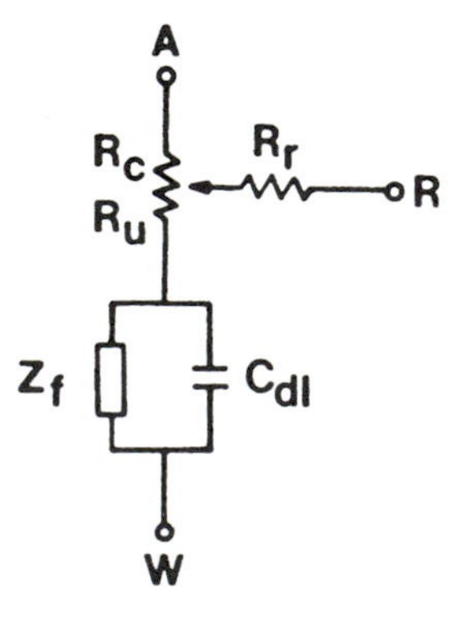


**Figure 9.1** Equivalent circuit of an electrochemical cell. A, Auxiliary electrode; R, reference electrode; W, working electrode; $R_c$, compensated resistance; $R_u$, uncompensated resistance; $R_r$, reference electrode impedance; $Z_f$, faradaic impedance; $C_{dl}$, double-layer capacitance.

## B. Low-Current Conditions

Low-current experiments present few problems with regard to cell construction. Unless there are chemical complications due to substances generated at the auxiliary electrode or released from the reference electrode probe, all three electrodes can be placed in a single compartment. Design of the "ideal" cell for accurate potential control is explained using the equivalent circuit shown in Figure 9.1. The total cell resistance $R_t$ is essentially the solution resistance between the working and auxiliary electrodes and is obviously dependent on the distance between the two electrodes, the electrode areas, and the solution conductivity. When a current i exists between the auxiliary and working electrodes, the potential at any point just outside of the working-electrode interface will be affected by the potential drop between the point and the auxiliary electrode (see Chap. 6). A symmetrical arrangement of the auxiliary and the working electrodes should be used where possible, to keep the potential and current density the same at all points along the working electrode surface.

Single-surface planar electrodes placed parallel to each other provide an ideal arrangement. Cylindrical (Pt wire) or spherical (HMDE) working electrodes placed in the center of a larger concentric auxiliary electrode (usually a metal screen or a wire loop) will be equally satisfactory for low-current applications in solutions of high conductivity. The $iR_t$ is small along any line from A to W. Variations from point to point along the working-electrode surface are therefore negligible. As indicated in Figure 9.1, the placement of the tip of the reference probe divides $R_t$ into two resistances, compensated and uncompensated. The reference probe tip *should* be placed close to the working electrode on a line *between* the working and auxiliary electrodes. Nevertheless, in very low-current experiments (common in analytical electrochemistry) the effect of electrode placement is usually negligible because the iR drop throughout the bulk solution may be less than 1 mV.

The potential drop in solution between the reference electrode probe tip and the auxiliary electrode, $iR_c$, is corrected for by the operation of the potentiostat (see Chap. 6 if this is not understood). The potential drop existing between the reference probe tip and the working electrode, $iR_u$, remains as an error in the accuracy of potential control. If the probe tip is too large or is placed too close to the working-electrode surface, it will interfere with mass transport of the electroactive material to the working electrode, an effect called "shielding" of the working electrode [1,2]. It is possible to correct for the uncompensated iR drop mathematically or by positive feedback electronic techniques, but the correction is rarely significant in low-current measurements (see Chap. 7).

Often the chemistry of an experiment dictates that the cell have special features. For example, *products generated at the auxiliary electrode or materials leaking from the reference probe will occasionally interfere with the reac-*

*tion at the working electrode*. In such cases, the auxiliary electrode and the reference electrode must be isolated from the working electrode compartment. The auxiliary electrode and the reference electrode are often isolated by a semi-permeable barrier (e.g., a glass frit, an ion-exchange membrane, "thirsty" glass, an asbestos fiber junction, a platinum glass seal) from the working electrode compartment. This restricts mass transport between the electrode compartments and yet permits electrical contact. For long-term experiments where interference is most likely, an intermediate solution compartment or salt bridge can be employed [3]. Use of strong electrolyte solutions and placement of the reference probe tip on the line of closest approach between the other two electrodes minimizes potential gradients along the working electrode surface.

The accuracy of working electrode potential control can be established by introducing an independent reference electrode into the cell, with the probe tip positioned about the same distance from the working electrode surface as the original reference electrode probe, and also on a line between working and auxiliary electrodes. The experimentally applied potential may then be compared with that measured between the working electrode (usually at real or virtual ground) and the independent reference electrode. This check is usually adequate for cases where the product $iR_u$ is small, such as is true in low-current experiments in strong electrolyte solutions. Correction of the working electrode potential error can be made by determining the magnitude of $iR_u$ experimentally or by instrumental compensation for $R_u$ (see Chap. 7).

## C. High-Current Conditions

High-current conditions can exist under constant-potential control when the working electrode is large, the concentration is high, the solution is in motion, or the potential is rapidly changed from one value to another (i.e., usually requiring high nonfaradaic current levels for double-layer charging). In order for the potentiostat to control the working-electrode potential, the current between the auxiliary and working electrodes must not overload the capability of the control amplifier at the applied potential.

A typical example would be a coulometric controlled-potential experiment, where initially a large current exists between the auxiliary and the working electrode (the load of the control amplifier). If a potentiostat is rated to have a maximum output of ±20 V at 1 A, it cannot supply more than 20 W of power [power (watts) = current (amperes) × potential (volts)]. If $R_t$ were 100 Ω, the potentiostat would *not* be able to control the potential of the working electrode at −2.0 V (or any other potential for that matter) if 0.5 A were demanded. At least 25 W ($I^2R_t$) of power would be required of the potentiostat for potential control to be maintained. As a result of our 5-W deficiency, the potential of the working electrode would be uncontrolled at a value less than the −2.0 V; less than 0.5 A, in fact, would pass through the cell.

Accurate potential control in an electrolysis is often necessary to limit the electrode reaction(s) to the one(s) of interest. Failure of the potentiostat in the preceding case could be solved by using a lower sample concentration or a smaller working electrode area in order to decrease the current demand. Decreasing the resistance $R_t$ between the auxiliary and working electrodes, by increasing the conductivity of the medium, might also lower the power ($I^2R_t$) required to operate the cell to a value within the potentiostat's capability. If a diffusion barrier is used to isolate the auxiliary electrode, it will usually contribute significantly to $R_t$. It might then be desirable to use a larger barrier or a material with a larger pore size. As should now be obvious, there are usually a number of ways to get around the problem of too much power demand.

The example above points out that *it is the potentiostat's voltage swing, not its current capability, which often limits its performance.* If we had used a 40-W potentiostat (±20 V at 2 A) the experiment might still have failed, whereas a 25-W model (±50 V at 0.5 A) could do the job.

Some attention must be paid to the speed at which a potentiostat can change the controlled potential from one value to another when using any of a variety of fast-sweep or potential-step experiments. The rise time of a potentiostat intended for high-current applications should be experimentally determined by measuring the time required to charge a realistic resistive-capacitive "dummy cell" following a known potential step. *Statement of an experimentally determined rise time without reference to the specific load driven is meaningless.* The experimental factors that limit the rise time of a potentiostat are complicated and have been examined in some detail in Chapter 7. The power of the potentiostat, the phase shift of the feedback signal with respect to the output signal of the control amplifier, the characteristics of the cell and sample, and the magnitude of $C_{dl}$ all influence the actual rise time of the interfacial potential excursion.

When the potential of the working electrode is stepped from one value to another, large amounts of current must be supplied in a very short time to charge the double-layer capacitance, $C_{dl}$ in Figure 9.1, before the final potential difference across $C_{dl}$ can be controlled and the faradaic current can commence. The apparatus and the experimental design must be matched to make maximum use of the rise time of the potentiostat. Experimentally, there is an advantage in using smaller potential steps, since large potential steps may be accompanied by large changes in the double-layer capacitance. Using the voltammogram in Figure 9.2 as an example, one would prefer to perform a potential step from an initial value B to a final value C rather than from A to C. Small working electrode areas are often desirable, because $C_{dl}$ is directly proportional to the area of the working electrode. For fast potential-step experiments, a working electrode area of about 0.01 $cm^2$ is often a reasonable compromise.

It is important to realize that a potentiostat's rise time using a dummy cell cannot be used as evidence for accepting a measurement made on a real sample.

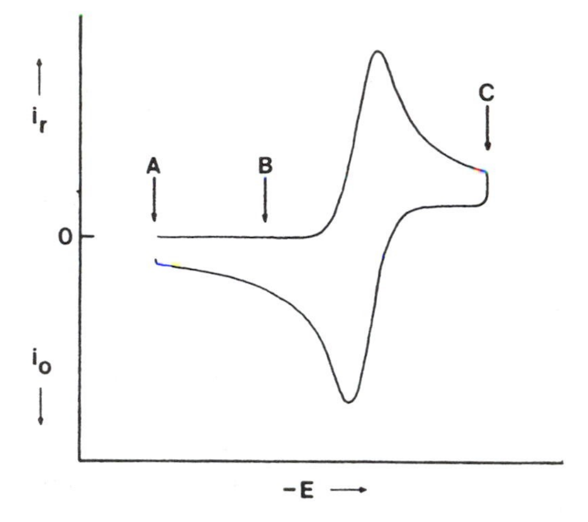


**Figure 9.2** Model cyclic voltammogram. A, B, initial applied potentials; C, step potential.

Moreover, establishing the optimum placement of the electrodes and their geometry ought to be performed while monitoring the effect these changes have on the response for the real sample. The time required for double-layer charging and therefore control of the working electrode potential should be determined by monitoring both the current and the potential in a potential-step experiment on a real sample. Reference electrode probe-tip geometry is particularly critical in the optimization of cell design for very fast potential-step experiments [1,2]; however, these experiments are rarely performed by analytical electrochemists.

## II. STATIONARY-SOLUTION EXPERIMENTS

Numerous electroanalytical techniques are performed on "quiet" or stationary solutions. The ideal stationary solution has no convective motion due to vibrations, mechanical stirring, motion of the electrode(s), temperature gradients, or density gradients. At long times only "natural convection" effects arising from the electrode reaction itself will contribute to nonideality.

Three popular cell designs are shown in Figure 9.3. These cells can be scaled to fit the available sample volume and the size and type of electrodes to be used. The cell may include a stationary semimicro working electrode [e.g., platinum wire, bead, or button; dropping mercury electrode (shown in Fig. 9.3C), hanging mercury drop electrode, carbon paste (shown in Fig. 9.3A), graphite, etc.], although simple modification could allow for rotating electrodes.

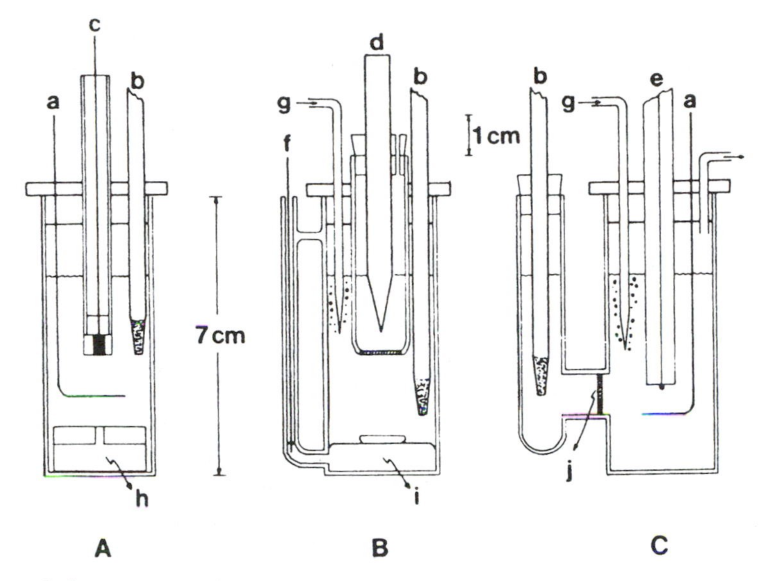


**Figure 9.3** Stationary solution voltammetry cells. (a) Platinum wire loop auxiliary electrode, (b) reference electrode or reference electrode probe tip, (c) carbon paste working electrode, (d) graphite auxiliary electrode, (e) dropping mercury electrode, (f) platinum wire contact to mercury pool working electrode, (g) nitrogen gas inlet tube, (h) magnetic stirrer, (i) mercury pool working electrode, (j) glass frit isolation barrier.

The auxiliary electrode shown in Figure 9.3A and C is a platinum wire loop concentric with the working electrode. A platinum wire or foil, a mercury pool, and carbon can also be used for this purpose. The graphite rods used for emission spectroscopy are useful as auxiliary electrodes. For optimum performance at short times, the position of the reference probe might have to be changed from that shown when using different auxiliary electrodes.

Physical isolation of the auxiliary electrode can be essential, unnecessary, or even detrimental to the experimental goal. It is important to recognize which situation exists for a given experiment. In general, isolation is essential for exhaustive coulometric experiments, unnecessary for nonexhaustive experiments (i.e., voltammetry), and detrimental to fast experiments (i.e., chronoamperometry). In the normal stationary-electrode voltammetry experiment, so little material is actually converted at the electrodes that the possibility for interference is very remote. When necessary, the auxiliary electrode can be easily isolated by using a glass frit as shown in Figure 9.3B. The sample and the auxiliary-electrode filling solution should be adjusted to the same level to minimize convection between the two compartments. The isolation of the auxiliary electrode historically has been achieved with an agar gel, but there is no advantage to this more tedious approach. Porous Vycor and ion-exchange membranes have been widely substituted for the glass frit with satisfactory results [2,4,5 and references therein]. Another type of auxiliary electrode can be used in cases where interferences from the product(s) of this electrode reaction are of concern in reductive experiments. Silver metal in contact with chloride electrolyte results in the formation of silver chloride on the surface of the electrode. This type of auxiliary electrode further ensures that contamination from the auxiliary compartment into the working-electrode compartment does not occur. The silver auxiliary electrode is polished at the conclusion of an experiment to provide a clean surface for subsequent experiments.

The reference-electrode probe shown in Figure 9.3 may be the reference electrode itself or a salt bridge connecting and isolating the sample solution from the reference electrode. As mentioned earlier, the primary reference-electrode filling solution must sometimes be isolated from the sample. Contamination of nonaqueous samples by water and denaturing of biological samples by metal ions in the filling solution are examples of experiments requiring an intermediate bridge. It is important to consider the stability of reference electrodes and salt bridges in work done in nonaqueous solvents. Fouling of either the reference electrode or the salt bridge junction can lead to potential errors and distorted voltammetric responses. A procedure for testing aqueous saturated calomel electrodes used in nonaqueous solvents employing impedance measurements has been described [6].

Several types of reference electrodes are convenient for use in analytical electrochemistry. The use of high-input-impedance operational amplifiers in the reference electrode inputs of potentiostats ensures that very low levels of current are drawn from the reference electrode (see Chap. 6). This permits the use of reference electrodes that do not have to contain a large number of redox equivalents in order to ensure a constant reference potential and are therefore very small. Three reference-electrode designs that are convenient for use in analytical electrochemistry are shown in Figure 9.4. Saturated calomel and silver-silver chloride (of various concentrations of chloride) are among the most common commercially available or conveniently fabricated reference electrodes.

Provision for nitrogen purging is shown in Figure 9.3B and C, although for many experiments employing the application of positive potentials, removal of dissolved oxygen is unnecessary. Techniques such as differential pulse polarography can be used without removal of oxygen when the redox potential of the reaction of interest is sufficiently positive or negative of the reduction potential of oxygen. More efficient removal of oxygen is achieved if a fritted glass dispersion tube (available from laboratory glass suppliers) is used; however, a small-diameter glass tube drawn out to a fine tip providing many small bubbles is usually adequate. In cases in which removal and exclusion of dissolved oxy-

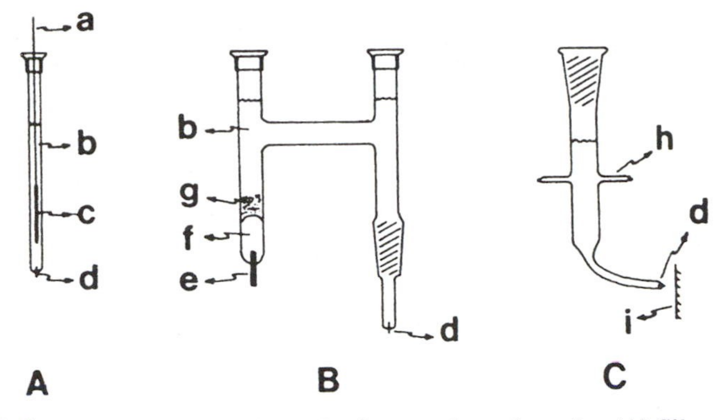


**Figure 9.4** Reference electrodes and reference electrode probe. (A) Silver-silver chloride reference electrode, (B) calomel reference electrode, (C) reference electrode probe. a, Silver wire, ca. 22 gauge; b, reference electrode filling solution, saturated or known concentration of KCl; c, silver chloride electrochemically precipitated on silver wire by anodization; d, reference electrode junction; e, Pt/glass seal; f, mercury; g, calomel; h, electrochemical cell body wall; i, working electrode surface.

gen is required, the bubbling of gas through the solution must be stopped prior to experiments employing quiet solutions. A slight positive pressure of inert gas is then maintained over the solution. This can be done by including another gas delivery tube in the cell cap positioned above the solution surface. The inert gas is alternatively passed through the dispersion or the delivery tubes. In some cases this can simply be accomplished by raising and lowering the inert-gas delivery tube above and into the solution.

The cell cap and cell body can be constructed of any suitable material that is inert to the sample. A snug, but not tight, press fit of the cap to the cell is desirable if glass is used for the cell since the diameter of ordinary glass tubing is imprecise and may break at the most inopportune moment. The cell body may be a beaker, test tube, glass tubing, any of a number of plastic containers, or may be machined out of Lucite, Delrin, Teflon, MACOR (machinable glass), or similar materials. The cell cap may be made of Lucite, Delrin, Teflon, rubber, neoprene, or the like. Cork borers may be used to make the necessary holes in the soft rubberlike cell caps, but freezing in liquid nitrogen or an acetone-dry ice bath allows the convenient use of an electric drill to bore necessary holes. Adjustment of the depth to which each part can be positioned in the sample solution is simple. A short length of the appropriate-diameter rubber tubing or an O-ring may be slipped around each part, providing a collar that can be adjusted for positioning. Microcells based on this design [7] can be used for sample volumes of 10 to 0.5 μL. Other microcells using similar design principles are useful for anodic stripping voltammetry on sample volumes down to as little as 5 μL [4,5], as well as for other voltammetric techniques [8,9].

Many variations of the design of the cells shown in Figure 9.3 have been reported. Reasonable cost, simple construction, flexibility, and ease of correct use have led to widespread acceptance of these general designs. These cell designs fit most aqueous and nonaqueous sample requirements where the presence of oxygen or water is not critical. When the removal and exclusion of these contaminants is required, special care must be taken to work in an inert atmosphere. A dry box (Chap. 19) or a cell that can be interfaced to a vacuum line (Chap. 18) may be required.

## III. CONVECTED-SOLUTION EXPERIMENTS

### A. Stirred Solutions

An exhaustive coulometric technique can be used both as an analytical tool and as a preparative tool. These two applications often require different cell designs. The present discussion is restricted to analytical coulometric cells. The design of electrochemical cells for preparative electrolysis has been treated in Chapter 22 and elsewhere [10].

Figure 9.5 (a more elaborate version of Fig. 9.3B) shows a versatile design for analytical controlled-potential coulometry [11]. The working electrode is a stirred mercury pool with provision for rapid and convenient exchange of spent sample and mercury through the stopcock. The same cell could be used with a platinum gauze working electrode, with the only change being provision for electrical contact through the cell cap. The auxiliary electrode is isolated by a porous glass tube and is located far enough above the surface of the working electrode to minimize potential gradients. Positioning of the reference probe tip is shown on the line of closest approach between the working and auxiliary electrodes but would have to be moved if a cylindrical platinum gauze electrode were used. A mechanical stirring paddle is included. The stirring rate should be adjusted to achieve maximum agitation of the mercury surface while keeping the agitation of the sample surface at a minimum to prevent splattering. The dimensions of the cell can be designed to hold sample volumes from about 20 to several hundred milliliters. Efficient cells (minimum volume, maximum electrode area) are desirable in controlled-potential coulometry to keep the time of the experiment short. A nitrogen gas dispersion tube is included, and the rate of nitrogen flow should be adjusted so that splattering is minimized. Several other controlled-potential electrolysis designs are given in the same reference [11]. A discussion of cell design philosophy is also included in this reference and in a previous paper [12].

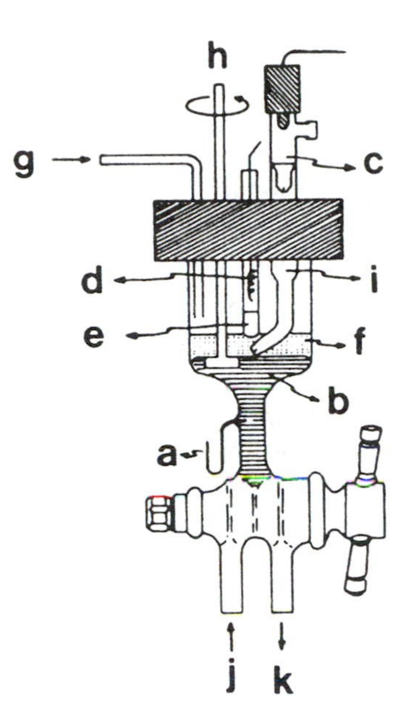


**Figure 9.5** Controlled-potential coulometry cell with a mercury pool working electrode. a, Platinum wire contact to mercury pool working electrode; b, mercury pool working electrode; c, reference electrode; d, auxiliary electrode; e, porous Vycor; f, sample solution; g, inert gas inlet; h, stirrer; i, reference electrode salt bridge; j, clean mercury; k, waste. [From Ref. 11, adapted with permission.]

## B. Rotating Electrodes

Rotating electrodes have been used to provide some degree of control of the rate of mass transport to the working electrode through control of the rate of electrode rotation. Wire, wire gauze, ring, and ring-disk electrodes are among the types of rotating electrodes that have found wide use in analytical and kinetic applications. Details of construction of some of these electrodes are found in Chapters 10, 11, and 14.

The cells shown in Figures 9.3 and 9.5 can be readily adapted for use in rotating-electrode experiments. Clamping the cell and top into position and drilling the cap to allow free rotation of the electrode shaft is required. Constant-speed rotating-electrode experiments usually have analytical objectives (e.g., amperometric titrations), and the cell volume may be any convenient size ranging upward from about 0.5 mL, as demonstrated in a study of sample volume requirements in rotating-electrode voltammetry [13].

Kinetic studies employing rotating-disk and rotating ring-disk electrodes require convenient and precise adjustment of the rotation speed from below 10 rotations per second (rps) to above 1000 rps. The high rates of rotation make wobble of the electrode a point of concern, and mating of the electrode shaft to the cell cap can require the use of a bearing joint. Commercial systems that avoid this problem are available. The rotating electrode should be positioned in the cell just below the solution surface to minimize convection in the solution due to electrode vibration. Kinetic experiments can require care with respect to both cell dimensions and electrode placement [14,15]. It should be recognized that the larger currents encountered in rotating-electrode experiments can result in depletion of the sample concentration if too many voltammetric scans are performed on a single sample. The use of cells containing large sample volumes (50–500 mL) can minimize this problem.

Rotating platinum screens have been used for electrogravimetric analyses. Commercial instruments employ two concentric cylindrical platinum screen electrodes with one or both electrodes rotating to increase convection. The cell itself is usually a beaker with a sample volume of about 150 mL. Typically, no cell top is used, and when running multiple analyses, there should be adequate ventilation to prevent accumulation of hydrogen. A simple operational amplifier circuit can be constructed for the instrument. The commercial instruments available are expensive given the simplicity of this type of experiment.

An interesting and clever variation on rotating electrodes is a cell that is itself rotated. Rotating mercury [16] and platinum [17] cells were initially developed and subsequently modified [18–23] to produce very practical cells for rapid electroanalytical measurements. These cells are designed for coulometric analysis on sample volumes of 2.0 mL. As the cell body rotates alternately at 1500 and 1800 rpm to enhance sample convection [22], the sample (and mercury when that is the working electrode) is held against the cell wall by centrifugal force in a thin film. This serves to optimize the contact of the sample volume with the working electrode surface. During operation of this cell, convective sample mixing permits rapid deoxygenation in a matter of about 20 s, and complete analysis at the microequivalent level in about 5 min.

### C. Flow-Through Cells

Flowing solution methods are particularly useful in on-line determinations and in providing a ready means of controlling mass transport in analytical and kinetic measurements. Cell designs for flowing solution analyses have used a wide variety of electrode materials, often seeking to take advantage of a particular electrode material's electrochemical characteristics or its ease of use in certain geometric configurations. Platinum wire [24–26], microconical platinum [27], platinum grid [26,28], hanging mercury drop [29], gold micromesh [30], graphite packing [31], reticulated vitreous carbon [32], tubular platinum [33–35], carbon [36–38], stainless steel [39], gold [40], mercury-covered platinum

[41,42], and mercury-covered graphite [43] were among the electrodes studied in flowing solution analyses. Also, flow cells have been employed as reagent generators for titrations [44,45] and for electrosynthesis [46]. Flow electrochemical cells that have been used with excellent success as detectors in liquid chromatography are described in Chapter 27.

The flow-cell design in Figure 9.6 is one of the many configurations that incorporate various types of working electrodes described earlier. In this example, reticulated vitreous carbon (RVC) is used as the working electrode, providing a large surface area and sample flow through the electrode itself. The reference electrode is isolated from the sample solution by a cation-exchange resin, and the internal filling solution of the reference electrode is continually replaced by flow. The cell body is machined conveniently from Plexiglas, and conventional connectors serve to mate the necessary input and output tubes to the cell body. The detection limits for this cell design are below $10^{-9}$ M [32]. Another flow-cell design provides for the use of rotated electrodes [47–49]. Increased sample convection is achieved, leading to high levels of current efficiency.

One major advantage of flowing solution methods is that the measurement is of steady-state currents. This simplifies the instrumentation and allows for very accurate and precise measurements at very low sample concentrations.

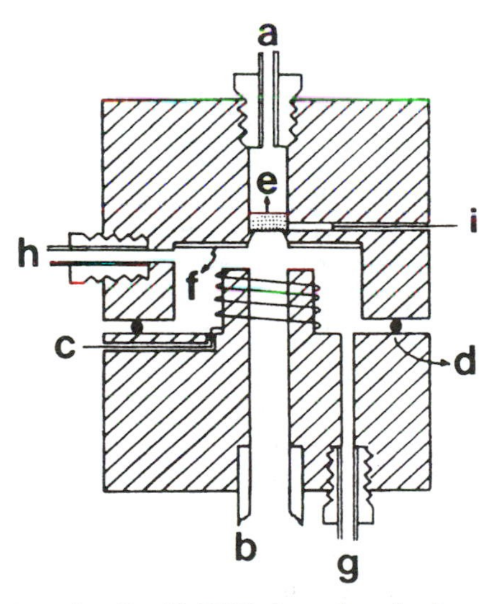


**Figure 9.6** Flow-through cell with RVC electrode and reference electrode. a, Sample solution inlet; b, sample solution outlet (Tygon); c, lead to reference electrode; d, O-ring; e, RVC electrode, one disk shown; f, cation exchange membranes; g, reference solution inlet; h, reference solution outlet; i, lead to working electrode. [From Ref. 32, adapted with permission.]

## IV. THIN-LAYER CELL DESIGN*

Problems in designing cells for semi-infinite diffusion techniques are well known to electrochemists. When the same considerations must be applied to a layer of solution less than 100 μm thick, the difficulties are significantly enhanced. Like most real-world components, thin-layer cells do not always conform to theory (Chap. 3), and design trade-offs are necessary to optimize performance in a particular application.

The thickness L of the layer ought to be uniform; thus the electrode must have a minimum surface roughness and be parallel to the opposing wall. Normally, the interstitial volume will have a juncture with bulk solution to accommodate the reference and/or auxiliary electrodes, or to permit filling. This is accomplished either at the edge of the thin layer or via a small orifice in the barrier wall. The cross-sectional area of this edge or orifice must be small relative to the electrode area in order to minimize diffusion of material in and out of the cell volume proper. The edge effects are minimized when the thin solution layer extends beyond the working electrode area. In general, it is best to keep the solution film horizontal to mitigate enhancement of edge effects by natural convection phenomena.

The rapid response theoretically inherent with thin-layer experiments is difficult to realize in practice because of the problem of achieving a uniform current-potential distribution. This is, of course, especially true for nonaqueous media. Except in steady-state twin-electrode experiments (see Chap. 3), current paths usually run parallel to the electrode surface. This most unfavorable situation can be rectified by use of a barrier wall that will effectively transmit current without significant transport of electroactive material during the experiment. This is a difficult property to achieve in practice while still maintaining the integrity of the thin-layer geometry. Porous glass ("thirsty" glass or porous Vycor) barriers have been used with some success; however, these frequently are more of a nuisance than the iR drop they purport to correct. The poor current distribution can make placement of the reference electrode more important than usual. Although "internal" reference probes are possible, they can result in overcompensation of the iR drop if placed in certain reaches of the cavity (e.g., above the working electrode in the sandwich cells discussed later).

The final cell design, as in semi-infinite electrochemistry, depends very much on the goals of the experimenter. For example, when one is interested in single-electrode coulometry for n value, concentration, or spectral measurements, the cell requirements are minimal. Potential scan experiments (voltammetry, potential scan coulometry, steady-state voltammetry) or experiments

---

*Written by Peter T. Kissinger, Department of Chemistry, Purdue University, and Bioanalytical Systems, Inc., West Lafayette, Indiana.

where transient response is important (e.g., kinetic studies) require the most grandiose thin-layer cells.

Cells with solid electrodes have incorporated a thin-layer volume in the form of either a slab or cylindrical shell. The latter have all been of fixed L value, whereas the former have been constructed with either fixed or variable thickness. A comprehensive review of thin-layer cells and applications has been published [50]. Since many of these are quite complex and will only be of interest to the specialist, only one approach is described in detail here.

Sandwich-type cells with fixed L values are easily constructed and have been widely employed for coulometric and spectroelectrochemical experiments. These designs have incorporated metal foils, a metal minigrid, deposited metal films, and semiconducting metal oxide electrodes [50–53]. A popular configuration uses glass slides and an inert spacer (e.g., Teflon) to form the sandwich. Cross sections of three models are shown in Figure 9.7. This specific approach originated with the minigrid cell of Murray et al. [54].

Figure 9.8 illustrates one practical scheme for using sandwich-type cells. The sample is introduced by capillary action from below or through a side arm from above. Solution contact to the reference and auxiliary electrodes is made

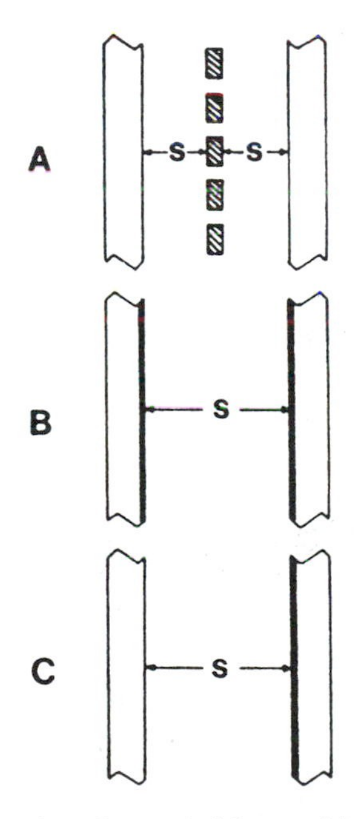


**Figure 9.7** Three configurations for sandwich-type thin-layer cells. (A) Minigrid suspended between two spacers. (B) Twin-electrode cell using metal films on glass. (C) Single-electrode cell, barrier plate and electrode plate. s, Sample solution.

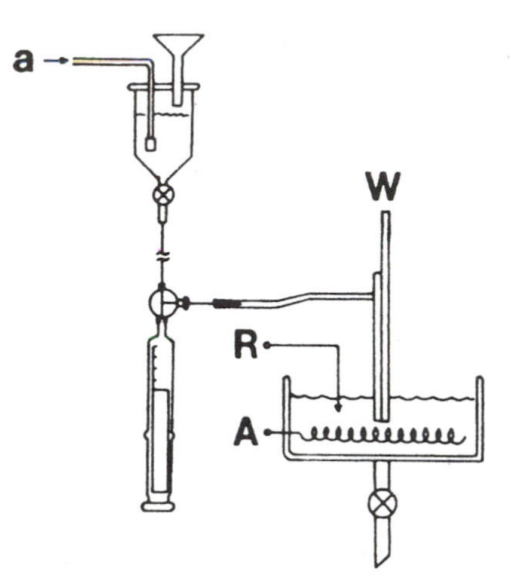


**Figure 9.8** Experimental arrangement for use of thin-layer electrode constructed with glass or quartz plates. W, Working electrode; R, reference electrode; A, auxiliary electrode; a, inert-gas inlet.

via the bottom edge of the cell in a small cup filled with supporting electrolyte. The thin-layer electrode itself (Fig. 9.9) consists of two glass microscope slides (an electrode plate and a barrier plate, two electrode plates, or two barrier plates with a central grid electrode) pressed to a thin (ca. 20–75 μm) spacer. Either

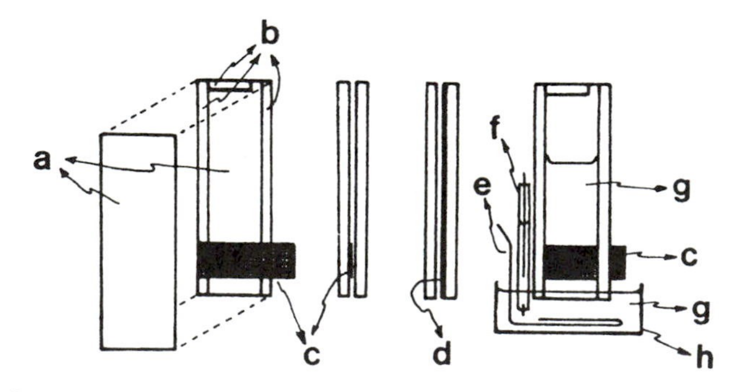


**Figure 9.9** Assembly of sandwich-type optically transparent thin-layer electrochemical cell. a, Glass or quartz plates; b, adhesive Teflon tape spacers; c, minigrid working electrode; d, metal thin-film working electrode, which may be used in place of (c); e, platinum wire auxiliary electrode; f, silver-silver chloride reference electrode; g, sample solution; h, sample cup. [Adapted with permission from T.P. DeAngelis and W.R. Heineman, *J. Chem. Educ.* *53*:594 (1976). Copyright 1976 American Chemical Society.]

clamps or epoxy cement can be used to hold the cell together. A spacer(s) of suitable geometry is carefully cut from stock film with a razor blade. The sandwich is then assembled and clamped tightly with a large spring-loaded paper clip, making sure that the spacer has no folds and the plates are properly oriented. Epoxy is lightly applied to three edges and the unit is placed in an oven or under an infrared lamp to facilitate curing of the resin. A large number of cells can be prepared in short order and at very little expense. Should convenient disassembly be necessary, a clamp would have to be devised. Two steel plates backed by rubber gaskets can be screwed together around the cell, or the entire cell can be machined from an engineering plastic [55].

Some attention must be paid to the electrode dimensions (see Fig. 9.9). The working electrode's lower edge should be close to the bottom of the cell plates to minimize iR-drop problems. The width of the working electrode in contact with the thin layer of solution should be small to minimize edge diffusion. As noted earlier, a vertical orientation is not desirable; however, it is convenient and compatible with the horizontal optical path of virtually all commercial spectrophotometers. Recommended sources of cell components (including minigrids) are listed in Table 9.1. Thin-layer cells for chromatographic detection and electron spin resonance spectroscopy are discussed in Chapters 27 and 29, and their application in optical studies is described in Chapter 3.

## V. CELLS FOR SPECTROELECTROCHEMISTRY

The coupling of electrochemical and spectroscopic techniques is finding increased use in analytical, kinetic, and mechanistic studies. The product of an electrode reaction or the product of a homogenous reaction between the electrogenerated species and some material (e.g., an enzyme) present in solution can be examined using optical absorption spectroscopy. Several approaches to joining electrochemical and optical absorption techniques have been taken. In one approach, the sample solution is continuously circulated from the working electrode to an optical cell where the absorbance is measured. A variety of designs have been described [56–60]. Possibly the simplest design employs a conventional cuvette as the electrochemical cell [58]. Classical kinetic experiments may also be performed on the electrogenerated product as long as the rate of solution mass transport is fast compared to the kinetic process under study, ensuring a uniform concentration gradient throughout solution. Determinations employing electrochemically generated titrants are also well suited to this approach.

Another approach is to employ a metal minigrid or a thin metal or semiconductor film for the working electrode (Chap. 11). These electrodes permit transmittance of the optical beam through the electrode and are readily adapted for use in thin-layer cells. Fast heterogeneous or homogeneous electron transfer reactions, or homogeneous chemical reactions initiated by the electrode

**Table 9.1** Electroanalytical Equipment and Accessories

| Source | Comments |
|---|---|
| Bioanalytical Systems, Inc.<br>2701 Kent Ave.<br>West Lafayette, IN 47906 | Instrumentation, electrodes, cells, accessories |
| Brinkman Instruments, Inc.<br>Cantiague Rd.<br>Westbury, NY 11590 | Hanging mercury drop electrodes |
| Buckbee-Mears Co.<br>245 E. 6th St.<br>St. Paul, MN 55101 | Minigrid electrodes, especially of gold, also nickel, copper, silver |
| Cypress Systems Inc.<br>P.O. Box 3931<br>Lawrence, KS 66046 | Instrumentation, electrodes, cells, accessories |
| Delta Technologies<br>13960 N. 47th St.<br>Stillwater, MN 55802-1234 | Indium oxide optically transparent electrodes on glass and quartz |
| Dionex Corporation<br>1228 Titan Way<br>P.O. Box 3603<br>Sunnyvale, CA 94088-3603 | Pulsed amperometric detector cells and instrumentation |
| Donnelly Corporation<br>414 E. 40th St.<br>Holland, MI 49423 | Indium oxide optically transparent electrodes on glass |
| EG&G Princeton Applied Research<br>P.O. Box 2565<br>Princeton, NJ 08543-2565 | Instrumentation, electrodes, cells, accessories |
| Electrosynthesis Co., Inc.<br>72 Ward Rd.<br>Lancaster, NY 14096-9779 | Instrumentation, electrodes, cells, accessories |
| ESCO Products<br>171 Oak Ridge Rd.<br>Oak Ridge, NJ 07438 | Optical components for spectroelectrochemical cells |
| Harrick Scientific Corp.<br>88 Broadway<br>Box 1288<br>Ossining, NY 10562 | Optical components for spectroelectrochemical cells |
| Pine Instrument Company<br>101 Industrial Dr.<br>Grove City, PA 16127 | Instrumentation, electrodes, cells, accessories rotating electrode equipment |

*Note*: Each year the August 15 issue of *Analytical Chemistry* contains extensive listing of suppliers of electrochemical instruments, electrodes and accessories, with current phone numbers.

reaction product, can best be followed via spectroelectrochemistry under semi-infinite diffusion conditions. The popular sandwich cell design that has evolved over many years in Kuwana's laboratory remains flexible and relatively simple to assemble [61–63]. Figure 9.10 illustrates a sandwich cell employing an optically transparent electrode (OTE). Some OTEs are available commercially with varying film conductivities and optical properties (see Table 9.1). In general, OTEs are fabricated by vapor deposition on the appropriate substrate as described in Chapter 11. For the cell shown in Figure 9.10, an RVC working electrode is used to effect rapid exhaustive electrolysis in a sandwich OTE cell especially designed for use in acquiring extended x-ray absorption fine structure (EXAFS) spectra for metal complex redox couples [64]. Mylar or Kapton optical windows provide x-ray transparency, and the assembly is simply held together with a clamp. A variety of OTE sandwich cell designs are described in the same paper [64]. Transmission spectroelectrochemical experiments can also be conducted under rather stringent experimental conditions. Other spectroelectrochemical cell designs using RVC that permit acquisition of data

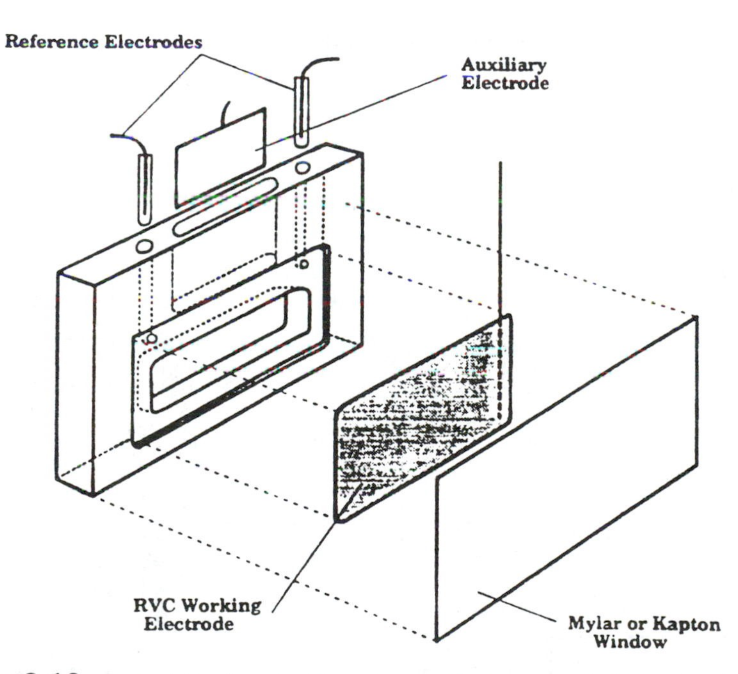


**Figure 9.10** Assembly of sandwich-type optically transparent electrochemical cell for extended x-ray absorbance fine structure (EXAFS) spectroelectrochemistry. Cell body is of MACOR; working electrode is reticulated vitreous carbon (RVC). [From Ref. 64, with permission.]

on room-temperature molten salt samples have been described [65]. In this work, the cell is positioned in a glove box so that stringent exclusion of oxygen and moisture can be maintained. The optical measurements are acquired using a microprocessor-controlled fiber-optic spectrophotometer.

The special properties of OTEs that permit the use of transmission spectroelectrochemical techniques are often at cross purposes with the acquisition of reliable electrochemical data. The desire to have the superior electrical properties of bulk conducting materials, and thereby reliable electrochemical data, together with the power of a coupled optical probe led groups to develop various diffraction and reflection approaches to spectroelectrochemistry. Light diffracted by a laser beam passing parallel to a planar bulk electrode can be used to significantly increase the effective path length and sensitivity in spectroelectrochemistry [66].

Normal reflection optics have been used to advantage with bulk electrode materials. Examples of this type of spectroelectrochemical cell are shown in Figure 9.11 [67]. Simple bifurcated fiber-optic waveguides are used to direct source light onto reflective bulk electrode surfaces and to collect the reflected light for transmission to a detector. This is a simple means for performing spectroelectrochemical experiments at bulk metal electrodes that cannot be as

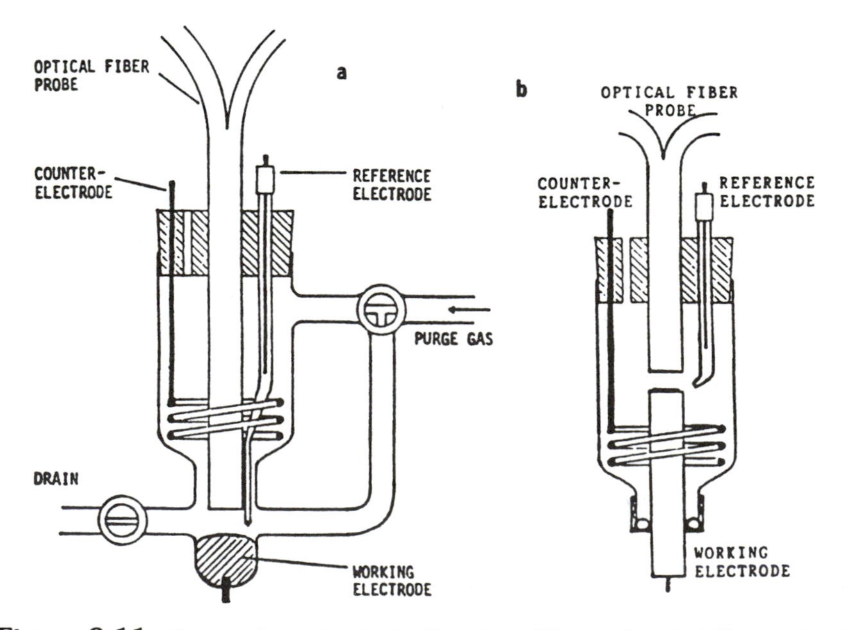


**Figure 9.11** Spectroelectrochemical cells using a bifurcated optical fiber probe: (a) mercury pool working electrode; (b) reflective metal working electrode. [From Ref. 67.]

conveniently fabricated as OTEs, such as silver [68], while also eliminating the cost and problems associated with some OTEs.

Spectroelectrochemical experiments have been extended to the infrared spectral region using simple optical and electrochemical cell arrangements. Figure 9.12 shows a spectroelectrochemical cell that permits acquisition Fou-

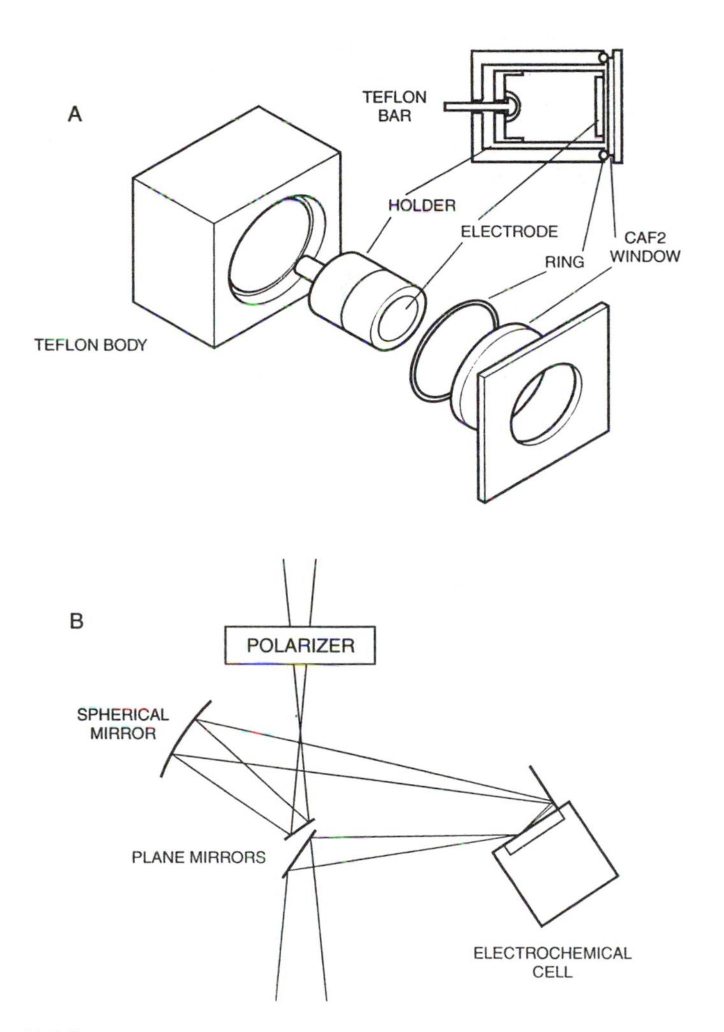


**Figure 9.12** Spectroelectrochemical cell for Fourier-transform infrared reflection absorption spectroscopy (FTIRRAS). (A) Cell components showing Teflon bar for controlling the sample path length; (B) retroreflection absorption optics for use with this cell. [From I.T. Bae, X. Xing, E.B. Yeager, and D. Scherson, *Anal. Chem. 61*:1164 (1989). Copyright 1989 American Chemical Society.]

rier-transform infrared reflection absorption spectra (FTIRRAS) of thin layers of aqueous electrolyte [69]. The bulk metal electrode is mounted on a Teflon holder that is used to control the path length between the electrode surface and the infrared-transparent calcium fluoride optical window. This simple arrangement permits acquisition of potential-dependent FTIRRAS of electrolyte layers with thicknesses down to 1 μm using infrared radiation polarized parallel to the plane of incidence on the electrode surface.

Structural information regarding the molecular orientation of adsorbates on electrodes can be obtained using a number of Raman spectroscopic probes. While these experiments are not routine to conduct, they are beginning to approach the simplicity of the FTIRRAS experiments just described. Figure 9.13

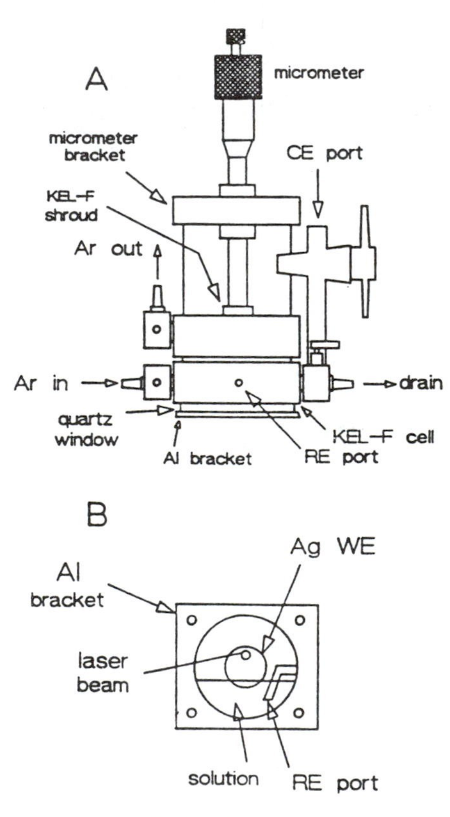


**Figure 9.13** Spectroelectrochemical cell for Raman spectroscopy studies using electrode emersion. (A) Top view of cell, micrometer adjusts cell path length; (B) front view with 90° rotation of reference electrode position. [From J.E. Pemberton and R.L. Sobocinski, *J. Electroanal. Chem. 318*:157 (1991). Copyright 1991 Elsevier Sequoia S.A., Lausanne.]

is an example of a Raman spectroelectrochemical cell that can be used to obtain in situ Raman spectra as well as Raman spectra on electrode surfaces that have been emersed [70]. As illustrated in Figure 9.12B, a part of the electrode surface can be emersed from sample solution permitting acquisition of Raman spectra of the electrode-solution interface in the absence of bulk solution interference by simply draining part of the sample solution. The micrometer shown in Figure 9.13A permits control of the optical path length facilitating careful rinsing of the sample compartment. The emersion procedure retains the molecular orientation of the solvent in contact with the electrode upon emersion and makes acquisition of Raman spectra of electrode-solution interfaces more general. Neither surface enhancement nor resonance enhancement is required to obtain Raman spectra of satisfactory signal-to-noise ratio.

Spectroelectrochemical cells that permit redox titrations of precious biological samples, require exclusion of oxygen, and allow acquisition of data from multiple spectroscopic domains have been described. A recent example of these cell designs combines electron paramagnetic resonance spectroscopy with UV-visible absorption spectroscopy [71] for studies of flavoproteins.

The design principles described in the preceding sections must also be considered in configuring an electrochemical cell for spectroscopic measurements. The need to minimize variations in current density, and therefore potential, along the surface of a disk electrode suggests the use of small-diameter OTEs. Physical placement of the auxiliary electrode and the reference electrode probe must be outside the optical beam path. Thin-metal-film OTEs are often not mechanically stable to large applied potentials, and good experimental technique must be followed to prevent destruction of the OTE. The working electrode lead should be connected last and disconnected first when starting and finishing experiments, respectively.

A variety of optically coupled electrochemical techniques have been recently advanced for addressing analytical and kinetic problems. The advantages and limitations of these different spectroscopic techniques as applied to particular problems have been described in other chapters in this book and in recent reviews [51–53,65].

## REFERENCES

1. J. E. Mumby and S. P. Perone, *Chem. Instrum. 3*:191 (1971).
2. B. D. Cahan, Z. Nagy, and M. A. Grenshaw, *J. Electrochem. Soc. 119*:64 (1972).
3. J. Lindberg, *J. Electroanal. Chem. 40*:265 (1972).
4. L. Huderova and K. Stulik, *Talanta 19*:1285 (1972).
5. K. Stulik and M. Stulikova, *Anal. Lett. 6*:441 (1973).
6. K. M. Kadish, S.-M. Cai, T. Malinski, J.-Q. Ding, and X. Q. Lin, *Anal. Chem. 55*:163 (1983).

7. G. Park, R. N. Adams, and W. R. White, *Anal. Lett. 5*:887 (1972).
8. M. Karolczak, R. Dreiling, R. N. Adams, L. J. Felice, and P. T. Kissinger, *Anal. Lett. 9*:783 (1976).
9. J. Wang and B. A. Frelha, *Anal. Chem. 54*:334 (1982).
10. H. Lund and P. Iverson, in *Organic Electrochemistry* (M. M. Baizer, ed.), Marcel Dekker, New York, 1973, pp. 165–249, and references therein.
11. J. E. Harrar and E. L. Pomernacki, *Anal. Chem. 45*:57 (1973).
12. J. E. Harrar and I. Shain, *Anal. Chem. 38*:1148 (1966).
13. B. Miller and S. Bruckenstein, *Anal. Chem. 46*:2033 (1974).
14. R. N. Adams, *Electrochemistry at Solid Electrodes*, Marcel Dekker, New York, 1969, pp. 80–92, and references therein.
15. W. J. Albery and M. L. Hitchman, *Ring-Disc Electrodes*, Clarendon Press, Oxford, 1971.
16. R. G. Clem, F. Jakob, D. H. Anderberg, and L. D. Ornelas, *Anal. Chem. 43*:1398 (1971).
17. R. G. Clem, *Anal. Chem. 43*:1853 (1971)
18. R. G. Clem, F. Jakob, and D. H. Anderberg, *Anal. Chem. 43*:918 (1971).
19. F. L. Stephens, F. Jakob, L. P. Rigdon, and J. E. Harrar, *Anal. Chem. 42*:765 (1970).
20. J. E. Harrar, *Anal. Chem. 35*:893 (1963).
21. R. G. Clem, *Ind. Res. 15*:50 (1973).
22. R. G. Clem, *LBL Nucl. Chem. Div. Annu. Rep.* 386 (1972).
23. R. Jakob and R. G. Clem, *MPI Appl. Notes 6*:9 (1971).
24. O. H. Muller, *J. Am. Chem. Soc. 69*:2992 (1947).
25. J. Jordan, *Anal. Chem. 27*:1708 (1955).
26. R. E. Sioda, *Electrochim. Acta 15*:783 (1970).
27. J. Jordan, R. A. Javich, and W. E. Ranz, *J. Am. Chem. Soc. 80*:3846 (1958).
28. R. E. Sioda, *Electrochim. Acta 13*:375 (1968).
29. G. Koster and M. Ariel, *J. Electroanal. Chem. 33*:339 (1971).
30. W. J. Blaedel and S. L. Boyer, *Anal. Chem. 45*:258 (1973).
31. R. E. Sioda, *Electrochim. Acta 13*:1559 (1968).
32. W. J. Blaedel and J. Wang, *Anal. Chem. 51*:799 (1979).
33. W. J. Blaedel, C. L. Olsen, and L. R. Sharma, *Anal. Chem. 35*:2100 (1963).
34. W. J. Blaedel and L. N. Klatt, *Anal. Chem. 38*:879 (1966).
35. W. J. Blaedel and S. L. Boyer, *Anal. Chem. 43*:1538 (1971).
36. W. D. Mason and C. L. Olson, *Anal. Chem. 42*:488 (1970).
37. W. D. Mason and C. L. Olson, *Anal. Chem. 42*:548 (1970).
38. W. D. Mason, T. D. Gardner, and J. T. Stewart, *J. Pharm. Sci. 61*:1301 (1972).
39. J. C. Bazan and A. J. Ariva, *Electrochim. Acta 9*:17 (1964).
40. D. B. Easty, W. J. Blaedel, and L. Anderson, *Anal. Chem. 43*:509 (1971).
41. T. O. Oesterling and C. L. Olson, *Anal. Chem. 39*:1543 (1967).
42. T. O. Oesterling and C. L. Olson, *Anal. Chem. 39*:1547 (1967).
43. W. R. Seitz, R. Jones, L. N. Klatt, and W. D. Mason, *Anal. Chem. 45*:480 (1973).
44. R. E. Meyer, M. C. Banta, P. M. Lantz, and F. A. Posey, *J. Electroanal. Chem. 30*:345 (1971).

45. H. S. Wroblowa and A. Saunders, *J. Electroanal. Chem. 42*:329 (1973).
46. D. Danly, *Organic Electrochemistry* (M. M. Baizer, ed.), Marcel Dekker, New York, 1973, pp. 911–924, and references therein.
47. J. Wang and M. Ariel, *Anal. Chim. Acta 99*:89 (1978).
48. B. Oosterhuis, K. Brunt, B. H. C. Westerink, and D. A. Doornbos, *Anal. Chem. 52*:203 (1980).
49. J. L. Anderson, *Anal. Chem. 51*:2312 (1979).
50. A. T. Hubbard and F. C. Anson, *Electroanalytical Chemistry*, Vol. 4 (A. J. Bard, ed.), Marcel Dekker, New York, 1970, pp. 129–214, and references therein.
51. T. Kuwana and W. R. Heineman, *Acc. Chem. Res. 9*:241 (1976). and references therein.
52. W. R. Heineman, *Anal. Chem. 50*:390A (1978), and references therein.
53. W. R. Heineman, F. M. Hawkridge, and H. N. Blount, *Electroanalytical Chemistry*, Vol. 13 (A. J. Bard, ed.), Marcel Dekker, New York, 1984, pp. 1–113, and references therein.
54. R. W. Murray, W. R. Heineman, and G. W. O'Dom, *Anal. Chem. 39*:1666 (1979).
55. C. W. Anderson, H. B. Halsall, and W. R. Heineman, *Anal. Biochem. 93*:366 (1979).
56. J. M. Sosonkin, A. N. Domarev, V. A. Subbotin, F. F. Lakomov, V. I. Kumantsov, and A. Y. Kaminskii, *Elektrokhimiya 8*:1168 (1972).
57. A. M. Gary, E. Pietmont, J. Roynette, and J. P. Schwing, *Anal. Chem. 44*:198 (1972).
58. P. Hobza, I. Nykl, and K. Uhlir, *Collect. Czech. Chem. Commun. 38*:994 (1973).
59. M. T. Stankovich, *Anal. Biochem. 109*:295 (1980).
60. J. L. Anderson and J. R. Kincaid, *Appl. Spectrosc. 32*:356 (1978).
61. T. Kuwana and N. Winograd, *Electroanalytical Chemistry*, Vol. 7 (A. J. Bard, ed.), Marcel Dekker, New York, 1974, pp. 1–78, and references therein.
62. F. M. Hawkridge, and T. Kuwana, *Anal. Chem. 45*:1021 (1973).
63. F. M. Hawkridge, J. E. Pemberton, and H. N. Blount, *Anal. Chem. 49*:1646 (1977).
64. L. R. Sharp, W.R. Heineman, and R.C. Elder, *Chem. Rev. 19*:705 (1990).
65. E. H. Ward and C. L. Hussey, *Anal. Chem. 59*:213 (1987).
66. P. Rossi, C. W. McCurdy, and R. L. McCreery, *J. Am. Chem. Soc. 103*:2524 (1981).
67. C.-H. Pyun and S.-M. Park, *Anal. Chem. 58*:25 (1986).
68. D. E. Reed and F. M. Hawkridge, *Anal. Chem. 59*:2334 (1987).
69. I. T. Bae, X. Xing, E. B. Yeager, and D. Scherson, *Anal. Chem. 61*:1164 (1989).
70. J. E. Pemberton and R.L. Sobocinski, *J. Electroanal. Chem. 318*:157 (1991).
71. K. E. Paulsen, A. M. Orville, F. E. Frerman, J. D. Lipscomb, and M. T. Stankovich, *Biochemistry 31*:11755 (1992).
72. R. J. Gale, ed., *Spectroelectrochemistry: Theory and Practice*, Plenum Press, New York, 1988.

# 10

# Carbon Electrodes

**Richard L. McCreery** *The Ohio State University, Columbus, Ohio*

**Kristin K. Cline** *Wittenberg University, Springfield, Ohio*

## I. INTRODUCTION

When one surveys the current state of electroanalytical research and the applications of electroanalysis, one concludes that solid electrodes in general and carbon in particular have joined mercury as widely used, practical electrode materials. Although solid electrodes are used in many applications in bioanalysis, analytical voltammetry, energy conversion, and electrosynthesis, their development has followed a complex path through many electrode materials, surface preparations, and experimental procedures. Historically, the dependence of analytically useful observables such as current and potential on sometimes poorly characterized interfacial phenomena often led to difficulties with solid electrode performance and reproducibility. The renewable surface of the dropping mercury electrode (DME) is the major reason for its reliability and popularity, despite its often serious limitations in potential range and its mechanical inconvenience. The headaches associated with surface irreproducibility on solid electrodes are countered by major advantages in electroanalysis and by large-scale industrial applications of high economic value. The mechanical stability, wide potential range, electrocatalytic activity, and versatility of solid electrodes make them more attractive than mercury for a variety of applications. Since solid electrodes began to be used in electroanalysis about forty years ago, analytical chemists have learned to control reproducibility and have devised a variety of surface treatments and derivatizations that enhance selectivity and sensitivity. In numerous cases, solid electrodes have permitted analytical and mechanistic applications that were not possible with mercury electrodes. The success in acquiring new chemical information combined with applications of major economic impact has provided much of the driving force for developing solid electrodes.

Carbon materials in the form of graphite, glassy carbon, carbon fibers, etc. have been important players in solid electrode development for several reasons. First, they are available in a variety of forms and are generally inexpensive. Second, the slow kinetics of carbon oxidation lead to a wide useful potential range, particularly in the positive direction. This characteristic is an important advantage over platinum and mercury, which exhibit significant or overwhelming background oxidation currents. Third, carbon has a rich surface chemistry, which can be exploited to influence reactivity. In particular, a wide variety of chemical derivatizations is possible on graphite and glassy carbon surfaces. Fourth, controlled variation of electron transfer kinetics and adsorption on carbon surfaces can be used to enhance analytical utility. Overall, these properties of carbon can be exploited to advantage provided the user is cognizant of the relationships between carbon materials properties, surface preparation, and electroanalytical behavior.

In the previous edition of this book, Dryhurst and McAllister described carbon electrodes in common use at the time, with particular emphasis on fabrication and potential limits [1]. There have been two extensive reviews since the previous edition, one emphasizing electrode kinetics at carbon [2] and one on more general physical and electrochemical properties [3]. In addition to greater popularity of carbon as an electrode, the major developments since 1984 have been an improved understanding of surface properties and structure, and extensive efforts on chemical modification. In the context of electroanalytical applications, the current chapter stresses the relationship between surface structure and reproducibility, plus the variety of carbon materials and pretreatments. Since the intent of the chapter is to guide the reader in using commonly available materials and procedures, many interesting but less common approaches from the literature are not addressed. A particularly active area that is not discussed is the wide variety of carbon electrodes with chemically modified surfaces.

The primary objective of the discussion that follows is to establish a basis for choosing and applying carbon electrodes for analytical applications. As with any electrode material or electroanalytical technique, the choice depends on the application; there is no "ideal" electrode for all situations. We first discuss the criteria that drive the chemist's choice of electrode or procedure. These criteria include background current, potential limits, and electrode kinetics, and may be considered dependent variables that are ultimately controlled by the properties of the carbon surface. Then we consider the independent variables that determine electroanalytical behavior. These include the choice of carbon material, surface roughness, cleanliness, etc. By considering the dependence of electroanalytical behavior on surface variables that the user can control, it should be possible to make rational choices of electrodes and procedures to lead to the desired analytical objective.

## II. PERFORMANCE CRITERIA

Before discussing particular carbon electrode materials, we should define the qualities cn which a choice of material will be based. These are the criteria that matter the most to the user, and the importance of each will vary with the application. For example, a carbon electrode to be used for detecting eluents from a liquid chromatograph should have a low background current and long stability, whereas an electrode used for studying redox mechanisms should usually exhibit fast electron transfer kinetics. The criteria relevant to carbon electrodes are conveniently classified into four types.

### A. Background Current and Potential Limits

The current observed for an amperometric technique in a blank electrolyte solution has several components, including double-layer charging, redox reactions of the electrode surface (e.g., surface oxide formation), and redox reactions of impurities in the electrolyte solution, particularly oxygen. Except to note that oxygen reduction is very inefficient on carbon, we ignore background currents originating in the solution and emphasize those dependent on the electrode itself. For voltammetric techniques, carbon electrodes exhibit the usual charging current, but its magnitude varies greatly with the carbon type. Several background voltammograms for carbon electrodes are shown in Figure 10.1, in all cases normalized for electrode area. Highly ordered pyrolytic graphite (HOPG), described later, has a very low background due to low interfacial capacitance (curve 1). Pure double-layer capacitance, $C_{dl}$, yields a constant current for a voltammetric experiment equal to the product of $C_{dl}$ and the scan rate, dE/dt. Apparent double-layer capacitance can vary from $<1\ \mu F/cm^2$ for highly ordered pyrolytic graphite to $>70\ \mu F/cm^2$ for much less ordered materials. Carbon materials also exhibit a nonclassical component to the voltammetric background, which increases the observed "capacitance" above that expected from double-layer effects alone. For comparison, Hg and Pt electrodes have typical double-layer capacitance values of 10–20 $\mu F/cm^2$, with corresponding voltammetric backgrounds. For well-behaved Hg or Pt surfaces in certain potential ranges, this capacitance behaves ideally, with only a slight dependence on scan rate. For partially or extensively oxidized glassy carbon (GC), the apparent capacitance can be much larger than 20 $\mu F/cm^2$ and varies significantly with scan rate. Most authors lump this nonclassical background together with the double layer, and express both as "capacitance." The excess current has been attributed to surface redox processes, ion adsorption, and porosity, and depends strongly on pretreatment, particularly oxidation. A typical background voltammogram for polished GC is shown in Figure 10.1, curve 2. Although it is fairly featureless, the background current is much larger than HOPG (curve 1) under the same conditions. Unlike Pt or Hg, GC exhibits no voltammetric features related to

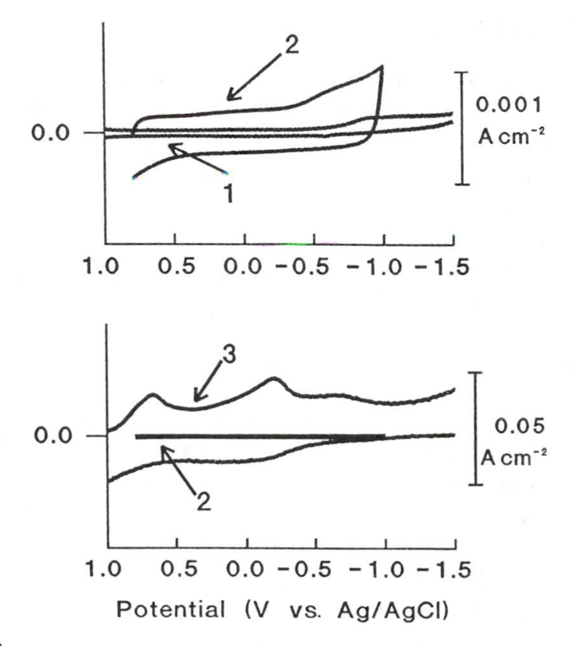


**Figure 10.1** Background currents (divided by electrode area) for (1) HOPG, (2) polished GC, and (3) polished GC following 10 ECP cycles (potential cycled from 0 to 2.2 V vs. Ag/AgCl in 0.1 M $H_2SO_4$). Electrolyte was 1 M KCl; scan rate was 2 V/s.

solvent oxidation or reduction, at least in the potential range of approximately ± 1.0 V vs. Ag/AgCl or SCE. In the lower plot of Figure 10.1, polished and oxidized GC are compared, and the large background of oxidized GC is demonstrated. The magnitude of this background current depends on oxidation conditions, but it is always larger than unoxidized GC.

Voltammetric potential limits are an important special case of the general issue of background current. In the simplest case, an electrode will exhibit only capacitive current over some potential range, with no redox reactions of either the electrode surface or the electrolyte. As the potential is scanned away from this double-layer region, eventually the surface or the solvent will begin to oxidize or reduce, until the current exceeds that from the redox process under study. The potential limit for an electrode/electrolyte combination is usually considered to be the potential where the voltammetric background current exceeds some arbitrary value. This value depends on scan rate, electrode area, etc., but nevertheless provides a useful comparison of electrodes and solvents.

For amperometric techniques at a fixed potential (e.g., liquid chromatography detectors), the classical charging current is zero since $dE/dt = 0$. There is still a background current, however, due to surface redox processes or slow changes in the surface microstructure or cleanliness. The background current is generally much lower than that for voltammetric techniques, but nevertheless affects the detection limits of the potentiostatic techniques. Obviously the potentiostatic background depends strongly on redox processes of the carbon surface, and carbon materials vary in their susceptibility to such processes.

## B. Electron Transfer Kinetics

The rate of electron transfer between an electrode surface and a redox system is most commonly discussed in terms of the heterogeneous electron transfer rate constant, k°, with units of cm/s [4]. The term k° is defined as the heterogeneous electron transfer rate constant at the formal potential, (as shown in Chap. 2), and is often expressed as $k_{s,h}$ in the literature. For a typical voltammetric experiment at a scan rate of 1 V/s or less, a reaction with a k° of ≥0.1 cm/s will appear to be electrochemically reversible. The k° can usually be determined only for single electron transfer reactions without complications of proton transfer, but even complex multistep redox processes involve one or more single electron transfer steps with accompanying k° values. Electron transfer kinetics are of fundamental importance to analytical and mechanistic electrochemistry for several reasons. First, the shape and magnitude of voltammograms depend on k°, as does the slope of a current versus concentration plot. Second, the range of useful scan rates is ultimately limited by k°, with higher scan rates and correspondingly short time scales requiring higher k°. An electrochemically reversible redox reaction will exhibit a voltammetric $\Delta E_p$ of 57/n mV, while systems with a small k° can exhibit $\Delta E_p$'s values greater than 1 V. Third, any electroanalytical technique requiring a mass transport limited current (e.g., liquid chromatography (LC) detection, chronoamperometry, controlled-potential electrolysis) must be conducted under conditions of fast electron transfer. Fourth, variations in k° with time, perhaps due to impurity adsorption or other electrode surface changes, can cause changes in current response, thus requiring recalibration or electrode renewal. Fifth, any attempts to study fast reactions of electrogenerated reactants depend on fast electron transfer to prepare the reactant of interest. Stated more generally, the use of voltammetry or chronoamperometry to probe reaction mechanisms depends on suitable electrode kinetics, particularly for fast reactions.

Carbon electrodes exhibit a wide range of electron transfer rates for benchmark redox systems, depending on carbon material and surface history. Two examples are shown in Figure 10.2, which compares two carbon surfaces with very different k° for $Fe(CN)_6^{3-/4-}$. In some cases, the variations in electrode kinetics have been particularly important to analytical applications. For example, carbon paste and carbon fiber electrodes have been used to monitor neurotransmitters in living animal brains [5,6]. The determination of catechol transmitters in the presence of relatively large amounts of interferents (e.g., ascorbate) de-

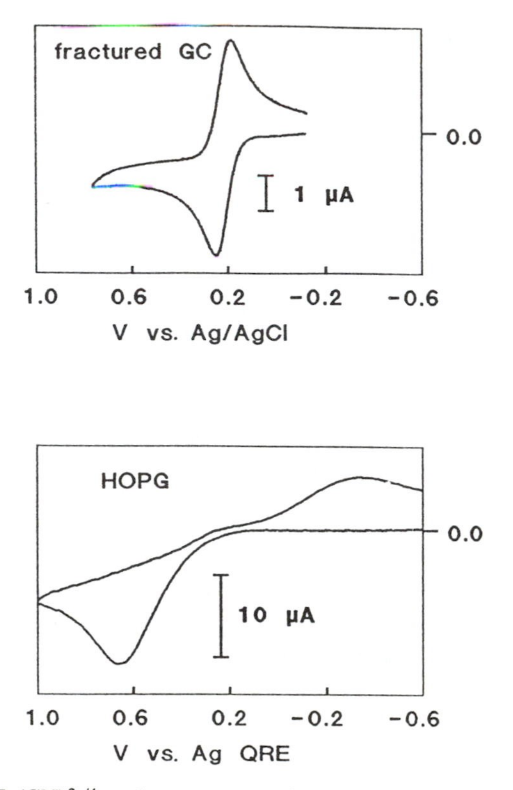


**Figure 10.2** $Fe(CN)_6^{3-/4-}$ voltammetry on glassy carbon (GC) fractured in solution, and on basal plane highly ordered pyrolytic graphite (HOPG). 1 mM $K_4Fe(CN)_6$ in 1 M KCl, scan rate = 0.2 V/s. $\Delta E_p$ for fractured GC voltammogram = 64 mV, corresponding to k° > 0.1 cm/s, $\Delta E_p$ for HOPG = 1005 mV, k° = 1 × $10^{-6}$ cm/s. Potential scale is relative to silver quasi-reference electrode.

pends on a combination of favorable kinetics and adsorption. In the course of discussing various carbon electrodes we will compare their electrode kinetics.

## C. Reproducibility and Stability

These parameters are largely self-explanatory, and generally refer to changes in electrode behavior with time or repeated surface preparation. Since the DME is reproducible to 1% or better, it serves as a standard to approach with solid electrodes. The variables in question generally include background current, analytical signal (e.g., voltammetric peak height), electron transfer rate, and electrode area.

### D. Adsorption

Although adsorption is usually undesirable in electroanalytical applications, there are several situations where it can be exploited to advantage. First, many surface modification schemes rely on adsorption to a solid electrode surface for binding of the modifier [7]. Second, adsorption can enhance an analytical response by preconcentrating an adsorbing analyte over a nonadsorber. The detection of micromolar levels of dopamine in the presence of millimolar ascorbate was already noted as an example. Third, adsorption can provide a measure of microscopic area, which can be significantly larger than the projected electrode area.

On the other hand, adsorption can have serious negative effects on analytical response. Adsorbed reactants will increase the current over that predicted from theory based on diffusion-controlled mass transport. Thus the usually powerful methods based on Nicholson/Shain voltammetric theory are seriously perturbed by adsorption, particularly at high scan rates. In addition, the adsorption of nonelectroactive impurities or reaction products can eventually deactivate the electrode, thus requiring electrode renewal.

While all electrodes are subject to adsorption of reactants, products, or impurities, carbon has special properties in this regard. The wide use of activated carbon for adsorption of a wide range of materials, organic and inorganic, is evidence of the likelihood of adsorption to carbon electrodes. As will be noted later, adsorption to carbon depends on electrode history, and adsorption effects can be an important issue when choosing carbon electrodes for a given application.

## III. CARBON ELECTRODE MATERIALS PROPERTIES

Section I identified the performance criteria that determine the suitability of a given electrode for an electroanalytical application. We now turn to the question of what aspects of the carbon determine its performance and electrochemical behavior. Since the structure of $sp^2$ carbon materials is more complex than that of pure metals like Pt, there are more structural variables that affect behavior. As a consequence, $sp^2$ carbon can vary widely in conductivity, stability, hardness, porosity, etc., and care must be taken to choose and prepare the carbon material for an electrochemical application. Before discussing particular carbon electrode materials, we first consider which structural variables affect the electrochemical observables discussed in Section II.

### A. Carbon Microstructure

Single-crystal graphite is inherently anisotropic, as shown in Figure 10.3. The ratio of the in-plane (a-axis) resistance to the c-axis resistance is about $10^{-4}$, with

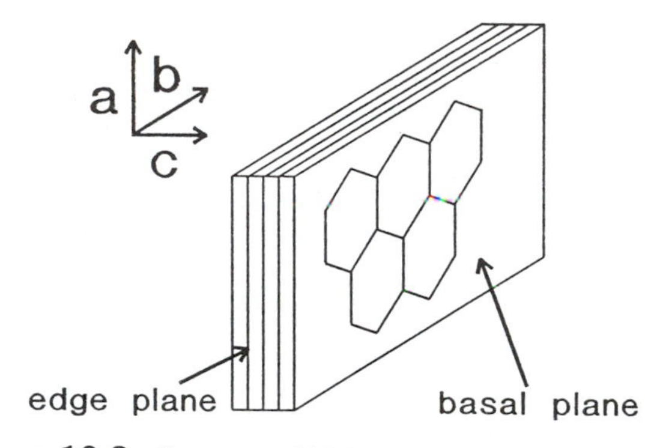


**Figure 10.3** Structure of highly ordered pyrolytic graphite.

the a-axis resistivity being about 25 times that of copper [2]. Parameters of importance are the a-axis and c-axis microcrystallite sizes, which vary from a few micrometers for ordered graphite down to tens of angstroms for glassy carbon and carbon black. The behavior of the material also depends on the interplanar spacing, which is 3.354 Å for crystals, up to 3.6 Å for glassy carbon or carbon black, and much larger for intercalation compounds. This parameter strongly affects the coupling between layers and as a consequence the electronic and spectroscopic behavior of the material.

The plane perpendicular to the c axis is known as the basal plane; that perpendicular to the a axis is usually referred to as edge plane. The electrochemical properties of the two faces differ greatly, as shown in Table 10.1. Clearly, one determinant of carbon electrode behavior is the relative density of edge and basal plane at the electrode surface. It is tempting to hypothesize that the observed electrochemical behavior of graphite is an area-weighted average of edge and basal properties [8], but this is probably an oversimplification. However, many of the effects of electrode treatments can be understood in terms of their

**Table 10.1** Anisotropic Electrochemical Properties of HOPG

| | Basal Plane | Edge Plane |
|---|---|---|
| Capacitance, $\mu F/cm^2$ | $<1.0$ | $\sim$50–70 |
| Adsorption of antraquinone disulfonate, $pMol/cm^2$ | $<1$ | $>150$ |
| k° for $Fe(CN)_6^{3-/4-}$, cm/sec | $<10^{-6}$ | 0.06–0.10 |

*Source*: Data from References 2, 9, and 10.

effects on the relative amounts of edge and basal regions. HOPG is very similar to single crystal graphite, with the same interplanar spacing but slightly greater rotational disorder.

Although HOPG and single-crystal graphite are examples of an informative limiting case, most carbon electrodes consist of randomly oriented $sp^2$ carbon, such as polycrystalline graphite with less oriented crystallites (e.g., spectroscopic graphite or pencil lead) or glassy carbon. These retain the parallel planes of condensed aromatic rings over short distances, but differ greatly from HOPG in long-range order.

## B. Surface Roughness

Except for a few special cases including metal single crystals and HOPG, no solid electrode surface is atomically smooth. Surface roughness affects electroanalytical measurements in several ways, depending on the scale of the roughness. The most obvious is roughness comparable to the diffusion layer thickness, which will affect the assumption of planar diffusion common to the majority of solid electrode experiments. For typical voltammetry or amperometry in the 0.01 to 10 s time scale, the diffusion layer thickness is about 2 to 70 μm thick, based on $\sqrt{Dt}$ (see Chapter 2 for the origin of $\sqrt{Dt}$). Thus a planar electrode assumption would require surface roughness below 1 μm for this time scale. For this case, the observed area will be the projected area of the electrode surface. At shorter times, the diffusion layer will track any roughness, yielding an apparently larger electrode area. These limiting cases are shown in Figure 10.4.

For electrodes with total dimensions smaller than $\sqrt{Dt}$, often called microvoltammetric electrodes (see Chapter 12), roughness is less of an issue compared to mass transport. Such electrodes are typically less than 10 μm in diameter and exhibit radial rather than planar diffusion. In most cases (the exception being fast voltammetry) the diffusion field is thicker than both the electrode diameter and its surface roughness, and the diffusion-limited response is unperturbed by roughness.

When one considers a distance scale much smaller than 1 μm, surface roughness also is an issue to observed electrode behavior. The ratio of the microscopic surface area to the projected electrode area is usually designated the roughness factor, and can vary from 1.0 to 5 or so for typical solid electrodes, or much higher for porous electrodes. Capacitance, surface faradaic reactions, adsorption, and electrode kinetics all depend on microscopic area. For example, double-layer capacitance increases with roughness such that the apparent capacitance ($C^{\circ}_{obs}$) is larger than the value for a perfectly flat electrode ($C^{\circ}_{flat}$) as shown in Equation 10.1:

$$C^{\circ}_{obs} = \sigma C^{\circ}_{flat} \tag{10.1}$$

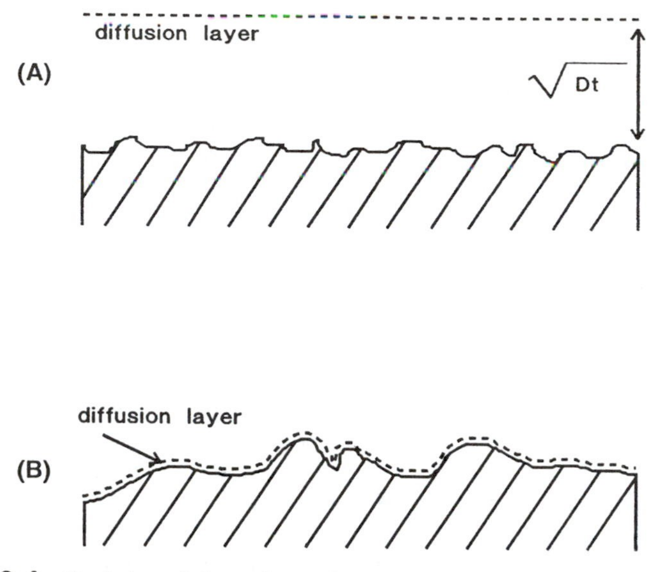


**Figure 10.4** Depiction of electrode roughness compared to diffusion layer thickness, $\sqrt{Dt}$. Dotted line indicates approximate boundary of diffusion layer, with (A) diffusion layer thickness greater than surface roughness, resulting in an observed area equal to the projected area, and (B) diffusion layer thickness on the order of surface roughness, resulting in a larger apparent electrode area.

where $\sigma$ is the roughness factor. Since capacitance is a major component in the voltammetric background, roughness can have a significant effect on electrode response. Recall that the diffusion-controlled analytical signal is usually proportional to *projected* electrode area, while the capacitive or surface faradaic background tracks the *microscopic* area.

## C. Surface Cleanliness

Since electroanalytical response ultimately depends on electron transfer between an electrode material and a solution species, any intervening layers or films are of obvious consequence. As noted earlier, carbon materials are prone to adsorption, and it is not trivial to prepare carbon surfaces that are not contaminated with adsorbed layers, usually of adventitious organic impurities. In order to avoid poor or irreproducible responses from carbon electrodes, the user must either find conditions where surface impurities have negligible or reproducible effects on the process of interest, or prepare the carbon surface in a way that avoids surface films. As is evident from the history of electroanalytical chemistry since the DME, surface cleanliness is a major issue with solid electrodes, no less so for carbon.

## D. Surface Oxidation

Both the background current and the interaction of carbon electrodes with solution redox systems can change when the electrode surface is oxidized. In many cases, surface oxidation is intentional, and numerous papers have appeared on electrochemical pretreatment (ECP) of carbon electrodes. In these cases, the surface is oxidized by various procedures in order to enhance electron transfer or adsorption. Surface oxidation may also be unintentional, resulting from air exposure, polishing, etc. In fact, it is quite difficult to prepare an oxide-free carbon surface, and typical polished surfaces have 5–15% oxygen on the surface (relative to carbon). Furthermore, the formation of surface oxides on carbon is usually chemically irreversible, so they are difficult to remove once formed. There are many reports proposing that surface oxides promote electron transfer, perhaps through a redox mediation mechanism where oxygen-containing surface redox groups transfer electrons to or from solution components. In some cases, known functional groups (e.g., quinones) were synthesized on the surface to promote a particular reaction, as shown in Figure 10.5 [11]. In other cases, oxides increase the voltammetric background, perhaps by ion association

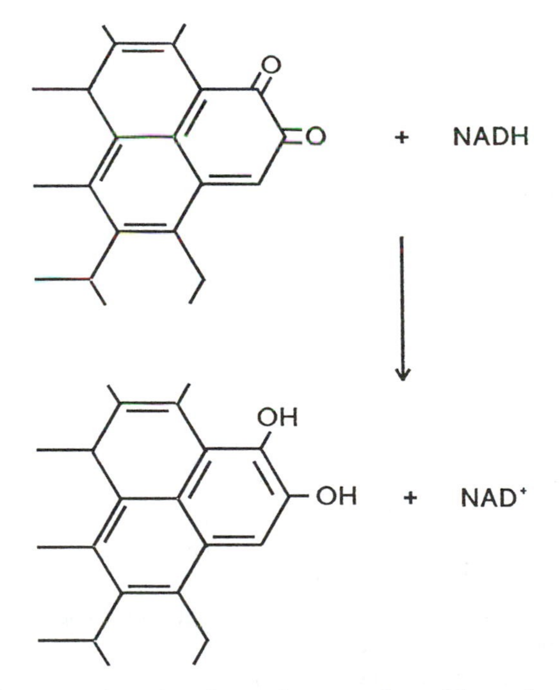


**Figure 10.5** Illustration of surface quinones and possible mechanism of electron transfer via catalytic proton and electron transfer from NADH. [See Ref. 11.]

or a surface faradaic reaction (recall Fig. 10.1). Clearly, surface oxides can be important contributors to electrochemical behavior, and are an issue in choosing and preparing carbon electrodes.

### E. Surface Heterogeneity

Any solid electrode, but particularly carbon, can have surface properties that vary spatially across the electrode plane. For example, an uneven coverage of organic impurities might lead to regions of fast and slow electron transfer kinetics. If the surface heterogeneity is spatially small compared to the diffusion layer thickness, one observes an electrochemical response that is a weighted average of the differing surface regions. The details of this phenomenon are fairly complex and often subtle [12,13], but the effect can be important to electroanalytical applications. Examples are considered in due course, and they include the behavior of carbon paste electrodes and carbon electrodes modified with catalysts.

## IV. COMMON CARBON ELECTRODE MATERIALS

Given the preceding overview of performance criteria and underlying electrode variables, we can now consider widely used carbon electrode types. They are grouped into four main classes: pyrolytic graphite, polycrystalline graphite, glassy carbon, and carbon fibers. This list neither is in historical order nor implies popularity, but rather is organized according to complexity of the carbon material, with the most ordered graphites first. Within each category we discuss four issues of importance to electroanalytical applications:

1. Origin and structure
2. Physical properties, including resistivity, porosity, hardness
3. Preparation and modification
4. Electroanalytical applications

### A. Pyrolytic Graphite (PG) and Highly Ordered Pyrolytic Graphite (HOPG)

#### *1. Origin and Structure*

PG is made by pyrolysis of light hydrocarbons onto a hot (ca. 800°C) stage, often followed by heat treatment to higher temperatures. Deposition occurs such that the a axis is parallel to the stage surface, and the process usually leads to many "islands" of graphite formation exhibiting a macroscopically rough surface. During high-temperature treatment often >2000°C, the carbon "graphitizes," meaning the $sp^2$ layers orient parallel to each other with 3.354 Å interlayer spacing. In addition, the crystallite sizes grow during graphitization, yielding an end product with several hundred angstrom crystallites. HOPG [14] is made from PG by pressure annealing in a hot press at ca. 3000°C and several kilobars, and is often referred to as pressure-annealed pyrolytic graphite (PAPG). The interlayer spacing is still 3.354 Å, but crystallite sizes are at least 1 μm (10,000 Å). HOPG has a smooth, shiny basal surface, while PG is mottled and dull. Due to the lower defect density in HOPG, it has a much higher ratio of c-axis to a-axis resistivity [2,14].

### 2. *Properties*

The dominant structural property of PG and HOPG is the long-range order of the graphitic layers. Since the anisotropy extends through the entire electrode piece (i.e., several cm) a typical electrode will have mostly basal or mostly edge plane exposed to the solution. Since PG has many more defects than HOPG, its basal surface will reflect "contamination" by a usually uncontrolled contribution of edge plane regions.

As shown in Table 10.2, both PG and HOPG have low a-axis resistance. Although the c-axis resistance is much higher, it is normally still too low to cause noticeable iR error in electroanalytical applications. Neither PG nor HOPG is porous, unlike common polycrystalline graphites such as "spectroscopic graphite." The roughness factor for HOPG basal plane is as close to 1.0 as any graphite, and except for defects, HOPG is atomically smooth. PG basal plane is much less so, in part because of the island structure, but it is still smoother than polycrystalline graphite. The edge planes of both HOPG and PG are qualitatively different from basal plane and are usually quite rough. The solvent will

**Table 10.2** Resistivities of Carbon Materials

| | $\rho$, $\Omega$ - cm[a] |
|---|---|
| HOPG, a axis | $4 \times 10^{-5}$ |
| HOPG, c axis | 0.17 |
| Pyrolytic graphite, a axis | $2.5 \times 10^{-4}$ |
| Pyrolytic graphite, c acis | 0.3 |
| Polycrystalline graphite[b] | $1 \times 10^{-3}$ |
| Glassy carbon[c] | $4 \times 10^{-3}$ |
| Carbon fiber[d] | $7.5 \times 10^{-4}$ |
| Platinum | $1.1 \times 10^{-5}$ |
| Copper | $1.7 \times 10^{-6}$ |

[a]The DC resistance of a carbon piece equals $\rho A/L$ where A is cross-sectional area and L is length along current path.
[b]Ultracarbon UF-4s.
[c]Tokai GC-20.
[d]Thornell P-55s 4K ($\rho$ varies significantly with fiber type).

not penetrate into the bulk material at the edges because of the close interplanar spacing and lack of porosity.

An ideal graphite crystal is a semimetal with some properties reminiscent of semiconductors [15–17]. There is no band gap, but there is a low density of electronic states near the Fermi level. The consequence of this electronic nuance is a small space charge capacitance in series with the usual double-layer capacitance [16]. Since capacitors in series exhibit a total capacitance lower than any of the components, the observed double-layer capacitance of a perfect graphite crystal basal plane is controlled by the low space charge capacitance of $<3$ $\mu F/cm^2$. HOPG basal plane is sufficiently ordered to exhibit electronic properties similar to an ideal crystal, and has an anomalously low observed capacitance of 1–3 $\mu F/cm^2$, with the variation caused by residual defects [9,17]. This low value is observed only for the basal plane, with the capacitance of HOPG (or PG) edge plane being $>50$ $\mu F/cm^2$. This unusual electronic effect is observed only for HOPG basal plane and is not an issue for most carbon electrodes. However, it is likely to be a major factor in controlling electrode kinetics on ordered graphite, and provides a physical basis for understanding observed rates of electron transfer [10].

*3. Preparation*

Due to their anisotropic structure, preparation of PG and HOPG electrodes differs from most other carbon electrodes. The weak interplanar bonding makes it possible to manually cleave PG and HOPG to expose basal surfaces. In the case of HOPG, common adhesive tape is adhered to the basal surface and lifted, thus removing several layers of graphite. The exposed surface is then mounted in a holder that defines the electrode area, usually made of Teflon. After recognizing that edge plane defects can differ greatly from basal plane in electrochemical properties (Table 10.1), it is clear that residual defects on cleaved basal plane may be important. In order to assure that the observed properties of HOPG are really those of the basal plane, some measure of defects is necessary. The capacitance is a useful semiquantitative indicator, since defects have much higher capacitance than basal plane. Thus a capacitance below 2 $\mu F/cm^2$ indicates a low defect surface. The ramifications of defects on HOPG basal plane are beyond the scope of this chapter, except to say that they must not be ignored if one is intending to study true basal plane behavior [9,10]. One must either reduce their density to an acceptable level or be convinced that they do not affect the phenomenon under study.

Since one starts with a less ordered material, cleaved PG basal plane has a much higher defect density. The bittersweet result is that the electrode behavior is more reproducible than that of HOPG, but one can never achieve low enough defect density to study true basal plane. Dryhurst has described the use of basal plane PG electrodes in detail [1], with the most common pretreatment being polishing of an epoxy-encapsulated PG piece. A generic mount for PG is shown in Figure 10.6A. The PG piece is housed in epoxy, with either the edge or basal plane exposed to the solution. Although basal plane PG has been used with success, electrodes based on glassy carbon and carbon paste have become both more popular and commercially available.

Edge plane HOPG and PG have been used as electrodes by mounting the graphite piece in a suitable holder with the edge exposed. After encapsulation in epoxy or Teflon such that only edge is exposed, the surface may be polished [8,18]. HOPG is rarely studied in this fashion, since one generally does not use a nearly perfect material to study its very imperfect edge. Edge plane pyrolytic graphite (EPG) electrodes are fairly common, however, because they combine the relatively high reactivity of edge plane with the low porosity of PG. In many cases, surface preparation of EPG electrodes is designed to maximize adsorption.

### 4. *Applications*

PG and HOPG materials play an important role in understanding carbon electrode structure and reactivity due to their anisotropic nature [2]. In some special cases, basal plane HOPG and EPG electrodes are of particular value in electroanalytical applications for both fundamental and pragmatic reasons. Examples are shown in Figures 10.7 to 10.9. Figure 10.7 shows the voltammetry

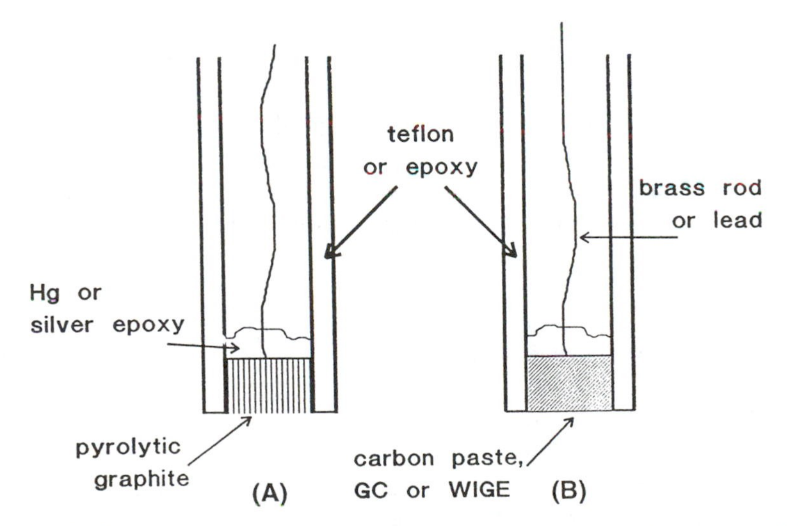


**Figure 10.6** Electrode construction (A) for pyrolytic graphite, where edge or basal plane orientation may be exposed to solution (edge is shown); (B) for GC, WIGE, or carbon paste.

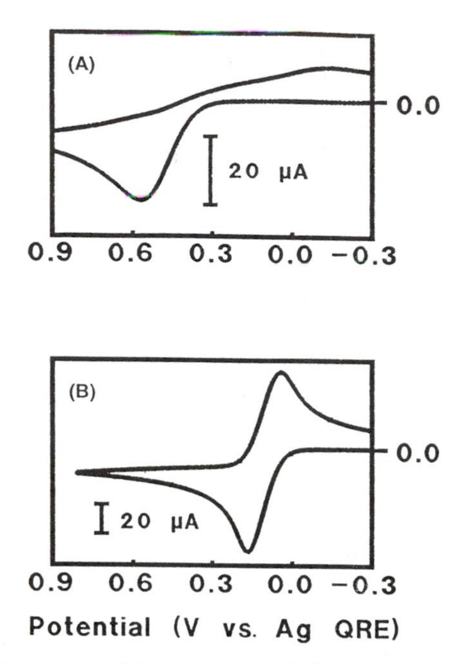


**Figure 10.7** Voltammetry of 2 mM $Co(phen)_3^{3+/2+}$ in 1 M KCl at (A) low edge plane defect density, $\Delta E_p$ =715 mV (k° = $8 \times 10^{-6}$ cm/s) and (B) high defect density HOPG, $\Delta E_p$ =126 mV (k° = $3 \times 10^{-3}$ cm/s). Scan rate = 0.2 V/s. [Adapted from Ref. 10.]

of $Co(phenanthroline)_3^{2+/3+}$ on basal HOPG for both low-defect and high-defect surfaces. Note the large difference in $\Delta E_p$ (and therefore k°) for the two surfaces. Figure 10.8 shows a kinetically fast ($Fe(phen)_3^{2+/3+}$) and slow (dihydroxyphenylacetic acid) system on basal plane HOPG. Note the very low background currents in both cases, due to the low capacitance of basal HOPG. Figure 10.9A shows voltammetry of 2,6-diamino-8-purinol on edge plane PG. The sharp peaks are indicative of strong adsorption, which is not observed on conventionally prepared glassy carbon (see Fig. 10.9B). Not only does adsorption to EPG permit observation of several redox couples, but the adsorbed species exhibit rapid electron transfer [18].

These examples notwithstanding, the unusual electronic properties of HOPG, the adsorption common to EPG, and the relative difficulty of handling anisotropic materials have prevented widespread practical applications. Polycrystalline graphite and glassy carbon exploit some of the same principles as HOPG and PG, but are more broadly applicable.

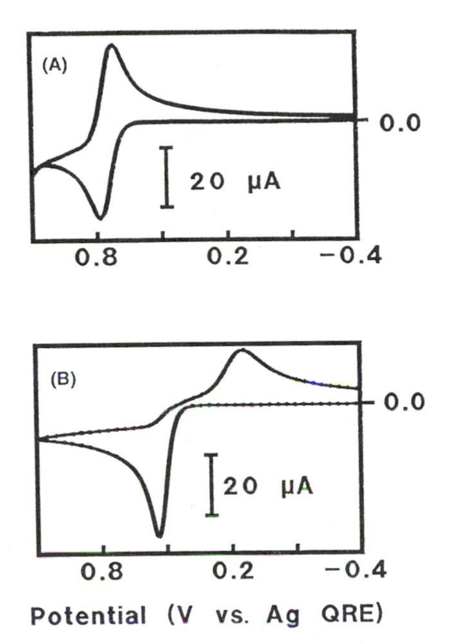


**Figure 10.8** Voltammetry illustrating variety of rates obtained for different systems at basal plane HOPG: (A) 2 mM $Fe(phen)_3^{3+/2+}$ in 1 M KCl ($\Delta E_p$ = 63 mV, k° = 0.07 cm/s); (B) 1 mM DOPAC in 0.2 M $HClO_4$ ($\Delta E_p$ = 403 mV). Scan rate = 0.2 V/s. [Adapted from Ref. 10.]

## B. Polycrystalline Graphite Electrodes

Throughout this chapter, the term polycrystalline graphite is used to denote microcrystalline graphitic materials with 3.354 Å interplanar spacing, but with randomly oriented a axes and small (50–500 Å) crystallites. The electrode materials in this group have a polycrystalline graphite structure in common, but differ substantially in the way the material is used. The carbon is usually graphitized, with microcrystallites of several hundred angstroms. It can be powdered graphite (e.g., Ultracarbon UCP-1-M), a solid with random crystallite orientation ("spectroscopic graphite"), graphitized carbon black, or pyrolyzed films. The key to practical use of these materials in electroanalytical chemistry is control of porosity.

### 1. *Origin and Structure*

Most polycrystalline graphite is made by heat treating high-molecular-weight petroleum fractions ("pitch") at high temperatures to bring about graphitization.

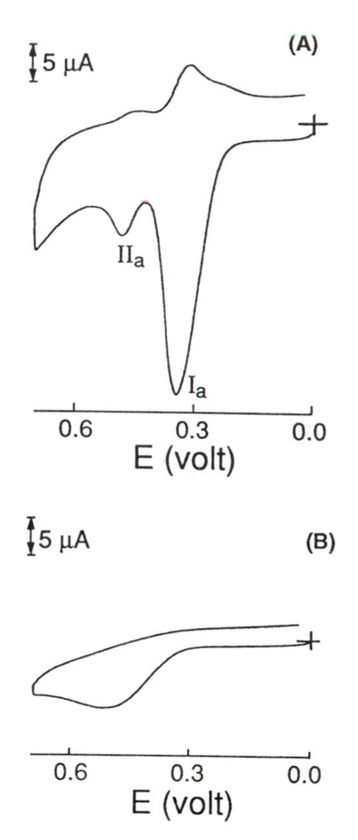


**Figure 10.9** Voltammetry of 0.2 mM 2,6-diamino-8-purinol in pH 7.0, 0.5 M phosphate buffer at: (A) pyrolytic graphite, resurfaced with 600 grit silicon carbide paper; (B) polished GC. Scan rate = 0.2 V/s. [Adapted from Ref. 18.]

In the case of spectroscopic rods, for example, heavy hydrocarbons are shaped and heated at about 600–800°C to form a "green body," then graphitized at ca. 2500°C. Alternatively, one may start with low molecular weight hydrocarbons to make powders such as carbon black. The term "graphite" is used to designate materials that have been subjected to high temperatures, and thus have aligned the $sp^2$ planes parallel to each other. "Graphitized carbon black" is a powder formed at relatively low temperatures (< 1000°C), then heated to above 2000°C. It is important to distinguish graphite materials from glassy carbon, which is made from high-molecular-weight polymers and does not form extended parallel graphitic planes.

### 2. *Properties*

A key property of polycrystalline graphite is porosity. The density of HOPG is 2.26 $g/cm^3$, while that of a typical spectroscopic graphite rod (Ultracarbon

UF-4S) is 1.8 g/cm$^3$. Since the interplanar spacing is the same for both materials, the difference is porosity; in this case the rod is about 21% void space. Any electrode made from powdered graphite will also contain void space. If nothing is done to reduce this inherent porosity, solution will enter the electrode material and large, usually uncontrolled background current will result. While polycrystalline graphite electrodes are useful in synthesis, often on an industrial scale, their porosity must be reduced for electroanalytical applications. With the exception of graphite thin films, all microcrystalline graphite electrodes involve some means to reduce porosity.

*3. Preparation*

**a. Carbon Paste Electrodes (CPE).** One means to control the porosity of microcrystalline graphite materials is to fill the pores with an inert liquid. In the case of carbon paste, a heavy hydrocarbon such as Nujol or hexadecane is mixed with powdered graphite (e.g., Ultracarbon UCP-1-M) in a composite that is typically 70% carbon by weight [1,19]. After thorough mixing, the paste is packed into a shallow well (typically 2–5 mm deep) in an inert holder (usually Teflon) with electrical contact at the back (Fig. 10.6B). It is important that the Teflon face of the electrode is flat and smooth, so that a flat, reproducible carbon paste surface results. The pasting liquid is chosen for inertness, low solubility in the solvent of interest, and low volatility. Heavier hydrocarbons or waxes yield stable electrodes with relatively slow electrode kinetics, while lighter liquids (e.g., hexane) yield more reactive surfaces with shorter lifetimes [20]. Ceresin wax yields a carbon paste that is compatible with a wide range of solvents used in electrochemistry [21]. The reactivity differences have been attributed to coating of the graphite particles by the hydrocarbon, with lighter liquids forming thinner films.

Once the paste is packed into the electrode body, the surface is "polished" on some mild abrasive such as an index card (historically an IBM computer card). Provided the face of the electrode holder is initially smooth, very reproducible electrode areas (less than 5% variation) and reactivities are attained. In fact, a major motivation for developing carbon paste electrodes was the ease of surface renewal and reproducibility.

The nature of the CPE is amenable to chemical modification by a variety of methods. CPEs are an example of the more general class of composite electrodes, in which chemically useful functionalities can be introduced during physical mixture of graphite and pasting liquids. Composite electrodes are considered in general terms later, but the specific case of modified CPEs is discussed briefly here.

To prepare a modified CPE, a modifying agent is either dissolved in the pasting liquid or physically mixed with the paste. In some cases modifier and liquid are dissolved in a common solvent such as hexane, which is later evaporated. Modifiers can be grouped into at least three different classes, depending on how they affect the interactions of solution species with the electrode. First, modifiers such as stearic acid introduce an anionic surface charge which affects local electrostatics. The resulting anionic surface promotes redox reactions of cations over anions [22]. Second, a modifier such as a zeolite [23] or complexing agent [24] may cause selective adsorption of molecules with particular sizes or charges. Zeolite Y in a CPE preconcentrated methyl viologen and $Cu^{2+}$ in the electrode, but excluded heptyl viologen and $Fe(CN)_6^{3-/4-}$ [25]. Third, modifiers such as metal particles [26] and redox mediators [27] can act as catalysts for normally slow redox reactions at the CPE. The major advantages of the CPE for such modification schemes include the ease of preparation and surface renewal, and the variety of possible modifying agents.

**b. Spectroscopic Graphite Electrodes.** As noted earlier, graphite can be formed in particular solid shapes, such as rods. Spectroscopic graphite is so named because of a low level of metallic impurities important to applications in emission spectroscopy. Such rods are useful for electrochemistry once the now familiar problem of porosity is addressed. A common approach is impregnation with ceresin or paraffin wax, leading to the wax-impregnated graphite electrode (WIGE) [1]. A suitable piece of graphite rod (e.g., 6 mm diameter, 1 cm long) is immersed in molten wax, often in a vacuum to remove air from the graphite pores. After cooling and mounting in an inert sheath (e.g., shrinkable Teflon tubing), the end of the rod may be polished before use. Because of its similarity of composition and composite nature, the WIGE has similar electroanalytical behavior to the CPE. However, the CPE is more convenient unless mechanical rigidity is required, and is currently more widely used. Except for applications as a substrate for mercury deposition preceding anodic stripping voltammetry, the WIGE has been largely superseded by the CPE and glassy carbon electrodes.

**c. Graphite Composite Electrodes.** A potentially valuable but more recent development is a variant of the CPE that leads to more rugged electrodes. Graphite powder is mixed with a suitable filler, then either physically or chemically bonded to form a conductive solid composite. An early example is the "kelgraf" electrode made by pressing powdered Kel-F and graphite [28,29]. At sufficient temperature and pressure, Kel-F is fluid enough to fill pores, and the resulting electrode exhibits good performance without wax or pasting liquid. Kelgraf has a wide range of solvent compatibility, since both Kel-F and graphite are quite inert. As noted in the next section, a composite electrode with exposed graphite particles surrounded by insulator can exhibit very low background currents of value to electroanalysis, particular in LC detectors.

Carbon composite electrodes have also been made by cross-linking a polymer after mixture with powdered graphite or carbon black [30]. In some cases, these electrodes incorporate modifiers such as zeolites to affect reactivity. As with the CPE, the electrode is polishable and relatively easily prepared and renewed.

### 4. *Applications*

An essential issue regarding applications of microcrystalline graphite electrodes is the effect of their composite nature on both analytical and background currents. The need to reduce porosity leads to an electrode surface consisting of graphite islands in a sea of electrochemically inert binder or filler. If the islands are small relative to $\sqrt{Dt}$ (they usually are), radial diffusion will occur toward the islands and there will be a locally high flux of redox-active species [2,12]. At long times, $\sqrt{Dt}$ usually exceeds the spacing between active particles, and the overall diffusion field becomes planar. Thus, for conventional experiments on a 0.1–10 s time scale, graphite composite electrodes exhibit diffusion behavior expected for a macroscopic homogeneous plane even though diffusion is nonplanar on a microscopic scale. A subtle consequence of this effect is the high local current density at the graphite particle compared to the average over the entire electrode. The electrode kinetics may not be fast enough to support this current density, leading to a decrease in the observed electron transfer rate. When the particle is also coated with a thin layer of pasting liquid or wax, the overall effect is reduced charge transfer reversibility.

A second consequence of the heterogeneity of graphite composites is reduced background current. Since the background originates in capacitive and surface faradaic currents on the particles themselves, a smaller particle area will yield smaller background. Stated more quantitatively, the background current from surface processes is proportional to the fraction of total electrode area that is graphite. This fraction is 20% for a typical kel-graf electrode, or only a few percent for the CPE. The large reduction in background current resulting from low active surface area in the CPE is a major reason for its effectiveness.

The overall picture of graphite composite electrodes that emerges from this discussion starts with a graphite solid or powder whose pores are filled with inert material. Solvent compatibility is determined by that of the filler, and can range from electrodes that are generally restricted to aqueous solutions (e.g., CPE with a light hydrocarbon) to essentially any solvent (kel-graf). Without special effort, the graphite particles are coated with thin layers of inert material, either residual hydrocarbon liquid or the inert solid that was smeared across the particle during polishing. The effects of this film may be minor, but generally it reduces electron transfer rates. Composites with a high surface density of graphite (e.g., WIGE) exhibit background currents that are similar to a homogeneous material such as glassy carbon, since most of the surface is electrochemically active. As the fractional coverage of graphite is reduced, however, the background current decreases. For cases where the current is diffusion-limited, this reduction in background is not accompanied by a reduction in the analytical signal.

**Table 10.3** Potential Limits for Carbon Electrodes

| Electrode | Electrolyte | Potential Limit vs. SCE, V[a] | |
|---|---|---|---|
| | | Positive | Negative |
| PG (basal) | 0.1 M HCl | 1.0 | -0.8 |
| Spectroscopic graphic | 0.1 M HCl | 1.19 | -0.46 |
| | 0.1 M HCl | 1.85 | -0.39 |
| WIGE | 0.1 M HCl | 1.0 | -1.3 |
| | Phosphate, pH 7.02 | 1.35 | -1.38 |
| CPE (Nujol) | 1 M HCl | $>1.0$ | -0.9 |
| CPE (ceresin wax paraffin oil) | 0.1 M $H_2SO_4$ | 1.7 | -1.2 |
| Glassy carbon[b] | 0.1 M HCl | 1.05 | -0.80 |
| | 0.05 M $H_2SO_4$ | 1.32 | -0.80 |

[a]Recall that these limits depend on scan rate and electrode area.
[b]Background is very dependent on condition and preparation.
*Source*: Adapted from Reference 1.

Potential limits for various carbon electrodes are shown in Table 10.3. As noted early in the chapter, Table 10.4 indicates the observed electron transfer rate constants for $Fe(CN)_6^{3-/4-}$. This redox system has long been used as a benchmark system for comparing electrode reactivity, even though thorough investi-

**Table 10.4** k° for $Fe(CN)_6^{3-/4-}$

| Electrode | Pretreatment | Electrolyte | k°, cm/s | Reference |
|---|---|---|---|---|
| HOPG, basal | Cleaved | 1 M KCl | $10^{-6}$ | 2 |
| HOPG, edge | Polished | 1 M KCl | ~0.08 | 2 |
| PG, basal | Cleaved | 0.5 M KCl | 0.002 | 2 |
| PG, basal | Ground | 0.5 M KCl | 0.007 | 2 |
| CPE (Nujol) | | 1 M KCl | $1 \times 10^{-4}$ | 2 |
| CPE (hexane) | | 1 M KCl | 0.003 | 2 |
| GC-20 | Polished | 1 M KCl | 0.002–0.06 | 2 |
| GC-20 | Ultraclean polish | 1 M KCl | 0.14 | 31 |
| GC-10 | High-speed polish | 1 M KCl | 0.007 | 32 |
| GC (LeCarbone) | High-speed polish | 1 M KCl | 0.098 | 33 |
| GC-20 | Fractured | 1 M KCl | 0.50–0.55 | 34 |
| GC-20 | Polished, laser-activated[a] | 1 M KCl | $>0.50$ | 34 |
| GC-30 | Fractured | 1 M KCl | $>0.50$ | 8 |
| GC-30 | Polished, laser-activated[a] | 1 M KCl | $>0.50$ | 8 |

[a]25 MW/cm$^2$.
*Source*: Adapted from Reference 2.

gation of its kinetics has revealed it to be far from simple [2]. Nevertheless, it is a redox couple whose kinetics are quite dependent on surface properties, and there is a large collection of kinetic data on different carbon surfaces.

Although the pasting liquid or binder generally decreases k° for $Fe(CN)_6^{3-/4-}$ compared to GC, there are trade-offs that may still favor the use of a composite electrode even with slower kinetics. For LC detection, for example, the difference between a k° of 0.001 cm/s and 0.1 cm/s may be unimportant, while the lower background current of the less reactive electrode may be essential. As always, the application dictates electrode choice.

## C. Glassy Carbon

Also known as vitreous carbon, glassy carbon has been the subject of intense electroanalytical research in the past 10 years. It is impermeable to liquids and gases, and thus porosity is not an issue, as it is with polycrystalline graphite. It is easily mounted, polishable, and compatible with all common solvents. These properties have led to widespread use in mechanistic electrochemistry, LC detection, and voltammetric analysis.

### *1. Origin and Structure*

Unlike PG or polycrystalline graphite, glassy carbon (GC) is made from a high-molecular-weight carbonaceous polymer, often polyacrylonitrile, phenol/formaldehyde resin, etc. When such polymers are heated to 600–800°C, most of the noncarbon elements are volatilized, but the polymer backbone is not degraded. Regions of hexagonal $sp^2$ carbon are formed during this treatment, but they are unable to form extensive graphitic domains without breaking the original polymer chain. After this initial heat treatment, the material is heated slowly under pressure to 1000°C (GC-10), 2000°C (GC-20), or 3000°C (GC-30). Even at 3000°C, only small graphitic domains (50–100 Å) are formed, and the material resembles a tangled mass of graphitic ribbons. GC is hard, isotropic on a scale larger than 100 Å, and about 1% as conductive as the a axis of HOPG (Table 10.2). GC density of 1.5 g/cm$^3$ implies a void volume of about 33%, but the voids are very small and unconnected, thus preventing liquid or gas penetration. The interplanar spacing in GC is about 3.48 Å, slightly larger than the 3.354 Å for HOPG. Although this difference is small, it is enough to perturb the electronic behavior of the material, and the disorder of GC appears to destroy the semimetal character observed for HOPG. In very general terms, GC is similar to polycrystalline graphite in composition, bonding, and resistance, but differs greatly in porosity, hardness, and mechanical properties, all due to the difference in structure caused by the origin of each material.

There are several manufacturers of GC, but Tokai is the most commonly used by electrochemists. GC-10 has somewhat irreproducible electrochemical properties, and GC-30 often contains small (a few μm) bubbles apparently formed during heat treatment. Thus Tokai GC-20 is the most commonly used material, and is available as rods, plates, or disks. Although cheaper than platinum, it is more expensive than spectroscopic graphite.

An unusual but often useful variant of glassy carbon is reticulated vitreous carbon (RVC) [35]. It is chemically similar to conventional GC, with a comparable preparation procedure, but the final material is very porous. The pores are large enough for easy penetration by electrolyte, to the point at which solvent readily flows through the RVC "solid." RVC is available with a range of pore sizes and densities, with a typical material being >90% pores. The material is easily cut with a cork borer or razor blade, but caution should be exercised to avoid exposure of skin or eyes to the hard, irritating dust. RVC is useful for experiments requiring high surface area, such as controlled-potential electrolysis. It has also been used as an optically transparent electrode [36] and for fabrication of microelectrode arrays by filling the pores with epoxy [37].

*2. Properties*

As noted earlier, the resistance of GC is low enough to be negligible in most electroanalytical applications. The unconnected pore structure prevents any bulk penetration of solvent, but there are reports of superficial entry of electrolyte ions into surface pores. The occasional pits that originate as gas bubbles during heat treatment cover a small fractional area, and do not appear to significantly affect electrochemical response. Surface roughness is a strong function of preparation technique, but roughness factors in the range of 1.3 to 3.5 have been reported [38–40]. Nodules with diameters of 50 to 300 nm have been observed with scanning electron microscopy (SEM) and scanning tunneling microscopy (STM), and contribute to surface roughness [39]. The size scale of roughness on polished surfaces is usually much smaller than $\sqrt{Dt}$ for most experiments, so planar diffusion conditions are readily achieved. The largest effect of roughness is an increase in microscopic area, particularly evident when measuring adsorption.

GC exhibits the "excess" apparent capacitance noted in general terms in Section II. For smooth, heat-treated GC, the observed capacitance is quite low, ca. 10–20 $\mu F/cm^2$ [41]. This value most closely approximates that of a classical double layer without any anomalous semimetal character of the electrode. For the vast majority of polished GC surfaces, the observed capacitance is larger, ranging from 30 to 70 $\mu F/cm^2$ [2,31]. Some of this increase is due to roughness (recall that $C^{\circ}_{obs}$ is proportional to microscopic area), but much is due to surface faradaic or adsorption processes. The background current on GC is generally larger than on graphite composites because the entire surface is active. Although the observed capacitance is generally larger on GC than on Pt, gross oxidation of carbon is kinetically slow, and the anodic range of GC is significantly more positive than Pt. This property has made GC an attractive electrode for studying oxidations, particularly in aqueous solution.

Since GC does not require the binders or fillers used in graphite composites, there is no residual binder on the active carbon surface. Thus GC is potentially more reactive toward electron transfer. The k° for $Fe(CN)_6^{3-/4-}$ ranges from <0.001 to >0.5 cm/s for GC in 1 M KCl, depending on surface pretreatment, as shown in Table 10.4.

*3. Preparation*

**a. Polishing.** Nearly all GC electrodes are polished at some point in their use, and in a majority of cases, polishing is the only surface preparation employed. Due to the sensitivity of k° on surface condition, a wide variety of polishing protocols has been developed [2,31]. The success of a given procedure is often assessed by measuring k° for $Fe(CN)_6^{3-/4-}$ and observing the background current. Although each lab has developed its own preferred procedure, some generalizations are available. Performance of polished GC depends on how it is mounted, and on the polishing materials and procedure.

A large number of GC electrodes, both commercial and custom, are encapsulated in Teflon or Kel-F. For example, Bioanalytical Systems press-fits a short GC rod into a Kel-F well (as in Fig. 10.6B), then polishes the exposed disk. In some cases GC is encapsulated in epoxy. For encapsulated electrodes, there is a danger of smearing a thin film of epoxy or Kel-F across the face of the GC. For example, Engstrom used XPS to demonstrate Teflon carryover during polishing of Teflon-encapsulated GC [42]. This can be a minor or negligible issue for numerous applications, but it can have a significant effect on experiments involving electrode kinetics or adsorption.

In order to avoid this issue altogether, many workers polish an unencapsulated GC disk before mounting in a suitable cell. For example, a 12-mm-diameter GC disk can be polished (or otherwise pretreated), then mounted in a Teflon cell. A Teflon washer placed on the polished GC surface then defines the electrode area, as shown in Figure 10.10.

Encapsulation of GC in glass is fairly uncommon, due to difficulty in making an adequate seal between the GC and glass. Apparently the two materials do not wet each other easily, and there can be some outgassing of the carbon during heating, thus interfering with the seal. Several suppliers (Cypress Systems, Electrosynthesis, Bioanalytical Systems) offer glass-encased carbon fiber electrodes that are very similar to GC. These are discussed in Section IV.D, as are epoxy-encased carbon fibers.

Whether the GC is mounted or not, the most common polishing procedure involves a series of abrasives used as slurries on a commercial fabric polishing cloth. The as-received electrode or GC piece is first sanded on 600 grit emery paper to remove gross surface defects and impurities. In most cases, each polishing step involves a circular motion for a few minutes, with the objective being uniform roughness over the electrode surface. The sandpaper is followed by

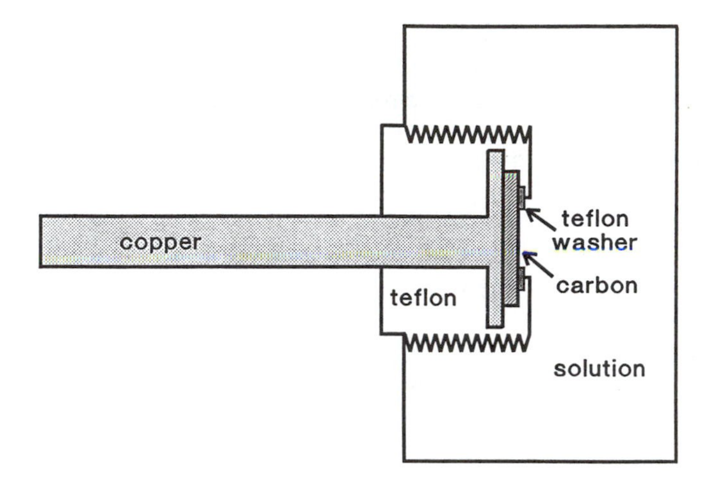


**Figure 10.10** Cell design for GC (or HOPG) disk with electrode area determined by Teflon washer or Viton O-ring.

slurries of alumina in water, of progressively smaller particle size. Most laboratories use prepackaged slurries (Buehler) and an adhesive-backed polishing cloth (Texmet from Buehler) on a glass plate. A few milliliters of the premixed slurry is placed on the cloth and a few minutes of hand polishing ensues. After thorough rinsing, often with sonication, the process is repeated with the next abrasive. Finally, the electrode is sonicated in purified water or electrolyte before immediate use. The most common sequence of alumina abrasives is 0.3 μm, 0.1 μm, 0.05 μm. Once the electrode has been polished with this series, it is generally necessary to repeat only the final, 0.05 μm polish (followed by sonication) between electrochemical experiments.

As an indication of the performance of this generic polishing procedure, the resulting observed k° for $Fe(CN)_6^{3-/4-}$ in 1 M KCl is in the range of $10^{-3}$–$10^{-2}$ cm/s [31]. For a wide range of electrochemical experiments, the electrode thus prepared is reproducible and well behaved. However, several important processes that are quite surface-sensitive (e.g., $Fe(CN)_6^{3-/4-}$ kinetics, ascorbate oxidation, and dopamine adsorption) will show significant variations with repeated polishing, and may depend strongly on the experimenter's care and the reagents used for polishing. For this reason, many variants of the generic procedure have been reported.

Of immediate concern is the purity of the materials, since carbon surfaces are prone to adsorption of trace impurities. Commercial alumina preparations often contain deagglomerating agents, which should be avoided. Some workers have used diamond dust preparations, which are often suspended in an organic paste, which can negatively affect electrode performance. To avoid these problems, dry alumina powders should be slurried with ultrapure water (e.g., Nanopure from Barnstead) before polishing, and commercial diamond slurries should be avoided. Based on $Fe(CN)_6^{3-/4-}$ rates, these simple steps significantly improve electrode kinetics, with k° typically equaling or exceeding 0.05 cm/sec. For exacting work, the polishing cloth can be avoided, and careful protection from contamination from one's hands should be provided. Hu, Karweik, and Kuwana achieved rates of 0.14 cm/s for $Fe(CN)_6^{3-/4-}$ for carefully polished GC [31], Thornton et al. observed 0.07 cm/s for mechanically polished GC [32], and Kamau et al. observed 0.098 cm/s following high-speed mechanical polishing [33]. Several procedures are compared in Table 10.4.

Although these polishing procedures may appear to be tedious and sometimes confusing, their importance depends on the application. When polished with homemade slurries and ultrapure water, GC will perform well for the majority of applications. Since part of the reason for choosing GC over graphite composites is avoidance of hydrocarbons, the straightforward polishing procedures will yield improved kinetics and surface behavior over the CPE or WIGE. For applications in which reproducible electrode area and stable background are of primary importance, conventionally polished GC will perform excellently. The more refined and sometimes more difficult polishing procedures are important when the process under study is quite sensitive to surface impurities. Examples include ascorbic acid oxidation, catechol adsorption, $O_2$ reduction, and more generally those systems where electrode kinetics and adsorption are important. In these cases, the extra effort required to maintain surface cleanliness during polishing is essential. This issue has also driven the development of alternative pretreatment procedures, which we discuss now. The methods come under the general label of *activation*, referring to an increase in electron transfer reactivity relative to a polished GC surface.

**b. Thermal Activation.** Heating GC by various methods has been shown to increase electron transfer rates and adsorption. The mechanism may be as simple as removal of surface impurities, but may also involve oxide removal or even structural rearrangement. Vacuum heat treatments (VHT) at $10^{-4}$ torr [43] or $< 10^{-7}$ torr [41] lead to large k° values for $Fe(CN)_6^{3-/4-}$, and greatly improved voltammetry for ascorbic acid. It appears that VHT exposes "pristine" carbon on the surface, and the reactions studied proceed rapidly on this surface because they are unimpeded by surface impurities. VHT leads to GC surfaces with excellent performance, although the procedure is quite slow and impractical for rapid, repeated use.

A very different but mechanistically similar procedure is exposure of the GC electrode to an intense laser pulse in situ [2,8,39]. A laser pulse (ca. 10 ns, 25 MW/cm$^2$, 1064 nm) rapidly heats the GC surface to temperatures calculated to be above 1000°C. There is little change to the surface morphology (as judged by SEM [34] and STM), but the observed rates for many redox systems increase greatly, sometimes by factors of more than 100. The most likely mechanism for laser activation of GC is thermal desorption of surface films, thus reducing inhibition of electron transfer. The attractions of laser activation include its repeatability, reproducibility, and applicability in situ.

**c. Electrochemical Pretreatment.** The acronym ECP is used here to denote an electrochemical pretreatment of carbon electrodes usually involving oxidation in aqueous electrolytes. Due in part to some surprising early results on the effect of ECP on dopamine and ascorbate voltammetry, ECP methods have been examined fairly vigorously. Researchers in the area sought both to enhance the selectivity and sensitivity of carbon electrodes, and to understand the relationship between ECP-induced surface changes and electrochemical reactivity. An example of ECP effects on initially polished GC is shown in Figure 10.11.

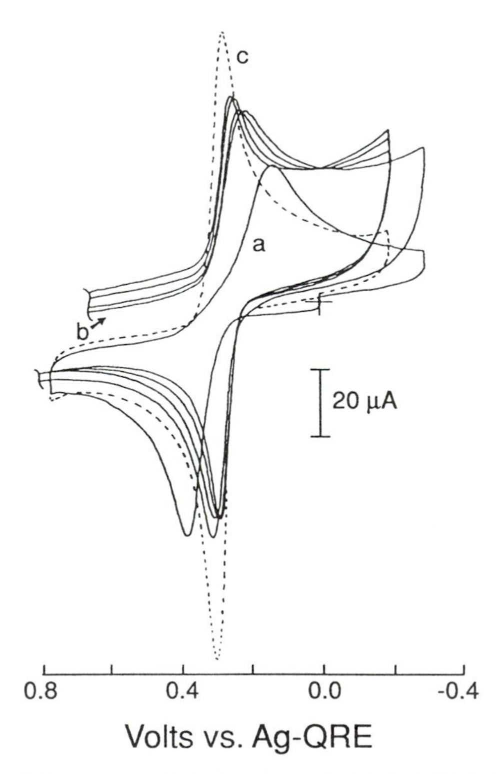


**Figure 10.11** Voltammetry of 0.45 mM catechol in 0.1 M $H_2SO_4$ at polished GC: (a) after polishing; (b) after first, second, fourth, and seventh activation cycles (cycled to 1.78 V vs. Ag QRE at 0.2 V/s); (c) dashed voltammogram was obtained 35 min after activation. [Adapted from Ref. 44.]

The decrease in $\Delta E_p$ and increase in peak current indicate both increased adsorption and increased electron transfer rate following ECP.

There exists a variety of ECP protocols, almost one per research lab. The procedures vary in potential waveform (dc, slow cyclic scans, high-frequency ac, etc.) and in medium (0.1 M $KNO_3$, 1 M $H_2SO_4$, neutral NaCl, etc.). Two well-studied procedures involve a 1.8 V dc potential (vs. SCE) in 0.1 M $KNO_3$ for 5 min, followed by −1.0 V for 1 min [42], or 0 to 2.0 V potential sweeps (200 mV/s) in 0.1 M $H_2SO_4$ [44]. The various procedures have quantitatively different effects on the electrode surface, but some qualitative generalizations are available. First, the surface oxygen/carbon ratio increases significantly during ECP. After extensive pretreatment, the surface approaches the composition of graphitic oxide, an oxygen-rich, nonconducting, generally nonstoichiometric compound. Second, ECP causes damage to the $sp^2$ carbon lattice, with an observable decrease in a-axis microcrystallite size and Raman-observable order [45]. Third, a surface film forms with time during ECP. For the case of potential sweeps in $H_2SO_4$, the film thickness can be measured by ellipsometry, and forms at an average rate of about 44 Å/scan. Fourth, ECP can remove surface layers of carbon, including adsorbed impurities, with beneficial effects on electrode performance. Fifth, ECP electrodes have substantially higher background currents compared to polished surfaces. Finally, ECP surfaces exchange ions with the solution, with a strong preference for cations over anions. In fact, adsorption and ion exchange of dopamine (a cation) is the main reason for the improved selectivity over ascorbate (an anion) in neurochemical analysis.

The model that emerges from these observations is based on formation of an oxide layer that is rich in oxygen-containing functional groups, some of which act as cation exchange sites; in some cases, enhanced electrode performance is due merely to removal of polishing debris or other surface impurities during anodization. More importantly, a variety of redox reactions are catalyzed by oxides, and these systems exhibit higher reactivity on the ECP surface. The ion-exchange and/or adsorption properties of the oxide layer can also preconcentrate cations, thus improving detection limits. This is an important issue for the carbon fibers discussed in Section IV.D. Since the ECP film is hydrated and very porous, the voltammetric background is high and can be a significant interference. As always, the user must assess the catalytic and adsorption advantages of ECP in light of possible problems with background current.

#### 4. *Applications*

GC is arguably the most common carbon electrode material in current use, and its applications are too extensive to list here in any comprehensive fashion. Examples are listed here to illustrate GC electrode performance, particularly for voltammetry. Figure 10.12 shows the effect of vacuum heat treatment on ascorbic acid oxidation on initially polished GC [41]. Note that the less positive peak

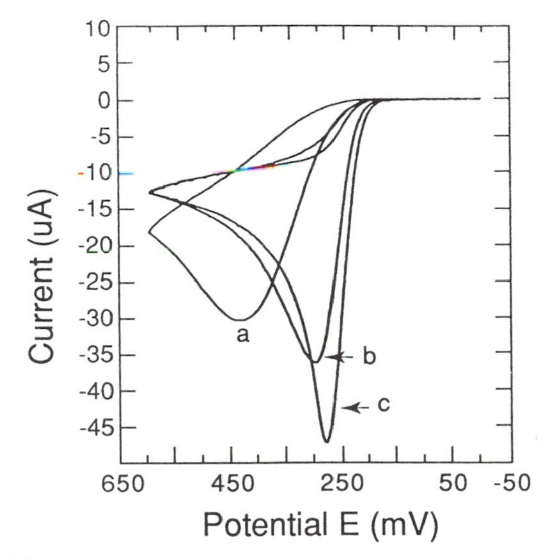


**Figure 10.12** Voltammetry of ascorbic acid at glassy carbon heat-treated at (a) 520°C, (b) 650°C, (c) 750°C. Scan rate = 0.1 V/s. [Adapted from Ref. 41.]

potential indicates more rapid oxidation kinetics. Figure 10.13 shows the effect of in situ laser activation on ascorbic acid and dopamine voltammetry [46]. The oxidation peaks of the two redox systems overlap on a polished surface, due to slow kinetics, but after laser activation kinetics are fast enough that both systems exhibit their thermodynamic potentials. In this case, the improved kinetics permit analytical discrimination of the two systems. Finally, as noted earlier, Figure 10.11 shows the effect of ECP on catechol oxidation on GC in 0.1 M $H_2SO_4$. With repeated cycling, $\Delta E_p$ decreases and the peak current increases.

## D. Carbon Fibers

Interest in microelectrodes, in vivo analysis, and carbon-reinforced structural materials has stimulated research on the electrochemical behavior of carbon fibers. Such fibers have diameters ranging from a few micrometers to about 60 μm, with the majority in the range of 5–15 μm. Although carbon fibers have a wide variety of structures and properties and are often less well characterized than GC or graphite, they have been used successfully in several important electroanalytical experiments.

### *1. Origin and Structure*

The majority of carbon fibers are made from petroleum pitch or polyacrylonitrile (PAN), with a heat treatment process similar to glassy carbon's [47]. The

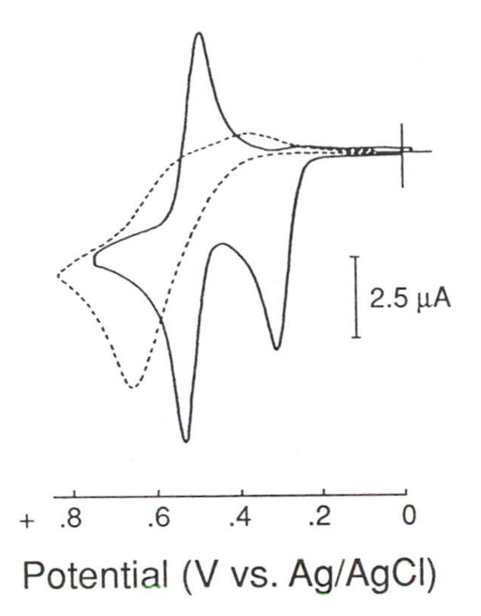


**Figure 10.13** Voltammetry of 1.0 mM dopamine and 1.0 mM ascorbic acid in 0.1 M $H_2SO_4$. Dashed line is polished GC; solid line is after three 25-MW/cm$^2$ laser pulses in situ. [Adapted from Ref. 46.]

carbon material is drawn out during curing to form filaments, with the process designed to orient the a axis along the fiber axis. The finished fiber has a cross section of the "onion," "radial," or "random" type, and the fiber end generally exhibits a high fraction of edge plane. Due to the importance of carbon fibers as structural materials, the literature is extensive.

For the case of ex-PAN fibers, it is tempting to consider them to be similar to glassy carbon, since their thermal history is similar. However, the orientation of the a axis during curing introduces a degree of anisotropy in the fibers not present in GC at dimensions greater than ca. 100 Å. For ex-pitch fibers, or the relatively recently developed chemical vapor deposited fibers, the degree of orientation can be quite high. In some cases, the cylindrical exterior of the fiber may be quite similar to HOPG basal plane, with low capacitance and relatively low electron transfer reactivity. While there is significant diversity among the many carbon fibers available, their structures fall between two extremes. At one end of the spectrum are nearly perfectly oriented onion fibers (Fig. 10.14A), like those made by catalytic chemical vapor deposition. Ordered fibers have high tensile strength and modulus, an HOPG-like surface along the exterior of the shaft, and a disordered end dominated by graphitic edge plane. At the other extreme are disordered fibers such as those made from PAN (Fig. 10.14B). These have lower tensile strength, more defects on the cylindrical wall, and less preferential orientation of the a axis along the fiber axis. Roughly speaking, oriented high-modulus fibers are HOPG-like, and low-modulus fibers are more

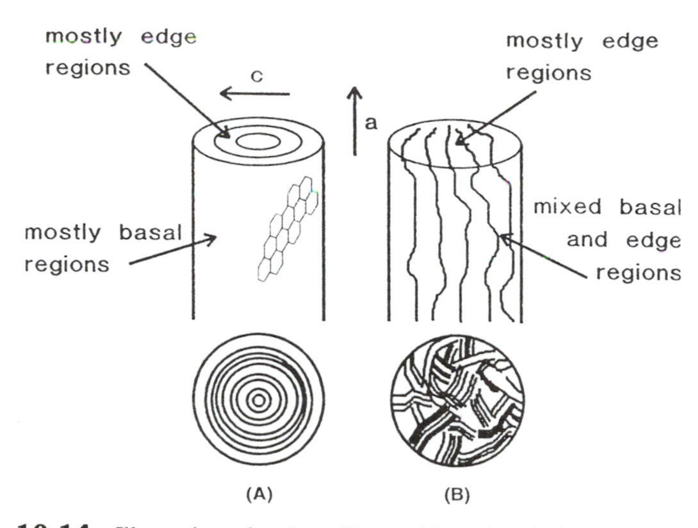


**Figure 10.14** Illustration of carbon fibers, side and end views: (A) ordered "onion" fiber; (B) disordered "random" fiber.

GC-like. However, this generalization should not be pushed too far, and in fact every fiber type has its own characteristics. There is no "standard fiber" with well-known electrochemical properties as yet, although some common types are Thornell P-55 from Union Carbide, and various diameter fibers from Avco Specialty Materials.

*2. Properties*

Carbon fibers have not been studied as extensively as GC or graphite, and in most cases the fiber is pretreated. Thus most of the electrochemical properties of fibers are discussed in the next section, Preparation. A few general points are useful here, dealing with size and resistance. Since the majority of carbon fibers are 5–15 μm in diameter, they will exhibit nonplanar diffusion under most conditions, whether they are used as disks or as cylinders. For example, $\sqrt{Dt}$ for a typical analyte ($D = 5 \times 10^{-6}$ cm$^2$/s) equals 2.2 μm at 10 ms. This is a significant fraction of a typical fiber diameter, so diffusion will become nonplanar even at short times. Thus any experiment lasting more than a few milliseconds will deviate from a response predicted for planar diffusion. Note that the deviation depends on whether the fiber end is used as a disk electrode or an exposed fiber is used as a cylinder, but quantitative theories have been presented for both cases [48].

A major advantage of carbon fiber electrodes is the possibility of using fast voltammetric scan rates (>100 V/s) due to small electrode area [48]. Even though internal resistance can be significant for carbon fibers (see later), the small disk area yields low iR error even at high scan rates. Fast voltammetry permits measurement of fast electron transfer rates (>0.1 cm/s) and enhances the current due to adsorbed reactants.

Although the resistivities of carbon fibers are comparable to GC and graphite, their small diameters can lead to nontrivial internal electrode resistance. For example, a 10 μm diameter fiber with a resistivity of $8 \times 10^{-4}$ Ω-cm and length of 1 cm has a resistance of 1 kΩ. In many cases, the small currents observed lead to negligible ohmic potential error, but resistance can be a problem for fast experiments where the current is larger. While resistance is not often a problem with carbon fiber electrodes, it is risky to ignore it. A related issue common to carbon fibers is the small electroanalytical current resulting from small electrode area. In the absence of contributions from the electronics, the signal-to-background ratio for fibers is at least as good as "macro" electrodes. In fact, nonplanar diffusion can significantly enhance the signal from solution redox species over capacitive or surface faradaic current. To exploit this advantage, however, care must be taken to reduce environmental or electronic noise below any contributions from the cell or electrodes. Several commercial potentiostats are designed to address this issue (Bioanalytical Systems, Cypress Systems, Ensman Instrumentation).

*3. Preparation*

Preparations of carbon fibers are similar to those of GC, with two important exceptions caused by the small fiber diameter. Fibers are mounted quite differently from GC, particularly for in vivo applications that require a small overall electrode diameter. In addition, fibers can experience much higher current densities during electrochemical pretreatment, which can qualitatively alter ECP effects.

**a. Polishing.** Although not studied extensively, conventional polishing of fiber electrodes is similar to that for GC. For obvious reasons, only carbon fiber electrodes with the disk exposed can be polished. Commercial fiber electrodes encased in glass are available (Bioanalytical Systems, Cypress), and homemade fiber electrodes are often encased in epoxy. The same issues of surface contamination apply, perhaps worsened by the small electrode area relative to the holder. Michael and Justice found that polished carbon fiber disk electrodes exhibited poor response for dopamine until they were lightly electrochemically pretreated [49]. Apparently ECP was necessary to remove surface films and to promote dopamine adsorption. An alternative was used by Wightman et al. to prepare fiber electrodes for in vivo voltammetry [50]. The fiber was sealed in a glass micropipette with a thin layer of epoxy. After cutting off a short length of fiber and glass, the end was briefly polished at an angle on a rotating wheel to expose an elliptical active surface. The carbon surface exhibited relatively fast kinetics for dopamine oxidation as well as good selectivity for dopamine over ascorbate. Apparently the polishing technique resulted in minimal impurity adsorption, thus promoting reactivity.

**b. Thermal Treatment.** Both vacuum heat treatment and laser activation have been employed on carbon fibers, but not yet extensively. Strein and Ewing used a small nitrogen laser to pretreat the exposed disk of an epoxy/glass-encapsulated, 11 μm diameter carbon fiber [51]. They observed significantly improved voltammetric wave shapes and increased electron transfer rates for several catechol derivatives. They attributed the changes to exposure of active carbon to which the catechols adsorb. Swain and Kuwana used vacuum heat treatment to pretreat carbon fibers, and observed a strong dependence of capacitance and kinetics on the presence of oxygen and previous electrochemical pretreatment [52]. As noted earlier, VHT leads to clean and often active surfaces, but is not practical for routine analytical applications.

**c. Electrochemical Pretreatment.** ECP is commonly used with carbon fiber electrodes, partly because there are few alternatives for activating the exposed cylindrical fiber. The first ECP procedures were developed for fibers after a remarkable improvement in sensitivity and selectivity for dopamine was discovered during fabrication of electrodes for in vivo analysis [53]. Due to quite different effects on electroanalytical behavior, ECP procedures for fibers are sometimes separated into "mild," with positive pretreatment potentials of 2.0 V or less, and "strong," with positive potentials often exceeding 2.5 V [2].

The effects of mild ECP on carbon fibers appear to be quite similar to that on "macro" GC electrodes. Anodization forms surface oxides and eventually a graphite oxide film. The oxide layer or film preferentially adsorbs or ion exchanges cations. In the case of in vivo analysis, this leads to enhanced sensitivity for cationic dopamine over the ascorbate anion [5,6,54]. There does not appear to be a standard procedure for mild ECP, but most workers alter the process to improve performance for a particular analytical target.

The effects of "strong" ECP on fibers differ from that on large GC disks, probably because of the high current densities present on the fiber during treatment. Such vigorous oxidation leads to an insulating oxide film over much of the fiber surface [55]. Cracks in this film comprising a small fraction of the total surface ($<10\%$) exhibit fast electron transfer activity. Thus the surface is composed of a sea of insulating oxide, which still adsorbs cations, with small islands of high electron transfer activity. The electrode exhibits reversible electron transfer to both anions and cations. The selective preconcentration of cations on the fiber surface is the main reason that such fibers are attractive for in vivo analysis of dopamine and related cations. The price of this advantage is comparatively slow response time due to the time required for preconcentration.

The effects of ECP on dopamine (DA) and ascorbate (AA) voltammetry at physiological pH are illustrated in Figure 10.15. For both DA and AA on untreated fiber (curves A1 and B1), slow kinetics cause a drawn-out oxidation

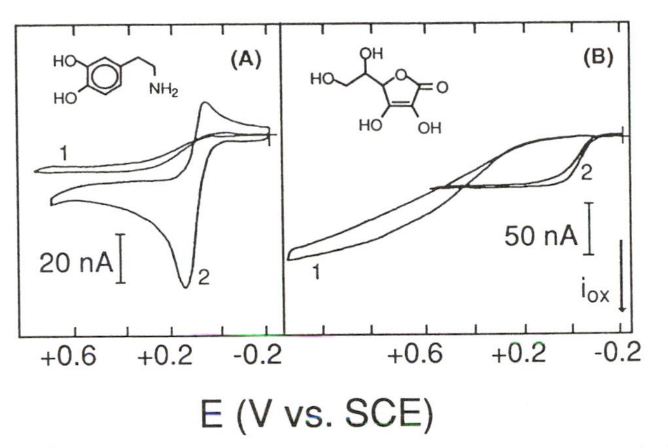


**Figure 10.15** Voltammetry at a cylindrical carbon fiber electrode (1) before and (2) after electrochemical pretreatment: (A) 0.1 mM dopamine and (B) 1.0 mM ascorbic acid, pH 7 solutions, scan rate = 0.1 V/s. [Adapted from Ref. 6.]

wave at potentials significantly positive of the thermodynamic potential. ECP greatly increases the response for DA (curve A2), not only because of faster charge transfer kinetics but also because of adsorption (a peak-shaped rather than sigmoidal wave for a microelectrode implies adsorption). For ascorbate, the wave shape indicates faster kinetics, but the wave height implies lower sensitivity following ECP (curve B2). Note that the concentration of AA is 10 times that of DA, but the wave heights are comparable. Thus ECP increased electron transfer reactivity, but also caused selective adsorption of cations.

## V. SELECTION OF CARBON ELECTRODES FOR ANALYTICAL APPLICATIONS

As noted in Section II, the performance criteria for an electrode include electron transfer kinetics, background current, reproducibility, adsorption properties, and pragmatic concerns such as ease of use and availability. It is clear that the carbon material and its pretreatment history strongly affect performance, mainly through effects on the electrode variables listed in Section III. While an application to a given problem will unavoidably involve some trial and error, the information in this chapter should provide a basis for rational selection of a carbon electrode material and preparation procedure. The user should first consider which performance criteria are most important to the application. For example, if reproducibility is more important than fast kinetics, carbon paste might be preferable to glassy carbon. Table 10.5 summarizes several properties of carbon electrodes and pretreatments, in generally qualitative terms. It

**Table 10.5** Features of Common Carbon Electrode Materials

| Carbon material | Pretreatment | Advantages | Problems | Comments | Availability[a] |
|---|---|---|---|---|---|
| HOPG, basal | Cleaved | Low background, well defined, very weak adsorption | Slow kinetics | Stronge dependence on defects | 2 |
| HOPG, or PG, edge | Polished | Easily renewed, strong adsorption | Variable impurities | | 2 |
| | Laser-activated or "roughened" | Strong adsorption fast kinetics | Moderately high background | | |
| Carbon paste | Lightly polished | Very low background, easily renewable, reproducible | Slow kinetics for some systems, limited solvent compatability | Widely used for aqueous oxidations, easy to modify | 1 |
| Graphite composites | Polished | Wide solvent compatability, increased signal/background current, easy chemical modification | Binder can coat carbon | Selectivity can be varied with chemical modification | 3 |
| Glassy carbon | Polished | Wide solvent compatability, easily prepared | Variable kinetics, background current | Probably the most common carbon electrode material | 1,2 |
| | Heat-treated | Wide solvent compatability, easily prepared, fast kinetics | Tedious renewal | | |
| | Laser-activated | Wide solvent compatability easily prepared, rapid in situ renewal | Expensive, optical window required in cell | Yields fast kinetics | 1,2 |
| | ECP | Sensitive, selective for cations, fast kinetics | High background current | | 1,2 |
| Carbon fibers | Polished | Small size, possibly very small volume | Variable carbon microstructure | Important for in vivo use | 1[b],2 |
| | ECP | Selectivity, preconcentration of cations | Slow response time | Important for in vivo use | 1[b],2 |

[a]1 = Commerically available as assembled electrode, 2 = materials commercially available, must be encapsulated or mounted before use, 3 = materials commercially available, some chemical or physical steps required before mounting.

[b]Encapsulated fibers are commercially available with large overall dimensions (ca. 5 μm), but fiber electrodes with small dimensions (<20 mm) must be custom made.

should not be considered a substitute for the details in the text, but does provide a quick overview. Finally, Table 10.6 lists the sources of materials and equipment mentioned in the text.

As a final note, the reader is reminded that the intent of this chapter is descriptive rather than comprehensive. Although glassy carbon and carbon paste are commonly used in electroanalysis, there are a variety of alternative carbon

**Table 10.6** Sources of Materials and Equipment

| Source | Materials/Equipment |
|---|---|
| Avco Speciality Materials Division<br>2 Industrial Ave.<br>Lowell, MA 01815 | Carbon fibers, including large (30–60 μm) diameter |
| Barnstead/Thermolyne Corp.<br>2555 Kerper Blvd.<br>PO Box 797<br>Dubuque, IA 52004-0797<br>(Also lab supply houses) | Nanopure water purification systems |
| Bioanalytical Systems, Inc.<br>2701 Kent Ave.<br>West Lafayette, IN 47906 | Conventional and microelectrodes, potentiostats, polishing kits, carbon paste electrodes |
| Buehler, Ltd.<br>41 Wankegan Rd.<br>Lake Bluf, IL 60044 | Polishing materials and equipment, alumina |
| Carbone of America<br>Ultracarbon Division<br>900 Harrison St.<br>Bay City, MI 48708 | Spectroscopic graphite, pyrolytic graphite, graphite powder |
| Cypress Systems, Inc.<br>PO Box 3931<br>Lawrence, KS 66046 | Conventional and microelectrodes, potentiostats |
| Electrosynthesis, Inc.<br>PO Box 430<br>East Amherst, NY 14015 | Electrodes, modified carbon materials, potentiostats, Tokai GC |
| Emerson-Cuming Inc.<br>Supplier: Ellsworth Adhesive Systems<br>3828 Royann Dr.<br>Fairview, PA 16415 | Eccobond epoxy for encapsulation |
| Ensman Instrumentation<br>4151 Broadway<br>Bloomington, IN, 47403 | Microelectrode potentiostat and bipotentiostat |
| ERG, Inc.<br>Lowell and 57th St.<br>Oakland, CA 94608 | Reticulated vitreous carbon |

(*continued*)

**Table 10.6** Continued

| | |
|---|---|
| Princeton Applied Research Corpation<br>CN 5206<br>Princeton, NJ 08543-5206 | Potentiostats, electrodes, carbon paste electrodes |
| Praxair/Union Carbide Corp.<br>12900 Snow Rd.<br>Parma, OH 44130 | Pyrolytic graphite, HOPG, Thornell fibers, carbon cloth |
| Tokai Carbon Co.<br>Supplier: Applied Industrial Materials Corp.<br>Park Ridge One Office Center<br>Ste. 200 Commerce Dr.<br>Pittsburgh, PA 15275<br>(Also supplied by Electrosynthesis, Inc.) | Glassy carbon rods, disks, plates |

materials that are beyond the scope of this chapter. The materials listed in Table 10.5 represent the overwhelming majority of carbon materials used in electroanalytical chemistry, but new candidates appear regularly. Some examples are vapor-deposited films, pyrolyzed thin films, and even fullerenes and doped diamond. There seems to be no limit to the diversity and versatility of carbon materials, and it is certain that electroanalytical performance will improve as the nature and preparation of such materials continues to be better understood.

## REFERENCES

1. G. Dryhurst and S. McAllister, in *Laboratory Techniques in Electroanalytical Chemistry* (P. T. Kissinger and W. R. Heineman, eds.), Marcel Dekker, New York, 1984.
2. R. L. McCreery, in *Electroanalytical Chemistry*, Vol. 17 (A. J. Bard, ed.), Marcel Dekker, New York, 1991, pp. 221–374.
3. K. Kinoshita, *Carbon: Electrochemical and Physicochemical Properties*, Wiley, New York, 1988.
4. A. J. Bard and L. R. Faulkner, *Electrochemical Methods*, Wiley, New York, 1980, Chap. 3.
5. R. N. Adams, *Anal. Chem. 48*:1126A (1976).
6. R. M. Wightman, L. J. May, and A. C. Michael, *Anal. Chem. 60*:769A (1988).
7. R. W. Murray, in *Electroanalytical Chemistry*, Vol. 13 (A. J. Bard, ed.), Marcel Dekker, New York, 1984, pp. 191–368.
8. R. J. Rice, N. Pontikos, and R. L. McCreery, *J. Am. Chem. Soc. 112*:4617 (1990).
9. M. T. McDermott, K. Kneten, and R. L. McCreery, *J. Phys. Chem. 96*:3124 (1992).

10. K. Kneten and R. L. McCreery, *Anal. Chem. 64*:2518 (1992).
11. D. C. S. Tse and T. Kuwana, *Anal. Chem. 50*:1315 (1978).
12. C. Amatore, J. M. Saveant, and D. Tessier, *J. Electroanal. Chem. 147*:39 (1983).
13. F. A. Armstrong, A. M. Bond, A. O. Hill, I. S. M. Psalti, and C. G. Zoski, *J. Phys. Chem. 93*:6485 (1983).
14. A. W. Moore, *Chem. Phys. Carbon 17*:233 (1981).
15. I. Spain, *Chemistry and Physics of Carbon*, Vol. 16 (P. L. Walker and P. A. Thrower, eds.), Marcel Dekker, New York, 1981, p. 119.
16. J. P. Randin and E. Yeager, *J. Electroanal. Chem. 36*:257 (1972).
17. H. Gerischer, R. McIntyre, D. Scherson, and W. Storck, *J. Phys. Chem. 91*:1930 (1987).
18. L. Bodalbhai and A. Brajter-Toth, *Anal. Chem. 60*:2557 (1988).
19. R. N. Adams, *Electrochemistry at Solid Electrodes*, Marcel Dekker, New York (1969).
20. M. Rice, Z. Galus, and R. Adams, *J. Electroanal. Chem. 176*:169 (1984).
21. K. L. Hardee and A. J. Bard, *J. Electrochem. Soc. 122*:739 (1975).
22. P. D. Lyne and R. D. O'Neill, *Anal. Chem. 62*:2347 (1990).
23. D. Rolison, *Chem. Rev. 90*:867 (1990).
24. S. V. Prabhu, R. P. Baldwin, and L. Kryger, *Anal. Chem. 59*:1074 (1987).
25. B. R. Shaw, K. E. Creasy, C. J. Lanczycki, J.A . Sargeant, and M. Tirhadc, *J. Electrochem. Soc. 135*:869 (1988).
26. J. Wang, N. Naser, L Angnes, H. Wu, and L. Chen, *Anal. Chem. 64*:1285 (1992).
27. L. Santos and R. P. Baldwin, *Anal. Chem. 58*:848 (1986).
28. D. E. Weisshaar, D. E. Tallman, and J. L. Anderson, *Anal. Chem. 53*:1809 (1981).
29. J. E. Anderson, J. B. Montgomery, and R. Yee, *Anal. Chem. 63*:653 (1991).
30. B. Shaw and K. Creasy, *Anal. Chem. 60*:1241 (1988).
31. I. Hu, D. H. Karweik, and T. Kuwana, *J. Electroanal. Chem. 188*:59 (1985).
32. D. C. Thornton et al., *Anal. Chem. 57*:150 (1985).
33. G. N. Kamau, W. S. Willis, and J. F. Rusling, *Anal. Chem. 57*:545 (1985).
34. N. Pontikos, Ph.D. Thesis, Ohio State University, 1992.
35. A. N. Strohl and D. J. Curran, *Anal. Chem.* 51:1050 (1979).
36. V. E. Norvell and G. Mamantov, *Anal. Chem. 49*:1470. (1977).
37. N. Sleszynski, J. Osteryoung, and M. Cartes, *Anal. Chem. 56*:130 (1984).
38. R. J. Rice and R. L. McCreery, *Anal. Chem. 61*:1638 (1989).
39. N. M. Pontikos and R. L. McCreery, *J. Electroanal. Chem. 324*:229 (1992).
40. C. W. Lee and A. J. Bard, *J. Electroanal. Chem. 239*:441 (1988).
41. D. T. Fagan, I. Hu, and T. Kuwana, *Anal. Chem. 57*:2759 (1985).
42. R. C. Engstrom and V. A. Strasser, *Anal. Chem. 56*:136 (1984).
43. K. J. Stutts, P. M. Kovach, W. G. Kuhr, and R. M. Wightman, *Anal. Chem. 55*:1632 (1985).
44. L. J. Kepley and A. J. Bard, *Anal. Chem. 60*:1459 (1988).
45. Y. W. Alsmeyer and R. L. McCreery, *Langmuir 7*:2370 (1991).
46. M. Poon and R. L. McCreery, *Anal. Chem. 58*:2745 (1986).
47. M. S. Dresselhaus, *Graphite Fibers and Filaments*, Springer-Verlag, Berlin, 1988.
48. R. M. Wightman and D. O. Wipf, in *Electroanalytical Chemistry*, Vol. 15 (A. J. Bard, ed.), Marcel Dekker, New York, 1989, pp. 267–353.

49. A. C. Michael and J. B. Justice, *Anal. Chem.* 59:405 (1987).
50. J. E. Baur, E. W. Kristensen, L. J. May, D. J. Wiedemann, and R. M. Wightman, *Anal. Chem. 60*:1268 (1988).
51. T. G. Strein and A. G. Ewing, *Anal. Chem. 63*:194 (1991).
52. G. M. Swain and T. Kuwana, *Anal. Chem. 64*:565 (1992).
53. F. G. Gonon, C. M. Fombarlet, M. J. Buda, and J. F. Pujol, *Anal. Chem. 53*:1386 (1981).
54. J.-X. Feng, M. Brazell, K. Renner, R. Kasser, and R. N. Adams, *Anal. Chem. 59*:1863 (1987).
55. P. M. Kovach, M. R. Deakin, and R. M. Wightman, *J. Phys. Chem. 90*:4612 (1988).

# 11

# Film Electrodes

**James L. Anderson** *The University of Georgia, Athens, Georgia*

**Nicholas Winograd** *The Pennsylvania State University, University Park, Pennsynlvana*

## I. INTRODUCTION

Electrodes based on films of conducting material have been making increasingly important contributions to electrochemical experiments in recent years, and the possibilities for future growth and novel applications are outstanding. The use of films as electrodes makes possible numerous experiments that would be difficult or impractical to implement with conventional bulk electrodes. Discussion here emphasizes either "thin" ($<5$ μm thick; usually quite a bit thinner) or "thick" ($>5$ μm; usually quite a bit thicker) film electrode materials, consisting of a conductor, either a continuous film or spatially patterned, most commonly deposited on a suitably prepared insulating substrate. Films consisting primarily of insulators are not considered here, except to the extent that they may be used to form patterned arrays or electrodes with special geometries. For example, films used to form chemically modified electrodes are discussed in Chapter 13.

The initial impetus for application of film electrodes for electrochemical experiments was the development of spectroelectrochemical experiments in which a light beam was directed through an optically transparent film electrode [1]. Today, however, an ever-increasing range of applications that rely on other properties as well is being developed. Of particular interest is the use of microlithographic technology to define a pattern, coupled with a method of deposition of conductor or insulator. For very high spatial resolution, various forms of vacuum deposition (e.g., evaporation or sputtering) are typically used. When spatial resolution is not so critical (spacing between features $>25$ μm) and thicker films are acceptable, screen printing of patterns is quite useful. Screen printing can also be useful for finer spacing if the device is prepared by layered ("sandwich") film deposition, with spacing defined by the film thickness rather than its lateral dimensions.

Electrodes or arrays of electrodes can be produced by these techniques with complex or fine patterns that would be very difficult to fabricate with conventional bulk electrode materials. Films can commonly be reproducibly formed in quite pure chemical form (with a caveat based upon interfering chemical effects that may arise from diffusion of adhesion-promoting intermediate layers through the thin film to the exposed surface). Film electrodes lend themselves to inexpensive fabrication of disposable electrodes, cells, or eventually complete integrated measurement devices in large quantities that would be economically infeasible with conventional materials, due to either the complexities of fabrication or the cost of materials.

Note particularly that film electrodes make it economically practical to use very expensive materials as electrodes because of the very small quantities required to form the film. For example, a platinum film of 300 nm thickness and 1 $cm^2$ cross-sectional area contains ~640 μg of platinum. A macroscopic wire electrode of the same surface area constructed with 26-gauge platinum wire (0.4 mm diameter, 7.9 cm long) contains ~210 mg of platinum, or 330 times as much platinum. At a price of $60/g, the cost of the platinum in the film ($0.04) is much lower than the cost of the wire ($12.80). Despite the great reduction in raw materials cost, the resulting electrode may not actually be proportionally less expensive than bulk platinum for small numbers of film electrodes, due to significantly greater fabrication costs. However, when produced on a large scale, costs of film electrodes may be dramatically lower.

It is likely that we will see increasingly numerous examples of such systems in applications as diverse as home clinical chemistry assay kits and remote sampling for environmental applications. The discussion here outlines general applications and some of the unique physical properties and fabrication considerations of film electrodes, and then focuses in greater detail on preparation and properties of widely used examples of metallic, carbon film, and semiconductor electrode materials.

## II. APPLICATIONS OF THIN-FILM ELECTRODES

Film electrodes can be conveniently prepared from numerous types of materials on a wide variety of substrates in a bewildering variety of configurations, with high purity and chemical and spatial reproducibility. Some applications rely on the convenience of preparing uniform electrochemical films on optical quality substrates for reflection spectrometry [2], or for electrodes in thin-layer cells, where flatness and parallelism of electrodes to tolerances of a few micrometers is necessary [3,4]. Film electrodes are useful for many other applications. In most of these applications, the thin dimension is perpendicular to the surface exposed to solution, and the film thickness is relevant only in considering optical light transmission when the electrode is used for spectroelectrochemistry [1], or in considering the electrical resistivity of the film [5,6]. However, the thin dimension is being increasingly used as part of the active portion of the electrode, that is, relying on the thinness of the film to define the smallest dimension in an ultramicroelectrode or array of ultramicroelectrodes [7–10]. Here, only the edge (including the thin dimension) of the electrode film may be exposed to solution.

Applications have been reported for photoelectrochemical experiments, for example, splitting of water [11], local generation of photoelectrodes by spatially selective laser excitation [12], and steady-state electrochemiluminescence at a band electrode array [13,14]. Band electrodes prepared from very thin films approaching molecular dimensions have been used to assess the limits of theory describing electrode kinetics at ultramicroelectrodes [9]. Spectroelectrochemical applications have been extensively reviewed [1]. In an intriguing approach, thin, discontinuous metal films have been prepared on a transparent semiconductor substrate; they are essentially transparent under conditions in which a continuous metal film containing the same quantity of metal would be expected to substantially absorb [15].

Film electrodes have been successfully produced with well-ordered surfaces containing a predominant crystal orientation by sputtering platinum on mica to produce an essentially Pt(111) surface [16]. Textured polycrystalline deposits have been produced with a large proportion of crystallites oriented along particular crystallographic axes by electrodeposition of copper/nickel superlattices with unique mechanical and chemical properties relative to the constituent metals [17]. Recent efforts to electrodeposit well-ordered compound semiconductors such as CdTe one atomic layer at a time essentially involve generation of a film electrode of radically different properties than the substrate electrode [18]. When a single crystal face is used as a substrate, thin films such as those formed by underpotential electrodeposition of copper on Au(111) electrodes can be studied with atomic resolution when evaluated by techniques such as scanning tunneling [19] or atomic force [20] microscopy, as well as conventional techniques such as x-ray [21,22] or electron diffraction [18].

Extremely thin, self-supporting film electrodes have been successfully produced. For example, polypyrrole films of ~120 nm thickness have been used successfully to grow microcrystallites, followed by removal of the film plus crystallites from the substrate for characterization of the crystallographic structure and orientation of the microcrystallites by transmission electron microscopy with selected area electron diffraction [23]. Thin-film electrodes have been applied to allow transmission of high-energy beams, including x-rays, for in situ extended x-ray absorption fine structure (EXAFS) or x-ray absorption near edge structure (XANES) spectroscopy at the electrode interface [21,22,24].

Film electrodes have been essential components of quartz microbalance studies of stoichiometry of many electrodeposition and dissolution experiments as well as polymer electrode characterization [25]. In the quartz microbalance, changes in mass are detected by measurement of changes in the resonant frequency of a quartz crystal oscillator as the mass adhering to the surface changes. The oscillation is feasible because thin-film metal electrodes (typically gold) applied to opposing faces of a piezoelectric quartz crystal serve both to induce the oscillation and to provide a site for electrochemical reaction.

Thin film electrodes have made feasible thermal jump kinetic measurements of extremely fast electrode reactions [26], by illuminating a thin electrode film by a very rapid laser pulse and monitoring the relaxation process on a nanosecond time scale. Thin films of silver have also been deposited on electrode materials such as carbon to enable surface-enhanced Raman spectroscopic investigation of surface-bound species [27].

Considerable activity has been evidenced in the area of thin-film electrodes utilized with the thin edge exposed to solution to serve as an ultramicroelectrode. Examples include ring electrodes, consisting of a very thin film deposited on a substrate oriented perpendicular to solution [7,8,28,29], and band ultramicroelectrodes, which have been used either singly [9,28,30,31] or in arrays of two or more band electrodes [10,14,30,32–35] to investigate reactions on a very small distance or time scale. In an interesting "microstep" variant, edge-based microelectrodes have been fabricated by conventional photolithography, in which an edge of a thin film patterned photolithographically in the plane of a substrate is exposed while the broad dimension of the film is overcoated with a photolithographically defined insulator [10]. In another variant, "vertically separated" interdigitated arrays have been fabricated with one set of electrodes deposited in one plane and covered with a thin $SiO_2$ insulating pattern, which is then covered by a second set of electrodes in a third plane, with the advantage that the distance between sets of electrodes can be minimized while allowing a higher number of interdigitated electrodes per unit length [35].

Great activity has also been evidenced in microlithographically fabricated arrays of microelectrodes, which are typically formed in one plane on an insulating substrate [7,8,13,34–45] for experiments involving either an array of electrodes held at a common potential [37,40,42,43], or an array of noninteracting electrodes held at two or more different applied potentials [42,44], or an array of interdigitated electrodes held at two different potentials [13,34,36,38,39,45–47]. Arrays have significantly better analytical detection limits than continuous electrodes of the same overall dimensions, due to enhanced mass transport fluxes that arise from an increase in the spatial dimensionality of mass transport due to the alternation of electrode zones with passive zones, which can serve as reservoirs of reactant to replenish reactant consumed at the active electrodes [38,40]. Interdigitated microelectrode arrays have also been used to enhance sensitivity and/or selectivity of analytical response by allowing redox cycling between adjacent electrodes [38].

Particularly intriguing applications of thin-film electrodes are in the area of practical devices that may make important technological contributions in the 21st century. Tin oxide optically transparent electrodes already play an important role in the operation of liquid crystal displays such as those found on digital watches and laptop computers. Additional applications that promise to have potential economic impact include electrochromic display devices, which rely on generation and removal of optically absorbing or reflecting materials at thin-film electrodes [48], and electrochemically tunable solar control windows [49], which could significantly cut energy costs in the 21st century by actively rather than passively controlling the transmission of solar radiation through windows as a function of light conditions and season.

## III. PROPERTIES OF FILM ELECTRODES

### A. Electrical Conduction

Film electrodes are generally fabricated from conducting or semiconducting materials, which may be deposited as a result of either a physical or a chemical process (or some combination) onto a suitable substrate, which is typically an insulator. Key factors governing the desired thickness of the film are the electrical resistivity ($\rho$) or conductivity ($\kappa = 1/\rho$) of the film material, which is a practical consideration in almost all cases, and the optical transparency or reflectance of the material if optical transmission or reflection is also desired. The optimum film thickness for an application involving both electrical and optical considerations will require a trade-off, since a decrease in resistivity (usually desirable) normally is also accompanied by a decrease in light transmission (undesirable for an optically transparent electrode).

The electrical resistivity ($\rho$) of a film electrode depends on numerous factors, including film structure and morphology (e.g., grain size, crystal texture, surface roughness, and continuity), the number of lattice defects and impurities, and film thickness. The resistivity of a film may differ significantly from the value for the bulk material. Resistivity generally increases as the number of scattering interactions of charge carriers increases, and is greatest in the thinnest films. While scattering can occur primarily at lattice defects in bulk materials, scattering at the surface of the film can become very important in thin films [5], since electrons encounter the surface increasingly frequently as the film becomes thinner. Resistivity will be higher than the bulk value as long as the ratio K of film thickness to the mean free path of conductors (electrons or holes) is not large.

When a film is very thin, it may not be continuous, and conduction is subject to the "percolation" effect, whereby charge migrates by hopping or tunneling between island sites [50,51]. Such a process is activation controlled, and such thin films do not obey Ohm's law. The activation energy can be decreased by the presence of an applied electric field, making development of a rigorous theory difficult. The resistivity can be expressed by the relationship [5]

$$\rho = C \exp[e^2/(4\pi\varepsilon\varepsilon_0 rkT)] \tag{11.1}$$

where C is a constant characteristic of the film, e is the electronic charge, $\varepsilon_0$ is the permittivity of free space, $\varepsilon$ is the dielectric coefficient of the substrate, r is the effective radius of the island site, k is Boltzmann's constant, and T is absolute temperature. Resistivity will be higher for substrates with smaller dielectric coefficients and for smaller island radii. It is clear that the temperature coefficient of resistance is negative, with resistance decreasing as temperature increases, due to the activation control, in contrast with the normal behavior for bulk conductors such as metals, but consistent with the behavior of semiconductors. In general, $\rho$ does not begin to approach a limiting value corresponding to the bulk value until the islands occupy a critical fraction of the total volume of the film. For spherical islands, the critical fraction is about 60% [52].

Once continuity is achieved, the ratio of the resistivity $\rho$ of the thin film to the bulk resistivity $\rho_0$ can be approximated in the limit that $K \ll 1$ (film much thinner than the mean free path) by the expression [5]

$$\rho/\rho_0 = -4/[3K \ln K] \tag{11.2}$$

which shows that resistivity normalized to the bulk value varies inversely with the logarithm of the ratio of film thickness to mean free path.

As film thickness becomes large relative to mean free path ($K \gg 1$), the ratio of resistivity to bulk resistivity asymptotically approaches unity and can be approximated by the relation [5]

$$\rho/\rho_0 = 1 + 3/(8K) \tag{11.3}$$

It is clear from Equation 11.3 that resistivity should approach within 10% of the bulk value when the film thickness exceeds about four times the mean free path. The better the conductor, the smaller the mean free path. Thus, the resistivity approaches the bulk value as the film thickness reaches typical values of 100–200 nm for metallic conductors, or perhaps as much as several micrometers for semiconductors, depending on the intrinsic or doped carrier density. For sufficiently thick metallic films with $K \gg 1$, the temperature coefficient of resistivity becomes positive, as bulk electron-phonon scattering becomes the primary contribution to resistivity [5]. Conduction in semiconductor films remains activation-limited, and retains a negative temperature coefficient. Figure 11.1 illustrates the dependence of resistivity on film thickness for sputtered

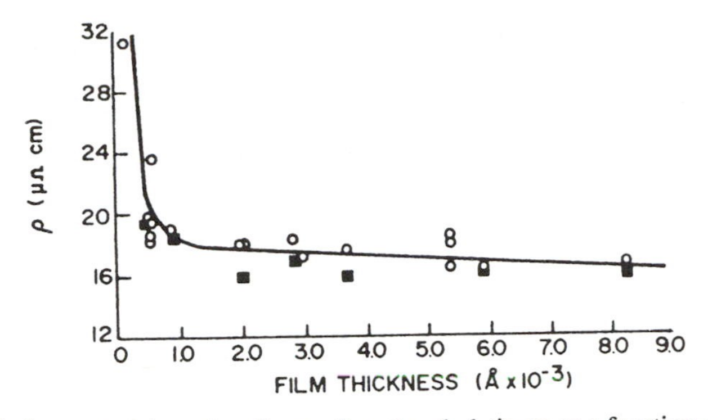


**Figure 11.1** Resistivity ρ in $\mu\Omega$-cm of sputtered platinum as a function of film thickness, in angstroms. The circle and square designations refer to the different substrates of glass and GaAs, respectively, that were used for the substrate. [From Ref. 53.]

platinum on two substrates [53]. The film on the substrate with a lower dielectric coefficient (glass) exhibits a larger resistance for thin films, consistent with the expression for the percolation-limited case.

It is frequently important to determine electrode resistance to account for any contributions to iR drop and circuit rise time due to the electrode itself. Two general strategies are available for this purpose. One strategy for accurately determining ρ is to use a four-point probe (see Fig. 11.2) for the measurement [6]. This device supplies a controlled current to be passed through the film via

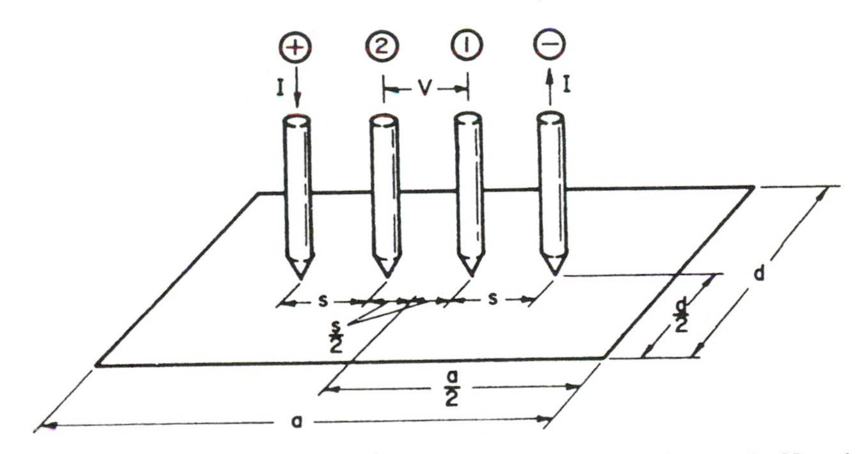


**Figure 11.2** Arrangement of a four-point probe on a rectangular sample. Note that the dimensions a, d, and s must be specified to calculate a specific value for the resistance. [From Ref. 9.]

the two outermost electrodes, and resistance is determined as the ratio of the voltage drop between the two inner probes to the current, thereby minimizing any errors associated with probe contact resistance. An alternative, more convenient but less accurate approach relies on the fact that the resistance R of a rectangular section of film of width w, length ℓ, and thickness t can be expressed:

$$R = \rho \ell/(wt) \tag{11.4}$$

If we measure the resistance directly with probes of width w equal to the distance ℓ between them to define a square, the expression reduces to

$$R = \rho/t \tag{11.5}$$

If the film is uniform in thickness and a square probe placement is used, $\rho/t$ (the sheet resistance) will be a constant for that film, and it is convenient to express the measured resistance in terms of the sheet resistance ($= \rho/t$), with units of ohms. Sheet resistance is usually expressed as ohms per square, since this approach yields a constant resistance for any square configuration of probes placed in contact with the film such that the probe width equals the separation distance, regardless of the size of the square.

## B. Optical Transparency

For use as an optically transparent electrode (OTE) for spectroelectrochemistry, the film optical transmission spectrum is important. As noted earlier, trade-offs must be made between maximizing transmittance and minimizing sheet resistance. Key parameters are film thickness (transmittance and resistance decrease with increasing thickness), charge carrier density in the conduction band (transmittance and resistance decrease with increasing carrier density), and the band gap. Metals generally exhibit strong absorption by the free electron charge carriers in the visible and infrared spectral regions, but may become transparent at wavelengths shorter than a critical edge value defined by the plasma frequency of the metal, usually in the ultraviolet region. For example, a 20-nm-thick platinum film exhibits an absorbance of ~0.5 at 500 nm. Sodium, which has a free electron concentration of $10^{22}$ electrons/cm$^3$, becomes transparent at wavelengths below 210 nm [54]. Optical excitation of valence-band electrons can also contribute to the spectrum. The color of gold films is attributed to the 5d → 6d transition at 470 nm.

Semiconductor electrodes, which have much lower charge carrier densities ($10^{13}$–$10^{19}$ carriers/cm$^3$), typically absorb in the infrared but exhibit much lower absorption by charge carriers than metals of comparable film thickness, and frequently show a transparency window in much of the visible spectrum due to a substantial band-gap energy, before absorbing again in the ultraviolet. For example, $SnO_2$ and ZnO, like many common semiconductor electrode materials, both have band gaps near 3 eV (414 nm), and are transparent in the visible but strongly absorbing in the ultraviolet (UV). In contrast, semiconductors with lower band gaps, such as Si (band gap 1.1 eV, 1140 nm) and germanium (band gap 1.5 eV, 1900 nm), are suitably transparent in the infrared but absorb strongly in the visible and UV regions.

Doping of intrinsic semiconductors with n-type (electron donor) or p-type (electron acceptor) impurities can increase the concentration of free charge carriers (electrons or holes) by several orders of magnitude. For example, $SnO_2$ is typically heavily doped with antimony or other impurities to form an n-type semiconductor with carrier densities as high as $10^{20}$–$10^{21}$ carriers/$cm^3$ for sheet resistances as low as 5–10 Ω/square. This compares to a carrier density of only $10^{19}$ for tin oxide films formed by chemical vapor deposition spraying of tin chloride onto a hot glass substrate without antimony doping. The free carriers in doped tin oxide absorb in the infrared region at wavelengths greater than ~2000 nm. The film is transparent in the visible, but again absorbs in the UV, with absorbance rising sharply above the intrinsic optical absorption edge at 3.7–3.8 eV (~330 nm). On the common substrate material glass, transmittance in the UV is limited by the absorption of the glass itself below 360 nm.

## C. Fabrication

Numerous film fabrication methods are available, depending on the film material. Table 11.1 summarizes some of the fabrication methods. General comments on substrate preparation and the various fabrication processes are presented in the order listed in the table. Applications to specific systems are summarized according to the electrode material type, including metals, carbon, and semiconductors. Carbon is sometimes classified as a semimetal, with properties intermediate between metals and semiconductors.

### *Substrate Preparation*

Preparation of an extremely clean substrate is an absolutely essential step in successful preparation of film electrodes. Neglecting this step is an excellent means of assuring poor quality or even unusable films. For complex devices produced by photolithography with very small feature size, even the most minute dust or particulate contamination can ruin a device. Thus, care for cleanliness and particle removal becomes an increasingly heroic enterprise as the feature size decreases.

The two most common substrates for thin film electrodes are various types of glass—soda-lime, Pyrex, and various forms of quartz or fused silica—and silicon wafers that have been treated to produce an insulating surface layer (typically a thermally grown oxide or nitride). Other possible substrates include mica, which can be readily cleaved to produce an ordered surface, and various ceramic materials. All of these materials can be produced in very flat, smooth

**Table 11.1** Fabrication Methods for Film Electrodes

| Fabrication method | Materials applicable | Comments |
|---|---|---|
| Vacuum evaporation | Metals, carbon, some semiconductors | Vacuum, simple, works best for metals with low melting points |
| DC sputtering | Metals, some semiconductors; some insulators (reactive sputtering) | Vacuum, works best with conductors, better than vapor evaporation for high-melting species, better adhesion than vapor evaporation, but rougher surface; reactive sputtering of metals in presence of oxygen or other reactive gas can deposit some insulators |
| RF sputtering | Metals, semiconductors, insulators | Radiofrequency oscillation breaks down the space charge that would inhibit direct dc deposition of insulators |
| Liquid film application | Metals, carbon/graphite, semiconductors, insulators | Typically organometallic resinates, deposited by painting, spin coating, etc., and reduced to final form (e.g., metal) when fired in a furnace or dried; carbon or graphite can be dispersed in a volatile vehicle, which evaporates to leave a conducting film; substrate must be able to withstand firing temperature |
| Screen printing | Metals, carbon/graphite, semiconductors, insulators | Apply liquid solution or suspension through fine mesh screen mask to form pattern; then convert film to desired form as above |
| Chemical vapor deposition | Metals, carbon (pyrolysis), semiconductors, insulators | Spray-heated substrate with reagents, which react to form the film on the substrate |
| Microlithographic pattern generation | All materials, positive resist, negative resist | Used to produce patterns, including arrays of electrodes, or complex electrode or cell geometries; can deposit film through photoresist patterned by a mask, or use photoresist to define areas of film to be stripped |

forms suitable for electrochemistry and reflectance spectrometry, and the materials that are transparent in the wavelength region of interest (particularly glass and mica) are also suitable for transmission spectroelectrochemistry. (The ceramic materials generally have rougher surfaces than the others, and may be less ideal for spectroelectrochemistry.) Some general aspects of surface preparation are common to all these surfaces, although some cleaning steps may be bypassed in some sequential processes for the silica wafers once the thermal oxide or nitride has been generated, since the surface preparation is carried out under clean conditions on a surface that has already been rigorously cleaned, or for the newly exposed surfaces of freshly cleaved mica.

Although many common cleaning procedures have been attempted, boiling in an aqueous detergent solution, followed by rinsing in triply distilled or high-quality deionized/distilled water, followed by an isopropanol vapor degreasing procedure, is usually quite successful for glass or other oxide surfaces. When the surface is clean, distilled water will readily run off the surface in a sheet, leaving no condensation or beading of droplets behind. This cleaning step should be carried out immediately before further processing steps. If photolithography is to be used, the sample should then be carefully oven-dried and immediately transferred under clean, dust-free conditions to the photolithographic resist-application stage for processing. If the film is to be deposited by a vacuum deposition process, the samples should be quickly transferred to the vacuum apparatus and pumped down to a pressure of $\sim 10^{-6}$ torr. A glow discharge with $\sim 50$ torr of Ar or residual air for about 10–20 min can provide a final cleaning in the vacuum [55], by applying a dc voltage of 1–5 kV at 50–200 mA between a metal block (typically aluminum) holding the substrate and a metal (e.g., aluminum) discharge electrode. This step can be carried out in either a vacuum evaporation or a sputtering apparatus that is appropriately set up. Commercial stand-alone plasma cleaning devices are also available for this step. However, storage in or transfer through air can quickly defeat the cleaning effect of these processes.

In many cases, particularly with evaporation and sputtering methods, adhesion of the deposited film to the substrate may be inherently poor. It is extremely desirable in such cases to deposit a thin layer of an intermediate material that has better adhesion to the substrate. Examples of appropriate adhesion layers are discussed after presentation of evaporation and sputtering techniques.

### *Vacuum Evaporation*

Vacuum evaporation is a widely used technology for deposition of a wide variety of materials, particularly in the coatings and electronics industries. A complete discussion of this technique may be found in the classic text by Holland entitled *Vacuum Deposition of Thin Films* [56]. The subject is also treated in numerous texts on microelectronics [5]. Discussion here focuses on metal deposition, since this is the case most commonly encountered in the preparation of film electrodes.

Deposition of metals by vacuum evaporation is normally carried out in a high-vacuum system ($10^{-5}$–$10^{-8}$ torr). The high vacuum is necessary to insure that the mean free path of evaporated particles is long enough to reach the substrate, but perhaps more importantly, to maintain cleanliness of the substrate.

Even in this pressure range, it must be understood that a monolayer of residual gas forms on the surface of the substrate within seconds.

Evaporation is most commonly initiated by melting the metal onto a conductive support, which may be a filament or boat, usually tungsten or tantalum, followed by further electrical heating of the metal to its melting point. For lower-melting metals, a concentrated source such as a conical wire basket is adequate. For higher-melting metals, such as platinum, it is advantageous to wrap the metal to be evaporated around a sinuous, multistranded filament such as tungsten. Alternatively, higher-melting metals, as well as carbon and other materials, can be successfully evaporated using electron beam heating techniques. Here, the metal to be deposited is placed in a water-cooled copper bucket and bombarded with high-energy electrons. The metal surface is rapidly heated and evaporates quickly. Thus, the electron beam method is particularly well suited to prepare very thick films.

*Sputtering*

Sputtering is closely related to the glow discharge cleaning process described earlier. An electrical discharge in a low-pressure gas such as argon at 10–100 mtorr is maintained between two electrodes, one of which is the material to be deposited (the target). When the target is made negative, with an accelerating voltage of 1–3 kV, it will be bombarded by energetic argon ions, which transfer momentum to the material and cause ejection of atoms or ions, which deposit with relatively high velocity on the substrate. Sputtering can be carried out using either a dc or an ac field. DC sputtering is sufficient for any conductive material to be deposited. For compactness, coplanar concentric electrodes are often used with a magnet to steer the bombarding ions (magnetron sputtering). In some cases, semiconducting or insulating materials can be deposited by reactive dc sputtering. Many metal oxides, for example, can be deposited by reactive sputtering, by sputtering the metal of interest in presence of a low pressure of oxygen or residual air. However, radiofrequency (rf) sputtering is needed when it is desired to use an insulator as a target material for sputter deposition, since the buildup of a space charge at an insulator prevents effective sputtering under dc conditions. DC sputtering is most commonly used.

Sputtering has some particular advantages and disadvantages relative to evaporation. Sputtered films generally have better adhesion to substrates than evaporated films, presumably because of the greater physical force with which they are deposited. Deposition can be carried out at very low temperatures. However, since sputtering is carried out at much higher pressures than evaporation, purity of the deposit is generally significantly poorer for sputtered than for evaporated films, with much higher inclusion of residual gaseous impurities. Resistivity of the films is also correspondingly higher for sputtered than for evaporated films. Deposition rates (which vary by element, but typically fall in the range of 2.5 nm/min) are also typically significantly slower for sputtering than for evaporation.

*Film Annealing*

Electrical and optical properties as well as adhesion of many metal films including Pt and Au can be markedly improved by annealing at temperatures up to 600°C in air [55]. These changes are probably manifestations of the coalescence of individual islands into a more continuous film. The effects are particularly pronounced for gold, where the resistivity drops by 30–50% and the optical transparency, as shown in Figure 11.3 [55], improves considerably. Similar effects are observed for Pt [55,57].

*Adhesion Layers*

Adhesion of metal films on glass or other oxide substrates can often be significantly improved by first depositing a thin layer of a different material on the glass. The best adhesion layers tend to be formed from metals (particularly transition metals) that readily form oxides. These adhesion-promoting metals can form a covalent bond to the oxide substrate lattice, while simultaneously alloying with the metal film to anchor the film to the substrate [58]. The most popular adhesion layers are Cr [10,39,43,59], Ti [13], W, Ti/W [58], Ta, Nb [42], V, and Zr. Evaporation or sputtering of only several layers of metals such as Cr, Ti, or Ta has improved the adhesion of Pt on glass without degrading the optical transparency or the electrochemical properties of the electrode [60]. Deposition of metal oxides, such as $Bi_2O_3$ and $PbO_2$, has improved the sticking of

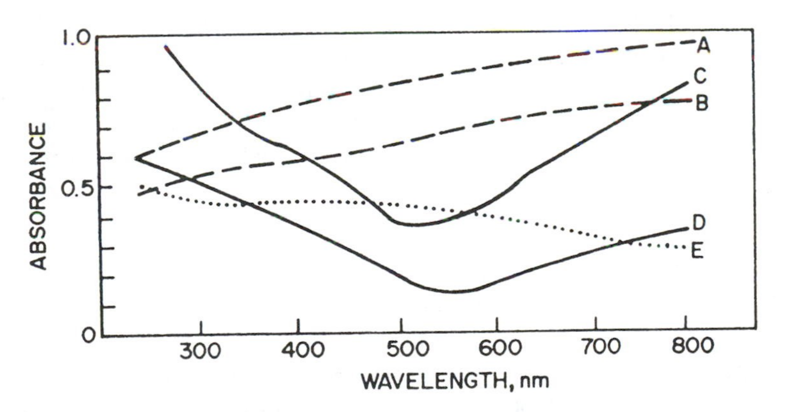


**Figure 11.3** Spectra of Pt and Au films. A, Pt on quartz, 11 Ω/□; B, Pt on quartz, 20 Ω/□; C, Au-bismuth oxide on quartz, 2.5 Ω/□; D, Au-bismuth oxide on quartz, 11 Ω/□, E, film of bismuth oxide on quartz. [From Ref. 55, with permission. Copyright 1970 American Chemical Society.]

Au films to glass [55]. Other combinations of undercoats have been suggested. These undercoats are critical for metal film electrodes deposited on glass, quartz, or thermally grown oxides on silicon, especially when operating in aqueous solution. The films invariably strip off during $H_2$ or $O_2$ evolution unless they are well bonded to the substrate. Gold patterns microlithographically deposited directly on thermally grown oxide without an adhesion layer could be removed by dragging a fingernail over them, yet functioned in nearly perfect quantitative accordance with theory for hours in a flow stream [40].

A critical problem with adhesion layers arises from grain boundary diffusion. Deposited films tend to be polycrystalline and granular. The electrochemistry of the adhesion film is frequently much less desirable than the electrochemistry of the primary film. Moreover, minute contamination of the primary metal film surface by adhesion components can dramatically degrade the electron transfer properties (e.g., electrochemical reversibility, as evidenced by cyclic voltammetric peak potential separation) of the film [58]. Thus it is essential that the adhesion layer is not exposed to solution. While the rate of diffusion of adhesion metals through the bulk of the primary layer is quite slow, grain boundary diffusion along the surfaces of grains is much faster. In many cases, the adhesion layer can seriously compromise the performance of the electrode. This is particularly a problem for chromium underlayers. Recently a codeposited Ti/W adhesion layer has been recommended as an alternative to chromium, with reportedly better adhesion and fewer interferences than Cr. A procedure was also described to recondition these electrodes to minimize interference by adhesion layer metals [58].

A procedure based on chemical affinity has also been reported, in which an oxygen-containing substrate such as glass is derivatized with a monolayer of an organofunctional silane compound which contains a functional group such as a thiol. The silane bonds the compound to the oxide substrate, and the thiol forms a chemical bond to metals, particularly gold. A gold film is subsequently vapor deposited upon the substrate to form a durably adherent film with significantly better electrochemical properties than a gold film with a chromium adhesion layer [61].

### *Liquid Film Application*

A relatively simple and inexpensive way to fabricate metal film electrodes is to paint a thin coating of various commercially available metal resinate inks (known as "Liquid Bright Gold," etc.) onto a suitable substrate, followed by firing in a reducing furnace at a temperature near the softening point of glass and a slow return to room temperature. These inks are widely used in the glass decoration industry. Such electrodes work satisfactorily for simple spectroelectrochemical applications [4]. The adhesion of these films is excellent. However, such electrodes contain many impurities (e.g., 4% of adhesion promoter Bi, with traces of Sn, Ag, and Pb) [4], and they are not well characterized. For many applications, they should afford satisfactory performance. Carbon or graphite films may be prepared by depositing an appropriate carbon-containing ink followed by 30 min in a drying oven at temperatures near 110°C [62], or by depositing a suspension of graphite powder in a solution containing 1.5% cellulose acetate in a 1:1 mixture of cyclohexanone and acetone, followed by overnight drying at room temperature [63].

*Screen Printing*

This approach has become increasingly popular recently for the fabrication of electrodes and complete cells for a number of diagnostic and other applications where a disposable, one-time-use electrochemical measurement is desired [35,62–66]. Masking patterns are photographically exposed and developed on an emulsion applied to a fine screen whose mesh is open in zones where ink is to be applied to the substrate but occluded in zones where ink is not to be applied. Where finer resolution is desired, an etched mask (more expensive than the screen mask) can replace the screen mask. Pastes or inks are then forced by means of a semiflexible squeegee blade through the masking screen onto the substrate to create the film electrode assembly. The finished assembly is then treated according to the type of electrode desired. In some cases (particularly metal electrodes), the inks are fired to form the electrode(s). In other cases (e.g., carbon films), the electrodes are ready as soon as the solvent has evaporated and the film has dried. The process of screen printing is described in greater detail elsewhere [5].

*Chemical Vapor Deposition*

This is a procedure in which volatile precursor compounds are transported in a carrier gas onto a substrate (typically heated), where they react to form the species to be deposited. For example, antimony-doped tin oxide films can be formed by spraying a substrate held at high temperatures (~500°C) with $SnCl_4$ and $SbCl_5$ in air and hydrolyzing the surface. Various references are available in journal articles [67] and the patent literature [68]. Films of $TiO_2$ on Ti with thicknesses of ~1 μm have been prepared by heating $Ti(OC_3H_7)_4$ at 80–85°C into a flowing $N_2$ carrier gas stream [69]. This gas jet is then combined with a similar water stream which hydrolyzes the $Ti(OC_3H_7)_4$. The resulting films are then annealed at 600°C at $10^{-5}$ torr to introduce oxygen vacancies in the n-type material, which considerably lowers its resistance. Other materials, such as $Fe_2O_3$ and $SnO_2$ films, have been prepared in a similar fashion [11]. A chemical deposition process that is particularly useful for production of carbon or graphite film electrodes is pyrolysis of a precursor carbon compound onto a substrate. This is discussed in the section on carbon electrode fabrication.

*Microlithography*

Microlithography is the process that has made possible most of the advances in microelectronics in the past few decades [70]. Microlithography also makes many new electrochemical experiments possible. The device to be fabricated is coated with a photosensitive polymeric coating, which is baked and then exposed to a light source through a mask or a sequence of masks. The masks are designed using computer-aided design, and are typically produced by covering a suitable metal film (e.g., chromium) on a high-quality glass plate with a negative photoresist, exposing and developing the resist, and then etching the metal to reveal the desired pattern [5]. In use, the mask is typically placed in intimate contact with the photoresist-covered substrate, and radiation (typically UV light) is passed through the mask to expose a pattern in the photoresist.

Most microelectronic facilities can readily achieve minimum feature sizes on the order of 10 μm or better. State-of-the art facilities can produce devices with feature sizes of a few tenths of a micrometer. The lower limit is still decreasing, but the effort required expands exponentially as feature size decreases. Frequently, large numbers of devices can be simultaneously fabricated on the same substrate and subsequently separated by scoring and breaking or sawing the substrate after all processing steps are complete. While complex multilayer devices are possible, most film electrode devices reported to date involve only one or two layers.

Masks may be positive or negative. Positive masks block light wherever an electrode feature is to be placed and transmit light elsewhere, and they appear identical in outline to the intended electrode pattern. Negative masks transmit light where electrode features will be and block light where electrode features will not be. The choice of mask type is dictated by the type of photoresist and processing that one wishes to use. Photoresists can also be classified as positive or negative. A positive photoresist is typically initially polymerized and relatively insoluble but can be depolymerized and made soluble where light strikes it. A negative photoresist is typically initially soluble but rendered insoluble at points that have been photopolymerized by exposure to light.

Typical processing sequences for both positive and negative resists are illustrated in Figure 11.4. The number and order of possible applications and removal of resist and active film materials are great. The general principles of microlithography have been extensively discussed elsewhere [70]. One strategy is to deposit a film of the desired electrode material (preceded and possibly followed by any desired adhesion layer), followed by spin-coating with a negative resist. The photoresist is then exposed through the mask and developed. The photopolymerized zones of the photoresist that remain after development protect the zones of the electrode film that are to be preserved, and the unpolymerized zones of the photoresist are washed away during development, exposing zones of the electrode material that are to be removed by an appropriate type

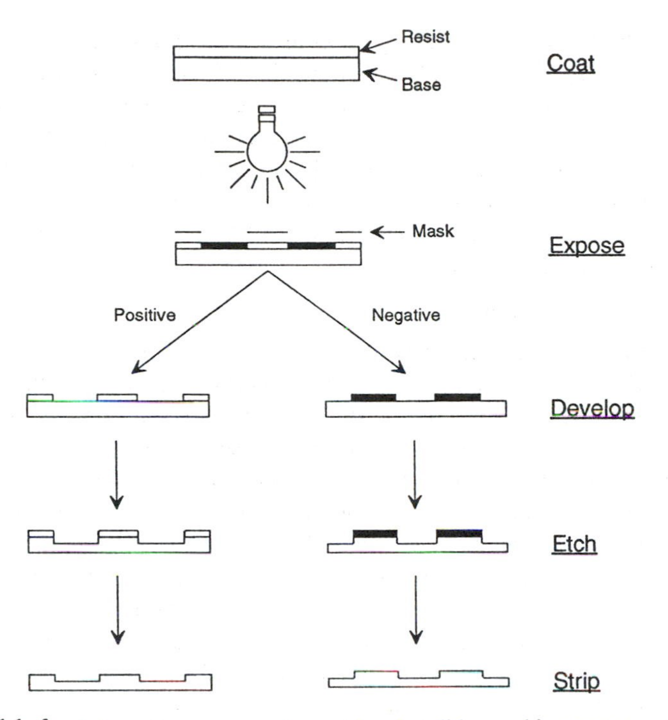


**Figure 11.4** Schematic representation of the photolithographic process sequence, for both positive and negative photoresists. [Adapted from Ref. 70.]

of etching process. For example, chemical dissolution [13,34,37,59], electrolytic dissolution [39,43], plasma etching [13,36,39], reactive ion plasma etching [10], or ion beam milling [13,36] can be used to remove undesired portions of the film. In some cases, such as ion beam milling, a negative resist may be applied over the electrode pattern to protect the electrode material from erosion by the ion beam and may then be stripped by a method such as plasma or reactive ion etching.

An alternative strategy is to deposit a positive photoresist to mask areas where electrode material is to be absent. Upon exposure through the mask and development, the base substrate will be exposed where the electrode components are to be deposited. The electrode materials are then typically deposited over both the bare substrate and the photoresist. A liftoff procedure is then used to remove both the photoresist and excess deposited electrode film material by swelling the photoresist polymer in a suitable solvent, leaving the electrode array behind [40,45,47] with the underlying substrate also exposed in regions where no electrode is present. Increasingly complex patterns of application of multiple resists may be utilized as the design becomes more complex and more masks and photoresist layers are used to create the device. A key concern is adhesion, not only of electrode material to the substrate but also of any overlying insulator materials to both electrode material and substrate. In some instances, developed and cured photoresists, such as polymethylmethacrylate [43], remain as insulating layers to define active electrode areas. Great care must be paid to durable adhesion in such cases. Detachment of the overlayer may cause catastrophic failure of the device.

*Robustness and Resurfacing of Devices*

A major consideration in utilizing thin-film electrodes is their robustness and durability. Films are typically at most a few micrometers thick. If the film electrode is fabricated so that its thinnest dimension is perpendicular to the solution interface and its broadest dimension is exposed to solution, all components of fabrication (substrate, adhesion layer, conductor layer, and any insulating overlayers) must be optimized to enable the electrode to survive. The conventional strategy of polishing to renew a bulk electrode is generally not available. The process of polishing would in many cases destroy the film or the structure of any pattern laid out in the film, due to the extreme thinness of the film. Any chemical or electrochemical cleaning treatments must be designed to avoid compromise of the integrity of any of the structural components of the device. (For example, with metal films such as gold or platinum, the accessible potential range may need to be restricted to avoid evolution of hydrogen or oxygen, which may peel the metal film from the substrate unless extremely robust adhesion treatments have been used.) Thus film electrodes of this design may frequently be relegated to one-shot experiments, and disposable devices may be the most practical approach.

Film electrodes based on exposure of the thinnest edge dimension to solution are potentially much more amenable to resurfacing or polishing, since the broadest dimension can be supported in a sandwich between two robust protecting insulator layers (e.g., microscope slides or plastic films). Removal of considerably more material is possible without destroying the device. However, even in this case polishing may be problematic, since smearing of adhesives or other materials of construction or the electrode material itself may seriously occlude or distort the already small and difficult to measure dimensions of the electrode, rendering quantitative evaluation of the fundamental response of the electrode more difficult. If polishing is to be attempted, better results will be obtained if the hardness of all materials of construction can be matched, to minimize differential removal of material during polishing, which can lead to undercutting or excessive exposure of the electrode material. For these reasons, cleavage of the device to expose a fresh surface may be a more reliable way to generate a new, clean surface.

Finally, although mechanical pressure applied between the outer support materials of the sandwich may stabilize the electrode device against failure more than if simply covered with an insulating polymer layer, a major failure mode of these devices is the result of inadequate sealing between materials. If solution can seep into fissures between layers, the device will be prone to catastrophic failure. The more layers, the greater the probability of failure, even if the probability of failure of a single layer is small. For this reason, photolithographic fabrication may be more practical than layered construction for fabrication of microelectrode arrays having large numbers of microelectrodes. While numerous microelectrode array designs can be readily conceived, the most difficult step typically involves solving the materials problems and design compromises needed to create a durable device.

## IV. METAL FILM ELECTRODES

Metal films can be prepared by numerous methods outlined earlier. Prominent among these are chemical deposition (e.g., reduction of metal resinate ink) of the metal onto the desired substrate, and vacuum evaporation and sputtering, generally in high vacuum. In principle, electroless deposition of the appropriate metal film would also be feasible, although examples of this approach for production of practical electrodes are scarce. Electrodeposition onto a conductive substrate is also feasible. Where optical transparency is important, the underlying conductor can be a semiconductor such as tin oxide, or a very thin metal film. Where optical transparency is not crucial, thicker conductor films or even bulk metal, carbon, or other conductors can be used as substrates for electrodeposition. Chemical methods, although quick and convenient, do not provide pure, reproducible films and therefore have had more limited application, except for Hg films. Vacuum evaporation and sputtering have been more widely applied. Since Pt and Au films have been most extensively employed as electrodes, examples of these films are discussed briefly here.

### A. Gold Film Electrode Fabrication

Gold films have been prepared by sputtering [28,33,38,44,71,72], vacuum evaporation [3,34,37,40,42,43,59,61], and liquid film application or screen printing [37,64,73]. Most of these films have been too thick for optical transparency (see later discussion). Gold films have been used in sandwich configurations with the edge exposed to solution [10,28,32,33,72] and in microlithographically defined arrays [10,34,37,38,40,42,47,59]. Microlithographically defined gold arrays have been fabricated by first depositing and developing a positive photoresist to define the electrode deposition geometry, followed by gold deposition, and then solvent swelling and liftoff of excess photoresist and its overlying metallization [38,40,42,47]. Microlithographically defined gold arrays have also been fabricated by first depositing a gold film, followed by overcoating with a negative photoresist, and either chemical etching with triiodide solution [10,34,37,59] or reactive ion plasma etching [10] to remove gold from unprotected zones, followed by stripping of the resist to expose the gold electrodes.

Chromium has frequently been used as an adhesion layer for gold deposition. It has been noted that gold films should be at least 200 nm thick to avoid interference by the chromium adhesion layer with surface electrochemistry [61]. Numerous etching procedures have been described to remove interfering chromium from the surface of gold electrodes. These include chemical etching with $Ce(SO_4)_2/HNO_3$ solution [10] or $K_3Fe(CN)_6/NaOH$ solution [59], and electrochemical dissolution etching [39,43]. Niobium has also been used as an adhesion layer for gold [42].

## B. Platinum Film Electrode Fabrication

Platinum films have been prepared by sputtering [13,16,28,30,36,45,58], vacuum evaporation [3,17,42], liquid film application [4], or screen printing [37], and by purchase of commercial thin (4 μm) foils [32]. Most of these films have been too thick for optical transparency (see later discussion). Platinum films have been used in sandwich configurations with the edge exposed to solution [28,30,32] and in microlithographically defined arrays [13,36,39,42,45]. Platinum has served as a substrate for a microelectrode array of pyrolytically deposited carbon [36]. Platinum electrodes have also been produced by electrodeposition onto other conductors. In one case, a gold microelectrode array was first fabricated by standard microlithographic processing, and then platinum was electrodeposited onto the gold film to produce the active electrode surface [39]. An advantage of the approach was the ability to achieve a narrower gap between interdigitated electrodes than feasible by microlithographic means alone, by varying the quantity of platinum electrodeposited.

Microlithographically defined platinum arrays have been fabricated by first depositing and developing a positive photoresist to define the electrode deposition geometry, followed by platinum deposition, and then solvent swelling and liftoff of excess photoresist and its overlying metallization [42,45,47]. Microlithographically defined platinum arrays have also been fabricated by first depositing a platinum film, followed sequentially by overcoating with a negative photoresist (to protect the pattern of platinum to be retained from the ion milling step), ion milling to remove platinum from unprotected zones [13], and plasma etching of the resist to expose the platinum electrodes.

Chromium [10,39], niobium [42], titanium [13], and 90% titanium/10% tungsten [58] have been used as adhesion layers for platinum deposition. Chemical etching procedures based on $H_2O_2$/ethylenediamine tetraacetic acid (EDTA)/$NH_4OH$ have been described for removal of interfering titanium [13] or titanium/tungsten [58] adhesion components from the surface of platinum electrodes. The procedure was shown to be effective for more than 24 h after treatment, as evidenced by cyclic voltammetric peak potential separations [58]. Titanium appeared to be a more serious contributor to surface contamination than tungsten.

## C. Optical and Electrochemical Properties

Pt acts like a neutral density filter through a wide wavelength range and can be used in the UV when the film is on quartz. Au has an optical window with maximum transmission at 540 nm. On quartz, Au can also be used in the UV when the conductivity is comparable to Pt. For the visible region, Au films with good transmission and with resistance as low as a few ohms per square can be prepared. Typical optical properties are shown in Figure 11.3 [55]. The electrochemical properties are similar to bulk Pt and Au, as seen in Figure 11.5 [74],

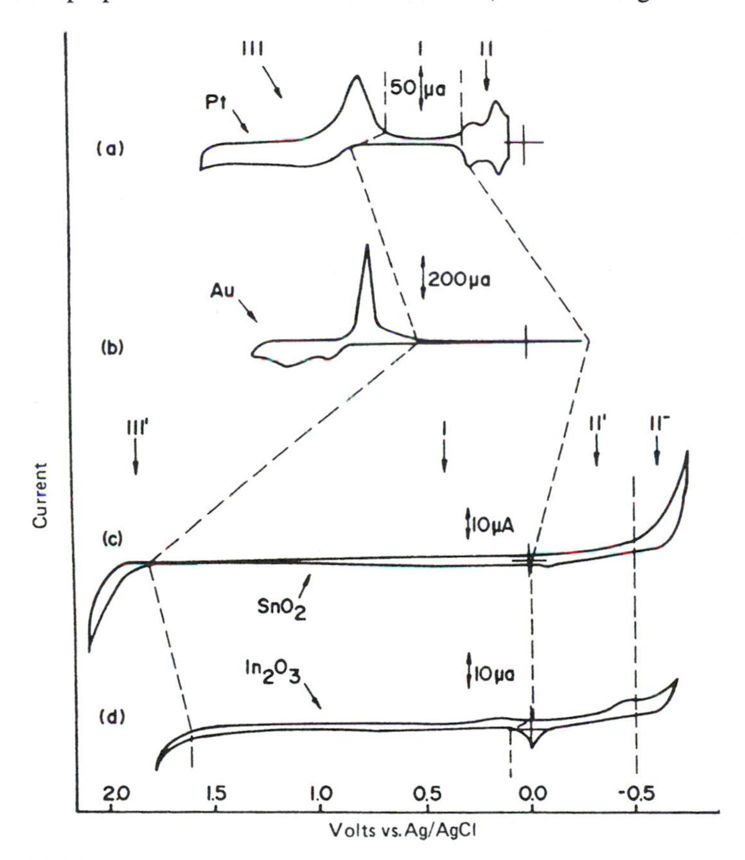


**Figure 11.5** Comparison of current-voltage curves for Pt, Au, $SnO_2$, and $In_2O_3$ electrodes in 1 N $H_2SO_4$ media. Potential scan rate was 87 mV/s. All electrodes were 1 $cm^2$ in geometric area. (a) Platinum thin film, (b) gold thin film, (c) $SnO_2$ thin film, (d) $In_2O_3$ thin film. I corresponds to the double-layer region; II indicates hydrogen reduction; III represents oxygen evolution. [From Ref. 74, with permission. Copyright 1976 American Chemical Society.]

except that the overpotential for hydrogen ion discharge is slightly less on a Pt film than for bulk Pt. Whether this activity is due to the "cleaner" surface or to the crystal structure of the Pt film is not clear, although it is known that the activity of platinum for hydrogen evolution varies with the crystal face, and preferential texture has been observed for vacuum-deposited thin platinum films [16]. Characteristics of these films in nonaqueous solutions have been evaluated by Osa and Kuwana [75]. The background potential limits are summarized in Table 11.2 [75].

An optically transparent, porous platinum film has been produced by photoelectrodeposition on an InP semiconductor substrate [15]. Polyester sheet covered with a thin film of sputtered gold has also proved suitable as an OTE [71]. When overcoated with a layer of $TiO_2$, these electrodes exhibited electrochemical behavior consistent with a microelectrode array, including cyclic voltammetric current plateaus instead of clearly defined peaks, although this feature was not recognized at the time [71].

## V. CARBON FILM ELECTRODES

### A. Carbon Film Electrode Fabrication

Carbon films and graphite films have been prepared by vacuum evaporation [28,42,76,77], pyrolysis [29,36,78–83], screen printing [46,62,63,65,66,84], and laser photoactivation of sites on a graphite or glassy carbon substrate [85]. Various pyrolytic processes have been successful, most based on the deposition of volatile precursor compounds. For example, methane can be pyrolyzed while

**Table 11.2** Background Potential Limits in Various Nonaqueous Solvents

| | Potential limits[a] | | |
|---|---|---|---|
| Solvents | Tin oxide-OTE | Pt-OTE | Au-OTE |
| Acetonitrile | +2.47 to -2.77 | +2.07 to -1.88 | -2.68 |
| DMF | +1.72 to -2.70 | +1.65 to -2.21 | +1.61 to -2.39 |
| Nitromethane | +2.48 to -1.08 | +1.42 to -1.15 | — |
| Dichloroethane | +2.32 to -2.18[b] | — | — |
| Propylenecarbonate | +2.52 to -2.66 | — | — |
| Monoglyme | +2.37 to -2.82[b] | — | — |
| Diglyme | +2.40 to -2.93[c] | — | — |

[a]Potentials reported vs. SCE; 0.1 M $Et_4NClO_4$ supporting electrolyte unless otherwise noted; scan rate 0.053 V/s, temperature 25°C; potentials are for current levels of 100 μA/cm$^2$.
[b]0.1 M $n$-$Bu_4NClO_4$.
[c]0.05 M $n$-$Bu_4NClO_4$.
*Source*: Ref. 75.

flowing past a suitable substrate such as glassy carbon at 1000°C [78], or a Macor glass-ceramic substrate at 926°C [79], or a quartz substrate heated in a Bunsen burner flame [29]. Ethylene can be catalytically reduced at 550°C on a nickel resinate film deposited on a glass substrate to form a film with properties similar to glassy carbon [80]. Higher-molecular-weight carbon compounds such as 3,4,9,10-perylenetetracarboxylic dianhydride (PTDA) can also be sublimed at low pressure (~0.01–0.1 torr) and then pyrolyzed to form a film with properties similar to glassy carbon, for example, on a quartz substrate at 850°C [81], or on a platinum substrate at 1000°C for preparation of a carbon-based microelectrode array [36]. A thin film of polyacrylonitrile has also been deposited on a quartz substrate and subsequently pyrolyzed at 1000°C to form a carbon film of thickness controllable by the initial polymer layer thickness [82]. Most of these films have been too thick for optical transparency (see later discussion).

Carbon films have been used in sandwich configurations with the edge exposed to solution [28] and in microlithographically defined arrays [36]. The complete process of microlithographically fabricating a carbon array, illustrated in Figure 11.6 [36], is quite complex. First an insulating layer of $SiO_2$ is thermally grown on a silicon wafer. Then a titanium adhesion layer is sputtered onto the wafer, followed without breaking vacuum by sputter deposition of a platinum film. The device is then transferred to a pyrolysis chamber for pyrolytic deposition of PTDA onto the platinum at 1000°C. The array is then spin-coated with a silicon-based positive photoresist, which is exposed through a mask and developed. Reactive ion etching in an oxygen plasma then removes undesired carbon to expose the platinum/titanium layer. The remaining photoresist and the exposed platinum/titanium layer are then ion-milled to expose the carbon array as well as the substrate. A 250-nm layer of $Si_3N_4$ is deposited by plasma chemical vapor deposition over the whole array, followed by application of a positive photoresist to mask areas of the array that are not to be exposed to solution, and the $Si_3N_4$ layer is reactive-ion-etched with a $CF_4$ plasma until the carbon layer is exposed, leaving $Si_3N_4$ flush with the surface of the carbon [36].

## B. Optical and Electrochemical Properties

A carbon OTE can be prepared by vacuum evaporating a thin carbon film on glass or quartz [77] with an electron beam evaporator. The Lebow Company (Goleta, CA) produces these films commercially. The electrochemical properties are similar to those of bulk graphite, and resistivity can be as low as twice that for bulk graphite in films as thin as 28 nm [77]. Optical transparency is good in the UV-visible range. An analogous carbon OTE on germanium has also been demonstrated in the infrared (IR) region [76].

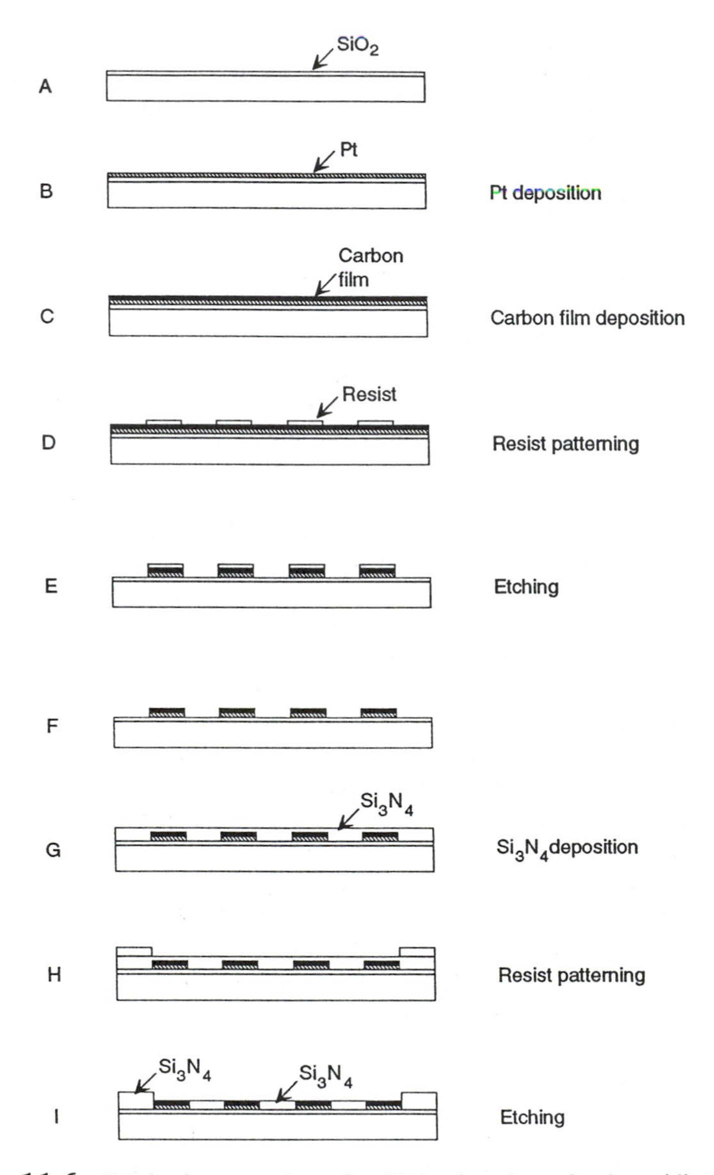


**Figure 11.6** Fabrication procedure of an IDA microelectrode. An oxidized silicon wafer (A) is coated with platinum (B). Carbon film is pyrolyzed on it (C). The substrate is coated with photoresist, which is exposed and developed (D); unnecessary portions are then removed by reactive ion etching (E). After removing the photoresist, an $Si_3N_4$ layer is deposited on the substrate (G). The substrate is coated with photoresist, which is exposed and developed (H); then the desired shape of the carbon electrode is exposed by reactive-ion etching (I). [Adapted from Ref. 36.]

## VI. SEMICONDUCTOR FILM ELECTRODES

A vast number of semiconductor materials have been tested for use as electrodes. Some of these are elucidated in Chapter 28, particularly with respect to their unique properties toward photochemically excited molecules, although most have been used as bulk single crystals and not as polycrystalline films. The two most common semiconducting film electrodes are $SnO_2$ and $InO_x$.

### A. Tin Oxide OTE

The optical and semiconducting properties of tin oxide coatings on glass and quartz plates have been known for a long time. These coatings have been available commercially under the trade names of Nesa and Electropane glass and were reputedly used in the early days of aviation for deicing windshields. Tin oxide coatings are also frequently used on oven windows to enable viewing of the interior with minimal heat loss to the outside due to absorption and reflection of infrared absorption. More recently, tin oxide has been deposited on plastic substrates such as polyester sheets [71]. Antimony has been the most commonly used dopant, and the resulting tin oxide is an n-type semiconductor. Commercial preparations are available from Libbey, Owen and Ford Glass Co. (Brackenridge, PA), Delta Technologies (Stillwater, MN), which can also supply the film on either glass or quartz, Watkins-Johnson Co. (Scotts Valley, CA), and other sources. The p-type material doped with fluoride is also available (Watkins-Johnson Co.).

Electron microscopic examination indicates a surface without any defined grain boundaries but having microcrystalline particles with mean size of $\sim 10$ nm. The commercial coatings are usually between 600 and 1000 nm thick, depending on the source and sample. The refractive index is 2.0 in the visible region of the spectrum. Because of the high refractive index and the thickness of these coatings, it is not surprising that the absorption spectrum, taken with the light beam perpendicular to the surface, exhibits an interference pattern. Typical spectra of n-type tin oxide on transparent substrates are shown in Figure 11.7. Coating thicknesses can be easily calculated from the interference maxima and a knowledge of the refractive index [86].

The doped tin oxide surface can be best described as a poor conductor in its behavior as an electrode. It can be used in aqueous solutions over a potential range quite similar to that of platinum. Linear sweep cyclic voltammetric scans [74] are shown in Figure 11.5 for $SnO_2$ in comparison to other film electrodes in acid media. Tin oxide is not recommended for use in highly alkaline solutions, since some surface degradation occurs. According to Laitinen and coworkers [87], oxygen or chlorine (in HCl media) evolution occurs without disruption of the surface. The positive potential extreme is quite high, on the order of 2 V, presumably since the Sn exists in its highest oxidation state, so that further oxidation, as is evident for Pt and Au, does not occur. At the negative

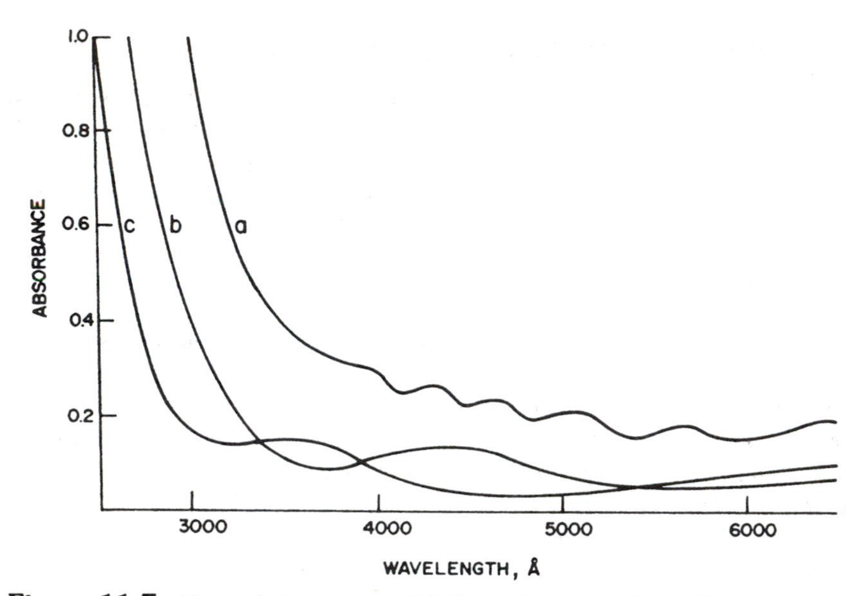


**Figure 11.7** Transmission spectra of $SnO_2$ coatings on various substrates: a, glass, 3 Ω/□; b, Vycor, 6 Ω/□; c, quartz, 20 Ω/□.

limit, hydrogen-ion discharge is accompanied by the concurrent reduction of the tin oxide to metallic tin, resulting in destruction of the electrode. In an intermediate region of potential, a charging current is observed by linear sweep cyclic voltammetry. This current is attributed to the charging and discharging of positive surface states and is most pronounced in chloride media. The amount of charge is several times greater than the estimated surface population of tin atoms.

The surface chemistry may also be affected by pH through groups that can be influenced by possible acid-base equilibria:

$$\sim SnOH \rightleftarrows SnO^- + H^+ \tag{11.6}$$

or

$$\sim SnOH \rightleftarrows Sn^+ + OH^- \tag{11.7}$$

In nonaqueous solutions, the usable potential range is large, on the order of 5 V. This potential range allows a wide variety of oxidations and reductions to be examined. Some typical values on different types of film electrodes are reported in Table 11.2 [75]. The electrolysis of several organic compounds in these solvents has been characterized. A general observation is that the tin oxide electrode appears more suited to studies of reduction processes than of oxidation processes in nonaqueous media. For example, the reversible oxidation of ferrocene at a Pt electrode is quite irreversible at tin oxide. This is presumably in large part due to the n-type semiconductor properties of tin oxide. The potential at which ferrocene is oxidized is quite positive of the flatband potential of $SnO_2$, resulting in a rather low population of electrons in the conduction band, and accordingly sluggish kinetics. Most reductions, on the other hand, give results similar to those obtained at Pt, since the conduction band is much more highly populated at more negative potentials, with the added advantage of a large accessible negative range.

### B. Others

Although $SnO_2$ has found the widest electroanalytical utility as a semiconducting electrode, other types have been investigated over the years. One promising type is the tin-"doped" indium-oxide-coated glass electrode manufactured by PPG (Barberton, OH) under the trade name of Nesatron. The electrochemical characteristics are similar to $SnO_2$ as shown in Figure 11.5, although the available potential range is somewhat smaller. The optical transmission is only slightly improved over $SnO_2$, so that Nesatron does not appear to possess any marked advantage over $SnO_2$.

Several other varieties of semiconducting electrodes have been produced by chemical vapor deposition, as noted earlier, including films of $TiO_2$ on Ti [69], $TiO_2$ on plastic substrates [71], and other materials, such as $Fe_2O_3$ and $SnO_2$ films [11]. Most of these other electrodes have been tested as electrodes for use in photoelectrochemically induced water splitting and have not found conventional electroanalytical utility.

Germanium plates have been employed as OTEs in the infrared region, although films have not been examined due to the high resistance. The surface also exhibits a limited usable potential range [88]. The electrochemical properties of germanium are substantially improved by vapor-depositing a thin carbon film on the surface [76].

## VII. TECHNIQUES OF CELL DESIGN BASED ON FILM ELECTRODES

A few special technical problems arise when attempting to mount film electrodes into conventional electrochemical cells due to difficulties in defining the electrode area and in making electrical connection to the active surface. These constraints usually result in cells that have been somewhat modified, although the same principles that were discussed in Chapter 9 still apply. In this final section we address some of these practical difficulties.

The film electrode should usually be partially masked to the minimum desired electrochemical area to reduce power requirements of the electronic equipment and to minimize any iR drop that may occur across the electrode surface. These problems are especially critical if short-time pulse experiments are being performed using electrodes with a significant resistance. The use of ultramicroelectrodes is particularly advantageous in this case [7,8]. Masking is conveniently accomplished using a rubber or elastomer (e.g., Viton) O-ring, which serves to define the electrode area and to provide a leaktight seal to the body of the cell. Electrical contact to the film electrode could be made using conducting silver paint or epoxy, although high resistance often results.

An excellent mechanical contact can be obtained by tightly compressing a thin copper or brass gasket concentrically around the O-ring between the electrode and the cell body. Large forces per unit area can be generated in this manner. Use of a second O-ring mounted concentric to the sealing O-ring but between the backing plate and the electrode provides strain relief and is very helpful to avoid fracture of fragile substrates such as glass. Alternatively, in some cases the electrode can be sealed to the cell with an adhesive such as epoxy, and a mechanical contact can be made to an exposed corner of the electrode surface. A typical cell arrangement [89] is shown in Figure 11.8.

A more sophisticated approach to this problem involves the use of integrated circuit technology, allowing the electrode to be directly press-fit onto the cell with an O-ring [90]. The electrical contact can be provided by evaporation of a thick film of gold onto the film electrode. The active area can then be defined by applying a photoresist which is exposed using a contact photographic mask of the shape of the desired electrode area. Subsequent removal of the exposed photoresist then allows etching of the gold by a concentrated KI solution, exposing the electrode area. Any gold that may be exposed to the electrochemi-

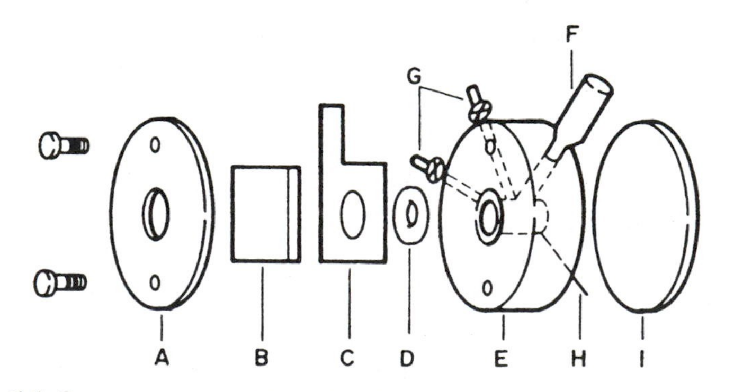


**Figure 11.8** Cell for use of thin-film electrodes. A, Aluminum retaining plate; B, OTE; C, copper foil; D, O-ring; E, Lucite body; F, glass salt bridge for reference electrode; G, Hamilton valves; H, auxiliary electrode; I, quartz disk. [From Ref. 89, with permission. Copyright 1972 American Chemical Society.]

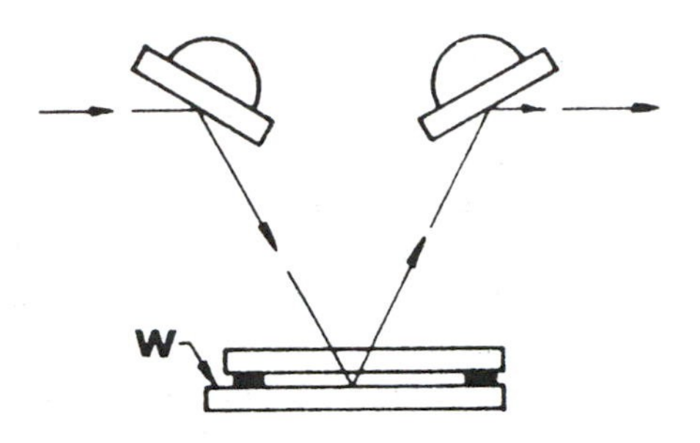


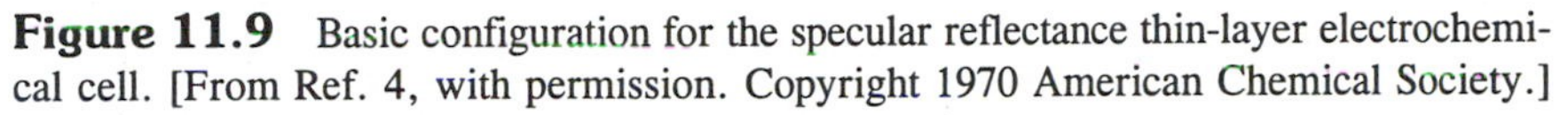

**Figure 11.9** Basic configuration for the specular reflectance thin-layer electrochemical cell. [From Ref. 4, with permission. Copyright 1970 American Chemical Society.]

cal solution can then be covered by evaporation of SiO or $MgF_2$ in vacuo or simply by coating the gold with an epoxy impervious to the solvent of interest.

Various cell designs have been proposed to accommodate film electrodes. Of special interest here is the development of a cell compatible with a vacuum manifold for use in studies of oxygen-sensitive systems [91]. In one design of a sandwich-type optically transparent electrochemical cell, the reference and auxiliary electrodes can be attached to vacuum valves on the cell body. The vacuum line attachment can be made to the remaining valve. This cell features a small volume of less than 2 ml and can keep the oxygen concentration below $5 \times 10^{-7}$ M for several days. Thin-layer electrochemical cells have relied on a very simple scheme [4] where a thick and thin-film electrode are separated by a spacer of a given thickness, perhaps several hundred micrometers. As shown in Figure 11.9 [4], a light beam can then be used for spectroscopic studies and the thin-layer cell can be operated in the conventional manner. A myriad of other cell designs exist in the literature—a list too long to describe—with some as simple to use as disposable cells with a film electrode held in place and sealed liquid-tight with epoxy. In general, the same rules apply for the design of these cells as those using foil electrodes.

## VIII. PROSPECTS FOR DISPOSABLE, INTEGRATED SENSOR SYSTEMS

Considerable progress has been made toward the goal of developing electrochemically based sensors based on disposable film electrodes or complete integrated cell or sensor systems. While some of the microlithographic techniques described earlier have the potential for production of practical, inexpensive devices in large number, very promising results have been reported for devices produced by screen printing and thick-film technology. For example, the commercially available ExacTech glucose sensor is a complete, pen-sized system designed for home measurement of blood glucose levels based on a ferrocene-mediated glucose oxidase enzyme-catalyzed reaction at a carbon electrode, fabricated by screen printing on a disposable polyvinyl chloride plastic strip [73]. A sensor for bacterial activity and spoilage of meat is under development, also based on a screen-printed carbon electrode [73]. A disposable microelectrode array based on laser-drilled holes in a dielectric layer covering screen-printed carbon electrodes is also commercially available. It has been modified by coelectrodeposition, with rhodium or platinum, of enzymes for determination of glucose or lactate [92]. A screen-printed graphite film containing cobalt phthalocyanine as a catalyst has been used to catalyze the enzymatic determination of reduced glutathione [63]. In an interesting twist, the screen-printed carbon electrical contact on the back side of a commercial ExacTech disposable test strip has been used as a substrate for mercury deposition and stripping voltammetry [65].

Down the road, we can expect integrated sensor systems to be developed using many of the concepts described in this chapter. Such systems may well incorporate complete integration of electronic control and output transducers into a single microelectronically integrated package also containing sensor input devices such as electrodes and cell packaging. Current trends in this area have been surveyed [73,93], and issues involved in integrating both sensor and electronic components in a single package have been discussed [94]. Methods for micromachining that may be helpful in designing and fabricating a complete practical sensor device have also been reviewed [95]. These ideas will be very valuable in designing the film-based integrated electrochemical sensor of the future. The future is very bright for film-based electrodes and the devices and applications that they make possible.

## REFERENCES

1. For reviews, see: (a) T. Kuwana and N. Winograd, in *Electroanalytical Chemistry*, Vol. 7 (A. J. Bard, ed.), Marcel Dekker, New York, 1974, pp. 1–78; (b) W. R. Heineman, F. M. Hawkridge, and H. N. Blount, *Electroanalytical Chemistry*, Vol. 13 (A. J. Bard, ed.), Marcel Dekker, New York, 1984, pp. 1–113; and references therein.
2. J. D. E. McIntyre, in *Advances in Electrochemistry and Electrochemical Engineering*, Vol. 9 (R. H. Mueller, ed.), Wiley, New York, 1973, pp. 61–166.
3. A. Yildiz, P. T. Kissinger, and C. N. Reilley, *Anal. Chem. 40*:1018 (1968). (A host of intriguing thin-layer cell applications are suggested, including the possibility of thin-layer electrochemical detectors for liquid chromatography, realized several years later by Kissinger and Adams.)
4. P. T. Kissinger and C. N. Reilley, *Anal. Chem. 42*:12 (1970).
5. M. Fogiel, *Modern Microelectronics, Basic Principles, Circuit Design, Fabrication Technology*, Research and Education Association, New York, 1972, pp. 94–100.

6. F. M. Smits, *Bell Syst. Tech. J. 37*:711 (1958).
7. M. Fleischmann, S. Pons, D. R. Rolison, and P. P. Schmidt, eds., *Ultramicroelectrodes*, Datatech Systems, Inc., Morganton, NC, 1987, and references therein.
8. R. M. Wightman and D. O. Wipf, in *Electroanalytical Chemistry, A Series of Advances*, Vol. 15 (A. J. Bard, ed.), Marcel Dekker, New York, 1989, pp. 267–369, and references therein.
9. R. B. Morris, D. J. Franta, and H. S. White, *J. Phys. Chem. 91*:3559 (1987).
10. M. Samuelsson, M. Armgarth, and C. Nylander, *Anal. Chem. 63*:931 (1991).
11. K. L. Hardee and A. J. Bard, *J. Electrochem. Soc. 123*:1024 (1976).
12. B. Miller, and J. M. Rosamilia, *J. Electrochem. Soc. 132*:2621 (1985).
13. C. E. Chidsey, B. J. Feldman, C. Lundgren, and R. W. Murray, *Anal. Chem. 58*:601 (1986).
14. J. E. Bartelt, S. M. Drew, and R. M. Wightman, *J. Electrochem. Soc. 139*:70 (1992).
15. A. Heller, D. E. Aspnes, J. D. Porter, T. T. Sheng, and R. G. Vadimsky, *J. Phys. Chem. 89*:4444 (1985).
16. H.-Y. Liu, F.-R. F. Fan, and A. J. Bard, *J. Electrochem. Soc. 132*:2666 (1985).
17. D. S. Lashmore and M. P. Dariel, *J. Electrochem. Soc. 135*:1218 (1988).
18. B. W. Gregory, D. W. Suggs, and J. L. Stickney, *J. Electrochem. Soc. 138*:1279 (1991).
19. T. Hachiya, H. Honbo, and K. Itaya, *J. Electroanal. Chem. 315*:275 (1991).
20. S. Manne, P. K. Hansma, J. Massie, V. B. Elings, and A. A. Gewirth, *Science 251*:183 (1991).
21. M. F. Toney and J. McBreen, *Interface 2(1)*:22 (1993).
22. A. J. Bard, H. D. Abruña, C. E. Chidsey, L. R. Faulkner, S. W. Feldberg, K. Itaya, M. Majda, O. Melroy, R. W. Murray, M. D. Porter, M. P. Soriaga, and H. S. White, *J. Phys. Chem. 97*:7147 (1993), and references therein.
23. N. S. Chong, M. L. Norton, and J. L. Anderson, *J. Electrochem. Soc. 138*:1263 (1991).
24. H. D. Dewald, *Electroanalysis 3*:145 (1991).
25. D. A. Buttry and M. D. Ward, *Chem. Rev. 92*:1355 (1992), and references therein.
26. J. F. Smalley, C. V. Krisnan, M. Goldman, S. W. Feldberg, and I. Ruzic, *J. Electroanal. Chem. 248*:255 (1988).
27. Y. W. Alsmeyer and R. L. McCreery, *Anal. Chem. 63*:1289 (1991).
28. K. R. Wehmeyer, M. R. Deakin, and R. M. Wightman, *Anal. Chem. 57*:1913 (1985).
29. Y.-T. Kim, D. M. Scarnulis, and A. G. Ewing, *Anal. Chem. 58*:1782 (1986).
30. T. Varco Shea and A. J. Bard, *Anal. Chem. 59*:2101 (1987).
31. J. T. McDevitt, R. W. Murray, and S. I. Shah, *J. Electrochem. Soc. 138*:1346 (1991).
32. J. A. Bartelt, M. R. Deakin, C. Amatore, and R. M. Wightman, *Anal. Chem. 60*:2167 (1988).
33. H. A. O. Hill, N. A. Klein, I. S. M. Psalti, and N. J. Walton, *Anal. Chem. 61*:2200 (1989).

34. D. G. Sanderson and L. B. Anderson, *Anal. Chem. 57*:2388 (1985).
35. O. Niwa, M. Morita, and H. Tabei, *Electroanalysis 3*:163 (1991).
36. H. Tabei, M. Morita, O. Niwa, and T. Horiuchi, *J. Electroanal. Chem. 334*:25 (1992).
37. W. Thormann, P. van den Bosch, and A. M. Bond, *Anal. Chem. 57*:2764 (1985).
38. A. Aoki, T. Matsue, and I. Uchida, *Anal. Chem. 62*:2206 (1990).
39. A. J. Bard, J.A. Crayston, G. P. Kittlesen, T. Varco Shea, and M. S. Wrighton, *Anal. Chem. 58*:2321 (1986).
40. L. E. Fosdick, J. L. Anderson, T. A. Baginski, and R. C. Jaeger, *Anal. Chem. 58*:2750 (1986).
41. G. P. Kittelsen, H. S. White, and M. S. Wrighton, *J. Am. Chem. Soc. 106*:7289 (1984).
42. R. S. Glass, S. P. Perone, and D. R. Ciarlo, *Anal. Chem. 62*:1914 (1990).
43. T. Hepel, and J. Osteryoung, *J. Electrochem. Soc. 133*:752 (1986).
44. T. Matsue, A. Aoki, E. Ando, and I. Uchida, *Anal. Chem. 62*:407 (1990).
45. O. Niwa, M. Morita, and H. Tabei, *Anal. Chem. 62*:447 (1990).
46. O. Niwa, Y. Xu, H. B. Halsall, and W. R. Heineman, *Anal. Chem. 65*:1559 (1993).
47. T. Horiuchi, O. Niwa, and H. Tabei, *Anal. Chem. 66*:1224 (1994).
48. K. Honda and A. Kuwano, *J. Electrochem. Soc. 133*:853 (1986).
49. C. B. Greenberg, *J. Electrochem. Soc. 140*:3332 (1993).
50. K. L. Chopre, L. C. Bobb, and M. H. Francombs, *J. Appl. Phys. 34*:1669 (1963).
51. C. A. Mengebauer and M. B. Webb, *J. Appl. Phys. 33*:74 (1962).
52. E. K. Sichel, J. I. Gittleman, and P. Sheng, *J. Electron. Mater. 11*:699 (1982).
53. S. P. Muraka, *Thin Solid Films 23*:323 (1974).
54. See, for example, C. Kittel, *Introduction to Solid State Physics*, Wiley, New York, 1971, p. 276.
55. W. Von Benken and T. Kuwana, *Anal. Chem. 42*:1114 (1970).
56. L. Holland, *Vacuum Deposition of Thin Films*, Wiley, New York, 1956.
57. R. B. Belser, *J. Appl. Phys. 28*:109 (1957).
58. M. Josowicz, J. Janata, and M. Levy, *J. Electrochem. Soc. 135*:112 (1988).
59. K. Yokoyama, E. Tamiya, and I. Karube, *Electroanalysis 3*:469 (1991).
60. J. O'M. Bockris and B. D. Cahan, *J. Chem. Phys. 50*:1307 (1969).
61. C. A. Goss, D. H. Charych, and M. Majda, *Anal. Chem. 63*:85 (1991).
62. X. Xing, T.-C. Tan, M. Shao, and C.-C. Liu, *Electroanalysis 4*:191 (1992).
63. S. A. Wring, J. P. Hart, and B. J. Birch, *Electroanalysis 4*:299 (1992).
64. N. A. Morris, M. F. Cardosi, B. J. Birch, and A. P. F. Turner, *Electroanalysis 4*:1 (1992).
65. J. Wang and B. Tian, *Anal. Chem. 64*:1706 (1992).
66. P. N. Bartlett and P. R. Birkin, *Anal. Chem. 65*:1118 (1993).
67. H. Kim and H. A. Laitinen, *J. Electrochem. Soc. 122*:53 (1975).
68. J. M. Mochel, Corning Glass Works, U.S. Patent 2,564,707 (1951).
69. K. L. Hardee and A. J. Bard, *J. Electrochem. Soc. 122*:739 (1975).
70. L. F. Thompson, C. G. Willson, and M. J. Bowden, eds., *Introduction to Microlithography*, ACS Symposium Series, Vol. 219, American Chemical Society, Washington, DC, 1983, and references therein.

71. R. Cieslinski and N. R. Armstrong, *Anal. Chem. 51*:565 (1979).
72. P. N. Bartlett, P. B. M. Archer, and S. K. Ling-Chung, *Sens. Actuators 19*:125 (1989).
73. A. P. F. Turner, *Sens. Actuators 17*:433 (1989).
74. N. R. Armstrong, A. Lin, M. Fujihira, and T. Kuwana, *Anal. Chem. 48*:741 (1976).
75. T. Osa and T. Kuwana, *J. Electroanal. Chem. 22*:389 (1969).
76. J. S. Mattson and C. A. Smith, *Anal. Chem. 47*:1122 (1975).
77. T. P. DeAngelis, R. W. Hurst, A. M. Yacynych, H. B. Mark, Jr., W. R. Heineman, and J. S. Mattson, *Anal. Chem. 49*:1395 (1977).
78. A. L. Beilby and A. Carlsson, *J. Electroanal. Chem. 248*:283 (1988) and references therein.
79. C. F. McFadden, L. L. Russell, P. R. Melaragno, and J. A. Davis, *Anal. Chem. 64*:1520 (1992).
80. R. A. Saraceno, C. E. Engstrom, M. Rose, and A. G. Ewing, *Anal. Chem. 61*:560 (1989).
81. A. Rojo, A. Rosenstratten, and D. Anjo, *Anal. Chem. 58*:2988 (1986).
82. J. Wang, A. Brennsteiner, A. P. Sylwester, and C. L. Renschler, *Electroanalysis 3*:505 (1991).
83. C. F. McFadden, P. R. Melaragno, and J. A. Davis, *Anal. Chem. 62*:742 (1990).
84. W. Lian and S. Wouters, *Sens. Actuators 17*:609 (1989).
85. K. D. Sternitzke and R. L. McCreery, *Anal. Chem. 62*:1339 (1990).
86. S. Tolansky, *Multiple Beam Interferometry of Surfaces and Films*, Clarendon Press, Oxford, England, 1948.
87. H. A. Laitinen, C. A. Vincent, and T. M. Bednarski, *J. Electrochem. Soc. 115*:1024 (1968).
88. D. E. Tallant and D. H. Evans, *Anal. Chem. 41*:835 (1969).
89. W. R. Heineman and T. Kuwana, *Anal. Chem. 44*:1972 (1972).
90. J. C. Holland, Ph.D. Thesis, Purdue University, W. Lafayette, IN, *Computer Controlled Internal Reflection Spectroelectrochemical Measurements of Fast Electron Transfer Reactions*, University Microfilms, No. 7926386, Ann Arbor, MI, 1979.
91. F. M. Hawkridge and T. Kuwana, *Anal. Chem. 45*:1021 (1973).
92. J. Wang and Q. Chen, *Anal. Chem. 66*:1007 (1994).
93. P. Bergveld, *Sens. Actuators 10*:165 (1986).
94. S. D. Senturia and R. L. Smith, *Sens. Actuators 15*:221 (1988).
95. G. Delapierre, *Sens. Actuators 17*:123 (1989).

# 12

# Microelectrodes

**Adrian C. Michael** *University of Pittsburgh, Pittsburgh, Pennsylvania*

**R. Mark Wightman** *The University of North Carolina at Chapel Hill, Chapel Hill, North Carolina*

## I. INTRODUCTION

Microelectrodes, defined in this chapter as electrodes of micrometer or smaller dimensions, have received considerable attention since the early 1980s in investigations by electroanalytical chemists. What can be done with these small electrodes? It should be obvious that they are not very useful for the generation of large amounts of material. On the other hand, it is clear that their small size should make them useful to probe the chemical composition of electroactive species in small spaces. Indeed, the use of true microelectrodes for amperometry was pioneered in the 1940s to measure oxygen concentrations inside biological tissues [1]. Today, microelectrodes have been developed that can be used inside the living brain to probe the dynamic concentrations of neurotransmitters, chemical substances that relay information between neurons [2]. Voltammetric electrodes can be fabricated with such a small size that they can actually be used to probe chemical events inside single biological cells [3].

The need to measure concentrations in very small volumes is not restricted to biological systems. For example, open tubular columns for liquid chromatographic separations offer the advantage of increased resolution, but because their internal diameters may be as small as 15 μm, the amount of material in the eluted peaks is very small. Thus, the use of these columns requires detectors that can be used with low concentrations in small volumes. Jorgenson and co-workers showed that this could be accomplished by the insertion of a 10-μm-diameter, cylindrical electrode made from a carbon fiber into the end of the column [4]. The close fit between the column wall and the fiber ensured that a large fraction of the eluting molecules were electrolyzed. When the electrochemical data were collected in a voltammetric mode, the resolved compounds could be classified according to their voltammetric properties. Another application of microelectrodes has been to examine the broadening of chromatographic peaks that occurs with conventional chromatographic columns (3.2 mm i.d.) across the axis of the column [5]. As solute bands pass through the column, the portion that interacts with the column wall experiences different forces than the portion that sees only the chromatographic packing material. The local band broadening that originates from these interactions can be readily resolved.

The use of microelectrodes mounted on piezoelectric micropositioners allows the local concentration of electroactive substances to be mapped in two dimensions. In this type of experiment, referred to as scanning electrochemical microscopy, the microelectrode is repetitively moved across a surface while kept at a constant height above the sample [6,7]. The current at each location is plotted against its x-y coordinate. In this way a three-dimensional image that shows local concentrations is generated. This technique has been used to examine chemical heterogeneity at the surface of minerals or biological samples, and can even be used to map local electrochemical events at the surfaces of larger electrodes.

In addition to their spatial resolution, several other features of microelectrodes make these devices useful to improve the quality of traditional electrochemical measurements. Depending on the experimental conditions, these features include improved temporal resolution of chemical and electrochemical reactions, much greater fluxes under diffusion-controlled conditions, and the possibility of the investigation of electrochemical processes in environments of high resistance. These advantages have made the use of microelectrodes widespread, so that today the use of microvoltammetric electrodes is indeed ubiquitous among electrochemists. In this chapter we examine ways to prepare and use these devices, the unique processes that govern voltammetric behavior at these electrodes, the properties that cause them to be useful for high-speed applications, and their use in resistive solvents. Although this chapter is a survey, several comprehensive books and reviews exist [8,9].

## II. CONSTRUCTION OF MICROELECTRODES

Techniques have been devised for constructing microelectrodes of various geometries, such as hemispheres, disks, cylinders, rings, and bands. In addition, because the electrodes themselves are smaller than their associated diffusion layers, arrays of closely spaced microelectrodes have also been of considerable practical and theoretical interest. In this section, construction of the more frequently employed microelectrode configurations is described. First, however, it should be pointed out that several commercial sources of microelectrodes now exist, and these may represent an economically viable alternative to the do-it-yourself approach for those who anticipate requiring only a few electrodes for their work. For work that involves harsh environments such as high pressures or corrosive conditions, where delicate electrodes may be subject to damage, home-built electrodes will probably be warranted.

The choice of which particular electrode geometry to use in any given situation obviously depends on the objectives of the experiment. If interpretation of the electrochemical results must involve comparisons with microelectrode theory for the extraction of heterogeneous or homogeneous rate constants, for instance, the electrode must satisfy the conditions specified in the development of the theory. Diffusion is the most commonly encountered mode of mass transport in microelectrode experiments, and each electrode geometry leads to a different solution to the relevant diffusion equations (see next section). Hence, each geometry has a unique current-voltage relationship. If, on the other hand, the experiment is aimed more at evaluation of the concentration of an electroactive species, which can be performed via normal calibration methods, the electrode geometry need not be so rigorously defined. In some situations, especially those concerned with measurements in small spaces, the dimensions of the insulating materials that surround the electrode proper may also be subject to restrictions.

## A. Microelectrode Construction with Wires and Fibers

Both disk-shaped and cylinder electrodes can be fashioned with metal microwires and carbon fibers. Two practical considerations that must be kept in mind when working with small wires and fibers are their resistivity and their delicacy. The former consideration demands that only short sections of the microwire or fiber be used for electrode construction, while the latter mainly represents a significant source of frustration for the novice electrode maker. A 5-$\mu$m-radius platinum wire at times seems to break just because you've looked at it! With a little practice, however, handling these small materials becomes straightforward.

Platinum and gold microwires can be sealed into soft glass by melting the glass with a Bunsen burner or an electrically heated coil of nichrome wire [10,11]. Soft glass is used because it has a lower softening temperature than most glasses and its thermal expansion and contraction behavior is well matched to that of the microwires. A short section of the microwire (about 2 cm) should first be attached to a hook-up wire (18 gauge single conductor) with conducting epoxy (e.g., Epo-Tek H20E, Epoxy Technology, Billerica, MA). The hook-up wire, with microwire attached, is then mounted in the lumen of a soft glass tube with the microwire protruding just a few millimeters from one end. The hook-up wire should first be secured to the glass tube with a drop of fast-setting epoxy before the tube is sealed, to avoid the possibility of placing any strain on the microwire. Finally, the end of the tube is melted to form a seal around the microwire. The quality of the seal should be inspected with a microscope to check for any signs of flaws or bubbles in the glass in the vicinity of the microwire. A large number of bubbles near the microwire is usually a result of overheating during the sealing step. Usually, only a millimeter or so of the microwire will actually be sealed in the glass.

To create a cylinder electrode, the protruding microwire need only be trimmed to the desired length. The practical advantage of a cylinder electrode is that while it has a micrometer-sized radius and thus exhibits the beneficial features of a microelectrode, its length gives rise to easily measured currents so that high-sensitivity potentiostats are not required. The drawbacks of cylinder electrodes are that they cannot be mechanically polished to refurbish their surfaces and they are physically unstable if their length is too great. Osteryoung has shown that the latter problem can be solved by anchoring both ends of the cylinder [12].

A disk electrode can be made by first grinding the sealed end of the glass tube with coarse sandpaper and subsequently polishing to a mirror finish with consecutively finer grades of abrasive. For disk electrodes, the quality of the seal between the microwire and the glass is very important [13]. Any cracks or crevices in the seal will prevent the electrode from exhibiting the properties of a polarizable interface. Rather, the electrode will exhibit an ohmic response ("ramping") in solution even in the absence of electroactive material. Such behavior can sometimes be eliminated by regrinding and repolishing, but often the electrode must be remade. Ideally, the electrode capacitance should be proportional to the electroactive area of the electrode. It is important to recognize, however, that there can be (at least) two sources of capacitance: one arising from the electrode-solution interface and the other arising from capacitive coupling between the inner conductor and the electrolyte solution [11]. Normally, the latter effect is too small to be noticed at large surface area electrodes, but it can become significant as the electrode area is reduced. Two construction methods minimize this effect. One is to keep to a minimum the amount of conductive epoxy that is used to connect the microwire and the hook-up wire, and the other is to place a grounded shield around the electrode body. Once the electrode tip is appropriately polished, the electrode is ready for use. Electrodeposition of mercury onto platinum disks has been used to form approximately hemispherical electrodes.

Carbon fibers are not particularly amenable to being heat-sealed into glass. Carbon fiber-based electrodes, therefore, are usually made with epoxy seals [12,14]. Wightman's group has long recommended the use of Epon 828 (Shell) with 14% *m*-phenylenediamine as hardener, which is resistant even to organic solvents. Both carbon-fiber disk and cylinder electrodes have been used extensively for measurements of electroactive neurotransmitters (e.g., dopamine) in living brain tissue [15]. This is an example of a situation where the entire electrode, including the surrounding insulator, must be of micrometer dimensions in order to minimize disruption of the biological environment in which the electrodes are employed. These in vivo style electrodes are made by placing a single carbon fiber in the lumen of a small glass tube (i.e., less than about 2 mm o.d.), which is then heated and pulled to a fine point. Although the glass tube collapses around the fiber during this process, a high-quality seal is not usually established. Epoxy is placed between the carbon and glass by capillary action, or epoxy can be injected into the barrel of the electrode with a small bore needle to backfill the electrode tip (care must be used to avoid insulating the carbon fiber with the latter method). When heated to about 70°C, Epon becomes very nonviscous and flows readily, but only for a short time after being cooled. Spurr low-viscosity embedding medium (Polysciences) can be used as an alternative. The electrode barrel can be filled with mercury to make contact between the carbon fiber and a hook-up wire. Again, a cylinder electrode is formed by simply trimming the fiber to the desired length. To form a disk electrode, however, a microelectrode beveler is required because these electrodes are too fragile to polish by hand. It should be noted that beveled carbon-fiber electrodes usually have an elliptical, rather than disk, shape.

## B. Microelectrode Construction with Foils and Films

In addition to disks and cylinders, microelectrodes with the geometry of bands and rings have also been constructed and used. Bands, like cylinders, exhibit the properties of microelectrodes by virtue of their narrow width, yet pass large currents due to their length. Ring electrodes, on the other hand, have been of interest because they can be constructed with very small surface areas.

Band electrodes have been constructed in several ways. Lithography has been used [16], especially for the construction of band arrays, but foils [17] and deposited films [18] can be used if lithography is not accessible (lithographic band arrays are commercially available). The thin metal film can be sandwiched with epoxy between a pair of microscope slides, and the edge can be ground and polished to expose the cross section of the metal. Thus, the electrode width is defined by the thickness of the foil or film used. Some care is required during the polishing steps due to the varying degrees of hardness of the materials used to construct the electrode. Overpolishing these electrodes can cause the epoxy and/or the electrode to become recessed between the harder microscope slides. This will inhibit diffusional access of electroactive species to the surface of the electrode and distort the current-voltage relationships.

Double- and triple-band electrodes have been constructed in a similar fashion by placing thin insulating films (e.g., Mylar) between the metal foils. These have been used in generator-collector experiments, in which one electrode (the collector) is used to monitor a product formed at the other electrode (the generator). This can be considered the microelectrode equivalent of the rotated ring-disk experiment. Here, the fact that the electrodes are smaller than their respective diffusion layers (see next section) is employed. The lithographically prepared band arrays, consisting of about eight individually addressable electrodes, can be used to probe the diffusion layer with several collectors, each at a different distance from the generator [19,20].

Ring electrodes have been built by depositing films on ceramic or glass rods [21], sealing the structure in a concentric glass tube, and grinding to expose the cross section of the film. Perhaps because the theoretical treatment of microring electrodes has proven somewhat more difficult than for other geometries, ring electrodes have been used less frequently than the electrodes mentioned earlier. Nevertheless, some interesting applications of rings have been reported. For example, Weber's group has constructed ring electrodes around the core of an optical fiber for electrochemical detection of photogenerated species [22]. Ewing's group has used the pyrolysis of methane to deposit a carbon ring onto the internal surface of a pulled glass capillary to produce an electrode small enough for intracellular electrochemical measurements [23].

## III. DIFFUSION AT MICROELECTRODES

### A. Diffusion Operators

The principles underlying the diffusion of molecules to a microelectrode are identical to those described earlier in this book for diffusion to macroelectrodes. A reaction occurring at the surface of the electrode produces a concentration gradient in solution in the vicinity of the electrode, which in turn gives rise to a diffusional flux. The diffusionally induced rate of change of concentration in solution is described, in general, by Fick's second law:

$$\frac{\partial C}{\partial t} = D \nabla^2 C \tag{12.1}$$

where C is the concentration of diffusing species, t is time, D is the diffusion coefficient, and $\nabla^2$ is the so-called Laplacian diffusion operator. The diffusion operator takes on a different form for each diffusion geometry such that Fick's second law has a unique solution for each electrode geometry. In Chapters 2 and 20, finite element diagrams are used to explain planar diffusion. We make use of the same approach to explain the nonplanar diffusion processes that take place at microelectrodes.

Consider the series of volume elements shown in Figure 12.1, which explains the terminology that will be used. Our task is to discover how the concentration in any given volume element changes with respect to time due to diffusion. The concentration change in any volume element, *i*, will be due to a

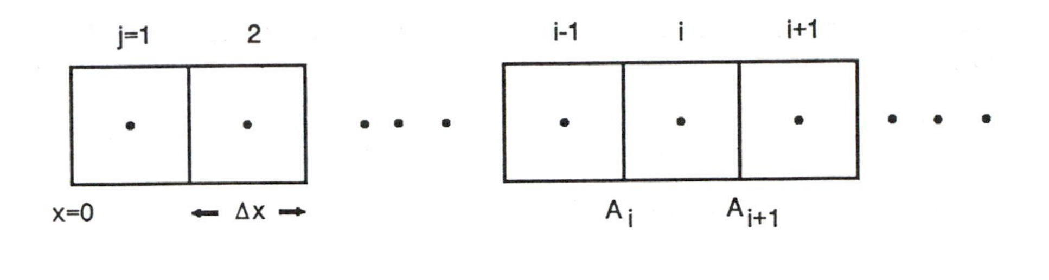


| CYLINDER: | SPHERE: |
|---|---|
| $x_i = \Delta x\,(i-½)$ | $x_i = \Delta x\,(i-½)$ |
| $A_i = 2\pi\,(i-1)\,\Delta x\, l$ | $A_i = 4\pi\,(i-1)^2\,\Delta x^2$ |
| $A_{i+1} = 2\pi\, i\,\Delta x\, l$ | $A_{i+1} = 4\pi\, i^2\,\Delta x^2$ |
| $v_i = 2\pi\, x_i\,\Delta x\, l$ | $v_i = 4\pi\, x_i^2\,\Delta x$ |

**Figure 12.1** Finite-element diagram for diffusion to electrodes. The coordinate, x, represents the direction normal to the surface of the electrode. The origin of x is at the central axis in the cylindrical or spherical case.

flux across the interfaces between the neighboring volume elements, $i - 1$ and $i + 1$. In general, this can be expressed by:

$$\frac{\Delta C}{\Delta t} = \frac{1}{v_i}\left[ DA_{i+1}\frac{(C_{i+1} - C_i)}{\Delta x} - DA_i\,\frac{(C_i - C_{i-1})}{\Delta x}\right] \tag{12.2}$$

which takes the difference between the flux in mol/s across each interface, and divides by the volume of the element, $v_i$, to give the desired units of concentration per time, mol/cm$^3$ s. For the case of planar diffusion $A_i = A_{i+1} = A$ and $v_i = \Delta x\ A$, so Equation 12.2 becomes

$$\frac{\Delta C}{\Delta t} = \frac{D}{\Delta x^2}\left[C_{i+1} - 2C_i + C_{i-1}\right] \tag{12.3}$$

which is the finite element form of Fick's second law for planar diffusion,

$$\frac{\partial C}{\partial t} = D\,\frac{\partial^2 C}{\partial x^2} \tag{12.4}$$

which was given before as Equation 2.11. To derive Fick's second law for cylindrical and spherical geometries one follows the same procedure using the appropriate entries in Figure 12.1 for $A_i$, $A_{i+1}$, and $v_i$. For example, for cylindrical geometry:

$$\frac{\Delta C}{\Delta t} = \frac{D}{\Delta x^2(2i-1)}\left[2i\,(C_{i+1}-C_i)-(2i-2)(C_i-C_{i-1})\right] \tag{12.5}$$

which can be rearranged to give

$$\frac{\Delta C}{\Delta t} = D\left[\frac{C_{i+1}-2C_i+C_{i-1}}{\Delta x^2} + \frac{1}{\Delta x\,(i-1/2)}\,\frac{C_{i+1}-C_{i-1}}{2\Delta x}\right] \tag{12.6}$$

which is the finite element form of Fick's second law for cylindrical geometry,

$$\frac{\partial C}{\partial t} = D\left(\frac{\partial^2 C}{\partial x^2} + \frac{1}{x}\frac{\partial C}{\partial x}\right) \tag{12.7}$$

Fick's second law for spherical geometry, generated in identical fashion, is

$$\frac{\partial C}{\partial t} = D\left(\frac{\partial^2 C}{\partial x^2} + \frac{2}{x}\frac{\partial C}{\partial x}\right) \tag{12.8}$$

The planar, cylindrical, and spherical forms of Fick's second law, and combinations of those forms, are sufficient to describe diffusion to most microelectrode geometries in use today. Just as was illustrated in Chapter 2, the appropriate form of Fick's second law is solved, subject to the boundary conditions that describe a given experiment, to provide the concentration profile information. The sought-after current-time or current-voltage relationship is then obtained by evaluating the flux at the electrode surface.

## B. Diffusion to Hemispherical and Disk Electrodes

For the purposes of considering diffusion at microelectrodes, it is convenient to introduce two categories of electrodes: those to which diffusion occurs in a linear fashion and those to which diffusion occurs in a nonlinear fashion. The former category consists of cylindrical and spherical electrodes. As shown schematically in Figure 12.2A, the lines of flux (i.e., the pathway followed by material diffusing to the electrode) are straight, and the current density is the same at all points on the electrode. Thus, the diffusion problem is one-dimensional (i.e., distance from the electrode surface) and involves solution of the appropriate form of Fick's second law, Equation 12.7 or 12.8, either by Laplace transform methods or by digital simulation (Chap. 20).

Chronoamperometry at a hemispherical electrode illustrates the consequences of diffusion geometry on the electrochemical response. The expression obtained by solution of Equation 12.8 for the chronoamperometric limiting

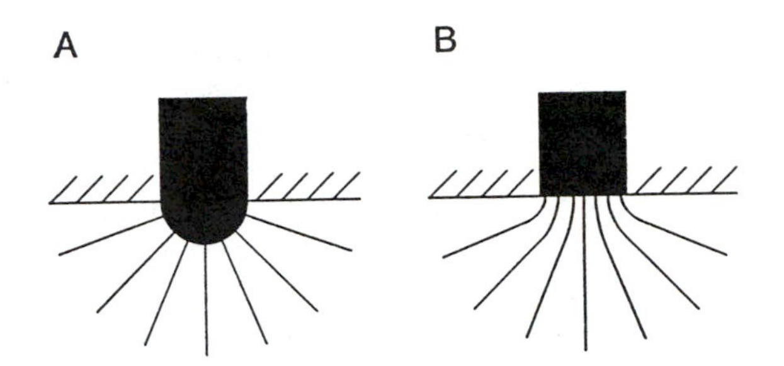


**Figure 12.2** (A) Cross section of a hemispherical or hemicylindrical electrode and the linear lines of flux to the electrode surface. (B) Cross section of a disk or band electrode and the nonlinear lines of flux.

current under conditions of semi-infinite linear diffusion at an electrode with radius $r_o$ is [24]

$$i_t = nFADC\left(\frac{1}{\sqrt{\pi Dt}} + \frac{1}{r_o}\right) \tag{12.9}$$

where the first term is identical to the Cottrell equation for macroplanar electrodes (Eq. 3.6) and the second, time-independent term is sometimes referred to as the "spherical correction." Thus, one consequence of spherical geometry is that at long experimental times, the current approaches a nonzero steady-state value of:

$$i_{t=\infty} = 2\pi nFr_oDC^* \tag{12.10}$$

whereas at a macroplanar electrode, the current density approaches a value of zero at long times. At very short times, however, the response of the hemispherical electrode is indistinguishable from that of the planar electrode, an important result for the interpretation of high-speed voltammetric methods discussed later in this chapter. From inspection of Equation 12.9, it is clear that when $(Dt)^{1/2} >> r_o$ (i.e., when the diffusion layer is larger than the electrode itself), steady-state (time-independent) behavior will be observed. The smaller the electrode, therefore, the sooner the steady-state condition is established. A hemispherical electrode with a radius of 5 $\mu$m will exhibit a steady-state response at times greater than about 250 ms in aqueous solutions where D is usually about $10^{-5}$ cm$^2$/s.

Phenomenologically speaking, the reason that a steady-state response is obtained with a hemispherical microelectrode is that the lines of flux converge on the electrode surface from all directions. Thus, the volume of the solution providing electroactive material to the electrode is very large in comparison to the electrode surface area. In fact, this can be seen in Figure 12.1 where the volume, $v_i$, of each finite element is constant in the planar case but increases with $x^2$ in the spherical case. The microelectrode is simply too small to deplete the surrounding solution of electroactive material, in contrast to a macroelectrode. With this explanation in mind, it should be clear that steady-state behavior will be exhibited by any electrode if *all* of its dimensions are smaller than $(Dt)^{1/2}$. So, by analogy to a hemisphere, a microdisk electrode should also exhibit steady-state behavior.

Disks fall into the second category of electrode, at which nonlinear diffusion occurs. The lines of flux to a disk electrode (Figure 12.2B) do not coincide with the simple geometries for which we derived Fick's second law, and the diffusion problem must therefore be expressed in two dimensions. Note that a line passing through the center of the disk and normal to the plane of the disk is a cylindrical axis of symmetry, so it is sensible to choose the radial distance from this axis as one of the coordinates for the problem. Diffusion along this radial coordinate, r, is described by Equation 12.7. Also, diffusion along the coordinate, x, normal to the plane of the electrode is described by Equation 12.4. Thus, the form of Fick's second law that must be solved for the disk is:

$$\frac{\partial C}{\partial t} = D\left(\frac{\partial^2 C}{\partial x^2} + \frac{\partial^2 C}{\partial r^2} + \frac{1}{r}\frac{\partial C}{\partial r}\right) \tag{12.11}$$

Several approaches to solving this expression for various boundary conditions have been reported [25,26]. The solutions are qualitatively similar to the results at a hemisphere: at very short times (i.e., when $(Dt)^{1/2} \ll r_o$), the Cottrell equation is followed, but at long times the current becomes steady-state. Simple analytical expressions analogous to the Cottrell equation for macroplanar electrodes or Equation 12.9 for spherical electrodes do not exist for disk electrodes. For the particular case of a disk electrode inlaid in an infinitely large, coplanar insulator, the chronoamperometric limiting current has been found to follow [27]:

$$\frac{i_t}{4nFr_oDC^*} = 0.7854 + 0.8862\tau^{-1/2} + 0.2146e^{-0.7823\tau^{-1/2}} \tag{12.12}$$

where $\tau = 4Dt/r_o^2$. The right-hand side of Equation 12.12 goes to unity as time goes to infinity, so the steady-state current value for the inlaid disk is $4nFr_oDC^*$. Note that the in-vivo-style electrodes mentioned earlier do not exactly follow Equation 12.12 because (a) they are not exactly disk-shaped, and (b) the glass capillary tip does not fit the definition of an infinitely large, coplanar insulator.

Thus, the behavior of hemispherical and disk electrodes with equivalent radii are analogous; that is, both show steady-state currents at large values of $(Dt)^{1/2}/r_o$. Again speaking phenomenologically, one could say that a molecule at a distance of $10r_o$ or so from the electrode surface, that is, at the outer reaches of the diffusion layer, has no idea if it is diffusing toward a disk or a hemisphere. However, at the surface of the disk electrode, a major difference with the hemisphere is found. The flux at the perimeter is much greater than at the center. For this reason, the current is said to be nonuniform at the disk.

The conclusions drawn from analysis of the chronoamperometric response of sphere and disk electrodes apply equally to other electrochemical techniques, such as cyclic voltammetry. The characteristic time, $t_c$, of a cyclic voltammetry experiment can be conveniently expressed by the reciprocal of the scan rate: RT/nFv. When $r_o >> (Dt_c)^{1/2}$, the voltammogram will appear as predicted for a macroplanar electrode (Chap. 3), and when $r_o << (Dt_c)^{1/2}$, the voltammogram will take on a sigmoidal shape given by:

$$\frac{i_{E_{app}}}{i_{ss}} = \frac{1}{1 + e^{(E_{app} - E^{\circ'})nF/RT}} \tag{12.13}$$

for a Nernstian reduction. Between the two limits, the shape of the voltammogram is dependent on $r_o/(Dt_c)^{1/2}$, as depicted in Figure 12.3. Usually, at least in conventional electrochemical experiments involving liquid electrolyte solutions, voltammograms of varying shape would arise either due to variations in the scan rate or the radius of the electrode in use. However, there are conditions where

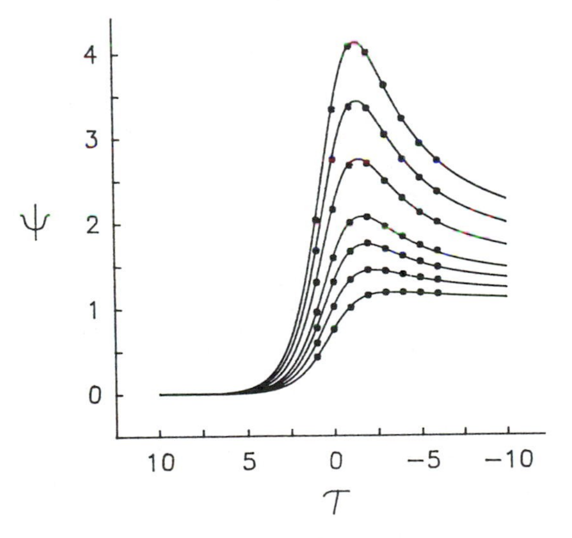


**Figure 12.3** Calculated voltammograms for reversible charge transfer reactions at microdisk electrodes for $r_o/(Dt_c)^{1/2}$ values of 10 (tallest curve), 8, 6, 4, 3, 2, and 1. Curves were calculated by digital simulation. Points are from Reference 25. [From Ref. 26, reproduced with permission of the copyright holder.]

the diffusion coefficient may change. For example, with a disk electrode with a diameter of 25 $\mu$m and a scan rate of 10 V/s, Bard's group observed the voltammogram of $Fe(bpy)_3^{+2}$ in $SO_2$ change from peak-shaped (i.e., planar diffusion) to sigmoidal (i.e., steady-state diffusion) upon conversion of the solvent from a liquid to a supercritical fluid by adjusting the temperature and pressure of the electrochemical cell [28].

## C. Diffusion to Cylinders and Bands

The equivalency found between the behavior of hemisphere and that of disk electrodes also exists between cylinder and band electrodes [29]. Diffusion to a cylinder electrode is linear and described by Equation 12.7, while diffusion to a band is nonlinear. A plane of symmetry passes through the center of the band and normal to its surface, so the nonlinear diffusion process can be broken down into two planar components, one in the direction parallel to the electrode surface, x, and the other in the direction perpendicular to the electrode surface, y. So Fick's second law for a band electrode is

$$\frac{\partial C}{\partial t} = D\left(\frac{\partial^2 C}{\partial x^2} + \frac{\partial^2 C}{\partial y^2}\right) \qquad (12.14)$$

which has also been solved under a variety of boundary conditions [30]. In contrast to the behavior of spheres and disks, cylinders and bands do not exhibit steady-state behavior [30,31]. Sometimes their response is referred to as quasi-steady-state, to highlight the fact that the current at a cylinder decays more slowly than the current at a macroelectrode (luckily, no one has yet coined the term quasi-ultramicroelectrode!). This occurs because the length of the cylinder usually exceeds $(Dt)^{1/2}$ on reasonable experimental time scales. Put in more mathematical terms, you can see from Figure 12.1 that the volume, $v_i$, of the individual finite elements increases linearly with x for cylindrical electrodes. So the behavior of cylinder and band electrodes is intermediate to that observed at either macroplanar or spherical electrodes.

Spatially resolved absorbance spectroelectrochemistry has been used to observe the concentration profile of an absorbing species generated at the surface of a cylinder electrode with a radius of 6 $\mu$m. The observed concentration profiles agreed very closely with those predicted by solution of Equation 12.7 by digital simulation [32]. As with the disk electrode, simple analytical expressions for the current-time and current-voltage relationships of cylinders and bands do not exist.

## D. Experimental Verification of Nonlinear Diffusion

As noted earlier, nonlinear diffusion to disks is predicted to produce nonuniform current densities across the electrode surface. Similarly, the current density at the edges of the band should be considerably higher than near the electrode center. Experimental verification of this prediction would, at least in part, provide evidence that the Fick's law expressions derived earlier do in fact describe the behavior of these microelectrodes. This has been achieved with double-band electrodes, that is, a pair of band electrodes positioned close enough together that their diffusion layers can overlap during an electrochemical experiment [18]. When the two electrodes are shorted together to form a single electrode, the current measured is less than the sum of the currents measured with each electrode operating independently. This is due to an effect called shielding, where the flux of material to each band is reduced due to the electrolysis occurring at the neighboring band. Moreover, the measured current is very similar to that predicted for a single band with a width equivalent to the sum of the widths of the two electrodes and the intervening spacer. This shows that most of the current does indeed flow at the edge of the band, because removing the center region of the band has little effect on the observed current. Electrochemical chemiluminescence (ECL) can also be used to observe the high current density at the edge of a disk electrode [33]. Oxidation of luminol in alkaline hydrogen peroxide produces ECL with an intensity proportional to the current density. The ECL intensity emitted from the edge of a disk electrode was measurably greater than that emitted from the center of the electrode, again confirming a nonuniform distribution of current.

### E. Practical Advantages Derived from Nonplanar (Convergent) Diffusion to Microelectrodes

The fact that diffusion to microelectrodes is convergent is the source of many of their advantageous properties. Convergent diffusion produces significantly higher current densities at microelectrodes than at their macroplanar counterparts, which leads to several advantages. For example, in any electrochemical experiment, an important concern is the magnitude of the charging current, since this may mask the desired faradaic current. In cyclic voltammetry, the charging current has a limiting value of $vC_d$. Since $C_d$ is proportional to the electrode area, the charging current decreases in proportion to the area of the electrode, while the faradaic current under steady-state conditions decreases in proportion to the radius. This means that the ratio of the faradaic current to the charging current is improved as the electrode size is decreased. In general, this advantage more than offsets any disadvantage derived from the small current flow that must be measured in microelectrode experiments. The small capacitance of microelectrodes also has great advantage in the application of pulsed voltammetric methods in which charging current rejection is enhanced because of the fast cell time constant [34].

With macroelectrodes, the duration of an experiment in stationary solutions is determined by the onset of distortion due to solution convection. Even in unstirred solutions, some convection arises from sources such as the vibration of the building, the air flow in the laboratory, or people moving about. Unless the experimental apparatus is carefully isolated from these vibrations, current contributions due to convection will be noticed once the diffusion current has decayed for a few seconds. This is less apparent at microelectrodes exhibiting steady-state behavior (i.e., spheres and disks), because the diffusion layer grows into the bulk of solution at a slower rate. This has proven useful for electrochemical detection in flowing solutions, such as liquid chromatography [35]. The current at cylinder and band electrodes is more time-dependent than at disks and hemispheres (see earlier discussion), so they are somewhat susceptible to convective effects. McCreery has measured convection at 6-$\mu$m-radius cylinder electrodes and found that, with the use of vibration isolation equipment, convection can be reduced sufficiently to allow diffusion-controlled experiments with durations of about 10–20 s [32].

A particularly nice feature of steady-state diffusion to hemisphere and disk electrodes is the theoretical simplification introduced by elimination of time as a variable. The steady-state behavior of a spherical electrode can be obtained by considering the expression

$$\frac{\partial^2 C}{\partial r^2} + \frac{2}{r}\frac{\partial C}{\partial r} = 0 \tag{12.15}$$

which has been solved for several combinations of heterogeneous and homogeneous kinetic complications by Oldham and colleagues [36]. They have also provided extensive steady-state analyses of different types of systems at microdisk electrodes [37]. For example, the determination of heterogeneous rate constants ($k_{s,h}$) under conditions where the current is time dependent normally involves variation of the sweep rate of a cyclic voltammogram. The sweep rates must be chosen such that the dimensionless kinetic parameter $k_{s,h}(RT/nFvD)^{1/2}$ has a value close to (or less than) unity. So, for systems with a larger value of rate constant, the required scan rate likewise must be increased. Under steady-state conditions, on the other hand, the relevant dimensionless kinetic parameter is $k_{s,h}r_o/D$, where $r_o$ is the radius of the electrode. Thus, the value of $k_{s,h}$ can be obtained by analyzing steady-state voltammograms obtained with several electrodes of different radii. Determination of a heterogeneous rate constant of 1 cm/s requires either a sweep rate of ~2500 V/s at a macroelectrode, or a microelectrode with a radius of ~1 $\mu$m [38]. Likewise, the relevant kinetic parameter for a first-order following homogeneous reaction is $kr_o^2/D$, and the rate constant can be evaluated under steady-state conditions by varying $r_o$ over an appropriate range. The main drawback of these steady-state methods is the rather small shifts in the half-wave potential (on the order of a few millivolts) that must be measured with considerable accuracy. In addition, each electrode used must exhibit highly reproducible behavior.

An alternative steady-state method for evaluating the kinetics of coupled homogeneous reactions is based on generator–collector experiments with arrays of microbands [39]. The diffusion layer at a band electrode can easily exceed the dimensions of the band itself, so a second electrode can be positioned within the diffusion layer. With a bipotentiostat, the potential of each electrode can be controlled independently (see discussion of rotated ring-disk electrodes in Chap. 3). Material generated electrochemically at the so-called generator electrode can diffuse to the adjacent collector electrode and be electrochemically detected there. If, during the diffusion process, the electrogenerated material undergoes a homogeneous reaction (to a material that is not electrochemically active at the potential of the collector electrode), the collector current will be decreased. The relevant dimensionless kinetic parameter for this experiment, therefore, is $kg^2/D$, where g is the thickness of the gap between the generator and collector(s). Theoretically, the collection efficiency (i.e., the ratio of the collector current, $i_c$, to the generator current, $i_g$) is predicted to follow

$$\frac{i_c}{i_g} = \frac{1}{2}\exp\left[-\left(\frac{k}{D}\right)^{1/2}\frac{\pi g}{2}\right] \tag{12.16}$$

for a single generator and collector when the collection efficiency is less than ~0.4 [40]. The latter criterion is readily established with appropriately sized electrodes and gaps. Because the experiment is carried out under quasi-steady-state conditions, interference from charging currents will be negligible. With a gap of 4.5 $\mu$m, a first-order rate constant of 100 $s^{-1}$ can be determined with a potential sweep applied to the generator at 0.1 V/s [40]. To determine such a fast rate constant by conventional cyclic voltammetry would require a range of sweep rates up to ~1000 V/s.

## IV. HIGH-SPEED CYCLIC VOLTAMMETRY

The small area of a microelectrode, with its proportionately low capacitance, allows its use at very short time scales compared to the time scale used with a classical voltammetric electrode. As we have seen earlier in this chapter, when microelectrodes are used at short time scales, the current follows the behavior expected for diffusion in one dimension. Thus, the development of high-speed voltammetric methods with microelectrodes was a logical step, and has greatly expanded the scope and capabilities of electrochemical techniques [41]. Rapid electrochemical methods allow evaluation of the larger rate constants of rapid heterogeneous and/or homogeneous reactions. For example, theories of heterogeneous electron transfer predict that $k_{s,h}$ values exceeding 10 cm/s are possible [42]. Evaluation of such a large rate constant by cyclic voltammetry requires a sweep rate in excess of 250 kV/s. Such high sweep rates, furthermore, allow electrochemical investigation of reactive intermediates with lifetimes in the nanosecond regime. In addition to kinetic studies, high-speed cyclic voltammetry has been extremely valuable for the study of the dynamics of rapid chemical events. In these studies, the voltammetric scans are repeated at regular intervals, and the concentration of an electroactive species is monitored by its current in each successive recording.

## A. Practical Considerations

Consideration of the equivalent circuit diagram of an electrochemical cell, such as that given in Figure 5.1, reveals the major limitation on the rate at which the potential of an electrode can be varied, namely, the time constant of the electrochemical cell, $R_uC_d$. When a potential sweep is applied across the cell, the nonfaradaic charging current that flows is described by [24]

$$i = \nu C_d + \left[ \left( \frac{E_i}{R_u} - \nu C_d \right) e^{-t/R_u C_d} \right] \tag{12.17}$$

Thus, at times less than about $3R_uC_d$, the potential drop across the double-layer capacitance will not be linearly dependent on the applied potential, and furthermore, once the charging current reaches the limiting value of $\nu C_d$, a consistent ohmic drop across the uncompensated resistance in the cell will be present. The time to reach the steady-state charging current, given by the exponential expression in Equation 12.17, becomes prohibitively large at macroelectrodes when the scan rate exceeds a few tens of volts per second. So, reducing the size of the electrode in order to minimize the double-layer capacitance allows rapid changes of the electrode potential to be employed, and cyclic voltammetry in the $10^5$ V/s range can be accomplished [11].

Even with small electrodes, the ratio of the charging current to the faradaic current will be large at elevated sweep rates because the former increases linearly with $\nu$ while the latter increases with $\nu^{1/2}$. Accurate charging current correction must therefore be devised. Normally, this is done by background-subtraction techniques (i.e., data is first obtained in the absence of the redox compound, then again in the presence of the redox compound, and the difference between the two data sets is used as the voltammogram). Explicit in the use of background subtraction is the idea that the charging and faradaic components of the current are independent. This assumption, however, breaks down at high sweep rates. Taking $E'$ as the potential difference across the electrode-solution interface, the double-layer charging current is given by [44]

$$i_c = C_d \frac{\partial E'}{\partial t} \qquad E' = E + R_u(i_f + i_c) \qquad E = E_i + vt \tag{12.17}$$

These equations show that a flow of current through the faradaic impedance perturbs E′ and $\partial E'/\partial t$, the effective scan rate. This in turn alters the charging current. This process, sometimes referred to as induced charging current, becomes significant when the charging current is large with respect to the faradaic current. Not only must the induced charging current be accounted for, but so also must be the effect on E′, which will again bear a nonlinear relationship to E while the faradaic current flows. These two sources of distortion of the observed current-voltage curve must be corrected post hoc with appropriately designed digital simulation algorithms.

For operation at high sweep rates, careful attention must be paid to the design of the potentiostat, the current amplifier, and the electrical connections to the electrodes. Any amplification stage in the circuitry must be designed to avoid temporal distortion of the voltammetric signal [43,44]. A rule of thumb is that RCnv must be less than 4 mV, where RC is the time constant of the amplifier, n is the number of electrons involved in the heterogeneous reaction, and v is the sweep rate. Any stray capacitance in the connections between the circuit and the electrodes will contribute to the time constant of the cell and the charging current because it will appear in parallel with the double-layer capacitance, and so must be avoided. This is mainly done by using as short a length as possible of shielded connecting wires. Circuit designs for ohmic drop compensation have been described that can be used in the $10^5$ V/s range [45]. The availability of ohmic compensation at high speed extends the upper sweep rate limit for electrodes of convenient dimensions (i.e., with radii of 1 $\mu$m or larger). This is important because electrode construction becomes more difficult as the electrode size decreases.

Although electrochemical cells with microelectrodes can be devised that will allow cyclic voltammetry in the MV/s range, additional sources of distortion make accomplishing this goal quite difficult. One problem is the performance of the operational amplifiers used in the potentiostat. At very large v, the open loop gain, the bandwidth, and the phase angle between the input and the output of the amplifier must be considered. Some of the distortion of the data by the potentiostat can be removed with deconvolution techniques [43]. Successful experiments have been reported at scan rates up to 5 MV/s. However, Amatore has pointed out that when such short time scales are probed, the dimensions of the diffusion layer are very similar to the dimensions of the diffuse region of the double layer [46]. Electrochemical theory extant to date has all been derived with the assumption that the diffusion and the diffuse layers can be treated separately (e.g., the Frumkin correction). The implications of the breakdown of this assumption on submicrosecond time scales are just beginning to receive theoretical consideration. Thus, for the time being at least, sweep rates on the order of a few megavolts per second can be regarded as both a practical and theoretical upper limit.

## B. Applications of High-Speed Cyclic Voltammetry

Carbon-fiber cylinder and disk electrodes have found extensive use for detection of catecholamine and indoleamine neurotransmitters and their metabolites. Carbon fibers are used for these measurements because they have been found to be less susceptible to surface degradation due to biofouling than noble metal electrodes, and their surface properties can be easily manipulated with electrochemical pretreatments so electrodes can be selectively optimized for particular purposes [47]. When used in conjunction with high-speed voltammetry, carbon fibers are found to be limited to sweep rates in the few hundreds of volts per second range because of the large amount of surface electrochemistry they exhibit. Nevertheless, carbon-fiber disk electrodes allow measurements to be performed on a millisecond time scale, which has proven extremely useful for studying the dynamic aspects of the release of neurotransmitters by cells in living tissues as well as in culture [48]. These events are extremely rapid, taking place on a subsecond time scale, so rapid analytical techniques are required.

The use of high-speed voltammetry for these in vivo studies is important not only from the perspective of temporal resolution, but also from the perspective of the selectivity of the measurement. Because the data are collected in a voltammetric mode, electrochemical identification of compounds observed in vivo can be obtained by comparison of the qualitative features of the voltammograms with those obtained from standards. The extracellular fluid of mammalian brain contains a variety of easily oxidizable compounds that could potentially interfere with the technique. The compound that is the cause for greatest concern is ascorbate (vitamin C), which is present in high concentrations and has a less positive E° than dopamine. Two possible mechanisms exist whereby ascorbate can interfere with the measurement of dopamine. One simply involves the direct oxidation of ascorbate at the electrode surface, and the other involves a solution electron transfer between the oxidized form of dopamine and ascorbate, which results in an enhanced dopamine signal. At a polished carbon-fiber electrode, the heterogeneous kinetics of ascorbate oxidation are sufficiently slow that at sweep rates of a few hundred volts per second, virtually no ascorbate oxidation is observed. Likewise, the homogeneous electron transfer with oxidized dopamine is also slow on these time scales [49]. So, with high-speed cyclic voltammetry, a highly selective dopamine measurement can be performed.

High-speed cyclic voltammetry in the $10^5$ V/s range has recently been applied to kinetic investigations of extremely rapid heterogeneous reactions and coupled homogeneous reactions. Consider, for example, the voltammetry of the ferrocene/ferricinium couple, which is often used as an example of a rapid outer-sphere electron transfer. When, on conventional time scales, a couple behaves reversibly, no kinetic information is available in the voltammetric data. However, at extremely high sweep rates, kinetic effects can in fact be found, and a $k_{s,h}$ of 3.1 cm/s has been found for the ferrocene/ferricinium couple. Several other electrochemical systems previously regarded as totally reversible have been investigated in this way.

Just as with heterogeneous events, operating on a time scale similar to that of coupled homogeneous events allows kinetic and mechanistic information to be obtained [41]. For example, halogenated hydrocarbons undergo cleavage of the carbon-halogen bond following electroreduction. The cleavage step is often rapid, so reoxidation of the anion-radical intermediate is not observed. This leaves open to question the existence of the intermediate, since the reduction and cleavage step could proceed in a concerted way. For aromatic halides, fast-scan voltammetry has been able to show the existence of the radical-anion intermediate. For example, 9-bromoanthracene shows a two-electron reduction to the parent hydrocarbon at low sweep rates, but at 20 kV/s a one-electron, reversible reduction to the radical anion is seen. Kinetic analysis yielded a rate constant of $6 \times 10^5$ $s^{-1}$ for the cleavage step. Saveant and coworkers have employed both fast cyclic voltammetry and fast potential-step experiments for investigation of the kinetics and mechanism of the formation of electropolymerized conducting polymers [50,51].

Perhaps one of the uses of voltammetry of most fundamental interest is the determination of E° values. However, this is not possible if the product of an electrode reaction is unstable. The presence of a following chemical reaction that consumes the product of an electrode reaction will shift the forward wave of the voltammogram from its reversible potential. If the following reaction is extremely fast, the forward wave may even be shifted so far from its E° that it takes on the appearance of a quasireversible process. One of the early recognized uses of high-speed cyclic voltammetry was to obtain E° values for unstable systems by operating at time scales shorter than that of the coupled chemical reactions [52]. In doing so, voltammetric peaks are observed on the reverse sweep of the cyclic voltammogram, enabling the E° value to be obtained from the average of the forward and reverse peak potentials. This technique was applied to a range of aromatic hydrocarbons and a linear correlation was found between the standard redox potentials and vertical-ionization potentials.

## V. OHMIC DROP AT MICROELECTRODES

A major advantage of microelectrodes is that electrochemical data is less distorted by ohmic drop than when recorded with electrodes of conventional size

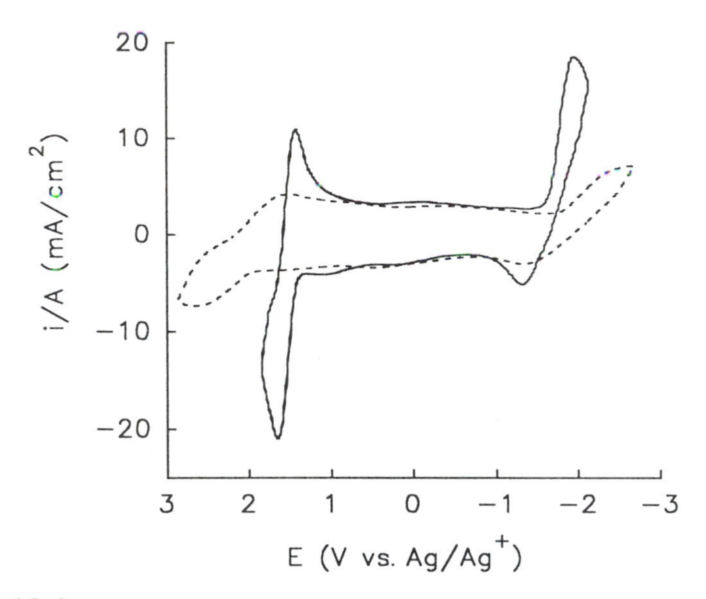


**Figure 12.4** Cyclic voltammograms of 1 mM diphenylanthracene in acetonitrile containing 0.01 M tetrabutylammonium hexafluorophosphate. Scan rate: 115 V/s. Solid line: voltammogram recorded at a disk electrode with a 14 $\mu$m radius. Dashed line: voltammogram recorded at a disk electrode with a 0.91 mm radius.

[8]. Consider the two normalized voltammograms shown in Figure 12.4, which were recorded with two different sizes of electrodes in acetonitrile containing 10 mM tetrabutylammonium hexafluorophosphate and 1 mM diphenylanthracene (DPA) at 115 V/s. The larger electrode has an area of 0.026 $cm^2$, while the other electrode has an area of $6 \times 10^{-6}$ $cm^2$ and thus can be viewed as a microelectrode. For each electrode, the value of $(Dt_c)^{1/2}$ is less than $r_o$, so peak-shaped voltammograms are obtained. For the voltammograms recorded at each electrode, the reversible oxidation of DPA to its radical cation and the quasi-reversible reduction to the corresponding radical anion can be seen. However, the voltammogram recorded with the smaller electrode is much more clearly defined. At the larger electrode, the voltammetric peaks are shifted to more extreme potentials, and the peak currents are not reached with the voltage limits employed. Because of this, the reverse waves for each process are barely apparent at the larger electrode. These results clearly indicate the greater distortion of voltammograms by ohmic drop found with larger electrodes.

## A. Resistance at Microelectrodes

As discussed earlier in this volume (Chaps. 6 and 7), ohmic drop is a consequence of the electrode current, the sum of the faradaic and residual currents, that flows through the solution. This current through a resistive media generates a potential that opposes the applied potential difference between the working and counter electrodes. At microelectrodes, the predominant cell resistance is from a region near the electrode. It arises from transport of ions in the region adjacent to the working electrode and has dimensions that are the same as that of the diffusion layer under steady-state conditions. At electrodes of larger dimensions, the resistance of the bulk solution between the working and counter electrodes can also contribute to ohmic drop. This source of resistance exerts a negligible effect with microelectrodes because the large volume of the bulk solution relative to that of the diffusion layer allows the current flowing to the counter electrode to be distributed broadly. The use of a three-electrode potentiostat, a traditional method for removal of ohmic drop, is generally not useful with microelectrodes because of the difficulty of placing the reference electrode so close to the working electrode surface. Furthermore, when a microelectrode is used under steady-state conditions, the presence of a reference electrode in the diffusion layer would perturb the voltammetric response.

The simplest specific example to examine is for a microelectrode with a spherical geometry [53]. The resistance (R) of a spherical microelectrode was given in Chapter 7. However, for a sphere whose radius is much smaller than the distance from the sphere to the counter electrode, the resistance is

$$R = \rho/4\pi r_o \tag{12.19}$$

where $\rho$ is the specific resistance of the solution adjacent to the electrode. Typical values for $\rho$ for some solutions of interest in electrochemistry are given in Table 12.1. If we consider ohmic drop under conditions where voltammograms are steady state, we can combine the expression for resistance with the expression for the diffusion-limited faradaic current at a sphere ($i_f = 4\pi nFr_oDC°$). Then it can be seen that the ohmic drop is independent of the electrode dimensions:

$$iR = \rho nFDC° \tag{12.20}$$

**Table 12.1** Specific Resistances ($\rho$) at 25°C

| Solution | $\varepsilon$[a] | $\rho$ (ohm-cm) |
|---|---|---|
| Water (1 M KCl) | 78 | 8.94 |
| Water (0.1 M KCl) | 78 | 77.6 |
| Acetonitrile (0.6 M TBAP[b]) | 37 | 37 |
| Dimethoxyethane (1.0 M TBAP) | 7.2 | 312 |
| Tetrahydrofuran (1.0 M TBAP) | 7.8 | 368 |

[a]Dielectric constant.
[b]Tetrabutylammonium perchlorate.
*Source*: Ref. 74.

Even in the solutions of highest resistance given in Table 12.1, the ohmic drop can be calculated to be less than 1 mV for a millimolar solution of electroactive species. In a solvent that does not promote ion pairing, the value of $\rho$ is, to a first approximation, inversely proportional to the supporting electrolyte concentration. Thus, the ohmic drop in steady-state voltammograms can be adjusted by changing either the concentration of the electrolyte or the electroactive species.

The independence of the ohmic drop on electrode dimensions is a general conclusion true for any electrode geometry as long as steady-state faradaic currents are obtained [54,55]. This is because the important parameter is the distribution of the current through solution in the vicinity of the electrode surface. For example, at a disk where $R = \rho/4r_o$ and the steady-state limiting faradaic current is given by $i = 4r_o nFDC°$, Equation 12.20 also describes the maximum ohmic drop. This advantage is also found with microband electrodes. They may have current in the microampere range for millimolar solutions when operated at near steady-state conditions, but show only modest ohmic drop. Under conditions where the diffusion layer is much smaller than that obtained at steady state, such as the fast-scan voltammetry discussed earlier, the current is proportional to the electrode area; thus, the ohmic drop is a function of the electrode dimensions. The equations that describe the resistances of various microelectrode geometries are available in the literature [53,54] and can be used to compute the ohmic drop for the particular experiment of interest, as long as there is an excess concentration of electrolyte relative to that of the electroactive species.

Advantages are also found with microelectrodes used at short time scales where time-dependent currents are obtained, as the example in Figure 12.4 shows [56]. Time-dependent currents are proportional to the electrode area. Thus, for the case of a sphere, combination of the expression for the time-dependent current with the resistance shows that the ohmic drop is directly proportional to the radius. Thus, ohmic drop is always minimized by the use of a smaller electrode.

## B. Classification of Measurements in Resistive Media

Several examples of the use of microelectrodes in highly resistive media exist. The first reported measurements were an examination of the reduction of aromatic hydrocarbons such as perylene in benzene containing tetrahexylammonium perchlorate [57]. Although this electrolyte is presumably in a quite associated state in benzene (or toluene [58]), it does impart sufficient conductivity for electrochemistry to be observed. In subsequent work, this result was confirmed and extended to other low-dielectric-constant solvents [59]. Even voltammetry in hexane has been shown to be possible with a microelectrode [60]. In this solvent, a mixture of fatty acids and bases was used to impart conductivity. Several reports have shown that the oxidation of ferrocene can be studied in acetonitrile with zero or very low concentrations of electrolyte [10,61–63]. In the experiments containing no purposely added electrolyte, it appears that acetonitrile contains sufficient ionic impurities that there is always some conductivity [64]. Other experiments have suggested the possibility of electrochemistry in media where ionic movement is restricted, such as at low temperatures [65] or in rigid polymers [66].

On the basis of these exploratory studies, three different classes of experiments can be described in which ohmic drop will play a role:

1. Media containing excess electrolyte whose ionic mobility is limited, that is solutions at low temperature or rigid polymeric solutions.
2. The use of solvents that favor association of electrolytes, for example, dimethoxyethane and tetrahydrofuran. This is the reason for their much higher specific resistances than the other solvents containing similar amounts of electrolyte (Table 12.1).
3. Solutions that contain a lower concentration of inert electrolyte compared to the concentration of the electroactive species. The extremes of this class range from the electrolysis of the solvent to the electrolysis of an electroactive species with no purposely added electrolyte.

The magnitude of the ohmic drop at a microelectrode can be evaluated quite readily for case 1 from a knowledge of the specific solution resistance (obtained from conductivity measurements such as in Table 12.1) and the expressions for the voltammetric current for the specific microelectrode employed. Case 2 is also straightforward if the free concentration of ions exceeds that of the electroactive species. However, the situation is somewhat more complicated for the third class. In this case, and in case 2 for fully associated electrolyte, migration as well as diffusion can affect the observed voltammetric signals. In all three cases, the situation may be further complicated by a change in structure of the double layer. However, this is ignored for now, and is considered in the section on very small electrodes.

## C. Effects of Migration

In a voltammetric experiment conducted with an excess of electrolyte, the current in solution is carried by the electrolyte, not by the electroactive couple. Therefore, when solving the appropriate mass transport equations, it is appropriate to evaluate only diffusion of the electroactive species. The flux of electrolyte, which certainly occurs, is not manifested in the amplitude or position of the voltammetric signal. However, when the concentration of electrolyte is lower than that of the electroactive couple, the effect of the potential gradient in solution on mass transport of the electroactive species must be considered. This form of mass transport is referred to as migration. To do this we must modify the expression for the flux (see Chap. 2) to include the effects of the potential gradient $d\phi/dr$. This gradient arises from the uneven distribution of ions in solution caused by the electrolysis [67]. The necessary expression, known as the Nernst-Planck equation, can be written for one-dimensional mass transport for a species i as:

$$J_i = -D_i\left[\frac{dC_i}{dr}\right] - \frac{z_iF}{RT}D_iC_i\left[\frac{d\phi}{dr}\right] \tag{12.21}$$

The equation must be written separately for each species in solution (the anion and cation of the electrolyte, the oxidized and reduced forms of the electroactive couple, and any ion accompanying the initial state of the electroactive substance). Note that the charge of each species (z) controls the direction of the migrational flux, and for a neutral molecule ($z = 0$), the term on the right disappears as expected. Simultaneous solution of the five equations and evaluation over the appropriate boundary conditions gives the current for conditions when migration of either (or both) members of the electroactive couple can occur.

As an example, consider a solution that contains an electroactive species present at a concentration 100 times greater than that of the completely dissociated electrolyte ($P^+M^-$). The electrochemical reaction to be considered is the one-electron, electrochemically reversible reduction of a cation ($O^+$) associated with an inert anion (if it is the same as the electrolyte anion, only four simultaneous equations have to be solved) to a neutral species (R). To further simplify the derivation, it will be assumed that the diffusion coefficients of all four species are equal, a reasonable approximation when dealing with small molecules. An experimental example of such a situation is the reduction of cobalticenium to cobaltacene in acetonitrile containing a very low concentration of tetrabutylammonium perchlorate [68]. At potentials corresponding to the limiting current, the neutral product is being formed at a rate determined by the flux of the charged reactant to the electrode. The flux of reactant is due to diffusion, as in the case of an excess of electrolyte, but also has a migrational component because the electrode process removes cations while the inert anion remains. Therefore, to maintain electroneutrality, more $O^+$ must accumulate in the diffusion layer. For the same reason, there will be migration of the electrolyte cation into the diffusion layer, although this will be negligible due to its low concentration relative to $O^+$, and migration of the electrolyte anion out of this region.

Since the fluxes of electrolyte ions are zero at the electrode surface ($r_o$), the Nernst-Planck equation for the anion becomes

$$\frac{dC_{M^-}(r_o)}{dr} = \frac{F}{RT}C_{M^-}(r_o)\left[\frac{d\phi(r_o)}{dr}\right] \tag{12.22}$$

Electroneutrality requires that the concentration of $M^-$ in the diffusion layer very closely equals that of $O^+$ and that $dC_{M-}/dr = dC_{O+}/dr$. (The term quasielectroneutrality is perhaps more appropriate; electroneutrality exists over dimensions large relative to the radius of the ion of interest, but not on much smaller dimensions [67].) Thus, we can write:

$$\begin{aligned} J_O + (r_\circ) &= -D_O + \left[\frac{dC_O + (r_\circ)}{dr}\right] - \frac{F}{RT} D_O + C_O + (r_\circ)\left[\frac{d\phi(r_\circ)}{dr}\right] \\ &= -2D_O + \left[\frac{dC_O + (r_\circ)}{dr}\right] \end{aligned} \tag{12.23}$$

From this result it can be seen that the flux is twice as large as in the absence of migration. Since for any electrode geometry the current is directly proportional to the flux, this means that the voltammetric current will be twice as large as that found under conditions of diffusional control. The transition between diffusional control and migration is illustrated in the voltammograms of cobaltocenium shown in Figure 12.5 (right panel), which were recorded at different

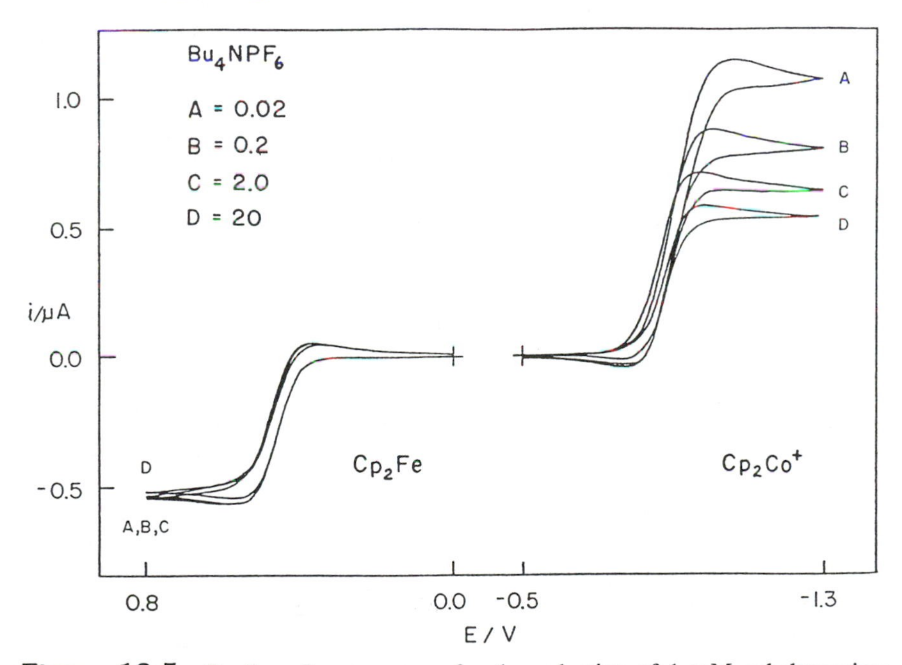


**Figure 12.5** Cyclic voltammograms for the reduction of 1 mM cobaltocenium ($Cp_2Co^+$) hexafluorophosphate and oxidation of 1 mM ferrocene ($Cp_2Fe$) in acetonitrile recorded at a band electrode (width = 4.6 $\mu$m) at a scan rate of 10 mV $s^{-1}$. The supporting electrolyte is tetrabutylammonium hexafluorophosphate at (A) 0.02 M, (B) 0.2 mM, (C) 2.0 mM, and (D) 20 mM. [From Ref. 68, reprinted with permission of the copyright holder.]

concentrations of electrolyte. Complete descriptions of the effect of migration on the limiting current have been developed [69,70], and in most cases experimental examples have verified these expressions. The case just considered yields the greatest percent increase in the limiting current when going from diffusional control to migrational/diffusional control for a one-electron process.

A common experimental situation is the electrolysis of a neutral species to form an ion. The one-electron oxidation of ferrocene to ferricenium is an example of such a process. Since the current arises from the flux of a neutral to the electrode, the diffusion-limited current is unaffected by lowering the electrolyte concentration (Fig. 12.5, left panel). However, evaluation of the diffusion/migration equations shows that the charged product of the electrochemical reaction is removed from the diffusion layer by ion migration while inert ions of opposite charge are drawn in by the process of migration, again to maintain electroneutrality [68].

The process of migration will cause the ohmic drop present in a voltammetric wave to differ from that calculated from the specific resistivity for experiments conducted under the conditions described in case 3 [53]. This is because the electroactive couple is carrying the majority of the current through solution, and the concentration of each is changing in the diffusion layer during the voltammogram. Consider what occurs in the diffusion layer during a potential scan through the voltammetric wave for the conversion of $O^+$ to R in the presence of a low concentration of electrolyte [71]. At potentials sufficiently positive that no current flows, the ionic composition of the diffusion layer, and thus its specific resistance, is the same as the bulk solution. However, since faradaic current is not flowing, there is no ohmic drop at this point. At the beginning of the voltammetric wave, the predominant cation ($O^+$) is consumed and the resistance will start to increase. This process will continue until the limiting current is reached, where the maximum resistance (and thus maximum ohmic drop) will be encountered.

For the electrolysis of a neutral species to an ion, the opposite process occurs. As the potential of the electrode approaches E°, ions are generated by the electroactive species and ions of opposite charge migrate into the diffusion layer. In this case the electrolysis generates a greater number of charged species in the diffusion layer, the region that provides the greatest contribution to ohmic drop. To examine this effect as a function of electrolyte concentration, changes in the value of the half-wave potential ($E_{1/2}$) can be examined [53,64]. The shift found at a spherical or hemispherical electrode for a one-electron oxidation of R in the presence of an inert electrolyte with anion $M^-$ is (assuming equal diffusion coefficients of the oxidized and reduced forms)

$$E^\circ - E_{1/2} = -\frac{RT}{F}\ln\left[1 + \frac{C_R}{8C_{M^-}}\right] \tag{12.24}$$

which predicts that the shift in $E_{1/2}$ will be linear with the logarithm of the electrolyte concentration at high values of $C_R/C_{M^-}$. In contrast, the use of the specific resistance to calculate the ohmic drop predicts a linear change in the $E_{1/2}$ value with this ratio. Experimentally, data for the oxidation of ferrocene are found to follow the behavior expected for migration [64,72].

## D. Experimental Examples

One condition where ohmic drop may be encountered is that in which the solution composition is varied to elucidate reactions in the area of organic electrochemistry. Consider, for example, the commercially important reaction that occurs during the reduction of acrylonitrile:

$$CH_2{=}CHCN + H_2O + e^- \rightleftharpoons \tfrac{1}{2}CN(CH_2)_4CN$$

This hydrodimerization reaction forms adiponitrile, which is a precursor for nylon. However, the two-electron reduction of this reactant results in propionitrile, an undesired product:

$$CH_2{=}CHCN + 2H_2O + 2e^- \rightleftharpoons CH_3CH_2CN$$

The products obtained depend on the solution conditions, and one commercial process to form adiponitrile employs an emulsion of acrylonitrile and an aqueous phosphate buffer containing various additives. Pletcher and coworkers [73] mimicked these reaction conditions and observed the limiting current at a microelectrode as a function of adiponitrile concentration. At low concentrations, the wave height observed corresponded to a two-electron process. However, at much higher concentrations ($>1$ M), conditions that more closely resemble the commercial process, the limiting current corresponded to a one-electron process (Fig. 12.6). Note that such high concentrations would preclude measurements with an electrode of conventional size because of ohmic drop caused by the high reactant concentrations. These investigators were also able to show that the hydrodimerization reaction requires the presence of tetraalkylammonium ions. Although the optimum conditions had previously been determined by empirical methods, this study nicely illustrates the quantitative results that can be obtained with microelectrodes.

Another experimental example of the effect of chemical reactions on voltammetry in resistive media is given by the reduction of tetracyanoquinodimethane (TCNQ) in acetonitrile. At high ionic strength, voltammograms of TCNQ have two one-electron waves of equal amplitude [75]. The waves correspond to the following electrochemical processes:

$$TCNQ + e^- \rightarrow TCNQ^-$$

$$TCNQ^- + e^- \rightarrow TCNQ^{2-}$$

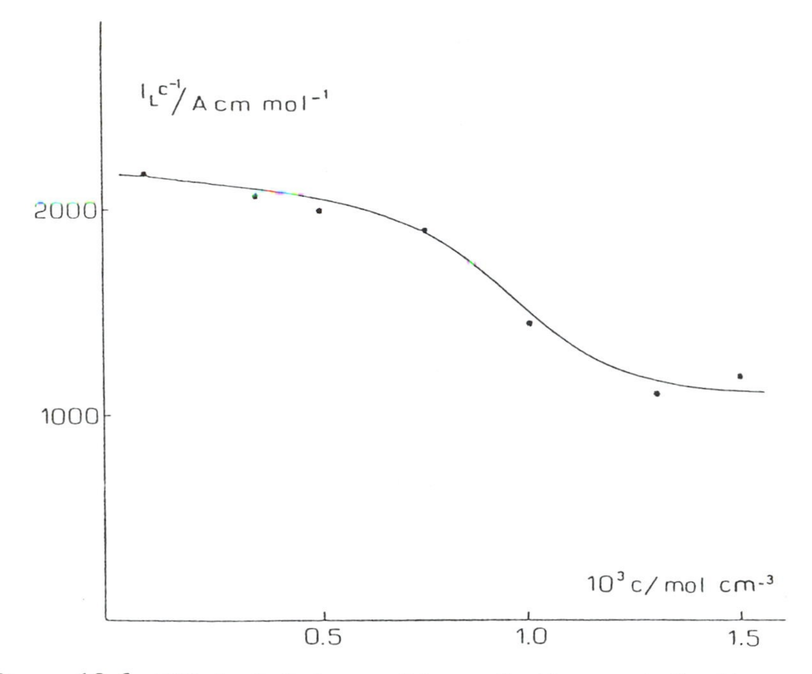


**Figure 12.6** Diffusion-limited current ($i_L$) normalized by concentration (c) versus concentration for the reduction of acrylonitrile in aqueous $Na_2HPO_4$ (15%) with $Bu_4NHSO_4$ (0.4%) at pH 7.0. Voltammograms were obtained at a mercury electrode ($A = 8 \times 10^{-7}$ cm$^2$) at a scan rate of 50 mV s$^{-1}$. [From Ref. 73, reproduced with permission of the copyright holder.]

As the concentration of electrolyte is lowered to a range equal to or less than the concentration of TCNQ, the first wave remains near its original height, but the second wave decreases in amplitude relative to the first (Fig. 12.7). The behavior of the first electrochemical wave is that expected for the reduction of a neutral, since the limiting current for such a process should be independent of migration. However, to explain the behavior of the limiting current of the second wave, the chemical processes in the diffusion layer must be considered.

When $TCNQ^{2-}$ is generated at the electrode surface, it is transported toward the bulk of solution. There, it can undergo the following reproportionation reaction:

$$TCNQ^{2-} + TCNQ \rightarrow 2TCNQ^{-}$$

The equilibrium constant can be calculated from the difference in standard reduction potentials ($\Delta E°$) for the two processes [$K = \exp(\Delta E°F/RT)$], and is

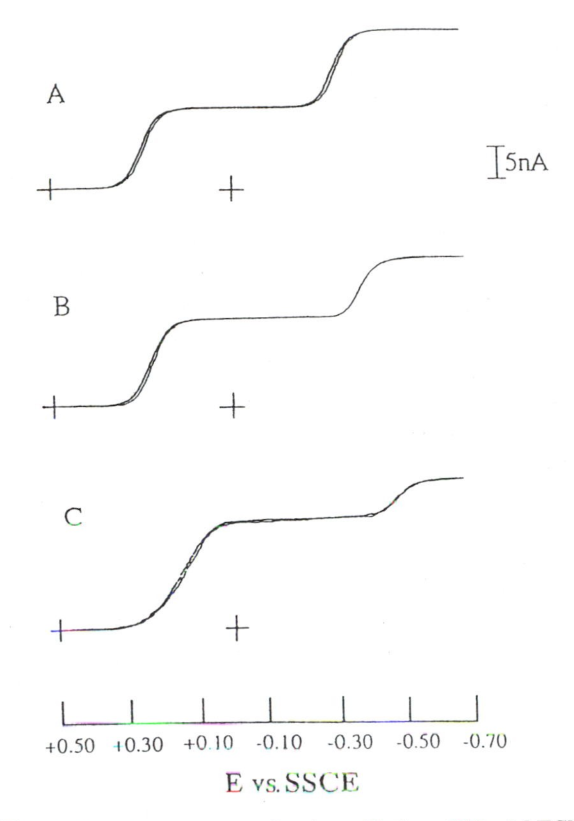


**Figure 12.7** Cyclic voltammograms for the oxidation of 10 mM TCNQ in acetonitrile at a 12.4-$\mu$m-radius Pt disk electrode. Concentration of the supporting electrolyte, tetrabutylammonium perchlorate: (A) 100 mM; (B) 1 mM; (C) 0 mM. The solutions were purged with $N_2$. [From Ref. 75, reproduced with permission of the copyright holder.]

greater than $10^9$ for this particular reaction. At high ionic strength, $TCNQ^-$ diffuses to the electrode, is reduced, and the effect of the solution reaction is not apparent. However, at low ionic strength, migration directs $TCNQ^-$ away from the electrode, resulting in less current at potentials corresponding to the second wave. The relative diminution in current at the second wave compared to the first can be used to estimate the rate constant of the reproportionation reaction. This effect may also be operant in some of the other systems that have been found to give unexplained behavior under conditions of low ionic strength [76].

Well-defined voltammograms can be obtained for the electrolysis of a solvent such as nitrobenzene with microelectrodes. The voltammogram is not distorted by ohmic drop because of migration of electrolyte into the diffusion-layer region. However, because the solvent and the electroactive species are the same, a complete description of the transport phenomena for the first wave requires consideration of factors other than diffusion and migration [77]. Interpretation of the second wave involves consideration of the reproportionation described earlier, but also may involve electron hopping in the solution [78].

There is, of course, a limit on the amount of solution resistance that can be overcome even with microelectrodes. Attempts have been made to perform electrochemistry in the gas phase [79] or in supercritical $CO_2$ [80] with microelectrodes. In each case, however, the conduction path was shown to be not through the bulk phase, but rather across the insulating surface between the microelectrode and the counter electrode. This mechanism enables electrochemical detection in highly unusual media for voltammetry and illustrates that only very small conduction pathways are required to obtain well-defined electrochemical behavior.

## VI. VERY SMALL ELECTRODES

Most of the results described earlier in this chapter were obtained with electrodes of micrometer dimensions. Much smaller electrodes, sometimes referred to as nanodes, are of considerable interest because of the possibility of probing chemical events with an electrode approaching molecular dimensions. However, such devices are considerably more difficult to fabricate with well-defined geometries. Furthermore, their use requires the measurement of very small currents because of the dependence of the current on the electrode dimensions. Submicrometer electrodes can be made from Wollaston wires [81], and even smaller electrodes can be made from etched wires sealed in an insulator [82–84]. Although not as small, several procedures have also been documented for fabrication of submicron carbon electrodes [85–89]. A micrograph of two styles of microcarbon electrodes adjacent to a biological cell is shown in Figure 12.8. Glass, wax, or electrochemically deposited films [90] serve as insulators and define the active region of the electrode. Typically, the electrode dimensions are characterized on the basis of the amplitude of the electrochemical signal by assuming an electrode geometry and using an electroactive system of known concentration and diffusion coefficient [91,92].

Band electrodes were the first to be fabricated with true nanometer dimensions [17,93]. Although small in only one dimension, the width, this controls the magnitude of the flux, while the finite length leads to easily measured currents. These electrodes can be fabricated with a smallest dimension as small as 2 nm by sealing deposited metal films between insulators. While voltammograms obtained with electrodes with widths greater than 50 nm follow the theory expected for diffusion-controlled mass transport, narrower electrodes show less

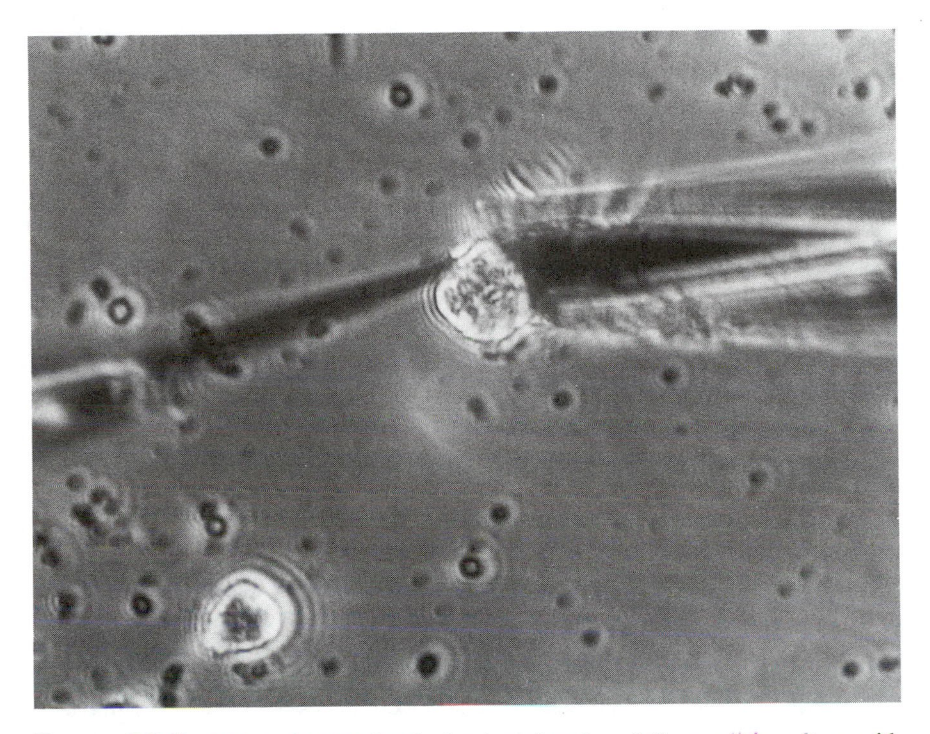

**Figure 12.8** Light micrograph of a bovine adrenal medullary cell in culture with etched and glass-encased carbon-fiber electrodes (r = 5 $\mu$m) placed adjacent to it. Magnification is 450×. [From Ref. 88, reproduced with permission of the copyright holder.]

current than expected. This has been attributed to two features: the failure of the equations to account for the finite size of the molecule that undergoes electron transfer, and the altered fluid density and viscosity near the surface of the electrode. Theory exists to evaluate each of these features [94], and these concepts should be considered when evaluating an electrode of any geometry when its size approaches molecular dimensions. Further study of these points has been hampered by the ability to evaluate the surfaces of such small structures.

Another important feature is the relative size of the diffusion layer with respect to the double layer. The diffusion-layer dimensions are proportional to the size of the electrode, and can approach the size of the double layer with electrodes of molecular dimensions [95]. This can also occur in other situations explored with microelectrodes. For example, in solutions of very dilute electrolyte, the diffuse double layer extends several nanometers into solution [62]. Alternatively, very fast cyclic voltammetry results in a very small diffusion layer, which may be of dimensions similar to the double layer [46]. In all these situtions, migration induced by the potential field inside the double layer will occur along with diffusion, and the measured currents will differ from those based only on diffusional fluxes. Unless all of these possibilities are accounted for and the electrode geometry is clearly defined, kinetic values obtained with very small electrodes are open to question [96].

As described in the introduction, submicrometer disk electrodes are extremely useful to probe local chemical events at the surface of a variety of substrates. However, when an electrode is placed close to a surface, the diffusion layer may extend from the microelectrode to the surface. Under these conditions, the equations developed for semi-infinite linear diffusion are no longer appropriate because the boundary conditions are no longer correct [97]. If the substrate is an insulator, the measured current will be lower than under conditions of semi-infinite linear diffusion, because the microelectrode and substrate both block free diffusion to the electrode. This phenomena is referred to as shielding. On the other hand, if the substrate is a conductor, the current will be enhanced if the couple examined is chemically stable. For example, a species that is reduced at the microelectrode can be oxidized at the conductor and then return to the microelectrode, a process referred to as feedback. This will occur even if the conductor is not electrically connected to a potentiostat, because the potential of the conductor will be the same as that of the solution. Both shielding and feedback are sensitive to the diameter of the insulating material surrounding the microelectrode surface, because this will affect the size and shape of the diffusion layer. When these concepts are taken into account, the use of scanning electrochemical microscopy can provide quantitative results. For example, with the use of a 30-nm conical electrode, diffusion coefficients have been measured inside a polymer film that is itself only 200 nm thick [98].

## VII. CONCLUSIONS

The features described in this chapter clearly show that microelectrodes have several considerable advantages over their conventional counterparts. By miniaturizing the electrode, one reduces the capacitance and the ohmic drop. This enables measurements to be made under previously prohibited conditions. The high flux found when the electrodes are used at steady state provides a new approach to the investigation of electrochemical phenomena. Their small size enables spatially heterogeneous chemical events to be probed. Together, these advantages open whole new domains for electrochemical investigation.

## ACKNOWLEDGMENT

Research support in this area by the National Science Foundation is gratefully acknowledged.

## REFERENCES

1. P. W. Davies and F. Brink, *Rev. Sci. Instrum. 13*:524 (1942).
2. R. N. Adams, *Prog. Neurobiol. 35*:297 (1990).
3. J. B. Chien, R. A. Wallingford, and A. G. Ewing, *J. Neurochem. 54*:633 (1990).
4. J. G. White and J. W. Jorgenson, *Anal. Chem. 58*:2992 (1986).
5. J. E. Baur, E. W. Kristensen, and R. M. Wightman, *Anal. Chem. 60*:2334 (1988).
6. R. C. Engstrom and C. N. Pharr, *Anal. Chem. 61*:1099A (1989).
7. A. J. Bard, F.-R. F. Fan, D. T. Pierce, P. R. Unwin, D. O. Wipf, and F. Zhou, *Science 254*:68 (1991).
8. R. M. Wightman and D. O. Wipf, in *Electroanalytical Chemistry*, Vol. 15 (A. J. Bard, ed.), 1989, Chap. 3.
9. *Microelectrodes: Theory and Applications, NATO ASI Series* (M. I. Montenegro and J. L. Daschbach, eds.), Kluwer Academic Publishers, Boston, 1991.
10. J. O. Howell and R. M. Wightman, *Anal. Chem. 56*:524 (1984).
11. D. O. Wipf, A. C. Michael, and R. M. Wightman, *J. Electroanal. Chem. 296*:15 (1989).
12. M. J. Nuwer and J. Osteryoung, *Anal. Chem. 61*:1954 (1989).
13. K. R. Wehmeyer and R. M. Wightman, *J. Electroanal. Chem. 196*:417 (1985).
14. P. M. Kovach, M. R. Deakin, and R. M. Wightman, *J. Phys. Chem. 90*:4612 (1986).
15. J. B. Justice, Jr., ed., *Voltammetry in the Neurosciences*, Humana Press, Clifton, NJ, 1987.
16. C. E. Chidsey, B. J. Feldman, C. Lundgren, and R. W. Murray, *Anal. Chem. 58*:601 (1986).
17. K. R. Wehmeyer, M. R. Deakin, and R. M. Wightman, *Anal. Chem. 57*:1913 (1985).
18. J. E. Bartelt, M. R. Deakin, C. Amatore, and R. M. Wightman, *Anal. Chem. 60*:2167 (1988).
19. S. Licht, V. Cammarata, and M. S. Wrighton, *Science 243*:1176 (1989).
20. S. Licht, V. Cammarata, and M. S. Wrighton, *J. Phys. Chem. 94*:6133 (1990).
21. W. Thormann and A. M. Bond, *J. Electroanal. Chem. 218*:187 (1987).
22. L. S. Kuhn, A. Weber, and S. G. Weber, *Anal. Chem. 62*:1631 (1990).
23. T. Abe, Y. Y. Lau, and A. G. Ewing, *J. Am. Chem. Soc. 113*:7421 (1991)
24. A. J. Bard and L. R. Faulkner, *Electrochemical Methods*, Wiley, New York, 1980.
25. K. Aoki, K. Akimoto, K. Tokuda, H. Matsuda, and J. Osteryoung, *J. Electroanal. Chem. 171*:219 (1984).
26. A. C. Michael, R. M. Wightman, and C. A. Amatore, *J. Electroanal. Chem. 267*:33 (1989).
27. D. Shoup, and A. Szabo, *J. Electroanal. Chem. 140*:237 (1982).
28. C. R. Cabrera, E. Garcia, and A. J. Bard, *J. Electroanal. Chem. 260*:457 (1989).
29. C. A. Amatore, B. Fosset, M. R. Deakin, and R. M. Wightman, *J. Electroanal. Chem. 225*:33 (1987).
30. M. R. Deakin, R. M. Wightman, and C. A. Amatore, *J. Electroanal. Chem. 215*:49 (1986)

31. S. Sujaritvanichpong, K. Aoki, K. Tokuda, and H. Matsuda, *J. Electroanal. Chem. 199*:271 (1986).
32. H. P. Wu and R. L. McCreery, *Anal. Chem. 61*:2347 (1989).
33. R. C. Engstrom, C. M. Pharr, and M. D. Koppang, *J. Electroanal. Chem. 221*:251 (1987).
34. Reference 9, see chapters by J. Osteryoung and by J. Osteryoung and M. M. Murphy.
35. L. J. Magee, Jr., and J. Osteryoung, *Anal. Chem. 62*:2625 (1990).
36. J. C. Myland, K. B. Oldham, and C. G. Zoski, *J. Electroanal. Chem. 193*:3 (1985).
37. A. M. Bond, K. B. Oldham, and C. G. Zoski, *J. Electroanal. Chem. 245*:71 (1988).
38. K. B. Oldham, C. G. Zoski, A. M. Bond, and D. Sweigart, *J. Electroanal. Chem. 248*:467 (1988).
39. T. Varco Shea and A. J. Bard, *Anal. Chem. 59*:2102 (1987)
40. B. Fosset, C. A. Amatore, J. E. Bartelt, A. C. Michael, and R. M. Wightman, *Anal. Chem. 63*:306 (1991).
41. R. M. Wightman and D. O. Wipf, *Acc. Chem. Res. 23*:64 (1990).
42. N. S. Hush, ed., *Reactions of Molecules at Electrodes*, Wiley, New York, 1971.
43. D. O. Wipf and R. M. Wightman, *Anal. Chem. 60*:2460 (1988).
44. C. P. Andrieux, D. Garreau, P. Hapiot, J. Pinson, and J. M. Saveant, *J. Electroanal. Chem. 243*:321 (1988).
45. C. Amatore and C. Lefrou, *J. Electroanal. Chem. 324*:33 (1992).
46. C. Amatore and C. Lefrou, *J. Electroanal. Chem. 296*:335 (1990).
47. P. M. Kovach, A. G. Ewing, R. L. Wilson, and R. M. Wightman, *J. Neurosci. Methods 10*:215 (1984).
48. R. M. Wightman, L. J. May, and A. C. Michael, *Anal. Chem. 60*:769A (1988).
49. M. A. Dayton, A. G. Ewing, and R. M. Wightman, *Anal. Chem. 52*:2392 (1980).
50. C. P. Andrieux, P. Audebert, P. Hapiot, M. Nechtschein, and C. Odin, *J. Electroanal. Chem. 305*:153 (1991).
51. C. P. Andrieux, P. Audebert, P. Hapiot, and J. M. Saveant, *J. Phys. Chem. 95*:10158 (1991).
52. J. O. Howell, J. M. Goncalves, C. Amatore, L.Klasinc, R. M. Wightman, and J. K. Kochi, *J. Am. Chem. Soc. 106*:3968 (1984).
53. K. B. Oldham, *J. Electroanal. Chem. 250*:1 (1988).
54. S. Bruckenstein, *Anal. Chem. 59*:2098 (1987).
55. K. B. Oldham, *J. Electroanal. Chem. 237*:303 (1987).
56. D. O. Wipf and R. M. Wightman, *Anal. Chem. 62*:98 (1990).
57. R. Lines and V. D. Parker, *Acta Chem. Scand. B. 31*:369 (1977).
58. L. Geng and R. W. Murray, *Inorg. Chem. 25*:3115 (1986).
59. J. O. Howell and R. M. Wightman, *J. Phys. Chem. 88*:3915 (1984).
60. L. Geng, A. G. Ewing, J. C. Jernigan, and R. W. Murray, *Anal. Chem. 58*:852 (1986).
61. R. M. Wightman, *Anal. Chem. 53*:1125A (1981).
62. A. M. Bond, M. Fleischmann, and J. Robinson, *J. Electroanal. Chem. 168*:299 (1984).

63. A. M. Bond and T. F. Mann, *Electrochim. Acta 32*:863 (1987).
64. B. D. Pendley, H. D. Abruna, J. D. Norton, W. E. Benson, and J. S White, *Anal. Chem. 63*:2766 (1991).
65. A. M. Bond, M. Fleischmann, and J. Robinson, *J. Electroanal. Chem. 180*:257 (1984).
66. J. F. Parcher, C. J. Barbour, and R. W. Murray, *Anal. Chem. 61*:584 (1989).
67. R. P. Buck, *J. Membr. Sci. 17*:62 (1984).
68. M. R. Deakin, R. M. Wightman, and C. A. Amatore, *J. Electroanal. Chem. 225*:49 (1987).
69. C. Amatore, B. Fosset, J. Bartelt, M. R. Deakin, and R. M. Wightman, *J. Electroanal. Chem. 256*:255 (1988).
70. J. C. Myland and K. B. Oldham, *J. Electroanal. Chem. 347*:49 (1993).
71. J. B. Cooper and A. M. Bond *J. Electroanal. Chem. 315*:143 (1991).
72. S. M. Drew, R. M. Wightman, and C. A. Amatore, *J. Electroanal. Chem. 317*:117 (1991).
73. M. I. Montenegro and D. Pletcher, *J. Electroanal Chem. 248*:229 (1988).
74. D. T. Sawyer and J. L. Roberts, Jr., *Experimental Electrochemistry for Chemists*, Wiley, New York, 1974, p. 188.
75. J. D. Norton, W. E. Benson, H. S. White, B. D. Pendley, and H. D. Abruna, *Anal. Chem. 63*:1909 (1991).
76. J. B. Cooper, A. M. Bond, and K. B. Oldham, *J. Electroanal. Chem. 331*:877 (1992).
77. R. B. Morris, K. F. Fischer, and H. S. White, *J. Phys. Chem. 92*:5306 (1988).
78. J. D. Norton, S. A. Anderson, and H. S. White, *J. Phys. Chem. 96*:3 (1992).
79. R. Brina, S. Pons, and M. Fleischmann, *J. Electroanal. Chem. 244*:81 (1988).
80. D. Niehaus, M. Philips, A. C. Michael, and R. M. Wightman, *J. Phys. Chem. 93*:6232 (1989).
81. M. Fleischmann, F. Lasserre, J. Robinson, and D. Swan, *J. Electroanal. Chem. 177*:97 (1984).
82. R. M. Penner, M. J. Heben, and N. S. Lewis, *Anal. Chem. 61*:1630 (1989).
83. B. D. Pendley and H. D. Abruna, *Anal. Chem. 62*:782 (1990).
84. C. Lee, C. J. Miller, and A. J. Bard, *Anal. Chem. 63*:78 (1991).
85. M. Armstrong-James, K. Fox, and J. Millar, *J. Neurosci. Methods 2*:431 (1980).
86. A. Meulemans, B. Poulain, G. Baux, L. Tauc, and D. Henzel, *Anal. Chem. 58*:2088 (1986).
87. R. A. Saraceno, and A. G. Ewing, *J. Electroanal. Chem. 257*:83 (1988).
88. K. T. Kawagoe, J. A. Jankowski, and R. M. Wightman, *Anal. Chem. 63*:1589 (1991).
89. T. G. Strein and A. G. Ewing, *Anal. Chem. 64*:1368 (1992).
90. K. Potje-Kamloth, J. Janata, and M. Josowicz, *Ber. Bunsen-Ges. Phys. Chem. 93*:1480 (1989).
91. J. E. Baur and R. M. Wightman, *J. Electroanal. Chem. 305*:73 (1991).
92. M. V. Mirkin, F.-R. F. Fan, and A. J. Bard, *J. Electroanal. Chem. 328*:47 (1992).
93. R. B. Morris, D. J. Franta, and H. S. White, *J. Phys. Chem. 91*:3559 (1987).
94. J. D. Seibold, E. R. Scott, and H. S. White, *J. Electroanal. Chem. 264*:281 (1989).

95. J. D. Norton, H. S. White, and S. W. Feldberg, *J. Phys. Chem. 94*:6772 (1990).
96. R. M. Penner, M. J. Heben, T. L. Longin, and N. S. Lewis, *Science 250*:1118 (1990).
97. J. Kwak and A. J. Bard, *Anal. Chem. 61*:1221 (1989).
98. M. V. Mirkin, F.-R. F. Fan, and A. J. Bard, *Science 257*:364 (1992).

# 13

# Chemically Modified Electrodes

**Charles R. Martin** *Colorado State University, Fort Collins, Colorado*

**Colby A. Foss, Jr.** *Georgetown University, Washington, D.C.*

Thus, when a substituent of interest is incorporated into an olefinic substance and the resulting compound allowed to react with the electrode surface, the substituent becomes connected to the surface. . . . By this means, ionic species have been tethered within the double layer region in order to probe the mechanisms of electrode reactions involving platinum complexes. . . . Alternatively, the electrochemical reactant itself can be connected to the electrode surface, allowing its reactivity to be observed as a function of charge, orientation, and structure, as described here.

R.F. Lane and A.T. Hubbard, *Journal of Physical Chemistry*, Vol. 77, p. 1401, 1973 [10]

If a method for securely anchoring such molecules could be found, advantage could be taken of the molecular structure to build surfaces with unique and widely varying properties. Indeed, the attached molecules could be used in the sense of chemical reagents to perform reactions in tandem with the electron transfer processes characteristic of chemically inert electrodes.

B.F. Watkins, J.R. Behling, E. Kariv, and L.L. Miller, *Journal of the American Chemical Society*, Vol. 97, p. 3549, 1975 [17]

We see this line of research as eventually leading to a wide array of chemically modified electrode surfaces with unusual analytical, chemical, catalytic and optical properties.

C.M. Elliott and R.W. Murray, *Analytical Chemistry*, Vol. 48, p. 1247, 1976 [97]

## I. INTRODUCTION

These quotes were chosen to introduce this chapter on "chemically modified electrodes" because they are from some of the earliest papers in the field and because they review the concepts and objectives of this research area. We learn that the field of chemically modified electrodes involves attaching specific molecules to the surfaces of conventional "inert" electrodes. We also discover the two major reasons for wanting to attach molecules to electrode surfaces. As explained by Lane and Hubbard, one objective is to obtain fundamental information about the mechanism of electron transfer at electrode surfaces. The second objective, as expressed by Watkins et al. and Elliott and Murray, is to impart to the electrode surface some chemical specificity not available at the unmodified electrode. For example, the modified electrode might catalyze a specific chemical reaction. Alternatively, the modified electrode might be able to recognize a specific molecule present in a contacting solution phase.

It is also of interest to note the age of these quotes. It has now been 20 years since the publication of the first papers on chemically modified electrodes. During these 20 years this concept has been the subject of intense research activity. Indeed, it is fair to say that chemically modified electrodes have been the most popular (important?) research area in electrochemistry during the previous two decades. There are now thousands of papers in the literature on this subject; fortunately, a number of authoritative reviews of this voluminous literature are available [1–8]. Because of the importance of this field, it is essential that students of modern electrochemistry have a working knowledge of chemically modified electrodes. The objective of this chapter is to provide this knowledge.

We first review methods for preparing chemically modified electrodes. This section is organized roughly according to the chronology of development of the various types of modified electrodes. We then discuss the fundamentals of electrochemical processes at chemically modified electrodes. Section IV then provides a discussion of methods used for characterizing chemically modified electrodes. Both electrochemical and spectroscopic methods are reviewed. The penultimate section describes potential applications of these devices. Applications in both fundamental and applied science are reviewed. We conclude with some highly biased observations about the impact of this field and where it is going.

## II. METHODS FOR PREPARING CHEMICALLY MODIFIED ELECTRODES

There are now over a half dozen demonstrated methods for preparing chemically modified electrodes. These range from simple chemisorption to electrosynthesis of a polymer film at an electrode surface. We review five methods here.

## A. Methods Based on Chemisorption

Chemisorption [9] is an adsorptive interaction between a molecule and a surface in which electron density is shared by the adsorbed molecule and the surface. Electrochemical investigations of molecules that are chemisorbed to electrode surfaces have been conducted for at least three decades. Why is it, then, that the papers that are credited with *starting* the chemically modified electrode field (in 1973) describe chemisorption of olefinic substances on platinum electrodes [10,11]? What is it about these papers that is different from the earlier work? The answer to this question lies in the quote by Lane and Hubbard at the start of this chapter. Lane and Hubbard raised the possibility of using carefully designed adsorbate molecules to probe the fundamentals of electron-transfer reactions at electrode surfaces. It is this concept of specifically tailoring an electrode surface to achieve a particularly desired goal that distinguishes this work from the prior literature on chemisorption, and it is this concept that launched the chemically modified electrode field.

Since the pioneering work of Lane and Hubbard, there have been numerous examples of using chemisorption to modify electrode surfaces. For example, Anson and his coworkers have investigated chemisorption of various aromatic systems onto carbon electrodes [12]. In this case, $\pi$-electron density is shared between the electrode and the adsorbate molecule. Examples of electroactive molecules that have been used to modify electrode surfaces via this approach are shown in Table 13.1 [8]. It is of interest to note that from the very beginning, there was considerable interest in modifying electrode surfaces with biochemical substances (Table 13.1). This is because such modified electrodes seemed to be likely candidates for use in electrocatalytic processes and biochemical sensors (see Section V).

Chemisorption requires direct contact between the chemisorbed molecule and the electrode surface; as a result, the highest coverage achievable is usually a monomolecular layer. This may be contrasted with several of the methods to be discussed later that allow the electrode surface to be covered with thick films (i.e., multimolecular layers) of the desired molecule. In addition to this coverage limitation, chemisorption is rarely completely irreversible. In most cases, the chemisorbed molecules slowly leach into the contacting solution phase during electrochemical or other investigations of the chemisorbed layer. For these reasons, electrode modification via chemisorption was quickly supplanted by other methods, most notably polymer-coating methods.

There has, however, been a recent "rebirth" of interest in using chemisorption to modify electrode surfaces. This rebirth is centered around the use of

**Table 13.1** Some Representative Electroactive Molecules That Have Been Chemisorbed onto Electrode Surfaces

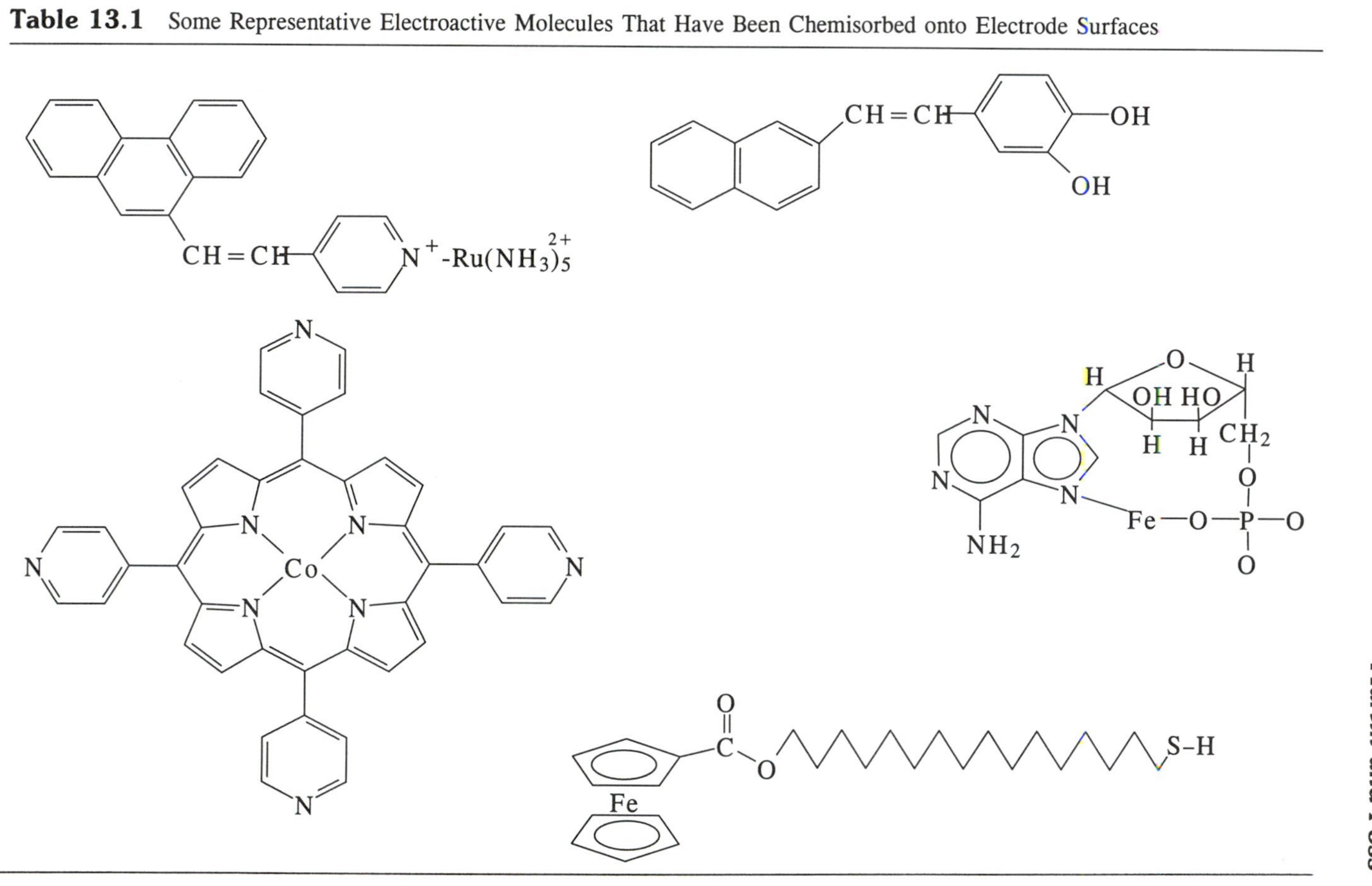

thiols, sulfides, and disulfides as chemisorption agents for derivatization of gold (and other) electrode surfaces [13]. In the case of an alkylthiol, this chemisorption reaction can be written as [14]

$$-\text{Au} + \text{R-SH} \longrightarrow -\text{Au-S-R} + 1/2\,\text{H}_2 \tag{13.1}$$

where Au represents a gold atom at the electrode surface and R is the alkyl substituent. The reaction is typically carried out by simply immersing the electrode into a dilute solution of the thiol. These derivatization agents are of tremendous current interest because this simple chemisorption process can yield densely packed and highly ordered monolayer films on metal surfaces and because the nature of the R group (Eq. 13.1) can be changed at will. These films have been called "self-assembled" monolayers.

## B. Methods Based on Covalent Bond Formation

Various methods were developed during the mid and late 1970s for forming covalent bonds between specific functional groups on the electrode surface and the molecule to be attached to the surface. The quintessential example involves reaction of a surface hydroxyl with a hydrolytically unstable silane [15]. This chemistry is illustrated in Equation 13.2, where M-OH represents a hydroxyl group on an electrode surface and R on the silane is the functional group that is to be attached to the electrode surface. Murray and his associates were the first to use this chemistry to modify electrode surfaces. This chemistry has since been used to attach an enormous number of functional groups to $SnO_2$, $RuO_2$, $TiO_2$, Pt, Au, and other electrode surfaces [15].

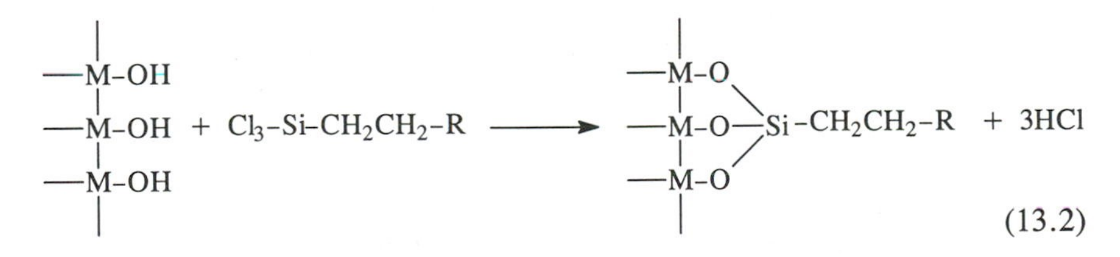


Equation 13.2 should be viewed as a "cartoon" version of the "silanization" reaction because while the silane is, in principle, capable of forming three covalent bonds to the surface, it is doubtful that all three actually form. Indeed, if traces of water are present, the silane will be hydrolyzed to form a siloxane polymer that will ultimately become covalently attached to the electrode surface. Because a polymer is formed, it is possible to achieve multimolecular layers at the electrode surface via this chemistry. Wrighton et al. have made extensive

use of such hydrolytically unstable silanes to prepare multilayer films containing numerous desired electroactive functionalities [16].

Various other surface chemistries have been used to attach chemical species to electrode surfaces. For example, Watkins et al. activated the carboxylic acid functionalities on carbon electrodes with thionyl chloride and then reacted this surface with amines [17]. Sagiv and his coworkers have recently invented a clever approach for monolayer-by-monolayer deposition of multilayer films based on organosilane chemistry [18]. Finally, Mallouk et al. have also developed a monolayer-by-monolayer approach for synthesizing well-ordered multilayer films [19]. Because of these interesting new synthetic strategies, covalent attachment of functional groups remains an attractive approach for modifying electrode surfaces.

## C. Coating Electrodes with Polymer Films

In 1978, Miller's group and Bard's group independently showed that chemically modified electrodes could be prepared by coating electrode surfaces with polymer films [20,21]. This has since proven to be the most versatile approach for preparing chemically modified electrodes. Indeed, until the recent rebirth of chemisorption and new covalent-attachment schemes (see earlier discussion), the polymer-film method had essentially supplanted all other methods for preparing chemically modified electrodes.

There are many reasons for the popularity of polymer films as electrode-modifying agents. First, it is easy to prepare multilayer films using the polymer route. Indeed, the quantity of polymer deposited (i.e., the thickness of the film) can be reliably and reproducibly varied. Polymer films can be made completely insoluble in the contacting solution phase; thus, loss of material from the electrode is not a problem. Many organic polymers have tremendous chemical stability (think how long it takes a plastic six-pack ring to degrade); hence, degradation of the film is usually not a problem. Polymers can be prepared in almost infinite variety. Thus, polymers that incorporate any desired electroactive chemical functionality can be synthesized. Finally, other types of functional groups can be easily added to the polymer; for example, ionic groups can be added to increase film conductivity.

An enormous number of polymers have been used to prepare chemically modified electrodes. Some examples are given in Table 13.2; Albery and Hillman provide a more extensive list [8]. As indicated in Table 13.2, these polymers can be divided into three general categories—redox polymers, ion-exchange and coordination polymers, and electronically conductive polymers. Redox polymers are polymers that contain electroactive functionalities either within the main polymer chain or in side groups pendant to this chain. The quintessential example is poly(vinylferrocene) (Table 13.2). The ferrocene groups attached to the polymer chain are the electroactive functionality. If fer-

**Table 13.2** Some Representative Polymer Molecules That Have Been Used to Modify Electrode Surfaces

Redox Polymers

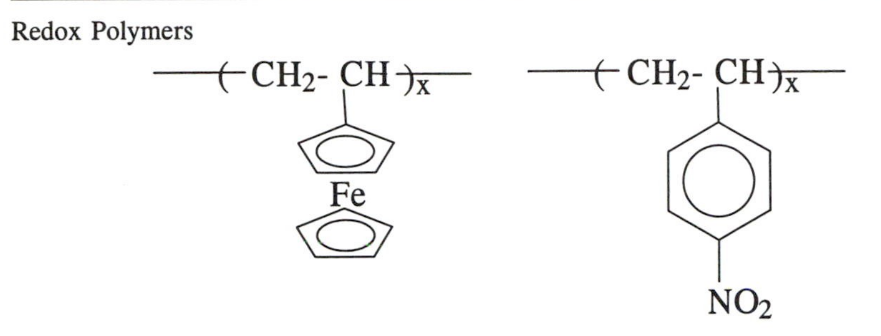


Ion-Exchange and Coordinating Polymers

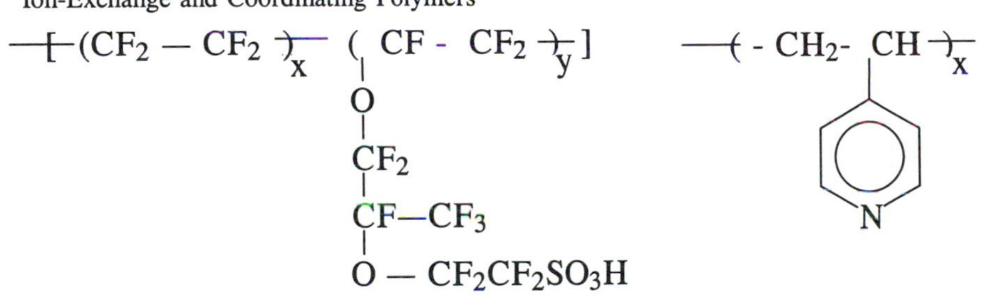


(Nafion)®

Electronically Conductive Polymers

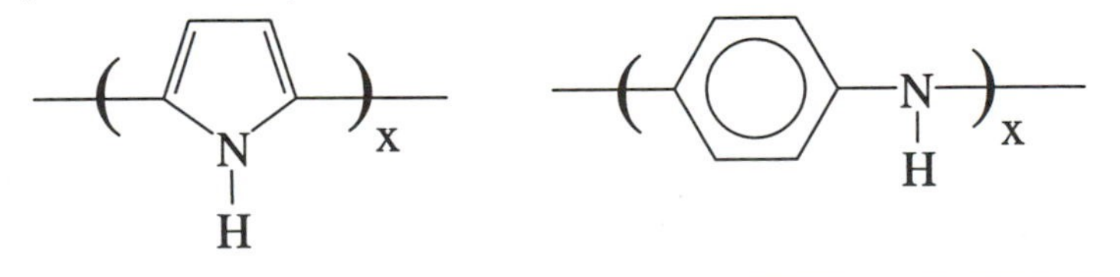

rocene is abbreviated Fc and ferricinium is abbreviated $Fc^+$, this oxidation process can be represented as follows:

$$Fc \rightarrow Fc^+ + e^- \quad (13.3)$$

We have more to say about this electrochemistry in the following section.

Ion-exchange and coordination polymers are not electroactive themselves, but can incorporate electroactive guest molecules. For example, Anson's group showed that films of poly(vinylpyridine) can incorporate *electroactive* coordinatively unsaturated metal complexes via coordination of the metal to the polymer-bound pyridine [22]. Likewise, ion-exchange polymers incorporate *electroactive* counterions via an ion-exchange reaction. The most extensively investigated polymer of this type is du Pont's perfluorosulfonate ionomer, Nafion® (see Table 13.2). Nafion® is a strong acid ion-exchange polymer. The proton can therefore be replaced with an electroactive cation ($M^{n+}$) via

$$n(\text{-}SO_3H)_{polym} + M^{n+}_{soln} \rightarrow [M^{n+}(\text{-}SO_3^-)_n]_{polym} + nH^+_{soln} \quad (13.4)$$

where the subscript "polym" denotes the polymer phase and the subscript "soln" denotes a contacting solution phase that contains the exchanging cation. A large number of electroactive cations can be incorporated into Nafion® films at electrode surfaces via this chemistry [23].

Nafion® film-coated electrodes have an interesting early history. Such electrodes are prepared by applying a solution of the polymer to the electrode surface. However, prior to 1982, only du Pont knew how to dissolve the polymer, and they gave solutions of this polymer to only two U.S. laboratories. These laboratories showed that this polymer was an extremely interesting and versatile material for preparing chemically modified electrodes. Unfortunately, no other laboratory could get its hands on the polymer solution! This situation changed dramatically when Martin's group developed a procedure for dissolving the film form of Nafion® [24]. Shortly thereafter, solutions of this polymer were marketed commercially. Since then, Nafion® film-coated electrodes have become the most extensively investigated chemically modified electrodes. Some examples of these electrodes are presented in Section V.

The third class of polymers used to prepare chemically modified electrodes is the electronically conductive polymers [25]. The polymer chains in this family of materials are themselves electroactive. For example, the polymer redox reaction for polypyrrole (Table 13.2) can be written as follows:

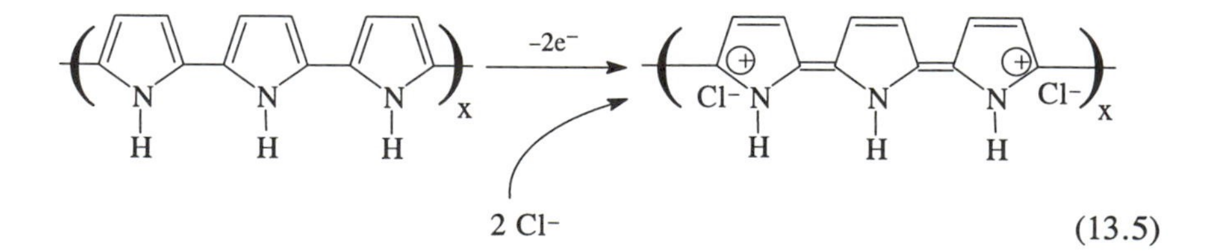


(13.5)

[The charge-balancing (discussed later) anions come from the contacting electrolyte phase.] Note that because the polymer is conjugated, the cationic site created upon oxidation is delocalized along the polymer chain. This may be contrasted with the case of a redox polymer, such as poly(vinylferrocene), where the positive charge created upon oxidation is localized within the ferrocene moiety. The delocalization in the conductive polymers causes these polymers to be electronic conductors (i.e., similar to metals) [25]. This unique class of materials has generated a tremendous amount of excitement during the last decade, and electrochemists have made important contributions to our understanding of these materials. In particular, Arturo F. Diaz at IBM has been instrumental in developing synthetic methods and exploring how structure affects conductivity in these materials [26].

A number of different methods can be used to prepare polymer film-coated electrodes. The simplest is to dip the surface to be coated into a solution of the polymer, remove the electrode from the solution, and allow the solvent to evaporate. While this method is simple, it is difficult to control the amount of material that ends up on the electrode surface. Alternatively, a measured volume of solution can be applied to the surface to be coated. This allows for accurate control of the amount of polymer applied. The polymer film may also be spin-coated onto the electrode surface. Spin-coating is used extensively in the semiconductor industry and yields very uniform film thicknesses.

Polymer films can also be electropolymerized directly onto the electrode surface. For example, Abruña et al. have shown that vinylpyridine and vinylbipyridine complexes of various metal ions can be electropolymerized to yield polymer films on the electrode surface that contain the electroactive metal complex (see Table 13.2) [27]. The electronically conductive polymers (Table 13.2) can also be electrosynthesized from the corresponding monomer. Again, a polymer film that coats the electrode surface is obtained [25]. Electropolymerized films have also been obtained from styrenic, phenolic, and vinyl monomers.

## D. Coating Electrode Surfaces with Inorganic Materials

Clays, zeolites, and other inorganic microcrystalline-structured materials have also been used to modify electrode surfaces [28–30]. These inorganic materials are of interest because they are ion exchangers, like ion-exchange polymers; however, unlike polymers, clays and zeolites can withstand high temperatures and highly oxidizing solution environments. Furthermore, these inorganic materials have well-defined microstructures. For example, clays have a sheetlike structure, and zeolites contain pores and channels of well-defined diameter. Zeolites have long been used as sorbants and catalysts; hence, it only seemed natural to explore the electrochemical properties of electrodes modified with these and related materials. Rollison and her coworkers were the first to explore zeolite-modified electrodes [29].

A considerable amount of work has also been done on an interesting family of transition metal hexacyanometalates, having the general formula $M_k^A[M^B(CN)_6]_x$ where $M^A$ and $M^B$ are transition metals with different formal oxidation numbers [30]. The quintessential member of this family is the material known as Prussian blue, $Fe_4[Fe(CN)_6]_3$. This material has been known since at least the 18th century [31] and, because of its intense blue color, has been used extensively as a pigment. This family of materials forms inorganic polymers that can be coated as thin films on electrode surfaces. These films have interesting electrochemical and optical properties. In particular, they display a property called electrochromism—the ability of a material to change its color upon a change in oxidation state. We discuss an electrochromic device, a "smart window," in the applications section (Section V) of this chapter.

Finally, Majda has investigated a novel inorganic membrane-modified electrode [32]. The membrane used was a microporous alumina prepared by anodizing metallic aluminum in an acidic electrolyte [33]. Majda et al. lined the pores of these membranes with polymers and self-assembled monolayers and studied electron and ion transfer down the modified pore walls to a substrate electrode surface [32]. Martin and his coworkers have used the pores in such membranes as templates to prepare nanoscopic metal, polymer, and semiconductor particles [34].

### E. Langmuir-Blodgett (LB) Methods

In the 1930s, Irving Langmuir and Katharine Blodgett invented a method for creating highly ordered monolayer films at the air/water interface and then transferring these films to substrate surfaces. The LB method entails the use of a molecule with a polar "head group" (e.g., a carboxylate) and a hydrophobic "tail" (e.g., an alkyl chain). When such amphiphilic molecules (called surfactants) are dispersed onto the surface of water, the head groups point down, because they are strongly solvated by water, and the hydrophobic tails point up. The LB method has recently been used to coat electrode surfaces with monolayer and, after multiple transfers, multilayer films [35]. In particular, surfactants with electroactive ions as the head group have been used, and the electrochemical and photoelectrochemical properties of the resulting films have been investigated [35]. The self-assembly and LB methods can, in principle, produce analogous structures on electrode surfaces. The self-assembly method seems, however, to be more versatile and, because of the formation of the chemisorptive chemical bond, should produce a more stable film.

## III. ELECTROCHEMISTRY AT CHEMICALLY MODIFIED ELECTRODES

The objective of this section is to give the reader the basics of how electrochemical reactions occur at chemically modified electrodes. We consider two simple limiting cases. The first is the case of an electroactive monolayer film attached to an electrode surface. The second is the case of an electroactive polymer film that is significantly thicker than a monolayer. Polymer films of this type have sometimes been called "multilayer" films because they can be conceptually divided into a collection of monolayers of the polymer stacked on top of each other. This is an artificial concept because there are no real layers (like the layers in plywood) within the polymer film; however, the concept can be useful in thinking about, and modeling, the electrochemistry of such films.

## A. Electrochemistry at a Monolayer Film-Coated Electrode

Let us assume that a gold-disk electrode has been coated with a "self-assembled" monolayer (discussed earlier) of the following thiol:

$$HS\text{-}(CH_2)_x\text{-}Fc \tag{13.6}$$

Clearly, this thiol can be used to attach ferrocene groups (Fc) to the Au electrode surface. If we abbreviate the surface-confined ferrocene group as -Fc, it should be possible to drive the following surface redox reaction:

$$\text{-}Fc \rightarrow \text{-}Fc^+ + e^- \tag{13.7}$$

This would be accomplished by immersing the chemically modified electrode, a reference electrode, and an auxiliary electrode into an appropriate electrolyte solution (e.g., 0.1 M $NaClO_4$ in acetonitrile). The potential difference between the modified electrode (the working electrode) and the reference would then be adjusted to a value appropriate to drive this reaction, using a commercially available potentiostat, and the resulting anodic current would be measured.

A schematic drawing of this surface-confined oxidation process is shown in Figure 13.1. Note that the film is represented as a rather disordered monomolecular layer of attached -Fc groups. Initially, all of these groups are in the unoxidized -Fc state (Fig. 13.1A); however, upon application of a suitably positive potential to the working electrode, these -Fc sites give up their electrons to the substrate electrode (Fig. 13.1B) until all of the sites have been converted to the corresponding oxidized $\text{-}Fc^+$ form (Fig. 13.1C). This electrochemical process could be driven using any of a variety of electrochemical methods. We consider the case of cyclic voltammetry (Chap. 3), since this is the most popular electrochemical method in use today.

A hypothetical cyclic voltammogram for the surface-confined redox reaction is shown in Figure 13.2. On the forward (positive-going) scan, an anodic wave is observed; this wave is associated with oxidation of the -Fc groups (Fig. 13.1). This wave rises to a peak and then decays to zero current at potentials positive of the peak. This points out the first difference between a surface-confined redox reaction and a redox reaction in which the electroactive species is dissolved in solution. If the ferrocene were dissolved in the electrolyte solution

A

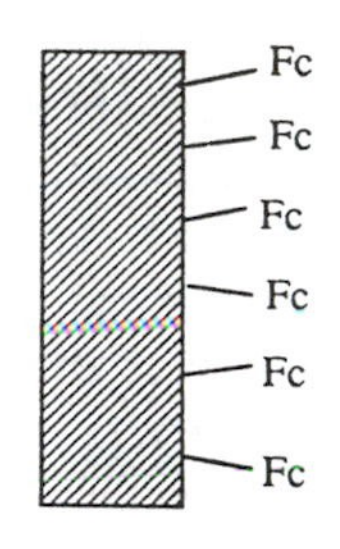


Electrode Solution

B

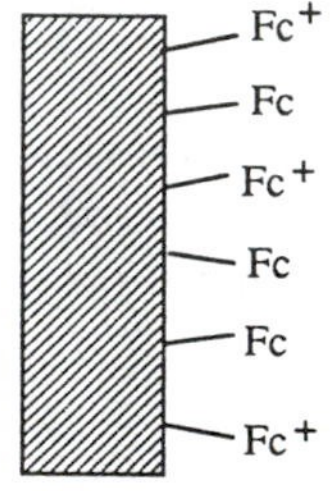


C

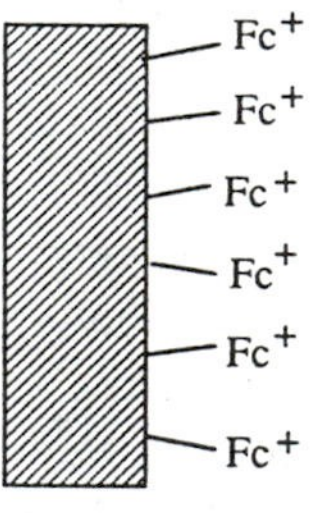


**Figure 13.1** Schematic diagram of the electrochemical oxidation of a monolayer of a surface-confined ferrocene (Fc) derivative: (A) before, (B) during, and (C) after oxidation.

(i.e., free Fc), the anodic current would gradually decrease at potentials positive of the peak (see Chap. 3). This gradual tailing of the current results because diffusion continuously brings fresh Fc to the electrode surface from the bulk solution. In the surface-confined case, there is no free Fc in the solution phase; therefore, once the -Fc has been completely oxidized, the current goes to zero. Finally, when the potential scan is reversed, a cathodic peak is observed;

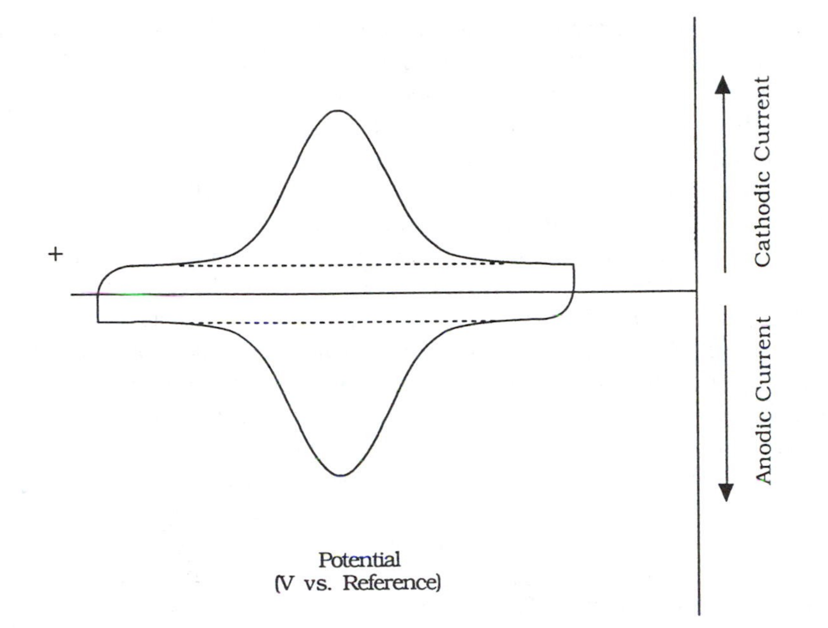


**Figure 13.2** Hypothetical cyclic voltammogram for the surface oxidation process illustrated in Figure 13.1.

this peak is associated with the reduction of the -$Fc^+$ generated on the forward scan.

## B. Electrochemistry at a "Multilayer" Film-Coated Electrode

Assume that a disk-shaped electrode (gold, platinum, carbon, etc.) has been coated with a film of poly(vinylferrocene) (Table 13.2). This can be accomplished by dissolving the polymer in chloroform, applying a drop of the solution to the electrode surface, and allowing the solvent to evaporate. The electrochemistry of the resulting polymer film-coated electrode can be investigated using the same electrochemical cell and equipment as described in the previous example.

The key feature of this polymer film-coated electrode is that it will always be thicker than the monolayer film considered in the previous example. Indeed, the film thickness can be controlled at will by varying the volume of solution applied and/or the concentration of polymer in the solution. The film thickness could be anywhere from tens of angstroms to hundreds of microns, or even thicker. This "multilayer film" situation is illustrated schematically in Figure 13.3A. Note that like the previous case, there are Fc groups sitting essentially right on the electrode surface. However, unlike the previous case, there are also Fc groups at distances removed from the electrode. The key questions that arise are: Can the Fc groups that are at distances quite removed from the electrode surface become electrochemically oxidized (Eq. 13.3)? And if so, by what mechanism does this oxidation occur?

The answer to the first question is yes—such multilayer films can be electrochemically oxidized and reduced using simple electrochemical experiments such as cyclic voltammetry. One mechanism by which such an oxidation process could occur is called "electron hopping." This mechanism was first proposed by Kaufman and Engler [36] and is illustrated schematically in Figure 13.3. Figure 13.3A shows the distribution of Fc sites in the polymer film before the electrochemical oxidation process is initiated. In Figure 13.3B the oxidation process has been initiated through appropriate control of the working electrode potential. Note that in complete analogy to the monolayer film case (Fig. 13.1), the first thing that happens is that the Fc sites sitting directly on the electrode surface become oxidized. This produces a layer of $Fc^+$ sites immediately adjacent to the electrode surface (Fig. 13.3B).

Initially, all the Fc sites further removed from the electrode surface are still in the unoxidized Fc state. Electrons can, however, "hop" from these distant Fc sites to the $Fc^+$ sites at the electrode surface (Fig. 13.3C). Electron hopping occurs via a well-known chemical process called "electron self-exchange" whereby the reduced half of a redox couple (Fc) simply gives its electron to an oxidized counterpart ($Fc^+$). This reaction can be written as

$$Fc_1^+ + Fc_2 \rightarrow Fc_1 + Fc_2^+ \qquad (13.8)$$

where the subscripts "1" and "2" are provided to show that the ferrocene molecule that was initially oxidized (subscript 1) ends up getting reduced, and the ferrocene molecule that was initially reduced (subscript 2) ends up getting oxidized.

As shown in Figure 13.3C, the result of this *first* electron hop is that a layer of reduced Fc sites is regenerated at the electrode surface. What will happen to these Fc sites? Clearly, they will give up their electrons to the electrode, regenerating a layer of $Fc^+$ at the electrode surface (Fig. 13.3D). Note the net result is that we now have two layers of $Fc^+$ sites near the electrode but a lot of Fc sites still in the bulk of the polymer film. What happens next? Two more electron hops and another electron transfer will occur to yield three layers of $Fc^+$ sites at the electrode surface (Fig. 13.3E). If we repeat this electron-hop/electron-transfer process many times, we will ultimately end up with a completely oxidized film (Fig. 13.3F).

The preceding discussion shows that a multilayer polymer film can be electrochemically oxidized or reduced via this process called electron hopping. There is a feature of this process, however, that we have not yet considered.

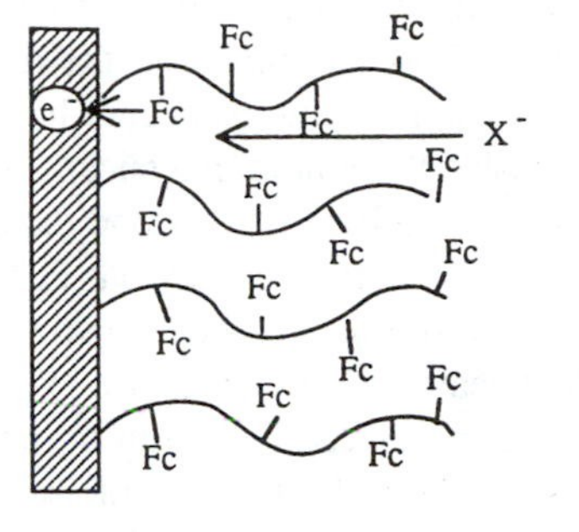


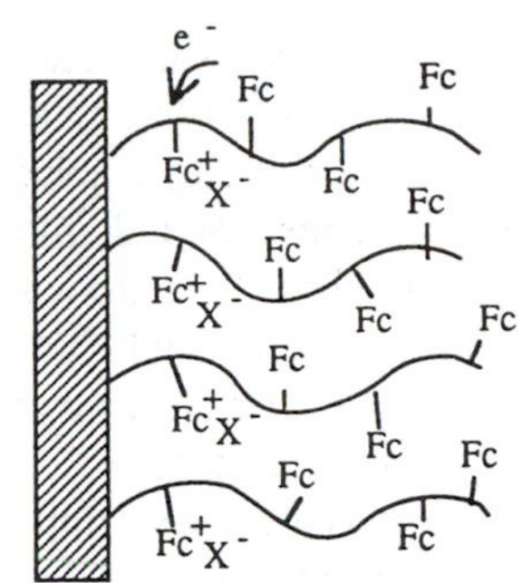


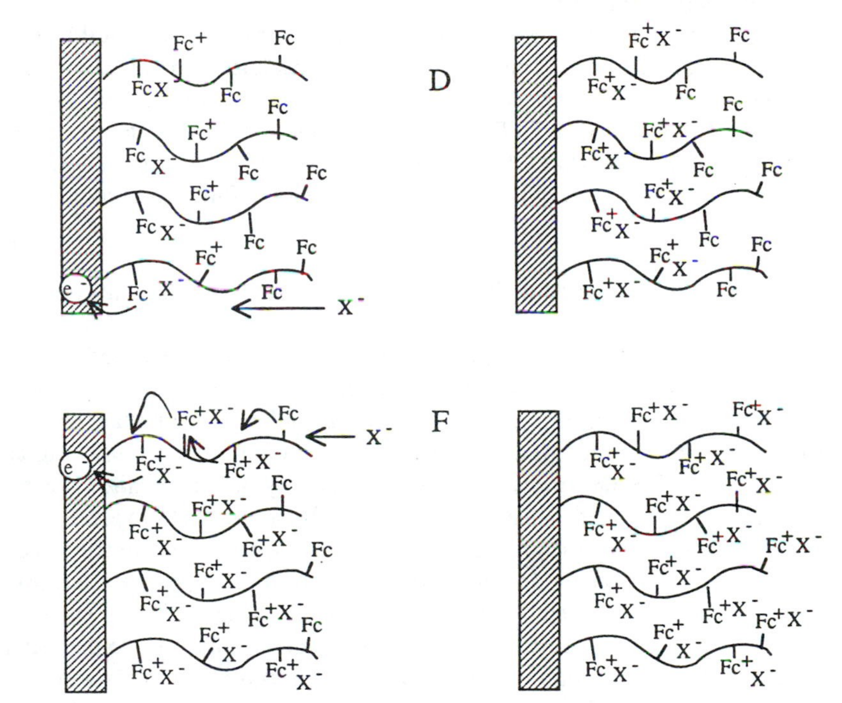


**Figure 13.3** Schematic diagram of the electrochemical oxidation of a multilayer film containing covalently attached ferrocene (Fc) sites. $X^-$ represents anions from the supporting electrolyte that neutralize the $Fc^+$ sites created. Parts A through F represent various times during the oxidation process.

In Figure 13.3, we show that every time an electron leaves an Fc to create an $Fc^+$ in the polymer film, an anion must simultaneously come into the film from the contacting electrolyte solution. This cotransport of anions occurs because all phases (e.g., the polymer film) have the desire to remain electrically neutral. (This is called the "electroneutrality principle." Lightning is a good example of the consequences of violating this basic principle.) Hence, when an $Fc^+$ is created in the film (by electron transfer to the electrode), an anion must also enter the film to ensure that the film remains electrically neutral (Fig. 13.3). Incidentally, you may be wondering how the contacting electrolyte-solution phase remains electrically neutral if anions are lost to the polymer-film phase. To answer this question, consider what is happening at the auxiliary electrode in the solution phase (Chap. 6).

Electron hopping is not the only mechanism by which a multilayer film containing an electroactive species can be electrochemically oxidized or reduced. Consider an electrode surface that has been coated with a film of the cation-exchange polymer Nafion® (Table 13.2). Let us assume that the electroactive cation $Fe^{3+}$ has been ion-exchanged into this Nafion® film (e.g., Eq. 13.4). In principle, the $Fe^{3+}$ in this film can be reduced via the following redox reaction:

$$Fe^{3+} + e^- \rightarrow Fe^{2+} \tag{13.9}$$

How will the $Fe^{3+}$ groups at sites distant from the electrode surface be reduced in this film?

It is important to point out that the ion-exchange polymer Nafion® is fundamentally different from poly(vinylferrocene). In poly(vinylferrocene), the electroactive Fc groups are *covalently attached* to the polymer chain and, therefore, cannot move through the polymer film (by either diffusion or migration; see Chap. 2). The electroactive groups in this polymer are "nailed down." In contrast, the $Fe^{3+}$ sites in the ion-exchange polymer are not covalently bound. Hence, they can, in principle, move through the polymer film in the same way that the charge-balancing anions moved through the poly(vinylferrocene) film in Figure 13.3. Hence, the $Fe^{3+}$'s in the portions of the Nafion® film removed from the electrode surface could simply diffuse through the film to the electrode surface where the reduction to $Fe^{2+}$ (Eq. 13.9) would occur. This would make the electrochemical reduction of the $Fe^{3+}$ at the Nafion® film-coated electrode completely analogous to the reduction of a solution of $Fe^{3+}$ at an uncoated electrode.

So, we have two possibilities for the case of the ion-exchange polymer film-coated electrode: reduction could occur by physical diffusion of the $Fe^{3+}$ through the film, or reduction could occur via electron hopping through the film. How can we know which process is operative? Electrochemists have devoted a considerable amount of research effort to answering this question. The answer clearly depends on the nature of the polymer, the extent of swelling of the polymer by solvent, the size of the electroactive counterion, and the rate of the relevant electron self-exchange reaction (Eq. 13.8). A simple-minded (first-order) answer to this question is as follows: If the magnitude of the diffusion coefficient for the electroactive ion in the polymer film is large and if the rate of the self-exchange reaction is low, physical diffusion will predominate. In contrast, if the diffusion coefficient for the electroactive ion is small and the self-exchange rate is high, electron hopping will predominate.

## IV. CHARACTERIZATION AND ANALYSIS OF CHEMICALLY MODIFIED ELECTRODES

We begin with the most routine characterization methods—electrochemical methods. We then discuss various instrumental methods of analysis. Such instrumental methods can be divided into two groups: ex situ methods and in situ methods. In situ means that the film on the electrode surface can be analyzed while the film is emersed in an electrolyte solution and while electrochemical reactions are occurring on/in the film. Ex situ means that the film-coated electrode must be removed from the electrolyte solution before the analysis. This is because most ex situ methods are ultra-high-vacuum techniques. Examples include x-ray photoelectron spectroscopy [37], secondary-ion mass spectrometry [38,39], and scanning or transmission electron microscopies [40]. Because ex situ methods are now part of the classical electrochemical literature, we review only in situ methods here.

### A. Electrochemical Methods

The first question an electrochemist might ask about a chemically modified electrode is: How much electroactive species is present in the film on the electrode surface? Cyclic voltammetry can provide an answer to this question.

Let us begin by considering the self-assembled monolayer film discussed in the previous section. We define the amount of electroactive, surface-confined -Fc in this film as $\Gamma_{Fc}$, which has units of moles of -Fc confined per square centimeter of gold electrode area. $\Gamma_{Fc}$ can be determined from the voltammogram shown in Figure 13.2. The x axis in this voltammogram is potential; however, since the potential is scanned at a constant rate ($\nu$ volts per second), any voltage, V, along the x axis can be converted to a corresponding time, t, via $t = V/\nu$. Hence, the x axis can be easily converted to a time axis. Since the y axis is current, this would make the area under the forward anodic peak the charge corresponding to the oxidation of the surface-confined -Fc. Let's call this charge or area $Q_{Fc}$. $Q_{Fc}$ can be easily obtained by electronic integration (which is available on many modern potentiostats) or by cutting the voltammogram from the paper it is recorded on and weighing. $\Gamma_{Fc}$ can then be calculated from $Q_{Fc}$ via Faraday's law,

$$\Gamma_{Fc} = Q_{Fc}/nFA \tag{13.10}$$

where F is Faraday's constant and A is the electrode area.

There is one caveat that should be mentioned. Note that both the anodic and cathodic peaks in Figure 13.2 sit on top of flat (ideally) background currents. These are the capacitive currents associated with double charging (see Chaps. 3 and 12). We do not want to include these capacitive currents in our determination of $\Gamma_{Fc}$. We want to integrate only the current associated with the oxidation of Fc, that is, the "Faradaic" current. We have delineated the area of the curve associated with the Faradaic current by extrapolating the background current (dashed line underneath the peak). The area we will use to determine $\Gamma_{Fc}$ is the area under the peak but above this dashed line.

The voltammogram shown in Figure 13.2 is for the "reversible" or "Nernstian" case. As discussed in Chapters 2 and 3, this means that at the scan rate employed, the rate of electron transfer is sufficiently high that the surface concentrations of -Fc and -$Fc^+$ are always at equilibrium with the applied electrode potential. For a surface-bound redox couple, the Nernstian case is characterized by a difference in potentials between the anodic and cathodic peaks ($\Delta E_{pk}$) of zero volts; that is, the cathodic wave sits right on top of the anodic wave (Fig. 13.2). It is worth mentioning that the formal potential (see Chaps. 2 and 3) for the Nernstian case is simply the potential of the anodic and cathodic peak.

If the rate of electron transfer is low (or the scan rate is too high), electron transfer will not be able to adjust the surface concentrations of -Fc and -$Fc^+$ to values that are at equilibrium with the applied potential (quasireversible or totally irreversible case, see Chap. 3). In this case, the anodic peak and the cathodic peaks will not be at the same potential; that is, $\Delta E_{pk}$ will be greater than zero volts. Kinetic information about the surface-bound redox couple can be obtained from such quasireversible or irreversible voltammograms. For example, methods for obtaining the standard heterogeneous rate constant (see Chap. 2) for the surface-confined redox couple have been developed [41,42].

Let us turn our attention now to a multilayer film-coated electrode. What information can electrochemical experiments provide about such multilayer films? First, we can again obtain the quantity of electroactive sites in the polymer film; however, whether this can be done using the simple voltammetric method outlined above depends on the film thickness and on the rate of charge transport in the film. If the multilayer film is thin and/or if the rate of charge transport is high, it may be possible to oxidize or reduce (i.e., electrolyze) all of the electroactive sites in the multilayer film during the voltammetric scan. If so, in analogy to Figure 13.2, the current at potentials beyond the voltammetric peaks will decay rapidly to zero. If this condition is satisfied, the quantity of electroactive sites in the film can be obtained from the areas under the peaks as per the monolayer film.

On the other hand, if the film is thick and/or if the rate of charge transport is low, only a fraction of the electroactive sites in the film will become electrolyzed during the voltammetric scan. The simplest way to think about this case is that the diffusion layer created at the electrode surface during the voltammetric scan extends only a fraction of the way into the film. If this is the case, the voltammetric currents at potentials beyond the peaks will show gradual diffusional tails exactly like those observed in a cyclic voltammogram for a redox-active molecule dissolved in a solution (see Chap. 3). The total quantity of electroactive sites in the film cannot be obtained from such voltammograms because only a fraction of the sites is electrolyzed during the scan. The total quantity of electroactive sites in such films can, however, be obtained using a simple coulometric method [43].

Scan rate clearly plays a role in determining whether all the electroactive sites in the film are electrolyzed during the voltammetric scan. If the scan rate is very low (e.g., 1 mV $s^{-1}$), then it might be possible to electrolyze the entire film during the scan, in which case a monolayer-type voltammogram (Fig. 13.2) would be obtained. In contrast, if the scan rate is high (e.g., 10 V $s^{-1}$), it is more likely that only those sites in close proximity to the electrode surface will be electrolyzed and a diffusional-type voltammogram will be obtained. In fact, it is often possible, through variation of the scan rate, to observe both monolayer-type (low scan rates) and diffusional-type (high scan rates) voltammograms for the same film.

A simple criterion has been developed for predicting when a monolayer-type and when a diffusional-type voltammogram might be observed [1]. Let us assume that the thickness of the polymer film in question is d cm and that the "apparent diffusion coefficient" associated with charge transport in this film is $D_{app}$ $cm^2/s^{-1}$; the apparent diffusion coefficient (discussed later) is a measure of the rate of charge transport in the electroactive polymer film. We can define the "time constant" for electrolysis of the film as $d^2/D$. (Note that $d^2/D$ has units of seconds.) We can define a time-scale parameter for the voltammetric scan as $RT/nF\nu$, where R is the gas constant, T is the temperature in degrees Kelvin, n is the number of moles of electrons transferred per mole of redox-active sites, F is the Faraday constant, and $\nu$ is the scan rate in V $s^{-1}$. (Note that $RT/nF\nu$ also has units of seconds.) If $RT/nF\nu \ll d^2/D$, then only a fraction of the redox-active sites in the film will be electrolyzed and the voltammetric wave will appear diffusional in nature. In contrast, if $RT/nF\nu \gg d^2/D$, then the entire film will be electrolyzed during the scan and a monolayer-type voltammogram will be obtained.

What is this parameter called the apparent diffusion coefficient, $D_{app}$? As indicated earlier, it is a measure of the rate of charge transport in a multilayer film at an electrode surface. What $D_{app}$ *physically* means depends on the type of film being investigated. If the electroactive species is free to diffuse through the multilayer film, and if this diffusional process controls the rate of charge transport, then $D_{app}$ is just the diffusion coefficient for the redox-active species in the polymer. Hence, in this case, $D_{app}$ is completely analogous to the diffusion coefficient for an electroactive molecule dissolved in a solution (see Chap. 2). In contrast, if charge transport in the film occurs by electron hopping, then $D_{app}$ is related to the rate of this electron hopping process. Electrochemists have measured $D_{app}$ values for all types of films on electrode surfaces. Various electrochemical methods have been used, including chronoamperometry (Chap. 3), chronocoulometry (Chap. 3), rotating-disk voltammetry (Chap. 3), microelectrode voltammetry (Chap. 12), and ac impedance methods (Chap. 5).

Finally, it is worth commenting on the shape of the voltammetric wave obtained for a multilayer film on an electrode surface. If the rate of electron transfer is high (Nernstian case) and if the criterion $RT/nFr \ll d^2/2D$ is satisfied, the film voltammogram should be completely analogous to a voltammogram for a redox species dissolved in solution, for example, diffusional anodic and cathodic peaks with $\Delta E_{pk} = 58/n$ mV (see Chap. 3). Such ideal waves are rarely observed for multilayer film-coated electrodes. For example, $\Delta E_{pk}$ values in excess of 58/n mV are almost always observed. This could be caused by film resistance or slow heterogeneous kinetics. Alternatively, it is possible that the film is chemically heterogeneous, and as a result, the local chemical environments "seen" by the electroactive species in the film are not the same. Digital simulations (see Chap. 20) have been useful in exploring these contributions to non-Nernstian behavior in multilayer films on electrode surfaces [43].

## B. Methods Based on Optical Spectroscopy

The electrochemical behavior of a modified electrode ultimately depends on structural details at the molecular level. For example, the molecular-level interaction between the redox site in the film and the solvent from the contacting solution phase might play an important role in the electrochemical response. Molecular-level details are often difficult to infer from electrochemical methods alone, but do lend themselves to spectroscopic analyses. In recent years there has been an explosion of new spectroscopic techniques for characterizing modified electrodes and the electrode-solution interface in general [44,45]. In this section, we review some of these "spectroelectrochemical" methods.

### *UV/Visible Transmission and Reflection Spectroscopy*

The simplest spectroscopic experiments are based on measurements of the amount of light transmitted by a sample. Such transmission-mode experiments are possible with chemically modified electrodes if the substrate electrode is transparent. In fact, tin oxide or indium tin oxide ("ITO") electrodes are quite transparent over the entire visible spectrum [46]. If the modifying layer contains visible-range chromophores, then examining potential-induced chemical changes in the film should be straightforward. For example, Elliott and Redepenning [47] coated a tin oxide electrode with a polymer that contained electroactive metal complex sites (Fig. 13.4A). The ruthenium complexes have seven oxidation states, ranging from +2 to −4. After establishing the potentials at which the oxidation states change (via cyclic voltammetry), these authors collected visible absorption spectra as a function of applied potential. Typical data are shown in Figure 13.4B.

Unfortunately, most electrode materials are not transparent, and it is therefore necessary to employ reflectance methods. The reflectance, R, is defined as the ratio of the intensity of reflected light to that of the incident light. If there are absorbing chromophores at the reflecting surface, then R will be attenuated. Reflectance data are typically processed by ratioing the measured reflectance R to the reflectance of some conveniently defined standard, $R_o$. $R_o$ might be the reflectance of a particular reference sample such as the uncoated electrode. Alternatively, $R_o$ might be the reflectance at the coated electrode at some reference potential, $E_{ref}$ [48]. In either case, a reflectance spectrum (analogous to an absorption spectrum) can be constructed by plotting log $(R/R_o)$ versus wavelength or energy. Alternatively, the differential reflectance $\Delta R/R = (R - R_o)/R_o$ can be plotted on the y axis [48].

Reflectance experiments are generally more complicated than transmittance-mode experiments. This is because the reflected intensity depends not only on the identity and concentration of chromophores in the film, but also on the angles of incidence and polarization of the incident light [49]. For example, light polarized with its electric field parallel to the reflecting surface is usually reflected more efficiently than light polarized perpendicular to the surface. Polarization is particularly important at metal electrodes; if an adsorbed molecule's transition dipole is aligned parallel to the metal surface, then light polarized in the plane of the surface will be "blind" to that particular absorption mode. This occurs because the incident light induces an image field in the metal that interferes destructively at the surface. Figure 13.5 demonstrates this effect quite dramatically [50].

### *Infrared Reflectance Spectroscopy*

Infrared spectroscopy can provide a great deal of information on molecular identity and orientation at the electrode surface [51–53]. Molecular vibrational modes can also be sensitive to the presence of ionic species and variations in electrode potential [51,52]. In situ reflectance measurements in the infrared spectrum engender the same considerations of polarization and incident angles as in UV/visible reflectance. However, since water and other solvents employed in electrochemistry are strong IR absorbers, there is the additional problem of reduced throughput. This problem is alleviated with thin-layer spectroelectrochemical cells [53].

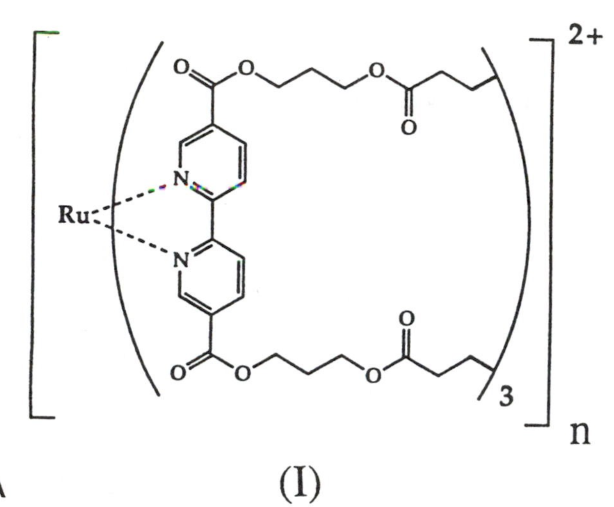

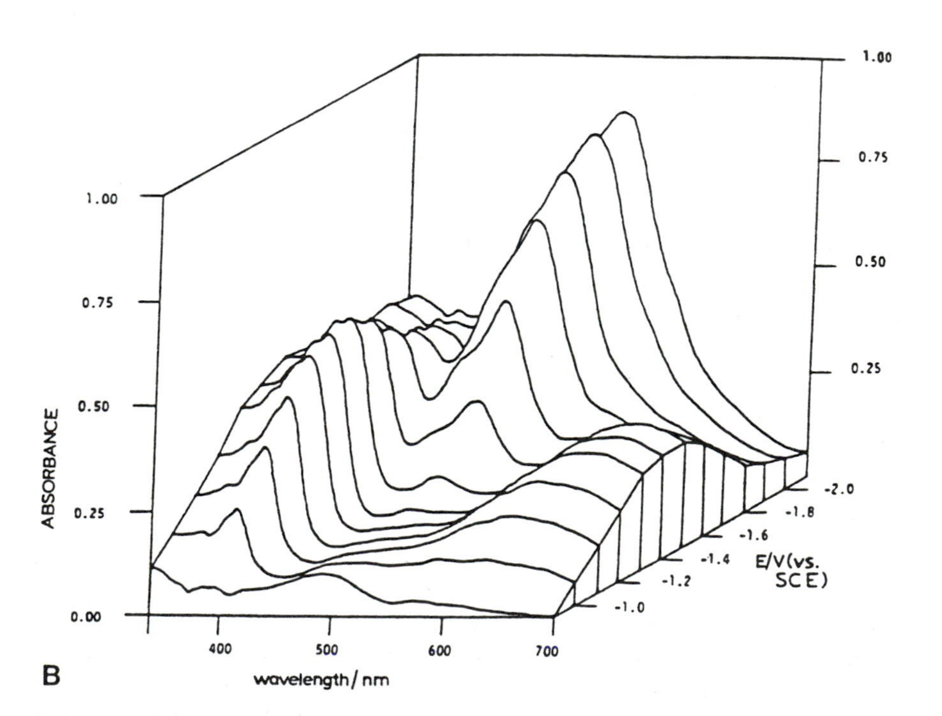


**Figure 13.4** (A) Structure of poly-(Ru(II) tris-[5,5′-dicarbo(3-acrylatoprop-1-oxy)-2,2′-bipyridine]). (B) UV/visible spectra of (I) thermally polymerized on $SnO_2$ transparent electrode as a function of applied potential. Film thickness ~0.2 μm. Potential measured against $Ag/Ag^+$ (0.1 M). [From Ref. 47.]

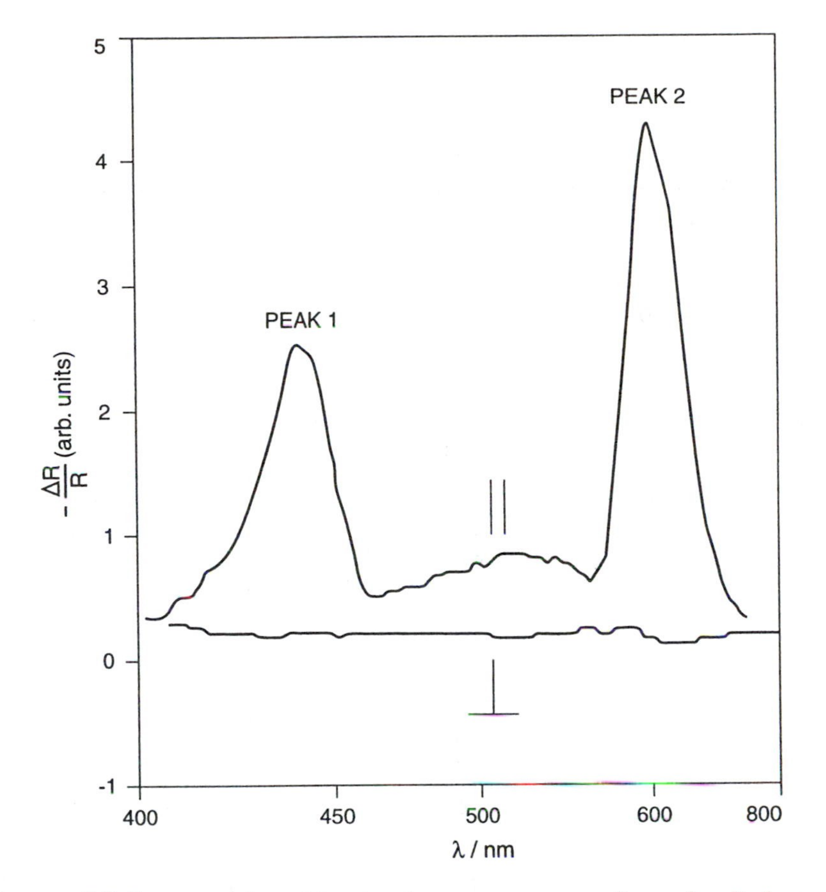


**Figure 13.5** Potential modulated reflectance spectrum of *p*-aminonitrobenzene (PANB) on platinum (solution phase 0.5 mM $Na_2SO_4$ + 0.05 mM PANB). Applied dc 0.44 V vs. SHE. Modulation amplitude ±50 mV. Modulation frequency 33 Hz. Incidence angle 65°. || signifies incident polarization parallel to incident plane and perpendicular to electrode surface. ⊥ signifies incident polarization perpendicular to incident plane (hence parallel to electrode surface). [From Ref. 50.]

A recent example of the application of in situ FTIR reflectance spectroscopy to modified electrodes can be found in the work of Korzeniewski and her group [54] on Pt electrodes coated with films of the polymer polyaniline. Figure 13.6 shows the reflectance spectrum (in ΔR/R units) of a polyaniline film-coated electrode as a function of applied potential. As the film becomes more oxidized, the band at ~1320 $cm^{-1}$ is shifted to higher energies, a result consistent with an increasing C-N bond strength upon oxidation [54]. The interac-

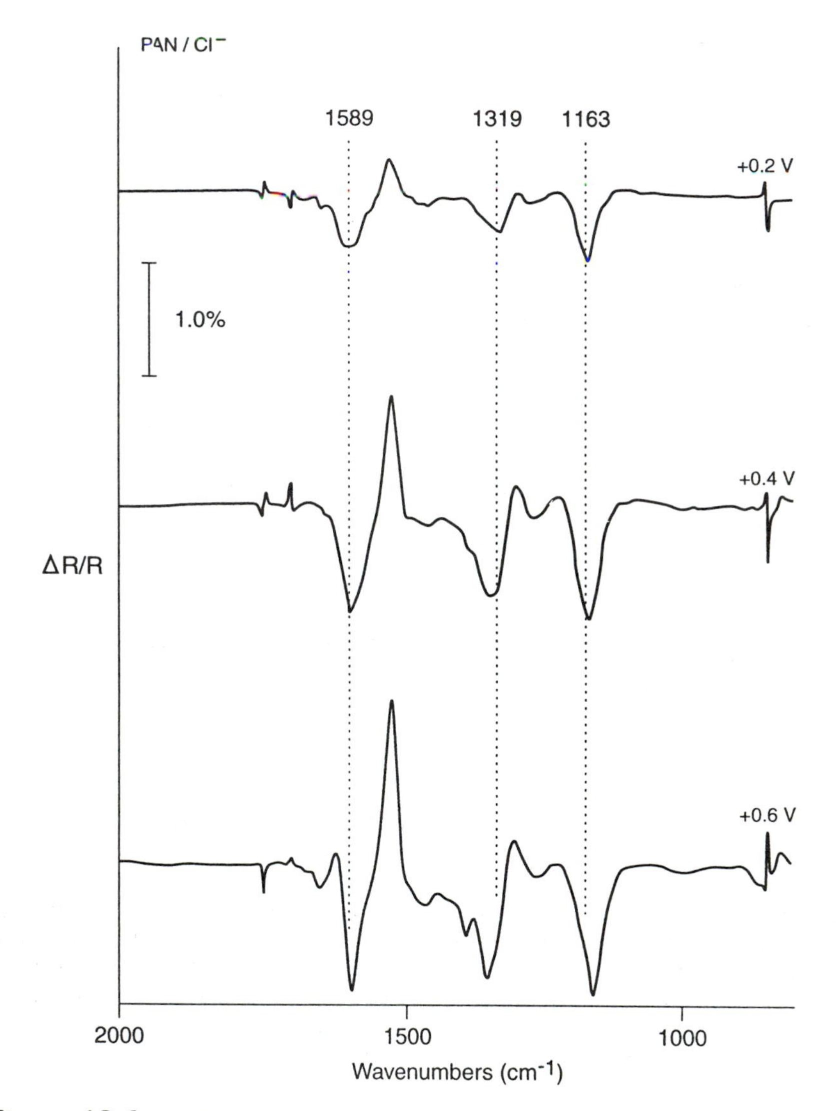


**Figure 13.6** In situ infrared reflectance spectra of polyaniline-modified electrodes in solution containing 0.10 M HCl. Spectra collected at applied potentials indicated at right of curves, and normalized according to Eq. 13.12 using $R_o$ collected at -0.1 V. Potentials measured against saturated calomel electrode. [From Ref. 54.]

tion between the dopant anion and the polymer was also investigated in this study. For example, molecular anions such as $ClO_4^-$ and $SO_4^{2-}$ are IR-active, and thus detectable in an FTIR reflectance experiment. Korzeniewski found that vibrational bands for anions with the polyaniline film were blue-shifted relative to the same anions in the bulk solution. This was ascribed to an ionic interaction between the dopant anions and the polyaniline film [54].

*Raman Spectroscopy*

Raman spectroscopy can offer vibrational information that is complementary to that obtained by IR. Furthermore, since the Raman spectrum reveals the "backbone" structure of a molecular entity [55], it is particularly useful in the examination of polymer film-coated electrodes. There are also some distinct advantages over in situ IR. For example, both the mid and far infrared spectral regions can be accessed with the same instrumental setup (in IR spectroscopy, these two regions typically require separate optics) [55]. Second, solvents such as water and acetonitrile are weak Raman scatterers; thus the solvent medium does not optically obscure the electrode surface as it does in an in situ IR experiment.

While signal loss due to the solvent may be low, it is also true that the Raman signal from the modified electrode surface may be weak, particularly for monolayer films. Hence, many workers exploit the surface-enhanced Raman scattering (SERS) effect [56–58]. Small metal particles, or protrusions on a roughened metal substrate, can amplify the incident electric field and lead to strong electromagnetic enhancements in the Raman signal of molecules adsorbed on their surfaces [57,58]; enhancements can range from three to six orders of magnitude. For Raman experiments employing excitation wavelengths in the visible spectrum, electromagnetic SERS enhancements occur on a few metals such as Ag, Au, and Cu [58].

The Raman scattering signal can also be enhanced if one chooses an excitation wavelength corresponding to an electronic transition of the molecule of interest. This resonance Raman effect can enhance the signal by two to six orders of magnitude [55]. Hence, exploiting both the surface enhancement and the molecular resonance leads to extremely low detection limits (e.g., picomolar and below).

Figure 13.7 shows an example of surface-enhanced resonance Raman (SERRS) of a modified electrode. Cotton and her group examined the dye Nile Blue A in solution, at a glassy carbon electrode, and at a roughened silver electrode [59]. By integrating currents in cyclic voltammetric measurements (discussed earlier), they found that multilayer films (50 to 100 monolayers) of the dye formed on the glassy carbon electrode. In contrast, monolayer films formed at the roughened silver electrode. In spite of the vast difference in coverage, the Raman signal from the roughened silver electrode is comparable in magnitude.

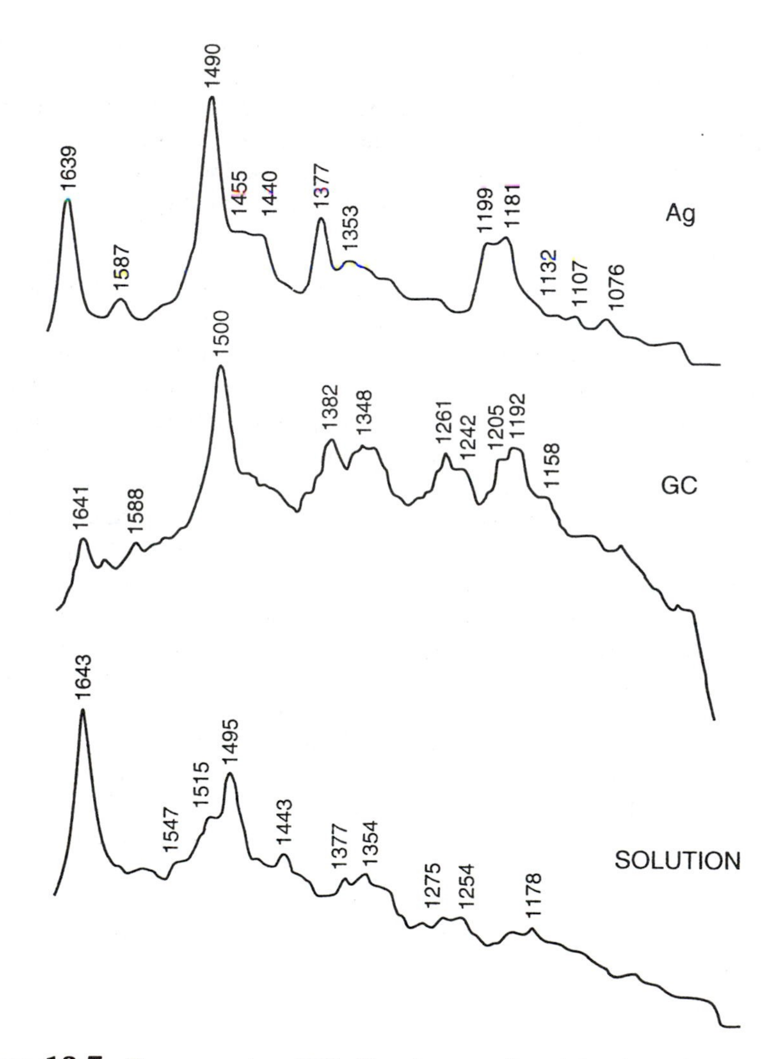


**Figure 13.7** Raman spectra of Nile Blue A on roughened Ag electrode, glassy carbon electrode (GC), and in solution; pH = 9.0. Excitation wavelength = 488 nm. Laser power for Ag, 10 mW; for GC, 200 mW; for solution, 180 mW. Raman shifts in $cm^{-1}$ shown above peaks. [From Ref. 59.]

*Ellipsometry*

We mentioned earlier that the intensity of reflected light depends strongly on the angles of incidence and polarization of the incident light. In fact, reflection can cause a change in polarization. For example, if the incident light is linearly polarized at some angle between 0 ( $\perp$ ) and 90 ( $||$ ) degrees, the reflected light may be *elliptically polarized* on account of the different degrees to which the $\perp$ and $||$ components are reflected. While UV/visible and infrared reflectance methods probe only reflected intensities, ellipsometry is a technique that measures the precise polarization state of the reflected light. The key advantage of ellipsometry is that one can determine both the thickness and the optical constants of a film [60–64]. A full treatment of this method is beyond the scope of this chapter. However, the reader is directed to a recent review by Collins and Kim [60] and the classic text by Azzam and Bashara [61].

*Nonlinear Optical Techniques*

A general objective in any in situ spectroscopic technique is to maximize the signal that arises specifically from the electrode surface. Nonlinear optical techniques such as second-harmonic generation (SHG) and sum frequency generation (SFG) are of interest because they involve optical signals that *by definition* can only arise at the electrode-solution interface [65].

Materials that have a nonzero second-order susceptibility will produce light at twice the incident frequency. The magnitude of this effect is small, and has been a practical consideration only since the advent of lasers. If the symmetry of a crystal or other medium is such that it has a center of inversion, no SHG effect will be observed. However, surfaces by their very nature break this inversion symmetry. Hence, an SHG signal may arise at the electrode-solution interface even though both bulk phases may be considered centrosymmetric [66]. The magnitude of the SHG signal is sensitive to surface conditions (e.g., electrode potential, ionic or molecular adsorption, etc.). Surface spectroscopy is also feasible since the SHG signal will be enhanced if either the incident frequency ($\omega$) or SHG ($2\omega$) corresponds to an electronic absorption of a surface species [66].

Most SHG studies involve incident energies in the visible or near-infrared spectrum. Infrared SHG studies are hindered by the current lack of sufficiently sensitive IR detectors. However, the sum frequency generation (SFG) technique allows one to obtain surface-specific vibrational spectra. In SFG, two lasers are focused on the sample surface, one with a fixed frequency in the visible and one with a tunable range of IR frequencies. The sample surface experiences the sum of these frequencies. When the frequency of the infrared component corresponds to a molecular vibrational mode, there is an increase in the total SHG signal, which is detected at the visible frequency [66]. The application of such techniques to in situ electrochemistry and modified electrodes is still in its infancy, but the outlook for nonlinear optical methods is certainly promising [66].

## C. In Situ X-Ray Methods

X-ray diffraction (XRD) is a routine method for determining crystal lattice parameters and molecular structure. The application of XRD to modified electrodes has been limited, particularly for actual molecular structure determination. First, such experiments presuppose a single-crystal electrode substrate. Second, the small amount of sample present in a thin film on an electrode surface means that the scattered intensities will be restrictively low, at least for commonly available x-ray sources [67]. However, if one is fortunate enough to have access to a synchrotron, such experiments are quite feasible. For details, the reader is directed to an excellent review by Toney and Melroy [68]. On the other hand, powder diffraction experiments with Cu or Mo $K_\alpha$ anode sources are straightforward, and can yield lattice-constant data in situ. For example, Ikeshoji and Iwasaki measured lattice constants for Prussian blue films (discussed earlier) on gold electrode surfaces [69].

A method that requires a synchrotron source but not single-crystal electrodes is extended x-ray absorption fine structure (EXAFS). Since the synchroton produces a continuum of x-ray energies, one can collect an absorption spectrum analogous to those obtained in UV or IR spectroscopy. However, since x-radiation corresponds to core level electronic transitions, the spectra are characterized by a sharp absorption "edge" rather than broad bands. At energies above the edge, oscillations in the absorbance occur because of backscattering of the ejected photoelectron off the nearest atom neighbors. The magnitude and period of the oscillations depend on the distance and identity of the nearest neighbors. By fitting the spectra to a model, it is possible to determine the distance and number of near neighbors [70]. Abruña and his group have used in situ EXAFS to examine platinum electrodes modified with polymers containing Ru(II) or Os(II) complex sites [71].

## D. Scanning Tunneling and Atomic Force Microscopy

Scanning tunneling (STM) was invented a decade ago by Binnig and Rohrer [72], and was first applied to the solid-liquid interface by Sonnenfeld and Hansma in 1986 [73]. Since then, there have been numerous applications of STM to in situ electrochemical experiments [74–76]. Because the STM method is based on tunneling currents between the surface and an extremely small probe tip, the sample must be reasonably conductive. Hence, STM is particularly suited to investigations of redox and conducting polymer-modified electrodes [76,77].

Atomic force microscopy (AFM) is one of many techniques that rely on a force interaction (e.g., electrostatic or magnetic) between the probe tip and the surface [78]. AFM relies on the attraction or repulsion forces that operate at atomic dimensions. Like STM, AFM may be done in situ, but offers an advantage in that the sample may be conducting or insulating. For example, Murray and his group used AFM to monitor the electrodeposition of poly(phenylene oxide) [79], and later to actually alter the polymer surface with the probe tip (Murray calls this "nanodozing") [80].

### E. Quartz Crystal Microbalance

Since the original work of Sauerbrey nearly four decades ago [81], the quartz crystal microbalance (QCM) has been applied in various contexts for the detection of mass changes at the nanogram level. The heart of the QCM method is a specially cut quartz crystal that oscillates at some resonant frequency when an alternating voltage is applied across its thickness. Adsorption of foreign material (i.e., atoms and molecules) on the surface of the crystal leads to minute but detectable changes in the resonant frequency. The change in frequency $\Delta f$ is related to the change in mass $\Delta m$ via the Sauerbrey equation, $\Delta m = \Delta f^2 C$, where C is a constant that contains the resonant frequency before mass addition, and the sheer modulus and density of the quartz crystal [82]. The first application of the QCM to in situ electrochemistry was made by Nomura and Iijima, who monitored silver deposition from micro- to nanomolar solutions of silver nitrate [83].

The electrochemical quartz crystal microbalance (EQCM) simply employs one of the two oscillator drive electrodes as the working electrode [82]. EQCM is particularly suited to modified electrode studies where oxidation or reduction of the film on the electrode surface causes ions to enter or leave the film [84,85]. For example, Varineau and Buttry used EQCM to monitor mass changes in poly(vinylferrocene) films on gold during concurrent cyclic voltammetric scans [84]. Figure 13.8 shows the simultaneous voltammetric and EQCM results. Upon oxidation, indicated by the anodic wave in curve A, the frequency of the quartz substrate decreases (curve B). This indicates that counteranions ($PF_6^-$) are entering the film as the Fc sites are oxidized to $Fc^+$. When the film is reduced (reverse scan), the frequency increases to its original value, indicating, as might be expected, that upon reduction the anions are expelled from the film.

## V. APPLICATIONS OF CHEMICALLY MODIFIED ELECTRODES

After twenty years of research, practical devices based on chemically modified electrodes are now finding their way into the commercial marketplace. We review several recent examples here. In addition, chemically modified electrodes have always been used as tools in fundamental scientific investigations. We discuss one very recent example.

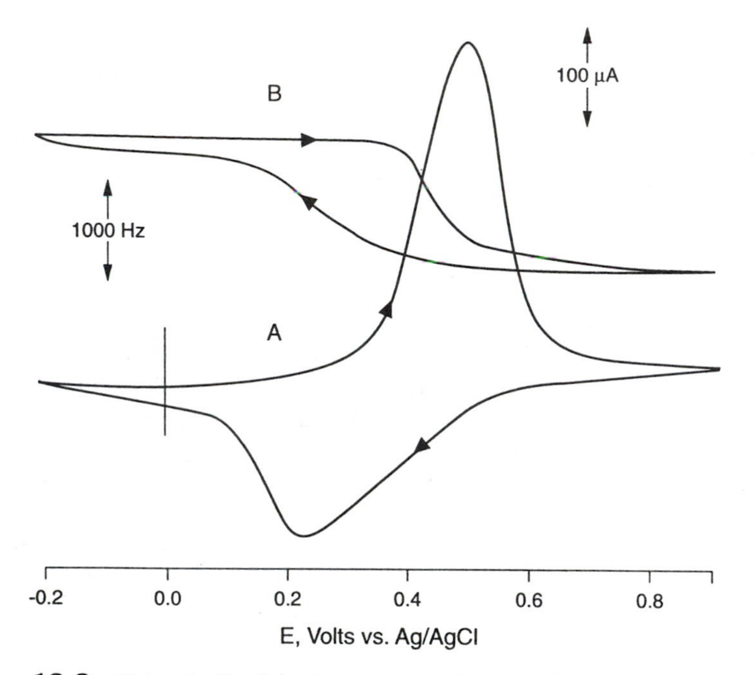


**Figure 13.8** Curve A: Cyclic voltammogram of polyvinylferrocene (PVF) on gold in 0.1 M $KPF_6$. Scan rate 10 mV $s^{-1}$. Curve B: EQCM frequency curve obtained simultaneously with Curve A. [From Ref. 84.]

## A. Chemical Sensors

A chemical sensor is a device that provides the concentration of a particular chemical species (called the analyte) in a sample solution. For example, your doctor might want to know the concentration of $Na^+$ in your blood, or an environmental scientist might want to know the concentration of DDT in a sample of river water. We have used these two examples in which a sensor might be useful to illustrate an important point—most sample solutions of "real world" interest are extremely complicated mixtures containing many different chemical components; your blood is a good example. The sensor must somehow recognize and act upon the analyte of interest and ignore all other chemical species in the sample solution. This is an extremely challenging problem.

One of the goals in the chemically modified electrode research area has been to develop new types of electrochemical sensors. Several review articles have recently been published on this subject [86,87]. Our intent is not to provide another review of this voluminous literature. Rather, we would like to introduce the reader to the concepts behind the use of chemically modified electrodes as electrochemical sensors. As we will see, the key to developing new sensors is building chemical sensitivity into the film covering the electrode surface so that only the analyte of interest is detected by the substrate electrode.

We begin by pointing out that this concept of covering an electrode surface with a chemically selective layer predates chemically modified electrodes. For example, an electrode of this type, the Clark electrode for determination of $O_2$, has been available commercially for about 30 years. The chemically selective layer in this sensor is simply a Teflon-type membrane. Such membranes will only transport small, nonpolar molecules. Since $O_2$ is such a molecule, it is transported to an internal electrolyte solution where it is electrochemically reduced. The resulting current is proportional to the concentration of $O_2$ in the contacting solution phase. Other small nonpolar molecules present in the solution phase (e.g., $N_2$) are not electroactive. Hence, this device is quite selective.

Research into chemically modified electrodes has led to a number of new ways to build chemical selectivity into films that can be coated onto electrode surfaces. Perhaps the simplest example is the use of the polymer Nafion® (see Table 13.2) to make selective electrodes for basic research in neurophysiology [88]. Starting with the pioneering investigations by Ralph Adams, electrochemists have become interested in the electrochemical detection of a class of amine-based neurotransmitters in living organisms. The quintessential example of this class of neurotransmitters is the molecule dopamine, which can be electrochemically oxidized via the following redox reaction:

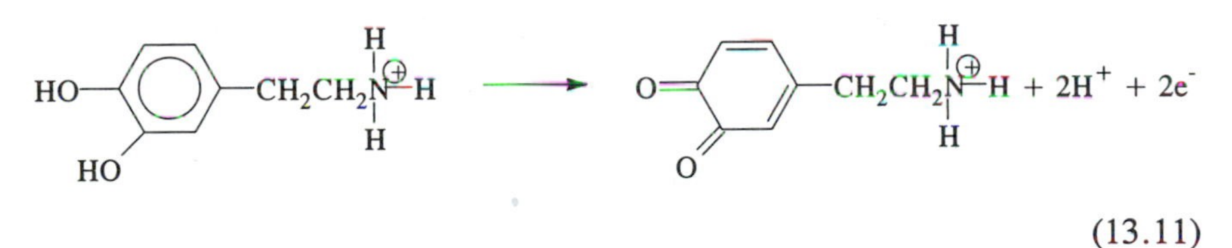


(13.11)

Electrochemists have shown that this molecule can be detected in the brains of living rats by surgically implanting electrodes into the rats' brains [88].

There is, however, a major problem with this analysis. The cerebral fluid analyzed also contains relatively high concentrations of ascorbate. Ascorbate is oxidized at roughly the same potential as dopamine. Hence, ascorbate interferes with the determination of dopamine in the in vivo electrochemical analysis. The solution to this problem, coating the surface of the electrode to be implanted with a thin Nafion® film, came out of a collaborative research effort between Martin's group and Adams' group [89]. Note that at physiological pH values, dopamine is a cation and ascorbate is an anion. Because Nafion® is a cation-exchange polymer, it transports dopamine but rejects ascorbate. Hence, the Nafion®-coated electrode provides the selectivity required for in vivo analysis of dopamine and other cationic neurotransmitters. The use of Nafion®-coated electrodes has since become standard procedure for such investigations.

Nafion® provides a very rudimentary form of chemical selectivity: selectivity based on the charge (cationic vs. anionic) of the analyte molecule. How can we build true molecule-recognition capability into a film that can be coated onto an electrode surface? One route that appears quite promising is to borrow from Mother Nature. Living systems have evolved a set of very selective chemical reagents called enzymes. An enzyme is a protein that recognizes and binds a specific molecule and then catalyzes some chemical transformation of that molecule. Because of this molecular-recognition capability, various highly successful enzymatic methods of chemical analysis have been developed [90]. From the very inception of the chemically modified electrode field, it seemed likely that enzymes could provide the molecule-recognition function that would lead to a new family of highly selective electrochemical sensors. Electrochemists have been vigorously pursuing this goal for 20 years.

The enzyme-based sensor that has received the most attention is the glucose sensor based on glucose oxidase. Glucose oxidase catalyzes the two-electron, two-proton oxidation of glucose (GU) to gluconolactone (GL). The electron/proton acceptor is the cofactor flavin adenine dinucleotide (FAD). In living systems, the electrons and protons collected from glucose by FAD are passed on to $O_2$ to generate hydrogen peroxide. This chemistry can be represented as follows:

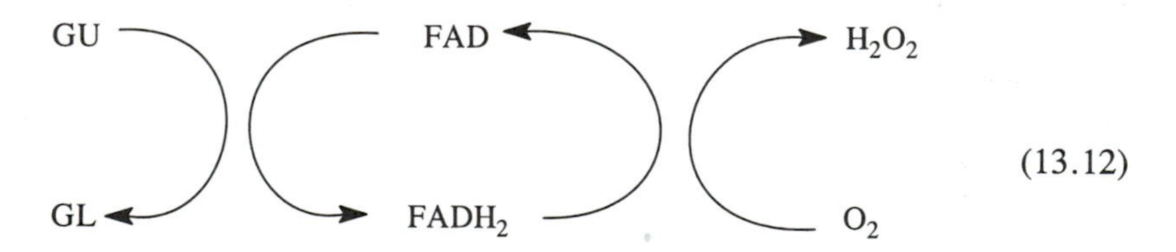


(13.12)

The most obvious way to incorporate this chemistry into an electrochemical sensor is to immobilize the enzyme onto an electrode surface and use this electrode to oxidize the hydrogen peroxide produced. An enzymatic sensor of this type was first prepared by Guilbault and Lubrano [91]. Numerous variations on this theme have since appeared, and sensors that employ this electrochemistry are now commercially available.

Detection of the enzymatic reaction (and therefore of glucose) via oxidation of $H_2O_2$ has some disadvantages. For example, a potential in excess of +0.7 V (vs. Ag/AgCl) must be applied to the substrate electrode in order to oxidize $H_2O_2$. If other oxidizable species are present in the sample (e.g., ascorbate, dopamine, etc.), these species will also be oxidized at this high positive potential. This creates the possibility for interference from these species. Hence, the molecular-recognition advantage of the enzyme will be lost. In 1984, Hill et al. proposed a simple solution to this problem [92]. Instead of using $O_2$ as the electron acceptor (Eq. 13.12), they employed an acceptor that could be reoxidized at lower potentials. This allows the sensor to be operated at lower potentials and thus decreases the possibility of interference from oxidizable species in the analyte solution.

Derivatives of ferrocene are most often used as electron acceptors for glucose sensors of this type. In this case, the electron transfer reactions can be written as shown in Equation 13.13:

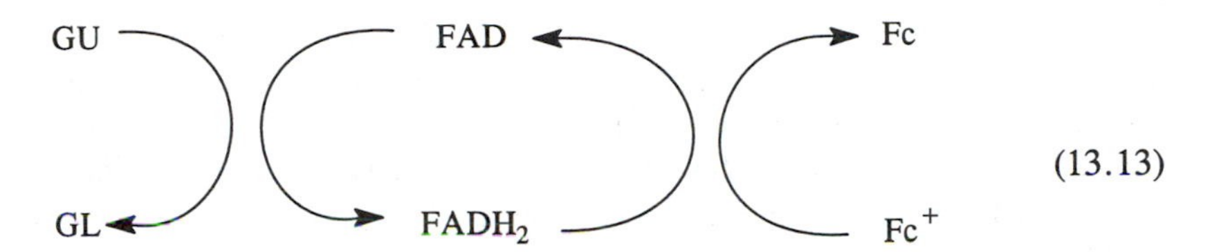


(13.13)

Glucose sensors based on this electrochemistry are now commercially available. Furthermore, it seems likely that this concept will soon be expanded to other types of enzyme-based sensors. Hence, sensor development is proving to be one of the great success stories of the chemically modified electrode research area.

## B. Energy-Producing Devices

There are two primary types of electrochemical energy-producing devices: batteries and fuel cells. Both of these devices convert chemical energy into electrical energy via electrochemical reactions. The difference between a battery and a fuel cell is that a battery contains all of the chemicals required for the energy-producing reaction within the device package. Hence, the advantage of a battery is that it is a completely self-contained energy-producing device. In contrast, a fuel cell does not store its chemical reactants within the device itself; the reactants are supplied from external tanks. Hence, the advantage of the fuel cell is that it will run continuously as long as it is supplied with the appropriate chemical fuels. Fuel cells have been used extensively to provide electrical power for the U.S. space program.

The concepts of modified electrodes have contributed tremendously to battery and fuel cell development. For example, a schematic of an interesting new type of fuel cell, the polymer electrolyte fuel cell, is shown in Figure 13.9. Hydrogen gas is supplied to the anode and is oxidized via

$$H_2 \rightarrow 2H^+ + 2e^- \tag{13.14}$$

Oxygen gas is supplied to the cathode and is reduced via

$$O_2 + 4H^+ + 4e^- \rightarrow 2H_2O \tag{13.15}$$

Note that hydronium ion is produced at the anode and consumed at the cathode. The purpose of the polymer membrane that separates these electrodes is to transport $H^+$ from the anode to the cathode. Hence, a cationically conductive ion-exchange polymer is used as the polymer electrolyte. The polymer Nafion® that has been discussed in so many other places in this chapter is often used (see Table 13.2).

The fuel cell in Figure 13.9 can be conceptually viewed as a combination of a *Nafion® film-coated cathode* and a *Nafion® film-coated anode*. Hence, the fuel cell is, in essence, a combination of two chemically modified electrodes. This idea is, in fact, more than just a concept, because electrochemical investigations of Nafion® film-coated electrodes have been used to obtain fundamental chemical and electrochemical information that is relevant to the operation of such devices [93]. For example, the kinetics of $O_2$ reduction in fuel cells can be investigated at such modified electrodes; the solubility and diffusion coefficient for $O_2$ in Nafion® and the proton conductivity of this membrane material can also be determined. Chemically modified electrodes have made analogous contributions to battery development.

## C. Electrochromic Devices

In Section II.D, we introduced the concept of electrochromism—the ability of a material to change color upon a change in its oxidation state. Electrochemists

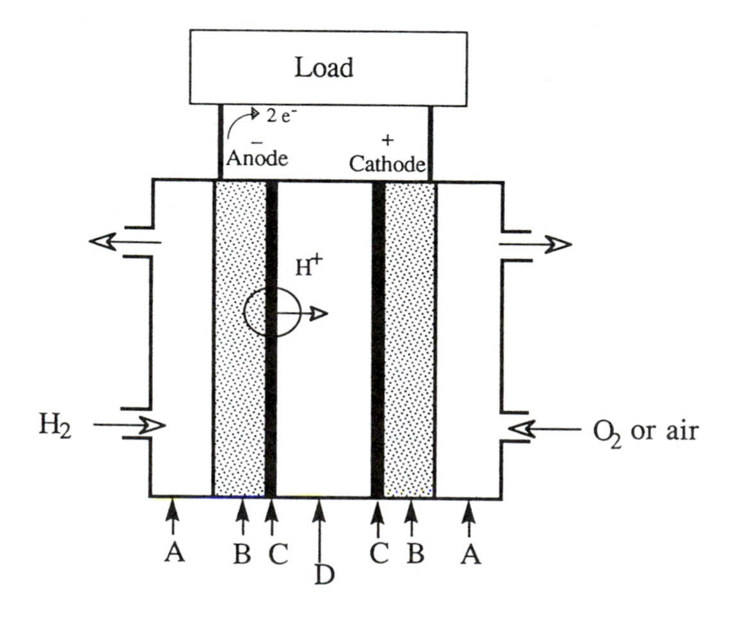


**Figure 13.9** Schematic diagram of a polymer electrolyte fuel cell: (A) gas manifolding, (B) porous graphite block, (C) active catalyst layer (dispersed Pt and Teflon binder), and (D) polymer electrolyte.

are interested in using electrochromism to make devices that change color on command. For example, let us assume that you are sitting in your car at the end of a rainstorm. During the storm the sky was dark, but now the clouds have cleared and it is extremely bright and sunny. Wouldn't it be nice if you could just turn a knob on your dashboard and tint your windows a little so that the sun would not hurt your eyes or interfere with your driving? This is the concept behind the "smart window," a concept being vigorously pursued by electrochemists at various companies around the world.

A general schematic for a smart window is shown in Figure 13.10. This device is, quite literally, two chemically modified electrodes sandwiched together. In this case, the films coating the electrode surfaces are electrochromic materials. A polymer electrolyte, analogous to that used in the fuel cell discussed earlier, is sandwiched between these two electrochromic material-coated electrodes. In a recent example of this concept by Habib and Maheswari of General Motors Research Laboratories [94], the cathodic electrochromic material was a tungsten oxide and the cathodic electrochromic material was the material Prussian blue, discussed in Section II of this chapter. It seems likely that electrochromic cells will soon find their way into the commercial marketplace.

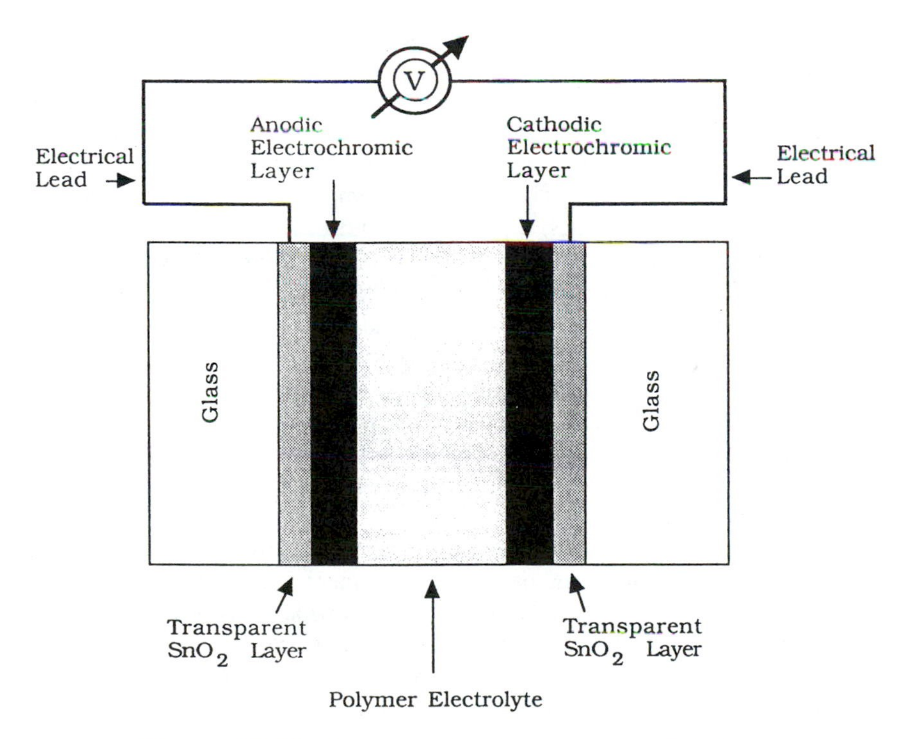


**Figure 13.10** Schematic diagram of a "smart window."

### D. Fundamental Chemistry

In addition to leading to new types of electrochemical devices, modified electrodes have been used as tools in fundamental scientific investigations. The objective of such investigations is simply to obtain fundamental scientific information. A good example is the use of modified electrodes to study the fundamentals of electron transfer (ET) reactions. For example, Christopher Chidsey has used self-assembled monolayers (alkanethiols) with terminal ferrocene functions to probe ET processes at the electrode-solution interface [95]. Because the electroactive sites are bound to the electrode, there is no need to separate kinetic and diffusional components of the measured current. Also, because the electrode potential can be varied, the driving force $\Delta G°$ for the reaction can be changed easily. (In homogeneous ET, one member of the donor/acceptor pair must be changed to change the driving force.) Hence, Chidsey is able to quantitatively evaluate Marcus' theory [96], which postulates a quadratic relation between $\Delta G°$ and the activation Gibbs energy $\Delta G^*$. Finally, the donor–acceptor distance can be varied simply by changing the length of the alkane group [95].

## VI. CONCLUSIONS

Chemically modified electrodes have been a dominant research theme in electrochemistry for two decades. As discussed in this chapter, many tangible benefits have accrued from this research effort. An important intangible benefit has also accrued. Through this research effort, electrochemists have raised the scientific profile of their research area and have helped break down barriers between traditional scientific disciplines. For example, because of research on chemically modified electrodes, a polymer scientist might now use such a device to synthesize an interesting new material. An experimental physicist might then conduct exotic spectroscopic measurements on this material, and a theoretical physicist might try to calculate the band structure of this material. Or because of this research effort, a physiologist might use a chemically modified electrode for in vivo investigations of a bioactive substance, or a medical doctor might use a modified-electrode sensor to help diagnose disease. Penetration of research ideas across traditional scientific boundaries is essential to modern science. The chemically modified electrode research area is a good example of such a borderless research topic.

What about the future? The chemically modified electrode field is now quite mature. A large fraction of the necessary fundamental work has already been done. Hence, further development of commercial applications of modified-electrode technology must dominate future research efforts. Since such product development research is applied in nature, much of this research should be done in the private sector and/or by chemical and electrochemical engineers. This is called technology transfer, and is the logical step for this next, extremely important phase of the overall research effort. So our question to the current and future generations of fundamentally oriented electrochemists is: Will another research topic ever galvanize our endeavors the way chemically modified electrodes have? If so, what is it?

## ACKNOWLEDGMENTS

The authors acknowledge the support of the Office of Naval Research and the Air Force Office of Scientific Research.

## REFERENCES

1. I. Rubinstein, in *Applied Polymer Analysis and Characterization*, Vol. II (J. Mitchell, Jr., ed.), Hanser, Munich, 1991.
2. A. Merz, in *Topics in Current Chemistry*, Vol. 152, Springer-Verlag, Berlin, 1990.
3. E. Barendrecht, *J. Appl. Electrochem. 20*:175 (1990).
4. H. D. Abruña, *Coordination Chem. Rev. 86*:135 (1988).
5. R. W. Murray, A. G. Ewing, and R. A. Durst, *Anal. Chem. 59*:379A (1987).
6. C. E. D. Chidsey and R. W. Murray, *Science 231*:25 (1986).
7. L. R. Faulkner, *Chem. Eng. News 62(9)*:28 (1984).
8. W. J. Albery and A. R. Hillman, *Annual Reports on the Progress of Chemistry*, Section C, Physical Chemistry, Vol. 78, p. 377, Royal Society of Chemistry, London, 1981.
9. A. W. Adamson, *Textbook of Physical Chemistry*, Academic Press, New York, 1973, pp. 664–670.
10. R. F. Lane and A. T. Hubbard, *J. Phys. Chem. 77*:1401 (1973).
11. R. F. Lane and A. T. Hubbard, *J. Phys. Chem. 77*:1411 (1973).
12. A. P. Brown and F. C. Anson, *Anal. Chem. 49*:1589 (1977).
13. L. H. Dubois and R. G. Muzzo, *Anal. Rev. Phys. Chem. 43*:437 (1992).
14. M. M. Walczak, C. Chung, S. M. Stole, C. A. Widrig, and M. D. Porter, *J. Am. Chem. Soc. 113*:2370 (1991).
15. R. W. Murray, *Acc. Chem. Res. 13*:135 (1980).
16. D. C. Bookbinder and M. S. Wrighton, *J. Electrochem. Soc. 130*:1080 (1983).
17. B. F. Watkins, J. R. Behling, E. Kariv, and L. L. Miller, *J. Am. Chem. Soc. 97*:3549 (1975).
18. L. Netzer and J. Sagiv, *J. Am. Chem. Soc. 105*:674 (1983).
19. H. Lee, L. J. Kepley, H.-G. Hong, and T. E. Mallouk, *J. Am. Chem. Soc. 110*:618 (1988).
20. L. L. Miller and M. R. Van de Mark, *J. Am. Chem. Soc. 100*:3223 (1978).
21. A. Merz and A. J. Bard, *J. Am. Chem. Soc. 100*:3222 (1978).
22. N. Oyama and F. C. Anson, *J. Am. Chem. Soc. 101*:739 (1979).
23. M. N. Szentirmay and C. R. Martin, *Anal. Chem. 56*:1898 (1984).
24. C. R. Martin, T. A. Rhodes, and J. A. Ferguson, *Anal. Chem. 54*:1639 (1982).

25. C. R. Martin and L. S. Van Dyke, in *Molecular Design of Electrode Surfaces* (R. W. Murray, ed.), John Wiley, New York, 1992.
26. A. F. Diaz and J. Bargon, in *Handbook of Conducting Polymers* (T. A. Skotheim, ed.), Marcel Dekker, New York, 1986, Vol. 1, pp. 81–116.
27. H. D. Abruña, P. Denisevich, M. Umana, T. J. Meyer, and R. W. Murray, *J. Am. Chem. Soc. 103*:1 (1981).
28. A. Fitch, *Clays and Clay Minerals 38*:391 (1990).
29. D. R. Rollison, *Chem. Rev. 90*:867 (1990).
30. K. Itaya, I. Uchida, and V. D. Neff, *Acc. Chem. Res. 19*:162 (1986).
31. Brown, J. *Philos. Trans. 33*:17 (1724).
32. C. A. Goss, C. J. Miller, and M. Majda, *J. Phys. Chem. 300*:377 (1991).
33. A. Despic and V. Parkhutik, in *Modern Aspects of Electrochemistry* (J. Bockris, R. E. White, and B. E. Conway, eds.), Plenum, New York, 1989, Chap. 6.
34. C. R. Martin, *Science*, (1994), pp. 266, 1961.
35. J. S. Facci, *Tech. Chem. 22*:119 (1992).
36. F. B. Kaufman and E. M. Engler, *J. Am. Chem. Soc. 101*:549 (1979).
37. R. Kötz, in *Spectroscopic and Diffraction Techniques in Interfacial Electrochemistry* (C. Gutiérrez and C. Melendres, eds.), Kluwer, Dordrecht, 1990, pp. 409–438.
38. F. Chao, J. L. Baudin, M. Costa, and P. Lang, *Makromol. Chem., Macromol. Symp. 8*:173 (1987).
39. P. C. Lacaze and G. Tourillon, *J. Chim. Phys. Phys.-Chim. Biol. 76*:371 (1979).
40. Y.-M. Tsou, H.-Y. Liu, and A. J. Bard, *J. Electrochem. Soc. 135*:1669 (1988).
41. A. T. Hubbard and F. C. Anson, in *Electroanalytical Chemistry*, Vol. 4 (A. J. Bard, ed.), Marcel Dekker, New York, 1970, pp. 129–214.
42. E. Laviron, *J. Electroanal. Chem. 101*:19 (1979).
43. C. R. Martin, I. Rubinstein, and A. J. Bard, *J. Am. Chem. Soc. 104*:4817 (1982).
44. H. D. Abruña, ed., *Electrochemical Interfaces: Modern Techniques for In-Situ Interface Characterization*, VCH, New York, 1991.
45. C. Gutiérrez and C. Melendres, eds., *Spectroscopic and Diffraction Techniques in Interfacial Electrochemistry*, Kluwer, Dordrecht, 1990.
46. N. Winograd and T. Kuwana, *Electroanalytical Chem. 7*:1 (1974).
47. C. M. Elliott and J. G. Redepenning, *J. Electroanal. Chem. 197*:219 (1986).
48. W. Plieth, in Ref. 45, pp. 223–260.
49. K. Ashley and S. Pons, *Trends Anal. Chem. 5*:263 (1986).
50. P. H. Schmidt and W. J. Plieth, *J. Electroanal. Chem. 201*:163 (1986).
51. K. Ashley, F. Weinert, M. G. Samant, H. Seki, and M. R. Philpott, *J. Phys. Chem. 95*:7409 (1991).
52. K. Ashley, F. Weinert, and D. L. Feldheim, *Electrochim. Acta 36*:1863 (1991).
53. (a) K. Ashley, *Spectroscopy 5*:22 (1990); (b) A. Bewick, K. Kunimatsu, and B. S. Pons, *Electrochim. Acta 25*:465 (1980); (c) S. M. Stole, D. D. Popenoe, and M. D. Porter, in Ref. 44, pp. 339–410.
54. D. Seeger, W. Kowalchyk, and C. Korzeniewski, *Langmuir 6*:1527 (1990).
55. H. H. Willard, L. L. Merritt, J. A. Dean, and F. A. Settle, *Instrumental Methods of Analysis*, Wadsworth, Belmont, CA, 1988.

56. M. Fleischmann, P. J. Hendra, and A. J. McQuillan, *Chem. Phys. Lett. 26*:163 (1974).
57. R. K. Chang and T. E. Furtak, eds., *Surface Enhanced Raman Spectroscopy*, Plenum, New York, 1982.
58. J. E. Pemberton, in Ref. 44, pp. 193–263.
59. F. Ni, M. Feng, L. Gorton, and T. M. Cotton, *Langmuir 6*:66 (1990).
60. R. W. Collins and Y.-T. Kim, *Anal. Chem. 62*:887A (1990).
61. R. M. A. Azzam and N. M. Bashara, *Ellipsometry and Polarized Light*, North Holland, Amsterdam, 1977.
62. J. O'M. Bockris, M. A. V. Devanathan, and A. K. N. Reddy, *J. Electroanal. Chem. 6*:61 (1963).
63. A. Redondo, E. A. Ticianelli, and S. Gottesfeld, *Mol. Cryst. Liq. Cryst. 160*:185 (1988).
64. Y.-T. Kim, R. W. Collins, K. Vedam, and D.L. Allara, *J. Electrochem. Soc. 138*:3266 (1991).
65. (a) G. L. Richmond, in Ref. 44, pp. 265–337; (b) Y. R. Shen, *Principles of Nonlinear Optics*, Wiley, New York, 1984.
66. Y. R. Shen, in Ref. 45, pp. 281–311.
67. J. Robinson, in Ref. 45, pp. 313–341.
68. M. F. Toney and O. R. Melroy, in Ref. 44, pp. 55–129.
69. T. Ikeshoji and T. Iwasaki, *Inorg. Chem. 27*:1123 (1988).
70. M. J. Albarelli, J. H. White, G. M. Bommarito, M. McMillan, and H. D. Abruña, *J. Electroanal. Chem. 248*:77 (1988).
71. H. D. Abruña, in Ref. 44, pp. 1–54.
72. G Binnig, H. Rohrer, and C. Gerber, *Phys. Rev. Lett. 49*:57 (1982).
73. R. Sonnenfeld and P. K. Hansma, *Science 232*:211 (1986).
74. K. Itaya, S. Sugawara, K. Sashikata, and N. Furuya, *J. Vac. Sci. Technol., A 8*:515 (1990).
75. K. Uosaki and H. Kita, *J. Vac. Sci. Technol., A 8*:520 (1990).
76. F. F. Fan and A. J. Bard, *J. Electrochem. Soc. 136*:3216 (1989).
77. J. H. Schott, C. P. Araña, H. D. Abruña, H. H. Petach, C. M. Elliott, and H. S. White, *J. Phys. Chem. 96*:5222 (1992).
78. D. Sarid and V. Elings, *J. Vac. Sci. Technol., B 9*:431 (1991).
79. C. A. Goss, J. C. Brumfield, E. A. Irene, and R. W. Murray, *Langmuir 8*:1459 (1992).
80. J. C. Brumfield, C. A. Goss, E. A. Irene, and R. W. Murray, *Langmuir 8*:2810 (1992).
81. G. Sauerbrey, *Z. Phys. 155*:206 (1935).
82. S. Bruckenstein and M. Shay, *Electrochim. Acta 30*:1295 (1985).
83. T. Nomura and M. Iijima, *Anal. Chim. Acta 131*:97 (1981).
84. P. T. Varineau and D. A. Buttry, *J. Phys. Chem. 91*:1292 (1987).
85. D. A. Buttry and M. D. Ward, *Chem. Rev. 92*:1355 (1992).
86. P. W. Stoecker and A. M. Yacynych, *Selective Electrode Rev. 12*:137 (1990).
87. J. Wang, *Electroanalysis 3*:225 (1991).
88. G. A. Gerhardt, A. F. Oke, G. Nagy, B. Moghaddam, and R. N. Adams, *Brain Res. 290*:390 (1983).

89. G. Nagy, G. A. Gerhardt, A. F. Oke, M. E. Rice, R. N. Adams, R. B. Moore III, M. N. Szentirmay, and C. R. Martin, *J. Electroanal. Chem. 188*:85 (1985).
90. G. G. Guilbault, *Enzymatic Methods of Analysis*, Pergamon, New York, 1970.
91. G. G. Guilbault and G. L. Lubrano, *Anal. Chim. Acta 64*:439 (1973).
92. A. E. G. Cass, G. Davis, G. D. Francis, H. A. O. Hill, W. J. Aston, I. J. Higgins, E. V. Plotkin, L. D. L. Scott, and A. P. F. Turner, *Anal. Chem. 56*:667 (1984).
93. A. Parthasarathy, S. Srinivasan, A. J. Appleby, and C. R. Martin, *J. Electrochem. Soc. 139*:2856 (1992).
94. M. A. Habib and S. P. Maheswari, *J. Electrochem. Soc. 139*:2155 (1992).
95. C. E. D. Chidsey, *Science 251*:919 (1991).
96. R. A. Marcus, *J. Chem. Phys. 43*:679 (1965).
97. C. M. Elliott and R. W. Murray, *Anal. Chem, 48*:1247 (1976).

# 14

# Mercury Electrodes

**Zbigniew Galus** *The University of Warsaw, Warsaw, Poland*

## I. INTRODUCTION

Mercury is widely used in the practice of electroanalytical chemistry, both for working electrodes and for reference electrodes (in the latter case usually as an electrode of the second kind).

The use of mercury is nearly ideal for working electrode construction for several reasons. Mercury has a large liquid range (–38.9 to 356.9°C at normal pressure), and therefore electrodes of various shapes may be easily prepared either in pure form or by deposition of mercury on the conducting support. The surface of such electrodes is highly uniform and reproducible if the mercury is clean.

The preparation of pure mercury is not difficult. Any metal with a standard potential more negative than that of mercury may be oxidized easily (with the exception of nickel, which forms a mercury intermetallic compound) by dispersing mercury into a solution of its salts acidified with $HNO_3$ and saturated with oxygen. Metals insoluble in mercury may be also removed this way, although the process may be slow. More effective in this respect is the separation of metal microcrystals by filtration. The elimination from mercury of metals more noble than itself (as well as less noble metals) is accomplished by distillation under reduced pressure. Usually such distillations are repeated several times. "Triple-distilled mercury" is commonly used for electrochemical purposes.

One of the most important reasons for the application of mercury to the construction of working electrodes is the very high overpotential for hydrogen evolution on such electrodes. Relative to a platinum electrode, the overpotential of hydrogen evolution under comparable conditions on mercury will be –0.8 to –1.0 V. It is therefore possible in neutral or (better) alkaline aqueous solutions to reduce alkali metal cations at mercury electrodes, giving relatively well-defined polarographic waves at potentials more negative than –2.0 V vs. SCE.

Using supporting electrolytes such as tetraalkylammonium salts, one may apply potentials as negative as –2.6 V vs. SCE in aqueous solutions, while in some nonaqueous systems even –3.0 V vs. SCE (aqueous) is accessible. Unfortunately, mercury electrodes have serious limitations in applications at positive potentials (with the exception of passivated mercury electrodes, which are described in Section VI), and this has led to extensive research in the development of solid metal and carbon electrodes. Oxidation of mercury occurs at approximately +0.4 V vs. SCE in solutions of perchlorates or nitrates, since these anions do not form insoluble salts or stable complexes with mercury cations. In all solutions containing anions that form such compounds, oxidation of the mercury proceeds at potentials less than +0.4 V vs. SCE. For example, in 0.1 M KCl this occurs at +0.1 V, in 1.0 M KI at –0.3 V, and so on.

The following types of mercury electrodes have been widely used for voltammetry: dropping mercury electrode (DME), hanging mercury drop electrode (HMDE), static mercury drop electrode (SMDE), streaming mercury electrode (SME), and mercury film electrode (MFE). We begin our discussion with a description of the construction and properties of the DME because this electrode has a long history and continues to be used for both analytical and fundamental studies.

## II. DROPPING MERCURY ELECTRODE

### A. Introduction

The DME is the essential component of all polarographic experiments. It was introduced by Kucera in 1903 [1]. Later, Heyrovský applied this electrode in the original polarographic apparatus [2]. Classically, the DME was formed from glass capillary tubing connected to a standtube of plastic or glass attached to a mercury reservoir (Fig. 14.1). If the level of mercury in the standtube of this device is sufficiently high with respect to the tip of the capillary, a small drop of mercury will form at the end of the capillary; the drop will eventually fall, and another drop will form. These drops, which have a nearly spherical shape (if their mass is not greater than about 15 mg and if the radius of the capillary is on the order of 0.05 mm), will fall at regular intervals. Electrolysis is carried out on the surface of the mercury drops. One of the most important advantages of the DME is continuous renewal of the electrode surface. The periodicity of this renewal is governed by the life of an individual drop.

Our description of the DME starts with a discussion of basic concepts involved in the formation of mercury drops. These considerations are subsequently used to define the capillary characteristics required for satisfactory elec-

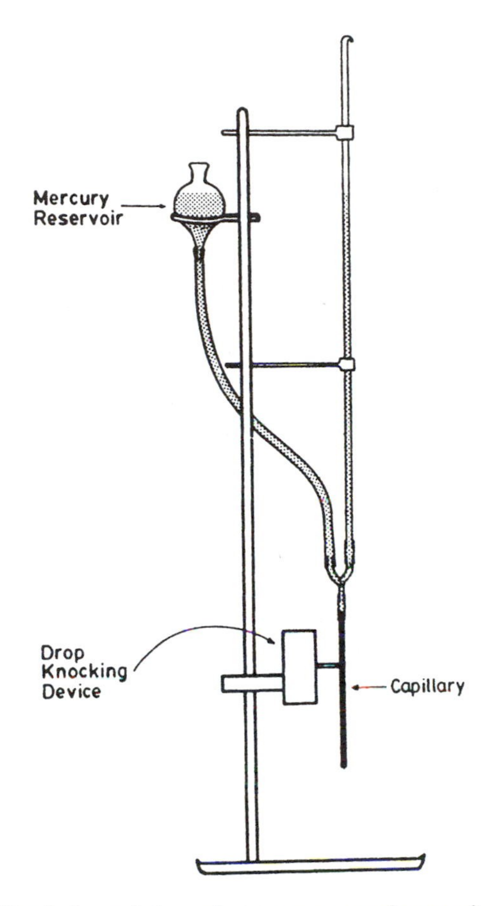


**Figure 14.1** Classical standtube and mercury reservoir assembly for use with dropping mercury electrodes.

trodes. This discussion is based partially on the classical monograph of Kolthoff and Lingane [3].

## B. General Discussion of the Flow of Mercury from Capillaries

The rate of flow of a liquid through a capillary can be defined by the Poiseuille equation. The volume V of liquid that flows in time t is given by

$$V = \frac{\pi r_c^4 P t}{8 \ell \eta} \tag{14.1}$$

where $r_c$ and $\ell$ are the internal radius and length of a capillary, respectively, $\eta$ is the viscosity of the flowing liquid, and P is the effective hydrostatic pressure (i.e., P is not exactly equal to the pressure created by the difference between the level of liquid in the reservoir and the lower end of the capillary). This equation may be strictly applied to the flow of a liquid that wets the walls of a capillary. This is not exactly true in the case of mercury flowing from glass capillaries; however, the resulting error is small.

Equation 14.1 may be easily rearranged to the form

$$m = \frac{Vd}{t} = \frac{\pi r_c^4 Pd}{8\ell\eta} \tag{14.2}$$

where d is the density of the liquid. In this rearrangement the following relationship was used:

$$mt = Vd \tag{14.3}$$

where m is the flow rate of mercury from the capillary (usually expressed as mg/s). Both mt and Vd express the total mass of mercury flowing for a given t and can easily be measured experimentally.

The effective pressure that acts on the mercury, P, is not exactly equal to the total pressure, $P_t$, which is given by the equation

$$P_t = h_t gd \tag{14.4}$$

where $h_t$ is the vertical distance between the end of the capillary and the level of the mercury in the reservoir, and g is the gravitational constant.

Mercury flows from the capillary in small drops. The size of these drops is determined by the interfacial tension at the mercury–solution interface. The internal pressure in the drop acting against $P_t$ is, according to Kucera, given by the relation

$$P_b = \frac{2\gamma}{r} \tag{14.5}$$

This internal pressure $P_b$, often called back pressure, is proportional to the interfacial tension $\gamma$ at the mercury–solution interface and is inversely proportional to r, the radius of the drop.

It follows from Equation 14.5 that the amount of mercury flowing into the solution as drops should be somewhat less than the amount of mercury that would flow from the same capillary at the same $h_t$ when dipped into a larger volume of mercury. Equation 14.5 also shows that the back pressure in a given solution changes with the growth of a drop and reaches the smallest value at the moment of detachment of a drop from the capillary. If one considers, for example, a capillary with an internal diameter of 0.06 mm placed into 3 M KCl at a potential of –0.6 V, the back pressure changes from 10.4 to 0.97 cm of mercury during the life of the drop. If the mercury column is lower than the first value, it is obvious that the outflow of mercury will not be observed.

The radius of the drop at an instant $t_1$ following drop birth may be easily calculated using Equation 14.3 and the expression for the volume of a sphere,

$$V = \frac{mt_1}{d} = \frac{4}{3}\pi r^3 \tag{14.6}$$

It is important to know the mean value of the back pressure over the lifetime of the drop, $t_d$. To obtain this value one can calculate the mean radius of the DME, $\bar{r}$:

$$\bar{r} = \frac{1}{t_d}\int_0^{t_d} r\, dt = \frac{3}{4}\left(\frac{3mt_d}{4\pi d}\right)^{1/3} \tag{14.7}$$

By combining Equations 14.7 and 14.5 with elimination of r, one obtains

$$\bar{P}_b = 4.31\frac{\gamma\, d^{1/3}}{(mt_d)^{1/3}} \tag{14.8}$$

Now the effective pressure may be expressed as

$$P = P_t - \bar{P}_b = h_t gd - 4.31\frac{\gamma\, d^{1/3}}{(mt_d)^{1/3}} \tag{14.9}$$

By introducing P given by Equation 14.9 into Equation 14.2, one obtains

$$m = \frac{\pi r_c^4 d}{8\ell\eta}\left[h_t gd - 4.31\frac{\gamma d^{1/3}}{(mt_d)^{1/3}}\right] \tag{14.10}$$

Using the following values for the constants,

$d = 13.53\ \text{g/cm}^3$, at 25°C

$\eta = 0.0152\ \text{dyn}\cdot\text{s/cm}^2$, at 25°C

$g = 980.6\ \text{cm/s}^2$

and taking $\gamma = 400$ dyn/cm, one can evaluate m according to Equation 14.11:

$$m = 4.64\times 10^9\,\frac{r_c^4}{\ell}\left[h_t V - \frac{3.1}{(mt_d)^{1/3}}\right] \tag{14.11}$$

If $r_c$ and $\ell$ are given in centimeters and $t_d$ in seconds, one obtains m in milligrams per second.

One should remember that the product $mt_d$ is constant for a given capillary at a given potential. This constancy follows from the equation

$$mt_dg = 2\pi r_c\gamma \qquad (14.12)$$

The left-hand side represents the weight of the falling drop, which is dependent on the radius of the capillary and the interfacial tension.

The correction term in Equation 14.10 is constant independent of $h_t$; however, in view of Equation 14.11, this term is quite significant. Assuming that $m = 2$ mg/s and $t_d = 4$ s (conditions that are very often met in practice), $h_t$ has to be corrected for 15.5 mm, which is quite significant in comparison to $h_t$, which sometimes may be as small as 150 mm.

The linear dependence between m and $h_t$ corrected for the back pressure has been found experimentally [4]. With a given capillary and the same $h_t$, m may be slightly different for different solutions because the back-pressure term will change from solution to solution due to variations in the interfacial tension. It follows from Equation 14.12 that the product $mt_d$ should be constant if $h_t$ is not changed. From Equations 14.10 and 14.12, it follows that $t_d$ is inversely proportional to $h_t$.

## C. Construction of the Classical Dropping Mercury Electrode

Capillaries suitable for DME construction are commercially available from many sources. They may also be prepared by heating and pulling ordinary capillary tubing with thick walls to shrink the bore, but this is rarely worth the trouble.

In choosing a capillary, one should remember that uniformity is a primary consideration. The capillary should have no irregularities or radial cracks extending from the orifice, because such capillaries often cause irregular limiting currents. Therefore, it is advisable to inspect each capillary with a magnifying glass before use. Although it is usual for analytical work to cut the tube normal to the capillary, this results in "shielding" the drop from the top side, since the diameter of the glass is typically 5–7 mm; a conical tip can be made to avoid shielding to some extent.

The length of the capillary used depends on its internal diameter and experimental objectives. Usually, the longest drop times should not exceed 7–8 s. In general, the capillary should give a drop time on the order of 3–4 s with $h_t$ equal to about 40 cm when the capillary is dipped into the solution. If drop times are too long, one should cut the capillary, remembering that a linear dependence exists between the length of the capillary and $t_d$.

The appropriate capillary length may also be estimated on the basis of simple calculations. Neglecting the correction term in Equation 14.11, one has for the mercury flow rate

$$m = 4.64 \times 10^9 \frac{r_c^4 h_t}{\ell} \tag{14.13}$$

By combining Equations 14.12 and 14.13 and introducing numerical constants, we have

$$t_d = 5 \times 10^{-7} \frac{\ell}{r_c^3 h_t} \tag{14.14}$$

Using this equation, one may calculate the length of the capillary necessary to have proper drop times at reasonable $h_t$.

For example, let us assume that the radius of the capillary is 0.04 mm and the shortest $t_d$ that we would like to have is 2 s at an $h_t$ of 80 cm. Introducing these data into Equation 14.14 gives $\ell$ = 20.5 cm. For such a capillary, if $h_t$ = 20 cm, the drop time is about 8.0 s. These calculated values are only approximate, since we did not consider the influence of the back pressure.

When cutting the capillary, one should remember that the end that enters the sample solution should be cut as close as possible to 90° to obtain the most reproducible currents [5,6]. The cut should also be smooth, but not mechanically polished. The other end of the capillary may be either sealed to a short length of glass tubing (several millimeters internal diameter) or connected directly to flexible tubing. The connection to the mercury reservoir is usually made using soft plastic tubing such as Tygon. The inside diameter of this tubing should be 5–7 mm and the connection to the capillary should be reinforced with wire. The plastic tube should not contain sulfur. Before use, the tube interior should be washed with a dilute solution of sodium or potassium hydroxide, rinsed with distilled water, and thoroughly dried.

The mercury reservoir usually has a volume of 100–200 cm$^3$. The platinum wire used to make contact to the DME may be dipped in the mercury reservoir or sealed in the standtube.

Once the capillary has been connected, the mercury reservoir should be filled with very pure mercury. For polarographic purposes, the triple-distilled mercury available commercially is usually satisfactory. If the mercury was previously used in voltammetric experiments, it should be chemically cleaned as indicated in Section I, and then twice or triply distilled. Due to the dangers associated with mercury vapors, it is normal practice to return dirty mercury to the supplier for commercial distillation.

When filling the apparatus with mercury, one should be especially careful to avoid entrapped air bubbles. The classical DME requires a stand that facilitates changing the position of the mercury reservoir. A centimeter scale is often included, making it convenient to determine $h_t$. Some early authors recommended an all-glass apparatus, but this seems inadvisable now that inert plastic materials are available. When a fixed glass reservoir is used, the drop time may be controlled by changing the gas pressure over the mercury. In any case, the polarographic apparatus should be protected against vibrations that might prematurely dislodge the drop.

## D. Maintenance of the DME and Determination of the Drop Time and Flow Rate of the Capillary

A capillary will work properly for a long time if special care is exercised both during experiments and when the electrode is not in use. An important rule of thumb is that solutions should never enter the capillary. The mercury reservoir should always be elevated and the flow of mercury observed before the capillary is placed in solution. When the polarographic work is finished, the DME should be withdrawn from the cell and transferred to a beaker containing distilled water (or a nonaqueous solvent, if used). Mercury should drop there for several minutes to wash the end of the capillary. Finally, the capillary should be carefully washed with distilled water.

There are several procedures for storing the DME after washing. Probably the best technique is to dry the capillary with filter paper and dip it into a vial of mercury. Then the reservoir should be lowered to decrease the flow of mercury. The DME may also be stored in distilled water; in this case, mercury should again be allowed to continue flowing very slowly.

After washing and drying the capillary, many experimenters simply lower the mercury reservoir enough to stop the flow of mercury, but this procedure can occasionally cause problems. The inevitable metal impurities in mercury can be air-oxidized at the capillary tip, and the oxidation products may change the parameters of the electrode (namely, m and $t_d$). There is also a possibility that a small amount of the sample solution may remain in the capillary and then evaporate, leaving a solid residue; this will also change the parameters of the electrode, and the residue may be very difficult to remove from the capillary.

If erratic capillary behavior is observed (this may be easily recognized from a decrease of current or irregularities in drop time), the tip may be cleaned by immersing it in a small beaker of 1:1 nitric acid (~6 M), followed by thorough rinsing. Mercury should, of course, flow from the DME throughout the procedure. This simple treatment may be insufficient when the residue is at some distance from the end of the capillary. In these cases the capillary may be removed from the apparatus and carefully washed out by using a vacuum to suck 1:1 hot nitric acid through the capillary. It may take a long time to remove the impurities. The capillary should then be rinsed with distilled water, dried, and remounted. Sometimes even this procedure is not effective, and it may be helpful to place the dry capillary in an oven for 24 h at approximately 450°C. For obvious reasons, such desperate measures are usually passed up in favor of replacing the capillary.

The parameters characterizing the DME may be determined in the following way. The capillary is placed in a solution of the supporting electrolyte. The mercury reservoir is adjusted to a given $h_t$ and a potential is applied, either a value related to experimental measurements or in the range –0.5 to –0.6 V vs. SCE. A stopwatch is used to measure the time for 10–20 drops to form, and the drop time, $t_d$, is obtained by dividing the elapsed time by the number of drops collected. Mass flow rate of mercury, m, can be measured by collecting drops for a measured time and then weighing the washed and dried mercury. Usual units are milligrams per second.

For accurate calculation of the area of the drop, a correction must be made for the area of the capillary by use of the formula

$$A = 2\pi r\left[r + \left(r^2 - r_c^2\right)^{1/2}\right] \tag{14.15}$$

The radius of the drop is calculated from m and $t_d$, and $r_c$ is measured with a microscope and ocular micrometer.

## E. Unusual Dropping Mercury Electrodes

Various DMEs have been developed [7] and they are useful in various experiments. Blunt-ended DMEs are, for example, not useful in experiments carried out at higher frequencies. The shielding effect mentioned earlier creates a nonuniform distribution of current along the surface of the drop, and spurious frequency dispersion effects appear.

To diminish this problem, a capillary with a sharp tip is used [8], which may be prepared by drawing out a normal capillary. To overcome the shielding effect significantly, the overall diameter of the capillary at its end should be considerably smaller than the mean diameter of the mercury drop. It should, however, be mentioned that work with this type of capillary is more difficult, and the production of such capillaries with a uniform drop time requires considerable practice.

Deviation of a capillary from a vertical position changes its drop time considerably. The horizontally positioned DME was originally used by Smoler [9]. Shielding and depletion effects are diminished to some extent, and current oscillations are much smaller. Probably the best performance is observed when the electrode position deviates 45° from the vertical; however, opinions about such electrodes vary [10]. In some experiments, electrodes with small mercury flow rates are helpful. For these electrodes, the probability of the appearance of polarographic maxima is diminished. When m is very low, the drop time is usually very long. In such cases, artificial regulation of the drop time should be used (see Sec. II.F).

Capillaries with small m and long drop time may be produced by restricting the internal diameter of a normal polarographic capillary at some distance from its end. Since the radius of the capillary at the end is not changed, the drop size should be practically identical at $t_d$ with that of the original nonrestricted capillary. Novotný et al. [11], using a 37-μm diameter for the main part of the capillary and a 170-μm diameter at the orifice, obtained a $t_d$ approaching 130 s. If the capillary were restricted at its end, the drop size would also be diminished. Electrodes of this type, however, may be easily blocked by small crystals of mercury salts or other impurities.

Anomalous effects are often observed due to solution entering the capillary at the instant the mercury drop falls. In ac measurements, this phenomenon leads to anomalous frequency dispersion [12]. In addition, the drop time becomes irreproducible. These effects may be diminished to a large extent by coating the internal wall of the capillary with a film of silicone [13]. A tip made of hydrophobic (solvophobic) material may also be attached to the glass capillary. For example, a polyethylene tip was used [14] to discriminate against the attack of fluorides on glass in the study of double-layer structure in the presence of fluorides. In another study, capillary tips were modified with commercial narrow-bore PTFE tubes to determine arsenic in basic solution [15]. This procedure is also used for the hanging mercury drop electrode discussed in Section III.

### F. Mechanical Control of Drop Life

Often, polarographic experiments are best carried out with a prescribed drop time. If not controlled, the drop time will change slightly as a function of potential, since surface tension is dependent on double-layer structure. Drop time can be controlled, for example, by a simple arrangement consisting of a glass rod sealed to the end of the polarographic capillary [16]. For modern techniques using digital timing circuitry (e.g., sampled dc and pulse polarography), it is most convenient to achieve a constant $t_d$ by dislodging the drop prior to its natural fall. Many systems are based on a solenoid-controlled hammer that periodically touches the capillary, detaching the mercury drop.

## III. HANGING MERCURY DROP ELECTRODE

Detection limits obtained with the DME are restricted by a relatively large background current due to the constantly changing electrode area. Also, a stationary electrode is required for stripping experiments and for mechanistic studies employing cyclic voltammetry. Thus, many applications require a stationary or hanging mercury drop electrode (HMDE).

## A. HMDE Suspended on a Metallic Contact

A simple HMDE was developed by Gerischer [17]. Using his approach, one or two mercury drops falling from the classical DME are collected on a miniature spoon and transferred to a small metal contact sealed in glass or plastic (Fig. 14.2). Platinum or gold-plated platinum wires are used. The length of the exposed contact wire is usually 0.1–0.5 mm and its diameter is about 0.5 mm.

Electrodes of this sort have many different chemical properties from pure mercury electrodes, because of the formation of a gold or platinum amalgam [18]. Normally, a drop is suspended just prior to an experiment, so this problem will be of no serious consequence. Nevertheless, since the solubility of these noble metals in mercury is about 0.05 M at room temperature [19], the concentration of gold or platinum in mercury may be quite significant on a longer time scale. In such cases, gold or platinum may form intermetallic compounds with several metals that are electrodeposited into the mercury [18].

## B. Capillary HMDE

The capillary version of the HMDE consists of a small mercury drop with radius usually not exceeding 1 mm, which hangs on a thin mercury thread in a glass capillary. The inner diameter of this capillary is about 0.15–0.20 mm, considerably larger than for the DME. An electrode of this type was used for the first time by Antweiler [20]; however, the preparation of a reproducible HMDE using his original model was very difficult. As pointed out by Kemula and Kublik [21], the suspended HMDE is more reproducible than the early design of the capillary HMDE.

In the most successful designs, the mercury is mechanically pushed out of the capillary. This type of electrode was constructed and used in the early experimental work of Randles and White [22]. Mercury was pushed from the

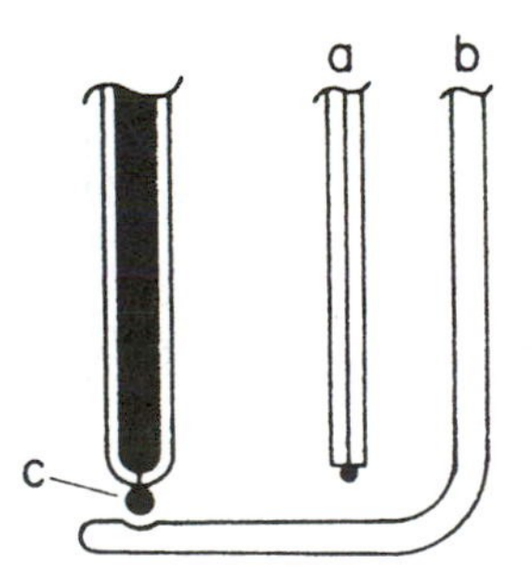


**Figure 14.2** HMDE as developed by Gerischer: (a) DME; (b) transfer cup; (c) HMDE.

capillary by turning a micrometer screw. The electrode of this type that is still used today was originally developed by Kemula and Kublik [23].

Many variations exist; the model in Figure 14.3 has been used for many years at the Department of Chemistry at Warsaw University. The glass part of this electrode is shown in Figure 14.3a. It is composed of thick-walled glass tubing with an internal diameter of 4 mm, approximately 5 cm long. This forms a mercury reservoir, which is sealed to a thick wall capillary. The internal diameter of this capillary is commonly 0.18 mm but may vary somewhat.

The length of the capillary will depend on the size of the electrolytic cell. A ground-glass joint on the mercury reservoir can provide a convenient means of supporting the electrode in a cell. A Teflon thermometer adapter also works well. The top of the reservoir resembles a glass ring that is flat and very smoothly polished. This part of the electrode is connected to the plastic head, which is composed of two parts (Fig. 14.3b and c). After the capillary is inserted through part b, part c is screwed tightly into part b. The end of the mercury reservoir should then be tightly sealed by the bottom of part c.

To provide a tight seal, a small plastic cylinder, a bit thicker than the internal diameter of the glass tubing, is extended from part c. This small cylin-

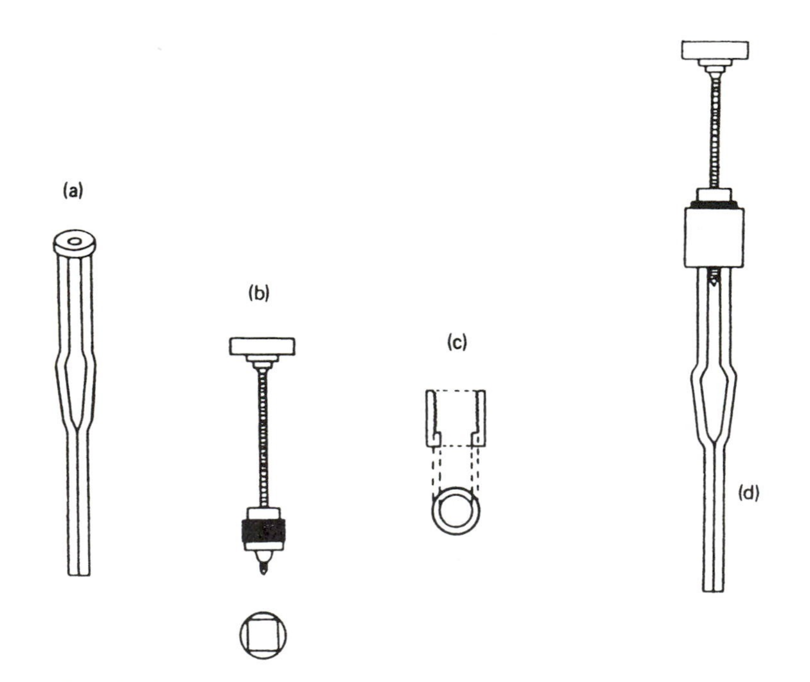


**Figure 14.3** Capillary HMDE: (a) reservoir and capillary (glass); (b) threaded plunger; (c) plastic collar; (d) assembled unit.

der (Fig. 14.3c) should enter into the reservoir tubing with a pressed fit when all parts of the HMDE are put together.

By turning the threaded steel piston (2 mm diameter) a fixed angle, for instance, 90° or 180°, mercury will be pushed out of the capillary and will hang as a drop from its end. It is important that the piston screw be set tight in the polyethylene part shown in Figure 14.3b.

In Figure 14.3d the HMDE is shown ready for use. To ensure that the electrode works properly and solution does not enter the capillary, special care should be taken to coat the internal part of the capillary with a hydrophobic silicone film. The capillary should be thoroughly cleaned before a new coating is applied. Sodium hydroxide (~2 M) should be used to remove any residual silicone. The capillary should then be washed with 3 M nitric acid, rinsed with distilled water, and thoroughly dried. Then a 5% solution of dichlorodimethylsilane in $CCl_4$ should be pulled through the capillary, followed by air. (Note: With the phaseout of chlorinated solvents due to atmospheric ozone depletion, toluene is often substituted for $CCl_4$.) After several hours at room temperature or about 15 min at 110°C, the capillary is ready for use. The silicone coating should be renewed from time to time, especially when the hanging drop tends to fall during an experiment. This happens when the silicone coating on the capillary wall is partly destroyed and solution enters between the glass and the mercury thread.

The mercury reservoir and the capillary should be completely filled with mercury. There must be no entrapped air! A special vessel useful in filling the HMDE was described by Kemula [24]. This vessel should be equipped with a ground joint corresponding to that of the capillary, and the end of the capillary should extend to within several millimeters of the bottom.

When the vessel is being filled with mercury, the end of the capillary should be covered when the vessel is placed in a horizontal position, as depicted in Figure 14.4. With all of the mercury in part b, the vessel should be connected to a vacuum pump, with a three-way stopcock to decrease the pressure in the vessel and the electrode to 0.01 mm.

After evacuation, the vessel should be tilted into the horizontal position shown in Figure 14.4 and the stopcock should be turned to disconnect the pump, and then turned further to admit air to the vessel. The pressure of air acting on the mercury in the vessel pushes it into the reservoir and the capillary of the HMDE. If the pressure was previously decreased to 0.01 mm, mercury now fills the reservoir and capillary completely. The HMDE is now ready for use. By turning the screw a fixed angle, a known amount of mercury is pushed out, forming a drop.

This type of electrode may be easily prepared in every laboratory. Modifications of this type of HMDE have been described by other authors [25]. The connection of the mercury reservoir to the plastic cap may be made in differ-

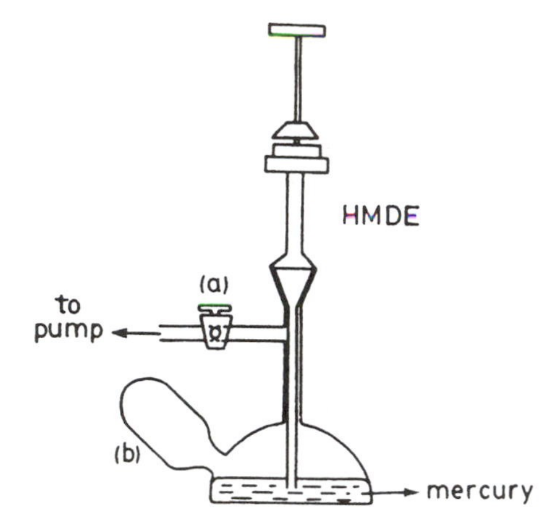


**Figure 14.4** Arrangement for filling the HMDE: (a) three-way stopcock, and (b) mercury compartment.

ent ways, but in every case the mercury reservoir must be closed tightly. Also, the tip of the glass capillary may be made from plastic (solvophobic) material, in order to prevent the solution from entering the capillary and the possible reaction between glass and components of the solution (fluorides). Some variations on the HMDE described earlier (available from several manufacturers) are equipped with micrometer screws for pressing out a well-defined amount of mercury.

### C. Sessile Mercury Drop Electrodes

Sessile mercury drop electrodes are obtained using J-shaped capillaries connected to the typical HMDE arrangement described earlier. The drop rests on the mercury thread rather than hanging from it. These electrodes are sometimes useful when working at very negative potentials. Then the classical HMDE may not be stable and the drop often falls off at the most inconvenient time. Sessile drops are more stable; however, they are not recommended for most experiments, because the area of such electrodes is not always well defined. Sessile mercury drop electrodes may also be prepared by placing the mercury drop on a small contact made of metal that is wetted by mercury [12,26].

## IV. STATIC MERCURY DROP ELECTRODE

The static mercury drop electrode (SMDE) was first introduced commercially in the late 1970s by EG&G Princeton Applied Research [27]. It utilizes a method of drop formation in which the mercury drop is dispensed rapidly and then allowed to hang stationary at the capillary tip. When used in a DME mode of operation, the drops can be repetitively formed and dislodged at desired time intervals for use in sampled dc polarography, pulse polarography, or differential pulse polarography. Alternatively, a drop can be maintained indefinitely for use as an HMDE.

When used for polarography experiments, the SMDE exhibits a fundamental difference in its current-time behavior compared to the conventional DME. Figure 14.5 shows the growth of drop area and the accompanying current profiles for an SMDE and a conventional DME. In the SMDE the drop is rapidly expanded to a given area, which is then maintained. This feature enables the current measurement in various polarographic techniques to be made on an electrode with a static area, eliminating the contribution of nonfaradaic current from changing surface area. By comparison, the conventional DME exhibits a continuously increasing surface area until the drop is dislodged. Thus a component of charging current due to this increasing surface area always exists.

The mechanical details of the original EG&G SMDE are illustrated in Figure 14.6a. Activation of the solenoid lifts the plunger, allowing mercury to flow from the large reservoir on top through the capillary to form a drop. The mercury flow rate is very rapid due to the relatively large bore of the capillary (0.15 mm). Drop size is controlled by the duration of solenoid activation. Solenoid deactivation stops the flow of mercury, resulting in a static drop. At the drop time selected, the mercury drop is mechanically dislodged from the capillary; a new drop is then grown by activation of the solenoid. A SMDE with similar features but based on a different principle was developed by Tesla Laboratorni Pristrojc in the Czech Republic (see for instance [28]). A renewable mercury microelectrode has also been made [29].

Metrohm and BAS have also introduced improved DME models capable of operating in the SMDE mode. The Metrohm electrode (Fig. 14.6b) has a needle valve and small-bore capillary. Much of it is pneumatically controlled. The BAS version (Fig. 14.6c) is called the controlled growth mercury electrode (CGME). It is based on the work of Kowalski, Osteryoung, and coworkers [30]. Its features include a low-resistance electrical contact to the mercury thread in the capillary via a stainless steel tube and a fast response valve. The fast valve has allowed unique experiments to be performed where precise control of mercury drop growth during the experiment is desirable [31–33]. The BAS (Fig. 14.7), EG&G Princeton Applied Research (Fig. 14.8), and Metrohm (Fig. 14.9) electrodes offer this easy and reproducible drop renewal in fully equipped cell stands.

## V. STREAMING MERCURY ELECTRODES

There are two types of streaming mercury electrodes (SMEs). The first type is used in the determination of the potential of zero charge, and the second was introduced for voltammetry by Heyrovský.

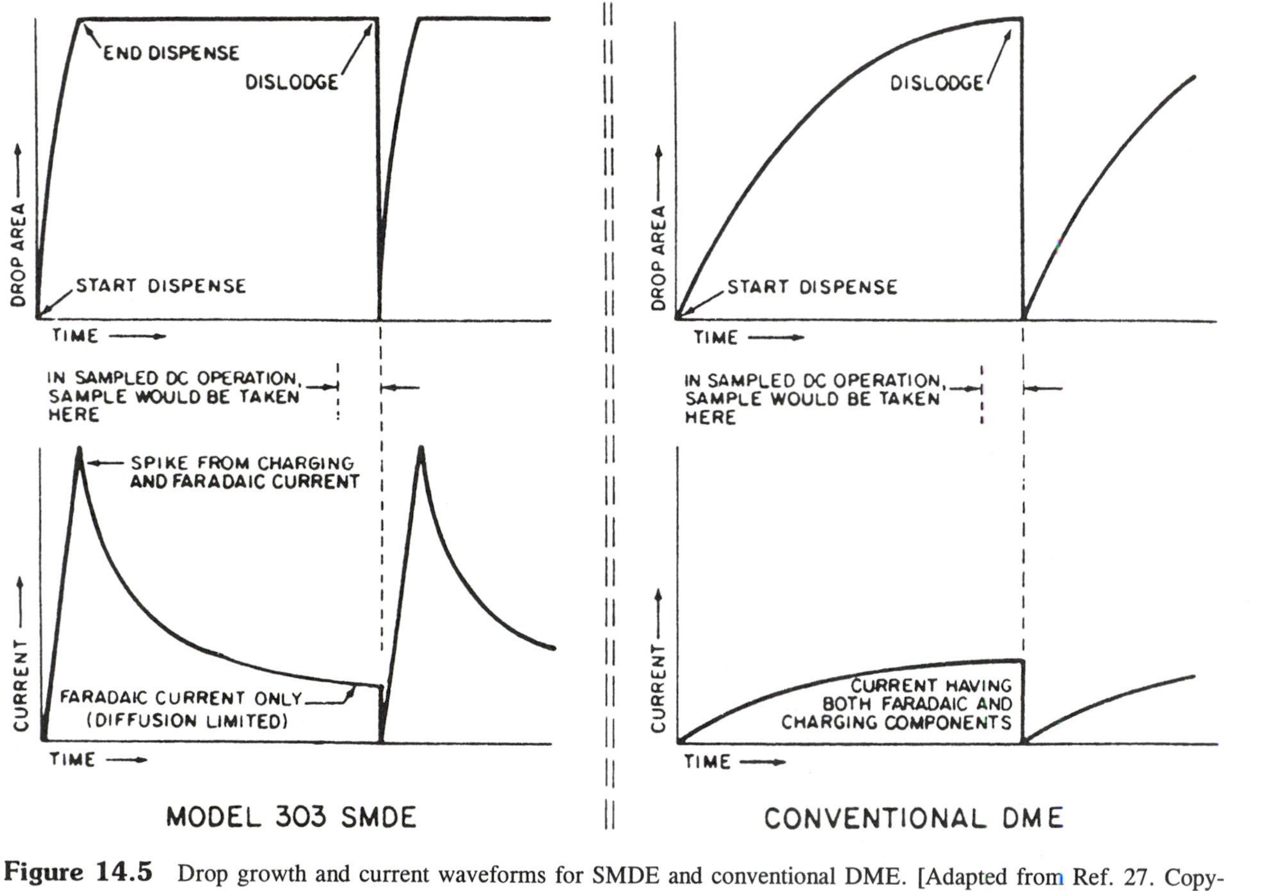


**Figure 14.5** Drop growth and current waveforms for SMDE and conventional DME. [Adapted from Ref. 27. Copyright 1979 International Scientific Communications, Inc.]

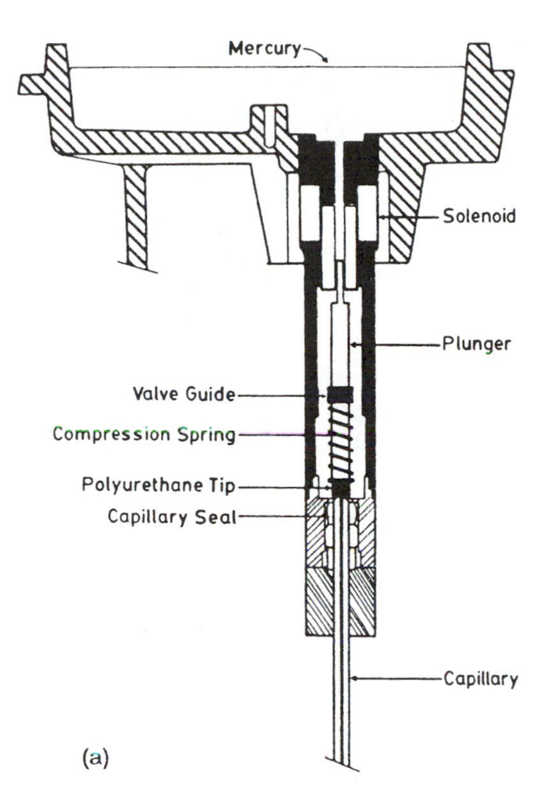


**Figure 14.6** Cross-sectional views of (a) the EG&G Princeton Applied Research model 303A SMDE; (b) Metrohm's electrode, featuring a needle valve and small-bore capillary; and (c) the BAS' controlled growth mercury electrode.

The rapid flow of mercury from a capillary will assume the potential of zero charge, which is then simply measured. This method was proposed at the end of the 19th century by Helmholtz, and the first experiments were carried out by Ostwald [34]. Later, a number of investigators adapted this approach [35–39]. This method is in wide use today in the determination of the zero charge potentials of mercury electrodes in different solutions (see for instance [40]).

The capillary used in such measurements is usually 1 cm long with an internal diameter of 80–100 μm. It should have a ragged end, which causes mercury to flow in several streams, resulting in as large a surface area as possible.

The potential of such streaming electrodes, placed in the sample solution, is dependent to some extent on the height of the mercury reservoir, $h_t$, going

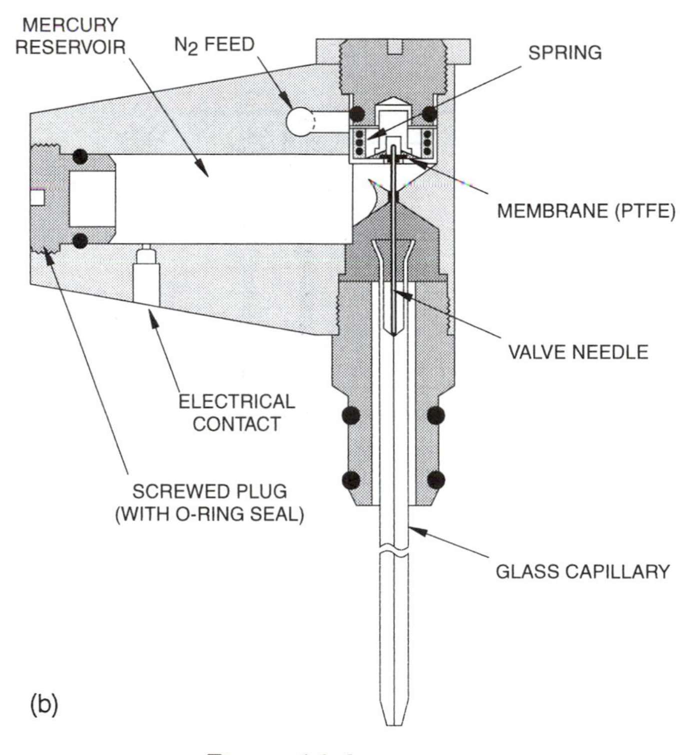


**Figure 14.6** Continued

through a maximum. The potential corresponding to this maximum is considered the zero charge potential.

For electrodes with the above parameters, $h_t$, when the electrode reaches a zero charge potential, is within the limits of 50–100 cm, depending on the type of solution.

In voltammetry it is occasionally necessary to have a mercury electrode with a constant but fresh mercury surface. Such an electrode was developed by Heyrovský [41–43] for "oscillopolarography." It is usually made of a thick-walled capillary (6–7 mm o.d.) with an internal diameter of 1 to 2 mm, but at its end the diameter decreases to 0.1 mm. Mercury flows from the capillary in one stream, which is usually directed upward, out of the solution. If the cylindrical shape of the mercury stream is maintained along its length, the surface of the electrode will be $A = 2\pi r_c \ell$ where $r_c$ is the radius of the capillary at its end and $\ell$ is the length of the continuous stream of mercury in solution. The rate of mercury flow is approximately a hundredfold greater than for a classical DME (about 0.2 g/s or 0.7 kg/h).

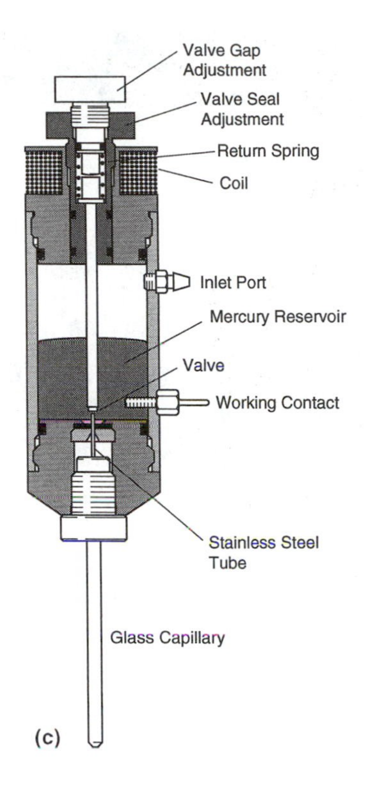


The streaming mercury electrode has been used in kinetic measurements [44,45]. Following introduction of the HMDE, streaming electrodes have become little more than historical curiosities. They are limited by the fact that the radius of the mercury stream and the flow rate are not exactly constant as a function of distance [46].

## VI. MERCURY FILM AND OTHER TYPES OF MERCURY ELECTRODES

In the last two decades, mercury film electrodes (MFEs) have been used frequently in electroanalytical practice. Using such electrodes, metal ions present in the solution in trace amounts may be determined with satisfactory accuracy by their reduction on the surface of MFEs, formation of relatively concentrated amalgam, and anodic oxidation from MFEs of the preconcentrated metal in a final step (see Chap. 24).

MFEs are also useful in hydrodynamic techniques, such as rotating-disk voltammetry (Chap. 3) and electrochemical detection for liquid chromatogra-

**Figure 14.7** The BAS controlled growth mercury electrode (CGME), in its fully equipped cell stand.

phy (Chap. 27). MFEs are manufactured by the deposition of mercury either on carbon (polished glassy carbon is preferred [47] due to its inertness and low porosity) or on a noble metal support. In the former case, a mercury film is produced in situ by electrodeposition on carbon of mercury ions introduced to the sample solution [47,48]. In fact, a mercury film produced in situ forms a layer composed of small droplets with a size dependent on the amount of mercury deposited and the deposition potential [49,50].

Since the distribution of mercury on a carbon support cannot be precisely controlled, MFEs should preferentially be used in analytical determinations, where experiments with a standard sample can be carried out under conditions

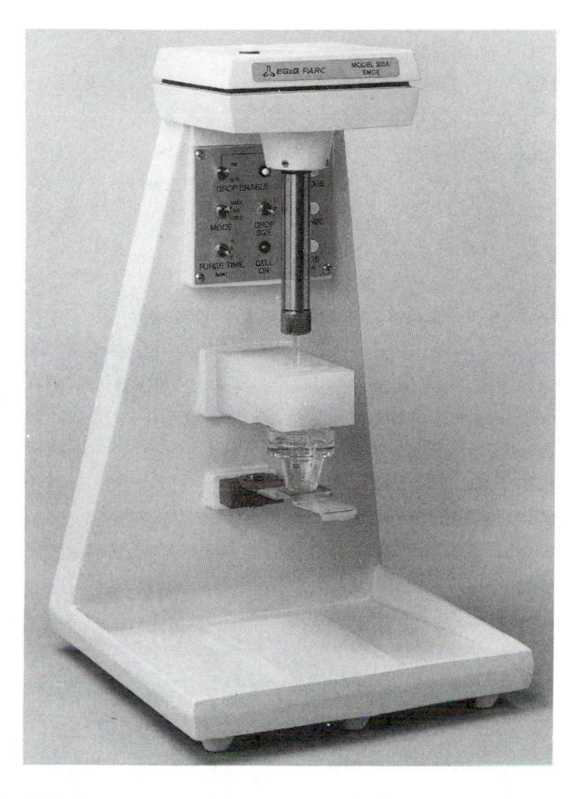


**Figure 14.8** The first commercial version of the static mercury drop electrode, the EG&G Princeton Applied Research model 303A, in fully equipped cell stand.

identical with those used in the determination. The advantage of this type of MFE is that the film of mercury is not contaminated by atoms of the material on which it is deposited. This is not the case when an MFE is formed on platinum, gold, silver, or other metal support, because each of these metals dissolves to some extent in mercury, and consequently, properties of the MFE may be slightly changed. However, a more homogeneous mercury film can be obtained on a metal support [51]. Various metal substrates have been studied, resulting in selection of iridium as the substrate of choice, partly because of its low solubility in mercury [52,53].

For instance, the stripping peaks of metals deposited in a contaminated mercury film may be changed by formation of intermetallic compounds. Compounds may form between the metals and the metal of a support that has dissolved in the film, or between the metals and the surface atoms of the metal

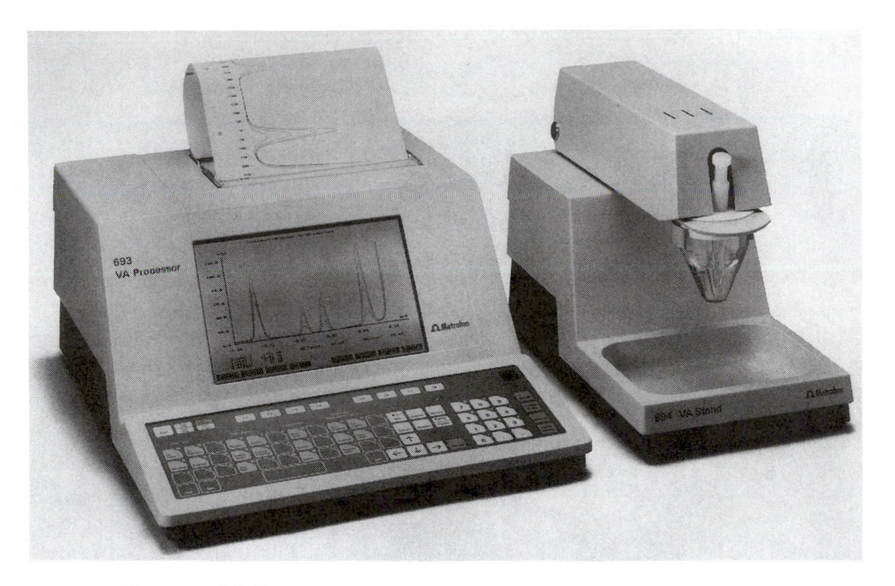


**Figure 14.9** The Metrohm 693 VA processor and VA stand.

support–mercury film interface. Such effects were observed when zinc and tin were introduced to the MFE formed on a gold or platinum support.

Also, the hydrogen evolution overpotential may be decreased when using this type of MFE. The mercury film thickness can be easily regulated by electrolytic deposition with a coulometric control. When the film is relatively thick, its thickness along the MFE surface may not be uniform. MFEs are not stable in time when the mercury film thickness is very low, due to the diffusion of mercury into the metallic support.

In recent years, mercury film ultramicroelectrodes have received wider application in electroanalytical practice [51,54,55]. Such electrodes are especially useful in analytical determinations, since they combine the features of ultramicroelectrodes (Chap. 12) with those of mercury film electrodes. In this case, the mercury can be deposited on carbon fibers, but many prefer a metallic support which is wetted by mercury. The solubility of the supporting metal in mercury should be low. Iridium [54] and silver [55], as well as platinum and nickel, have been used as supporting metals. Surprisingly, even gold fibers wetted by mercury have been very successfully used as electrodes in microchromatography and capillary electrophoresis detectors (Chap. 27).

Various types of mercury pool electrodes have been used [56–63]. These electrodes are prepared by placing a suitable amount of mercury in containers of desired geometry equipped with metal contacts. Suitable holders may be produced from Teflon, or if made of glass, their surfaces should be made hydrophobic with a water-repellent material such as silicone. Such electrodes are usually used as working electrodes for slow experiments in which the spherical effect on the HMDE is too great. Large mercury pools formed in the bottom of a cell are used for separation of some metals, purification of electrolytes, or exhaustive electrolysis for coulometric or synthetic purposes.

The mercury thread electrode (MTE) has recently been introduced [64]. The mercury is contained in a short, thin-walled dialysis tube (Fig. 14.10). The

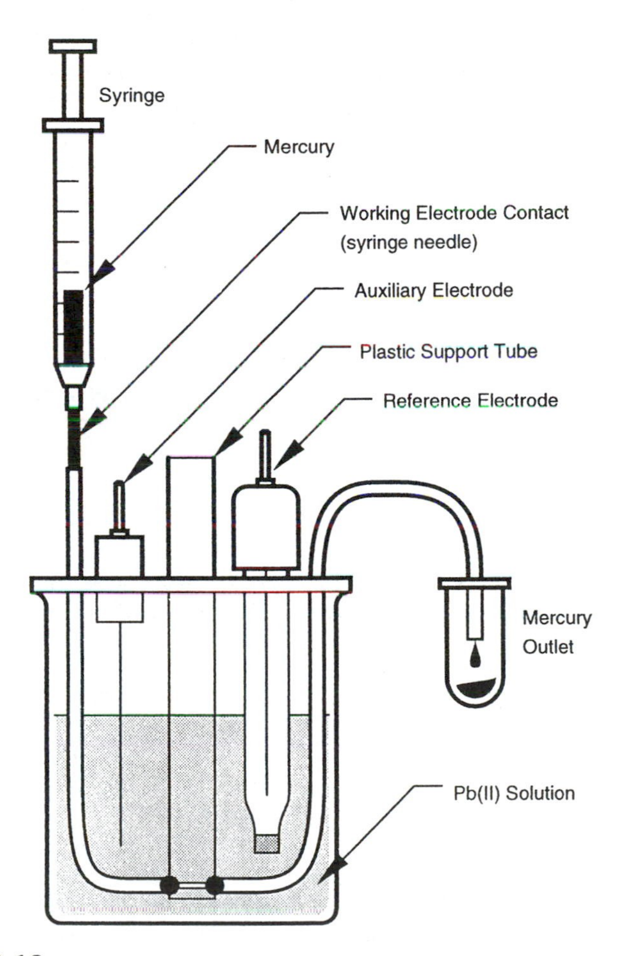


**Figure 14.10** Mercury thread electrode design and electrochemical cell arrangement (BAS).

dialysis tube prevents the electrode surface from being fouled by large organic molecules. Waste disposal is simplified since the mercury is collected at the outlet and does not contaminate the sample solution.

An effort has been made to apply mercury electrodes at positive potentials. Such electrodes are constructed from mercury covered by a thin layer of insoluble mercury compound. This electrode was used for the first time by Kuwana and Adams [65,66]. In their experiments mercury was covered by a thin layer of calomel produced by the oxidation of mercury in chloride solutions. *N,N'*-Dimethyl-*p*-phenylenediamine and ferrocyanide were oxidized at these passivated electrodes.

Using the HMDE covered by calomel, Kemula and coworkers studied the oxidation of several other substances [67]. Kublik studied the oxidation of various ions at the HMDE covered by a film of HgO [68]. These electrodes may be used at even more positive potentials than platinum electrodes in the same solutions. Although the mechanism of oxidation at these electrodes is very interesting, platinum and carbon electrodes are more useful in practical work since their properties and stability are not as dependent on the nature of the sample. When using such electrodes at positive potentials, some current due to oxidation of mercury is inevitable. This limits the application of passivated mercury electrodes to rather concentrated sample solutions.

This chapter has focused on the history, principles, and construction of working electrodes that utilize mercury. Several other chapters in this volume consider applications of such electrodes to both fundamental studies and quantitative analysis.

## REFERENCES


1. B. Kucera, *Ann. Phys. 11*:529 (1903).
2. J. Heyrovský, *Chem. Listy 16*:256 (1922); *Philos. Mag. J. Sci. 45*:303 (1923).
3. I. M. Kolthoff and J. J. Lingane, *Polarography*, Vol. 1, Interscience, New York, 1952.
4. L. Meites, *J. Am. Chem. Soc. 73*:2035 (1955).
5. I. M. Kolthoff and G. J. Kahan, *J. Am. Chem. Soc. 64*:2553 (1942).
6. O. H. Miller, *J. Am. Chem. Soc. 66*:1019 (1944).
7. I. M. Kolthoff and Y. Okinaka, in *Progress in Polarography* (P. Zuman and I. M. Kolthoff, eds.), Vol. 2, Wiley-Interscience, New York, 1962, p. 357.
8. D. C. Grahame, *J. Am. Chem. Soc. 68*:301 (1946).
9. I. Smoler, *Chem. Listy 47*:1667 (1953); I. Smoler, *Collect. Czech. Chem. Commun. 19*:238 (1954); I. Smoler, *J. Electroanal. Chem. 6*:465 (1963); E. Scarano, M. Forina, M. G. Bonicelli, and M. Mioscu, *J. Electronal. Chem. 21*:279 (1969).
10. L. Meites, *Polarographic Techniques*, 2nd ed., Wiley-Interscience, New York, 1965, p. 75; L. Meites, *J. Am. Chem. Soc. 73*:3724 (1951).
11. L. Novotný, I. Smoler, and J. Kuta, *Collect. Czech. Chem. Commun. 48*:963 (1983).

12. G. C. Barker, *Anal. Chim. Acta 18*:118 (1958).
13. G. Tessari, P. Delahay, and K. Holub, *J. Electroanal. Chem. 17*:69 (1968).
14. L. Novotný, J. Kuta, and I. Smoler, *J. Electroanal. Chem. 88*:161 (1978).
15. I. G. R. Gutz, L. Agnes, and J. J. Pedrotti, *Anal. Chem. 65*:500 (1993).
16. E. M. Skobets and N. S. Kavetskii, *Zavod. Lab. 15*:1299 (1949).
17. H. Gerischer, *Z. Phys. Chem. 202*:302 (1953).
18. W. Kemula, Z. Galus, and Z. Kublik, *Bull. Acad. Pol. Sci. Cl. III 7*:613 (1959); W. Kemula, Z. Galus, and Z. Kublik, *Rocz. Chem. 33*:1431 (1959); W. Kemula, Z. Galus, and Z. Kublik, *Bull. Acad. Pol. Sci. Cl. III 7*:723 (1959).
19. Solubility Data Series, *Metals in Mercury* (C. Hirayama, Z. Galus, and C. Guminski, eds.), Vol. 25, Pergamon Press, New York, 1986, pp. 330 and 369.
20. H. J. Antweiler, *Z. Elektrochem. 44*:831 (1938).
21. W. Kemula and Z. Kublik, in *Advances in Analytical Chemistry and Instrumentation* (C. N. Reilley, ed.), Vol. 2, Wiley-Interscience, New York, 1963, p. 123.
22. J. E. B. Randles and W. White, *Z. Elektrochem. 59*:666 (1955).
23. W. Kemula and Z. Kublik, *Rocz. Chem. 30*:1005 (1956); *Anal. Chim. Acta 18*:104 (1958).
24. W. Kemula, *Microchem. J. 11*:54 (1966).
25. J. J. Vogel, *J. Electroanal. Chem. 8*:82 (1964); H. Specker and H. Trub, *Z. Anal. Chem. 186*:123 (1962); F. von Sturm and M. Ressel, *Z. Anal. Chem. 186*:63 (1962); Z. Kowalski, *Rocz. Chem. 35*:365 (1961); Z. Kowalski, *J. Electroanal. Chem. 21*:P9 (1969); see also R. Neeb, *Inverse Polarographie und Voltammetrie*, Akademie-Verlag, Berlin, 1969.
26. R. Neeb, *Z. Anal. Chem. 171*:321 (1959).
27. W. M. Peterson, *Am. Lab. 11*(12):69 (1979).
28. *Pristroje pro Elektrochemii*, Laboratorni Pristroje, Prague, 1983.
29. L. Novotný, *Electroanalysis 2*:257 (1990).
30. Z. Kowalski, K. H. Wang, R. A. Osteryoung, and J. Osteryoung, *Anal. Chem. 59*:2216 (1987).
31. Z. Kowalski, *Analyst 113*:15 (1988).
32. Z. Kowalski and J. Migdalski, in *Contemporary Electroanalytical Chemistry* (A. Ivaska, A. Lewenstam and R. Sara, eds.), Plenum Press, New York, 1990, p. 149.
33. Z. Kowalski and J. Migdalski, *Electroanalysis 4*:915 (1992).
34. W. Ostwald, *Z. Phys. Chem. 1*:583 (1887).
35. F. Paschen, *Ann. Phys. 39*:43 (1890); *40*:36 (1890); *43*:568 (1891).
36. D. C. Grahame, E.M. Coffin, J. I. Cummings, and M. A. Poth, *J. Am. Chem. Soc. 74*:1207 (1952).
37. D. C. Grahame, R. P. Larsen, and M. A. Poth, *J. Am. Chem. Soc. 71*:2978 (1949).
38. B. B. Damaskin, N. B. Nikolaeva-Fedorovich, and A. N. Frumkin, *Dokl. Akad. Nauk SSSR 121*:129 (1958).
39. R. Payne, *J. Phys. Chem. 69*:4113 (1965); *70*:204 (1966).
40. E. Arndt, J. Bogus, J. Dojlido, I. Osinska, and P. K. Wrona, *Pol. J. Chem. 56*:307 (1982).
41. J. Heyrovský and J. Forejt, *Z. Phys. Chem. 193*:77 (1943).
42. J. Heyrovský, *Chem. Listy 40*:222 (1946).
43. J. Heyrovský, F. Sorm, and J. Forejt, *Collect. Czech. Chem. Commun. 12*:11 (1947).

44. J. Koryta, *Collect. Czech. Chem. Commun. 19*:433 (1954).
45. J. Koryta, *Collect. Czech. Chem. Commun. 20*:1125 (1955).
46. J. R. Weaver and R. W. Parry, *J. Am. Chem. Soc. 76*:6258 (1954); *78*:5542 (1956).
47. T. M. Florence, *J. Electroanal. Chem. 27*:273 (1970).
48. G. E. Bately and T. M. Florence, *J. Electroanal. Chem. 55*:23 (1974).
49. D. N. Hume and J. M. Carter, *Chem. Anal. (Warsaw) 17*:747 (1972).
50. M. Štulikova, *J. Electroanal. Chem. 48*:33 (1973).
51. R. M. Wightman and D. O. Wipf, in *Electroanalytical Chemistry* (A. J. Bard, ed.), Vol. 15, Marcel Dekker, New York, 1989, p. 267.
52. S. P. Kounaves and J. Buffle, *J. Electroanal. Chem. 216*:53 (1987).
53. S.P. Kounaves and J. Buffle, *J. Electroanal. Chem. 239*:113 (1988).
54. J. Golas, Z. Galus, and J. Osteryoung, *Anal. Chem. 59*:389 (1987).
55. P. Krasinski and Z. Galus, *J. Electroanal. Chem. 346*:135 (1993).
56. R. Kalvoda, *Collect. Czech. Chem. Commun. 22*:1390 (1957).
57. R. Kalvoda, *Anal. Chim. Acta 18*:132 (1958).
58. I.M. Kolthoff, J. Jordan, and S. Prager, *J. Am. Chem. Soc. 76*:5221 (1954).
59. I. M. Kolthoff and J. Jordan, *J. Am. Chem. Soc. 77*:3215 (1955).
60. A. G. Stromberg and V. E. Gorodovykh, *Zavod. Lab. 26*:46 (1960).
61. C. A. Streuli and W. D. Cooke, *Anal. Chem. 25*:1691 (1953).
62. C. A. Streuli and W. D. Cooke, *J. Phys. Chem. 57*:824 (1953).
63. D. J. Rossie and W. D. Cooke, *Anal. Chem. 27*:1360 (1955).
64. H. G. Jayaratna, *Anal. Chem. 66*:2985 (1994).
65. T. Kuwana and R. N. Adams, *J. Am. Chem. Soc. 79*:3610 (1957).
66. T. Kuwana and R. N. Adams, *Anal. Chim. Acta 20*:51, 60 (1959).
67. W. Kemula, Z. Kublik, and J. Taraszewska, *J. Electroanal. Chem. 6*:119 (1963).
68. Z. Kublik, Habilitation thesis, University of Warsaw, 1968.

# 15

# Solvents and Supporting Electrolytes

**Albert J. Fry** *Wesleyan University, Middletown, Connecticut*

## I. INTRODUCTION

Almost every electrochemical experiment is carried out in a medium consisting of a solvent and supporting electrolyte. Many such combinations have been used by electrochemical experimenters; as a glance at the Appendix will show, selection of a solvent and electrolyte has been the subject of much discussion in the literature. Yet the problem is not intrinsically complicated; a handful of solvent–electrolyte combinations would probably suffice for the majority of electrochemical applications. For this reason, a few particularly good solvent systems (the term "solvent system" is used here to describe the medium consisting of both solvent and supporting electrolyte) are recommended here, in lieu of an extended but uncritical listing of all such systems. The systems to be recommended have been chosen from among many that have been reported. However, because the novice will certainly want to develop a feel for what makes a "good" solvent system, the first part of the chapter discusses desirable criteria, as well as the ways in which electrochemical results may depend on the nature of the particular solvent–electrolyte combination employed. Because special situations frequently do require special solutions, Section III describes some less common solvents and electrolytes that have been found useful in special situations. Finally, the Appendix offers a guide to further readings of a more comprehensive nature.

### A. Solvent Selection Criteria

There is no universal solvent, and even for a given application one rarely finds an ideal system. One must factor some informed guesswork into one's choice of solvent and electrolyte. In order to optimize conditions for an electrode reaction, one must consider how its chemical and electrochemical features, for example, the chemical properties of expected intermediates, might be affected by the solvent or electrolyte. A solvent is chosen such that its merits outweigh its disadvantages in a particular situation. A good solvent system for one type of experiment may be wholly unsuitable for other applications. The most important properties that the ideal solvent system ought to possess are *electrochemical inertness* (stability over a wide range of potentials), *high electrical conductivity*, *good solvent power, chemical inertness*, *availability in pure form* or *ease of purification*, and *low cost*, although others must sometimes be considered. We now examine each of these in detail.

*Electrochemical Inertness*

The solvent system should not undergo any electrochemical reaction over a wide range of potentials from very positive (strongly oxidizing) to very negative (strongly reducing). The potential at which electrochemical reaction of the solvent system (either the supporting electrolyte or the solvent itself) commences is known variously as the *solvent breakdown potential*, *solvent decomposition potential*, or *solvent background limit*. Certain systems can be used over a very wide potential range; one of the best for this purpose is a solution of tetrabutylammonium hexafluorophosphate in acetonitrile, which exhibits anodic and cathodic breakdown limits of +3.4 and -2.9 V (vs. SCE), respectively [1]. In practice, however, one rarely needs both extended positive and negative ranges in a single set of experiments; solvents are frequently used that exhibit either a very negative or very positive decomposition potential, but not both. Thus, both nitrobenzene [2] and methylene chloride [3] have been used for studies of oxidation processes because they are hard to oxidize, notwithstanding the fact that they are quite easily reduced. Conversely, liquid ammonia and methylamine are good solvents for electrochemical reductions, although poor for oxidations [4]. It is reported that potentials as positive as +4.5 V (vs. SCE) can be reached in liquid $SO_2$ [5].

The potential limit is frequently set by the electrochemical behavior of the supporting electrolyte, not the solvent. For example, dipolar aprotic solvents such as acetonitrile (AN), dimethylformamide (DMF), and dimethyl sulfoxide (DMSO) are very difficult to reduce; the reduction of alkali metal ions to the corresponding metals (ca. -2 V vs. SCE) is the potential-limiting process in such solvents, and by use of the more difficult-to-reduce tetraalkylammonium salts one can reach considerably more negative potentials. The decomposition potentials of tetraalkylammonium ions at mercury themselves become more negative as the alkyl groups become larger ($Me_4N^+$: -2.65 V; $Bu_4N^+$: -2.88 V) [6]. One can frequently extend the accessible cathodic range by using a large tetraalkylammonium ion, but double-layer and adsorption effects can negate this generalization [6,7]. DMF and DMSO are not suitable as solvents for anodic studies because they are fairly easily oxidized, but as noted earlier, AN is very difficult to oxidize, and in this solvent the decomposition potential of the supporting electrolyte is usually the limiting process. At very positive potentials, oxidation of the anion of the supporting electrolyte is often the limiting process. Not only should easily oxidized ions such as bromide and iodide not be used for anodic studies, but even nitrate and perchlorate salts should be avoided when one wishes to operate at very positive potentials; at these potentials tetrafluoroborate and hexafluoroborate salts are preferred because of their stability to oxidizing conditions [1].

Electrochemical reaction of the solvent or supporting electrolyte need not be undesirable. Sometimes one can take advantage of such behavior by using the product of such a reaction as a chemical reagent. For example, one can generate bromine by anodic oxidation of bromide ion or lithium by cathodic reduction of lithium ion. There can be real advantages to electrochemical generation of reagents this way instead of using the bulk reagent, for example, avoiding the need to handle dangerous substances and being able to add the reagent in regulated amounts by careful control of the electrolysis current [8].

*Electrical Conductivity*

In order to support passage of an electric current, the solvent system should have low electrical resistance. This implies that the solvent should have a moderately high dielectric constant ($\geq$10); the prevalence of ion pairing and even multiple ionic association in less polar solvents results in relatively low ionic mobility and conductance. The fact that nonpolar solvents tend not to be very good solvents for salts in the first place makes it even harder to contemplate using them. A number of the most common solvents for organic electrochemistry have satisfactory dielectric constants (DMF: 36.7; AN: 37.5; DMSO: 46.7), and their electrolytic solutions have acceptably high conductances [9–11]. Solvents of substantially lower polarity than these, however, have been used successfully despite their high electrical resistance (e.g., tetrahydrofuran, THF [12], and methylene chloride [3], whose dielectric constants are 7.3 and 8.9, respectively [13]). Lower conductivity can be tolerated more readily in low-current applications such as voltammetry than in preparative electrolyses. Although voltammograms can be badly distorted by uncompensated resistance under conditions of high electrical resistance, the distortion can be largely corrected by using high electrolyte concentrations (not always possible in resistive solvents, which tend to be poor solvents for salts) and by computational correction of the voltammogram [11]. A new element has entered the scene in recent years with the development of small-diameter (diameter $\leq$10 $\mu$m) microelectrodes [14]. These can be used in highly resistive media [11,15–19], including the absence of supporting electrolyte [20], and in fact even in the gas phase [21].

*Good Solvent Power*

An electrochemical solvent must be able to dissolve a wide range of substances at acceptable concentrations. In general, this means that electrolytes must be soluble at least to the extent of 0.1 M, while the electroactive substance must be soluble to the extent of 0.1–1 mM for application of the various electroanalytical techniques, and higher concentration for preparative electrolyses. Salts in which either the cation or anion, and preferably both, are large, such as tetraalkylammonium perchlorates, hexafluorophosphates, and tetrafluoroborates, generally exhibit the greatest solubility in organic solvents. The latter two are becoming more popular than perchlorates because of safety hazards with perchlorates, especially in acidic media. Ethyltributylammonium tetrafluoroborate is quite soluble in nonpolar solvents [12], as is tetrabutylammonium hexafluorophosphate.

*Chemical Inertness*

The solvent should not react with the electroactive material nor with intermediates or products of the electrode reaction under investigation. No one solvent can reasonably be expected to be unreactive toward all of the many kinds of reactive species that can be generated electrochemically. When selecting a solvent, one does have to make some informed guesses about likely intermediates and products, based on chemical experience. For example, many organic and anodic processes generate cationic, that is, electrophilic species [22]; one would therefore not use a nucleophilic solvent if one wants such intermediates to be long-lived. This is equally true for organic and inorganic species. Acetonitrile is frequently used in anodic studies because of its high conductivity and wide accessible potential range, but it is moderately nucleophilic; it can react with organic intermediates and replace ligands on inorganic intermediates. Attempts to characterize anodically generated intermediates are therefore frequently carried out in a less nucleophilic solvent such as methylene chloride or one of the "superacid" solvents to be discussed in Section II.A. Conversely, reduction of organic substrates frequently forms anionic intermediates, and one must therefore avoid the use of electrophilic solvents, in particular protic substances, to avoid acid-base reactions between such intermediates and the solvent. Finally, one would like the solvent to be reasonably stable, so that purification, preparation, and storage of standard solutions will present no major problems.

*Liquid Range*

The solvent should exhibit a convenient liquid range; that is, it should be neither too low-boiling nor too high-boiling. The former makes for potential safety and fire hazards and difficulty in storing solutions, particularly when degassing them, and the latter makes purification by distillation more difficult. On the other hand, the definition of convenient liquid range depends on the particular experiment involved, in particular the temperature at which it is to be carried out. High-boiling solvents are preferred for experiments at higher temperatures, whereas low-boiling solvents are better for low-temperature electrochemistry (they retain favorable viscosity properties at low temperatures). Other factors to be considered are (a) whether (and how) the solvent must be removed and solutes recovered at the end of the experiment, and (b) the particular purification and transfer procedures to be used in the experiment.

*Miscellaneous Considerations*

Solvents and electrolytes should also be inexpensive, nontoxic, and nonflammable. The latter two characteristics are not well satisfied by most organic solvents, but with reasonable safety precautions and reasonable ventilation they can be used routinely without incident. Another solvent property, viscosity, may be of importance on occasion. High viscosities are useful when one wishes to extend the time interval over which mass transport occurs purely by diffusion, such as for potential-step experiments, but a low-viscosity solvent is preferred when efficient mass transport is required, as in preparative electrolyses.

## II. RECOMMENDED SOLVENTS AND ELECTROLYTES

### A. Anodic Electrochemistry

Solutions of inorganic salts in water meet many of the preceding desirable solvent criteria, and have been used for innumerable electrochemical studies. Much inorganic electrochemistry has been carried out in water with little difficulty. Organic and organometallic compounds either are not very soluble in water or are reactive toward it; this has stimulated interest in the use of nonaqueous solvents in electrochemistry. Ethanol and methanol are good organic solvents, inexpensive, readily available in high purity, dissolve salts readily, and in general have much to commend them for electrochemical purposes. On the other hand, alcohols are relatively easily oxidized and are highly nucleophilic. For this reason, aprotic organic solvents are more commonly used, and of these, the most commonly used are acetonitrile and *N,N*-dimethylformamide. *Acetonitrile* (AN) is probably the most widely used solvent for organic anodic electrochemistry, and we believe that it is the preferred solvent for most such purposes. It is a good solvent for organic compounds and many inorganic and organic salts, has a high dielectric constant and a convenient liquid range (mp –45.7°C, bp 81.6°C), and is transparent in the ultraviolet region. (For purification of AN, see Sec. VI.) Recommended electrolytes include tetrabutylammonium hexafluorophosphate (TBAHFP) and tetrafluoroborate (TBATFB), both of which are readily obtained in pure form (Sec. VI), have high solubilities in AN, and are stable to very positive potentials. These salts, particularly the hexafluorophosphate, are rather insoluble in water and can often be separated from reaction products by evaporation of the AN followed by extraction with a nonpolar solvent, chromatography, or volatilization of the desired product. Sometimes separation of the electrolyte from an organic product presents difficulties, however. An inorganic electrolyte (e.g., $NaBF_4$, $LiClO_4$, or $NaClO_4$) may be necessary in such cases.

Acetonitrile is rather nucleophilic; this is a disadvantage when the electrochemical reaction produces a reactive electrophilic intermediate, or when the substrate is an inorganic or organometallic species containing a replaceable ligand. In this case, solutions of TBATFB or TBAHFP in methylene chloride have frequently been used as alternatives, although the latter solutions are reported to decompose slowly [23]. Meyer, however, has recently found that 1,2-difluorobenzene is a better solvent than methylene chloride [24]. Other choices, although chemically more reactive than the preceding halogenated solvents, include trifluoroacetic acid (no electrolyte needed) [25], liquid sulfur dioxide [26], an aluminum chloride-alkali halide melt [27], or trifluoromethanesulfonic acid [28]. Anodically generated reactive electrophilic intermediates are also relatively long-lived at low temperatures [29] or in very dry AN [30,31].

### B. Cathodic Electrochemistry

Water and alcohols are often useful in inorganic electrochemistry. Dimethylformamide (DMF), acetonitrile, and dimethyl sulfoxide (DMSO) are all excellent solvents for reductive electrochemistry of organic compounds. TBAHFP and TBATFB are good electrolytes for use in these solvents. Despite their rather high boiling points (156°C and 189°C, respectively), *DMF and DMSO are particularly good solvents for cathodic electrochemistry*. DMF is fairly easily purified (Sec. VI.B) and DMSO is in fact fairly pure as received. Both are excellent solvents for both organic and inorganic solvents. Water (which is always present in polar aprotic solvents) is a poor proton donor in DMSO and DMF and can be reduced by appropriate procedures (Sec. VI.A). Probably the most widely used organic solvents for cathodic electrochemistry are DMF and acetonitrile. DMF has a number of desirable properties, but when wet is prone to slow decomposition, liberating dimethylamine. Decomposition is much less of a problem in dry DMF, which can be prepared by vacuum distillation from calcium hydride, followed by treatment with 3-Å molecular sieve [32]. Acetonitrile would be our primary recommended solvent for cathodic electrochemistry except for problems related to its purification. Very pure, dry AN can be obtained by proper treatment, but this can be tedious [33]. The efficacy of a number of simpler procedures in the literature for drying AN has been disputed [34].

## III. SOME OTHER SOLVENTS

Many solvents and electrolytes have been used in electrochemical applications. In the preceding sections we suggested a few solvents for general use. It may happen, however, that none of these is acceptable for a particular application. There are many exhaustive discussions of solvent systems available in the literature, which we shall not duplicate here (see the Appendix). A few solvents deserve special mention here, because each has a property likely to be useful in certain special situations.

### A. Low-Temperature Electrochemistry

Low-temperature applications (Chap. 16) present special problems. With most solvents, solvating power decreases (this is a particular problem with supporting electrolytes, which have to be soluble in concentrations of 0.1 M or higher for conventional electrochemistry) and viscosity increases as the temperature is lowered. Both effects result in higher solution resistance at low temperatures. Reilley and Van Duyne [29] and, more recently, Evans [11] and Murray [15] and their coworkers have examined a variety of solvent combinations and have found several that suffer the least degradation of solvent properties with decreasing temperature. Reilley and Van Duyne recommended butyronitrile for use with conventional electrodes at temperatures as low as −105°C. The development of microelectrodes in recent years has been a boon to those interested in low-temperature electrochemistry, since these electrodes can function well in highly resistive media (Chap. 12). Evans recommended acetonitrile for operations down to −40°C and butyronitrile to −80°C for high-speed low-temperature voltammetry at microelectrodes. Murray found a number of solvent mixtures that are suitable for voltammetry at ultra-low temperatures: 1:1 butyronitrile (BN):ethyl bromide (with 0.2 M TBAP as electrolyte) is usable to −148°C; 1:1:1:1 isopentane: methylcyclopentane:BN:ethyl bromide to −158°C; and remarkably, 1:1 BN:ethyl chloride, in which voltammetric data could be obtained at temperatures as low as −185°C! The use of microelectrodes also permits voltammetric measurements in frozen solvent [16] and in solids [17]. Gosser has carried out voltammetry in frozen DMSO with conventional electrodes, but has suggested that the electrochemistry actually occurs in pools of liquid solvent within the frozen medium [18].

### B. Solvated Electrons

Solvated electrons are readily prepared in hexamethylphosphoramide (HMPA), ammonia, and methylamine [19]. Lithium chloride is the preferred electrolyte for this application.

### C. Nonpolar Media

Nonpolar media are of interest as electrochemical solvents because they permit the investigator to observe the properties of polar intermediates undistorted by interaction with the solvent medium, but they present major problems because of their high electrical resistance. This is compounded by the poor solubility of most electrolytes in such solvents. The high electrical resistance causes severe signal distortion in low-current experiments (voltammetry) and resistive heating (which can be so severe as to cause solvent boiling) in high-current experiments (preparative-scale electrolyses). As mentioned in the preceding section, the use of microelectrodes can largely mitigate the resistance problem in voltammetry in such media.

There are hydrocarbon-like media that can permit voltammetric studies. It turns out that molten naphthalene and its 1-chloro and 1-methyl derivatives are highly electrically conductive and good solvents for organic salts at elevated temperatures (1.0 M TBAP solutions can be prepared) [35]. Well-formed ac and cyclic voltammograms can be obtained at 150°C. These media were found useful for substrates such as phthalocyanines, which are completely insoluble in all common solvents near room temperature. A barely explored idea in nonaqueous electrochemistry is that of incorporating the functions of solvent and supporting electrolyte into a single substance, that is, a liquid or low-melting organic salt. A solution of tetrabutylammonium tetrafluoroborate (TBATFB) in toluene separates into two liquid phases; the upper is pure toluene, and the lower is a novel "liquid electrolyte" whose composition corresponds to three molecules of toluene and one of TBATFB [36]. Fry and Touster examined the properties of this substance as a voltammetric solvent and found them to be remarkably similar to those of DMF, AN, or DMSO [37]. Tetrahexylammonium benzoate, a liquid at room temperature, has been found suitable for electrochemical use [38]. Better candidates might be found among the many low-melting organic salts now available [39].

## IV. SOLVENT- AND ELECTROLYTE-DEPENDENT PHENOMENA

Electrochemical reactions are unavoidably influenced by the nature of the solvent and supporting electrolyte. We have already alluded several times to the possibility of chemical reaction of electrochemically generated intermediates with the solvent. One must always be on guard against this possibility. The nature of the problem is exemplified by a study of the electrochemical reduction of iodobenzene in DMF containing tetraethylammonium bromide [40]. A variety of experiments demonstrated that the highly basic and nucleophilic phenyl anion is generated in this reaction. Benzene is the major product; presumably it is formed by attack by the phenyl anion on traces of moisture in the solvent. When the water content in the solvent was lowered, benzene was still found as the major product, but now it was found to be formed by proton attack upon the cation of the supporting electrolyte (Hoffmann elimination). When the electrolyte was changed to lithium bromide to prevent Hoffmann elimination, phenyl anion proceeded to react with the solvent by a combination of proton abstraction and nucleophilic addition! This is a reminder that carbanions generated in dipolar aprotic solvents such as DMF are especially reactive; they cannot bind covalently to tetraalkylammonium counterions, and even ion pairing to the counterion is minimal in such solvents because of their high dielectric constants. Electrochemically generated phenyl anions might perhaps be longer lived in a less reactive solvent (e.g., THF) with a metal salt as electrolyte. The anodic counterpart of this problem has already been mentioned: the fact that anodically generated cations readily attack nucleophilic solvents such as acetonitrile, mandating the use of *electrophilic* solvents when it is desired to extend the lifetime of such cations. All of the special solvent systems mentioned at the end of Section II meet this criterion.

The solvent and electrolyte can affect an electrode process in a number of ways other than by reacting chemically with the electrolysis intermediates or products. The rate of heterogeneous electron transfer from the electrode to the electroactive substance can be affected by the structure of the electrical double layer at the electrode surface, which in turn is dependent on the nature of the solvent and supporting electrolyte [41]. This isn't the only role the electrolyte can play when incorporated into the double layer. Acidic cations, such as triethylammonium and guanidinium ions, can protonate short-lived intermediates generated at the electrode surface before such intermediates can escape into bulk solution [42]. Similarly, preferential solvation of small ions by one constituent of a mixture can raise the local concentration of this substituent at the electrode surface if the ion is in the double layer (anions at potentials positive of the potential of zero charge [PZC], cations at potentials negative of PZC). Examples of both cathodic [42] and anodic [43] electrochemistry thought to be affected this way have been reported. A third type of surface electrolyte effect has been reported, in which optically active products have been obtained from electrode reactions in which an optically active supporting electrolyte has been used [44]. The degree of asymmetric induction obtained in this fashion has generally been small. Reasons for this have been suggested [45].

The most common electrochemical effects exerted in bulk solution are related to association (solvation, ion-pairing, complex formation, etc.) with the electroactive substance or electrochemically generated intermediates [4,19]. The importance of solvation can be gauged by comparing calculated and measured values of the parameter $\Delta E_{1/2}$ (defined as the difference, in volts between the half-wave potentials of the first and second polarographic waves) exhibited by polycyclic aromatic hydrocarbons (PAH) in dipolar aprotic solvents [46,47]. It can be shown that $\Delta E_{1/2}$ is related to the equilibrium constant for disproportionation of the aromatic radical anion into neutral species and dianion, that is,

$$2Ar^{\dot{-}} \rightleftharpoons ArH + ArH^{2-}$$

The computed value of $\Delta E_{1/2}$ by a reliable computational method [48] is ~5 V [49]. The experimental values are generally only about one-tenth of this (0.4–0.6 V) in DMF, AN, or DMSO, showing that the disproportionation equilibrium is significantly perturbed by solvation [46,47,49]; the computed value is essentially a gas-phase, solvent-free value. The disproportionation equilibrium depends on the nature of the supporting electrolyte, demonstrating that ion pairing is operative even in solvents as polar as these, although it appears to involve only the dianion, not the radical anion. In low-dielectric-constant solvents, where ion pairs and even higher aggregates are prevalent, or with electrolytes (such as lithium salts) that form strong covalent bonds [50], association between the electrolyte and charged intermediates plays an even greater role in affecting measured potentials. In fact, in the specific case of radical anion disproportionation under discussion here, $\Delta E_{1/2}$ can even become negative; that is, the two one-electron waves in polar media coalesce into a single two-electron wave [49,51].

## V. EXPERIMENTAL PROCEDURES

### A. Obtaining Dry Solvent

Water is the most ubiquitous impurity in organic solvents. Considering how sensitive many reactions are to the presence of water, the literature on drying solvents is surprisingly full of contradictions and misinformation. Anyone doubting this statement is referred to a remarkable series of papers by Burfield [32,34,52,53]. The Introduction sections to these papers will provide a healthy dose of reality to those who feel they can rely solely on their chemical intuition when deciding how to purify their organic solvents. Burfield developed a highly accurate method for determining the water content of organic solvents and then used it to assess the dryness of a number of common organic solvents, each of which was dried by a variety of methods. (The analytical method is based upon treatment of the anhydrous solvent with tritiated water, followed by drying and analysis of the solvent for residual tritium by scintillation counting [54].) Burfield made a number of useful discoveries in the course of this work, including the following: (a) a number of common laboratory drying procedures are almost totally ineffective with some solvents; (b) a reagent that is good for drying one solvent may be rather poor with another; and (c) probably the best general method for drying most solvents is by the use of molecular sieves, assuming they are used properly, that is, (1) activating 3-Å molecular sieves for 15 h at 300°C, (2) allowing them to cool in a $P_2O_5$ dessicator, (3) treating the solvent with activated sieve (5–10% w/v) for 6 h to 7 days, and (4) allowing the solvent to stand over a second batch of fresh molecular sieve. This procedure was shown to reduce water levels in the following solvents to the indicated levels: DMF: ≤1.5 ppm; AN: ≤0.6 ppm; DMSO: ≤10 ppm. A series of experiments in which the water content of DMF standing over molecular sieve was monitored by gas chromatography confirmed the fact that several days are necessary to achieve the lowest water levels [55]. An important qualification must be added at this point, since it was not considered by Burfield: His analysis assumes that the only solvent impurity is water. This is frequently not the case. DMF contains readily detectable quantities of dimethylamine, DMSO contains traces of dimethyl sulfide, AN can contain acrylonitrile, THF and other ethers contain peroxides, etc. Thus it is frequently necessary to subject the solvent to a preliminary treatment to remove such impurities and lower its water content before drying. (Commercial DMF is frequently ≥0.1 M in water, for example.) Peroxides can usually be removed by passing the solvent through a column of activated alumina.

Dipolar aprotic solvents are generally hygroscopic, and contact with the atmosphere should be avoided after drying. For critical applications, the solvent should be distilled into the electrochemical cell [56] or in a dry box, or passed through a column containing the dry agent repeatedly in a sealed system [57]. Simpler and almost as satisfactory alternatives include drying the solvent over a molecular sieve either in a dropping funnel attached to the cell, followed by direct addition of the solvent to the cell [55], or in a separate flask, followed by solvent transfer using cannula techniques [58]. Supporting electrolytes are frequently hygroscopic and should be dried before use; this is readily done in an Abderhalden apparatus using heat and vacuum, or for work requiring scrupulous drying, by heating the electrochemical cell containing the electrolyte in vacuo before adding the solvent.

Water in organic solvents can be determined by a variety of methods. One of the fastest involves gas chromatographic analysis on columns packed with Porapak Q or N [59,60].

## B. Solvent Purification Procedures

In view of the more extended discussions presented in the references cited in the Appendix, we present here only selected procedures for purification of materials for general use, and only for those advocated in Section II, since the large majority of applications will be satisfied by one or another of these.

*Acetonitrile (AN)*

Many purification procedures have been reported for AN. Indeed, there is probably more disagreement over just how to dry and purify this material than any other electrochemical solvent. The confusion, or rather, disagreement among authors arises in part because there are different commercial processes for production of AN, each generating different impurities. Common impurities are water and acrylonitrile. Kuwana and coworkers [61] surveyed a variety of procedures for purification of AN, and concluded that the best procedure involves simply passing the solvent twice through a column packed with alumina that has been dried at 400°C. Parker has claimed that successive treatments with highly activated alumina reduce the water content below 1 ppm [31]. We have found that this procedure works well for general use as long as the alumina is replaced regularly (when voltammetric currents begin to increase). However, the activated alumina procedure is relatively expensive when large amounts of solvent are required. Two treatments with 3 Å activated molecular sieves reduce the water content to ~0.6 ppm [52]. The 4-Å sieves are much less efficient; they afford material of 500 ppm water content [52]. A procedure consisting of distillation from $P_2O_5$ or $B_2O_3$, followed by treatment with molecular sieves, should also reduce the water content below 10 ppm [34]. A recent four-step procedure has been claimed to afford very high-purity acetonitrile [33]. If only a small amount of AN is needed, one may use spectrophotometric-grade AN, after drying over molecular sieve or activated alumina.

*Dimethylformamide (DMF)*

The principal impurities in commercial DMF are water, dimethylamine, and formic acid. Vacuum distillation from calcium hydride, followed by successive treatments with activated molecular sieve, affords material of water content 1.5 ppm [32].

*Dimethyl Sulfoxide (DMSO)*

This solvent is fairly pure as received. The only impurities are water and traces of dimethyl sulfide. Dimethyl sulfide can be removed by a preliminary vacuum distillation, or by bubbling an inert gas through the solution for 10 to 20 min before use. The water content can be reduced to 10 ppm by sequential treatment with two batches of 3 Å molecular sieves (activation of the sieve at 500°C for 16 h in an inert atmosphere has been advocated [62], but activation for 15 h at 300°C should suffice [52]). Calcium hydride and a number of other basic reagents have been advocated as drying agents, but in fact all of these are ineffective [32]. Tetrabutylammonium hexafluorophosphate exhibits good solubility in THF.

*Tetrahydrofuran (THF)*

The primary contaminants in commercial THF are water and organic peroxides (THF is notoriously prone to peroxide formation and should not be exposed to air after purification). Molecular sieves reduce the water content to 28 ppm [52] but do not destroy peroxides. Modern organic laboratories typically purify THF by distillation from sodium benzophenone ketyl under a nitrogen atmosphere, which has the virtue of destroying both peroxides and water [63]. This is the method of choice for obtaining high-quality THF.

*Methylene Chloride*

This solvent is commercially available in a high state of purity. It can be dried simply by allowing it to stand over molecular sieve for 7 days, after which the water content is less than 0.1 ppm [52].

*Ethanol*

The solvent is available commercially in pure form. Distillation from magnesium ethoxide [64] affords fairly dry material ($H_2O$ content = 50 ppm), but a better job can be done: $H_2O$ content = 18 ppm can be obtained by allowing the alcohol to stand over powdered 3-Å molecular sieve [52]. It is important that the molecular sieve be powdered; the more commonly found beads only reduce the water content to 99 ppm.

## C. Preparation and Purification of Electrolytes

A number of studies on the relative merits of different salts as supporting electrolytes, including measurements of the resistance of various solvent–electrolyte combinations at different electrolyte concentrations, have been reported [9–11]. Several electrolytes appear to be especially useful for general work.

*Tetrabutylammonium hexafluorophosphate* (TBAHFP, $Bu_4NPF_6$) is a good general-purpose electrolyte. It is soluble in a wide variety of solvents, including even liquid sulfur dioxide [26] and methylene chloride (it is reported to decompose after a few hours in the latter solvent [23,24] but in the author's laboratory we have found solutions of this salt to be quite stable in DMF), is only slightly hygroscopic, exhibits a wide potential range, and is readily prepared by the following procedure [64]. Tetrabutylammonium bromide (100 g, 0.27 mol) is dissolved in the minimum amount of reagent acetone. To this is added, with stirring, an acetone solution of ammonium hexafluorophosphate (50 g, 0.31 mol). It is important to maintain at least a 5% excess of hexafluorophosphate ion. The precipitate of ammonium iodide is removed by filtration. Water is then added slowly to precipitate TBAHFP, which is removed by filtration and washed well with water to remove all extraneous ammonium salts. The crude product can be reprecipitated from a 5% solution of ammonium hexafluorophosphate to insure complete removal of iodide. The white solid is recrystallized from ethanol-water and dried *in vacuo* at 100°C for 10 h. The yield is ~80%.

*Tetrabutylammonium tetrafluoroborate* (TBATFB, $Bu_4NBF_4$) is another excellent electrolyte for general use. Like other tetrafluoroborate salts, it is somewhat more soluble in both organic compounds and water than TBAHFP, and is more stable than TBAHFP in methylene chloride. House and coworkers have described procedures for the synthesis of TBATFB and other tetraalkylammonium tetrafluoroborates [9]. Thus, a solution of 8.4 g (0.025 mol) of tetrabutylammonium bromide in a minimum amount of water (18 mL) is treated with 3.6 mL (0.026 mol) of aqueous 48–50% tetrafluoroboric acid ($HBF_4$). The resulting mixture is stirred for a few minutes and the crystalline TBATFB is collected by filtration, washed with ice-cold water until the washings are neutral, and dried. The crude salt, 6.3 g (79%), can be dissolved in ethyl acetate and precipitated with pentane to afford 6.0 g (75%) of the salt, which can then be dried as with TBAHFP.

## APPENDIX: OTHER LITERATURE

One of the most comprehensive reviews on solvent use and purification was prepared by C.K. Mann [13]. Many solvents are discussed, and for each particular solvent, suitable supporting electrolytes and reference electrodes are presented. Lund and Iversen [65], among others [66], have also surveyed the use, purification, and properties of solvents. There are overlaps among all of these reviews and the present one, but each has its own special flavor and the different viewpoints make for instructive comparisons. There are also a number of articles devoted to specific solvents, such as DMSO [62,67], DMF [68], methylene chloride [69], and pyridine [70].

As mentioned earlier, tetraalkylammonium hexafluorophosphate and tetrafluoroborate salts are replacing the corresponding perchlorates for general use. There does not appear to be any reason to use the perchlorates in view of the potential safety (explosion) hazard that they present [42,71,72]. A number of methods for preparation of hexafluorophosphate and tetrafluoroborate salts have been reported [9–11,57,73].

Tetraalkylammonium tosylates [74] and trifluoromethanesulfonates [72] are also excellent electrolytes. Although tetraalkylammonium ions are favored as the cations for supporting electrolytes because of their wide potential range, other cations are sometimes used for special applications—for example, methyltriphenyl phosphonium, whose tosylate is freely soluble in methylene chloride, and other fairly nonpolar solvents [74] or metal ions (lithium salts tend to have the best solubility in organic solvents) where undesirable reactions of the tetraalkylammonium ion might occur [13,75]. The properties of many electrolytes suitable for nonaqueous use have been surveyed [76].

## ACKNOWLEDGMENTS

Financial support by the National Science Foundation and the State of Connecticut under its High Technology program are gratefully acknowledged. This article was written during a sabbatical at the University of California at Santa Barbara under the hospitality of Prof. R. Daniel Little. Joseph Leonetti provided some key references on the drying properties of molecular sieves.

## REFERENCES


1. M. Fleischmann and D. Pletcher, *Tetrahedron Lett.* 6255 (1968).
2. L. Marcoux, P. Malachesky, and R. N. Adams, *J. Am. Chem. Soc. 89*:5766 (1967).
3. O. Hammerich and V. D. Parker, *J. Am. Chem. Soc. 96*:4289 (1974).
4. (a) A. J. Bard and A. Demortier, *J. Am. Chem. Soc. 95*:3495 (1973); (b) W. H. Smith and A. J. Bard, *J. Am. Chem. Soc. 97*:5203 (1975); (c) O. R. Brown, Specialist Periodical Reports in Electrochemistry, Vol. 4, Chemical Society, London, 1974, p. 55ff; (d) R. A. Benkeser and S. J. Mels, *J. Org. Chem. 35*:361 (1970).
5. E. Garcia, J. Kwak, and A. J. Bard, *Inorg. Chem. 27*:4377 (1988).
6. A. J. Fry and R. L. Krieger, *J. Org. Chem. 41*:54 (1976).
7. S. R. Missan, E. I. Becker, and L. Meites, *J. Am. Chem. Soc. 83*:58 (1961).
8. J. Janata and H. B. Mark, Jr., in *Electroanalytical Chemistry*, Vol. 3 (A. J. Bard, ed.), Marcel Dekker, New York, 1969, p. 1.
9. H. O. House, E. Feng, and N. P. Peet, *J. Org. Chem. 36*: 2371 (1971).
10. K. M. Kadish, J. Q. Ding, and T. Malinski, *Anal. Chem. 56*:1741 (1984).
11. W. J. Bowyer, E. E. Engelman, and D. H. Evans, *J. Electroanal. Chem. 262*:67 (1989).
12. T. J. Curphey, C. W. Amelotti, T. P. Layloff, R. L. McCartney, and J. H. Williams, *J. Am. Chem. Soc. 91*:2817 (1969).
13. C. K. Mann, in *Electroanalytical Chemistry*, Vol. 3 (A. J. Bard, ed.), Marcel Dekker, New York, 1969, p. 57.
14. R. M. Wightman, in *Electroanalytical Chemistry*, Vol. 15 (A. J. Bard, ed.), Marcel Dekker, New York, 1989, p. 267.
15. J. T. McDevitt, S. Ching, M. Sullivan, and R. W. Murray, *J. Am. Chem. Soc. 111*:4528 (1989).
16. A. M. Bond, M. Fleischmann, and J. Robinson, *J. Electroanal. Chem. 180*:257 (1984).
17. W. Gorski and J. A. Cox, *J. Electroanal. Chem. 323*:163 (1992).
18. D. K. Gosser and Q. Huang, *J. Electroanal. Chem. 267*:333 (1989).
19. (a) L. A. Avaca and A. Bewick, *J. Electroanal. Chem. 41*:405 (1973); (b) O. R. Brown and P. O. Stokes, *J. Electroanal. Chem. 57*:425 (1974).
20. A. M. Bond, M. Fleischmann, and J. Robiasa, *J. Electroanal. Chem. 168*:299 (1984).
21. J. Ghoroghchian, S. Pons, and M. Fleischmann, *J. Electroanal. Chem. 317*:101 (1991).
22. A. J. Fry, in *Synthetic Organic Electrochemistry*, 2nd ed., John Wiley and Sons, New York, Chap. 8.
23. R. S. Drago, M. J. Desmond, B. B. Corden, and K. A. Miller, *J. Am. Chem. Soc. 105*:2287 (1983).
24. T. R. O'Toole, J. N. Younathan, B. P. Sullivan, and T. J. Meyer, *Inorg. Chem. 28*:3923 (1989).
25. (a) O. Hammerich, N. S. Moe, and V. D. Parker, *Chem. Commun.* 156 (1972); (b) U. Svanholm and V. D. Parker, *Tetrahedron Lett.* 471 (1972).

26. (a) L. L. Miller and E. A. Mayeda, *J. Am. Chem. Soc. 92*:5818 (1970); (b) D. Ofer, R. M. Crooks, and M. S. Wrighton, *J. Am. Chem. Soc. 112*:7869 (1990).
27. V. R. Koch, L. L. Miller, and R. A. Osteryoung, *J. Org. Chem. 39*:2416 (1974).
28. P. Bernhard, U. Diab, and A. Ludi, *Inorg. Chim. Acta 173*:65 (1990).
29. R. P. Van Duyne and C. N. Reilley, *Anal. Chem. 44*:142,153,158 (1972).
30. U. Svanholm and V. D. Parker, *J. Am. Chem. Soc. 96*:1234 (1974).
31. O. Hammerich and V. D. Parker, *Electrochim. Acta 18*:537 (1973).
32. D. R. Burfield and R. H. Smithers, *J. Org. Chem. 43*:3966 (1978).
33. H. Kiesele, *Anal. Chem. 52*:2230 (1980).
34. D. R. Burfield, K. H. Lee, and R. H. Smithers, *J. Org. Chem. 42*:3060 (1977).
35. R. H. Campbell, G. A. Heath, G. T. Hefter, and R. C. S. McQueen, *Chem. Commun.* 1123 (1983).
36. C. J. Pickett, *Chem. Commun.* 323 (1985).
37. A. J. Fry and J. Touster, *J. Org. Chem. 51*:3905 (1986).
38. C. G. Swain, A. Ohno, D. K. Roe, R. Brown, and T. Maugh II, *J. Am. Chem. Soc. 89*:2648 (1967).
39. (a) J. E. Gordon, *J. Am. Chem. Soc. 87*:1499,4347 (1965); (b) J. E. Gordon, *J. Org. Chem. 30*:2760 (1965); (c) W. T. Ford, *J. Org. Chem. 38*:3916 (1973); (d) Y. Chauvin, B. Gilbert, and I. Guibard, *Chem. Commun.* 1715 (1990).
40. (a) J. W. Sease and R. A. de la Torre, *J. Am. Chem. Soc. 101*:1687 (1979); (b) R. Alvarado, M. A. thesis, Wesleyan University, 1975.
41. D. M. Mohilner, in *Electroanalytical Chemistry*, Vol. 1 (A. J. Bard, ed.), Marcel Dekker, New York, 1966, pp. 241–409.
42. (a) A. J. Fry and R. G. Reed, *J. Am. Chem. Soc. 94*:8475 (1972); (b) R. Breslow and R. F. Drury, *J. Am. Chem. Soc. 96*:4702 (1974). The latter authors employed guanidinium perchlorate as electrolyte. Those contemplating use of this substance should be aware that it is an explosive and was surveyed for military use by Nernst in 1915: see K. Mendelssohn, *The World of Walther Nernst*, University of Pittsburgh Press, Pittsburgh, PA, 1973, p. 90. A more stable anion (tetrafluoroborate or hexafluorophosphate) is strongly advised.
43. (a) K. Nyberg, *Chem. Commun.* 774 (1969); (b) L. Eberson and B. Olofsson, *Acta Chem. Scand. 23*:2355 (1969).
44. L. Horner and D. Degner, *Electrochim. Acta 19*:611 (1974).
45. Ref. 22, Chap. 4, pp. 121–122.
46. A. J. Fry, L. L. Chung, and V. Boekelheide, *Tetrahedron Lett.* 445,1354 (1974).
47. A. J. Fry, J. A. Simon, M. Tashiro, T. Yamato, R. Mitchell, T. W. Dingle, R. V. Williams, and R. Mahedevan, *Acta Chem. Scand. 37B*:445 (1983).
48. M. J. S. Dewar and C. de Llano, *J. Am. Chem. Soc. 91*:789 (1969).
49. A. J. Fry, C. S. Hutchins, and L. L. Chung, *J. Am. Chem. Soc. 97*:591 (1975).
50. Ref. 22, p. 114.
51. H. Lehmkuhl, S. Kintopf, and E. Janssen, *J. Organomet. Chem. 56*:41 (1973).
52. D. R. Burfield, G. H. Gan, and R. H. Smithers, *J. Appl. Chem. Biotech. 28*:23 (1978).
53. (a) D. R. Burfield and R. H. Smithers, *J. Org. Chem. 48*:2420 (1983); (b) D. R. Burfield and R. H. Smithers, *J. Org. Chem. 46*:629 (1981).
54. D. R. Burfield, *Anal. Chem. 48*:2285 (1976).

55. J. W. Sease and R. C. Reed, personal communication.
56. A. J. Bard, *Pure Appl. Chem. 25*:345 (1971).
57. (a) H. Kiesele, *Anal. Chem. 53*:1952 (1981); (b) R. Lines, B. S. Jensen, and V. D. Parker, *Acta Chem. Scand. 32B*:510 (1978).
58. A. I. Vogel, *Textbook of Practical Organic Chemistry*, 5th ed., Longman, Essex, England, 1989, pp. 122–125.
59. M. Walter and L. Ramaley, *Anal. Chem. 45*:165 (1973).
60. L. C. Portis, J. C. Roberson, and C. K. Mann, *Anal. Chem. 44*:294 (1972).
61. T. Osa and T. Kuwana, *J. Electroanal. Chem. 22*:389 (1969).
62. J. N. Butler, *J. Electroanal. Chem. 14*:89 (1967).
63. A. J. Gordon and R. A. Ford, *The Chemist's Companion*, Wiley, New York, 1972, p. 439.
64. J. A. Ferguson, *Interface 6*:2 (1970).
65. H. Lund, in *Organic Electrochemistry*, 2nd ed. (M. M. Baizer and H. Lund, eds.), Marcel Dekker, New York, 1983, pp. 192–215.
66. (a) J. A. Riddick and W. B. Bunger, *Organic Solvents*, 3rd ed., Wiley, New York, 1970; (b) Ref. 63, p. 429ff.
67. (a) T. R. Koch and W. C. Purdy, *Talanta 19*:989 (1972); (b) T. B. Reddy, *Pure Appl. Chem. 25*:459 (1972).
68. (a) D. S. Reid and C. A. Vincent, *J. Electroanal. Chem. 18*:427 (1968); (b) C. D. Ritchie and G. H. Megerle, *J. Am. Chem. Soc. 89*:1447 (1967); (c) M. Bréant and G. Demange-Guérin, *Bull. Soc. Chim. Fr.* 2935 (1969).
69. D. Coutagne, *Bull. Soc. Chim. Fr.* 1940 (1971).
70. D. A. Hall and P. J. Elving, *Anal. Chim. Acta 39*:141 (1967).
71. *Handbook of Laboratory Safety*, 2nd ed., CRC Press, Cleveland, 1971, pp. 268–269.
72. K. Rousseau, G. C. Farrington, and D. Dolphin, *J. Org. Chem. 37*:3968 (1972).
73. (a) L. Byrd, L. L. Miller, and D. Pletcher, *Tetrahedron Lett.* 2419 (1972); (b) N. S. Moe, *Acta Chem. Scand. 19*:1023 (1965).
74. M. M. Baizer, *J. Electrochem. Soc. 111*:215 (1964).
75. J. L. Webb, C. K. Mann, and H. M. Walborsky, *J. Am. Chem. Soc. 92*:2042 (1970).
76. C. J. Janz and R. P. T. Tompkins, *Nonaqueous Electrolytes Handbook*, Vols. I and II, Academic Press, New York, 1972, 1974.

# 16

# Electrochemical Studies at Reduced Temperature

**Dennis H. Evans** *University of Delaware, Newark, Delaware*

**Susan A. Lerke** *DuPont Merck Pharmaceutical Company, Wilmington, Delaware*

## I. INTRODUCTION

The vast majority of electrochemical measurements are made at a single temperature, in fact, often at the prevailing temperature of the laboratory. The latter practice is convenient and perfectly justified if qualitative information is the goal of the experiment. However, when quantitative interpretations take priority, temperature control becomes mandatory, and often it is desirable to make measurements at a variety of controlled temperatures.

In this chapter, we first discuss the principal motivations for doing electrochemistry at other than room temperature and attempt to delineate the type of chemical information that can be obtained from such measurements. The emphasis is on measurements at reduced temperatures, though the principles apply to high-temperature electrochemistry as well.

In the second part, examples of electrochemical studies at low temperatures will be given, followed by a discussion of some of the practical aspects of doing electrochemistry under such conditions. Some of the techniques and procedures differ considerably from those of high-temperature electrochemistry, and, in fact, this field of low-temperature measurements has been given the moniker *cryo-electrochemistry*.

In keeping with the spirit of this publication, the coverage is not exhaustive. Instead, only key references to the most critical information are cited.

## II. MOTIVATIONS FOR VARIATION OF THE TEMPERATURE IN ELECTROCHEMICAL STUDIES

### A. Thermodynamic Information

$$\Delta G = -nFE_{rxn} \tag{16.1}$$

The electromotive force (emf) of an electrochemical cell, $E_{rxn}$, is related to the free-energy change of the cell reaction, $\Delta G$, by Equation 16.1, in which n is the number of electrons involved in the cell reaction and F is the Faraday constant. We consider here a cell comprising a reference electrode and a working electrode at which a redox reaction of interest occurs. The entire cell is at a single temperature (*isothermal operation*) and $\Delta G$ is the free-energy change of the entire cell reaction, including both reference and working half-reactions, at that temperature. If all species are at unit activity, the measurement gives $\Delta G°$, the standard free-energy change of the cell reaction.

This is all that can be learned from measurements at a single temperature. However, if $E_{rxn}$ is measured at a series of temperatures, the entropy change for the cell reaction, $\Delta S$, can be calculated from Equation 16.2. Knowing both $\Delta G$ and $\Delta S$, we may also compute the enthalpy change from $\Delta H = \Delta G + T \Delta S$.

$$\Delta S = nF\left(\frac{\partial E_{rxn}}{\partial T}\right)_p \tag{16.2}$$

Thermodynamic parameters for many reactions have been evaluated and tabulated in this way. Nevertheless, there has been a quite natural desire to obtain thermodynamic information about single half-reactions, in particular the reaction occurring at the working electrode. In the strictest sense, it is impossible to separate the thermodynamic parameters for a cell reaction into those of its component half reactions. Nevertheless, a procedure has been advanced and evaluated that accomplishes this separation with negligible errors [1].

This simple procedure is based on the use of a *nonisothermal* cell in which the reference electrode is maintained at a constant temperature while the temperature of the working electrode (and the solution containing the solutes involved in the half reaction) is varied. Basically, the contribution of the reference electrode to the cell emf remains constant while that of the other half-reaction changes. Thermal potentials develop along the salt bridge connecting the two half-cells and the metal contact to the working electrode. It was the demonstration that these thermal potentials were negligible that popularized this technique and led to the wide acceptance of data so obtained.

The most frequently evaluated parameter is the entropy change of the half reaction, $\Delta S_{hr}$, which is given by Equation 16.3:

$$\Delta S_{hr} = nF\left(\frac{\partial E^{\circ\prime}}{\partial T}\right)_p \tag{16.3}$$

Here the reversible formal potential, $E^{\circ\prime}$, of the redox couple of interest has been measured at a series of temperatures versus a reference electrode maintained at constant temperature. Values of $\Delta S_{hr}$ have been interpreted in terms of differences in water–ligand hydrogen bonding [1] and have been compared to the entropy change for spin crossover in iron complexes [2].

This is an appropriate point at which to comment on the common practice of evaluating the formal potential from voltammetric measurements. When a reversible response is obtained in voltammetry, what is actually measured is the reversible half-wave potential, $E_{1/2}$, which (except for hydrodynamic voltammetry) is related to the formal potential by a term involving the diffusion coefficients of the oxidized and reduced forms of the half-reaction, $D_O$ and $D_R$, respectively.

$$E_{1/2} = E^{\circ\prime} + \frac{RT}{nF}\ln\frac{D_R^{1/2}}{D_O^{1/2}} \tag{16.4}$$

In cyclic voltammetry, for example, $E_{1/2}$ is, to a good approximation, equal to the potential midway between the cathodic and anodic peak potentials of a reversible voltammogram. Normally, $D_O \approx D_R$ so the term involving the diffusion coefficient is small (a few millivolts) and $E_{1/2}$ is an accurate measure of $E^{\circ\prime}$, but in some instances the difference is significant; the diffusion coefficients should be measured and a correction applied.

When determining $\Delta S_{hr}$ by measuring the temperature dependence of $E_{1/2}$, the assumption is less demanding, as it is only required that

$$\frac{\partial(D_O/D_R)}{\partial T} = 0 \tag{16.5}$$

It is also common to measure by voltammetry the thermodynamic properties of purely chemical reactions that are in some way coupled to the electron transfer step. Examples include the determination of solubility products, acid dissociation constants, and metal–ligand complex formation constants for cases in which precipitation, proton transfer, and complexation reactions affect the measured formal potential. Also in these instances, studies at variable temperature will afford the thermodynamic parameters of these coupled chemical reactions.

## B. Kinetic Information

*Mass Transport*

The most common rate phenomenon encountered by the experimental electrochemist is mass transport. For example, currents observed in voltammetric experiments are usually governed by the diffusion rate of reactants. Similarly, the cell resistance, which influences the cell time constant, is controlled by the ionic conductivity of the solution, which in turn is governed by the mass transport rates of ions in response to an electric field.

The dependence of the diffusion coefficient upon temperature affords the activation energy for the diffusion process, $E_{a,dif}$:

$$E_{a,dif} = R \frac{\partial \ln(D)}{\partial T} \tag{16.6}$$

Though one may learn something about the diffusion process by determining activation energies, such motivations usually do not underlie current research. Rather, one needs to be aware of the effects of temperature on transport in order to avoid errors and to interpret properly the experimental data. As temperatures decrease, diffusion coefficients decrease, giving reduced voltammetric currents. At the same time, solution resistance will increase, causing greater possibility of error due to uncompensated iR drop. Also, when using hydrodynamic voltammetry (e.g., the rotating disk electrode), lower temperatures will lead to higher kinematic viscosities with concomitantly lower limiting currents.

*Rates of Heterogeneous Electron Transfer Reactions*

There has been keen interest in determination of activation parameters for electrode reactions. The enthalpy of activation for a heterogeneous electron transfer reaction, $\Delta H^*_{ex}$, is the quantity usually sought [3,4]. It is determined by measuring the temperature dependence of the rate constant for electron transfer at the formal potential, that is, the standard heterogeneous electron transfer rate constant, $k_s$. The activation enthalpy is then computed by Equation 16.7:

$$\Delta H^*_{ex} = R \frac{\partial \ln(k_s)}{\partial T} \tag{16.7}$$

A principal focus of research has been to compare the experimental enthalpies with the predictions of modern electron transfer theory [5–7].

*Rates of Chemical Reactions Coupled to Electron Transfer*

One of the most popular reasons for conducting electrochemical studies at low temperature is to investigate chemical reactions that are coupled to the primary electrode reaction. An example is the EC mechanism (Chap. 2) or variants thereof:

$$A + e \rightleftarrows B; \quad B \xrightarrow{k_1} C \qquad (16.8)$$

Here, the electrode reaction is followed by a first-order irreversible chemical reaction in solution that consumes the primary product B and forms the final product C. The rate of this chemical reaction can be measured conveniently with cyclic voltammetry, double-potential-step chronoamperometry, reverse pulse voltammetry, etc. However, this is only true if the half-life of B is greater than or equal to the shortest attainable time scale of the experiment.

As discussed in Chapter 12, microelectrode techniques allow the easy determination of the rates of reaction of species with half-lives on the order of 10 μs, which can be extended in certain cases to submicrosecond half-lives. Nevertheless, it is sometimes better to lower the rate of the chemical reaction by lowering the temperature; this is probably the most common motivating factor for such studies.

As an example, consider the effect of temperature on a first-order rate constant for a process with an activation energy, $E_a$, that is considered to be independent of temperature. Table 16.1 summarizes the calculated effect of temperature on the half-life of a first-order reaction, based on Equation 16.9 and the assumption that the preexponential factor, A, is independent of temperature.

$$k_1 = A \exp\left(\frac{-E_a}{RT}\right) \qquad (16.9)$$

**Table 16.1** Effect of Temperature on the Rate of a First-Order Chemical Reaction as a Function of Activation Energy[a]

| Temperature | | $\log(k_{1,T}/k_{1,298})$ | | | |
|---|---|---|---|---|---|
| K | °C | $E_a$ = 2 kcal/mol | $E_a$ = 6 kcal/mol | $E_a$ = 10 kcal/mol | $E_a$ = 14 kcal/mol |
| 298 | 25 | 0 | 0 | 0 | 0 |
| 278 | 5 | -0.11 | -0.32 | -0.53 | -0.74 |
| 258 | -15 | -0.23 | -0.68 | -1.14 | -1.59 |
| 238 | -35 | -0.37 | -1.11 | -1.85 | -2.59 |
| 218 | -55 | -0.54 | -1.61 | -2.69 | -3.77 |
| 198 | -75 | -0.74 | -2.22 | -3.70 | -5.19 |
| 178 | -95 | -0.99 | -2.97 | -4.94 | -6.92 |
| 158 | -115 | -1.30 | -3.90 | -6.50 | -9.10 |
| 138 | -135 | -1.70 | -5.10 | -8.50 | -11.9 |
| 118 | -155 | -2.24 | -6.71 | -11.2 | -15.7 |
| 98 | -175 | -2.99 | -8.98 | -15.0 | -21.0 |

[a]From Equation 16.9

Note that the rates of reactions with substantial activation energies can be profoundly suppressed by lowering the temperature, but of course, the rates of reactions with small activation energies are not strongly affected. Unfortunately, those are the very reactions that are likely to be fast at room temperature, so the beneficial effect of reducing the temperature is least prominent for fast reactions where assistance is most sorely needed. Nevertheless, it is found that the rates of many reactions can be moved into a convenient voltammetric time scale by even rather modest reductions in temperature.

### C. Electrochemistry in Novel Media

Other low-temperature studies have been motivated by the desire to characterize and understand processes occurring in unusual media. For example, the use of liquid ammonia [8–10] and liquid sulfur dioxide [11–13] naturally requires reduced temperatures unless high pressures are used, as is done for electrochemistry in supercritical fluids [14]. Frozen media are interesting systems in terms of mass transport phenomena and microstructural effects. Examples include glasses of acetonitrile and acetone [15], frozen dimethyl sulfoxide solutions [16,17], and the solid electrolyte $HClO_4 \cdot 5.5\ H_2O$ [18–20].

## III. EXAMPLES OF THE STUDY OF THE RATES OF COUPLED CHEMICAL REACTIONS

Studies of the effect of temperature on the rates of coupled chemical reactions are characterized by a great deal of complexity and subtlety, so we have chosen to emphasize this type of application in this presentation of examples of experimental techniques and interpretation of data.

### A. Cleavage of Radical Anions of Diiron Carbonyl Dimers

Electrochemical reduction of $[Fe(CO)_2(\eta^5\text{-}Cp)]_2$ is an overall two-electron process at room temperature with the monomeric anion as the final product [21]. When the temperature is lowered, the radical anion formed by addition of one electron to the dimeric starting material can be detected (Fig. 16.1). Cyclic voltammetry is employed to obtain qualitative information about the reaction mechanism, and the solvent, propionitrile, is an excellent polar medium for low-temperature studies.

At room temperature, the cyclic voltammogram features an irreversible two-electron reduction (peak a) with oxidation of the monomeric anion evident on the return sweep (peak b). As the temperature is lowered, a new peak (peak c) grows in on the positive-going scan. This peak, due to oxidation of the dimeric radical anion to starting material, grows at the expense of peak b until the latter is entirely absent at –76°C. These results are qualitatively in accord with an ECE mechanism (Chap. 2) as shown in Equations 16.10 to 16.12 where the monomeric fragment is represented by $ML_2Cp$.

$$[ML_2Cp]_2 + e \rightleftarrows [ML_2Cp]_2^- \tag{16.10}$$

$$[ML_2Cp]_2^- \xrightarrow{k_1} ML_2Cp + ML_2Cp^- \tag{16.11}$$

$$ML_2Cp + e \rightleftarrows ML_2Cp^- \tag{16.12}$$

Not considered here is a solution-electron-transfer reaction that occurs in this DISP1 scheme, which was fully discussed in the original paper.

The appearance of peak c at –54°C (Fig. 16.1) and 250 mV/s suggests that $k_1$ is on the order of 10 $s^{-1}$ at this temperature. In order to obtain a quantitative determination of the rate constant, double-potential-step chronoamperometry (DPSC) was employed (Fig. 16.2). The potentials were selected so that forward reaction 16.10 occurred during the first step and the reverse reaction took place after the switching time $\tau_{exp}$; that is, any current seen during the second step will be due to oxidation of the short-lived radical anion. Clearly, the radical anion is detected (Fig. 16.2A), and the data analysis (illustrated in Fig. 16.2B) is fully consistent with the ECE-DISP1 mechanism. Similarly good agreement was found at other values of $\tau_{exp}$ and other concentrations of reactant. The rate constant was evaluated over the range of –23 to –78°C and the activation energy was determined.

Double-potential-step chronoamperometry does not require precise control of potential. It is only necessary that the potential during each step be at a value for which the desired reaction occurs at the mass-transport-limited rate. Thus, it was not necessary to correct for solution iR drop in these experiments as it would have been if cyclic voltammetry had been selected as the method for quantitative evaluation of the rate constant.

Another voltammetric method, rotating ring-disk electrode (RRDE) voltammetry, was applied to study the same reaction and excellent agreement was found with the DPSC results. The studies were conducted at temperatures as low as –78°C and at rotation rates between 500 and 6000 rpm. Unlike DPSC, analysis of the RRDE data required knowledge of both the kinematic viscosity of the solution and the diffusion coefficient of the reactant at each temperature, so it was necessary to measure these quantities to complete the analysis.

## B. Reactions Preceding Electron Transfer: Conformational Change in trans-1,2-Dibromocyclohexane

The electrochemical reduction of *vic*-dibromides is a highly irreversible two-electron process in which two bromide ions are expelled leaving the correspond-

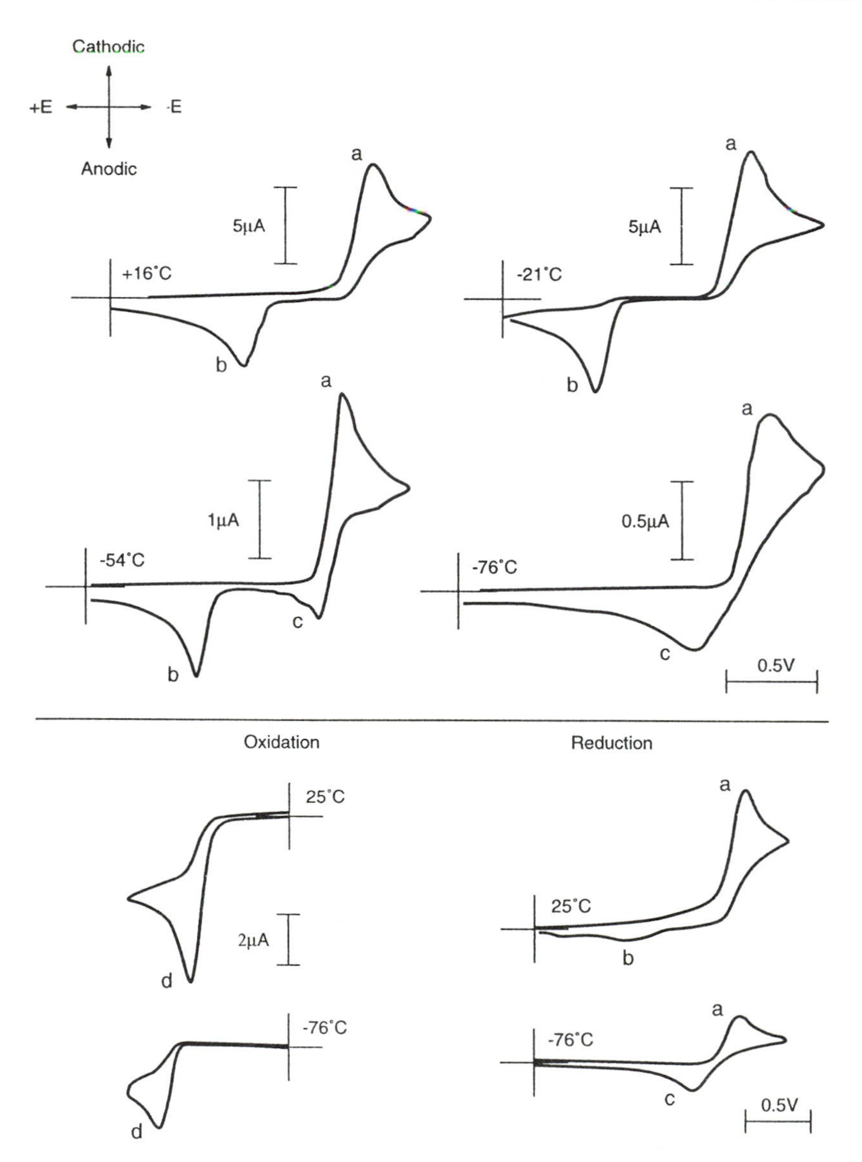


**Figure 16.1** Cyclic voltammetry of $[Fe(CO)_2(\eta^5\text{-}Cp)]_2$ in 0.1 M $Bu_4NPF_6$/ propionitrile. Top panel: +16 and –76°C (250 mV/s, 5.4 mM $[Fe(CO)_2(\eta^5\text{-}Cp)]_2$); –21°C (300 mV/s, 8.9 mM $[Fe(CO)_2(\eta^5\text{-}Cp)]_2$); –54°C (250 mV/s, 6.0 mM $[Fe(CO)_2(\eta^5\text{-}Cp)]_2$). Bottom panel: oxidation (left) and reduction (right) voltammograms for 3.3 mM $[Fe(CO)_2(\eta^5\text{-}Cp)]_2$ at room temperature (100 mV/s) and –76°C (400 mV/s). The electrode area is 0.0032 cm$^2$; the potential axis origin is at –0.3 V vs. SCE. [Reprinted with permission from E.F. Dalton, S. Ching, and R.W. Murray, *Inorg. Chem. 30*:2642 (1991). Copyright 1991 American Chemical Society.]

ing olefin. The effective rate constant for this electrode reaction is very dependent upon the dihedral angle, θ, between the two carbon-bromine bonds, with θ = 0° or 180° having the highest rates (lowest overpotential) and an angle of 90° giving the minimum rate (highest overpotential).

*trans*-1,2-Dibromocyclohexane exists in two rapidly interconverting conformations, one with axial bromine atoms (aa conformation; θ ≈ 180°) and one with equatorial substituents (ee conformation; θ ≈ 60°) [22,23].

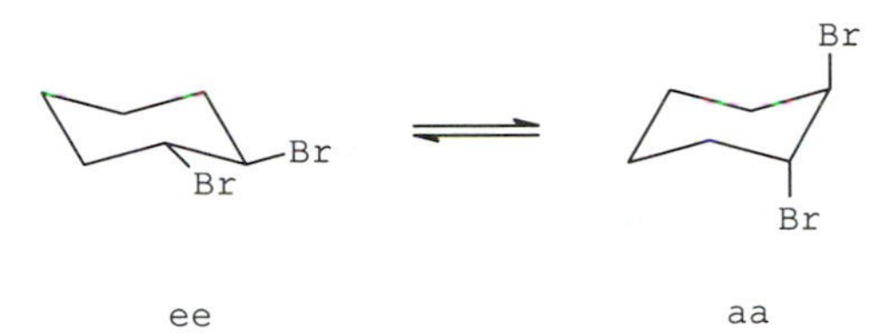


Thus, it would be expected that this compound would show two reduction peaks by cyclic voltammetry: the first process should be reduction of aa and the second reduction of ee. However, at room temperature, only a single reduction peak is seen because the rate of interconversion of the two forms is so rapid that the entire population of reactant molecules near the electrode is reduced via the more easily reduced conformation, aa.

As the temperature is lowered to -60°C (Fig. 16.3A), the single peak appearing near –2.3 V is joined by a smaller peak at more negative potentials. At still lower temperatures (Fig. 16.3B,C), the second peak grows at the expense of the first, though their relative heights do not continue to change below -80°C.

The qualitative interpretation is that the aa form is reduced near –2.3 V and the ee at more negative potentials. In the low-temperature limit, the two peak heights simply reflect the equilibrium populations of the two conformations that are present in comparable amounts. The fact that the first peak is much larger than the second at higher temperatures is explained by the conversion of significant amounts of ee to the aa form, which can then be reduced at the first peak.

Clearly, the data contain information about both the equilibrium constant and the rate constants for the conformational interconversion. In this instance, the quantitative analysis was based upon the cyclic voltammetric data. The points in Figure 16.3 are the background-corrected experimental data, and the curves were computed by digital simulation with values of the equilibrium and rate constants selected to achieve best agreement with the experimental data. A given set of parameters was found to account for the data at a variety of scan rates, a necessary condition if the kinetic model is to be judged adequate.

Because cyclic voltammetry was chosen as the method for quantitative evaluation of the kinetic parameters, close attention to the effects of solution

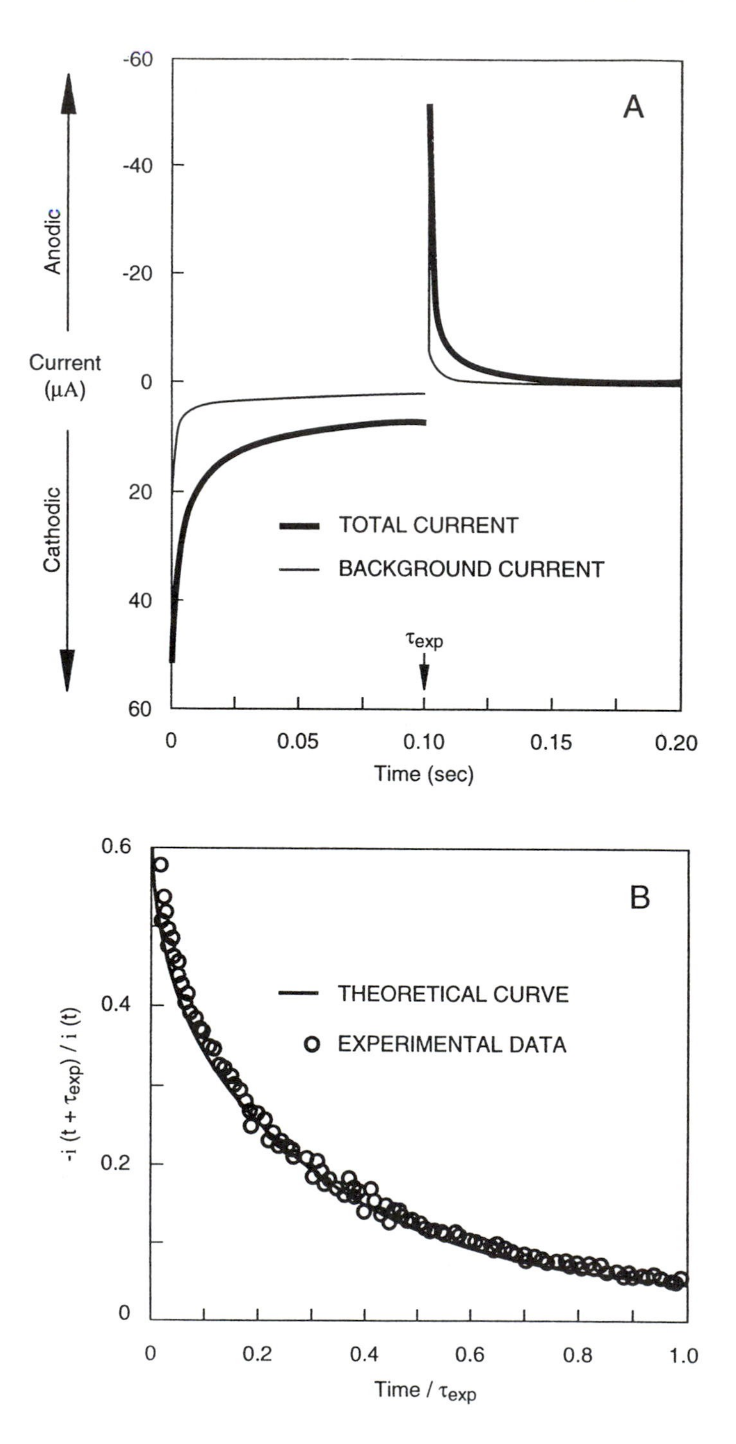
-60
-40
-20
0
20
40
60
Anodic
Current
(μA)
Cathodic
A
TOTAL CURRENT
BACKGROUND CURRENT
τexp
0
0.05
0.10
0.15
0.20
Time (sec)
0.6
0.4
0.2
0
B
THEORETICAL CURVE
EXPERIMENTAL DATA
-i (t + τexp) / i (t)
0
0.2
0.4
0.6
0.8
1.0
Time / τexp

resistance was required. Positive-feedback iR compensation was employed. In order to determine the effective resistance between the working electrode and the Luggin capillary of the reference electrode, a compound that presumably undergoes a reversible reduction (nitrobenzene) was studied and sufficient compensation was applied to achieve a reversible peak separation at each temperature. The amount of resistance so compensated was taken as the effective solution resistance.

## C. Reduction of 2,3-Dinitro-2,3-Dimethylbutane

In this example, the rate of a fast cleavage reaction was determined by a combination of low-temperature and fast-scan voltammetry [24]. Upon insertion of an electron into a *vic*-dinitro compound, the resulting radical anion undergoes rapid C–N bond cleavage to form nitrite ion and a β-nitroalkyl radical, which in turn undergoes further reduction.

Unlike the previous examples, a microdisk electrode was employed in order to minimize the iR drop encountered at fast sweep rates with macroelectrodes. The reduction of 2,3-dinitro-2,3-dimethylbutane at –30°C and 1000 V/s (Fig. 16.4) features a first reduction peak at –1.0 V due to the formation of the corresponding radical anion. Reversal after this peak produces a peak for oxidation of the radical anion to the starting material. As shown by digital simulation, a negligible amount of the radical anion decomposed under these conditions. When the negative-going scan was extended, a second reduction peak was observed. This corresponded to further reduction of the radical anion and was the height predicted by digital simulation if a negligible amount of radical anion had been lost. The product of the one-electron reduction of the radical anion is short-lived and could not be detected in these experiments; that is, the second reduction peak was always irreversible.

Data at smaller scan rates and higher temperatures allowed determination of the rate constant for cleavage of nitrite from the radical anion of this and other *vic*-dinitro compounds. In addition to the use of microelectrodes to reduce iR drop, another interesting feature of this study was the observation of the effect

---

**Figure 16.2** Double-potential-step chronoamperometry of 4.6 mM $[Fe(CO)_2(\eta^5\text{-}Cp)]_2$ in 0.1 M $Bu_4NPF_6$/propionitrile at –43°C. Step time $\tau_{exp}$ = 0.1 s. (A) Current transients for potential step from –0.8 to –1.9 V in blank electrolyte solution (thin line) and with added $[Fe(CO)_2(\eta^5\text{-}Cp)]_2$ (dark line). (B) Ratio of experimental currents $-i(t + \tau_{exp})/i(t)$ for $0 < t < \tau_{exp}$ versus normalized time $(t/\tau_{exp})$ compared to theory (solid line) for $k_{obs}$ = 10.5 $s^{-1}$. Electrode area = 0.0032 $cm^2$. [Reprinted with permission from E.F. Dalton, S. Ching, and R.W. Murray, *Inorg. Chem. 30*:2642 (1991). Copyright 1991 American Chemical Society.]

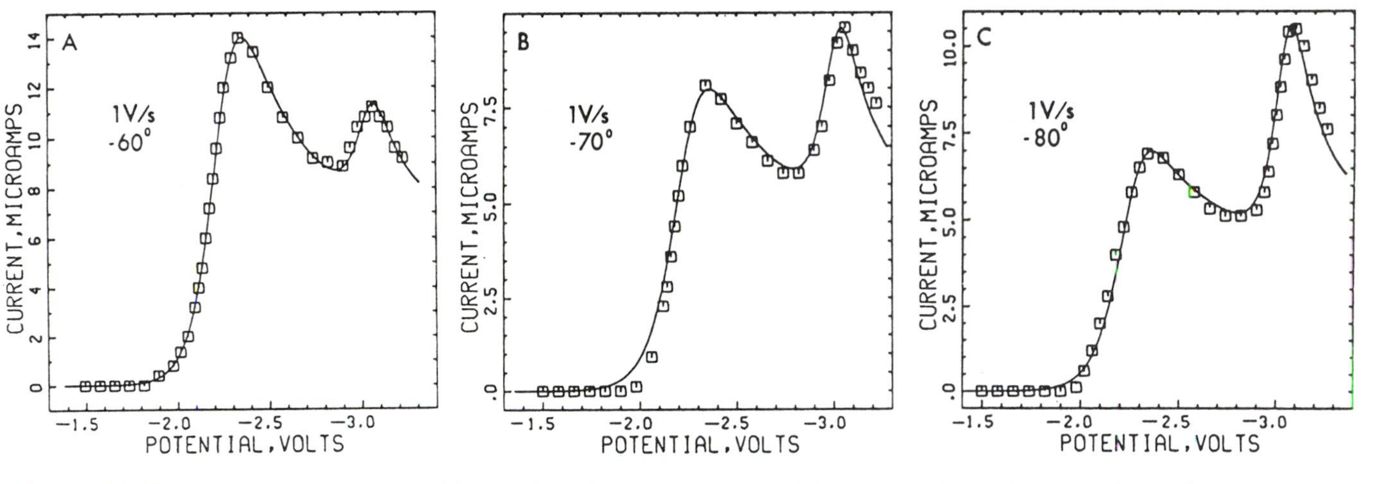


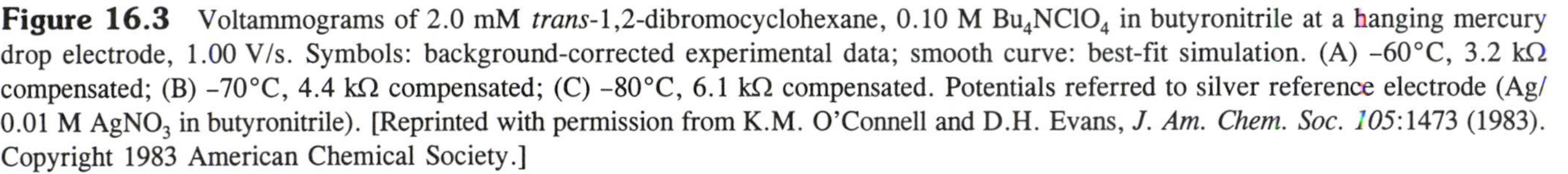
**Figure 16.3** Voltammograms of 2.0 mM *trans*-1,2-dibromocyclohexane, 0.10 M $Bu_4NClO_4$ in butyronitrile at a hanging mercury drop electrode, 1.00 V/s. Symbols: background-corrected experimental data; smooth curve: best-fit simulation. (A) −60°C, 3.2 kΩ compensated; (B) −70°C, 4.4 kΩ compensated; (C) −80°C, 6.1 kΩ compensated. Potentials referred to silver reference electrode (Ag/0.01 M $AgNO_3$ in butyronitrile). [Reprinted with permission from K.M. O'Connell and D.H. Evans, *J. Am. Chem. Soc.* *105*:1473 (1983). Copyright 1983 American Chemical Society.]

of temperature on the rates of both homogeneous chemical reactions and heterogeneous electron transfer reactions. The large separation (about 0.4 V) between cathodic and anodic peaks in Figure 16.4 is due to the strong suppression of the electron transfer rate constant at –30°C.

## D. Transition from Microelectrode to Macroelectrode Diffusion Regimes

Our last example does not involve the rate of a chemical reaction, but instead, the effect of temperature on diffusion rates [25]. One of the motivations for using microelectrodes as in the previous example is to allow fast experiments without appreciable iR drop. When used in the opposite extreme of very small scan rates, microdisk electrodes produce steady-state voltammograms that have the same sigmoidal shape as dc polarograms and RDE voltammograms (cf. Chap. 12).

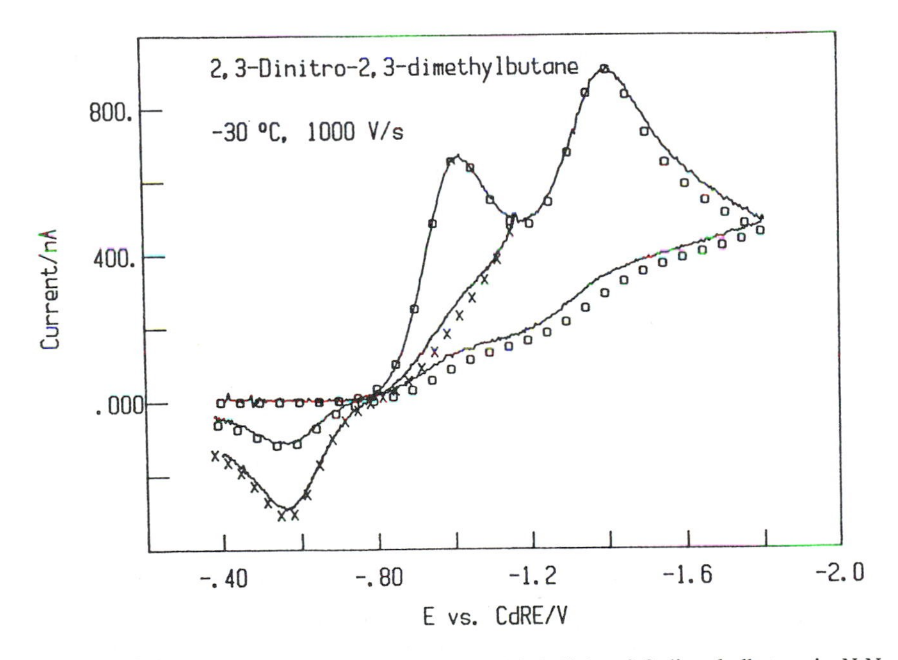


**Figure 16.4** Cyclic voltammogram of 4.5 mM 2,3-dinitro-2,3-dimethylbutane in *N,N*-dimethylformamide/0.20 M $Bu_4NPF_6$ at a 25-μm-diameter mercury electrode. Curves: experimental voltammograms after subtraction of background current. Points: digital simulations. Potentials referred to cadmium reference electrode [cadmium amalgam/$CdCl_2$ (sat'd) in DMF]. [Reprinted with permission from W.J. Bowyer and D.H. Evans, *J. Org. Chem.* *53*:5234 (1988). Copyright 1988 American Chemical Society.]

The steady state will be approached when the diffusion-layer thickness becomes much larger than the radius of the disk, roughly speaking, when $(Dt)^{1/2}/r \gg 1$, where D is the diffusion coefficient, t is the duration of the experiment, and r is the electrode radius. In the opposite extreme, where $(Dt)^{1/2}/r \ll 1$, a normal peak-shaped response will be obtained that is characteristic of planar diffusion with little contribution of diffusion from the solution at the periphery of the disk.

Reduction of the solution temperature allows transition from steady-state to peak-shaped response simply by way of the marked diminution of D at low temperatures. Figure 16.5 shows slow-scan cyclic voltammograms obtained at two microdisk electrodes as a function of solution temperature. Between −120 and −140°C there is a particularly clear transition for the 25-μm-diameter electrode as the diffusion-layer thickness becomes less than the disk radius. Also illustrated here is the immense decrease in the limiting currents that is seen over this range of temperatures due to the 100-fold decrease in D.

The solvent used in this work was a mixture of butyronitrile and ethyl chloride. The solvent system was found to produce usable data at temperatures as low as −170°C. The resistivities are high, however, so fast experiments would undoubtedly be difficult.

## IV. PRACTICAL ASPECTS OF ELECTROCHEMICAL STUDIES AT LOW TEMPERATURES

### A. Cells

Fortunately, it turns out that the general cell design that is suitable for room temperature measurements (cf. Chap. 9) will suffice at low temperatures as well. Glass is the preferred material of construction. There are two general cell types: *dip-type* cells that are designed to be immersed in a bath of coolant, and *jacketed* cells whose contents are cooled by passage of coolant fluid through the jacket.

The dip-type cell is often constructed with a slightly elongated cell body to facilitate immersion of the working electrode compartment in the coolant [23]. For jacketed cells, it is common to construct two jackets, an outer evacuated jacket (with feedthroughs for the coolant) to improve temperature stability, and an inner jacket through which the coolant is pumped [23,25]. Single-jacketed cells have also been used. In one instance [7], a glass cell was inserted into a Teflon jacket through which chilled nitrogen gas was passed. Jacketed cells have also been used within a glove box with cooling by nitrogen gas that is chilled outside the box and enters the box through a modified entrance port [26].

Detailed drawings of low-temperature cells, including cells for liquid ammonia solvent [8], have been published [27].

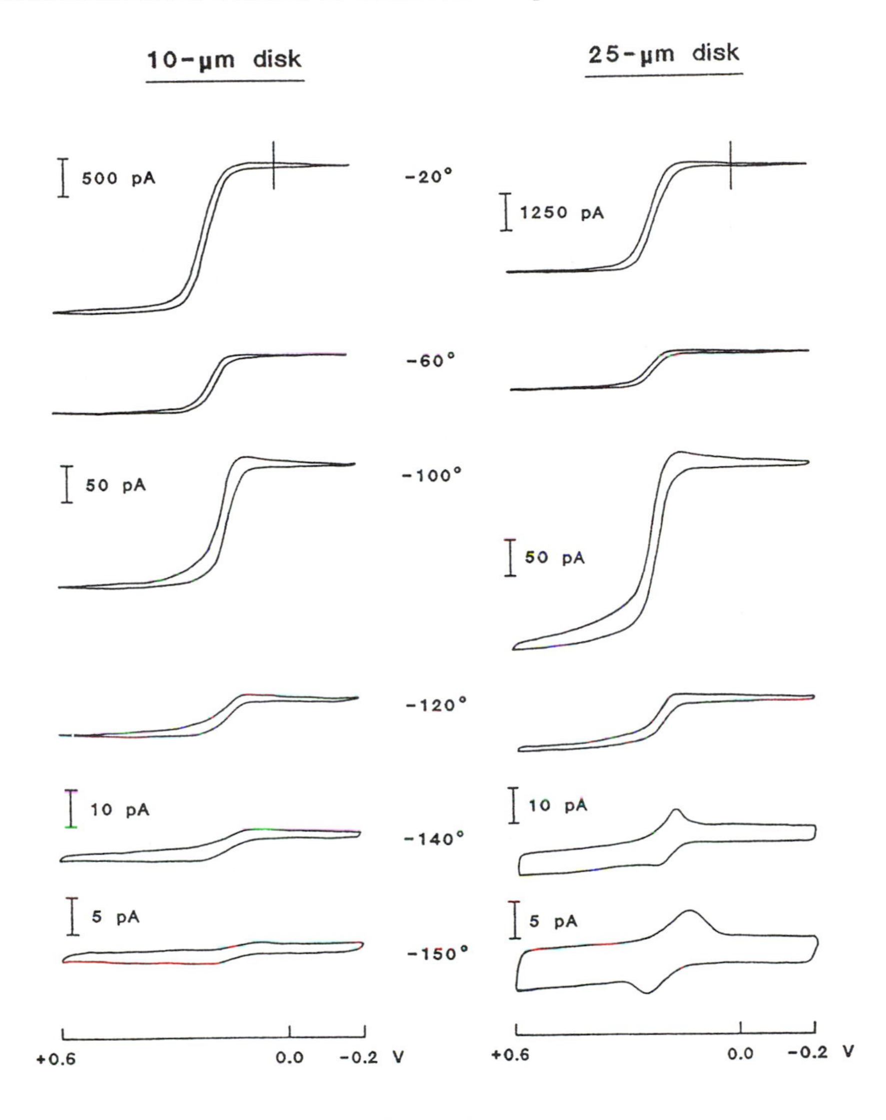


**Figure 16.5** Variable-temperature cyclic voltammograms (50 mV/s) recorded at 10- (left panel) and 25-μm (right panel) Pt microelectrodes for 2 mM $Cp^*_2Fe$ in 0.2 M $Bu_4NPF_6$, 1:2 butyronitrile:ethyl chloride. [Reprinted with permission from S. Ching, J.T. McDevitt, S.R. Peck, and R.W. Murray, *J. Electrochem. Soc. 138*:2308 (1991). Copyright 1991 The Electrochemical Society, Inc.]

## B. Cooling Systems

*Immersion*

As already mentioned, there are two general approaches to cooling the cell, immersion in the coolant and pumping coolant through the cell jacket. The simplest approach [21,27] for immersion is to use standard slush baths or salt–ice mixtures that are available for temperatures down to –160°C [28]. Crude but effective control of temperature can be achieved by cooling the cell in liquid nitrogen followed by slow warm-up in the vapor above the boiling liquid [5].

The bath temperature can also be controlled with an immersion cooler that has been found to be practical for temperatures down to about –80°C [23]. Suggested coolant fluids are methanol [5,23] and 2-propanol [11]. High-power coolers for temperatures as low as –100°C are manufactured by FTS Systems, Inc., Stone Ridge, NY. There are many manufacturers of refrigerated baths for less demanding temperatures.

*Jacketed Cells*

The aforementioned refrigerated baths are often equipped with circulation pumps so that the chilled fluid can be passed through the jacket of a nearby electrochemical cell. The tubing connecting the circulator to the cell should be insulated to minimize heat loss. With such a circulator, experiments have been conducted at temperatures as low as –20°C [29,30], and there appears to be no reason that lower temperatures can not be achieved.

The most common coolant for very low-temperature studies is chilled nitrogen gas [7,23,25–27]. Typically, pressurized nitrogen gas is passed through a coil of copper tubing immersed in liquid nitrogen and then to the cell. Relatively effective temperature control can be achieved by manual adjustment of the nitrogen flow rate so that a steady-state temperature is achieved in which the rate of cooling equals the rate of warming of the cell by its surroundings. Control to within ±0.5°C down to –120°C can be achieved [25] with this technique, though a patient operator is essential. Better temperature control is achieved by passing the prechilled nitrogen over a resistance heater that is controlled in a feedback mode using the signal from a thermocouple positioned in the cell solution [7,27]. Thermal stability of ±1°C over a period of more than an hour can be realized as well as achievement of a new temperature within 3–4 min [7]. In a slightly different approach, a commercial voltammetry cell (EG&G Princeton Applied Research, Princeton, NJ; cell for 303 static mercury drop electrode) was modified with a Kel-F jacket through which chilled nitrogen gas was passed. The flow rate of the gas, and hence the cell temperature, was controlled in feedback mode by a small heater immersed directly in liquid nitrogen in a dewar flask connected to the cell [30].

## C. Measurement of Temperature

The actual temperature of the solution in the vicinity of the working electrode should be measured. It is not wise to assume that the solution temperature equals that of the bath, particularly when using a circulating refrigerated bath, because the temperature of the coolant will rise on passage from the bath to the cell. Low-temperature thermometers can be used in principle, but the almost universal choice is a thermocouple. Many commercial units are available with digital output and control features. The thermocouple junction can be coated with Teflon and inserted directly into the cell.

The spatial uniformity of temperature in the cell is difficult to determine, and we are not aware of a careful study of this problem. In most experiments, it is the temperature of the electrode–solution interface or that of the diffusion layer that is relevant. A possible "internal thermometer" could be created by measuring a temperature-sensitive voltammetric function, for example, the peak separation in the cyclic voltammogram of a reversible reaction, which is $2.22RT/F$. The resolution is not likely to be outstanding, but such a technique would probably allow detection of serious differences between the thermocouple reading and the actual temperature of the electrode–solution interface.

## D. Electrodes

The working electrodes found to be useful at room temperature can also be used at low temperature. There are no special constraints. Platinum is probably the most widely used, simply because it is the most common electrode material for room-temperature work. A mercury electrode can also be employed, either as a hanging mercury drop electrode (HMDE) [23], a thin mercury film on a solid support, or an amalgam [26]. The HMDE was reported to extend the range of usable potentials to somewhat more negative values than found with platinum [23]. Of course, below −39°C, the HMDE is actually a solid electrode; however, no detectable change in its voltammetric behavior is noted at the phase transition.

By far the most common mode of operation is *nonisothermal*, in which the reference electrode is maintained (usually) at room temperature; hence, there are no special requirements for the reference electrode. The temperature gradient occurs along the salt bridge, which usually contains the solvent and electrolyte that are present in the low-temperature portion of the cell.

## E. Solvents and Supporting Electrolytes

The first question that usually arises when considering a low-temperature experiment is the choice of solvent and electrolyte. The freezing point of water limits the lowest practical temperatures for studies in fluid aqueous solution to values near 0°C. Of course, the freezing point of concentrated aqueous salt solutions is considerably depressed so such systems could be used as fluid electrolytes at lower temperatures, but this is not commonly done.

It is much more popular to use nonaqueous solvents for low-temperature studies. There are two motivations, the more common of which is the desire to make measurements down to the lowest temperature possible using a solvent/electrolyte system compatible with the chemical properties of the substances to be studied. In other instances, the purpose of the experiments is to study the effect of solvent on a temperature-sensitive parameter (e.g., a heterogeneous electron-transfer rate constant [5]), so a variety of solvents is sought in which low-temperature measurements can be made.

A summary of some of the solvent/electrolyte systems that have been used for low-temperature electrochemistry is given in Table 16.2. The "lowest tem-

**Table 16.2** Solvent and Electrolyte Systems for Low-Temperature Electrochemistry[a]

| Solvent | Electrolyte | Lowest Temp. (K) | Ref. |
|---|---|---|---|
| Acetone | $TEAClO_4$ (sat'd) | 195 | 36 |
| | 0.30 M $TEAPF_6$ | 198 | 7 |
| | 0.10 M $TBAClO_4$ | 223 | 42 |
| Acetonitrile | 0.30 M $TBAPF_6$ | 258 | 7 |
| | 0.10 M $TBAClO_4$ | 228 | 42,46 |
| | 0.10 M $TEAClO_4$ | 253 | 29 |
| | 0.10, 0.30, and 0.60 M $TBAPF_6$ and $TBAClO_4$ | 233 | 45 |
| Benzonitrile | 0.10 M $TEAClO_4$ | 273 | 48 |
| | 0.10 M $TBAClO_4$ | 273 | 48 |
| | 0.10 M $TOAClO_4$ | 273 | 48 |
| Butyronitrile | 0.30 M $TBAPF_6$ | 198 | 7 |
| | 0.10 M $TBAPF_6$ | 195 | 21 |
| | 0.10 M $TBAClO_4$ | 168 | 2,22,27,41 |
| | 0.30 M $TBAClO_4$ | 143 | 23 |
| | 0.10, 0.30, and 0.60 M $TBAPF_6$ and $TBAClO_4$ | 193 | 45 |
| 1,2-Dichloroethane (ethylene chloride) | 0.10 M $TBAClO_4$ | 243 | 48 |
| Dichloromethane (methylene chloride) | 0.50 M $TBAClO_4$ | 183 | 32 |
| | 0.10 M $TBAPF_6$ | 183 | 32 |
| | 0.10 M $TBAClO_4$ | 183 | 33,48 |
| | 0.10 M TBACl | 183 | 34 |
| 1,2-Dimethoxyethane (ethylene glycol dimethyl ether; glyme) | $TBAPF_6$ | 213 | 49 |

**Table 16.2** Continued

| Solvent | Electrolyte | Lowest Temp. (K) | Ref. |
|---|---|---|---|
| *N,N*-Dimethylformamide | 0.10 M $TBAClO_4$ | 230 | 31,39 |
| | 0.10 M $TEAClO_4$ | 213 | 40,43 |
| | 0.30 M $TBAPF_6$ | 223 | 24,44 |
| | 0.10, 0.30, and 0.60 M $TBAPF_6$ and $TBAClO_4$ | 213 | 45 |
| Ethanol | 0.10 M $LiClO_4$ | 190 | 5 |
| | 0.50 M $LiClO_4$ | 170 | 27 |
| Methanol | 0.10 M $LiClO_4$ | 190 | 5 |
| | 1.0 M $LiClO_4$ | 183 | 27 |
| 1-Propanol | 0.10 M $LiClO_4$ | 220 | 5 |
| 2-Propanol | 0.10 M $LiClO_4$ | 193 | 27 |
| Propionitrile | 0.30 M $TBAPF_6$ | 198 | 7 |
| | 0.10 M $TBAPF_6$ | 195 | 21 |
| | 0.10 M $TBAClO_4$ | 173 | 27 |
| Pyridine | 0.10 M $TEAClO_4$ | 243 | 48 |
| | 0.10 M $TBAClO_4$ | 243 | 48 |
| | 0.10 M $TOAClO_4$ | 243 | 48 |
| Tetrahydrofuran | 0.10 M $TBAClO_4$ | 208 | 37 |
| | 0.20 M $TBAClO_4$ | 195 | 38 |
| | 0.10 M $TBAPF_6$ | 231 | 21 |
| Mixed solvents: | | | |
| $CH_2Cl_2/CF_3CO_2H/(CF_3CO)_2O$ (20:1:1 by volume) | 0.10 M $TBABF_4$ | 195 | 35 |
| $CH_2Cl_2/CH_3CN$ (1:1 by volume) | 0.50 M $TBAPF_6$ | 223 | 52 |
| Butyronitrile/propionitrile (69:31 wt %) | 0.10 M $TBAClO_4$ | 155 | 27 |
| Butyronitrile/ethyl bromide (1:1 by volume) | 0.20 M $TBAClO_4$ | 128 | 47 |
| Butyronitrile/ethyl chloride (1:1 by volume) | 0.20 M $TBAClO_4$ | 88 | 47 |
| Butyronitrile/ethyl chloride (1:2 by volume)[b] | 0.20 M $TBAPF_6$ | 100 | 25 |
| Acetonitrile/toluene (1:5.4 by volume) | 0.10 M $TBAPF_6$ | 263 | 50 |
| *N,N*-Dimethylformamide/toluene (2:3 by volume) | 0.10 M $TEAPF_6$ | 185 | 51 |
| Ethyl chloride/THF/2-MeTHF (2:0.88:0.12 by volume) | 0.60 M $LiBF_4$ | 100 | 53 |

[a]TEA = tetraethylammonium; TPA = tetra-*n*-propylammonium; TBA = tetra-*n*-butylammonium; TOA = tetra-*n*-octylammonium.

[b]Volume ratio calculated for room temperature. Volume of butyronitrile measured at room temperature. Volume of ethyl chloride measured at −78°C. See this reference for other useful solvent mixtures.

perature" reported is not necessarily the lowest practical temperature for studies with a particular system. Rather, it is simply the lowest temperature at which measurements were made in the work cited.

It is clear that a variety of solvents commonly used in electrochemistry is available for low-temperature studies. Particularly noteworthy are the solvent mixture butyronitrile/ethyl chloride, which can be used down to about 100 K [25,47], and the inclusion of the low-polarity cosolvent, toluene, to enhance the solubility of a substrate that is insoluble in many polar solvents, in this case the fullerene, $C_{60}$ [50,51]. When low solution resistance is a priority and only moderately low temperatures are needed (above ca. −50°C), polar solvents such as acetonitrile and *N,N*-dimethylformamide are preferred.

## F. Addressing the Problem of Solution Resistance

One of the first observations one is likely to make when carrying out low-temperature voltammetric measurements is that the effects of solution iR drop that may have been scarcely noticeable at room temperature are suddenly alarmingly pronounced. The iR drop, of course, is governed by the cell current and the effective resistance between the working electrode and the Luggin capillary of the reference electrode. For example, the solution resistance for an embedded circular disk electrode of radius r with a distant reference electrode is given by Equation 16.13, where ρ is the resistivity of the solution.

$$R = \rho/4r \tag{16.13}$$

Thus, as the resistivity increases, the iR drop can become excessive. This problem is somewhat ameliorated by the fact that diffusion coefficients also decrease with decreasing temperature so that mass-transport-controlled currents will be smaller. However, this effect is not large enough to offset significantly the deterioration in response due to increased resistance.

There are four general responses to the problem of solution resistance. First, if only qualitative information is sought in the experiment, a certain amount of iR error can be tolerated, perhaps 100 mV. Second, electronic compensation of solution resistance can be applied, and this is often quite successful and will allow accurate data to be obtained even with macroelectrodes. Nevertheless, problems of potentiostat stability and signal distortion must be addressed.

A third approach is to use microelectrodes for low-temperature studies. The iR drop is intrinsically lower with microelectrodes, and it is often possible to obtain accurate current-potential curves at low temperatures even at rather large scan rates. The reason for this is that the current (when diffusion from the periphery of a disk electrode is negligible) is proportional to the area of the electrode; that is, it is proportional to $r^2$. Thus (considering Eq. 16.13), iR is proportional to r and the iR error decreases smoothly as smaller and smaller disk electrodes are used. Similar predictions pertain to microelectrodes of some other geometries (e.g., hemispheres).

When using microelectrodes to obviate resistance problems, it is convenient to develop a procedure to determine what conditions are required to reduce the error to an acceptable level. The results of such a procedure applied to disk electrodes are shown in Figure 16.6 [45]. In this and the remaining discussion, the technique of cyclic voltammetry is considered, as it is one of the most widely used voltammetric methods. The region of practical working conditions of electrode radius and scan rate is defined by the area set off by lines A, B, and C.

Above and to the right of line A, the contribution of edge diffusion will be less than 3%. This is convenient when quantitative interpretation of the data is attempted, because the calculation of the effects of edge diffusion are somewhat more complicated than planar diffusion. Below and to the left of line B, the iR error will be less than 3 mV. Finally, it is desirable to work in the region to the left of line C where distortion of the shape of the voltammogram by the RC time constant of the cell is not significant. Figure 16.6 predicts that

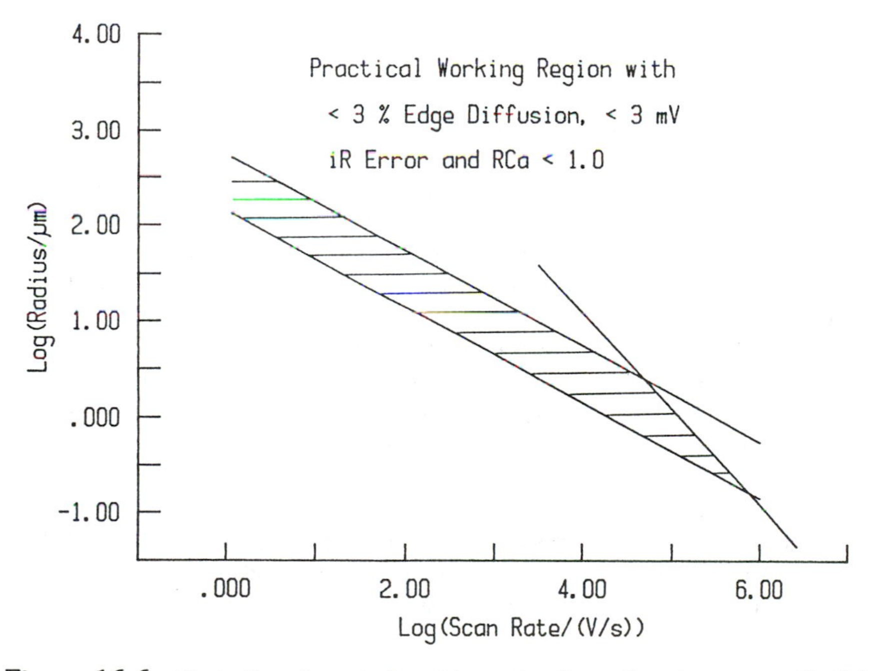


**Figure 16.6** Illustration of practical working region for cyclic voltammetry with disk electrodes. $D = 5 \times 10^{-6}$ cm$^2$/s, $C^* = 1$ mM, $\rho = 100\ \Omega$ cm, $C = 20\ \mu$F/cm$^2$, temperature = $-50$°C. [Reprinted with permission from W.J. Bowyer, E.E. Engelman, and D.H. Evans, *J. Electroanal. Chem. 262*:67 (1989). Copyright 1989 Elsevier Sequoia S.A.]

scan rates up to about $10^5$ V/s could be used for micrometer-sized disk electrodes. Diagrams analogous to Figure 16.6 can be constructed for the values of D, ρ, and C (the capacitance per unit area) that pertain to a given experimental system [45] (note error in Eq. 1 of Ref. 45, where the right-hand side should be divided by 4).

The fourth approach involves post facto analysis to account for the effects of solution iR drop. The iR drop can be that obtained without any electronic resistance compensation, or, on the other hand, the effect to be analyzed may be the residual iR drop resulting from partial compensation. In either case, the goal is to account for the effect of solution resistance in a quantitative manner.

Two general approaches have been used in low-temperature studies. In the first, the uncompensated resistance, electrode capacitance, diffusion coefficient, and kinetic and thermodynamic parameters describing the electrode reaction are incorporated in a master model, which is treated (usually by some form of digital simulation) to calculate the expected voltammetric response for comparison with experiment [7,49].

The second approach corrects the experimental voltammogram by subtracting the charging current and correcting the observed potentials for iR drop. The charging current is usually measured experimentally under the same conditions as the experiment using solvent and electrolyte alone (background voltammogram). Two more subtle corrections are also applied. The iR drop causes the effective scan rate in cyclic voltammetry to vary with time; that is, it is not constant like the apparent, applied scan rate. This in turn affects the actual charging current, so a correction must be applied to the background. Finally, the effective scan rate is inserted into the model used to calculate the voltammetric response (digital simulation), which is then compared with the corrected voltammogram [45].

Both approaches have been found to account adequately for the effects of solution iR drop in the analysis of voltammetric data. It is apparent that it is necessary to know the resistivity of the solution for the solvent, electrolyte, and temperature used in the experiments. Low-temperature resistivities of various electrolytes in acetone [7], acetonitrile [7,25,45], butyronitrile [7,25,45], *N,N*-dimethylformamide [7], and propionitrile [25,45], as well as butyronitrile/ethyl chloride and butyronitrile/ethyl bromide mixtures [25], have been published. Reference to these works prior to selection of solvent and electrolyte could save the inconvenience of determining the resistivity of another system that might be selected.

## ACKNOWLEDGMENT

This work was supported by the National Science Foundation, grant CHE9100281.

## REFERENCES

1. E. L. Lee, R. J. Cave, K. L. Guyer, P. D. Tyma, and M. J. Weaver, *J. Am. Chem. Soc. 101*:1131 (1979).
2. K. M. Kadish, K. Das, D. Schaeper, C. L. Merrill, B. R. Welch, and L. J. Wilson, *Inorg. Chem. 19*:2816 (1980).
3. M. J. Weaver, *J. Phys. Chem. 80*:2645 (1976).
4. M. J. Weaver, *J. Phys. Chem. 83*:1748 (1979).
5. A. J. Baranski, K. Winkler, and W. R. Fawcett, *J. Electroanal. Chem. 313*:367 (1991).
6. W. R. Fawcett and M. Opallo, *J. Electroanal. Chem. 331*:815 (1992).
7. L. K. Safford and M. J. Weaver, *J. Electroanal. Chem. 331*:857 (1992).
8. A. Demortier and A. J. Bard, *J. Am. Chem. Soc. 95*:3495 (1973).
9. W. H. Smith and A. J. Bard, *J. Am. Chem. Soc. 97*:5203 (1975).
10. C. Amatore, J. Badoz-Lambling, C. Bonnel-Huyghes, J. Pinson, J. M. Savéant, and A. Thiébault, *J. Am. Chem. Soc. 104*:1979 (1982).
11. L. A. Tinker and A. J. Bard, *J. Am. Chem. Soc. 101*:2316 (1979).
12. L. A. Tinker and A. J. Bard, *J. Electroanal. Chem. 133*:275 (1982).
13. J. G. Gaudiello, P. R. Sharp, and A. J. Bard, *J. Am. Chem. Soc. 104*:6373 (1982).
14. R. M. Crooks, F.-R. F. Fan, and A. J. Bard, *J. Am. Chem. Soc. 106*:6851 (1984).
15. A. M. Bond, M. Fleischmann, and J. Robinson, *J. Electroanal. Chem. 180*:257 (1984)
16. D. K. Gosser and Q. Huang, *J. Electroanal. Chem. 267*:333 (1989).
17. D. K. Gosser and Q. Huang, *J. Electroanal. Chem. 278*:399 (1990).
18. U. Stimming and W. Schmickler, *J. Electroanal. Chem. 150*:125 (1983).
19. U. Frese, T. Iwasita, W. Schmickler, and U. Stimming, *J. Phys. Chem. 89*:1059 (1985).
20. M. Cappadonia and U. Stimming, *J. Electroanal. Chem. 300*:235 (1991).
21. E. F. Dalton, S. Ching, and R. W. Murray, *Inorg. Chem. 30*:2642 (1991).
22. A. J. Klein and D. H. Evans, *J. Am. Chem. Soc. 101*:757 (1979).
23. K. M. O'Connell and D. H. Evans, *J. Am. Chem. Soc. 105*:1473 (1983).
24. W. J. Bowyer and D. H. Evans, *J. Org. Chem. 53*:5234 (1988).
25. S. Ching, J. T. McDevitt, S. R. Peck, and R. W. Murray, *J. Electrochem. Soc. 138*:2308 (1991).
26. S. A. Lerke, Ph.D. thesis, University of Delaware, Newark, DE (1992).
27. R. P. Van Duyne and C. N. Reilley, *Anal. Chem. 44*:142 (1972).
28. A. J. Gordon and R. A. Ford, *The Chemist's Companion*, John Wiley and Sons, New York, 1972, pp. 451–452.
29. (a) R. A. Petersen and D. H. Evans, *J. Electroanal. Chem. 222*:129 (1987); (b) A. G. Gilicinski and D. H. Evans, *J. Electroanal. Chem. 267*:93 (1989).
30. T. Nagaoka and S. Okazaki, *Anal. Chem. 55*:1836 (1983).
31. G. Farnia, F. Marcuzzi, G. Melloni, and G. Sandonà, *J. Am. Chem. Soc. 106*:6503 (1984).
32. N. J. Stone, D. A. Sweigart, and A. M. Bond, *Organometallics 5*:2553 (1986).
33. A. M. Bond and D. A. Sweigart, *Inorg. Chim. Acta 123*:167 (1986).

34. P. O'Brien and D. A. Sweigart, *J. Chem. Soc., Chem. Commun.*:198 (1986).
35. S. F. Nelsen, D. L. Kapp, R. Akaba, and D. H. Evans, *J. Am. Chem. Soc. 108*:6863 (1986).
36. A. M. Bowden, J. A. Bowden, and R. Colton, *Inorg. Chem. 13*:602 (1974).
37. M. G. Richmond and J. K. Kochi, *Inorg. Chem. 25*:656 (1986).
38. K. M. Kadish, C.-L. Yao, J. E. Anderson, and P. Cocolios, *Inorg. Chem. 24*:4515 (1985).
39. G. Farnia, G. Sandonà, F. Marcuzzi, and G. Melloni, *J. Chem. Soc., Perkin Trans. 2*:247 (1988).
40. G. Farnia, J. Ludvík, G. Sandonà, and M. G. Severin, *J. Chem. Soc., Perkin Trans. 2*:1249 (1991).
41. S. F. Nelsen, L. Echegoyen, E. L. Clennan, D. H. Evans, and D. A. Corrigan, *J. Am. Chem. Soc. 99*:1130 (1977).
42. S. F. Nelsen, E. L. Clennan, and D. H. Evans, *J. Am. Chem. Soc. 100*:4012 (1978).
43. D. H. Evans and R. W. Busch, *J. Am. Chem. Soc. 104*:5057 (1982).
44. W. J. Bowyer and D. H. Evans, *J. Electroanal. Chem. 240*:227 (1988).
45. W. J. Bowyer, E. E. Engelman, and D. H. Evans, *J. Electroanal. Chem. 262*:67 (1989).
46. Y. Zhang, D. K. Gosser, P. H. Rieger, and D. A. Sweigart, *J. Am. Chem. Soc. 113*:4062 (1991).
47. J. T. McDevitt, S. Ching, M. Sullivan, and R. W. Murray, *J. Am. Chem. Soc. 111*:4528 (1989).
48. D. Dubois, G. Moninot, W. Kutner, M. T. Jones, and K. M. Kadish, *J. Phys. Chem. 96*:7137 (1992).
49. J. Mortensen and J. Heinze, *Angew. Chem. Int. Ed. Engl. 23*:84 (1984).
50. Q. Xie, E. Pérez-Cordero, and L. Echegoyen, *J. Am. Chem. Soc. 114*:3978 (1992).
51. Y. Ohsawa and T. Saji, *J. Chem. Soc., Chem. Commun.*:781 (1992).
52. G. A. Heath and R. G. Raptis, *Inorg. Chem. 30*:4106 (1991).
53. S. J. Green, D. R. Rosseinsky, and M. J. Toohey, *J. Chem. Soc., Chem. Commun.* 325, 1994.

# 17

# Electroanalytical Chemistry in Molten Salts

**Charles L. Hussey** *The University of Mississippi, University, Mississippi*

## I. INTRODUCTION

Molten salts or ionic liquids (also referred to as fused salts by some authors) were among the very first media to be employed for electrochemistry. In fact, Sir Humphrey Davy describes electrochemical experiments with molten caustic potash (KOH) and caustic soda (NaOH) [1] as early as 1802! A wide variety of single molten salts and molten salt mixtures have been used as solvents for electroanalytical chemistry. These melts run the gamut from those that are liquid well below room temperature to those melting at more than 2000°C. The former present relatively few experimental challenges, whereas the latter can present enormous difficulties. For example, commercially available Teflon- and Kel-F-shrouded disk electrodes and Pyrex glass cells may be perfectly adequate for electrochemical measurements in ambient temperature melts such as the room-temperature chloroaluminates, but completely inadequate for use with molten sodium fluoroaluminate or cryolite (mp = 1010°C), which is the primary solvent used in the Hall–Heroult process for aluminum electrowinning.

This chapter is aimed at the inexperienced researcher who desires to carry out electroanalytical measurements in molten salts and seeks introductory information about the experimental details associated with the use of these solvents. It is intended to complement the chapters appearing elsewhere in this volume that discuss conventional molecular solvents and supporting electrolytes and various electroanalytical techniques.

### A. Classification of Melts

It is important to recognize that not all inorganic melts are ionic. In fact, a great many substances are categorized as molten salts that do not deserve this distinction.

Lovering and Gale [2] classify melts into the following convenient categories: (a) *ionic melts* that are completely ionized, such as molten KCl and NaBr; (b) *polyanionic melts* that are completely ionized, such as molten $NaNO_3$ and $K_2SO_4$; (c) *network/glass melts* that are partially ionized, such as $K_3PO_4$; (d) *molecular melts* such as $AlCl_3$ that exhibit almost no ionization in the liquid state; and (e) *hydrate melts* that are essentially very concentrated aqueous solutions of salts, such as $Ca(NO_3)_2 \cdot 4H_2O$. Thus, just as molecular solvents encompass a wide range of polarities, melts exhibit a wide range of "ionicities."

The classification system described earlier is limited to the simplest kinds of individual melts and is not intended to include mixtures. However, molten mixtures of these different classes of compounds are often more practical solvents than the melts of the individual compounds, due to their much lower melting points and other favorable properties, and this system of classification can usually be extended to these mixtures. For example, the very popular molten LiCl–KCl eutectic mixture is simply a binary ionic melt, whereas molten $NaNO_3$–$KNO_3$–$LiNO_3$ is a ternary polyanionic melt. Interestingly, the equimolar molten mixture of the simple ionic salt NaCl (a) and the molecular compound $AlCl_3$ (d) produces a simple polyanionic salt melt (b) composed of $Na^+$ and $AlCl_4^-$ ions!

The melts and melt mixtures that are discussed in this chapter are limited to those classified in categories (a) and (b) because these intrinsically conductive ionic melts are the most widely used electrochemical solvents. For the purpose of this chapter, low-melting salts are defined as those melting below 100°C, moderate-melting salts encompass those melting below 300°C, and all others are considered to be high-melting salts (>300°C).

## B. Molten Salts as Electrochemical Solvents

As electrochemical solvents, ionic melts possess certain advantages and disadvantages relative to molecular solvents. Among the advantages are high electrical conductivity, low or negligible vapor pressure, and good thermal stability. Many of the molten salts and molten salt mixtures that fall into categories (a) through (c) can be rendered almost completely anhydrous and aprotic. Some molten salt systems possess adjustable acid–base properties analogous to the variable Bronsted acidity of aqueous solutions. As a general rule, molten salts are good solvents for inorganic compounds. In fact, they may be the *only* solvents available for some materials. For example, $Al_2O_3$, which is practically insoluble in all conventional molecular solvents, dissolves readily in molten $Na_3AlF_6$ [3].

In terms of disadvantages, most inorganic molten salts are liquid at temperatures considerably above room temperature, necessitating that provisions be made for heating and thermostatting the electrochemical cell. Many exhibit rather limited electrochemical windows. In addition, some melts are corrosive to the usual cell and electrode materials and reactive with atmospheric moisture. The former attribute mandates careful selection of cell and electrode construction materials, whereas the latter necessitates the use of glove-box techniques (see Chap. 19) or gas-tight cells. In general, molten salts are poor solvents for most organic and organometallic solutes because many of these materials are thermally labile. The room-temperature chloroaluminates and other organic salt melts are an exception to this rule.

## II. MOLTEN SALT SYSTEMS

### A. Alkali Carbonates

Molten carbonates are of interest because of their applications as electrolytes in molten salt fuel cells. The preparation, handling, physical and electrochemical properties, and important applications of molten alkali carbonates have been described at length [4,5]. $Li_2CO_3$, $Na_2CO_3$, and $K_2CO_3$ and various mixtures of these salts are the carbonates most frequently used as electrochemical solvents. The major impurities in alkali carbonates are alkali hydroxides and oxides produced through hydrolysis and dissociation:

$$M_2CO_3 + H_2O \rightarrow 2MOH + CO_2 \tag{17.1}$$

$$M_2CO_3 \rightarrow M_2O + CO_2 \tag{17.2}$$

Purification of these salts and mixtures of these salts involves conversion of the adventitious hydroxide and oxide ion impurities to carbonate ion. This can be accomplished by using the procedure recommended by Maru [4]: (1) drying the salt or salt mixture for at least 24 h at 200°C, (2) melting the salt or salt mixture slowly under an atmosphere of dry, high-purity $CO_2$, and (3) passing $CO_2$ through the melt for an extended period. Contamination by atmospheric moisture must be avoided during the preparation and utilization of these materials in order to avoid the reaction shown in Equation 17.1. Thus, careful experimentation with these melts necessitates the use of a gas-tight cell or an inert-atmosphere glove box. It is also desirable to maintain an atmosphere of $CO_2$ above these melts in order to suppress the reactions shown in Equations 17.1 and 17.2.

Two different reference electrodes are commonly used for electrochemical measurements in molten carbonates: the oxygen electrode and the carbon monoxide electrode.

$$CO_2 + \tfrac{1}{2}O_2 + 2e^- \rightleftharpoons CO_3^{2-} \tag{17.3}$$

$$2CO_2 + 2e^- \rightleftharpoons CO_3^{2-} + CO \tag{17.4}$$

Physically, these gas reference electrodes consist of an alumina tube containing carbonate melt with a gold wire current collector. Provisions are made for purging the melt in the tube with the appropriate gas mixture. Complete details about these reference electrodes are summarized in the review by Selman and Maru [5].

Molten carbonates display variable acid–base behavior according to the Lux–Flood modification of the Lewis definition of acidity,

$$\text{Acid} + O^{2-} \rightleftharpoons \text{base} \tag{17.5}$$

This "oxoacidity" is described by $pO^{2-} = -\log a(O^{2-})$, where $a(O^{2-})$ is the oxide ion activity [6]. Not surprisingly, the anodic and cathodic potential limits of carbonate melts are very dependent upon the oxoacidity. Anodic decomposition of the melt is assumed to arise from the oxidation of oxide provided by the thermal decomposition of $CO_3^{2-}$:

$$O^{2-} \rightarrow \tfrac{1}{2}O_2 + 2e^- \tag{17.6}$$

The anodic evolution of oxygen may also occur by the direct oxidation of $CO_3^{2-}$:

$$CO_3^{2-} \rightarrow CO_2 + \tfrac{1}{2}O_2 + 2e^- \tag{17.7}$$

In the case of the ternary eutectic $Li_2CO_3$–$Na_2CO_3$–$K_2CO_3$ at 605°C saturated with pure $CO_2$ ($pO^{2-} = 6$), the anodic limit is about 0.27 V vs. the oxygen electrode; however, when saturated with $Li_2O$ ($pO^{2-} = 0$) this melt is reported to exhibit an anodic limit of only –0.23 V [5]. The cathodic limit of ternary eutectic carbonate melts with $pO^{2-} = 2$ to 6 is about –1.9 to –2.1 V [5]. The reduction process produces elemental carbon according to the reaction

$$CO_3^{2-} + 4e^- \rightarrow C(s) + 3O^{2-} \tag{17.8}$$

Gold is the only material that is not subject to appreciable corrosion in molten carbonates, and it is the material of choice for constructing working electrodes.

## B. Halides

White [7] gives a general review that includes information about the preparation and purification of a variety of alkali and alkaline earth halide melts. Information about fluorides can also be found in an earlier review article by Bamberger [8]. Many different halide melts have been used as solvents for electrochemistry, and a complete discussion of all of these melts is outside the scope of this chapter. However, two systems that have generated continuous interest over the years are the LiCl–KCl eutectic (58.8–41.2 mol%, mp = 352°C) [9] and the LiF–NaF–KF (46.5-11.5-42.0 mol%, mp = 454°C) also known as FLINAK [10]. The former is an electrolyte commonly used in thermal batteries, whereas the latter molten salt is of interest for refractory metal plating.

An excellent procedure for purification of the LiCl–KCl melt has been summarized by Levy and Reinhardt [11]. To apply this procedure the melt is maintained under an atmosphere of dry HCl gas at 400°C while HCl is passed through the melt to convert hydroxide impurities to $H_2O$:

$$MOH + HCl \rightarrow MCl + H_2O \tag{17.9}$$

This is followed by the passage of $Cl_2$ gas at 600°C for 48 h in order to oxidize any remaining $OH^-$ to $O_2$:

$$2OH^- + 2Cl_2 \rightarrow O_2 + 2HCl + 2Cl^- \quad (17.10)$$

Residual HCl and/or $Cl_2$ is then removed by passing dry argon gas through the melt at 400°C for 16 h.

FLINAK is purified by treatment with the HF released by ammonium bifluoride ($NH_4HF_2$); the HF converts oxide impurities in the melt to $H_2O$ [7]. In this purification procedure, the fluoride salt mixture is combined with 15 wt% $NH_4HF_2$ and heated to about 500°C in a graphite crucible. The molten mixture is poured into a platinum container and heated to 750°C. Hydrogen is passed through the molten mixture for approximately 2 days. Further purification can be achieved by controlled-potential electrolysis at an applied potential of about 3 V between a tungsten cathode and glassy carbon anode.

White [7] lists some of the various redox systems that have been employed as reference electrodes in molten halides. Albeit inconvenient, the $Cl_2(g)/Cl^-$ couple is the thermodynamic reference system of choice in the LiCl-KCl eutectic [12]. However, a reference electrode consisting of a silver wire dipping into a solution of AgCl in molten LiCl–KCl (i.e., the $Ag^+/Ag$ couple) is far more convenient [11,13]. This mixture is usually isolated from the bulk melt in a Pyrex tube with a closed end. No porous diaphragm is needed because the Pyrex glass becomes conductive at high temperatures. The recommended reference electrode for LiF–NaF–KF is the $NiF_2$(satd)/Ni couple isolated from the melt with a fluoride ion conductive $LaF_3$ membrane [14] as modified by Bamberger [8]. The Pt(II)/Pt couple is also used as a reference by some workers [12,15].

At a platinum electrode, highly purified FLINAK has a voltammetric window extending from about +1.5 to −2.0 V vs. the nickel reference electrode [7]. The positive limits of the alkali halide melts discussed herein arise from the oxidation of halide ions, whereas the negative limits are due to reduction of the alkali metal ions. Because chloride ion is substantially easier to oxidize than fluoride ion, the potential window of the LiCl–KCl melt is approximately 1.5 V smaller than that for FLINAK.

Plambeck [16] summarizes the important physical properties of the molten LiCl-KCl eutectic such as density, viscosity, and electrical conductivity and gives references to some other sources of this data. Tabulated physical property data for FLINAK are scarce but are available in experimental articles.

Molten alkali chlorides like LiCl–KCl can be handled in Pyrex containers if the melt is anhydrous and the temperature is kept below the softening point of the glass. Fluoride melts present a greater challenge because fluoride ion attacks glass and some ceramic materials. Bamberger [8] recommends a welded nickel container with a molybdenum liner for work with molten fluorides. Most workers maintain halide melts under an inert gas such as high-purity argon in order to exclude atmospheric moisture. Alkali fluorides, fluorine, and hydrogen fluoride are very toxic; caution and good judgment must be used when handling these materials.

## C. Haloaluminates

A haloaluminate melt results from the combination of an aluminum halide and an appropriate halide salt. A list of the different haloaluminate melts that have been used for electrochemistry is quite large, and it is impossible to discuss each system in a brief introductory chapter of this nature. Thus, only a few of the more popular systems are covered in this section. The commonly used haloaluminates can be divided into three broad categories depending upon their melting temperatures: (a) mixtures of aluminum chloride and organic chloride salts that are liquid at or below room temperature, (b) moderate-melting (less than ~300°C) aluminum chloride–alkali chloride mixtures, and (c) high-temperature aluminum fluoride–alkali fluoride melts such as cryolite. Examples from each of these categories are discussed next.

*Room-Temperature Chloroaluminates*

Comprehensive reviews describing the preparation, purification, and physical and electrochemical properties of these melts have been published [17–20]. The most popular systems are mixtures of $AlCl_3$ with either 1-(1-butyl)pyridinium chloride (BupyCl) or 1-methyl-3-ethylimidazolium chloride (MeEtimCl). These systems are very versatile solvents for electrochemistry because they are stable over a wide temperature range. In many ways they can be considered to be a link between conventional nonaqueous solvent/supporting electrolyte systems and conventional high-temperature molten salts.

Good-quality $AlCl_3$ (actually $Al_2Cl_6$) can be prepared by direct synthesis from high-purity aluminum metal and HCl gas [18]. Alternately, commercial aluminum chloride can be purified by sublimation in sealed tubes or under continuous vacuum [20]. As of this writing, the best quality commercial $AlCl_3$ is marketed by Fluka Chemie AG, Ronkonkoma, NY. The BupyCl salt is prepared by refluxing together stoichiometric quantities of 1-chlorobutane and pyridine [18]. MeEtimCl is synthesized by combining freshly distilled 1-methylimidazole with a stoichiometric excess of chilled ethyl chloride in a heavy-wall glass pressure vessel. This mixture is heated under pressure to about 60°C for several days and stirred vigorously [21]. Both BupyCl and MeEtimCl are purified by precipitation from dry acetonitrile with ethyl acetate or diethyl ether. Exposure of the $AlCl_3$–BupyCl and $AlCl_3$–MeEtimCl melts to moisture leads to the formation of both proton- and oxide-containing impurities. Various strategies for removing these impurities have been summarized [20].

The usual reference electrode for use in these melts is the Al(III)/Al couple that results when high-purity aluminum wire is immersed in either 66.7–33.3 or 60.0–40.0 mol% melt. This reference melt is separated from the bulk melt with a

Pyrex tube that is closed on one end by a fine-porosity frit. The latter melt is preferred because changes in the composition of this melt lead to smaller changes in the reference electrode potential. The formal potential of the ferrocenium/ferrocene couple is located at 0.250 V vs. the Al(III)/Al couple in the former melt [22].

Room-temperature chloroaluminate melts exhibit adjustable Lewis acidity, defined by the chloride ion concentration or "chloroacidity" [6]. Acidic melts result when a molar excess of aluminum chloride is combined with the organic chloride salt (>50 mol% $AlCl_3$), and basic melts are obtained when an excess of the salt is mixed with $AlCl_3$ (<50 mol% $AlCl_3$). The chloroacidity of these melts is well described by the following equilibrium reaction:

$$2AlCl_4^- \rightleftharpoons Al_2Cl_7^- + Cl^- \qquad (17.11)$$

The equilibrium constant for Equation 17.11 is almost the same for both melts and is slightly greater than $1 \times 10^{-17}$ at 40°C [23]. Like the carbonate systems described earlier in this chapter, the potential range of these ionic solvents is highly dependent on acidity. The electrochemical potential window of acidic melt extends from about 2.1 to 0 V vs. the 60.0–40.0 mol% reference electrode, whereas the basic melt window extends from 0.8 to −1.9 V in the case of $AlCl_3$–MeEtimCl and from 0.8 to −1.0 V for $AlCl_3$–BupyCl. The positive and negative limits of acidic melt are due to the oxidation of $AlCl_4^-$ and to the reduction of $Al_2Cl_7^-$ to aluminum metal, respectively. For basic melts, these limits correspond to the oxidation of chloride ion and the reduction of the organic cation. The $MeEtim^+$ ion is considerably more resistant to reduction than the $Bupy^+$ ion; in fact, the latter is reduced by aluminum metal in basic melt. "Neutral" melts (exactly 50 mol% $AlCl_3$) exhibit the largest electrochemical window; they display the positive limit of an acidic melt and the negative limit of a basic melt. However, they are considerably more difficult to prepare than acidic and basic melts.

Comprehensive composition-dependent phase equilibria and physical property data is available for the $AlCl_3$–MeEtimCl system [24]. Only a smattering of such data is available for the $AlCl_3$–BupyCl melt; the available data are summarized in reviews [17,20]. Because room-temperature chloroaluminates react vigorously with moisture, most researchers prefer to work with these molten salts in a high-quality glove box, necessitating the use of techniques described in Chapter 19. These molten salts can be studied in standard Pyrex electrochemical cells with Teflon- or Kel-F-shrouded working electrodes; however, bear in mind that these electrodes usually develop leaks when heated to 50°C or more. These melts attack epoxy glue and some plastics.

### *Alkali Metal Chloroaluminates*

The binary mixture of $AlCl_3$ and NaCl is the alkali metal chloroaluminate most commonly used as an electrochemical solvent. The preparation, purification, and general properties of this and several related inorganic chloroaluminate systems have been discussed in reviews [7,18,25]. References to methods for the preparation and purification of $AlCl_3$ are presented in the preceding section. Inexpensive, high-quality NaCl is available commercially. Organic contaminants can be removed from the NaCl by heating it to its melting point in a small furnace. The NaCl-saturated and equimolar $AlCl_3$–NaCl melts can be further purified by constant-current electrolysis between high-purity aluminum electrodes, but the resulting melts must be filtered through a glass frit to remove aluminum particulate matter before use.

The reference electrode recommended for use in the $AlCl_3$–NaCl melt is the Al(III)/Al couple that is obtained by placing an aluminum wire in a tube containing NaCl(satd) melt (49.8–50.2 mol% $AlCl_3$–NaCl at 175°C). The reference electrode tube is terminated at one end with a Pyrex frit [26] or a thin Pyrex membrane [27]. A platinum wire quasi-reference electrode is used by some researchers [28].

Besides Equation 17.11, additional equilibria govern the chloroacidity of the $AlCl_3$-NaCl melt:

$$3Al_2Cl_7^- \rightleftharpoons 2Al_3Cl_{10}^- + Cl^- \quad (17.12)$$

$$2Al_3Cl_{10}^- \rightleftharpoons 3Al_2Cl_6 + 2Cl^- \quad (17.13)$$

The equilibrium constants for these reactions are $8.9 \times 10^{-8}$ (Eq. 17.11), $1 \times 10^{-7}$ (Eq. 17.12), and $1 \times 10^{-14}$ (Eq. 17.13) at 175°C [29]. As a consequence of Equation 17.13, there is a significant vapor pressure of $Al_2Cl_6$ over acidic $AlCl_3$–NaCl melt, making experimentation with this ionic solvent difficult. This is not a problem for the equimolar and NaCl(satd) melts.

The useful potential window of the equimolar $AlCl_3$–NaCl melt extends from about 2.2 to 0 V vs. the Al(III)/Al couple in NaCl(satd) melt. Plambeck [30] summarizes the physical properties of the equimolar (mp = 151°C) and 63–37 mol% eutectic melts. Pyrex cells are satisfactory for use with the $AlCl_3$–NaCl melt. Teflon has also been used in these melts, but it slowly decomposes. Experimentation with the alkali metal chloroaluminates requires the use of an inert-atmosphere glove box or a gas-tight cell.

*Alkali Metal Fluoroaluminates*

The alkali metal fluoroaluminates are of considerable interest because they constitute the electrolyte in the Hall–Heroult electrolytic process used to produce aluminum. Several reviews describe the use of fluoroaluminates for the electrolytic production of aluminum [31,32]. The majority of electrochemical studies involving fluoroaluminates have been conducted in the $AlF_3$–NaF melt, especially the 25.0–75.0 mol% composition or $Na_3AlF_6$ (mp = 1010°C). $Na_3AlF_6$ is also referred to as "cryolite." In practice, electrochemical measurements are rarely conducted in pure molten cryolite; instead such measurements are almost invariably carried out in cryolite containing substantial amounts of dissolved alumina ($Al_2O_3$) and other additives, including excess $AlF_3$.

$AlF_3$ is typically purified by sublimation in a corundum crucible with a platinum lid in a vertical vacuum furnace [32]. The bottom of the crucible is heated to 1000°C and the temperature of the platinum lid becomes about 750°C. As a result, the $AlF_3$ is evaporated from the crucible and deposited on the inside of the platinum cover. The alkali metal fluorides are typically recrystallized from their melts in platinum crucibles; the impure core of the solidified melt is discarded [32].

A number of different gas, oxide, and metal reference electrodes have been used in molten cryolite. The Al(III)/Al couple produced by placing molten aluminum in a sintered corundum or boron nitride tube is the traditional reference electrode [33]. However, a $CO_2$/C electrode has also been used. This electrode consists of a graphite-covered copper tube; $CO_2$ gas is passed through the tube and allowed to bathe the graphite surface while it is immersed in the melt [32].

The electrochemical window of pure molten cryolite has not been expressly stated, but a voltammogram of purified cryolite recorded at a graphite working electrode exhibits very little residual current over the range of potentials extending from 0.4 to −1.9 V vs. a nickel wire quasi-reference electrode [7]. Physical property data for molten cryolite and phase equilibria for the $AlF_3$–NaF melt system have been summarized [31,32]. The extremely high temperature of cryolite places severe constraints on the materials that can be used for cells. Platinum and boron nitride are the materials of choice.

## D. Hydroxides

The majority of the published electrochemical studies involving hydroxides have been conducted in the NaOH-KOH eutectic (49.0–51.0 mol%, mp = 170°C). This eutectic mixture can be used at temperatures considerably below those of the individual salts; pure NaOH and KOH melt at 318 and 400°C, respectively. Water, oxygen, and carbon dioxide are the major impurities in molten hydroxides. The various procedures for removing these contaminants have been summarized [34]. Water can be eliminated by passing dry inert gas over the melt when it is heated to about 450 or 500°C; heating the melt to these temperatures also leads to the thermal decomposition of peroxide and superoxide contaminants [34]:

$$O_2^{2-} \rightarrow O^{2-} + \tfrac{1}{2}O_2(g) \tag{17.14}$$

$$2O_2^{-} \rightarrow O_2^{2-} + O_2(g) \tag{17.15}$$

These contaminants result from the reaction of hydroxide ion with adventitious water and oxygen. Contamination by $CO_2$ does not seem to cause difficulties, and usually no effort is made to remove $CO_2$ from hydroxide melts.

The various reference electrodes that have been used in hydroxide melts are discussed by Claes and Glibert [34], who conclude that there is really not a suitable reference electrode for these melts. In practice, voltammetric measurements conducted in the NaOH–KOH eutectic are usually referenced to a platinum wire quasi-reference electrode. In turn, this electrode is related to the negative limit of the solvent, which corresponds to the reduction of the alkali metal cations [35].

Molten hydroxides display acid–base behavior defined by the equilibrium

$$2OH^- \rightleftharpoons H_2O + O^{2-} \qquad (17.16)$$

where $H_2O$ is the acid and $O^{2-}$ is the base [6]. This equilibrium is analogous to the autodissociation of water. The acidity of molten hydroxides is conveniently described by either $pH_2O = -\log a(H_2O)$ or $pO^{2-} = a(O^{2-})$. As is the case for many of the other molten salts described in this chapter, the useful negative and positive limits of the molten hydroxides are dependent upon the melt acidity. Exactly neutral melts exhibit a useful potential range extending from 0 to about 2.1 V vs. the reference system described in the previous paragraph; the positive limit arises from the oxidation of hydroxide ions to superoxide ions [36]. At a platinum electrode, acidic melts exhibit a voltammetric reduction wave at about 0.4 V due to the reduction of water, whereas basic melts show a wave at approximately 1.4 V due to the oxidation of oxide ion [36]. Obviously, large concentrations of either $H_2O$ or $O^{2-}$ will foreshorten the electrochemical window of the melt. Molten hydroxides are exceedingly corrosive. A Pyrex glass cell can be used with these melts, but the area contacted by the melt must be protected with Teflon or some other relatively inert material, such as glassy carbon [36]. These melts must be shielded from atmospheric moisture and oxygen.

## E. Nitrates

Molten nitrates are of technological interest as electrolytes for metal anodization. In addition, a large number of academic electrochemical studies have been carried out in these melts. According to Plambeck [37], the $LiNO_3$–$KNO_3$ eutectic (43–57 mol%, mp = 132°C), equimolar $NaNO_3$–$KNO_3$ (50.0–50.0 mol%, mp = 228°C), and the ternary eutectic $LiNO_3$–$NaNO_3$–$KNO_3$ (30.0–17.0–53.0 mol%, mp = 125°C) are the nitrate melts most often used as electrochemical solvents. Purification of nitrate melts is difficult because the main contaminants are $H_2O$ and the $NO_2^-$ ion, and both are difficult to remove. The latter results from the thermal decomposition reaction [38]

$$2NO_3^- \rightarrow 2NO_2^- + O_2 \qquad (17.17)$$

This reaction is difficult to avoid, but it can be minimized by working with these melts at the lowest practical temperature. Some authors suggest purging the melt with $O_2$ or $NO_2$ in order to oxidize the $NO_2^-$ to $NO_3^-$, but these procedures may lead to the production of still more impurities such as hydroxide, superoxide, and hydrogen ion [39]. By far, the best strategy is to use the highest quality starting materials available and to subject these materials to careful recrystallization before use. The freshly recrystallized materials are combined in the correct proportions, melted under vacuum, and filtered through a glass frit before use.

The recommended reference electrode for use in molten nitrates is the Ag(I)/Ag couple isolated from the bulk melt by a glass membrane, asbestos wick, or porous ceramic plug. Ag(I) is introduced into the melt by dissolving $AgNO_3$. A number of different Ag(I) concentrations have been used in this reference electrode; furthermore, different Ag(I) concentrations are often used with different melts [37]. For a recent example, see Miles et al. [40].

Some relevant physical property data for nitrate melts are summarized in two chapters by Plambeck [37]. The useful electrochemical window of equimolar $NaNO_3$–$KNO_3$ melt extends from approximately +1.0 to −1.6 V vs. the 0.07 molal Ag(I)/Ag reference electrode. The negative limit of nitrate melt corresponds to the reduction of $NO_3^-$ to $NO_2^-$. It is strongly emphasized that nitrates are *powerful oxidants*; contact with aluminum and its alloys, cyanides, organic materials, and other easily oxidized materials is to be avoided [41]! Pyrex glass cells are adequate for ordinary electrochemical experimentation, but investigations involving the oxygen electrode in nitrate melt revealed unwanted participation by $SiO_2$ extracted from the glass [39].

### F. Organic Salts

This class includes molten salts with organic cations, organic anions, or both, as opposed to the molten salts described earlier that are derived strictly from inorganic ions. The room-temperature haloaluminate melts described in Section II.C can be classified as organic salts. Several reviews covering organic molten salts other than the room-temperature haloaluminates are available [17,42–44]. Surprisingly, except for the room-temperature haloaluminates, a great majority of the common organic salts are not especially low-melting, and most of those that do melt at low temperature are syrupy, poorly conductive, rather useless liquids as far as the electrochemist is concerned. Most of the interest in organic salts stems from their use as solvents for organic synthetic reactions; however, on occasion some of these salts have been used for electrochemistry. Table 17.1 lists a few of the organic salts that have been employed as electrochemical solvents; the references listed in the table should be consulted for information about the preparation and physical properties of these salts.

### G. Thiocyanates

Electroanalytical studies involving thiocyanate melts are largely restricted to the alkali metal salts. Investigations have been carried out in both the NaSCN–KSCN eutectic (26.3–73.7 mol%, mp ~130°C) and in pure KSCN (mp = 177°C) [50, 51]. In view of their relatively low melting points, both the binary eutectic and the

**Table 17.1** Examples of Organic Molten Salts Used as Electrochemical Solvents

| Molten salt | Liquidus temperature (°C) | Potential window (V)/ working electrode | Reference electrode | Ref |
|---|---|---|---|---|
| $N_{2226}B_{2226}$[a] | −78[b] | ~2/Pt | aq AgCl/Ag | [45] |
| $[C_2H_5]_3NHCuCl_2$ | 0 | <1/Pt | Cu(II)/Cu | [46] |
| $[C_6H_{13}]_4NOCOC_6H_5$[c] | <25 | ~1.5/Pt | aq AgCl/Ag | [48] |
| $CF_3COOK–CF_3COONa$ (65–35 mol%) | 116 | 4.9/Pt | Ag(I)/Ag[d] | [49] |
| $CH_3COOLi–CH_3COONa–CH_3COOK$ (20.0–35.0–45.0 mol%) | 187 | >1/DME | Ag(I)/Ag[d] | [49] |

[a]Triethyl-*n*-hexylammonium triethyl-*n*-hexylboride.
[b]Glass transition temperature; no crystallization has been observed.
[c]Tetra-*n*-hexylammonium benzoate hemihydrate.
[d]In molten $LiNO_3–KNO_3$.

single salt are very attractive. Alkali metal thiocyanates do not appear to require any significant pretreatment; Potter and Schiffrin [52] simply dried the high-quality salts in a vacuum oven before use. Thiocyanate melts are compatible with Pyrex, but under certain circumstances platinum and gold dissolve in these melts [51]. In addition, these melts must be protected from oxygen in order to avoid decomposition.

Plambeck [51] discusses the different reference electrodes that have been used in thiocyanate melts. A reference electrode consisting of AgCl/Ag in KCl-saturated $LiNO_3–KNO_3$ eutectic was employed in early research [53]; however, an $Ag_2S/Ag$ reference electrode was used in more recent work [54]. The electrochemical window of the eutectic melt is approximately 2.0 V. The positive limit is due to the oxidation of thiocyanate ion to parathiocyanogen, and the negative limit corresponds to the reduction of thiocyanate to sulfide and cyanide [51]. Both Kerridge [50] and Plambeck [51] provide references to physical property data for these salts.

## H. Other Systems

The molten salt systems listed earlier are among those most frequently used as solvents for electroanalytical chemistry. This list is by no means inclusive because in many cases there are "subspecies" of these melt systems that have not been covered that enjoy equal or only slightly less popularity than the systems that were discussed. In addition, there are a number of molten salts used for electrochemistry that have not achieved the level of popularity of the systems described in the preceding sections. Molten alkali sulfates, for example, $Li_2SO_4–K_2SO_4$ eutectic (80.0–20.0 mol%, mp = 535°C) [55,56], are included in this category.

## III. APPARATUS AND TECHNIQUES

### A. Cells and Electrodes

The high temperatures and corrosive environments associated with some molten salts place exacting requirements on the types of materials that can be used in the fabrication of electrochemical cells. For those molten salts requiring the use of containers more robust than Pyrex glass, there are a number of relatively inert high-temperature nonmetallic materials to choose from: alumina, boron nitride, fused silica, glassy carbon, graphite, mullite, and stabilized zirconia. The review by White [7] lists the compatibility of some of these materials. Except for the requirement for special construction materials, electrochemical cells designed for use with moderate- and high-melting molten salts are in concept no different from those employed with low-temperature molten salts and conventional solvents (see Chap. 9); that is, provisions must be made for working, auxiliary, and reference electrodes and for sparging/blanketing the cell contents with an inert gas. Also, in molten salt cells it is customary and sometimes necessary to include auxiliary electrodes for preelectrolysis of the melt and to make provisions for a thermocouple or thermistor temperature sensor. Additional constraints are placed on the design of the cell if the molten salt solvent is moisture- and oxygen-sensitive and/or it exhibits a significant vapor pressure. Electroanalytical experiments with moisture- and oxygen-sensitive salt melts that possess relatively low vapor pressures can be carried out directly inside a glove box (see Chap. 19 and the review by Bartak [57]). However, if the vapor pressure of the molten salt is high, then it is necessary to work in a completely sealed cell. The review by Lantelme et al. [58] gives diagrams of some common cell designs. A simple, but very effective vacuum-tight Pyrex glass cell for electrochemical measurements with high-vapor-pressure, moderate-melting salts is shown in Figure 17.1. A typical cell used for experiments with molten fluorides is illustrated in Figure 17.2.

Suggested reference electrodes were discussed in conjunction with each of the molten salts listed in Section II. However, reviews describing the various reference electrodes that have been used in molten salts should be consulted for additional details [58,59]. The material compatibility and the design of the working electrodes to be used in molten salt electrochemical experiments are both critical to the success of such experiments. The list of working electrodes that have been used in molten salts is extensive and includes dropping metal electrodes based on liquid bismuth, lead, and mercury and solid electrodes constructed from glassy (vitreous) carbon, gold, iridium, palladium, platinum, pyrolytic graphite, molybdenum, nickel, silver, tungsten, and other materials. Some of the different working electrode designs that have been employed in molten salts are reproduced in Figure 17.3. It can be quite difficult to fabricate solid working electrodes with well-defined surface areas and diffusional characteristics (e.g., disk electrodes) for use in high-temperature molten salt systems. This is due to the lack of thermally robust insulating materials with expansion coefficients that are compatible with the materials just listed. As

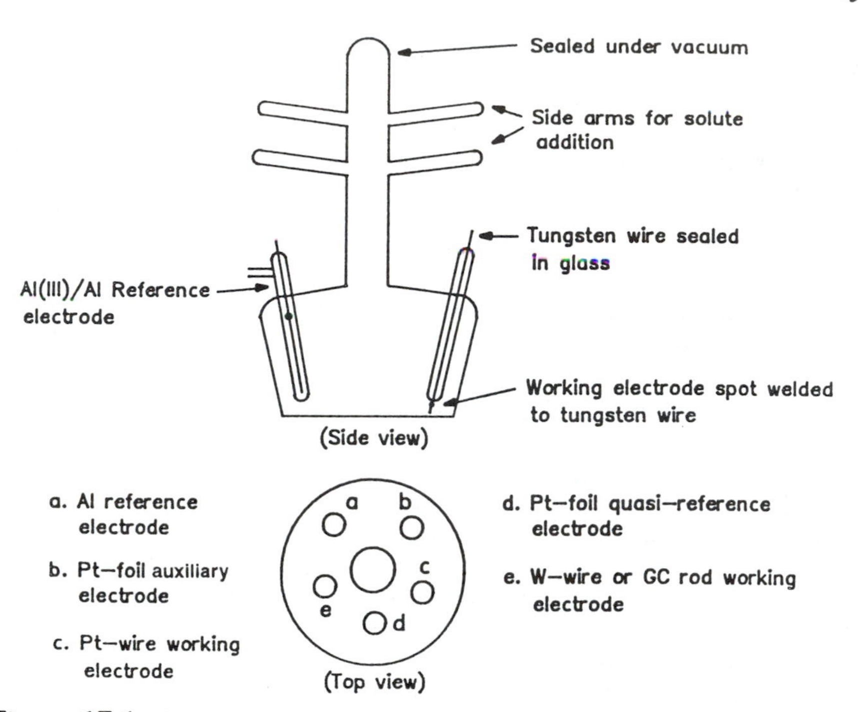


**Figure 17.1** Gas-tight Pyrex glass electrochemical cell for use with moderate-melting molten salts. [Courtesy of Prof. G. Mamantov, University of Tennessee.]

discussed in Section II.C, Teflon-shrouded disk electrodes such as those available from Pine Instruments Co., Grove City, PA, are perfectly suitable for use in room-temperature melts provided that they are not heated to more than about 50°C. For experiments with higher melting salts, it is possible to fabricate insulated disk electrodes by sealing tungsten wire into uranium glass, platinum wire into soft glass, and glassy carbon rod into Pyrex glass [11] and exposing the cross section of the rod or wire with a cut on a glass saw. A very elegant example of the latter is the Pyrex glass/glassy carbon ring-disk electrode fabricated from glassy carbon rod and Pyrex tubing by Phillips et al. [60]. Naturally, the electrodes just described can only be used at temperatures well below the softening point of the glass used in their construction. It is not always possible to construct insulated disk electrodes for use in very-high-temperature molten salts; the investigator must often be satisfied with a metal flag electrode suspended in the melt by a thin wire. A platinum-flag electrode is easily constructed by spot-welding a length of Pt wire to the edge of a suitably sized piece of Pt foil. The wire can be shielded by encasing it in a thin ceramic tube. As of this writing, microelectrodes have not been employed extensively in molten salts; however, a 25mm tungsten electrode has been used to study metal deposition in the LiCl–KCl eutectic at 400°C [61].

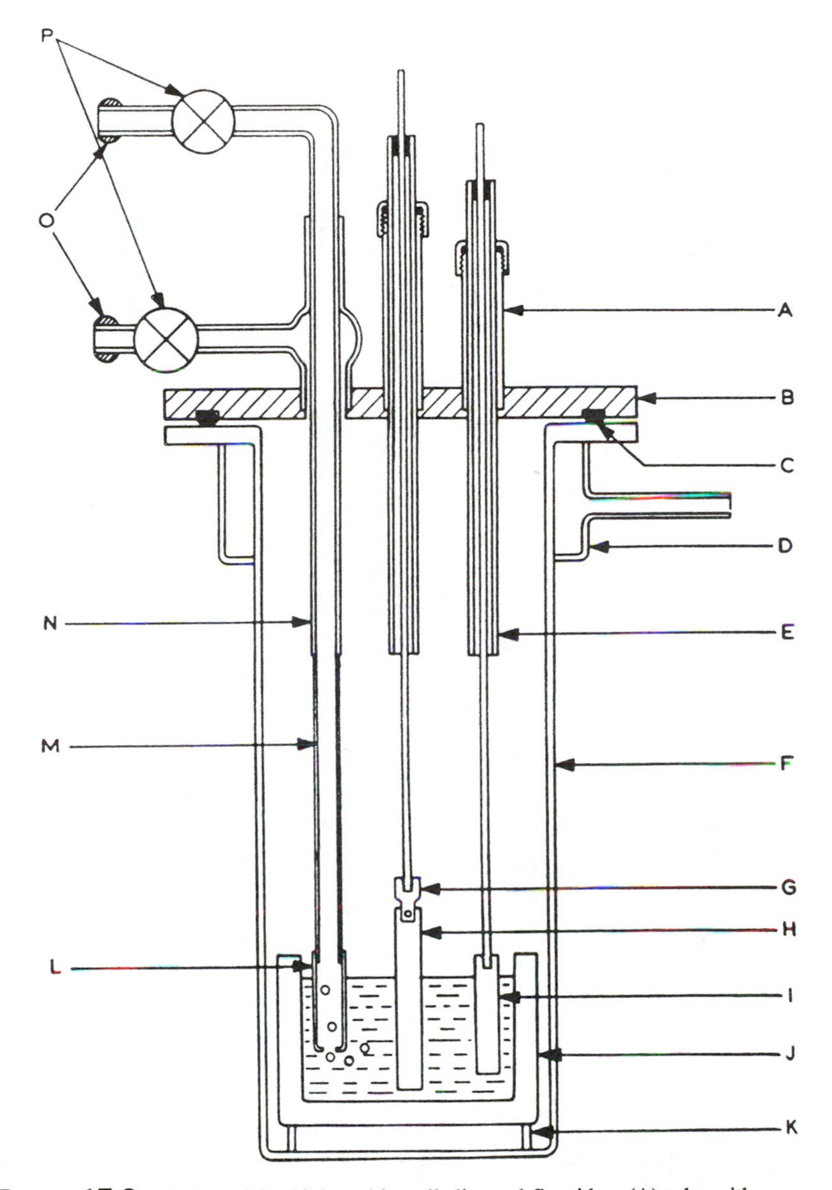


**Figure 17.2** Cell used for high-melting alkali metal fluorides: (A) tube with screw cap, (B) brass head, (C) Viton O-ring, (D) cooling jacket, (E) Pyrex glass support tube, (F) cell body made from nickel or Inconel, (G) stainless-steel connector block, (H) graphite or tungsten cathode, (I) graphite anode, (J) graphite or glassy carbon crucible, (K) alumina base, (L) molybdenum or graphite sparging tube, (M) stainless steel tube, (N) Pyrex glass gas inlet tube, (O) ball joints, (P) vacuum valves. [From Ref. 7, with permission.]

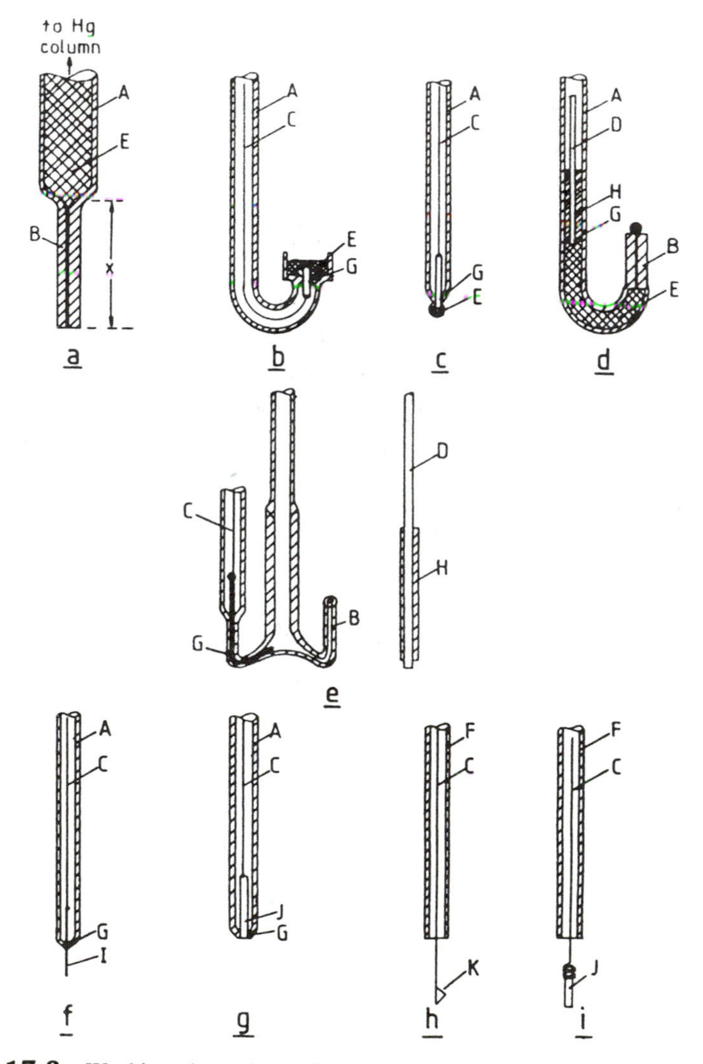


**Figure 17.3** Working electrode configurations: (a) capillary for dropping metal, (b) microcup, (c) hanging drop, (d) metal drop formed by plunger, (e) alternate design of (d), (f) wire, (g) disk, (h) flag, (i) rod. The capital letters in these drawings refer to (A): Pyrex glass sheath, (B): glass capillary, (C): contact wire, (D): metal contact, (E): liquid metal, (F): ceramic sheath, (G): glass/metal seal, (H): plunger, (I): thin metal wire, (J): metal rod, and (K): metal flag. [From Ref. 58, with permission.]

## B. Electroanalytical Techniques

Many of the electroanalytical techniques that are routinely employed in conventional solvents, such as, chronoamperometry, chronocoulometry, chronopotentiometry, coulometry, cyclic (stationary electrode) voltammetry, rotating electrode voltammetry, and pulse voltammetry, have also been applied to molten salts. Some of these techniques are discussed next with special attention to their employment in molten salts. References to noteworthy examples appearing in the literature are included. Background information about these techniques is available elsewhere in this book.

### *Chronoamperometry*

The theory and practice of chronoamperometry (CA) or large-amplitude potential-step experiments have been covered in detail in Chapter 3 of this volume, and the application of CA techniques to molten salts has been discussed at length [62]. The behavior of the faradaic current with time following the application of a potential step for an uncomplicated, diffusion-controlled reaction is governed by the Cottrell equation (Eq. 3.6). In molten salts, chronoamperometry has been used primarily to measure diffusion coefficients and to produce current–potential curves. The latter are obtained by recording current–time transients at different potentials and then plotting the current measured at the same time on each transient versus the potential. Because the diffusion layer at the electrode surface is ordinarily destroyed by stirring the solution before the collection of a new transient, the resulting current-potential curve is quite similar in appearance to a normal-pulse polarogram (vide infra). An example of such a current–time curve is shown in Figure 17.4 for the oxidation of Ti(II) to Ti(III) and Ti(IV) in the $AlCl_3$–NaCl melt. The advantage of this method for producing current–potential curves is that it can be effected at solid electrodes by using very simple instrumentation.

A more elaborate version of the chronoamperometry experiment is the symmetrical double-potential-step chronoamperometry technique. Here the applied potential is returned to its initial value after a period of time, t, following the application of the forward potential step. The current–time response that is observed during such an experiment is shown in Figure 3.3(B). If the product produced during a reduction reaction is stable and if the initial potential to which the working electrode is returned after t is sufficient to cause the diffusion-controlled oxidation of the reduced species, then the current obtained on application of the reverse step, $i_r$, is given by [63]

$$-i_r = nFAD_o^{1/2}C_o^{\circ}\pi^{-1/2}[1/(t - \tau)^{1/2} - 1/t^{1/2}] \tag{17.18}$$

For an uncomplicated reaction, the ratio of the current resulting from the forward step, $i_f$, to that produced during the reverse step is [63]

$$-i_r/i_f = [t_r/(t_r - \tau)]^{1/2} - (t_f/t_r)^{1/2} \tag{17.19}$$

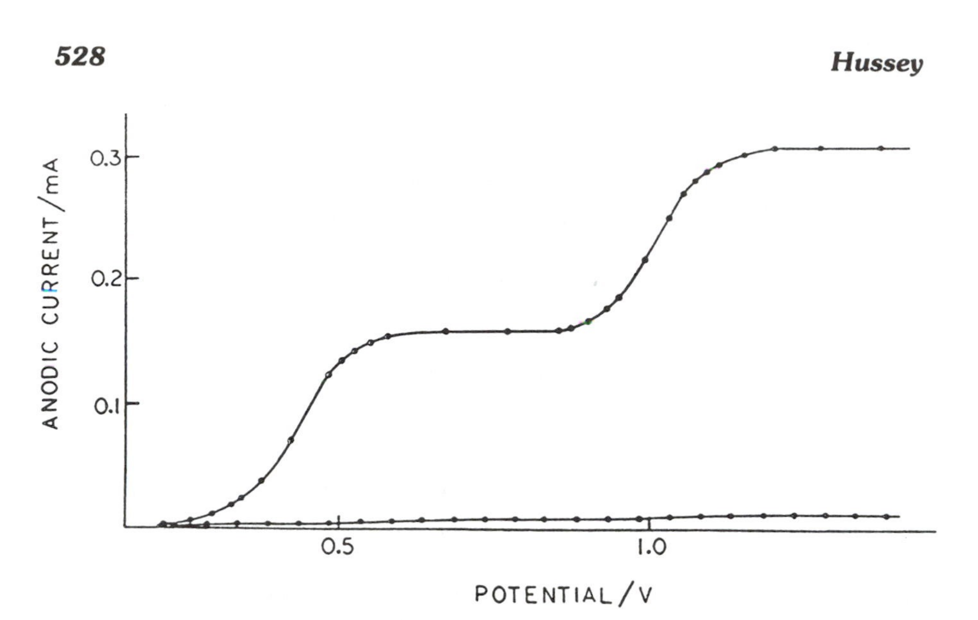


**Figure 17.4** Current–potential curves constructed from chronoamperometric current–time data recorded at a platinum electrode in a 3.34 × $10^{-2}$ *M* solution of Ti(II) in the 65–35 mol% $AlCl_3$–NaCl melt at 185°C. [From K. W. Fung and G. Mamantov, *J. Electroanal. Chem. 35*:27 (1972), with permission.]

where $t_f$ and $t_r$ are times selected in such a way that $t_r = t_f + \tau$. A plot of $-i_r/i_f$ versus $t_r/\tau$ can be used to assess the stability of the reduced species and to obtain kinetic data about homogeneous chemistry coupled to the electrode reaction [63]. Some typical results for the $[TaCl_6]^{-/2-}$ electrode reaction in $AlCl_3$–NaCl(satd) melt are given in the article by Laher et al. [64]. The chronocoulometric variant of the double-potential-step experiment has been utilized to study adsorption in molten salts [65].

*Chronopotentiometry*

Chronopotentiometry is the principal controlled-current method applied to molten salts. The theory and practice of single-step and current-reversal chronopotentiometry are presented in Chapter 4 of this volume. A typical current excitation signal and the resulting potential response are illustrated in Figure 4.3. For a single-current-step experiment involving a simple, diffusion-controlled reaction, the applied current is related to the transition time, t, by the Sand equation (Equation 4.6), whereas the potential-time response is described by Equation 4.4. Historically, chronopotentiometry has been a very popular technique for investigating electrode reactions in molten salts, and some investigators continue to employ it. Jain et al. [66] have reviewed the applications of chronopotentiometric techniques to molten salts. Chronopotentiometry has been used to study adsorption and metal deposition reactions, to unravel the mechanisms of complex electrode reactions, and to measure diffusion coefficients in molten salts.

*Stationary Electrode Voltammetry*

Linear and cyclic sweep stationary electrode voltammetry (SEV) play preeminent diagnostic roles in molten salt electrochemistry as they do in conventional solvents. An introduction to the theory and the myriad applications of these techniques is given in Chapter 3 of this volume. Examples of the linear and cyclic sweep SEV current–potential responses expected for a reversible, uncomplicated electrode reaction are shown in Figures 3.19 and 3.22, respectively. The important equation of SEV, which relates the peak current, $i_p$, to the potential sweep rate, $\nu$, is the Randles-Sevcik equation [67]. For a reversible system at some temperature, T, this equation is

$$i_p = 0.4463(n^{3/2}F^{3/2}/R^{1/2}T^{1/2})AD^{1/2}C^o\nu^{1/2} \tag{17.20}$$

The symbols appearing in this equation are defined in Chapter 3. At 25°C, Equation 17.20 simplifies to Equation 3.18. (However, in molten salt experiments it is extremely rare for this equation to be employed at such a low temperature!) The reversibility of an experimental system is often judged by comparing the experimental values of $|E_p - E_{p/2}|$, $E_{p/2} - E_{½}$ and $\Delta E_p$ to the theoretical values calculated from these equations in the case of a reversible redox system. Therefore, it is important to point out that these parameters are temperature dependent and that they *increase* with increasing temperature. The complete theoretical expressions for these parameters are given in Equations 17.21 to 17.23, respectively [67].

$$\left|E_p - E_{p/2}\right| = 2.2RT / nF \tag{17.21}$$

$$E_{p/2} - E_{1/2} = 1.09RT / nF \tag{17.22}$$

$$E_{pa} - E_{pc} = \Delta E_p \approx 2.3RT / nF \tag{17.23}$$

A cyclic voltammogram of the $[FeBr_4]^{-/2-}$ electrode reaction at a Pyrex glass/glassy carbon electrode in molten $AlBr_3$–1-methyl-3-ethylimidazolium bromide at 60°C is displayed in Figure 17.5.

*Pulse Voltammetry*

In addition to the traditional SEV techniques discussed earlier, various pulse voltammetric techniques have been employed at solid electrodes in molten salts, especially in the room-temperature haloaluminate melts. Numerous pulse techniques have been devised, and some of the more common examples of this family of voltammetric methods are described in Chapters 3 and 5 of this volume. However, the application of these methods to molten salts is limited primarily to large amplitude pulse voltammetry (LAPV), differential-pulse voltammetry (DPV), and, more recently, reverse normal-pulse voltammetry (RNPV). The application of LAPV and

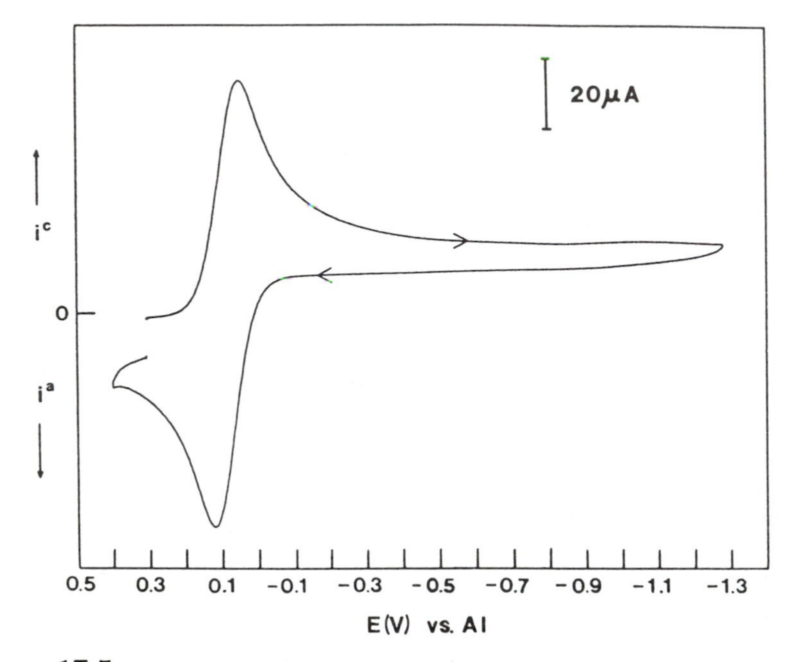


**Figure 17.5** Cyclic voltammogram of a 15.2 mM solution of Fe(III) at a Pyrex glass/glassy carbon electrode in the 49.5–50.5 mol% $AlBr_3$–1-methyl-3-ethylimidazolium bromide melt at 60°C. The sweep rate was 50 mV $s^{-1}$. [From I.-W. Sun, J. R. Sanders, and C. L. Hussey, *J. Electrochem. Soc. 136*:1415 (1989), with permission.]

DPV to molten salts is discussed in a recent review [62]. The excitation waveform used and the current–potential response expected during a LAPV experiment in a molten salt are shown in Figure 3.31. For DPV, the excitation waveform used is similar to that shown in Figure 5.7; the resulting current-potential response takes a derivative shape similar to that shown in Figure 5.2(c). Because this technique is normally carried out in molten salts at solid electrodes rather than at dropping metal electrodes, renewal of the electrode diffusion layer between pulses in DPV is difficult to achieve, except by imposing a long waiting period between pulses. Therefore, some investigators recommend that DPV experiments be conducted at slowly rotated electrodes.

In the RNPV technique (see Ref. 68 and references therein), which is a derivative of LAPV, the initial potential is set to a value where the electrode reaction of interest occurs, causing an accumulation of the reaction products. Next, an analysis pulse similar to that used in LAPV is applied so as to oxidize or reduce the reaction products as appropriate [68]. The RNPV technique is well suited for studying reactions complicated by adsorption or those producing products that foul the electrode surface, because the potential program can be adjusted so as to electrochemically "clean" the electrode surface between analysis pulses. An example of a current-potential curve resulting from the NRPV oxidation of Ti(II) to Ti(III) and Ti(IV) at a Pt electrode in a room-temperature chloroaluminate melt is shown in Figure 17.6. The pulse train used to produce curve (a) in Figure 17.6 is depicted in Figure 17.7. The Ti(III) species resulting from the oxidation of Ti(II) in this melt adsorbs on the Pt surface, causing the normal pulse voltammogram to be poorly defined. By using the NRPV technique, it is possible to observe well-defined waves for both the Ti(II)/Ti(III) and Ti(III)/Ti(IV) reactions.

*Rotating Electrode Voltammetry*

Rotating-disk electrode voltammetry (RDEV) has been employed extensively in moderate- and low-melting salts. An introduction to the theory and practice of rotating electrode methods is given in Chapter 3 of this volume, and the application of these techniques to molten salts has been reviewed [62]. The RDEV current-potential response expected for a reversible reaction in a molten salt is similar to that shown in Figure 3.38, provided that equal concentrations of O and R are present. Equation 3.48 (Levich equation) describes the relationship between the convective mass transport limited current and the angular velocity of a rotating disk

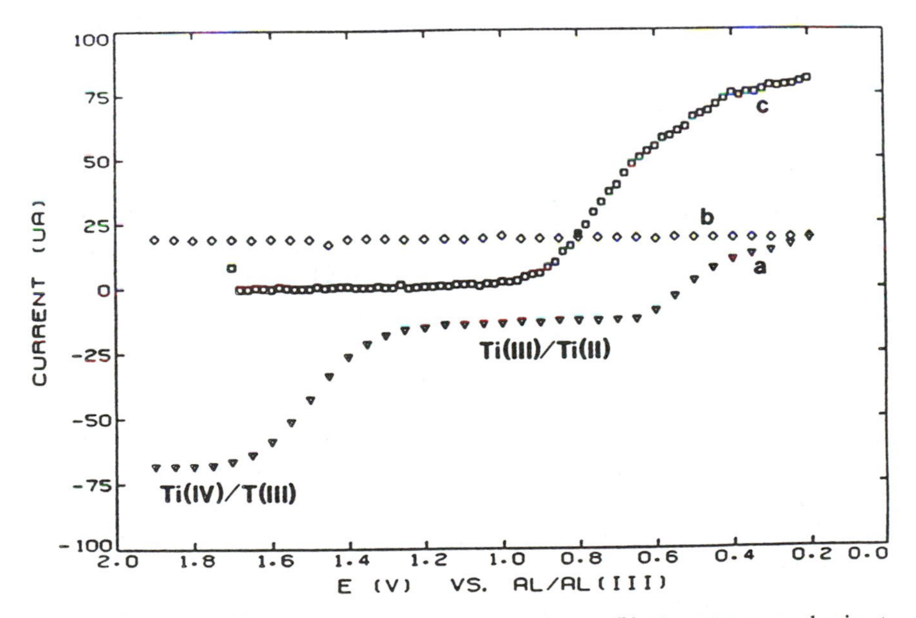


**Figure 17.6** (a) Reverse normal pulse voltammogram, (b) current measured prior to the analysis pulse in (a), and (c) normal large-amplitude pulse voltammogram at a platinum electrode for a 13.4 m*M* solution of Ti(IV) in the 60–40 mol% $AlCl_3$–1-methyl-3-ethylimidazolium chloride melt at ~25°C. [From Ref. 68, with permission.]

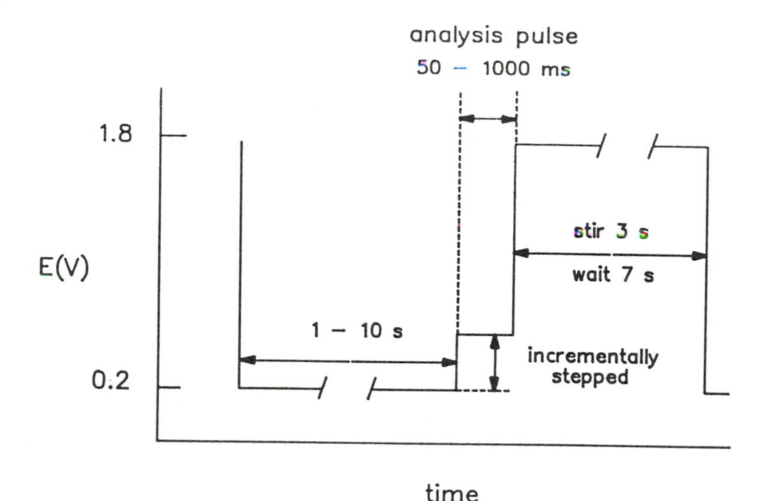


**Figure 17.7** Reverse normal-pulse voltammogram pulse train used to produce wave (a) in Figure 17.6.

electrode. From this expression, it is obvious that the calculation of diffusion coefficients from RDEV limiting-current data requires knowledge of the kinematic viscosity of the solvent. An extensive compilation of absolute viscosity and density data that can be used to calculate the kinematic viscosity is available [69].

As discussed in Section III.A, it is difficult to fabricate shrouded disk electrodes that are stable in high-temperature molten salts, and this factor above all has conspired to limit the employment of RDEV to the moderate- and low-melting systems. In addition, because some finite clearance must be provided around the rotating electrode shaft at the point where it enters the electrochemical cell, it is very difficult to fabricate an RDEV cell that can be hermetically sealed for use in the open atmosphere or for use with molten salt systems that exhibit high vapor pressures. A very elegant hermetically sealed, magnetically driven RDEV cell designed for use under these conditions is shown in Figure 17.8. RDEV experiments with low-vapor-pressure salts are usually conducted in open vessels inside an inert-atmosphere glove box. Rotating ring-disk electrode voltammetry (RRDEV) has also been used in molten salts (see Phillips and Osteryoung. [70]), but is not widely employed owing to the difficulty of constructing thin-gap ring-disk electrodes that are stable at elevated temperatures.

*Controlled-Potential Coulometry*

Controlled-potential coulometry (CPC) provides a direct measure of the number of electrons transferred in an electrode reaction. It is widely employed in molten salts for this purpose. Background information about CPC is given in Chapter 3 of this

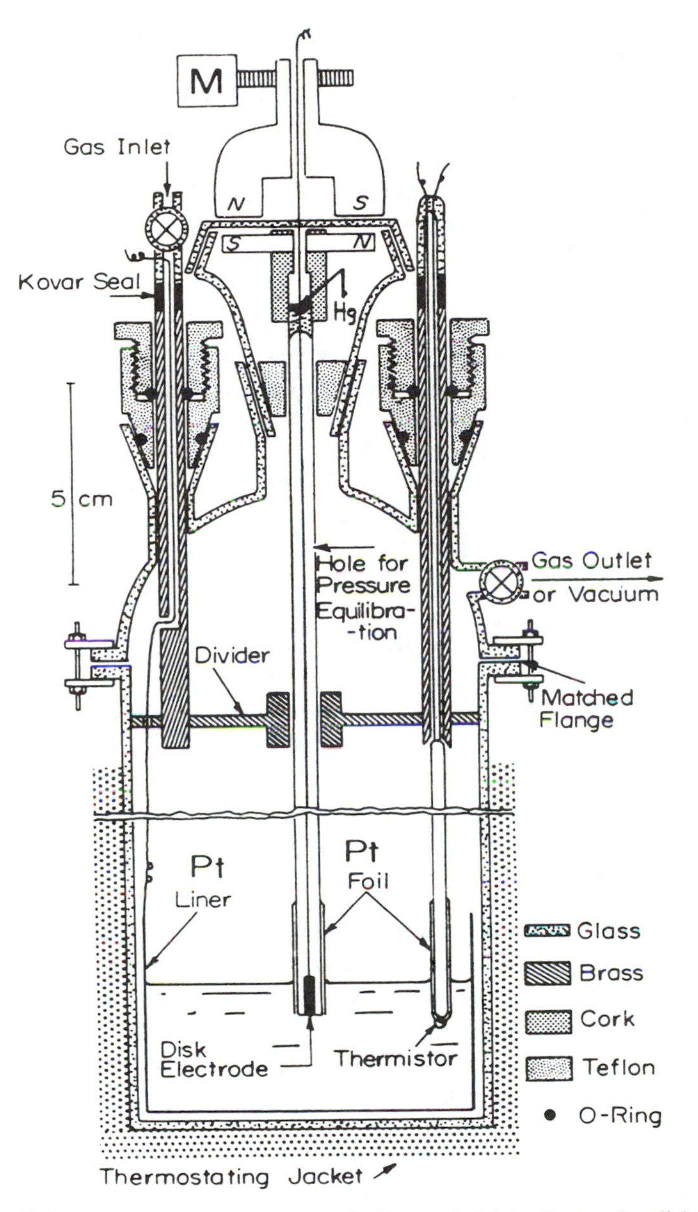


**Figure 17.8** Gas-tight cell with a magnetically coupled drive for rotating disk electrode voltammetry in moderate-melting salts. [From P. G. Zambonin, *Anal. Chem. 41*:868 (1969), with permission.]

volume, and the application of this electroanalytical technique to molten salts has been discussed [62]. Controlled-potential electrolysis ordinarily proceeds rapidly in highly conductive molten salt media. Faraday's law (Eq. 3.48) relates the number of electrons transferred in an electrode reaction, n, to the charge passed and the number of moles of substance undergoing electrolysis. Equation 3.48 can be recast in terms of the concentration of the electroactive species, $C^\circ$, and the volume of the solution, V, undergoing electrolysis:

$$n = Q/FVC^\circ \tag{17.24}$$

CPC experiments are most easily carried out if current integration equipment (such as a digital coulometer) is available. If the background current in the molten salt solution being electrolyzed is significant, then it is necessary to either carry out a blank experiment on an equal volume of pure solvent or to employ some type of digital background correction during the actual experiment.

In the event that current integration equipment is not available, it is still possible to conduct a quality coulometry experiment by observing the decay of the electrolysis current with time as suggested by MacNevin and Baker [71]. The electrolysis current can be monitored by using a simple strip-chart recorder or a personal computer (PC) equipped with a simple analog-to-digital (A/D) conversion board. Equation 3.49 can be rewritten as

$$i = i_0 e^{-kt} \tag{17.25}$$

where $k = AD/V\delta$. In logarithmic form, Equation 17.25 becomes

$$\ln i = \ln i_0 - kt \tag{17.26}$$

whereas the integration of Equation 17.25 yields

$$Q = i_0/k \tag{17.27}$$

The value of $i_0$ is determined from the intercept of a linear plot of ln i versus t, and Q is calculated from $i_0$ by using Equation 17.27. If the plot (ln i vs. t) is not linear, then the reaction under study may be complicated by slow coupled homogeneous chemistry. In this event this method may not be applicable for measuring Q, but the shape of the plot gives diagnostic information. The only stringent requirement for using this method to determine Q and hence n is that the stirring of the cell contents be stable and reproducible for the duration of the experiment; that is, $\delta$ must be fixed at all times so that k is indeed constant. This method has been employed in the author's laboratory with excellent results.

*A Precautionary Note About Convective Mass Transport*

Great care must be taken in the design of experiments employing the techniques discussed earlier that are normally carried out in quiescent solutions, in order to avoid unintentional convective mass transport or "convective stirring." This mode of transport results from nonuniform heating of the cell and/or its contents. It is caused by density gradients that stir the solution in the cell, destroying the expected quiescent conditions. Convective stirring is further exacerbated by the low viscosities of some moderate- and high-melting molten salts, which rival those of organic solvents like acetone and acetonitrile. The necessity of taking into account the contributions of convective stirring can not be overstated. In fact, it is possible to find examples in the molten salt electrochemistry literature wherein simple deviations from the expected mode of mass transport arising from convective stirring during CA experiments were ascribed to an elaborate ECE mechanism!

The best way to gauge the influence of convective stirring on prospective electroanalytical experiments such as chronoamperometry or cyclic voltammetry is to conduct preliminary experiments with a well-behaved, reversible electrode reaction employing the same experimental setup to be used for examination of the system of interest. Increases in the expected constant values of $it^{1/2}$ with increasing t or $i_p/v^{1/2}$ with decreasing $v$, respectively, signal the point at which convective transport becomes important.

## C. Thin-Layer Spectroelectrochemistry

The application of thin-layer spectroelectrochemical techniques to the study of electrode reactions in molten salts is becoming increasingly popular. Experiments have been conducted in both the ultraviolet–visible and infrared spectral regions. Background information about thin-layer spectroelectrochemical techniques is given in Chapter 3. In addition, Norvell and Mamantov [72] have reviewed the application of spectroelectrochemical techniques to molten salts.

The capillary action, microscope slide, thin-layer cells (see Fig. 1 in Ref. 73) that are commonly used in conventional solvents are inadequate for experiments with molten salts because they are open to the atmosphere, difficult to heat, and constructed with materials like epoxy glue that are attacked by many molten salts. Gas-tight Pyrex glass cells that have been constructed specifically for ultraviolet–visible and infrared experiments in moderate-temperature molten salts are shown in Figures 17.9 and 17.10, respectively. Figure 17.11 shows a Teflon cell designed for use with room-temperature chloroaluminate molten salts and other corrosive, low-vapor-pressure liquids. Experiments with the latter cell are typically conducted inside a glove box; the cell is optically accessed through fiber-optic waveguides passing through the glove box wall. Platinum screen and 100 ppi reticulated vitreous carbon (RVC) with path lengths of 1 mm or less are the most common materials used for constructing optically transparent electrodes (OTEs). In principle, it is possible to use a gold minigrid OTE, but the useful potential range of gold is small in some molten salts.

Two of the more common spectroelectrochemical techniques that have been employed for investigations in molten salts are chronoabsorptometry and spectro-

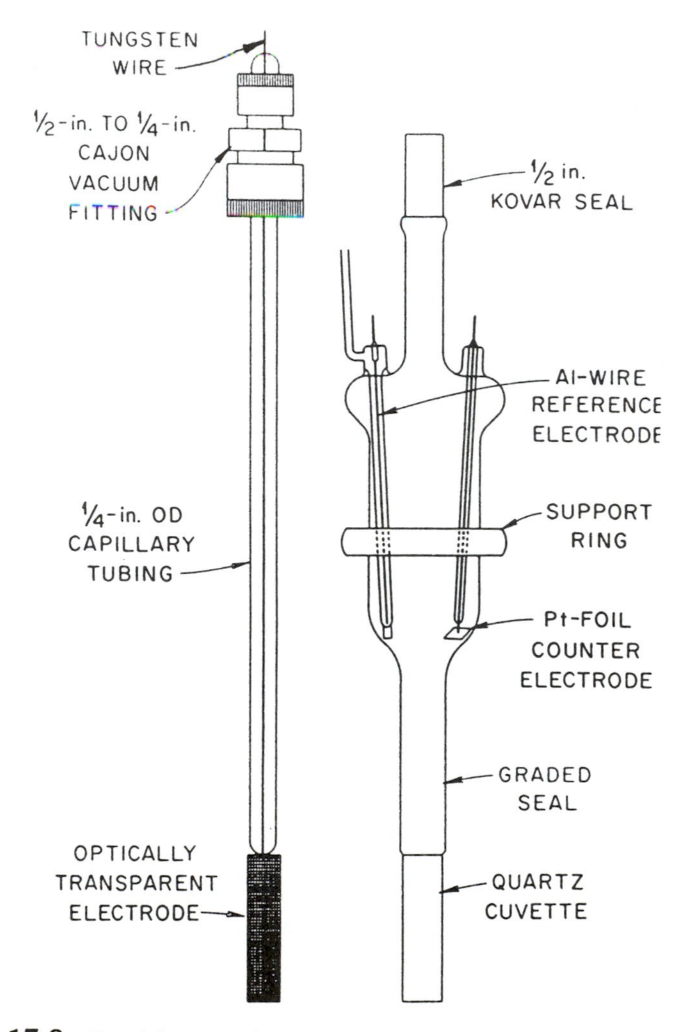


**Figure 17.9** Gas-tight transmission cell for UV-visible spectroelectrochemistry in moderate-melting salts. [From G. Mamantov, V. E. Norvell, and L. Klatt, *J. Electrochem. Soc.* *127*:1768 (1980), with permission.]

potentiometry. An example of the latter experiment that was conducted in the Teflon cell shown in Figure 17.11 is illustrated in Figure 17.12. The inset shows a Nernst plot that was constructed from the absorbance-potential data resulting from this experiment.

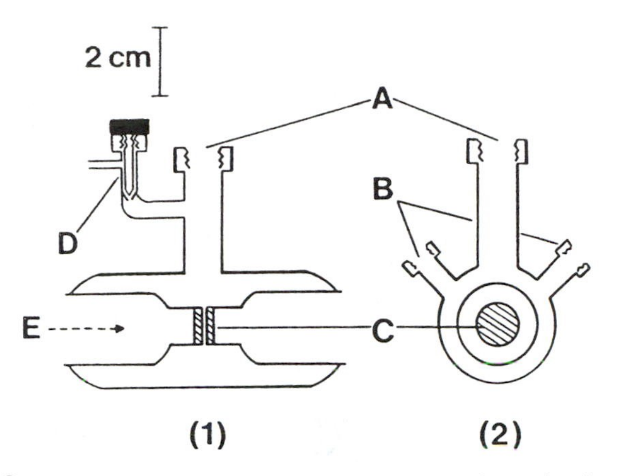


**Figure 17.10** Gas-tight transmission cell for IR spectroelectrochemistry in moderate-melting salts: (A) optically transparent electrode (OTE) port, (B) reference electrode and auxiliary electrode ports, (C) Si windows, (D) vacuum valve, (E) light path. [From P. A. Flowers and G. Mamantov, *J. Electrochem. Soc. 136*:2944 (1989), with permission.]

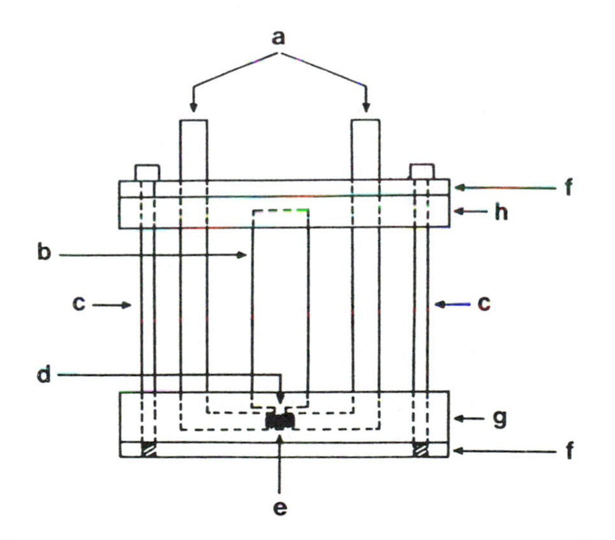


**Figure 17.11** Transmission spectroelectrochemistry cell designed for use with room-temperature haloaluminate melts and other moisture-reactive, corrosive liquids. (a) Auxiliary electrode and reference electrode compartments, (b) quartz cuvette containing the RVC-OTE, (c) brass clamping screw, (d) passageway between the separator and OTE compartment, (e) fritted glass separator, (f) Al plate, (g) lower cell body (Teflon), (h) upper cell body (Teflon). This cell is normally used inside a glove box and is optically accessed with fiber optic waveguides. [From E. H. Ward and C. L. Hussey, *Anal. Chem. 59*:213 (1987), with permission.]

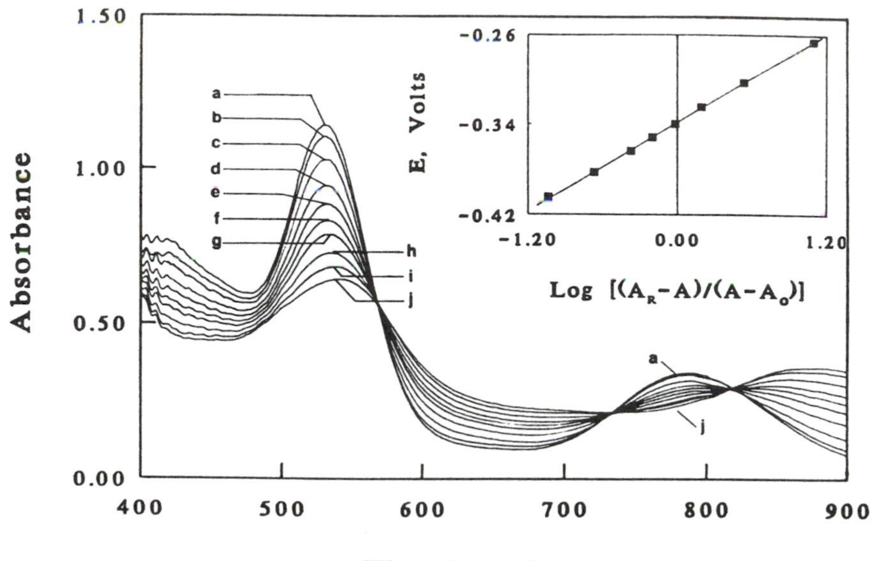


**Figure 17.12** Spectropotentiostatic experiment conducted with a 6.08 m*M* solution of $[Re_3Cl_{12}]^{3-}$ in the 49.0–51.0 mol% $AlCl_3$–1-methyl-3-ethylimidazolium chloride melt at 40°C using the cell shown in Figure 17.11. Applied potentials (V): (a) open circuit, (b) -0.266, (c) -0.303, (d) -0.325, (e) -0.340, (f) -0.352, (g) -0.364, (h) -0.383, (i) -0.405, (j) -0.550. Inset: Nernst plot constructed from the spectra in this figure. [From S. K. D. Strubinger, I.-W. Sun, W. E. Cleland, and C. L. Hussey, *Inorg. Chem. 29*:993 (1990), with permission.]

## ACKNOWLEDGMENT

Financial support for the preparation of this chapter was supplied by the National Science Foundation through grants CHE-9016632 and CHE-9411165.

## REFERENCES

1. C. J. Brockman, *J. Chem. Educ. 4*:512 (1927).
2. D. G. Lovering and R. J. Gale, in *Molten Salt Techniques*, Vol. 1 (D. G. Lovering and R. J. Gale, eds.), Plenum, New York, 1983, p. 4.
3. G. A. Wolstenholme, in *Molten Salt Technology* (D. G. Lovering, ed.), Plenum, New York, 1982, Chap. 2.
4. H. C. Maru, in *Molten Salt Techniques*, Vol. 2 (R. J. Gale and D. G. Lovering, eds.), Plenum, New York, 1984, Chap. 2.
5. J. R. Selman and H. C. Maru, in *Advances in Molten Salt Chemistry*, Vol. 4 (G. Mamantov and J. Braunstein, eds.), Plenum, New York, 1981, pp. 159–390.

6. B. L. Tremillon, in *Molten Salt Chemistry, An Introduction and Selected Applications*, NATO ASI Series C: Mathematical and Physical Sciences, Vol. 202 (G. Mamantov and R. Marassi, eds.), Reidel, Dordrecht, 1987, pp. 279–303.
7. S. H. White, in *Molten Salt Techniques*, Vol. 1 (D. G. Lovering and R. J. Gale, eds.), Plenum, New York, 1983, Chap. 2.
8. C. E. Bamberger, in *Advances in Molten Salt Chemistry*, Vol. 3 (J. Braunstein, G. Mamantov, and G. P. Smith, eds.), Plenum, New York, 1975, Chap. 4.
9. J. R. Selman, D. K. DeNuccio, C. J. Sy, and R. K. Steunenberg, *J. Electrochem. Soc. 124*:1160 (1977).
10. H. W. Jenkins, G. Mamantov, and D. L. Manning, *J. Electrochem. Soc. 117*:183 (1970).
11. S. C. Levy and F. W. Reinhardt, *J. Electrochem. Soc. 122*:200 (1975).
12. H. A. Laitinen and J. W. Pankey, *J. Am. Chem. Soc. 81*:1053 (1959).
13. J. O'M. Bockris, G. J. Hills, D. Inman, and L. Young, *J. Sci. Instr. 33*:438 (1956).
14. H. R. Bronstein and D. L. Manning, *J. Electrochem. Soc. 119*:125 (1972).
15. R. A. Bailey, E. N. Balko, and A. A. Nobile, *J. Inorg. Nucl. Chem. 37*:971 (1975).
16. J. A. Plambeck, in *Encyclopedia of Electrochemistry of the Elements*, Vol. 10 (A. J. Bard, ed.), Marcel Dekker, New York, 1976, Chap. 2.
17. C. L. Hussey, in *Advances in Molten Salt Chemistry*, Vol. 5 (G. Mamantov and C. B. Mamantov, eds.), Elsevier, Amsterdam, 1983, pp. 185–230.
18. R. J. Gale and R. A. Osteryoung, in *Molten Salt Techniques*, Vol. 1 (D. G. Lovering and R. J. Gale, eds.), Plenum, New York, 1983, Chap. 3.
19. R. A. Osteryoung, in *Molten Salt Chemistry, An Introduction and Selected Applications*, NATO ASI Series C: Mathematical and Physical Sciences, Vol. 202 (G. Mamantov and R. Marassi, eds.), Reidel, Dordrecht, 1987, pp. 329–364.
20. C. L. Hussey, in *Chemistry of Nonaqueous Solvents: Recent Advances* (G. Mamantov and A. I. Popov, eds.), VCH, New York, 1994, pp. 227–275.
21. J. S. Wilkes, J. A. Levisky, R. A. Wilson, and C. L. Hussey, *Inorg. Chem. 21*:1263 (1982).
22. Z. J. Karpinski, C. Nanjundiah, and R. A. Osteryoung, *Inorg. Chem. 23*:3358 (1984).
23. Z. J. Karpinski and R. A. Osteryoung, *Inorg. Chem. 23*:1491 (1984).
24. A. A. Fannin, D. A. Floreani, L. A. King, J. S. Landers, B. J. Piersma, D. J. Stech, R. L. Vaughn, J. S. Wilkes, and J. L. Williams, *J. Phys. Chem. 88*:2614 (1984).
25. C. R. Boston, in *Advances in Molten Salt Chemistry*, Vol. 1 (J. Braunstein, G. Mamantov, and G. P. Smith, eds.), Plenum, New York, 1971, Chap. 3.
26. J. Phillips and R. A. Osteryoung, *J. Electrochem. Soc. 124*:1465 (1977).
27. K. Tanemoto, R. Marassi, C. B. Mamantov, Y. Ogata, M. Matsunaga, J. P. Wiaux, and G. Mamantov, *J. Electrochem. Soc. 129*:2237 (1982).
28. B. Gilbert, G. Mamantov, and K. W. Fung, *Inorg. Chem. 14*:1802 (1975).
29. H. A. Hjuler, A. Mahan, J. H. von Barner, and N. J. Bjerrum, *Inorg. Chem. 21*:402 (1982).
30. J. A. Plambeck, in *Encyclopedia of Electrochemistry of the Elements*, Vol. 10 (A. J. Bard, ed.), Marcel Dekker, New York, 1976, Chap. 8.
31. J. Thonstadt, in *Advances in Molten Salt Chemistry*, Vol. 6 (G. Mamantov, C. B. Mamantov, and J. Braunstein, eds.), Elsevier, Amsterdam, 1987, pp. 73–126.

32. D. Bratland, in *Molten Salt Techniques*, Vol. 3 (D. G. Lovering and R. J. Gale, eds.), Plenum, New York, 1987, Chap. 3.
33. R. Piontelli, *Ann. N.Y. Acad. Sci. 79*:1025 (1960).
34. P. Claes and J. Glibert, in *Molten Salt Techniques*, Vol. 1 (D. G. Lovering and R. J. Gale, eds.), Plenum, New York, 1983, Chap. 4.
35. P. Claes, Y. Dewilde, and J. Glibert, *J. Electroanal. Chem. 250*:327 (1988).
36. P. Tilman, J. P. Wiaux, C. Dauby, J. Glibert, and P. Claes, *J. Electroanal. Chem. 167*:117 (1984).
37. J. A. Plambeck, in *Encyclopedia of Electrochemistry of the Elements*, Vol. 10 (A. J. Bard, ed.), Marcel Dekker, New York, 1976, Chaps. 6 and 7.
38. D. A. Nissen and D. E. Meeker, *Inorg. Chem. 22*:716 (1983).
39. D. Inman and D. G. Lovering, in *Comprehensive Treatise of Electrochemistry*, Vol. 7 (B. E. Conway, J. O'M. Bockris, E. Yeager, S. U. M. Khan, and R. E. White, eds.), Plenum, New York, 1983, Chap. 9.
40. M. H. Miles, D. A. Fine, and A. N. Fletcher, *J. Electrochem. Soc. 125*:1209 (1978).
41. A. Rahmel, in *Molten Salt Technology* (D. G. Lovering, ed.), Plenum, New York, 1982, Chap. 10.
42. J. E. Gordon, in *Techniques and Methods of Organic Chemistry*, Vol. 1 (D. B. Denny, ed.), Marcel Dekker, New York, 1969, Chap. 3.
43. P. Tissot, in *Molten Salt Techniques*, Vol. 1 (D. G. Lovering and R. J. Gale, eds.), Plenum, New York, 1983, Chap. 6.
44. R. M. Pagni, in *Advances in Molten Salt Chemistry*, Vol. 6 (G. Mamantov, C. B. Mamantov, and J. Braunstein, eds.), Elsevier, Amsterdam, 1987, pp. 211–346.
45. W. T. Ford, *Anal. Chem. 47*:1125 (1975).
46. J. R. Silkey and J. T. Yoke, *J. Electrochem. Soc. 127*:1091 (1980).
47. C. G. Swain, A. Ohno, D. K. Roe, R. Brown, and T. M. Maugh, *J. Am. Chem. Soc. 89*:2648 (1967).
48. R. Dallenbach, O. John, and P. Tissot, *J. Appl. Electrochem. 9*:643 (1979).
49. G. G. Bombi, M. Fiorani, and C. Macca, *J. Chem. Soc., Chem. Commun.* 455 (1966).
50. D. H. Kerridge, in *Advances in Molten Salt Chemistry*, Vol. 3 (J. Braunstein, G. Mamantov, and G. P. Smith, eds.), Plenum, New York, 1975, Chap. 5.
51. J. A. Plambeck, in *Encyclopedia of Electrochemistry of the Elements*, Vol. 10 (A. J. Bard, ed.), Marcel Dekker, New York, 1976, Chap. 11.
52. R. J. Potter and D. J. Schiffrin, *J. Electroanal. Chem. 206*:253 (1986).
53. R. E. Panzer and M. J. Schaer, *J. Electrochem. Soc. 112*:1136 (1965).
54. R. J. Potter and D. J. Schiffrin, *Electrochem. Acta 30*:1285 (1985).
55. J. A. Plambeck, in *Encyclopedia of Electrochemistry of the Elements*, Vol. 10 (A. J. Bard, ed.), Marcel Dekker, New York, 1976, Chap. 5.
56. A. J. B. Cutler, in *Molten Salt Techniques*, Vol. 1 (D. G. Lovering and R. J. Gale, eds.), Plenum, New York, 1983, Chap. 5.
57. D. E. Bartak, in *Molten Salt Techniques*, Vol. 3 (D. G. Lovering and R. J. Gale, eds.), Plenum, New York, 1987, Chap. 6.
58. F. Lantelme, D. Inman, and D. G. Lovering, in *Molten Salt Techniques*, Vol. 2 (R. J. Gale and D. G. Lovering, eds.), Plenum, New York, 1984, Chap. 5.

59. N. Q. Minh and L. Redey, in *Molten Salt Techniques*, Vol. 3 (D. G. Lovering and R. J. Gale, eds.), Plenum, New York, 1987, Chap. 4.
60. J. Phillips, R. J. Gale, R. G. Wier, and R. A. Osteryoung, *Anal. Chem. 48*:1266 (1976).
61. R. T. Carlin and R. A. Osteryoung, *J. Electrochem. Soc. 136*:1249 (1989).
62. G. Mamantov, C. L. Hussey, and R. Marassi, in *Characterization of Electrodes and Electrochemical Processes* (R. Varmi and J. R. Selman, eds.), Wiley, New York, 1991, Chap. 10.
63. A. J. Bard and L. R. Faulkner, *Electrochemical Methods, Fundamentals and Applications*, Wiley, New York, 1980, p. 181.
64. T. M. Laher, L. E. McCurry, and G. Mamantov, *Anal. Chem. 57*:500 (1985).
65. E. Temmerman and R. A. Osteryoung, *J. Electroanal. Chem. 64*:1 (1975).
66. R. K. Jain, H. C. Gaur, E. J. Frazer, and B. J. Welch, *J. Electroanal. Chem. 78*:1 (1977).
67. A. J. Bard and L. R. Faulkner, *Electrochemical Methods, Fundamentals and Applications*, Wiley, New York, 1980, Chap. 6.
68. R. T. Carlin, R. A. Osteryoung, J. S. Wilkes, and J. Rovang, *Inorg. Chem. 29*:3003 (1990).
69. G. J. Janz, *J. Phys. Chem. Ref. Data, 17*:Suppl. 2 (1988).
70. J. Phillips and R. A. Osteryoung, *J. Electrochem. Soc. 124*:1465 (1977).
71. W. M. MacNevin and B. B. Baker, *Anal. Chem. 24*:986 (1952).
72. V. E. Norvell and G. Mamantov, in *Molten Salt Techniques*, Vol. 1 (D. G. Lovering and R. J. Gale, eds.), Plenum, New York, 1983, Chap. 7.
73. W. R. Heineman, B. J. Norris, and J. F. Goelz, *Anal. Chem. 47*:79 (1975).

# 18

# Vacuum-Line Techniques

**Vladimir Katovic** *Wright State University, Dayton, Ohio*

**Michael A. May** *Central State University, Wilberforce, Ohio*

**Csaba P. Keszthelyi** *Louisiana State University, Baton Rouge, Louisiana*

## I. INTRODUCTION

When doing electrochemistry with compounds that are sensitive to trace amounts of moisture or oxygen present in nonaqueous solvents, it is necessary to use vacuum electrochemical techniques. The rigorous exclusion of water is particularly important when the starting material is sensitive to moisture, since it can decompose rapidly to hydrolysis products and make the electrochemical measurements erroneous. The products of electrode reactions are often highly reactive species and are usually present at concentrations low enough to make small amounts of water and oxygen critically important. However, specialized vacuum-line techniques might not be required if the starting material is not sensitive to hydrolysis or if only qualitative measurements are done, for it is possible to remove moisture in an electrochemical cell by adsorption onto activated alumina [1]. It has been demonstrated that vacuum technique alone is not sufficient for the complete removal of adsorbed water from the cell [2]. In this case, a special vacuum electrochemical cell equipped with an internal drying column can be used to recirculate the solvent–electrolyte solution over activated alumina to remove the last traces of water. Moisture can be introduced into the electrochemical system via the solvent, the electrolyte, the electrode assembly, and the glass surfaces. Using vacuum-line techniques, the concentrations of water and oxygen inside the cell can usually be reduced to acceptable levels. Further advantages of using vacuum-line techniques include thorough solution degassing and effective long-term isolation of the solution from the atmosphere. In carefully dried and deoxygenated solvents, the usable "potential window" can be increased, since $H^+$ reduction is avoided in dry nonaqueous solvents.

The subject of vacuum-line technique is approached here from the perspective of the novice user who may wish to construct a system in collaboration with a glassblower. The literature on vacuum technology [3–11] and glassblowing [12–17] is vast and there are many commercial firms that specialize in it. Within the past decade, newer components such as Teflon–glass needle valves, O-ring seals, and oil diffusion pumps have been introduced into vacuum systems.

## II. VACUUM LINE

Figure 18.1 is a diagram of a vacuum line that has been used for nonaqueous electrochemical studies. Major components of the system include a mechanical pump, an oil diffusion pump, a liquid nitrogen cold trap, a glass manifold with multiple ports, Teflon vacuum valves, vacuum gauges, and an inert gas introduction port. Starting from atmospheric pressure, the ultimate vacuum of a clean, leak-tight system can usually be achieved by overnight pumping. With a properly constructed system, it is possible to achieve a "high vacuum" of $10^{-5}$–$10^{-6}$ torr. Due to its polar nature, water is strongly adsorbed onto glass surfaces. The removal of adsorbed water from the surface of glass can be accelerated by carefully heating the system with a heatgun under dynamic vacuum (i.e., continuous pumping).

The main vacuum manifold is almost always constructed from borosilicate glass, and its internal diameter is usually >25 mm. The manifold could have multiple ports, several of which are equipped with 4-mm Teflon–glass valves for a variety of uses. Any ground-glass joints (between the manifold and the auxiliary ports) should be lightly greased with high-vacuum silicone or Apiezon H grease, the latter being more resistant to high temperature (Apiezon: Fisher Scientific Co., Pittsburgh, PA). When grease cannot be tolerated in the system, high-vacuum O-ring joints are an excellent choice to connect standard taper glass joints together.

In the author's laboratory, two mechanical pumps are used for convenience to evacuate the system. One pump (number 1) is attached to the oil diffusion pump outlet with a combination of glass tubing and heavy wall rubber or Tygon tubing (Nalgene Co., Rochester, NY). The diffusion pump inlet is connected to a liquid nitrogen trap via a 35/25 O-ring ball joint. The liquid nitrogen trap is then connected to the main vacuum manifold through a 12-mm Teflon vacuum valve. The other pump (number 2) is directly attached to the vacuum manifold through a special purge valve, which has a check valve for use under vacuum, and is used for "roughing" the vacuum line in isolation from the diffusion pump. Inert gas can be introduced into the main vacuum manifold through this purge valve. This type of valve has largely replaced the older mercury manometer valve. The inert gas source is connected to the vacuum manifold using copper

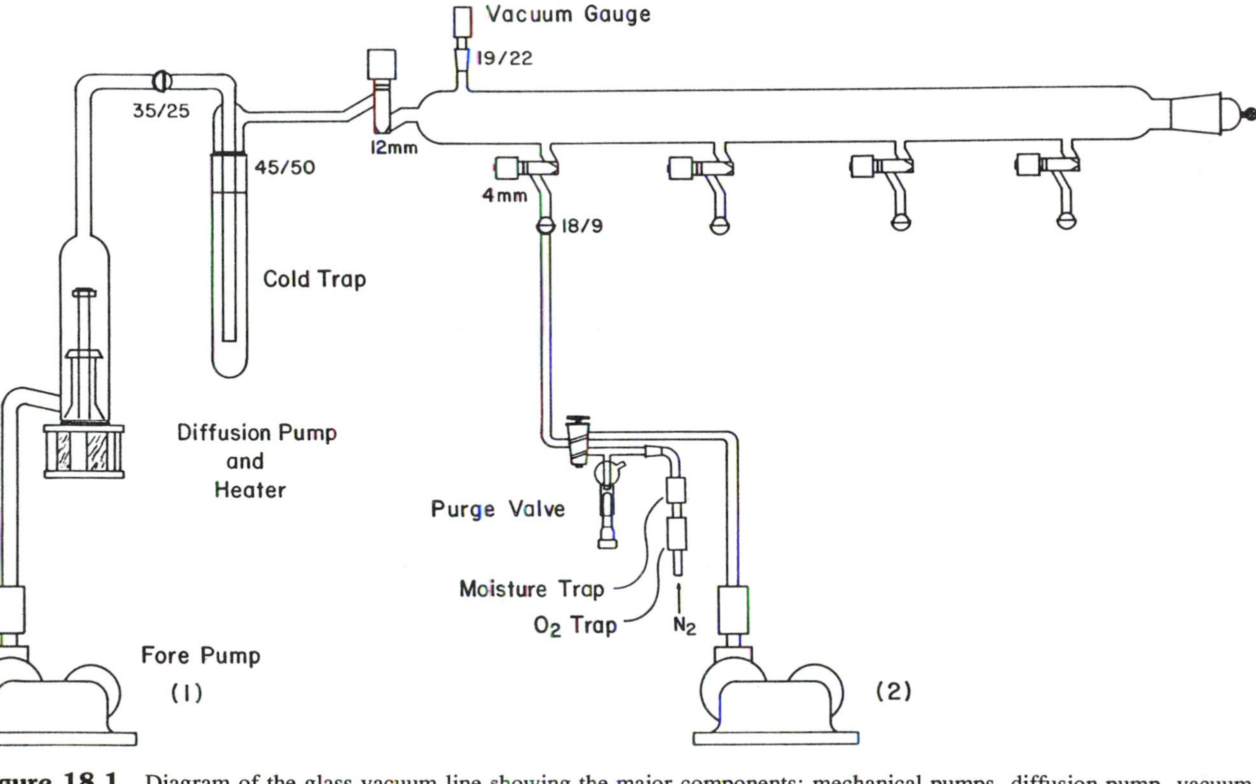


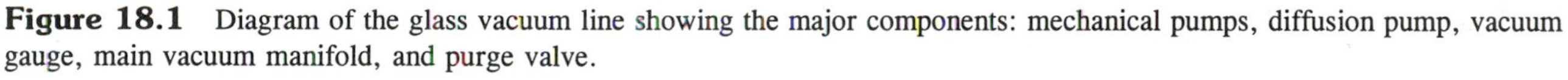
**Figure 18.1** Diagram of the glass vacuum line showing the major components: mechanical pumps, diffusion pump, vacuum gauge, main vacuum manifold, and purge valve.

tubing and Swagelok connectors (Swagelok Co., Solon, OH). Commercially obtained inert gas is available in many grades of purity, for example, prepurified $N_2$ (99.998% pure) or ultrapure $N_2$ (<5 ppm total $H_2O$ and $O_2$). For more rigorous work, the inert gas can be further purified by first passing it through oxygen and moisture traps such as Ridox catalyst (or BTS catalyst) and 4-Å molecular sieves (Ridox: Fisher Scientific Co.; BTS Catalyst: Fluka Chemical Corp., Ronkonkoma, NY).

The general operation of the vacuum line is briefly outlined. Initially the system is "roughed" down using the mechanical pump. Shortly after turning the mechanical pump on, the trap between the manifold and the pump should be filled with liquid nitrogen. It should be noted that liquid oxygen (which can explosively react with hydrocarbons) can condense in the trap if the system is not first evacuated before filling the cold trap with liquid nitrogen. Depending upon the quality of the mechanical pump, a fore-vacuum of $10^{-2}$–$10^{-4}$ torr can be attained in several hours. At this stage, the oil diffusion pump can be activated by turning on the oil heater. Once high vacuum is established in the system ($10^{-5}$–$10^{-6}$ torr), external glassware can be attached to the vacuum manifold through the ports. The diffusion pump is then isolated from the manifold so that auxiliary glassware can be preevacuated using only a mechanical pump (number 2). An obvious alternative to using two mechanical pumps is to use only one mechanical pump and add a bypass connection (with a valve) between the mechanical pump and the manifold.

The use of the vacuum line and vacuum electrochemical cells can be very time-consuming. For example, a cyclic voltammogram (CV) itself can be obtained in just 2 min. However, the preparation leading up to that step can take many hours (or days). The only poor expenditure of time spent on the vacuum line is when rigorous vacuum conditions are forsaken in one of the many experimental steps. An example would be the addition of solids to a solution through the top opening of a vacuum electrochemical cell, even under a positive inert gas pressure. Once such a means of addition is used, the purpose of vacuum-line technique is compromised, and one might just as well use a glove box for sample preparation.

### A. Pumps

Vacuum pumps are the active components of any vacuum line, and many types exist depending upon the pumping needs. The pumping speed, S, in L $s^{-1}$ of a given pump is one measure of its efficiency, and it represents the volume of gas pumped past a plane per unit time at constant temperature [5]:

$$S = S_0 [1 - (p_0 / p)]$$

where $S_0$ is the intrinsic pumping speed at the lowest attainable pump pressure

$p_0$, and p is the vacuum system pressure. This equation is an approximation valid for many types of laboratory pumps over a broad pressure range. Note that pumping speed decreases at low system pressure. In practice, $S_0$ is typically experimentally measured, and S is then calculated.

The most commonly used mechanical pump for the vacuum line is the rotary type (an internal rotor rotates about a stator, with alternate cycles of gas intake and exhaust). The moving pump components are immersed in oil for lubrication and also to seal the pump from backstreaming of the gas exhaust. Such a pump with a single stage can achieve a vacuum on the order of $10^{-1}$ torr, whereas a two-stage mechanical pump can reach $10^{-4}$ torr. Thus, a two-stage mechanical pump alone is sometimes sufficient for electrochemical experiments. Pumping speeds for mechanical pumps generally range from 0.2 to 150 L $s^{-1}$. "Direct drive" pumps are often preferred over the older belt-driven types since they are smaller and quieter. Since volatile solvents such as acetonitrile and dichloromethane can readily contaminate the pump oil, the quality (i.e., the composition and vapor pressure) of the oil directly affects the ultimate vacuum. For this reason, solvents should never be directly pumped without a proper cold trap; also, the pump oil should be periodically replaced (degraded pump oil is usually a dark color). For reasons of safety, it is good laboratory practice to always connect the pump exhaust to a well-ventilated fume hood with large-diameter tubing.

An all-glass diffusion pump is an excellent addition to the mechanical pump if pressures on the order of $10^{-5}$–$10^{-6}$ torr are required for the vacuum line. The diffusion pump inlet is connected to the system side of the vacuum line, and the exhaust outlet is attached to a mechanical forepump. Two types of diffusion pump working fluids are commonly used, mercury and oil. Both types operate by boiling the working fluid and directing the vapor through a nozzle, where the vapor stream collides with gas molecules to be pumped, and thus transfers the gas molecules to the diffusion pump outlet and the forepump. The working fluid vapor then recondenses and returns to the boiler reservoir. The pumping speed for a small, single-stage, all-glass oil diffusion pump ranges from about 5 to 30 L $s^{-1}$. The magnitude of the vacuum is regulated by the boiling rate of the working fluid, which in turn is controlled by the heater. During operation, diffusion pumps are usually cooled by heat exchange with water or air. Special water flow monitors can be commercially obtained that shut the diffusion pump oil heater off in the event of a water leak (Instruments for Research and Industry, Cheltenham, PA). Like mechanical pumps, the efficiency and pumping speed of diffusion pumps can be increased by the use of multiple stages. Oil diffusion pumps are currently preferred over mercury, due to the toxic effect of mercury vapor. Diffusion pump oils are characterized by very low vapor pressures, for example, $10^{-9}$–$10^{-10}$ torr for Dow Corning 705 silicone oil (Dow Corning Corp., Midland, MI). One disadvantage with all diffusion pumps is backstreaming of the pump oil vapor into the vacuum system, although this is not usually a major problem for electrochemical studies.

Turbomolecular pumps could, in principle, also be used to achieve a high ultimate vacuum on the vacuum line (e.g., $10^{-7}$–$10^{-9}$ torr). A turbomolecular pump is essentially a series of jet turbines that spin at $>10^4$ rpm and collide with gas molecules, forcing them toward the pump outlet. It can pump most gases efficiently (with the exception of $H_2$) if the outlet pressure is below $10^{-2}$ torr. Thus, the outlet to a turbomolecular pump is connected to a mechanical forepump. The main advantages of modern turbomolecular pumps are their high pumping speeds (~60 to 5000 L $s^{-1}$), high gas compression ratios, and the fact that they are very clean; that is, they do not backstream oil or mercury vapor. Turbomolecular pumps have not replaced diffusion pumps on glass vacuum lines for reasons of cost.

## B. Vacuum Gauges

The pressure regimes normally encountered on the preparative vacuum line range from $10^{-1}$ to $10^{-7}$ torr. A wide variety of vacuum gauges is commercially available. For the preparative vacuum line, three types of gauges are commonly used: MacLeod gauges, thermal conductivity gauges, and ionization gauges. Very often it is necessary to use two or more gauges to measure the entire pressure range of interest.

The MacLeod gauge has been used on vacuum lines for a long time, and it remains popular today since its use gives a direct measure of pressure [18]. It operates on the principle of compressing a large, known volume of gas at (unknown) low pressure into a known smaller volume. This compressed small volume of gas then has a pressure high enough to be measured using a manometer. The sensitivity of MacLeod gauges depends upon the size of the working fluid reservoir. Compact swivel-type MacLeod gauges are available that require only 4–8 mL of mercury or oil (Gilmont Instruments, Inc., Barrington, IL), and can measure vacuum from $10^0$ to $10^{-3}$ torr. These convenient swivel gauges are commonly used to measure rough vacuum and to calibrate electronic vacuum gauges. Larger scale MacLeod gauges can measure pressure down to $10^{-5}$ torr, although they can require up to 0.5 L of mercury. However, condensable gases (such as water, ammonia, carbon dioxide, and pump oil vapor) can lead to erratic pressure readings when compressed.

A thermal conductivity gauge uses a constant electric current to heat an element whose temperature is a linear function of gas pressure over a limited range. The temperature is typically measured with a thermocouple. In the popular Pirani gauge, a single metal filament is substituted for a thermocouple, and filament resistance is monitored [19]. The range of pressures detected by thermal conductivity gauges is ~$10^{-1}$–$10^{-4}$ torr, which makes them useful for measuring moderate vacuum. Thermal conductivity gauges are often calibrated against MacLeod gauges for a particular gas of interest, since thermal conductivity gauge response is dependent on the gas composition and gauge design.

Ionization gauges [20] are available in two basic types, "cold" and "hot" cathode. The cathode of the cold-cathode (Penning) gauge operates near room temperature [21,22]. During gauge operation, a high voltage is applied between the anode and the cathode, which creates an electric "cold-cathode" discharge. In the magnetron (or "Redhead") gauge, a magnetic field is also superposed onto the same region of space in which the electric discharge occurs. The magnetic field causes electrons (emitted from the cathode) to move in long spiral orbits, thus increasing the probability of gas ionization. The advantages of the Penning gauge are its wide range of pressure sensitivity ($10^{-2}$–$10^{-7}$ torr), its resistance to "burnout" (it has no delicate filament or grid electrode), and its moderate resistance to chemical fouling. Since they are robust, cold-cathode gauges are frequently used for measuring high vacuum on glass vacuum lines.

The "hot" cathode (or thermionic) gauge is a common type of ionization gauge, and it can be used to measure high or ultrahigh vacuum. Hot-cathode gauges were originally designed like triode vacuum tubes, which have filament, grid, and collector electrodes. A filament was resistively heated to a temperature sufficient to boil off a high flux of free electrons. The free electrons were accelerated toward a positively biased grid, which caused electron ionization of the neutral gas particles. The positive ions were collected by a physically large, negatively biased collector. Most modern thermionic gauges are based on the Bayard–Alpert design [23]. The Bayard–Alpert gauge has a thin collector wire surrounded by a grid electrode coil. This clever design reduced the high background current generated when free electrons emitted x-rays that struck the collector electrode. The usable pressure range of the Bayard-Alpert gauge is between $10^{-3}$ and $10^{-10}$ torr, above which the filament will burn out. More complex thermionic gauges have also been designed for measuring pressure as low as $10^{-13}$ torr [24].

An electronic vacuum gauge is typically positioned directly next to the cold-trap valve, as shown in Figure 18.1. It is also possible to use two vacuum gauges for the rapid isolation of vacuum leaks. In Figure 18.1, for example, the first gauge could be positioned between the 12-mm valve and the cold trap, and the second gauge positioned at the most distant end of the vacuum manifold. In order to protect a vacuum gauge from chemical contamination, it should be isolated from the vacuum manifold by a high-vacuum valve.

## C. Vacuum Valves

Large-bore valves in the main evacuation path allow for a more rapid pumpdown of the vacuum system. Traditionally, glass stopcock valves were used on high-vacuum lines. Their main drawback is the requirement of valve lubrication with vacuum grease, which can contaminate the solvent of an electrochemical system. Currently, high-vacuum Teflon needle valves have largely replaced glass valves due to their chemical inertness. Several types of Teflon vacuum valves, sealed by Teflon and elastomeric O-rings, are commercially available. The O-rings can be made of a variety of materials, such as Viton, PTFE, silicone, butyl rubber, or ethylene-propylene elastomer. An ultimate vacuum of $\sim 10^{-7}$ torr can be achieved with the standard O-rings. An ultrahigh vacuum of $10^{-9}$ torr can be obtained using special bakeable vacuum valve plugs.

A typical Teflon vacuum valve consists of a threaded outer glass body and a threaded Teflon plug. The plug consists of a solid Teflon or glass stem, a threaded Teflon bushing, and a knurled handle. The stem is made of Teflon or glass and is sealed to the valve body by Teflon and/or O-rings. Figure 18.2 is a diagram of a typical high-vacuum Teflon valve. In general, the valve stem seal consists of several O-ring-to-glass seals, whereas the seal at the valve stem seat can be either Teflon-to-glass or O-ring-to-glass. Such vacuum valves require no grease for the seal at the valve seat, with the advantage that high-purity electrochemical solvents and electrolytes are not contaminated by grease. For maximum chemical inertness, the valve stem seat seal should be Teflon-to-glass; however, the O-ring-to-glass valve-stem seal is often preferred for variable-temperature experiments (above the brittleness temperature of the O-ring). For experimental considerations, the valve-stem seat seal is superior to the valve stem lateral seal for maintaining a static vacuum.

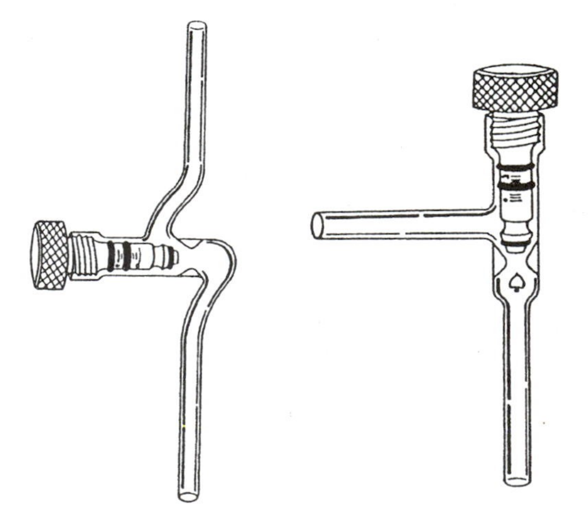

**Figure 18.2** High-vacuum Teflon needle valves, showing "straight through" and "right angle" designs. [Reproduced with permission of the copyright owner, Ace Glass, Inc.]

As stated earlier, the valve stem lateral seal consists of several O-ring-to-glass seals, and it also represents the major source of vacuum leakage. Sometimes a leaky valve can be visually identified by poor O-ring contact or uneven O-ring contact with the outer glass body. Lightly greased O-rings allow for a higher compression seal at the expense of possible solvent contamination with the grease. Some valve designs provide mechanical support for the O-rings on the valve stem with a precision-machined Teflon "wiper."

## D. Vacuum-Line Connections to Auxiliary Glassware

Auxiliary glassware is usually connected to the vacuum line by any of three types of connections: ball-and-socket joints, tapered joints, and threaded glass connectors. The most common type of connection is a ground-glass ball-and-socket joint, which allows for moderate mechanical flexibility. A ball-and-socket joint is pictured in Figure 18.3, and it can be secured together by a spring clamp. Tapered glass joints are used when mechanical flexibility is not desirable, for example, in the connection between the vacuum gauge and the main vacuum manifold. Both the ball-and-socket and the tapered glass joint can be sealed either by vacuum grease or by O-rings, with the same considerations previously discussed. A typical threaded glass connector is shown in Figure 18.4, and it consists of a threaded plastic bushing mated to a threaded outer glass tube. The inner tube (i.e., the second tube to be connected) can be made of glass, plastic, or metal, and fits inside the bore of the plastic bushing. The seal on the threaded connector is achieved by means of an O-ring at the end of the plastic bushing, as shown in Figure 18.4. Connectors such as those made by Ace Glass are very useful for quickly connecting different materials together on the vacuum line, such as metal tubing to glass tubing (Ace Glass Co., Vineland, NJ). Adapters are also available for the direct connection of a Swagelok metal connector to the bushing of the glass connector.

A vacuum adapter is used to connect two different sizes (or types) of joints together, and it often has a vacuum valve between its two ends (see Fig. 18.5). For example, the glass joints of the vacuum manifold and solvent reservoirs are

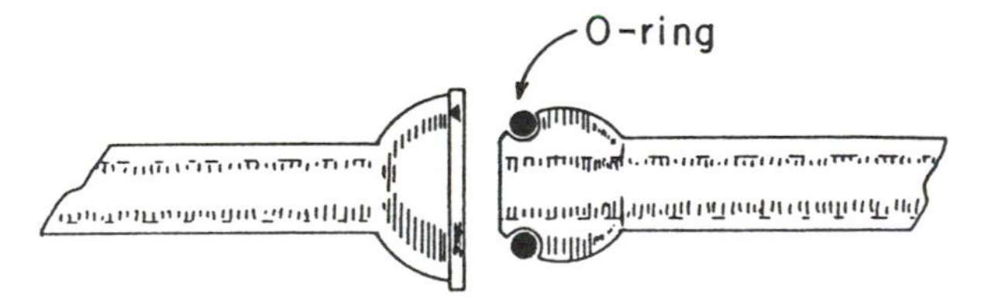


**Figure 18.3** Ground-glass ball-and-socket joint with grooved O-ring seal. The physical contact between the socket and ball is minimized by the use of the O-ring, and the flexibility of the joint is increased. Lubricant is not essential to hold a vacuum.

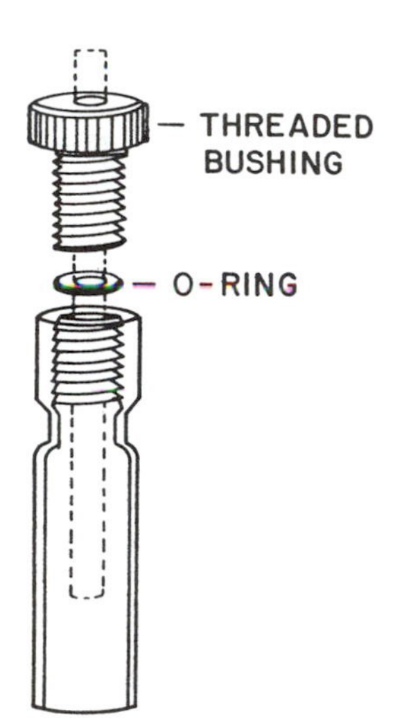


**Figure 18.4** Threaded glass connection that consists of a threaded outer tube, a threaded bushing, and an O-ring.

usually of different size, but are conveniently connected together with the appropriate adapter.

## III. ELECTROCHEMICAL GLASSWARE FOR THE VACUUM LINE

### A. Solvent Storage Containers

Solvents and electrolytes used for electrochemistry frequently require special purification. Commercially "dry" solvents can be further dried in multiple stages

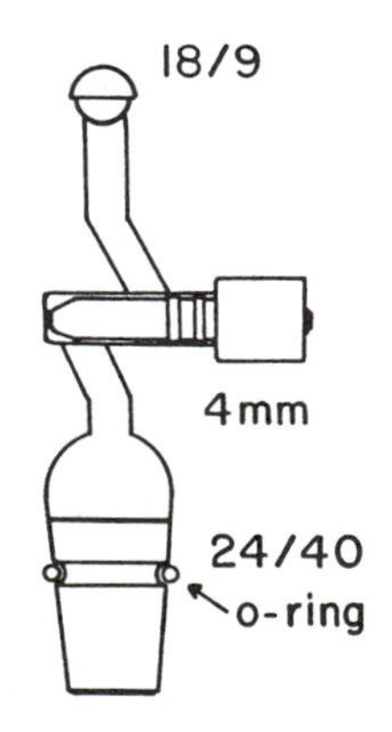


**Figure 18.5** Vacuum glass adapter for the connection of two glass joints of different size. It is typically used for connecting glassware to the main vacuum manifold.

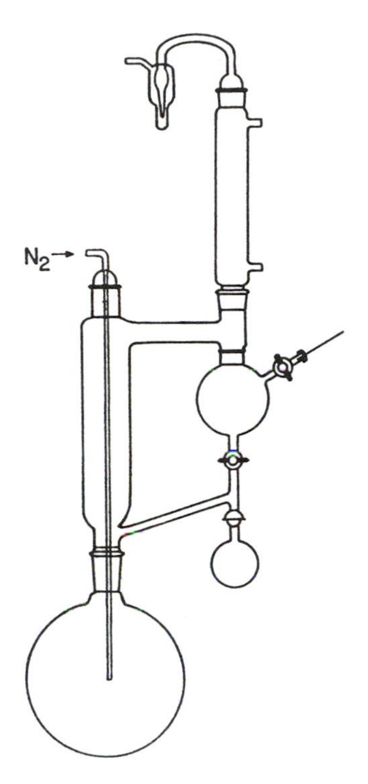


**Figure 18.6** Distillation apparatus for the purification of solvents.

to reduce the water concentration to an acceptable level (for a more complete treatment see Chaps. 15 and 19). Volatile solvents can be first purified by reflux over a drying agent under dry nitrogen gas. For example, the solvent can be distilled in an apparatus similar to that of Figure 18.6. The solvent is placed in the main reflux flask, to which an appropriate drying agent is added. (Caution: It is extremely important to use a drying agent that is chemically compatible with the solvent in order to avoid the possibility of explosion.) Table 18.1

**Table 18.1** Drying Agents for Nonaqueous Solvents Used in Electrochemistry

| Solvent | Drying agent | Ref. |
|---|---|---|
| Acetonitrile | Alumina or calcium hydride | 25–27 |
| Tetrahydrofuran | Lithium or aluminum hydride | 28 |
| | or sodium benzophene | 29 |
| Methylene chloride | Alumina | 30 |
| Dimethyl sulfoxide | Molecular sieves | 31 |
| Dimethylformamide | Calcium hydride or alumina | 1,32–34 |
| Pyridine | Molecular sieves or KOH | 35 |

contains a list of drying agents useful for several common solvents used for vacuum electrochemistry.

Once the solvent is distilled, it is stored over a drying agent under vacuum in a glass solvent reservoir. Subsequent storage of the solvent should preserve its quality. Two principal types of storage containers are usually used, depending on the solvent volatility. Low-volatility solvents such as propylene carbonate or *N,N*-dimethylformamide (DMF) can be transferred to the vacuum electrochemical cell by gravity. A suitable device for the gravity transfer of solvents is shown in Figure 18.7 [36].

Volatile solvents such as tetrahydrofuran, acetonitrile, and methylene chloride can be conveniently transferred by vacuum distillation from a special solvent storage container that contains a Teflon needle valve and a ball-and-socket joint. Such a solvent storage container is shown in Figure 18.8. In operation, the appropriate drying agent is first added to the empty solvent container. If molecular sieves are used as the drying agent, they can be activated in the solvent container by heating the bottom of the glass container with a "sand bath" at ~250°C overnight under dynamic vacuum. After that, the storage container valve should be closed and then the heat turned off so that the molecular sieves can cool down. Once activated, high-purity solvent can be vacuum-distilled into the glass storage container. The original container is next evacuated (to remove

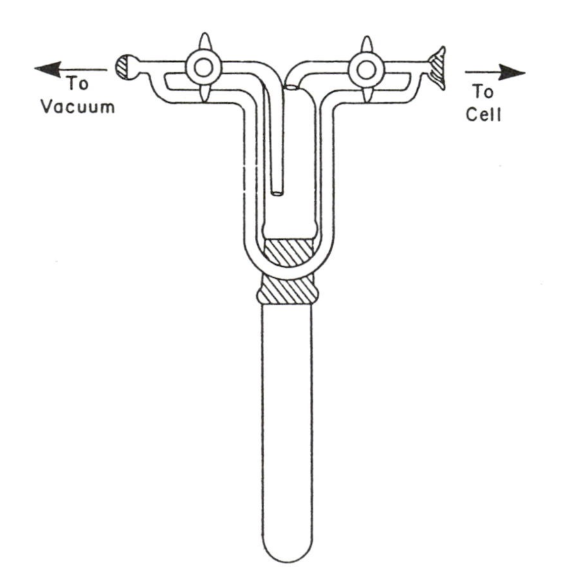


**Figure 18.7** Storage container for the gravity transfer of low-volatility solvents under vacuum. [From Ref. 36, with permission.]

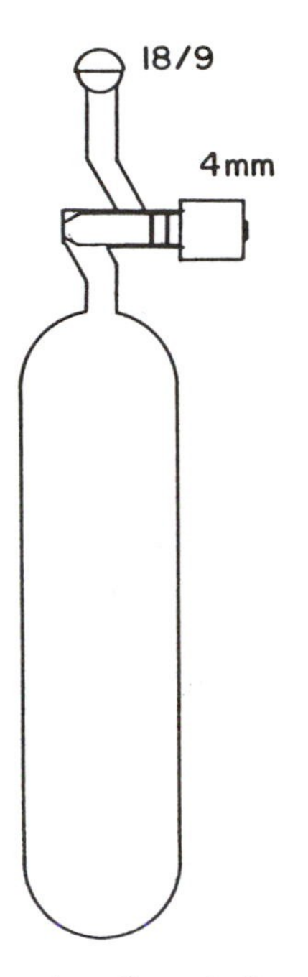


**Figure 18.8** Glass storage container for volatile solvents. It is used for the storage and the vacuum distillation of solvents.

dissolved gases) while the solvent continues to be frozen. The solvent storage container (i.e., the receiver flask) is cooled with liquid nitrogen. The dynamic vacuum to the vacuum manifold is shut off by closing the main vacuum valve. The original container and the solvent storage container are opened to each other under static vacuum, and the original container is warmed with tepid water. Since the solvent storage container is still cooled with liquid nitrogen, solvent is transferred by its vapor pressure from the original container to the storage container.

## B. Vessels for Solution Preparation and Degassing

Dissolved oxygen and other gases can be removed from the solvent by the freeze–pump–thaw (FPT) method, or by the expansion of gases above the solvent into an evacuated container on the vacuum line. It is customary to freeze–pump–thaw the solvents before vapor transfer from the storage container to the vacuum electrochemical cell. Such FPT cycles should be done as soon as the solvent is brought into contact with the drying agent. The FPT method involves three basic steps (i.e., one cycle), which are often repeated:

1. With the Teflon vacuum valve of the solvent container closed (and the vacuum line under dynamic vacuum), the contents are thoroughly frozen by placing a dewar holding liquid nitrogen around the solvent storage container. The liquid nitrogen level should not reach the Teflon vacuum valve (it should be at least 1 in below the valve).
2. With the liquid nitrogen bath in place, the Teflon valve to the vacuum line is carefully opened and the solvent container is evacuated for 10–30 min. Because it is advantageous to go through several FPT cycles, step 2 should not be unnecessarily prolonged for any one cycle. During the first cycle, the vacuum gauge should be closed to the vacuum line.
3. After evacuation of the frozen solvent, the Teflon stopcock is closed and the frozen solvent is thawed. This is accomplished by quickly placing tepid water in a large beaker in contact with the outside of the solvent container. Allowing the frozen solvent to thaw by itself can result in breakage of the glass container. Even with proper procedure, the solvent container can break if the solvent has a large thermal expansion coefficient. Thus a large container of acetonitrile should be subjected to FPT cycles only if it has a round bottom and is less than half full. FPT cycles can be repeated until the pressure over the frozen solvent approaches the base pressure of the vacuum line.

The removal of oxygen by the expansion of gases above the solvent on a vacuum line is usually done directly in the vacuum electrochemical cell [37]. This method is much faster than the FPT method and allows one to monitor the amount of dissolved oxygen by electrochemical means.

Electrochemical cells with elaborate interior construction are not well suited for FPT cycles. A convenient experimental arrangement for degassing solutions is to use a special glass vessel for the FPT cycles that allows for gravity transfer of the solution into the electrochemical cell after degassing. Figure 18.9 shows suitable glassware for the FPT cycles [38]. Once the solution is degassed, the vessel can be rotated into the upright position (flipped 180°) and the solution gravity-transferred. The physical position of such a preparation vessel on the vacuum line would be at the end, followed by the electrochemical cell itself.

When a volatile solvent is to be vapor-transferred from a large storage container to make up a solution, it is feasible to omit the FPT step in the final preparation. To make up an exact volume, a calibrated preparation vessel should be used. The major disadvantage of using an intermediate vessel for the preparation of solutions is the increased possibility of contamination. Contamination by stopcock grease can be avoided by using O-ring ball joints and Teflon vacuum valves. An alternative arrangement is to use a special vacuum electrochemical cell equipped with a sidearm for FPT cycles (see Sec. III.C).

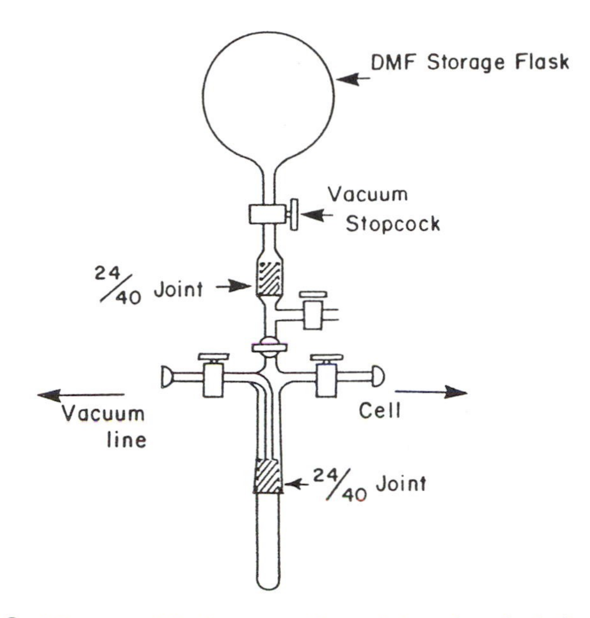


**Figure 18.9** Glass vessel for the preparation and degassing of solutions by the freeze–pump–thaw method. The degassed solution is gravity-transferred to the electrochemical cell by rotating the container 180°. [From Ref. 38, with permission.]

## C. Vacuum Electrochemical Cells

There are numerous designs for vacuum electrochemical cells, ranging from very simple to extremely complex. In operation, the vacuum electrochemical cell parts are first cleaned, washed with solvent, and dried in an oven at 200°C. The hot cell parts should then be quickly assembled and evacuated on the vacuum line for several hours or overnight. Once the electrochemical cell has been pumped down, it should be closed off and transferred to a dry box, where the air-sensitive sample and the electrolyte can be added to it. Alternatively, the solid electrolyte could be added into the electrochemical cell before assembly.

It is possible to degas solutions in simple vacuum electrochemical cells such as that shown in Figure 18.10 by the FPT method [39]. Note that the cell has three electrodes (working, auxiliary, and reference) and consists of two glass pieces. Such cells have been widely used for preliminary work, and for elevated-temperature studies with molten scintillator dyes as solvents [40]. The main advantage of a simple electrochemical cell is the minimal time required for pumpdown of the sample and solution. The vacuum electrochemical cell shown in Figure 18.11 has several additional features, including an attached sidearm chamber for FPT cycles, a microelectrode as the working electrode, and a macroelectrode for bulk electrolysis [41,42].

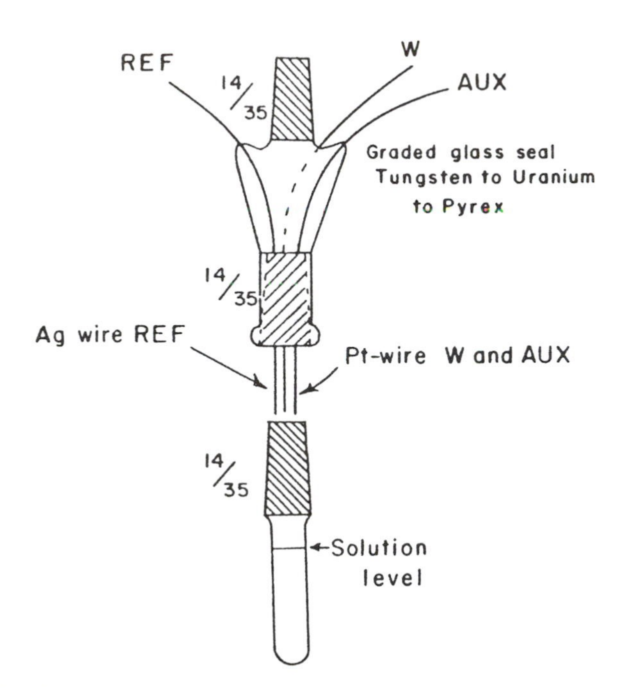


**Figure 18.10** Simple vacuum electrochemical cell that is suitable for direct freeze–pump–thaw of the solution. It is very useful for rapid exploratory experiments under vacuum. [From Ref. 39.]

For quantitative electrochemical work, some experimental means for controlling sample concentration is desirable. Typically, the delivery of known volumes of solutions involves the use of some type of buret or graduated ampule which can be attached to the electrochemical cell. Figure 18.12 shows a rather flexible electrochemical cell system that incorporates a dropping mercury working electrode [26]. For example, semiquantitative concentration studies can be done by the addition of reagents from ampule C′. Sample could also be added to the cell by gas-tight syringe, which would result in more precise volumetric delivery (at the expense of increased amounts of $H_2O$ and $O_2$). Another useful feature of the system shown in Figure 18.12 is the capability to remove samples during bulk electrolysis. Microliter-sized samples can be removed by syringe through a rubber septum in sidearm C′ and analyzed by gas chromatography/mass spectroscopy (GC/MS) to separate and identify the electrolysis products. A Solv-Seal container can be positioned so that samples from the bulk electrolysis compartment can be removed for (noninvasive) ultraviolet-visible (UV/Vis), electron spin resonance (ESR), or nuclear magnetic resonance (NMR) spectroscopy, and then reintroduced into the same compartment for further electrolysis.

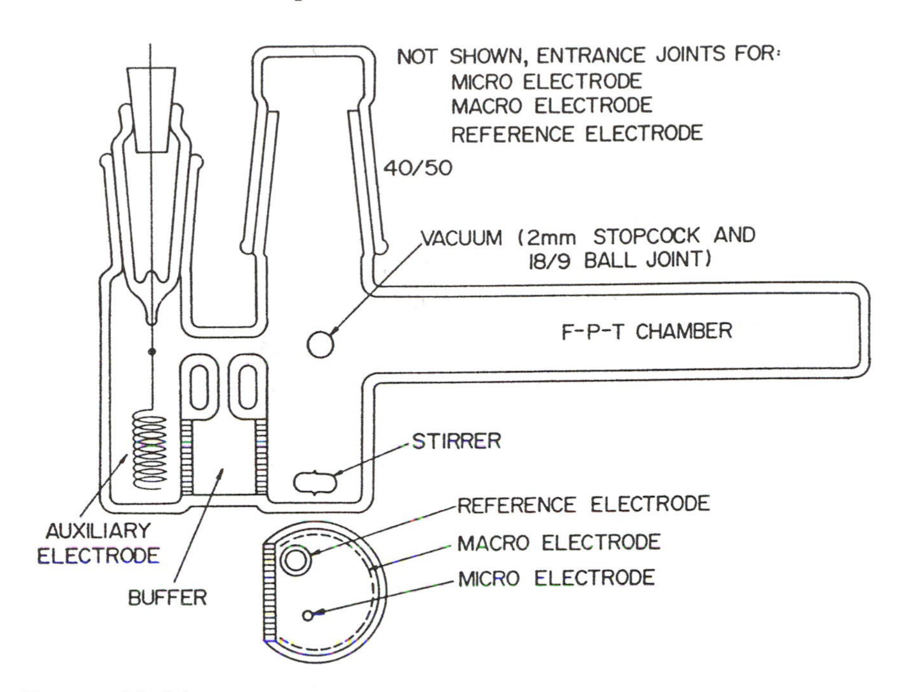


**Figure 18.11** Vacuum electrochemical cell with a freeze–pump–thaw chamber. [From Ref. 47, with permission.]

For the electrochemistry of extremely air-sensitive compounds, it is necessary to incorporate an internal $Al_2O_3$ drying column into the vacuum electrochemical cell. With such a specialized cell, it is feasible to reduce the water concentration in the nonaqueous solution to less than $10^{-5}$ *M*. Nondried commercial solvents typically contain $\sim 10^{-3}$ *M* water concentration. One example of this type of vacuum electrochemical cell shown in Figure 18.13 incorporates an internal, aluminum oxide drying column that allows purification of the solvent–electrolyte solution by repeated filtration of the solution through the alumina at low temperature ($\sim 0°C$). Solvent is transferred into the cell from a storage vessel through a rubber septum by means of a steel needle. The cold solution is passed through the alumina column by a simple 90° rotation of the cell. Generally, two cycles of filtration were adequate to reduce the moisture concentration to the required level. For example, the working "potential window" of a 0.2 *M* $NBu_4BF_4/CH_3CN$ solution in this cell ranged from +1.80 to -2.60 V (vs. Ag/0.01 *M* $Ag^+$) before filtration, and the potential window ranged from +2.70 to -3.30 V after filtration.

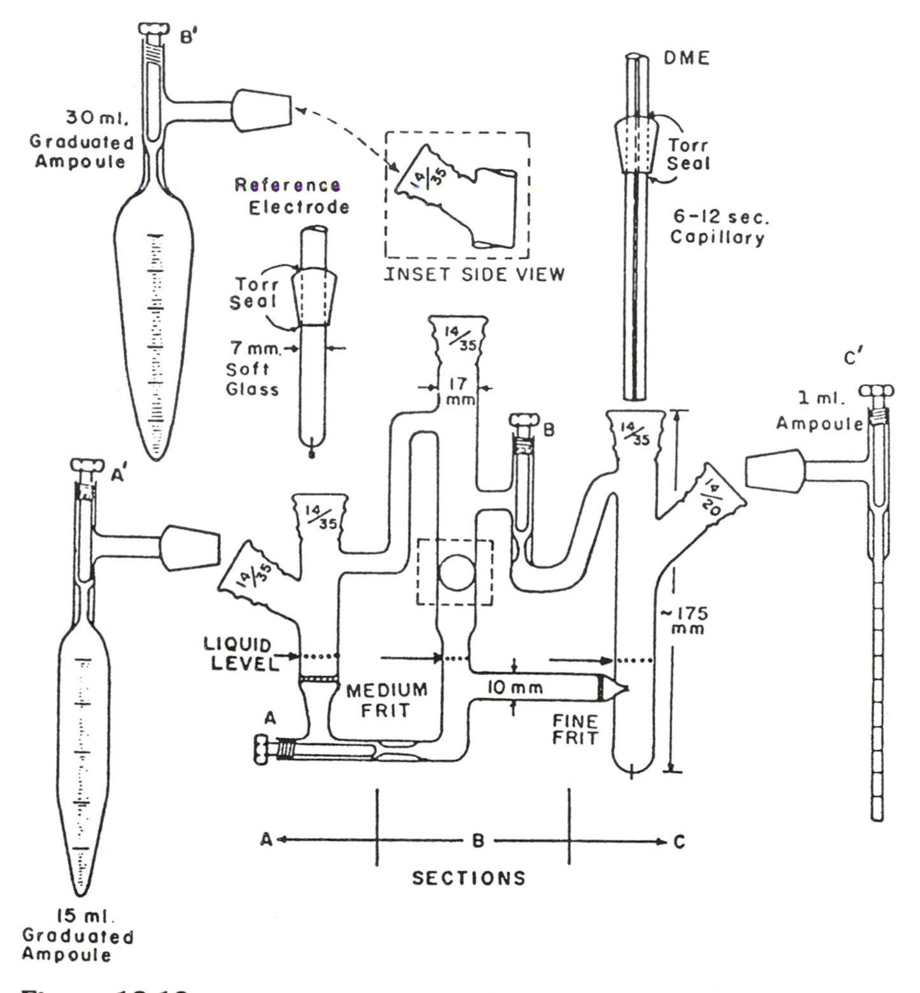


**Figure 18.12** Vacuum electrochemical cell and assorted glassware that allows for concentration-dependent electrochemical studies using a dropping mercury electrode. [From Ref. 26, with permission.]

Another vacuum electrochemical cell, shown in Figure 18.14, incorporates both an internal alumina drying column [37] and also a specialized sample loading device [43]. This sample loading device is extremely advantageous, because air-sensitive solids can be directly added to the solvent–electrolyte solution after it has been "superdried." The cell is used in the following manner: The vacuum electrochemical cell parts are cleaned, washed with solvent, and dried in a

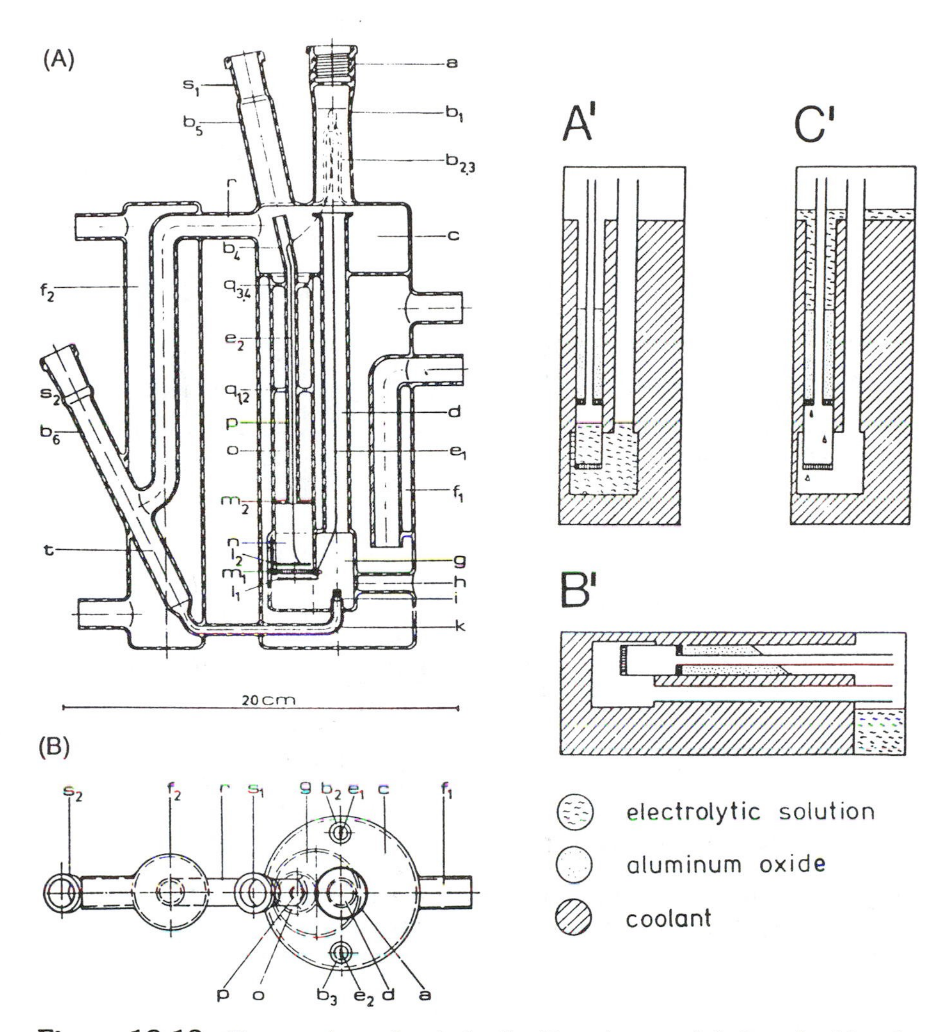


**Figure 18.13** Vacuum electrochemical cell with an integrated drying tube (o) and water-cooled jackets (f1, f2) from (A) a front view and (B) a top view. Schematic representation of the drying operation is shown in A′, B′, and C′. The cell is filled with aluminum oxide and electrolyte solution in A′. The solution is transferred into the cell by a 90° rotation in B′. After back-rotation, the solution flows into the electrode compartment, passing through the cooled alumina drying tube in C′. [From Ref. 2, with permission.]

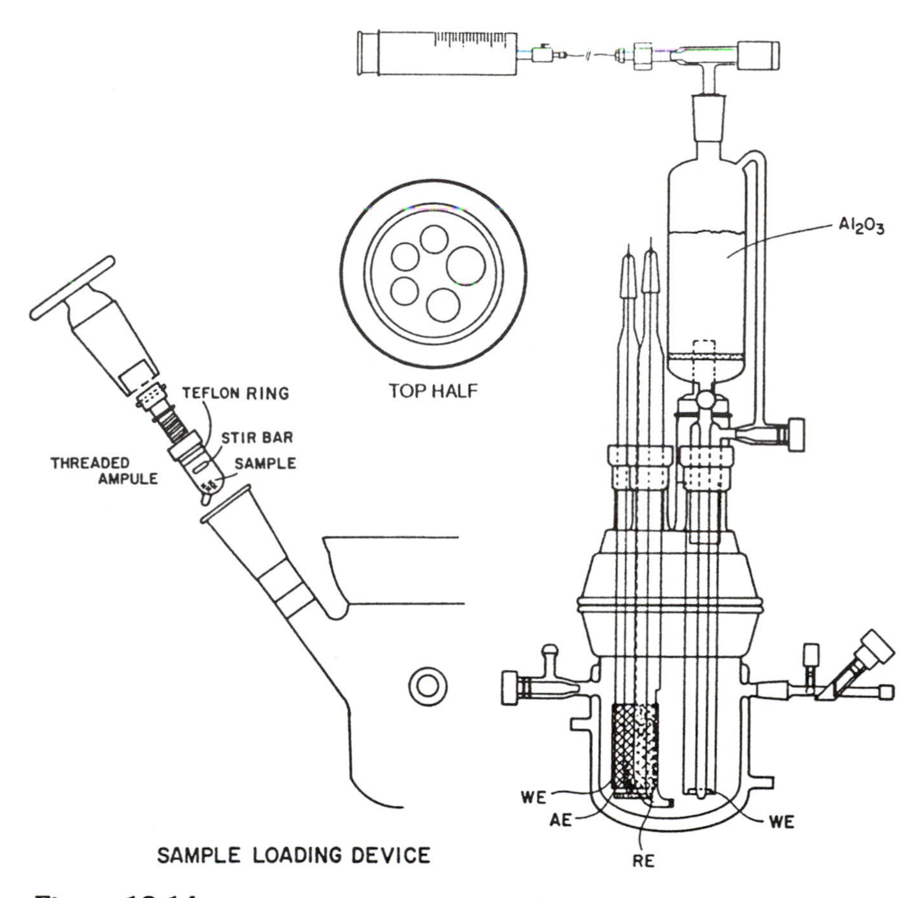


**Figure 18.14** Vacuum electrochemical cell with upper and lower halves, electrodes, sample loading device, solvent introduction port, and alumina column. [From Ref. 37, with permission.]

vacuum oven at 160°C; after all hot parts of the cell are assembled, the solid electrolyte, an internal moisture indicator (DPA: 9,10-diphenylanthracene), and a Teflon stir bar are added to the cell; the sample loading device, previously filled in a dry box with a known amount of the sample, is inserted into its appropriate cell port; the alumina column is installed in the cell top, and hot, previously activated (~550°C) neutral alumina is poured into it; the cell and the alumina column are then evacuated for several hours, during which time the alumina is heated to 170°C; and purified, dry solvent is transferred from the solvent container into the cell by way of an entry port equipped with a rubber septum and a steel needle. Oxygen can be removed from the solution at room temperature on the vacuum line by expansion of the dissolved gases into an evacuated 1-L flask [44]. Removal of the oxygen can be monitored with cyclic voltammetry by observing the disappearance of the $O_2$ to $O_2^-$ reduction peak at –0.98 V vs. normal hydrogen electrode (NHE). Moisture is removed from the solvent by passing the oxygen-free solution, cooled to 0°C, over the activated alumina column. This is accomplished by drawing the solution into the top of the alumina column with the attached syringe. Removal of water is monitored with cyclic voltammetry by observing the stability of the electrogenerated radical cation $DPA^{+\bullet}$ and the dication $DPA^{2+}$. In the presence of traces of water, $DPA^{2+}$ is not stable, and the CV does not display a cathodic peak due to the reduction of $DPA^{2+}$ to $DPA^{+\bullet}$ (see Fig. 18.15, trace A). After water is removed from the solution by repeatedly passing the solution over alumina, $DPA^{2+}$ becomes stable, and the cyclic voltammogram (CV) displays the coupled cathodic peak on the reverse scan (see Fig. 18.15, trace B). The solid sample can then

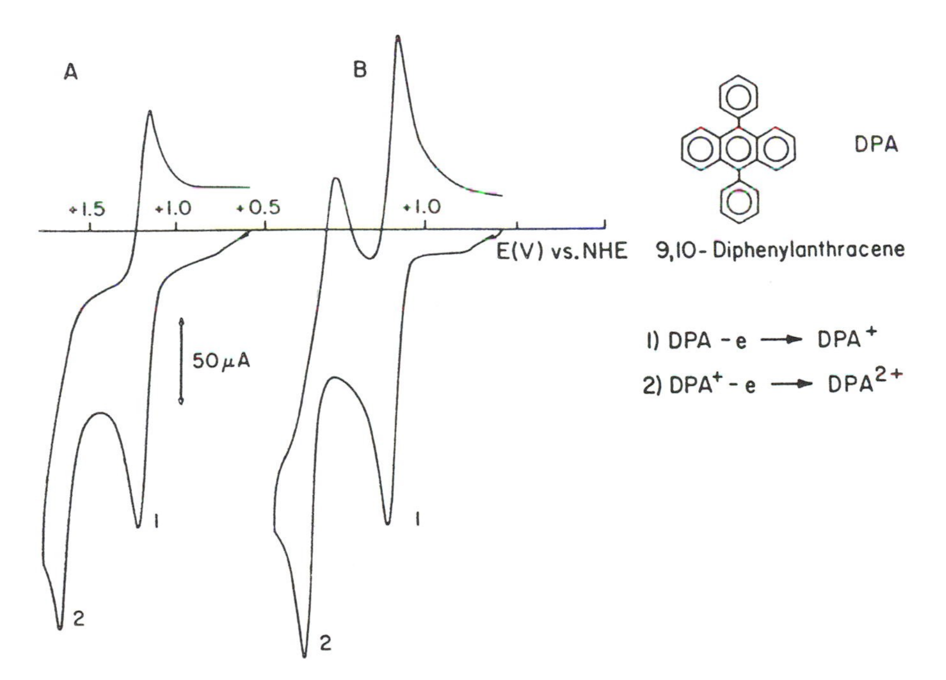


**Figure 18.15** Cyclic voltammograms of 9,10-diphenylanthracene (DPA) in 0.1 M $Bu_4NPF_6/CH_3CN$, (A) before drying, and (B) after contact with activated alumina. [From Ref. 37, with permission.]

be transferred into the solvent–electrolyte solution by means of the sample loading device. This electrochemical cell also incorporates a large working electrode for bulk controlled-potential electrolysis. The cell has been used to study the reduction mechanism of niobium and tantalum halides in "superdry" acetonitrile [37].

A vacuum spectroelectrochemical cell that also contains an optically transparent thin-layer electrode (OTTLE) is shown in Figures 18.16 and 18.17. The cell can function either as a spectroelectrochemical cell employing an OTTLE or as an electrochemical cell for voltammetric measurements. This low-volume cell is useful for UV/Vis spectral studies in nonaqueous solvents when the reduction product is sensitive to traces of molecular oxygen present in the solvent. The cell is physically small enough to fit inside the sample compartment of the spectrophotometer. The performance of such a cell was evaluated from visible spectroscopy and coulometry of methyl viologen in propylene carbonate [45].

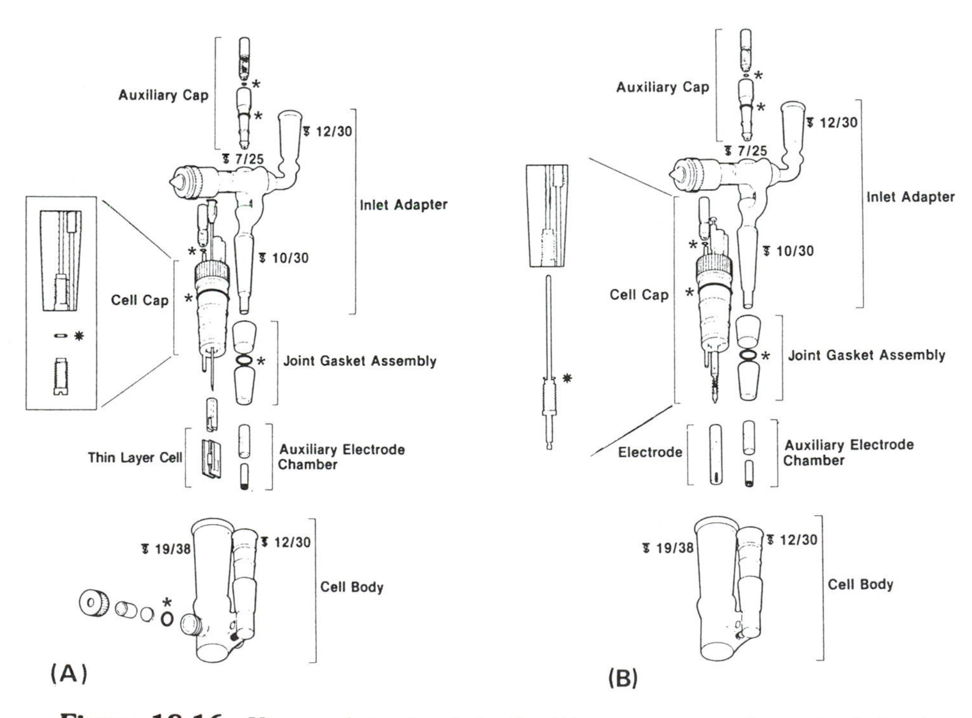


**Figure 18.16** Vacuum electrochemical cells: (A) vacuum spectroelectrochemical cell that contains an optically transparent thin-layer electrode (OTTLE) and (B) electrochemical cell assembly. [From Ref. 45, with permission.]

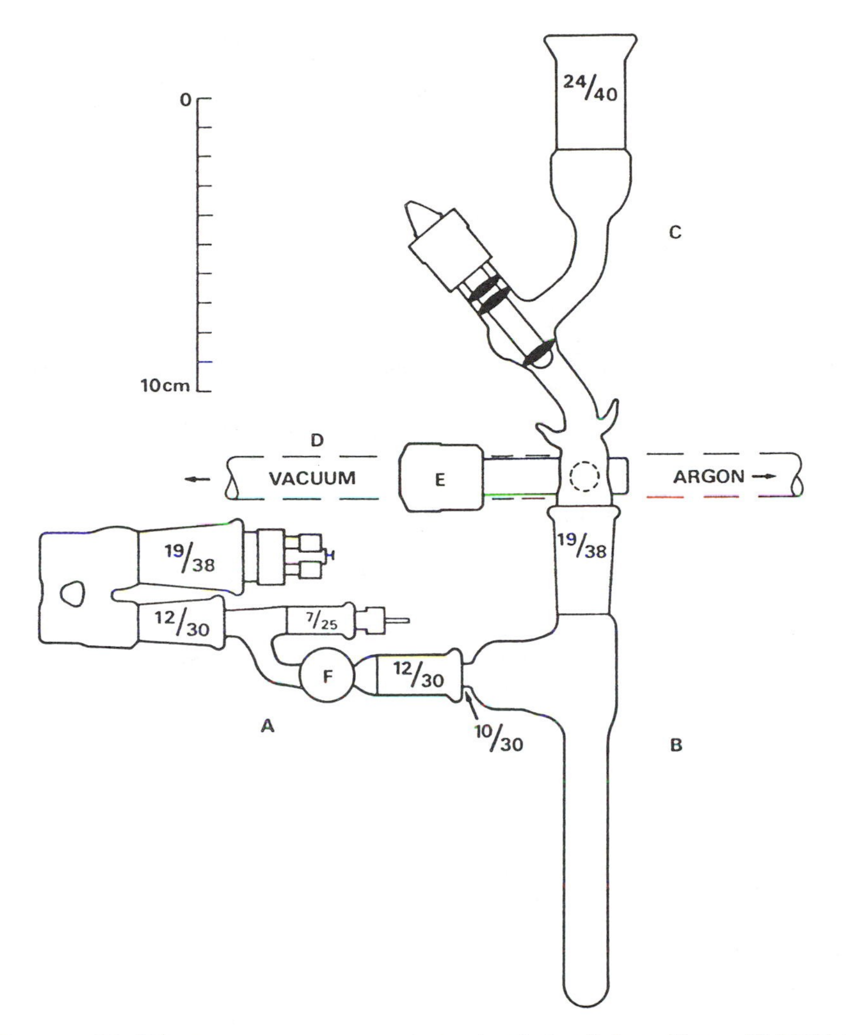


**Figure 18.17** The vacuum spectroelectrochemical cell from Figure 18.16(A) is shown attached to (B) the solution ampule and (C) the solvent header, which in turn is attached to (D) the main vacuum line. [From Ref. 45, with permission.]

## REFERENCES

1. O. Hammerich and V. D. Parker, *Electrochim. Acta 18*:537 (1973).
2. H. Kiesele, *Anal. Chem. 53*:1952 (1981).
3. J. F. O'Hanlon, *A User's Guide to Vacuum Technology*, 2nd ed., John Wiley and Sons, New York, 1989.

4. P. H. Plesch, *High Vacuum Techniques for Chemical Syntheses and Measurement*, Cambridge University Press, Cambridge, England, 1989.
5. D. P. Shoemaker, C.W. Garland, and J.W. Nibler, *Experiments in Physical Chemistry*, 5th ed., McGraw-Hill, New York, 1989, Chap. 17.
6. D. F. Shriver, *The Manipulation of Air-Sensitive Compounds*, 2nd ed., John Wiley and Sons, New York, 1986.
7. R. R. Lapelle, *Practical Vacuum Systems*, McGraw-Hill, New York, 1972.
8. N. T. M. Dennis and T. A. Heppell, *Vacuum System Design*, Chapman and Hall, London, 1968.
9. P. A. Redhead, J.P. Hobson, and E.V. Kornelsen, *The Physical Basis of Ultra-high Vacuum*, Chapman and Hall, London, 1968.
10. K. Diels and R. Jaeckel, *Leybold Vacuum Handbook*, Pergamon Press, Oxford, 1966.
11. A. Roth, *Vacuum Sealing Techniques*, Pergamon Press, Oxford, 1966.
12. J. E. Hammesfahr and C.L. Stong, *Creative Glass Blowing*, W. H. Freeman, San Francisco, 1968.
13. *Laboratory Glass Blowing with Corning's Glasses*, Bulletin B-72, Corning Glass Works, Corning, NY, 1969.
14. R. Barbour, *Glassblowing for Laboratory Technicians*, Pergamon Press, Oxford, 1968.
15. F. L. Wheeler, *Scientific Glassblowing*, Interscience, New York, 1963.
16. W. E. Barr and V. J. Ahorn, *Scientific and Industrial Glass Blowing and Laboratory Techniques*, Instruments Publishing Company, Pittsburgh, PA, 1959.
17. A. J. B. Robertson, *Laboratory Glass Working for Scientists*, Academic Press, New York, 1957.
18. W. Gaede, *Ann. Phys. 41*:289 (1913).
19. M. von Pirani, *Dtsch. Phys. Gesell. 8*:24 (1906).
20. D. E. Buckley, *Proc. Natl. Acad. Sci. U.S.A.* 2:683 (1916).
21. W. Gaede, *Zh. Tekh. Fiz. 15*:664 (1934).
22. F. M. Penning and K. Nienhuis, *Philips Tech. Rev. 11*:116 (1949).
23. R. T. Bayard and D. Alpert, *Rev. Sci. Instrum. 21*:571 (1950).
24. W. J. Lange, *Phys. Today 25*(8):40 (1972).
25. T. Osa and T. Kuwana, *J. Electroanal. Chem. 22*:389 (1969).
26. J. L. Mills, R. Nelson, S.G. Shore, and L.B. Anderson, *Anal. Chem. 43*:157 (1971).
27. H. Kiesele, *Anal. Chem. 52*:2230 (1980).
28. H. O. House, E. Feng, and N.P. Peet, *J. Org. Chem. 36*:2371 (1971).
29. A. J. Gordon and R. A. Ford, *The Chemist's Companion*, John Wiley and Sons, New York, 1972.
30. O. Hammerich and V. D. Parker, *J. Am. Chem. Soc. 96*:4289 (1974).
31. J. N. Butler, *J. Electroanal. Chem. 14*:89 (1967).
32. B. S. Jensen and V. D. Parker, *J. Am. Chem. Soc. 97*:5211 (1975).
33. D. M. LaPerriere, W. F. Carroll, Jr., B. C. Willet, E. C. Torp, and D. G. Peters, *J. Am. Chem. Soc. 101*:7561 (1979).
34. R. Lines, B. S. Jensen, and V. D. Parker, *Acta Chem. Scand. 32B*:510 (1978).

35. D. A. Hall and P. J. Elving, *Anal. Chim. Acta 39*:141 (1967).
36. L. R. Faulkner, H. Tachikawa, and A. J. Bard, *J. Am. Chem. Soc. 94*:691 (1972).
37. J. R. Kirk, D. Page, M. Prazak, and V. Katovic, *Inorg. Chem. 27*:1956 (1988).
38. J. T. Bowman, Ph.D. dissertation, University of Texas at Austin, 1970.
39. N. E. Tokel, C. P. Keszthelyi, and A. J. Bard, *J. Am. Chem. Soc. 94*:4872 (1972).
40. C. P. Keszthelyi and A. J. Bard, *J. Electrochem. Soc. 120*:241 (1973).
41. A. J. Bard, *Pure Appl. Chem. 25*:379 (1971).
42. A. Demortier and A. J. Bard, *J. Am. Chem. Soc. 95*:3495 (1973).
43. J. D. L. Holloway, F. C. Senftleber, and W. E. Geiger, Jr., *Anal. Chem. 50*:1010 (1978).
44. R. Battino, M. Banzhof, M. Bogan, and E. Wilhelm, *Anal. Chem. 43*:806 (1971).
45. E. A. Blubaugh and L. M. Doane, *Anal. Chem. 54*:329 (1982).

# 19

# Electrochemistry in the Dry Box

**Steven N. Frank** *Texas Instruments, Dallas, Texas*

**Su-Moon Park** *The University of New Mexico, Albuquerque, New Mexico*

## I. INTRODUCTION

Controlled atmospheres are employed in electrochemistry whenever oxygen, water, or other constituents of the air may interfere with the reaction under study. The problem is basically twofold: to supply and contain an atmosphere of suitable composition, and to exclude air. For our present purposes it is sufficient to confine the discussion to inert atmospheres: that is, gases devoid of oxygen and water (and sometimes nitrogen). The use of high-purity argon, helium, or nitrogen is most common.

Much progress has been made in the last few decades in making high-purity inert gases available at reasonable cost. The design of enclosed working spaces to contain these inert atmospheres and exclude air has resulted from the demands of high-technology areas such as metallurgy, semiconductor physics, and atomic energy [1]. The enclosed working space most often used in the research lab is the dry (or glove) box.

The interest in nonaqueous solvents [2,3] has presented the electrochemist with a number of new problems in solvent manipulation. Since most anhydrous solvents are quite hygroscopic, they should not be exposed to the laboratory atmosphere unnecessarily. The time it takes to pour or pipette from one container to another can be sufficient to pick up significant quantities of water. In many cases even minute quantities of water cannot be tolerated, since they will often participate in the reaction mechanism. Small amounts can also significantly reduce the available potential range. Often, nonaqueous solvents also suffer from the disadvantage of being too volatile to be conveniently deoxygenated by the usual method of bubbling an inert gas through the solvent. Electrochemists have typically circumvented these problems by using the vacuum-line techniques discussed in Chapter 18. Vacuum-line techniques, however, suffer from a number of disadvantages, not the least of which is that they are often very time-consuming. Using a dry box, on the other hand, can often shorten the time required to prepare for an experiment to about an hour; also, there is much less chance of breaking very expensive cells, such as those often used on a vacuum line.

The advantages of using a dry box over vacuum techniques may be summarized as follows:

1. There is a considerable saving of time.
2. Dry-box procedures are often simpler than vacuum-line techniques.
3. A high quality of oxygen and water removal is more easily maintained. Attainment of $10^{-4}$ torr vacuum may require several hours for large cells, but it is a simple matter to maintain routinely less than a 1 ppm level of oxygen (we have successfully maintained less than 0.1 ppm oxygen as determined with a Chemtrix model 30 oxygen meter) in the inert atmosphere of the dry box.
4. There is less danger of breaking important, expensive cells.
5. Solvent volume can be conveniently measured to any degree of accuracy desired. A high accuracy in weighing materials may be obtained by placing a balance in the dry box.
6. The handling of sublimable materials is easier than with vacuum techniques.

There are, however, several disadvantages of using a dry box for electrochemistry:

1. Every liquid has to be thoroughly deoxygenated before being transferred to the dry box.
2. Reference electrodes requiring aqueous solutions, such as SCE or Ag/AgCl,$Cl^-$ electrodes, cannot be used for nonaqueous work. This disadvantage can be overcome by using a nonaqueous Ag/$AgNO_3$ electrode and calibrating it against an SCE, or by using a silver wire "pseudo" reference.
3. It is difficult to carry out temperature studies.
4. The work space is rather limited.
5. So-called "inert" gases are not always inert. A vacuum may be better.
6. A good dry box can be very expensive to install and maintain.

## II. CHOOSING AN INERT-ATMOSPHERE SYSTEM

In choosing an inert-atmosphere system, one should select a system commensurate with the job at hand. It makes no sense to invest thousands of dollars in a sophisticated dry box when an inexpensive glove bag might serve just as well.

The most inexpensive and simplest system is the glove bag, which is basically a plastic bag open at one end with glove inserts. The materials to be used are placed in the bag in the collapsed condition. Holes are cut in the plastic for making electrical connections and so on, and then sealed with tape or other suitable material. The bag is then purged with inert gas for several minutes and the open end is sealed with the gas continuing to flow.

The glove bag has several advantages. It is economical and available in a variety of sizes. However, although a glove bag is versatile, it is difficult to maintain the quality of the atmosphere. At best the atmosphere can be no better than the inert gas used. The highest quality tanked gases usually have in excess of 10 ppm impurity, and this establishes a lower limit on the impurity level. Glove bags also tend to leak and there is a large diffusional area (most plastics are somewhat permeable to air). The purging procedure is usually not very efficient at removing all of the air initially trapped in the bag. The impurity level of the atmosphere in the glove bag is therefore usually much higher than that in the gas supply. When a large number of experiments are anticipated that require an inert atmosphere, the glove bag is not very convenient. It is ideal for the occasional experiment in which the systems being investigated are only moderately sensitive to water and oxygen. There are a number of systems somewhere between the glove bag and the conventional dry box. Nevertheless, in our experience these systems offer no significant advantages over the glove bag.

Dry boxes vary widely in design and in the materials used in construction. Prices range from a few thousand dollars for economy models to over $25,000 for more sophisticated units. All dry boxes contain a working space that incorporates glove ports, gas connections, electrical connections and lighting, and a viewing window. In addition, most boxes have an antechamber or transfer box for moving material in and out of the working area without disturbing the integrity of the dry box atmosphere. Two kinds of transfer chambers are available: an evacuation chamber and a gas purging chamber. In an evacuation chamber the samples are placed in the chamber, and the chamber is evacuated. Gas of the same composition and purity as in the working area is then introduced and the dry box is opened. With a purging chamber, gas, usually from the same cylinders as the supply gas for the box, purges air from the chamber. The evacuation transfer chamber is preferred since it does a more complete job of removing air and adsorbed water and saves the cost of the purging gas.

The inert atmosphere for the box may be supplied continuously with a bleed-off valve, or recirculated through purification trains that remove oxygen and water. Since all boxes leak to some extent (1 $f^3$/day is typical), recirculation through a purification train is desirable. Figure 19.1 shows a block diagram for a typical dry box with the recirculation option. Most trains can maintain oxygen and water to less than 1 ppm. If economics is a consideration, scientists can

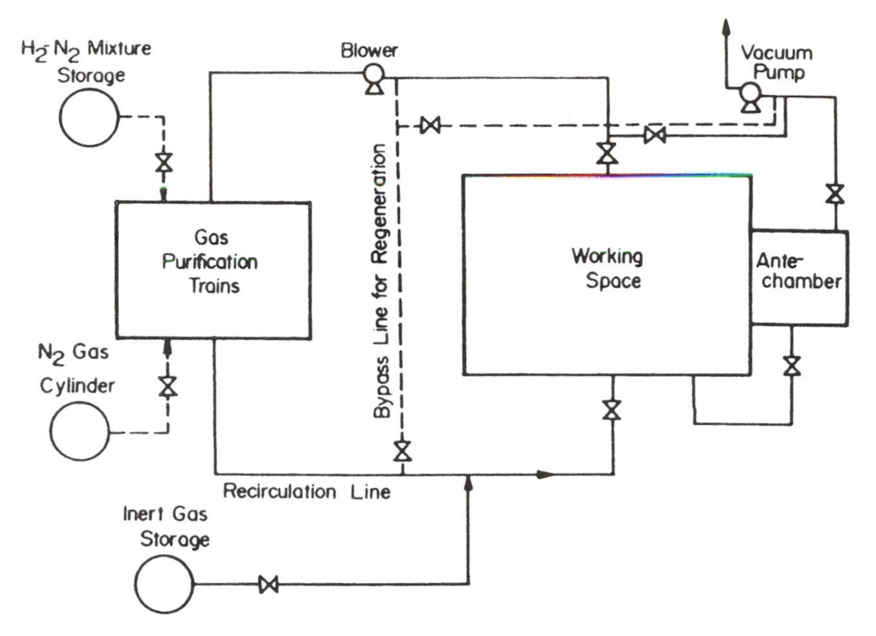


**Figure 19.1** Block diagram of dry box with inert-gas recirculation through purification train. The $N_2$ and $N_2$–$H_2$ mixtures are used during regeneration of the purification train.

construct their own purification trains [1], although they would be much less convenient than the commercial product.

Systems that permit automatic regulation of the gas pressure in the box within present limits are preferable to those that do not. This feature can prevent accidents that may result from over- or underpressurizing the box. All-metal boxes with glass or plastic windows (Fig. 19.2) are generally better than all-plastic boxes. Metal boxes offer the advantages of durability, resistance to chemical attack, and increased ease of modification.

## III. EXPERIMENTAL PROCEDURES

### A. Basic Dry-Box Procedures and Considerations

Although there are several classes of dry boxes as described earlier, we limit our discussion to the considerations involved in using a fairly sophisticated dry box equipped for gas recirculation through a purification train and including an evacuation transfer chamber. The dry box will also provide for automatic control of the gas pressure within the box. These features make a dry box especially well suited for carrying out high-quality nonaqueous electrochemistry with a minimum of effort.

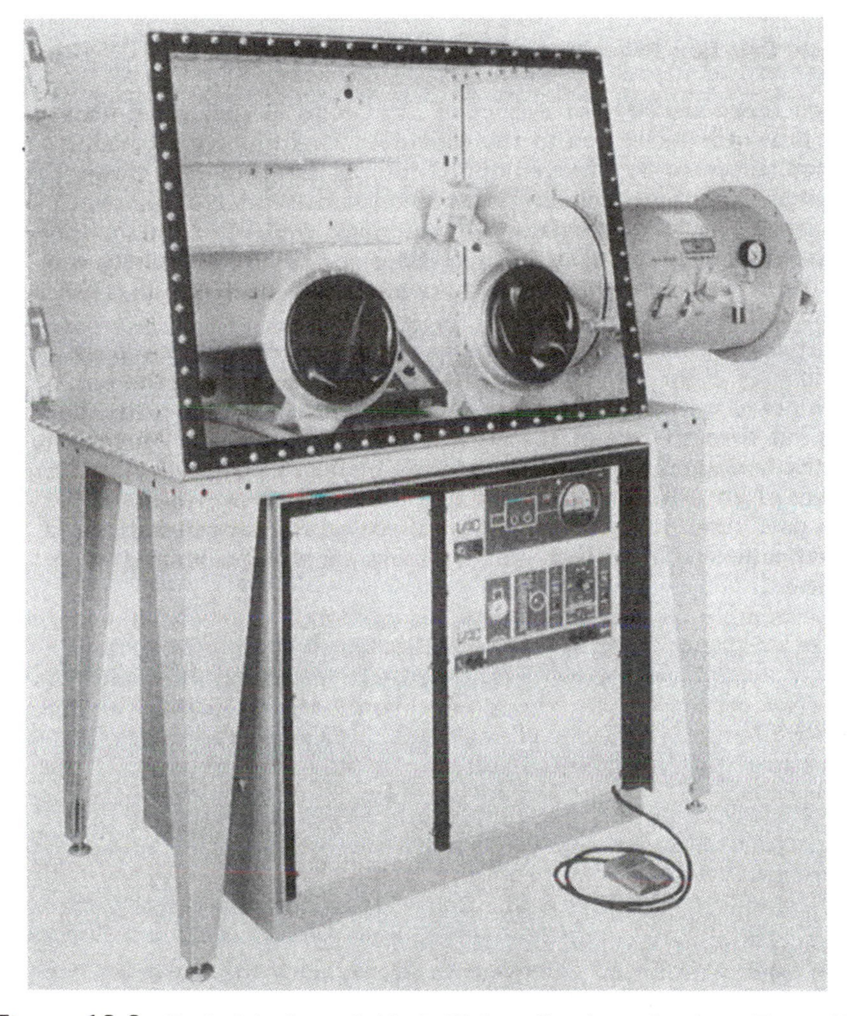

**Figure 19.2** Typical dry box suitable for high-quality electrochemistry. The purification train, recirculation unit, and automatic pressure control system are housed in the unit beneath the dry box. [Courtesy of Vacuum/Atmospheres Corporation, Hawthorne, CA.]

Initial purging of the dry box is readily carried out by passing approximately eight volumes of the inert gas through the box at positive pressure (the equivalent of about 2 in of water) with the gas flowing out one of the service ports of the box. During the purge, the box should be recirculating in order to purge the purification train of air. While the box is still under positive pressure, the service port should be sealed. After three or four regenerations of the purification train, a high-quality inert atmosphere should be established.

Immediately following the initial assembly (and at any other time leaks are suspected), the box and purification train should be tested for leaks. A quick test for leaks is to pressurize the dry box until the gloves stick straight out. The gloves should remain in this position for several hours if no leaks are present. If leaks are indicated, testing is most easily accomplished while the box is pressurized. If the inert atmosphere is helium, the preferred method is the helium "sniff" test. All joints, welds, and connections should be checked. In the absences of a helium-sensitive probe or if the inert atmosphere is other than helium, the bubble method may be used: a small amount of soapy water is placed on leak-prone welds and joints and the appearance of any bubbles is noted.

Glassware and other material to be introduced into the box should be as clean and dry as possible. This saves in the time required to pump down the antechamber, minimizes the possibility of contaminating the dry box, and increases the time between regenerations of the purification train. Simple glassware that is clean and dry should be pumped down for only 20–30 min. Large electrochemical cells containing many frits and compartments should be pumped down for several hours. By removing glassware directly from a drying oven and placing it immediately in the antechamber, the efficiency of the pump-down procedure is considerably enhanced, since desorption of adsorbed material is greatly enhanced at elevated temperatures. The efficiency of oxygen and water removal is also increased if the antechamber is evacuated a second time after filling it with inert gas from the working area of the dry box (via the appropriate plumbing). Before pumping down the transfer box, one should make sure that the door between the transfer box and the working area is well sealed, or a rapid depletion of the supply gas for the dry box will result. When possible we have found it convenient to pump down complicated equipment such as stirring motors or cells overnight.

Most solid chemicals can be introduced without any difficulty after they have been thoroughly dried. Supporting electrolytes such as tetra-*n*-butylammonium perchlorate (TBAP) are usually vacuum-dried, at elevated temperatures if possible, for 24–48 h and can be pumped down in the antechamber in the bottle with the top removed. Sublimable compounds should not be pumped down in the antechamber, for obvious reasons. These materials should be placed in specially designed sample containers having two small stopcocks. The containers can be purged with a dry inert gas for several minutes (Fig. 19.3). The design should be such as to permit evacuation of the surrounding atmosphere. Screw caps are especially convenient for sealing the top opening of the vials. As an alternative, metal pipes or cylinders sealed at one end with screw caps and O-rings and stainless or brass valves for purging make excellent containers for introducing sublimable compounds.

Solvents present special problems in handling since they must be dried and deoxygenated before being introduced into the dry box. Solvent manipulation

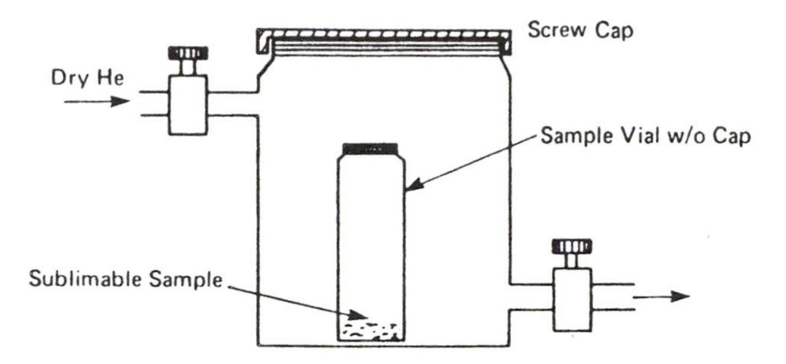


**Figure 19.3** Design of apparatus for introducing sublimable materials into a dry box equipped with an evacuation transfer chamber.

best illustrates how vacuum-line techniques complement dry-box techniques. For this reason and because the published procedures [3] seem unduly complicated and time-consuming, we present the procedure for purifying (drying) the widely used nonaqueous solvent acetonitrile and the introduction of the thus purified solvent (less than 0.1 m$M$ in $H_2O$) into the dry box.

The design of the equipment used in our laboratory for the purification is shown in Figure 19.4. Spectrograde acetonitrile is successively distilled three times from $P_2O_5$ under vacuum at room temperature, with the final distillation

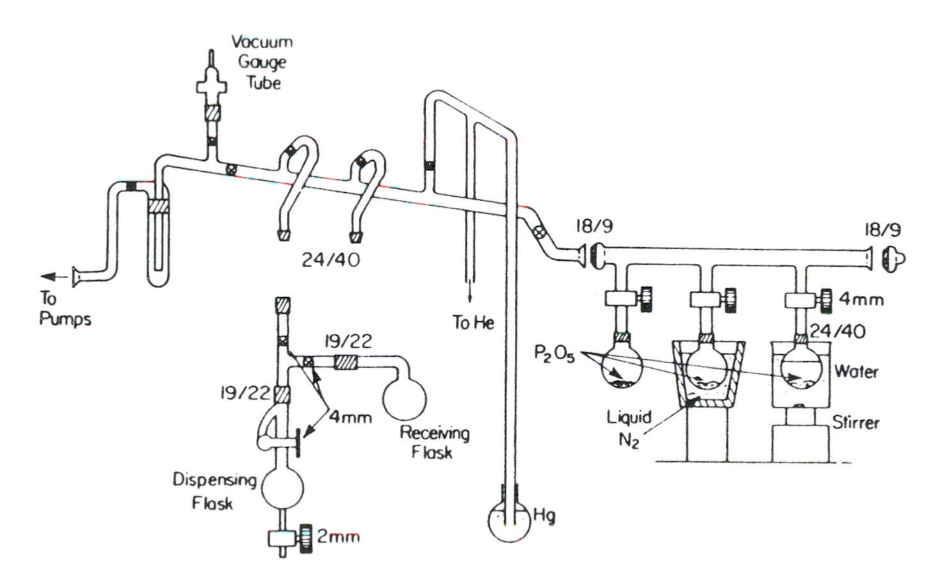


**Figure 19.4** Apparatus for vacuum distillation of acetonitrile. For more details, see Figure 18.1.

being into the receiving flask. The purified and dry acetonitrile is degassed in this flask by performing one or two freeze–pump–thaw (FPT) cycles. Only one or two FPT cycles are necessary since the solvent has already been pumped down to $10^{-5}$ torr three times during the distillation. The acetonitrile is now transferred to the final holding and dispensing flask with the aid of a little positive helium pressure in the receiving flask, obtained by passing some helium onto the vacuum line (see Chap. 18 for more details). Before placing the dispensing flask in the antechamber and evacuating, the stopcocks should be tied to the flask to prevent them from popping out. It is important to consider similar possibilities whenever any container that is not to be evacuated is placed in the transfer chamber.

Enough solvent should always be prepared to last for some time, since solvent preparation may be time-consuming and inconvenient. Any liquid, be it solvent or solute, should be degassed and dried (*not* always with $P_2O_5$) before transferring it to the working area. Solvents such as *N,N′*-dimethylformamide and methyl sulfoxide, which cannot be purified by vacuum distillation, should be degassed more thoroughly by three or four FPT cycles.

What equipment should always be in the dry box? We have found that placing a large plastic sheet on the bottom of the box facilitates cleaning. A small brush and adhesive tape for cleaning up solid chemical spills may prove useful. The arrangement of the equipment is important since the working area is rather limited. Usually, several ring stands, a magnetic stirring motor, a buret stand, a pair of tongs, several clamps, and an automatic balance are necessary. A large plastic syringe attached to a rubber tube is useful for pipetting. Tongs with rubber tubing over the tips are especially convenient for handling glassware when using the balance. A small spatula is necessary for transferring solid materials. A box of wipes (these should be pumped down in the vacuum chamber for at least 12 h to avoid introducing oxygen and water into the dry box) and weighing paper are also useful, as is silicone vacuum grease should one wish to remove apparatus or solutions containing air-sensitive compounds from the dry box.

For electrochemical experiments inside the box, one should be certain that there are enough electrical connections and plugs. Few if any of the commercial dry boxes will have enough of these, and they will have to be installed. Connections to the outside world are a source of leaks, and considerable care should be exercised to ensure that the connectors are well sealed. These connectors, called feed throughs, are available from vendors of vacuum-related products. By having enough connections, two or three people can work simultaneously. For example, we have installed two connectors with six contacts on each. Long color-coded wires to the connectors are recommended to avoid confusion.

When working in the box, a slightly less than atmospheric pressure is preferred inside the box, since it makes handling objects with the gloves easier. When not in use, however, the pressure in the box should be higher than atmospheric to minimize leaking into the dry box.

Sprinkling a little talc or baby powder in the gloves facilitates putting on and taking off the gloves. It also helps absorb moisture, as the hands and arms tend to perspire when they have been in the gloves for a few minutes. Better yet, wearing a pair of disposable surgical gloves used with a small amount of talc will greatly aid insertion and removal of hands from the dry-box gloves. The gloves should be inspected periodically and replaced when leaks or holes are found.

Always double check to make sure the inner door between the working area and the antechamber is closed before opening the outer door to the antechamber. It is surprising how often this simple task is forgotten.

The presence of oxygen can be detected by exposing the filament of a broken light bulb to the inert atmosphere. Under ideal conditions the filament should not burn out for several weeks. Unfortunately, the oxygen transients that sometimes result upon introducing materials into the box can exceed 15 ppm. Although the quality of the atmosphere is rapidly reestablished within a few minutes, the filament may have already burned out.

Oxygen can also be detected chemically. Tetrakis(dimethylamino)ethylene forms a luminescent compound with oxygen and is very sensitive to trace amounts of oxygen [4]. One drawback of this compound is that it has a high affinity for silicone stopcock grease, which it deteriorates quite rapidly. A small bottle of diethylzinc mixed with heptane is also useful. The absence of a vapor cloud indicates less than 5 ppm of oxygen. (Caution: This material is pyrophoric; use extreme care in handling.)

Probably the best method of oxygen measurement is an oxygen meter, which uses an amperometric oxygen sensor such as the Clark electrode [5]. The amperometric oxygen electrode is placed in the box and connected to the external meter via the aforementioned electrical connections. When placing the oxygen probe in the dry box, it should be placed in a vacuum-tight container when pumping down the antechamber. Otherwise, the pressure differential will rupture the electrode membrane. Periodically, the electrode will have to be removed and the filling solution replenished. No significant water contamination results from using this electrode. Several instrument vendors offer oxygen meters.

Titanium tetrachloride will indicate 10 ppm or more of water by the appearance of a vapor cloud. Hygrometers are also available, but the expense hardly seems justified. If significant quantities of oxygen are detected, it is safe to assume that the dry box is also contaminated with water. When oxygen is detected, the gas purification train should be regenerated.

Since the gas purification trains consist of chemicals such as manganous oxide, calcium, or copper to remove oxygen and molecular sieves to remove water, they eventually become saturated and deactivated and must be regenerated. It is best to regenerate the train on a regular basis before its effectiveness decreases. Regeneration usually consists of heating the train while purging with nitrogen and nitrogen–hydrogen mixture for 4–12 h.

The vacuum pump should be serviced regularly. Whenever the pump oil seems contaminated, it should be replaced with new oil.

## B. Electrochemistry in the Dry Box

As we have pointed out previously, oxygen and water concentrations can be kept at extremely low levels with a properly maintained purification train. In fact, contamination by water is much more easily controlled in a dry box than on a vacuum line. This may result in part from the number of operations and manipulations necessary to use a vacuum line. The oxidation of the cation radical of thianthrene to the dication illustrates this point. The dication is very electrophilic and is rapidly attacked by any nucleophiles (e.g., water). The electrochemistry of the dication in solutions prepared in the dry box (with acetonitrile as the solvent purified as described earlier) is reversible if a little care is taken in preparing the solvent and in drying the glassware. It is more difficult to obtain such reversible behavior when solutions are prepared on a vacuum line.

Transient electrochemical measurements such as cyclic voltammetry are well suited for the dry box. Solvents, supporting electrolytes, and any compounds to be used can be kept in the box. Only the cell and electrodes need be introduced for experiments. Solution preparation usually takes only 20–30 min. The electrodes can be connected to external apparatus as described earlier. Measurements of voltammetric current functions should be avoided at slow sweep rates because vibrations due to the various motors and pumps result in some convection. In fact, theoretical measurements using any technique that assumes mass transport by diffusion should not be attempted in the dry box if the time of data acquisition is greater than 2 or 3 s (e.g., sometimes chronopotentiometry and chronocoulometry). In these cases the solutions may be prepared in the dry box and the cells made airtight (we recommend using cells with ground-glass joints) with silicone grease and removed from the dry box. A considerable saving of time is realized by preparing the solutions in the dry box instead of on a vacuum line. The dry box offers similar advantages when performing controlled-potential electrolysis.

Conventional polarography with a dropping mercury electrode can conveniently and routinely be carried out in the dry box, although due to height limitations it may be necessary to use a lower mercury head than is possible outside the box.

Dry boxes are especially advantageous for rotating ring-disk electrode (RRDE) experiments [6]. Because the motor can be placed in the dry box, one is saved the expense of a rather expensive RRDE cell and sealed motor and the attendant problems of working outside the dry box.

Electrogenerated chemiluminescence (ECL) experiments [7] are difficult to perform in the dry box due to the need to exclude light from all external sources. Placing a black cloth over the dry box window is usually not sufficient unless the intensity of the ECL is very high. It may be necessary to use a monochromator in these experiments. This is especially unfortunate in the case of ECL at an RRDE, since the complexity of cell and motor assembly is greatly increased. Nevertheless, the dry box is very useful for performing initial ECL experiments, perhaps to establish whether or not ECL occurs.

Dry boxes also offer significant advantages over vacuum line techniques in the study of mixed-solvent systems. It is difficult to mix two different solvents accurately on a vacuum line. In a dry box, the procedure is rendered almost trivial. Definite advantages of a dry box can also be envisioned for electrochemistry in low-temperature molten salts, which are often very hygroscopic.

For those who wish to work with the old model Kemula hanging mercury drop electrode, some problems may be encountered when evacuating the antechamber (separation of the mercury column in the capillary and retraction of the mercury column up the capillary after the drop is dislodged). To prevent these problems, the electrode should be clamped in an upright position with the tip of the electrode submerged in a small mercury pool. This electrode also suffers the disadvantage that a vacuum is necessary for filling and it therefore cannot be refilled in the dry box. The new model Kemula electrode (available from Brinkmann Instruments and Princeton Applied Research) and the hanging mercury drop electrode available from Beckman Instruments can be evacuated in the antechamber and filled in the dry box without any problems.

Aqueous reference electrodes should not be used since they cannot be evacuated in the antechamber. In any case there would be a serious possibility of contaminating the dry box with water. A silver wire in contact with a solution of silver can be dissolved in many solvents being used in the dry box. The electrode can later be calibrated against an aqueous SCE. For many experiments a bare silver wire will serve satisfactorily, although the potential of this electrode will not be constant.

With the high solution resistance that may be present with nonaqueous solvents and the long electrode connections recommended earlier, some impedance problems may result. We have found the severity of these problems to be highly dependent on the potentiostat being used [e.g., we have had far fewer problems when using the PAR (Princeton Applied Research) model 173 than the more expensive PAR 273 electrochemical system]. To remedy this difficulty, one might consider placing a voltage follower in the dry box to minimize the lead length to the reference electrode.

A note of caution is needed if one plans to use open electrochemical cells in the dry box. Because the rate of recirculation of the atmosphere is rather high in most dry boxes (one complete change of atmosphere every few minutes), evaporation of most organic solvents is rapid. Open containers such as beakers are recommended only for a short period during solution preparation.

## APPENDIX

A reasonable place to begin one's search for a dry box is the Lab Guide issue of *Analytical Chemistry*, which appears yearly. When selecting a box the criteria set forth in the text should be considered. Consideration should also be given to the needed physical dimensions of the box. For those on a tight budget, one of the plastic or fiberglass dry boxes may be satisfactory. If purification trains or a vacuum transfer chamber are not listed as options, the manufacturer should be consulted to see if the dry box can easily be modified to accommodate them. We recommend buying a complete system (which would include all pumps and motors) from one manufacturer when possible. Two manufacturers that make especially good systems are Kewaunee Scientific Corp., Adrian, MI, and Vacuum/Atmospheres Company, Hawthorne, CA.

Glove bags may be purchased from Instruments for Research and Industry ($I^2R$), Cheltenham, PA.

## REFERENCES

1. P. A. F. White and S. E. Smith, *Inert Atmospheres in Chemical, Metallurgical, and Atomic Energy Industries*, Butterworth, Washington, DC, 1962.
2. (a) C. K. Mann and K. N. Barnes, *Electrochemical Reactions in Nonaqueous Systems*, Marcel Dekker, New York, 1970; (b) M. M. Baizer and H. Lund, *Organic Electrochemistry*, 2nd ed., Marcel Dekker, New York, 1983.
3. C. K. Mann, in *Electroanalytical Chemistry*, Vol. 3 (A.J. Bard, ed.), Marcel Dekker, New York, 1969.
4. (a) R. L. Purett, J. T. Barr, K. E. Rapp, C. T. Bahner, J. D. Gibson, and R. H. Lafferty, Jr., *J. Am. Chem. Soc. 72*:3646 (1950); (b) J. P. Paris, *Symposium on Chemiluminescence*, Durham, NC, March–April, 1965.
5. (a) E. Gneiger and H. Forstner, eds., *Polarographic Oxygen Sensors*, Springer-Verlag, New York, 1983; (b) M. L. Hitchman, *Measurement of Dissolved Oxygen*, Wiley, New York, 1978.
6. (a) W. J. Albery, *Trans. Faraday Soc. 62*:1915 (1966); (b) W. J. Albery and S. Bruckenstein, *Trans. Faraday Soc. 62*:1920 (1966); (c) W. J. Albery and S. Bruckenstein, *Trans. Faraday Soc. 62*:1932 (1966); (d) W. J. Albery, S. Bruckenstein, and D. C. Johnson, *Trans. Faraday Soc.. 62*:1938 (1966); (e) W. J. Albery and S. Bruckenstein, *Trans. Faraday Soc. 62*:1964 (1966); (f) W. J. Albery and S. Bruckenstein, *Trans. Faraday Soc. 62*:2584 (1966); (g) W. J. Albery and S. Bruckenstein,

*Trans Faraday Soc.* *62*:2569 (1966); (h) W. J. Albery, *Trans. Faraday Soc.*. *63*:1771 (1967).

7. (a) L. R. Faulkner and A. J. Bard, in *Electroanalytical Chemistry*, Vol. 10 (A. J. Bard, ed.), Marcel Dekker, New York, 1977; (b) S.-M. Park and D. A. Tryk, *Rev. Chem. Intermed.* *4*:43 (1981).

# 20

# Digital Simulation of Electrochemical Problems

**J. T. Maloy** *Seton Hall University, South Orange, New Jersey*

## I. INTRODUCTION

This chapter originates from a set of notes developed by the author for the members of Allen Bard's research group at the University of Texas in the late 1960s. Based entirely upon the pioneering research of Steve Feldberg [1–3], this work began as a how-to guide for those who wanted to learn how to write the computer code to perform digital simulations of electrochemical phenomena. These notes then circulated underground for more than a decade while the first edition of this text was in preparation. During that period, numerous practitioners of the art used these notes, or materials based upon them [4], as an introduction to digital simulation. It was in this pedagogical milieu that the version found in the first edition was written.

Many advances in digital simulation have taken place since the publication of the first edition. Some of these advances have occurred in hardware through the development of the personal computer. Others have taken place by the development of commercial software that will perform specific simulations or will create a computer environment (e.g., a spreadsheet) that will allow one to do simulations without having to write a computer program. Finally, there have been theoretical advances where newer implicit algorithms are used to solve the necessary partial differential equations more efficiently than is possible using the more intuitive explicit methods described herein.

This revision does not attempt to take many of these recent advances into account, even though some of them are cited in this chapter. Rather, it continues to provide a rigorous foundation for writing programs that will perform explicit finite difference simulations. In learning how to do this, the reader develops an appreciation of the method and, more importantly, its limitations.

Not the least of these results from the absence of an analytical solution to almost any problem of consequence, except in the limits of the boundary value problem. An appreciation of this limitation and how one may minimize its impact is essential to the utilization of digital simulations in any application.

## II. THE FINITE DIFFERENCE REPRESENTATION OF FICK'S LAWS

Any discussion of the digital simulation of problems involving diffusion begins with a consideration of the combined form of Fick's laws [5],

$$\frac{\partial}{\partial t} C(x,t) = D \frac{\partial^2}{\partial x^2} C(x,t) \tag{20.1}$$

for the concentration C(x,t), a function of position and time, where D is the diffusion coefficient. In terms of its finite difference definition, this law may be written

$$\begin{aligned} &\lim_{t \to 0} \frac{C(x,t+\Delta t) - C(x,t)}{\Delta t} = \\ &\lim_{\Delta x \to 0} \frac{D}{(\Delta x)^2} \{[C(x+\Delta x,t) - C(x,t) - [C(x,t) - C(x-\Delta x,t)]\} \end{aligned} \tag{20.2}$$

This may be readily rearranged to yield

$$\begin{aligned} &C(x,t+\Delta t) - C(x,t) = \Delta C(x,t) = \\ &\lim_{\substack{\Delta x \to 0 \\ \Delta t \to 0}} \frac{D\Delta t}{(\Delta x)^2} [C(x+\Delta x,t) - 2C(x,t) + C(x-\Delta x,t)] \end{aligned} \tag{20.3}$$

Two substitutions are necessary to obtain the diffusion algorithm. First, division of both members of Equation 20.3 by C, the bulk concentration, transforms all concentration to fractional values:

$$f(x,t) \equiv \frac{C(x,t)}{C} \tag{20.4}$$

Second, the definition of $D_M$, the model diffusion coefficient, relates $\Delta x$, $\Delta t$, and $\Delta D$:

$$D_M = \lim_{\substack{\Delta x \to 0 \\ \Delta t \to 0}} \frac{D\Delta t}{(\Delta x)^2} \tag{20.5}$$

Thus the fundamental equation used in the simulation of diffusion is obtained by substituting Equations 20.4 and 20.5 into Equation 20.3:

$$f(x,t) = D_M[f(x+\Delta x,t) - 2f(x,t) + f(x-\Delta x,t)] \tag{20.6}$$

To use Equation 20.6 to calculate the fractional concentration change resulting from a diffusion process, one partitions the medium (the solution) into an as yet unspecified number of volume elements of thickness $\Delta x$ and requires that $f(x,t)$ be uniform within each element. In the simplest case, these elements would form a linear array originating at a planar surface (the electrode) of area A located at $x = 0$. Therefore, the volume of each element will be $A\ \Delta x$, as represented in Figure 20.1. Each element may be assigned a serial number, J, as indicated in this figure, so that any distance x is given as

$$x = (J - 1)\ \Delta x \tag{20.7}$$

provided that the electrode is placed in the center of the first volume element. [This placement is particularly convenient in modeling the potential-step experiment; other techniques (e.g., chronopotentiometry) are sometimes more easily treated by placing the electrode at the exterior face of the first volume element. This choice is arbitrary but not without consequences, and is discussed further in this chapter.]

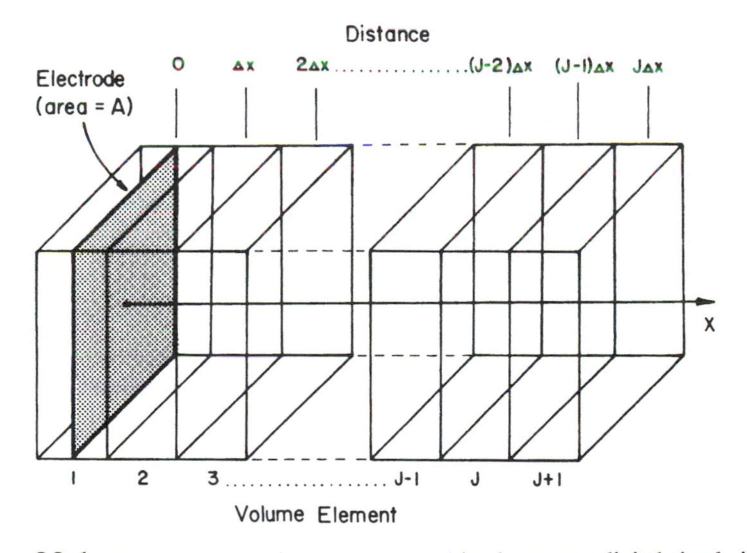


**Figure 20.1** Model volume element array used in elementary digital simulations of electrochemical problems. Note that the planar electrode has been placed in the middle of the first volume element in this model.

With this model in mind, one may abbreviate Equation 20.6 as

$$\Delta f(J) = D_M[f(J+1) - 2f(J) + f(J-1)] \tag{20.8}$$

to obtain an expression for the change in fractional concentration occurring in the Jth element at any time due to diffusion. In this form, $D_M$ is seen as the proportionality constant governing flux between adjacent elements containing different fractional concentrations. Applied successively to each element in the array, Equation 20.8 predicts the new fractional concentration,

$$f'(J) = f(J) + \Delta f(J) \tag{20.9}$$

that will occupy each element after diffusion has occurred during the interval $\Delta t$.

This basic method of treating diffusion requires

1. The assignment of a value to $D_M$.
2. The assurance that $\Delta t$ and $\Delta x$ approach zero.
3. The establishment of initial conditions ($t = 0$) and boundary conditions ($x = 0$, $x \to \infty$) so that concentration gradients may be obtained.
4. The expression of time, distance, and material flux in experimentally accessible dimensionless parameters, instead of the more ambiguous computer parameters (e.g., volume element J) indicated earlier.

## III. THE MODEL DIFFUSION COEFFICIENT: DEFINING $\Delta t$ AND $\Delta x$

Although the assignment of a value to $D_M$ is arbitrary, it is by no means unrestricted. From its definition (Equation 20.5) it is clear that $D_M$ must be positive. To find the upper limit for $D_M$, one need only substitute the Einstein definition [6] of the diffusion coefficient,

$$D = \lim_{\substack{\Delta y \to 0 \\ \Delta t \to 0}} \frac{(\Delta y)^2}{2\Delta t} \tag{20.10}$$

(where $\Delta y$ is the distance a diffusing particle travels during $\Delta t$) into Equation 20.5 to obtain

$$D_M = \lim_{\substack{\Delta x \to 0 \\ \Delta y \to 0}} \frac{(\Delta y)^2}{2(\Delta x)^2} \tag{20.11}$$

Since $\Delta y$ may approach zero in the limit but $\Delta x$ may only approach some small finite value in the finite difference method, it is clear that $(\Delta y)^2 \leq (\Delta x)^2$. In other words, particle diffusion between nonadjacent volume elements is forbidden in this model because $\Delta y$ may never exceed $\Delta x$. Thus the maximum acceptable value for $D_M$ is obtained when $\Delta y$ equals $\Delta x$, and the condition

$$\tfrac{1}{2} \geq D_M > 0 \tag{20.12}$$

is a requirement of the technique.

Since the assignment of a value is arbitrary, it would appear that setting $D_M$ equal to its maximum value would tend to minimize $\Delta x$ with respect to $\Delta t$. This is true, but setting $D_M$ equal to 0.50 in the simulation of a simple potential step to diffusion limiting conditions results in simulated currents that oscillate between accurate and very inaccurate representations of the Cottrell equation [7]. These oscillations are especially severe during the early stages of the simulation when relatively large currents are predicted, but they become insignificant sources of error in the predicted current as the number of iterations (each of duration $\Delta t$) is increased. Because the extent of oscillation decreases with decreasing $D_M$, Feldberg has used $D_M = 0.45$ to obtain reliable simulations of current–time behavior [3]. Unfortunately, the prediction of any behavior requiring the numerical integration of simulated current–time data may not be accurate when $D_M$ is this large because of the accumulation of large errors in current during the first few iterations. Marked improvements in the accuracy of simulated charge–time [8,9] or current semi-integral–time [10] behavior are obtained if smaller values of $D_M$ are used; a satisfactory value that has been used in these simulations of potential-step behavior is $D_M = 0.2184$.

Differences in real diffusion coefficients may also be taken into account using these methods. Because $\Delta x$ and $\Delta t$ would be the same for two different diffusing species A and B, it follows from Equation 20.5 that

$$\frac{D_{MA}}{D_{MB}} = \frac{D_A}{D_B} \tag{20.13}$$

Thus different rates of diffusion may be simulated by defining $D_{MA}$ and $D_{MB}$ in the same proportion as the actual coefficients $D_A$ and $D_B$. The conditions given in Equation 20.12 hold for both $D_{MA}$ and $D_{MB}$, however.

As indicated earlier, Equation 20.8 is applied successively to each element in the array of volume elements to simulate material transfer during the interval $\Delta t$. To define a $\Delta t$ unit, one must select some known time ($t_k$) in the physical experiment and partition that known time into L equal intervals for the purpose of the simulation. Thus

$$\Delta t \equiv \frac{t_k}{L} \tag{20.14}$$

and increasing L, the number of iterations representing a known time in the physical experiment, causes $\Delta t$ to approach zero. (The known time parameter

$t_k$ may change as different electrochemical techniques are simulated. For example, in the simulation of a double-potential-step experiment, $t_k$ is usually taken to be $t_f$, the interval between the two potential steps. In the simulation of a constant-current experiment, $t_k$ may be taken as $\tau$, the chronopotentiometric transition time; given a known current, the value of L may be calculated during the execution of the simulation rather than being defined. In some simulations, time-related quantities may be used to represent $t_k$. Thus $\Delta E/v$ may be used as the known time in potential sweep simulations, while $\omega^{-1}$ may be taken as $t_k$ in the simulation of rotating disk or ring-disk behavior.)

Real times may be related to $t_k$ through Equation 20.14. Thus, if Equation 20.8 has been applied to all the volume elements in the array K times so that diffusion has occurred for a time

$$t = (K - \tfrac{1}{2})\,\Delta t \tag{20.15}$$

($\frac{1}{2}$ is subtracted from K because time is measured midway through each iteration), the real time t may be related to the known time $t_k$ by substituting Equation 20.14 into Equation 20.15:

$$\frac{t}{t_k} = \frac{K - \frac{1}{2}}{L} \tag{20.16}$$

This provides a dimensionless representation of time throughout the simulation.

Because $\Delta x$, $\Delta t$, and $D_M$ are related by Equation 20.5, the specification of $D_M$, L, and $t_k$ results in the definition of $\Delta x$.

$$\Delta x \equiv \left(\frac{Dt_k}{D_M L}\right)^{1/2} \tag{20.17}$$

Thus, increasing L causes both $\Delta t$ and $\Delta x$ to approach zero. It is important to note, however, that $\Delta x$ decreases with $L^{1/2}$; therefore, once reasonable accuracy has been obtained by increasing L to a large value (*ca.* 1000), little is to be gained by increasing L any further.

## IV. ESTABLISHING INITIAL AND BOUNDARY CONDITIONS

The establishment of initial conditions is straightforward: Usually, an initial relative concentration equal to 1.0 is set in each theoretical volume element. It is possible that the concentration array might be initialized with some preestablished concentration profile other than bulk concentration throughout, but these cases do not arise frequently.

In the simulation of diffusion-limited semi-infinite linear diffusion, the establishment of the electrode boundary condition is equally straightforward: The concentration at the electrode surface is set equal to zero. If the electrode is placed in the middle of the first volume element in the model envisioned, the concentration in that element is set equal to zero to achieve the diffusion-limited condition. If the electrode is placed at the exterior face of the first volume element in the model under consideration, the concentration in the first element must be adjusted to one-third of that in the second element to maintain zero electrode surface concentration (see Fig. 20.2).

The simulation of other electrochemical experiments will require different electrode boundary conditions. The simulation of potential-step Nernstian behavior will require that the ratio of reactant and product concentrations at the electrode surface be a fixed function of electrode potential. In the simulation of voltammetry, this ratio is no longer fixed; it is a function of time. Chronopotentiometry may be simulated by fixing the slope of the concentration profile in the vicinity of the electrode surface according to the magnitude of the constant current passed. These other techniques are discussed later; a model for diffusion-limited semi-infinite linear diffusion is developed immediately.

In the case of semi-infinite linear diffusion, the concentration of electroactive species A will be maintained at zero at the electrode surface if the diffu-

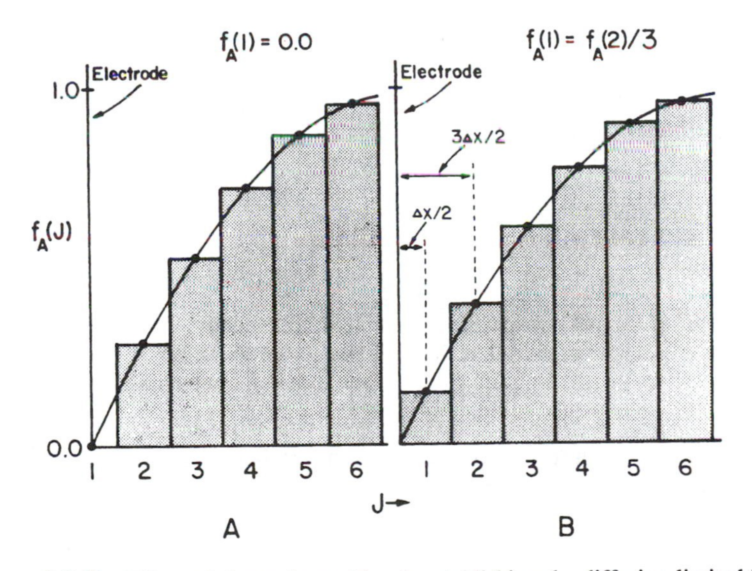


**Figure 20.2** Effect of electrode position in establishing the diffusion-limited boundary condition. (A) The electrode has been positioned in the middle of the first volume element so that zero concentration at the electrode surface may be readily maintained. (B) The electrode has been placed at the exterior edge of the first volume element so that zero surface concentration can be obtained only by maintaining the concentration in the first element at one-third of that in the second element.

sion-limited conditions exist. Thus the electrode boundary condition for species A may be written

$$f'_A(1) = f_A(1) = 0 \qquad (20.18)$$

For the electrode reaction

$$A \rightarrow B$$

the electrode boundary condition for the product B is not as straightforward. It must be written

$$f'_B(1) = f_B(1) + D_{MA}f_A(2) - D_{MB}[f_B(1) - f_B(2)] \qquad (20.19)$$

where the second term of the right member accounts for the A converted to B by the electrode reaction, and the third term accounts for the diffusion of B out of the first volume element.

At the other boundary, bulk concentration of A must be maintained at some finite distance from the electrode, while the concentration of B will be zero at the same point. This distance may be regarded as the diffusion layer thickness. In terms of the simulation, the establishment of the semi-infinite boundary condition requires the determination of the number of volume elements making up the diffusion layer. This will be a function of the number of time iterations that have taken place up to that point in the simulation. At any time in the physical experiment, the diffusion layer thickness is given by $6(Dt)^{1/2}$. This rule of thumb may be combined with Equation 20.7 to calculate $J_d$, the number of volume elements in the diffusion layer:

$$J_d = \frac{6(Dt)^{1/2}}{\Delta x} + 1 \qquad (20.20)$$

Direct substitution of Equations 20.16 and 20.17 into Equation 20.20 yields

$$J_d = 6\sqrt{D_M(K - \tfrac{1}{2})} + 1 \cong 3\sqrt{2K} \qquad (20.21)$$

Thus the number of volume elements required to model the diffusion layer increases with the square root of the iteration number, and the maximum number of volume elements required in any simulation of this sort is given by

$$J_{max} = 3\sqrt{2K_{max}} \qquad (20.22)$$

where $K_{max}$ is the serial number of the last time iteration attempted. To maintain the semi-infinite boundary condition, then, one needs only to maintain bulk concentration of electroactive species A in the $J_{max}$th volume element, while holding the product concentration of B at zero at the same point.

These observations bring up an interesting situation that occurs when the diffusion layer is of some fixed thickness d. This occurs when one models thin-layer cell behavior or membrane diffusion. In this case, $\Delta x$ is defined

$$\Delta x = \frac{d}{J_{max} - 1} \tag{20.23}$$

because one partitions the known thickness d into ($J_{max}$ - 1) volume elements. Since d is constant, it is possible to define a known time in terms of the diffusion coefficient,

$$t_k = \frac{d^2}{D} \tag{20.24}$$

These definitions may be combined with Equations 20.5 and 20.14 to yield

$$D_M = \frac{(J_{max} - 1)^2}{L} \tag{20.25}$$

This result is generally applicable to any situation in which a fixed-width diffusion medium is partitioned into ($J_{max}$ - 1) volume elements. In this case, the selection of $J_{max}$ fixes the magnitude of the model diffusion coefficient $D_M$. The constraints of Equation 20.12 still apply, however. Therefore, one must resist the temptation to increase the size of $J_{max}$ (without limit) to reduce the size of $\Delta x$. The maximum value of $J_{max}$ selected must be such that $D_M$ is still less than ½.

## V. DIMENSIONLESS PARAMETERS

The concept of a dimensionless representation of time was introduced in Equation 20.16. This was obtained by substituting the definition of $\Delta t$ into Equation 20.15 and rearranging the result so that physical quantities appeared on one side of the equation and model parameters appeared on the other. Since the model parameters are all dimensionless, this method assures that the arrangement of physical quantities obtained will also be dimensionless. This procedure (of substituting in the definitions of $\Delta t$, $\Delta x$, and $D_M$ and rearranging the result) is followed in this chapter in all assignments of dimensionless parameters.

The dimensionless distance parameter may be found by substituting the definition of $\Delta x$ into Equation 20.7. In the case of semi-infinite linear diffusion, Equation 20.17 is used to define $\Delta x$; this substitution yields

$$x = (J - 1)\left(\frac{Dt_k}{D_M L}\right)^{1/2}$$

which may be rearranged to obtain a dimensionless distance parameter,

$$\frac{x}{(Dt_k)^{1/2}} = \frac{J-1}{(D_M L)^{1/2}} \tag{20.26}$$

Thus the ambiguous "box number" may be transformed to a meaningful dimensionless distance $[x/(Dt_k)^{1/2}]$ by applying Equation 20.26. In the case of thin-layer diffusion, a different expression for $\Delta x$ is used (Eq. 20.23). When this is substituted into Equation 20.7 and rearranged, a different dimensionless distance parameter is obtained:

$$\frac{x}{d} = \frac{J-1}{J_{max}-1} \tag{20.27}$$

Dimensionless current representations also exist. If the electrode is placed at the center of the first volume element in the model, this current will be proportional to the material flux into the first element from the second. For electroactive species A, this flux is given by $D_{MA}$ $[f_A(2) - f_A(1)]$, where $f_A(J)$ is the fractional concentration of A in the Jth element. This fractional flux may be converted to moles of flux through multiplication by C (bulk concentration) and A $\Delta x$ (volume-element volume, assuming a planar electrode of area A). Appropriate electrochemical conversion and the recognition that this material flux occurs during the interval $\Delta t$ yield the current expression

$$i(t) = \frac{nFAC\ \Delta x D_{MA}[f_A(2) - f_A(1)]}{\Delta t}$$

Substitution of the definition of $\Delta x$ and $\Delta t$ (Eq. 20.14 and 20.17) and rearrangement yields the appropriate dimensionless current parameter

$$\frac{i(t)t_k^{1/2}}{nFACD_A^{1/2}} = \left(\frac{L}{D_{MA}}\right)^{1/2} D_{MA}[f_A(2) - f_A(1)] \tag{20.28}$$

Thus $(L / D_{MA})^{1/2}$ is the appropriate factor used to convert an ambiguous material flux to a meaningful current term. In the simulations reported next, the dimensionless current parameter is given by Z(K). If diffusion-limited conditions are obtained, the relative concentration of electroactive species in the first element is zero. Thus the form most frequently used is

$$Z(K) = \frac{i(t)t_k^{1/2}}{nFACD_A^{1/2}} = \left(\frac{L}{D_{MA}}\right)^{1/2} D_{MA} f_A(2) \tag{20.29}$$

In the case of thin-layer diffusion,

$$Z(K) = \frac{i(t)d}{nFACD_A} = \frac{L}{J_{max}-1} D_{MA} f_A(2) \tag{20.30}$$

by substituting Equations 20.24 and 20.25 into Equation 20.29.

Other dimensionless parameters, particularly those related to the study of kinetics, are obtained in a similar manner. These are discussed later.

## VI. A SAMPLE BASIC PROGRAM

It is now possible to write a computer program that will simulate the current–time behavior that occurs following a single step to diffusion-limiting conditions. The solution to this problem (the Cottrell equation) is well known and there is no need to perform this simulation to obtain any new electrochemical information. However, it is instructive to see how the principles developed earlier may be used to write a program and how the results of that simulation may be verified by comparing them to known results. Any additional programming will be left as an exercise to the reader.

The annotated BASIC program appears in Figure 20.3. This program is suitable for up to 1000 iterations; therefore, 1001 elements are reserved in the DIM statement for Z and T, the dimensionless current and time parameters. This fixes the maximum number of volume elements necessary to represent the diffusion layer according to Equation 20.22. Two volume element arrays (old and new) are reserved for each of the two diffusing species (A and B). New concentrations are calculated using the diffusion algorithm (Eq. 20.8), and these become old concentrations just before the next time iteration. Values for L, $D_{MA}$, and $D_{MB}$ are entered by the user during execution [lines 110, 130, and 150]. The old concentration arrays are then initialized [$f_A(J) = 1$ and $f_B(J) = 0$ throughout] and the first time iteration takes place. All of the A in the first element is transformed to B at the midpoint of this time iteration; this exhaustive electrolysis results in "unit" current [SQR(L/DMA)] flow. Each subsequent iteration begins at line 500. Electrode boundary conditions are established and the maximum number of volume elements is calculated for each iteration [line 530] in order to save computer time. The diffusion algorithm is applied to each element in the array [lines 550 and 560], and the dimensionless time and current are calculated [lines 600 and 610] at the end of each iteration. The new arrays become old arrays [lines 750–780] and the process is repeated until a maximum number of time iterations, L, have transpired [line 800]. The dimensionless current–time curve is then printed to the screen [lines 900–970].

Selected results obtained with this program appear in Table 20.1. The exact value for Z(K) as obtained from the Cottrell equation is shown here also. [The Cottrell equation may be written

$$i(t) = \frac{nFACD^{1/2}}{\pi^{1/2}t^{1/2}} = \frac{nFACD^{1/2}}{\pi^{1/2}t^{1/2}}\left(\frac{L}{K - \frac{1}{2}}\right)^{1/2}$$

by employing Equation 20.16. Rearrangement of this expression results in the determination of an exact expression for Z(K):

```
100 REM  COTTRELL.BAS - SIMULATES COTTRELL CURRENT; UP TO 1000 ITERATIONS
110 REM                    SET UP CONCENTRATION AND CURRENT-TIME ARRAYS
120 DIM FAOLD(136), FBOLD(136), FANEW(136), FBNEW(136), Z(1001), T(1001)
130 INPUT "ENTER A VALUE FOR L, THE NUMBER OF ITERATIONS [UP TO 1000]: "; L
140 PRINT
150 INPUT "ENTER THE MODEL DIFFUSION COEFFICIENT OF A [UP TO 0.5]: "; DMA
160 PRINT
170 INPUT "ENTER THE MODEL DIFFUSION COEFFICIENT OF B [UP TO 0.5]: "; DMB
180 PRINT:PRINT:PRINT
200 REM                      INITIALIZE VOLUME ELEMENT ARRAYS FOR TIME = 0
210 FOR J=1 TO 136
220 FAOLD(J)=1!
230 FBOLD(J)=0!
240 NEXT J
300 REM              FIRST ITERATION: CALCULATE FIRST CURRENT-TIME POINT
310 K=1
320 FAOLD(1)=0!
330 FBOLD(1)=1!
340 T(1)=.5/L
350 Z(1)=SQR(L/DMA)
400 REM                   START OF ALL SUBSEQUENT ITERATIONS [K=2 TO K=L]
410 K=K+1
420 REM                    ELECTRODE BOUNDARY CONDITIONS FOR K=2 TO K=L
430 FANEW(1)=0!
440 FBNEW(1)=FBOLD(1)+DMA*FAOLD(2)-DMB*(FBOLD(1)-FBOLD(2))
500 REM              CALCULATE MAXIMUM VALUE OF J FOR GIVEN VALUE OF K
510 JMAX=3*SQR(2*K)+1
520 REM     NEW CONCENTRATIONS CALCULATED BY DIFFUSION ALGORITHM - EQ. 8
530 FOR J=2 TO JMAX
540 FANEW(J)=FAOLD(J)+DMA*(FAOLD(J+1)-2*FAOLD(J)+FAOLD(J-1))
550 FBNEW(J)=FBOLD(J)+DMB*(FBOLD(J+1)-2*FBOLD(J)+FBOLD(J-1))
560 NEXT J
600 REM                   CURRENT-TIME BEHAVIOR CALCULATED - EQ. 27; EQ. 16
610 T(K)=(K-.5)/L
620 Z(K)=DMA*FAOLD(2)*SQR(L/DMA)
700 REM          TRANSFORM NEW ARRAYS TO OLD ARRAYS FOR NEXT ITERATION
710 FOR J=1 TO JMAX
720 FAOLD(J)=FANEW(J)
730 FBOLD(J)=FBNEW(J)
740 NEXT J
800 REM                      STARTS NEXT ITERATION IF K DOES NOT EXCEED L
810 IF K=>L THEN 900 ELSE 400
900 REM               PRINT TIME - CURRENT DATA TO MONITOR IN TABULAR FORM
910 PRINT TAB(6) "K" TAB(20) "T(K)" TAB(40) "Z(K)"
920 PRINT TAB(6) "-" TAB(20) "----" TAB(40) "----"
930 PRINT
940 FOR K=1 TO L
950 PRINT TAB(5) K TAB(19) T(K) TAB(39) Z(K)
960 NEXT K
999 END
```

**Figure 20.3** Sample BASIC program.

**Table 20.1** Comparison of Simulated and Exact Current–Time Data Obtained from the Sample Program

| K | $T(K)=t/t_k$ | $Z(K) = \frac{i(t)t_k^{1/2}}{nFACD^{1/2}}$ (simulated) | $Z_c(K) = \frac{i(t)t_k^{1/2}}{nFACD^{1/2}}$ (exact) | Percent error |
|---|---|---|---|---|
| 1 | 0.0005 | 47.14 | 25.23 | 86.8 |
| 5 | 0.0045 | 8.684 | 8.410 | 3.26 |
| 10 | 0.0095 | 5.962 | 5.788 | 3.01 |
| 50 | 0.0495 | 2.548 | 2.536 | 0.47 |
| 100 | 0.0995 | 1.793 | 1.789 | 0.22 |
| 500 | 0.4995 | 0.7987 | 0.7983 | 0.05 |
| 1000 | 0.9995 | 0.5645 | 0.5643 | 0.04 |

$$Z_c(K) = \frac{i(t)t_k^{1/2}}{nFACD^{1/2}} = \frac{1}{\pi^{1/2}}\left(\frac{L}{K - \frac{1}{2}}\right)^{1/2} \tag{20.31}$$

which may be compared with Z(K) obtained in the simulation to give some idea of the reliability of the simulation at any point.]

A comparison of the simulated and exact values of Z(K) in Table 20.1 reveals that most of the error associated with the simulated current occurs during the first few iterations. Once K exceeds 50, however, the error in the simulated current is less than 1%, and if K exceeds 500, the error drops below 0.1%. This is well within the uncertainty associated with typical electrochemical measurements. The worst error occurs during the first iteration when Z(1) is set equal to $\sqrt{L/D_{MA}}$ . Since this choice is somewhat arbitrary, it may be eliminated by setting K = 1 in Equation 20.31 and using

$$Z(1) = \sqrt{\frac{2L}{\pi}} \tag{20.32}$$

in its place.

Typical concentration profiles obtained using a program similar to that given here are shown in Figure 20.4. The agreement between the results of the simulation (solid lines) and the exact solution (points) is obvious.

## VII. CHRONOCOULOMETRY

The dimensionless current parameter Z(K) may be shown as a function of the dimensionless time parameter $t/t_k$. This is illustrated in Figure 20.5. The mathematical form of this function is not obtained; instead, the function is evaluated

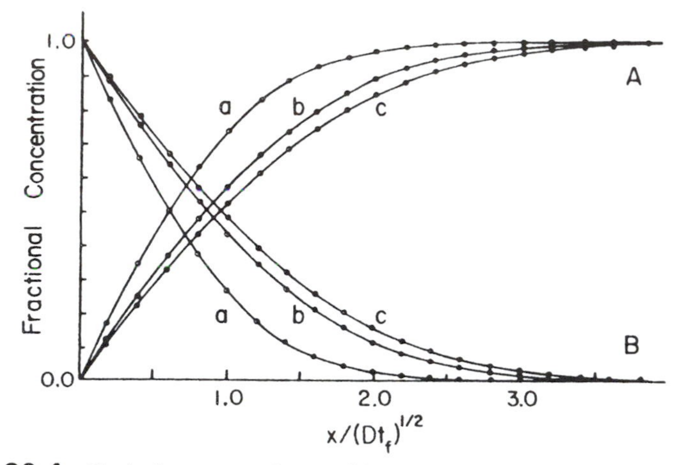


**Figure 20.4** Typical concentration profiles obtained using digital simulation. The conditions are identical to those employed in the sample program where the electrode reaction is A$\pm$ne $\rightarrow$ B. The lines represent the results of the simulation while the points show the exact solution obtained using partial differential equations. For a, b, and c the values of $t/t_f$ employed were 0.40, 0.80, and 1.00, respectively.

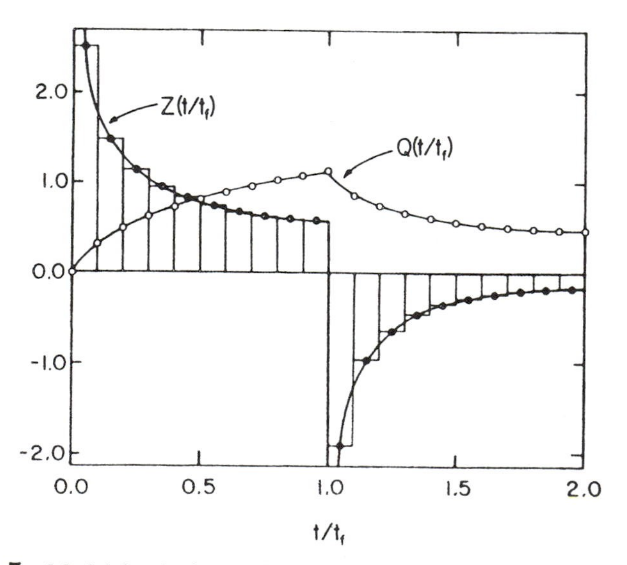


**Figure 20.5** Model for the integration of the dimensionless current. Note that while the dimensionless current $[Z(t/t_f) = i(t)_f^{1/2}/nFACD^{1/2}]$ is sampled at the midpoint of each item iteration, the dimensionless charge $[Q(t/t_f) = Q(t) / nFACD^{1/2}t_f^{1/2}]$ is evaluated at the end of each iteration.

at $t/t_k = 0.5/L$ and $L^{-1}$ intervals thereafter. This immediately suggests a method for obtaining the integral of the dimensionless current by summing the areas of the rectangular strips that make up the area under the curve. While $Z(t/t_k)$ is evaluated at the midpoint of each strip, the integral of $Z(t/t_k)$ will always be obtained at a time corresponding to the right edge of the strip (at integer numbers of time iterations).

Formally then,

$$\sum_{K=1}^{K=K_t} Z\left(\frac{K-\frac{1}{2}}{L}\right)L^{-1} = \int_0^{K_t/L} Z\left(\frac{t}{t_k}\right) d\left(\frac{t}{t_k}\right) = \int_0^{t/t_k} \frac{i(t/t_k)t_k^{1/2}}{nFACD^{1/2}} d\left(\frac{t}{t_k}\right) \tag{20.33}$$

where $K_t$ is the iteration number corresponding to time t. Since

$$Q(t) = \int_0^t i(t)\,dt$$

an appropriate change in the variable of integration results in

$$\sum_{K=1}^{K=K_t} Z\left(\frac{K-\frac{1}{2}}{L}\right)L^{-1} = \frac{Q(t)}{nFACD^{1/2}t_k^{1/2}} \tag{20.34}$$

Thus the sum of the areas of the rectangular strips making up $Z(t/t_k)$ form a dimensionless representation of the total charge passed at that time. In the discussions that follow, the dimensionless charge parameter that makes up the right member of Equation 20.34 is referred to (albeit ambiguously) as $Q(t/t_k)$ or $Q(K)$.

Dimensionless current data generated by the sample program just given may be integrated to obtain $Q(t/t_k)$. Some results are shown in Table 20.2 for selected values of $t/t_k$. These results may be compared with the expected value of $Q_c(K)$ obtained from the integration of the Cottrell equation:

$$Q_c(K) = \frac{\int_0^t (nFACD^{1/2}/\pi^{1/2}t^{1/2})\,dt}{nFACD^{1/2}t_k^{1/2}} = \frac{2}{\pi^{1/2}}\left(\frac{t}{t_k}\right)^{1/2} = \left(\frac{4K}{\pi L}\right)^{1/2} \tag{20.35}$$

This comparison is not nearly as encouraging as that in Table 20.1; even after 1000 iterations the simulated chronocoulometry is more than 2% in error. It is encouraging, however, that the absolute error is essentially constant after 10 iterations. This indicates that most of the error is introduced during the early stages of the simulation when the absolute error in the current is quite large. Therefore, some sort of fix-up at these early stages of the simulation is neces-

**Table 20.2** Comparison of Simulated and Exact Charge–Time Data Obtained from the Results of the Sample Program

| K | $T(K) = t/t_k$ | $Q(K) = \frac{Q(t)}{nFACD^{1/2}t_k^{1/2}}$ (simulated) | $Q_c(K) = \frac{Q(t)}{nFACD^{1/2}t_k^{1/2}}$ (exact) | Error | Percent error |
|---|---|---|---|---|---|
| 1 | 0.001 | 0.047 | 0.036 | 0.011 | 32.0 |
| 5 | 0.005 | 0.099 | 0.080 | 0.019 | 24.6 |
| 10 | 0.010 | 0.134 | 0.113 | 0.021 | 18.8 |
| 50 | 0.050 | 0.275 | 0.252 | 0.023 | 8.99 |
| 100 | 0.100 | 0.380 | 0.357 | 0.023 | 6.49 |
| 500 | 0.500 | 0.821 | 0.798 | 0.023 | 2.90 |
| 1000 | 1.000 | 1.152 | 1.128 | 0.023 | 2.09 |

sary if the accuracy of simulated charge–time curves is to approach that of simulated current–time curves. An immediate improvement may be obtained by substituting the exact value for Q(1) as determined from Equation 20.35:

$$Q(1) = \sqrt{\frac{4}{\pi L}} \tag{20.36}$$

Additional improvement may be noted if the simulated current–time curve is rendered more exact through the judicious selection of $D_{MA}$. This may be done by noting that unit relative concentration of A exists in the second volume element at the start of the second time iteration. Therefore, Z(2) as calculated in the simulation will be $\sqrt{LD_{MA}}$. This may be set equal to $L[Q_c(2) - Q_c(1)]$, as obtained from Equation 20.35 and solved for $D_{MA}$. Thus

$$\sqrt{LD_{MA}} = L\left[\left(\frac{8}{\pi L}\right)^{1/2} - \left(\frac{4}{\pi L}\right)^{1/2}\right]$$

and

$$D_{MA} = \left[\frac{2}{\pi^{1/2}}(\sqrt{2} - 1)\right]^2 = 0.2184\ldots \tag{20.37}$$

This value of $D_{MA}$ will render Q(2) to be exact, at no loss in accuracy for subsequent values of Z(K). This may be seen in Table 20.3, which shows values of Z(K) and Q(K) obtained using this value of $D_{MA}$. With these two most serious sources of error in Q(K) eliminated, errors no greater than 0.3% can be expected in the simulation of chronocoulometry if values of L in the vicinity of 1000 are employed.

**Table 20.3** Simulated Current–Time and Charge–Time Data Obtained from the Modified Sample Program[a]

| K | $T(K)=t/t_k$ [defined for Q(K)] | Z(K) | Percent error in Z(K) | Q(K) | Percent error in Q(K) |
|---|---|---|---|---|---|
| 1 | 0.001 | 25.23 | 0.00 | 0.036 | 0.00 |
| 5 | 0.005 | 8.555 | 1.72 | 0.080 | 0.63 |
| 10 | 0.010 | 5.840 | 0.90 | 0.114 | 0.77 |
| 50 | 0.050 | 2.540 | 0.16 | 0.253 | 0.54 |
| 100 | 0.100 | 1.790 | 0.06 | 0.358 | 0.43 |
| 500 | 0.500 | 0.7984 | 0.02 | 0.800 | 0.21 |
| 1000 | 1.000 | 0.5644 | 0.01 | 1.130 | 0.14 |

[a] $Z(1) = Z_c(1)$; $Q(1) = Q_c(1)$; $D_{MA} = 0.2184. \ldots$

## VIII. OTHER NERNSTIAN ELECTRODE BOUNDARY CONDITIONS

One of the more common electrochemical methods treated by digital simulation techniques is double-potential-step chronoamperometry [8,10]. In this method a product B is electrogenerated from a reactant A under diffusion-limited conditions for a period $t_f$ (the conditions of the sample program given earlier). The electrode potential is then reversed to a point such that B is converted back to A under diffusion-limited conditions also. The current so obtained gives a measure of the amount of B remaining in the vicinity of the electrode following electrogeneration and provides a very good measure of the stability of B to homogeneous reaction within the diffusion layer. Therefore, this method is suited to the study of homogeneous kinetics.

To model this method using digital simulation techniques, one need only change the electrode boundary conditions after some predetermined number of time iterations (representing $t_f$) have taken place. The electrode boundary conditions become

$$f'_A(1) = f_A(1) + D_{MB}f_B(2) - D_{MA}[f_A(2) - f_A(1)] \tag{20.38}$$

and

$$f'_B(1) = f_B(1) = 0 \tag{20.39}$$

which may be compared with Equations 20.18 and 20.19. The dimensionless current expression becomes

$$Z(K) = -\left(\frac{L}{D_{MA}}\right)^{1/2} D_{MB}f_B(2) \tag{20.40}$$

by comparison with Equation 20.29; note that currents generated under these conditions are opposite in sign to those obtained following the initial potential step. Current passed during this second half-cycle may be compared with the current at some known time during the first half-cycle to get some measure of the stability of B. Methods for simulating the instability of B are discussed later.

A different set of boundary conditions is obtained during the second half-cycle if the potential of the electrode is stepped to a value such that both B and A are converted to a third oxidation state (D) of the initial substance under diffusion-limited conditions. (A situation like this is encountered in the study of electrogenerated chemiluminescence when an aromatic hydrocarbon is first reduced to its anion radical, and then both the parent and the anion radical are oxidized to the cation radical at the same electrode.) In this case the electrode boundary conditions are

$$f'_A(1) = f_A(1) = f'_B(1) = f_B(1) = 0 \tag{20.41}$$

and

$$f'_D(1) = f_D(1) + D_{MA} f_A(2) + D_{MB} f_B(2) - D_{MD}[f_D(2) - f_D(1)] \tag{20.42}$$

The dimensionless current now results from the flux of two different materials,

$$Z(K) = \left(\frac{L}{D_{MA}}\right)^{1/2} [n_A D_{MA} f_A(2) + n_B D_{MB} f_B(2)] \tag{20.43}$$

where $n_A$ and $n_B$ now take into account the differing number of faradays transferred per mole of A and B. In general, the application of the principles just developed to other multicomponent systems is equally straightforward.

In the event that the electrode potential is stepped to a level that does not bring about diffusion-limiting conditions, the relative concentration of A and B at the electrode surface must be calculated. If the reaction

$$A + ne \rightleftarrows B$$

is reversible, the ratio of the electrode surface concentration of A to that of B is given by the Nernst equation

$$\theta = \frac{[A]_{x=0}}{[B]_{x=0}} = \frac{f_A(1)}{f_B(1)} = \exp\left[\frac{nF}{RT}(E - E^{o\prime})\right] \tag{20.44}$$

where all quantities have their usual electrochemical significance. Given the necessary potential parameters, one may calculate $\theta$. This may be used in the simulation to calculate a new electrode surface concentration ratio from any set of surface concentrations:

$$\theta = \frac{f'_A(1)}{f'_B(1)} = \frac{f_A(1) - \Delta f_A(1)}{f_B(1) + \Delta f_A(1)} \tag{20.45}$$

The change of fractional surface A concentration (which is proportional to the current) may be obtained by rearranging Equation 20.45:

$$\Delta f_A(1) = \frac{f_A(1) - \theta f_B(1)}{\theta + 1} \tag{20.46}$$

In the simulation it is probably best to allow diffusion to occur at the electrode surface before using Equation 20.46 to adjust the surface concentrations:

$$\begin{aligned} f'_A(1) &= f_A(1) + D_{MA}[f_A(2) - f_A(1)] \\ f'_B(1) &= f_B(1) + D_{MB}[f_B(1) - f_B(2)] \end{aligned} \tag{20.47}$$

These new surface concentrations of A and B may then be used to calculate $\Delta f_A(1)$, which may be used to achieve the desired concentration ratio,

$$\begin{aligned} f''_A(1) &= f'_A(1) - \Delta f_A(1) \\ f''_B(1) &= f'_B(1) + \Delta f_A(1) \end{aligned} \tag{20.48}$$

The current is calculated from the quantity $\Delta f_A(1)$ also,

$$Z(K) = -\left(\frac{L}{D_{MA}}\right)^{1/2} \Delta f_A(1) \tag{20.49}$$

rather than from the flux into the first element. This same general procedure is used to simulate linear sweep voltammetry (when $\theta$ is a functional of time).

## IX. HOMOGENEOUS KINETICS

The real power of digital simulation techniques lies in their ability to predict current–potential–time relationships when the reactants or products of an electrode reaction participate in some intervening chemical reaction. These "kinetic complications" often result in a fairly difficult differential equation (when combined with the conditions for diffusion or convection encountered in electrochemical problems) that resists solution by ordinary means. Through simulation, however, the effect of any number of chemical steps may be predicted. In practice, it is best to limit these predictions to cases where the reactants and products participate in one or two rate-determining steps; each independent step adds another dimensionless kinetics parameter that must be varied over the range of experimental significance in order to predict the effect of that step. The method for obtaining these dimensionless kinetic parameters will be discussed in this section; their use in the simulation is also illustrated.

Consider the electrode reaction

$$A \pm ne \rightleftarrows B \tag{20.50}$$

where the product B undergoes subsequent chemical reaction. This may be a simple first-order (or pseudo-first-order) decomposition

$$B \xrightarrow{k_1} C \tag{20.51}$$

where C may be electroactive (as in the ECE mechanism) or may not be electroactive (as in the EC mechanism) at the potential at which B is formed. Equation 20.50 represents a reaction taking place at the electrode surface; its effect on electrode boundary conditions in the simulation has already been discussed. Equation 20.51 represents a homogeneous reaction that occurs in each volume element of the diffusion layer array. In order to account for the effect of Equation 20.51 on each element of the array, one must find an expression for the fractional change in the concentration of B due to the chemical reaction. This is obtained from the rate law for Equation 20.51,

$$-\frac{d[B]}{dt} = \frac{d[C]}{dt} = k_1[B]$$

by dividing through by the bulk concentration and rewriting the expression in finite difference form:

$$-\frac{\Delta f_B(J)}{\Delta t} = \frac{\Delta f_C(J)}{\Delta t} = k_1 f_B(J)$$

Rearrangement and combination with Equation 20.14 yields

$$-\Delta f_B(J) = \Delta f_C(J) = k_1 \Delta t f_B(J) = \frac{(k_1 t_k)}{L} f_B(J) \tag{20.52}$$

Introduced as a factor in the last term of Equation 20.52 is the dimensionless first-order kinetics parameter $k_1 t_k$. Variations in this parameter will allow the effect of any first-order reaction to be simulated.

In the simulation of the EC mechanism, a double potential step may be employed; these boundary conditions have been discussed previously. Diffusion would be modeled in the usual way. Following the diffusion step, however, the concentration in each element of the resulting array is adjusted according to Equation 20.52 using the input parameter $k_1 t_k$:

$$f'_B(J) = f_B(J)\left(\frac{(k_1 t_k)}{L}\right)$$
$$f'_C(J) = f_C(J) + \frac{(k_1 t_k)}{L} f_B(J) \tag{20.53}$$

This process is repeated throughout the simulation, and the current obtained at some known time during the first half-cycle is compared with that obtained at some time during the second. The simulation is then repeated and a new value is used for the kinetics parameter $k_1 t_k$; another comparison of currents is made. The variation of the current ratio with values of $k_1 t_f$ when displayed graphically is known as a working curve. Typical working curves are shown in Figure 20.6. These working curves may be compared with experimental data in an attempt to elucidate the mechanism and evaluate the rate constant $k_1$.

The simulation of the ECE mechanism may also employ the double-potential-step technique, but a working curve can be constructed from single-potential-step data also. This is because some of the current that passes, as A is converted to B, is due to the electrolysis of C, the decomposition product of B. The greater the decomposition rate of B, the more current flows, approaching the rate of

$$A \pm (n_A + n_C)e \rightleftarrows D$$

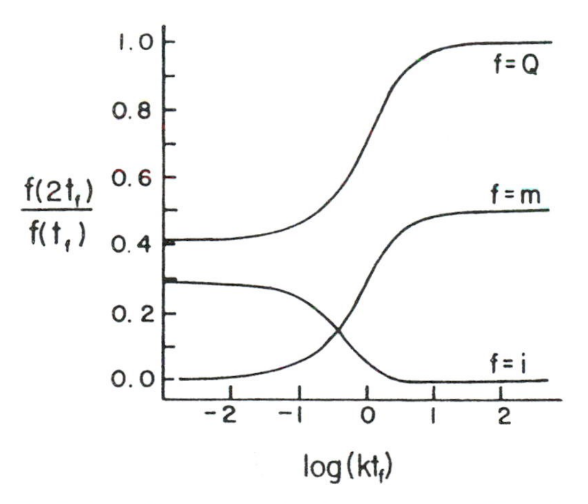


**Figure 20.6** Typical working curves for the first-order EC mechanism. The curve shown for f = i shows the behavior of the current ratio $[i(2t_f)/i(t_f)]$ as a function of $kt_f$. That given for f = Q shows the charge ratio, while that given for f = m shows the current semi-integral ratio over the same range. [From Ref. 10.]

in the limit of fast kinetics (where D is the product of the electrode reaction of C). The treatment of the electrode boundary condition in this case is along the same lines as the treatment of multicomponent systems given earlier:

$$
\begin{aligned}
f'_A(1) &= f_A(1) = 0 \\
f'_B(1) &= f_B(1) + D_{MA} f_A(2) - D_{MB}[f_B(1) - f_B(2)] \\
f''_B(1) &= f'_B(1)\left[1 - \frac{(k_1 t_k)}{L}\right]
\end{aligned}
\tag{20.54}
$$

$$
\begin{aligned}
f'_C(1) &= 0 \\
f'_D(1) &= f_D(1) + D_{MC} f_C(2) - D_{MB}[f_B(1) - f_B(2)] \\
f''_D(1) &= f'_D(1) + \frac{(k_1 t_k) f'_B(1)}{L}
\end{aligned}
\tag{20.55}
$$

Equations 20.54 and 20.55 result from the fact that the ongoing chemical reaction occurs at the electrode surface also; since this does take place at the electrode, however, the decomposition product of B is immediately converted to D. This flux must be accounted for in the determination of the dimensionless current parameter for that iteration,

$$
Z(K) = \left(\frac{L}{D_{MA}}\right)^{1/2}\left[D_{MA} f_A(2) + D_{MC} f_C(2) + \frac{(k_1 t_k) f'_B(1)}{L}\right] \tag{20.56}
$$

The treatment of second-order kinetics is equally feasible. Given the chemical reaction

$$
B + A \xrightarrow{k_2} D
$$

one may write the rate law

$$
-\frac{d[B]}{dt} = -\frac{d[A]}{dt} = \frac{d[D]}{dt} = k_2[A][B]
$$

and the corresponding finite difference representation

$$
-\frac{\Delta f_B(J)}{\Delta t} = -\frac{\Delta f_A(J)}{\Delta t} = \frac{\Delta f_D(J)}{\Delta t} = k_2 C f_A(J) f_B(J) \tag{20.57}
$$

where C represents bulk concentration of A. This may be rearranged with Equation 20.14 to yield

$$-\Delta f_B(J) = -\Delta f_A(J) = \Delta f_D(J) = (k_2 C\ \Delta t) f_A(J) f_B(J)$$
$$= \frac{(k_2 C t_k)}{L} f_A(J) f_B(J) \tag{20.58}$$

which shows $k_2Ct_k$ as the proper dimensionless kinetics parameter for a second-order reaction. This would be supplied as an input parameter, and kinetically adjusted concentrations would be determined as indicated previously:

$$\begin{aligned} f'_A(J) &= f_A(J)\left[1 - \frac{(k_2Ct_k)f_B(J)}{L}\right] \\ f'_B(J) &= f_B(J)\left[1 - \frac{(k_2Ct_k)f_A(J)}{L}\right] \\ f'_D(J) &= f_D(J)\frac{(k_2Ct_k)f_A(J)f_B(J)}{L} \end{aligned} \tag{20.59}$$

No special treatment like Equation 20.54 is necessary in the calculation of the surface condition, even if D is assumed electroactive, if diffusion-limited conditions are obtained; since $f_A(1) = 0$ under these circumstances, no conversion of B to D is possible in the first volume element.

Dimerizations, which yield nonlinear partial differential equations if treated rigorously, may be handled with ease using digital simulation techniques. In this case the reaction

$$2B \xrightarrow{k_2} D$$

yields the rate law

$$-\tfrac{1}{2}\frac{d[B]}{dt} = \frac{d[D]}{dt} = k_2[B]^{2D}$$

which provides the finite difference form used in the simulation,

$$-\Delta f_B(J) = 2\ \Delta f_D(J) = 2\frac{(k_2Ct_k)}{L} f_B(J)^2 \tag{20.60}$$

This furnishes the method for determining the kinetically adjusted concentrations of B and D:

$$\begin{aligned} f'_B(J) &= f_B(J)\left[1 - \frac{2(k_2Ct_k)f_B(J)}{L}\right] \\ f'_D(J) &= f_D(J)\frac{(k_2Ct_k)f_B(J)^2}{L} \end{aligned} \tag{20.61}$$

One admonition is necessary concerning the maximum kinetics parameter that may be employed in any simulation. As is evident in Equations 20.53, 20.59, and 20.61, if too large a value is used for $k_1t_k$ or $k_2Ct_k$, a negative concentration may result. Since this is impossible, any values of these kinetics parameters that yield negative concentrations may be regarded as too large for treatment by these methods. In the case of first-order kinetics, the maximum rate that can be handled occurs when

$$k_1t_k = L \tag{20.62}$$

Any values of $k_1t_k$ in excess of this will produce erroneous results. Similarly, the upper limit of $k_2Ct_k$ is approached when

$$k_2Ct_k = L \tag{20.63}$$

(or L/2 in the case of Eq. 20.61). Values higher than this should be attempted only when some check for negative concentrations is built into the simulation. It is of interest to note that in the case of a second-order reaction, this restriction defines the rate of diffusion control for the purpose of the simulation:

$$k_d = \frac{L}{Ct_k} \tag{20.64}$$

Substituting in "typical" electrochemical values for L, C, and $t_k$, one obtains $k_d \cong 10^8$ L/(mol s). Comparison of this result with the equation of Osborne and Porter [11] shows that diffusion control in the simulation occurs at a rate two orders of magnitude less than predicted by accepted theory. In other words, one cannot distinguish any difference between the effects of second-order rate constants in the vicinity of $10^8$ L/(mol s) from those in the vicinity of $10^{10}$ L/(mol s) by using digital simulation.

## X. PARAMETRIC SUBSTITUTIONS

The definition of a second time-dependent variable (such as $k_1t_k$ or $k_2Ct_k$) within a simulation permits one to have some flexibility in the way other time-dependent quantities are displayed. For example, in the case where no kinetic complications are introduced, one has no choice but to reference all times with respect to some known time in the physical experiment (Eq. 20.16). With the introduction of $k_1t_k$ or $k_2Ct_k$, however, experimental times may be referenced with respect to $k_1^{-1}$ or $(k_2C)^{-1}$. That is,

$$\frac{t}{t_k} k_1t_k = k_1t = k_1t_k \frac{K - \frac{1}{2}}{L} \tag{20.65}$$

Thus one may obtain kt by multiplying the quantity previously referred to as the dimensionless time by $k_1t_f$, the dimensionless rate constant. This is particularly useful in constructing the single-potential-step working curve for the ECE mechanism mentioned earlier. This parametric substitution allows the experimental time to be rendered dimensionless by the inverse of the rate constant instead of by some known time $t_k$.

Similarly, the dimensionless current may be expressed in terms of $k_1^{-1}$ instead of $t_k$. That is,

$$\frac{i(t)t_k^{1/2}}{nFACD^{1/2}}(k_1t_k)^{-1/2} = \frac{i(t)}{nFACk_1^{1/2}D^{1/2}} = \frac{Z(K)}{\sqrt{k_1t_k}} \tag{20.66}$$

This parametric substitution allows the current to be rendered dimensionless by a quantity that is independent of the time of the physical experiment (all the factors of $nFACk_1^{1/2}D^{1/2}$ are constant for a given physical system). A typical working curve for the ECE mechanism obtained using these parametric substitutions is illustrated in Figure 20.7.

It must be emphasized that no new information is obtained through these parametric substitutions. Rather, the information that comes from the simula-

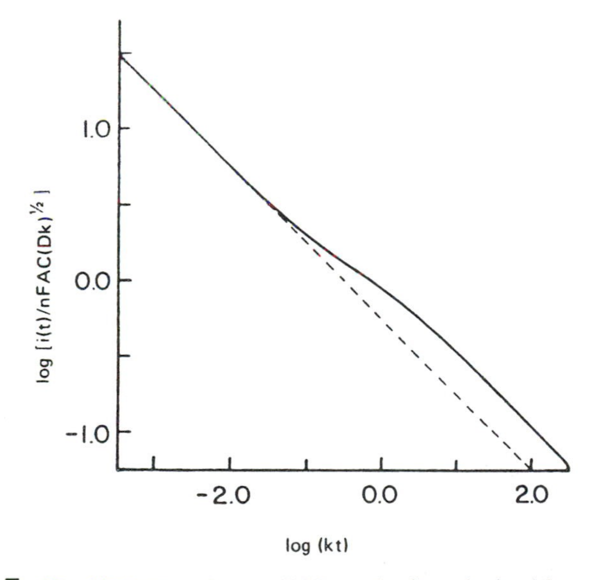


**Figure 20.7** Working curve for the ECE mechanism obtained by parametric substitution. Note that this represents the current–time behavior over five time decades. The slope of the dashed line is −1/2, and the separation between it and the parallel working curve immediately above it is 0.301 logarithmic units.

tion is presented in an alternative form that may be more understandable or more readily compared with experimental data. Nor are the parametric substitutions discussed herein the only ones possible. The variety of possible substitutions is greatly increased when two different kinetic processes are considered in the model. The consideration of these possibilities is left as an exercise to the reader.

## XI. ELECTROGENERATED CHEMILUMINESCENCE

No discussion of digital simulation would be complete without some mention of the methods employed in treating electrogenerated chemiluminescence (ECL). In the most common statement of the problem, an aromatic hydrocarbon (A) is alternately oxidized and reduced at a single electrode to produce its radical anion and cation species (B and D). These species, in turn, react within the diffusion layer to regenerate the hydrocarbon.

$$B + D \xrightarrow{k_2} 2A \tag{20.67}$$

The boundary conditions and the necessary kinetic considerations have been introduced previously and are not considered here. However, some fraction ($\phi_{ECL}$) of the radical ions reacting according to Equation 20.67 produces a molecule of A in an excited state that goes on to produce a photon. The sum of this photon flux originating in the diffusion layer may be related to I(t), the intensity of ECL emission.

To define an ECL intensity parameter, one first notes that the photon flux from an individual volume element is given by $\phi_{ECL}(A\ \Delta x)C(k_2Ct_k)f_B(J)f_D(J)/L$. The total ECL intensity during any iteration is given by

$$I(t) = \frac{\phi_{ECL} A\ \Delta x C(k_2 C t_k)}{\Delta t\ L} \sum_J f_B(J) f_D(J) \tag{20.68}$$

where the sum is taken over all volume elements. Substitution of Equations 20.14 and 20.17 into Equation 20.68 and rearrangement yields a dimensionless representation of ECL intensity,

$$I(t) = \frac{\phi_{ECL} A\ \Delta x C(k_2 C t_k)}{\Delta t\ L} \sum_J f_B(J) f_D(J) \tag{20.69}$$

where I(t) is expressed in einsteins/s. This dimensionless intensity factor may be rendered independent of $t_k$ through an appropriate parametric substitution [division of both members of Eq. 20.69 by $(k_2Ct_k)^{1/2}$]. Numerous examples of digital simulations of ECL problems exist in the literature [1,2,12–14].

## XII. CHRONOPOTENTIOMETRY

As indicated previously, in the simulation of chronopotentiometric behavior, one writes the computer program to calculate L, the number of iterations associated with $\tau$, the transition time. This is achieved by defining a constant current $i_c$ such that

$$Z(K) = \frac{i_c \tau^{1/2}}{nFACD^{1/2}} = \left(\frac{L}{D_M}\right)^{1/2} \delta = \text{constant} \tag{20.70}$$

by analogy with Equation 20.28. The fractional concentration $\delta$, a defined constant, is simply subtracted from $f_A(1)$ and added to $f_B(1)$ during each iteration to simulate constant-current flow:

$$\begin{aligned} f'_A(1) &= f_A(1) + D_{MA}[f_A(2) - f_A(1)] - \delta \\ f'_B(1) &= f_B(1) - D_{MB}[f_B(1) - f_B(2)] + \delta \end{aligned} \tag{20.71}$$

Because the transition time corresponds to that time at which the concentration of A reaches zero at the electrode surface, a record is kept of the number of iterations required to achieve this end. This represents L in Equation 20.70 and, combined with input parameters $D_M$ and $\delta$, allows the computation of Z(K) at the conclusion of the simulation.

During the course of the simulation, the most important variables are the electrode surface concentrations of A and B because they determine E(t). These may be calculated directly by placing the electrode in the center of the first volume element in the model as in previous simulations. In this case, however, it turns out to be more straightforward to place the electrode at the exterior edge of the first volume and to calculate the electrode surface concentrations by extrapolating the concentration profiles to $x = 0$. This is illustrated in Figure 20.8. The extrapolation is made easier by the fact that one boundary condition in constant-current electrolysis requires that the concentration gradient at the electrode surface be constant:

$$\frac{\partial C(0,t)}{\partial x} = \frac{i_c}{nFAD} = \text{constant}$$

Written in finite difference form this becomes

$$\frac{\Delta f_A}{\Delta x} = \frac{i_c}{nFACD} = \text{constant} \tag{20.72}$$

Equation 20.72 may be solved for $\Delta f_A$, the change in fractional A concentration over one volume element, and combined with Equation 20.17:

$$\Delta f_A = \frac{i_c \ \Delta x}{nFACD} = \frac{i_c \tau^{1/2}}{nFACD^{1/2}} (D_M L)^{-1/2} = \frac{\delta}{D_M} \tag{20.73}$$

Inspection of Figure 20.8 reveals that the distance of extrapolation is one-half of a volume element. Therefore, the electrode surface concentration is given by

$$f_A(x=0) = f_A(1) - \frac{\delta}{2D_M}$$
$$f_B(x=0) = f_B(1) + \frac{\delta}{2D_M} \tag{20.74}$$

These surface concentrations may be used in the determination of L or in the calculation of E(t) by means of the Nernst equation.

Some comments about the size of the input parameter δ are necessary. If δ is too large, not enough iterations will take place before the transition time to make the simulation accurate. If δ is too small, a great deal of computer time will be wasted reaching the transition point. A useful technique is to estimate the number of iterations that one would like to represent the transition time. One may then use this estimate to calculate δ by means of the Sand equation:

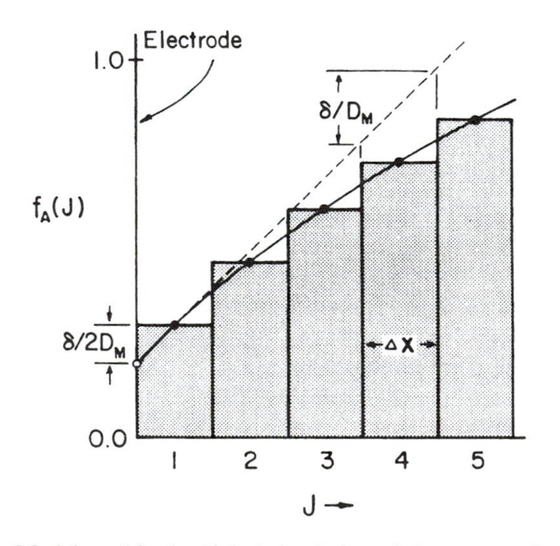


**Figure 20.8** Model used in the digital simulation of chronopotentiometric behavior. A constant amount (δ) is removed from the contents of the first element following the diffusion step. The electrode surface concentration is determined by extrapolating the line of known slope through the point representing first element concentration to the electrode surface.

$$\frac{i_c \tau^{1/2}}{nFACD^{1/2}} = \frac{\pi^{1/2}}{2} = \left(\frac{L'}{D_M}\right)^{1/2} \delta \tag{20.75}$$

where L′ represents the estimated value for L. Rearrangement yields

$$\delta = \sqrt{\frac{\pi D_M}{4L'}} \tag{20.76}$$

In kinetically uncomplicated simulations of chronopotentiometric behavior, the value of L that is calculated during the simulation is found to agree with L′ within a fraction of one iteration out of a thousand. The corresponding simulated potential-time behavior is generally accurate within 0.1 mV.

## XIII. LINEAR SWEEP AND CYCLIC VOLTAMMETRY

One can also use these methods to model linear sweep and cyclic voltammetry. If one wishes to consider only the effect of homogeneous chemical kinetics upon the resulting Nernstian voltammogram, one need only modify the electrode boundary condition to generate surface concentrations of A and B that vary according to the Nernst equation in the potential scan. In practice, however, most voltammetry also depends on heterogeneous electron transfer kinetics; thus, the Nernstian boundary condition is the exception rather than the rule. This combination of homogeneous kinetics, heterogeneous kinetics, and mass transport often results in a sequence of events taking place on vastly different time scales. These "stiff" problems present a computational challenge that is beyond the scope of this work.

To see how one might incorporate the effect of a changing electrode potential into the boundary value problem, let us consider the simulation of Nernstian linear sweep voltammetry. One first selects an appropriate potential scan range

$$\Delta E = E_1 - E_2 \tag{20.77}$$

where $E_1$ and $E_2$ are the initial and final potentials (vs. $E^{\circ\prime}$ for the electrode reaction) to be achieved, with $E_1$ the more positive of the two. The combination of $\Delta E$ with $\nu$, the potential scan rate, yields the known time parameter for the simulation,

$$t_k = \frac{\Delta E}{\nu} \tag{20.78}$$

For a potential scan in the negative direction then,

$$E = E_1 - \nu t = E_1 - \nu \frac{K - \frac{1}{2}}{L} t_k = E_1 - \nu \frac{K - \frac{1}{2}}{L} \Delta E \tag{20.79}$$

so that the potential scan range is seen to have been partitioned into L equal parts.

Equation 20.79 may be combined with Equation 20.44 to calculate

$$\theta = \frac{f_A(1)}{f_B(1)} = \exp\left[\frac{nF}{RT}\left(E_1 - \frac{K - \frac{1}{2}}{L} \Delta E - E^{\circ\prime}\right)\right] \tag{20.80}$$

where $\theta$ now depends on K. This time-dependent $\theta$ may be used to calculate $\Delta f_A(1)$ according to Equation 20.46, which in turn may be used to calculate the dimensionless current according to Equation 20.49. In this case, however, the dimensionless current is given by

$$Z(K) = \frac{i(t)\nu^{1/2}}{nFACD^{1/2}(\Delta E)^{1/2}} \tag{20.81}$$

in view of Equation 20.78. Examples of simulated linear sweep voltammetry have been reported in the literature [15].

Since the publication of the first edition, a number of authors have addressed those issues of mass transport and competitive heterogeneous and homogeneous kinetics that make the simulation of voltammetric phenomena so difficult [16–20]. Some of these [16,17] have employed the explicit finite difference method that has been used in this chapter. Others have used nonlinear spatial elements and implicit finite difference methods [18–20] to handle the "stiff" problems that sometimes result when one takes both homogeneous and heterogeneous kinetics into account in a digital simulation. A particularly good discussion of the differences between explicit and implicit methods from proponents of the implicit method may be found in the current literature [21]. An impressive exposition of the explicit method in voltammetry has also been presented [22]. Each of these approaches is even represented with a commercial software package that allows the user to investigate the virtues of each method without having to write a program. A copy of CVSIM, Gosser's explicit finite difference software, is included with his text [22]. Rudolph's DigiSim™ [23], which employs the implicit finite difference method, is available commercially through Bioanalytical Systems, Inc. (West Lafayette, IN). Readers who believe they might be able to investigate complex competitive kinetics using cyclic voltammetry are encouraged to compare the results obtained from each of these software products with their experimental results.

## XIV. SIMULATION OF ROTATING DISK HYDRODYNAMICS

The study of rotating disk electrode behavior provides a unique opportunity to develop a model that predicts the effect of diffusion and convection on the current. This is one of the few convective systems that have simple hydrodynamic equations that may be combined with the diffusion model developed herein to produce meaningful results. The effect of diffusion is modeled exactly as it has been done previously. The effect of convection is treated by integrating an approximate velocity equation to determine the extent of convective flow during a given $\Delta t$ interval. Matter, then, is simply transferred from volume element to volume element in accord with this result to simulate convection. The whole process repeated results in a steady-state concentration profile and a steady-state representation of the current (the Levich equation).

In rotating-disk hydrodynamics, an approximate equation describing the velocity of the fluid in the vicinity of the electrode is given by

$$v_x = \frac{dx}{dt} = -0.51\sqrt{\frac{\omega^3}{\nu}}x^2 \tag{20.82}$$

where $\omega$ is the angular velocity of the rotating electrode, $\nu$ is the kinetic viscosity of the solution, and x is measured on an axis perpendicular to the rotating disk. Integration of Equation 20.82 yields

$$\int_{x_2}^{x_1} \frac{dx}{x^2} = \frac{1}{x_2} - \frac{1}{x_1} = -0.51\sqrt{\frac{\omega^3}{\nu}} \int_{t_1}^{t_2} dt = -0.51\sqrt{\frac{\omega^3}{\nu}}\ \Delta t \tag{20.83}$$

where $x_2$ represents the position of a given volume element of fluid initially (at $t_1$) and $x_1$ represents the position of that same fluid after the interval $\Delta t$. Equation 20.83 may be solved for $x_2$ to obtain an expression for the initial position of the fluid volume element that resides at $x_1$ at the conclusion of the $\Delta t$ interval:

$$x_2 = \frac{x_1}{1 - 0.51\sqrt{\omega^3/\nu}\ \Delta t x_1} \tag{20.84}$$

Substitution of Equations 20.14 and 20.26 into Equation 20.84 results in

$$J_2 - 1 = (J_1 - 1)\Bigg/\left(1 - 0.51\sqrt{\frac{\omega^3 D t_k^3}{\nu}}\ \frac{J_1 - 1}{\sqrt{D_M L^3}}\right) \tag{20.85}$$

where $J_1$ and $J_2$ are volume element numbers corresponding to $x_1$ and $x_2$. Examination of Equation 20.85 reveals a simplification that occurs if

$$t_k \equiv \left(\frac{\nu}{D}\right)^{1/2} \omega^{-1} \tag{20.86}$$

for this simulation; Equation 20.85 then becomes

$$J_2 - 1 = \frac{J_1 - 1}{1 - 0.51(J_1 - 1)/\sqrt{D_M L^3}} \tag{20.87}$$

Thus given the volume element $J_1$, one may calculate the serial number of that volume element $J_2$ that supplies material to element $J_1$ during one iteration. In theory, one then simply transfers the contents of element $J_2$ to element $J_1$ in order to simulate convection.

Of course, it is not quite that simple. Given an integer value for $J_1$, it is quite unlikely that an integer will be obtained for $J_2$. Thus some sort of interpolation scheme must be devised to obtain the relative concentrations of the species of interest at noninteger values of $J_2$; these concentrations are then transferred to element $J_1$. In a typical program, this process would take place after the diffusion step in any iteration but before any kinetic perturbations are applied in that iteration. Care must be exercised to transfer to interior elements first so that one does not write over a meaningful value in the array.

Finally, it should be noted that the choice of $t_k$ indicated in Equation 20.86 provides the proper hydrodynamic expression for Z(K). The steady-state value of Z(K) obtained by this method (with L = 1000) is 0.61. That is,

$$\lim_{K \to \infty} Z(K) = \frac{i_{lim} \nu^{1/6}}{nFACD^{5/6} \omega^{1/2}} = 0.61 \tag{20.88}$$

This expression is identical to the Levich equation within the reliability of the simulation.

## XV. SIMULATION OF ROTATING RING DISK BEHAVIOR

The methods developed for rotating hydrodynamics are easily extended to ring-disk electrodes. This extension is facilitated by the fact that one needs only to consider radial convection in addition to linear connection and diffusion to extend disk methods to ring disk problems. This does require some model development, however, because mass transport is now permitted in two directions, x and r. Thus a linear volume element array is no longer sufficient, and a two-dimensional volume element array is required. The linear axis of this array is identical to that developed above for linear diffusion and convection. Thus linear distances are still expressed in volume element units,

$$x = (J_x - 1)\Delta x \tag{20.89}$$

Radial distances are expressed in a similar fashion:

$$r = (J_r - 1)\ \Delta r \tag{20.90}$$

where $\Delta r$ is defined in terms of $r_c$, the radial distance to the outermost edge of the ring electrode,

$$\Delta r = \frac{r_c}{J_c - \frac{1}{2}} \tag{20.91}$$

where $J_c$ is the serial number associated with the outermost ring volume element (*ca.* 100). Other radial dimensions such as $r_a$, the radius of the disk, and $r_b$, the radial distance to the innermost edge of the ring, may be expressed in terms of Equations 20.90 and 20.91.

The approximate equation describing radial convection in rotating-disk and ring-disk hydrodynamics is

$$v_r = \frac{dr}{dt} = 0.51\sqrt{\frac{\omega^3}{\nu}}xr \tag{20.92}$$

This may be integrated just as Equation 20.82 was to obtain an expression for the position of a radial fluid volume element at the end of a $\Delta t$ interval with respect to its position at the start of the interval:

$$\ln\frac{r_2}{r_1} = 0.51\sqrt{\frac{\omega^3}{\nu}}x\ \Delta t \tag{20.93}$$

where $r_1$ represents the position at the start of the interval and $r_2$ the position at the end. Substitution of Equations 20.14, 20.17, 20.86, 20.89, and 20.90 into Equation 20.93 yields

$$\ln\frac{J_{r_2} - 1}{J_{r_1} - 1} = 0.51\frac{J_x - 1}{\sqrt{D_M L^3}} \tag{20.94}$$

This may be rearranged to permit one to calculate the radial position of a given fluid volume element at the start of an iteration given the linear and radial position of that element at the conclusion of the iteration:

$$J_{r_1} - 1 = \frac{J_{r_2} - 1}{\exp[0.51(J_x - 1)/\sqrt{D_M L^3}\,]} \tag{20.95}$$

Equation 20.95 represents the algorithm for simulating radial convection. Fractional concentrations, now a function of $J_x$ and $J_r$, found in the element

$(J_x, J_{r_1})$ are transferred to the element $(J_x, J_{r_2})$ according to Equation 20.95. Since $J_{r_2}$ need not be an integer, some interpolation scheme must be developed to calculate $f_A(J_x, J_{r_2})$ and $f_B(J_x, J_{r_2})$, but this is accomplished easily. Some care must be exercised to transfer concentrations to outermost radial elements first to maintain the proper radial concentration gradient. And, of course, all volume elements must be subjected to the same treatment for linear convection and linear diffusion that was developed in the preceding section. The digital simulation technique does permit one to handle this fairly complicated problem of linear diffusion, linear convection, and radial convection with relative ease. Of course, the effects of homogeneous kinetics may be superimposed on this problem of mass transport by using the methods developed previously. Examples of simulations that predict the effect of kinetic perturbation on rotating ring-disk electrode behavior have appeared in the literature [2,24–28].

## XVI. THE STEADY-STATE ASSUMPTION

Many problems involving competitive reaction kinetics may be treated by invoking the steady-state assumption within the digital simulation; this has been done in at least two instances [29–34]. The first of these involves the development of a model for enzyme catalysis in the amperometric enzyme electrode [29–31]. In this model, the enzyme E is considered to be immobilized in a diffusion medium covering an electrode that is operated at a fixed potential such that the product (P) of enzyme catalysis is electroactive under diffusion-controlled conditions. (This model has also served as the basis for the simulation of the voltammetric response of the enzyme electrode [35].) The substrate (S) diffuses through the medium that contains the immobilized enzyme and is catalyzed to form P by straightforward enzyme kinetics:

$$S + E \underset{k_2}{\overset{k_1}{\rightleftarrows}} ES \xrightarrow{k_3} P \tag{20.96}$$

The membrane containing the immobilized enzyme is handled by partitioning it into a specified number of volume elements so that Equations 20.23 and 20.24 are valid in this model. While the concentration of each species may vary from element to element, the steady-state assumption (d[ES]/dt = 0) may be invoked independently for each volume element. This results in the definition of the Michaelis–Menton constant, $K_M$:

$$K_M = \frac{k_2 + k_3}{k_1} = \frac{[E][S]}{[ES]} \tag{20.97}$$

The rate of product formation (and substrate loss) is given by

$$\frac{d[P]}{dt} = k_3[ES] \tag{20.98}$$

If $C_E$ is taken as the total enzyme concentration in each volume element so that

$$C_E = [E] + [ES] \tag{20.99}$$

a differential rate law may be formulated by combining Equations 20.96–20.98,

$$\frac{d[P]}{dt} = \frac{d[S]}{dt} = \frac{k_3 C_E}{(K_M/[S]) + 1} \tag{20.100}$$

In finite difference form this becomes

$$\frac{\Delta f_P(J)}{\Delta t} = -\frac{\Delta f_S(J)}{\Delta t} = \frac{k_3 C_E/C}{(K_M/C)/f_S(J) + 1} \tag{20.101}$$

Combination of Equation 20.101 with Equations 20.14 and 20.24 and rearrangement yields

$$\Delta f_P(J) = \Delta f_S(J) = \frac{k_3 C_E d^2/DK_M}{1 + (C/K_M) f_S(J)} \times \frac{f_S(J)}{L} \tag{20.102}$$

Thus the invocation of the steady-state assumption results in a rate law involving two independent dimensionless parameters: $C/K_M$, which compares the bulk substrate concentration with the Michaelis constant for the enzyme, and $k_3C_Ed^2/DK_M$, which in effect compares the rate of catalysis with the rate of diffusion. These two parameters may be varied to show their effect on the detected current, and the reader is directed to the cited papers to view the results of this simulation. What is intended here, however, is a specific example of the use of the steady-state assumption in the development of an atypical rate expression (Eq. 20.102).

The derivation of a similarly atypical rate expression is required for the simulation of the electrochemical behavior encountered in electrohydrodimerization studies. In these studies, the variation of the bulk concentration of the olefin (e.g., ethyl cinnamate, diethyl furmarate) reveals that there is a concentration dependence to the reaction order associated with the dimerization of the electrogenerated radical ion [33]. This variation in apparent reaction order with concentration can only be attributed to a two-step mechanism [25] involving two independent rate or equilibrium processes. A mechanism that meets this criterion and appears to fit the electrochemical data is the "preequilibrium" mechanism [36] in which the electrogenerated radical ions first engage in an equilibrium dimerization before the rate-determining ring closure of the dimer takes place. Symbolically, this mechanism may be written:

At $E_1$: $\quad R + e \longrightarrow R^{\bar{\cdot}}$ (20.103)

Chemical equilibrium: $\quad 2R^{\bar{\cdot}} \overset{K}{\rightleftarrows} R_2^{\,2-}$ (20.104)

Chemical reaction: $\quad R_2^{\,2-} \xrightarrow{k} X$ (20.105)

At $E_2$: $\quad R^{\bar{\cdot}} \longrightarrow R + e$ (20.106) where X is the cyclic product formed and $E_1$ and $E_2$ are potentials necessary to bring about the diffusion-controlled reaction of R and $R^{\bar{\bullet}}$, respectively. This mechanism appears to be first order if the cyclization is rate-determining or second order if the radical ion equilibrium is rate-determining. Intermediate reaction orders may be obtained depending on the thermodynamic extent of Equation 20.104. The rate law for this mechanism may be derived by combining the equilibrium constant expression for Equation 20.104,

$$K = \frac{[R_2^{2-}]}{[R^{\bar{\bullet}}]^2} \tag{20.107}$$

with the material balance equation for $R^{\bar{\bullet}}$ and $R_2^{2-}$:

$$C_{R^{\bar{\bullet}}} = [R^{\bar{\bullet}}] + 2[R_2^{2-}] \tag{20.108}$$

(where $C_{R^{\bar{\bullet}}}$ is the total concentration of the product of the electrode reaction exclusive of any product of the homogeneous chemical reaction) to yield

$$K = \frac{[R_2^{2-}]}{(C_{R^{\bar{\bullet}}} - 2[R_2^{2-}])^2} \tag{20.109}$$

Solution of Equation 20.109 yields

$$[R_2^{2-}] = \frac{\left(\sqrt{8KC_{R^{\bar{\bullet}}} + 1} - 1\right)^2}{16K} \tag{20.110}$$

which may be substituted into the first-order rate law equation for the reaction given in Equation 20.105 to obtain a rate law for the mixed mechanism:

$$\frac{d[X]}{dt} = k[R_2^{2-}] = \frac{k\left(\sqrt{8KC_{R^{\bar{\bullet}}} + 1} - 1\right)^2}{16K} \tag{20.111}$$

The finite difference form of this rate law may be represented by

$$\frac{\Delta f_X(J)C}{\Delta t} = \frac{k\left[\sqrt{8KCf_{R^{\bar{\bullet}}}(J) + 1} - 1\right]^2}{16K} \tag{20.112}$$

or

$$\Delta f_X = \frac{(kt_f)\left[\sqrt{8(KC)f_{R^{\bar{\bullet}}}(J) + 1} - 1\right]^2}{16(KC)L} \tag{20.113}$$

where C is the bulk concentration of parent olefin, L is the number of time iterations representing step time $t_f$, and $f_X(J)$ and $f_{R^{\bar{\bullet}}}(J)$ are the fractional concentrations of product X and $R^{\bar{\bullet}}$ at any time or position within the diffusion layer.

The application of this rate law to the simulation of electrochemical behavior requires two dimensionless input parameters $kt_f$ and KC. When these are supplied, three-dimensional chronoamperometric or chronocoulometric working surfaces [34] are generated. These working surfaces both indicate first-order behavior when KC is large and second-order behavior when KC is small. Intermediate values of KC produce the variable reaction orders between one and two that are observed experimentally when the bulk olefin concentration is varied. Appropriate curve fitting of the experimental i(t,C) data to the simulation results in the evaluation of k and K; details appear in the referenced work.

## XVII. BEYOND THE BASICS

Many advances in the field of digital simulation have taken place since the publication of the first edition. As indicated previously, these are beyond the scope of (and outside the space allotted for) this introduction. Reference has already been made to more comprehensive treatments that discuss advances such as the implicit finite difference method in greater detail [18–21]. Another noteworthy technique involves the variation of the volume element size ($\Delta x$) within the space array according to the position of the volume element within the array. This causes a local variation in $D_M$ in accord with Equation 20.5. This method has been used by Joslin and Pletcher [37] and Feldberg [38] to achieve rapid simulations of electrochemical problems by varying element size in an exponential manner with distance from the electrode; this describes the events taking place in the vicinity of the electrode in greater detail than those taking place within the bulk. Accurate representations of the current may be achieved within only a few time iterations. Thus the technique may be suited for "stiff" problems—those in which two events are taking place on vastly different time scales. For example, if one wishes to model the complete electrochemical transport problem (involving diffusion, convection, and migration) through the solution of the Nernst–Planck equation [4] to obtain a transient expression for the total current due to double-layer charging and a faradaic process, one must address the fact that migration in the vicinity of the electrode takes place on a much more rapid time scale than diffusion from the bulk. Because of this, early simulations of this process [39,40] have resulted in steady-state solutions without achieving good transient results. Variations in volume element size in the vicinity of the electrode may permit the simulation of the combined effects of diffusion and migration, so that it appears to be possible to treat the transient problem in this manner [41]. Another example of the utility of this method lies in the simulation of the complex diffusion problems encountered in polymer film electrodes [42]. These problems are complicated by the fact that diffusion coefficients within the polymer film are vastly different than those within the solution phase; thus transport takes place at significantly different rates on either side of the solution-membrane interface. By employing different volume-element sizes on either side of the membrane, we have been able to model the combined effects of solution and membrane transfer in the polymer-film-covered rotating-disk electrode [43]. Without resorting to this method, the simulation of this problem is quite time-consuming.

Another development that appears to be forthcoming is an attempt to eliminate "working curves" by fitting the results of simulations to specific functional forms. Examples of this have already appeared in the literature [2,13,14,29,44]. Equations obtained in this manner permit one to analyze experimental data functionally, even though these functions are not rigorously derived. Even if the functions so obtained only fit the simulation results over a limited range, they may still be quite useful in predicting experimental results falling within that region. More developments along these lines are anticipated. While falling outside the electrochemical boundaries of this work, this concept has already been extended to the moment analysis of chromatograms obtained by finite difference methods [45–48].

## REFERENCES

1. S. W. Feldberg, *J. Am. Chem. Soc. 88*:390 (1966).
2. S. W. Feldberg, *J. Phys. Chem. 70*:3928 (1966).
3. S. W. Feldberg, in *Electroanalytical Chemistry*, Vol. 3 (A.J. Bard, ed.), Marcel Dekker, New York, 1969, p 271.
4. L. R. Faulkner and A. J. Bard, *Electrochemical Methods*, Wiley, New York, 1980.
5. A. Fick, *Ann. Phys. (Pogg.) 94*:59 (1855).
6. A. Einstein, *Z. Electrochem. 14*:235 (1908).
7. F. G. Cottrell, *Z. Phys. Chem. 42*:385 (1902).
8. W. V. Childs, J. T. Maloy, C. P. Keszthelyi, and A. J. Bard, *J. Electrochem. Soc. 118*:874 (1971).
9. R. D. Grypa, M. S. thesis, West Virginia University, 1974.
10. R. J. Lawson and J. T. Maloy, *Anal. Chem. 46*:559 (1974).
11. A. D. Osborne and G. Porter, *Proc. R. Soc., Ser. A 284*:9 (1965).
12. S. A. Cruser and A. J. Bard, *J. Am. Chem. Soc. 91*:267 (1969).
13. J. T. Maloy, K. B. Prater, and A. J. Bard, *J. Am. Chem. Soc. 93*:5959 (1971).
14. L. R. Faulkner, *J. Electrochem. Soc. 122*:1190 (1975).
15. R. P. Van Duyne and C. N. Reilley, *Anal. Chem. 44*:153 (1972).
16. D. K. Gosser, Jr. and P. H. Rieger, *Anal. Chem. 60*:1159 (1988).
17. D. Britz, *Digital Simulation in Electrochemistry*, 2nd ed., Springer-Verlag, Berlin, 1988.

18. B. Speiser, *Comput. Chem. 14*:127 (1990).
19. M. Rudolph, *Electroanal. Chem. 314*:13 (1991).
20. M. Rudolph, *Electroanal. Chem. 338*:85 (1992).
21. M. Rudolph, D. P. Reddy, and S. W. Feldberg, *Anal. Chem. 66*:586A (1994).
22. D. K. Gosser, Jr., *Cyclic Voltammetry: Simulation and Analysis of Reaction Mechanisms*, VCH, New York, 1993 (includes CVSIM software).
23. DigiSim™ is a registered trademark of Bioanalytical Systems, Inc., West Lafayette, IN.
24. K. B. Prater and A. J. Bard, *J. Electrochem. Soc. 117*:207 (1970).
25. K. B. Prater and A. J. Bard, *J. Electrochem. Soc. 117*:335 (1970).
26. K. B. Prater and A. J. Bard, *J. Electrochem. Soc. 117*:1517 (1970).
27. K. B. Prater, in *Computers in Chemistry and Instrumentation*, Vol. 2 (J. S. Mattson, H. B. Mark, Jr., and H. C. MacDonald, Jr., eds.), Marcel Dekker, New York, 1972, Chap. 8.
28. J. T. Maloy, in *Computers in Chemistry and Instrumentation*, Vol. 2 (J. S. Mattson, H. B. Mark, Jr., and H. C. MacDonald, Jr., eds.), Marcel Dekker, New York, 1972, Chap. 9.
29. L. D. Mell and J. T. Maloy, *Anal. Chem. 47*:299 (1975).
30. L. D. Mell and J. T. Maloy, *Anal. Chem. 48*:1597 (1976).
31. F. R. Shu and G. S. Wilson, *Anal. Chem. 48*:1679 (1976).
32. R. D. Grypa and J. T. Maloy, *J. Electrochem. Soc. 122*:377 (1975).
33. R. D. Grypa and J. T. Maloy, *J. Electrochem. Soc. 122*:509 (1975).
34. B. M. Bezilla, Jr. and J. T. Maloy, *J. Electrochem. Soc. 126*:579 (1979).
35. J. Randriamahazaka and J.-M. Nigretto, *Electroanalysis 5*:221 (1993).
36. E. L. King, *J. Chem. Educ. 51*:186 (1974).
37. T. Joslin and D. Pletcher, *J. Electroanal. Chem. 49*:171 (1974).
38. S. W. Feldberg, *J. Electroanal. Chem. 127*:1 (1981).
39. S. W. Feldberg, *J. Phys. Chem. 74*:87 (1970).
40. D. Laser and A. J. Bard, *J. Electrochem. Soc. 123*:1828, 1837 (1976).
41. S. W. Feldberg, personal communication, 1982.
42. P. J. Peerce and A. J. Bard, *J. Electroanal. Chem. 112*:97 (1980).
43. Work done in collaboration with J. Leddy and A. J. Bard, 1983.
44. R. Bezman and L. R. Faulkner, *J. Am. Chem. Soc. 94*:3699 (1972).
45. B. M. Bezilla, L. Pyles, and J. T. Maloy, 1981 Pittsburgh Conference on Analytical Chemistry and Applied Spectroscopy, Atlantic City, NJ, Abstract 197.
46. S. A. Kevra, D. L. Bergman, and J. T. Maloy, *J. Chem. Educ.*, *71*:1023 (1994).
47. D. L. Bergman, Ph.D. Thesis, Seton Hall University, 1995.
48. S. A. Kevra, Ph.D. Thesis, Seton Hall University, 1994.

# 21

# Evaluation of Mechanisms of Organic Reactions

**M. Dale Hawley** *Kansas State University, Manhattan, Kansas*

## I. INTRODUCTION

The applications of various electrochemical methods to the study of several organic electrode processes are examined in this chapter. Because space limitations permit only the salient features of each electrochemical system to be presented, the discussion of much of the original data, the experimental procedures, and the instrumentation is abbreviated. The reader is encouraged to consult the primary literature when more information is desired.

The goal of the electrochemical studies is the elucidation of the sequence of electron transfer and chemical reactions that occur at or near the electrode surface. Although no single experimental plan can be devised that will suffice for all studies, answers to the following questions provide much of the information needed to describe a redox pathway:

1. Is the redox pathway chemically reversible or are solution reactions coupled to the electron transfer?
2. How many electrons are involved in each of the observed redox processes?
3. Is each redox process kinetically or diffusion controlled?
4. What are the reaction kinetics of each of the kinetically controlled processes?
5. What are the intermediates in the electrode reaction?
6. Is the redox behavior affected by an added substrate (e.g., an electroinactive proton donor), a change in the concentration of the electroactive species, a component of the solvent–electrolyte system, or the electrode material?
7. What are the products and their yields?

The selection of the experimental techniques that will be used in the mechanistic studies is usually based on the instrumentation available to the investigator, the kinetics of the heterogeneous and homogeneous chemical reactions, and the responses of intermediates and final products to various analytical probes. A potentiostat and a waveform generator providing cyclic voltammetric, chronoamperometric, and coulometric capabilities are essential to the performance of most electrochemical studies. In addition, when the identification of intermediates and the determination of their reaction kinetics are necessary for the elucidation of the reaction mechanism, capabilities of performing experiments at subambient temperatures and in the millisecond and submillisecond time domains are often necessary. The latter experiments will require appropriate data acquisition devices (e.g., laboratory digital computer and/or digital oscilloscope) and, possibly, ultramicroelectrodes. Although spectroscopic studies at optically transparent electrodes were not used in the present examples, spectroscopic methods have considerable utility and should be used whenever appropriate.

## II. REDUCTION OF *p*-CHLOROBENZONITRILE [1]

Although the one-electron reduction of nitrobenzene to its radical anion in dipolar aprotic solvents is a classical example of a chemically reversible redox couple, the reductions of many organic compounds are chemically irreversible. The redox behavior of *p*-chlorobenzonitrile is typical of those systems in which the initial electrode product undergoes rapid, irreversible chemical reaction to give another reducible species.

### A. Cyclic Voltammetry

Nearly all electrochemical studies of organic systems are initiated by a cyclic voltammetric examination of the redox behavior (Chap. 3, Sec. III.B). From several cyclic voltammograms obtained at widely separated scan rates, one not only can determine which redox processes are chemically reversible, but also can often identify intermediates and final products and ascertain the relationships among them.

In the case of *p*-chlorobenzonitrile (refer to Fig. 21.1):

A. The potential scan is initiated at –0.8 V in the negative-going direction.
B. The one-electron reduction of *p*-chlorobenzonitrile here

$$p\text{-ClC}_6\text{H}_4\text{CN} + e^- \rightleftarrows p\text{-ClC}_6\text{H}_4\text{CN}^{\cdot-} \quad (21.1)$$

is followed by the rapid loss of chloride ion from the unstable radical anion, the reduction of the *p*-cyanophenyl radical at the electrode surface at the applied potential

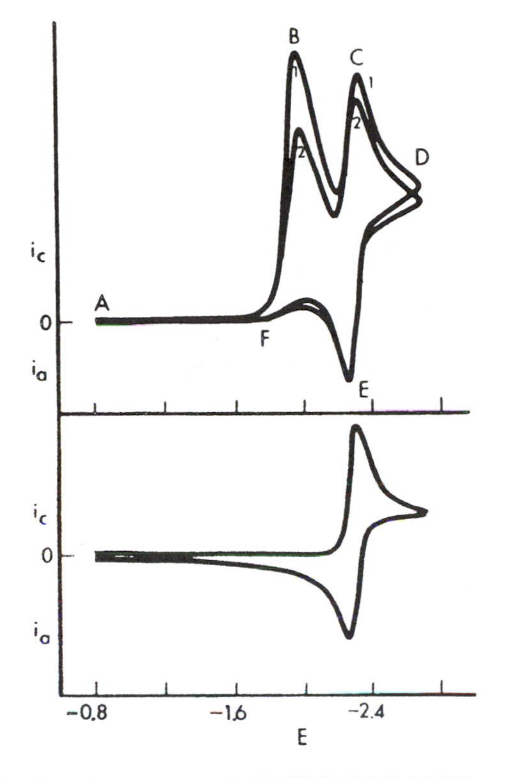


**Figure 21.1** Cyclic voltammograms in DMF–0.1 M $Et_4NClO_4$ at a scan rate of 80.6 mV/s on a planar platinum electrode: (top) $2.15 \times 10^{-3}$ M *p*-chlorobenzonitrile; (bottom) $1.44 \times 10^{-3}$ M benzonitrile. Numbers 1 and 2 represent cycles 1 and 2, respectively. [From Ref. 1, reprinted with permission.]

$$p\text{-}ClC_6H_4CN^{\dot{-}} \xrightarrow{\text{fast}} p\text{-}NCC_6H_4^{\cdot} + Cl^- \quad (21.2)$$

$$p\text{-}NCC_6H_4^{\cdot} + e^- \rightarrow p\text{-}NCC_6H_4^- \quad (21.3)$$

and the abstraction of a proton from a component of the solvent system, HA, by the *p*-cyanophenyl anion

$$p\text{-}NCC_6H_4^- + HA \rightarrow C_6H_5CN + A^- \quad (21.4)$$

C. Benzonitrile, the product of these coupled chemical reactions, is reduced reversibly to its radical anion here.

$$C_6H_5CN + e^- \rightleftarrows C_6H_5CN^{\dot{-}} \quad (21.5)$$

D. The direction of the potential scan is now reversed . . .

E. . . . and on the positive-going sweep, benzonitrile is regenerated here

$$C_6H_5CN^{\cdot-} \rightleftarrows C_6H_5CN + e^- \tag{21.6}$$

F. The absence of an anodic peak here indicates that $p$-$ClC_6H_4CN^{\cdot-}$ decomposes rapidly on the time scale of this experiment.

Because only one stable electroactive product arises from the reduction of $p$-chlorobenzonitrile, the interpretation of the cyclic voltammetric behavior is relatively straightforward. The fact that electron attachment to halogen-containing compounds frequently results in the cleavage of the carbon–halogen bond immediately suggests benzonitrile as a possible product. The similarity between the cyclic voltammetric behavior of benzonitrile and that of the $p$-chlorobenzonitrile reduction product supports this prediction (see Fig. 21.1).

## B. Chronoamperometry

Although cyclic voltammetry is eminently well suited for the qualitative investigation of complex electrode processes, it is less satisfactory for the quantitative study of reaction kinetics. The shape of the current–voltage curve, and hence its interpretation, may be seriously affected by the rate of the follow-up chemical reaction, the rate of heterogeneous electron transfer, the charging of the electrical double layer, uncompensated resistance, and inadequate potential control. These problems are much less important with potential-step techniques, such as chronoamperometry, chronocoulometry, and chronoabsorptometry, and accordingly, these methods are usually preferred by the author for the quantitative study of reaction rates (Chap. 3, Sec. II). Moreover, whereas each current–voltage curve gives only one value of the rate constant, the single-potential-step experiment gives as many values as the number of times the current is sampled. With the potentiostats and data acquisition equipment that are currently available at rather moderate cost, studies of electrochemically generated intermediates that have half-lives ranging from a millisecond or less to tens of seconds can be performed readily.

The single-potential-step chronoamperometric technique was selected here for the study of diffusion versus kinetic control for the formation of benzonitrile.

A. In this experiment (refer to Fig. 21.2), the potential is stepped abruptly from here, where $p$-chlorobenzonitrile is electroinactive, to . . .

B. . . . here, where $p$-chlorobenzonitrile and benzonitrile are reduced concurrently and the concentration of each species is effectively zero at the electrode surface.

C. Note that the chronoamperometric $it^{1/2}/CA$ value is constant throughout the accessible time range. These data indicate that the rate of benzonitrile formation is controlled only by the diffusion rate of $p$-chlorobenzonitrile, a

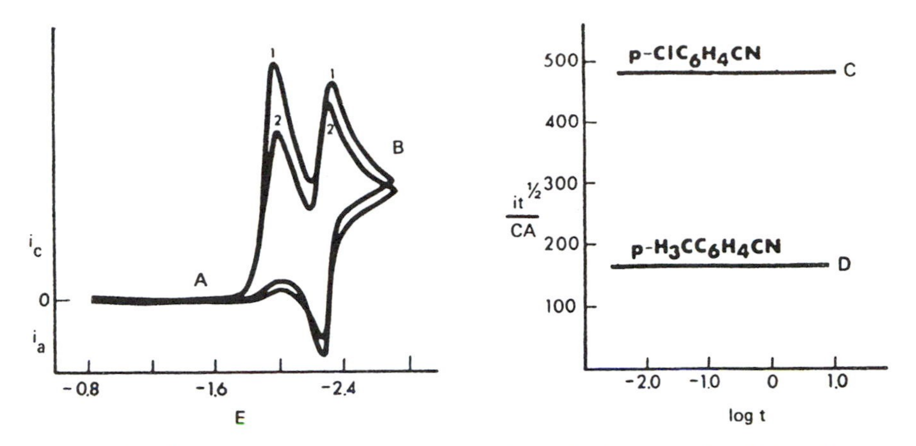


**Figure 21.2** (Left) Cyclic voltammogram of *p*-chlorobenzonitrile. (Right) Chronoamperometric data for the reduction of *p*-chlorobenzonitrile (top curve) and *p*-tolunitrile (bottom curve). [From Ref. 1, reprinted with permission.]

conclusion that further requires the half-life of any intermediate in the electrochemical dechlorination reaction to be less than the shortest observation time, which in this case is 1 ms.

D. The chronoamperometric results can also be used to ascertain the number of electrons involved in the formation of benzonitrile from *p*-chlorobenzonitrile. In order to translate the chronoamperometric data into a meaningful n value, a compound is selected that has a diffusion coefficient very similar to that of *p*-chlorobenzonitrile and that gives a stable, known product upon electroreduction. Tolunitrile, which satisfies these criteria, is known to be reduced to its radical anion at a diffusion-controlled rate. Since this one-electron process gives a value of 168 $\mu A \cdot s^{1/2} \cdot M^{-1} \cdot cm^{-2}$ for $it^{1/2}/CA$, the corresponding value of 480 $\mu A \cdot s^{1/2} \cdot M^{-1} \cdot cm^{-2}$ for the reduction of *p*-chlorobenzonitrile to benzonitrile anion radical must represent an overall three-electron process. When we subtract the one electron that is required to reduce benzonitrile to its radical anion from this total, we immediately conclude that two electrons are involved in cleavage of the carbon–chlorine bond in *p*-chlorobenzonitrile. A scheme that is consistent with these data is described by Equations 21.1 to 21.6.

## C. Capture of the Intermediate *p*-Cyanophenyl Radical

An intermediate in an electrode process often can be studied directly by electrochemical and spectroscopic methods, or its existence can be inferred from product and kinetic analyses. In the present example, the intermediacy of the *p*-cyanophenyl radical can be demonstrated by a trapping experiment (refer to Fig. 21.3).

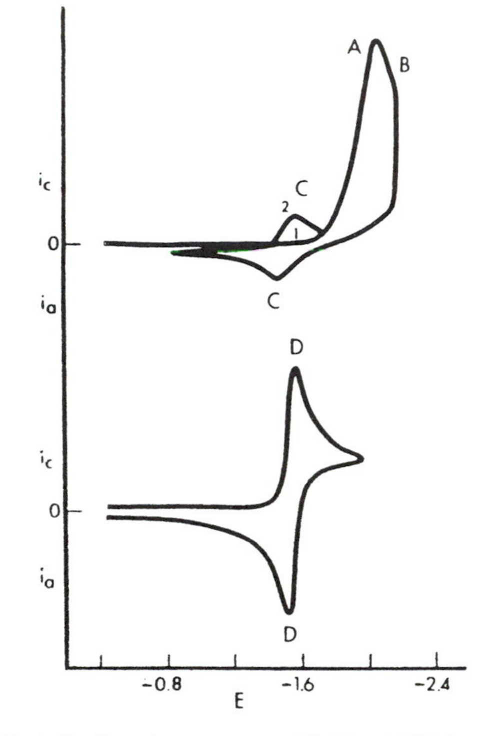


**Figure 21.3** (Top) Cyclic voltammogram of $3.75 \times 10^{-3}$ M *p*-chlorobenzonitrile in DMF–0.2 M $Et_4NClO_4$. (Bottom) Cyclic voltammogram of $1.80 \times 10^{-3}$ M terephthalonitrile in DMF–0.1 M $Et_4NClO_4$. The working electrode in both experiments is platinum; the scan rate is 80.6 mV/s. [From Ref. 1, reprinted with permission.]

When the cathodic switching limit is adjusted to –2.2 V:

A. The only cathodic peak that is observed on the first negative-going sweep is due to *p*-chlorobenzonitrile.

B. The sweep is interrupted and the potential held for approximately 30 s at the cathodic switching limit. A small fraction of the short-lived *p*-cyanophenyl radical intermediate is captured by cyanide ion, the trapping agent, to give . . .

C. . . . terephthalonitrile radical anion (Eq. 21.7), a species that is oxidized reversibly (Eq. 21.8) at this potential when the potential scan is resumed in the positive-going direction. Note that the reduction of terephthalonitrile to its radical anion is not seen until the second negative-going potential sweep is made.

$$p\text{-NCC}_6\text{H}_4^{\cdot} + \text{CN}^- \rightarrow p\text{-NCC}_6\text{H}_4\text{CN}^{\cdot -} \quad (21.7)$$

$$p\text{-NCC}_6\text{H}_4\text{CN}^{\cdot -} \rightleftarrows p\text{-NCC}_6\text{H}_4\text{CN} + e^- \quad (21.8)$$

D. An authentic sample of terephthalonitrile gives this cyclic voltammetric behavior.

## D. Coulometry and Product Identification

Numerous examples could be cited in which two or more suspected products of an electrode process exhibit similar electrochemical behavior. In other instances, the species that is stable during the time required to complete the cyclic voltammetric experiment may undergo a slow chemical reaction to give the product that is isolated. These problems arise sufficiently frequently that the identification of products and the determination of the product distribution are required.

The procedure employed in many laboratories involves exhaustive controlled-potential electrolysis of the parent species and the subsequent identification of the products and their yields by standard analytical procedures. In the case of *p*-chlorobenzonitrile, exhaustive electrolysis at a potential between the first and second cathodic waves (–2.08 V; see Fig. 21.1) gave the anticipated n value ($n_{exptl} = 2.02 \pm 0.05$). Cyclic voltammetric examination of the electrolyzed solution immediately after electrolysis indicated a nearly quantitative yield of benzonitrile (95%); a gas-chromatographic analysis of the same solution showed that benzonitrile was formed in 85% yield. No other product was detected by either of these methods.

# III. OXIDATION OF ADRENALINE [2]

*p*-Chlorobenzonitrile and adrenaline, our second example, both give electrode products that are unstable with respect to subsequent chemical reaction. Because the products of these homogeneous chemical reactions are also electroactive in the potential range of interest, the overall electrode reaction is referred to as an ECE process; that is, a chemical reaction is interposed between electron transfer reactions. Adrenaline differs from *p*-chlorobenzonitrile in that (1) the product of the chemical reactions, leucoadrenochrome, is more readily oxidized than the parent species, and (2) the overall rate of the chemical reactions is sufficiently slow so as to permit kinetic studies by electrochemical methods. As a final note before the experimental results are presented, the enzymic oxidation of adrenaline was known to give adrenochrome. Accordingly, the emphasis in the work described by Adams and co-workers [2] was on the preparation and study of the intermediates.

## A. Cyclic Voltammetry

The cyclic voltammetric behavior of adrenaline and several other catecholamines was studied as a function of pH. In strongly acidic media (refer to Fig. 21.4), extensive protonation of the amine precludes cyclization of the adrenalinequinone on the time scale of this experiment.

A. As a result, the oxidation of adrenaline to adrenalinequinone is a chemically reversible process under these reaction conditions:

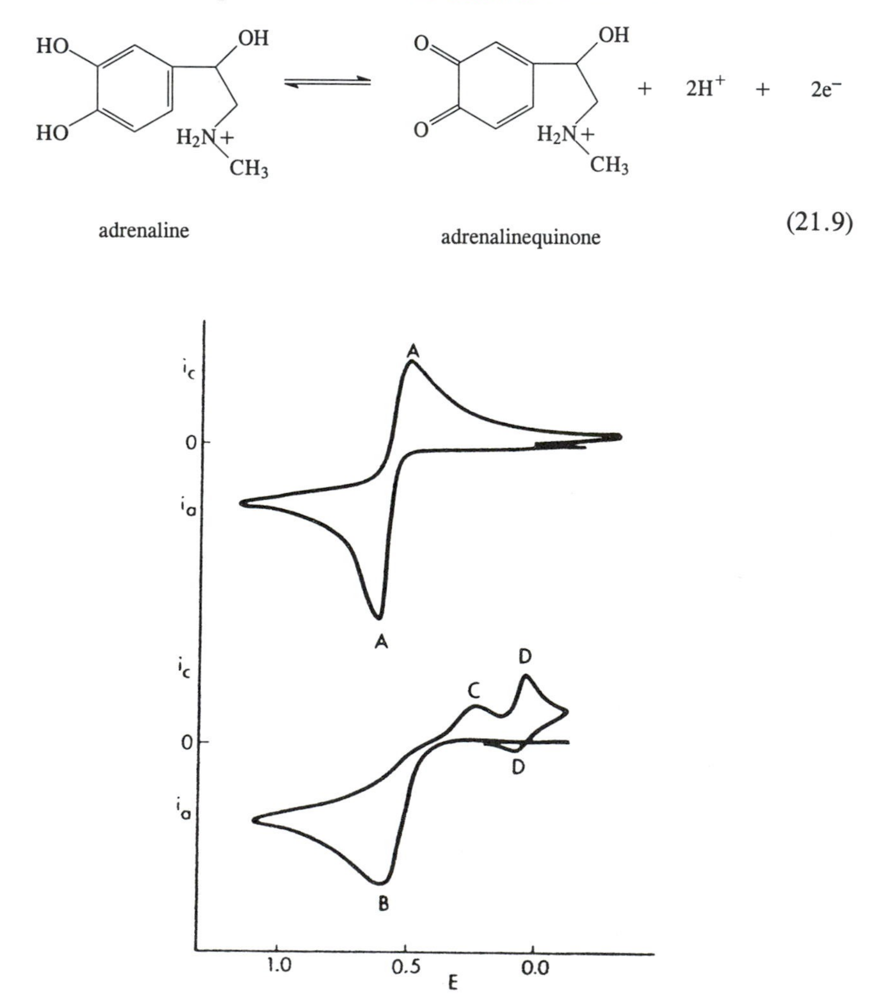


**Figure 21.4** (Top) Cyclic voltammogram of adrenaline in 0.29 *M* $HClO_4$ at a scan rate of 55 mV/s. (Bottom) Cyclic voltammogram of adrenaline at pH 3.0 at a scan rate of 278 mV/s. [From Ref. 2, reprinted with permission.]

B. At pH 3.0, though, deprotonation of the conjugate acid of adrenaline-quinone (Eq. 21.10) and the subsequent cyclization of the unprotonated form of adrenalinequinone (Eq. 21.11) become important.

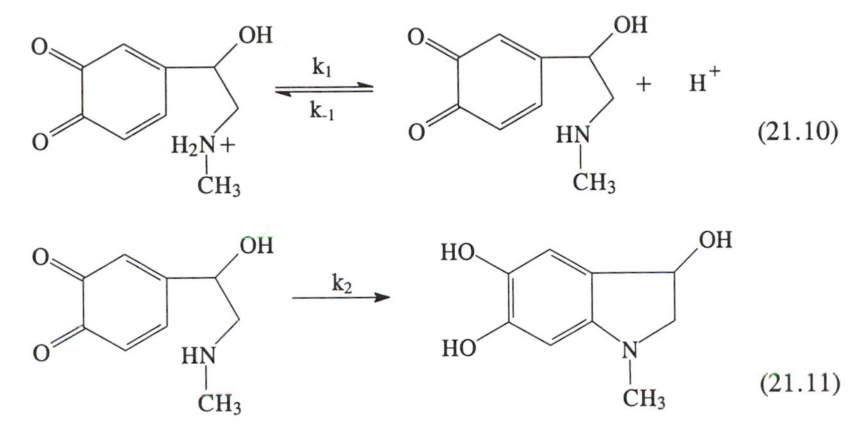


Because the product of these reactions, leucoadrenochrome, is more readily oxidized than adrenaline, a homogeneous electron transfer reaction (Eq. 21.12) involving leucoadrenochrome and unreacted adrenaline-quinone ensues. In addition to the formation of adrenochrome as the final product, adrenaline is regenerated by the solution redox reaction. The reader should be satisfied that the overall transformation of adrenaline to adrenochrome requires four electrons.

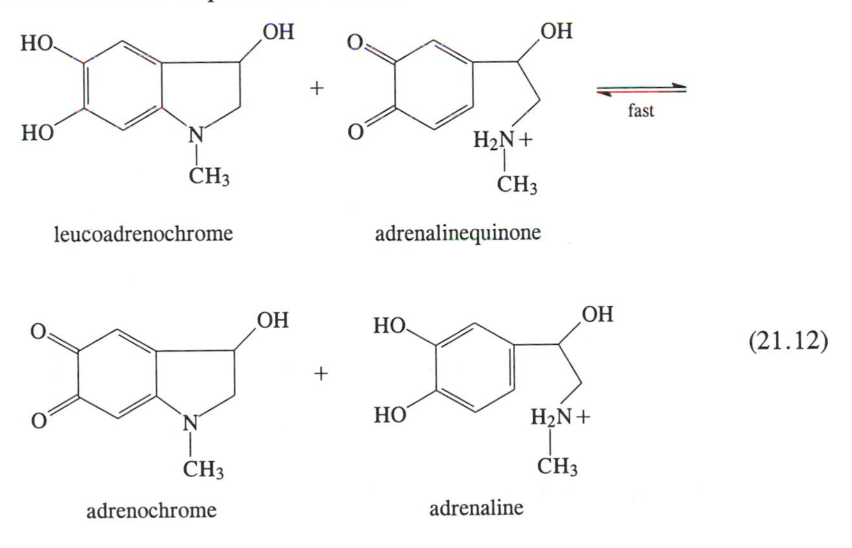

C. The cyclization of adrenalinequinone and the subsequent homogeneous electron transfer reaction cause the peak height for the reduction of unreacted adrenalinequinone on the reverse, negative-going sweep to be reduced, and . . .

D. . . . a redox couple to emerge at a less positive potential for the chemically reversible reduction of adrenochrome to leucoadrenochrome:

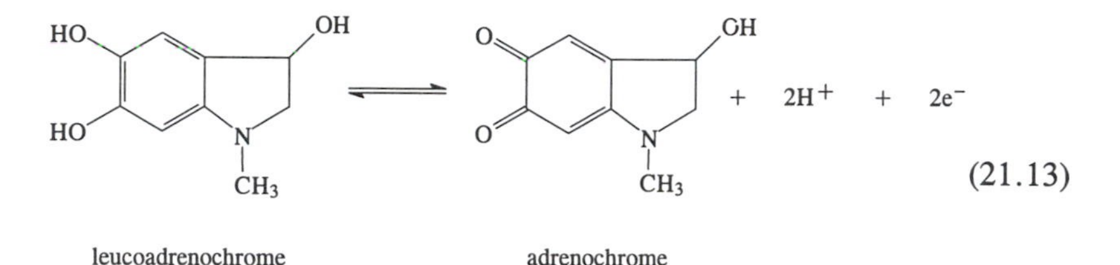


(21.13)

Note that the anodic peak due to the oxidation of leucoadrenochrome to adrenochrome near 0 V is not seen until the second positive-going potential sweep is made. The voltage separation between the anodic and cathodic peaks for the oxidation of adrenaline (peak B, Fig. 21.4, bottom) and the reduction of adrenalinequinone (peak C) is large when compared to most chemically reversible redox couples. However, this behavior is typical of many quinone-hydroquinone systems on a carbon paste surface at intermediate values of pH.

## B. Kinetics of the Solution Reactions

Kinetic studies of ECE processes (sometimes called a DISP mechanism when the second electron transfer occurs in bulk solution) [3] are often best performed using a constant-potential technique such as chronoamperometry. The advantages of this method include (1) relative freedom from double-layer and uncompensated iR effects, and (2) a new value of the rate constant each time the current is sampled. However, unlike certain large-amplitude relaxation techniques, an accurately known, diffusion-controlled value of $it^{1/2}/CA$ is required of each solution before a determination of the rate constant can be made. In the present case, diffusion-controlled values of $it^{1/2}/CA$ corresponding to $n = 2$ and $n = 4$ are obtained in strongly acidic media (i.e., when kt can be made small) and in solutions of intermediate pH (i.e., when kt can be made large), respectively. The experimental rate constant is then determined from a dimensionless working curve for the proposed reaction scheme in which the apparent value of n ($n_{app}$) is plotted as a function of kt.

No single working curve suffices for all first-order ECE processes. The equilibrium constant of the solution redox reaction has a marked effect upon the current–time behavior and, accordingly, an important influence on the shape of the working curve. Experimentally, the value of the equilibrium constant is calculated using the peak potentials determined by cyclic voltammetry. The actual working curve is then obtained by the digital simulation of the diffusion, kinetics, and electron-transfer processes according to Feldberg's procedure [4].

Once the appropriate working curve has been determined, either of two methods may be used for the calculation of the rate constant. In the method that was employed in the original study, each experimental value of $n_{app}$ gives a corresponding value of kt from the working curve. If the limiting values of $it^{1/2}/CA$ have been determined correctly, then a plot of kt versus t should yield a straight line with a zero intercept (Fig. 21.5). The second and preferred method involves fitting the experimental $n_{app}$ versus log t plot to the working curve ($n_{app}$ vs. log kt) until the best match is obtained. A value of k is then calculated directly.

Distinction among the several working curves for ECE processes is sometimes difficult, especially if data are obtained over a relatively restricted range of $n_{app}$. Although the use of pen-and-ink recording limited the dynamic range in the present example, current instrumentation usually permits acquisition of results over several decades of time. The full capability of the instrumentation should be utilized in order to minimize errors in interpretation.

The method for determining the observed rate constant for cyclization of adrenalinequinone is illustrated in Figure 21.5. The potential is stepped . . .

A. . . . from here, where adrenaline is electroinactive, to . . .

B. . . . here, where the concentrations of adrenaline and the cyclization product, leucoadrenochrome, are effectively zero at the electrode surface. Kinetically controlled values of $it^{1/2}$ are divided by the diffusion-controlled value of $it^{1/2}$ when $kt = 0$. (The diffusion-controlled value of $it^{1/2}$ may be obtained either at small values of t at this pH or in a strongly acidic medium where k is small.) This ratio, multiplied by 2, gives $n_{app}$. Each experimental value of $n_{app}$ gives . . .

C. . . . a corresponding value of kt from this working curve. Values of kt are then plotted . . .

D. . . . as a function of t to give the observed rate constant. Measurement of the rate constant at several values of pH suggests . . .

E. . . . a rate law of the form $k = k_1k_2/(k_{-1}[H^+] + k_2)$. The solid curve is calculated from this equation using $k_{-1} = 10^{10}\ M^{-1}\ s^{-1}$, $k_2 = 10^4\ s^{-1}$, and $K = k_1/k_{-1} = 10^{-9}$.

## IV. OXIDATION OF α-TOCOPHEROL [5–7]

The elucidation of organic electrode processes frequently requires studies in which the chemical environment for the electrochemical and accompanying

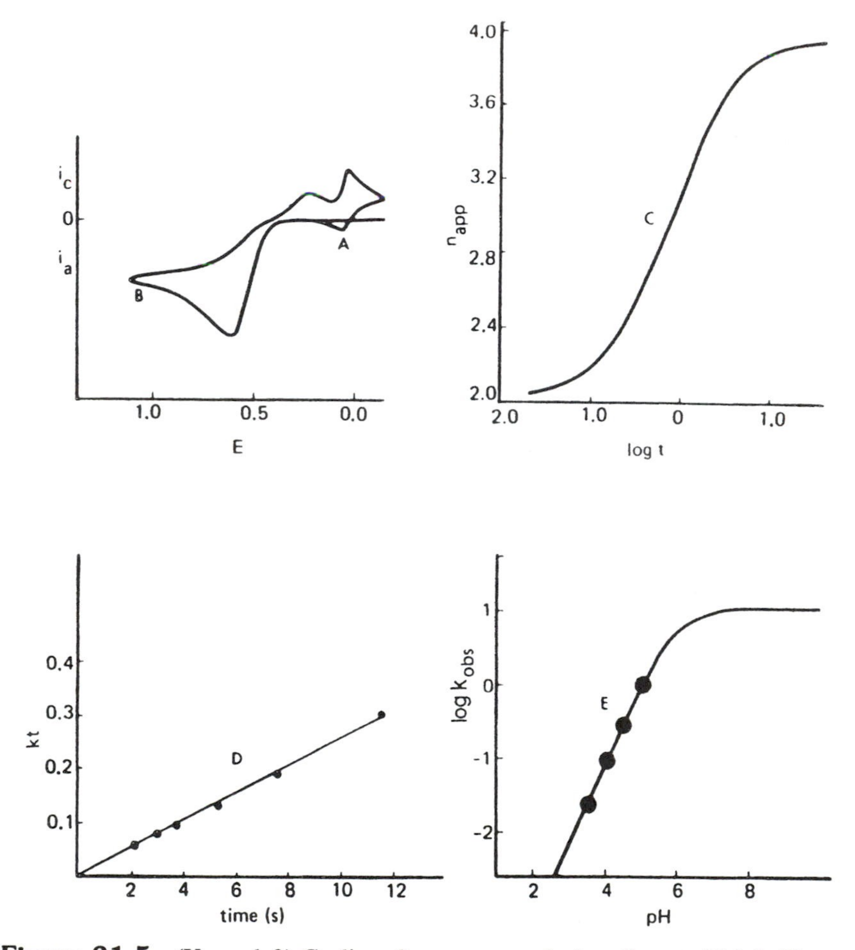


**Figure 21.5** (Upper left) Cyclic voltammogram of adrenaline at pH 3.0. The scan rate is 0.278 V/s. (Upper right) Working curve for this ECE process. (Lower left) Chronoamperometric data for the cyclization of adrenalinequinone at pH 3.5. (Lower right) Observed rate constant for the cyclization of adrenalinequinone as a function of pH. [From Ref. 2, reprinted with permission.]

chemical reactions is varied extensively. The studies involving the oxidation of $\alpha$-tocopherol illustrate how the effects of pH, compositions of the solvent and the buffer system, and the variation of an electroinactive reagent concentration might be used to deduce the principal electrochemical and chemical steps in an electrode process.

## A. Kinetics of the Hydrolysis Reaction (Fig. 21.6)

A. The cyclic voltammetric oxidations of a-tocopherol (Ia) and its model compound, 2,2,5,7,8-pentamethyl-6-hydroxychroman (Ib), are chemically reversible in acetonitrile–0.1 *M* $Et_4MClO_4$:

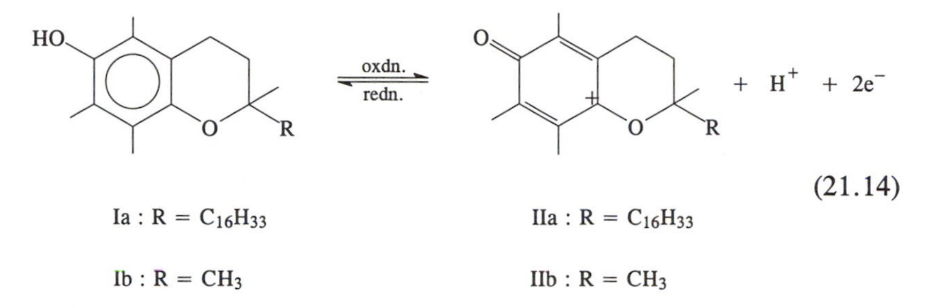


B. However, when water is added as a nucleophile, hydrolysis of the carbonium ion II occurs to give III, which is electroinactive:

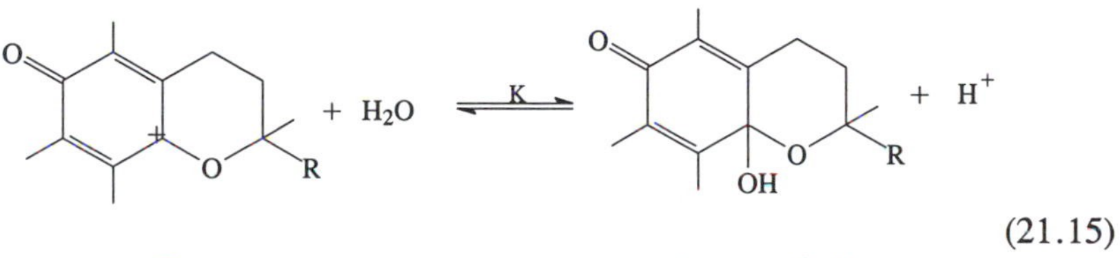


IIIb : R = $CH_3$

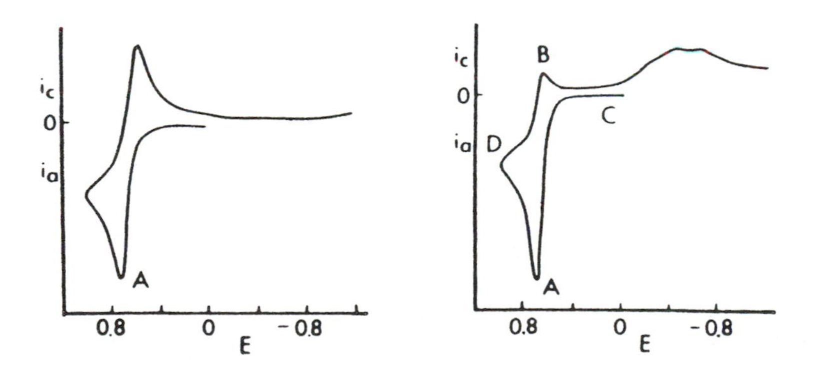


**Figure 21.6** (Left) Cyclic voltammogram of α-tocopherol in acetonitrile with 0.1 *M* $Et_4NClO_4$ as the supporting electrolyte. The scan rate is 83.5 mV/s. (Right) Cyclic voltammogram in acetonitrile–0.1 *M* $Et_4NClO_4$ to which 200 m*M* water has been added. The scan rate is 83.5 mV/s. [From Ref. 5.]

C. The kinetics of the hydrolysis reaction were studied by double-potential-step chronoamperometry. The potential is stepped from here, where I is electroinactive, . . .

D. . . . to here, where the oxidation of I occurs at a diffusion-controlled rate. After a prescribed electrolysis time t, the potential is stepped back to the initial potential, C. Only the reduction of the unreacted carbonium ion II occurs at this potential. The value of the pseudo-first-order rate constant, $k_1$, is then determined from a dimensionless working curve that relates the ratio of the cathodic and anodic currents to $k_1t$. Details for the construction of the working curves (each ratio of $t_r/t$ requires a different working curve) and their subsequent use may be found in the literature [8].

Results of the electrochemical measurements of the hydrolysis rate constant with 60 m*M* added water are shown in Figure 21.7 (left). The pseudo-first-order rate constant is determined from the slope of the plot of $k_1t$ as a function of the forward electrolysis time, t. From a series of these measurements at different concentrations of water, a second plot (Fig. 21.7, right) is obtained from which a value of 2.5 $M^{-1}$ $s^{-1}$ is calculated for the second-order rate constant. Note that this "dry" acetonitrile contains approximately 25 m*M* water.

## B. Ring Opening of III and the Formation of α-Tocopherylquinone

Inspection of Figure 21.6 (right) will show that three small cathodic peaks are present near -0.2, -0.4, and -0.6 V. The first of these peaks arises from the

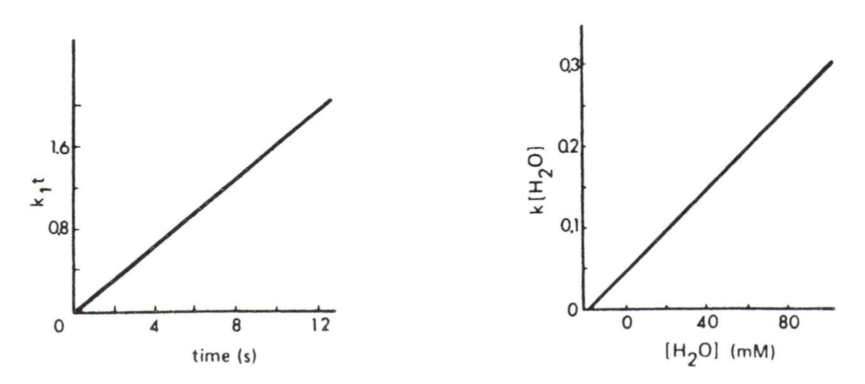


**Figure 21.7** (Left) Plot of $k_1t$ versus t for the hydrolysis of the carbonium ion prepared by the electrochemical oxidation of 2,2,5,7,8-pentamethyl-6-hydroxychroman. The solution contains 60 m*M* $H_2O$ and 0.1 *M* $Et_4NClO_4$ supporting electrolyte. (Right) Plot of $k[H_2O]$ vs. $[H_2O]$ for the hydrolysis of the carbonium ion prepared by the electrochemical oxidation of 2,2,5,7,8-pentamethyl-6-hydroxychroman. The solution is 1.1 m*M* in the chromanol and 0.1 *M* in $Et_4NClO_4$ supporting electrolyte. [From Ref. 5.]

reduction of the protons that are formed during the electrooxidation of α-tocopherol, while the two more-negative peaks arise from the stepwise reduction of α-tocopherylquinone to its hydroquinone. The formation of the quinone involves ring opening of the electroinactive III, a reaction that can be studied conveniently in a water–acetonitrile (75 vol%–25 vol%) mixture (Fig. 21.8).

A. The reaction of the intermediate carbonium ion II with water in this solvent system is rapid, causing the oxidation of I to be . . .

B. . . . chemically irreversible at this scan rate. In addition, the ring-opening reaction,

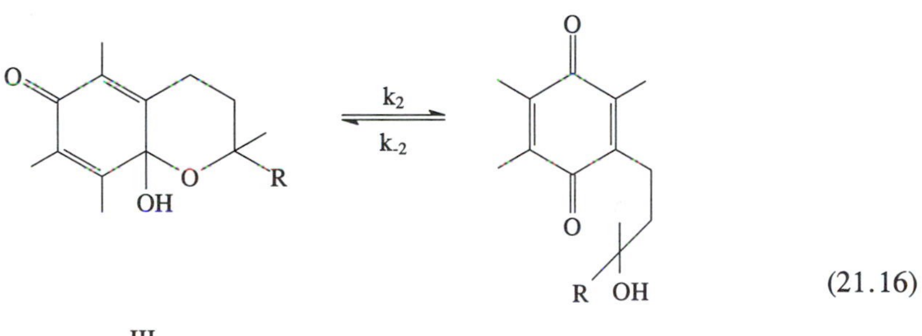


(21.16)

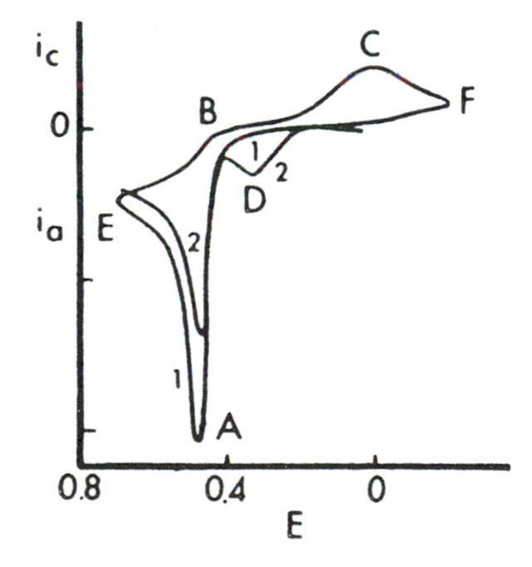


**Figure 21.8** Cyclic voltammogram of 2,2,5,7,8-pentamethyl-6-hydroxychroman in water-acetonitrile (75:25, v/v). The solution is 0.95 m*M* in the chromanol and 0.87 *M* in perchloric acid. The scan rate is 0.167 V/s. The numbers 1 and 2 denote the first and second cycles, respectively. [From Ref. 5.]

C. . . . which occurs relatively rapidly in a strongly acidic medium, affords the quinone, IV, which is reduced at this potential to its hydroquinone:

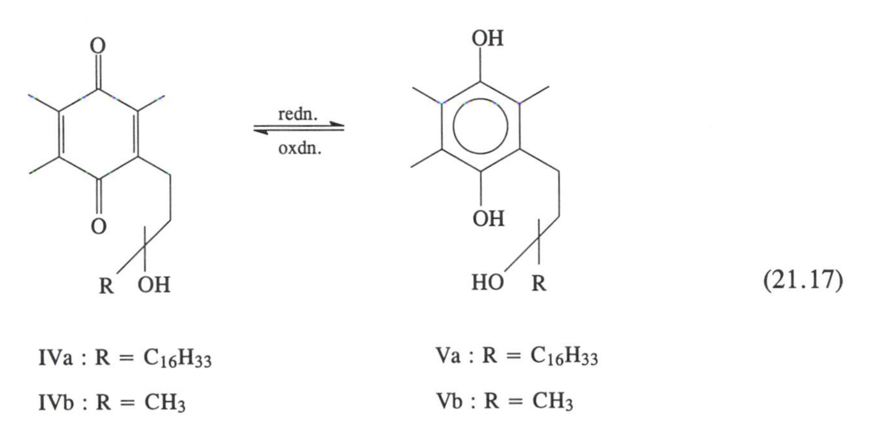


(21.17)

D. Note that the oxidation of hydroquinone near 0.3 V is not observed until the second positive-going sweep is made. Also note the rather large separation in peak potentials of the quinone–hydroquinone redox couple.

The ring-opening reaction (Eq. 21.16) occurs on the time scales of these electrochemical experiments in the pH range 2–7. Because the hydrolysis of carbonium ion II remains rapid in this pH range in this solvent system, the kinetics of the ring-opening reaction can be ascertained by following the rate at which quinone IV is formed. Again, double-potential-step chronoamperometry was used for the study. The potential is stepped from here, where I is electroinactive, . . .

E. . . . to here, where the oxidation of I occurs at a diffusion-controlled rate. After time $\tau_f$, the potential is jumped to . . .

F. . . . here. Any quinone IV formed by the ring-opening reaction is reduced to its hydroquinone at this potential. In this work, the cathodic current was sampled at times $\tau_r$ that were equal to 0.1, 0.2, 0.3, 0.5, and 1.0 times $\tau_f$. From the ratio of the cathodic current to the anodic current at each value of $\tau_r/\tau_f$, a corresponding value of $k\tau_f$ is obtained from an appropriate dimensionless working curve. Although the rate constant in this work was determined from a plot of $k\tau_f$ versus $\tau_f$ (Fig. 21.9), the value of k could also be obtained by fitting the plot of the experimental data, $i_c/i_a$ (at $t = \tau_f$), to the dimensionless working curve that corresponds to that value of $\tau_r/\tau_f$. Each dimensionless working curve, which has $i_c/i_a$ (at $t = \tau_f$) calculated as a function of $k\tau_f$, is obtained by the digital simulation of the processes described by Equations 21.14–21.17. The rate constant is then obtained from an average of the best fits of the data to the several working curves.

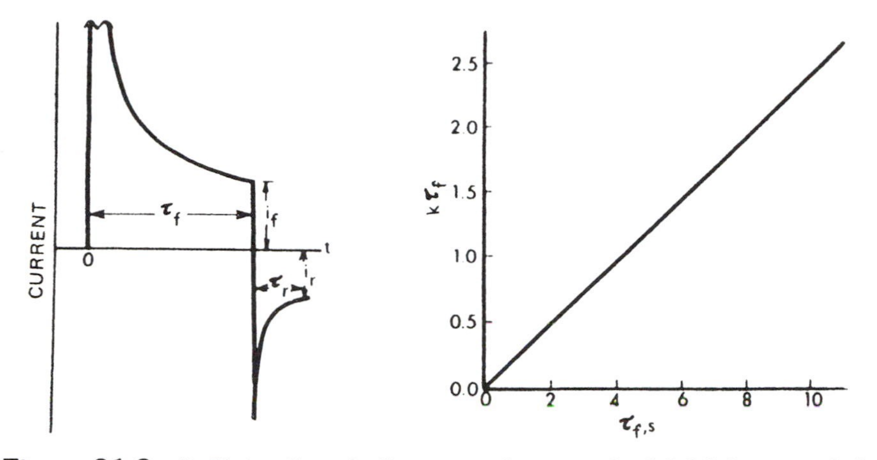


**Figure 21.9** (Left) Anodic–cathodic current–time curve for 2,2,5,7,8-pentamethyl-6-hydroxychroman (Ib) showing the method of current and time measurements for the double-potential-step chronoamperometric method. The $\tau_r/\tau_f$ ratio shown here is 0.3. (Right) Plot of $k\tau_f$ vs. $\tau_f$ for the ring opening of IIb in 75 vol% water–25 vol% acetonitrile: pH 3.99; 0.5 *M* chloroacetic acid, 0.5 M sodium chloroacetate. [From Ref. 6, reprinted with permission.]

### C. Effect of pH on the Ring-Opening Reaction

Numerous measurements of the rate constant in the pH range 2–7 and various buffer concentrations give the results shown in Figure 21.10. Catalytic constants for the ring opening of IIIb at 23° are $H^+$, $4 \times 10^2$ $M^{-1}$ $s^{-1}$; $OH^-$, $1 \times 10^7$ $M^{-1}$ $s^{-1}$; $CH_3COO^-$, $1.6 \times 10^{-1}$ $M^{-1}$ $s^{-1}$; and intrinsic, $4 \times 10^{-3}$ $s^{-1}$. The solid curve in Figure 21.10 is constructed using these data.

## V. CONCLUDING REMARKS

Errors in the formulation of the reaction pathway will invariably arise. Their numbers can be minimized, however, if product studies are performed carefully and electrochemical data are analyzed critically. Predicted changes in electrochemical behavior when a proton donor or acceptor is added, the concentration of the electroactive species is altered, etc. should be examined and confirmed. It should also be clear from the examples in this chapter that one electrochemical technique is seldom capable of yielding all information that is needed to specify the electrode process. A variety of techniques, including nonelectrochemical methods, should be standard fare for the electroanalytical chemist.

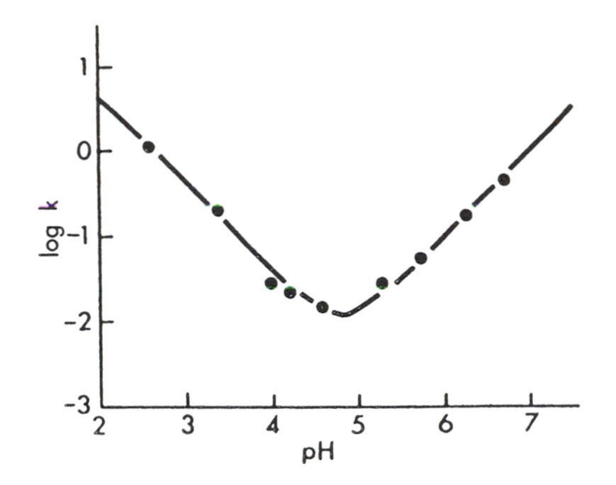


**Figure 21.10** Plot of log k, where k is the first-order rate contant, versus pH for the ring opening of 8*a*-hydroxy-2,2,5,7,8-pentamethyl-6-chromanone (IIIb) at zero ionic strength and zero buffer concentration. [From Ref. 6, reprinted with permission.]

## REFERENCES

1. D. E. Bartak, K. J. Houser, B. C. Rudy, and M. D. Hawley, *J. Am. Chem. Soc. 94*:7526 (1972).
2. M. D. Hawley, S. V. Tatawawadi, S. Piekarski, and R. N. Adams, *J. Am. Chem. Soc. 89*:447 (1967).
3. C. Amatore and J. M. Saveant, *J. Electroanal. Chem. 85*:27 (1977).
4. M. D. Hawley and S. W. Feldberg, *J. Phys. Chem. 70*:3459 (1966).
5. M. F. Marcus and M. D. Hawley, *Biochim. Biophys. Acta 201*:1 (1970).
6. M. F. Marcus and M. D. Hawley, *J. Org. Chem. 35*:2185 (1970).
7. M. F. Marcus and M. D. Hawley, *Biochim. Biophys. Acta 222*:163 (1970).
8. W. M. Schwarz and I. Shain, *J. Phys. Chem. 69*:30 (1965).

# 22

# Electroorganic Synthesis

**Eberhard Steckhan** *Rheinische Friedrich-Wilhelms-Universität Bonn, Bonn, Germany*

## I. WHAT IS ELECTROORGANIC SYNTHESIS [1]?

Electrochemistry provides a very interesting and versatile means for the selective reduction or oxidation of organic compounds. By addition (reduction) or removal (oxidation) of electrons, organic substrates are activated. Thus, starting from neutral compounds, radical anions (reduction) or radical cations (oxidation) are formed (Scheme 22.1). Therefore, organic electrochemistry may be considered as the chemistry of radical ions. As compared with redox reagents, the selectivity of the electrochemical redox process can be determined by the selection of the electrode potential (see below). Furthermore, as the electrons are not bound to a reagent, no waste from used-up reagents is produced. The ecologically and economically problematic waste treatment can be avoided to a large extent.

From the retrosynthetic point of view, electrochemical redox reactions are an easy way to accomplish the principle of redox-umpolung (polarity reversal) [2]. As can be seen in Scheme 22.1, oxidation of an electron-rich neutral compound will lead to an electrophilic cation radical, or starting with a nucleophilic anion, anodic oxidation may lead to an electrophilic cation. In the mirror image reductions, an electron-poor neutral compound is transformed to a nucleophilic anion radical, or an electrophilic cation will end up as a nucleophilic anion.

## II. ADVANTAGES AND DISADVANTAGES OF ELECTROORGANIC REACTIONS

The **advantages** of electroorganic reactions may be summarized as follows:

- The *selectivity* of the reaction can be influenced by the applied electrode potential (potential-controlled reaction), which, in contrast to redox reagents, can be varied continuously.

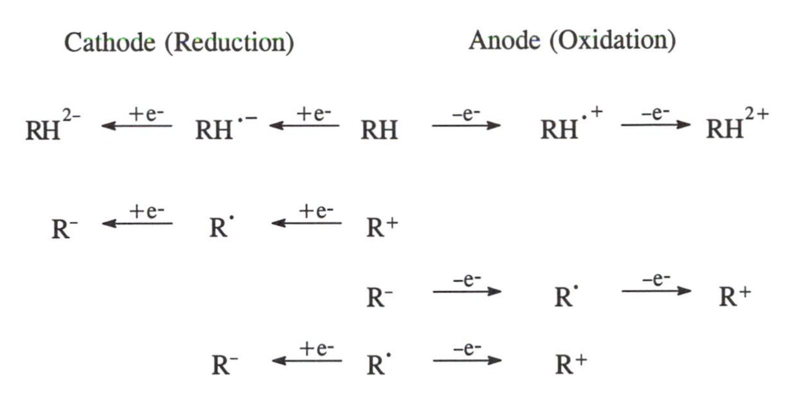


**Scheme 22.1** Electrochemically accessible reactive intermediates.

- The *reaction rate* can be controlled by the current density.
- The *turnover degree* can be fixed by the charge consumption.
- The *electrode material* and the *electrolyte composition* can be used as parameters for controlling the selectivity and reaction rate.
- The reaction conditions are generally very *mild*. Usually, electrolysis is performed at ambient temperature and pressure. The rate of the electron transfer is not influenced very much by the temperature. Thus, lowering the reaction temperature will not diminish the electron transfer rate but may lower the tendency for chemical side reactions.
- The electrons are *reagent free*. Therefore, the pollution of the environment by spent reagents can be avoided.

However, the **disadvantages** of electrochemical reactions must also be taken into account:

- For electrochemical reactions, special reactors (*electrolysis cells*) are required. Fortunately, today a large number of electrolysis cells are commercially available for many types of electrochemical reactions. For laboratory use or small- and medium-scale technical applications, no cell construction is necessary.
- Electrochemical reactions are in principle *heterogeneous*. They take place at the electrode/electrolyte phase boundary. Therefore, for technical applications, special care must be taken in optimizing the space–time yield. This can be done by using a high electrode surface area to volume ratio by application of electrodes with very high specific surface area (carbon felt, reticulated vitreous carbon, packed bed, or fluidized bed electrodes). In other cases, a mediator system, which acts as an electron shuttle between electrode and substrate, may accelerate the reaction rate by shifting the rate-determining

redox reaction into the homogeneous solution (*indirect electrolysis*) [3] (see Sec. IV.F).

- The recovery of the *supporting electrolyte* may create problems and can be costly. This problem can be diminished by using an electrolysis cell with a very small gap between the electrodes, so that very low concentrations of supporting electrolyte may be used.

## III. TECHNICALLY INTERESTING PROCESSES

Some reaction types that have proved valuable not only for the laboratory but also for industrial applications are presented here:

- Anodic substitution in the aromatic nucleus, the side chains of aromatic compounds, and the allylic position of double bonds.
- Anodic coupling of carboxylates, carbanions, or CH-acidic methylene compounds.
- Cathodic hydrodimerization of active olefins and carbonyl compounds.
- Anodic dehydrodimerization and additive dimerization of electron-rich olefins.
- Anodic substitution in the alpha-position to hetero atoms.
- Cathodic substitution of halide atoms by electrophiles like carbon dioxide, aldehydes, or protons.
- Electrochemical regeneration of expensive but highly selective chemical redox agents (indirect electrolysis). This is very attractive, if the application of such reagents in stoichiometric amounts is excluded because of economical and ecological reasons.

Some selected examples, which have been developed in industry, illustrate the technical interest in electroorganic synthesis.

### A. Anodic Oxidations

Several industrial processes make use of anodic substitution in the side chain of alkyl aromatic compounds. Thus, aromatic aldehydes or aldehyde dimethyl acetals are generated at graphite electrodes in methanol. A typical example is the formation of 4-methoxybenzaldehyde (anisaldehyde) dimethyl acetal starting from 4-methoxytoluene [4]:

$$CH_3O{-}C_6H_4{-}CH_3 \xrightarrow[CH_3OH/KF]{\text{C anode}} CH_3O{-}C_6H_4{-}CH(OCH_3)_2 \quad (22.1)$$

Anodic aromatic nuclear substitution is performed to introduce a hydroxy function leading to phenolic compounds. In the first step, acetoxylation or trifluoroacetoxylation occurs usually at carbon electrodes using the acid as solvent [5]:

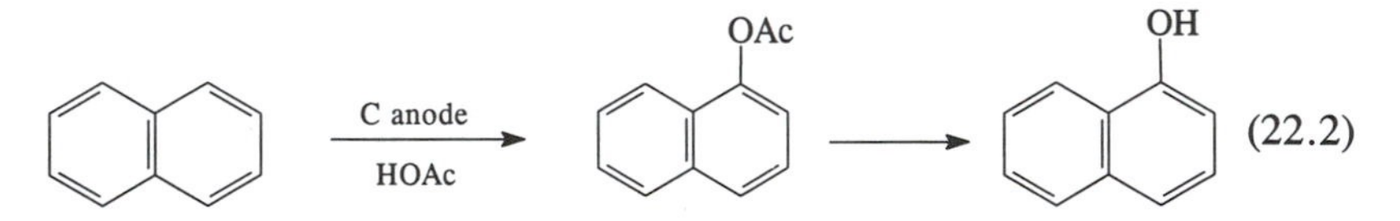


Quinones can be synthesized by anodic nuclear aromatic substitution, if high-valent chromium(VI) acts as an electrochemically generated and regenerated chemical oxidant in acidic solution (indirect electrolysis) [6]:

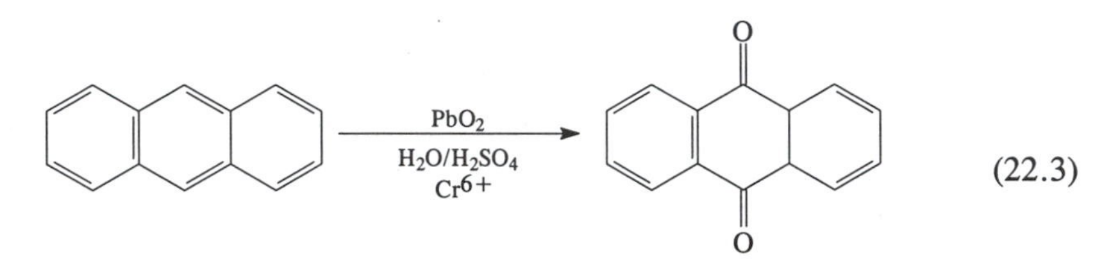


Anodic addition to an electron-rich heteroaromatic compound is used to transform furan to 2,5-dimethoxy-2,5-dihydrofuran, a valuable synthetic intermediate. Again, an indirect electrochemical process occurs. The bromide ion as redox catalyst is electrochemically oxidized to give bromine, which then acts as chemical oxidant for furan [7]:

Furan $\xrightarrow[\text{KBr, } CH_3OH]{\text{C anode}}$ 2,5-dimethoxy-2,5-dihydrofuran ($CH_3O$, O, $OCH_3$) (22.4)

Symmetrical diesters are easily prepared by Kolbe oxidation of half esters via decarboxylative radical dimerization [8]:

$$CH_3O{-}\underset{\underset{O}{\|}}{C}{-}(CH_2)_4{-}COO^- \xrightarrow{\text{Pt-anode}} CH_3O{-}\underset{\underset{O}{\|}}{C}{-}(CH_2)_8{-}\underset{\underset{O}{\|}}{C}{-}OCH_3 \quad (22.5)$$

A phosgene-free synthesis of alkylisocyanates makes use of the indirect electrochemical oxidation in the alpha-position to nitrogen of formamides. Bromide in methanol solution acts as the redox catalyst, which, presumably, is oxidized to the methyl hypobromite [9]:

$$R{-}CH_2{-}NH{-}CHO \xrightarrow[\text{KBr, } CH_3OH]{\text{C anode}} R{-}CH_2{-}NH{-}\underset{\underset{O}{\|}}{C}{-}OCH_3 \xrightarrow[-CH_3OH]{\Delta} R{-}CH_2{-}N{=}C{=}O \quad (22.6)$$

The anodic oxidation of perfluoro propene is applied in industry to synthesize perfluoro propene oxide at $PbO_2$ anodes [10]:

$$CF_3{-}CF{=}CF_2 \xrightarrow[\text{AcOH, HNO}_3\text{, HF}]{\text{PbO}_2\text{ anode}} CF_3{-}\overset{O}{CF{-}CF_2} \qquad (22.7)$$

In several cases, the oxidation of primary alcohols to carboxylic acids is desired as a technical process. Direct and indirect electrochemical oxidations have been developed for this purpose.

The direct anodic oxidation of 1,4-butynediol at $PbO_2$ anodes in sulfuric acid yields the desired acetylenic dicarboxylic acid [11]:

$$HO{-}CH_2{-}C{\equiv}C{-}CH_2{-}OH \xrightarrow[\text{H}_2\text{SO}_4\text{, H}_2\text{O}]{\text{PbO}_2\text{ anode}} HOOC{-}C{\equiv}C{-}COOH \qquad (22.8)$$

For the oxidation of glucose to Ca gluconate, an indirect electrooxidation via the intermediate formation of sodium hypobromite is effective [12]:

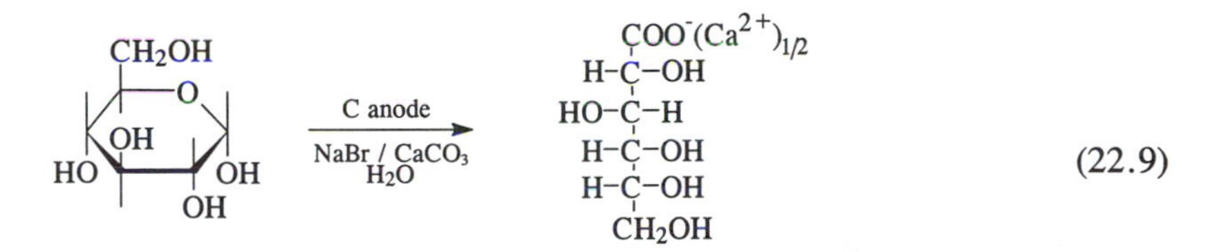


In the synthesis of vitamin C, the oxidation of diacetone L-sorbose to diacetone 2-keto-L-gulonic acid proceeds at an Ni-anode in the presence of hydroxide. Under these conditions, the nickel hydroxide surface is anodically transformed to NiOOH, the "nickel peroxide," which acts as chemical oxidant via hydrogen atom abstraction. Thus, a chemically modified redox-active electrode acts as a heterogeneous redox catalyst [13]:

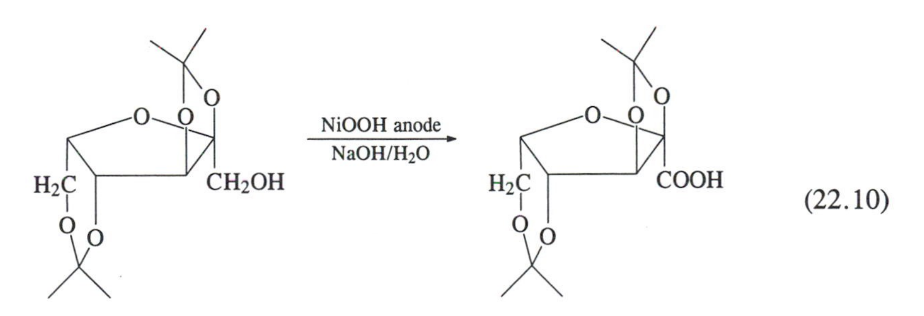


Fluorination of hydrocarbons can only be performed electrochemically in HF solution at nickel electrodes. A typical example for this technically extremely important process is [14]:

$$CH_3{-}(CH_2)_n{-}SO_2Cl \xrightarrow[\text{Ni anode}]{\text{HF}} CF_3{-}(CF_2)_n{-}SO_2F \qquad (22.11)$$

In a different way, perfluorinated functionalized compounds can be synthesized using electrogenerated peroxodisulfuryldifluoride, which starts an additive teleomerization with tetrafluoroethylene [15]:

$$2FSO_3H \xrightarrow[-2H^+]{\text{anode}} FSO_2{-}O{-}O{-}SO_2F \xrightarrow{n(CF_2{=}CF_2)}$$

$$FSO_2{-}O{-}(CF_2{-}CF_2)_n{-}O{-}SO_2F \longrightarrow F{-}\underset{\underset{O}{\|}}{C}{-}(CF_2{-}CF_2)_{n-2}{-}\underset{\underset{O}{\|}}{C}{-}F \qquad (22.12)$$

Metal alkyls can be generated by anodic oxidation of a grignard reagent at a sacrificial metal anode. Tetraethyl lead is generated in this manner [16]:

$$4\ RMgCl + Pb^o \xrightarrow{\text{Pb-anode}} R_4Pb + 4\ MgCl^+$$

$$4\ MgCl^+ \xrightarrow{\text{cathode}} 2\ Mg^o + 2\ MgCl_2 \qquad (22.13)$$

$$4\ Mg^o + 4\ RCl \longrightarrow 4\ RMgCl$$

## B. Cathodic Reductions

The hydrogenation of heteroaromatic compounds can easily be performed at lead cathodes in sulfuric acid electrolytes. Examples are the formation of piperidine or hexahydrocarbazole [17]:

$$\text{pyridine} \xrightarrow[H_2O,\ H_2SO_4]{\text{Pb cathode}} \text{piperidine (NH)} \qquad (22.14)$$

Similarly, phthalic acid anhydride can be reduced to dihydrophthalic acid, if dioxane is present as cosolvent [18]:

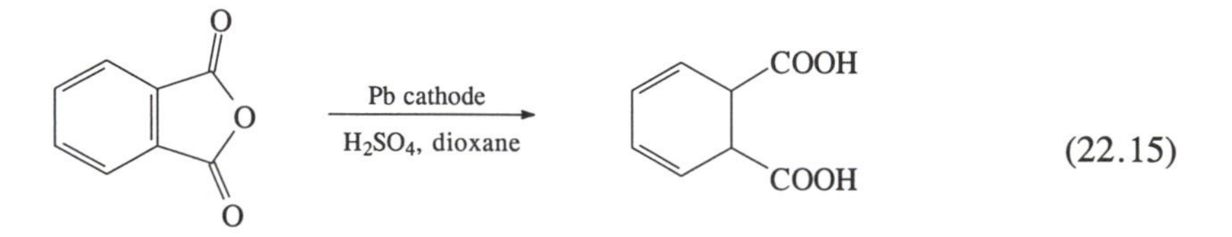


(22.15)

Starting from nitrobenzene, *p*-aminophenol can be generated at lead cathodes via the intermediate hydroxyl amine, which rearranges under the reaction conditions. This reaction is also unprecedented by other methods [19]:

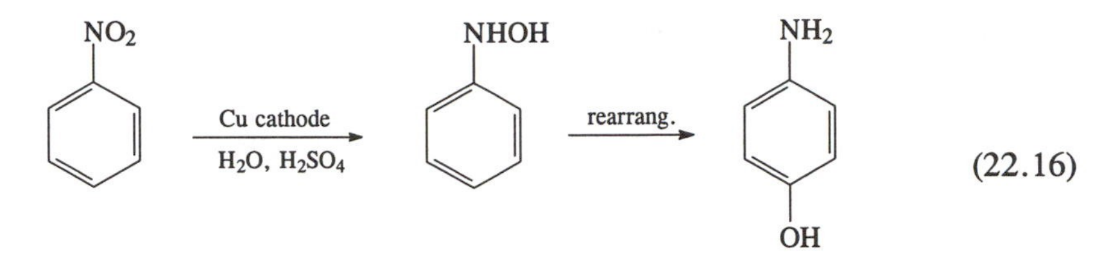


(22.16)

Cathodic hydrodimerization has been applied to a large number of electron-poor olefins. The most prominent example is the hydrodimerization of acrylonitrile to give adipodinitrile. This so-called "Monsanto process" has a production capacity of about 300,000 tons per year worldwide [20]:

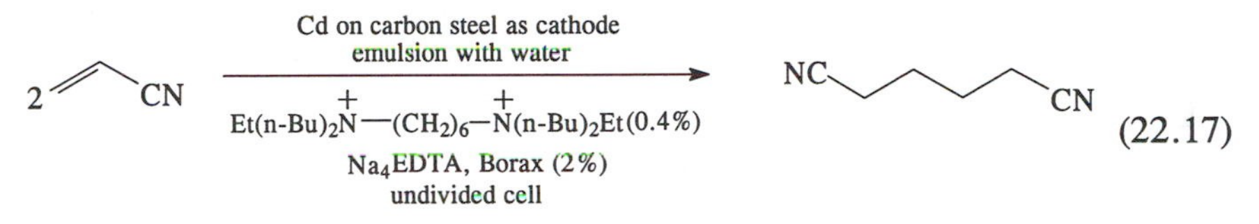


(22.17)

Cathodic dehalogenation has been applied for the reduction of dichloroacetic acid to monochloroacetic acid. Thus, in the synthesis of chloroacetic acid, the always formed dichloroacetic acid can be recycled to give the desired product and chlorine, which is used again for the chlorination [21]:

$$CHCl_2\text{—}COOH \xrightarrow[Pb^{2+},\ 2H^+]{\text{C cathode}} CH_2Cl\text{—}COOH + HCl$$

$$2HCl \xrightarrow{\text{anode}} Cl_2 + 2H^+ \qquad (22.18)$$

The cathodic substitution of halides has been applied in the synthesis of anti-inflammatory agents. Arylpropionic acids are formed by reductive dehalogenation of the corresponding benzylic halides in the presence of carbon dioxide as electrophile [22]:

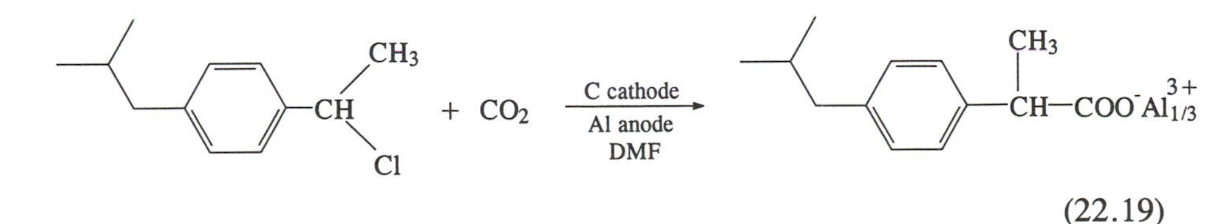


(22.19)

## IV. CLASSIFICATION OF ELECTROORGANIC REACTIONS

Electrochemical reactions can be classified by their reaction mechanisms.

## A. Electrochemical Substitution

*Substitution at the Anode*

Substitution of an electrophile (E) by a nucleophile (Nu):

$$R{-}E + Nu^- \xrightarrow[-2e^-]{} R{-}Nu + E^+$$

A typical example is the nuclear acetoxylation of an aromatic compound [23]:

$$R{-}C_6H_4{-}H + AcO^- \xrightarrow{-2e^-} R{-}C_6H_4{-}OAc + H^+ \qquad (22.20)$$

This is exemplified in α-naphthol formation (Eq. 22.2) or in the synthesis of *p*-methoxybenzonitrile from 1,4-dimethoxybenzene (hydroquinone dimethyl ether) [24]:

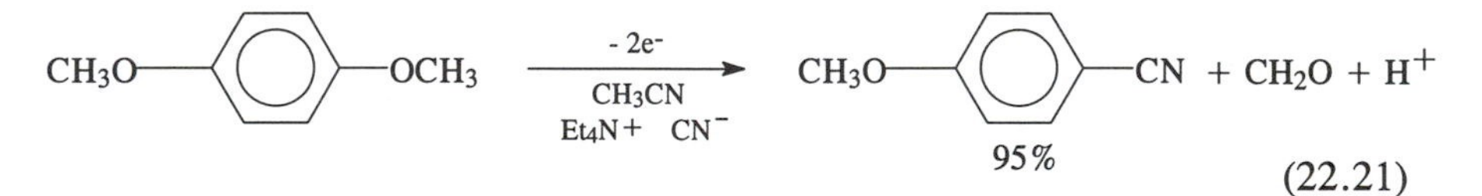


(22.21)

Side-chain substitutions of alkyl aromatic compounds are applied in industry (Eq. 22.1).

Anodic substitution can also occur easily in the α-position to heteroatoms like nitrogen or oxygen. Thus, the indirect electrochemical oxidation of ethylene glycol dimethyl ether in methanol using tris(2,4-dibromophenyl)amine as redox catalyst leads to the formation of 2-methoxyacetaldehyde dimethylacetal [25]:

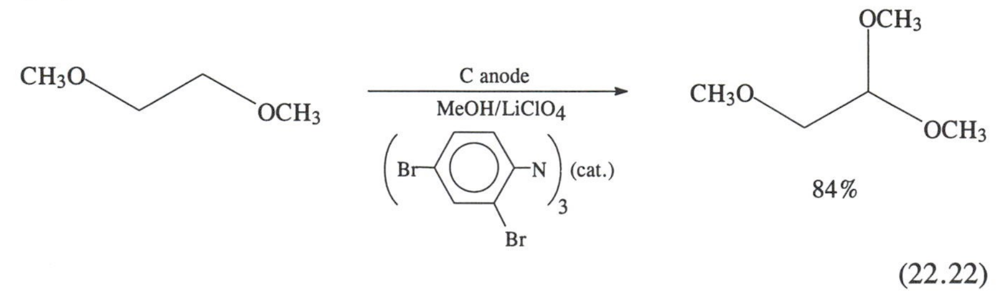


(22.22)

Especially the α-methoxylation or acetoxylation of amides like *N*-acyl and *N*-alkoxycarbonyl derivatives of amines, amino acids, and peptides leads to the synthetically very useful *N*,*O*-acetals, which are powerful amidoalkylation reagents [26]:

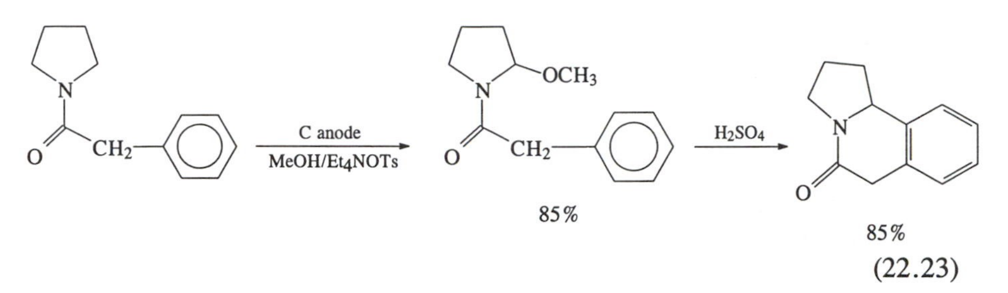


(22.23)

In the case of *N*-acylated amino acid derivatives, the oxidation has to take place via an indirect electrochemical process using chloride ions as redox catalysts. In methanol as solvent, methylhypochlorite is formed as oxidant, which yields, via *N*-chlorination, HCl elimination, and methanol addition, the α-methoxylated amino acids as amidoalkylation reagents [27]:

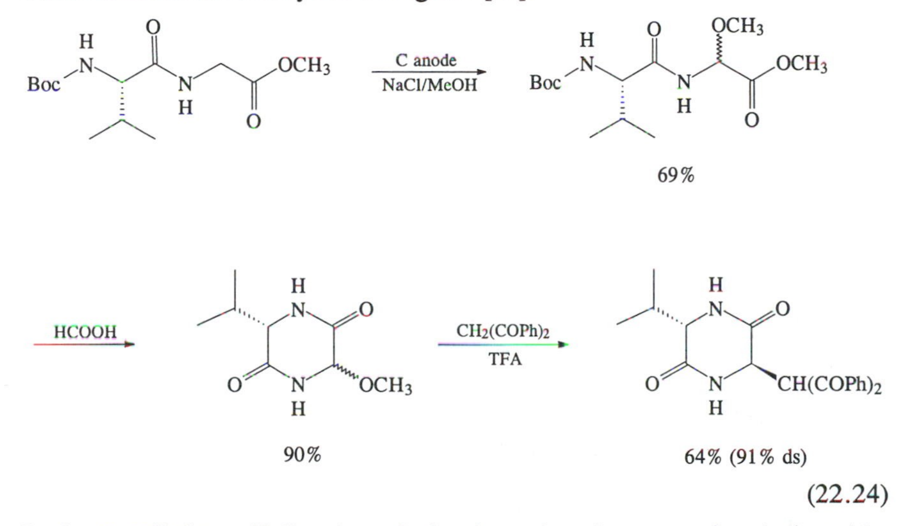


(22.24)

In the so-called non-Kolbe electrolysis, the carboxylate group is substituted by a nucleophile like methanol. This is especially effective if heteroatoms like nitrogen or oxygen are situated in the α-position. A number of amino acid derivatives have thus been transformed into the already mentioned *N,O*-acetals as effective amidoalkylation reagents [28]:

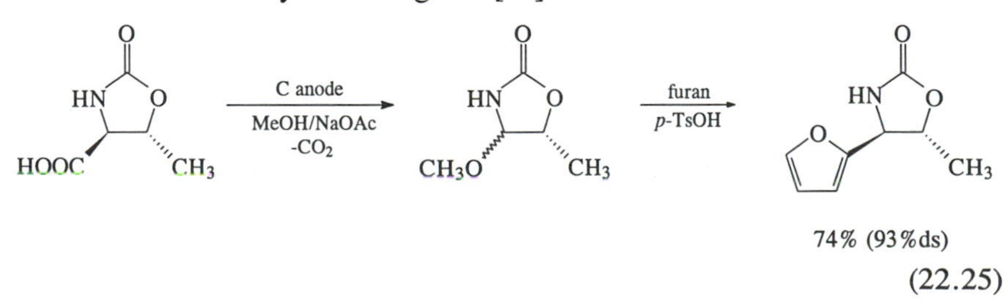


(22.25)

*Substitution at the Cathode*

Substitution of a nucleophile (X) by an electrophile (E):

$$R{-}X + E^+ \xrightarrow[+2e^-]{} R{-}E + X^-$$

A typical example is the substitution of a halide ion by an electrophile like a carbonyl compound [29]:

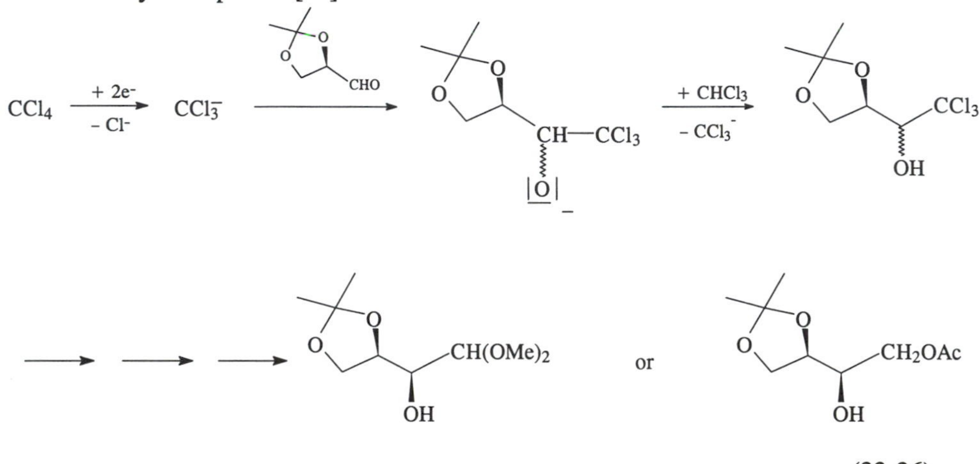


(22.26)

In other cases, carbon dioxide has been used as an electrophile, such as in the synthesis of arylpropionic acids (Eq. 22.19).

In an indirect electrochemical process, vitamin $B_{12}$ has been used as catalyst for the cathodic alkylation of $\alpha,\beta$-unsaturated carbonyl compounds by alkyl iodides as exemplified in the formation of (+)-*exo*-brevicomine [30]:

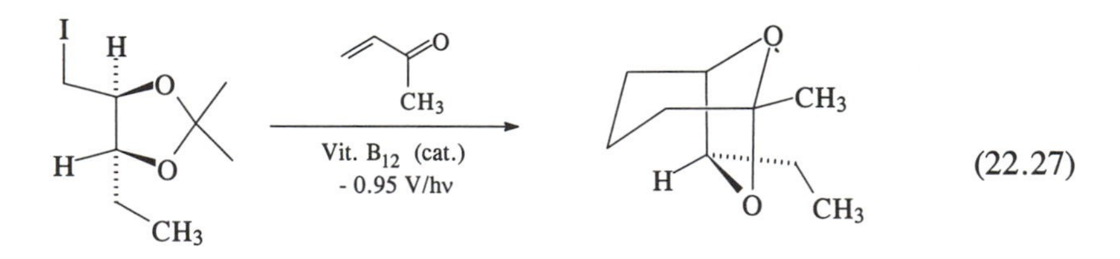


(22.27)

## B. Electrochemical Addition

*Addition at the Anode*

Addition of two nucleophiles to an electron-rich double-bond:

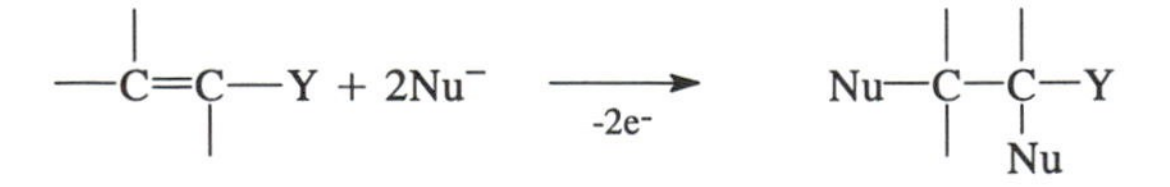

A typical example is the dimethoxylation of conjugated dienes [31]:

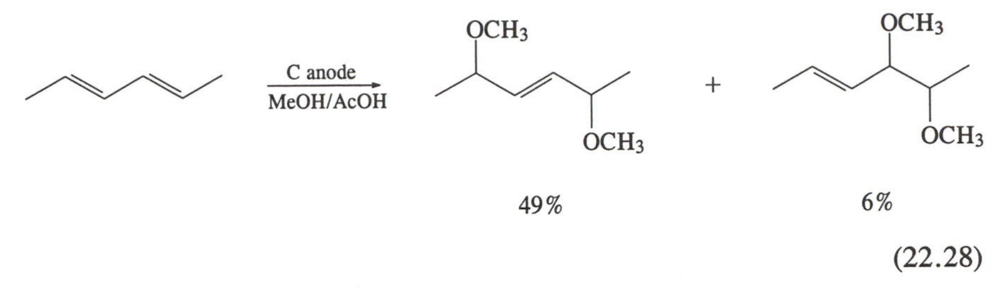


(22.28)

Another example has been given in the dimethoxylation of furan (Eq. 22.4).

*Addition at the Cathode*

Addition of two electrophiles to an electron-poor C-C-double bond (C=C) or a C-heteroatom double bond (C=X):

$$-\overset{|}{C}=X + 2E^+ \xrightarrow[+2e^-]{} E-\overset{|}{\underset{|}{C}}-XE$$

The electrochemical hydrogenation of double bonds can be performed either electrocatalytically at Raney nickel, palladium, or platinum modified electrodes [32] or via electron transfer under Birch conditions to the intermediate anion radical [33]. Examples are given in the dihydrogenation of phthalic acid (Eq. 22.15) and the hydrogenation of heteroaromatic compounds (Eq. 22.14).

Another typical example is the addition of two protons to a carbonyl group [34]:

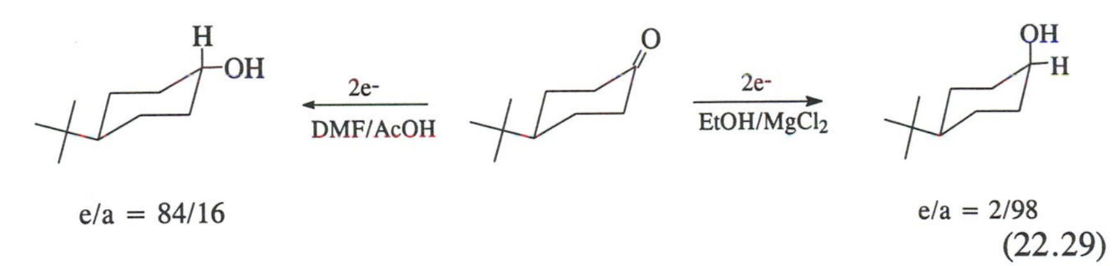


(22.29)

## C. Electrochemical Elimination

*Elimination at the Anode*

Elimination of two electrophiles:

$$-\underset{E}{\overset{|}{C}}-\underset{E}{Y} \xrightarrow[-2e^-]{} \rangle C=Y + 2E^+ \quad Y=O, CH_2 \quad E^+=H^+, CO_2, R_3Si^+$$

Typical examples are bisdecarboxylations [35] and decarboxylative desilylations [36]:

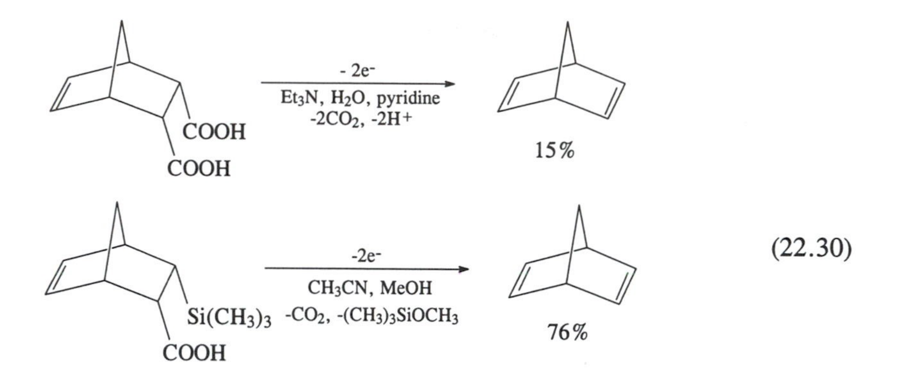


*Elimination at the Cathode*

Elimination of two nucleophiles:

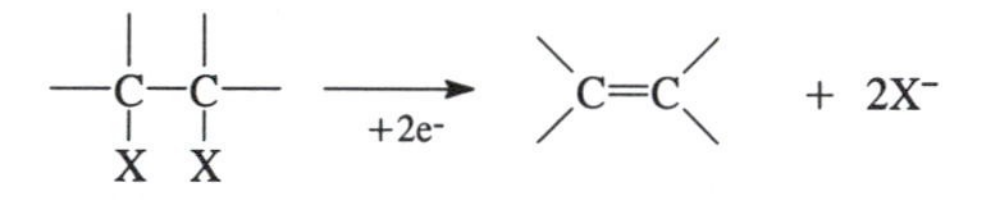


A typical example is the dehalogenation of 1,2-dihalides. The elimination of 1,2-dibromides occurs stereospecifically in the form of an *anti*-elimination. Thus, the *meso*-dibromide yields the (*E*)- and the *d,l*-dibromide yields the (*Z*)-alkene selectively [37]:

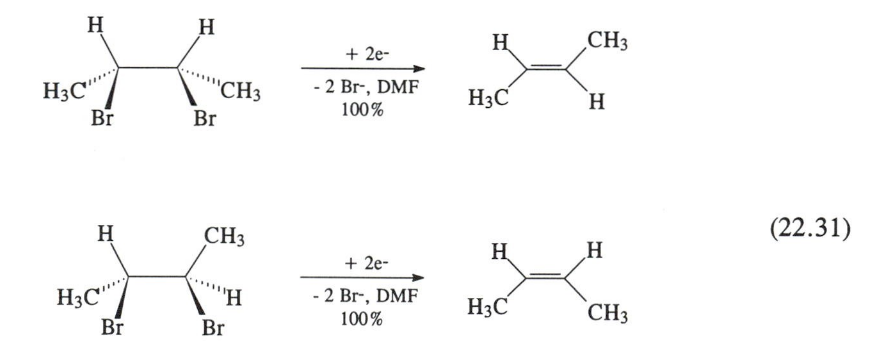


Even highly strained cyclic olefins can thus be obtained [38]:

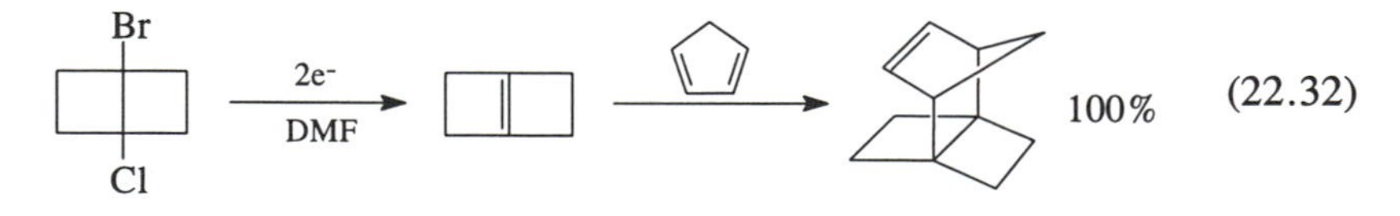

The 1,1- and 1,3-eliminations can also be performed cathodically.

## D. Electrochemical Coupling

*Coupling at the Anode*

Coupling of two electron-rich components like alkenes, enol acetates, enol ethers, carbanions, or carboxylates:

$$\text{Type 1} \quad 2RH \xrightarrow[-2e^-]{} R{-}R + 2H^+ \qquad \text{via cation radicals}$$

$$\text{Type 2} \quad 2R^- \xrightarrow[-2e^-]{} R{-}R \qquad \text{via radicals}$$

A typical example for type 1 is the dimerization of electron-rich olefins like aryl olefins, enol ethers, enol acetates, conjugated dienes, or enamino esters [39]:

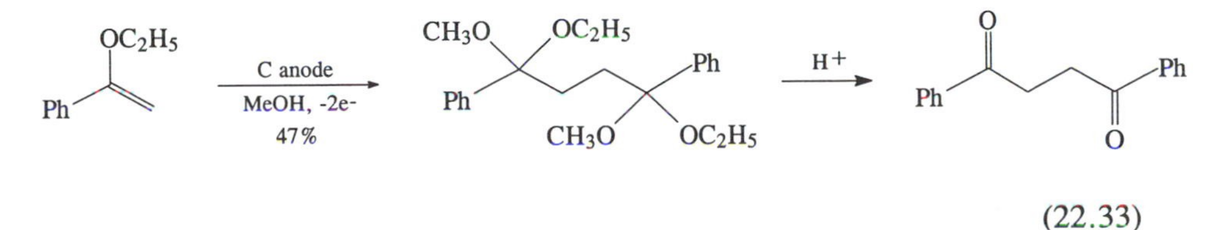


(22.33)

A retrosynthetic analysis leads to one synthon with reversed polarity and another with normal polarity:

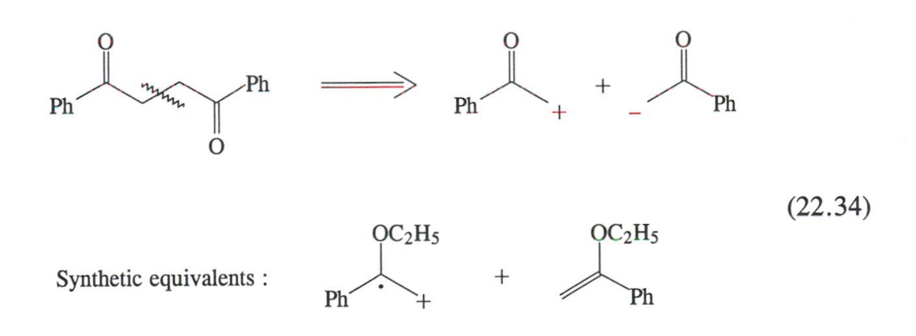


(22.34)

A typical example for type 2 is the Kolbe dimerization of carboxylates [40] as exemplified in the dimerization of half esters like adipates to give sebacates (see Eq. 22.5). Alkatrienes, sex attractants of Lepidoptera, were prepared by mixed Kolbe electrolysis starting from linolenoic acid [41]:

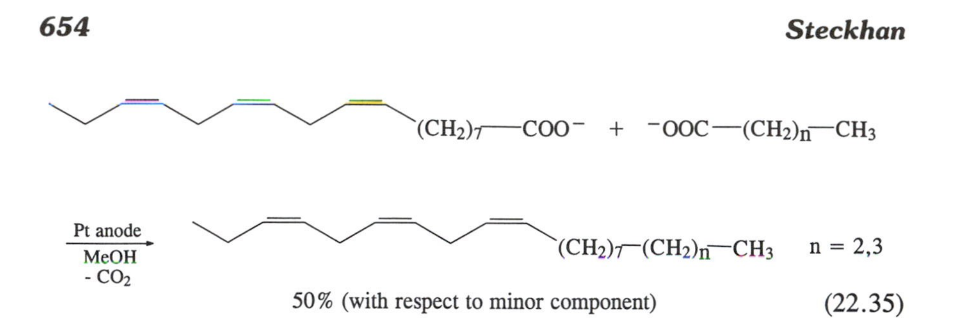


(22.35)

*Coupling at the Cathode*

Coupling of two electron-poor components:

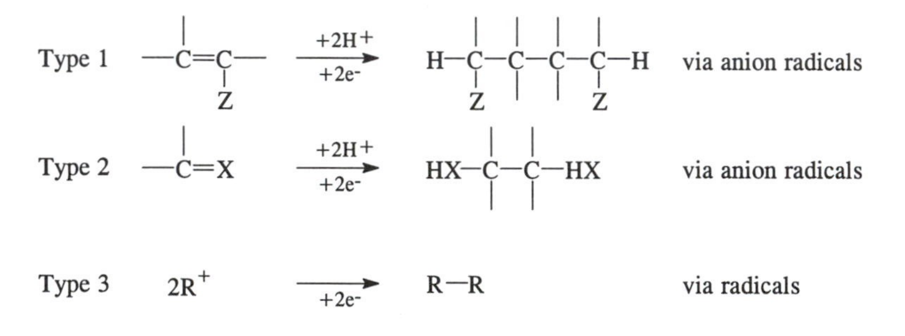


A typical example for type 1 is the electrohydrodimerization of electron-poor olefins like acrylonitrile (Eq. 22.17). The electrohydrodimerization can also be performed intramolecularly as shown in Equation 22.36 for an unsaturated diester, leading predominantly to the trans isomer [42]:

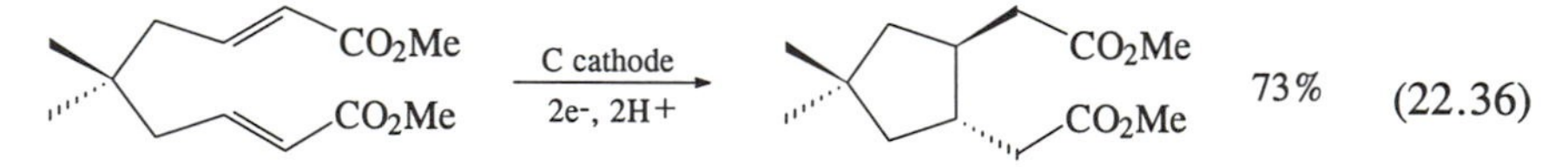


(22.36)

A typical example for type 2 is the pinacolization of acetone [43]:

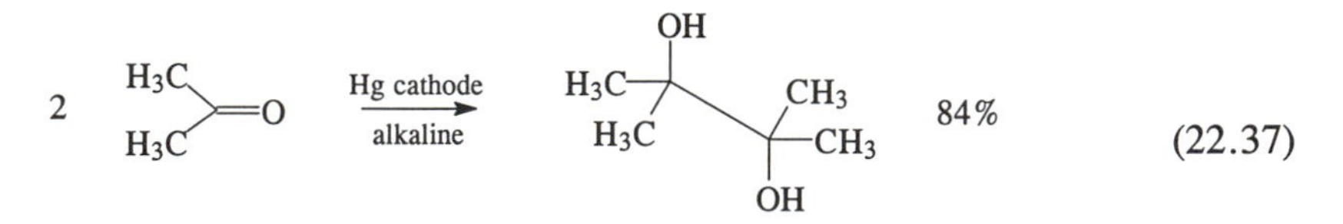


(22.37)

It has been observed that in the presence of Cr(III), the pinacolization is strongly favored over alcohol formation [44]:

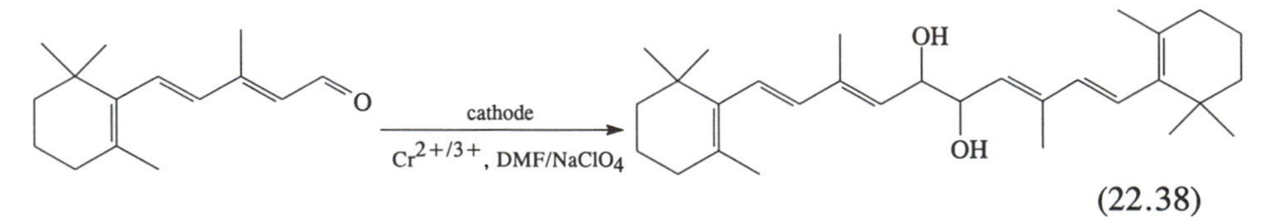


(22.38)

A typical example for type 3 is the dimerization of the tropylium ion [45]:

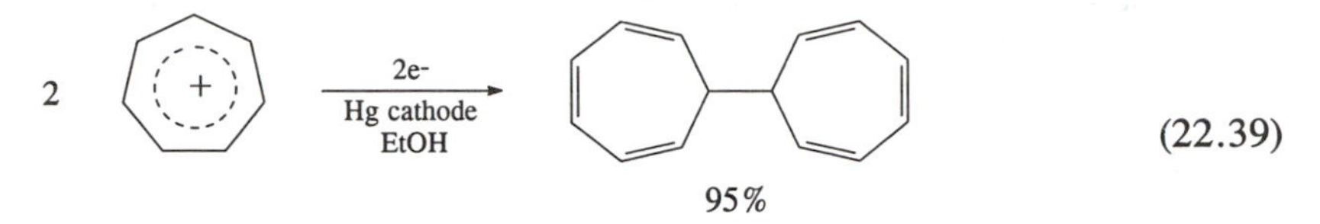


(22.39)

Retrosynthetic analysis of the adiponitrile synthesis leads to one synthon with reversed polarity and another with normal polarity:

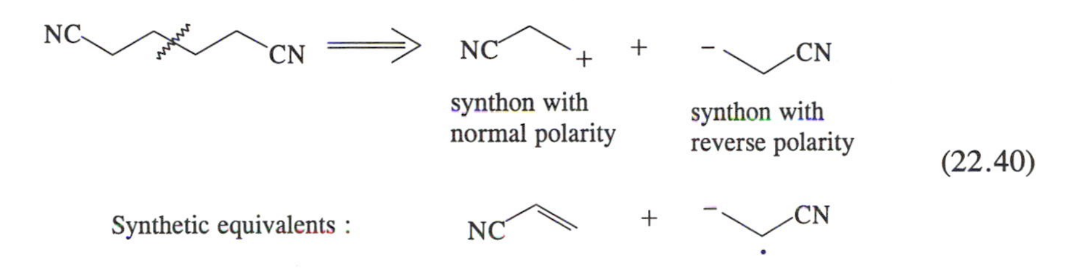


(22.40)

## E. Electrochemical Cleavage

*Cleavage at the Anode*

Type 1 $\quad R{-}X \xrightarrow[-e^-]{} R{-}X^{\cdot+} \longrightarrow R^+ + \tfrac{1}{2}X_2$

Type 2 $\quad RCH_2X \xrightarrow[-e^-]{} RCH_2X^{\cdot+} \xrightarrow[-H^+]{} R\text{-}\dot{C}H\text{-}X$

$\xrightarrow[e^-]{} RCH{=}X^+ \xrightarrow[H_2O]{} RCH{=}O + HX + H^+$

Typical examples for type 1 are the anodic cleavages of two carbon–sulfur bonds in 1,3-dithianes [46] or dithiolanes [47]. This reaction is especially effective if performed under the conditions of indirect electrolysis using triarylamine cation radicals as regenerable oxidative mediators [47]:

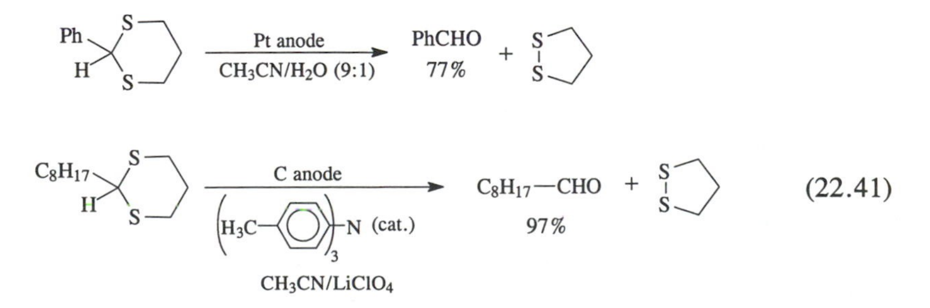


(22.41)

Similarly, thiol esters may be cleaved [48].

Typical examples for type 2 are the cleavage of the carbon–oxygen bond in benzylic ethers (X = OR) either directly or indirectly via triarylamine cation radicals [49] (Eq. 22.42), the cleavage of the carbon–nitrogen bond in amines (X = $NR_2$) [50] (Eq. 22.43), and the cleavage of the carbon–sulfur bond in sulfides (X = SR) [51] (Eq. 22.44):

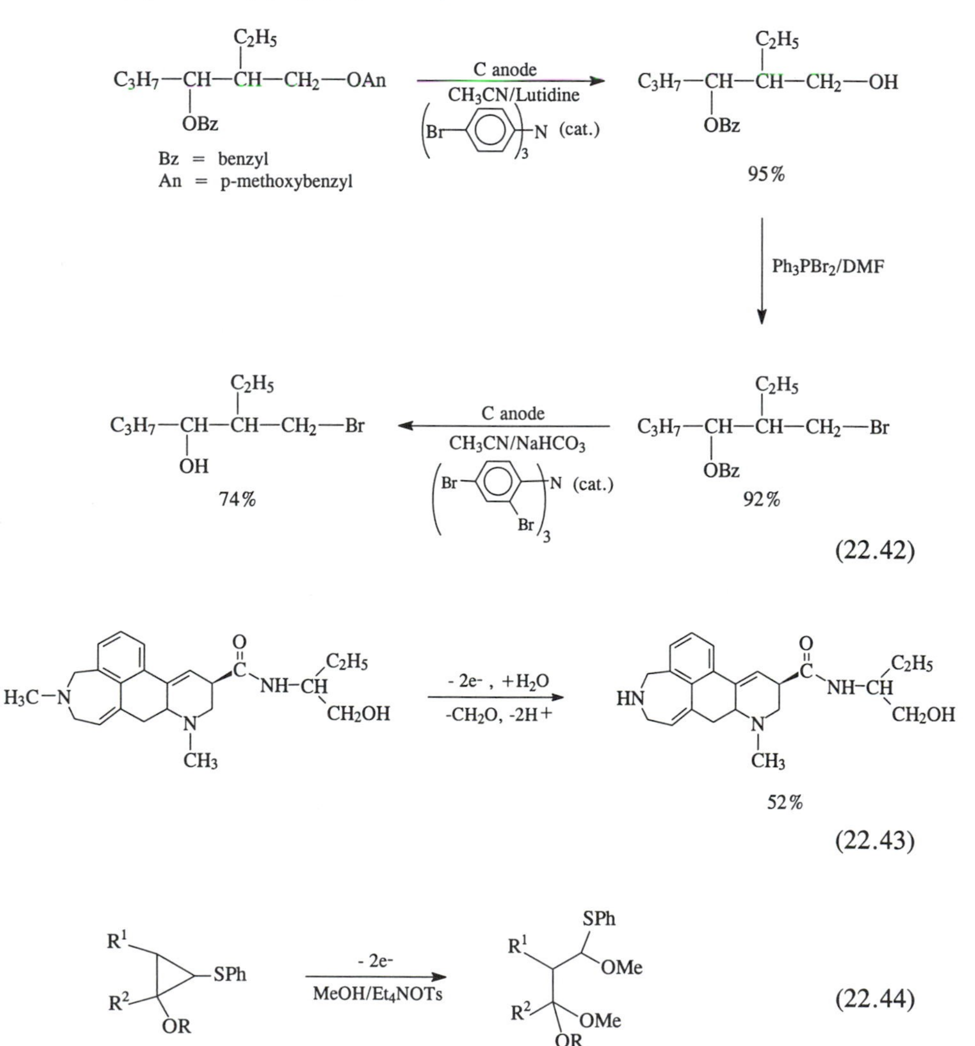


These reactions have been used for the deblocking of protecting groups. In the case of amines, the formation of metabolites is possible because the anodic oxidation of amines equals the microsomal oxidation to a large extent.

*Cleavage at the Cathode*

$$\mathrm{R{-}X} \xrightarrow[+\,e^-]{} \mathrm{R{-}X^{\cdot -}} \longrightarrow \mathrm{R^{\cdot}} + \mathrm{X^-}$$
$$\mathrm{R^{\cdot}} \xrightarrow{+\,e^-} \mathrm{R^-}$$

In the cathodic cleavage, halogen, phosphonium, ammonium, nitrile, thioalkyl, or sulfonate groups can act as leaving groups and are replaced by hydrogen. Selectivity between different leaving groups can be obtained by potentiostatic reduction (Eq. 22.45). The application to deprotection has been used successfully [52]. The selectivity in the cathodic cleavage of phosphonium salts was used very early for the formation of optically active phosphine and phosphonium ions [53]. The carbon sulfur bond in aliphatic diphenyldithioacetals, α-carbonyldiphenyldithioacetals, and α-carbonylketene dimethyldithioacetals can be cleaved cathodically in the presence of tetrabutylammonium hydrogen sulfate (Eq. 22.46) [54]:

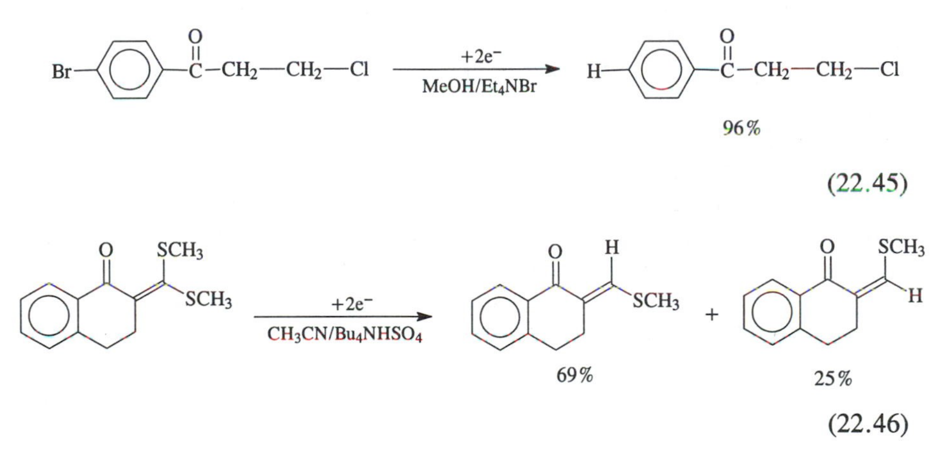


Another important classification of electroorganic reactions is obtained by dividing them into those in which the substrate undergoes direct electron transfer with the electrode (direct electrolysis) and those in which an additional compound (redox catalyst, mediator) transfers the redox equivalents between the substrate and the electrode (indirect electrolysis).

## F. Indirect Electrochemical Reactions

In indirect electrochemical reactions [3], the heterogeneous reaction between the substrate and the electrode is replaced by a homogeneous redox reaction in

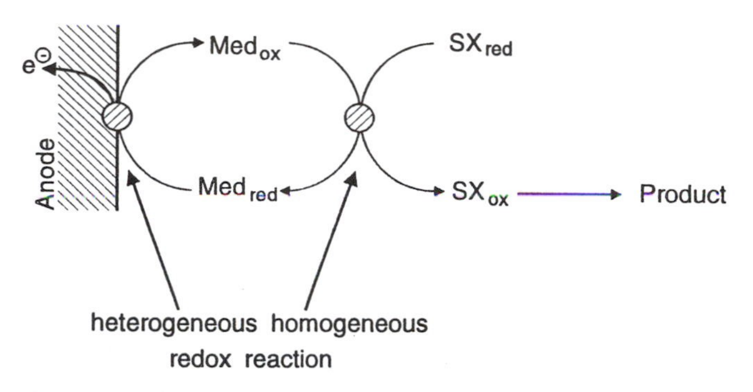


**Figure 22.1** Schematic representation of an indirect electrochemical process (given for an oxidation; $Med_{ox}$, $Med_{red}$: oxidized and reduced forms of the redox catalyst = mediator; $S_{ox}$, $S_{red}$: oxidized and reduced forms of the substrate).

solution between the substrate and an electrochemically activated and regenerated redox catalyst (see Fig. 22.1).

Two types of redox catalysts are used in indirect electrolyses [3]: (1) pure, outer-sphere, or non-bonded electron transfer agents, and (2) redox reagents that undergo a homogeneous chemical reaction that is intimately combined with a redox step. This may be called inner-sphere electron transfer or bonded electron transfer mechanism.

In type 1, the mediator catalyzes the electron exchange between the electrode and the substrate. Compared to direct electrolyses, overvoltages may be reduced and passivation of electrode surfaces may be avoided. If the homogeneous electron transfer equilibrium is followed by a fast and irreversible chemical step, mediators can be used with redox potentials that are up to 600 mV lower than the standard potential of the substrate (electron transfer against the standard potential gradient). Also, slow heterogeneous electron transfers at the electrode surface can be replaced by fast homogeneous electron transfers in solution. This is especially valuable if large biomolecules are the substrates (see Sec. IV.G). Typical examples of such electron transfer mediators are stable radical anions or dianions of mostly aromatic or heteroaromatic systems [3] or the stable radical cations of substituted triarylamines [3,55]. Examples of this type are given in Equations 22.22, 22.41, and 22.42.

In type 2, the homogeneous redox reaction of the electrogenerated and regenerated redox catalyst consists of a chemical reaction. For oxidations, these reactions may be hydride ion or hydrogen atom abstraction, oxygen transfer, or an intermediate complex or bond formation. For reductions, hydride or carbanion transfer from a metal complex is often observed. In all these cases, very large potential differences between the standard potential of the substrate and the redox catalyst may be overcome. The selectivity can be very high and may be tuned by the proper selection of the metal ions and the ligands in the case of metal complexes as redox catalysts. All chemical steps involved must be very fast for an efficient redox catalytic system. Most of the inorganic redox catalysts, especially transition metal complexes, act in this way. For oxidations, not only transition metal salts or complexes but also halide ions are applied. Examples are given in Equations 22.3, 22.4, 22.6, 22.9, 22.10, 22.24, and 22.27.

## G. Electroenzymatic Reactions

Electroenzymatic reactions are not only important in the development of amperometric biosensors. They can also be very valuable for organic synthesis. The enantio- and diasteroselectivity of the redox enzymes can be used effectively for the synthesis of enantiomerically pure compounds, as, for example, in the enantioselective reduction of prochiral carbonyl compounds, or in the enantioselective, distereoselective, or enantiomer differentiating oxidation of chiral, achiral, or *meso*-polyols. The introduction of hydroxy groups into aliphatic and aromatic compounds can be just as interesting. In addition, the regioselectivity of the oxidation of a certain hydroxy function in a polyol by an enzymatic oxidation can be extremely valuable, thus avoiding a sometimes complicated protection–deprotection strategy.

As is the case with all redox enzymes, freely dissociated (NADH, NADPH) or enzyme-bound (FAD, PQQ) cofactors are necessary to shuttle the redox equivalents from the enzyme to the substrate. For synthetic applications, therefore, the main problem is usually the cofactor regeneration. Until now, this problem has been generally solved by the application of a second enzymatic reaction (enzyme-coupled regeneration). Thus, besides a production enzyme for the transformation of the substrate into the product, a regeneration enzyme together with its cosubstrate is necessary to reactivate the cofactor of the production enzyme. The additionally formed coproduct then has to be separated from the desired product [56]. As these systems can be complicated and not stable enough for synthetic purposes, it would be extremely interesting to use the reagent-free electrochemical procedure to regenerate the cofactors. In the case of the enzymatic oxidation with $NAD^+$-dependent alcohol dehydrogenases, direct electrochemical oxidation of NADH has been applied successfully. By using high-surface-area carbon electrodes, like reticulated vitreous carbon or especially carbon felt, the otherwise slow NADH oxidation can be performed effectively. As a model system, the production of D-gluconic acid from β-D-glucose, catalyzed by glucose dehydrogenase, could be performed using the direct electrochemical regeneration of the cofactor $NAD^+$ at a flow-through graphite felt anode [57] (Fig. 22.2).

The deracemization of lactate in principle is also interesting. In this process, pyruvate is reduced at the cathode to racemic lactate which is reoxidized to pyruvate by the cheap L-lactate dehydrogenase combined with the anodic

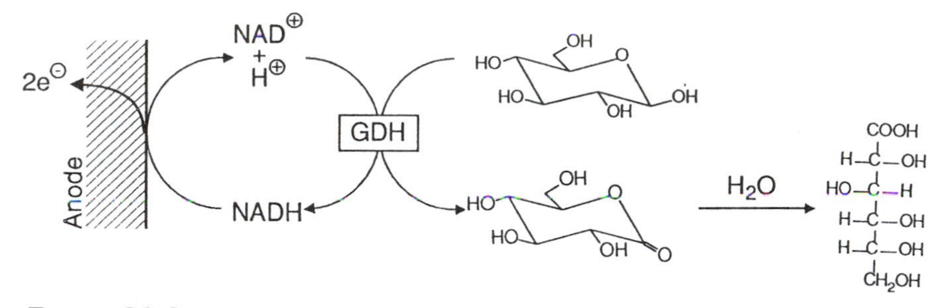


**Figure 22.2** Electroenzymatic oxidation of β-D-glucose with direct electrochemical $NAD^+$ regeneration (GDH = glucose dehydrogenase).

regeneration of the oxidized cofactor $NAD^+$. As D-lactate is not accepted by the enzyme it is accumulated during the process [58] (Fig. 22.3).

The electron transfer between NADH and the anode may be accelerated by the use of a mediator. Synthetic applications have been described for the oxidation of primary and secondary alcohols to aldehydes and ketones catalyzed by yeast alcohol dehydrogenase (YADH) and the alcohol dehydrogenase from *Thermoanaerobium brockii* (TBADH) with indirect electrochemical regeneration of $NAD^+$ and $NADP^+$, respectively, using the tris(3,4,7,8-tetramethyl-1,10-phenanthroline) iron(II/III) complex as redox catalyst [59].

For enzymatic reductions with NAD(P)H-dependent enzymes the electrochemical regeneration of NAD(P)H always has to be performed by indirect electrochemical methods. Direct electrochemical reduction, which requires high overpotentials, always leads to larger or smaller amounts of enzymatically inactive NAD dimers generated in a one-electron transfer reaction. If one-electron transfer redox catalysts are used as mediators, the same problem occurs.

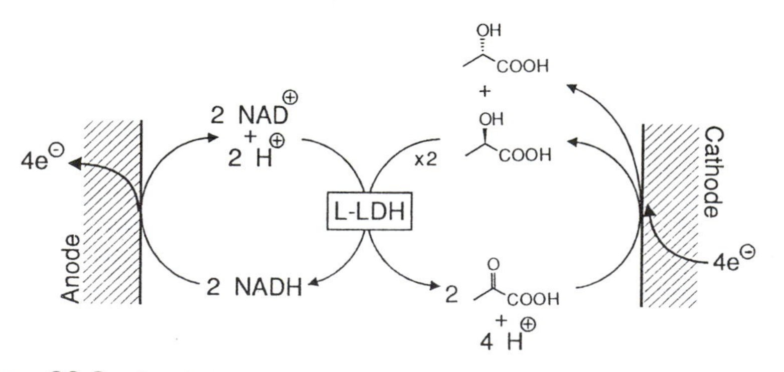


**Figure 22.3** Coupled electrochemical process for the formation of D-lactate from pyruvate (L-LDH = L-lactate dehydrogenase).

The dimerization can only be prevented if an additional regeneration enzyme is applied, which must be able to accept two electrons in two steps from the mediator and then to transfer one electron pair to $NAD(P)^+$. This is the case if ferredoxin reductase [60], lipoamide dehydrogenase [61], enoate reductase [62], or the so-called "VAPOR" enzymes (viologen-accepting pyridine nucleotide oxidoreductases) [63] are used as regeneration enzymes in combination with methyl viologen (*N*,*N*-dimethyl-4,4-bipyrinium dichloride) as a one-electron mediator. Instead of the isolated enzymes, resting cells of the microorganisms containing both the production enzyme and the cofactor-regenerating enzyme can be used successfully [61]. To be able to regenerate NADP(H) by an indirect electrochemical procedure without the application of a second regeneration enzyme system, the redox catalyst must fulfill three conditions:

1. The active redox catalyst must transfer two electrons in one step or a hydride ion.
2. The electrochemical activation of the catalyst must be possible at potentials less negative than –0.9 V vs. SCE. At more negative potentials, the direct electrochemical reduction of $NAD(P)^+$ will lead to NAD dimer formation.
3. The active form of the catalyst must transfer the electrons or the hydride ion to $NAD(P)^+$, but not directly to the substrate. Otherwise, this nonenzymatic reduction will lead to low chemoselectivity and/or low enantioselectivity.

Systems that fulfill these conditions are substituted or unsubstituted (2,2′-bipyridyl)(pentamethylcyclopentadienyl)rhodium complexes. Electrochemical reduction of these complexes at potentials between –680 mV and –840 mV vs. SCE leads to the formation of rhodium hydride complexes. Strong catalytic effects observed in cyclic voltammetry and preparative electrolyses are indicating a very fast hydride transfer from the complex to $NAD(P)^+$ under formation of the starting complex, as shown in the following reaction scheme [64]:

$$[Cp^*Rh(bpy)L]^{2+} + e^- \rightleftharpoons [Cp^*Rh(bpy)L]^+$$

$$[Cp^*Rh(bpy)L]^+ \longrightarrow [Cp^*Rh(bpy)]^+ + L$$

$$[Cp^*Rh(bpy)L]^+ + e^- \rightleftharpoons Cp^*Rh(bpy) + L$$

$$[Cp^*Rh(bpy)]^+ + e^- \rightleftharpoons Cp^*Rh(bpy)$$

$$[Cp^*Rh(bpy)L]^+ + [Cp^*Rh(bpy)]^+ \rightleftharpoons Cp^*Rh(bpy) + [Cp^*Rh(bpy)L]^{2+}$$

$$Cp^*Rh(bpy) + H^+ \longrightarrow [Cp^*Rh(bpy)H]^+$$

$$[Cp^*Rh(bpy)H]^+ + NAD(P)^+ + L \longrightarrow [Cp^*Rh(bpy)L]^{2+} + NAD(P)H$$

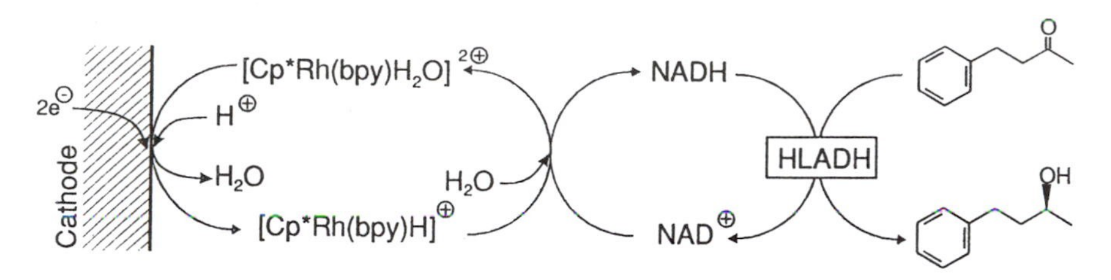


**Figure 22.4** Electroenzymatic reduction of 4-phenyl-2-butanone catalyzed by HLADH with in situ indirect electrochemical regeneration of NADH.

This system has been successfully applied to the in-situ electroenzymatic reduction of pyruvate to D-lactate using the NADH-dependent D-lactate dehydrogenase or the reduction of 4-phenyl-2-butanone to (*S*)-4-phenyl-2-butanol using the NADH-dependent horse liver alcohol dehydrogenase (HLADH) with high enantioselectivity (Fig. 22.4) [65].

In the field of enzymatic oxidations especially, the class of flavoenzymes with bound FAD as cofactor is interesting for synthetic applications. The regeneration of the oxidized FAD within the enzyme can be performed by oxygen. In this case, however, hydrogen peroxide is formed, which drastically diminishes the enzyme stability and activity, rendering it unsuitable for synthesis.

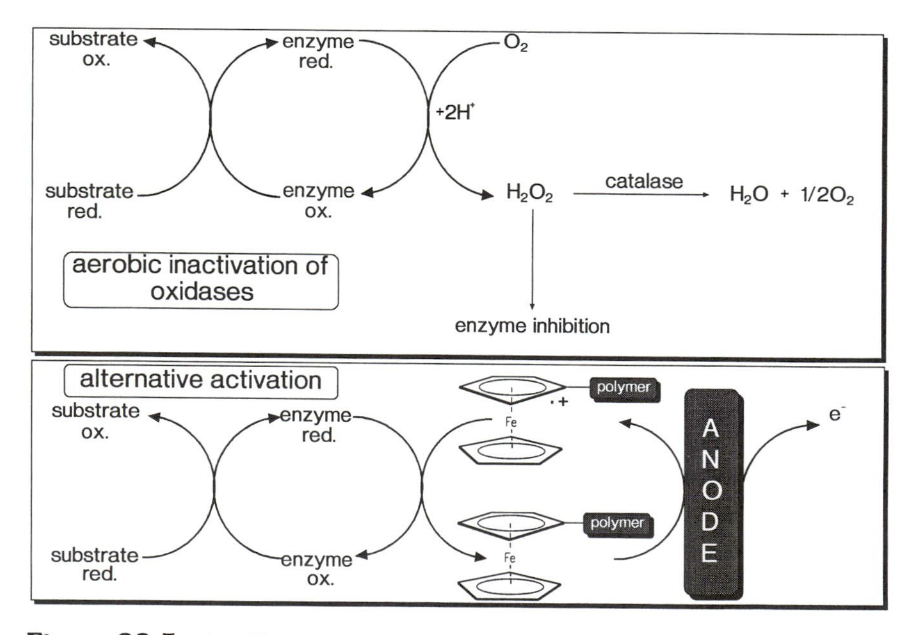


**Figure 22.5** Aerobic versus anaerobic electrochemical activation of FAD-dependent oxidases.

Even the presence of large amounts of catalase to destroy the hydrogen peroxide very often does not lead to a system that is stable enough for the application in synthesis. Therefore, an anaerobic reactivation of flavoenzymes and other similar oxidizing enzymes is necessary. The most simple anaerobic oxidation would be the electrochemical procedure. However, as pointed out earlier, the large biomolecules do not undergo fast enough electron exchange with the electrode. For analytical purposes, this limitation can be solved by the application of "promoter" molecules, which are adsorbed at the electrode surface. For preparative applications, the most effective way is the indirect electrochemical procedure. As mediators, ferrocenes are most effective with flavoenzymes (Fig. 22.5).

A first application using ferroceneboronic acid as mediator [66] has been described for the transformation of *p*-hydroxytoluene to *p*-hydroxybenzaldehyde, which is catalyzed by the enzyme *p*-cresolmethyl hydroxylase (PCMH) from *Pseudomonas putida*. This enzyme is a flavocytochrome containing two FAD and two cytochrome prosthetic groups. To develop a continuous process using ultrafiltration membranes to hold back the enzyme and the mediator, water-soluble polymer-bound ferrocenes have been applied as mediators for the application in batch electrolyses (Fig. 22.6) or in combination with an electrochemical enzyme membrane reactor (Fig 22.7) [67].

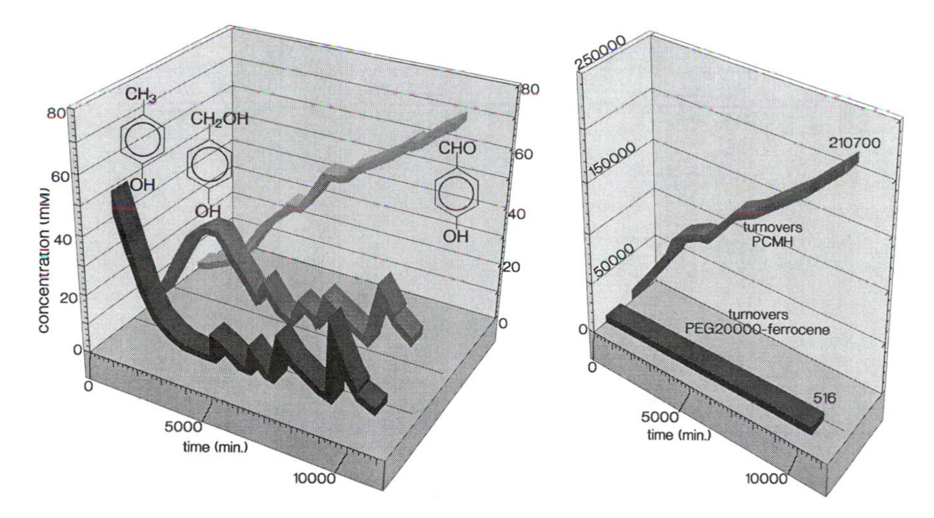


**Figure 22.6** Electroenzymatic oxidation of *p*-cresol under catalysis by PCMH in a long-time batch electrolysis. PCMH: 16 U = 5.6 nmol; PEG-20000 ferrocene: 0.51 mmol = 9.45 μmol ferrocene; starting concentration of *p*-cresol 41.25 m*M* = 0.66 mmol; additions of substrate after 4140 min (0.0925 mmol), 5590 min (0.0784 mmol), 6630 min (0.184 mmol), 11253 min (0.371 mmol), in 10 mL Tris/HCl-buffer of pH 7.6; divided cell: Sigraflex anode, 26 $cm^2$.

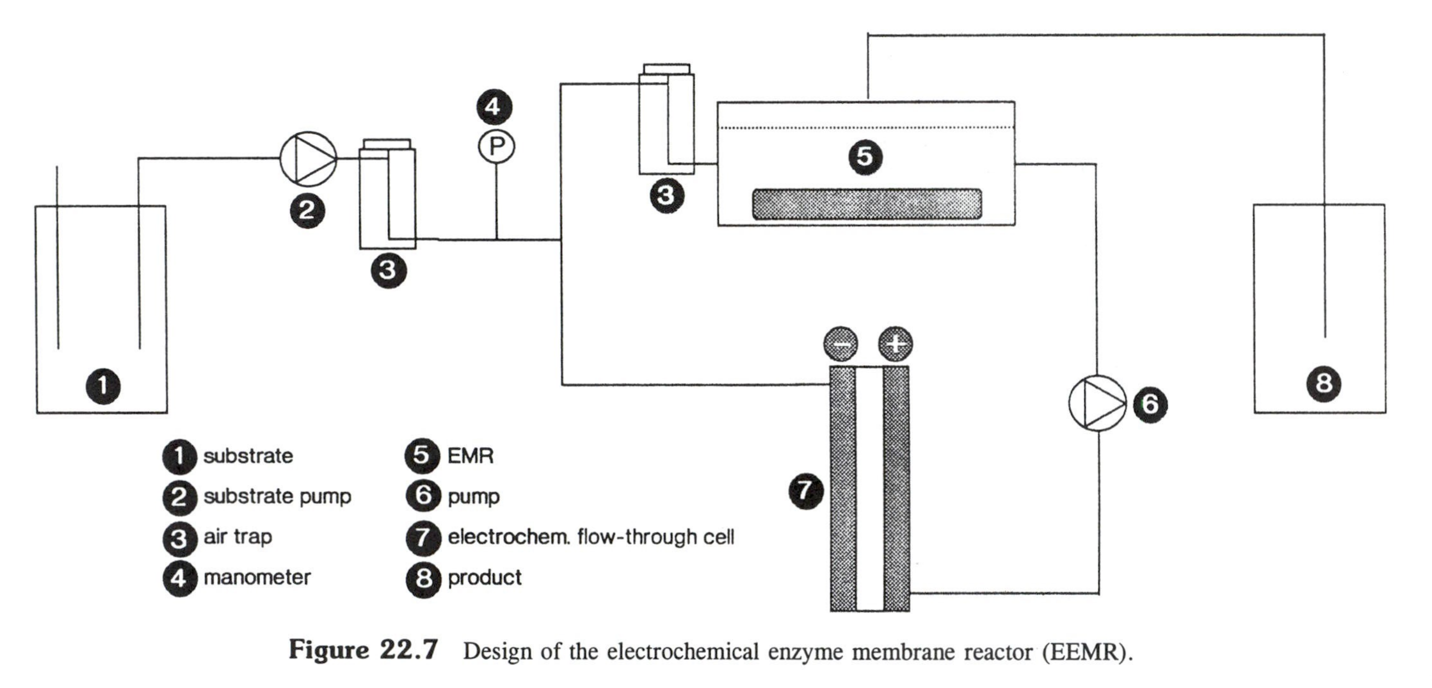


**Figure 22.7** Design of the electrochemical enzyme membrane reactor (EEMR).

Here, an electrochemical flow-through cell using a carbon-felt anode is combined with an enzyme–membrane reactor. The residence time is adjusted by the flow of the added substrate solution. The off-flow of the enzyme membrane reactor only contains the products *p*-hydroxybenzaldehyde and *p*-hydroxybenzylalcohol. By proper adjustment of the residence time and the potential, total turnover of the *p*-hydroxytoluene can be obtained. In a 14-day run, the enzyme underwent 500,000 cycles and the polymer-bound mediator, which was present in a higher concentration than the enzyme, more than 500 cycles. At the end the system was still active.

## V. EXPERIMENTAL FACTORS AND TECHNIQUES

The following experimental factors have to be taken into account for the synthetic application of an electrochemical reaction:

- The choice of the electrolysis procedure either working at controlled potential (potentiostatic or controlled-potential electrolysis, cpe) or at a fixed current density (galvanostatic)
- The type of electron exchange, which can take place either by direct heterogeneous electron transfer between the electrode and the substrate or by using an indirect process, in which a redox catalyst mediates the exchange of redox equivalents between the electrode and the substrate
- The redox potential applied at the working electrode, if controlled-potential electrolysis is selected
- The current density, if galvanostatic conditions are selected
- The charge consumption
- The composition of the electrolyte
- The type of electrode material
- The type of electrolysis cell, divided or undivided, flow-through or beaker-type

### A. Electrolysis Cells

The main features of electrolysis cells as used in electroanalytical applications are described in Chapter 9. For application in the laboratory and in industry, some other aspects are important. These aspects are discussed here.

For larger scale application, especially in industry, one should try to reach the following three goals:

1. To keep the energy consumption as low as possible, the distance between working and auxiliary electrode should be minimized. The energy consumption is the product of the cell voltage, $U_C$, and the current. The cell voltage is given by the following equation:

$$U_C = E_A + E_C + iR$$

$E_A$ and $E_C$ are the applied potentials at the electrodes, while iR is the sum of the ohmic drops at the resistances of the electrolyte and, for example, of the diaphragm. The electrolyte resistance is linearly dependent on the electrode distance. Therefore, this distance should be small and a diaphragm should be avoided (undivided cell), if possible. These conditions are best fulfilled by capillary gap cells with gaps of 0.1–0.2 mm. In this case, the concentration of the supporting electrolyte can also be very small.

2. The electrode surface should be as large as possible to reach a high space-time yield.
3. If possible, the cell should be undivided to minimize the construction cost and also the energy consumption (see goal 1). The application of a controlled reaction at the auxiliary electrode taking place at low potential allows for the use of undivided cells in many cases. For oxidations, the cathodic process at the auxiliary electrode may be a proton reduction under formation of hydrogen. For reductions, the anodic process may be the oxidation of formate or oxalate under production of carbon dioxide [68] or the dissolution of sacrificial anodes [69] (see also Sec. V.B).

For laboratory cells, minimizing the energy consumption and optimizing the space–time yield are not as important. It is more important that the different reaction parameters like electrode materials, diaphragms, and the working potentials can be varied easily. For electrolyses under potential control, a three-electrode construction has to be used, which is schematically shown in Figure 22.8.

Any direct current source may be used as a current source (CS), in connection with a voltmeter with high input impedance, to control the applied potential at the working electrode. The applied potential can be measured via the reference electrode (R) using a voltmeter with a high input resistance (V). If a potentiostat is available, it will automatically control the working potential.

*Undivided Laboratory Cells*

Figure 22.9 shows a very simple beaker-type glass cell with parallel plate electrodes and a cooling mantle. Mass transport is performed by a magnetic stirrer. A Teflon stopper is used to hold the connections for the electrodes.

A better mass transport is obtained with a construction originally designed as a photoreactor (Fig. 22.10). In this case, a magnetic Teflon stirrer with inlet and outlet holes sucks in the electrolyte from above so that the solution is circulating efficiently. The working electrode is best constructed as a cylinder with a coaxial auxiliary electrode.

If a large space–time yield is also required in the laboratory application, a capillary gap cell (disc-stack) is a good choice (Fig. 22.11). Only the upper-

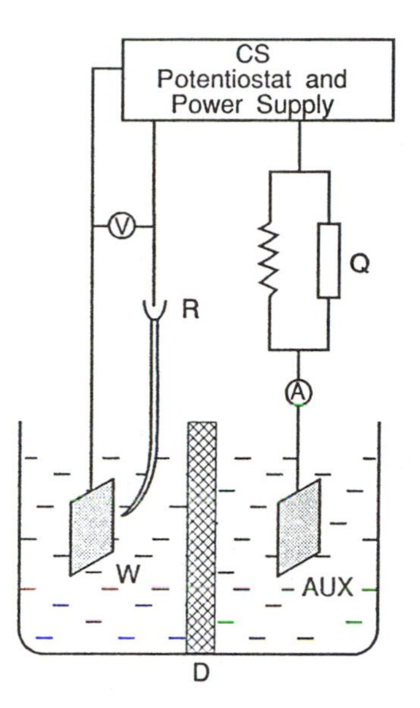


**Figure 22.8** Schematic representation of an electrolysis cell (three-electrode construction: W = working electrode; AUX = auxiliary electrode; R = reference electrode; Q = coulometer; D = diaphragm, if necessary; CS = current source; A = amperometer; V = voltmeter).

and lower-most electrodes are electrically connected. The other electrodes are polarized in the electrical field and act as bipolar electrodes [70]. As spacers between the electrodes, thin plastic (Teflon) strips of 0.3 to 0.5 mm thickness are used.

A special construction, called Swiss-Roll cell [13] (Fig. 22.12), is used for the application of the NiOOH electrode in the oxidation of primary alcohols to carboxylic acids.

With many of the laboratory cells, some problems occur because of complex cell geometries. They may cause an uneven distribution of the working potential at different points of the electrode surface and a poor and sometimes irreproducible mass transport, which is difficult to describe mathematically. These problems can most easily be avoided by the use of flow-through cells with parallel plate electrode geometry. With these, the results are easily reproducible, the selectivity can be better controlled, the electrode geometry guarantees the same working potential at all points of the electrode surface, the mass transport can be reproduced and described easily, larger amounts of product may be produced, and the cell is a good model for scale-up purposes. Parallel-plate flow-through cells can either be used as undivided or divided cells (discussed later) without changing the design.

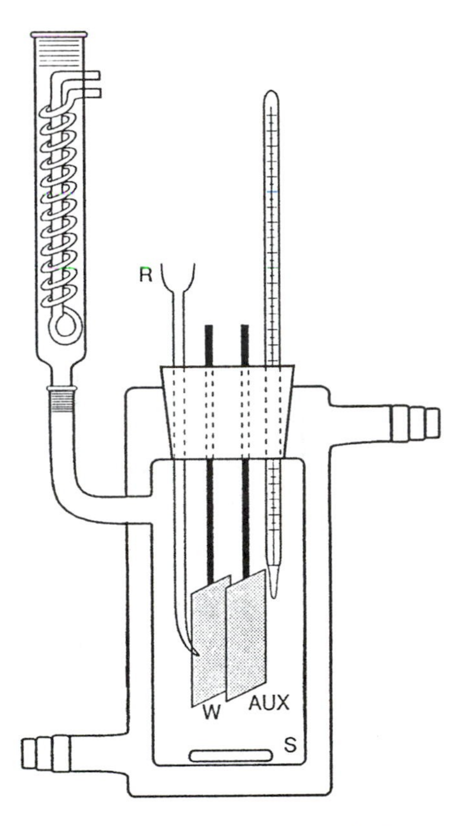


**Figure 22.9** Undivided beaker-type laboratory cell (S = magnetic stirrer; W = working electrode; AUX = auxiliary electrode; R = reference electrode).

*Divided Laboratory Cells*

Similarly to the beaker-type glass cell shown in Figure 22.9, a divided cell can be constructed using a cylindrical ceramic diaphragm or a glass cylinder closed with a glass frit (Fig. 22.13).

Divided parallel plate flow-through cells are especially advantageous for laboratory use because they can easily be constructed from two Teflon cell body halves according to Figure 22.14. In such a construction, ion-exchange membranes can easily be applied for the separation of the two cell compartments. For anodic oxidations, cation exchange membranes like Nafion are most often used. For cathodic reactions, anion exchange membranes are applicable. Asbestos or microporous membranes can also be used as separators.

The flow-through system is schematically shown in Figure 22.15. Similar constructions can be purchased from a number of commercial suppliers. All

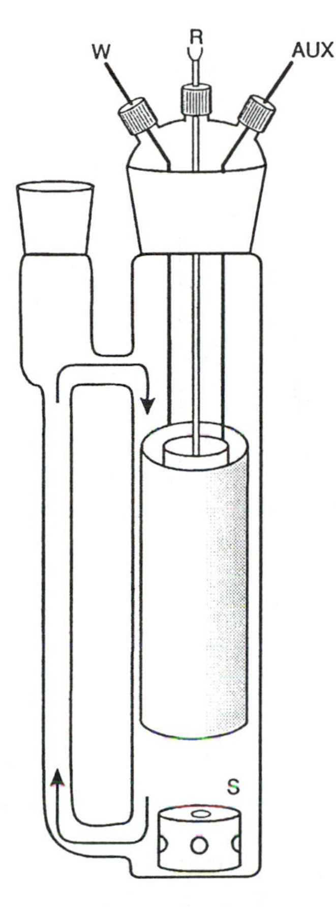


**Figure 22.10** Undivided laboratory cell with circulating electrolyte (S = magnetic stirrer; W = working electrode; AUX = auxiliary electrode; R = reference electrode).

these constructions can also be used as undivided cells. However, these cells are usually not constructed for the introduction of a reference electrode.

## B. Reaction Conditions and Their Influence on the Reaction Pathway

*The Redox Potential*

The knowledge of the redox potential of the substrate is necessary to develop a process for its selective electrochemical transformation. It allows the selection of an electrolyte that is stable at the applied working potential (see next subsection). It also gives information about the possibility of transforming one functionality selectively within a multifunctional substrate. Current–voltage

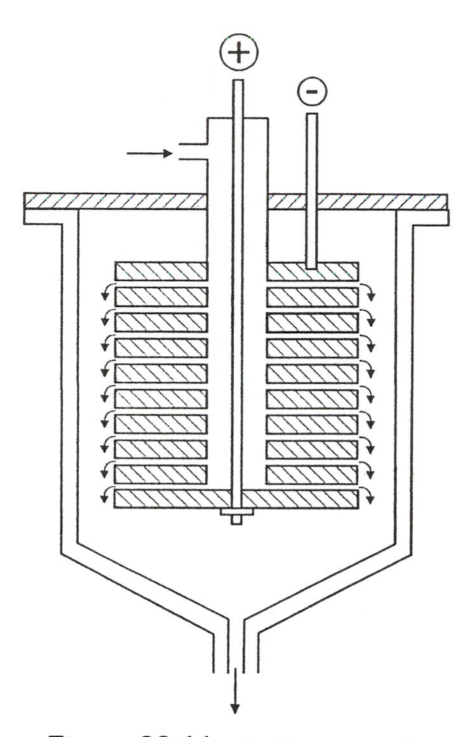


**Figure 22.11** Capillary gap cell.

curves, especially cyclic voltammograms, supply the desired information on the redox potentials. For this purpose, one should start with the recording of the current–voltage curve of the electrolyte alone (background) using the electrode material that was also selected for the preparative electrolysis. The addition of the substrate to the electrolyte will then show if the substrate can be electrochemically transformed in the potential window of the electrolyte. Running several continuous potential scans will give information on possible electrode passivation. It will show up as a strong decrease of the current (i.e., the peak current) in cyclic voltammetry. If passivation is observed, preparative scale electrolyses will not lead to an appreciable product formation. Therefore, one should study different electrolytes and electrode materials by analytical methods until nonpassivating conditions are found. One should take into account that chemically or electrochemically irreversible redox potentials have no thermodynamic meaning because they are influenced by kinetics. They only indicate the reducibility or oxidizability of the substrate under the given conditions. Reversible redox potentials are correlated to the standard potentials. However, for electrochemical synthesis, a follow-up reaction to the product of interest is

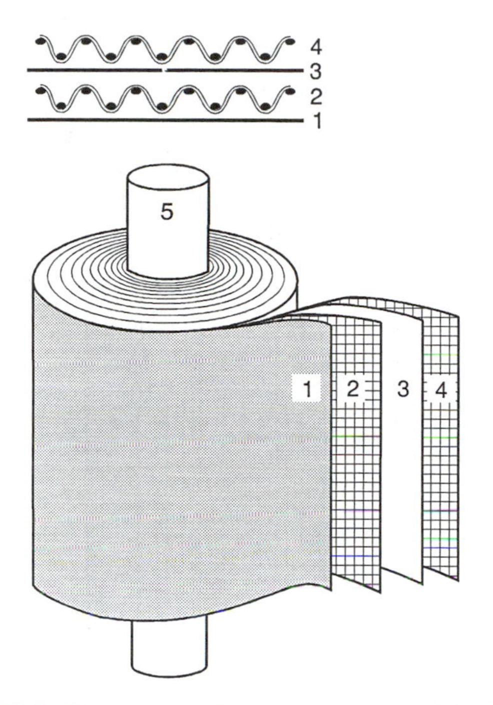


**Figure 22.12** "Swiss-Roll cell" for the applications of NiOOH electrodes (1 = steel net cathode; 2 + 4 = polypropylene net; 3 = nickel net anode; 5 = current feeder).

the goal. Therefore, an electrochemically stable redox pair is usually not desirable. More detailed information about the application of cyclic voltammetry, especially in the elucidation of electroorganic mechanisms, can be found in Chapter 21.

*Influence of the Electrolyte*

The solvents and supporting electrolytes are discussed in Chapter 15. Therefore, only some aspects with regard to electroorganic syntheses are discussed here. For this application, the following conditions have to be fulfilled by the electrolyte:

1. It should have a high conductivity so that low cell voltages are possible at low concentrations of the supporting electrolyte. For this purpose, solvents of high dielectric constant are desirable.
2. The supporting electrolyte salts should be totally dissociated in the applied solvent. The dissociation of salts of larger ions is usually more readily obtained than that of smaller ions.

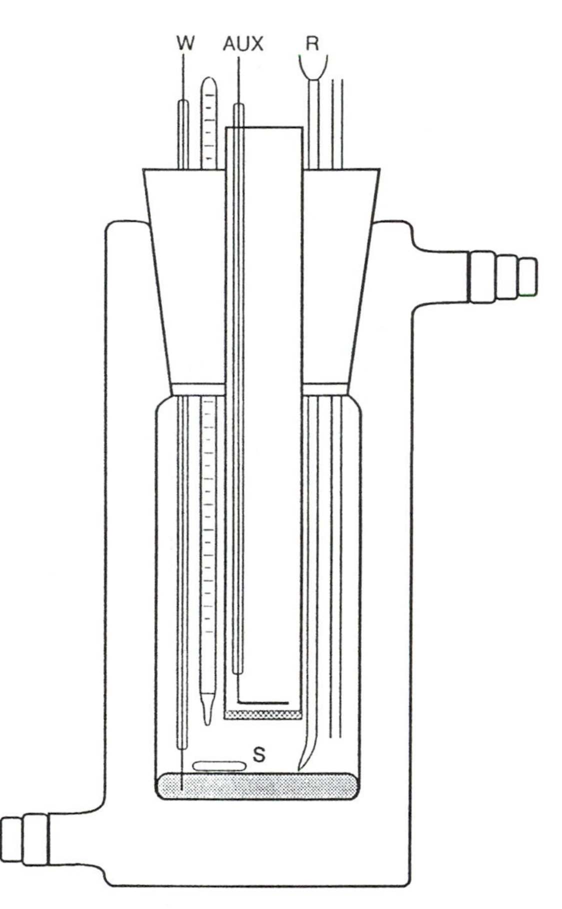


**Figure 22.13** Divided beaker type laboratory cell (S = magnetic stirrer; W = working electrode; AUX = auxiliary electrode; R = reference electrode).

3. The solubility of the components in the solvent must be sufficient. To improve the solubility, cosolvents can be used. Another possibility is the application of a two-phase system or an emulsion in the presence of phase-transfer catalysts. A two-phase system also has advantages in product isolation and continuous electrolysis procedures. A typical example is the synthesis of *p*-methoxy benzonitrile by anodic substitution of one methoxy group in 1,4-dimethoxybenzene by the cyanide ion (Eq. 22.21). The homogeneous cyanation system (acetonitrile, tetraethylammonium cyanide) [24] can be efficiently replaced by a phase-transfer system (dichloromethane, water, sodium cyanide, tetrabutylammonium hydrogen sulfate) [71].

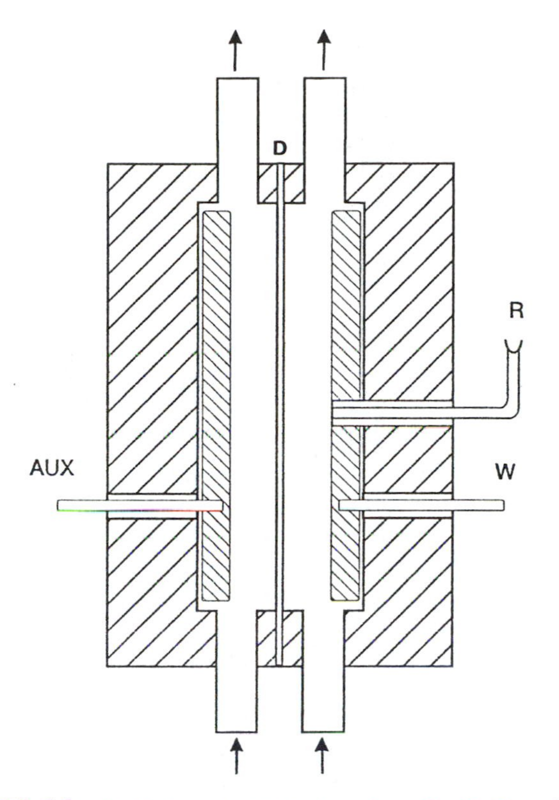


**Figure 22.14** Divided parallel-plate flow-through cell (D = diaphragm; W = working electrode; AUX = auxiliary electrode; R = reference electrode).

4. The stability of the electrolyte system at the applied potential or current density has to be high enough to insure high current yields and to prevent electrolyte destruction. The stability of an electrolyte can be determined by measuring the foot potential in current-voltage curves. Limiting values can be found in most of the books on electroorganic chemistry [1,72].
5. For application in industry, the work-up of the product containing electrolyte must be easy and efficient. With capillary gap cells, the concentration of a supporting electrolyte may be so low that its recovery is not necessary from the economical standpoint.
6. Other properties of the solvent-supporting electrolyte may influence the reaction directly. For example, in aprotic solvents, the cathodic limiting potential can be much more negative. However, under these conditions,

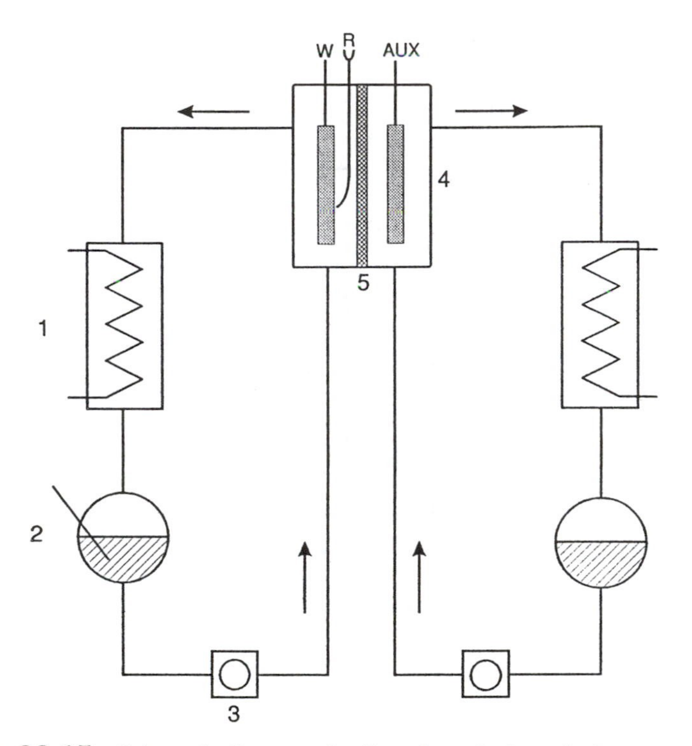


**Figure 22.15** Schematic diagram of a flow-through electrolysis system (1 = heat exchanger; 2 = containers; 3 = pumps; 4 = Teflon body; 5 = diaphragm or ion exchange membrane; W = working electrode; AUX = auxiliary electrode; R = reference electrode).

many compounds and electrogenerated intermediates tend to polymerize. Much cleaner reductions can often be obtained if small amounts of a proton donor are present to trap the reactive anionic intermediates. The nucleophilicity or electrophilicity of the electrolyte also has to be taken into account. Oxidations very often run more smoothly and give cleaner products if at least small amounts of a nucleophile are present to trap cationic intermediates. In the case of reductions, it can be very helpful to trap the intermediates by the addition of electrophiles like trimethylsilyl chloride, alkyl halides, carbonyl compounds, or carbon dioxide.

The electrolyte may also influence the product distribution. This can be demonstrated in the case of the side-chain oxidation of alkyl aromatics [73]:

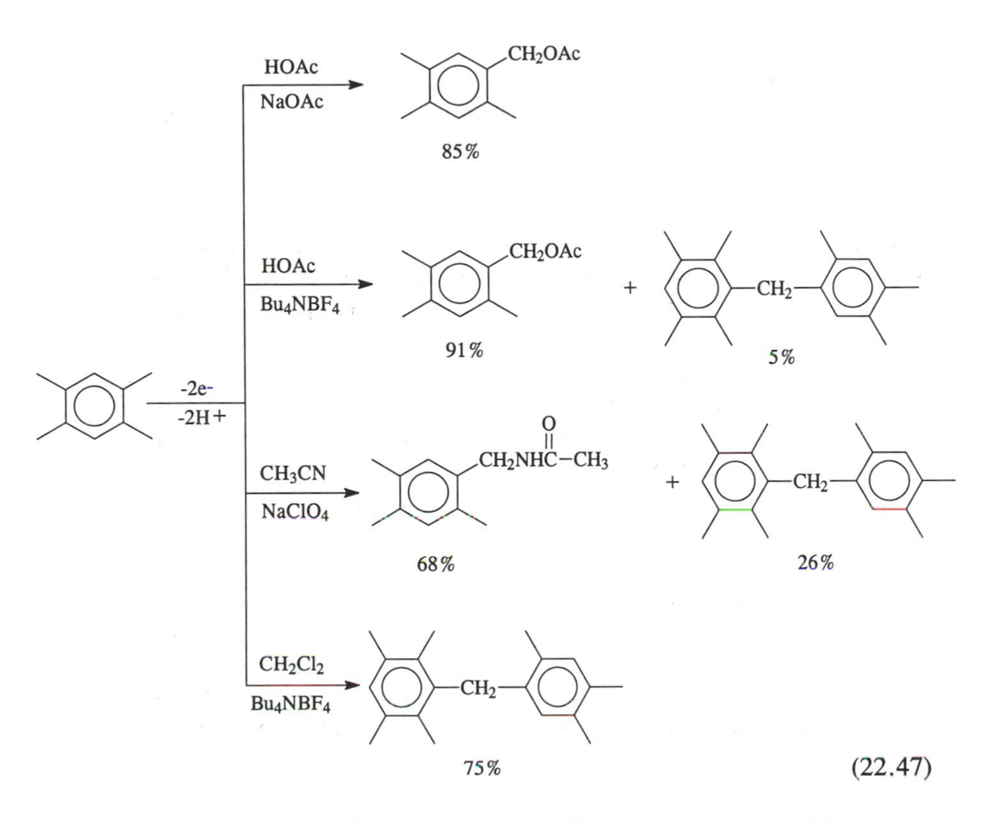


(22.47)

## C. Influence of the Electrode Material

*Cathodes*

In aqueous media, the cathodes should have high enough hydrogen overvoltage to prevent hydrogen generation. The following list presents the hydrogen overpotentials of some common electrode materials in 0.5 *M* aqueous sulfuric acid:

| | Pd | Fe | Pt | Ni | Sn | Pb | Zn | Hg |
|---|---|---|---|---|---|---|---|---|
| Hydrogen overpotential (V) | 0.00 | 0.08 | 0.09 | 0.21 | 0.53 | 0.64 | 0.70 | 0.78 |

In nonaqueous, aprotic media, the hydrogen overvoltage does not play any role. Commonly used cathode materials are lead, aluminum, copper, titanium, steel, and carbon of all kinds.

*Anodes*

The main problem with anodes is their possible corrosion. The most commonly used anode materials are platinum, lead dioxide, magnetite, and all kinds of carbon, such as graphite, glassy carbon, carbon cloth, and carbon felt. Carbon felt is especially valuable as a material with very large surface area. The nickel peroxide electrode is especially valuable for alcohol and amine oxidations [13].

The electrode material can also influence the product distribution, as shown in the Kolbe electrolysis of carboxylates. With platinum anodes, the Kolbe dimerization of the intermediate radicals predominates strongly (Eq. 22.5). At carbon anodes, however, further oxidation to the carbenium ion (non-Kolbe reaction or Hofer–Moest reaction) becomes the main pathway (Eq. 22.25).

In the anodic olefin oxidation, the electrode material plays an important role as well. The oxidation of enolethers at platinum anodes mainly leads to the dimethoxylation product [74], while the application of graphite anodes favors the additive dimerization (Eq. 22.33) [39].

The electrode materials can act in very different ways. In the simplest case, they only act as electron transfer agents. However, sometimes the electrode material can behave as a reactant or as a heterogeneous redox catalyst, as in the case of the NiOOH anode [13]. Electrodes like lead dioxide may act as oxygen transfer agents. Platinum-group metals or high-surface-area nickel or nickel alloys can behave as hydrogen transfer catalysts via adsorbed hydrogen [32]. Another possibility is the formation of organometallic intermediates with accompanying dissolution of the electrode material (sacrificial anodes [56]). The adsorption of intermediates is also a function of the electrode material. All these effects of the electrode materials will strongly influence the outcome of a reaction.

## D. Controlled-Potential or Controlled-Current Electrolyses

In general, controlled-current electrolyses need less expensive equipment. Only a controlled-current source in combination with a coulomb integrator is necessary. Therefore, in industry, electroorganic reactions are always performed at a fixed current density. In the laboratory, it is advisable to start with controlled-potential electrolyses using a potentiostat and a three-electrode electrolysis cell (Fig. 22.8). In this way, the reaction can be controlled at the redox potential of the substrate determined analytically, and the selectivity of the process can be studied at different potentials. After determination of the selectivity controlling factors, it is usually possible to change over to current control by proper selection of the current density and the concentration of the substrate. Using a continuous process, the concentration can be fixed at the desired value. Thus, selectivity can also be obtained under these conditions.

## E. Indirect vs. Direct Electrochemical Reactions

In direct electrochemical reactions, the substrate undergoes a heterogeneous redox reaction at the electrode surface within the Helmholtz layer. The reactive intermediate (i.e., an ion radical) thus formed undergoes the chemical follow-up reaction to the product in the reaction layer. The steep concentration gradients near the electrode are an important factor in the chemical follow-up reaction of the electrogenerated intermediates. Second-order reactions of the intermediates can thus be obtained at high current densities. In indirect electrochemical reactions [3] (see also Sec. IV.F), the heterogeneous reaction between the electrode and the substrate is replaced by a homogeneous redox reaction between the substrate and an electrochemically activated redox catalyst (mediator) in solution (see Fig. 22.1). First-order reactions of the electrogenerated intermediates may be favored under these conditions. In addition, overpotentials for the reaction of the substrate at the electrode may be avoided and reactions may be accelerated. If electrode passivation causes problems for direct electrolyses, an indirect pathway could be an alternative, because the electrode reaction of the substrate can thus be prevented. In many other cases, the selectivity properties of the redox catalysts can be used profitably in indirect processes. On the other hand, direct electrochemical reactions are less complicated, because no additional catalysts must be added and separated.

## REFERENCES

1. Books on organic electrosynthesis: (a) *Organic Electrochemistry, An Introduction and a Guide*, 3rd ed. (H. Lund and M.M. Baizer, eds.), Marcel Dekker, New York, 1991; (b) F. Beck, *Elektroorganische Chemie*, Verlag Chemie, Weinheim, 1974; (c) L. Eberson, Electroorganic Synthesis, in *Modern Synthetic Methods*, Vol. 2 (R. Scheffold, ed.), Salle und Sauerländer, Frankfurt, 1980; (d) S. D. Ross, M. Finkelstein, and E. J. Rudd, *Anodic Oxidation*, Academic Press, New York, 1975; (e) *Techniques of Chemistry*, Vol. V, Parts I and II, Technique of Electroorganic Synthesis (N. L. Weinberg, ed.), Wiley, New York, 1974/1975; (f) T. Shono, *Electroorganic Chemistry as a New Tool in Organic Synthesis*, Springer-Verlag, Berlin, 1984; (g) S. Torii, *Electroorganic Syntheses, Methods and Applications*, Part I, Oxidations, Kodansha, Tokyo, 1985.
2. (a) H. J. Schäfer, *Kontakte (Darmstadt)* 2:17 (1987) and *3*:37 (1987); (b) H. J. Schäfer, *Angew. Chem. 93*:978 (1981); *Angew. Chem. Int. Ed. Engl. 20*:911 (1981).
3. (a) E. Steckhan, *Topics Curr. Chem. 142*:1 (1987); (b) E. Steckhan, *Angew. Chem. 98*:681 (1986); *Angew. Chem. Int. Ed. Engl. 25*:693 (1986).
4. D. Degner, *Topics Curr. Chem. 148*:20 (1988).
5. D. Degner, in *Techniques of Chemistry* Vol. V, Part III, Technique of Electroorganic Synthesis (N. L. Weinberg and B. V. Tilak, eds.), J. Wiley, New York, 1982, p. 277.
6. C. Jackson and A. T. Kuhn, in *Industrial Electrochemical Processes* (A.T. Kuhn, ed.), Elsevier, Amsterdam, 1971, p. 515.
7. (a) N. Clauson-Kaas, F. Limborg, and K. Glens, *Acta Chem. Scand. 6*:531 (1952); (b) H. Tanaka, Y. Kobayasi, and S. Torii, *J. Org. Chem. 41*:3482 (1976); (c) D.

Degner, in *Techniques of Chemistry*, Vol. V, Part III, Technique of Electroorganic Synthesis (N. L. Weinberg and B. V. Tilak, eds.), Wiley, New York, 1982, p. 273.
8. (a) L. Eberson, in *Chemistry of Carboxylic Acids and Esters* (S. Patai, ed.), Wiley, London, 1969, p. 53; (b) J. H. P. Utley, in *Techniques of Chemistry*, Vol. V, Part I, Technique of Electroorganic Synthesis (N.L. Weinberg, ed.), Wiley, New York, 1974, p. 793; (c) D. Degner, *Topics Curr. Chem. 148*:1 (1988).
9. D. Degner, *DECHEMA-Monographien*, Vol. 112, VCH, Weinheim, 1989, p. 375.
10. H. Millauer, *Chem.-Ing.-Tech. 52*:53 (1980).
11. D. Degner, in *Techniques of Chemistry*, Vol. V, Part III, Technique of Electroorganic Synthesis (N. L. Weinberg and B. V. Tilak, eds.), Wiley, New York, 1982, p. 265.
12. (a) C. Jackson and A. T. Kuhn, in *Industrial Electrochemical Processes* (A. T. Kuhn, ed.), Elsevier, Amsterdam, 1971, p. 516; (b) C. G. Pink and D. B. Summers, *Trans. Am. Electrochem. Soc. 74*:625 (1938); (c) H. S. Isbell, H. L. Frush, and F. J. Bates, *Ind. Eng. Chem. Res. 24*:375 (1932).
13. (a) P. Seiler and P.M. Robertson, *Chimia 36*:305 (1982); (b) H.J. Schäfer, *Topics Curr. Chem. 142*:101 (1987); (c) P. M. Robertson, P. Berg, H. Reimann, K. Schleich, and P. Seiler, *J. Electrochem. Soc. 130*:591 (1983).
14. (a) A. J. Rudge, in *Industrial Electrochemical Processes* (A. T. Kuhn, ed.), Elsevier, Amsterdam, 1971, p. 71; (b) J. H. Simons, *J. Electrochem. Soc. 95*: 47,53,55,59,64 (1949); (c) M. Sander and W. Blöchel, *Chem.-Ing.-Tech. 37*:7 (1965); (d) H. Millauer, *DECHEMA-Monographien*, Vol. 112, VCH, Weinheim, 1988, p. 429.
15. H. Millauer, *DECHEMA-Monographien*, Vol. 112, VCH, Weinheim, 1988, p. 429.
16. (a) M. Fleischmann, D. Pletcher, and C.J. Vance, *J. Electroanal. Chem. 29*:325 (1971); (b) R. Galli and F. Oliviani, *J. Electroanal. Chem. 25*:331 (1970).
17. M. Ferles, A. Silhankowa, and D. Doskocilowa, *Coll. Czech. Chem. Commun. 34*:1976 (1969).
18. D. Degner, in *Techniques of Chemistry*, Vol. V, Part III, Technique of Electroorganic Synthesis (N. L. Weinberg and B. V. Tilak, eds.), Wiley, New York, 1982, p. 259.
19. G. Trümpler and P. Rüetschi, *Helv. Chim. Acta 36*:1649 (1953).
20. (a) D. E. Danly, *Chem. Ind. (London) 1979*:439; (b) M. M. Baizer, *Chem. Ind. (London) 1979*:435; (c) D. E. Danly, *Hydrocarbon Process. 60*:161 (1981); (d) M.M. Baizer, *Chem. Tech. 10*:161 (1980); (e) D.E. Danly, *Chem. Tech. 10*:302 (1980); (f) D. E. Danly and C. R. Campbell, in *Techniques of Chemistry*, Vol. V. Part III, Technique of Electroorganic Synthesis, J. Wiley, New York, 1982, p. 283; (g) D. Pletcher, *Industrial Electrochemistry*, Chapman and Hall, London, 1982, p. 153–166.
21. (a) S. Dapperheld, *DECHEMA-Monographien*, Vol. 112, VCH Weinheim, 1988, p. 317; (b) S. Dapperheld, *Chem. Ind. 1990 (10)*:85.
22. (a) G. Silvestri, S. Gambino, G. Filardo, and A. Gulotta, *Angew. Chem. 98*:978 (1984); *Angew. Chem. Int. Ed. Engl. 23*:979 (1984); (b) J.F. Fauvarque, C. Chevrot, A. Jutand, M. Francois, and J. Perichon, *J. Organomet. Chem. 264*:273 (1984); (c) J. Perichon, *Nouv. J. Chim. 5*:621 (1981).

23. L. Eberson and K. Nyberg, *Acc. Chem. Res. 6*:106 (1973).
24. S. Andreades, E.W. Zahnow, *J. Am. Chem. Soc. 91*:4181 (1969).
25. K. D. Ginzel, E. Steckhan, and D. Degner, *Tetrahedron 43*:5797 (1987).
26. (a) S. D. Ross, M. Finkelstein, and R. C. Petersen, *J. Am. Chem. Soc. 86*:2745 (1964); *J. Org. Chem. 31*:128 (1966); *J. Am. Chem. Soc. 88*:4657 (1966); (b) T. Shono, *Tetrahedron 40*:824 (1984); (c) T. Shono, *Electroorganic Chemistry as a New Tool in Organic Synthesis*, Springer-Verlag, Berlin, 1984, pp. 63–77; (d) E. Steckhan, in *Organic Electrochemistry: An Introduction and a Guide* (H. Lund and M.M. Baizer, eds.), 3rd ed., Marcel Dekker, New York, 1991, pp. 601–607.
27. (a)T. Shono, Y. Matsumura, and K. Inoue, *J. Org. Chem. 48*:1388 (1983); (b) A. Papadopoulos, J. Heyer, K.-D. Ginzel, and E. Steckhan, *Chem. Ber. 122*:2159 (1989); (c) A. Papadopoulos, B. Lewall, E. Steckhan, K.-D. Ginzel, F. Knoch, and M. Nieger, *Tetrahedron 47*:563 (1991); (d) K.-D. Ginzel, P. Brungs, and E. Steckhan, *Tetrahedron 45*:1691 (1989).
28. (a) C. Herborn, A. Zietlow, and E. Steckhan, *Angew. Chem. 101*:1392 (1989); *Angew. Chem. Int. Ed. Engl. 28*:1399 (1989); (b) T. Nishitani, H. Horikawa, T. Iwasaki, K. Matsumoto, I. Inoue, and H. Miyoshi, *J. Org. Chem.47*:1706 (1982); (c) P. Renaud and D. Seebach, *Synthesis 1986*:424;(d) D. Seebach, R. Charczuk, C. Gerber, and P. Renaud, *Helv. Chim. Acta 72*:401 (1989).
29. (a) T. Shono, H. Ohmizu, and N. Kise, *Tetrahedron Lett. 23*:4801,1609 (1982); (b) F. Karrenbrock and H. J. Schäfer, *Tetrahedron Lett.* 1521 (1978); (c) T. Shono, H. Ohmizu, S. Kawakami, S. Nakano, and N. Kise, *Tetrahedron Lett. 22*:871 (1981); (d) T. Shono, N. Kise, and T. Suzumoto, *J. Am. Chem. Soc. 106*:259 (1984); (e) F. Karrenbrock, H. J. Schäfer, and I. Langer, *Tetrahedron Lett.* :2915 (1979); (f) H. J. Schäfer and H. Claus, *Recent Advances in Electroorganic Synthesis* (S. Torii, ed.), Elsevier, Amsterdam, 1989, p. 227.
30. R. Scheffold, S. Albrecht, R. Orlinski, H.-R. Ruf, P. Stamouli, O. Tinembart, L. Walder, and C. Weymuth, *Pure Appl. Chem. 59*:363 (1987).
31. (a) H. Baltes, E. Steckhan, and H. J. Schäfer, *Chem. Ber. 111*:1294 (1978); (b) T. Shono, I. Nishiguchi, and M. Okawa, *Chem. Lett.* :573 (1976).
32. (a) K. Junghans, *Chem. Ber. 107*:3191 (1974); (b) T. Matsue, T. Yansada, H. Takahashi, and T. Osa, *Bull. Chem. Soc. Jpn. 59*:3690 (1986).
33. K. Junghans, *Chem.Ber. 106*:3465 (1973).
34. J. P. Coleman, R. J. Holman, and J. H. P. Utley, *J. Chem. Soc., Perkin Trans. 2*:879,884 (1976).
35. H. H. Westberg and H. J. Dauben, Jr., *Tetrahedron Lett.* :5123 (1968).
36. D. Hermeling and H. J. Schäfer, *Angew. Chem. 96*:238 (1984); *Angew. Chem. Int. Ed. Engl. 23*:233 (1984).
37. (a) J. Casanova and H. R. Rogers, *J. Org. Chem. 39*:2408 (1974); (b) H. Lund and E. Holboth, *Acta Chem. Scand. B 30*:895 (1976); (c) U. Husstedt and H. J. Schäfer, *Tetrahedron Lett. 22*:623 (1981).
38. J. Casanova and H. R. Rogers, *J. Org. Chem. 39*:3803 (1974).
39. (a) R. Engels, H.J. Schäfer, and E. Steckhan, *Liebigs Ann. Chem.* :204 (1977); (b) H. Baltes, E. Steckhan, and H.J. Schäfer, *Chem. Ber. 111*:1294 (1978); (c) D. Koch, H. J. Schäfer, and E. Steckhan, *Chem. Ber. 107*:3640 (1974); (d) D. Koch

and H. J. Schäfer, *Angew. Chem. 85*:264 (1973); *Angew. Chem. Int. Ed. Engl. 12*:245 (1973).
40. H. J. Schäfer, *Topics Curr. Chem. 152*:91 (1990).
41. H. J. Bestmann, K. Roth, K. Michaelis, O. Vostrowsky, H. J. Schäfer, and R. Michaelis, *Liebigs Ann. Chem.* :417 (1987).
42. L. Moens, M. M. Baizer, and R. D. Little, *J. Org. Chem. 51*:4497 (1986).
43. H. Kita, S. Ishikura, and A. Katayama, *Electrochim. Acta 20*:441 (1975).
44. D. W. Sopher and J. H. P. Utley, *J. Chem. Soc., Perkin Trans 2*:1361 (1984).
45. R. W. Murray and M. L. Kaplan, *J. Org. Chem. 31*:962 (1966).
46. Q. N. Porter and J. H. P. Utley, *J. Chem. Soc., Chem. Commun.* :255 (1978).
47. M. Platen and E. Steckhan, *Tetrahedron Lett. 21*:511 (1980); *Chem. Ber. 117*:1679 (1984).
48. (a) E. Steckhan, *Topics Curr. Chem. 142*:51 (1987); (b) N. Schultz, S. Töteberg-Kaulen, S. Dapperheld, J. Heyer, M. Platen, K. Schumacher, and E. Steckhan, in *Recent Advances of Electroorganic Synthesis* (S. Torii, ed.), Kodansha, Tokyo, 1987, p. 127.
49. (a) L. L. Miller, J. F. Wolf, and E. A. Mayeda, *J. Am. Chem. Soc. 93*:3306 (1971), *94*:6812 (1972); (b) R. F. Garwood, N. ud Dim, and B. C. L. Weedon, *J. Chem. Soc., Perkin. Trans. 1*:2471 (1975); (c) S. M. Weinreb, G. A. Epling, R. Comi, and M. Reitano, *J. Org. Chem. 40*:1356 (1975); (d) W. Schmidt and E. Steckhan, *Angew. Chem. 90*:717 (1978), *91*:850 (1979), *91*:851 (1979); *Angew. Chem. Int. Ed. Engl. 17*:673 (1978), *18*:802 (1979).
50. (a) T. Shono, *Electroorganic Chemistry as a New Tool in Organic Synthesis*, Springer-Verlag, Berlin, 1984, p. 54; (b) T. Shono, T. Toda, and N. Oshino, *J. Am. Chem. Soc. 104*:2639 (1982), *Drug Metab. Dispos. 9*:481 (1981).
51. S. Torii, T. Inokuchi, and N. Takahashi, *J. Org. Chem. 43*:5020 (1978).
52. (a) V. G. Mairanovsky, *Angew. Chem. 88*:283 (1976), *Angew. Chem. Int. Ed. Engl. 15*:281 (1976); (b) M.I. Montenegro, *Electrochim. Acta 31*:607 (1986).
53. L. Horner, *Pure Appl. Chem. 9*:225 (1964).
54. N. Schultz-von Itter and E. Steckhan, *Tetrahedron 43*:2475 (1987).
55. (a) W. Schmidt and E. Steckhan, *Chem. Ber. 113*:577 (1980); (b) S. Dapperheld, E. Steckhan, K.-H. Grosse-Brinkhaus, and T. Esch, *Chem. Ber. 124*:2557 (1991).
56. W. Hummel and M.-R. Kula, *Eur. J. Biochem. 184*:1 (1989)
57. (a) A. Fassouane, J.M. Laval, J. Moiroux, and C. Bourdillon, *Biotechnol. Bioeng. 35*:935 (1990); (b) J. Bonnefoy, J. Moiroux, J.M. Laval, and C. Bourdillon, *J. Chem. Soc., Faraday Trans. 1 84*:941 (1988).
58. A.-E. Biade, C. Bourdillon, J.-M. Laval, G. Mairess, and J. Moiroux, *J. Am. Chem. Soc. 114*:893 (1992).
59. J. Komoschinski and E. Steckhan, *Tetrahedron Lett. 29*:3299 (1988).
60. M. Ito and T. Kuwana, *J. Electroanal. Chem. 32*:415 (1971).
61. R. DiCosimo, C.-H. Wong, L. Daniels, and G.M. Whitesides, *J. Org. Chem. 46*:4622 (1981).
62. (a) H. Simon, J. Bader, H. Günther, S. Neumann, and J. Thanos, *Angew. Chem. 97*:541 (1985); *Angew. Chem. Int. Ed. Engl. 24*:539 (1985); (b) I. Thanos, J. Bader, H. Günther, S. Neumann, F. Krauss, and H. Simon, *Methods Enzymol.* (K. Mosbach, ed.) *136*:302 (1987).

63. (a) H. Günther, A. S. Paxinos, M. Schulz, C. van Dijk, and H. Simon, *Angew. Chem. 102*:1075 (1990); *Angew. Chem. Int. Ed. Engl. 29*:1053 (1990); (b) S. Nagata, R. Feicht, W. Bette, H. Günther, and H. Simon, *Eur. J. Appl. Microbiol. Biotechnol. 26*:263 (1987).
64. (a) E. Steckhan, S. Herrmann, R. Ruppert, E. Dietz, M. Frede, and E. Spika, *Organometallics 10*:1586 (1991); (b) E. Steckhan, M. Frede, S. Herrmann, R. Ruppert, E. Spika, and E. Dietz, *DECHEMA-Monographien*, Vol. 125, VCH, Weinheim, 1992, pp. 723–752.
65. (a) R. Ruppert, S. Herrmann and E. Steckhan, *Tetrahedron Lett. 28*:6583 (1987); (b) E. Spika and E. Steckhan, unpublished.
66. H. A. O. Hill, B. N. Oliver, D. J. Page, D. J. Hopper, *J. Chem. Soc., Chem. Commun.* :1469 (1985).
67. (a) E. Steckhan, M. Frede, S. Herrmann, R. Ruppert, E. Spika, and E. Dietz, *DECHEMA-Monographien*, Vol. 125, VCH, Weinheim, 1992, pp. 712–752; (b) M. Frede and E. Steckhan, *Tetrahedron Lett. 32*:5063 (1991); B. Brielbeck, M. Frede, and E. Steckhan, *Biocatalysis 10*:49 (1994).
68. R. Engels, C. J. Smit, and W. J. M. van Tilborg, *Angew. Chem. 95*:502 (1983), *Angew. Chem. Int. Ed. Engl. 22*:492 (1983).
69. G. Silvestri, S. Gambino, and G. Filardo, in *Recent Advances in Electroorganic Synthesis* (S. Torii, ed.), Kodansha, Tokyo, 1987, p. 287; also see Ref. 22.
70. F. Beck, *Elektroorganische Chemie*, Verlag Chemie, Weinheim, 1974, p. 123.
71. L. Eberson, B. Helgee, *Acta Chem. Scand. B 29*:451 (1975).
72. L. Eberson, *Modern Synthetic Methods*, Vol. 2, Otto Salle Verlag, Frankfurt, 1980, p. 40.
73. K. Nyberg, *J. Chem. Soc. Chem. Commun.* :774 (1969); *Acta Chem. Scand. 24*:1609 (1970).
74. B. Belleau and Y. K. Au-Yang, *Can. J. Chem. 47*:2117 (1969).

# 23

# Instructional Examples of Electrode Mechanisms of Transition Metal Complexes

**William E. Geiger** *University of Vermont, Burlington, Vermont*

This chapter concerns the study of electrode reaction mechanisms of inorganic and organometallic complexes. The emphasis is on proper use of experimental measurables from cyclic voltammetry for diagnosis of common mechanisms such as E, EC, CE, and ECE reactions. We employ the standard designation of electron transfer (et) reactions as E, and other chemical reactions as C. In practice, mechanistic studies make use of an array of electrochemical and other physical and chemical methods, but space limitations restrict our attention to the powerful and versatile technique of cyclic voltammetry (CV). If necessary, the reader may review the fundamentals of this technique in Chapter 3.

## I. ELECTRODE MECHANISMS

In Chapter 21, Hawley has formulated a series of questions about the mechanism of an electrode reaction. Complete diagnosis of the mechanism includes knowledge of the electrode reaction products and the sequential steps (E and/or C) by which they are formed. If a chemical reaction follows rapidly upon an electron transfer, the new (secondary) product may be produced close to the electrode, and may be subject to further electrochemistry. If the secondary products are formed slowly, after the primary electrolysis product has diffused away from the electrode, their formation will ordinarily not influence the electrode mechanism, except in bulk electrolysis. We limit our treatment to reactions occurring on the CV time scale, approximately 20 s to 10 ms for routine technology. Ultramicroelectrode technology (Chap. 12) extends the short-time limit to below 1 μs.

For simplicity, we restrict our model reactions to reductions, although all statements and conclusions can be adjusted for consideration of oxidations.

## A. Diagnostic Parameters

Electrode mechanisms are usually determined by evaluating trends in theoretically significant parameters with variables such as the time scale and temperature of the experiment and the concentration of the analyte. CV studies typically monitor two current-related parameters, namely,

Current function $X = (\text{const})\ i_p/C^o v^{1/2}$
Current ratio $i_{rev}/i_{fwd}$

and three potential-related parameters, namely

Peak potential $E_p$
Peak (potential) separation $\Delta E_p = E_{pa} - E_{pc}$
Peak breadth $\delta E_p = E_p - E_{p/2}$

all of which are defined in Chapter 3.* Reference will be made to these parameters in the examples that follow later in this treatment.

## B. Qualitative Observations: How Useful?

Suppose that an initial CV scan at a scan rate, $v$, of 0.1 V/s shows cathodic and anodic waves that appear to be coupled, as in Figure 23.1. This is often taken as qualitative evidence of a simple E mechanism (Eq. 23.1), and the experimenter might try to (1) determine the chemical reversibility† of the couple from the value of $i_{rev}/i_{fwd}$, (2) calculate a formal potential ($E_{1/2} \approx E^{\circ\prime}$) from the average of the cathodic and anodic peak potentials, and (3) determine the n value from the height of the cathodic peak current and/or the peak potential separation, $\Delta E_p$. This thinking would be premature, since other mechanisms, not requiring a stable product B, could also give rise to a CV with this general morphology. Figure 23.1 shows that virtually the *same* electrode response may be obtained *at a single scan rate* for E, EC, or more complicated mechanisms under some circumstances.

$$A + ne^- \xrightarrow{k_s} B \qquad E^{\circ\prime} \tag{23.1}$$

*The current function is defined as $\Psi$ in Chapter 3. $C^o$ is the bulk concentration of the electroactive species.

†As noted in Chapter 2, a chemically reversible couple is one in which both forms of the redox couple are stable (i.e., persistent) on the time scale of the experiment.

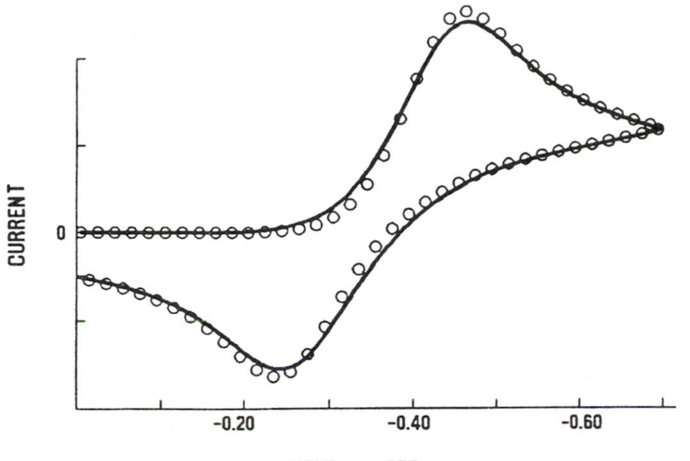


**Figure 23.1** CV waveforms computed for EC,E (circles) or E (line) mechanisms. Scan rate = 0.10 V/s. Parameters employed: EC,E mechanism (Eq. 23.2–23.4), $E^{\circ\prime}_{A/B}$ = −0.40 V, $k_s$ = 3 × $10^{-3}$ cm/s; $E^{\circ\prime}_{C/D}$ = −0.30 V, $k_s$ = 3 × $10^{-3}$ cm/s; $k_f$ = 10 $s^{-1}$, $k_b$ = 1 × 10 $^{-3}$ $s^{-1}$. E mechanism (Eq. 23.1), $E^{\circ\prime}$ = −0.35 V, $k_s$ = 1 × $10^{-3}$ cm/s. In all cases, $\alpha$ = 0.5.

The solid line of Figure 23.1 gives the calculated trace for Equation 23.1 with a moderate heterogeneous charge-transfer rate (a quasi-reversible E system). The circles give the calculated response on the basis of Equations 23.2–23.4, in which the initial product B reacts rapidly to give X, which is oxidized on the return sweep. The anodic wave therefore arises from the electroactive *decomposition product* (X) of B. The cathodic (forward) scan is therefore an EC process and the reverse scan an E process, so that the overall cyclic mechanism might be referred to as "EC,E" (Eq. 23.2–23.4).

$$A + e^- \longrightarrow B \tag{23.2}$$

$$B \underset{k_b}{\overset{k_f}{\rightleftarrows}} X \quad \text{(fast)} \tag{23.3}$$

$$X \rightleftarrows Y + e^- \tag{23.4}$$

To dramatize the possible similarity of the responses for E compared to EC,E mechanisms, our calculations employed similar $E^{\circ\prime}$ values for the A/B and X/Y couples. This is not, however, an unusual situation for electrode products, considering that many may be closely related in structure to the reactant (e.g., as isomers). Obviously, the CV result of Figure 23.1 alone does not provide sufficient information to prove the cathodic fate of A.

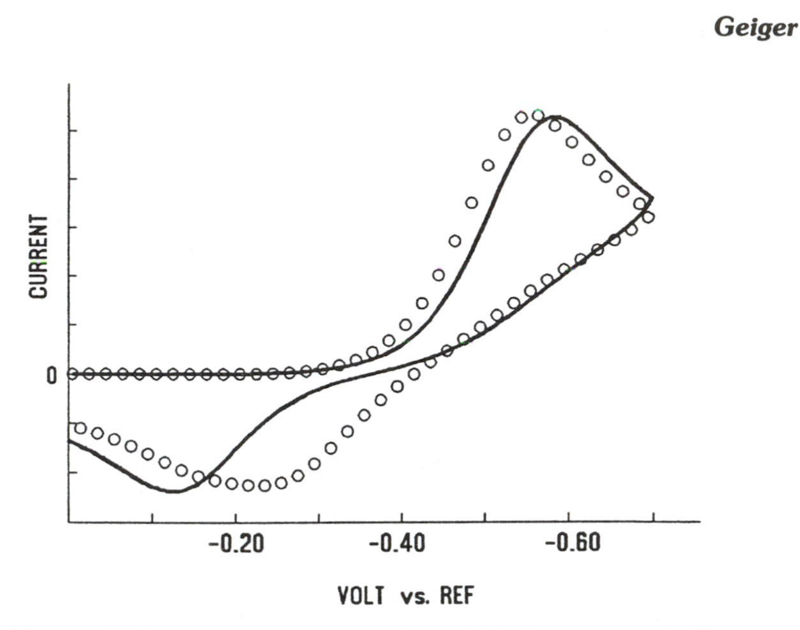


**Figure 23.2** Same parameters as Figure 23.1. Scan rate = 10 V/s.

The differences between the E and EC,E responses are easily probed, however, by viewing how the CV wave and its diagnostic parameters change with scan rate. Such changes are indicated in Figure 23.2, which shows the CV curves calculated for the two mechanisms at a higher scan rate, $\nu = 10$ V/s. The shorter experiment time is sufficient to essentially outrun the decomposition of B; the anodic current now arises primarily from oxidation of B, rather than X, for both mechanisms at this $\nu$. That is, the EC,E sequence has been reduced to an E mechanism on this time scale. The anodic wave for B is at more positive potentials in the simple E mechanism (solid line) because the $k_s$ value assumed for Equation 23.1 was lower than that for Equation 23.2. In addition to demonstrating that a single CV response is insufficient to confirm a mechanistic hypothesis, this example shows the importance of matching theory and experiment over a *range* of scan rates.

## II. OBTAINING HIGH-QUALITY DATA

Since the data to be evaluated must be of high integrity (accuracy), we now consider the most common experimental errors in voltammetry and make recommendations on how to minimize or otherwise deal with these errors.

Good electrochemical technique starts with high-purity solvents, supporting electrolytes, and inert atmospheres, as well as clean working electrode surfaces. Aspects of these factors are treated elsewhere in this book, so our discussion assumes impurity-free experimental conditions.

We consider three factors, namely: (1) residual currents, (2) ohmic losses, and (3) mass transport variations.

### A. Residual Currents

Residual currents, also referred to as background currents, are the sum of faradaic and nonfaradaic currents that arise from the solvent/electrolyte blank. Faradaic processes from impurities may be practically eliminated by the careful experimentalist, but the nonfaradaic currents associated with charging of the electrode double layer (Chap. 2) are inherent to the nature of a potential sweep experiment. Equation 23.5 describes the relationship between this charging current $i_{cc}$, the double-layer capacitance $C_{dl}$, the electrode area A, and the scan rate $\nu$:

$$i_{cc} = A\, C_{dl}\, \nu \qquad (23.5)$$

Equation 23.6 gives the approximate relationship between the charging current and the peak faradaic current, $i_p$ (the target quantity), for a test compound of concentration $C^o$ (m*M*) assuming normal double-layer capacitances [1],

$$i_{cc}/i_p \approx 10^{-5}\, \nu^{1/2}/C^o \qquad (23.6)$$

so that when $C^o = 1$ m*M* and $\nu$ is 0.1 V/s, the charging current is only about 0.3% of $i_p$, a favorable value. This value rises to 100% for $C^o = 0.1$ m*M* and $\nu = 100$ V/s. Discrimination against charging current is therefore favored by low sweep rates and high analyte concentrations. The latter comes at a price, however, namely, the higher ohmic error (Sec. II.B), which is associated with a larger faradaic current.

Peak currents are often measured with the baseline extrapolation method reproduced in Figure 23.3, which reasonably suits the needs of this single-point ($i_p$) analysis. If a more in-depth analysis of the wave shape is desired, the extrapolation method is inadequate, and a point-by-point subtraction of the background current from the total current with analyte present is necessary. The latter approach is greatly aided by the widespread interfacing of voltammetric equipment with computers. Typically, scans of the electrolyte blank recorded over an appropriate range of potentials and scan rates are digitally subtracted from those in which the test compound is present. Although somewhat tedious, background subtraction allows reasonably accurate [2] measurement of the entire faradaic wave. It does not, however, eliminate ohmic distortions from either the faradaic current *or* the charging current now subtracted from the visual display. Figure 23.4 gives an example of a CV scan before and after background sub-

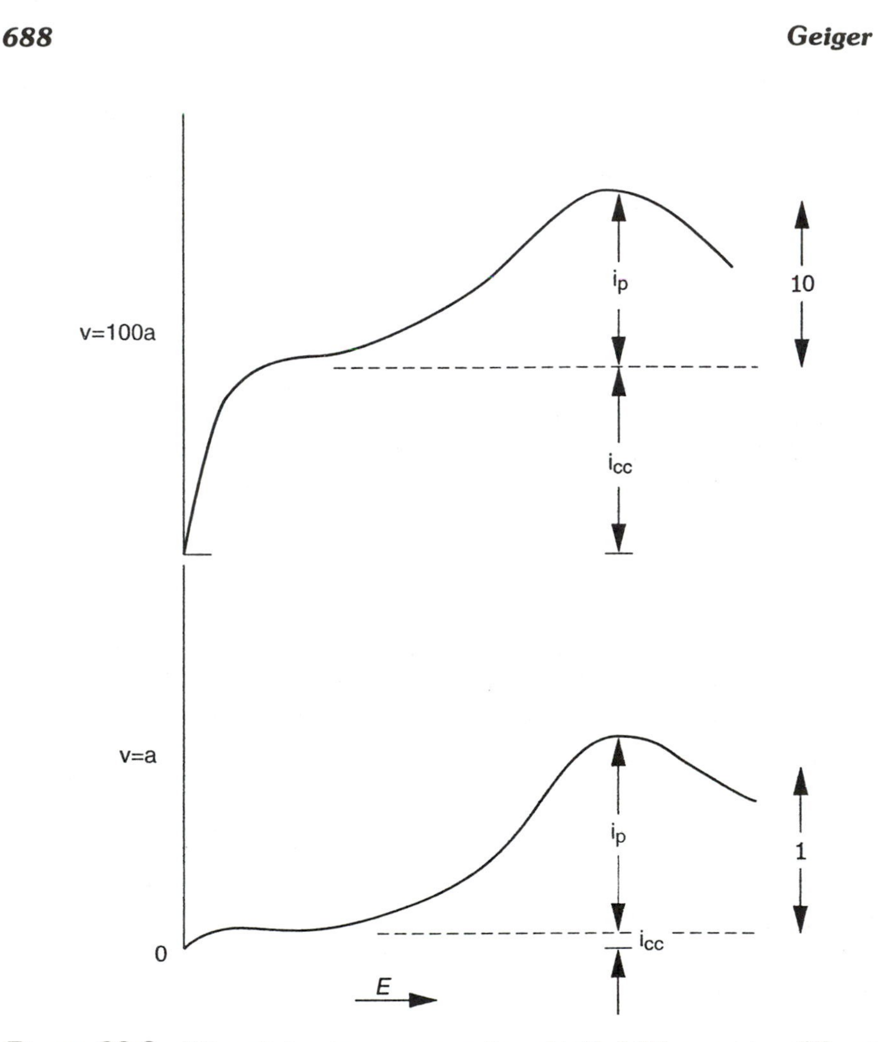


**Figure 23.3** Effect of charging currents on foward half of CV scan at two different scan rates, assuming that $C_{dl}$ is independent of v. Top trace is at 100 times the scan rate and 1/10 the current sensitivity of the bottom trace. The faradaic peak current, $i_p$, is indicated. Adapted from Ref. 1, p. 221, with permission.

traction. The unusual shape of the background-corrected CV wave arises from a coupled chemical reaction as discussed in Sec. IV.C.

## B. Ohmic Losses

Whereas large charging current contributions are found only under fairly demanding conditions (low analyte concentrations, very high v), ohmic distortions (iR losses) are almost always present. A full chapter in this book is devoted to

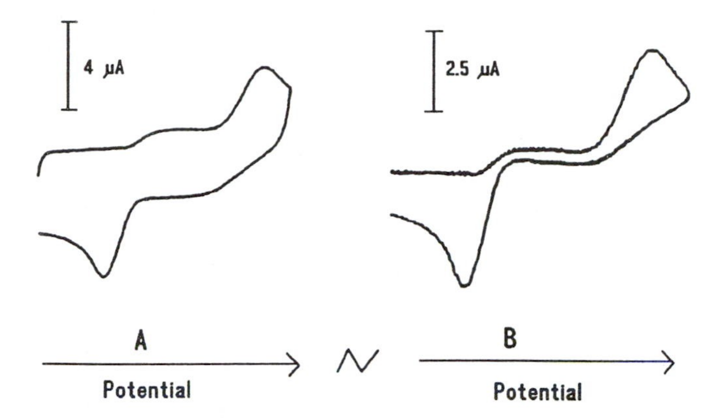


**Figure 23.4** CV scans of 0.7 m*M* $Cp_2Cr_2(CO)_6$ in $CH_2Cl_2$/0.1 *M* $Bu_4NPF_6$ at T = 243 K, ν = 100 V/s: (A) raw data; (B) faradaic component after background subtraction. Adapted from study in Ref. 18.

minimization of resistance effects. Our treatment assumes that residual ohmic effects are present in the data in spite of the experimenter's best efforts, an assumption almost certainly true for poorly conductive electrolytes involving $CH_2Cl_2$, dimethoxyethane, and tetrahydrofuran, which are popular for the study of inorganic and organometallic redox couples.

Equation 23.7 shows that the actual potential of the electrode, $E_{true}$, differs from that of the apparent potential, $E_{app}$, by the amount of the ohmic product, $iR_u$. $R_u$ is the uncompensated resistance and cathodic current is defined as positive:

$$E_{true} = E_{app} + iR_u \tag{23.7}$$

The ohmic error varies at different points on the CV wave, reflecting the inherent variations of current along the wave. All five diagnostics listed on p. 684 may be affected by ohmic losses. In unfavorable situations, ohmic distortions of the CV response may be severe. To reinforce this point, we consider some data on X and $\Delta E_p$ for the oxidation of ferrocene at a Pt electrode (area = 0.086 $cm^2$) in $CH_2Cl_2$/0.1 M $Bu_4NPF_6$ with 0.2 < ν < 160 V/s (Fig. 23.5). In the absence of ohmic distortions, both parameters undergo little change with ν since the couple $Cp_2Fe^{0/+}$ exhibits fast electron transfer kinetics between a stable reactant and product. Only a slight increase of the $\Delta E_p$ value from the Nernstian value of ~60 mV is anticipated at higher ν, and the current function is expected to be practically unchanged. As Figure 23.5 demonstrates, however, ohmic losses produce a large increase in the experimentally determined peak potential separation, $\Delta E_p$, and a large decrease in the measured current

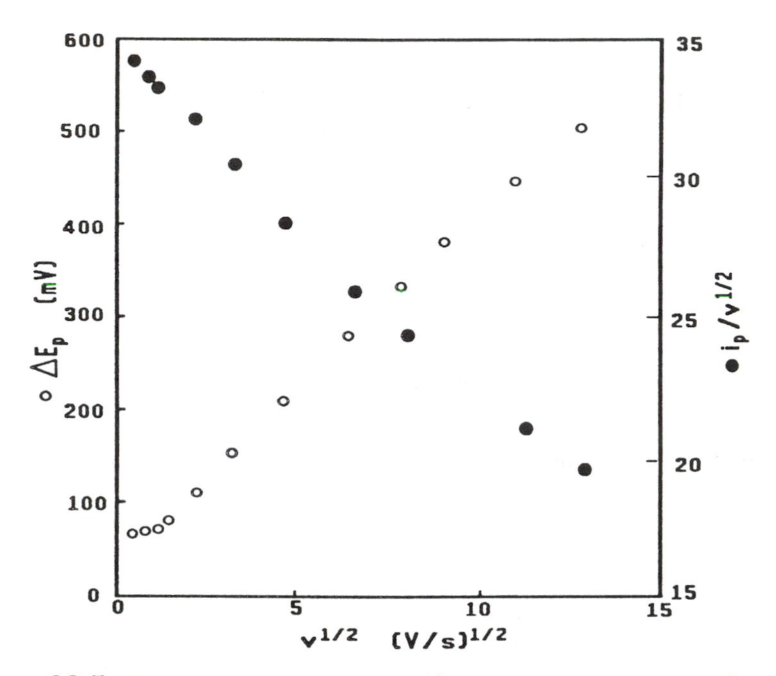


**Figure 23.5** Plots of current function, $i_p/v^{1/2}$ (●) and $\Delta E_p$ (○), against $v^{1/2}$ for oxidation of 0.5 mM $Cp_2Fe$ in $CH_2Cl_2$/0.1 *M* $Bu_4NPF_6$ at Pt disc of area 0.086 $cm^2$, showing the effect of uncompensated resistance on the peak current and peak separation.

function, X, over a 1000-fold increase in sweep rate. The fact that the changes are linear with $v^{1/2}$ implicates $iR_u$ errors, given that the current is proportional to $v^{1/2}$ for diffusion-controlled systems.

Ohmic effects render $E_{pc}$ more negative, $E_{pa}$ more positive, $\Delta E_p$ and $\delta E_p$ larger, and X smaller than the true values. Since experimental approaches to elimination of $iR_u$ errors are not foolproof (see Chap. 7), the presence of ohmic distortions should be tested by measurements on a Nernstian couple such as ferrocene/ferrocenium under conditions identical to those used to probe the test compound. In principle, errors in the measured CV parameters for a *test* compound can be eliminated by referencing its responses to those of the Nernstian *standard*. Note that this approach is accurate only if the *current level* of the standard, rather than its *concentration*, is equal to that of the test compound, since the diffusion coefficients of the two species may appreciably differ.

## C. Mass Transport Variations

Excluding adsorption, there are three sources of mass transport of reactant from solution to the electrode surface: migration, convection, and diffusion. The first of these is practically eliminated by the addition of an excess of supporting electrolyte. Elimination of the second is also necessary when employing equations based on diffusion to quantitatively describe the observed currents. Even nominally unstirred solutions, however, are subject to convection under some experimental conditions, and we return to this point later. First, though, we briefly consider the problem of nonlinear (or radial) diffusion.

In evaluating the dependence of X on v for CV studies at a planar disk electrode, the mass transport mechanism is generally assumed to be that of linear diffusion (Fig. 23.6, left). At long times (slow scan rates), however, radial diffusion may also be effective in bringing reactant to the electrode (Fig. 23.6, right) and in so doing result in mixed (linear and radial) diffusion and an increase in current. The contributions to X from the two processes are easily calculated knowing the scan rate and the radius of the electrode [3]. Radial diffusion may be safely ignored when $4Dt/r^2 \ll 1$, where D is the reactant's diffusion coefficient, t is the time elapsed in the CV sweep, and r is the radius of the electrode. When r is very small, as with an ultramicroelectrode, this limit is reached at rather high scan rates. With conventional electrodes ($r > \sim 2$ mm), however, less than 5% error is introduced by ignoring radial diffusion down to sweep rates of ca. 0.05 V/s [4].

Comparable or larger errors are introduced by unwanted *convective* mass transport. Convection is caused by physical motion of the solution, sometimes purposefully introduced for techniques such as rotating electrode voltammetry. When a quiet solution is desired, however, convective errors may arise at longer experiment times (slow scan rates) from mechanical vibrations of the solution. Convection is a particular problem for cells inside inert-atmosphere boxes, on which fans and vacuum pumps may be operative. Convection raises the current

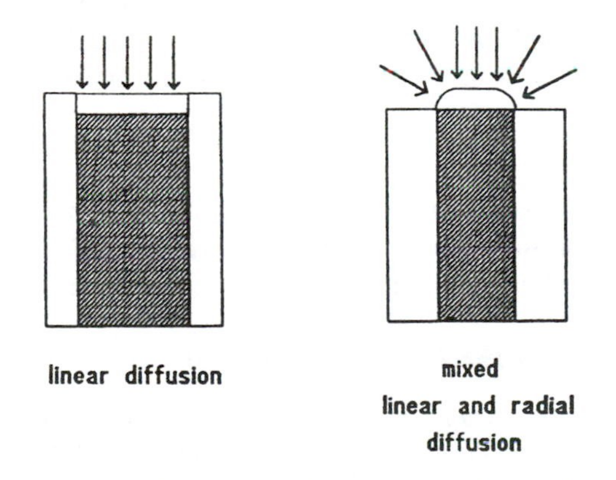


**Figure 23.6** Depiction of (left) linear diffusion and (right) combined linear and radial (or "edge") diffusion to electrodes.

**Table 23.1** Typical Observed Current Functions for the One-Electron Reduction of $Cp_2Rh^+$ in $CH_3CN$/0.1 *M* $Bu_4NPF_6$ at a Pt Disk Electrode of ~2 mm Diameter

| Scan rate, ν (V/s) | Current function, X (μA m$M^{-1}$ $s^{-1/2}$) |
|---|---|
| 0.050 | 11.2 |
| 0.065 | 10.9 |
| 0.100 | 10.5 |
| 0.130 | 10.2 |
| 0.260 | 10.1 |
| 0.390 | 10.0 |
| 0.500 | 10.0 |
| 1.00 | 10.2 |
| 5.00 | 10.0 |

*Note*: Conc. = $4 \times 10^{-4}$ *M*, temp. = ambient, $E^{\circ\prime}$ = −1.81 V vs. Fc, cell housed in vacuum atmosphere drybox.

by an amount that cannot be treated theoretically under the hydrodynamically indeterminate conditions of the experiment.

Table 23.1 reproduces data affected by breakdown of linear diffusion for the one-electron reduction of $Cp_2Rh^+$ at a Pt disk electrode (r = 2 mm). At sweep rates above about 0.1 V/s, the current function is essentially constant, consistent with the simple one-electron reaction of Equation 23.8. The increase of over 10% in X that is observed at lower scan rates arises from the breakdown of linear diffusion, rather than from additional reactions coupled to Equation 23.8. The radial diffusion contribution is less than 3% [5], with convection accounting for most of the additional mass transport.

$$Cp_2Rh^+ + e^- \rightleftarrows Cp_2Rh \qquad (23.8)$$

One can deal with mixed mass transport by ratioing the X value of the test compound to that of an internal standard such as $Cp_2Fe^{0/+}$, for which the electrode mechanism is known to be independent of scan rate at very low ν.

## D. A Comment on the Current Function

X is often the key parameter for the determination of the apparent number of electrons of the reaction. Two aspects of this quantity, namely, its absolute value and its changes with scan rate, should be considered. The dependence of X on ν is used next as a *relative* probe to *changes* in the apparent n value with time.

The *absolute* value of X is equally important, for it gives the n value directly through

$$X = (\text{const})i_p/v^{1/2}C^o = k_{RS}\ A\ n^{3/2}D^{1/2} \tag{23.9}$$

in which D is the diffusion coefficient of the test compound, A is the area of the electrode, and $k_{RS}$ is the Randles-Sevĉik constant [6]. However, when D is also unknown (the usual case), there is little to be gained from knowing the absolute value of X. Therefore, X is usually measured relative to that of a reference couple having known values of n and D. If the diffusion coefficients of the reference and test compounds are similar, an estimate of the number of electrons transferred in the reaction of the test compound is obtained.

Compounds differing in charge, shape, or molecular weight may have markedly different diffusion coefficients. For example, D values in $CH_3CN$ for the metal sandwich complexes ferrocene and $(C_6Me_6)_2Ru^{2+}$ are reported as $2 \times 10^{-5}$ cm$^2$/s and $0.7 \times 10^{-5}$ cm$^2$/s, respectively [7].

## III. MECHANISTIC STUDIES

In general, electrode mechanisms are best probed by a variety of electrochemical, spectroelectrochemical, and other techniques, combined to bear on a single problem. Seldom does CV, or any other method for that matter, give unequivocal answers concerning electrode behavior. In the following treatments, we seek to establish how diagnostics for CV are consistent with certain mechanisms, while recognizing that supporting evidence from other methods usually plays a key role in the investigations.

### A. Parametric Analysis or Digital Simulations?

Use of the five current- or potential-related parameters given on p. 684 forms the focus of the rest of this chapter. One should recognize that these five parameters are calculated from a limited number of points on the CV trace, notably those at the peak and half-peak positions. The rest of the trace is ignored in the treatment. In contrast, digital simulation techniques (Chap. 20) provide an approach in which the *entire* trace is calculated for a particular mechanism and compared to the experimental curve for consistency and agreement. The latter approach is obviously more powerful, and such calculations are increasingly accessible through the use of personal computers and fast computational algorithms [8]. Much can still be accomplished, however, without resorting to full theoretical simulations; examination of the five parameters will usually be a good starting point, even with systems for which full simulations are later intended. The main limitation in either approach involves the integrity of the experimental data.

## IV. EXAMPLES OF SELECTED MECHANISMS

### A. The E Mechanism

The E mechanism is given by

$$A + ne^- \rightleftarrows B \tag{23.10}$$

The five principal diagnostics, X, $i_{rev}/i_{fwd}$, $E_{pc}$, $\Delta E_p$, and $\delta E_p$, of a Nernstian system are all independent of $\nu$ in the absence of $iR_u$ distortions. Numerically, $\Delta E_p$ is ~60/n mV and $\delta E_p$ is 57/n mV at 298 K. The peak separation $\Delta E_p$ has a slight dependence on the switching potential of the scan, and may vary from 57 to 61 mV for (n = 1) Nernstian systems [9].

The formal potential of the couple, $E^{\circ\prime}$, is often calculated from Equation 23.11, which is valid for a Nernstian system.

$$E^{\circ\prime} = (E_{pc} + E_{pa})/2 \tag{23.11}$$

*What Is Implied for a Nernstian System?*

It is commonly, but incorrectly, thought that observation of a Nernstian wave implies that no significant structural changes accompany the electron transfer reaction. In fact, many redox processes are known in which Nernstian or near-Nernstian behavior is observed for systems that undergo structural rearrangements and/or making and breaking of bonds in the electrochemical reaction [10]. A more correct statement is that a Nernstian response implies that any structural changes or coupled chemical reactions are *fast and reversible* on the electrochemical time scale. We will see an example of this later in our consideration of a CE mechanism.

*Non-Nernstian Systems*

$$A + ne^- \underset{k_b}{\overset{k_f}{\rightleftarrows}} B$$

When the heterogeneous electron transfer (ET) between the electrode and the molecule in solution is slow, dramatic changes may be exhibited in the CV diagnostic criteria. Modest decreases in $k_f$ and $k_b$ (a "quasireversible" system) cause a modest increase in $\Delta E_p$, the change of which with $\nu$ may be used to calculate $k_s$, the standard heterogeneous ET rate ($k_s = k_f = k_b$ at $E = E^{\circ\prime}$). As the ET rates become smaller, the waves are more likely to reflect the influence of the transfer coefficient, $\alpha$, on the wave shape. The forward and reverse branches are shaped as mirror images only if $\alpha = 0.5$. If $\alpha < 0.5$, the cathodic branch is broader; the converse is true if $\alpha > 0.5$ (Fig. 23.7).

Yet smaller $k_s$ values give rise to "irreversible" waves. The distinction between reversible, quasireversible, and irreversible processes depends on the

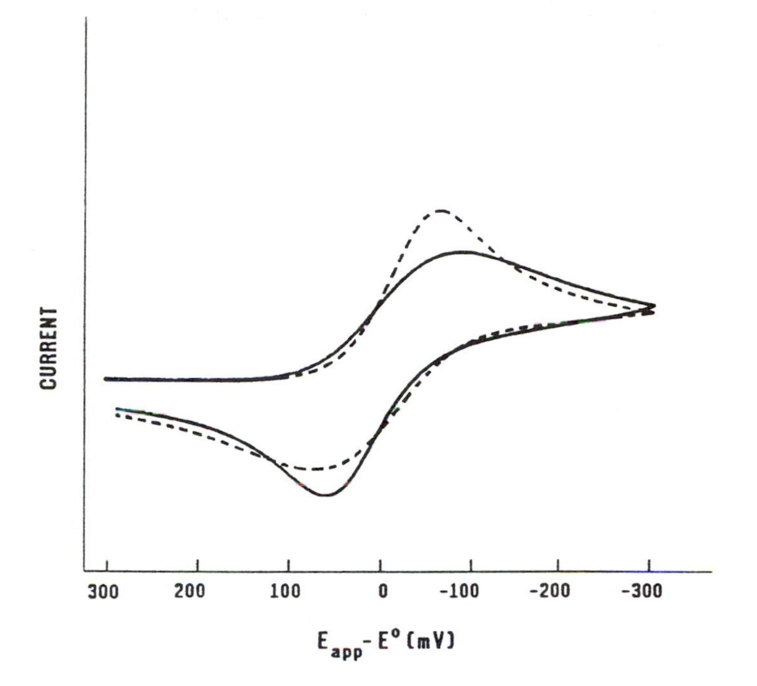


**Figure 23.7** Computed CV waveforms for two quasireversible $1e^-$ couples with different $\alpha$ values. Dashed line, $\alpha = 0.70$; solid line, $\alpha = 0.30$. $k_s = 0.1$ cm/s, $\nu = 1$ V/s.

time scale of the experiment, determined in CV by the scan rate. Because early usage of these terms related to reversibility on the polarographic time scale of several seconds, it became common to call couples with $2 \times 10^{-2}$ cm/s $< k_s < 3 \times 10^{-5}$ cm/s quasireversible and those with $k_s < 3 \times 10^{-5}$ cm/s irreversible. A non-Nernstian system may behave, however, as either, depending on $\nu$. Irreversible waves have their own set of diagnostic criteria, adherence to which fixes the redox system as being at the irreversible limit. In CV scans, couples with $\Delta E_p$ values greater than 200 mV may generally be treated as obeying irreversible charge-transfer kinetics. In-depth treatments of non-Nernstian systems are available [11].

*Example: Determination of $k_s$ from $\Delta E_p$ Measurements*

The $k_s$ value of a quasireversible system is conveniently determined by monitoring the peak separation of the CV wave as a function of sweep rate, using the results of calculations by Nicholson [12] that relate $\Delta E_p$ to the dimensionless parameter $\psi$, through Equation 23.12:

**Table 23.2** Relationship of the Dimensionless Kinetic Parameter $\Psi$ to CV Peak Potential Separations[a]

| $\Psi$ | $n\,\Delta E_p$ (mV) |
|---|---|
| 20 | 61 |
| 7 | 63 |
| 6 | 64 |
| 5 | 65 |
| 4 | 66 |
| 3 | 68 |
| 2 | 72 |
| 1 | 84 |
| 0.75 | 92 |
| 0.50 | 92 |
| 0.35 | 121 |
| 0.25 | 141 |
| 0.10[a] | 212[a] |

*Note*: There is a slight dependence of $\Delta E_p$ on switching potential. Values given are for switching potential 112.5/n mV beyond $E_p$. Other conditions: $\alpha = 0.5$; T = 298 K.
[a]Values for larger values of $\Delta E_p$ can be obtained graphically by noting that the plot of log $\Psi$ vs. $\Delta E_p$ is linear at low $\Psi$ values.
*Source*: Ref. 12.

$$\Psi = k_s/(\pi a D_A)^{1/2} \tag{23.12}$$

in which $a = nF\nu/RT = 39.1\ V^{-1}\ \nu$ at 298 K. Equation 23.12 assumes equal diffusion coefficients for species A and B, and an $\alpha$ value close to 0.5. Table 23.2 gives the calculated dependence of $\psi$ on $n\,\Delta E_p$. From an individual measurement of $\Delta E_p$ at a given $\nu$, one computes a value of $k_s/D_A{}^{1/2}$. Independent measurement of $D_A$ is required to obtain $k_s$.

Some results from our laboratory are given in Tables 23.3A and 23.3B. These measurements refer to a nickelacarborane complex, $[Ni(C_2B_9H_9Me_2)_2]^{m-}$, in which both the m=1/m=2 couple, involving Ni(III)/Ni(II), and the m=2/m=3 couple, involving Ni(II)/Ni(I), are quasireversible. These particular data are chosen because they illustrate results compromised by $iR_u$ effects.

The data on the Ni(III)/Ni(II) couple, (Table 23.3A) suggest the presence of ohmic errors, since the apparent $k_s/D^{1/2}$ values steadily decrease at higher sweep rates. The minimum ohmic error occurs at the lowest $\nu$, for which the current is minimum. If the true value of $k_s/D^{1/2}$ were 80 (the measured value at the lowest $\nu$), the expected $\Delta E_p$ at $\nu$ = 68 V/s would be only 88 mV, significantly lower than the observed value of 101 mV. The difference of 13 mV

**Table 23.3A** Sample Calculation of $k_s$ for Ni(III)/Ni(II) Couple of $[Ni(C_2B_9H_9Me_2)_2]^{-/2-}$ in $CH_3CN$/0.3 *M* $Bu_4NPF_6$ from CV Peak Separations at Hanging Hg Drop Electrode, T = ambient, $E^{\circ\prime}$ = -0.92 V vs. Fc[a]

| Scan rate (V/s) | $a^{1/2}$ | $\Delta E_p$ (mV) | $\Psi^a$ | $k_s/D^{1/2}$ ($s^{-1/2}$) |
|---|---|---|---|---|
| 11.3 | 21.0 | 71 | 2.15 | 80 |
| 22.6 | 28.0 | 76 | 1.54 | 76 |
| 33.9 | 36.5 | 84 | 1.00 | 65 |
| 54.2 | 42.0 | 90 | 0.80 | 60 |
| 56.5 | 47.3 | 96 | 0.66 | 55 |
| 67.8 | 51.5 | 101 | 0.57 | 52 |

[a]Values read from working curve constructed from data of Table 23.2

arises from $iR_u$ loss. An experimenter seeing such a trend in $k_s$ with $\nu$ should seek better minimization of uncompensated resistance. A simple average of the results at different $\nu$ gives $k_s$ = 0.42 cm/s (since $D_A = 4.1 \times 10^{-5}$ $cm^2/s$), a value that, if reported, should be given as a lower limit to $k_s$.

Ohmic error is *not* obvious in data on the Ni(II)/Ni(I) couple for this molecule (Table 23.3B), since a systematic decrease in the determined values of $k_s/D^{1/2}$ is not found. With a $k_s$ value of 0.025 cm/s, this couple has $\Delta E_p$ values that are much larger than those of the Ni(III)/Ni(II) couple, and errors of, for example, 10 mV are small when compared to the inherent $\Delta E_p$ arising from sluggish electron transfer kinetics. In this system, the slow electron transfer kinetics probably arise from molecular structure changes concomitant with the Ni(II)/Ni(I) electron transfer. A good discussion of the influence of ohmic errors on determination of $k_s$ values in various nonaqueous media is available [13].

**Table 23.3B** Calculation of $k_s$ for Ni(II)/Ni(I) Couple, $E^{\circ\prime}$ = -2.38 V vs. Fc with Other Conditions as in Table 23.3A

| Scan rate (V/s) | $a^{1/2}$ | $\Delta E_p$ (mV) | $\Psi^a$ | $k_s/D^{1/2}$ ($s^{-1/2}$) |
|---|---|---|---|---|
| 0.045 | 1.33 | 74 | 1.75 | 4.1 |
| 0.170 | 2.58 | 88 | 0.86 | 3.9 |
| 0.840 | 5.70 | 116 | 0.37 | 3.7 |
| 2.30 | 9.40 | 148 | 0.24 | 4.0 |
| 5.70 | 14.8 | 185 | 0.15 | 3.9 |
| 11.3 | 21.0 | 209 | 0.10 | 3.8 |

[a]Values read from working curve constructed from data of Table 23.2

### B The EE Mechanism

$$A + e^- \rightleftarrows B \qquad E_1^{\circ\prime} \tag{23.13}$$

$$B + e^- \rightleftarrows Z \qquad E_2^{\circ\prime} \tag{23.14}$$

This discussion is limited to situations in which both electron transfer rates are fast, since a high degree of complexity is introduced if one or more of these reactions is quasireversible.

Assuming Nernstian behavior for both processes of the EE reaction scheme, symmetrically shaped waves arise that are relatively easy to interpret. If $E_2^{\circ\prime}$ is well negative of $E_1^{\circ\prime}$ ($E_2^{\circ} << E_1^{\circ\prime}$), two separate waves are observed, in which $i_{pc(2)}$ is measured from extrapolation of the current from $i_{pc(1)}$ (see Fig. 23.8). Each individual wave obeys the diagnostics listed in Section IV.A for the simple E mechanism.

When the value of $\Delta E^{\circ\prime}$ ($= E_2^{\circ\prime} - E_1^{\circ\prime}$) is sufficiently small, the waves merge and either the $\Delta E_p$ value or the breadth of the cathodic wave ($\delta E_p$) may be used to estimate $E_2^{\circ\prime} - E_1^{\circ\prime}$, the difference between the formal potentials of the reactions [14]. Figure 23.9 shows the result for $\Delta E^{\circ\prime} = -0.05$ V, wherein the measured $\Delta E_p$ value is 72 mV. It is important to recall that this treatment assumes no contribution to peak spreading from either slow electron transfer or

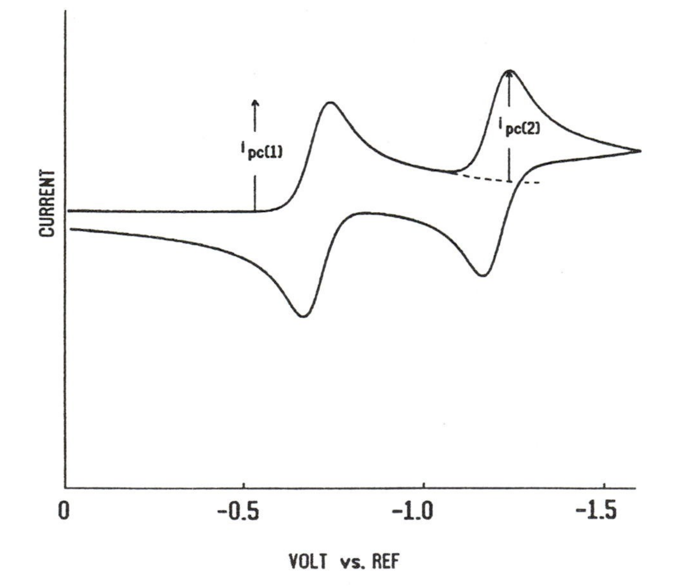


**Figure 23.8** Computed responses for an EE mechanism (Eq. 23.13 and 23.14) with $k_s(1) = k_s(2) = 1.0$ cm/s; $\alpha(1) = \alpha(2) = 0.5$; $E_1^{\circ\prime} = -0.70$ V, $E_2^{\circ\prime} = -1.20$ V; $\nu = 1$ V/s.

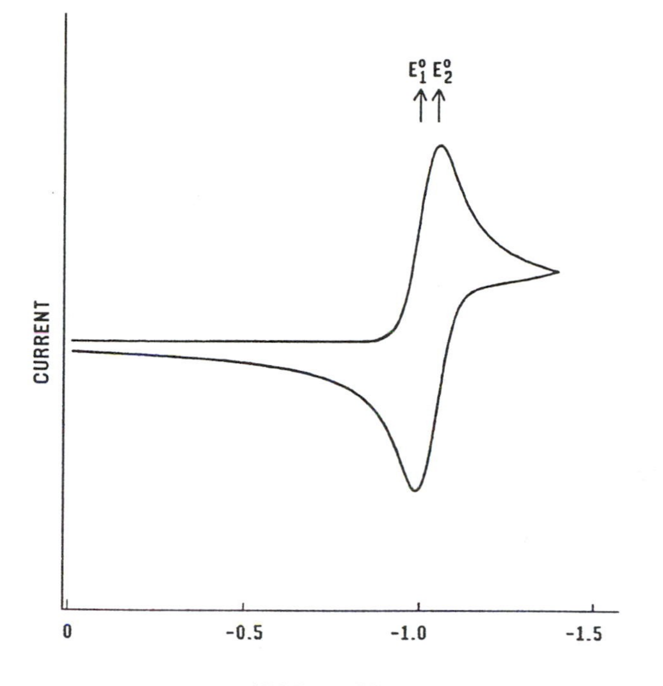


**Figure 23.9** EE mechanism with all parameters same as Figure 23.8, except that $E_1^{\circ\prime} = -1.00$ V, $E_2^{\circ\prime} = -1.05$ V.

from $iR_u$ loss. Careful measurements at different $\nu$ are required to establish the correctness of these assumptions.

When the second electron transfer is thermodynamically more facile than the first, $\Delta E^{\circ\prime}$ ($= E_2^{\circ\prime} - E_1^{\circ\prime}$) is positive, and a single wave is found. At large values of $\Delta E^{\circ\prime}$, the wave has the characteristics of a "two-electron" process, with $\Delta E_p = 30$ mV and $\delta E_p = 28.5$ mV.

Examples of EE processes abound in transition metal electrochemistry. One that conveniently shows the effect of $\Delta E^{\circ\prime}$ on wave shapes is that involving the two-step reduction of the sandwich complexes, where M = Rh or Ir:

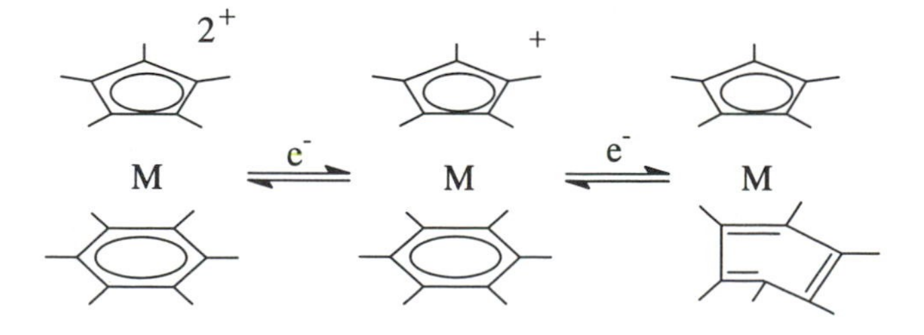

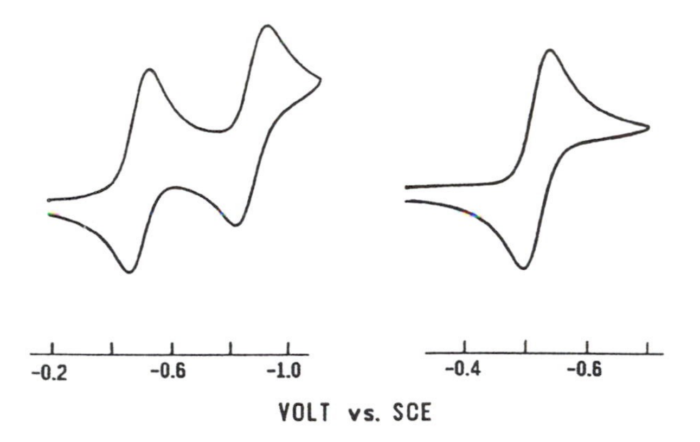


**Figure 23.10** Examples of EE mechanisms for metal sandwich compounds. (Left) Two separate 1e⁻ waves for Rh(III)/(Rh(II)/Rh(I) processes of $(C_5Me_5)Rh(C_6Me_6)^{2+/+/0}$, $E_1^{\circ\prime} = -0.50$ V, $E_2^{\circ\prime} = -0.87$ V. (Right) Single "2e⁻" wave for Ir(III)/Ir(I) couple of $(C_5Me_5)Ir(C_6Me_6)^{2+/0}$, $E^{\circ\prime} = -0.55$ V (Ref. 15).

When M = Rh, $\Delta E^{\circ\prime} \approx -0.20$ V, and two separate waves are observed (Fig. 23.10, left). When M = Ir, $\Delta E^{\circ\prime} \approx +0.30$ V, and a single wave is seen that has properties approaching those of a two-electron Nernstian process. These systems deviate slightly from ideality owing to quasireversibility of the second electron transfer, as evidenced by the somewhat larger peak separation in the second wave of the rhodium complex [15].

One useful commodity calculated from such measurements is the equilibrium constant for disproportionation (Eq. 23.15):

$$2B \rightleftarrows A + Z \qquad K_{disp} = [A][Z]/[B]^2 \tag{23.15}$$

through the relationship (at 298 K)

$$\log K_{disp} = 16.6\,(E_2^{\circ\prime} - E_1^{\circ\prime}) \tag{23.16}$$

A molecule will rarely take up or release a large number of electrons without bond disruptions. An exceptional example involves the reduction of an Fe(III) tris-bipyridyl complex, in which six electrons are added stepwise to the complex. The chemical reversibility of the processes may be followed by cyclic voltammetry (Fig. 23.11).

### C. The CE Mechanism

$$Y \underset{k_b}{\overset{k_f}{\rightleftarrows}} A + e^- \rightleftarrows B \tag{23.17}$$

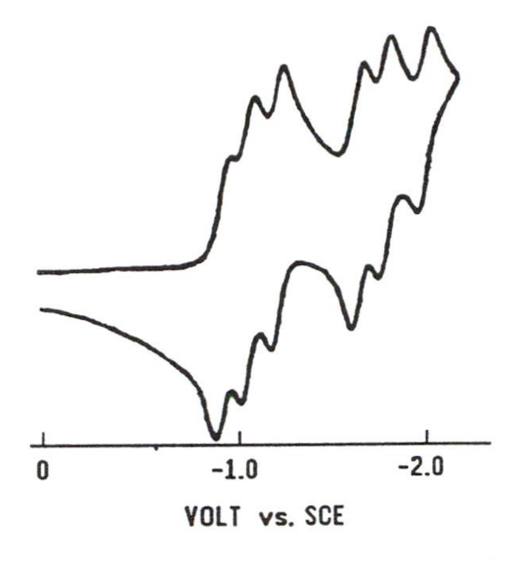


**Figure 23.11** Six successive $1e^-$ couples for the reduction of 1 m*M* Fe(4,4′-bis(ethoxycarbonyl)-2,2′-bipyridyl)$_3$ in DMF. Reprinted with permission from C.M. Elliott and E.J. Hershenhart, *J. Am. Chem. Soc. 104*:7519 (1982).

Here we introduce a reversible reaction coupled to the reactant, A, and assume that Y is not electroactive at the potential at which A is reduced. The manner in which the diagnostics respond to this coupled reaction depends in a complex way on the position of the equilibrium ($K_{eq} = k_f/k_b = [A]/[Y]$) and the values of the rate constants compared to the CV sweep rate [16]. For illustrative purposes, we pick a case in which $K_{eq}$, $k_f$, and $k_b$ are all moderate, and note the effect of $\nu$ on the current function, X, of compound A.

If the value of $K_{eq}$ is chosen for convenience as 1, the concentration of A in the bulk of solution is one-half its formal value. The current function for A, X(A), is *one-half* the value that would have been observed had no preceding reaction been present, *if* the equilibration is slow on the CV time scale. This "frozen equilibrium" or "fast scan" limit occurs when $k_f/a \ll 1$, where $k_f$ is defined in Equation 23.17 and $a = nF\nu/RT$ ($= 38.9\ V^{-1}\ \nu$ at 298 K). In this limit, X(A) is independent of sweep rate, since the reactant flux to the electrode is diffusion limited. A faster equilibration rate (larger $k_f$) may increase the observed value of X(A) by allowing significant conversion of Y to A *during* the scan. As the applied potential enters the range in which A is reduced, the reactant A is depleted near the electrode, resulting in a left-to-right stress on Equation 23.17 and conversion of Y to A. This results in an increase of X(A) owing to the additional A supplied by the "reservoir" of Y. An increase in current function at lower scan rates is an important diagnostic for this mechanism. An experimental example will follow later.

In the "fast reaction" or "slow scan" limit, $k_f/a >> 1$, all Y is converted to A during the scan, and X(A) has the same value that would have been found had there been no preceding reaction.

The three kinetic ranges of high, moderate, and low values of $k_f/a$ are all observed in the equilibration and reduction of the 17-electron complex $CpCr(CO)_3$ ($Cp = \eta^5\text{-}C_5H_5$). In this system (Eq. 23.18 and drawing), the monomer $CpCr(CO)_3$ corresponds to A and the metal-metal bonded dimer $Cp_2Cr_2(CO)_6$ to Y in Equation 23.17. The monomer is reduced to the 18-electron monoanion ($E^{\circ\prime} = -0.82$ V vs. Fc), and the dimer is electroactive only at much more negative potentials.

$$\underset{Y}{\tfrac{1}{2}Cp_2Cr_2(CO)_6} \underset{k_b}{\overset{k_f}{\rightleftharpoons}} \underset{A}{CpCr(CO)_3} + e^- \overset{E^{\circ\prime}}{\rightleftharpoons} \underset{B}{CpCr(CO)_3^-} \qquad (23.18)$$

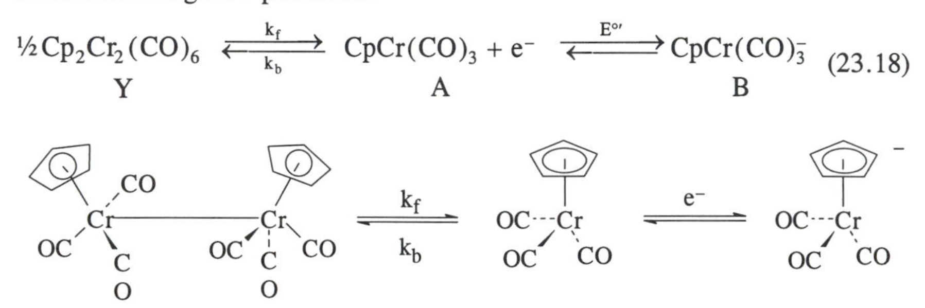


The dimer is strongly favored at formal concentrations in the millimolar range, with the monomer constituting only about 10% of the mass balance [17] at ambient temperatures. Surprisingly, though, the reduction wave for the monomer reaches approximately the full X value expected for a one-electron reactant when slow CV sweeps are employed. As shown in Figure 23.12, the CV wave has all the characteristics of a nearly Nernstian process. Looking only at the CV data, one might conclude that the electrode reaction is a simple E process and that any monomer/dimer equilibrium lay completely in favor of the monomer. In fact, the deceptively simple CV response is obtained because the experiment falls in the slow scan limit, $k_f/a >> 1$, in which monomer forms rapidly from dimer during the scan. Indications of the presence of the preceding equilibrium are found at lower temperatures or concentrations and higher sweep rates.

At reduced temperatures, $k_f$ is diminished, and the system is no longer in the slow scan limit. The revealing diagnostic is that of the change of X with $v$ (Table 23.4). At higher sweep rates, X[$CpCr(CO)_3$] falls off considerably, finally reaching a constant value when $v > 100$ V/s, as the fast scan limit is reached. The constant X value in this limit is reflective of the amount of equilibrated $CpCr(CO)_3$ in the bulk of solution.

The X value in the fast scan limit may be used to estimate the value of $K_{eq}$, and its increases at slower sweep rates may be used to estimate the monomerization/dimerization rate constants [16]. Full digital simulations may also be used

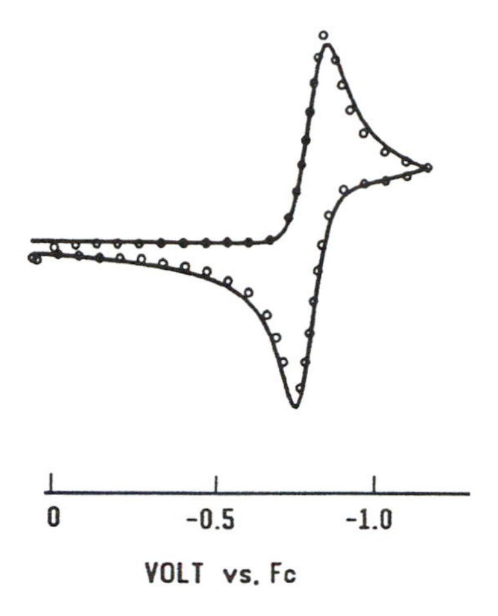


**Figure 23.12** Comparison of experiment (circles) for 0.9 m*M* $Cp_2Cr_2(CO)_6$ in $CH_2Cl_2$/0.1 *M* $Bu_4NPF_6$, ν = 0.1 V/s, T = 293 K with theory (line) for a $1e^-$ Nernstian couple. Taken from Ref. 18.

to extract the equilibrium and rate constants. In the present example, the latter were preferable because they allowed fitting of both the monomer wave at −0.8 V and the (irreversible) dimer reduction wave at more negative potentials (Fig. 23.13) [18]. The latter is not explicitly included in Equation 23.18.

## D. The EC Mechanism

$$A + e^- \rightleftarrows B \xrightarrow{k_c} Z \tag{23.19}$$

Equation 23.19 describes a commonly encountered electrode mechanism, namely electron transfer followed by an irreversible reaction of the primary electrode

**Table 23.4** Current Function, X, in Relative Units, for Reduction of $CpCr(CO)_3$ in $CH_2Cl_2$ at Pt Electrode, T = 240 K, Formal Concentration of $CpCr(CO)_3$ = 7 × $10^{-4}$ *M*.

| ν (V/s) | X |
|---|---|
| 1.0 | 3.6 |
| 5.0 | 2.4 |
| 20 | 1.7 |
| 100 | 1.1 |
| 125 | 1.0 |
| 150 | 1.0 |

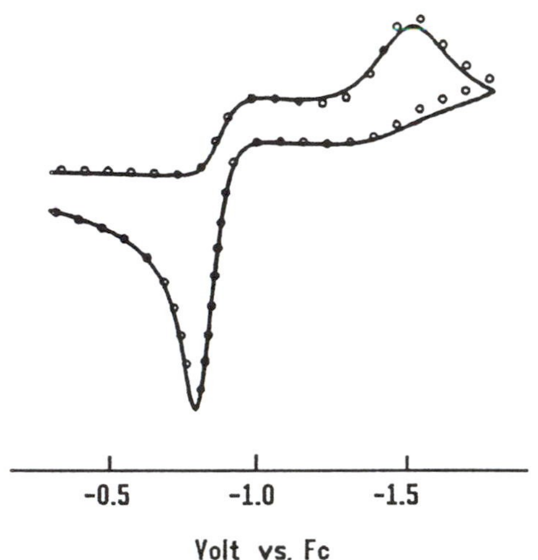


**Figure 23.13** Comparison of experiment (circles) with theory (line) for CV response of 0.7 m*M* $Cp_2Cr_2(CO)_6$ in $CH_2Cl_2$/0.1 *M* $Bu_4NPF_6$, T = 243 K, ν = 100 V/s, with experimental points corrected for charging current and uncompensated resistance (see Ref. 18). Cathodic features are those for reduction of the monomer $CpCr(CO)_3$ (ca. −0.9 V) and the dimer (ca. −1.5 V). Reduction of the dimer furnishes $CpCr(CO)_3^-$, which is reoxidized at ca. −0.75 V.

product, B, forming the secondary product, Z. When Z is not electroactive, the quantitative treatment of this mechanism is relatively simple.

As the value of $k_c/a$ goes from small to large, the expected shift in the cathodic peak potential with scan rate goes from zero to 30 mV per 10-fold increase in sweep rate. This diagnostic is difficult to apply, however, in light of the fact that ohmic errors produce shifts in the same direction.

The more convenient CV diagnostic is the current ratio, $i_{rev}/i_{fwd}$, which is unity when $k_c/a$ is small (fast scan or slow reaction limit) and less than unity when $k_c/a$ is large enough to diminish the amount of B available for re-electrolysis on the reverse sweep. The current ratio is measured conveniently by the empirical method of Nicholson [19], which requires only the measurement of the values $i_{p(fwd)}$, $i_{p(rev)}$, and $i_\lambda$ (Fig. 23.14), where λ designates the switching point. These quantities are then used in Equation 23.20 to obtain the current ratio:

$$i_{rev}/i_{fwd} = i_{p(rev)}/i_{p(fwd)} + 0.48\, i_\lambda/i_{p(fwd)} + 0.086 \tag{23.20}$$

The current ratio in turn is used to calculate the apparent rate constant for the follow-up reaction of B → Z. If a first-order or pseudo-first-order rate constant,

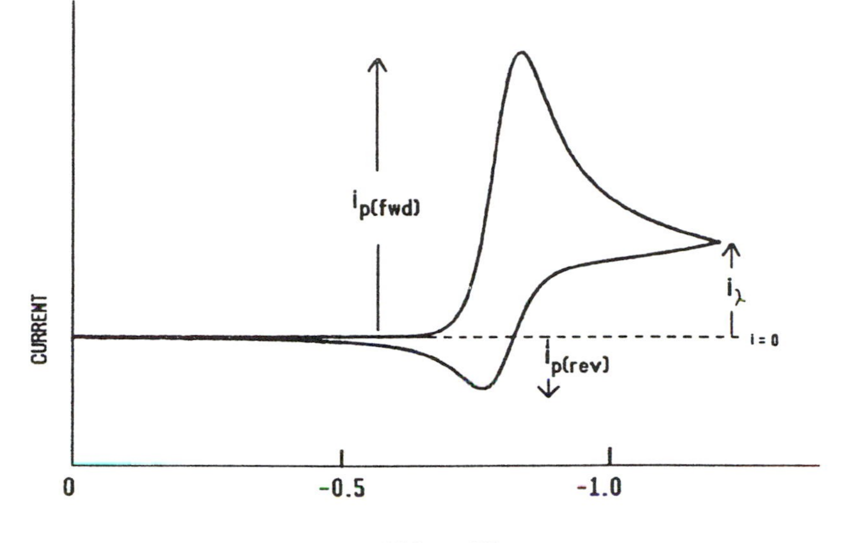


**Figure 23.14** Theoretical waveform for an EC mechanism (Eq. 23.20). Parameters: $k_s$ = 1 cm/s, $E^{\circ\prime}$ = -0.80 V, $k_c$ = 0.2 $s^{-1}$, $\nu$ = 1 V/s. The values to be used in Eq. 23.20 are depicted.

$k_{obs}$, is assumed, a working curve of $i_{rev}/i_{fwd}$ vs. $k_{obs}\tau$ constructed from the data in Table 23.5 is used to obtain $k_{obs}$. Here $\tau$ is the time in seconds required to scan from the $E_{1/2}$ ($\sim E^{\circ\prime}$) value to the switching potential in an individual scan; $\tau$ therefore changes with $\nu$.

Each individual CV scan yields an estimate of $k_{obs}$. In practice, the scan rate is usually varied in order to obtain a range of values of $i_{rev}/i_{fwd}$ from about 0.5 to 0.9, and the resulting $k_{obs}$ values are averaged. Alternatively, the set of $k_{obs}\tau$ values may be plotted against $\tau$ and the rate constant $k_{obs}$ obtained from the slope. The graphical treatment more readily reveals systematic errors.

**Table 23.5** CV Data (from Ref. 20) on the Oxidation of 0.2 m*M* Cd(Hg) in the Presence of 30 m*M* CaEDTA, 29 m*M* $Ca(NO_3)_2$, at pH 8.0, T = 298 K

| $\nu$ (V/s) | $i_c/i_a$ | $\tau$ (s) | $k_{obs}\tau$ | $k_{obs}$ ($s^{-1}$) |
|---|---|---|---|---|
| 1.49 | 0.51 | 0.094 | 0.92 | 9.8 |
| 2.22 | 0.61 | 0.063 | 0.62 | 9.8 |
| 3.27 | 0.69 | 0.043 | 0.41 | 9.5 |

Two literature examples serve to illustrate these procedures. The first involves rates of competitive ligand exchange reactions and relates to a single reaction condition, wherein $i_{rev}/i_{fwd}$ ratios are measured at a series of different scan rates. The second involves ligand substitution of an organometallic diiron complex, and relates to a set of different reaction conditions (nucleophile concentrations), wherein the $i_{rev}/i_{fwd}$ ratio is measured for each concentration at a single scan rate.

Kuempel and Schaap studied the relative rates of complexation of $Cd^{2+}$ and $Ca^{2+}$ with the tribasic anion of EDTA ($HY^{3-}$) in the scheme [20]:

$$\begin{array}{ccc} Cd \rightleftharpoons Cd^{2+} + 2e^- + & HY^{3-} & \xrightarrow{k_{obs}} CdY^{2-} + H^+ \\ & {\scriptstyle Ca^{2+}}\downarrow & \\ & CaHY^- & \end{array} \tag{23.21}$$

The EC mechanism is that of Equation 23.21, the horizontal line in the scheme, encompassing the oxidation of Cd to $Cd^{2+}$ (E) followed by complexation of $Cd^{2+}$ by $HY^{3-}$ (C). The electrochemical reactant, $Cd^{2+}$, was cleverly furnished by a cadmium amalgam working electrode, which upon oxidation gave $Cd^{2+}$ at the electrode for reaction with $HY^{3-}$. $Ca^{2+}$ added to the bulk solution served to compete with $Cd^{2+}$ for available $HY^{3-}$.

Typical data are those reported for three scan rates under the experimental conditions given in Table 23.5. An average $k_{obs}$ of 9.7 $s^{-1}$ was obtained. Note that in this case as well as in the following example, $k_{obs}$ is a pseudo-first-order rate constant and is equal to $k_c[Z]$, where Z ($HY^{3-}$ in this case) is the reagent reacting with B ($Cd^{2+}$).

A modified approach was used to obtain the rate of the cleavage of the dinuclear monocation $[Cp_2^*Fe_2(CO)_2(\mu\text{-}CO)_2]^+$ (drawing) by $CH_3CN$ (Eq. 23.22), in which Cp* = $\eta^5\text{-}C_5Me_5$ [21]:

$$[Cp_2^*Fe_2(CO)_2(\mu - CO)_2]^+ + CH_3CN \xrightarrow{k_{obs}} [Cp^*Fe_2(CO)_2(NCCH_3)]^+ + Cp^*Fe(CO)_2 \tag{23.22}$$

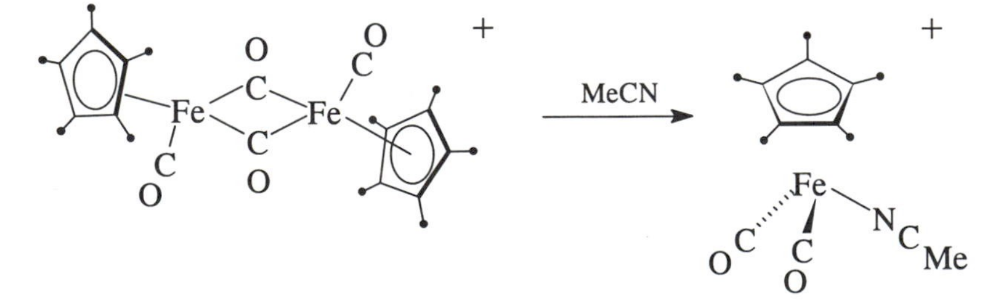

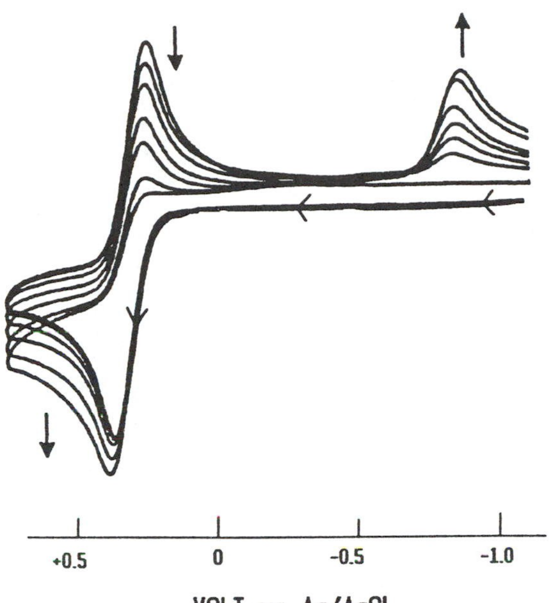


**Figure 23.15** CV scans of 1 mM $[Cp_2^*Fe_2(CO)_2(\mu\text{-}CO)_2]$ in $CH_2Cl_2$ containing $CH_3CN$ in increasing amounts, from 0 to 65.9 M (see Table 23.6 and Ref. 21); $\nu$ = 0.20 V/s in each scan.

Figure 23.15 shows the CV scans obtained after successive additions of $CH_3CN$ to a 0.98 m*M* solution of $Cp_2^*Fe_2(CO)_2(\mu\text{-}CO)_2$ in $CH_2Cl_2$. The one-electron oxidation at approximately +0.4 V is that of the reactant diiron complex. Equations 23.23 (E) and 23.22 (C) constitute the EC mechanism. The reduction wave at -0.85 V is that of the ultimate reaction product, $Cp^*Fe(CO)_2(NCCH_3)^+$, and although it is a useful indicator of the reaction product, it is not used in the calculations to obtain $k_{obs}$.

$$[Cp_2^*Fe_2(CO)_2(\mu - CO)_2] \rightleftarrows [Cp_2^*Fe_2(CO)_2(\mu - CO)_2]^+ + e^- \quad (23.23)$$

The data in Figure 23.15 were all obtained at the scan rate of 0.25 V/s. It is readily observed that at this single $\nu$, the chemical reversibility of the wave at +0.4 V decreases with increasing concentration of $CH_3CN$. Table 23.6 gives the raw data derived from these scans ($\tau$ = 1.75 s in all cases). Noting that $k_{obs} = k_c[CH_3CN]$, it is possible to calculate a value for $k_c$ at each concentration of acetonitrile.

**Table 23.6** CV Data for Oxidation of 0.98 m*M* $Cp_2^*Fe_2(CO)_2(\mu - CO)_2$ in $CH_2Cl_2$/0.1 *M* $Bu_4NPF_6$ at Ambient Temperature with Various Added Amounts of $CH_3CN$

| $[CH_3CN]$ (m*M*) | $i_c/i_a$ | $k_{obs}t$ | $k_{obs}$ | $k_c$ ($s^{-1}$) |
|---|---|---|---|---|
| 0 | 1.02 | — | — | — |
| 0.20 | 1.03 | — | — | — |
| 0.40 | 0.98 | — | — | — |
| 2.0 | 0.96 | (0.05) | — | — |
| 4.2 | 0.83 | (0.24) | (0.14) | (57) |
| 8.3 | 0.71 | 1.40 | 0.23 | 48 |
| 12.8 | 0.61 | 0.62 | 0.35 | 48 |
| 32.2 | 0.45 | 1.25 | 0.71 | 39 |
| 65.9 | 0.33 | >2 | >1.1 | >30 |

*Note*: All scans at $v = 0.25$ V/s, with $\tau = 1.75$ s.
*Source*: Data from Mann et al. [21 and personal communications].

Not all of the data sets in Table 23.6 are equally useful. At the lowest concentrations of $CH_3CN$ (the first four rows in Table 23.6), the deviation of $i_{rev}/i_{fwd}$ ($i_c/i_a$) from the unity value of a chemically reversible system is small, leading to large uncertainties in the calculation of $k_{obs}\tau$. Additionally, when $[CH_3CN] = 4.2$ m*M*, although the value of $i_{rev}/i_{fwd}$ (0.83) is in the sensitive range, the ratio of ~4.3 for $[CH_3CN]/[Cp_2^*Fe_2(CO)_2(\mu - CO)_2]$ is too low to fulfill the requirements of pseudo-first-order kinetics. Therefore, the apparent value of $k_c$ (57 $s^{-1}$) derived from this condition is erroneously high.

The last row entry in Table 23.6 is also problematic, since the value of $i_{rev}/i_{fwd}$ (0.33) is too low to be reliably measured by the Nicholson method; the values of $k_{obs}\tau$ and $k_c$ derived from the last row should therefore be viewed as lower limits. It is the data at the three $CH_3CN$ concentrations of 8.3, 12.8, and 32.2 m*M* that are most appropriate for evaluation of the rate constant, and the resultant values in Table 23.6 are consistent with rate = $45[CH_3CN]\ s^{-1}$.

## E. The ECE Mechanism

$$A + e^- \rightleftarrows B \qquad E_1^{o\prime} \tag{23.24}$$

$$B \xrightarrow{k_c} Z \qquad C \tag{23.25}$$

$$Z + e^- \rightleftarrows Z^- \qquad E_2^{o\prime} \tag{23.26}$$

Manifold possibilities exist for this scheme, depending on the relative values of $E_1^{o'}$ and $E_2^{o'}$, the value of $k_c/a$, and the possibility that the product of the chemical reaction may be produced in either its oxidized (Z) or reduced ($Z^-$) state. Our discussion is therefore limited in scope. Furthermore, we use examples and models in which all the electrochemical reactions are Nernstian one-electron processes.

*Case: $E_2^{o'} < E_1^{o'}$*

In this case the product wave ($Z/Z^-$) appears at a potential negative of the reactant wave (A/B). When the chemical reaction rate is moderate compared to the sweep rate, $k_c$ is measured conveniently from the $i_{rev}/i_{fwd}$ values of the A/B wave, clipped at the appropriate potential (Fig. 23.16), by treating the data in the same fashion as in Section IV.D (the EC case). An alternate approach would be to employ the ratio of $i_{pc2}$ to $i_{pc1}$ (Fig. 23.16), which ranges from zero to unity as $k_c/a$ goes from large to small values. Figure 23.17 displays the working curve for $i_{pc2}/i_{pc1}$ as a function of $k_c\tau$, where $\tau$ is the time required to scan

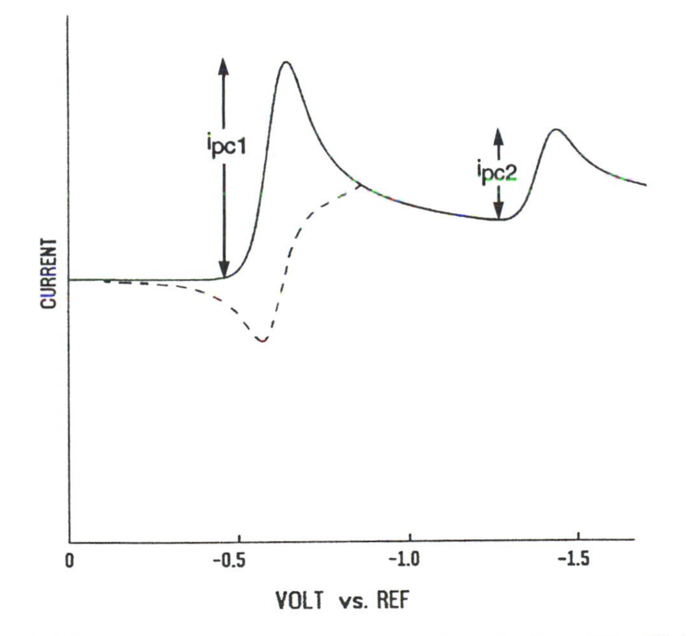


**Figure 23.16** Computed waveforms for the ECE mechanism of Eq. 23.24–23.26, in which $E_2^{o'} < E_1^{o'}$. Solid line, cathodic scan including both the reduction of A and Z; dashed line, CV scan of couple A/B alone. Parameters: $E_1^{o'} = -0.60$ V, $E_2^{o'} = -1.40$ V, $k_s(1) = k_s(2) = 1$ cm/s, $k_c = 0.25\ s^{-1}$, $\nu = 0.075$ V/s.

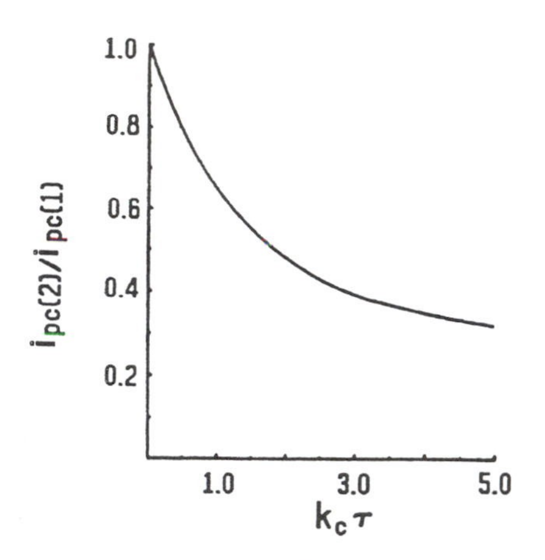


**Figure 23.17** Ratio of $i_{pc}(2)/i_{pc}(1)$ for ECE mechanism of Eq. 23.24–23.26 for different values of $k_c\tau$, where $k_c$ is the intervening chemical rate constant of Eq. 23.25 and $\tau$ is the time (in seconds) required to scan from $E_{1/2}(1)$ to $E_{1/2}(2)$. Taken with permission from Ref. 22.

from $E_1^{\circ\prime}$ to $E_2^{\circ\prime}$ [22]. The rate constant is evaluated from the values of $k_c\tau$ at different sweep rates in a fashion analogous to that described in Section IV.D for $k_{obs}$.

When $k_c/a$ is large, the first wave is chemically irreversible. This is illustrated (Fig. 23.18) by the reduction of nitroprusside ion, $[Fe(CN)_5NO]^{2-}$ in nonaqueous media [23], in which the reduction at approximately −1.3 V (Eq. 23.27) is followed by fast loss of cyanide (Eq. 23.28) to give a complex reducible at about −1.7 V (Eq. 23.29).

$$[Fe(CN)_5NO]^{2-} + e^- \rightleftarrows [Fe(CN)_5NO]^{3-} \qquad E_1^{\circ\prime} \tag{23.27}$$

$$[Fe(CN)_5NO]^{3-} \xrightarrow{k_c} [Fe(CN)_4NO]^{2-} + CN^- \tag{23.28}$$

$$[Fe(CN)_4NO]^{2-} + e^- \rightleftarrows [Fe(CN)_4NO]^{3-} \qquad E_2^{\circ\prime} \tag{23.29}$$

Moderate increases in $v$ give no new qualitative features to the CV traces. The peak potential of the first wave moves negative with increasing sweep rate (the expected magnitude is 30 mV per decade increase in $v$, reducing to the EC result). $E_{pc}$ for the second wave is essentially unchanged with $v$, implying that the $[Fe(CN)_4NO]^{2-/3-}$ couple is nearly Nernstian.

*Case:* $E_2^{\circ\prime} > E_1^{\circ\prime}$

If the product, Z, is reducible at a potential more positive than that required to reduce the reactant, A, a single wave is observed on the forward scan, near the

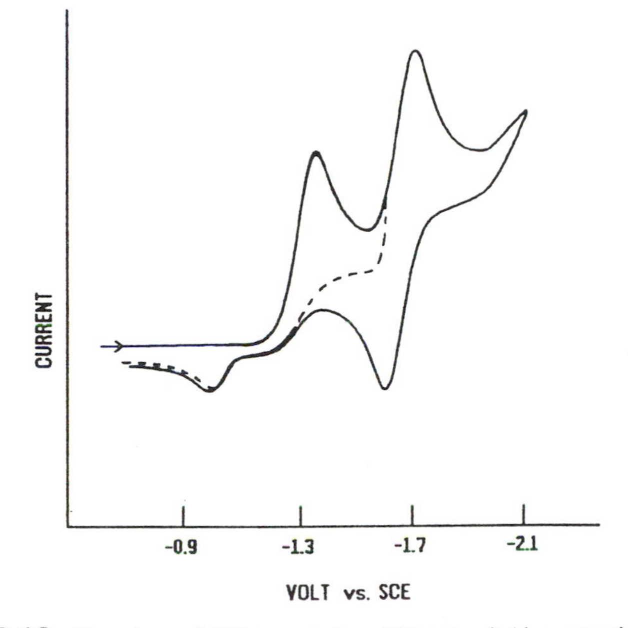


**Figure 23.18** Experimental CV scans to two different switching potentials for a complex reducing by the ECE mechanism of Eq. 23.24–23.26 with a rapid chemical reaction (Eq. 23.25). The system is $[Fe(CN)_5NO]^{2-}$ in $CH_2Cl_2$ (Eq. 23.27–23.29), and the anodic feature at approximately −1.0 V arises from a product not referred to in Eq. 23.27–23.29. $\nu$ = 0.08 V/s. Reprinted from Ref. 23 with permission.

potential of the A/B couple. The anodic current for the reoxidation of B is diminished owing to the partial or whole disappearance of B through Equation 23.25. The cathodic current (for A) is enhanced proportional to the amount of Z produced by Equation 23.25. The current function of A, $X_A$, is usually the most conveniently applied diagnostic for this mechanism. We now consider examples in two different $k_c/a$ regimes.

**$k_c \approx a$.** In this example, the rate constant $k_c$ is sufficient to produce substantial but not complete conversion of B to Z at scan rates in the range of 0.05 to 1.0 V/s. Equations 23.30–23.32 give the ECE sequence, in which the ring-addition intermediate $CpCo(\eta^4\text{-}C_5H_5CH_2CN)$ is the postulated (but unproven) product of the reaction of the cobaltocene anion (see drawing) with acetonitrile when the latter is used as the electrochemical solvent:

$$Cp_2Co + e^- \rightleftarrows Cp_2Co^- \qquad E^{\circ\prime} = -1.90V \qquad (23.30)$$

$$Cp_2Co^- + CH_3CN \xrightarrow{k_{obs}} CpCo(\eta^4 - C_5H_5CH_2CN) \quad (Z) \qquad (23.31)$$

$Z + e^- \rightarrow Z^-$ $E^{\circ\prime} > -1.90V$ (23.32)

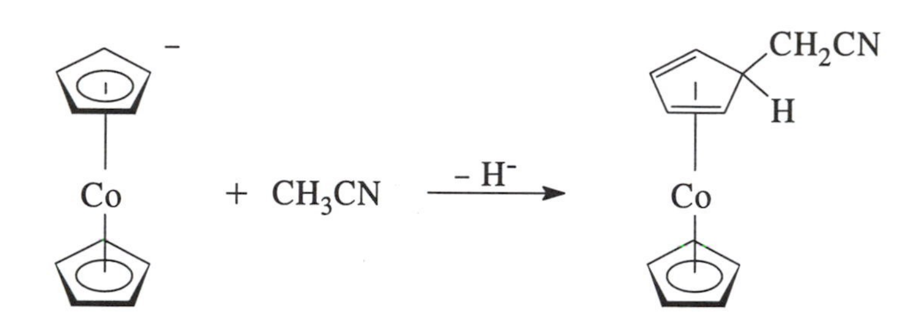


When $\nu > 1$ V/s, $k_{obs}/a$ is small and the system behaves as a one-electron E process, with the current function independent of scan rate. At lower $\nu$, Equation 23.31 contributes additional electroactive material and X(A) increases. Figure 23.19 demonstrates the increase in CV peak height and decrease in $i_a/i_c$ that accompanies the onset of the effect of the chemical reaction. The change of $X_A$ with $\nu$ may be used to calculate $k_{obs}$ if the current function that would have been obtained in the absence of the chemical reaction is known. Fortunately, the latter is often accessible from data either in the high scan rate limit or by making measurements under different chemical conditions (e.g., by changing the concentration of a reagent with which B is reacting). One must also know the total number of electrons transferred in the two E steps, but this is often deduced from bulk coulometry experiments.

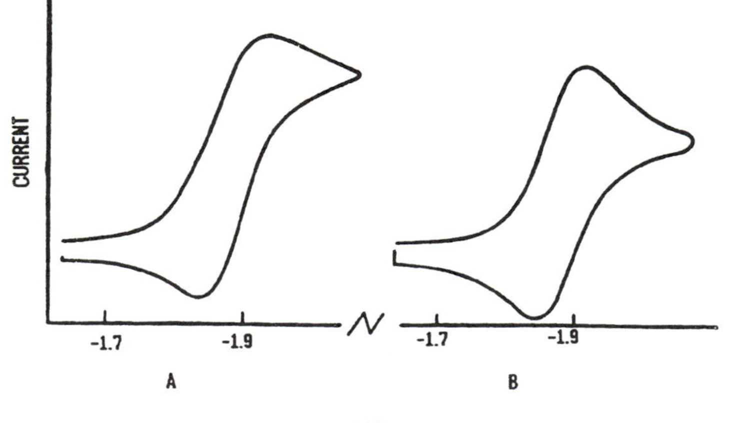


**Figure 23.19** CV scans of the $Cp_2Co^{0/-}$ wave in $CH_3CN$/0.1 *M* $Bu_4NPF_6$ at Hg electrode, with current axis normalized for scan rate by dividing by $\nu^{1/2}$. (Left) $\nu = 0.05$ V/s; (right) $\nu = 0.50$ V/s.

In the present example, the reactant A, cobaltocene, also undergoes a diffusion-controlled one-electron oxidation for the process $Cp_2Co^{+/0}$ at a more positive potential, $E^{\circ\prime} = -0.9$ V. X for $Cp_2Co^{+/0}$ is independent of ν over the recorded range of scan rates, and provides a reference for the limiting value of X for $Cp_2Co^{0/-}$. As seen in Figure 23.20, the current function for $Cp_2Co^{0/-}$ decreases smoothly with ν to a limiting value that is the same as that of the uncomplicated E process of $Cp_2Co^{0/+}$.

Nicholson and Shain [24] have shown that when the same number of electrons is transferred in each step of the ECE mechanism, Equation 23.33 holds, in which $X_c$ and $X_D$ are the current functions in the presence and absence, respectively, of the intervening chemical reaction ($X_D$ is the hypothetical value of the current function under diffusion-controlled conditions, i.e., if the C step were not present):

$$\frac{X_c}{X_D} = \frac{0.400 + k_c/a}{0.396 + 0.469(k_c/a)} \tag{23.33}$$

A separate estimate of $k_c$ is obtained for each experimental scan rate. One obtains an average of $k_c = 0.16\ s^{-1}$ from this data, which may then be used with

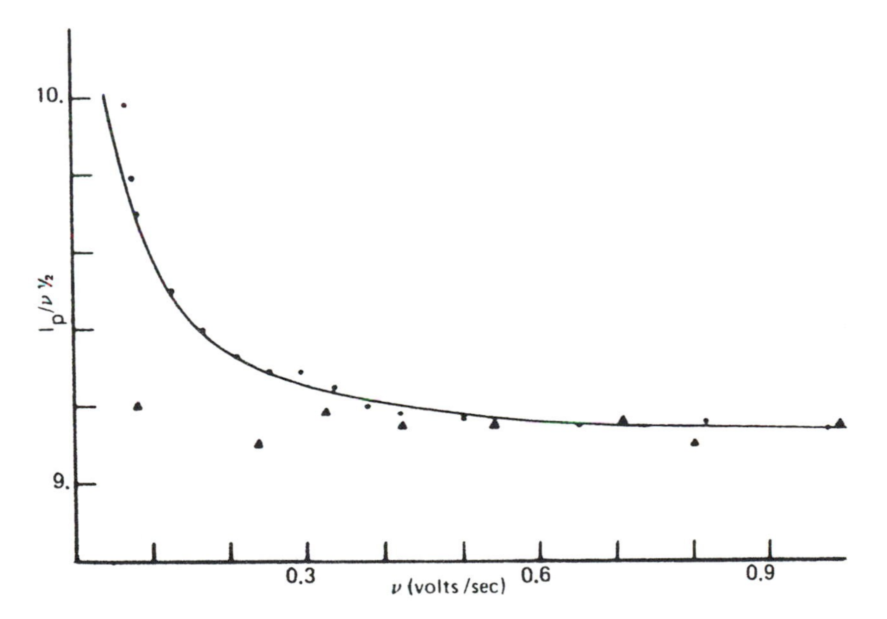


**Figure 23.20** Plots of cathodic current functions against ν for the 1e⁻ reduction of $Cp_2Co^{+/0}$ (triangles) and the ECE reduction of $Cp_2Co$ (Eq. 23.30–23.32) in $CH_3CN$. The solid line gives the theoretical values, assuming $k_{obs} = 0.16\ s^{-1}$ (Eq. 23.31).

Equation 23.33 to calculate a theoretical curve for the current function enhancement over the spread of ν recorded. The solid line in Figure 23.20 shows that the experimental enhancements in X for $Cp_2Co^-$ under the ECE conditions of Equations 23.30–23.32 adequately track those predicted by theory if one treats $k_c$ as a first-order rate constant.

**$k_c$ >> a.** The foregoing example involved an intervening chemical reaction of moderate rate giving a product that is irreversibly reduced. It is instructive to observe, at least qualitatively, some possible effects if the chemical step of Equation 23.25 were rapid and reversible.

For this case, we refer to literature reports concerning the reduction of *cis*-$(CO)_2Mn(\eta^2\text{-dppe})_2^+$, in which dppe = bis(diphenylphosphino)ethane [25]. This molecule has a cathodic wave of two-electron height at approximately −1.8 V vs. SCE (Fig. 23.21, peak 1). The major electrolysis product was identified as $(CO)_2Mn(\eta^2\text{-dppe})(\eta^1\text{-dppe})^-$, in which one of the dppe ligands is partially dissociated from the metal, as shown.

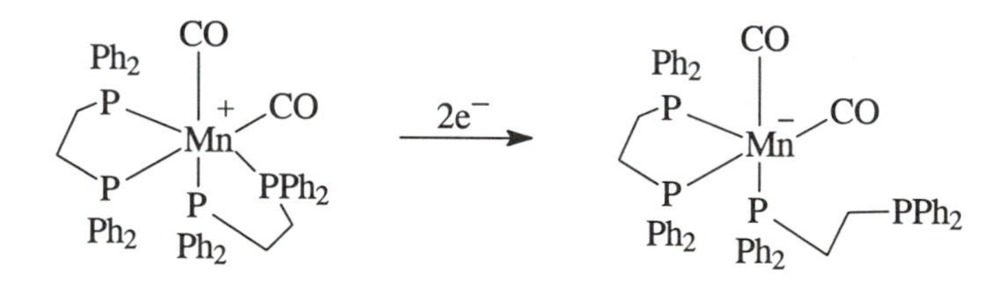


The overall ECE process of Equations 23.34–23.36 was implicated, with the intervening chemical reaction (Eq. 23.35) being the partial dissociation of one dppe ligand:

$$(CO)_2Mn(\eta^2-\text{dppe})_2^+ + e^- \rightleftarrows (CO)_2Mn(\eta^2-\text{dppe})_2 \qquad E_{pc} = -1.84\text{ V} \quad (23.34)$$

$$(CO)_2Mn(\eta^2-\text{dppe})_2 \xrightarrow{k_c} (CO)_2Mn(\eta^2-\text{dppe})(\eta^1-\text{dppe}) \quad (23.35)$$

$$(CO)_2Mn(\eta^2-\text{dppe})(\eta^1-\text{dppe}) + e^- \rightleftarrows (CO)_2Mn(\eta^2-\text{dppe})(\eta^1-\text{dppe})^- \qquad E_2^{o\prime} = -1.45\text{ V} \quad (23.36)$$

Peaks 2 and 3 of Figure 23.21 are those of the one-electron couple of the rearranged product (Eq. 23.36). Peak 3 is detected by a second negative-going scan initiated immediately after the first CV cycle has returned near to its beginning point.

There are two important differences between the overall scheme for this Mn complex and that discussed previously for the cobaltocene anion. One is that $k_c$ is very large in the former case, estimated as $> 10^6\ s^{-1}$ by the authors [25],

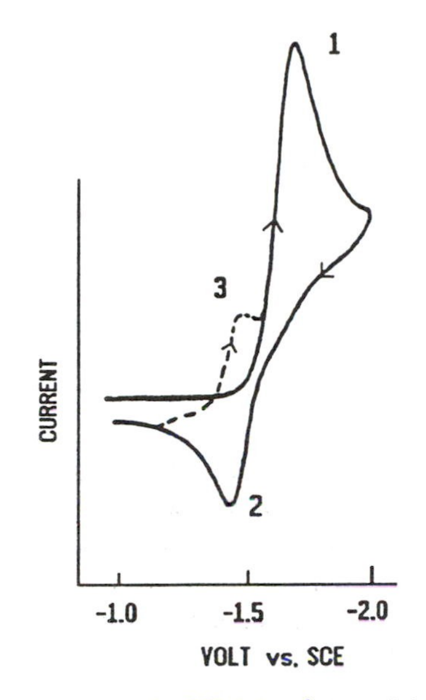


**Figure 23.21** CV scans of 5 m*M* $(CO)_2Mn(\eta^2\text{-dppe})_2^+$ in THF, $\nu$ = 0.5 V/s. The three features correspond to (1) reduction of $(CO)_2Mn(\eta^2\text{-dppe})_2^+$ (Eq. 23.34) and (2 and 3) oxidation and rereduction of $(CO)_2Mn(\eta^2\text{-dppe})(\eta^1\text{-dppe})^-$ (Eq. 23.36). Reprinted with permission from Ref. 25.

so that X for wave 1 is essentially independent of $\nu$ and yields no quantitative information about $k_c$ (the lower limit of $k_c$ was estimated from the observed chemical irreversibility of wave 1 at very high sweep rates). Second, the reversible electron transfer reaction of the product, Equation 23.36, allows the determination of the $E^{\circ\prime}$ value of the secondary reaction product, through observation of waves 2 and 3.

## V. COMMENTS ON SECOND-ORDER HOMOGENEOUS REACTIONS

To this point, we have considered only the electron transfer reactions that occur between the electrode and soluble substrates, that is, heterogeneous processes. In most cases, heterogeneous electron transfer reactions are sufficient to account completely for the shapes and diagnostic responses of voltammetric curves. It is well known, however, that in the solution layer adjacent to the electrode, second-order electron transfer reactions occur between electrolysis products and reactants. There is a growing body of information showing that under some circumstances these homogeneous electron transfer reactions present a more facile electron transfer pathway than do the heterogeneous reactions and therefore must be specifically treated to account for the experimental behavior [26,27]. This situation often arises when the heterogeneous electron transfer processes are slow, or when very fast chemical reactions are involved. An example arises from the aforetreated reduction of *cis*-$(CO)_2Mn(\eta^2\text{-dppe})_2^+$ [25].

Labeling the two isomers of the Mn system by the hapticity (number of ligating atoms) of the dppe ligand, the relevant homogeneous reaction to consider is that of Equation 23.37:

$$[\eta^2,\eta^2]^+ + [\eta^2\eta^1]^- \underset{k_b}{\overset{k_f}{\rightleftharpoons}} \eta^2\eta^2 + \eta^2\eta^1 \qquad (23.37)$$

The equilibrium constant, $k_f/k_b$, for Equation 23.37 is determined by the $E^{\circ\prime}$ values of Equations 23.34 and 23.36 and will be >> 1 only if $E_2^{\circ\prime} < E_1^{\circ\prime}$.

It should be seen that Equation 23.37 allows a pathway for the reduction of the reactant, $(CO)_2Mn(\eta^2\text{-dppe})_2^+$, separate from that of the heterogeneous pathway of Equation 23.34. Quantification of the extent to which the homogeneous cross-reaction contributes to the redox process almost always requires digital simulations that attempt to fit the shape of the whole CV curve. Such studies allowed Kuchynka and Kochi (25), for example, to obtain values for the rate constants of Equation 23.37 ($k_f = 200\ M^{-1}\ s^{-1}$, $k_b = 20\ M^{-1}\ s^{-1}$), and to account for minute details observed in continuous-cycle CV curves for this system.

*Concluding Comments*

Cyclic voltammetry has gained widespread usage as a probe of molecular redox properties. I have indicated how this technique is typically employed to study the mechanisms and rates of some electrode processes. It must be emphasized that adherence of the CV responses to the criteria diagnostic of a certain mechanism demonstrates consistency between theory and experiment, rather than proof of the mechanism, since the fit to one mechanism may not be unique. It is incumbent upon the experimenter to bring other possible experimental probes to bear on the question. These will often include coulometry, product identification, and spectroelectrochemistry.

## ACKNOWLEDGMENTS

I wish to thank Drs. Dale Hawley and Marcin Majda for helpful comments on this manuscript, and Dr. Kent Mann for providing the raw data that allowed calculation of the chemical rate constant for Equation 23.22. I am grateful to the National Science Foundation for research support during this period.

## REFERENCES


1. A. J. Bard and L. R. Faulkner, *Electrochemical Methods*, John Wiley and Sons, New York, 1980, p. 220.
2. Charging current subtraction assumes that the electrode double-layer capacitance does not change greatly in the presence of the test compound. See D.O. Wipf, E. W. Kristensen, M. R. Deakin, and R. M. Wightman, *Anal. Chem. 60*:306 (1988).
3. M. A. Dayton, J. C. Brown, K. J. Strutts, and R. M. Wightman, *Anal. Chem. 52*:946 (1980).
4. M. C. Longmire, M. Watanabe, H. Zhang, T. T. Wooster, and R. W. Murray, *Anal. Chem. 62*:747 (1990).
5. If a value of $D = 2.5 \times 10^{-5}$ $cm^2/s$ is used for $Cp_2Rh^+$ [N. El Murr, W. E. Geiger, J. E. Sheats, and J. D. L. Holloway, *Inorg. Chem. 18*:1443 (1979)], then $4Dt/r^2 = 0.005$ for a 1-V scan. The resulting mixture of linear and radial diffusion may then be obtained from the working curve in Ref. 4.
6. R. N. Adams, *Electrochemistry at Solid Electrodes*, Marcel Dekker, New York, 1969, pp. 124 ff.
7. D. T. Pierce and W. E. Geiger, *J. Am. Chem. Soc. 114*:6063 (1992) and Ref. 2.
8. M. Rudolph, D. P. Reddy, and S. W. Feldberg, *Anal. Chem. 66*:589A (1994).
9. Ref. 1, pp. 228–230.
10. (a) W. E. Geiger in *Progress in Inorganic Chemistry* (S. J. Lippard, ed.), Vol. 33, John Wiley and Sons, New York, 1985, p. 275; (b) D. H. Evans and K. M. O'Connell in *Electroanalytical Chemistry* (A. J. Bard, ed.), Vol. 14, Marcel Dekker, New York, 1986, p. 113.
11. Ref. 1, pp. 222–231, and references therein.
12. R. S. Nicholson, *Anal. Chem. 37*:1351 (1965).
13. D. F. Milner and M. J. Weaver, *Anal. Chim. Acta 198*:245 (1987).
14. (a) D. Polcyn and I. Shain, *Anal. Chem. 38*:370 (1966); (b) D.E. Richardson and H. Taube, *Inorg. Chem. 20*:1278 (1981).
15. W. J. Bowyer and W. E. Geiger, *J. Am. Chem. Soc. 107*:5657 (1985).
16. (a) D.D. Macdonald, *Transient Techniques in Electrochemistry*, Plenum Press, New York, 1977, pp. 195–201; (b) W. E. Geiger in *Inorganic Reactions and Methods* (J. J. Zuckerman, ed.), VCH, Deerfield Beach, FL, 1986, pp. 128–133.
17. S. J. McLain, *J. Am. Chem. Soc. 110*:643 (1988).
18. T. C. Richards and W. E. Geiger, *Organometallics 16*:4494 (1994).
19. R. S. Nicholson, *Anal. Chem. 38*:1406 (1966).
20. J. R. Kuempel and W. B. Schaap, *Inorg. Chem. 7*:2435 (1968).
21. J. R. Bullock, M. C. Palazzotto, and K.R. Mann, *Inorg. Chem. 30*:1284 (1991).
22. K.-D. Plitzko, B. Rapko, B. Gollas, G. Wherle, T. Weakley, D. T. Pierce, W. E. Geiger, R. C. Haddon, and V. Boekelheide, *J. Am. Chem. Soc. 112*:6545 (1990).
23. W. L. Bowden, P. Bonnar, D. B. Brown, and W. E. Geiger, *Inorg. Chem. 16*:41 (1977).
24. R. S. Nicholson and I. Shain, *Anal. Chem. 37*:179 (1965).
25. D. J. Kuchynka and J. K. Kochi, *Inorg. Chem. 28*:855 (1989).
26. D. H. Evans, *Chem. Rev. 90*:739 (1990).
27. T. C. Richards and W. E. Geiger, *J. Am. Chem. Soc. 116*:2028 (1994).

# 24

# Electrochemical Preconcentration

**Joseph Wang** *New Mexico State University, Las Cruces, New Mexico*

## I. WHY PRECONCENTRATION?

The quantitation of trace and ultratrace components in complex samples of environmental, clinical, or industrial origin represents an important task of modern analytical chemistry. In the analysis of such dilute samples, it is often necessary to employ some type of preconcentration step prior to the actual quantitation. This happens when the analyte concentration is below the detection limit of the instrumental technique applied. Besides its main enrichment objective, the preconcentration step may serve to isolate the analyte from the complex matrix, and hence to improve selectivity and stability.

Electrolytic deposition represents an efficient way for the enrichment and isolation of trace components [1]. It has the advantage that little or no reagent addition is required, thus minimizing contamination risks. The electrochemical preconcentration scheme can be coupled with a range of instrumental techniques. Such combinations are described in the following sections.

## II. STRIPPING ANALYSIS

Stripping analysis is the best-known analytical method that incorporates an electrolytic preconcentration step [2–5]. The technique couples the advantages of extremely low detection limits ($\sim 10^{-10}$–$10^{-11}$ *M*), multielement and speciation capabilities, suitability for on-line and in situ measurements, and low cost.

Stripping techniques involve two discrete steps: a preconcentration step and a stripping step. In the preconcentration step, the target analyte is accumulated onto or into the working electrode. In the stripping step, the accumulated material is oxidized or reduced back into the solution. The response, recorded during this step, is proportional to the concentration of that analyte in or on the electrode, and thus, in the sample solution. There are different versions of stripping analysis, each dependent upon the nature of its preconcentration and stripping steps.

### A. Anodic Stripping Voltammetry

Anodic stripping voltammetry (ASV) is the most common version of stripping analysis. It involves the reduction of a metal ion to the metal (which usually dissolves in mercury, i.e., amalgam formation), as the preconcentration step:

$$M^{n+} + Hg + ne \rightarrow M(Hg) \tag{24.1}$$

Preconcentration is followed by a positive-going potential scan to cause reoxidation of the species in the amalgam:

$$M(Hg) \rightarrow M^{n+} + Hg + ne \tag{24.2}$$

The amalgamated metals are thus stripped out of the electrode in an order that is a function of each metal standard potential, and give rise to anodic peak currents that are measured. Figure 24.1 displays the potential-time sequence used in ASV, along with the resulting voltammogram.

For the deposition step, the working electrode is maintained at a potential cathodic (by at least 0.4 V) of the standard potential of the least easily reduced ion to be determined. Forced convection (via rotation, stirring, or flow) is usually used to facilitate the deposition step. Quiescent solutions may be employed in connection with ultramicroelectrodes. The deposition time required is dependent on the sample concentration, with 1- to 10-min periods usually being sufficient for measurements in the range of $10^{-7}$ *M* to $1 \times 10^{-9}$ *M*. Because only a small fraction of the metal ions is deposited, it is essential that all experimental parameters be as reproducible as possible during a series of measurements.

The concentration of the metal in the mercury electrode after a given preconcentration period is given by Faraday's law:

$$C_{Hg} = \frac{i_l t_d}{nFV_{Hg}} \tag{24.3}$$

where $i_l$ is the limiting current for the deposition of the metal, $t_d$ the deposition period, and $V_{Hg}$ the volume of the mercury electrode. In practice, the preconcentration period results in a nonuniform (parabolic) distribution of the metal concentration within the mercury electrode (e.g., Fig. 24.2).

At the end of the preconcentration period, the forced convection is stopped and a 15-s rest period is employed to establish a uniform concentration (within the mercury) and to ensure that the subsequent stripping step is performed under quiescent conditions.

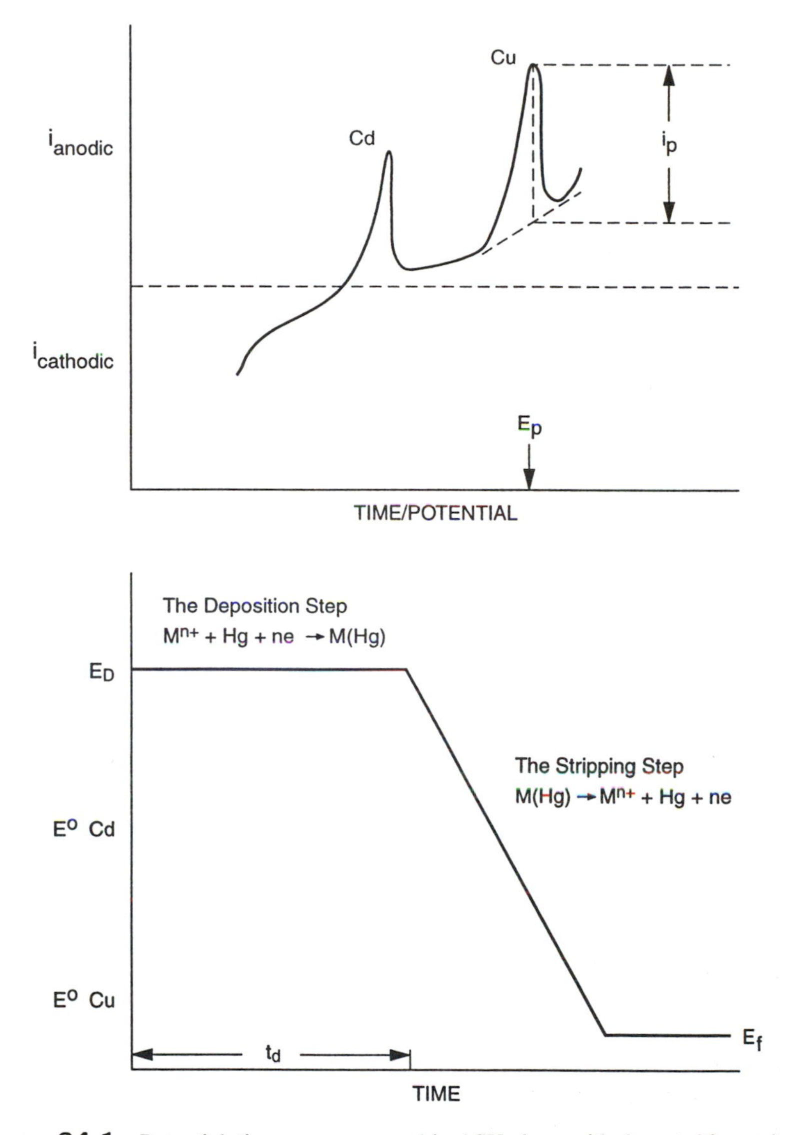


**Figure 24.1** Potential–time sequence used in ASV along with the resulting stripping voltammogram.

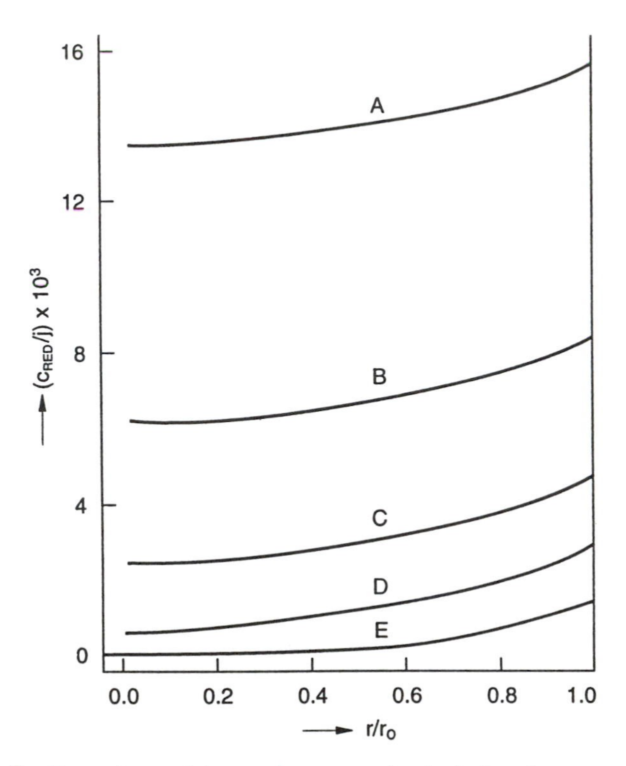


**Figure 24.2** Dependence of the metal concentration in the hanging mercury drop electrode on the distance from the center of the drop for different preconcentration times: (A) 240, (B) 120, (C) 60, (D) 30, and (E) 10 s. [From I. Shain and J. Lewinson, *Anal. Chem. 33*:187 (1961), reproduced with permission.]

Various voltammetric waveforms can be employed during the stripping step, including linear scan, differential pulse, square-wave, staircase, or alternating-current operations. The differential pulse and square-wave modes are usually performed at the hanging mercury drop electrode, while linear scan stripping is usually performed in connection with the mercury film electrode.

For reversible stripping reactions, the applied potential controls the concentration at the mercury–solution interface (according to the Nernst equation). Because of the rapid depletion of all the metal from thin mercury films, the stripping behavior at these electrodes follows a thin-layer behavior. The peak current for the linear scan operation at thin mercury film electrodes is thus given by

$$i_p = \frac{n^2F^2vA\ell C_{Hg}}{2.7\ RT} \tag{24.4}$$

where v is the scan rate, and A and $\ell$ are the film area and thickness, respectively. For thicker films, semi-infinite linear diffusion predominates and the peak current is proportional to the square root of the scan rate.

At the hanging mercury drop electrode, not all the metal is stripped during the potential scan. The peak current for differential pulse stripping voltammetry at the hanging mercury drop electrode is proportional to the pulse amplitude $\Delta E$ [6]:

$$i_p = kn^2r(\Delta E)U^{1/2}t_dC^o \tag{24.5}$$

where r is the drop radius, U the stirring rate (during deposition), and $C^o$ the bulk concentration of the metal. However, relatively small amplitudes (25–50 mV) are chosen to maintain the peak sharpness. The square-wave stripping mode offers a similar compensation against the charging current contribution, but also a significant speed advantage and a better immunity against dissolved oxygen. Such ability to conduct trace metal determinations in the presence of dissolved oxygen (e.g., Fig. 24.3) is attributed to the fast scanning ability. Simply, oxygen that is depleted from the surface at the end of the deposition step cannot be replenished during the fast square-wave scanning. Dual working-electrode techniques, such as subtractive ASV or anodic stripping with collection, offer effective compensation of the background current, but require more complex instrumentation.

It is often desirable to perform the stripping step in a medium other than the sample matrix. In particular, the medium exchange step can minimize certain interferences caused by various sample components. For example, the resolution between neighboring stripping peaks can be improved by performing the stripping step after an exchange to an appropriate complexing medium.

## B. Potentiometric Stripping Analysis

Potentiometric stripping analysis (PSA) is another attractive version of stripping analysis [7]. The preconcentration step in PSA is the same as for ASV; that is, the metal is electrolytically deposited (via reduction) onto the mercury electrode (usually a film). The stripping, however, is done by chemical oxidation, for example, with oxygen or mercuric ions present in the solution:

$$M(Hg) + \text{oxidant} \rightarrow M^{n+} + Hg \tag{24.6}$$

Alternately, it is possible to strip the metal off by applying a constant anodic current through the electrode. The potential of the electrode, when monitored as a function of time, gives an experimental curve, analogous to a normal redox titration curve (Fig. 24.4), which contains the quantitative and qualitative information. A sudden change in the potential occurs when all the metal deposited in the electrode has been depleted from the surface. The time required to

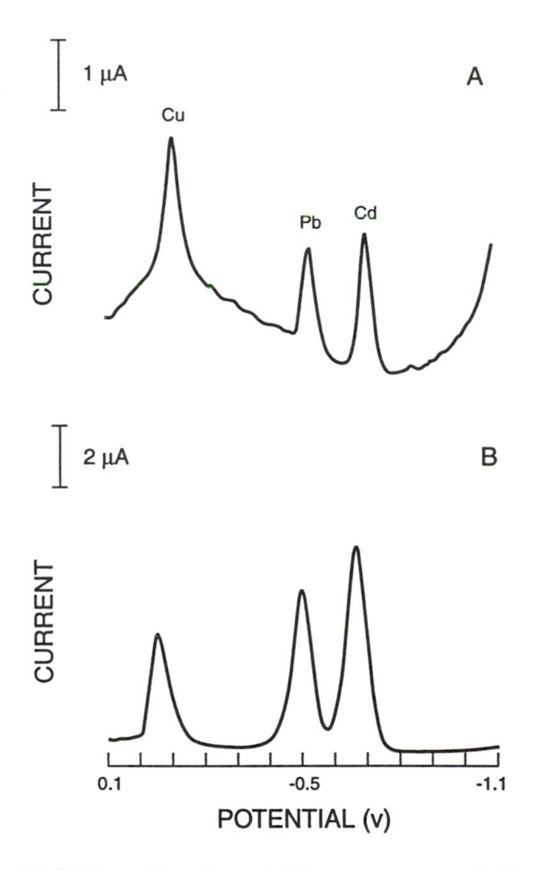


**Figure 24.3** (A) Differential pulse and (B) square-wave stripping voltammograms for a nondeaerated solution containing 50 μg/L (ppb) Cd, Pb, and Cu ions. Two-minute deposition at a mercury film electrode held at −1.1 V. Acetate buffer solutions (0.1 *M*, pH 4.5).

reach the "equivalence point" is proportional to the bulk concentration of the metal ion:

$$t \propto C_{M^{n+}} t_d / C_{ox} \qquad (24.7)$$

where $C_{ox}$ is the concentration of the oxidant. For constant-current PSA, the stripping time is inversely proportional to the stripping current. The qualitative identification relies on measuring the reoxidation potential, as determined by the Nernst equation for the amalgamated metals. A more convenient peak-shaped response (of dt/dE vs. E) is provided by modern PSA instrumentation (through derivatization of the sigmoidal signal). The use of PSA circumvents serious

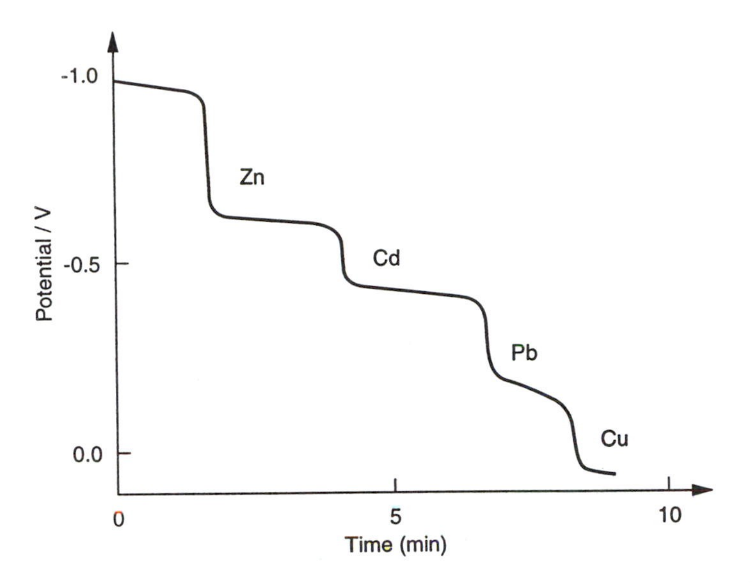


**Figure 24.4** Potentiometric stripping analysis of a solution containing $1.5 \times 10^{-6}$ M $Zn^{2+}$, $Cd^{2+}$, $Pb^{2+}$, and $Cu^{2+}$, using 3-min deposition and mercury as an oxidant.

interferences characteristic of ASV (e.g., from oxygen or organic surfactants), and hence can simplify the sample pretreatment step. For example, lead and cadmium have been determined in whole blood after simple dilution with hydrochloric acid [8].

## C. Cathodic Stripping Voltammetry

Cathodic stripping voltammetry (CSV) involves anodic (oxidative) deposition of an insoluble film of material on the electrode; it is subsequently stripped off during a negative-going potential sweep [9]. Most CSV applications rely on the anodic accumulation of sparingly soluble mercury compounds on a mercury surface:

$$A^{n-} + Hg \underset{\text{Stripping}}{\overset{\text{Deposition}}{\rightleftarrows}} HgA + ne \tag{24.8}$$

Hence, CSV can be viewed as the "mirror image" of ASV. CSV is best suited for the determination of inorganic anions (e.g., halides, cyanide) or organic sulfur compounds (e.g., penicillins, thiols) that form insoluble salts with the electrode material. In addition to mercury, silver electrodes may be advantageous for the determination of anions that form insoluble silver salts.

## D. Adsorptive Stripping Voltammetry

Adsorptive stripping voltammetry (AdSV) has been demonstrated to be a very sensitive method for numerous trace metals, the quantitation of which by conventional (electrolytic) stripping procedures is complicated or not feasible [10–12]. The method uses the formation of an appropriate metal chelate, followed by its controlled interfacial accumulation onto the working electrode; then the adsorbed metal chelate is reduced by application of a negative-going potential scan (Fig. 24.5). The reduction can proceed through the metal or ligand of the complex. The resulting adsorptive stripping response reflects the corresponding adsorption isotherm, as the surface concentration of the analyte ($\Gamma$) is proportional to its bulk concentration. The most frequently used isotherm is that of Langmuir:

$$\Gamma = \Gamma_m\left(\frac{BC}{1 + BC}\right) \tag{24.9}$$

where $\Gamma_m$ is the surface concentration corresponding to a monolayer coverage, and B the adsorption coefficient. Hence, calibration plots often exhibit deviations from linearity at "high" concentrations ($>10^{-7}$ *M*) corresponding to a full

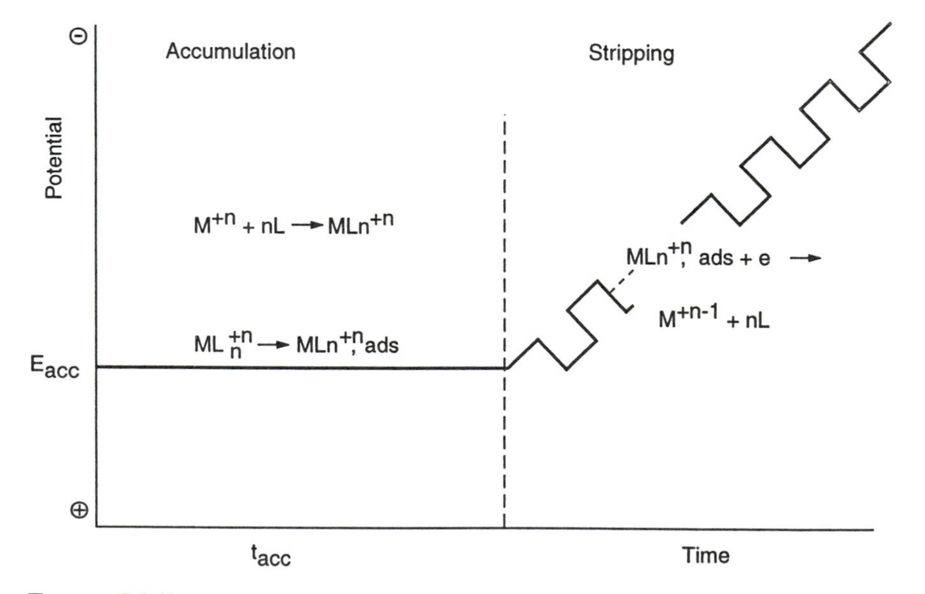


**Figure 24.5** Steps in adsorptive stripping voltammetric measurements of a metal ion, based on the formation, adsorptive accumulation, and reduction of its surface-active complex.

surface coverage. The surface-confined complex may also be quantitated by a chronopotentiometric approach, utilizing a constant cathodic current.

The adsorptive accumulation scheme results in a very effective preconcentration, allowing ultrasensitive measurements (down to the $10^{-10}$ *M* level) following short adsorption times. To attain such high sensitivity, it is essential to optimize operational variables (e.g., pH or accumulation potential) favoring strong adsorption. This scheme greatly extends the scope of stripping analysis toward important metals, such as uranium, chromium, vanadium, iron, molybdenum, palladium, aluminum, and yttrium (Table 24.1). For a detailed list of adsorptive stripping procedures, the reader is referred to a recent review article [23]. In addition, many surface-active organic compounds of biological, pharmaceutical, and environmental significance can be interfacially accumulated and measured at ultratrace levels (Table 24.2). Figure 24.6 displays voltammograms for nanomolar concentrations of the anticancer drug daunorubicin following different preconcentration periods (0–300 s). In the case of large biological macromolecules, the adsorptive accumulation results in conformational changes that facilitate the electron transfer between the redox center and the electrode. The adsorptive accumulation can also be exploited for trace measurements of nonelectroactive surfactants (e.g., detergents), via the subsequent tensammetric (nonfaradaic) response.

Ultrasensitive adsorptive stripping procedures, based on the unique coupling of catalytic and adsorption processes, have been developed recently [33–35]. Such dual amplification strategy results in remarkably low detection limits of $10^{-12}$ *M*. Note, for example, the large (μA) hydrogen catalytic currents observed

**Table 24.1** Representative Adsorptive Stripping Procedures for Measurements of Trace Metals

| Metal | Complexing agent | Detection limit, n*M* | Reference |
|---|---|---|---|
| Aluminum | 1,2-Dihydroxyanthraquinone-3-sulfuric acid | 1.0 | 13 |
| Cobalt | Nioxime | 0.006 | 14 |
| Chromium | Diethylenetriamine pentaacetic acid | 0.4 | 15 |
| Iron | Solochrome violet RS | 0.7 | 16 |
| Molybdenum | Oxine | 0.1 | 17 |
| Nickel | Dimethylgloyoxime | 0.1 | 18 |
| Tin | Tropolone | 0.2 | 19 |
| Uranium | Cupferron | 0.4 | 20 |
| Vanadium | Catechol | 0.1 | 21 |
| Yttrium | Solochrome violet RS | 1.4 | 22 |

**Table 24.2** Representative Adsorptive Stripping Procedures for Measurements of Organic Compounds

| Analyte | Electrolyte | Detection limit | Reference |
|---|---|---|---|
| Adriamycin | Acetate buffer | $1 \times 10^{-8}$ *M* | 24 |
| Diazepam | Acetate buffer | $5 \times 10^{-9}$ *M* | 25 |
| Dionoseb | Britton–Robinson buffer | $4 \times 10^{-10}$ *M* | 26 |
| Folic acid | Sulfonic acid | $1 \times 10^{-10}$ *M* | 27 |
| Nitro-containing pesticides | Britton–Robinson buffer | $5 \times 10^{-10}$ *M* | 28 |
| Progesterone | Sodium hydroxide | $2 \times 10^{-10}$ *M* | 29 |
| Riboflavin | Sodium hydroxide | $2.5 \times 10^{-11}$ *M* | 30 |
| Streptomycin | Sodium hydroxide | $7 \times 10^{-10}$ *M* | 31 |
| Trichlorobiphenyl | Britton–Robinson buffer | $1 \times 10^{-8}$ *M* | 32 |

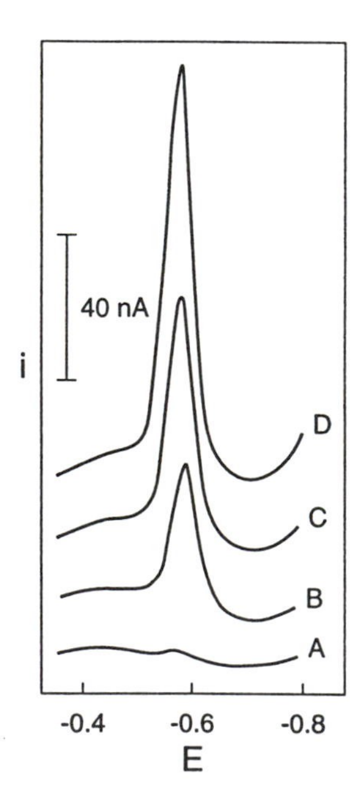


**Figure 24.6** Adsorptive stripping voltammograms for $4 \times 10^{-8}$ *M* daunorubicin following different preconcentration times: (A) 0, (B) 90, (C) 180, and (D) 300 s. [From J. Wang, M. S. Lin, and V. Villa, *Analyst 112*:1303 (1987), reproduced with permission.]

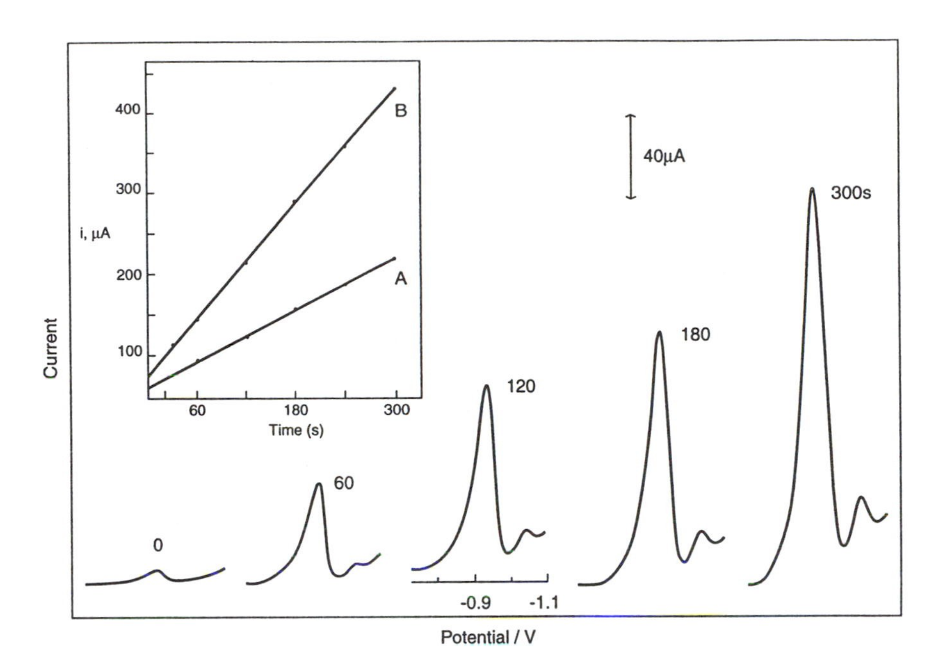


**Figure 24.7** Catalytic-adsorptive stripping voltammograms for 0.2 μg/L (ppb) platinum after different preconcentration times: 0, 60, 120, 180, and 300 s. Inset: Current vs. preconcentration time plots for (A) 0.2 and (B) 0.6 μg/L platinum. [From Ref. 33, reproduced with permission.]

in Figure 24.7 for subnanomolar concentrations of platinum in the presence of formazone.

## E. Preconcentration/Voltammetry at Modified Electrodes

Specific reactions at chemically modified electrodes (CMEs) can also be exploited for the preferential accumulation of target analytes onto electrodes prior to the voltammetric quantitation of the surface-bound species [10,36–38]. The rich chemistry that can be used for preconcentration at these molecularly tailored electrodes offers exciting prospects for trace analysis. The reactivity obtained by proper choice of the surface modifier can greatly enhance the selectivity and sensitivity of voltammetric devices. The preconcentrating agent is commonly incorporated into the surface by mixing it within carbon paste matrices or by forming derivatives of polymeric films. The major requirements for a successful use of modified electrodes as preconcentrating surfaces include selective accumulation, prevention of surface saturation, and a convenient regeneration of a "fresh" (analyte-free) surface.

Most popular schemes used to collect analytes are based on coordination reactions and electrostatic attraction. Common examples include the accumulation of nickel onto dimethylglyoxime-containing surfaces [39], the uptake and voltammetry of mercury on a diphenylcarbazide-carbon paste electrode [40], the use of surface-bound crown ethers for the collection and measurements of lead [41], or of trioctylphosphine oxide for the preconcentration of uranium [42], and the utility of polyelectrolyte-coated electrodes for the electrostatic collection of counterionic reactants [43,44]. Bioaccumulation through binding to surface-bound microorganisms [45] or biocatalytic processes [46] can also offer the desired sensitivity and selectivity enhancements.

## F. Electrodes for Stripping Analysis

A proper choice of the working electrode is crucial for the success of the stripping operation. In particular, mercury electrodes (Chap. 14) play a central role in the success of stripping analysis. The two mercury electrodes most commonly used are the hanging mercury drop electrode (HMDE) and the mercury film electrode (MFE). Particularly attractive for stripping work are automatic HMDEs or rotating mercury-coated glassy carbon surfaces. The MFE can be formed on glassy carbon either by preliminary deposition or by in situ plating (after adding $\sim 10^{-5}$ $M$ mercuric ion to the analyte solution [47]). The resulting film is typically nonhomogeneous and consists of a collection of fine mercury droplets. Despite its nonhomogeneity, the MFE offers higher sensitivity (larger surface area-to-volume ratio) and improved resolution of neighboring peaks (Fig. 24.8). It is also more robust and suitable for field operation. In contrast, the HMDE alleviates intermetallic and saturation effects, possesses a higher hydrogen overvoltage (as needed for measurements of zinc), and offers reproducible renewal of the surface. The HMDE is preferred for adsorptive stripping and cathodic stripping work, while the MFE is commonly used in potentiometric stripping analysis. Coverage of the MFE with permselective coatings, such as the Nafion [48] or Kodak AQ [49] ionomers, has been extremely useful for protecting this electrode against surfactants common in clinical or environmental matrices. Disposable, screen-printed electrodes for "one-shot" applications have also been described [50].

Bare electrodes (particularly of gold and carbon) have been used for the determination of trace metals with oxidation potentials positive of that of mercury (e.g., Ag, Se, As, Au, and Te), and for adsorptive stripping measurements of oxidizable organic compounds. Gold film electrodes (on glassy carbon substrates) have been particularly useful for trace measurements of mercury.

The introduction of ultramicroelectrodes (described in Chap. 12) holds great promise for stripping measurements [51–53]. In particular, microelectrodes offer enhanced deposition efficiency (and hence efficient preconcentration from qui-

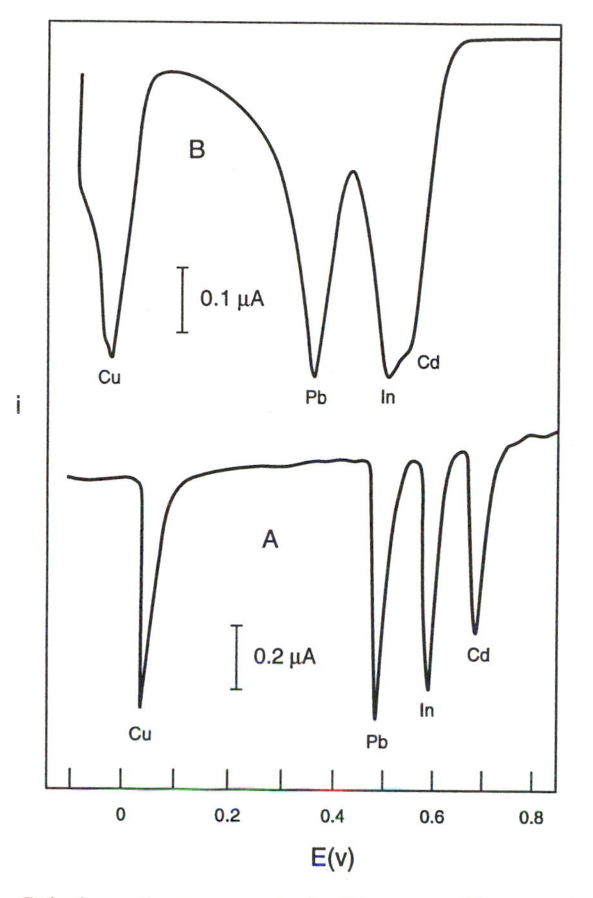


**Figure 24.8** Stripping voltammograms at the (A) mercury film and (B) hanging mercury drop electrodes for a solution containing $2 \times 10^{-7}$ *M* $Cd^{2+}$, $In^{3+}$, $Pb^{2+}$, and $Cu^{2+}$. Deposition for (A) 5 and (B) 30 min. [From Ref. 47, reproduced with permission.]

escent solution), and work in low-ionic-strength solutions or assays of microliter samples. The immunity against ohmic drops is particularly attractive for ultratrace work. By eliminating the need for adding a supporting electrolyte, possible contamination by metal impurities is minimized, and changes that may affect metal speciation studies are avoided [54]. Mercury-coated microelectrodes have been prepared using a variety of geometries and substrates. Particularly useful have been mercury-plated carbon fiber microcylinder [53] and iridium microdisk [55] electrodes. Arraylike composition electrodes, coated with mercury, couple the enhanced preconcentration efficiency with large (collective) stripping currents [56].

## G. Stripping Flow Systems

Numerous studies have illustrated the advantages of combining stripping analysis with flow systems [57,58]. In particular, the coupling of stripping schemes with flow injection systems provides the attractive features of enhanced sample throughput, small sample volumes, improved reproducibility and versatility, and medium-exchange capability [59,60]. The theoretical aspects of flow injection stripping analysis have been treated [61]. Automated flow systems based on adsorptive stripping voltammetry [62] and potentiometric stripping analysis [63] have been developed for continuous monitoring and discrete analysis, respectively. The latter offers a convenient way to perform the medium-exchange procedure, as illustrated in Figure 24.9. Computerized stripping flow systems have been applied (aboard a small vessel) for in situ marine surveys [64]. Thin-layer and wall-jet flow cells are usually employed in connection with stripping flow measurements. Specially designed flow cells have been useful for reducing intermetallic and oxygen interferences [65]. On-line matrix isolation or enrichment can be accomplished via the use of an appropriate flow-through column [66,67].

## H. Interferences

Stripping analysis can be complicated by the presence of organic surfactants that adsorb on the mercury electrode and block the deposition. Depressions in the stripping peaks of interest may also occur from the formation of intermetallic

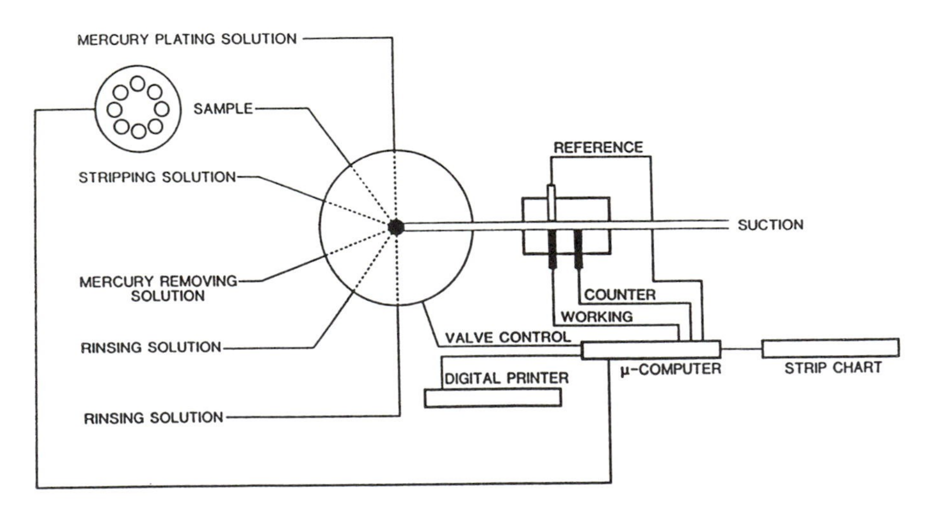


**Figure 24.9** Schematic of a computerized flow potentiometric stripping analyzer. [From Ref. 63, with permission.]

compounds between metals (e.g., Cu–Zn, Cu–Cd) deposited in the mercury electrode. Another problem is the resolution of overlapping peaks (e.g., Tl/Cd, Sn/Pb, Bi/Cu) associated with the similarity in their potentials. Many strategies are available to successfully circumvent these problems [2–4], including the UV destruction of surfactants, the preferential formation of another intermetallic compound, the selective masking of an interfering metal, or a judicious choice of the deposition potential.

In addition, the reliability of the stripping data strongly depends on the degree to which contamination can be minimized. Hence, as in other trace analytical methods, all principles of good laboratory practice (glassware cleanliness, sample collection and storage, etc.) must be observed to obtain high accuracy and low detection limits.

## I. Metal Speciation Studies

An important advantage of stripping techniques is that the speciation (physicochemical forms) of the metals can also be determined [68]. By adjustment of the solution and deposition conditions, it is possible to divide metal species (in a water sample) into groups. The fraction determined in conventional ASV work involves the free metal ion and the labile complexes (which dissociate at the applied deposition potential). The lability of metal species has been related to toxicity. Stripping-speciation experiments, including titrimetric stripping methods and the construction of pseudopolarograms, have been designed to study trace metal complexation parameters (e.g., coordination number, dissociation constant) and for measuring the complexation capacity of the sample.

The adsorptive stripping approach provides different speciation information (compared to ASV) based on ligand competition experiments. In this case, the fraction of metal measured involves that forming a complex (with the foreign ligand) in the time between the ligand addition and the start of the adsorptive accumulation. This is a function of the concentration of the added ligand, the stability constant of complexes formed, the concentration of natural ligands, and the stability constants of the natural complexes. Redox speciation can also be obtained, since any oxidation state can be accumulated (rather than the metallic state in the case of ASV).

## J. Applications

While early ASV schemes were limited to about a dozen amalgam-forming metals, the recent development of adsorptive stripping procedures has greatly expanded the scope of stripping analysis to more than 40 trace metals.

Numerous applications of stripping analysis to many relevant environmental, clinical, and industrial problems have been reported. Some typical examples include the determination of uranium or titanium in natural waters [34,62], flow injection/stripping determinations of thallium in soils [69], assays of blood and urine for their lead or cadmium [8,70] levels, and determination of zinc and lead in human hair [71], platinum in blood [72], iron in wine [73], tin in fruit juices [74], mercury in fish [75], cobalt in zinc electrolyte [76], and antimony in gunshot residues [77]. Particularly noticeable is the major role of stripping analysis in oceanographic studies (including in situ marine investigations). Such a role is attributed to the ability of the technique to handle saline solutions and to its excellent (p*M*–n*M*) detection limits. Advances in in situ stripping assays of natural waters have been reviewed [78].

The compact nature of its instrumentation makes stripping analysis particularly attractive for field monitoring (since other trace metal techniques are not noted for their mobility and low power needs). Hence, increased use of on-line stripping analysis for in situ automated environmental surveillance or real-time monitoring of industrial processes is anticipated. In addition, the development of disposable stripping electrodes [48] and compact hand-held analyzers [79] should lead to unique applications, including self-testing of drinking water or mass screening of children's blood samples for lead levels. Stripping analysis is now a remarkably sensitive technique, applicable to the measurements of numerous analytes. Its utility is expanding rapidly and will continue to do so in the future.

## III. ELECTROCHEMICAL PRECONCENTRATION FOR SPECTROSCOPIC ANALYSIS

Spectroscopic analysis can also benefit from a preceding electrochemical preconcentration. In particular, such coupling has been widely used for minimizing matrix interferences in atomic absorption spectroscopy (AAS). For example, lead, nickel, and cobalt have been determined in seawater with no interferences from the high sodium chloride content [80]. By adjusting the deposition potential and the pH, it is possible to obtain information on the oxidation and complexation states of the metal ions present [81].

The electrodeposition/AAS approach is accomplished by plating the target metals on metal wires of high melting point (e.g., W or Ir), or on various graphite surfaces. The electrode is then transferred to the AAS instrument, where the metal is determined by either flame or electrothermal atomization [80,81]. A flow system with a replaceable graphite tube electrode (Fig. 24.10) offers effective and reproducible electrodeposition and permits in situ preconcentration (which minimizes errors arising from contamination or adsorption). Improved selectivity and sensitivity have also been reported for the coupling of controlled-potential electrolysis with inductively coupled plasma atomic emission spectroscopy, using reticulated vitreous carbon [82] and hanging mercury drop [83] electrodes. Besides atomic spectroscopy, electrolytic preconcentrations have been

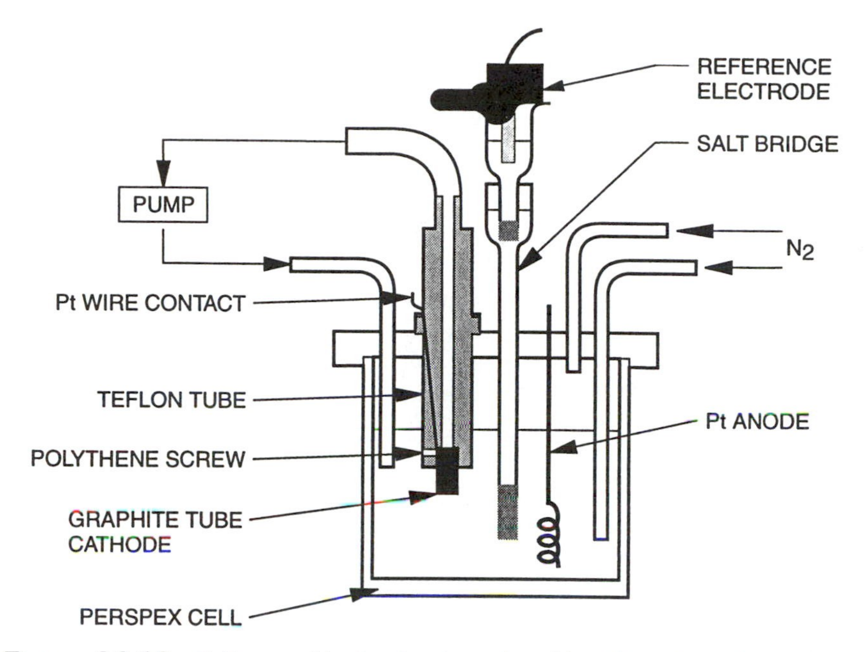


**Figure 24.10** Cell asssembly for the electrodeposition–electrothermal atomization technique. [From Ref. 81, with permission.]

successfully coupled with neutron activation analysis [84], x-ray fluorescence [85], and Mössbauer spectroscopy [86].

## REFERENCES

1. R. Sioda, G. Batley, W. Lund, J. Wang, and S. Leach, *Talanta 33*:421 (1986).
2. J. Wang, *Stripping Analysis*, VCH, Deerfield Beach, FL, 1985.
3. F. Vydra, K. Stulik, and E. Julakova, *Electrochemical Stripping Analysis*, J. Wiley and Sons, New York, 1976.
4. T. R. Copeland and R. K. Skogerboe, *Anal. Chem. 46*:1257A (1974).
5. T. M. Florence, *J. Electroanal. Chem. 168*:207 (1984).
6. W. Lund and D. Onshus, *Anal. Chim. Acta 86*:109 (1976).
7. D. Jagner, *Analyst 107*:593 (1982).
8. J. Christensen, L. Kryger, and N. Pind, *Anal. Chim. Acta 136*:39 (1982).
9. T. M. Florence, *J. Electroanal. Chem. 97*:219 (1979).
10. J. Wang, *Electroanalytical Chemistry*, Vol. 16 (A. J. Bard, ed.), Marcel Dekker, New York, 1989.
11. C. M. G. van den Berg, *Anal. Chim. Acta 250*:265 (1991).
12. R. Kalvoda and M. Kopanica, *Pure Appl. Chem. 61*:97 (1989).

13. C. M. G. van den Berg, K. Murphy, and J. Riley, *Anal. Chim. Acta 188*:177 (1986).
14. J. R. Donat and K. Bruland, *Anal. Chem. 60*:240 (1988).
15. J. Golimowski, P. Valenta, and H.W. Nürnberg, *Fresenius Z. Anal. Chem. 322*:315 (1985).
16. J. Wang and J. S. Mahmoud, *Fresenius Z. Anal. Chem. 327*:789 (1987).
17. C. M. G. van den Berg, *Anal. Chem. 57*:1532 (1985).
18. B. Pihlar, P. Valenta, and H. W. Nürnberg, *Fresenius Z. Anal. Chem. 307*:337 (1981).
19. J. Wang and J. Zadeii, *Talanta 34*:909 (1987).
20. J. Wang and R. Setiadji, *Anal. Chim. Acta 264*:205 (1992).
21. C. M. G. van den Berg, *Anal. Chem. 56*:2383 (1984).
22. J. Wang and J. Zadeii, *Talanta 33*:321 (1986).
23. M. Paneli and A. Voulgaropolous, *Electroanalysis 5*:355 (1993).
24. E. Chaney and R. Baldwin, *Anal. Chem. 54*:2556 (1982).
25. R. Kalvoda, *Anal. Chim. Acta 162*:197 (1984).
26. M. Pedrero, F. de Villena, J. Pingarron, and L. Polo, *Electroanalysis 3*:419 (1991).
27. D. B. Luo, *Anal. Chim. Acta 189*:277 (1986).
28. H. Benadikova and R. Kalvoda, *Anal. Lett. 17*:1519 (1984).
29. J. Wang, P. Farias, and J. Mahmoud, *Anal. Chim. Acta 171*:195 (1984).
30. J. Wang, D. B. Luo, P. Farias, and J. Mahmoud, *Anal. Chem. 57*:158 (1985).
31. J. Wang and J. S. Mahmoud, *Anal. Chim. Acta 186*:31 (1986).
32. N. Lam and M. Kopanica, *Anal. Chim. Acta 161*:315 (1984).
33. J. Wang, J. Zadeii, and M. S. Lin, *J. Electroanal. Chem. 237*:281 (1987).
34. K. Yokoi and C. M. G. van den Berg, *Anal. Chim. Acta 245*:167 (1991).
35. J. Wang, J. Lu, and Z. Taha, *Analyst 117*:35 (1992).
36. A. Guadalupe and H. Abruña, *Anal. Chem. 57*:142 (1985).
37. J. Wang, B. Greene, and C. Morgan, *Anal. Chim. Acta 158*:15 (1984).
38. K. Kalcher, *Electroanalysis 2*:419 (1990).
39. R. Baldwin, J. Christensen, and L. Kryger, *Anal. Chem. 58*:1790 (1986).
40. Z. Navratilova, *Electroanalysis 3*:799 (1991).
41. S. Prabhu, R. Baldwin, and L. Kryger, *Electroanalysis 1*:13 (1989).
42. K. Lubert, M. Schnurrbusch, and A. Thomas, *Anal. Chim. Acta 144*:123 (1982).
43. N. Oyama and F. C. Anson, *J. Electrochem. Soc. 127*:247 (1980).
44. M. Szentirmay and C. Martin, *Anal. Chem. 56*:1898 (1984).
45. J. Gardea-Torresdey, D. Darnall, and J. Wang, *Anal. Chem. 60*:72 (1988).
46. J. Kulys, N. Cenas, G. Svirmicka, and V. Svirmickiene, *Anal. Chim. Acta 138*:19 (1982).
47. T. M. Florence, *J. Electroanal. Chem. 27*:273 (1970).
48. B. Hoyer, T. M. Florence, and G. Bately, *Anal. Chem. 59*:1609 (1987).
49. J. Wang and Z. Taha, *Electroanalysis 2*:383 (1990).
50. J. Wang and T. Baomin, *Anal. Chem. 64*:1706 (1992).
51. K. Wehmeyer and R. M. Wightman, *Anal. Chem. 57*:1989 (1985).
52. A. Baranski, *Anal. Chem. 59*:662 (1987).
53. J. Wang, P. Tuzhi, and J. Zadeii, *Anal. Chem. 59*:2119 (1987).

54. J. Wang and J. Zadeii, *J. Electroanal. Chem. 246*:297 (1988).
55. R. De Vitre, M. Tercier, M. Tsacopoulos, and J. Buffle, *Anal. Chim. Acta 249*:419 (1991).
56. J. Wang, A. Brennsteiner, L. Angnes, A. Sylwester, R. LaGasse, and N. Bitsch, *Anal. Chem. 64*:151 (1992).
57. J. Wang, *Am. Lab.15*(7):14 (1983).
58. H. Gunasingham and B. Fleet, *Electroanalytical Chemistry*, Vol. 16 (A. J. Bard, ed.), Marcel Dekker, New York, 1989.
59. J. Wang, H. Dewald, and B. Greene, *Anal. Chim. Acta 146*:45 (1983).
60. M. Luque de Castro and A. Izquierdo, *Electroanalysis 3*:457 (1991).
61. J. Wang and H. Dewald, *Anal. Chim. Acta 162*:189 (1984).
62. J. Wang, R. Setiadji, and S. Morton, *Electroanalysis 4*:161 (1992).
63. D. Jagner, *Trends Anal. Chem. 2*:53 (1983).
64. A. Zirino, S. Lieberman, and C. Clavell, *Environ. Sci. Technol. 12*:73 (1978).
65. J. Wang and H. Dewald, *Anal. Chem. 55*:933 (1983).
66. W. Kubiak, J. Wang, and D. Darnall, *Anal. Chem. 61*:468 (1989).
67. B. Daih and H Huang, *Anal. Chim. Acta 258*:245 (1992).
68. T. M. Florence, *Analyst 111*:489 (1986).
69. Z. Lukaszewski and W. Zembrzuski, *Talanta 39*:221 (1992).
70. D. Jagner, M. Josefson, S. Westerlund, and K. Aren, *Anal. Chem. 53*:1406 (1981).
71. C. Liu and K. Jiao, *Anal. Chim. Acta 238*:367 (1990).
72. O. Nygren, G. Vaughan, T. M. Florence, G. Morrison, I. Warner, and L. Dale, *Anal. Chem. 62*:1637 (1990).
73. J. Wang and S. Mannino, *Analyst 114*:643 (1989).
74. S. Mannino, *Analyst 108*:1257 (1983).
75. J. Golimowski and I. Gustavsson, *Fresenius Z. Anal. Chem. 317*:481 (1984).
76. A. Bobrowski and A. M. Bond, *Electroanalysis 3*:157 (1991).
77. N. Konaunur and G. van Loon, *Talanta 24*:184 (1977).
78. M. Tercier and J. Buffle, *Electroanalysis 5*:187 (1993).
79. C. Yarnitzky, J. Wang, and B. Tian, *Talanta*, in press.
80. G. E. Bately and J. Matousek, *Anal. Chem. 49*:2031 (1977).
81. G. E. Bately and J. Matousek, *Anal. Chem. 52*:1570 (1980).
82. D. Ogaram and R. Snook, *Analyst 109*:1597 (1984).
83. M. Habib and E. Salin, *Anal. Chem. 57*:2055 (1985).
84. H. B. Mark, Jr., *J. Pharm. Belg. 25*:367 (1970).
85. B. Vassos, R. Hirsch, and H. Letterman, *Anal. Chem. 45*:792 (1973).
86. F. Berlandi and H. Mark, Jr., *Nucl. Appl. 6*:409 (1969).

# 25

# Controlled-Current Coulometry

**David J. Curran** *University of Massachusetts at Amherst, Amherst, Massachusetts*

## I. INTRODUCTION

The idea of in situ generation of known amounts of chemical reagents is an intriguing one that has occupied the attention of many chemists. Electrochemists, in particular, have been notably successful in this regard. The basis for this is, of course, Faraday's law of electrolysis, which states that the quantity of electricity passed (coulombs) is directly proportional to the amount of chemical reaction (equivalents) that takes place at an electrode. As we have seen in Chapter 3, a mathematical expression for this concept is readily apparent:

$$Q = \int_0^t i \, dt = (F)(\text{equivalents}) \tag{25.1}$$

where Q is the number of coulombs, i the current in amperes, t the time in seconds, and F the Faraday. This equation is fundamental to both controlled-potential and controlled-current coulometry. If the current is constant and the time for electrolysis is measured, Equation 25.1 may be written

$$\text{equivalents} = Q/F = it/F \tag{25.2}$$

From these equations, it is seen that the experimental variables in a controlled-current coulometric experiment are current and time, and it is possible to identify the following components of an appropriate apparatus: an electrolysis cell, a current source, a method of measuring elapsed time (or a method of measuring coulombs), and a switching arrangement to control experimental variables. Electrochemical experiments using controlled-current methods are widespread and include titrimetry, kinetic studies, process stream analysis, and others (see Chap. 4).

## II. COULOMETRIC GENERATION OF REAGENTS

### A. Constant Current

*Electrochemical Considerations*

Let the half-reaction at the generator electrode be the following:

$$\text{Precursor} \pm ne \rightleftharpoons \text{reagent} \qquad (25.3)$$

To use Faraday's law for the purpose at hand, the generator electrode reaction must be limited to the one of interest to assure that the charge consumed by the desired electron transfer reaction is essentially equal to the charge applied to the cell. Difficulties arise because it is not possible to control independently both the current and the electrode potential, and because only the current applied to the cell can be known directly in the experiment. A knowledge of the current–voltage curves for the precursor and the supporting electrolyte will provide valuable information for the design of practical experiments.

For controlled-current work, the cell requires an auxiliary electrode in addition to the generator electrode. Frequently, a separate compartment is used for the auxiliary electrode to prevent mass transport between the two electrodes (see Chap. 9). The addition of a reference electrode will permit current–voltage curves to be obtained under conditions identical to those prevailing during the constant-current experiment. The limiting current $i_l$ of this hydrodynamic voltammogram will be directly proportional to the concentration of precursor in the electrolyte. If a series of precursor solutions were prepared and the current–voltage curves obtained for each solution, a family of curves would result in which the limiting current would become progressively smaller as the concentration of the precursor decreased. Such a family of curves for a reversible system can be represented by a single curve by normalizing the axes of the graph as shown in Figure 25.1, where the vertical axis is expressed as the ratio of the current at any point on the current–voltage curve to the limiting current, $i/i_l$, and the horizontal axis is plotted as $E - E_{1/2}$. Only faradaic current is considered; it is assumed that there is no background current.

If a constant current i equal to 60% of the limiting current found for the most concentrated solution of precursor (i.e., $i = 0.6\, i_l$) is applied to the cell, then the ratio $i/i_l$ is 0.6 and is represented by line A of Figure 25.1. The potential of the electrode at the time the constant current is applied would be the potential at the intersection of line A with the curve. Now, consider a second precursor solution which is only 90% as concentrated as the first one. The new limiting current, $i_l'$, would be nine-tenths of the original limiting current, or $i_l' = 0.9 i_l$. However, in a constant-current experiment, the applied current would not change, so the new ratio, $i/i_l'$, is larger than the former ratio and can be calculated as

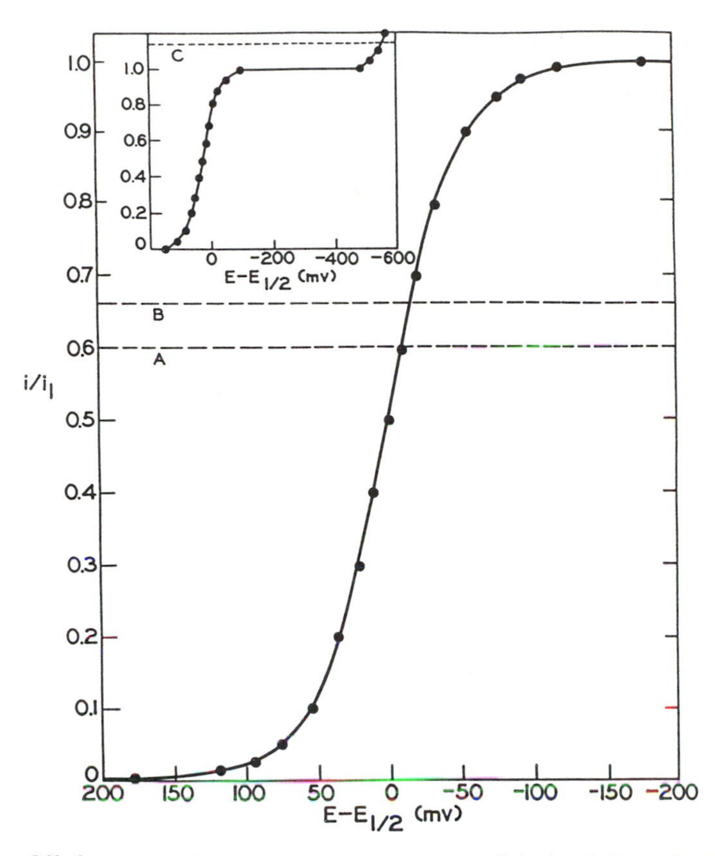


**Figure 25.1** Normalized current–voltage curve. Calculated from the equation $E = E_{1/2} + (RT/nF) \ln (i_l - i)/(i)$, where $i_l$ is the limiting current.

$$\frac{i}{i_l'} = \frac{0.6i_l}{0.9i_l} = 0.667 \qquad (25.4)$$

This ratio is shown by line B in the figure, and the electrode potential has a slightly more negative value at the new intersection. As the solution of precursor becomes more dilute, the potential of the generator electrode continues to shift negatively. As long as the applied current is smaller than the limiting current corresponding to the prevailing concentration of precursor, the electrode potential is governed by the precursor half-reaction and the electrode process is limited to a single redox reaction. If the applied current is greater than the limiting current, the electrode potential is forced to shift until a second process occurs to help carry a portion of the current. This condition is shown in the insert of

Figure 25.1. Line C represents a current ratio greater than 1, and the precursor reaction alone is unable to support the current applied to the cell. This intolerable situation is usually avoided by using a high concentration of precursor.

The current efficiency of an electrode reaction refers to the fraction of the current applied to the cell that actually contributes to the faradaic reaction of interest. The magnitude of the applied current relative to any residual (background) current $i_r$ that might exist is therefore of great importance. In any real cell, there will be background current that can arise from any or all of the following: oxidation or reduction at the working electrode of chemical impurities in solution, adsorption processes at the electrode, and charge or discharge of the electrical double layer at the electrode. A discussion of these processes as they relate to constant-current coulometric titrations is found in the work by Meites [1]. In practice, the magnitude of the residual current at any potential can be determined from a residual current–voltage curve obtained in the constant-current cell containing only the supporting electrolyte. The percent current efficiency is defined as

$$\%\text{C.E.} = \frac{100(i_{\text{applied}} - i_r)}{i_{\text{applied}}} \tag{25.5}$$

In general, the residual current is a function of the electrode potential. However, the dependence is often slight over a limited range of potential. Many applications of constant-current coulometry are performed in a manner in which the reagent generated from the precursor is consumed by a chemical reaction that regenerates the precursor. The concentration of precursor is thereby nearly constant and the potential of the generator electrode remains nearly constant during electrolysis. A measurement of the residual current at the potential of the generator electrode will serve for the calculation of the current efficiency.

The applied current must be 1000 times the residual current to achieve a current efficiency of 99.9%. In many cases, a ratio of applied current to a residual current of $10^3$ is reasonable for applied currents down to about 10 $\mu$A using generator electrode areas of about 0.1 $cm^2$. Currents in excess of a few hundred milliamperes are seldom used in constant-current coulometry because the solubility limit of the precursor is reached and/or the experiment may be over too quickly to permit accurate measurement of the time. Heating effects ($i^2R$) are also a problem when high currents are used.

*Instrumentation*

A constant-current source can be constructed from a series combination of a battery and a resistance network (see Chap. 6). Such an arrangement is shown in Figure 25.2a. The current with the load shorted is equal to $E/R_{adj}$. Viewed as an electrical network, a constant-current source should have a very high

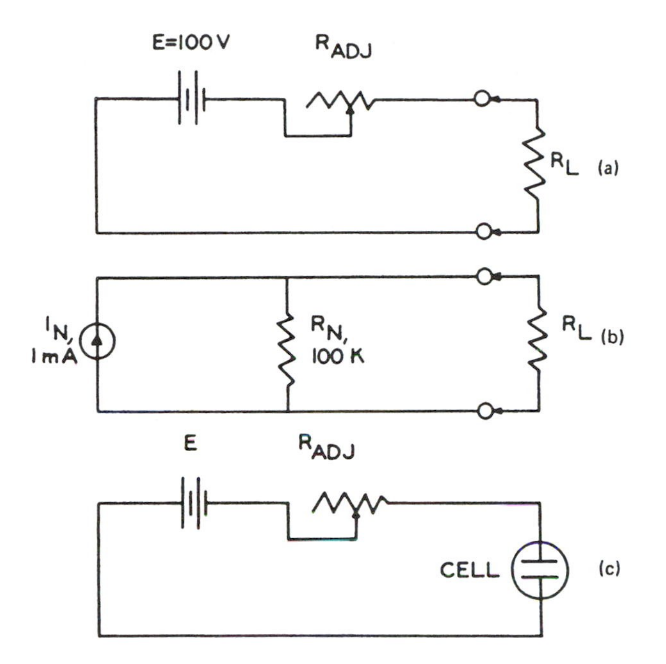


**Figure 25.2** Constant-current source using a battery and series resistor. (a) Dummy (resistor) load; (b) Norton's equivalent circuit for part a; (c) electrolysis cell as the load.

internal resistance. This may be seen from the Norton's equivalent circuit for Figure 25.2a. Let $R_{adj}$ be set to 100 kΩ and E to 100 V. The Norton circuit is shown in Figure 25.2b. Since the load $R_L$ is placed in parallel with the Norton's equivalent resistance, it is clear that the 1 mA of current will only pass through the load resistor if $R_N$ is very much larger than the load. If $R_L$ were 100 Ω, the load current would be 0.999 mA, which is acceptably close to 1.000 mA. When the load is an electrochemical cell, the situation is more complex. The circuit is shown in Figure 25.2c. Let $E_{app}$ be the voltage applied to the cell. From Kirchhoff's voltage law,

$$E = iR_{adj} + E_{app} \tag{25.6}$$

Now the applied voltage is the sum of the back electromotive force (emf) of the cell, $E_{back}$, and the iR drop through it, or

$$E_{app} = E_{back} + iR_{cell} \tag{25.7}$$

The back emf, or galvanic voltage, is the difference between the anode and cathode potentials, each of which is the sum of a reversible potential given by the Nernst equation, and an activation overpotential term $\eta$, which serves in the case of the cathode to shift the potential toward more reducing values and in the case of the anode toward more oxidizing potentials. Thus

$$\begin{aligned} E_{back} &= E_{anode} - E_{cathode} \\ &= E_a^{o\prime} + \frac{RT}{nF}\ln\frac{[O_1]}{[R_1]} + \eta_a - E_c^{o\prime} - \frac{RT}{nF}\ln\frac{[O_2]}{[R_2]} - \eta_c \end{aligned} \tag{25.8}$$

where redox couple 1 has been arbitrarily chosen as potential-determining at the anode. Substituting Equations 25.7 and 25.8 in Equation 25.6 yields

$$E = iR_{adj} + iR_{cell} + (E_a^{o\prime} - E_c^{o\prime}) + (\eta_a - \eta_c) + \frac{RT}{nF}\ln\frac{[O_1][R_2]}{[R_1][O_2]} \tag{25.9}$$

Although E may be constant, $R_{cell}$, the overvoltage term, and the logarithmic term are subject to change during electrolysis and the current must therefore change. One recourse is to choose a supply voltage large enough that any changes can be neglected. A better solution is to monitor the current and make adjustments to correct for any changes. Operational amplifiers (OAs) are ideally suited for this purpose (see Chap. 6).

A typical OA circuit for controlled-current coulometry is shown in Figure 25.3. If no significant amplifier input current is drawn, the cell current is identical to $i_{in}$. The latter is constant if the constant-voltage source and $R_{adj}$ are stable. The OA will automatically adjust the voltage applied to the cell to maintain the cell current equal to the input current. Perhaps the greatest source of instability is temperature change. Constant-voltage sources in integrated circuit packages that are quite stable to temperature changes are commercially available. An

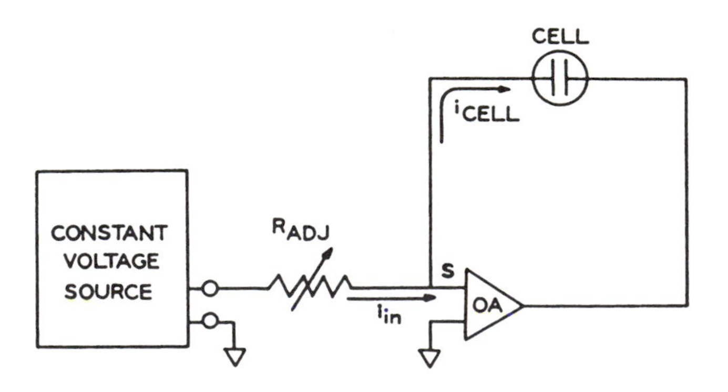


**Figure 25.3** Operational amplifier constant-current source. [Adapted from J. Janata and H.B. Mark, Jr., *Electroanal. Chem. 3*:11 (1969); reprinted by courtesy of Marcel Dekker, Inc.]

example is the Analog Devices model AD 2702L (Analog Devices Inc., One Technology Way, P.O. Box 9106, Norwood, MA 02062-9106), which delivers $\pm$ 10.000 $\pm$ 0.0025 V at 0–10 mA and has a temperature coefficient of 3 ppm/°C. Metal film resistors have temperature coefficients better than ±0.1% and their use is recommended.

Assuming a 10-V output from the constant-voltage source, the values of $R_{adj}$ shown in Table 25.1 were calculated for currents covering the range 1–100 mA. Some appreciation of the heating problem can be gained from the last two columns of the table. Resistors that should have a much higher wattage rating than the power to be dissipated should be used with either a heat sink or good access to air. The accuracy of the components is not as important as their stability, since the current can be calibrated, although this may not be necessary with high-quality components. In any case, it is entirely possible to generate currents that approach ±0.01% in stability, accuracy, and reproducibility. Inexpensive (< $1) commercially available integrated-circuit operational amplifiers are of a quality to match these performance figures.

*Timers*

In the early literature, time measurements were made with electromechanical timers. Lingane [2] has discussed these in detail. However, these devices have been replaced by timers based on digital logic circuits. An arrangement of a timer in an apparatus for controlled-current coulometry is shown in Figure 25.4. The double-pole, double-throw switch can be manually or automatically actuated. The actuating signal might be derived from some concentration-dependent transducer associated with the cell, such as a redox indicator electrode or a photodiode. The current is switched between the cell and a dummy resistor $R_D$, which has about the same resistance as that of the cell. In this way, the current source is working into a similar electrical load at all times.

Digital timers of excellent quality that are based on crystal oscillators are available. The resonant frequency of such oscillators is temperature dependent and subject to aging effects, but the state of the art is highly developed. Crystal-controlled clock oscillators having a maximum error of ±1 ppm at 25°C are commercially available from a number of manufacturers. Some units are tem-

**Table 25.1** Heat Generated in Resistors

| i (mA) | $R_{adj}$ (kΩ) | Power (W) | Time to generate 1 cal of heat (s) |
|---|---|---|---|
| 1 | 10 | 0.01 | 24 |
| 10 | 1 | 0.1 | 2.4 |
| 100 | 0.1 | 1.0 | 0.24 |

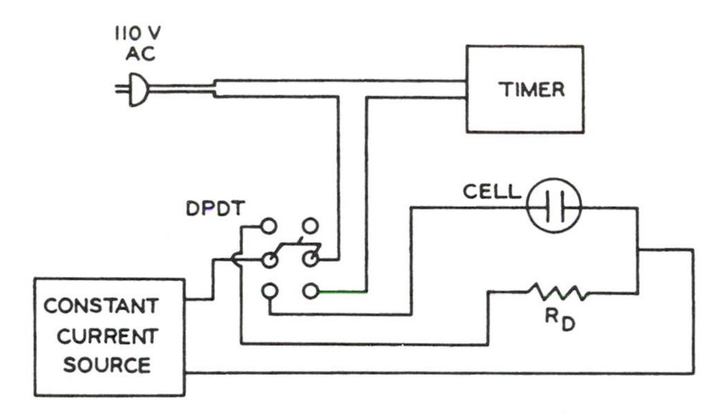


**Figure 25.4** Block diagram of an apparatus for controlled-current coulometry. A double-throw, double-pole switch is used to turn the timer and electrolysis current on and off. [Adapted from G.W. Ewing, in *Am. Lab.* *13*(6):16–22 (1981). Copyright 1981 by International Scientific Communications, Inc.]

perature compensated and have better stability than the figure quoted earlier, with aging rates of only 1 ppm per year. Overcontrolled crystal oscillators can have aging rates of $1 \times 10^{-9}$ per day and a temperature variation in the output of only $1 \times 10^{-9}$ over a 50°C temperature range. Such oscillators, used in conjunction with digital counting circuits, provide an excellent method for the time interval measurement in constant-current work. With adequate switching, timing to the nearest microsecond is routinely possible, resolution to six digits is not unusual, and very high accuracy can be achieved.

A block diagram of a counting circuit arranged for time interval measurements is shown in Figure 25.5. Individual digital circuits will not be discussed

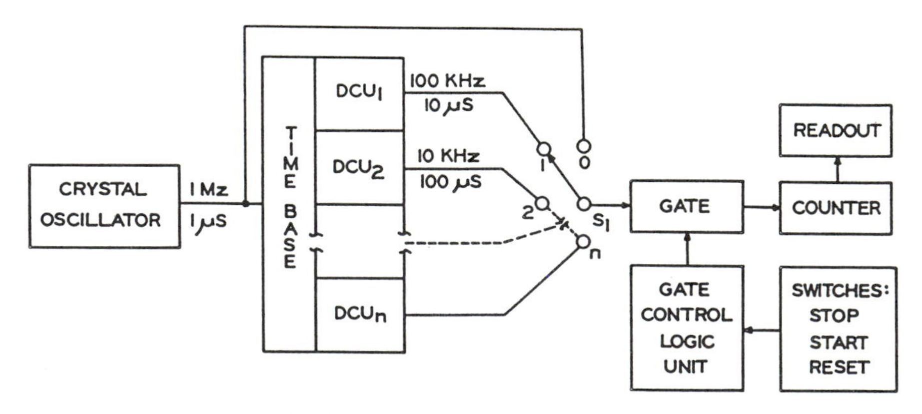


**Figure 25.5** Block diagram of a solid-state timer.

here; the reader is referred to several references [3–6]. A general understanding of digital time interval measurements can be gained from the block diagram in Figure 25.5. Pulses from the crystal oscillator are sent directly to a gate and to a time base (scalar) unit which consists of decade counting units (DCUs). Each DCU is a divide-by-10 circuit, and the oscillator output frequency is sequentially divided by 10 up to n times.

In the time domain, if the oscillator (clock) puts out a pulse every microsecond, the output of the first DCU is a 10-μs pulse train, and so on through n decade counting units. When the gate is open, pulses are passed on to the counter, where they are summed. The gate is controlled by a logic signal generated in the gate control logic unit. Manual switches may initiate this signal. A numerical readout displays the count accumulated by the counter. Switch $S_1$ serves as a range switch. With a six-digit counter and readout, 999,999 counts or 0.999999 s can be accumulated in the readout with $S_1$ in position 0. Switch position 1 corresponds to a maximum readout of 9.99999 s, and so forth, to switch position n, where $0.999999 \times 10^n$ s can be accumulated. It might seem that the gate control logic unit could be eliminated and the gate operated directly by the switches. This is not feasible when it is realized that one count could correspond to 1 μs and manual switching is not possible in so short a time. Flip-flops or other circuits are triggered by the manual switches to provide the control signals to the gate. Once the time measurement has begun, it can be terminated automatically or manually.

The basic time interval counter is more useful for coulometry if a preset capability is added to it. Such counters are commercially available and are also readily constructed from digital circuits. This counter will automatically stop counting at a preselected time. In addition, a logic-level signal is available to indicate the state of the counter: for example, logic level HI for counting and logic level LO for not counting. The change in logic level occurs at the time counting stops and can be used in conjunction with a logic-operated solid-state switch to turn the cell current on and off. This provides a direct way to switch the cell current in 1 μs or less.

A block diagram of a complete apparatus for constant-current coulometry using an operational amplifier constant-current source, solid-state switch, and digital timing is shown in Figure 25.6. The solid-state switch occupies a central position in the diagram to emphasize that it is a key element in the circuit design because it provides the link between the analog and digital parts of the circuit. The actuating circuit is logic-operated, or gated, while the switching circuit operates on the analog current. Solid-state switches are available from a number of manufacturers and can be obtained in a variety of configurations such as single-pole, single-throw (SPST), DPST, DPDT, and others. Make-before-break and break-before-make switching action is provided, and units compatible with nearly all logic families (TTL, CMOS, etc.) are manufactured.

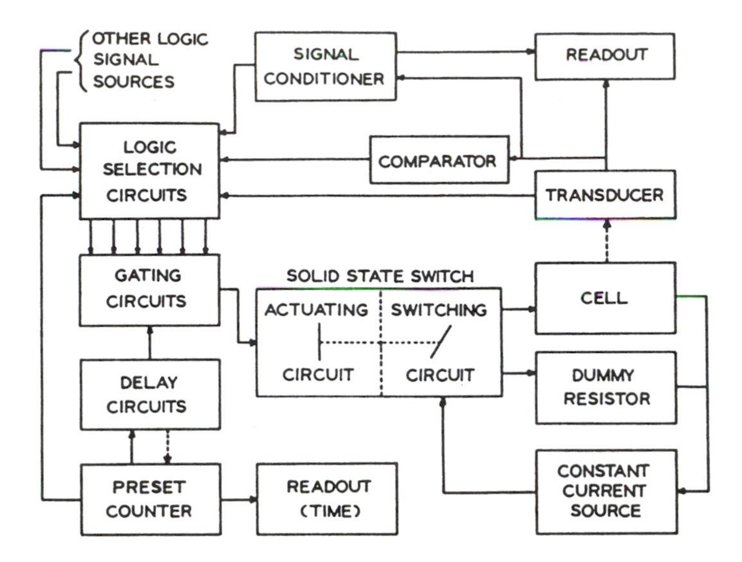


**Figure 25.6** Block diagram of an all-solid-state apparatus for controlled-current coulometry.

Switching times are typically less than 1 μs, and the complementary metal oxide semiconductor (CMOS) switches can be obtained with switching times of less than 10 ns. A useful introduction to analog switches can be found in Reference 7, and the subject is discussed further in Reference 8. The electrolysis cell, dummy resistor, and current source appear in the lower right-hand side of Figure 25.6. Some concentration-dependent transducer may be associated with the cell, such as a potentiometric indicator electrode or a phototransistor. Its output can be presented on a readout, either directly or after signal conditioning (amplification, waveshaping, etc.). Its output can also be used to switch the cell current on and off. In principle, the transducer output could be digital, in which case the output would be sent to a logic selection network, where a choice would be made of which logic signal or signals would be sent to the gating circuits and on to the analog switch. It is more likely that the transducer output is an analog signal. In this case, a comparator circuit can be used for the analog-to-digital conversion, or perhaps signal conditioning is required, which will have both an analog output for the readout and a digital output to operate the switch.

The block diagram in Figure 25.6 represents a modern implementation of the classical apparatus illustrated in Figure 25.4. The solid-state approach has the advantage of having the capability of using the time variable for an active role in the experiment. The output of the preset counter can be used to activate the current switching circuit through the logic selector and gates. This is a necessity for subsecond electrolysis times. Furthermore, the counter can also be used to trigger delay circuits, which can override all other signals to the gates controlling the input to the actuating circuit of the solid-state switch. Once the delay circuits are activated, they also control the counter. In this way, a "waiting" period can automatically be introduced into the electrolysis experiment. One might, for example, want to wait for equilibrium to be established or an induction period in a kinetics experiment. The system is versatile; a logic signal from any other source can be used to activate the solid-state switch. Electrolysis times from several microseconds to thousands of seconds are readily measured with this apparatus. Jaycox and Curran have described a coulometric titrator based on these design considerations [9].

*Current-to-Frequency Converters for Measuring Charge*

The basis of the preceding discussion is Equation 25.2, in which the integration operation is accomplished by simply calculating the product of the time of electrolysis and the constant current. More versatility is introduced if the constant-current source is programmable and the integration operation is approached as suggested in Equation 25.1. A satisfactory method for the latter is to perform the integration operation by current-to-frequency conversion (IFC) followed by counting. Current-to-frequency converters are available in integrated circuit packages such as the Analog Devices model AD 652, which has a maximum frequency output of 2 MHz, full scale. The input can be 0 to 10 V, 0 to –10 V, or ±5 V. The full-scale calibration error is typically 0.25 to 0.5% and the linearity as a percent of maximum output is 0.005 to 0.05. The output is a pulse train whose frequency is proportional to the input current.

Integration in a digital mode is accomplished by counting the output pulses of the IFC over a set period of time. This can be understood from a brief discussion of the charge balance type of IFC. The charge produced by the input current is compared with the charge in a standard charge packet. When the charges are equal, a pulse containing that charge appears at the output of the IFC. When the number of pulses is summed by a counter over a given period of time, the counter reading is the total charge. The time interval is controlled by logic START and logic STOP signals applied to a gate controlling the counter. This method of integration is superior to the classical analog operational amplifier integrator in terms of accuracy, precision, and resolution, and has replaced it where high performance is demanded. Coulometers based on IFCs have been described by Stock [10] and Huffman [11].

Programmable constant-current sources can be constructed from programmable operational amplifiers or programmable power supplies. These devices coupled with the IFC produce a powerful approach to controlled-current coulometry.

## III. COULOMETRIC TITRATIONS

### A. Introduction

A cell for internal generation of titrant is easily constructed. Often it is necessary to remove dissolved oxygen, and a closed vessel equipped with a gas dispersion tube and a vent is provided. Nitrogen is usually used for deoxygenation. The working electrode compartment is separated from the counterelectrode compartment by a glass frit. Materials for the working electrode and counterelectrode can be any of those commonly used in electrochemistry: mercury, platinum, gold, and carbon in its various forms, as well as active electrode materials such as silver. A magnetic stirring bar is used to mix the titrant with the solution. If silver ion and chloride ion can be tolerated, the two-compartment cell can be eliminated and a silver–silver chloride counter electrode can be placed directly into the solution to be titrated. A method of endpoint detection must be provided.

Cells that fit the preceding description have been used since coulometric titrations were first reported. Although the electrodes used are typically of fairly large area, the cells usually have a large volume-to-electrode-area ratio. Since it is the amount of titrate present that determines the time for electrolysis at a given current, small-volume cells have the advantage of shorter titration times and higher sample throughput. Ruttinger and Spahn [12] have devised a cell that has a working electrode chamber volume of only 10 to 11 mL and an electrode area of 4.6 $cm^2$. It features a magnetically coupled centrifugal pump that drives solution through channels that contain biamperometric and photometric flow-through detectors. Titration times of 10 to 100 s were achieved for both direct and back titrations.

### B. Primary and Secondary Processes

In a controlled-current coulometric titration, either the electrogenerated reagent reacts with an unknown substance in a stoichiometric or reproducible manner, or the unknown itself reacts directly at an electrode. The latter, called a primary coulometric titration, is rarely found in practice because the concentration of unknown continuously decreases during the electrolysis and it is difficult to maintain 100% current efficiency. The former approach, known as secondary coulometric titration, is used most of the time.

### C. Titration Efficiency

If the reaction at the generator electrode is given by Equation 25.10a and the titration reaction by Equation 25.10b,

$$ox_2 + ne \rightleftharpoons red_2 \tag{25.10a}$$

$$ox_1 + red_2 \rightleftharpoons red_1 + ox_2 \tag{25.10b}$$

where $ox_1$ is the titrate and $red_2$ is the titrant, the electrode potential of couple 1 must be more positive than the electrode potential of couple 2 (IUPAC sign convention) for the titration reaction to be thermodynamically feasible. There are interesting consequences of this in terms of the electrochemistry of the coulometric titration, and they can be seen by referring to the insert of Figure 25.1. Initially, the solution contains two oxidants. The titrate, $ox_1$, is more easily reduced than the precursor of the titrant, $ox_2$, and therefore the first wave shown in the figure corresponds to the reduction of the titrate, while the second wave (only the foot of the wave is shown) corresponds to the reduction of the precursor. If the current applied to the cell is represented by line C, it is clear that most of the current is initially carried by the titrate. The current efficiency is less than 100% for both couple 1 and couple 2. However, the charge passed is the same whether the titrate is reduced directly at the electrode or reduced by the titrant, $red_2$. Thus 100% *titration efficiency* can be achieved, although the current efficiency is less than 100%. In fact, it is possible for the current efficiency of the precursor reaction to be less than 100% due to background and have the titration efficiency maintained at an acceptable level when a large percentage of the titrate reacts directly at the electrode. Suppose that 85% of the titrate undergoes direct electroreduction with 100% current efficiency and the remaining 15% is titrated by the titrant, which is generated with 90% current efficiency. The titration efficiency is $0.85 + (0.9 \times 0.15) = 0.985$. An error of 1.5% might be entirely acceptable in some cases.

## D. Internal and External Generation

The common technique in coulometric titrations is to generate the titrant at an electrode placed physically within the coulometric electrolysis cell. This approach sometimes fails. An obvious case is when an electroactive material present in the sample undergoes a reaction at the generator electrode concurrently with the precursor. It is also possible to foul the generator electrode with surface-active materials such as proteins. These problems can be eliminated by generating the titrant in a stream of electrolyte which flows into the titration vessel. The operation becomes entirely analogous to that of a buret, and some means of controlling the flow rate and stopping the flow are usually required [13]. It is physically impossible for the titrate to undergo direct electrochemical reaction, and the titrant generation reaction must proceed at 100% current efficiency. Lingane has discussed a number of cell designs for external titrant generation [14], and much of the information on electrochemical detectors in Chapter 27 is useful in this regard.

## E. Endpoint Detection

Some method of signaling is required to indicate when the amount of titrant generated is equivalent to the amount of unknown present, and all of the endpoint detection methods used in volumetric titrimetry are, in principle, applicable to coulometric titrations. A list that covers most of the published coulometric titration procedures is given in Table 25.2. It is beyond our scope here to describe any of these in detail because each of these methods is a subject for discussion in its own right. Discussions of the equations for a number of types of titration curves are found in texts by Lingane [15], Butler [16], and Laitinen and Harris [17].

Several practical features relating to endpoint detection are worthy of comment. As mentioned earlier, in most coulometric titration cells, the ratio of solution volume to generator electrode area is large. The solution must be stirred to mix the titrant with the solution, and local excesses of titrant are common, particularly in the vicinity of the endpoint. The endpoint detection system is easily fooled by this nonequilibrium condition, and false detector signals arise. Such signals are seldom mistaken for equilibrium conditions since they do not persist, but advantage can be taken of them to introduce endpoint anticipation into the titration. A false signal corresponding to the endpoint condition would shut off the generator current for a short time to allow the system to reach equilibrium. To enhance this effect, the endpoint detector is placed in a position where it is bathed by the solution coming directly from the generator electrode. At the same time, it is necessary with electrometric endpoint detection methods to avoid placing the indicator electrodes in the current path between the generator electrodes, since interaction between the electrode pairs can occur.

Different experimental approaches are possible with the same endpoint detection method. For example, the titration curve can be plotted and the endpoint determined graphically. First and second derivative curves can be plotted or the derivatives obtained electronically. Another approach is to titrate to a predetermined endpoint signal. This technique is very useful with coulometric titrations, and many examples, especially those involving potentiometric endpoint detection, are found in the literature. The most widely applicable way

**Table 25.2** Selected Endpoint Detection Methods for Coulometric Titrations

| | |
|---|---|
| Visual indicators | Potentiometric |
| Colorimetric | Amperometric |
| Spectrophotometric | Biamperometric |
| Pressurometric | Conductometric |

of selecting the endpoint potential is to acquire the data for the complete titration curve with a known sample. The time required to reach the endpoint is known beforehand, and examination of the titration curve will reveal the potential corresponding to this time. Several advantages are realized from the use of a predetermined endpoint signal with coulometric titrations: (1) the supporting electrolyte can often be pretitrated to the predetermined potential to remove reactive impurities, (2) samples can be added successively to the same supporting electrolyte, (3) coulombic error due to double-layer charging or discharging may be minimized, and (4) the ease of automation is enhanced. With respect to the latter, endpoint detection becomes operationally a matter of a null comparison measurement, and the operational amplifier comparator is nicely suited to this task [18]. The third advantage listed occurs because the potential of the generator electrode is the same at the beginning and at the end of the titration, but this will be true only if the structure of the double layer remains unchanged (i.e., surfactant, for example, must be absent). The pretitration technique is widely applicable but can be subject to error. In the titration of a weak acid, the presence of unsuspected weak acids of different acid strengths from the sought-for acid would lead to an incorrect pretitration procedure. Also, it is not generally true that endpoint detection methods are independent of the initial concentration of analyte, and the dilution produced by the successive addition of samples may be enough to shift the correct endpoint from the predetermined one. In any event, the correctness of the pretitration technique can be checked by determining known amounts of the substance to be determined.

Using a titrator based on the concepts described in Figure 25.6, Jaycox and Curran [9] were able to titrate as little as 26 ng of Mn(VII) automatically with electrogenerated Fe(II) to a precision and accuracy of 2% relative using potentiometric endpoint detection to a preset endpoint potential. The same apparatus, operated in a manual mode, was used to titrate V(V) solutions as dilute as 2.1 ppb, for which the total elapsed electrolysis time to reach the endpoint was 20.5 ms. Cadwgan and Curran [19] described a single-beam fluorescence detector for use with the same titrator. Acid was titrated with electrogenerated base in the presence of dichlorofluorescein, a fluorescent indicator. Only the conjugate base form of this indicator fluoresces. At a solution pH equal to the $pK_a$ of the indicator, the ratio of acidic to basic forms is 0.5, and slight changes in pH will produce large changes in the emission light intensity. An arbitrary null point pH was chosen where the indicator was present primarily as the conjugate base. This was accomplished by adding small volumes of dilute HCl and/or KOH. The photodiode output voltage at the null point pH was set on the reference side of the comparator. A sample of acid was added to the cell with a microburet and the solution was titrated back to the null point pH. Amounts of hydrogen ion, as $KH(IO_3)_2$, from 1284 to 12.84 nEq were automatically determined with errors of a few percent in most cases and precision was ±0.5 and ±8.8% relative, respectively. It was estimated that addition of the smallest samples to the cell produced a change in hydrogen ion concentration of 0.31 ppb for a solution of about pH 5. This corresponded to a pH change of 0.013 unit, which would be difficult to determine with an expanded-scale pH meter. Very small changes in pH can be measured with this fluorescence detector, and the working pH region can be selected by choosing a fluorescent indicator of the appropriate $pK_a$.

With the same coulometric titrator, de Soto Perera and Curran [20] were able to titrate as little as 24.2 ppb of As(III) (or $3.22 \times 10^{-7}$ *M* solution) with electrogenerated bromine using thin-layer hydrodynamic biamperometric endpoint detection. A sandwich-type thin-layer cell containing two identical platinum electrodes (see Chaps. 3, 9, and 27) was placed in a closed loop attached to the coulometric cell. Solution was pumped through the flow loop at a rate of about 16 mL/min with a peristaltic pump and the solution in the coulometric cell was stirred magnetically. The potential difference across the thin-layer platinum electrodes was 170 mV and the titrations were performed incrementally. Plots of thin-layer cell current versus time of electrolysis were reverse L-shaped, since a response is obtained for this titration only in the presence of excess bromine. Solutions about 800 ppb in As(III) were determined with a relative error and precision of 0.1%. The method was also applied to the coulometric titration of some phenolic steroids with electrochemically generated bromine [21]. The titration curves were recorded continuously, and endpoints were obtained by graphical extrapolation of the linear segments.

With programmable constant-current sources, control of the current can be based on the output of a transducer that reflects the concentration of some species in solution as suggested in Figure 25.6. Not only can the current be switched on and off, but the magnitude of the current can be controlled. Early in the titration the current is large, but as the equivalence point is approached, the current becomes progressively smaller to avoid overshoot. This technique is particularly attractive in conjunction with the current-to-frequency converter, because the latter automatically takes into account the change in the magnitude of the current. A very elegant application of the controlled-current method has been presented by Zetlmeisl and Laurence [22] for the trace determination of chloride. An instrument is described that takes the signal from an indicator electrode (proportional to the log of the chloride concentration) and uses it to force the current applied to the cell to decay exponentially. Overshoot of the equivalence point is avoided, and since at the equivalence point the current is effectively zero, the result is the same as titrating to a preset endpoint and the same advantages pertain.

## F. High-Precision and Accuracy Titrations

The point has been made that the current and time variables (or the charge passed) in a coulometric titration can be measured with high precision and accuracy. This suggests that coulometric titrations are useful for assay of high-purity materials. Equation 25.1 shows that the result of such an assay will also depend on the value of the Faraday. In 1973, the International Union of Pure and Applied Chemistry defined the value of the Faraday as the product of Avogadro's number and the charge on the electron, or 96,484.56 coulombs/equivalent [23]. Thus, the result of an assay by coulometric titration depends only on the fundamental physical quantities current, time, and mass, as pointed out by Yoshimori in an excellent review of the subject [24]. Lingane made the same observation earlier and commented to the effect that such a titration is free of the need for primary standard compounds and standardized titrants and eliminates the uncertainties these bring to a titration procedure [25]. These features were recognized even earlier at the National Bureau of Standards (now the National Institute of Standards and Technology, NIST), and the field was pioneered in a paper by Taylor and Smith in 1959 [26]. Subsequent work at NIST and other laboratories around the world developed assay methods for a variety of compounds such as acids and bases, halides, arsenic trioxide, and others, often with results with uncertainties of 0.005% relative or better.

Some of the difficulties in carrying out such work are interesting and perhaps not obvious. Consider, for example, the hypothetical case that the only source of error in an assay by coulometric titration is in the determination of mass, and that this error should be no more than 0.001%, or one part in 100,000 parts (10 ppm). For a 1-g sample, the weighing must be made to 10 μg or better. This is not possible using the usual semimicro analytical balance; one would have to use a good microbalance. Another problem in the weighing process is moisture. According to Yoshimori, the water content of dried reagents is often greater than 10 ppm [24]. Another problem is the loss of sought-for constituent via the frits of the titration cell. To overcome this problem, the cell design at NIST incorporates one or more intermediate compartments between the generator anode and cathode compartments; these are flushed into the working electrode compartment when the titration is close to the equivalence point. Consider a 1-g sample again; take the molecular weight of the compound to be 100 g/mol, $n = 2$ equivalents per mole, and a current of 200 mA. The time for complete electrolysis is 2.68 h. Furthermore, the procedure followed reduces the current in the vicinity of the equivalence point so the electrolysis time is longer. Rinsing of compartments lengthens the assay time. All of this adds up to a considerable period of time required for an assay, and it is not surprising that the work at NIST continues. Virtually the entire process has been automated under computer control as reported by Pratt [27].

Other features and details on high-purity assays can be found in the review by Yoshimori [24].

## G. Titrants

*Survey*

Many titrants have been electrogenerated, most of them in aqueous solution. A list is given in Table 25.3, which was compiled from biannual reviews appearing in *Analytical Chemistry* between 1968 and 1984 [28–36], from the text by Lingane [2], and from the later original literature. A number of the common volumetric titrants are included as well as titrants that are unstable or difficult

**Table 25.3** Electrogenerated Titrants

| | | |
|---|---|---|
| A. Oxidants | | |
| Bromine | Cerium(IV) | Thallium(III) |
| Hypobromite ion | Permanganate ion | Vanadium(IV) |
| Bromine monochloride | Chromium(VI) | Vanadium(V) |
| Iodine | Silver(II) | Lead tetraacetate |
| Chlorine | Manganese(III) | Oxygen |
| Iron(III) | Mercury(II) | Quinone |
| Ferricyanide ion | Gold(II) | Silver(I) |
| B. Reductants | | |
| Iron(II) | Titanium(III) | Vanadium(IV) |
| Iron(II)-EDTA | Copper(I) | Vanadium(III) |
| Ferrocyanide ion | Hexacyanomanganese(I) | Uranium(III) |
| Molybdenum(V) | Tin(II) | Iodide ion |
| Octacyanomolybdenum(V) | Chromium(II) | Uranium(IV) |
| C. Precipitants and complexing agents | | |
| Silver(I) | Copper(II) | Iodide ion |
| Mercury(I) | Lanthanum(III) | Fluoride ion |
| Lead(II) | Mercury(II) | Sulfide ion |
| Bismuth(III) | Ferricyanide ion | Thiocyanate ion |
| Aluminum(III) | Ferrocyanide ion | EDTA |
| D. Bases | | |
| Aqueous and nonaqueous media | | |
| E. Acids | | |
| Aqueous and nonaqueous media | | |
| F. Miscellaneous | | |
| Tungsten(II) in LiCl–KCL eutectic | | |
| Karl Fischer reagent | | |

*Source*: See text.

to prepare volumetrically. In aqueous solution the strongest oxidant is silver(II) ($E° \approx 2$ V) and the strongest reductant is uranium(III) ($E° = -0.63$ V). The former is thermodynamically capable of oxidizing water and the latter similarly capable of being oxidized by hydrogen ion. The electrogeneration of these titrants requires that the kinetics of the other reactions be slow. Bromine is a moderately strong oxidizing agent ($E° = 1.065$ V) and its electrogeneration is quite free of side reactions in an aqueous solvent. It can be easily generated with 100% current efficiency at a platinum electrode and is a titrant of major importance because of its reactions in both inorganic and organic chemistry. It has wide scope in the determination of organics because substitution reactions, addition reactions, and others are possible. Specific examples of its application to the determination of pharmaceuticals are given in Chapter 26. Another important titrant is Karl Fischer reagent, which is used for the direct titration of water and indirectly for the determination of many organics in a stoichiometric or quantitative manner [37].

In early work, iodine was electrochemically generated from an iodide salt in anhydrous methanol containing sulfur dioxide and pyridine. The titration reaction is

$$H_2O + C_6H_5N \cdot I_2 + C_6H_5N \cdot SO_2 + CH_3OH + C_6H_5N \rightarrow 2C_6H_5N \cdot HI + C_6H_5NH \cdot SO_4CH_3 \tag{25.11}$$

Later studies have resulted in the elimination of the pyridine. For example, the work of Scholz used a reagent of KI, diethanolamine and sulfur dioxide in methanol [38]. Spohn and co-workers [39] have devised a new cell for coulometric Karl Fischer titrations that features a small cell volume, complete filling of the cell, very small sample volumes (about 85 μL) injected by a slider valve arrangement that also provides a second pathway for the injection of fresh electrolyte or iodine standards or other solutions, mixing by means of a centrifugal pump arrangement, and biamperometric flow-through detection, the flow being generated by the pump arrangement. The latter produces high sensitivity in the detection process. Sample throughput is impressive with 100 titrations possible in 1 h.

## H. Uranium Electrochemistry

When hydrogen ion is involved in a half-reaction, such as $ox + mH^+ + ne^- \rightleftharpoons red$, the potential is pH-dependent, as the following Nernst equation shows.

$$E = E° + \frac{RT}{nF} \ln \frac{a_{ox}}{a_{red}} - \frac{m}{n}(0.0591)\text{pH} \tag{25.12}$$

At unit activities of the oxidant and reductant, the potential depends only on pH; the slope of the line for a plot of potential versus pH is governed by the ratio m/n. Potential–pH diagrams are a concise means to display the redox properties of a system. We will take uranium as an example. The +6, +5, +4, and +3 oxidation states are known in aqueous solution. The determination of +6 uranium by coulometric titration has been investigated by many workers and the lower oxidation states have all been used as coulometric titrants. Hydrolyzed uranium species exist in a noncomplexing solution, but the chemistry is simplified considerably if the discussion is limited to solutions more acidic than about pH 4. Some of the half-reactions to be considered are listed next with E° vs. NHE:

$$+0.063 \qquad UO_2^{2+} + e^- \rightleftharpoons UO_2^+ \tag{25.13}$$

$$+0.246 \qquad UO_2^{2+} + 4H^+ + 2e^- \rightleftharpoons U^{4+} + 2H_2O \tag{25.14}$$

$$+0.429 \qquad UO_2^+ + 4H^+ + e^- \rightleftharpoons U^{4+} + 2H_2O \tag{25.15}$$

$$-0.635 \qquad U^{4+} + e^- \rightleftharpoons U^{3+} \tag{25.16}$$

The standard potentials given are those cited or calculated by Farrington and Lingane [40]. The only species of importance that is not shown in these half-reactions is $UOH^{3+}$. The predominant form of U(IV) in solutions more acidic than pH 1 is the simple aquated ion, but in the pH region 1–4, $UOH^{3+}$ predominates. Thus the following half-reactions also need to be considered:

$$UO_2^{2+} + 3H^+ + 2e^- \rightleftharpoons UOH^{3+} + H_2O \tag{25.17}$$

$$UO_2^+ + 3H^+ + e^- \rightleftharpoons UOH^{3+} + H_2O \tag{25.18}$$

$$UOH^{3+} + e^- \rightleftharpoons U^{3+} + OH^- \tag{25.19}$$

The potential–pH diagram shown in Figure 25.7 summarizes the information contained in these seven half-reactions. At any pH more acidic than about 1.7, the U(VI),U(V) couple is analogous to the Cu(II),Cu(I) couple in that the lower oxidation state is the stronger oxidant and therefore U(V) is expected to disproportionate to U(VI) and U(IV). The diagram also predicts that U(VI) should be directly reduced to U(IV) in solutions more acidic than pH 1.7 but to U(V) in more basic solutions. However, the diagram contains no information about any overpotentials for these reactions. A fairly large overpotential is required for the +6,+4 reaction, while the +6,+5 couple behaves reversibly. As a result, reduction directly to U(IV) occurs in solutions more acidic than about pH 1. Solutions of U(V) are readily oxidized by dissolved oxygen, and air must be excluded, but deaeration is not necessary for U(IV) solutions. For many years, the strongest reductant in analytical use has been Cr(II) (E° = –0.41 V).

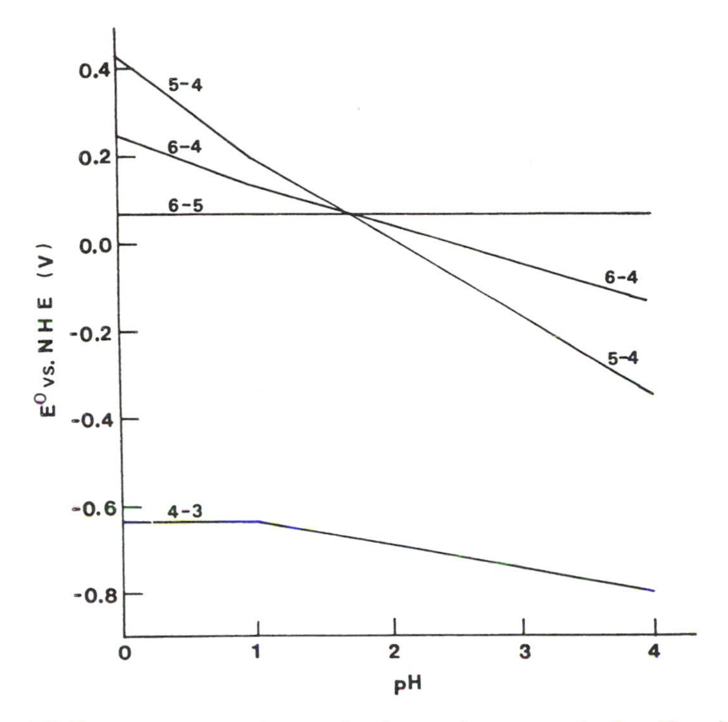


**Figure 25.7** Potential–pH diagram for the uranium system in the pH region 0–4. [Adapted from Ref. 2, p. 585.]

The +4, +3 uranium couple has a potential at least 0.2 V more negative than the +3, +2 chromium couple, and U(III) is a strong reductant indeed.

## I. Coulometric Titration of Uranium(VI)

Uranium(VI) is a rather weak oxidant and its direct determination requires a strong reductant. Lingane and co-workers developed methods using electrogenerated titanous ion, first at a mercury cathode [41], and later at platinum electrodes [42]. In the latter case, a supporting electrolyte 8 *M* in sulfuric acid and 0.6 *N* in titanyl sulfate was recommended. At this acidity, U(VI) is reduced to U(IV). To overcome the slow rate of reaction, a small amount of Fe(II) was added to the solution. The Fe(III) produced by the reaction of Fe(II) with U(V) is reduced by Ti(III). Biamperometric endpoint detection using platinum electrodes was used. Quantities of uranium from 8 to 225 mg were determined with an average error of less than +0.05%. Taylor and Marinenko later modified the procedure, using a 9 *M* $H_2SO_4$ supporting electrolyte, 1 *M* in titanyl ion (see Ref. 43 and the reference cited therein). They adopted the procedure of titrating beyond the endpoint, waiting for equilibrium, and then generating excess Ti(III) in small increments. Using amperometric endpoint detection, the endpoint was found from the slope of the titration curve beyond the endpoint. One-gram samples of uranium metal were determined with a standard deviation of 0.08%. All ions reducible by Ti(III) interfere. Lingane and Kennedy applied their procedure to the titration of mixtures of uranium and vanadium. The standard potential of the $VO_2^+$,$VO^{2+}$ couple is +1.000 V and that of the $VO^{2+}$,$VO^{3+}$ couple is +0.361 V. The former is sufficiently positive of the U(VI),U(V) potential that V(V) titrated first. The V(IV) so formed is titrated simultaneously with the U(VI). The first endpoint yields the amount of vanadium present and the amount of uranium is obtained by difference, using the second endpoint. Solutions containing 10–100 ppm of vanadium and 180–1000 ppm of uranium were determined with an average error close to ±0.2% for both elements.

Other constant-current coulometric titrations of U(VI) have been indirect. Carson reduced U(VI) to U(IV) by passing the solution of the +6 ion containing 2 *M* HBr through a lead-amalgam reductor column directly into an air-free titration vessel [44]. The column was washed with a 3 *M* HBr solution containing ferric bromide. The latter is reduced to Fe(II). Titrant was then generated at a platinum anode to oxidize the U(IV) to U(VI). The titration was carried out at 95°C to increase the rate of the reaction. It is not clear if electrogenerated bromine or Fe(III), or some combination of both, is the actual titrant. Lingane argues in favor of Fe(III) on the basis of formal potentials and relative concentrations [45]. Carson recommended constant-current potentiometric endpoint detection, which involves passage of a small constant current through a pair of identical indicator electrodes, with the potential difference between them being measured. This technique is often appropriate for titration reactions involving irreversible couples. Seven-milligram amounts of uranium were titrated with a precision and accuracy of ±0.3%, and 30-μg amounts were determined to ±6%. Several approaches have been devised to avoid the endpoint detection difficulties due to the irreversibility of the U(VI),U(IV) couple, but like Carson's work, they all involve an initial reduction step to convert U(VI) to U(IV). Furman and co-workers [46] reported that a Jones reductor (zinc amalgam) was unsatisfactory because a variable amount of U(III) was produced. A 3% cadmium amalgam was inconvenient because it is a liquid. A solid amalgam containing about 26% cadmium was finally used. Malinowski [47] developed a coulometric finish to a method described by Pszonicki [48]. Uranium(VI) is reduced primarily to U(III) by metallic aluminum in the presence of cadmium ion. The U(III) is oxidized to U(IV) by the addition of phosphoric and hydrochloric acids. This oxidation by protons should also be applicable with the Jones reductor. Goode and co-workers used the Ti(III) reaction to reduce U(VI) [49]. The excess Ti(III) was removed by reaction with a solution of sulfamic and nitric acids. In the work by both the Furman and Goode groups, the U(IV) was converted back to U(VI) by reaction with ferric ion, and the equivalent amount of Fe(II) so produced was determined by coulometric generation of Ce(IV). The overall scheme is presented in the following equations:

$$U(VI) + 2e^- \xrightarrow{\text{chemical reductant}} U(IV) \tag{25.20}$$

$$U(IV) + 2Fe(III) \rightleftharpoons U(VI) + 2Fe(II) \tag{25.21}$$

$$Ce(III) \rightleftharpoons Ce(IV) + e^- \text{ (platinum or gold anode)} \tag{25.22}$$

$$Ce(IV) + Fe(II) \rightleftharpoons Ce(III) + Fe(III) \tag{25.23}$$

The procedure of Goode et al. was developed to provide a highly precise determination of uranium and it succeeded admirably, since a precision of ±0.004% was achieved. The accuracy was reportedly better than 0.01%, but the method does require an accurately known solution of Ti(III). The Malinowski method offers another alternative. The U(IV) is oxidized by addition of an amount of solid primary standard $K_2Cr_2O_7$ that is in excess by a few milligrams. Ferric chloride is added to the solution and the excess dichromate is determined by reaction with electrogenerated Fe(II). Excellent results were found for the analysis of $U_3O_8$ obtained from NBS. A precision of ±3 parts in $10^5$ was reported and an error of 7 parts in $10^4$.

The initial reduction of U(VI) shown in Equation 25.20 would be simplified operationally if the metallic phase were eliminated. Davies and Gray [50] demonstrated that this step could be carried out homogeneously with excess Fe(II) in a strong phosphoric acid solution that also contained nitric and sulfamic acids. The excess Fe(II) was oxidized to Fe(III) by reaction with the $HNO_3$ after addition of molybdate catalyst, and the resulting $HNO_2$ was removed by reaction with the sulfamic acid, with the reaction producing $N_2$. This feature was necessary to prevent nitrite catalysis of the nitric acid oxidation of Fe(II) and U(IV). Goldbeck and Lerner [51] later modified the method to incorporate a coulometric step. A solution of vanadium(IV) was added to the uranium solution following the conversion of excess Fe(II) to Fe(III), and vanadium(V) was electrogenerated at gold or platinum electrodes. The endpoint was detected potentiometrically. Because of the possibility of V(IV) reacting with U(IV) to yield V(III) and U(VI), it was not clear whether the electrogenerated V(V) was being used to titrate U(IV) or V(III), but this was of no consequence since the same number of equivalents of V(V) would be required for the titration in either case. Solutions placed in the electrolysis cell typically contained about 100 mg of uranium, and the determinations were precise and accurate to better than 0.1%. The method has been automated (a serious consideration in view of the radioactivity involved) [52] and scaled down to 5- to 40-mg samples [53]. Chitnis et al. [54] have developed a method that combines features of the Malinowski and Goldbeck-Lerner procedures. Instead of adding V(IV) to the reduced U(VI) solution, these authors add a known excess of $K_2Cr_2O_7$ to oxidize the U(IV) and titrate the dichromate left with Fe(II) electrogenerated from the Fe(III) present in solution.

## J. Nonaqueous Titration of Organic Bases

Some of the reasons for considering coulometric titrations in nonaqueous solvents are that many organic compounds are not soluble in water, metals can exist in oxidation states that are not found in water, and advantage can be taken of the acidity or basicity of the solvent to improve the basic or acidic strength of a base or acid, respectively.

It is now well known that a difficulty in the titration of organic bases with electrogenerated hydrogen ion at a platinum electrode is that many bases are oxidized at the generator electrode, leading to a negative error in the determination. Two interesting solutions to this problem have been developed by workers in Yugoslavia. The first involves anodic oxidation of organic compounds in nonaqueous solvents to products, one of which is protons. For example, Mihajlovic and co-workers have generated protons in propylene carbonate with a sodium perchlorate supporting electrolyte by anodic oxidation of esters of gallic acid such as methyl through butyl gallates, phenols, and ascorbic acid [55]. Organic nitrogen bases such as puridine and piperidine were titrated. Recoveries ranged from 99.6 to 100.3% leading to the conclusion that the oxidation of the organics was quantitative. In other work, similar results were found in acetonitrile, acetic anhydride–acetic acid mixtures, nitromethane, and sulpholane as solvents [56].

The second method involves using hydrogen or deuterium dissolved in a palladium anode as a source of anodically generated protons or deuterium ions. The Yugoslavian group has pioneered the work in this area. Mihajlovic and co-workers determined a number of organic nitrogen bases in acetone, methyl ethyl ketone, methyl isobutyl ketone, cyclohexanone, acetic anhydride, and 1:6 acetic acid–acetic anhydride mixture as solvents [57]. Sodium acetate and potassium hydrogen phthalate were also determined. Visual endpoint detection with methyl red indicator was used in the ketone solvents and malachite green in the others.

Potentiometric endpoint detection was also used with a palladium indicator electrode saturated with hydrogen gas. Sodium perchlorate was used as the supporting electrolyte. Current–voltage curves showed that the hydrogen in the palladium electrode was oxidized at potentials about 1 V less positive than any other component of the solution. Current efficiencies ranged from 99.9 to 100.0% using visual indicators, and precision was typically about ±0.5% relative. Results using potentiometric endpoint detection ranged from 99.9 to 100.2% current efficiency for the generation of protons. Precision was slightly improved with the potentiometric detection. The palladium was charged with hydrogen gas by electrolysis of water acidified with sulfuric acid. According to the authors, a palladium plate with a volume of 1 $cm^3$ held enough hydrogen gas to generate hydrogen ions for more than 100 h using a current of 10 mA. Other work involved the coulometric titration of bases in propylene carbonate using hydrogen/Pd and deuterium/Pd electrodes [58], and the coulometric titration of bases in ketone solvents using deuterium/Pd electrodes [59].

Coulometric determination of organics by redox chemistry in nonaqueous solvents has been discussed by Abdullin and Budnikov [60] with particular emphasis on work at their own institution. Conditions for the coulometric generation of titrants by anodic dissolution of metals and from salts are summarized along with the conditions for the generation of halogens. Applications include the titration of sulfur-, nitrogen-, and oxygen-containing compounds, as well as determinations based on electrophilic substitution reactions.

## K. Use of Computers

Advantages of coulometric over volumetric titrations include (1) elimination of the need for standard solutions, (2) the use of unstable reagents, (3) better accuracy and precision, (4) ease of adding small increments of titrant, and (5) ease of automation. Certainly the computer can contribute to automation. An example is the automatic determination of calcium and magnesium with electrogenerated ethylenediamine tetraacetic acid (EDTA) [61]. An IBM PC-type computer was equipped with a 14-bit, 16-channel analog-to-digital conversion board and a relay board. An autosampler and an autoburet for delivery of sample were interfaced to the computer via an RS-233C interface. The relays were used to control a commercial galvanostat. A program was written in Quick-BASIC that provided control, data acquisition, and data processing. Endpoint detection was potentiometric. The method was applied to the determination of water hardness and Ca and Mg in soils. The results agreed well with volumetric titrations of the same samples.

A very interesting example is the automatic carbon and hydrogen determination reported by Houde et al. [62]. The sample is flash combusted in a stream of oxygen, and the water produced is frozen in a trap, while the $CO_2$ is absorbed by a solution of 0.2 *M* tetrabutylammonium bromide in a mixture of 90 mL of dimethyl sulfoxide (DMSO), 10 mL of $H_2O$, and 3 mL of ethanolamine, contained in a coulometric titration cell. Oxides of nitrogen are removed from the gas train by an $MnO_2$ packing. It was suggested that the $CO_2$ reacts with the primary amine to yield the corresponding carbamic acid, which is then titrated with electrogenerated base. The cold trap containing the water is heated and the water vapor passed over solid carbonyldiimidazole, which results in a quantitative yield of $CO_2$ that is then determined as discussed earlier. A combination of a dual water trapping line, two titration cells, and a multiport valve allowed the carbon from one sample and the hydrogen from the previous one to be simultaneously determined.

The entire process was placed under computer control using a Hewlett-Packard 9825A desktop computer with 15K of memory. The only operator time required was for weighing the samples, which were then placed in an automatic sample tray. The computer was interfaced to the Mettler balance, so the sample weight and identification could be stored in memory. The computer controlled the oxygen pressure, the temperature of the combustion furnace, heating and cooling of the traps, operation of the valve, and the coulometric cells, the latter through a commercial coulometer and potentiometric titration to a preset neutralization point. Data were received by the computer and the results after calculation were printed out as percent carbon and hydrogen. The sample tray held 23 samples and 4 h was required to complete the analyses. Two sets of 23 samples, were run during a working day and a third set was started at the end of the day and left to run overnight. Results were precise to 0.1% for carbon and 1% for hydrogen. This work clearly shows the control capabilities of the computer.

A second major area of computer use is interactive control of the experiment. Harrar and Pomernacki [63] describe a programmable constant-current source based on a Kepco model 40-0-5B power supply (Kepco, Flushing, NY). Digital control signals are converted to analog with a 14-bit digital-to-analog converter (DAC); the analog signal determines the magnitude of the current. The current range available was 0–500 mA and the digital resolution was 30 μA. The current can be set either by a computer or by front-panel switches. Further details of the circuits can be obtained from the original article. The stability of this current source is impressive. After initial calibration and zeroing, the measured current changed from 500.01 mA to 500.16 mA over a time span of 1½ years. The percent change for the same time period for a current of about 12 mA was +0.06. No recalibrations or zero adjustments were made. The computer used was a Digital Electronics Corporation PDP-8/e.

Earlier, the work of Zetlmeisl and Laurence [22] was described. With their instrument, the current decayed exponentially during the titration. Control of the current via an algorithm can be done with the computer. An example of this is found in the work of Earle and Fletcher [64]. Their titrator was based on the Intel 8008, an early 8-bit microprocessor. For acid–base titrations, the applied current was reduced linearly with the difference in pH, ΔpH, between the measured value and the endpoint pH. An algorithm compared ΔpH with a set of rate functions specified in units of mA/pH. The magnitude of the current was then computed by multiplying ΔpH by the rate function. This process was repeated every 65.536 ms. The coulombs passed were computed for each of these time intervals and summed until the endpoint was reached. The result was then used by the microprocessor to calculate the concentration of the unknown. The maximum initial current was 10.24 mA and the minimum was 0.1 mA or 10% of the initial current, whichever was larger. A delay time between 0.1 and 100 s could be selected as a wait period to ensure that equilibrium had been established at the endpoint.

The titrator was designed for process stream analyses, and a unique ceramic slider valve was developed for sampling the stream. A similar valve was used to deliver supporting electrolyte containing the titrant precursor to the cell. The sliders were solenoid driven. The solution in the working compartment of the cell was stirred by a magnetically coupled impeller driven by a motor. All of these peripherals were under computer control. A digital readout was used to display data and readout from the titrator, and keyboard entry of operating parameters was provided. Among the acids and bases titrated were HCl, NaOH, trishydroxymethylaminomethane (THAM), and potassium hydrogen phthlate (KHP), all in a supporting electrolyte of 0.2 *M* $Na_2SO_4$. Solutions of the latter two were 0.01 M and the former two were 0.1, 0.01, and 0.001 *M*. The results of several hundred titrations of each of these solutions were precise and accurate to ±0.2 relative. The titrator was also capable of consecutive endpoint determinations and an astonishing number of alkalinity determinations, (about 300,000 titrations) of tap water was made. All results obtained agreed with a standard method to within ±0.5%.

A weakness of coulometric titrations is poor selectivity. This could be remedied by using a separation process prior to the coulometric finish. Hahn and co-workers have made interesting progress in this regard using several approaches to the separation, and computer control of the experiment. The separation step must be quantitative to retain the absolute nature of the coulometric titration. In one approach, ion exchange columns [65] were coupled to a coulometric cell similar to that described earlier [39]. Two modes of operation were used; in the first, the column was used to remove possible interfering ions, but the analyte of interest was not exchanged. In the second, the ion of interest was exchanged and eluted from the column with a small volume of solution, which effected a concentration of the ion of interest. Urea was determined in the presence of ammonium ion using a cation exchange resin and titration with electrogenerated hypobromite ion. Successful titrations were achieved with ammonium ion in excess of urea by a factor as large as about 100. Experiments demonstrating the enrichment mode were performed using ammonium ion or nitrate ion and a cation exchanger or anion exchanger, respectively. Both the separation step and the titration step were under computer control.

In another paper, adsorbable organohalogen compounds (AOX) were adsorbed on activated charcoal, which, after treatment to remove inorganic halogens, was pyrolyzed at about 1000°C and the HCl formed was dried online and determined by titration with electrogenerated silver ion [66]. The computer used was the IBM AT type. In another approach to the separation step, membranes were used to separate gases [67,68]. GORETEX, a microporous PTFE membrane, was used and analyses of ammonia and sulfur dioxide were studied.

As a final illustration of the use of computers in coulometric titrations, a method of automatic dilution of samples developed by Ruzicka, Christian, and co-workers is discussed. Flow injection (FI) involves the injection of a solution of analyte into a flowing carrier stream contained in narrow-bore tubing (e.g., 0.5 mm diameter) where controlled dispersion of the injected solution takes place [69]. However, some techniques of FI have incorporated a gradient chamber into the flow line, particularly where stopped-flow methods of FI are involved [70]. The gradient chamber can be viewed as a mixing chamber. With the proper injection and gradient chamber volumes, the flow can be stopped at a particular time (delay time) and the entire sample zone, which now occupies a larger volume than that which was injected, can be trapped in the gradient chamber. At other delay times, a portion of the sample zone can be trapped in the mixer along with a volume of carrier solution. The mixing chamber is an exponential dilutor and as such will capture the same fraction of the sample volume for a given delay time. If the delay time is varied for a given injection volume, various degrees of dilution can be achieved. It is necessary to characterize the dilution factor for the mixing chamber experimentally. With this basis, Ruzicka, Christian, and co-workers have developed flow-injection coulometric titrations [71], and have applied them to the determination of bromine number, which for olefinic compounds is a measure of unsaturation [72]. The mixing chamber is equipped with a stirring bar, electrodes for the coulometry, and light pipes for spectrophotometric endpoint detection. Using commercial interface hardware and an IBM XT or a Toshiba 3200X computer, the entire process and data acquisition were under computer control. Additional details concerning the type of pumps and other information can be obtained from the original articles.

## REFERENCES

1. L. Meites, *Polarographic Techniques*, 2nd ed., Wiley-Interscience, New York, 1965, pp. 550–554.
2. J. J. Lingane, *Electroanalytical Chemistry*, 2nd ed., Interscience, New York, 1958, pp. 511–513.
3. H. V. Malmstadt, C. G. Enke, and S. R. Crouch, *Electronics and Instrumentation for Scientists*, Benjamin-Cummings, Menlo Park, CA, 1981.
4. T. P. Sifferlan and V. Vartanian, *Digital Electronics with Engineering Applications*, Prentice-Hall, Englewood Cliffs, NJ, 1970.
5. R. E. Simpson, *Introductory Electronics for Scientists and Engineers*, 2nd ed., Allyn and Bacon, Inc., Boston, 1987.

6. P. Horowitz and W. Hill, *The Art of Electronics*, 2nd ed., Cambridge University Press, Cambridge, 1989.
7. Ref. 6, pp. 140–143; pp. 904–911.
8. Ref. 5, pp. 246–251; pp. 278–284.
9. L. B. Jaycox and D. J. Curran, *Anal. Chem. 48*:1061 (1978).
10. J. T. Stock, *Microchem. J. 30*:92 (1984).
11. E. W. D. Huffman, Jr., *Microchem. J. 22*:567 (1977).
12. H. H. Ruttinger and U. Spohn, *Anal. Chim. Acta 202*:75 (1987).
13. D. D. DeFord, C. J. Johns, and J. N. Pitts, *Anal. Chem. 27*:938,941 (1951).
14. Ref. 2, pp. 522–528.
15. Ref. 2, Chaps. V–VII.
16. J. N. Butler, *Ionic Equilibria: A Mathematical Approach*, Addison-Wesley, Reading, MA, 1964.
17. H. A. Laitinen and W. E. Harris, *Chemical Analysis*, 2nd ed., McGraw-Hill, New York, 1975.
18. Ref. 3, pp. 74–76.
19. G. E. Cadwgan, Jr, and D. J. Curran, *Microkhim. Acta* 461 (1977).
20. M. A. de Soto Perera and D. J. Curran, *Anal. Chim. Acta 119*:251 (1980).
21. Ref. 20, p. 263.
22. M. J. Zetlmeisl and D. F. Laurence, *Anal. Chem. 49*:1557 (1977).
23. IUPAC, "Status of the Faraday Constant as an Analytical Standard," in *Information Bulletin, Appendices on Provisional Nomenclature, Symbols, Units, and Standards*, No. 35, IUPAC Secretariat, Oxford (1974).
24. T. Yoshimori, *Rev. Anal. Chem. 6*:13 (1982).
25. Ref. 2, p. 487.
26. J. K. Taylor and S. W. Smith, *J. Res. Natl. Bur. Std. 63A*:153 (1959).
27. K. W. Pratt Abstract No. AN4L28, 204th National Meeting, American Chemical Society, Washington, DC, August, 1992.
28. A. J. Bard, *Anal. Chem. 40*:64R (1968).
29. A. J. Bard, *Anal. Chem. 42*:22R (1970).
30. D. G. Davis, *Anal. Chem. 44*:79R (1972).
31. D. G. Davis, *Anal. Chem. 46*:21R (1974).
32. J. T. Stock, *Anal. Chem. 48*:1R (1976).
33. J. T. Stock, *Anal. Chem. 50*:1R (1978).
34. J. T. Stock, *Anal. Chem. 52*:1R (1980).
35. J. T. Stock, *Anal. Chem. 54*:1R (1982).
36. J. T. Stock, *Anal. Chem. 56*:1R (1984).
37. D. M. Smith and J. Mitchell, Jr., *Aquametry*, 2nd ed., Wiley-Interscience, New York, 1984.
38. E. Scholz, *Fresenius J. Anal. Chem. 303*:203 (1980).
39. U. Spohn, M. Hahn, H. H. Ruttinger, and H. Matschiner, *Fresenius J. Anal. Chem. 333*:39 (1989).
40. G. C. Farrington and J. J. Lingane, *Anal. Chim. Acta 60*:175 (1972).
41. J. J. Lingane and R. T. Iwamoto, *Anal. Chim. Acta 13*:465 (1955).
42. J. H. Kennedy and J. J. Lingane, *Anal. Chim. Acta 18*:240 (1958).

43. G. W. C. Milner and G. Phillips, in *Electroanalytical Chemistry*, Vol. 10 (H. W. Nürnberg, ed.), Wiley, New York, 1974, p. 78.
44. W. N. Carson, *Anal. Chem. 25*:466 (1953).
45. Ref. 2, pp. 543–544.
46. N. H. Furman, C. H. Bricker, and R. V. Dilts, *Anal. Chem. 25*:442 (1953).
47. J. Malinowski, *Talanta 14*:263 (1967).
48. L. Pszonicki, *Talanta 13*:403 (1966).
49. G. C. Goode, J. Harrington, and W. T. Jones, *Anal. Chim. Acta 37*:445 (1967).
50. W. Davies and W. J. Gray, *Talanta 11*:1203 (1964).
51. C. G. Goldbeck and M. W. Lerner, *Anal. Chem. 44*:594 (1972).
52. K. Lewis, Report, 1980, *NBL-294*, 14 pp. Available from the National Technical Information Service; *Chem. Abstr. 93*:1250484 (1980).
53. W. G. Mitchell and K. Lewis, *NBS Spec. Publ. (U.S.) 589*:140–146 (1980); *Chem. Abstr. 93*:125139d (1980).
54. R. T. Chitnis, S. G. Talnikar, and A. H. Paranjare, *J. Radioanal. Chem. 59*:15 (1980).
55. R. Mihajlovic, V. Vajgand, Lj. Jaksic, and M. Manetovic, *Anal. Chim. Acta 229*:287 (1990).
56. R. Mihajlovic, V. Vajgand, and Z. Simic, *Anal. Chim. Acta 265*:35 (1992).
57. R. P. Mihajlovic, Lj. V. Mihajlovic, V. J. Vajgand, and Lj. N. Jaksic, *Talanta 36*:1135 (1989).
58. R. P. Mihajlovic, Lj. N. Jaksic, and V. Vajgand, *Talanta 39*:1587 (1992).
59. R. P. Mihajlovic, V. J. Vajgand, and R. M. Dzudovic, *Talanta 38*:673 (1991).
60. I. F. Abdullin and G. K. Budnikov, *Russ. J. Anal. Chem. 47*:735 (1992).
61. A. Cladera, A. Caro, J. M. Estela, and V. Cerda, *Water, Air, Soil Pollut. 59*:321 (1991).
62. M. Houde, J. Champy, and R. Furminieux, *Microchem. J. 24*:300 (1979).
63. J. E. Harrar and C. L. Pomernacki, *Chem. Instrum. (N.Y.) 7*:229 (1976).
64. W. E. Earle and K. S. Fletcher III, *Chem. Instrum. (N.Y.) 7*:101 (1976).
65. M. Hahn, C. Voigt, H. H. Ruttinger, and H. Matschiner, *Fresenius J. Anal. Chem. 344*:311 (1992).
66. M. Hahn, H. H. Ruttinger, H. Matschiner, and N. Lenk, *Fresenius J. Anal. Chem. 340*:22 (1991).
67. U. Spohn, M. Hahn, H. Matschiner, G. Ehlers, and H. Berge, *Fresenius J. Anal. Chem. 332*:849 (1989).
68. M. Hahn, H. H. Ruttinger, and H. Matschiner, *Fresenius J. Anal. Chem. 343*:269 (1992).
69. J. Ruzicka and E. H. Hansen, *Flow Injection Analysis*, 2nd ed., Wiley, New York, 1988.
70. G. D. Christian and J. Ruzicka, *Anal. Chim. Acta 261*:11 (1992).
71. R. H. Taylor, J. Ruzicka, and G. D. Christian, *Talanta 39*:285 (1992).
72. R. H. Taylor, C. Winbo, G. D. Christian, and J. Ruzicka, *Talanta 39*:789 (1992).

# 26

# Electrochemistry in Pharmaceutical Analysis

**Marvin A. Brooks and Eric W. Tsai** *Merck Research Laboratories, West Point, Pennsylvania*

## I. INTRODUCTION

Analytical chemistry plays a critical role in the development of a compound from its synthesis to its marketing as part of a drug formulation. The instrumental methods most commonly used for quantitation in a pharmaceutical laboratory fall into four basic categories: chromatographic, spectrophotometric, electrochemical, and radiometric analysis. Among these four, electrochemical analysis is the least often employed. This is historically attributable to a lack of trained personnel and dependable commercial instrumentation. The relative neglect of electroanalytical chemistry is indeed unfortunate since many problems of pharmaceutical interest can easily be solved with a high degree of accuracy and precision employing this approach. It is the purpose of this chapter to examine the various problems presented to the analytical pharmaceutical chemist and to explore the possible role of electroanalytical chemistry in solving these problems.

### A. Development of a Drug

Analytical methods are required to assure the purity of starting materials and can be used to monitor a synthetic process to determine if it has reached completion. After synthesis, analysis is required to determine percent yield and to separate, isolate, and quantitate by-products of the reaction.

Analytical research receives the drug after synthesis and characterizes its physicochemical parameters to ensure that all pharmacological testing is done with pure material and to establish specifications so that future production lots will yield reproducible in vivo response. These specifications will include the purity of drug, identification and quantitation of other products of synthesis and degradation, the p$K_a$ of the drug, and physical characterizations such as thermal stability, optical isomers, surface properties, particle size, and crystalline form. These studies will include the stability and solubility of the drug in aqueous media as a function of pH.

Upon completion of the analytical profile, the drug is released for preliminary pharmacological and toxicological testing in laboratory animals. At this stage of development, the drug is usually administered in a simple preliminary formulation by parenteral and oral routes, to determine its pharmacological activity, its relative safety, and dose-response characteristics. During toxicological testing, blood, urine, and tissue samples are collected and analyzed to correlate both pharmacological response and toxicity to the blood concentration. These data aid in the assessment of adequate absorption, dose-response behavior, metabolic effects such as accumulation, enzyme induction, or attainment of steady-state conditions, intra- and interspecies variations, and the relationship of drug concentration to symptomatology [1].

In addition to toxicological studies, various in vitro studies are performed to determine if any physicochemical factors such as solubility, dissolution or release rate, permeability, gastrointestinal degradation, and rapid biotransformation will affect the drug's oral absorption characteristics after administration of a given dose in a pharmaceutical dosage form [2]. Assays must be devised to assess the solubility and dissolution rate of drugs as a function of pH within the physiological pH 1–8 range at 37°C, and to evaluate the drugs' permeability characteristics across the intestinal mucosa. The utility of these in vitro studies has been reported by Kaplan [3]. Metabolic transformation is evaluated in vitro by incubating the drug in a liver microsomal enzyme fraction (9000 × g supernatant) at 37°C, which was separated from the liver homogenates [4]. Analytical methodology is needed for these studies to quantitate the rate of biotransformation and to isolate and quantitate the metabolites formed.

The design of a suitable dosage form at a clinically effective and safe dosage is based on the data obtained from the pharmacological and toxicological studies together with biopharmaceutical data gained from the in vitro studies. Such dosage forms are evaluated for bioavailability following oral administration to laboratory animals at a relatively high dose (typically 10 mg/kg body weight in the dog). The bioavailability, rate, and extent of absorption of this formulation are assessed by comparing its blood, urine, and fecal elimination profile to that obtained from an intravenous and oral solution of the drug at the same dosing level [3]. The dosage form exhibiting optimal bioavailability is used for dosing animals in toxicity studies in several animal species, in which blood, urine, and tissue concentrations of drug and/or metabolites are monitored. If the drug is shown to be safe, it is then ready for limited testing in humans. An Investigational New Drug (IND) application is filed with the Food and Drug Administration (FDA); it describes the drug formulation (list of components, quantitative composition), the analytical procedures employed to establish and maintain standards of identity, strength, and purity, and all the preclinical information pertaining to the administration of the drug to laboratory animals.

With approval from the FDA, the drug proceeds into phase 1 clinical testing, where it is administered to humans to determine its safe dosage range, pharmacological activity, side effects, metabolism, and distribution, absorption, and elimination characteristics. From these studies, samples of blood, urine, and feces are collected and analyzed to assess the preliminary bioavailability and pharmacokinetic profile in humans. (Pharmacokinetics is defined as the study of the kinetics of absorption, distribution, metabolism, and excretion of drugs and other endogenous substances.) During phase 2 and phase 3 testing of a drug for safety and effectiveness in diagnosis, treatment, or prophylaxis of a given disease or condition, biological fluids are collected to obtain comprehensive pharmacokinetic information and to correlate such data with activity of the drug. At the completion of the clinical trials, all the data are accumulated about the drug from its initial synthesis to its clinical testing and submitted to the FDA as a New Drug Application (NDA) for approval of the drug as a marketable product.

Subsequent to marketing, constant quality control monitoring of the product from lot to lot must be maintained to provide a reproducible product. To guarantee this reproducibility in the dosage form, it is necessary to develop assays that will analyze reagents, bulk chemicals, and intermediates that go into making the dosage form, and also to obtain and analyze representative samples of the dosage form.

In summary, it is apparent that many analytical techniques at varying levels of sensitivity are required to solve analytical-pharmaceutical problems. Analysis of reagents, bulk chemicals, intermediates, and the final dosage form requires rapid, uncomplicated assays of relatively low sensitivity that (due to the large number of samples involved) can be readily automated. These assays use titrations, spectrophotometric techniques, or liquid chromatography (LC). Analysis of by-products from synthesis and degradation products, the assay of drugs from in vitro testing, and the assay of drugs in biological fluids from high-dose ($\geq$10 mg/kg body weight) animal studies require assays of intermediate detection limit (typically about 1.0 $\mu$g/mL of sample) that are fairly specific to be able to distinguish possible metabolites or breakdown products from the parent drug. Procedures that satisfy these requirements include gas chromatography employing a flame ionization detector, LC employing an ultraviolet detector, and thin-layer scanning densitometry. Thin-layer chromatography (TLC) followed by elution of the compounds from the plate and spectrophotometric, spectrofluorometric, or voltammetric quantitation is often used.

The most demanding of the pharmaceutical analytical assays are those that require the measurement of drugs and their metabolites in biological fluids in humans following the administration of low doses of drugs. These assays must be of high specificity and capable of determining drugs and their metabolites with subnanogram detection limit, and the urinary excretion of drugs with a detection limit of 50–100 ng/mL urine. Gas-liquid chromatography (GLC) with electron capture, nitrogen flame ionization or mass spectrometric detection, LC with ultraviolet or fluorescence detection, and radioimmunoassay are the most useful procedures in measuring these blood levels. Voltammetry, coupled with TLC or LC to ensure specificity, has been shown to be extremely useful in measuring blood levels and urinary excretion of drugs following low doses. This is a particularly valuable technique for the determination of polar metabolites that cannot be readily derivatized for gas chromatography.

## B. Electroanalytical Methods

The choice of the appropriate technique to solve an analytical–pharmaceutical problem is often controlled by the sample matrix, and the amount of preparation that is required before the analytical measurement can be made. If the material is an organic bulk substance or intermediate, it is usually possible to dissolve the compound in a suitable solvent and analyze the sample by a spectrophotometric, chromatographic, or electrochemical technique. A dosage form may be dissolved in an aqueous medium, filtered, and analyzed directly by an electrochemical or LC technique or extracted into an organic solvent and analyzed chromatographically or spectrophotometrically. A biological specimen cannot usually be analyzed directly (an exception is discussed in the voltammetric techniques section) but rather requires selective extraction and cleanup, separation of the parent drug from any metabolites and/or other substances present, and a sensitive and specific detection system for accurate quantitation. A detailed discussion of each step required for the analysis of drugs in biological fluids has been reported by de Silva [5].

The final selection of the analytical technique is dependent on the intrinsic physicochemical properties of the pure compound or its easily formed derivatives. Thus compounds or their derivatives should be processed through an analytical screening program to examine their spectrophotometric, chromatographic, and electrochemical properties that might be amenable to chemical analysis. Based on the signal response per nanogram or microgram of a compound, a projected limit of sensitivity per milliliter or gram of sample is calculated. Many analytical procedures can be eliminated at this stage of method development, due to their inability to provide adequate sensitivity.

The use of spectrophotometric and chromatographic techniques has been extensive throughout the pharmaceutical literature, with little or no emphasis on electrochemical techniques. This discussion now turns to the potential use of electroanalytical chemistry to solve analytical–pharmaceutical problems.

In the analytical screen described earlier, the four electrochemical techniques—voltammetry (i.e., direct-current, alternating-current, fast-scan, and pulse polarography along with fast-scan, pulse, square-wave, and stripping voltammetry at solid electrodes), coulometry (constant-current and -potential), potentiometric titrations (including the use of ion-specific electrodes), and amperometric titrations—are most commonly employed. Each of these electroanalytical techniques and its practical range of usefulness is presented in Figure 26.1. Voltammetric methods can be employed in a wide concentration range and can be used for the analysis of drugs in dosage forms at low sensitivities, and for the analysis of drugs in biological fluids at highest sensitivities. Constant-current coulometry possesses intermediate sensitivity and can determine drugs in biological fluids following only relatively high dosages. The techniques of constant-potential coulometry, amperometry, and potentiometry are restricted by their limits of sensitivity to the measurement of drugs in dosage forms and bulk materials.

In the analysis of drugs in dosage forms, electroanalytical chemistry has been shown to be an exceptional method, and quite often superior to classical wet methods and spectrophotometric methods. For the most part this is due to the great ease of sample preparation and lack of interferences from incipients in the dosage form. Dosage forms are typically pulverized (if necessary), dis-

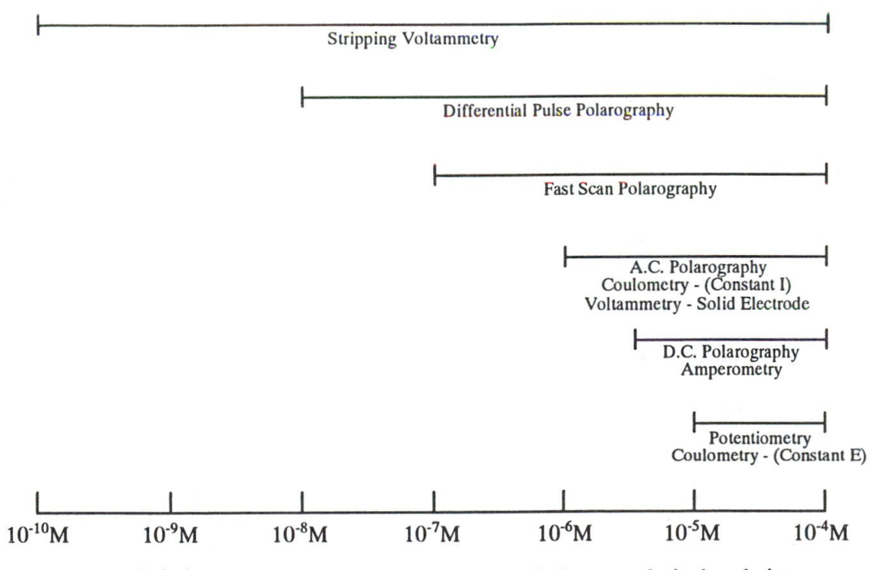


**Figure 26.1** Range of practical usefulness of electroanalytical techniques.

solved in an aqueous solvent, filtered, and analyzed directly by one of the four electroanalytical methods. Coulometric methods have been shown to be particularly useful since they are "absolute" methods requiring no chemical standards. Fast-scan and differential pulse polarography and, more recently, stripping voltammetry have become extremely useful for the measurement of drugs in biological fluids and tissues. This can be directly attributed to the availability of highly sensitive commercial instrumentation that yields easily interpretable data for routing quantitation at low levels. The establishment of absolute specificity is the major limitation of these two highly sensitive polarographic techniques. In an in vivo study, the assay must be able to measure the compound administered without any interferences from the biological matrix or any metabolic products. To ensure specificity, TLC separation is often coupled with voltammetric measurements. These TLC separation steps often introduce time-consuming and tedious spotting, developing, and elution procedures into the electrochemical assays. To avoid some of the problems associated with TLC, a liquid chromatograph can be coupled directly to an electrochemical detector. This approach has been demonstrated to be useful for a wide number of pharmaceutical problems (see Chap. 27).

As with any analytical procedure, the reproducibility and accuracy of an electroanalytical assay must be verified for the intact compound and/or a chemical derivative used for quantitation. Methods developed should be uncomplicated for use on a routine basis by the general scientific community.

This discussion centers on coulometric and voltammetric electroanalytical procedures to demonstrate their utility in solving analytically oriented pharmaceutical problems in the various stages of drug development. The examples chosen are by no means a comprehensive review for each technique. Since discussions of the pharmacological activity are not presented, several texts are recommended [6,7]. The purpose of this chapter is to suggest the types of problems that can be solved using electrochemical techniques. However, it should be noted that although most problems in pharmaceutical analysis are not amenable to an electrochemical approach, there are a large number of examples for which electrochemistry should be the method of choice.

## II. COULOMETRIC METHODS

### A. Controlled-Potential Coulometry

Controlled-potential coulometry (Chap. 3, Sec. IV.F) has been used in pharmaceutical analysis for determination of pure compounds and compounds in their dosage forms, and in studies to determine the nature of electrode reactions. Controlled-potential coulometry as an analysis tool has exhibited its greatest utility in the measurement of inorganic ions and nitro and halogenated compounds [8–11]. The applications to pharmaceutical problems are few, which attests to both the unfamiliarity of the technique and its lack of specificity.

The amine oxides of atropine, brucine, 3-bromopyridine, scopolamine, strychnine, and chlordiazepoxide have been assayed at a level of $8 \times 10^{-5}$ *M* in solvents of varying pH by this technique using a mercury pool cathode [12]. Solvents utilized were sulfuric acid–potassium sulfate systems of varying pH and an acetic acid–sodium acetate system of pH 4.63. Recoveries ranged from 98 to 100% for the six compounds investigated at $5 \times 10^{-5}$ M. Smaller samples could not be used because of weighing errors, and larger samples led to unacceptably long coulometric reduction periods. Pyridine *N*-oxide, 2,6-lutidine *N*-oxide, and quinoline *N*-oxide gave meaningless recoveries since their amine oxide reduction overlapped the catalytic hydrogen wave.

An assay for the determination of vitamin $K_3$ (2-methyl-1,4-naphthoquinone) by a combination of both constant-potential and constant-current coulometry has been reported [13]. The assay requires the two-electron reduction of the compound to the corresponding hydroquinone at a mercury pool electrode (E = -0.60 V vs. SCE) in acetate buffer, pH 5.9, followed by the coulometric titration of the reduction product with electrogenerated Ce(VI). This method is preferable to the standard method requiring preliminary reduction to the hydroquinone by zinc dust in acid medium, followed by titration with standard Ce(IV) solution. It is capable of low-level determination (1–2 mg) of this vitamin in pharmaceuticals, biological fluids, and foods.

The *N*-substituted phenothiazines, which are employed as antipsychotic drugs, have been measured by controlled-potential coulometric oxidative analysis in sulfuric acid at the rotating platinum wire mesh electrode [14]. The compounds could be quantitatively oxidized to a cation radical or to a sulfoxide by selection of suitable acid concentrations and applied potentials. Electroreduction of the free radicals occurs at approximately +0.25 V vs. SCE on a platinum electrode. In the case of the sulfoxides, a single two-electron reduction step occurred at approximately -0.95 V vs. SCE on a mercury pool cathode. The determinations showed good reproducibility, and an accuracy of about ±0.1% was obtained for sample concentrations of $10^{-3}$ *M* or greater. The compounds examined were chlorpromazine, chlorpromazine sulfoxide, promethazine, promazine, triflupromazine, trifluoperazine, prochlorperazine, and thioridazine (see Fig. 26.2). The phenothiazines have also been examined by chronopotentiometry (Chap. 4, Sec. I.B) at a flow-through tubular graphite electrode in 0.1 $N$ $H_2SO_4$ electrolyte. Using this methodology, drugs were quantitated in tablet formulations in the range of 2.5–100 mg/tablet based upon the relationship between transition time ($\tau$) and concentration [15].

The oxidation of ascorbic acid at a platinum electrode in potassium acid phthalate buffer (pH 6) has been reported at +1.09 V vs. SCE [16]. Under these

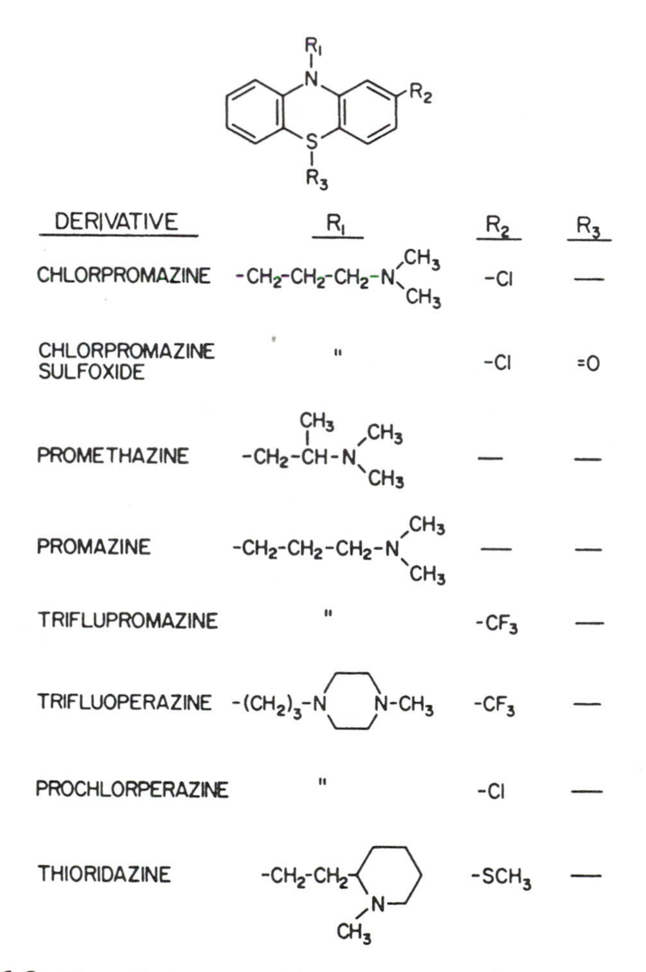


**Figure 26.2** Phenothiazines assayed by controlled-potential coulometry. [From Ref. 14, reprinted with permission.]

conditions, 15–100 mg of ascorbic acid can be determined with an error of ±0.7 mg with no interference from citric, succinic, and tartaric acids.

Controlled-potential coulometry is quite often used as an ancillary technique to polarography to calculate the number of electrons involved in a polarographic reduction. The determination of an n value for some antibacterial aminoacridines using this technique demonstrated that with the exception of 5-aminoacridine, all other substituted aminoacridines undergo a one-electron reduction that corresponds to the first step of the current-voltage curve [17]. This fact substantiated the formation of an intermediate stable semiquinone radical, which is of great importance in the relation to the free-radical mechanism of bacteriostasis by the aminoacridines. Analysis was performed in 0.5 *M* potassium sulfate at a mercury pool cathode using 1–5 mg of sample.

The coulometric method was also applied to the determination of the number of electrons involved in the two-step polarographic reduction of chlordiazepoxide [18]. The reduction of a $2.5 \times 10^{-5}$ *M* solution of the compound in 0.1 N $H_2SO_4$ yielded n values of 1.97 and 4.06, respectively, at a mercury pool cathode employing working electrode potentials of −0.45 V and −0.90 V vs. Ag/AgCl wire reference electrode.

These experiments clearly demonstrated that two electrons are involved in each of the two polarographic steps, which correspond to the reduction of the amine oxide and the 4,5-azomethine functional group.

The mechanism of reduction of an *N*-oxide, the antibiotic myxin, 6-methoxy-1-phenazinol 5,10-dioxide, has also been examined [19]. Below pH 3, the protonated species of myxin is reduced in a single four-electron step to 6-methoxy-1-phenazinol, whereas between pH 3 and 9, two two-electron steps are observed that are most likely due to the formation of an intermediate, which is in turn reduced to 6-methoxy-1-phenazinol. At pH values greater than 9, a single wave is observed, apparently due to the four-electron reduction of the anion to 6-methoxyphenazinol. The postulated mechanism is shown in Figure 26.3.

## B. Constant-Current Coulometry (Coulometric Titration)

Coulometric titrations can be conveniently carried out in the concentration range 1 μg–100 mg/mL for practically any titration that can be done volumetrically. Since current and time can be determined easily and accurately, the titrations are generally very accurate and precise (Chap. 25). In the microgram range, coulometric titrations show clear advantages over volumetric techniques, in that they eliminate problems of reagent stability and standardization. However, at concentrations greater than 100 mg/mL coulometric titrations have problems, such as high currents needed for completing the titration in a reasonable amount of time and the extremely high molar strength of generating electrolytes needed to sustain 100% current efficiency. In such cases, dilution of the sample is required.

An advantage of the technique is the use of an electrical standard to replace chemical standards and the problems associated with their preparation and stability. The coulometric titration also permits the generation of reagents such as copper(I) or bromine, which are difficult to employ as standard solution, or others such as silver(II) or chlorine, which are virtually impossible to use in any other way. A disadvantage of the coulometric titration is its lack of specificity.

Coulometric titrations have been employed for redox, acid–base, precipitation, and complexation titrations of organics and inorganics in both aqueous

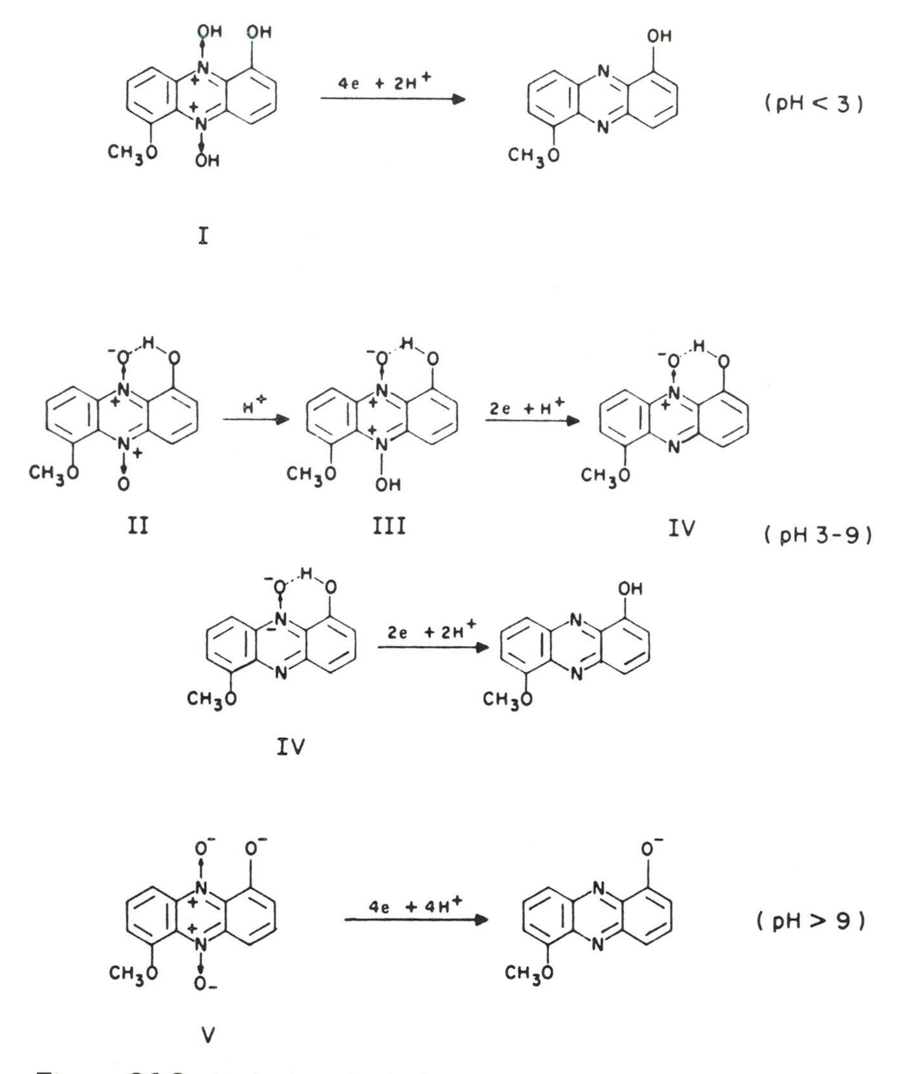


**Figure 26.3** Mechanism of reduction of the antibiotic myxin. [From Ref. 19.]

and nonaqueous solvents [8–11]. The basic theory [20] and applications [21] to the analysis of pharmaceuticals have been described. The present discussion is divided into compounds that are determined by generation of a halogen and those determined by generation of other reagents. Due to the sensitivity restrictions of the technique, very few examples have been reported for the use of coulometric titrations of drugs in biological materials. Most research has been directed toward use of the titration technique to determine bulk materials and drugs in dosage forms.

*Generation of Halogens*

The halogens are the most widely employed oxidation–reduction titrants for coulometric titrations. Bromine can be easily generated with 100% current efficiency from a solution consisting of dilute acid and 0.2 *M* sodium bromide. Bromine is generated at the platinum anode, while hydrogen is generated at a platinum cathode that is isolated in a fritted glass tube. Chlorine, a much stronger oxidizing agent than bromine, can be generated in solutions of acids and chloride salts; however, strict control of reagent concentration is required to guarantee a current efficiency of 100%. Iodine, the weakest oxidizing agent, can be generated from iodide solutions ranging from strongly acidic to approximately pH 8 with 100% current efficiency.

The titration of pharmaceuticals employing electrogenerated halogens has been extensively reported by Kalinowska and Kalinowski and their associates. Their work has been reviewed by Kalinowska [22], in which the compounds analyzed were classified according to the electrogenerated halogen used. Isoniazid, cyclobarbital, hexobarbital, folic acid, and the salts of *p*-aminosalicylic acid were determined with electrogenerated chlorine. Reserpine was reported to consume 4 mol of electrogenerated chlorine and was determined in the range 100–300 μg in dosage forms. Other compounds determined in pure form or in dosage forms in the range 0.2–1 mg by electrogeneration of chlorine included caffeine, theobromine, theophylline, carbutamide, benzylpenicillin, phenylbutazone, and methionine. The coulometric titrations of bulk anabolic steroids [23], levodopa, methyldopa and carbidopa [24], sulfothiourea [25], primaquine [26] and butemtanide and furosemide [27], sulfa drugs [28,29], and 6-mercaptopurine [30] in dosage forms using electrogeneration of chlorine and a methyl orange or methyl red indicator endpoint have been described. The indicator serves as an excellent endpoint, since its rate of reaction with the electrogenerated chlorine is significantly slower than the analytes. Procaine, pontocaine, thiopental, isoniazid, ethylmorphine, codeine, quercetin, *p*-aminobenzoic acid, oxytetracycline, chlorotetracycline, sulfonamides, sulfanilic acid, and phenylbutazone have been titrated with electrogenerated $Br_2$ from an acidic solution of a bromide salt. The electrogeneration of the excess hypobromate ion in alkaline solution of bromide salts was reported for the back-titration of methylthiouracil with $As_2O_3$ and $KMnO_4$ in pure form and tablets. Bromine titration was also used for the titration of benzylpenicillin after alkaline hydrolysis. Electrogenerated iodine was employed for the coulometric titration of thiamine in capsules and suspensions (3–9 mg), sodium noraminopyrine methanesulfonate, ascorbic acid, and benzylpenicillin after alkaline hydrolysis and phenoxymethylpenicillin. Most of the work of these authors employs four platinum electrodes:

two for generation and two for a biamperometric endpoint determination. Kalinowski has also reported assays for the indirect coulometric titrations of papaverine [31] and strychnine [32]. These methods [31,32] require the precipitation of the compound to be assayed with excess Reinecke salt (ammonium tetrathiocyanodiammonochromate) followed by titration of the excess salt with electrogenerated chlorine. Coulometric titrations of paracetamol [33], pyrazolone derivatives [34], and propyphenazone [35] in pharmaceutical formulations with electrogenerated bromine, utilizing methyl orange indication, have been reported.

Procarbazine [*N*-isopropyl-α-(2-methylhydrazino)-*p*-toluamide] used in the treatment of Hodgkin's disease has been determined with electrogenerated bromine [36]. The assay is not specific in the presence of most intermediates and decomposition products, since the electrogenerated bromine not only oxidizes the hydrazo group but also substitutes onto the phenyl ring. A report described the coulometric iodination of the same compound. This approach is more specific in that only oxidation of the hydrazo moiety to the azo group occurred with no ring substitution [37]. Iodine was generated from 0.5 *M* KI in a buffered sodium bicarbonate solution (pH 8.4). Determination of procarbazine hydrochloride in dosage forms was evaluated. The average for eight determinations was 51.31 mg per capsule (102.6% of claim, based on 50 mg). The standard deviation was ±0.17 mg per capsule or ±0.3%.

An automated constant-current coulometric system employing electrogenerated iodine for assay of ascorbic acid or sodium ascorbate has been reported [38]. Using this apparatus, 25 samples of 30 mg could be determined in 2.5 h with an accuracy and precision of ±0.3%. The automated system demonstrates accuracy and precision that are equivalent to or exceed the manual USP method, is significantly more rapid, and eliminates the need for preparation, standardization, and storage of titrant.

Considerable attention has been given to the coulometric determination of sulfa drugs with electrogenerated bromine. Sixteen different sulfa drugs and two anesthetics, benzocaine and tetracaine, have been assayed directly with electrogenerated bromine [39]. Other workers have found that these compounds react too slowly with bromine and describe methods where the drug is first reacted with excess bromine, and then the excess bromine is determined coulometrically with Cu(I) [40], or employing arsenic trioxide in a coulometric back titration [41]. Coulometric back titrations utilizing excess thiouracil with electrogenerated iodine and hydroquinone with electrogenerated bromine have been reported for the analysis of biologically active thiols including cysteine, 2-thiouracil, 6-mercaptopurine, and 6-thioguanine [42].

Seven phenothiazines have been titrated with electrogenerated bromine in the range 0.1–3 mg with an average error of less than 1% [43]. The assay utilizes the oxidation of the phenothiazine in two one-electron steps to a cation-free radical and sulfoxide, respectively.

The analysis of protaglandin $F_{2\alpha}$ [44] and phenothiazine [45] content of pharmaceutical preparations with triangle-programmed bromimetric titration in flowing systems has been described. Using this technique a sample solution stream and chemical reagent stream are mixed and the reaction is monitored biamperometrically.

The uses of constant-current coulometry for the determination of drugs in biological fluids are few, basically due to sensitivity restriction. Monforte and Purdy [46] have reported an assay for two allylic barbituric acid derivatives, sodium seconal and sodium sandoptal, with electrogenerated bromine as the titrant and biamperometry for endpoint detection. Quantitative bromination required an excess of bromine; hence back titration with standard arsenite was performed. The assay required the formation of a protein-free filtrate of serum with tungstic acid, extraction into chloroform, and sample cleanup by back extraction, followed by coulometric titration with electrogenerated bromine. The protein precipitation step resulted in losses of compound due to coprecipitation. The recoveries of sodium seconal and sodium sandoptal carried through the serum assay were approximately 81 and 88%, respectively. Samples in the concentration range 7.5–50 μg/mL serum were analyzed by this procedure.

*Miscellaneous Reagents*

Other typical reagents generated for coulometric titrations are hydrogen and hydroxyl ions, redox reagents such as ceric, cuprous, ferrous, chromate, ferric, manganic, stannous, and titanous ions, precipitation reagents such as silver, mercurous, mercuric, and sulfate ions, and complex-formation reagents such as cyanide ion and EDTA [8–10].

The coulometric titration of weak bases with electrogenerated $H^+$ has been successful for nicotinamide, sodium salicylate, metronidazole, sodium veronal, isonicotinic acid, and other compounds, in 0.1 *M* sodium perchlorate in a mixture of acetic acid and acetic anhydride (1:6) [47]. The endpoint of the titration was noted photometrically using a malachite green indicator. Approximately 5 mg of each compound was determined with an average deviation of ±0.3%.

The most commonly employed oxidation reagent for the determination of organics is the ceric ion. Employing a generating reagent of 3 *M* sulfuric acid and 0.05 *M* cerous sulfate, ceric ion was generated at a platinum electrode for the determination of phenothiazines [43,48]. The same compounds were also determined using electrogenerated manganic ion [43]. Results for the titrations of phenothiazines employing electrogenerated bromine (discussed earlier), ceric ion, and manganic ion demonstrated that quantities of 0.1–3 mg could be determined with an average error of less than ±1%. Vitamin $K_3$ (2-methyl-1,4-naphthoquinone) was also determined by the same authors using electrogenerated ceric ion [13]. The quinone was first coulometrically reduced at a mercury pool electrode or with zinc and then titrated with electrogenerated ceric ion.

Complex formation coulometric titrations of pharmaceuticals have been reported employing electrogenerated mercuric ion and silver ion. Thiols and disulfides, such as methimazole, cysteine, glutathione, and thiopentone sodium, were titrated with mercuric ion generated by oxidation of a mercury pool electrode [49]. A study reports the coulometric determination of barbiturates, including the sodium salts of phenobarbital, barbital, sandoptal, and secobarbital, with $Hg^{2+}$ employing a biamperometric endpoint detection system [50]. The compounds were determined in a range of approximately 0.2–2 $\mu$Eq in a supporting electrolyte of 0.5 *M* $NaClO_4$ or $KNO_3$ in 20% acetone. Structural analysis of the coulometrically formed product of sodium phenobarbital indicated that a polymeric compound containing continuous Hg–N linkages was formed. Titrations with electrogenerated silver and potentiometric endpoint determination have been utilized to assay the alkylated derivatives of procaine, procainamide, lidocaine, butanilicaine, and tetracaine with a silver electrode [51] and halogen-containing drugs (e.g., diphenylhydramine hydrochloride and procaine hydrochloride) with or without combustion with an amalgamated gold electrode [52]. Precipitation determination of organic bases and alkaloids as their tetraphenylborates has been described [53]. Two to ten milligrams of these compounds, including papaverine, atropine, pilocarpine, thiamine, sparteine, yohimbine, procaine, strychnine, tetracaine, eserine, quinine, and brucine, was precipitated with an alkaline sodium tetraphenylborate agent and separated from the excess reagent. The precipitates were dissolved in a solution of 0.4 *M* sodium nitrate in 30% v/v acetic acid. The tetraphenylborate content of the precipitate was determined with $Ag^+$ ion generated from a silver anode.

## III. VOLTAMMETRIC METHODS

Whereas potentiometric and coulometric titration methods are fairly universal and can be applied to a wide variety of problems, voltammetric methods are distinctly limited to easily reducible or oxidizable compounds. On the other hand, voltammetric methods are clearly the most useful for low-level quantitation and have a wide linear dynamic range (see Fig. 26.1). Voltammetry is defined here as electrolysis which is limited by the mass transport rate at which molecules move from the body of the solution to the electrode. The current is followed as a function of the potential applied using three-electrode cells of 1–50 mL working volume. For most reductions, the working electrode is a dropping mercury electrode (DME) or a hanging mercury drop electrode (HMDE). Solid electrode voltammetry, which is typically employed for anodic (oxidation) processes, can use a rotating platinum electrode (RPE), a wax-impregnated graphite electrode (WIGE), a carbon paste electrode (CPE), or a glassy carbon electrode (GCE). The commonly used voltammetric methods are polarography (dc, fast-scan, and differential pulse), linear sweep voltammetry, cyclic voltammetry, hydrodynamic voltammetry, differential pulse voltammetry, square-wave voltammetry, and stripping voltammetry, which involves a preconcentration step followed by a voltammetric measurement.

Since the voltammetric techniques demonstrate a large linear dynamic range ($10^{-3}$–$10^{-8}$ *M*), any one method can readily be applied to the analysis of pharmaceuticals in bulk and in dosage forms. Solid electrode voltammetry (which excludes polarography) has an inherent detection limit of approximately $10^{-4}$–$10^{-5}$ *M* and is consequently limited to these applications. Small-amplitude polarographic techniques (ac and pulse) can be applied to the measurement of impurities in bulk and dosage forms, and to the assay of drugs and metabolites in biological fluids following high-dose administration when concentrations are at least $10^{-6}$ *M* (1 μg/mL sample). Only the techniques of fast-scan polarography, differential pulse polarography, and stripping voltammetry with limits of detection of approximately 100–200, 10–20, and < 10 ng/mL of biological fluids, respectively, are capable of determining drugs in biological fluids following therapeutic administration on a routine basis. Levels as low as nanograms per milliliter of pharmaceuticals have been measured by coupling a liquid chromatograph with a low-volume electrochemical detector (LCEC) (see Chap. 27).

The appeal of the voltammetric methods of analysis for organic compounds is attributable to their simplicity and rapidity. An extremely large number of organic compounds are either directly reducible at the DME or oxidizable at a solid electrode or can be readily derivatized chemically to yield electroactive derivatives. The functional groups that show excellent voltammetric properties include the nitro, nitroso, quinone, azo, azoxy, azomethine, activated carbonyls, and activated double bonds [54]. For the most part, the voltammetric activity of a compound can be deduced from an examination of its functional groups. However, to ensure that electrochemical activity of a compound is not missed, the analytical screen should include voltammetric scans of the compound in aqueous supporting electrolytes (it is convenient to use 0.1 *M* HCl, 0.1 *M* NaOH, and buffers with pH values of 3, 5, 7, 9, and 11) using both the DME and glassy carbon electrode. If the compound is insoluble in aqueous media, then organic cosolvents such as alcohols, dioxane, acetonitrile, dimethyl sulfoxide, and dimethylformamide must be employed. Occasionally, pure nonaqueous solvents of high dielectric constant containing tetraalkyl salts, such as dimethylformamide, acetonitrile, dimethyl sulfoxide, pyridine, and glacial acetic acid, may be required to attain highly negative potentials (greater than −1.8 V vs. SCE) without supporting electrolyte decomposition. Generally, nonaqueous polarography is not successful at the microgram level because of impurities in the supporting electrolytes or solvents that tend to distort or mask the voltammetric peak. The reproducibility of determinations in nonaqueous media is also problematic at these low levels. In the event that no electroactivity is discovered in aqueous or nonaqueous media, derivatization methods can be used to transform the compound into an electroactive moiety. Derivatizations most commonly used include nitration, nitrosation, condensation, addition, substitution, oxidation, hydrolysis, and complex formation.

The number of voltammetric assays far outnumbers all other electroanalytical assays reported for pharmaceuticals [55–66]. Recent reviews have discussed the use of modified carbonaceous materials as voltammetric sensors [67], the use of adsorptive voltammetry for pharmaceutical analysis [68], fast-scan voltammetry and cyclic voltammetry of pharmaceuticals in flow-injection analysis [69], and voltammetric on-line analysis of biological importance [70]. In this section, the applications of voltammetric assays are discussed in reference to the various stages of drug development. Particular emphasis is placed on the determination of drugs in biological fluids, an important area that has recently come of age through the advent of commercial instrumentation capable of performing pulse polarography, stripping voltammetry, and LCEC.

### A. Synthesis and Characterization of Pure Compound

Voltammetry can be employed to monitor a reaction in progress or to measure the percentage yield at completion. An example of this type of study was the use of dc polarography to monitor the formation of *N*-desmethyldiazepam from the condensation reaction of 5-chloro-2-aminobenzophenone with ethyl glycinate [71]. The *N*-desmethyldiazepam was subsequently methylated to yield diazepam, the compound contained in the tranquilizer Valium (see Fig. 26.4). Polarography was performed on an aliquot of the condensation reaction in a McIlvaine buffer (pH 3.8) to determine the amount of *N*-desmethyldiazepam formed by simultaneous measurement of the decrease in wave height for the carbonyl in the benzophenone and the increase in wave height for the 4,5-azomethine in the *N*-desmethyldiazepam. The two analytical waves were separated by 200 mV. A similar procedure was described for monitoring the synthesis of chloramphenicol [72]. A dc polarographic assay was used to measure three nitrofunction intermediates in the presence of each other.

Pulse polarography has been used to characterize the products of the photochemical oxidation of 3,4-dihydro-1-isoquinolineacetamide to 1-isoquinolineacetamide (see Fig. 26.5) [73]. After photochemical oxidation, the polarographic activity that the compound had previously demonstrated in acid and neutral supporting electrolytes was destroyed. This was in agreement with the development of fluorescence activity due to the formation of the unsaturated derivative.

The determination of quinicine, the oxidative decomposition product of quinine, has been reported by polarography [74]. Quinicine was determined in the presence of a 1000-fold excess of quinine in quinine salts and injectibles. The assay was rendered specific for quinicine in the presence of quinine by the

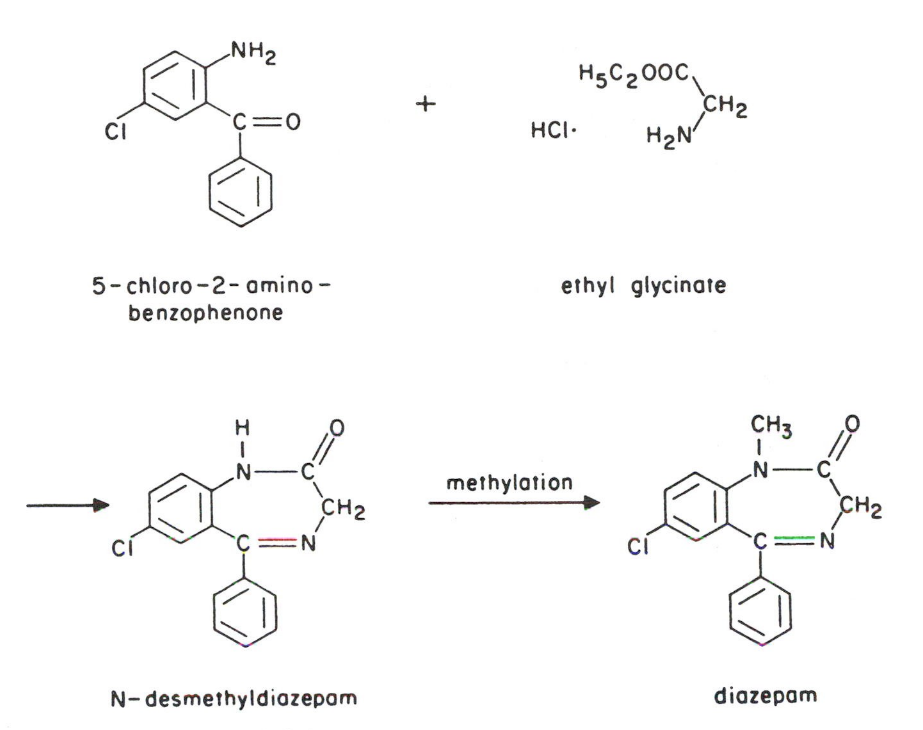


**Figure 26.4** Synthesis of diazepam. [From Ref. 71.]

reduction of the carbonyl group, which is present only in quinicine. Polarography ($E_{1/2}$ = −0.59 V vs. SCE) was performed in 0.2 *M* acetate buffer (pH 4.4) on a solution of 10 μg/mL of quinicine. Various penems, structurally closely related carbapenems antibiotics, and the reaction intermediate 4-acetoxyazeitidin-2-one has been studied by direct-current, sampled dc, differential pulse polarography, and cyclic voltammetry [75]. The relative standard deviation of the electroanalytical method was reported to be 2.5%.

## B. Analytical Research

Voltammetric research on pharmaceutical compounds is quite extensive. Therefore, this section attempts to demonstrate the wide variety of molecules that can be determined by voltammetric techniques with selected examples.

Chloramphenicol, a broad-spectrum antibiotic, has probably received more attention in the polarographic literature than any other pharmaceutical. The aromatic nitro group is quite easily reduced. Studies [76] employing polarography, cyclic voltammetry, and constant-potential coulometry have suggested that

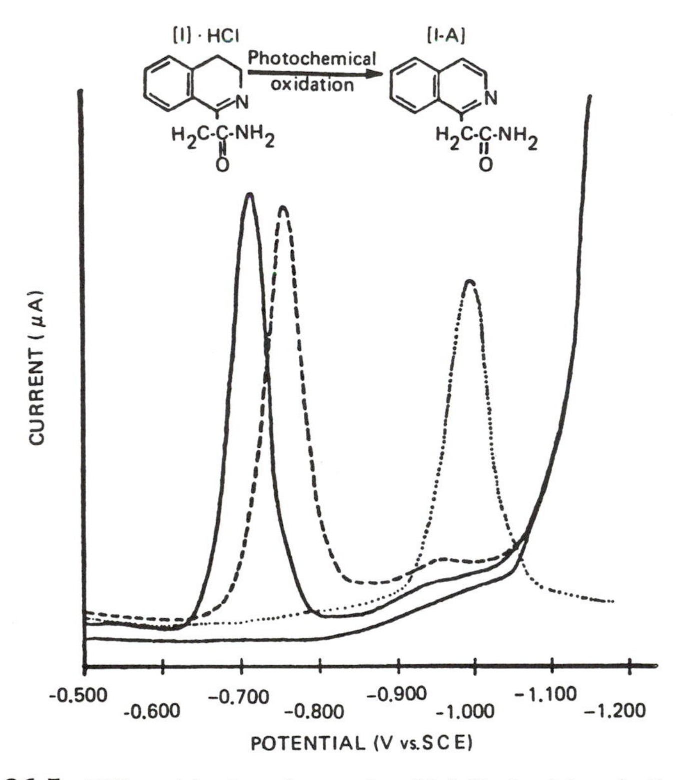


**Figure 26.5** Differential pulse polarography of 3,4-dihydro-1-isoquinolineacetamide [I]·HCl and isoquinoleacetamide. Solid line, 1.0 *N* HCl; dashed line, 0.1 *N* HCl; dotted line, 1.0 *M* KCl/1.0 *M* $PO_4^{3-}$ (pH = 7.0). [From Ref. 73, reproduced with permission.]

the drug undergoes a slow two-electron reduction followed by a fast two-electron reduction of hydroxylamine. At more negative potentials, the hydroxylamine group is further reduced to the arylamine. In pH 4.7 acetate buffer with 0.003% decylamine present, the drug produces a well-defined four-electron polarographic wave, which can be used for quantitation over the range from 0.3 to 600 μg/mL.

The 1,4-benzodiazepine class of pharmaceuticals (see Fig. 26.6), which are in widespread clinical use today, all possess intrinsic electrochemical activity that renders them ideal for voltammetric studies. Senkowski et al. were the first to perform a comprehensive study of this class of compounds by polarography [77] and demonstrated that the 1,4-benzodiazepines show a wave for the two-electron reduction of the 4,5-azomethine "electrophore" in a supporting electrolyte of 0.1 *M* HCl containing 20% methanol. Secondary polarographic waves

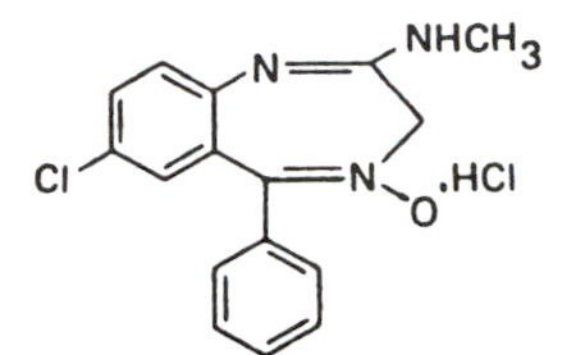

Chlordiazepoxide HCl

Diazepam

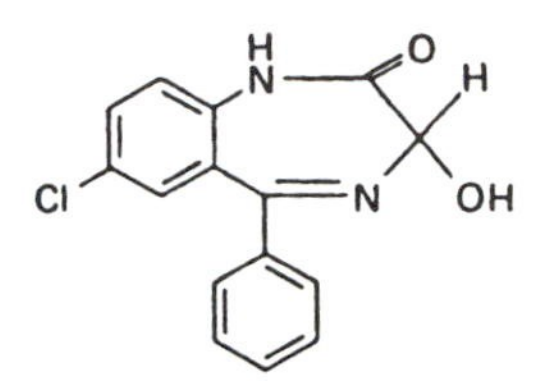

Oxazepam

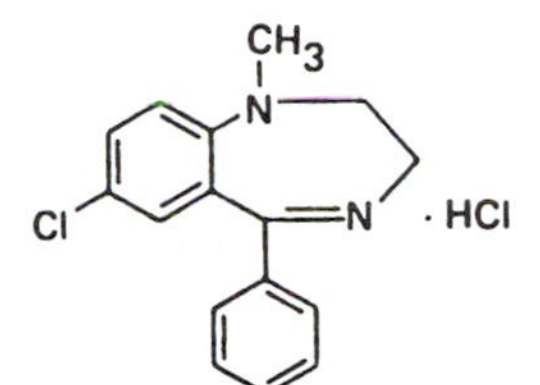

Medazepam HCl

Flurazepam HCl

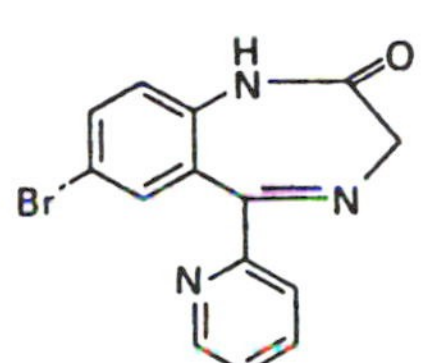

Bromazepam

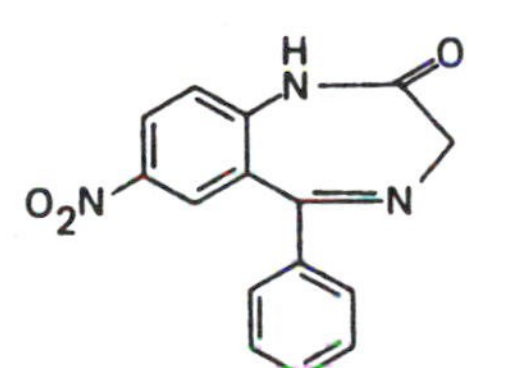

Nitrazepam

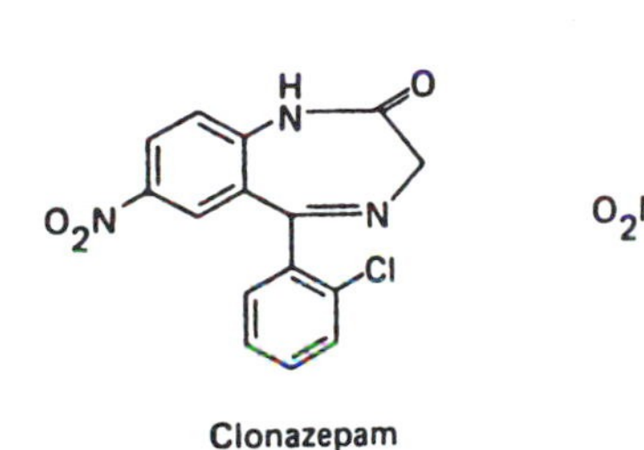

Clonazepam

Flunitrazepam

**Figure 26.6** Structural formulas of some 1,4-benzodiazepines.

were attributed to the reduction of the 1,2-azomethine and $N_4$ oxide in chlordiazepoxide, and to the reduction of the nitro group in nitrazepam. Factors that influenced the azomethine reduction, such as pH or substituents, were also reported. Another study of the polarographic activity of oxazepam demonstrated that the compound in acidic media (pH $< 6$) showed a four-electron step corresponding to the simultaneous reduction of the 4,5-azomethine bond and reductive cleavage of the hydroxyl group in the 3-position of the 1,4-diazepine ring [78]. The acid–base equilibrium and polarographic reduction mechanism of six 1,4-benzodiazepine derivatives [79] and some of their metabolic products [80] have been reported. The general range of analytical utility for the compounds was reported as $10^{-4}$–$10^{-7}$ *M*. Brooks et al. have studied the effect of substituent groups and the ease of reduction of the 4,5-azomethine on 1,4-benzodiazepine derivatives related to chlordiazepoxide, medazepam, and *N*-desmethyldiazepam in pH 3.0 and 7.0 phosphate buffer as supporting electrolytes [81]. A pulse polarogram of the acetyl derivative of chlordiazepoxide is shown in Figure 26.7. A good Hammett correlation was obtained for all three series of compounds. Polarographic data were also obtained for two groups of 1,4-benzodiazepines structurally related to medazepam and *N*-desmethyldiazepam with heterocyclic substituents in the 5-position.

An investigation of the dc and ac polarographic responses of 24 pharmaceuticals in an aprotic solvent system has been described [82]. Barbiturates, salicylates, corticosteroids, and alkaloids were examined in acetonitrile-tetrabutylammonium perchlorate. The majority of the compounds yielded one-electron reversible waves in this solvent; they were suitable for analytical pur-

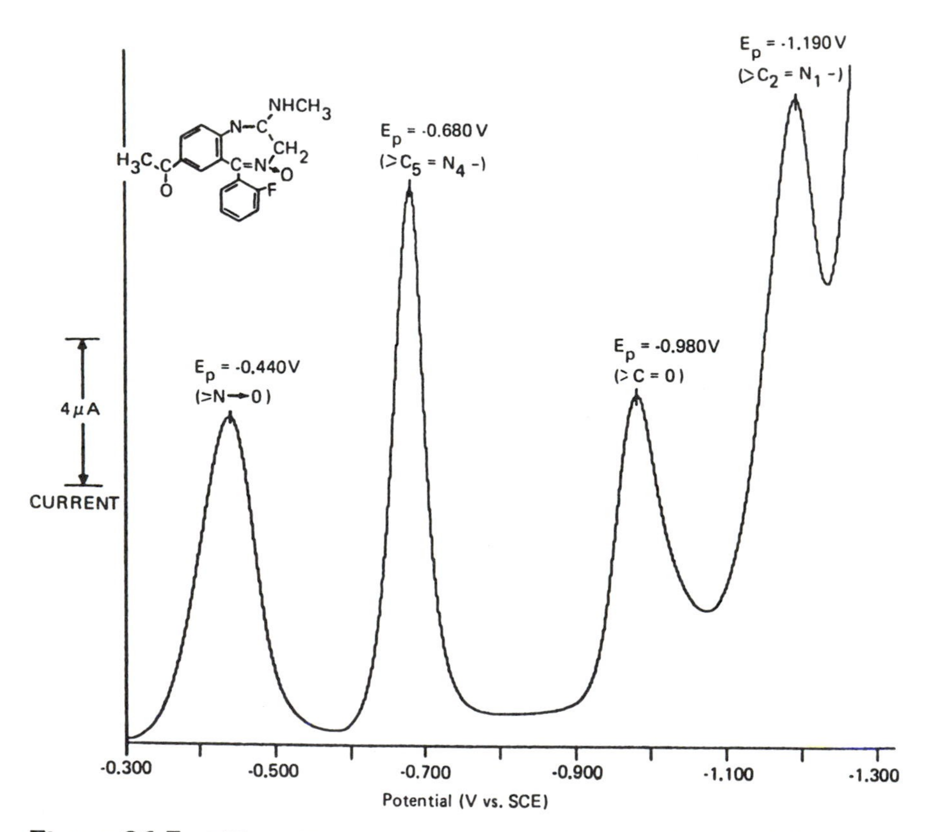


**Figure 26.7** Differential pulse polarography for the 7-acetyl analog of chlordiazepoxide. [From Ref. 81.]

poses, in contrast to the corresponding aqueous solution responses, which were irreversible or nonexistent. Data presented for each compound included the dc half-wave potential, fundamental harmonic ac peak potential, the potential of the second-harmonic ac minimum, and the apparent n value for each polarographic technique. The detection limit of dc and fundamental harmonic ac polarography was typically $5 \times 10^{-5}$ *M*, while that for the second harmonic ac polarography was one order of magnitude lower.

The polarography of steroids has received considerable attention [55]. A comparative study of ac and dc polarography as a method of analysis for several $\Delta^4$-3-ketosteroids, including testosterone, methyltestosterone, and progesterone, in a supporting electrolyte of pH 1.3 buffer and tetrabutylammonium iodide in 50% ethanol was described [83]. Results showed that the lowest practical detection limit for both techniques was $3.3 \times 10^{-5}$ *M*, and that ac polarography was more precise. A combined polarographic/thin-layer chromatographic (TLC) assay has been described for the separation and determination of some corticoids in mixtures [84]. The method has the distinct advantage of employing elution of the TLC plate directly into the polarographic cell with a specially designed suction apparatus (see Fig. 26.8), thus avoiding time-consuming filtration, separation, and evaporation procedures.

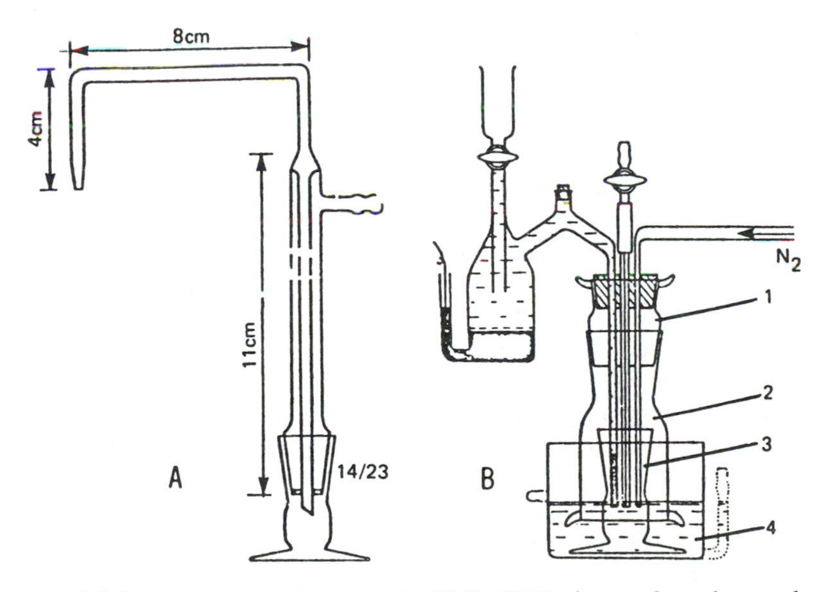


**Figure 26.8** (A) Suction equipment for TLC. (B) Equipment for polarography of small volumes: 1, upper part of ground joint; 2, mantle of the ground joint; 3, electrolysis cell; 4, water container sealing the inner space. [From Ref. 84.]

The dc polarography of a large group of pteridines in pH 2.0 phosphate buffer has been reported [85]. Three groups of pteridines (2,4,7-triaminopteridines, 7-substituted-4-amino-2-aryl-5-pteridine carboxamides, and 4,7-diamino-2-aryl-5-pteridine carboximides) demonstrated polarographic reduction that was found to occur in two steps, to be pH-dependent, and to involve two electrons per step.

Other dc and ac polarographic assays have been reported for warfarin [86], tetramethylthiuram disulfide employing the reduction of the S–S bond [87], and hexachlorophene using reductive cleavage of the halogram atom [88]. Another report dealt with dc polarographic assay of the *p*-fluoro-substituted butyrophenone derivatives triperidol, fluanisone, and haloperidol [89].

The classical dc polarography of vitamins A, $B_1$, $B_2$, $B_6$, $B_{12}$, and C, nicotinamide, tocopherols, and naphthoquinones has been reviewed [55]. Other studies have examined in detail the cyclic voltammetry of vitamin $B_{12}$ employing rapid-scan voltammetry at the DME [90] and the HMDE [91]. Vitamin $B_{12}$ is complexed with trivalent cobalt ion at the heterocyclic nitrogen atoms. As a result of the complexation, a catalytic hydrogen wave is formed for the compound. In addition to the catalytic wave, a wave corresponding to the reduction of the trivalent cobalt to the monovalent state is observed.

Derivatization techniques such as nitration, nitrosation, and others have commonly been employed for the analysis of pharmaceuticals. Benactyzine [92,93], antipyrine [94], 4-hydroxybenzoic acid ester [95], phenacetin [96], phenobarbital [97], diphenylhydantoin [97], meprobamate [92], glutethimide [98], and phenothiazines [99] have been determined as their nitro derivatives. Nitrosation has been employed to convert phenothiazine to an electroactive derivative, which could be determined with a sensitivity limit to 50 ppm [100]. Chlorpromazine has been assayed as its sulfoxide by cathode ray polarography after oxidation with saturated aqueous bromine [101]. Phenelzine was determined at a level of $4 \times 10^{-4}$ *M* as an acetone derivative using a 4% acetone solution containing 0.01% gelatin [102]. Pulse polarographic determination of sulfonamides, including sulfanilamide, sulfathiazole, sulfaguanidine, and sulfacetamide as their diazo derivatives, was described [103]. The compounds are reacted with sodium nitrate in acid solution, sulfamic acid is added, the solution is made basic, and the coupling is completed with 1-naphthol to yield the polarographically active azo dye. The lower limit of detection was about $5 \times 10^{-8}$ for sulfanilamide, sulfathiazine, and sulfacetamide, and about $1.2 \times 10^{-7}$ *M* for sulfaguanidine (see Fig. 26.9). These values are approximately twofold greater than those of spectrophotometric techniques. Epicillin [104] and ampicillin [105] have been assayed by dc or pulse polarography as the 2-hydroxy-3-phenyl-6-methylpyrazine derivative following acid hydrolysis in the presence of formaldehyde. A method has been described for the determination of a new fotemustine antineoplastic using adsorptive stripping voltammetry with differential pulse polarography for

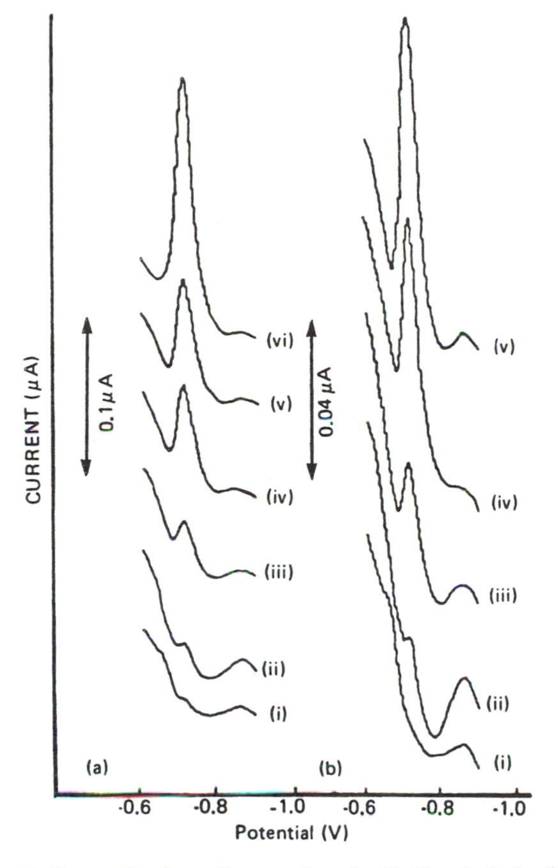


**Figure 26.9** Differential pulse polarography of sulfathiazole derivatives; sensitivities: (a) 0.5 μA, (b) 0.2 μA; concentrations ($\times 10^{-8}$ M): i, 4; ii, 12; iii, 20; iv, 32; v, 40; and vi, 80. [From Ref. 103.]

the voltammetric measurement [106]. The method requires the derivatization of the analyte(s) by means of diazotization and coupling reactions.

The voltammetric examination of the oxidation of phenothiazines in strong acid using various solid electrodes [14,107–110] has been reported by several groups. Fourteen phenothiazine tranquilizers were found to yield regular, reproducible, well-defined anodic voltammetric waves [107] employing the gold electrode. The report describes the effect of pH, concentration, and temperature on the position and size of the waves as well as the effects of substituents in the 10- and 2-positions. The quantitative oxidation of this series of compounds to a free radical or to a sulfoxide at the rotating platinum electrode has been described [14,108]. The results indicated that in 9 *N* sulfuric acid, two successive one-electron oxidations occurred, involving a stable semiquinone free-radical intermediate and a sulfoxide.

The oxidative voltammetric behavior of a series of phenothiazines in a variety of aqueous and nonaqueous supporting electrolytes using a silicon-rubber-based graphite electrode [109,110] and a platinum and gold rotating-disk electrode [111,112] has been described. The same silicon-rubber graphite electrode was also used for determining other pharmaceuticals, such as methyldopa, phenacetin, *p*-aminosalicylic acid, and amidopyrine [109]. The voltammetric determinations of phenylbutazone at the glassy carbon electrode [113] and morphine at platinum and graphite electrodes [114] have also been described.

Recent studies describe the use of cyclic voltammetry in conjunction with controlled-potential coulometry to study the oxidative reaction mechanisms of benzofuran derivatives [115] and bamipine hydrochloride [116]. The use of fast-scan cyclic voltammetry and linear sweep voltammetry to study the reduction kinetic and thermodynamic parameters of cefazolin and cefmetazole has also been described [117]. Determinations of vitamins have been studied with voltammetric techniques, such as differential pulse voltammetry for vitamin $D_3$ with a rotating glassy carbon electrode [118,119], and cyclic voltammetry and square-wave adsorptive stripping voltammetry for vitamin $K_3$ (menadione) [120].

Stripping voltammetry has recently received a great deal of interest due to its high sensitivity and rapid direct assay (see Sec. III.F). Examples using stripping voltammetry for the trace-level determination have included assays for penicillins [121], diltiazem (calcium channel blocker) [122], mitomycin C (quinone-containing antitumor antibiotic) [123], captopril (hypotensive drug) [124], and naftazone (for the treatment of traumatic or spontaneous hemorrhages) [125]. Detection limits for these studies were reported to be $10^{-9}$ to $10^{-11}$ *M*.

## C. Stability Testing

Voltammetric techniques are quite useful for the direct measurement of the stability of some pharmaceuticals in aqueous solution. Depending on the compounds involved, voltammetry can be used to monitor either the decomposition of a drug or the formation of a decomposition product. This type of study is typically performed by applying a potential to the working electrode that is on the voltammetric plateau for the starting material or end product to be monitored. The resulting current versus time data provide the desired kinetic information.

The polarographic reductions of cephaloridine, cephalothin, and cephalosporin C each yield one wave that is both pH and concentration dependent [126]. The reduction of these compounds and a related derivative, 3-(5-methyl-1,3,4-thiadiazol-2-ylthiomethyl)-7-[2-(3-sydnone)acetamido]-3-cephem-4-carboxylic acid-sodium salt [**I**], have been described [127]. The reduction of **I** yields two waves. The first is believed to be the two-electron reductive elimination of the

3-position substituent, and the second is believed to be the six-electron reduction of the sydnone acetamido 7-position substituent, common to all cephalosporin C derivatives. The first wave can be used to follow the acid, base, and β-lactamase degradation of **I**, which involves the loss of the 3-position substituent.

Polarography has also been applied to the determination of ethacrynic acid in the presence of its principal degradation product, a dimer [128]. Ethacrynic acid contains an electroactive, α,β-unsaturated ketone with $E_{1/2} = -1.28$ vs. SCE in pH 8 phosphate buffer, whereas the dimer degradation product does not have this functional group. By monitoring the decrease in wave height of the parent compound, the stability of ethacrynic acid in dosage forms was evaluated.

The hydrolysis of azomethine (imine) bonds is a fairly common mechanism for the degradation of pharmaceuticals. The acid hydrolysis of thiacetazone [129] and *p*-chlorobenzaldoxime (PCBO) [130] can easily be monitored employing polarographic reduction of the azomethine in each of the compounds. The latter compound is extremely interesting in that it was inactive as a muscle relaxant when administered orally, but active when administered parenterally. The data obtained in this study showed that PCBO was hydrolyzed to *p*-chlorobenzaldehyde at the pH of the stomach and thus accounted for the loss of activity following oral dosing of the drug.

Flurazepam and two of its metabolites have demonstrated reversible hydrolysis of the 4,5-azomethine [131]. In acid solution, this reaction may be monitored employing pulse polarography to measure the decrease in concentration of the 4,5-azomethine of the 1,4-benzodiazepine or increase in carbonyl concentration of the "open" benzophenone (see Fig. 26.10).

## D. Assays of Dosage Forms

The voltammetric analysis of drugs in pharmaceutical products is by far the most common use of electrochemistry for analytical–pharmaceutical problems. As a rule, many of the active constituents of formulations (in contrast to excipients) can be readily oxidized or reduced. Thus sample preparation usually consists of dissolving out the active ingredient from the particular formulation with a suitable solvent (nonaqueous, if necessary) and performing direct analysis on an aliquot of this solution. The specificity of the method is usually excellent because the compound can be identified by its voltammetric redox potential. Examples of polarographic determination of organic compounds in pharmaceutical preparations include many classes of drugs: tranquilizers, sedatives, hypnotics, antibiotics, steroids, antihistamines, diuretics, muscle relaxants, anticoagulants, and others.

A rapid polarographic determination of the antihistamine chlorpheniramine maleate in tablets based on the two-electron reduction of the compound at –0.7

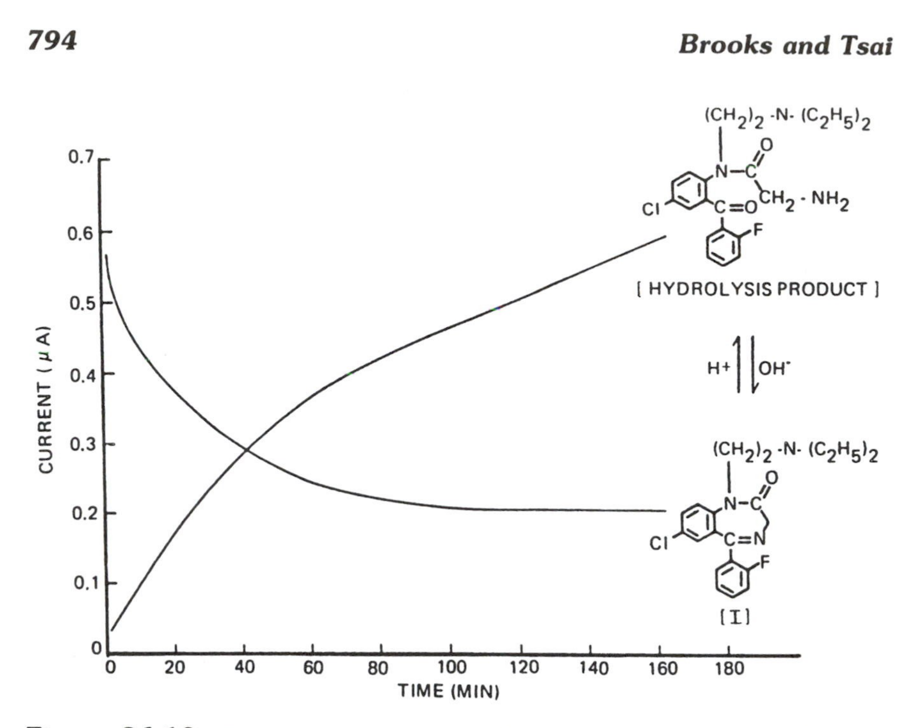


**Figure 26.10** Time-course plot of the hydrolysis of flurazepam in 0.1 *N* $H_2SO_4$ determined by differential pulse polarography. [From Ref. 131.]

V vs. Ag/AgCl in acid has been described [132]. The assay requires the extraction of the compound from a tablet into the 0.5 *M* sulfuric acid, dilution, and direct polarographic assay of an aliquot of this solution. The assay is accurate, sensitive, and, since it does not require the removal of insoluble matter, more rapid than the spectrophotometric procedures.

Disulfiram [bis(diethylthiocarbamoyl)disulfide] used in the treatment of alcoholism can be assayed directly by pulse polarography in an aliquot of a solution of a ground tablet dissolved in ethanol–acetate buffer (pH 4.5) [133]. A mechanism for the electrode process was proposed involving the reaction of the disulfiram with the mercury drop to form an insoluble mercuric salt, which then underwent reduction at the electrode surface.

The development of a simple fast-scan polarographic method for the determination of the $\Delta^4$-3-ketosteroid flurandrenolone in pharmaceutical preparations has been reported [134]. The polarographic peak due to the reduction of the carbon–fluorine bond is measured in ointments and creams to determine concentrations as low as 0.01% w/w. Pulse polarographic procedures have been described for progesterones [135], $\Delta^4$-3-ketosteroids [136], hydrocortisone [137], and flucytosine [138] in pharmaceutical preparations. Recent studies have illustrated the use of fast-scan differential pulse polarography (electroreduction) for the determination of some sulfonamides in tablets [139]. A submicromolar assay concentration was reported with acceptable assay precision. Square-wave polarography has been applied to the determination of 5-chloro-7-iodo-8-quinolinol and its homologues in tablet dosage forms [140]. The method was reported to be as simple, sensitive, and reliable as LC and gas chromatography–mass spectrometry. Cyclic voltammetry (although less sensitive) has also become popular in the assay of dosage form because of its simplicity and reproducibility. Quantitation of lorazepam (tranquilizers) tablets by cyclic voltammetry has been reported with results comparable to LC [141].

The determination of vitamins in pharmaceutical preparations continues to receive considerable attention. The voltammetric oxidation of vitamin A at a carbon paste electrode in the presence of vitamin E, a potential source of error in the assay, has been described [142,143]. Other assays involve the polarographic determination of niacinamide [144–146], menadione (vitamin $K_3$) [147], riboflavin (vitamin $B_2$) [148], thiamine, riboflavin, and nicotinamide in multivitamin preparations [149], and multivitamins [150].

A large volume of literature can be found describing the use of solid electrodes for the study of oxidative voltammetric behavior of drugs in dosage forms. Among the voltammetric methods described, differential pulse voltammetry is the most frequently used technique for quantitation. Quantitative determination of chlorpromazine and thioridazine [151], paracetamol [152], sumatriptan succinate [153], and famotidine [154] in tablets has been demonstrated. A detection limit of $10^{-7}$ *M* and relative standard deviation of 2–3% were achieved. A simple, rapid, and accurate method for the simultaneous determination of ascorbic acid, caffeine, and paracetamol in various formulations such as tablet and cough syrup has been reported with results in good agreement with LC [155]. The method is sensitive and can be applied for routine applications. A simple linear sweep voltammetric method has been developed for the quantitation of heroin in illicit dosage forms [156]. The results obtained compared favorably with spectrophotometric, chromatographic, and radiometric techniques.

Stripping voltammetry has also been applied to the quantitation of the drug in formulations. A sensitive and precise method using square-wave adsorptive stripping voltammetry has been developed for the determination of sulphaquinoxaline in veterinary formulations [157]. The differential pulse adsorptive stripping voltammetric determination of midazolam in injectable formulations as a method for quality control has been demonstrated [158].

The use of a tubular carbon electrode (TCE) for the electrochemical oxidative determination of ascorbic acid [159], L-dopa [160], and methyldopa [161] in dosage forms has been described. The flow system, electrode assembly, and electrochemical instrumentation required for these assays are shown in Figure 26.11. The method is based on continuous analysis in flowing streams

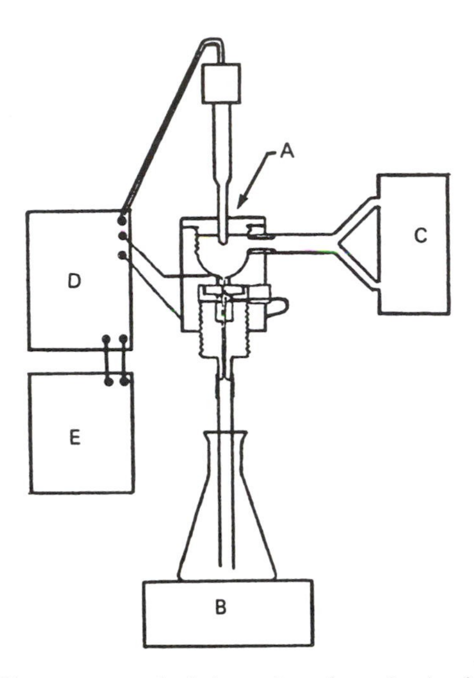


**Figure 26.11** Flow system and tubular carbon electrode: A, electrode assembly; B, magnetic stirrer; C, pump; D, polarograph; E, recorder. [From Ref. 159, reproduced with permission.]

by oxidation of the compound of interest at the TCE. Comparisons to colorimetric methods show this procedure to require fewer manipulative steps and to give comparable precision and accuracy. In the determination of ascorbic acid [159], no interferences were noted by antioxidants due to their lower concentration (usually 1:10,000) and greater oxidation potential. No interferences were noted from other vitamins in multivitamin preparations. Although the process was not automated in these studies, it is suggested that the method may be readily incorporated into automated or semiautomated systems because it employs continuous analysis on a flow stream of sample.

Lorazepam, a member of the 1,4-benzodiazepine class of compounds, has been determined in dosage forms using a sensitive and versatile automated procedure capable of performing voltage-scanning polarographic analysis [162]. The system interfaces a polarograph with a Technicon continuous-flow system to permit the analysis of samples at a high rate with a broad selection of electrode configurations. Fifteen tablet or capsule formulations of lorazepam per hour can be analyzed at the 0.5-mg level with a standard deviation of ±1.4%.

Sulfenazone shows two polarographic waves on reduction at the DME due to reduction in the pyridazine nucleus and the antipyrine substituent of the molecule. An automated procedure has been described for the determination of the compound after extraction using dc or ac polarography ($E_{1/2}$ = −1.5 V vs. Ag/AgCl) at a rapidly dropping DME (0.1–0.9 s) in a 140-$\mu$L Plexiglas cell [163]. The supporting electrolyte is phosphate buffer (pH 8.4) employing a Tween-80 surface-active agent. The system is capable of determining 60 samples per hour. The automatic and simultaneous determination of methylesculetol derivatives by polarography and of rutine by colorimetry in pharmaceutical preparations has been described using a modification of the equipment used for the determination of sulfenazone [164].

## E. In Vitro Studies

Voltammetric studies have been used to correlate pharmacological activity of a group of compounds with minor changes in their structures. A correlation of the polarographic behavior of 1,2-dimethyl-3-arylpyrazolinium salts and their antidepressant activity has recently been reported [165]. Results indicated that making substituent changes at the 3-aryl position would change the $E_{1/2}$ consistent with a Hammett type of relationship. A good correlation of $E_{1/2}$ values and antidepressant activity was also demonstrated. The authors conclude that apart from other factors that might affect the total drug-receptor interaction, the antidepressant activity of these salts is associated with a redox effect.

In similar studies, the reduction potentials of 1,2-benzothiazole derivatives in DMF were correlated to antimycotic activity [166], and those of 1,3,5,7-substituted-1,3-dihydro-2H-1,4-benzodiazepin-2-ones and benzodiazepine-2-thiones in dimethylformamide (DMF)-water solutions were correlated to inhibition of orientation reactions and to protection from electric shock [167].

Dissolution rate testing is an important aspect of the development of a drug. It can serve as a guide to formulation development and may be correlated to the absorbability potential of some poorly soluble drugs. A review of the role of in vitro dissolution testing in the development of a drug has been published [168]. Voltammetric methods are ideal for these studies since they are selective and the electrode used to monitor the rate of dissolution may be placed directly in the dissolution apparatus, obviating the need for repetitive sampling. In this experimental setup, a constant potential (on the reduction/oxidation plateau of the analyte) is applied to the working (sensor) electrode, and current–time are accumulated to indicate percentage of drug dissolved.

The silicon-rubber-based graphite electrode has been employed as an appropriate voltammetric sensor to measure the rates of dissolution of several oxidizable pharmaceutical compounds [169]. The recirculating dissolution measuring arrangement with this sensing electrode and an Ag/AgCl reference are shown in Figure 26.12. A peristaltic pump continuously samples the system in the thermostatically controlled dissolution vessel. An example of this system was demonstrated for the dissolution of aminopyrine tablets and promethazine tablets. A similar design experimental setup using a dropping mercury electrode in a flow cell has recently been applied to the determination of drug dissolution of chlordiazepoxide, trimethoprim, ornidazole, and isoniazid using polarography [170]. The dissolution rate of L-dopa in 0.1 *M* HCl has been determined employing a tubular carbon electrode [160]. A 1-L sample was pumped through a filter unit to the TCE and returned to the beaker. After setting the electrode potential at +0.9 V vs. SCE and establishing a baseline current, a tablet was added and the limiting current was recorded as a function of time. Another dissolution device has been described [171], in which voltammetric sensors (graphite electrode and membrane-coated mercury electrode) were utilized with a flow-through method to determine the dissolution profiles of tolperisone hydrochloride in enteric-coated tablets. A differential pulse voltammetric method using a glassy carbon electrode sensor has been applied to monitor the dissolution profiles of salbutamol tablets [172]. This procedure showed the advantage of continuously monitoring the concentration of the active ingredient in the standard dissolution cell without the need for withdrawing aliquots for measurement (as in LC).

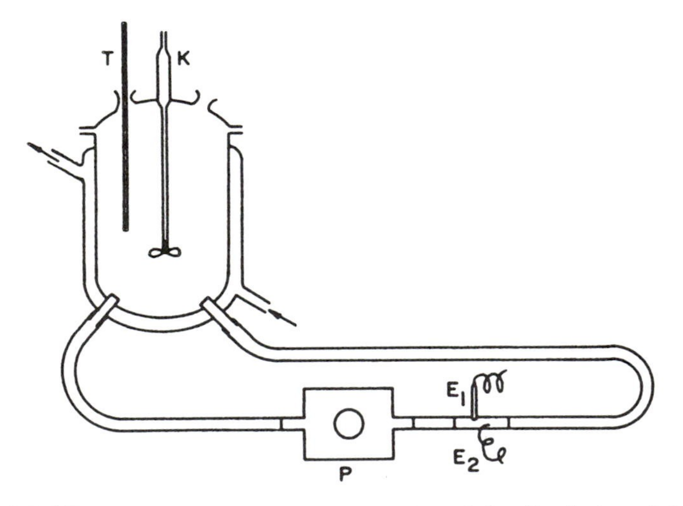


**Figure 26.12** Measuring setup for investigation of the dissolution of drugs from pharmaceutical preparations: T, thermometer; K, stirrer; P, peristaltic pump; $E_1$, silicone-rubber-based electrode; $E_2$, silver–silver chloride reference electrode. [From Ref. 169.]

The in vitro everted sac technique has been used for screening the permeability parameters of drug substances associated with their in vivo absorbability [173]. The time of onset of permeability across the sac and the cumulative amount of drug transferred per unit concentration of drug in mucosal solution were shown to relate to potential permeability-limited absorption characteristics of a drug substance. The system utilizes a 10-cm segment of rat ileum, which is everted, cannulated, and suspended in a solution of drug in pH 7.4 Krebs–Ringer bicarbonate buffer. The amount of drug passing across the intestinal mucosa is then determined by serial sampling at various time intervals. The solutions removed from the everted sac have been analyzed directly polarographically for the determination of the permeability characteristics of diazepam [174]. The polarographic assay has an advantage of not requiring the extraction of the compound from the buffer to remove interfering UV-absorbing substances.

The pharmacological response exerted by a chemitherapeutic agent may be related to the concentration of unbound drug in the plasma water. The extent to which a drug is bound to plasma albumin may influence its distribution, onset of activity, and rate of metabolism and excretion. A study has investigated the degree of binding to serum albumin for four series of 1,4-benzodiazepine homologues employing scintillation radiometry and pulse polarography for quantitation [175]. The procedure consisted of placing drug and plasma in a dialysis membrane sac and suspending this sac in an isotonic buffer solution for a fixed interval of time at 37°C. After completion of the incubation, the solutions both inside and outside the dialysis membrane were removed, made alkaline, and extracted with ether, the residue of which was analyzed by pulse polarography in 0.1 *M* HCl or pH 7.0 phosphate buffer. The pulse method was rapid and sensitive (~0.5 μg/mL of solution) and allowed a large number of compounds to be determined under the same conditions, in contrast to gas chromatography, which would have required specific conditions for each compound.

## F. Determination of Drugs in Biological Fluids

The use of voltammetric methods for the determination of drugs in biological fluids has only become practical due to the advent of commercial instrumentation capable of measuring concentration with limits of $10^{-7}$–$10^{-10}$ *M* [176,177]. The techniques of differential pulse polarography, fast-scan polarography, stripping voltammetry, and liquid chromatography with amperometric detection (LCEC) are the only techniques capable of routinely determining drugs at levels of less than $10^{-6}$ *M* (typically 10 μg/mL of biological fluids). Solid electrode voltammetry does not have sufficient sensitivity to measure levels below 10 μg/mL of biological fluids unless microcells (≤10 μL) are used in conjunction with preconcentration. The use of LCEC to measure drugs in biological fluids is discussed in Chapter 27.

The earlier works using polarography for the determination of drugs in biological fluids employed direct analysis of the biological material. Determinations of 2-ethyl-4-thioureidopyridine, for example, were performed directly in plasma without any additional supporting electrolyte or buffering [178]. The dc polarographic determination of chlordiazepoxide [179] and diazepam [180] in plasma, and urine diluted with the appropriate supporting electrolyte, was suggested by Oelschlager et al.

There are two basic shortcomings of these direct assays: (1) they are usually only sufficiently sensitive to measure the drug after the ingestion of an overdosage, and (2) they are inherently nonspecific in that compounds of similar structure, metabolites, and biological materials may interfere and thus yield erroneous results. In order for polarographic assays to yield analytically meaningful results for the determination of drugs in biological fluids, selective extraction, cleanup, and separation of the parent drug from any metabolites and other substances present must be employed prior to polarographic assay. Such extraction procedures are routine in bioanalytical laboratories, so they do not result in serious inconvenience. After the administration of many drugs it is quite common to find high percentages of the dose excreted in urine. For example, if 50% of a 50-mg dose is recoverable in the urine in a 1000-mL volume over an excretion period of 24 h, the resulting concentration of a drug in the urine sample would be 25 $\mu$g/mL. Pulse polarography is ideally suited for measurements in this concentration range.

A pulse polarographic method to determine dantrolene sodium and its major metabolites in urine after ethyl acetate extraction has been reported [181]. The ethyl acetate is brought to a residue and the dantrolene plus the total extractable metabolites are analyzed for reduction of the azomethine linkage at −0.86 V in a DMF–acetate buffer (pH 4.0). The nitro compounds are simultaneously determined in the same media as dantrolene equivalents from the reduction of the nitro group at −0.26 V (see Fig. 26.13). The difference between the two determinations represents the metabolites not containing the nitro group. Levels as low as 0.1 $\mu$g/mL can be determined for either functional group.

Differential pulse polarography has been employed for the determination of trimethoprim and its *N*-oxide metabolites in urine of humans, dogs, and rats [182]. By combining a TLC separation of a urinary extract with polarography, a high degree of specificity was obtained to denote quantitatively the preferred metabolic *N*-oxidation of each species. The pulse polarographic determination in pH 3.0 phosphate buffer is particularly specific because each compound can be distinguished by a peak due to the N-oxide and azomethine reductions (see Fig. 26.14).

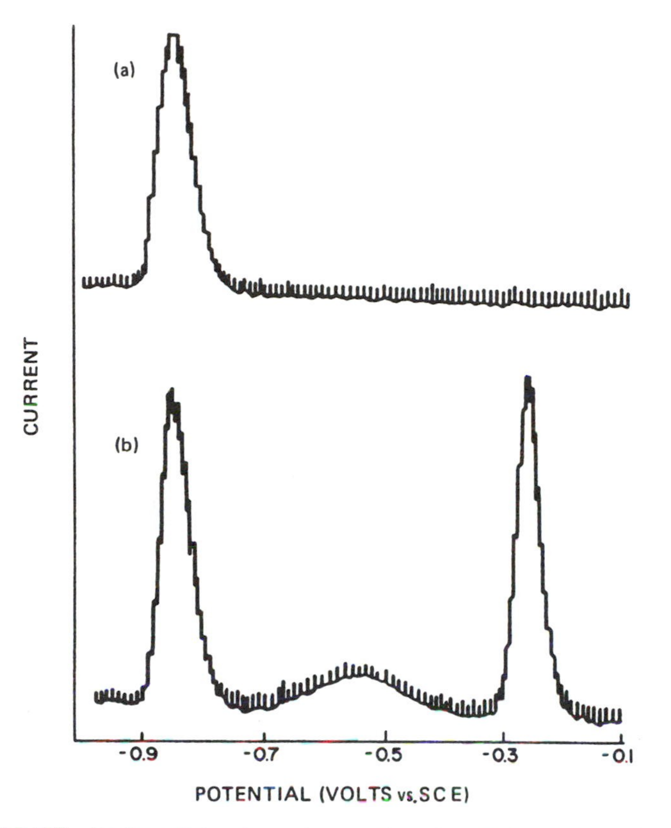


**Figure 26.13** Differential pulse polarography of (a) reduced dantrolene, and (b) dantrolene and nonreduced metabolites. [From Ref. 182, reproduced with permission.]

Polarographic methods have been extremely useful for the determination of the urinary excretion of the 1,4-benzodiazepines. An assay that employs selective solvent extraction and acid hydrolysis of diazepam and its major metabolites, *N*-desmethyldiazepam and oxazepam, to their respective benzophenones has been employed to measure the urinary excretion of diazepam [183]. A pulse polarographic assay has been reported that will measure the urinary excretion of bromazepam following a single 12-mg dose [184]. The assay employs selective extraction of bromazepam and the 2-amino-5-bromobenzoylpyridine metabolite from the deconjugated metabolites, 3-hydroxybromazepam and 2-amino-3-hydroxy-5-bromobenzoylpyridine, into separate diethyl ether fractions. The residues of the respective extracts are dissolved in phosphate buffer (pH 5.4) and analyzed by pulse polarography, which yields two distinct

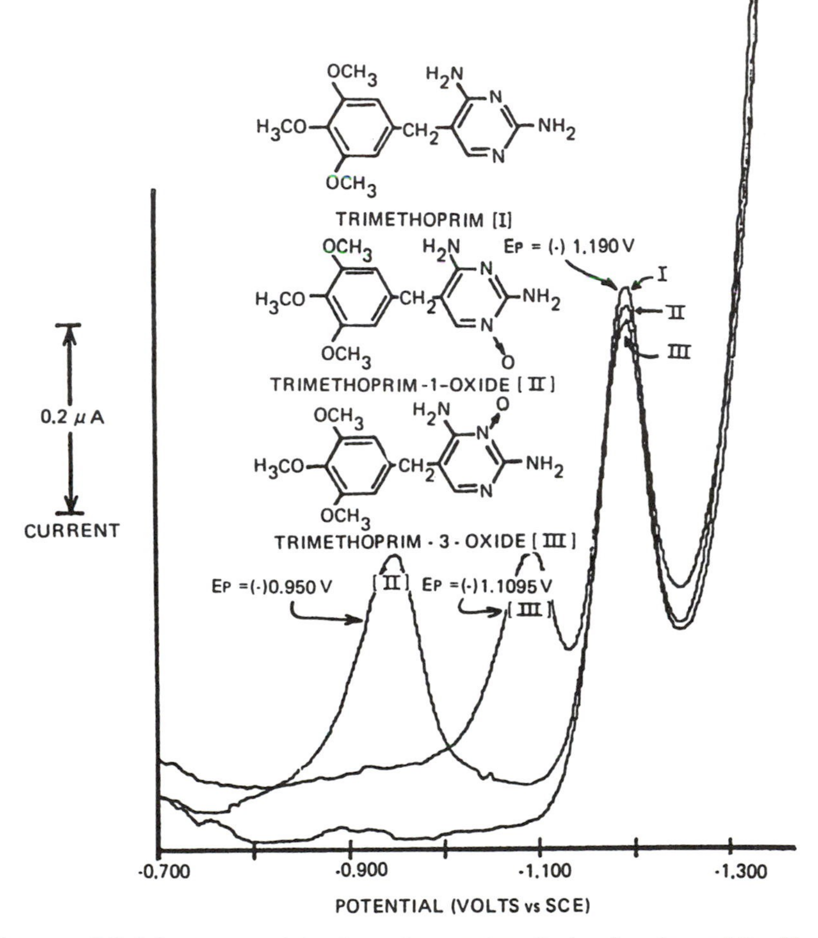


**Figure 26.14** Differential pulse polarography of trimethoprim and its *N*-oxide metabolites in 1 *M* phosphate buffer (pH 3). [From Ref. 184, reproduced with permission.]

well-resolved reduction peaks for the 4,5-azomethine functional group of the benzodiazepine-2-one and for the carbonyl function group of the benzoylpyridine in each fraction (see Fig. 26.15). Pulse polarographic assays have also been reported for the determination of the urinary excretion of 7-chloro-1,3-dihydro-5-(2′-chlorophenyl)-2*H*-1,4-benzodiazepine-2-one [185], clonazepam [186], and flurazepam [131]. The assay for flurazepam involves the determination of the three major metabolites of flurazepam employing a combination of selective extraction, TLC, and pulse polarography and was used to measure the urinary

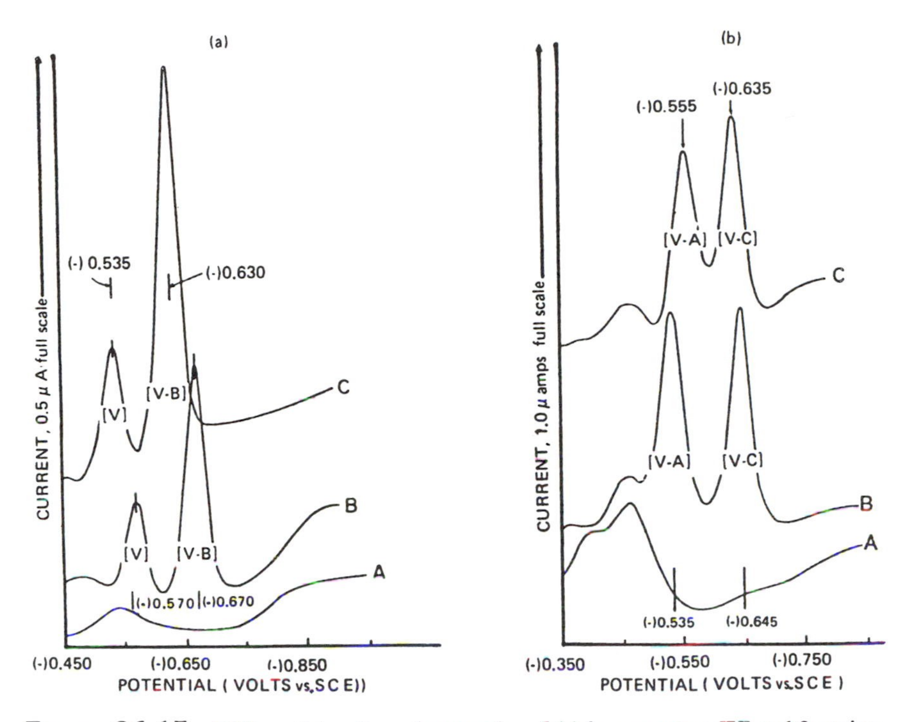


**Figure 26.15** Differential pulse polarography of (a) bromazepam [**V**] and 2-amino-5-bromobenzoylpyridine [**V-B**], and (b) 3-hydroxybromazepam [**V-A**], and 2-amino-3-hydroxy-5-bromobenzoylpyridine [**V-C**] in 1.0 *M* pH 5.5 phosphate buffer. A, control urine blank; B, authentic standard mixture; C, authentic compounds recovered from urine. [From Ref. 185, reproduced with permission.]

excretion following a 90-mg dose. A direct assay with a detection limit of 5 μg/mL that employs 0.1 mL of urine diluted with phosphate buffer (pH 7.0) is also described as a rapid means of measuring total benzodiazepines to confirm ingestion of flurazepam. The pulse assays [131,184–187] for the 1,4-benzodiazepine-2-ones typically have a detection limit of 0.1–0.2 μg/mL, which is more than sufficient to determine the urinary excretion of these compounds. A sensitive and selective differential pulse voltammetric measurement of tricyclic antidepressants in urine samples using lipid-coated glassy carbon electrodes has been reported [188]. The deliberate modification of electrodes with a coating excludes potential interference while allowing transport of the analyte. Besides the excellent reproducibility and high sensitivity (limit of detection of $10^{-8}$ *M*), the technique shows great promise in demonstrating the applicability to direct measurements of the drugs in urine samples.

The use of polarographic assays for the determination of drugs in blood is the most demanding on the detection limitations of the technique. Differential pulse polarography, stripping voltammetry, and LCEC are the only electrochemical methods currently available for routine determination of drugs below 1.0 μg/mL of blood.

The determination of diazepam in plasma with detection limits of 0.03–0.2 μg/mL by cathode ray [189] and pulse polarography [174,190] has been described. The first reported pulse assay [174] differs from the other polarographic assays [189,190] in that a more polar solvent is employed to ensure quantitative extraction, followed by TLC separation and determination of diazepam and its major blood metabolite, N-desmethyldiazepam, to ensure specificity.

The pulse polarographic determination of chlordiazepoxide and its metabolites in plasma has also been described [191]. Chlordiazepoxide and its metabolites are extracted from serum buffered to pH 9.0 followed by a TLC separation, elution, and final quantitation in 0.5 M $H_2SO_4$. The detection limit of the assay of 0.05–0.1 μg of each compound per milliliter of serum using a 2-mL sample is sufficiently useful for pharmacokinetic studies.

Pulse polarography has also been used to measure dantrolene in plasma [181] and trimethoprim [192] and nitroimidazoles in blood [193,194]. The overall recovery of the trimethoprim assay was reported to be 81.7 ± 6.3% (SD) with a detection limit of 0.5–0.75 μg/mL of blood. There was no interference from the sulfamethoxazole, which is administered simultaneously. Glibornuride [195], phenobarbital, and diphenylhydantoin [97] have all been determined as their nitro derivatives after extraction from blood. The recovery of phenobarbital and diphenylhydantoin from blood was 72.3 ± 6.5% (SD) and 76.6 ± 2.3% (SD), respectively, with a detection limit of 1–2 μg/mL. A modified assay [97] for the determination of both compounds in blood with TLC separation was also described. A differential pulse polarography determination of cephalosporin antibiotics in human serum samples has recently been described [196].

Some work has demonstrated that pulse voltammetry at the carbon paste electrode can be applied to the assay of drugs in body fluids [197,198]. Methods based on the electrooxidation of theophylline [197] and acetaminophen [198] were described at concentrations of approximately 5–10 μg/mL plasma.

Pulse polarographic studies have been described using a microcell of 0.5 mL capacity, which analyzed two 1,4-benzodiazepines, with the lowest detection limit reported to date being 10–20 ng/mL of blood [199]. Detailed construction of the cell and electrode assembly was also described (shown in Fig. 26.16). Further miniaturization of this type of three-electrode cell is not practical; hence further increases in sensitivity will have to rely on electrochemical detector flow cells of microliter capacity such as those used in conjunction with liquid chromatography (see Chap. 27).

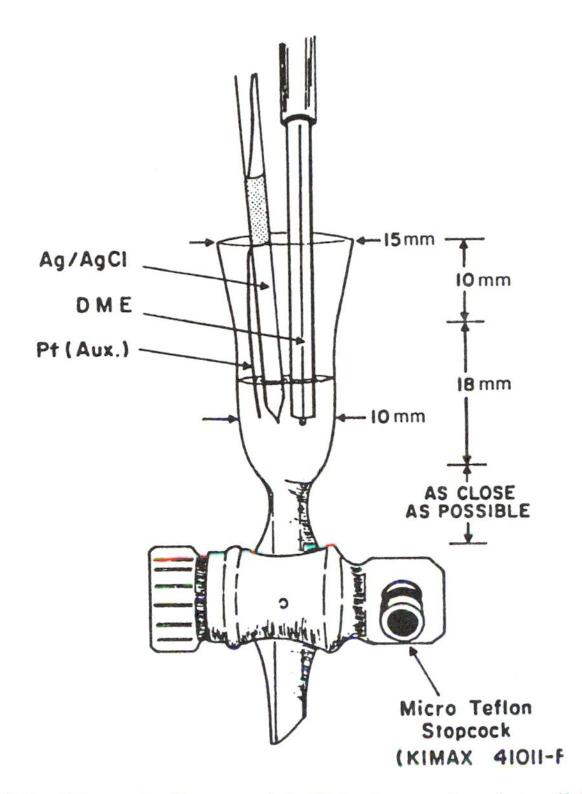


**Figure 26.16** Schematic diagram of the 0.5-mL-capacity microcell for pulse polarographic studies. [From Ref. 199, reprinted with permission.]

Stripping voltammetry is one of the most sensitive analytical techniques and has received a great deal of attention in recent years. The technique consists of two stages: a preconcentration step and voltammetric measurement. A brief review of the principles and instrumentation of stripping analysis has been reported [200]. Because of the high sensitivity and better selectivity of the technique, it is extremely useful for the analysis of drugs in biological fluids. A large volume of publications has recently been published describing the stripping analysis of a wide range of pharmaceutical compounds. Antihypertensive drugs such as reserpine, rescinnamine, and hydralazine in urine samples were assayed by differential pulse stripping measurement at trace levels [201]. Other studies report the determination of daunorubicin (antitumor drug) in urine samples [202], 5-fluorouracil (anticancer drug) in serum samples [203], amethopterine (antineoplastic drug) in serum samples [204], nifedipine (calcium antagonist) in human serum samples [205], and bentazepam (anxiolytic drug) in urine samples [206]. The sensitivity limit was typically $10^{-8}$ to $10^{-10}$ *M* in these investigations.

Modified solid electrodes have also been employed in the stripping analysis to enhance assay selectivity by utilizing size exclusion, preferential uptake, or specific complexation effects. A cellulose acetate-modified glassy carbon electrode has been applied for the adsorptive stripping voltammetric analysis of chlorpromazine and trimipramine in urine samples. The detection limit was found to be $2.5 \times 10^{-11}$ *M* [207]. A bentonite-modified carbon paste electrode was used to analyze flunitrazepam in serum and urine samples with a detection limit of 0.04 μg/mL [208]. A lipid-modified glassy carbon electrode was used to assay marcellomycin in urine samples [209]. With the high sensitivity and enhanced selectivity demonstrated by modification of the working electrode, stripping voltammetry has shown great utility in complex media. From the authors' perspective, it is predictable that additional highly specific and sensitive applications should be expected in this area of research in the near future.

## REFERENCES

1. C. B. Coutinho, J. A. Cheripko, T. Crews, B. H. Min, and A. C. Levy, Abstract, Society of Toxicology Meeting, Williamsburg, VA, March 9–13, 1975.
2. S. A. Kaplan, *Drug. Metab. Rev. 1*:15 (1972).
3. S.A. Kaplan, in *Current Concepts in the Pharmaceutical Sciences: Dosage Form Design and Bioavailability* (J. Swarbrick, ed.), Lea & Febiger, Philadelphia, 1974, pp. 1–30.
4. M. A. Schwartz and E. Postma, *Biochem. Pharmacol. 17*:2443 (1968).
5. J. A. F. de Silva, in *Current Concepts in the Pharmaceutical Sciences: Biopharmaceutics* (J. Swarbrick, ed.), Lea & Febiger, Philadelphia, 1970, pp. 203–264.
6. J. R. DiPalma, *Basic Pharmacology in Medicine*, McGraw-Hill, New York, 1982.
7. W. C. Cutting, T. Z. Csaky, and B. A. Barnes, *Cutting's Handbook of Pharmacology: The Actions and Uses of Drugs*, 6th ed., Appleton-Century-Crofts, Norwalk, CT, 1984.
8. K. Abresch and I. Claassen, *Coulometric Analysis*, Franklin Publishing, Englewood, NJ, 1966.
9. G.W.C. Milner and G. Phillips, *Coulometry in Analytical Chemistry*, Pergamon Press, New York, 1967.
10. W.C. Purdy, in *Electroanalytical Methods in Biochemistry*, McGraw-Hill, New York, 1965, pp. 223–276.
11. G. Patriarche, *Contribution de l'analyse coulométrique*, Éditions Arscia S.A., Bruxelles, 1963.
12. R.W. Janssen and C. A. Discher, *J. Pharm. Sci. 60*:798 (1971).
13. G. Patriarche and J. J. Lingane, *Anal. Chim. Acta 49*:241 (1970).

14. F. H. Merkel and C. A. Discher, *Anal. Chem. 36*:1639 (1964).
15. R. Parkash, H. O. Gupta, and J. Dutt, *Collect. Czech. Chem. Commun. 56*:1833 (1991).
16. K. S. V. Santhanam and V. R. Krishnan, *Anal. Chem. 33*:1493 (1961).
17. F. P. Wilson, C. G. Butler, P. H. B. Ingle, and H. Taylor, *J. Pharm. Pharmacol. 12*:220T (1960).
18. E. Jacobsen and T. V. Jacobsen, *Anal. Chim. Acta 55*:293 (1971).
19. J. J. Donahue and S. Oliveri-Vigh, *Anal. Chim. Acta 63*:415 (1973).
20. M. L. Girard, F. Rousselet, P. Levillain, and H. Fouye, *Ann. Pharm. Fr. 26*:535 (1968).
21. K. Stulik and J. Zyka, *Chemist-Analyst 55*:120 (1966).
22. Z. E. Kalinowska, *Pharmazie 22*:1 (1967).
23. K. Nikolic, *Pharmazie 44*:350 (1989).
24. K. Nikolic and M. Medenica, *Il Farmaco 45*:1037 (1990).
25. K. I. Nikolic and R. Velasevic, *Acta Polon. Pharm. 44*:319 (1987).
26. A. Berka, K. Nikolic, and K. Velasevic, *Acta Polon. Pharm. 47*:7 (1990).
27. K. I. Nikolic and K. Velasevic, *J. Pharm. Belg. 44*:387 (1989).
28. K. Nikolic and M. Medenica, *Acta Polon. Pharm. 47*:11 (1990).
29. K. Nikolic and M. Medenica, *J. Pharm. Biomed. Anal. 9*:199 (1991).
30. K. I. Nikolic and K. R. Velasevic, *J. Pharm. Belg. 43*:455 (1988).
31. K. Kalinowski and A. Olech, *Acta Pol. Pharm. 25*:171 (1968); *Chem. Abstr. 69*:46099 (1968).
32. K. Kalinowski, *Chem. Anal. 15*:277 (1970); *Chem. Abstr. 73*:38596U (1971).
33. K. I. Nikolic and K. R. Velasevic, *Acta Polon. Pharm. 42*:209 (1985).
34. K. Nikolic, K. Velasevic, and M. Medenica, *Acta Pharm. Jugosl. 34*:177 (1984).
35. K. Nikolic and K. Velasevic, *Acta Pharm. Jugosl. 35*:41 (1985).
36. H. Beral and V. Stoicescu, *Pharm. Zentralhalle 108*:469 (1969).
37. S. Oliveri-Vigh, J. J. Donahue, J. E. Heveren, and B. Z. Senkowski, *J. Pharm. Sci. 60*:1851 (1975).
38. S. A. Moros, C. M. Hamilton, J. E. Heveren, J. J. Donahue, and S. Oliveri-Vigh, *J. Pharm. Sci. 64*:1229 (1975).
39. M. L. Girard, F. Rousselet, H. Fouye, and P. Levillain, *Ann. Pharm. Fr. 27*:173 (1967).
40. S. Ebel and S. Kalb, *Arch. Pharm. (Weinheim, Ger.) 307*:2 (1974).
41. R. L. Charles and A. M. Knevel, *J. Pharm. Sci. 54*:1678 (1965).
42. T. Pastor and J. Barek, *Mikrochim. Acta 1*:407 (1989).
43. G. Patriarche, *Mikrochim. Acta* 950 (1970).
44. Z. Feher, G. Nagy, K. Toth, and E. Pungor, *Analyst (London) 104*:560 (1979).
45. Z. Feher, I. Kolbe, and E. Pungor, *Fresenius J. Anal. Chem. 332*:345 (1988).
46. J. R. Monforte and W. C. Purdy, *Anal. Chim. Acta 52*:433 (1970).
47. V. J. Vaygand and R. Mihaylovic, *Talanta 16*:1311 (1969).
48. G. J. Patriarche and J. J. Lingane, *Anal. Chim. Acta 49*:25 (1970).
49. C. A. Mairesse-Ducarmois, J. L. Vandenbalak, and G. J. Patriarche, *J. Pharm. Belg. 28*:300 (1973); *Anal. Abstr. 26*:3470 (1974).
50. J. R. Monforte and W. C. Purdy, *Anal. Chim. Acta 52*:25 (1970).

51. K. Nikolic, S. Vladimirov, D. Zivanov-Stakic, and K. Velasevic, *Acta Polon. Pharm. 44*:438 (1987).
52. H. Kitamura, T. Miyahara, K. Yamamoto, and K. Narita, *Microchem. J. 41*:156 (1990).
53. G. J. Patriarche and J. J. Lingane, *Anal. Chim. Acta 37*:455 (1967).
54. P. Zuman, *Organic Polarographic Analysis*, Pergamon Press, Elmsford, NY, 1964.
55. P. Zuman and M. Brezina, in *Progress in Polarography*, Vol. 2 (P. Zuman and I. M. Kolthoff, eds.), Wiley-Interscience, New York, 1962, pp. 687–701.
56. W. C. Purdy, in *Electroanalytical Methods in Biochemistry*, McGraw-Hill, New York, 1965, pp. 148–174.
57. R.J. Gayan, in *Oscillographic Polarography in Methods in Pharmacology*, Vol. 2: *Physical Methods* (C. F. Chignell, ed.), Appleton-Century-Crofts, New York, 1972, pp. 443–463.
58. J. E. Page, *J. Pharm. Pharmacol. 4*:1 (1952).
59. H. Hoffmann and J. Volke, in *Advances in Analytical Chemistry and Instrumentation*, Vol. 10: *Electroanalytical Chemistry* (H. W. Nurnberg, ed.), Wiley, New York, 1974, pp. 287–334.
60. M.R. Smyth and W. F. Smyth, *Analyst (London) 103*:529 (1978).
61. W. F. Smyth, *Polarography of Molecules of Biological Significance*, Academic Press, London, 1979.
62. W. F. Smyth and A. D. Woolfson, *J. Clin. Pharm. Ther. 12*:117 (1987).
63. G. J. Patriarche and J.-C. Vire, *Anal. Chim. Acta 196*:193 (1987).
64. P. M. Bersier, *J. Pharm. Biomed. Anal. 1*:475 (1983).
65. P. M. Bersier and J. Bersier, in *Critical Reviews in Analytical Chemistry*, Vol. 16 (W. L. Zielinski, ed.), CRC Press, Boca Raton, FL, 1985, pp. 15–128.
66. W. F. Smyth, in *Critical Reviews in Analytical Chemistry*, Vol. 18 (W. L. Zielinski, ed.), CRC Press, Boca Raton, FL, 1987, pp. 155–208.
67. J.-M. Kauffmann, M. P. Prete, J.-C. Vire, and G. Patriarche, *Fresenius J. Anal. Chem. 321*:172 (1985).
68. W. F. Smyth in *Electrochemistry, Sensor and Analysis* (M. R. Smyth and J. G. Vos, eds.), Elsevier Science, Amsterdam, 1985, pp. 29–36.
69. I. J. Nagels, G. Mush, and D. L. Massart, *J. Pharm Biomed. Anal. 12*:1479 (1989).
70. W. F. Smyth, I. Ivaska, J. S. Burmitz, I. E. Davidson, and Y. Vaneesorn, *J. Electroanal. Chem. 128*:459 (1981).
71. S. Arizan, R. Simionovici, and V. Voinov, *Pharmazie 24*:746 (1969).
72. D. Dumanovic, J. Volke, and R. Jovanovic, *J. Assoc. Off. Anal. Chem. 54*:884 (1971).
73. J. A. F. de Silva, N. Strojny, and N. Munno, *J. Pharm. Sci. 62*:1066 (1973).
74. M. Girard and F. Rousselet, *Ann. Pharm. Fr. 20*:109 (1962); *Anal. Abstr. 9*:3872 (1962).
75. P. M. Bersier, J. Bersier, G. Sedelmeier, and E. Hungerbubler, *Electroanalysis 2*:373 (1990).
76. K. Fossdal and E. Jacobsen, *Anal. Chim. Acta 56*:105 (1971).

77. B. Z. Senkowski, M. S. Levin, J. R. Urbigikit, and E. G. Wollish, *Anal. Chem. 36*:1991 (1964).
78. H. Oelschlager, J. Volke, G. T. Lim, and U. Bremer, *Arch. Pharm. Berl. 303*:364 (1970).
79. J. M. Clifford and W. F. Smyth, *Z. Anal. Chem. 264*:149 (1973).
80. J. Barrett, W. F. Smyth, and J. P. Hart, *J. Pharm. Pharmacol. 26*:9 (1974).
81. M. A. Brooks, J. J. Bel Bruno, J. A. F. de Silva, and M. R. Hackman, *Anal. Chim. Acta 56*:105 (1971).
82. A. L. Woodson and D. E. Smith, *Anal. Chem. 42*:242 (1970).
83. J. L. Spahr and A. M. Knevel, *J. Pharm Sci. 55*:1020 (1966).
84. J. Hakl, *J. Electroanal. Chem. 11*:37 (1966).
85. M. Lapidus and M. E. Rosenthale, *J. Pharm. Sci. 55*:555 (1966).
86. T. Komura, S. Morishita, C. Aikawa, and Y. Ueda, *Jpn. Anal. 18*:943 (1969); *Anal. Abstr. 19*:4374 (1970).
87. M. J. D. Brand and B. Fleet, *Analyst (London) 95*:1023 (1970).
88. E. Jacobsen and T. Rojahn, *Anal. Chim. Acta 61*:320 (1972).
89. J. Volke, L. Wasilewska, and A. Ryvolova-Kejharova, *Pharmazie 26*:399 (1971).
90. S. L. Tackett and J. W. Ide, *J. Electroanal. Chem. 30*:510 (1971).
91. P. G. Swetik and D. G. Brown, *J. Electroanal. Chem. 51*:433 (1974).
92. V. Vorel, J. Prokes, and V. Dolezal, *Soudni Lek. 5*:49 (1961); *Anal. Abstr. 9*:3419 (1962).
93. J. Vachek, *Cesk. Farm. 10*:187 (1961); *Anal. Abstr. 8*:4340 (1961).
94. H. Oelschlager and D. Hamel, *Arch. Pharm. Berl. 302*:847 (1969).
95. S. Tammilehto and M. Perala, *Pharm. Acta Helv. 46*:351 (1971).
96. H. Oelschlager, *Arch. Pharm. Berl. 296*:7 (1963).
97. M. A. Brooks, J. A. F. de Silva, and M. R. Hackman, *Anal. Chim. Acta 64*:165 (1973).
98. A. Danek and H. Strozik, *Diss. Pharm. Warsz. 18*:519 (1966); *Anal. Abstr. 15*:429 (1968).
99. A. G. Dumortier and G. J. Patriarche, *Z. Anal. Chem. 264*:153 (1973).
100. Y. I. Tur'yan, T. V. Merkyvkova, and O. V. Bogdanova, *Zh. Anal. Khim. 25*:384 (1970); *Anal. Abstr. 21*:1455 (1971).
101. G. S. Porter, *J. Pharm. Pharmacol., Suppl. 16*:24T (1964).
102. L. Schlitt, M. Rink, and M. vonStackelberg, *J. Electroanal. Chem. 13*:10 (1967).
103. A. G. Fogg and Y. Z. Ahmed, *Anal. Chim. Acta 70*:241 (1974).
104. J. A. Squella and L. J. Nunez-Vergara, *Acta Pharm. Suec. 16*:339 (1979).
105. A. G. Fogg and N. M. Fayad, *Anal. Chim. Acta 113*:91 (1980).
106. M. M. Diaz De Guerenu, R. J. Barrio, A. Arranz, and J. F. Arranz, *J. Pharm. Biomed. Anal. 10*:481 (1992).
107. P. Kabaskalian and J. McGlotten, *Anal. Chem. 31*:431 (1959).
108. F. H. Merkle and C. A. Discher, *J. Pharm. Sci. 53*:620 (1964).
109. E. Pungor, Z. Feher, and G. Nagy, *Magy. Kem. Foly. 77*:298 (1971); *Anal. Abstr. 22*:350 (1972).
110. E. Pungor, Z. Feher, and G. Nagy, *Acta Chim. Hung. 70*:207 (1971); *Anal. Abstr. 23*:747 (1972).

111. E. Bishop and W. Hussein, *Analyst 109*:229 (1984).
112. N. Zimova-Sulcova, I. Nemec, K. Waisser, and H. L. Kies, *Microchem. J. 32*:33 (1985).
113. H. K. Chan and A. G. Fogg, *Anal. Chim. Acta 109*:341 (1979).
114. B. Proksa and L. Molnar, *Anal. Chim. Acta 97*:149 (1978).
115. B. Gallo-Hermosa, J.-M. Kauffmann, G. J. Patriarche, and G. G. Guilbault, *Anal. Lett. 19*:2011 (1986).
116. I. Biryol, M. Kabasakaloglu, and Z. Senturk, *Analyst 114*:181 (1989).
117. E. Munoz, L. Camacho, and J. L. Avila, *Analyst 114*:1611 (1989).
118. J. H. Mendez, A. S. Perez, M. D. Zamarreno, and M. L. H. Garcia, *J. Pharm. Biomed. Anal. 6*:737 (1988).
119. M. D. Zamarreno, A. S. Perez, J. H. Mendez, and F. A. M. Dominguez, *J. Pharm. Biomed. Anal. 7*:1213 (1989).
120. J.-C. Vire, A. A. El Maali, G. H. Patriarche, and G. D. Christian, *Talanta 35*:997 (1988).
121. U. Forsman, *Anal. Chim. Acta 146*:71 (1983).
122. J. Wang, P. A. M. Farias, and J. S. Mahmoud, *Analyst 111*:837 (1986).
123. J. Wang, M. S. Lin, and V. Villa, *Anal. Lett. 19*:2293 (1986).
124. P. Passamonti, V. Bartocci, and F. Pucciarelli, *J. Electroanal. Chem. 230*:99 (1987).
125. M. Khodari, J.-C. Vire, G. J. Patriarche, and M. A. Ghandour, *Anal. Lett. 23*:1873 (1990).
126. I. F. Jones, J. E. Page, and C. T. Rhodes, *J. Pharm. Pharmacol., Suppl. 20*:45S (1968).
127. D. A. Hall, *J. Pharm. Sci. 62*:980 (1973).
128. E. M. Cohen, *J. Pharm. Sci. 60*:1702 (1971).
129. Y. Asah, *Chem. Pharm. Bull. Jpn. 11*:930 (1963); *Anal. Abstr. 11*:4514 (1964).
130. E. R. Garrett, *J. Pharm. Sci. 51*:410 (1962).
131. J. A. F. de Silva, C. V. Puglisi, M. A. Brooks, and M. R. Hackman, *J. Chromatogr. 99*:461 (1974).
132. E. Jacobsen and K. Hogbert, *Anal. Chim. Acta 71*:157 (1974).
133. D. G. Prue, C. R. Warner, and B. T. Kho, *J. Pharm. Sci. 61*:249 (1972).
134. M. J. Heasman and A. J. Wood, *J. Pharm. Pharmacol., Suppl. 23*:176S (1971).
135. L. G. Chatten, R. N. Yadav, S. Binnington, and R. E. Moskalyk, *Analyst (London) 102*:323 (1977).
136. R. N. Yadav and F. W. Teare, *J. Pharm. Sci. 67*:436 (1978).
137. E. Jacobsen and B. Korvald, *Anal. Chim. Acta 99*:255 (1978).
138. F. W. Teare, R. N. Yadav, and M. Spino, *J. Pharm. Sci. 67*:1642 (1978).
139. M. Kotoucek, J. Ruzickova, and I. Cechova, *Mikrochim. Acta II*:109 (1989).
140. K. Hasebe, N. Osanai, M. Taga, and J. G. Osteryoung, *Anal. Sci. 5*:275 (1989).
141. J. Rosas, J. M. Pinilla, and L. Hernandez, *Il Farmaco 45*:353 (1990).
142. S. S. Atuma and J. Linquist, *Analyst (London) 98*:886 (1973).
143. S. S. Atuma, J. Linquist, and K. Lundstrom, *Analyst (London) 99*:683 (1974).
144. A. Y. Taira, *J. Assoc. Off. Anal. Chem. 57*:910 (1974).
145. E. Jacobsen and K. B. Thorgersen, *Anal. Chim. Acta 71*:175 (1974).

146. J. M. Moore, *J. Pharm. Sci. 58*:1117 (1969).
147. A. M. Mayo and O. Nudelman, *J. Assoc. Off. Anal. Chem. 56*:1460 (1973).
148. B. Breyer and T. Biegler, *J. Electroanal. Chem. 1*:453 (1959/60).
149. M. E. Schertel and A. J. Sheppard, *J. Pharm. Sci. 60*:1070 (1971).
150. J. Ballantine and A. D. Woolfson, *J. Pharm. Pharmacol. 32*:353 (1980).
151. N. Zimova, I. Nemec, and J. Zima, *Talanta 33*:467 (1986).
152. I. Navarro, D. Gonzalez-Arjona, E. Roldan, and M. Rueda, *J. Pharm. Biomed. Anal. 6*:969 (1988).
153. K. Sagar, J. M. F. Alvarez, C. Hua, M. R. Smyth, and R. Munden, *J. Pharm. Biomed. Anal. 10*:17 (1992).
154. J. A. Squella, C. Rivera, I. Lemus, and L. J. Nunez-Vergara, *Mikrochim. Acta I*:343 (1990).
155. O. W. Lau, S. F. Luk, and Y. M. Cheung, *Analyst 114*:1047 (1989).
156. J. R. Barreira Rodriguez, V. C. Diaz, A. C. Garcia, and P. T. Blanco, *Analyst 115*:209 (1990).
157. J. J. Berzas, J. Rodriguez, J. M. Lemus, and G. Castaneda, *Anal. Chim. Acta 273*:369 (1993).
158. S. Kir, A. N. Omar, and A. Temizer, *Anal. Chim. Acta 229*:145 (1990).
159. W. D. Mason, T. D. Gardner, and J. T. Stewart, *J. Pharm. Sci. 61*:1301 (1972).
160. W. D. Mason, *J. Pharm. Sci. 62*:999 (1973).
161. J. T. Stewart, H. C. Loo, and W. D. Mason, *J. Pharm. Sci. 63*:954 (1974).
162. L. F. Cullen, M. P. Brindle, and G. J. Paperiello, *J. Pharm. Sci. 62*:1708 (1973).
163. A Cinci and S. Silvestri, *Farmaco, Ed. Prat. 27*:28 (1972).
164. P. B. Brunai, A. Cinci, and S. Silvestri, *Farmaco, Ed. Prat. 27*:89 (1972).
165. N. M. Omar and N. A. El-Rabbat, *Can. J. Pharm. Sci. 9*:57 (1974).
166. V. Riganti and G. Spini, *Farmaco Ed. Sci. 28*:243 (1973).
167. A. V. Bogatskii, S. A. Andronati, V. P. Gul'tyai, Y. I. Vikhlyaev, A. F. Galatin, Z. I. Zhilina, and T. A. Klygul, *J. Gen. Chem. USSR 41*:1364 (1971).
168. C. D. Lathia and U. V. Banakar, *Drug Dev. Ind. Pharm. 12*:71 (1986).
169. Zs. Feher, G. Nagy, K. Toth, and E. Pungor, *Analyst 99*:699 (1974).
170. M. R. Hackman and M. A. Brooks, *J. Pharm. Sci. 67*:842 (1978).
171. I. Kolbe and Z. Feher, *Anal. Chem. Symp. Ser. 18*:341 (1984).
172. K. A. Sagar, M. R. Smyth, and R. Munden, *J. Pharm. Biomed. Anal. 11*:533 (1993).
173. S. A. Kaplan and S. Cotler, *J. Pharm. Sci. 61*:1361 (1972).
174. M. A. Brooks, J. A. F. de Silva, and M. R. Hackman, *Am. Lab. 5*(12):23 (1973).
175. R. W. Lucek and C. B. Coutinho, *Mol. Pharmacol. 12*:621 (1976).
176. J. A. F. de Silva and M. A. Brooks, in *Drug Fate and Metabolism*, Vol. 2 (E. Garrett and J. Hirtz, eds.), Marcel Dekker, New York, 1978, pp. 1–48.
177. M. A. Brooks, in *Analytical Chemistry Symposia Series* (W. F. Smyth, ed.), Vol. 2: *Electroanalysis in Hygiene, Environmental, Clinical and Pharmaceutical Chemistry* (Proceedings of Conference), Elsevier Scientific, New York, 1980, pp. 287–298.
178. P. O. Kane, *Nature 183*:1674 (1959).
179. H. Oelschlager, *Arch. Pharm. Berl. 296*:396 (1963).

180. H. Oelschlager, J. Volke, and E. Kurek, *Arch. Pharm. Berl. 297*:431 (1964).
181. P. L. Cox, J. P. Heotis, D. Polin, and G. M. Rose, *J. Pharm. Sci. 58*:987 (1969).
182. M. A. Brooks, J. A. F. de Silva, and L. D'Arconte, *J. Pharm. Sci. 62*:1395 (1973).
183. R. Dugal, G. Caille, and S. F. Cooper, *Union Med. Can. 102*:2491 (1973).
184. J. A. F. de Silva, I. Bekersky, M. A. Brooks, R. E. Weinfield, W. Glover, and C. V. Puglisi, *J. Pharm. Sci. 63*:1440 (1974).
185. J. A. F. de Silva, I. Bekersky, and M. A. Brooks, *J. Pharm. Sci. 63*:1943 (1974).
186. J. A. F. de Silva, C. V. Puglisi, and N. Munno, *J. Pharm. Sci. 63*:520 (1974).
187. M. A. Brooks and J. A. F. de Silva, *Talanta 22*:849 (1975).
188. J. Wang, T. Golden, M. Ozsoz, and Z. Lu, *Bioelectrochem. Bioenerg. 23*:217 (1989).
189. D. J. Berry, *Clin. Chim. Acta 32*:235 (1971).
190. E. Jacobsen, T. V. Jacobsen, and T. Rojahn, *Anal. Chim. Acta 64*:473 (1973).
191. M. R. Hackman, M. A. Brooks, J. A. F. de Silva, and T. S. Ma, *Anal. Chem. 46*:1075 (1974).
192. M. A. Brooks, J. A. F. de Silva, and L. M. D'Arconte, *Anal. Chem. 45*:263 (1973).
193. J. A. F. de Silva, N. Munno, and N. Strojny, *J. Pharm. Sci. 59*:201 (1970).
194. M. A. Brooks, L. D'Arconte, and J. A. F. de Silva, *J. Pharm. Sci. 65*:1 (1976).
195. J. A. F. de Silva and M. R. Hackman, *Anal. Chem. 44*:1145 (1972).
196. B. Ogorevc and S. Gomiscek, *J. Pharm. Biomed. Anal. 9*:225 (1991).
197. J. W. Munson and H. Abdine, *Talanta 25*:221 (1978).
198. J. W. Munson and H. Abdine, *J. Pharm. Sci. 67*:1775 (1978).
199. M. A. Brooks and M. R. Hackman, *Anal. Chem. 47*:2059 (1975).
200. J.-C. Vire, J.-M. Kauffmann, and G. J. Patriarche, *J. Pharm. Biomed. Anal. 7*:1323 (1989).
201. J. Wang, T. Tapia, and M. Bonakdar, *Analyst 111*:1245 (1986).
202. J. Wang, M. S. Lin, and V. Villa, *Analyst 112*:1303 (1987).
203. A. J. M. Ordieres, M. J. G. Gutierrez, A. C. Garcia, and P. T. Blanco, *Analyst 112*:243 (1987).
204. T. R. I. Cataldi, A. Guerrieri, F. Palmisano, and P. Zambonin, *Analyst 113*:869 (1988).
205. R. J. B. Diez-Cabellero, L. L. De La Torre, J. F. A. Valentin, and A. A. Garcia, *Talanta 36*:501 (1989).
206. L. Hernandez, P. Hernandez, E. Lorenzo, C. Gonzalez, and I. Gonzalez, *Fresenius J. Anal. Chem. 336*:222 (1990).
207. J. Wang, M. Bonakdar, and M. M. Pack, *Anal. Chim. Acta 192*:215 (1987).
208. L Hernandez, P. Hernandez, M. H. Blanco, and E. Lorenzo, *Analyst 113*:1719 (1988).
209. O. Chastel, J.-M. Kauffmann, G. J. Patriarche, and G. D. Christian, *Anal. Chem. 61*:170 (1989).

# 27

# Electrochemical Detection in Liquid Chromatography and Capillary Electrophoresis

**Susan M. Lunte and Craig E. Lunte** *The University of Kansas, Lawrence, Kansas*

**Peter T. Kissinger** *Purdue University, and Bioanalytical Systems, Inc., West Lafayette, Indiana*

## I. INTRODUCTION

### A. Liquid Chromatography/Electrochemistry: The State of the Art

Liquid chromatography with electrochemical detection (LCEC) is in widespread use for the trace determination of easily oxidizable and reducible organic compounds. Detection limits at the 0.1-pmol level have been achieved for a number of oxidizable compounds. Due to problems with dissolved oxygen and electrode stability, the practical limit of detection for easily reducible substances is currently about 10-fold less favorable. As with all detectors, such statements of the minimum detectable quantity must be considered only with the proverbial grain of salt. Detector performance varies widely with the analyte and the chromatographic conditions. For example, the use of 100-$\mu$m-diameter flow systems can bring attomole detection limits within reach, but today this is not a practical reality.

The modern interest in electrochemical detectors for liquid chromatography was stimulated by the recognition that this technique was ideal for the study of aromatic metabolism in the mammalian central nervous system. Most of the papers published during the past 20 years have focused on applications of the LCEC technique to neurochemical problems. Since the first commercial detectors became available in 1974, a number of other areas of application have been explored as well. The trade publication *Current Separations* provides a useful overview of current applications [1]. The basic concepts of LCEC have also been reviewed in several publications [2–5].

Liquid chromatography (LC) and hydrodynamic electrochemistry are, for the most part, very compatible technologies that in combination yield important advantages for a number of trace determinations. In order of decreasing importance, the three major advantages are selectivity, low detection limits, and modest cost. The use of modern LC for residue determinations requires a selective detector with a rapid response time, wide dynamic range, and low active dead volume ($<20\ \mu L$). Because electrochemistry is a surface technique, small-volume transducers ($<1\ \mu L$) can easily be constructed that meet these criteria. Figure 27.1 illustrates the basic concept. A redox reaction supplies (or removes) electrons to (or from) an electronic conductor (the electrode). The resulting current is proportional to the concentration of reactant passing through the cell.

Direct-current amperometry (the measurement of electrochemical current in response to a fixed electrode potential) continues to be the most widely used finite-current electrochemical technique. Popular applications include endpoint

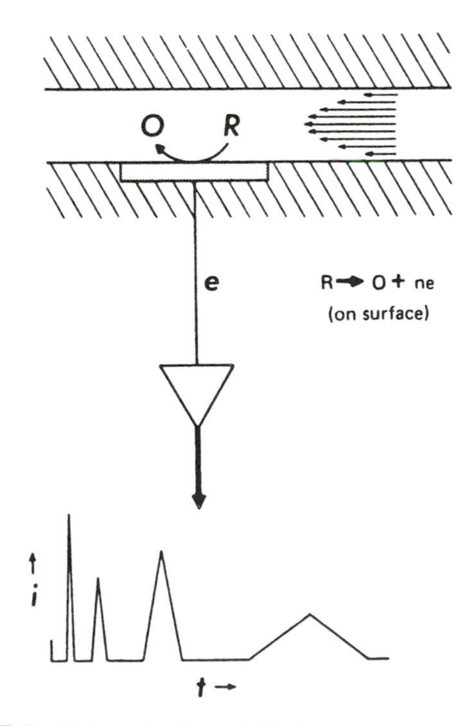


**Figure 27.1** Schematic view of thin-layer amperometric detection.

detection in volumetric and coulometric titrations, and the measurement of oxygen activity in gases and liquids. In addition, amperometry can be used to measure the activity of redox enzymes or the concentration of substrates for redox enzymes [6]. In the latter case, the enzyme itself is used as a reagent in the solution, trapped behind or in a membrane, or even covalently bound to the electrode surface. These types of detectors have been used in LC as well as in on-line biosensors.

It is now clear that electrochemical measurements can often have significant advantages over the classical spectroscopic approaches. Amperometry can be more specific; therefore, lower detection limits are often feasible. Because electrochemical detectors do not require optical carriers, they can be much less expensive than UV absorption or fluorescence detectors. This is especially true when one considers that electrochemical detectors are inherently tunable without the need for such things as monochrometers or filters. On the other hand, there can be significant problems with reliability, and, more often than not, there is a lack of acceptance by chemists weaned on Beer's law. Amperometric methods in biochemistry are just beginning to be commercialized, and it is now almost certain that they will come into widespread use.

Another recent development is the advent of pulse amperometry in which the potential is repeatedly pulsed between two (or more) values. The current at each potential or the difference between these two currents ("differential pulse amperometry") can be used to advantage for a number of applications. Similar advantages can result from the simultaneous monitoring of two (or more) electrodes poised at different potentials. In the remainder of this chapter it will be shown how the basic concepts of amperometry can be applied to various liquid chromatography detectors. There is not one universal electrochemical detector for liquid chromatography, but, rather, a family of different devices that have advantages for particular applications. Electrochemical detection has also been employed with flow injection analysis (where there is no chromatographic separation), in capillary electrophoresis, and in continuous-flow sensors.

## B. Fundamental Principles

As described in Chapter 3, amperometry is normally carried out in stirred (or flowing) solutions or at a rotated electrode. Current is measured as the compound undergoes an oxidation or reduction at the electrode held at a fixed operating potential (strictly speaking, the potential difference between the electrode and the solution is held constant). Hydrodynamic voltammetry, a steady-state technique from which amperometry is derived, is used to select the operating potential. Thermodynamic data (e.g., tables of $E°$ values), although quite valuable for some purposes, are often insufficient for determining the operating potential for most amperometric methods. Electrode kinetics, which depend on a number of factors (electrode material, choice of electrolyte, etc.), also play a significant role in practice. Often a greater potential is required for amperometry than would be predicted on the basis of equilibrium thermodynamics.

LCEC is a special case of steady-state hydrodynamic chronoamperometry. In LCEC, the concentration changes as the chromatographic zones flow past the detector. The electrode is operated in the limiting current region for the eluted compounds, even though the concentration varies as the zones enter and leave the detector compartment. It is important to note that the volume of solution in the active region of the typical electrochemical detector ($\sim 1$ μL or even less) is very small compared to the volume occupied by the typical chromatographic zone ($\sim 0.1$–1 mL).

The resolution of the column provides much of the selectivity in LCEC; therefore, the practical limitations of amperometry are circumvented to a large extent. Nevertheless, amperometry is more often than not used to improve the selectivity of an LC method. Compounds that oxidize or reduce at low potentials can be detected with great selectivity.

Adjustment of the electrode potential and the use of complex excitation functions (pulse amperometry) can improve the selectivity of amperometric detection. However, amperometry in standard solutions is one thing and amperometry in complex samples is quite another. In the real world, direct amperometric detection without prior cleanup is useless in most applications. Proteins, lipids, and industrial surfactants can contribute to electrode fouling and coating, often rendering the electrode inoperative. This problem has been overcome to a large degree by (1) the addition of membranes over the electrode surface, which can exclude molecules on the basis of permeability, (2) the use of various extraction techniques to "clean up" the sample, and (3) the introduction of a liquid chromatography column prior to the amperometric transducer to separate the desired component from possible surface contaminants or other reactive compounds that might interfere. With these points in mind, amperometry can be a very reliable detection approach for even the most complex mixtures.

## II. TRANSDUCER DESIGN

### A. Electrode Materials

The choice of electrode material is more critical in LCEC than in the usual electroanalytical experiment, primarily due to the mechanical ruggedness and long-term stability required. "Carbon paste" (an admixture of graphite powder and a dielectric material) remains a useful choice as an electrode material for LCEC. While carbon paste can be used in nonaqueous solvents if formulated using high-molecular-weight waxes or various polymers (e.g., Kel-F or epoxies) as the dielectric binder, vitreous ("glassy") carbon has been generally found to be more satisfactory. Carbon pastes are the first "composite electrodes" and there is a possibility that they will reemerge in alternate forms. Vitreous carbon has excellent mechanical and electrical properties and is relatively free of impurities. It can be brought to a high surface polish using standard metallographic techniques. Besides its solvent resistance, glassy carbon has a more favorable background current at negative potentials than the usual carbon paste formulations, making it possible to carry out more difficult reductions.

It is not yet possible to make detailed comparisons between different electrode materials, because organic electron transfer reactions at surfaces are not well understood at a molecular level. Reactions that are favorable (fast, reversible) at one surface may be unfavorable at another surface that appears to have similar properties. Perhaps the most important characteristic of an electrode for hydrodynamic amperometry is the background current and noise at a potential where the compound of interest reacts at a mass transport-limited rate. The fact that one electrode gives a lower background current than another at a given potential is not by itself indicative of a better electrode. The electron transfer rates for the analyte(s) of interest might also be lower and the resulting signal-to-noise ratio could be much worse. At the present time, it would appear that glassy carbon and composites such as carbon paste have advantages in given situations. The superior solvent resistance of glassy carbon will in many cases decide the issue.

A number of other carbon materials have been used for electrochemical detection [6]; however, at this point none of these appear to have a clear advantage over the electrodes described earlier. Nevertheless, these alternative materials certainly do "work" and there is little doubt that we will continue to see additional entries as the search for the ideal electrode continues. The chapter on carbon electrodes by McCreery and Kneten (Chap. 10) is a good place to review the fundamental issues.

Mercury is the electrode material of choice for many electrochemical reductions and some unique oxidations (see Chap. 14). We have explored the use of both small mercury pools and amalgamated gold disks in thin-layer amperometry. Other workers have used pools in a capillary tube [7] and amalgamated platinum wire [8]. In 1979, Princeton Applied Research introduced a unique approach based on their model 303 static mercury drop electrode (see Sec. II.F). Our laboratories and MacCrehan et al. [9] have focused on the use of amalgamated gold disks. This approach results in an inexpensive, easily prepared, and mechanically rigid electrode that can be used in conventional thin-layer cells (Sec. II.C) of the type manufactured by Bioanalytical Systems.

Metal electrodes other than mercury have been used for a few inorganic LCEC applications, but appear to be unsuitable for most organic applications.

Copper electrodes have been used to determine amino acids and carbohydrates [10]. Metal oxide electrodes (including thin-film semiconductors) show some promise, but nothing of substance has yet been published with regard to LCEC. Pulsed amperometric detection (PAD) takes advantage of metal oxides formed in situ. This approach is discussed later.

It would appear certain that the most important need in LCEC is the development of improved electrode materials. It may be possible in the near future to design an electrode that will give superior performance for certain classes of compounds. Modifying electrode surfaces by covalent attachment of various ligands or electron-transfer catalysts (including enzymes) can provide the key to better amperometric devices for all sorts of analytical purposes. Research in the area of chemically modified electrodes (CMEs) has been reviewed (see Chap. 13) [6,11]. Those interested in improving the performance of electrochemical detectors would do well to study these developments in detail.

## B. Mobile-Phase Limitations

All detectors place limitations on mobile-phase composition to some degree; however, in electrochemical detection one must be conscious of the fact that a complex surface reaction is involved that depends on both the physical and chemical properties of the medium. Considerable effort may be required to optimize an LCEC determination. Both column and detector must be considered together. In fact, the entire chromatograph may be considered part of the electrochemical cell. Fortunately, it is possible to make some generalizations regarding chromatographic mobile phases compatible with the electrochemical detector. For all practical purposes, nonpolar organic solvents cannot be used because of their inability to support a significant ionic strength, necessary for conductivity. This seemingly rules out classical "normal-phase" separations on silica or alumina columns, although there have been publications to the contrary.

The majority of LC separations performed today are reversed-phase, including ion-exchange and ion-pair methods. The mobile phases are typically aqueous solutions, with or without organic modifiers (e.g., methanol, acetonitrile, tetrahydrofuran). Retention is altered by adjusting the modifier concentration, the pH, or the ionic strength, or by adding ion-pairing agents. In some cases temperature control can also be very useful. Mobile phases of this type are generally ideal for electrochemistry, which requires that the mobile phase be able to carry an ionic current, be chemically and electrochemically inert in the desired potential range, have a convenient liquid temperature range, and dissolve the analyte of interest.

When buffered solutions are used, they are usually in the moderate- to low-ionic strength range (0.010–0.10 *M*) to provide adequate conductivity while minimizing the contribution to the background current of impurities found in the buffer reagents. Trace impurities found in the mobile phase are especially a problem when reductive electrochemical detection techniques are employed. Very minute amounts of electrochemically reducible metal ions, which apparently cannot be totally eliminated during the manufacturing process, become a major source of baseline problems.

Electrolysis of the mobile phase prior to the LC pump or injection valve can be an effective way to minimize such problems and may be an essential procedure when very low detection limits are required. Good reagents and pure solvents are clearly the best solution to this problem.

An even greater problem in reductive LCEC determinations is the removal of dissolved oxygen from the mobile phase. As in dc polarography, dissolved oxygen can cause serious problems. If it is not removed from the mobile phase and sample prior to injection, and precautions are not made to prevent oxygen from reentering the LC system after deoxygenating, the utility of this detection scheme is greatly limited. Even at relatively low working electrode potentials, very large background currents are produced by the reduction of oxygen; however, the severity of the problem depends on the electrode material (mercury or carbon).

Totally nonaqueous mobile phases [such as acetonitrile, dimethylformamide (DMF), and dimethyl sulfoxide (DMSO)] can be used with electrochemical detection but require salts such as tetrabutylammonium hexafluorophosphate or tetrafluoroborate. Using these totally nonaqueous solvents can give a distinct advantage when attempting to detect very difficult to oxidize or reduce compounds. Many times the potential limits in a particular electrochemical system are determined by solvent breakdown (e.g., $H^+$ reduction, $H_2O$ oxidation). In aqueous solutions, the limits are from about $+1.2$ to $-1.2$ V, depending on the electrode. In dry acetonitrile, with either of the salts mentioned earlier, the limits can be extended from $+3$ to $-3$ V. Many reactions inaccessible in water become facile at high potentials; examples include oxidation of alcohols and aromatic hydrocarbons, and reduction of activated alkenes. The major problem in using these reactions for trace determinations is the difficulty in eliminating water from the mobile phase. Water must be removed to the ppm level to avoid major background current problems. In addition, the higher impedance of nonaqueous electrolytes makes noise pickup more of a problem than for water. These problems are not easily solved, but the overall utility of electrochemical detection could be greatly expanded by the advent of nonaqueous LCEC.

### C. Thin-Layer Cells

The most popular electrochemical detectors to date have been based on the amperometric conversion of analyte in a cross-flow thin-layer cell. The basic functioning of this mode of detection is depicted schematically in Figure 27.2.

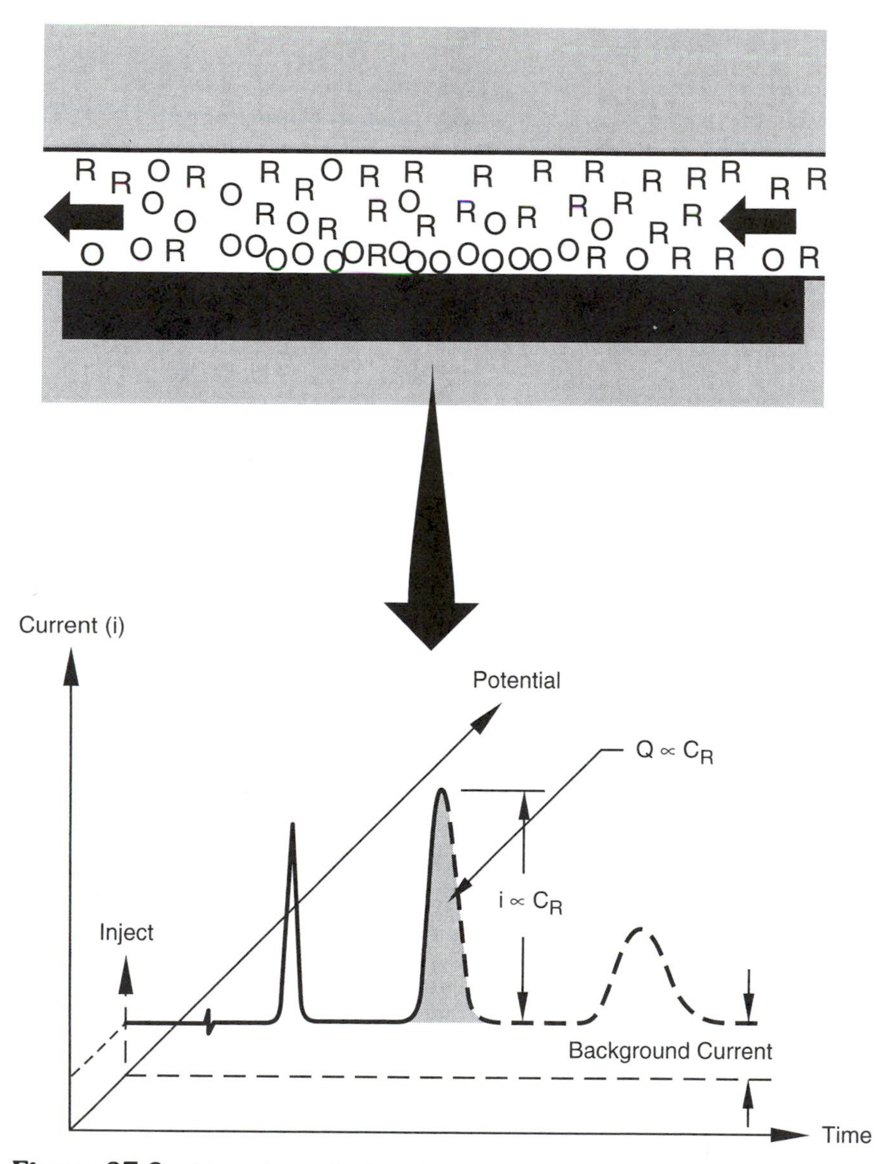


**Figure 27.2** At any instant in time, the current in an LCEC experiment reflects the rate of conversion of reactant (R) to product (O).

As illustrated, an electrochemically active substance passes over an electrode held at a potential sufficiently great (positive or negative) for an electron transfer (either oxidation or reduction) to occur. An amperometric current is produced that is proportional to the concentration of analyte entering the thin-layer cell.

As the concentration changes as a function of time due to elution of the analyte from the chromatographic column, so does the current; thus the usual chromatographic trace is recorded when current is displayed as a function of time. Thin-layer cells have excellent flow characteristics and a high ratio of electrode surface area to solution volume. In contrast to tubular electrodes (discussed later), the thin-layer geometry is suitable for essentially all electrode materials and provides for convenient surface polishing and chemical modification. Thin-layer cells can be used to react a negligible amount of analyte or all of it, depending on the mass transport efficiency of the cell (related to electrode area, channel thickness, and flow rate). All faradaic electrochemistry ultimately depends on Faraday's law:

$$Q = nFN \tag{27.1}$$

where the number of coulombs (Q) passed is directly proportional to the number of moles (N) converted to product in the chromatography zone, number of electrons (n) involved in the reaction, and the Faraday constant ($F = 9.65 \times 10^4$ C/eq). The instantaneous current is given by $i = dQ/dt = nF\,dN/dt = 9.65 \times 10^4 \times$ (equivalents converted per second). A current of $10^{-11}$ A (easily measured in practice) therefore corresponds to a conversion rate of $10^{-16}$ eq/s. The detection limit actually achieved will of course depend on the chromatographic efficiency, the detector conversion efficiency, and the baseline noise. Unfortunately, these three are not independent.

Amperometric detectors can operate over a range of conversion efficiencies from nearly 0% to nearly 100%. From a mathematical point of view, a classical amperometric determination (conversion of analyte is negligible) is one where the current output is dependent on the cube root of the linear velocity across the electrode surface as described by Levich's hydrodynamic equations for laminar flow. Conversely, the current response for a cell with 100% conversion is directly proportional to the velocity of the flowing solution. While the mathematics describing intermediate cases is quite interesting, it is beyond the scope of this chapter.

The question arises as to what conversion efficiency is best suited for electrochemical detection for LC. At present, low-efficiency detectors have the advantage of being easier to construct and have lower detection limits. To obtain 100% conversion, the electrode must have a large surface area. This can be achieved in practice only at the expense of a larger dead volume and residence time. Such cells are difficult to construct. The electrodes are usually beds made from conducting particles or fibers to increase the surface area-to-volume ratio (see Sec. II.E). While 100% conversion does give an absolute measurement for those compounds completely resolved chromatographically, this is often not a significant advantage for LC. As the electrode surface area is increased to improve efficiency, each added increment contributes proportionately less to the total amount converted but approximately equally to the background current from solvent, electrolyte, and electrode breakdown. Detector noise increases with the increasing background current developed by these large electrode areas, and the signal-to-noise ratio decreases. Although less material is converted at smaller electrodes, lower detection limits are attainable. It should, however, be obvious that extremely low conversion efficiencies are also not desirable because the current becomes extremely small and noise from the electronics will take over to defeat us. There does not appear to be an optimum cell geometry that will hold up for all applications.

The iR drop commonly associated with thin-layer electrochemical cells has not been a significant problem for most LCEC applications because aqueous mobile phases of relatively high ionic strength have been used. For the commercial version of this cell, a linear range from 1 pg to 1 $\mu$g injected is easily achieved for the typical separation of small molecules (typically MW of 200) with capacity factors between about 2 and 5. Obviously, many factors can influence the detector response in a given case, including volume flow rate, channel thickness, chromatographic efficiency, ionic strength, applied potential, electron transfer rate constant, temperature, and background current (chromatographic "baseline"). When thin-layer detectors are used in nonaqueous solvents, the iR-drop problem can be very severe due to the poor conductivity. In addition, severe iR problems can exist when operating at extremes of potential (positive and negative) where the background current is very great (often far greater than the chromatographic peak heights). Although the background current is always "bucked out" by the electronics and is therefore not "seen" on the chart recorder, it can be the major contributor to the iR drop.

In thin-layer cells with a downstream auxiliary electrode, the current between the working and auxiliary electrodes passes along the thin-layer channel, including that portion adjacent to the electrode (Fig. 27.3A). Therefore, the potential between the working electrode and the solution is not uniform across the face of the electrode. The potential difference at the downstream edge of the electrode will be closest to the value controlled by the potentiostat, whereas the upstream potential may be insufficient to oxidize or reduce the compounds of interest! The best solution to this problem is to position the auxiliary electrode opposite the working electrode so that the charge passes perpendicular to the working electrode surface (Fig. 27.3B). This geometry ensures both that the uncompensated iR drop will be extremely small (due to the thinness of the layer) and that the potential will be uniform at all points on the working electrode surface. Moving the electrodes in relation to each other certainly impacts on the physical properties of the cell as an electronic circuit element. A more detailed treatment of the iR problem requires rather complex mathematics; however, the influence on the design of practical detector cells is important and can be easily understood from a qualitative point of view.

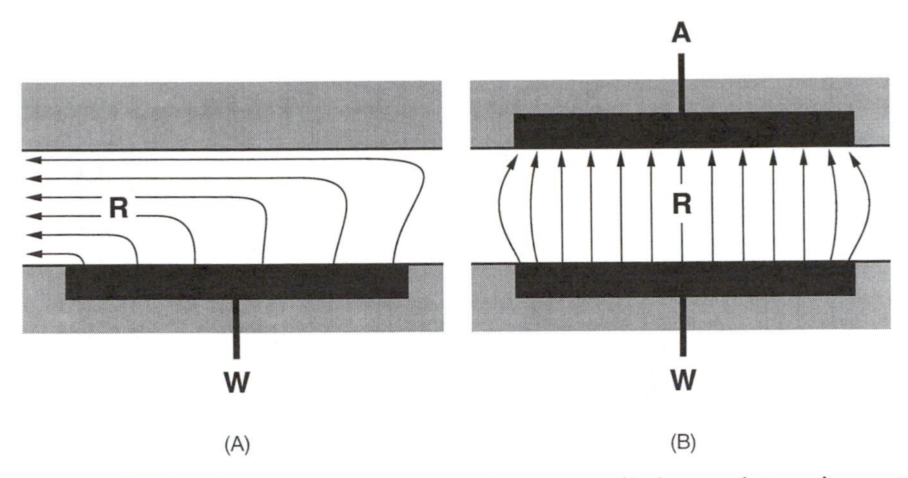


**Figure 27.3** Thin-layer experiments are subject to large iR drops under certain conditions (A). Placement of the auxiliary electrode across the channel from the working electrode (B) minimizes this problem.

Figure 27.4 illustrates five cross-flow cell geometries that we have investigated for a number of years and are currently using for various applications. Figure 27.4A illustrates the original geometry employed for LCEC experiments with both the reference (R) and auxiliary (A) electrodes mounted downstream. This design is relatively inexpensive, but can suffer from iR-drop problems as described earlier. It is important to recognize that for this design the three-electrode system does not compensate for any significant solution resistance, but does have the advantage of not drawing current through the reference electrode. At low currents the auxiliary electrode is not needed and the cell can be operated with the reference electrode serving as a classical counter electrode (as in two-electrode polarography or Clark amperometric oxygen electrodes). This cell has the disadvantage that it is not suitable for collection of fractions (especially radiolabeled metabolites from metabolism studies).

Figure 27.4B illustrates a simple modification in which a planar auxiliary electrode is placed opposite the working electrode. The thin-layer channel serves as a salt bridge to the reference electrode, but since no current passes along the thin layer, there is no resulting iR drop in the solution. In most instances the transit time for molecules along the electrode face is short compared to the diffusion time across the thin layer of solution. Therefore, no interference occurs between chemical events taking place at the working and auxiliary electrodes. Some attention should be paid to this problem with large electrode areas, low flow rates, and very thin gaskets, in which case the geometry shown

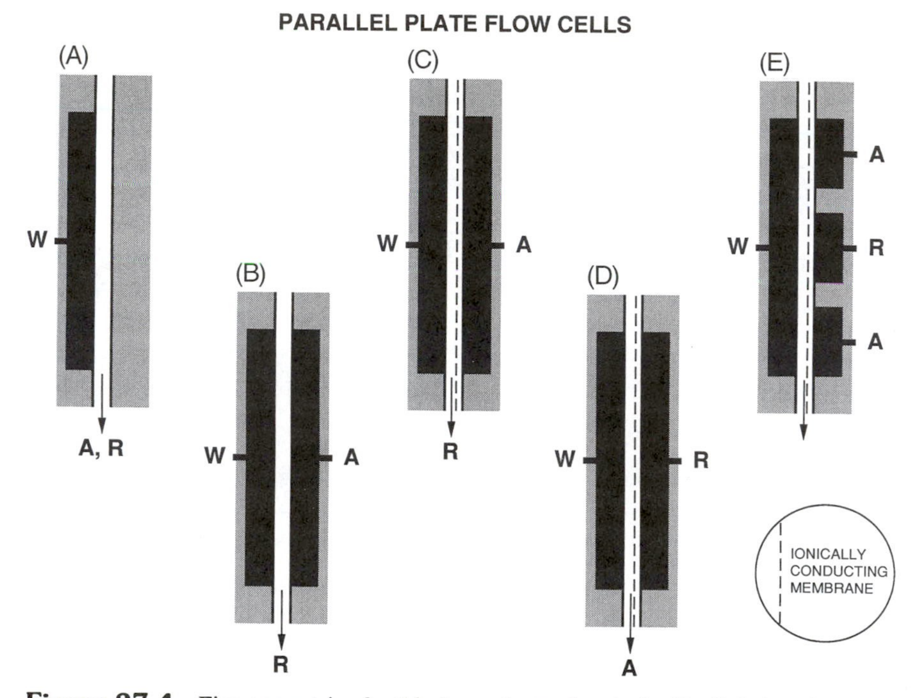


**Figure 27.4** Five geometries for thin-layer electrochemical cells. Relative placement of the working (W), reference (R), and auxiliary (A) electrodes is shown.

in Figure 27.4C can be advantageous. The problem here is that cell construction and maintenance becomes awkward when a porous diffusion barrier must be incorporated in the cell to isolate the auxiliary electrode.

Placing the reference electrode probe opposite the working electrode in the wall of the thin channel can best be accomplished by use of an ionic junction constructed from a porous material that permits ions to migrate but blocks the flow of electrolyte solution (Fig. 27.4D). Porous Vycor glass ("thirsty" glass, Corning 7930 glass) or ion exchange membranes (DuPont Nafion 811) are useful for this purpose. With the potential-sensing element across the stream from the working electrode, the uncompensated solution resistance is small (but not uniform) and the auxiliary electrode can be placed downstream with relatively little ill effect. The compensated resistance, however, is quite large and the current must pass along the layer of solution adjacent to the electrode.

Configurations 27.4B and C are superior on fundamental grounds because the current density across the electrode will not be significantly influenced by iR drop, and the potential of the auxiliary electrode will remain low due to the low impedance between A and W.

Configuration 27.4E is perhaps most satisfactory from a fundamental point of view because the uncompensated solution resistance and the compensated resistance are both small. As in B, the current density across the electrode will not be significantly influenced by iR drop. This design also provides for convenient collection of fractions. In summary, configurations 27.4A, B, and E are the most practical approaches to rugged cross-flow cells capable of routine use for many months. With the geometry illustrated by B the linear range of the detector extends to over six orders of magnitude even in mobile phases with poor conductivity. In general, configuration 27.4A will work for most trace applications, and 27.4B or E should be used for those cases where either a high background current or high resistance causes significant iR problems. Figure 27.5 illustrates early commercially available detectors of type 27.4B that are marketed by Bioanalytical Systems. Figure 27.6 illustrates a more recent cross-flow design with the added feature of parallel or series working electrodes. A historical perspective on the development of these cells has been reported [12].

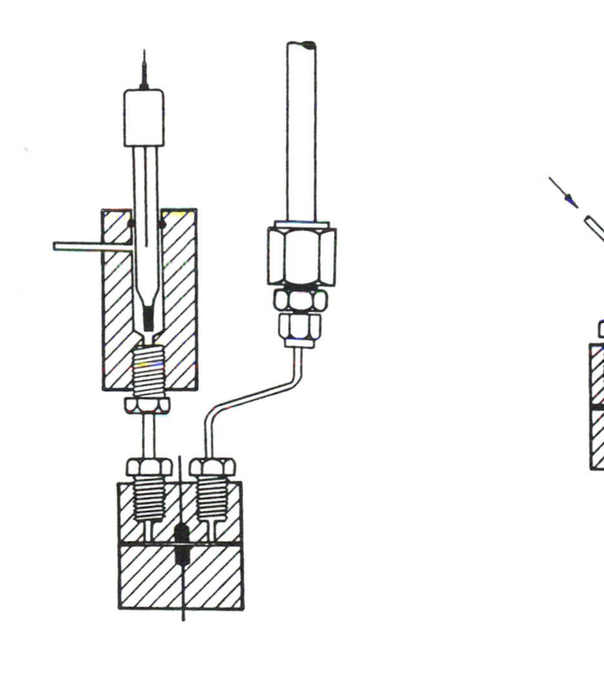

**Figure 27.5** Commercial thin-layer LCEC detectors. [Courtesy of Bioanalytical Systems, Inc.]

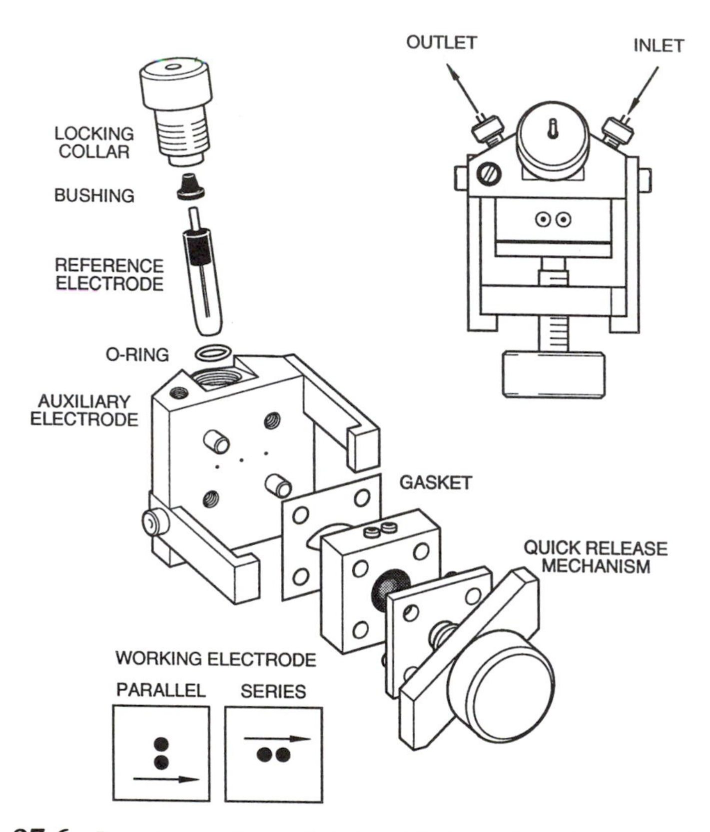


**Figure 27.6** Recent cross-flow cell design. Working electrodes may be in parallel or series.

While thin-layer cells have most commonly been used with flow parallel to the electrode surface as described earlier, several detectors have employed a radial-flow geometry and operate with flow entering from a jet perpendicular to and centered on the electrode surface as illustrated in Figure 27.7. This is intended to reduce dead volume and provide more effective mass transport to the electrode surface. The cell illustrated acts as an end fitting for a microbore LC column. Thin-layer cells with a radial-flow (vs. cross-flow) geometry give superior performance at lower flow rates [13]. While conventional LC columns operate at $\sim 1$ mL/min, it is not uncommon to use microbore columns at 10 $\mu$L/min, a hundredfold lower flow. It is important not to confuse these cells with the wall-jet concept. Here the orifice is very small and close to the working electrode. The cell is very thin and the wall-jet hydrodynamics are blocked since there are two walls.

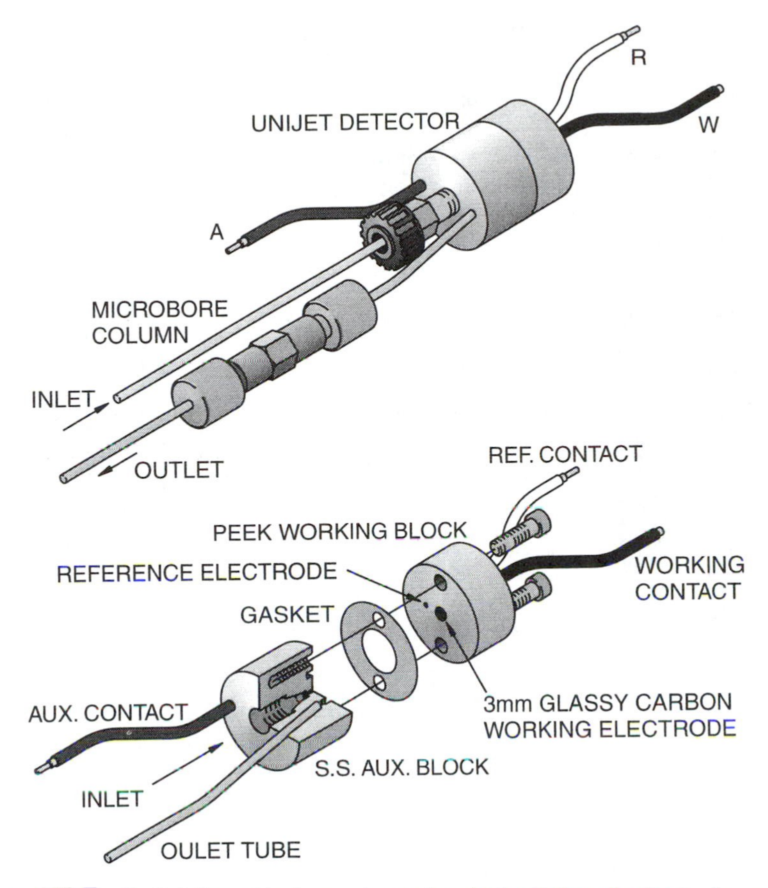


**Figure 27.7** Radial-flow thin-layer electrochemical cell for microbore chromatography.

## D. Tubular Electrodes

Flow-through electrochemical detectors based on a cylindrical geometry have been used for many years for a variety of purposes. For years Walter Blaedel and his co-workers at the University of Wisconsin stimulated interest in so-called tubular electrodes, although several other groups carried out pioneering work as well. Figure 27.8 illustrates three tubular electrode configurations. All three geometries have been used for LCEC applications. Electrochemists like open tubular electrodes (Fig. 27.8A) because the mass-transport mathematics for the cylindrical geometry is relatively easily solved. The problem with this approach to LCEC detectors is that the surface of the electrode is relatively difficult to polish and modify (compared to planar thin-layer cells), some electrode mate-

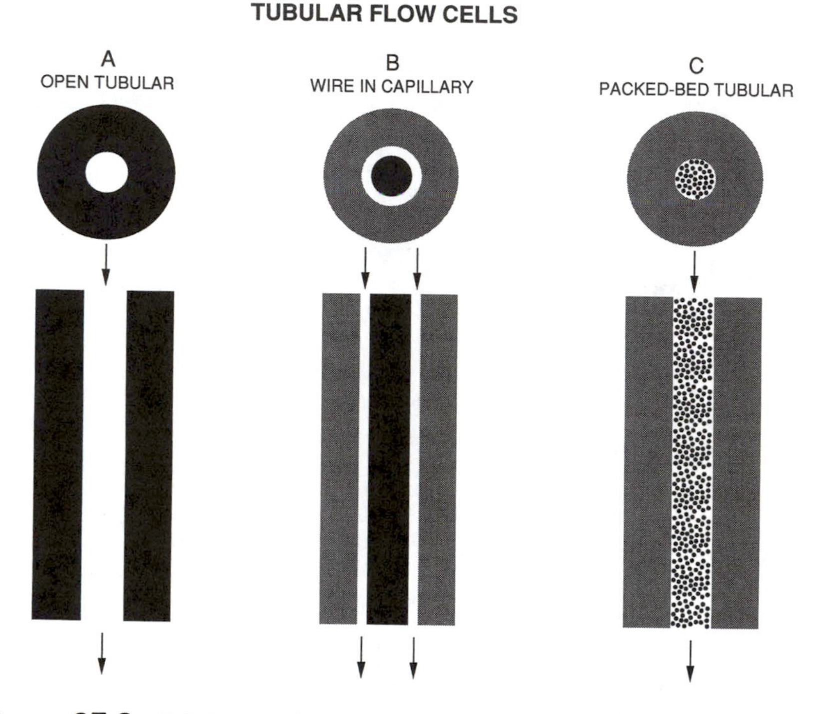


**Figure 27.8** Tubular working electrodes for hydrodynamic electrochemistry: (A) open tubular electrode; (B) wire electrode in capillary; (C) packed-bed tubular electrode.

rials cannot be easily used (e.g., mercury or carbon paste), and it is not possible to use a large surface area-to-volume ratio without encountering a severe iR drop (since the auxiliary and reference electrodes cannot be easily placed in an optimum position). In many cases these problems are not particularly significant and satisfactory results have been obtained.

The wire in a capillary geometry (Fig. 27.8B) has been used in two forms by electrochemists. In one case the capillary is an impervious diffusion barrier, and in the other case a capillary is used that is capable of transporting ions in a radial direction. In the latter situation, the auxiliary electrode can be positioned in a solution outside the capillary in order to minimize uncompensated iR drop. Porous Vycor glass (Corning 7930 glass) and ion exchange membranes (DuPont Nafion) have been used for the capillary, but in both cases construction difficulties prohibit the production of reliable devices capable of withstanding a variety of mobile phases. A wire placed in ordinary glass or in a hole drilled in a chemically resistant plastic is not quite as elegant, but will work very well for most low-current LCEC applications. Johnson and his group at Iowa State have reported a very simple cell that functions with a typical conversion efficiency of 35% using a platinum wire electrode [14]. While centering the wire in the capillary would help the theoreticians, this is not necessary for a practical device and greatly simplifies construction of the detector cell. This type of electrode is commonly used in open tubular liquid chromatography and capillary electrophoresis, where a small electrode is needed in order to detect effluent eluting from 10–100 μm i.d. fused silica capillaries.

## E. Packed-Bed Electrodes

Tubular packed-bed electrodes are extremely useful for electrosynthetic purposes, and adapting them for LCEC detection has been a worthwhile endeavor (Fig. 27.8C). Johnson and his students have also pioneered in this area. As for the wire in a capillary, one can distinguish between electrodes packed in an impervious tube (usually metal or graphite) and packed in a porous ionic conductor (usually porous Vycor). If the particles in the bed are small (typically metal chips or graphite powder), high efficiency can be achieved in a relatively short bed (a few millimeters) at typical LC flow rates, as long as the current is low (always true for trace analysis). Under these conditions an impervious bed works fine and the auxiliary and reference electrodes can be conveniently placed downstream (as shown in Fig. 27.4 for a thin-layer cell). If a high material throughput is necessary (higher concentration at a higher linear velocity), it is essential to operate the cell with current perpendicular rather than parallel to the axis to minimize iR drop. In such cases a porous tube is necessary. An example of a flow cell of this type is shown in Figure 27.9. This cell has been used for synthetic purposes and to remove electroactive contaminants from LC mobile phases. The ESA Company pioneered the commercial use of small porous carbon electrodes for high conversion efficiency LCEC. Their detectors have been especially well received for multiple electrode experiments (e.g., 16 working electrodes).

Packed-bed electrodes need not be cylindrical. Takata and Muto reported on a rectangular design some years ago in which a bed of carbon fibers was used for the electrode material [15]. While a number of innovative applications were reported, for the reasons described earlier, cells of this type do not provide detection limits competitive with those that can be achieved using more conventional amperometric detectors.

## F. Dropping Mercury Electrodes

While various dropping mercury electrodes (DMEs) have long been proposed as LC detectors, it is fair to say that they have never been put to significant practical use for this purpose. Kemula innovated this area in 1952 [16]. As the efficiency of LC columns improved, it became essential to reduce the volume

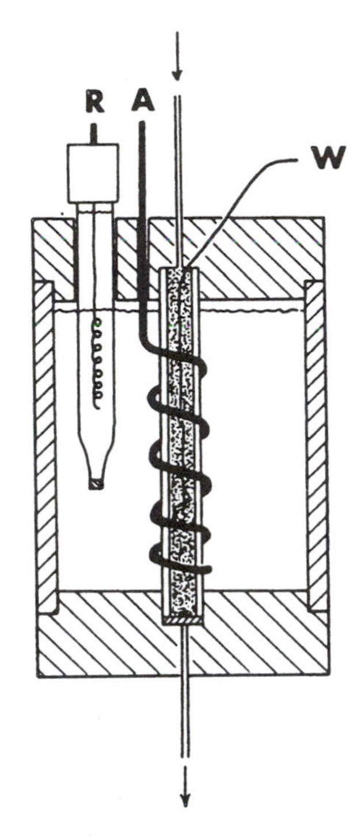


**Figure 27.9** Flow-through packed-bed coulometric reactor.

of DME-based detector cells and significant early work was published by Koen et al. [17], Stillman and Ma [18], and Fleet and Little [19], among others. Several papers have appeared that advocate the use of a rapidly dropping DME using a nonvertical (or even horizontal) capillary. Michel and Zatka were apparently the first to try this [20], and subsequent reports appeared from the Frei group in Amsterdam [21,22] and from the Kemula group in Warsaw [23]. The active volumes of these devices are adequately small for modern liquid chromatography.

Unfortunately, the very concept of a dropping mercury electrode is fraught with problems of both a fundamental (high double-layer charging due to changing surface area) and a practical nature (clogged capillaries, mercury toxicity). At this date, the minimum detectable quantities achieved with DMEs in flow streams are not competitive with stationary electrodes. Nevertheless, the renewable surface and the high hydrogen overpotential of pure mercury are decided advantages that may give the DME a competitive edge for some applications.

## G. Multiple-Electrode Detectors

While most practical LCEC experiments will continue to be carried out with transducers incorporating a single working electrode, it is a relatively simple matter to monitor the current at several working electrodes simultaneously. The electronics needed to accomplish this are very straightforward (Chap. 6) and in principle any number of electrodes could be used. In practice, the use of more than two electrodes is rather awkward and expensive. Figure 27.10A schematically depicts a thin-layer cell with two working electrodes in parallel, with a corresponding auxiliary electrode across the channel and a reference electrode downstream. Using this arrangement, one can monitor the amperometric current at two different applied potentials. This is perfectly analogous to the "dual-wavelength UV absorption detector," and the output is typically plotted using a dual-pen recorder. This arrangement is particularly useful when one wishes to quantitate a very easily oxidized (or reduced) substance in the presence of others which react at higher energies. The selectivity for the easily reacted substance can thus be excellent and the other compounds detected as well. The parallel dual-electrode arrangement can also provide good qualitative information in that the ratio of the response at each electrode will be characteristic of

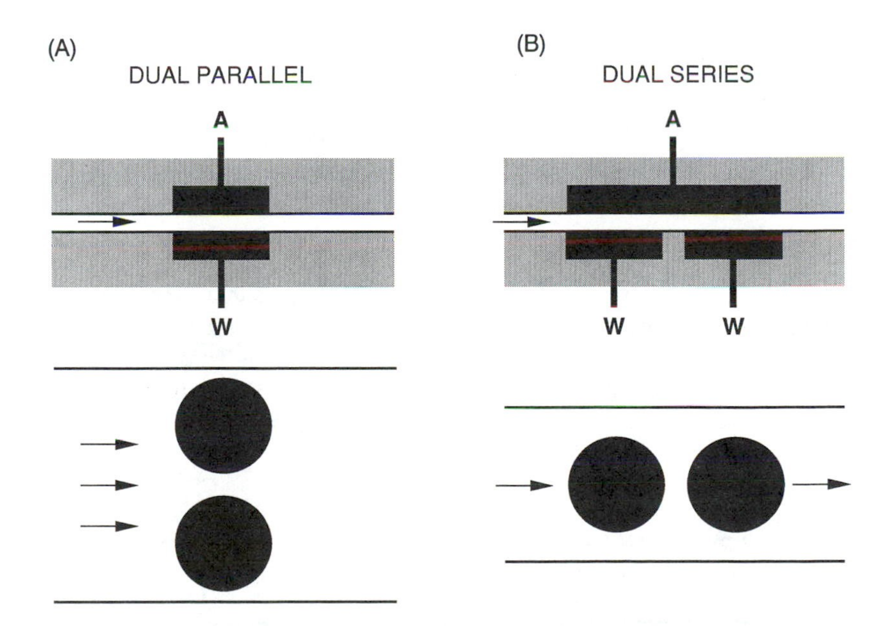


**Figure 27.10** (A) Dual parallel and (B) dual series thin-layer amperometric transducers.

a particular hydrodynamic voltammogram (just as the absorbance ratio at two wavelengths is characteristic of a given UV spectrum).

Figure 27.10B illustrates another arrangement, the series dual-electrode transducer. This device is used in the same manner as the classical ring-disk electrode (see Chap. 3). The products of the upstream electrode are monitored at the downstream electrode. This can be useful to enhance selectivity when the electrolysis products can be detected in a more favorable potential region than was necessary to carry out the original reaction. In addition, some electrochemical reactions are "chemically irreversible" and some are "chemically reversible," and therefore the dual-electrode arrangement can eliminate the response to irreversible processes while enhancing the response to reversible reactions. Both the parallel and series arrangements provide most of the advantages of differential pulse operation without the disadvantage of enhancing double-layer charging current. Instrumentation for LCEC based on these ideas is available from Bioanalytical Systems, and unique cells consisting of porous electrode reactors in series are available from the ESA Company.

Figure 27.11 illustrates a third dual-electrode arrangement that permits enhancement of the response by reversible redox cycling. Many more electrons are therefore transferred than would be the case with a single electrode, and the current is amplified dramatically. This concept does not work well with conventional LC columns because the volume flow rate is too large to permit a significant number of redox cycles. Nevertheless, the concept is certainly interesting, and, as reversed-phase capillary columns are developed, it may well have some practical value. A detailed treatment of multiple-electrode LCEC has been published [24].

## III. WAVEFORMS

Electrochemical detectors can be broadly classified as either amperometric or voltammetric. An amperometric detector is one in which the potential applied to the detecting electrode is held constant and the resulting current is measured as a function of time. A voltammetric detector is one in which the applied potential is varied with time and the current response is measured as a function

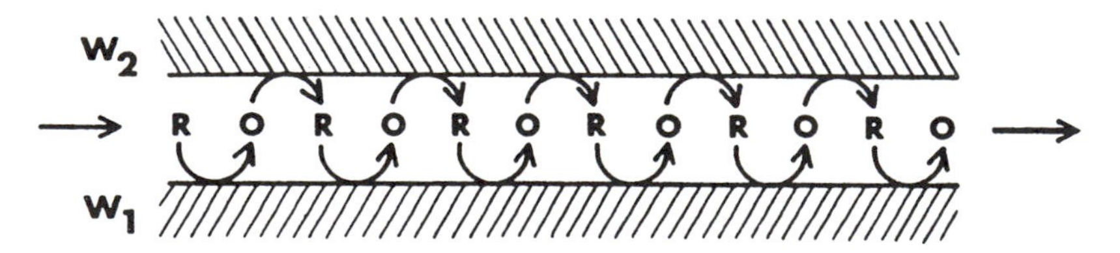


**Figure 27.11** Parallel opposed thin-layer amperometric transducer.

of both time and applied potential. The majority of reports on applications of LCEC and all commercially available electrochemical detectors deal with the amperometric type. Several designs of voltammetric detectors for liquid chromatography have been described in the literature, but thus far there have been few applications [25].

However, voltammetric detectors offer several advantages over amperometric detectors. By providing a complete voltammogram for each chromatographic peak, greater selectivity and more qualitative information are obtained. A voltammetric detector can be thought of as the electrochemical equivalent of a photodiode array absorbance detector. The major drawback to voltammetric detectors is that much higher detection limits are achieved compared to those of amperometric detection. This is the result of large background currents associated with scanning the applied potential. With the electrochemical flow cells currently used for chromatographic detection, this background current can be three to four orders of magnitude larger than the current response from the analyte. As LCEC is typically used because of the low detection limits it offers, the higher detection limits of voltammetric detection have been viewed as unacceptable.

## A. Voltammetric Detection

For voltammetry to be useful for chromatographic detection, the experiment must be designed such that the electrochemistry and chromatography are mutually compatible. The voltammetric scan must cover a potential window sufficient to obtain useful voltammetric information. If the potential window is too narrow, nothing is gained relative to simple dual-electrode amperometric detection. The voltammetric scan must also be sufficiently rapid that several scans can be obtained for each chromatographic peak. If the voltammetric scan is too slow, chromatographic resolution will be lost. For a conventional chromatographic separation in which a peak width is 30–60 s at a flow rate of 1 mL/min, these factors require a potential window of approximately 1 V and a scan rate of at least 1 V/s. Any voltammetric technique to be used for chromatographic detection must meet these minimal requirements and address the problem of the background current.

### *Square-Wave Voltammetry*

Most of the early voltammetric detectors for liquid chromatography were based on modifications of the dropping mercury electrode (DME) or hanging drop mercury electrode (HDME). With these cells, a square-wave potential scan was used because this waveform discriminates against the background current. Using square-wave voltammetry, the background current is minimized by subtracting the currents of the forward and reverse steps. Because mercury is easily oxidized, these detectors are limited to compounds that can be reduced. Examples of compounds detected by LC with square wave voltammetric detection include nitrosamines, nitroaromatics, quinones, triazine, and several metal species.

*Staircase Voltammetry*

Other waveforms have been evaluated for use with solid electrodes, particularly carbon. The staircase waveform gives the best results for a combination of reasons. First, each potential step of the staircase is identical in height, leading to more constant background currents. This background can be very effectively eliminated after data acquisition by background subtraction. Second, the potential steps are much smaller for staircase voltammetry than for any of the other waveforms. Because of this, the magnitude of the background current is also smaller. Third, because only a forward step is taken, the scan rate is twice that for techniques requiring a backstep. This point can become critical in maintaining the chromatographic resolution while performing voltammetry. Staircase voltammetry is currently the most commonly used waveform for voltammetric detection of chromatographic effluents. Detection limits of 1 $\mu M$ for various catecholamines have been achieved.

*Coulostatic Detection*

Instead of directly scanning the potential using a potentiostat, it is possible to indirectly achieve a voltammetric scan by coulostatic detection [26]. The potential of an electrode and the charge on that electrode are related according to

$$V = Q/C_{dl} \tag{27.2}$$

where V is the potential change that results from injection of Q coulombs of charge to an electrode with a double-layer capacitance, $C_{dl}$. Using this relationship, the potential can be changed through addition of small charge pulses through a coulostat. If a faradaic reaction occurs, this removes some of the charge and the voltage therefore decays. The average faradaic current, $i_f$, can be obtained from Equation 27.3,

$$i_f = V'C_{dl}/t \tag{27.3}$$

where V′ is the decay in potential and t is the time interval of the step. A large contribution to the background current with scanned potential methods is the charging of the electrode double-layer capacitance; because the coulostatic approach charges this directly, much of the background current is eliminated. A detection limit of 5 $\mu M$ was obtained for acetaminophen using coulostatic detection. Coulostatic detection does not appear to have widespread appeal, mostly because much more sophisticated electronics is required.

*Microelectrodes*

Several different types of electrodes have been employed for voltammetric detection. As stated earlier, the background current is directly related to the surface area of the electrode. By decreasing the size of the electrode, the amount of background current is diminished (Chap. 12). As the electrode becomes very small, the background current decreases faster than the faradaic current from the analyte. Jorgenson's group has used this phenomenon to design an electrochemical cell using a carbon fiber electrode for voltammetric detection [27]. Using a staircase waveform with the fiber electrode, a detection limit of $10^{-7}$ $M$ was found for both hydroquinone and catecholamines.

*Dual-Electrode Thin-Layer Cell*

The background current can also be eliminated through the use of a series configuration dual-electrode cell. In this scheme, the potential of the upstream electrode is scanned while the potential of the downstream electrode is held constant. The downstream electrode is used to monitor the redox reaction occurring at the upstream voltammetric electrode without the background current from scanning the potential. Because one electrode is operated voltammetrically and the other amperometrically, this detection scheme has been termed voltammetric–amperometric detection [28,29]. The downstream electrode can be used either to monitor the products of the upstream reaction, termed collection, or to monitor excess reactant, termed shielding. These processes are illustrated in Figure 27.12. A detection limit of 35 n$M$ was found for gentisic acid using voltammetric-amperometric detection. This detection scheme has also been used with a coulostatic system for the upstream electrode.

*Advantages of Voltammetric Detection*

The most obvious advantage of voltammetric detection relative to amperometric detection is that a voltammogram (current vs. potential) is obtained for each peak with a single chromatographic injection. To obtain the same information with an amperometric detector is a tedious process requiring multiple chromatographic runs and changes in the detection potential between runs. Voltammetric detection results in a three-dimensional array of current versus time and potential as shown in Figure 27.13. A voltammogram for a given chromatographic peak is obtained by plotting the current versus potential data at one time (the retention time of the compound). The voltammograms of the peaks from Figure 27.13 are shown in Figure 27.14. These voltammograms can be used to aid in peak identification by comparing the voltammograms for standard and sample peaks. When these data are used in addition to comparing the retention times, peak identities can be assigned with a much higher degree of certainty.

Because of the more extensive data array obtained by voltammetric detection, other data manipulation techniques to improve selectivity or sensitivity can also be devised. One example is to measure the difference in response at two different potentials in order to gain selectivity for compounds that are more difficult to oxidize. Compounds that are easily oxidized will give the same response at both potentials, and the difference signal will therefore be zero. Only

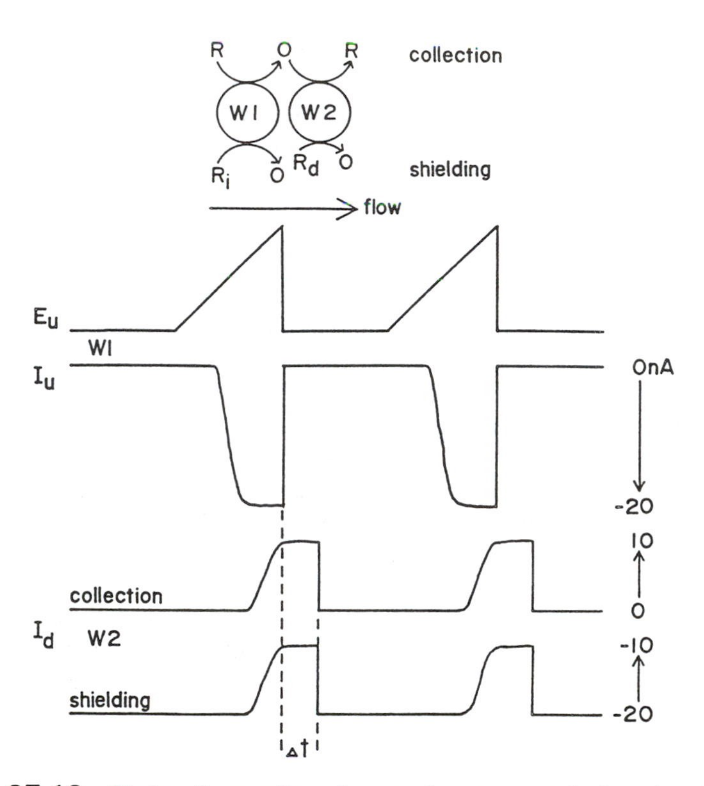


**Figure 27.12** Timing diagram for voltammetric-amperometric detection. W1, upstream electrode; W2, downstream electrode; $E_u$, upstream (W1) potential; $I_u$, upstream (W1) current; $I_d$, downstream (W2) current; t, time delay between electrodes. [Reprinted with permission from Ref. 25.]

compounds that are oxidizable at the high but not the low potential will give a response. A second application is to increase the resolution of the separation. Compounds that are not resolved chromatographically may be resolved voltammetrically.

## B. Pulsed Amperometric Detection (PAD)

Most aliphatic amines and alcohols are considered to be nonelectroactive. The reason for this is that the product of the oxidation adsorbs to the electrode surface, fouling the electrode. Therefore, most reactions of these compounds at noble metal electrodes have been transient and not amenable to direct amperometric detection. In voltammetry experiments, electrodes are cleaned between experiments by electrochemical or chemical treatment to restore the electrode response.

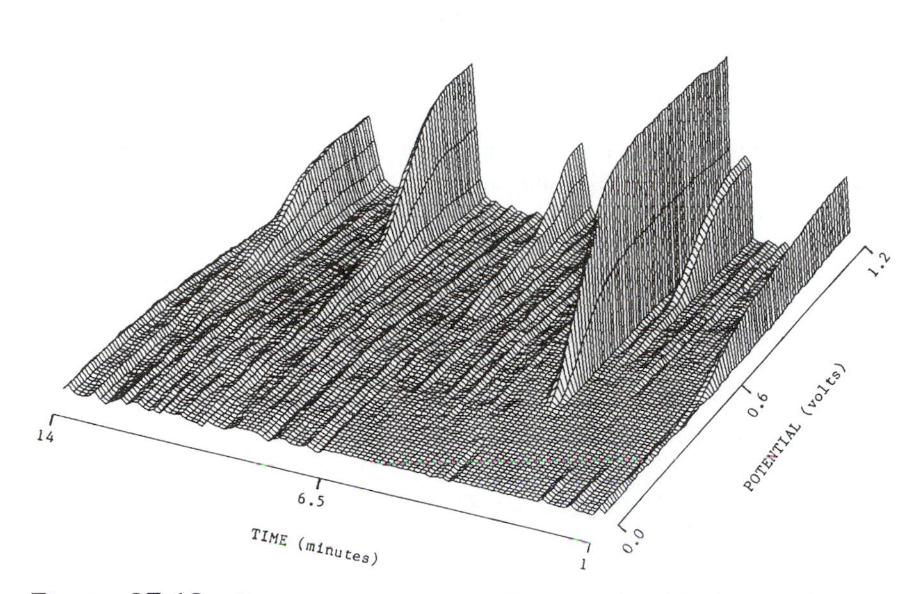


**Figure 27.13** Chromatovoltammogram of a phenolic acid mixture using voltammetric–amperometric detection. Conditions: scan rate, 2.0 V/s; W2 potential, −0.2 V vs. Ag/AgCl; flow rate, 1.0 mL/min. Peak identities (in order of elution): gentisic acid, 80 pmol; vanillic acid, 130 pmol; caffeic acid, 220 pmol; *p*-coumaric acid, 520 pmol; ferulic acid, 410 pmol; sinapic acid, 580 pmol. [Reprinted with permission from Ref. 25.]

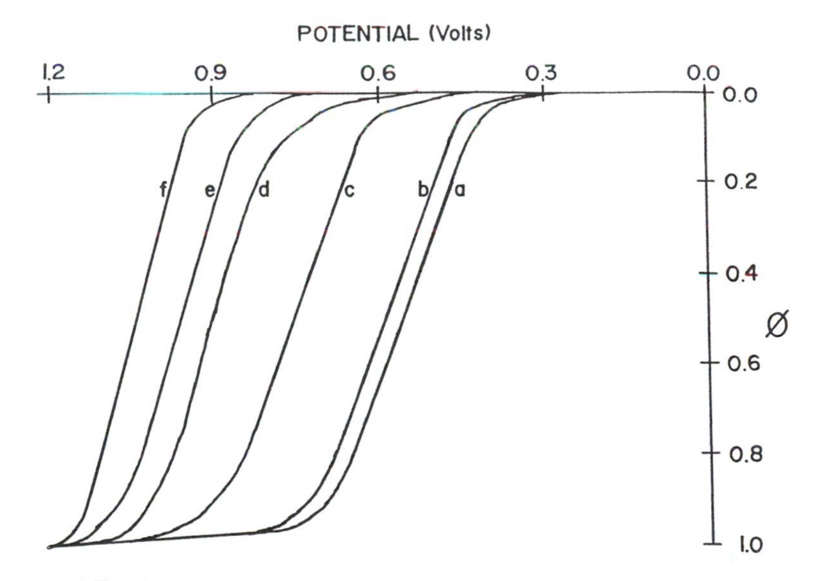


**Figure 27.14** Voltammograms of the chromatographic peaks extracted from the chromatovoltammogram of Fig. 27.12. Compounds: a, caffeic acid; b, gentisic acid; c, sinapic acid; d, ferulic acid; e, vanillic acid; f, *p*-coumaric acid. $\phi$ is the normalized current response. [Reprinted with permission from Ref. 25.]

Dennis Johnson and co-workers pioneered the area of pulsed electrochemical detection (PAD) [30,31]. In this technique, electrode fouling is circumvented through the use of a triple-pulse waveform. Alcohols, carbohydrates, amines, and sulfur compounds can all be detected using this technique. Alcohols and carbohydrates are determined by direct oxidation at the regenerated electrode surface, while amines and sulfur compounds undergo an oxide-catalyzed oxidation.

Figure 27.15 shows the residual voltammetric response for a gold rotated-disk electrode in 0.1 *M* NaOH. The part of the cyclic voltammogram indicated by A is due to the formation of the gold oxide. Part B is the evolution of oxygen, C is the reduction of the oxide produced at the electrode surface, and D is the reduction of oxygen. Aldehydes and reducing carbohydrates are detected in the range indicated by I and alcohols and nonreducing sugars are detected

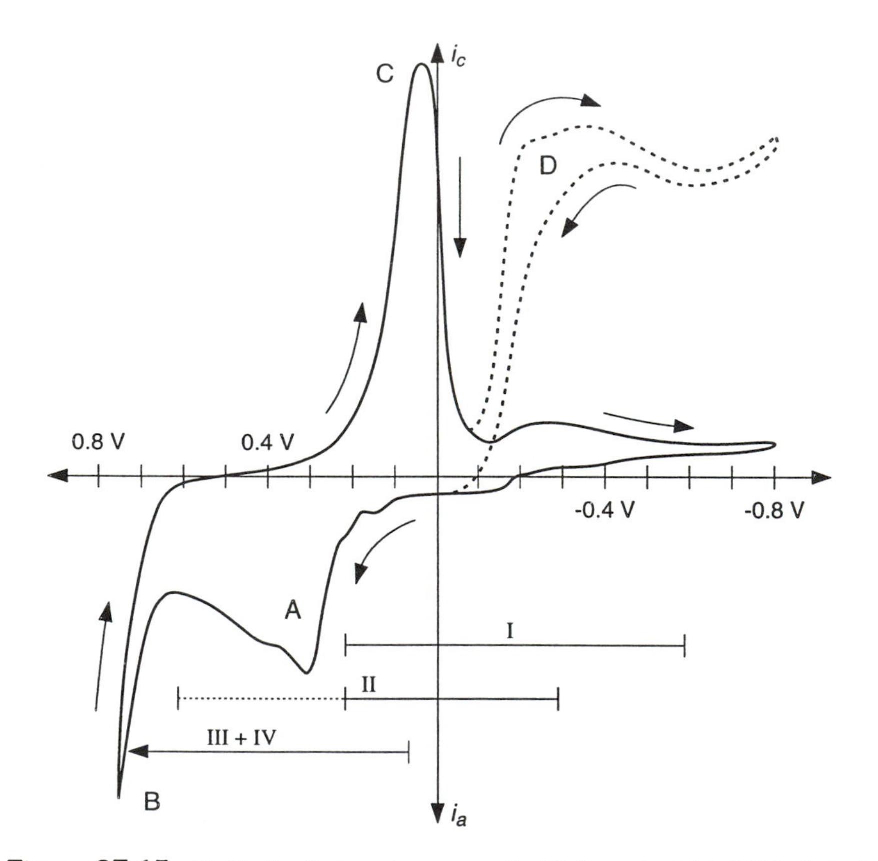


**Figure 27.15** Residual voltammetric response (i – E) for an Au rotated-disk electrode in 0.1 *M* NaOH. [Reproduced with permission from Ref. 31.]

in region II. In both cases, the anodic response is due to direct oxidation at the oxide-free electrode surface. Amines and sulfur compounds are adsorbed to the electrode surface at potentials of less than -0.1 V and are detected in region III through an oxide-catalyzed oxidation. The positive potential employed for the detection of these compounds is limited by oxygen evolution, which occurs at potentials above +0.6 V.

For the liquid chromatographic detection of carbohydrates and alcohols, a pulsed waveform of the type illustrated in Figure 27.16A is employed; the technique is called pulsed amperometric detection (PAD). In this case, the potential is stepped to E1, where oxidation of the compound takes place. The current is sampled for a short time at the end of the pulse (16.7 ms), where charging current is at a minimum. The potential is then stepped to E2, where the electrode undergoes oxidative cleaning. Last, the oxide-free surface is regenerated at potential E3.

In the case of oxide-catalyzed detection (amines and sulfur compounds), the baseline signal is due to a combination of double-layer charging and the formation of surface oxide. Therefore, the current does not decay as quickly as it does at a clean electrode surface. If a small detection period is employed (as described earlier), negative peaks can result because of the initial inhibition of oxide formation by the adsorbed analyte. Positive peaks can be obtained by increasing the sampling time to more than 150 ms.

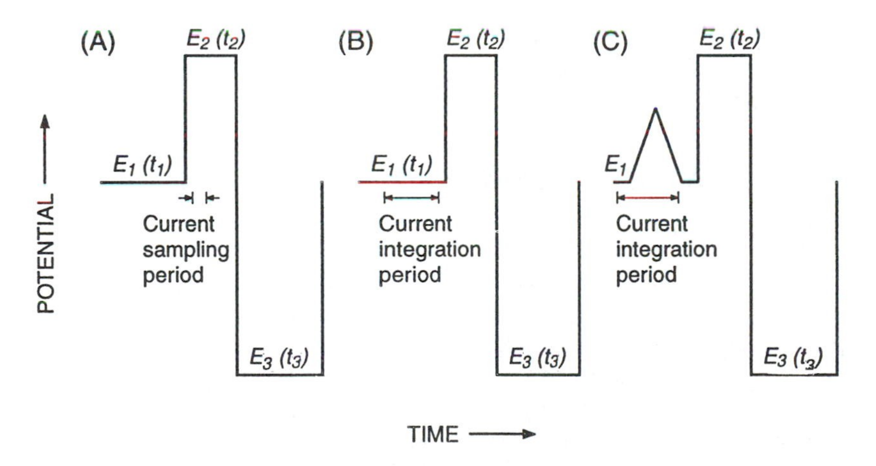


**Figure 27.16** Potential–time (E – t) waveforms. Processes: $E_1$, anodic detection; $E_2$, oxidative cleaning; $E_3$, cathodic reactivation. Waveforms are (A) pulsed amperometric detection (PAD) with a short current sampling period, (B) PAD with a long current integration period, and (C) integrated PAD with a long integration period. [Reproduced with permission from Ref. 31.]

Figure 27.16B shows the method known as pulsed coulometric detection (PCD). In this case, the current is integrated over a longer period and the time period is an integral number of 16.7-ms segments with typical total integration times of greater than 200 ms. The use of this type of waveform eliminates the most common electrical interference (60 Hz sinusoidal) encountered in pulsed electrochemical detection, and thereby increases the detection limits for most compounds.

The most recent addition to the PAD family is illustrated in Figure 27.16C. This is integrated PAD, which has an advantage over PAD for the determination of amines and sulfur compounds. The potential is scanned during E1 in the region of the oxide-catalyzed detection. The anodic charge for the formation of the oxide (A in Fig. 27.15) is counterbalanced by the corresponding cathodic charge obtained for the dissolution of the oxide (C in Fig. 27.15). This waveform eliminates background due to baseline drift caused by the gradual change of electrode area as it goes through the waveform thousands of times. Integrated PAD has no real advantage over PAD for the detection of carbohydrates and alcohols since these are not oxide-catalyzed reactions.

Most liquid chromatographic experiments performed with PAD employ alkaline mobile phases or use postcolumn addition of base to get the electrode at the appropriate pH for the formation of the oxide. The exceptions to this are the detection of carbohydrates and alcohols in acidic media and the detection of sulfur compounds. The oxidation of carbohydrates and alcohols is not oxide catalyzed, and since they exhibit a stronger adsorption to platinum than gold, they can be determined under acidic conditions. Sulfur compounds are adsorbed at oxide-free surfaces, and the kinetics for detection are favorable even at pH values below 7.

## IV. APPLICATIONS REVIEW

### A. Introduction

LCEC systems are used for a wide variety of applications, many of which have been published [1,32]. Space does not permit a thorough review; however, it is possible to generalize by considering the classes of compounds that have most frequently been studied: phenols, aromatic amines, thiols, quinones, and nitro compounds. By analogy with liquid chromatography with UV detection (LCUV), it is frequently desirable to obtain an electrochemical "spectrum" of a compound to assess its suitability for LCEC. Cyclic voltammetry (CV) is the electrochemical equivalent of spectroscopy (Chap. 3). It is useful to carry out CV experiments in several possible mobile phases, since electrochemical reactions can be very dependent on the medium.

## B. Oxidative Applications

*Phenols*

Phenolic substances are for the most part readily oxidized at a graphite electrode. The oxidation potentials for phenols vary widely with structure, and some (hydroquinones and catechols) are far more readily oxidized than others (cresols). Many compounds of biological interest (catecholamines, pharmaceuticals, plant phenolics) [32] and industrial interest (antioxidants, antimicrobials, agricultural chemicals) [33] are phenolic, and trace determination based on LCEC is now quite popular.

*Aromatic Amines*

Like phenols, aromatic amines are oxidized at a graphite electrode over a wide range of oxidation potentials. Some compounds (phenylenediamines, benzidines, and aminophenols) are ideal candidates due to their very low oxidation potentials; numerous applications have been developed.

*Thiols*

Thiols ("sulfhydryls" or "mercaptans") are very easily oxidized to disulfides in solution, but this thermodynamically very favorable redox reaction occurs very slowly at most electrode surfaces (e.g., glassy carbon). Therefore, LCEC methods for thiols usually depend on the unique behavior of these compounds at a mercury electrode surface at about +0.10 V (a very low potential). The reaction involves formation of a stable complex between the thiol and mercury. In a formal sense, the mercury, not the thiol, is oxidized. This approach has been used to determine the amino acid cysteine, the tripeptide glutathione, and the pharmaceuticals penicillamine and captopril. Allison and Shoup have developed a unique approach whereby both thiols and their corresponding disulfides can be determined in a single chromatographic experiment [34] employing a series dual-electrode detector. Besides thiols, many other sulfur-containing compounds are good candidates for LCEC.

*Miscellaneous Oxidizable Compounds*

A number of unique substances have been studied by oxidative LCEC. Being an excellent reducing agent, ascorbic acid is easily detected with excellent selectivity in very complex samples (food products, biological fluids, multivitamin products). Similarly, uric acid is readily detected in biological materials. The important enzyme cofactor NADH is readily oxidized at carbon electrodes. Indoles play a role in many biochemical processes and have been used as tagging reagents for electrochemical detection. Other heterocycles of pharmaceutical interest, including phenothiazines, imipramine, and pterins, are also uniquely suited to LCEC analysis.

## C. Reductive Applications

*Quinones*

Quinones are among the best-behaved organic compounds to undergo redox reactions in aqueous solutions. There are a reasonably large number of synthetic and natural products containing the quinone moiety, and many are excellent candidates for selective determination by LCEC. Unfortunately, some of the most important of these compounds are extremely hydrophobic due to the presence of long hydrocarbon side chains, and are therefore quite difficult to study by reversed-phase LC.

*Nitro Compounds*

Nitrobenzene was one of the first organic compounds studied by classical polarography in the 1920s. Thus, it is not surprising that nitro (and nitroso) compounds have been among the compounds most extensively investigated by both organic and analytical electrochemists. Aromatic nitro and nitroso compounds are very readily reduced at both carbon and mercury electrodes, but other compounds such as nitrate esters, nitramines, and nitrosamines are often good candidates as well. Many pharmaceuticals [32], explosives [35], agricultural chemicals [36], and important industrial intermediates fall into these classes, and a number of LCEC methods have been developed. Often, the selectivity is extremely good in biological and environmental samples because the nitro group is rare in nature and few other organic compounds are as easily reduced.

*Miscellaneous Reducible Compounds*

A number of organometallic compounds show promise for LCEC study, and a few have already been examined in detail (especially mercury alkyls) [9]. More highly conjugated organic compounds such as $\alpha,\alpha$-unsaturated ketones and imines are occasionally good candidates, but at this time UV detectors frequently outperform electrochemical detectors for such systems. At this writing there have been only a few reported LCEC studies of metal ions or metal complexes. Perhaps the major reason for this is that very little modern LC has been carried out on them using any detector. It is difficult to compete with atomic spectroscopy techniques for the determination of most elements, but as "speciation" becomes more important, it seems likely that more LCEC methods will be developed for metal complexes.

## D. Pre- and Postcolumn Reactions for LCEC

The limitations of electrochemical detection can be overcome in some cases by the use of pre- or postcolumn reactions. There are numerous examples of these reactions, and only a few are covered here. Several reviews that cover this area more thoroughly have been published [37,38].

*Precolumn Derivatization*

Precolumn derivatization is the most common mode of derivatization for LCEC. The ideal derivatizing reagent has the following properties: (1) the reagent reacts rapidly and quantitatively; (2) the reaction is selective for a specific functional group; (3) the reagent itself is not electroactive, but yields a product with a low redox potential; (4) the product undergoes an oxidation or reduction reaction that involves more than one electron; (5) the product is stable; and (6) the product is easily separated by reversed-phase liquid chromatography (RP-LC).

Reagents containing the aromatic nitro group have frequently been used to derivatize amines, aldehydes, ketones, and carboxylic acids to improve their characteristics for determination by absorption spectroscopy. The same or closely related reagents have been used to provide an electrochemically reactive handle for many of these same compounds [37,39]. Unfortunately, the popularity of these reagents has been limited by the necessity of using reductive detection, which requires rigorous deoxygenation of the liquid chromatographic system. Phenolic compounds, ferrocene, and aromatic amines have been employed as tags for oxidative detection. However, very few derivatizing reagents employing these functional groups possess all of the properties of an ideal derivatizing reagent for LCEC. In most cases, the reagent itself is electroactive and must be separated from the analytes of interest. In many other cases, a hydrolysis product is formed that is also electroactive and can interfere with the separation.

One group of tagging reagents that best approximates the properties of the ideal reagent is the aromatic dialdehydes. These compounds are electrogenic (that is, an electroactive product is produced from a nonelectroactive reagent). The aromatic dialdehydes form a Schiff base with primary amines, which then yields an isoindole product in the presence of a nucleophile. Both orthophthalaldehyde (OPA) and naphthalenedialdehyde (NDA) have been employed for the detection of primary amine analytes using *t*-butylthiol and cyanide, respectively, as nucleophiles. The products of both reactions are easily oxidized (+750 mV) and easily separated by RP-LC. These reagents have been used for numerous applications in the neurosciences. A common application is the determination of aspartate and glutamate in microdialysis samples [40,41] (see Fig. 27.17).

*Postcolumn Reactions*

Postcolumn reactions in LCEC can take several forms, most of which are illustrated in Figure 27.18. These include transformation of the analyte by chemical, electrochemical, photochemical, and enzymatic methods.

*Chemical Derivatization*

Postcolumn chemical reactions in LCEC are less common than in optical detection methods, since both the derivatization chemistry and the electrochemis-

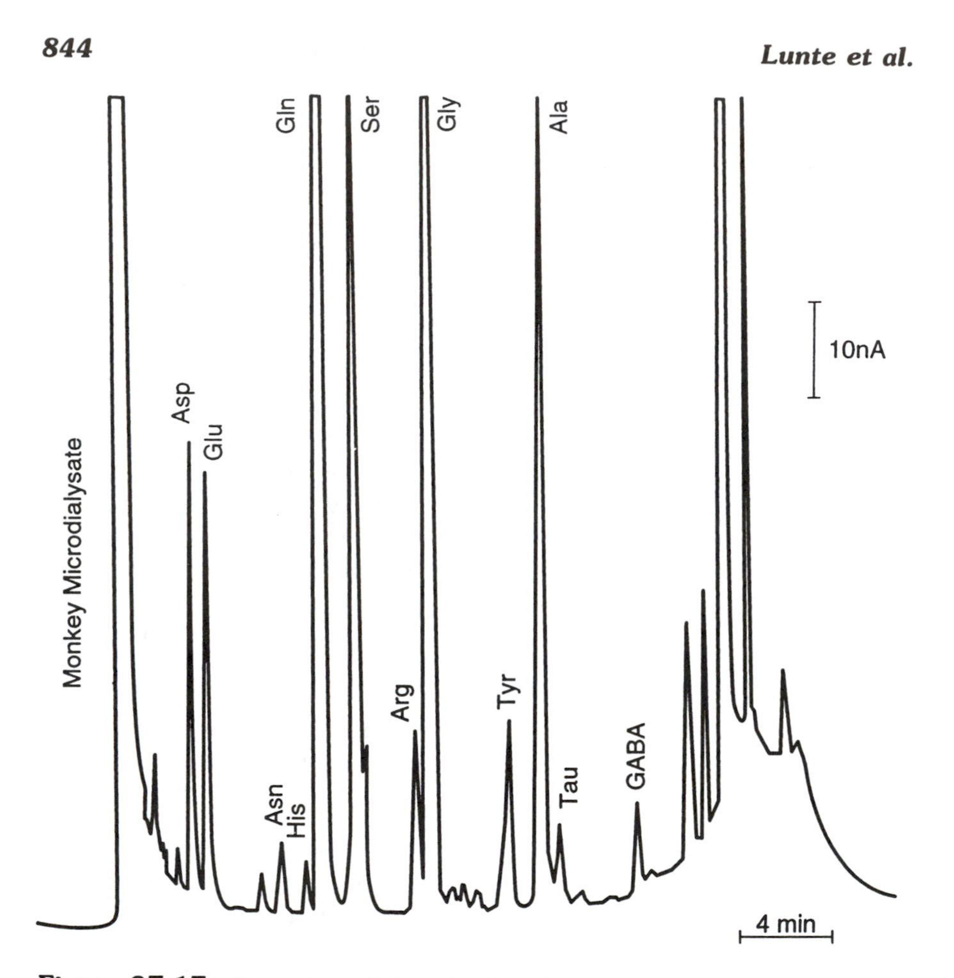


**Figure 27.17** Separation and determination of amino acids by reversed phase gradient LCEC of isoindole derivatives. The sample was obtained from an awake monkey using a microdialysis sampling probe to collect amino acids from the extracellular fluid of the brain.

try of the product must be considered in designing the reaction scheme. A successful example of postcolumn reaction is the detection of peptides using the classical biuret reaction [42]. In this scheme, Cu(II) reacts with peptides to produce Cu(II)–peptide complexes, which can be oxidized at a relatively modest potential. Selectivity can be further increased if a dual electrode system is employed. Cu(II)–peptides can be oxidized to Cu(III)–peptides, which can be detected reductively at a second electrode. The advantage of this method over

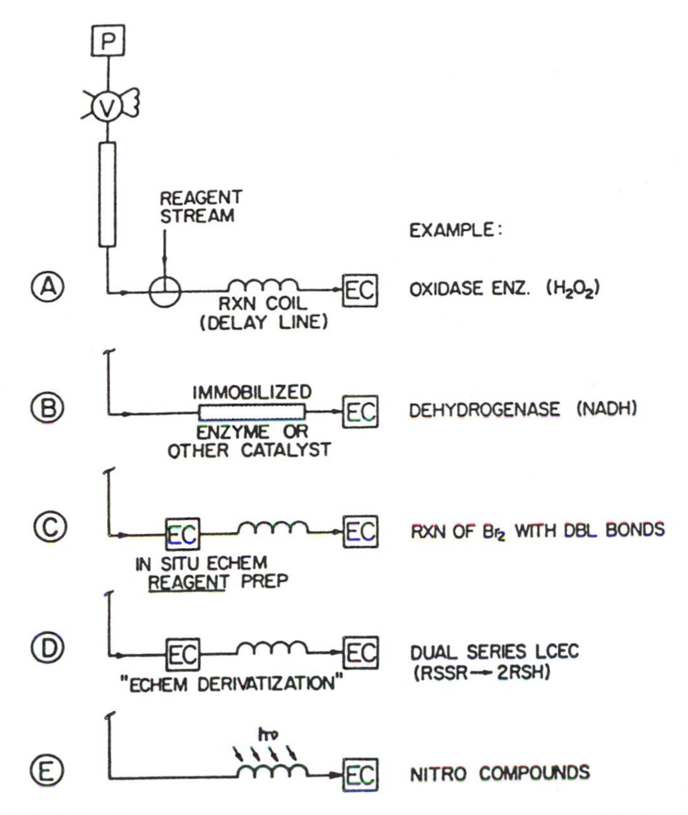


**Figure 27.18** Common configuration for postcolumn reactors with electrochemical analysis. (A) LC–chemical reaction–EC. Postcolumn addition of a chemical reagent (for example, $Cu^{2+}$ or an enzyme). (B) LC–enzyme–LC. Electrochemical detection following postcolumn reaction with an immobilized enzyme or other catalyst (for example, dehydrogenase or choline esterase). (C) LC–EC–EC. Electrochemical generation of a "derivatizing reagent." The response at the second electrode is proportional to analyte concentration (for example, production of $Br_2$ for detection of thioethers). (D) LC–EC–EC. Electrochemical derivatization of an analyte. In this case a compound of a more favorable redox potential is produced and detected at the second electrode (for example, detection of reduced disulfides by the catalytic oxidation of Hg). (E) LC–hν–EC. Photochemical reaction of an analyte to produce a species that is electrochemically active (for example, detection of nitro compounds and phenylalanine). Various combinations of these five arrangements have also been used. [Reprinted with permission from Bioanalytical Systems, Inc.]

several precolumn derivatization methods for amines is that peptides that do not contain primary amine groups can also be detected.

*Electrochemical Derivatization*

Dual-electrode LCEC is very useful for the selective detection of chemically reversible redox couples. In this case, two electrodes are placed in series (Fig. 27.10B). The first electrode acts as a "generator" to produce an electroactive species that is detected more selectively downstream at the second electrode, which is set at a more analytically useful potential. One excellent example of the use of a dual-electrode detector for electrochemical derivatization is the detection of disulfides [34]. In this case, the first electrode is used to reduce the disulfide to the corresponding thiol. The thiol is then detected by the catalytic oxidation of mercury, described earlier. Because of the favorable potential employed at the second electrode, the selectivity and sensitivity of this method are extremely high. In addition, thiols can be distinguished from disulfides by simply turning off the generator electrode.

Electrochemistry can also be used to generate an electroactive reagent (such as bromine) that will react with an analyte of interest. In this case, a reduction in response at the second electrode indicates that some of the reagent has been used up. This approach has been used with bromine for the detection of thioethers and prostaglandins [43].

*Photochemical Derivatization*

Light has been employed to generate electroactive species from nonelectroactive compounds. Among the substances that can be analyzed using this technique are organonitro compounds, organothiophosphates, and α-lactams [44]. A recent example is the detection of phenylalanine. Phenylalanine itself is not electroactive at an analytically useful potential. However, when it is exposed to light, a species is produced that can be detected at +0.60 V vs. Ag/AgCl. This method has been extended to the detection of proteins and peptides containing phenylalanine [45].

*Enzymatic Reactions*

Numerous postcolumn enzymatic reactors have been designed for LCEC. Enzymes can be used to produce an electroactive compound from the analyte of interest or, alternatively, to generate an electroactive species that is proportional to analyte concentration. An example of the latter is the detection of acetylcholine [46]. In this case, acetylcholinesterase is used to convert acetylcholine to choline. The resulting choline is reacted with choline oxidase to produce hydrogen peroxide. The amount of hydrogen peroxide produced is directly proportional to the initial concentration of acetylcholine. Detection limits are in the 100 femtomole range.

## V. CAPILLARY ELECTROPHORESIS/ELECTROCHEMISTRY

Capillary electrophoresis (CE) is a relatively new technique employed for the separation of charged molecules. The apparatus employed is fairly simple. It consists of a fused silica capillary (usually 5–100 μm i.d.) filled with buffer. A potential (usually 10–40 kV) is applied across the capillary using a high-voltage power supply. Due to the ionized silanol groups on the capillary surface, an electroosmotic flow is produced in the capillary, which causes all compounds, regardless of charge, to elute at the cathode. Generally, the order of elution is cations, neutrals, and anions. Compounds are separated based on their differences in electrophoretic mobility.

Spectroscopic detection techniques (UV, fluorescence) are the most common methods of detection employed in CE. UV detection, although the simplest method of detection to adapt to CE, suffers from a loss of sensitivity due to the extremely small pathlengths involved in CE. Laser-induced fluorescence detection is much more sensitive, but is limited by the number of wavelengths available for excitation. In addition, this technique is very expensive to implement and maintain. Electrochemical detection has several advantages for CE [47]. Since electrochemical detection is based on a reaction at the electrode surface, the cell volume can be very small without loss of sensitivity. The concentration-based limits of detection for capillary electrophoresis with electrochemical detection (CEEC) are comparable to those of LCEC.

One of the advantages of CE, which is also a disadvantage, is the very small volume needed for analysis. Typical injection volumes are 1–10 nL. This ability to inject small volumes has been shown to be useful in the analysis of single cells and microdialysis samples. Although the concentration-based detection limits of LCEC and CEEC are similar, detection limits based on actual mass analyzed are much lower for CEEC. Typical mass limits of detection for an electroactive compound are in the attomole range.

### *Detector Design*

The primary challenge for electrochemical detection in capillary electrophoresis is decoupling the high voltage and current associated with the CE separation from the small voltages and currents being applied and measured in the electrochemical cell. Two approaches, both originating from Ewing's group at Pennsylvania State University, have been implemented. The first approach (on-column detection) involves the use of a conducting coupler to isolate the high current generated during the electrophoretic separation from the picoamperes of current generated at the electrode surface by the redox process [48]. This method of detection is illustrated in Figure 27.19. The working electrode is generally a carbon fiber electrode placed in the end of the separation capillary. The coupler, positioned over a small crack in the capillary, is placed in the buffer res-

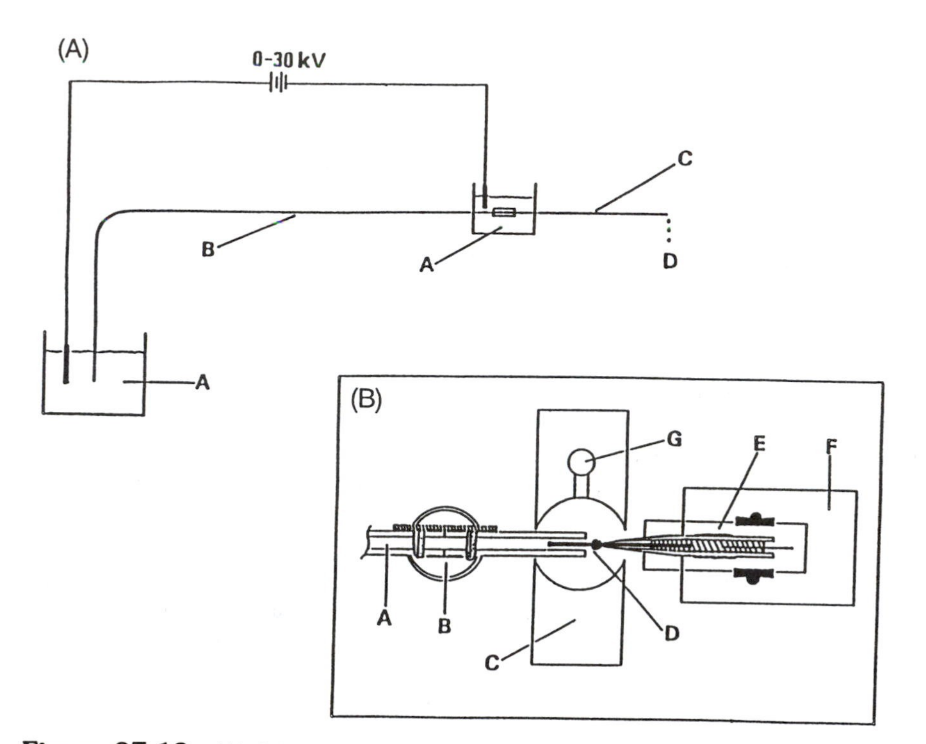


**Figure 27.19** (A) Schematic of the CE system using the porous glass coupler. A, CE buffer reservoirs; B, separation capillary; C, detection capillary; D, eluting buffer droplets. (B) Top view of the porous glass and electrochemical cell. A, column; B, porous glass coupler; C, Plexiglas block; D, carbon fiber working electrode; E, microscope slide; F, micromanipulator; G, reference electrode port. [Reproduced with permission from Ref. 48.]

ervoir, which is connected to ground. If the resistance of the coupler is low, the majority of the current associated with the electrophoretic separation will be sent to ground and will thus minimize the background noise at the detection electrode. This approach is particularly useful for separations employing larger diameter capillaries (50–100 μm), where the currents associated with the separation are in the microampere range. Several different types of couplers have been employed, including porous glass, Nafion tubing, porous carbon, and cellulose acetate.

The second approach (end-column detection) is best suited for capillaries of 25 μm or less [49]. This mode of detection is illustrated in Figure 27.20. In this case, the electrode is placed at the end of the capillary (but not inside) and no coupler is employed. A 70-cm, 5-μm-i.d. capillary filled with a zwitterionic

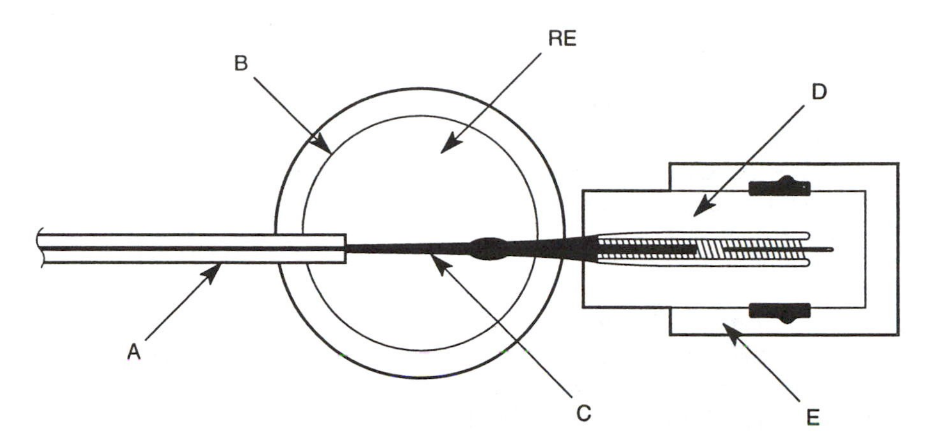


**Figure 27.20** Schematic drawing of CE with end-column amperometric detection: A, capillary; B, cathodic buffer reservoir and electrochemical cell; C, carbon fiber electrode; D, electrode assembly, E, micromanipulator; RE, reference electrode. [Adapted with permission from Ref. 49.]

buffer has a resistance on the order of $10^{12}\ \Omega$. Virtually all the voltage associated with the separation is dropped across the capillary. Therefore, an amperometric detector can be employed, provided that the working electrode is outside of the glass capillary.

*Electrode Materials*

Most electrode materials that are employed in LCEC can also be used for CEEC. The most commonly employed working electrode is a carbon fiber. Carbon fibers come in many different sizes and can also be etched to smaller diameters. Common applications of CEEC with carbon fiber electrodes are the detection of catecholamines in single neuronal cells and amino acids in brain microdialysis samples following derivatization with NDA/CN.

Other types of electrode materials can also be employed. The use of a gold wire allows the use of pulsed amperometric detection in conjunction with CE. Charged carbohydrates have been detected at the micromolar level by CE-PAD [50]. It is anticipated that this technique will be applied to the determination of glycopeptides in the near future. A gold fiber amalgamated with mercury has also been employed for the determination of thiols [51]. Since many important biological thiols (cysteine, glutathione) are charged, CE provides an excellent method of separation. Copper wire has been employed for the detection of both carbohydrates and amino acids by CEEC [52,53]. Again, concentration detection limits for these species are in the micromolar range. The use of chemically modified electrodes as detectors for CEEC has also been demonstrated. A working electrode based on cobalt phthalocyanine was developed by O'Shea and Lunte [54]. This electrode consists of cobalt phthalocyanine mixed with carbon paste and placed inside a fused silica capillary. The electrode has a diameter much larger than the internal diameter of the separation capillary; it is placed at the end of the separation capillary to form a "wall-jet" microelectrode. This electrode can be used for detection of a number of analytes including thiols, carbohydrates, and organic acids.

*Applications*

Recent applications of CEEC have been reviewed [47,55,56]. The types of compounds analyzed by CEEC are virtually identical to those determined by LCEC. However, CEEC has an advantage in the separation of charged species and in the analysis of very small samples. Many of the applications of CEEC have been targeted toward the analysis of catecholamines in single cells. The dopamine content of a single neuron can be analyzed by this technique [57]. Microdialysis sampling is another sampling method that generates very small, protein-free samples that are amenable to analysis by CEEC [58].

Capillary electrophoresis has found use in the biotechnology industry for structural analysis of recombinant proteins. The high resolving power of CE for charged analytes makes it a powerful tool for the analysis of tryptic digests. Therefore, many of the techniques given here, such as the determination of thiols, carbohydrates, and amino acids, will be employed for this purpose.

## VI. CONCLUSIONS

Electrochemical detectors for liquid chromatography have reached a level of maturity in that thousands of these devices are used routinely for a variety of mundane purposes. Nevertheless, the technology is advancing rapidly in several respects. Multiple electrode and voltammetric detectors have been developed for more specialized applications. Small-volume transducers based on carbon fiber electrodes are being explored for capillary and micropacked columns. Recently, electrochemical detection has also been coupled to capillary electrophoresis [47]. Finally, new electrode materials with unique properties are likely to afford improved sensitivity and selectivity for important applications.

It is important for the reader to recognize that there is no single optimal electrochemical detector, but, rather, a whole family of devices that can be adapted to solve a variety of problems. To make the best choice, there is no substitute for a little knowledge of the principles of electrochemical instrumentation. Even more important is an understanding of the chemistry involved when a redox-active molecule ventures into the high-field region near an electrode–solution interface.

## VII. PERFORMANCE CRITERIA FOR ELECTROCHEMICAL DETECTORS

The terms "sensitivity" (or "response factor") and "detection limit" (or "minimum detectable quantity") are frequently confused and used in a manner that is impossible to interpret. It is now generally accepted that sensitivity should refer to output/input. In the case of a chemical determination, this is generally a measured signal (current, voltage, absorbance, etc.) divided by an amount (weight in grams, or moles of substance). For electrochemical detection, sensitivity for a given substance under given chromatographic conditions is described in terms of "peak current per injected equivalents" (nA/nEq or nA/nmol for a one-electron transfer). Other expressions, such as nA/ng, are convenient for routine use, but make it more difficult to compare results under different conditions. For thorough work, the sensitivity in terms of "peak area (coulombs) per injected equivalents" is a useful measure because it permits direct calculation of the conversion efficiency of the cell. Sensitivity by itself does not reveal anything about the minimum detectable quantity; it is also necessary to evaluate the baseline noise. When comparing two sets of conditions, the one

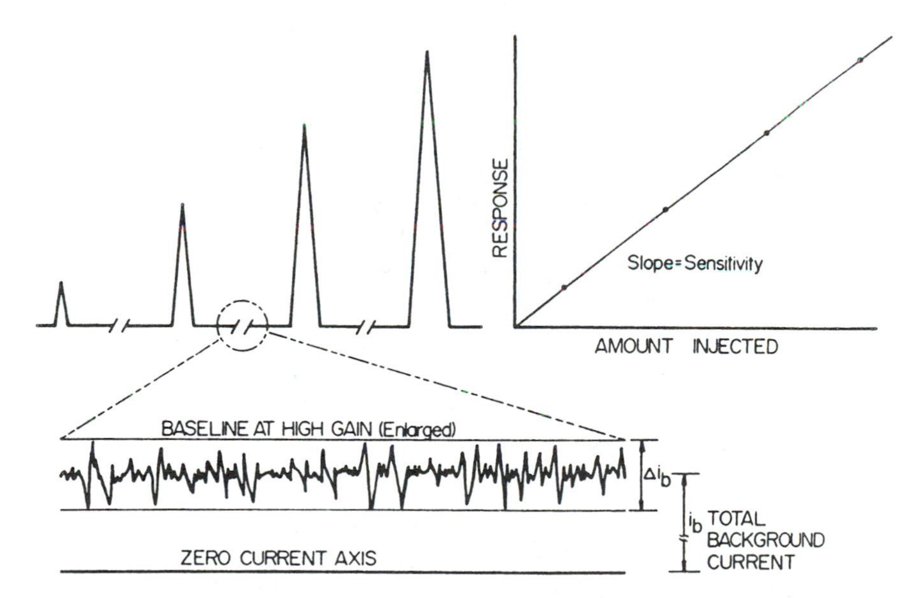


**Figure 27.21** Establishing detection limits in LCEC or FIA requires measurement of both the response slope (sensitivity) and noise. Sensitivity alone does not tell the story.

giving the highest sensitivity does not necessarily result in the lowest detection limit.

It is useful to evaluate the peak-to-peak baseline noise, $\Delta i_b$, over a time period about 10 times the width of the chromatographic peak. The frequency content of the noise relative to that of the chromatographic peak is very important since frequencies much higher or lower can usually be removed. The high frequencies are easily filtered out and the very low frequencies are, in effect, baseline drift, which is eliminated when the peak height is quantitated. The amount injected that would give a peak height equal to $\Delta i_b$ is clearly below the limit of usefulness. An amount that would give a peak height of five times $\Delta i_b$ is useful under many circumstances. It is recommended that LCEC users report both sensitivity for each peak of interest and $\Delta i_b$. With this information it is possible to predict a reasonable detection limit for each substance being determined. Figure 27.21 illustrates the procedure required to obtain the necessary data.

## REFERENCES

1. *Current Separations*, BAS Press, West Lafayette, IN.
2. S. G. Weber, in *Detectors for Liquid Chromatography* (E. S. Yeung, ed.), John Wiley & Sons, New York, 1986, p. 229.
3. D. C. Johnson, S. G. Weber, A. M. Bond, R. M. Wightman, R. E. Shoup, and I. S. Krull, *Anal. Chim. Acta 180*:187 (1986).
4. R. E. Shoup, *High Performance Liquid Chromatography 4*:91 (1986).
5. P.T. Kissinger and R. E. Shoup, *J. Neurosei. Methods 34*:3 (1990).
6. M. D. Ryan and J. Q. Chambers, *Anal. Chem. 64*:79R (1992).
7. D. L. Rabenstein and R. Saetre, *Anal. Chem. 49*:1036 (1977).
8. R. C. Buchta and L. J. Papa, *J. Chromatogr. Sci. 14*:213 (1976).
9. W. A. MacCrehan, R. A. Durst, and J. M. Bellama, *Anal. Lett. 10*:1175 (1977).
10. P. Luo, F. Zhang, and R. P. Baldwin, *Anal. Chim. Acta 244*:169 (1991).
11. J. A. Cox, R. K. Jaworski, and P. J. Kulesza, *Electroanalysis 3*:869 (1991).
12. P. T. Kissinger, *Electroanalysis 4*:359 (1992).
13. C. Bohs, M. Linhares, and P. T. Kissinger, *Current Separations 12*:4 (1994).
14. R. C. Koile and D. C. Johnson, *Anal. Chem. 51*:741 (1980).
15. Y. Takata and G. Muto, *Anal. Chem. 45*:1864 (1973).
16. W. Kemula, *Roca. Chem. 26*:281 (1952).
17. J. G. Koen, J. F. K. Huber, H. Poppe, and G. den Boef, *J. Chromatogr. Sci. 8*:192 (1970).
18. R. Stillman and T. S. Ma, *Mikrochim. Acta* 491 (1973).
19. B. Fleet and C. J. Little, *J. Chromatogr. Sci. 12*:747 (1974).
20. L. Michel and A. Zatka, *Anal. Chim. Acta 105*:109 (1979).
21. H. B. Hanekamp, P. Bos, and R. W. Frei, *J. Chromatogr. 186*:489 (1979).
22. H. B. Hanekamp, W. H. Voogt, P. Bos, and R. W. Frei, *J. Liq. Chromatogr. 3*:1205 (1980).

23. W. Kutner, J. Debowski, and W. Kemula, *J. Chromatogr. 191*:47 (1980).
24. D. A. Roston, R. E. Shoup, and P. T. Kissinger, *Anal. Chem. 54*:1417A (1982).
25. C. E. Lunte, *LC-GC 7*:493 (1989).
26. T. A. Last, *Anal. Chim. Acta 155*:287 (1983).
27. J. G. White, R. L. St. Claire III, and J.W. Jorgenson, *Anal. Chem. 58*:293 (1986).
28. C. E. Lunte, T. H. Ridgway and W. R. Heineman, *Anal. Chem. 59*:761 (1987).
29. C. E. Lunte, J. F. Wheeler, and W. R. Heineman, *Anal. Chim. Acta 200*:101 (1987).
30. D. C. Johnson and W. R. LaCourse, *Electroanalysis 4*:367 (1992).
31. D. C. Johnson and W. R. LaCourse, *Anal. Chem. 62*:589A (1990).
32. D. M. Radzik and S. M. Lunte, *Crit. Rev. Anal. Chem. 20*:317 (1989).
33. P. T. Kissinger, K. Bratin, W. P. King, and J. R. Rice, *ACS Symp. Ser. 136*:57 (1980).
34. L. A. Allison and R. E. Shoup, *Anal. Chem. 55*:8 (1983).
35. K. Bratin and R. C. Briner, *Current Separations 2*:1 (1980).
36. P. T. Kissinger, K. Bratin, G. C. Davis, L. A. Pachla, *J. Chromatogr. Sci. 17*:137 (1979).
37. I. S. Krull, C. M. Selavka, C. Duda, and W. Jacobs, *J. Liq. Chromatogr. 8*:2845 (1985).
38. S. M. Lunte, *TRACS 10*:97 (1991).
39. W. A. Jacobs and P. T. Kissinger, *J. Liq. Chromatogr. 5*:669 (1982).
40. L. A. Allison, G. S. Mayer, and R. E. Shoup, *Anal. Chem. 56*:1089 (1984).
41. J. A. Karn, T. Mohabbat, R. D. Greenhagen, C. J. Decedue, and S. M. Lunte, *Current Separations 11*:57 (1992).
42. A. M. Warner and S. G. Weber, *Anal. Chem. 61*:2664 (1989).
43. W. Th. Kok, J. J. Halvax, V. H. Voogt, U. A. Brinkman, and R. W. Frei, *Anal. Chem. 57*:2580 (1985).
44. I. S. Krull and W. R. LaCourse, in *Reaction Detection in Liquid Chromatography* (I. S. Krull, ed.), Marcel Dekker, New York, 1986, p. 303.
45. I. Dou, J. Mazzeo, and I. S. Krull, *Biochromatography 5*:74 (1990).
46. P. C. Gunaratna and G. S. Wilson, *Anal. Chem. 62*:402 (1990).
47. P. D. Curry, C. E. Engstrom-Silverman and A. E. Ewing, *Electroanalysis 3*:587 (1991).
48. R. A. Wallingford and A. G. Ewing, *Anal. Chem. 59*:1762 (1987).
49. X. Huang, R. N. Zare, S. Sloss, and A. G. Ewing, *Anal. Chem. 63*:189 (1991).
50. T. J. O'Shea, W. R. LaCourse, and S. M. Lunte, *Anal. Chem. 65*:948 (1993).
51. T. J. O'Shea and S. M. Lunte, *Anal. Chem. 65*:247 (1993).
52. L. A. Colon, R. Dadoo, and R. N. Zare, *Anal. Chem. 65*:476 (1993).
53. T. J. O'Shea and S. M. Lunte, *Anal. Chem. 66*:307 (1994).
54. C. E. Engstrom-Silverman and A. G. Ewing, *Microcol. Sep. 3*:141 (1991).
55. S. M. Lunte and T. J. O'Shea, *Electrophoresis 15*:79 (1994).
56. Y. L. Li, *TRACS 11*:325 (1992).
57. T. M. Olefirowicz and A. G. Ewing, *Anal. Chem. 62*:1872 (1990).
58. T .J. O'Shea, M. Telting-Diaz, S. M. Lunte, C. E. Lunte, and M. R. Smith, *J. Chromatogr. 652*:377 (1993).

# 28

# Photonic Electrochemistry

**Andrew B. Bocarsly** *Princeton University, Princeton, New Jersey*

**Hiroyasu Tachikawa** *Jackson State University, Jackson, Mississippi*

**Larry R. Faulkner** *University of Illinois at Urbana–Champaign, Urbana, Illinois*

## I. INTRODUCTION

Photonic electrochemistry, taken in its most general sense, involves the intimate interaction of light with electrochemical processes. Thus, processes in which illumination of the electrode–electrolyte interface produces charge-transfer events as well as electrochemical reactions that produce light as a product (electrochemiluminescence, ECL) fall in this category. A third related area is the electroanalytical detection of transient species formed by a photochemical process which takes place in solution. Techniques such as spectroelectrochemistry are excluded from consideration, since they utilize photons as a nonperturbing probe of purely electrochemical processes.

Our understanding of the interaction of light with a metal substrate to generate free charge carriers dates back to Einstein's description of the photoelectric effect, in which illumination of a metal substrate with photons having energy greater than the work function of the metal causes the ejection of electrons. This experiment takes on an electrochemical context if the metal involved is immersed in an electrolyte, in which case the photoelectrons can interact with dissolved species to carry out a charge-transfer reaction. Although the photoelectric effect itself is historically well established, the application to electrochemistry is relatively recent. A second application of photonic electrochemistry is the interaction of light with a semiconductor electrode–electrolyte interface. Photocurrents generated in such systems, which are referred to as photoelectrochemical cells (PEC), can be substantial; thus, the photoresponse of semiconductor–electrolyte interfaces is of interest in terms of both optical (i.e., solar) energy conversion and our physical understanding of semiconductor quantum structure. In addition, photoelectrochemical cells offer interesting practical prospects with respect to the microfabrication of semiconductor structures and remediation of pollutants from contaminated water sources. A third type of interaction of light with an electrochemical environment is the photogalvanic cell, in which a species dissolved in the electrolyte (or chemisorbed on the electrode surface) undergoes a photochemical process that generates a current that is collected at an electrode. The electrode utilized may be constructed of either a metal or a semiconducting material. Recently semiconducting photogalvanic cells have been shown to offer high efficiency for the conversion of visible optical energy to electricity.

Electrochemiluminescence typically involves the electrochemical generation of species that undergo homogeneous reactions to generate electronically excited dissolved compounds. In certain cases, emission can then be observed as these molecular species relax to the ground state. As such, ECL in some sense represents the opposite process of that observed in a photogalvanic cell.

## II. PHOTOELECTROCHEMISTRY—SEMICONDUCTOR/ ELECTROLYTE INTERFACES

### A. Semiconductor Solid-State Physics

While many of the standard electroanalytical techniques utilized with metal electrodes can be employed to characterize the semiconductor–electrolyte interface, one must be careful not to interpret the semiconductor response in terms of the standard diagnostics employed with metal electrodes. Fundamental to our understanding of the metal–electrolyte interface is the assumption that all potential applied to the back side of a metal electrode will appear at the metal electrode surface. That is, in the case of a metal electrode, a potential drop only appears on the solution side of the interface (i.e., via the electrode double layer and the bulk electrolyte resistance). This is not the case when a semiconductor is employed. If the semiconductor responds in an ideal manner, the potential applied to the back side of the electrode will be dropped across the internal electrode–electrolyte interface. This has two implications: (1) the potential applied to a semiconducting electrode does not control the electrochemistry, and (2) in most cases there exists a "built-in" barrier to charge transfer at the semiconductor–electrolyte interface, so that, electrochemical reversible behavior can never exist. In order to understand the radically different response of a semiconductor to an applied external potential, one must explore the solid-state band structure of the semiconductor. This topic is treated at an introductory level in References 1 and 2. A more complete discussion can be found in References 3, 4, 5, and 6, along with a detailed review of the photoelectrochemical response of a wide variety of inorganic semiconducting materials.

Solid-state physics depicts a semiconductor as being energetically composed of a series of closely spaced, low-lying, mainly filled orbitals, known as the valence band, and a series of closely spaced, high-energy, mainly vacant orbitals, the conduction band, separated by a forbidden zone known as the band gap (see Fig. 28.1) [6]. The band gaps of several semiconductors that have been extensively investigated as photoelectrodes are given in Table 28.1. The orbitals within each band are delocalized in nature. Thus, a single crystal of a semiconductor is best described as a single molecule, within traditional chemical nomenclature! The interorbital spacing within a band is so small in energy that orbital to orbital transitions occur by thermal activation. Thus, the primary optical characteristics of a semiconductor (including its color) are established by the energy required to promote electrons across the band gap, from the valence band to the conduction band.

In addition to energy, the semiconductor band gap is characterized by whether or not transfer of an electron from the valence band to the conduction band involves changing the angular momentum of the electron. Since photons do not have angular momentum, they can only carry out transitions in which the electron angular momentum is conserved. These are known as direct transitions. Momentum-changing transitions are quantum-mechanically forbidden and are termed indirect (see Table 28.1). These transitions come about by coupling

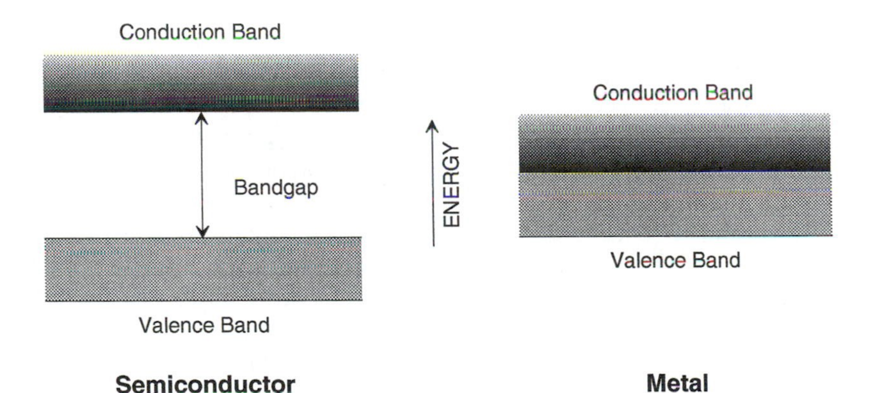


**Figure 28.1** The electronic structure of a solid can be described in terms of a band model in which bonding electrons are primarily found in a low-energy valence band, while conduction is typically associated with antibonding or nonbonding high-energy orbitals known as the conduction band. In the case of a semiconductor (left), these two bands are separated by a quantum-mechanical forbidden zone, the band gap. Excitation of electrons from the valence band to the conduction band gives rise to the bulk optical and electronic properties of the semiconductor. In the case of a metal (right), the conduction band and valence band overlap, giving rise to a continuum of states.

**Table 28.1** Bandgaps of Common Semiconductors

| Semiconductor | Bandgap[a] (eV) (transition type)[b] | Onset wavelength[c] (nm) | Bulk color | Dopant type |
|---|---|---|---|---|
| Si | 1.11 (I) | 1100 | Black | p or n |
| InP | 1.28 (D) | 960 | Black | p or n |
| GaAs | 1.35 (D) | 915 | Black | p or n |
| CdTe | 1.50 (D) | 820 | Black | p or n |
| CdSe | 1.74 (D) | 710 | Black | n only |
| $\alpha$-$Fe_2O_3$ | 2.2 (I) | 560 | Orange-red | primarily n |
| GaP | 2.24 (I) | 550 | Orange | p or n |
| CdS | 2.4 (D) | 515 | Yellow | n only |
| $TiO_2$ | 3.0 (D) | 412 | Pale yellow/ white | n only |
| $SrTiO_3$ | 3.2 (D) | 386 | White | n only |
| ZnO | 3.2 (I) | 386 | White | n only |
| $SnO_2$ | 3.6 (D) | 340 | White | n only |

[a]At 300 K.
[b]Transitions that conserve momentum are direct transitions (D); those that involve a change in momentum are indirect transitions (I).
[c]Absorption edge of the bandgap transition.

the absorption of a photon with a phonon (a quantum of solid-state lattice vibration). The net effect is that direct semiconductor transitions give rise to extreme absorptivity very close to the band gap energy, while indirect transitions are associated with a mild increase in absorptivity as one moves to energies greater than the band gap energy.

In general, whether an electrode is a metal or a semiconductor, to sustain a current, electrons must populate the conduction band levels, since these empty delocalized orbitals are needed to allow the electron to be physically transported through space (i.e., from orbital to orbital). In the case of a metal, the band gap is zero; therefore, the valence band and conduction band orbitals overlap (see Fig. 28.1). Thermal excitation of electrons from the valence band to the conduction band orbitals provides for a conductivity pathway. The relatively large band gap of a semiconductor, however, precludes thermal population of the conduction band by valence band electrons. However, optical excitation of the semiconductor with photons of energy greater than or equal to the band gap energy can be employed to produce conduction. Excitation of an electron generates not only an excited conduction band electron but also an empty orbital in the valence band. This empty orbital can undergo a self-exchange process with a filled valence band orbital to effectively "move" the empty orbital through the semiconductor lattice. The "movement" of an empty photoexcited orbital is physically equivalent to the motion of a positively charged particle through a semiconductor lattice. These empty orbitals are referred to as holes ($h^+$). Thus, a photoinduced current in a semiconductor can be thought of as a flow of conduction band electrons in one direction and an opposing flow of valence band holes. Holes can be given pseudoattributes similar to the properties of electrons, such as charge, mass, velocity, etc.

Simple promotion of electrons across the semiconductor band gap is insufficient to generate a photocurrent, since nonradiative relaxation of excited electrons to the ground state is quite facile in the solid state. This deactivation pathway can be overcome by carrying out a thermal charge-transfer reaction between the semiconductor and a second phase to generate a charge separating electric field at the semiconductor surface (prior to photoexcitation). This interfacial electric field can be used to spatially separate photoinduced electron–hole pairs, thereby preventing their nonradiative recombination. The necessary interfacial electric field can be developed by placing the semiconductor surface in intimate contact with any medium having a free energy different from that of the semiconductor. From an electrochemical point of view, the most interesting contacting phase is an electrolyte.

## B. The Ideal Semiconductor-Electrolyte Interface

As shown in Figure 28.2, the contacting of a semiconductor surface by an electrolyte containing an electroactive species having a well-defined redox potential causes (in the dark) the transfer of a limited number of electrons between the semiconductor and the electrolyte. This brings the free energy difference between the semiconductor (the semiconductor's Fermi level) and the electrolyte (the redox potential) to zero, establishing equilibrium. The direction of electron flow during this process will depend on the relative values of the Fermi level and the redox potential. In theory, both the free energy of the semiconductor and the free energy of the electrolyte should change in order to establish the equilibrium condition. However, since there exists a tremendous excess of charge carriers in the electrolyte (i.e., the number of electroactive molecules) compared to the number of charge carriers in the semiconductor, the redox potential of the electrolyte is virtually unaffected by this process. Rather, the semiconductor Fermi level shifts to the electrolyte redox potential.

In order to understand this process, the nature of the Fermi level must be considered. Within the nomenclature of physics, the Fermi level is defined as the energy at which there is a 50% probability of finding an electron in a metal—what a chemist would call the highest filled molecular orbital (i.e., the top of the valence band). In an *intrinsic* semiconductor, a material having a completely filled valence band and a completely empty conduction band, this energy level

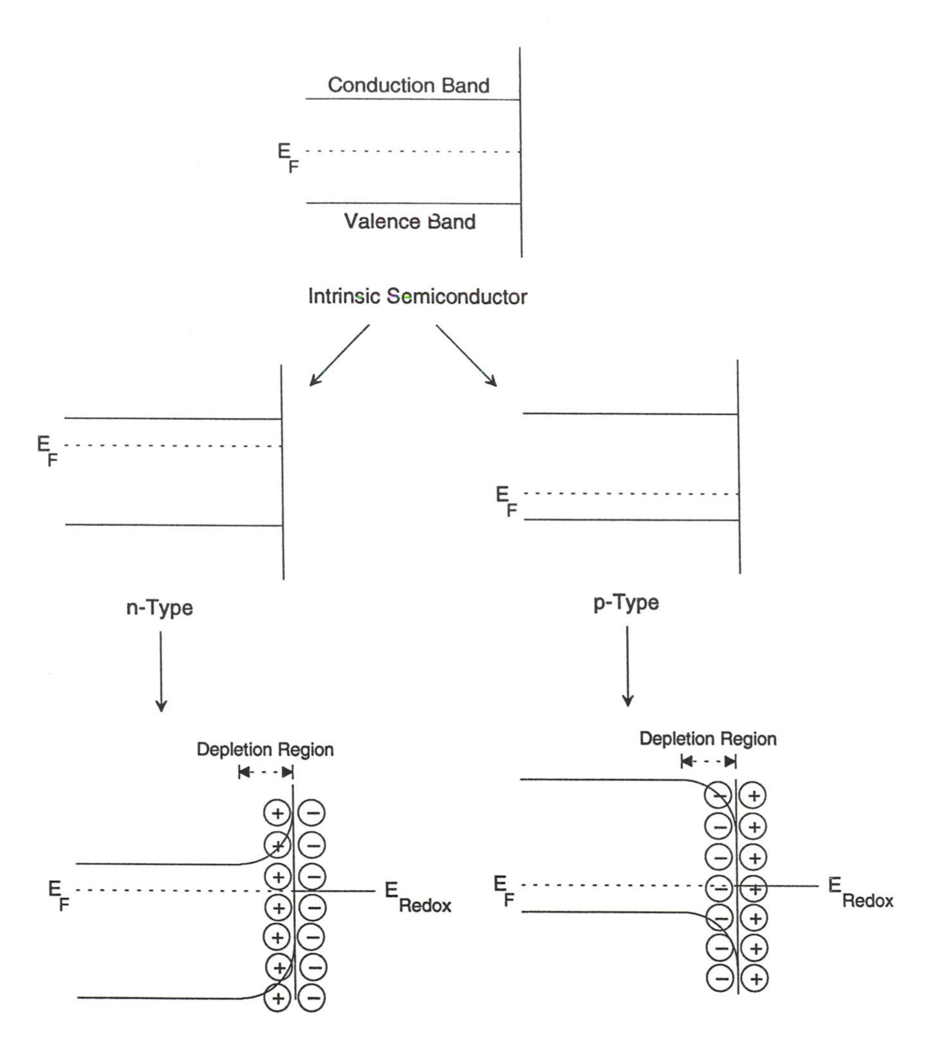


**Figure 28.2** An intrinsic semiconductor (top) has a Fermi level that lies in the band gap halfway between the conduction and valence bands. Doping the semiconductor n-type adds electrons to the conduction band, causing the Fermi level to shift higher in energy to a position just below the conduction band edge. Doping a semiconductor p-type adds holes to the valence band, lowering the free energy of the semiconductor and moving the Fermi level to slightly above the valence band edge. In either case, if the semiconductor is contacted with an electrolyte, a dark charge-transfer reaction occurs to bring the semiconductor–electrolyte interface into equilibrium. This establishes a depletion region and related electric field at the semiconductor interface. The band edges in the depletion region bend due to the electric field. In the case of an n-type material, the field is established such as to propel holes in the semiconductor to the interface, while a p-type material yields band bending that causes electrons to move through the interface.

mathematically lies at the center of the bandgap (Fig. 28.2). Of course, no electrons can be found at this energy, since no orbitals exist there. Thus, while retaining its thermodynamic meaning in a semiconductor, the physical picture of the Fermi level is lost. However, using Equation 28.1 the physical significance of the Fermi level can be understood as a solid-state analog of an electrolyte's redox potential,

$$\Delta G = -nFE_F \tag{28.1}$$

where n, the stoichiometric number of electrons, is set equal to 1, F is Faraday's constant, and $E_F$ is the Fermi level. As such, it is appropriate to equate redox potentials and Fermi levels. The Fermi level of a given semiconductor can be adjusted by adding dopants to the semiconductor lattice. These dopants either donate or accept charge (depending on electronegativity). Typical dopant concentrations are very low, generating from $10^{17}$ to $10^{19}$ charges per cubic centimeter. If a dopant that is less electronegative than the semiconductor is employed, a small population of high-energy electrons will be added to the semiconductor (conduction band) and the Fermi level will lie close to the conduction band edge (in the band gap). Such semiconductors are known as *n-type*. The electrons in such materials are referred to as the *majority carriers*, while holes are the *minority carriers*. When placed in contact with an electrolyte, electrons typically move from the n-type semiconductor to the solution to establish the equilibrium condition. On the other hand, addition of a highly electronegative dopant to the semiconductor will withdraw electrons from the semiconductor valence band, shifting the Fermi level to an energetic position near the valence band edge (see Fig. 28.2) and making holes the majority carrier. In this case, the semiconductor is referred to as *p-type*, and establishing an electron-transfer equilibrium typically involves moving charge from the electrolyte to the p-type semiconductor. In either case, an electric field, known as the *space charge* or *depletion region*, is established at the semiconductor interface due to a deficiency of majority carriers (Fig. 28.2). This layer typically penetrates 100–10,000 Å into the semiconductor.

At the n-type interface, the electric field generated causes photogenerated conduction band electrons to move into the bulk of the semiconductor, to the back metal contact, and into the external circuit. The valence band holes access the semiconductor interface due to the influence of the interfacial electric field (Fig. 28.2). Thus, redox species can be oxidized by the excited n-type semiconductor. These materials act as photoanodes. On the other hand, the electric field in a p-type material is reversed in potential gradient; therefore, excited electrons move to the semiconductor surface, while holes move through the semiconductor to the external circuit (Fig. 28.2). These materials are photocathodes. The presence of an electric field at the semiconductor–electrolyte interface is usually depicted by a bending of the band edges as shown in Figure 28.2. Electrons "roll" down the potential gradients indicated by the band bending, while holes "float" up such gradients.

At first glance, band bending appears to suggest that the energetics of the semiconductor surface are fixed while the bulk energetics are changing; the opposite is true. Note that the distance of the band edges with respect to the Fermi level, which is the thermodynamic reference point, is invariant in the bulk of the semiconductor, but changes in the depletion region. Thus, the energetics of the semiconductor are only perturbed in the near-surface region, as is expected for an interfacial charge-transfer reaction involving a relatively insulating material. In order to obtain efficient electron–hole separation, the photoexcited charge carriers must reside in the space charge region. This can occur via two processes: the excited carriers may be generated in the space charge region if a photon is absorbed in this spatial area, or electron–hole pairs may be generated in the bulk of the semiconductor and diffuse into the space charge region. The latter process is relatively inefficient due to the rapidity of excited-state nonradiative decay. Since the photoelectrochemical cell is a front surface device (i.e., the light-absorbing surface is coincident with the space charge region), most of the excited carriers are generated in the region of high band bending. However, as the optical energies employed are decreased to the minimum energy needed to sustain the band-to-band transition, the absorptivity of the semiconductor decreases and charge carriers can be formed in the semiconductor bulk.

Photoexcitation of the semiconductor is expected to decrease the degree of band bending, since promotion of electrons to the conduction band must raise the free energy moving the Fermi level to more negative potentials on an electrochemical scale. At sufficiently high light intensities, the bands will totally flatten out. Under these conditions, the semiconductor cannot support a photocurrent, due to rapid electron–hole recombination; however, a photopotential can still be measured. As shown in Figure 28.3a, in the case where a semiconductor is immersed in an electrolyte having a well-established redox potential (i.e., both components of the redox couple are present) and a second electrode that is in thermodynamic equilibrium with the redox couple is employed, the observed photovoltage is given by

$$V_{photo} = E_F - E_{Redox} \qquad (28.2)$$

This value is maximized when $E_F$ reaches the *flatband potential*. Thus, a plot of incident light intensity versus open-circuit electrode potential is expected to saturate at the flatband potential, thereby allowing identification of this potential. A typical experiment of this sort is shown in Figure 28.3b for an n-CdSe electrode immersed in a ferri/ferrocyanide electrolyte.

For an ideal semiconductor–electrolyte interface, the energetics of the band edges at the semiconductor-electrolyte interface are held constant by boundary

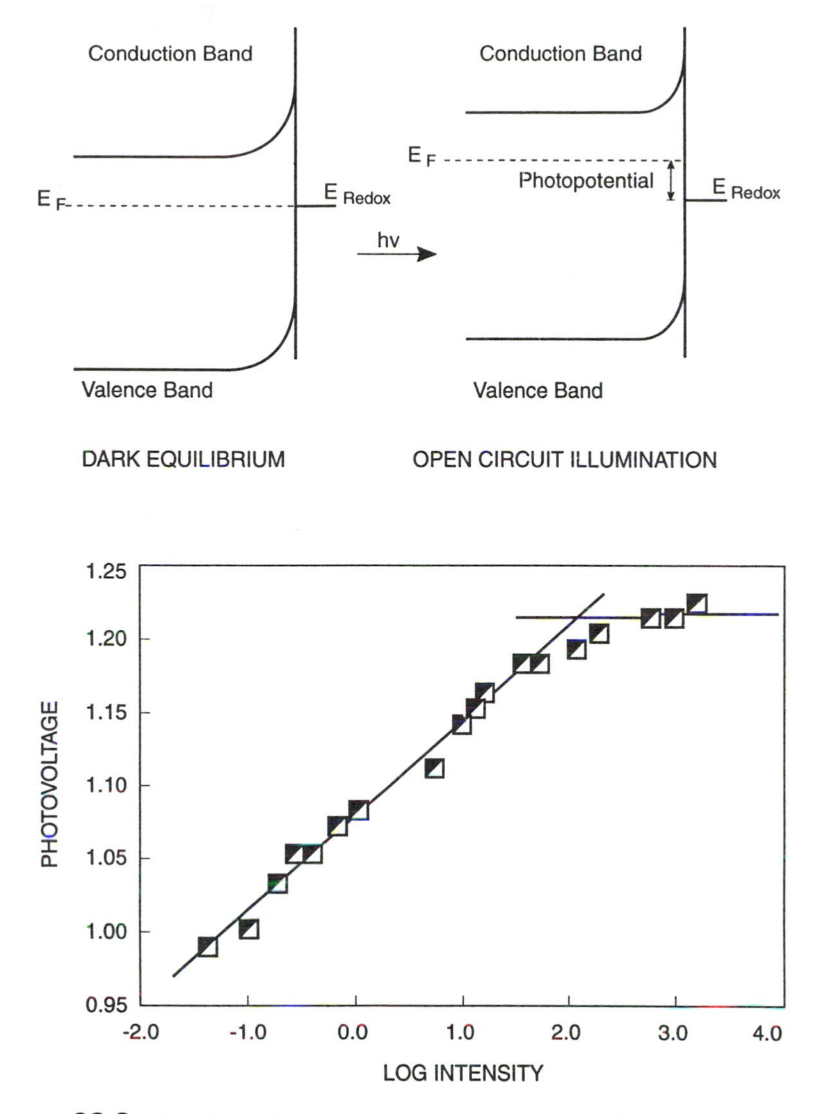


**Figure 28.3** The flatband potential of a semiconductor can be established by measuring the photopotential of the semiconductor as a function of illumination intensity. In the dark (left), the semiconductor Fermi level and the redox potential of the electrolyte are equal, providing an equilibrium condition. However, illumination of the semiconductor (right) generates charge carriers that separate the Fermi level and the redox potential. The difference in these two parameters is the observed photovoltage as shown for an n-CdS electrode immersed in a ferri/ferrocyanide electrolyte (bottom). The measured photovoltage is observed to saturate at the flatband potential. In this case, a value of –0.2 V vs. SCE is obtained. Note that the photovoltage response yields a linear functionality at low light intensity with saturation behavior occurring as the flatband potential is approached.

constraints. Additionally, the position of the band edges in the semiconductor bulk is related to the Fermi level by Equation 28.3, and is only dependent on the fractional dopant concentration of the semiconductor:

$$E_c = E_F + kT \ln\left(\frac{n}{N_c}\right) \tag{28.3}$$

where $E_c$ is the energy of the conduction band edge, n is the concentration of donor states per cubic centimeter, and $N_c$ is the density of states in the semiconductor ($N_c \approx 10^{20}$ $cm^3$). Thus, from the measured flatband potential and a knowledge of the semiconductor band-gap energy, the energetic position of the band edges at the semiconductor surface can be determined. The band-gap energy can be obtained by photoaction spectroscopy. This technique involves monitoring the photopotential or photocurrent of a photoelectrochemical cell as a function of excitation wavelength. The onset of photocurrent (potential) is taken as the band gap. Using the data provided in Figure 28.3 ($E_F$ = −1.2 V vs. SCE) and the observed band gap of 1.7 eV, it can be seen that for the n-CdSe/ $[Fe(CN)_6]^{4-/3-}$ cell the conduction band edge is at −1.3 V vs. SCE (based on Eq. 28.3, assuming n ≈ $10^{18}$ $cm^3$), and the valence band edge lies at +0.4 V vs. SCE.

In addition to the use of open-circuit photopotentials, the variation in interfacial capacitance with electrode potential can be utilized to determine the flatband potential as well as the semiconductor dopant concentration. A discussion of the capacitance-potential response of the semiconductor–electrolyte interface is beyond the scope of this text. The reader is referred to Reference 7 for a more complete discussion of this subject.

The position of a semiconductor's Fermi level is sensitive not only to dopant concentration and illumination intensity but also to application of an external potential, as is the case for a metal electrode. Note, however, that although shifting the Fermi level varies the energy of the highest filled metal orbital, and thus the energy at which electrons or holes react at the metal–electrolyte interface, varying the potential of an ideal semiconductor only affects the energy of the Fermi level, and through that the degree of band bending. The extent of band bending determines the efficiency of photoinduced electron–hole separations, it does not affect the energy of the band edges at the semiconductor surface. Under ideal conditions, electrons and holes only react at the energy of the band edge, since nonradiative transfer from an orbital high in the band to the band edge occurs on a rapid time scale. Thus, the oxidizing energetics of an ideal, illuminated n-type semiconductor or the reducing potential of an ideal, illuminated p-type semiconductor is totally determined by the fixed interfacial band edge positions and independent of the electrode potential.

From this discussion, it can be seen that while control over semiconductor potential does not provide control over the interfacial energetics, it does allow one to control the efficiency with which electron-hole pairs are allowed to recombine or alternatively undergo interfacial charge transfer. This process is usually discussed in terms of a quantum yield for electron flow, $\Phi_e$, as defined in

$$\Phi_e = \left( \frac{i/F}{\text{incident photons/s}} \right) \tag{28.4}$$

where i is the current (hence $i/F$ = moles of electrons in the external circuit per second). Unlike the normal definition of quantum yield, the present definition does not consider photons absorbed; rather, photons incident on the photoelectrochemical cell are monitored. As such, cell reflectivity is viewed as an inherent inefficiency. A given photoelectrochemical cell does not have a single quantum yield that characterizes it. All cells will produce $\Phi_e = 0$ at the flatband potential, with $\Phi_e$ rising to a maximum as band bending increases. This effect typically saturates ~300 mV past the flatband potential, at which point the cell current is maximized. The exact potential of $\Phi_e^{max}$ depends on the detailed kinetics of interfacial charge transfer, in terms of both electron–hole dynamics within the semiconductor and the classic electrochemical dynamics of heterogeneous charge transfer.

As illustrated in Figure 28.4 for an n-type semiconductor, the dynamics of semiconductor heterogeneous charge-transfer processes are most easily ascertained experimentally by observation of the dark and illuminated steady-state current-potential response of the semiconductor. This voltammetric experiment is typically carried out with a potential scan rate of ~10 mV/s generating conditions similar to those found in a solid electrode polarography experiment. At the flatband potential (Region II), electron–hole pairs cannot be separated and thus no current flows independent of illumination. Moving the electrode potential positive of this point (Region III) introduces band bending. In the dark, this establishes a barrier to charge transfer. The electrode is blocking with respect to anodic current. However, illumination of the electrode under these conditions generates valence band holes that are accelerated through the semiconductor interface by the space charge field. As the band bending is increased, the observed anodic current increases, until additional band bending does not improve charge separation efficiencies and current saturation is observed. If one returns to the flatband potential and now scans the electrode potential in the negative direction, into Region I, the bands are forced to invert, generating a majority carrier *accumulation layer* in the space charge region. Under these conditions, an excess of electrons exists in the interfacial region of an n-type semiconductor (an enrichment of holes would occur in a p-type material). Note that the

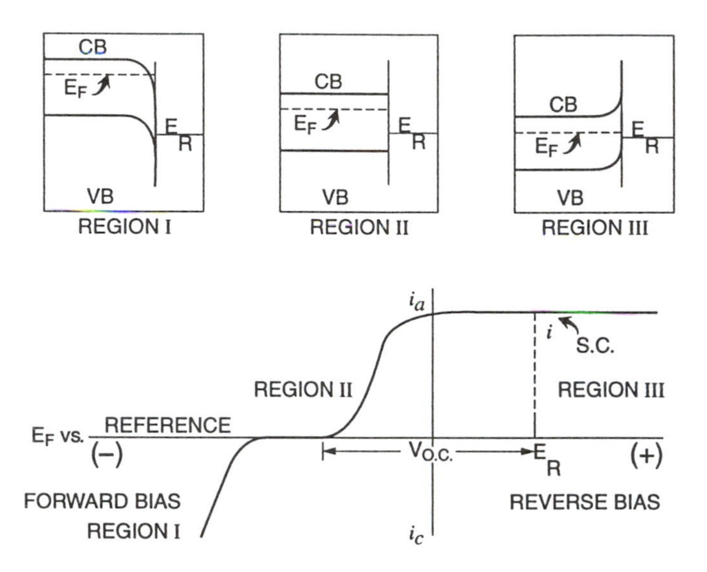


**Figure 28.4** The current–potential response of an n-type semiconductor can be divided into three regions. In region I, an accumulation layer exists which allows electrons to be injected into the electrolyte. Thus, the electrode behaves like a dark cathode. In region II, the flatband potential for the electrode is encountered and the electrode is blocking with respect to interfacial charge transfer. In region III, a depletion layer exists. In the dark, the electrode is blocking with respect to charge transfer; however, illumination of the electrode allows valence band holes to be injected into the electrolyte. Thus, the electrode acts as a photoanode. The maximum open-circuit photopotential of an electrode can be obtained from the current–potential curve by noting the potential between photocurrent onset and the redox potential of the electrolyte. The short-circuit photocurrent for the cell is given when the electrode potential is equal to the redox potential of the electrolyte. The fill factor of the cell can be computed based on the percentage of the box defined by the open-circuit photovoltage and the short-circuit current filled by the actual current–voltage curve.

Fermi level under these conditions lies above the conduction band edge in the accumulation region. The conduction band orbitals that fall below the Fermi level are filled with electrons (in the dark). In addition, the direction of band bending causes these electrons to move toward the semiconductor–electrolyte interface. If a reducible electroactive species is present in the electrolyte with a potential positive of the highest filled conduction band orbital, a cathodic current will flow. The mechanism of current flow is similar to that observed at a metal electrode in that moving the Fermi level more negative causes higher energy orbitals to be populated. Thus, n-type semiconductors are not only photoanodes, but dark cathodes. Similar reasoning leads to the conclusion that p-type materials behave as photocathodes and dark anodes.

The current-potential response of a p-type material, p-GaP, is shown in Figure 28.5. A 1 *M* HCl electrolyte has been employed. This electrode acts as an anode at sufficiently positive potentials in the dark (not shown). As expected, the electrode is totally blocking in the dark, at negative potentials indicative of the formation of a depletion layer. The flatband potential for this cell is measured to be about –0.2 V vs. SCE. Under illumination at potentials negative of this potential, photoexcited conduction band electrons migrate to the electrode surface leading to an observed photocathodic current, associated with the reduction of protons to $H_2$. The thermodynamic redox potential for this process is –0.24 V vs. SCE, and thus a *thermodynamic underpotential* of 40 mV is obtained. In this case, most of the light energy is consumed in overcoming the activation energy for reduction of protons to hydrogen at the GaP interface. Photogenerated valence band holes move into the electrode bulk and through the external circuit. Increasing the band bending to ~300 mV ($E_F$ = –0.1 V vs. SCE) maximizes the charge separation capability of the electrode leading to saturation of

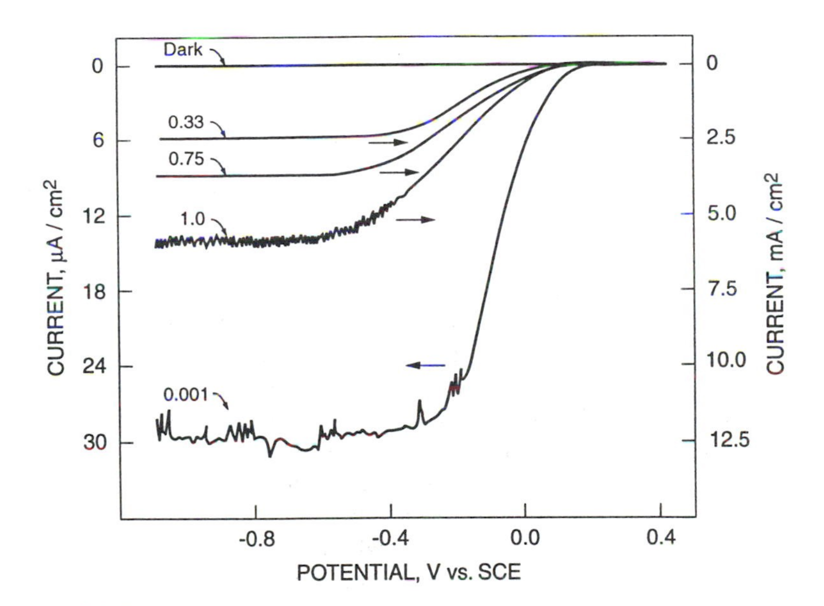


**Figure 28.5** Current–potential curves for p-GaP under low- to moderate-intensity illumination; a 1 *M* NaCl (pH = 1) electrolyte is employed. Illumination is from a 200-W high-pressure mercury lamp filtered with neutral density filter. Intensity is relative to the full lamp output. The $H_2/H^+$ redox potential is –0.3 V vs. SCE in this cell. Thus, this cell yields approximately 400 mV of open-circuit photovoltage. Note that increased illumination increases both the saturation photocurrent and the onset potential. Although the photocurrent is increased at higher light intensities, a calculation of the quantum yield for electron flow indicates that this parameter decreases with increased light intensity.

the photocurrent. Figure 28.5 also demonstrates the effect of light intensity on a potentiostatted photoelectrochemical cell. Under these conditions the semiconductor Fermi level is totally controlled by the potentiostat and is unaffected by the illumination level. The photocurrent is observed to scale with the light intensity. The effect is nonlinear, indicating that $\Phi_e$ decreases with increasing light intensity. This may be due to enhanced charge recombination as the effective "concentration" of photoinduced carriers increases, or due to the onset of limitations associated with heterogeneous charge transfer as the carrier concentration is increased.

The photocurrent onset potential is often taken as the flatband potential, since the measurement of the flatband potential is typically only good to $\pm 100$ mV and the onset of photocurrent is often observed with less than 100 mV of band bending. This practice is dangerous, however, since the onset potential is actually the potential at which the dark cathodic current and the photoanodic current are equal. Even though in the case of the p-GaP illustration, the observation of an anodic current and a photocathodic current are separated by several hundred millivolts, in many systems these two currents overlap. In those cases, the relationship between the flatband potential and the onset potential becomes unclear.

In addition to the quantum yield for charge transfer, several other parameters are useful in evaluating a semiconductor's current–potential response. These parameters allow one to quantify the output characteristics of a photoelectrochemical cell as well as the kinetically limiting factors associated with the semiconductor–electrolyte interface. The *fill factor* is a measure of the "squareness" of the current-potential profile and is indicative of the heterogeneous charge-transfer kinetics. In order to define this term, the thermodynamically important positions of the current–potential response must be identified. When the semiconductor Fermi level is equal to the electrolyte redox potential of the electrolyte (and thus the Fermi level of a reversible counterelectrode), no bias voltage is present and a short-circuit condition exists. The cell's open-circuit voltage (i.e., the point at which only a photopotential exists, and no current flows) is obtained as the difference between the short-circuit potential and the photocurrent onset potential. An idealized current–voltage response, devoid of all charge-transfer kinetics, can then be defined as having a photocurrent that jumps from zero to its maximum value instantaneously at the open-circuit potential. A rectangle of fixed area is defined running from the onset potential to the short-circuit potential (see Fig. 28.5). This rectangle represents the active potential region of a photoelectrochemical cell. For a p-type semiconductor at potentials negative of this region, the electrode acts as a dark cathode and thus photoconversion does not occur. Even under optimized conditions, real electrodes do not yield this type of behavior, since the kinetics of charge separation, recombination, and transport influence the current–voltage response. As a result, a trivially small degree of band bending does not lead to maximum charge separation. Instead, the efficiency with which carriers undergo interfacial charge transfer is observed to increase as the band bending increases. The fill factor is the fraction of the ideal rectangle that actually lies under the current–voltage curve. As such, it represents the impact of carrier kinetics on the thermodynamic response of a photoelectrochemical cell. Improved carrier kinetics lead to an increase in the slope of the current–voltage curve, and an increase in the portion of the thermodynamically defined rectangle that is filled.

In addition to yielding information about semiconductor charge-transfer dynamics, the fill factor parameterizes the efficiency with which the photoelectrochemical cell can be expected to convert optical energy to electricity. The practical value of a photoelectrochemical cell is usually evaluated by its maximum conversion efficiency. The energy conversion efficiency is defined as

$$\eta = \left(\frac{\text{electrical power output}}{\text{optical power input}}\right) - \frac{i \times V_b}{\text{optical watts incident}} \tag{28.5}$$

where $V_b$ is the bias voltage ($= |E_{el} - E_{sc}|$; $E_{el}$ = electrode potential; $E_{sc}$ = short-circuit potential) and i is the current measured at $V_b$. As the electrode potential is varied from the open-circuit voltage to the short-circuit voltage, $V_b$ decreases while i increases, and thus a plot of $\eta$ versus potential goes through a maximum. The *maximum power point* is found near the knee of the current–voltage curve.

## C. Real (Nonideal) Semiconductor–Electrolyte Interfaces

### *Photodecomposition Processes*

Although several single-crystal, wide-band gap semiconductors provide electrochemical and optical responses close to those expected from the ideal semiconductor–electrolyte model, most semiconducting electrodes do not behave in this manner. The principal and by far overriding deviation from the behavior described in the previous section is photodecomposition of the electrode. This occurs when the semiconductor thermodynamics are such that thermal or photogenerated valence band holes are sufficiently oxidizing to oxidize the semiconductor lattice [8,9]. In this case, kinetics routinely favor semiconductor oxidation over the oxidation of dissolved redox species. For example, irradiation of n-CdX (X = S, Se, or Te) in an aqueous electrolyte gives rise exclusively to semiconductor decomposition products as indicated by

$$CdX \xrightarrow[H_2O]{h\nu} Cd^{2+}_{(aq)} + X^0_{(s)} + XO_y^{n-} \tag{28.6}$$

Likewise, irradiation of an n-Si electrode in an aqueous electrolyte efficiently forms an insulating $SiO_x$ interface [3]. Reductive decomposition of semiconductors is rarely observed, although it has been claimed that certain p-type III–V semiconductors in the phosphide family decompose to form $PH_3$.

Several successful schemes have been developed to overcome semiconductor photodecomposition. The initial successful approach to this problem was based on the addition of a redox-active species to the electrolyte that could compete with autooxidation of the semiconductor. Thus, addition of ~1 M sulfide to an aqueous electrolyte was found to totally suppress the photodecomposition of n-CdS-based cells [11–13]. In this case, oxidation of the sulfide to a polysulfide ($Sn^{2-}$) is observed. The polysulfide can be reversibly reduced at the dark counterelectrode, giving rise to a cell that undergoes no net chemical change, but is capable of converting optical energy to chemical energy. Similarly, addition of $Se^{2-}$ to n-CdSe-based cells and $Te^{2-}$ to n-CdTe-based cells provides for the stable generation of photocurrent. In these systems a kinetic advantage is achieved by using a reagent which both undergoes a two-electron oxidation and chemically interacts with the electrode interface. The latter conclusion was demonstrated by the observation that the surfaces of n-CdSe photoanodes that were utilized in a $S^{2-}$ containing electrolyte were converted to $CdS_xSe_{1-x}$ [14,15]. It has also been found that chalconide-based electrolytes (i.e., Group 16 anions) are effective at stabilizing certain III–V semiconductor interfaces [16]. In several instances, apparent outer sphere charge-transfer reagents have also been observed to yield some degree of stability to semiconductor interfaces. Most notable is the observation that ferrocyanide stabilizes certain semiconducting interfaces. Although this was initially though to be due to the successful competition of outer sphere kinetics with semiconductor autooxidation, Bocarsly has demonstrated that cyanometalates react with the $Cd^{2+}$ ions generated by the decomposing semiconductor to form an ultrathin ($\leq 1$ μm) interfacial layer of $[CdFe(CN)_6]^{2-}$ [17]. Stabilization is obtained via a redox reaction between photogenerated holes in the semiconductor and the interfacial cyanometalate layer.

A second approach to semiconductor stabilization is the utilization of an electrolyte in which semiconductor photodecomposition products cannot form. Thus, in the case of n-Si, Lewis has noted that employment of a rigorously anhydrous nonaqueous electrolyte eliminates the possibility of interfacial oxide formation [18]. However, the fact that subnanomolar concentrations of water are sufficient to generate surface oxides makes the application quite difficult. Semiconductors that undergo decomposition to metal ions can likewise be stabilized by using a low-dielectric-constant nonligating electrolyte. Both organic liquids and solid-state ion conductors have been employed for this purpose. Unfortunately, such electrolytes are at best high resistance, and thus observable photocurrents are minimal. However, a hybrid approach in which a nonaqueous electrolyte (which may contain water) is coupled with a fast outer sphere redox reagent has proved successful in some cases. For example, Wrighton demonstrated that a methanol/TBAP electrolyte containing ferrocene provided n-Si electrodes with a good degree of stability, while allowing for the production of a sizable photocurrent [19].

A more molecular approach to the photocorrosion problem has been to modify the electrode surface chemically with a stabilizing species. The controlled growth of layers of $M_x[CdFe(CN)_6]$ (M = alkali cation) on n-CdX electrodes is an example of this approach. The concept was first introduced by Wrighton, who demonstrated that n-Si electrodes could be stabilized by the covalent attachment of a silylferrocene layer to the electrode surface. This was accomplished by the reaction of a hydrolytically unstable silane such as trichlorosilylferrocene with a single crystal silicon surface, etched to produce a surface containing hydroxyl functionalities. Under these conditions, two types of reactions occurred to tether the reagent to the electrode and to generate a multilayered (polymeric) interfacial structure, illustrated by the following:

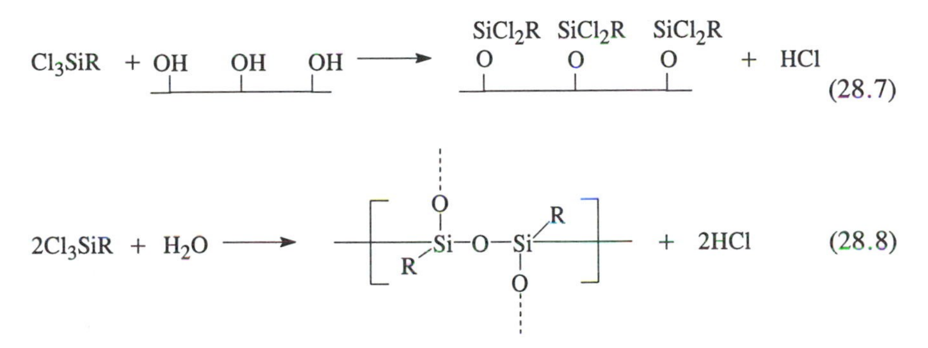


Electrode stabilization is produced by two effects associated with interfacial siloxane formation: The hydrophobic siloxane polymer provides a physical barrier limiting solvation of the electrode decomposition products, and the polymer generates a high effective concentration of the redox moiety (R) at the electrode interface, enhancing the overall interfacial charge-transfer rate. Key to this approach is the presence of a stable surface oxide, since dissolution of this oxide would lead to undercutting of the siloxane interface, physically separating the semiconductor from the protective polymer. Thus, although this approach has been demonstrated to be of utility for the stabilization of Si, Ge, and GaAs electrodes, there is still a small amount of semiconductor decomposition. As indicated earlier for CdX semiconductors in a ferrocyanide electrolyte, one approach to this problem is to utilize one or more of the electrode decomposition products to build the protective overlayer. Using this strategy, transient decomposition processes simply lead to the growth of the protective interface. As in the case of siloxane-derivatized interfaces, a high concentration of a redox-active species must be present in the interfacial structure to establish a kinetic competition between reversible oxidation of the interfacial material and autooxidation of the semiconductor. In the $CdX/M_x[CdFe(CM)_6]$ systems, the ferrocyanide unit present at $\sim 6\ M$ serves this purpose [20,21].

One of the key advantages of surface modification, independent of the synthetic approach, is that the stabilized electrode can be operated with a variety of electrolyte solvents and redox couples. For example, the insolubility of ferrocene in water precludes using this reagent to stabilize n-Si in an aqueous electrolyte; however, silylferrocene-derivatized n-Si photoanodes are relatively stable in aqueous electrolytes [22]. In addition to placing stabilizing species on the electrode surface, this same approach can be employed to attach electrocatalysts to the semiconducting electrode. For example, Wrighton's group has placed polymer-bound alkylviologen species on p-Si surfaces [23]. This cationic surface structure could be impregnated with $[PtCl_6]^{2-}$ via an ion-exchange reaction. Once in place, the platinum complex was electroreduced to platinum metal. Upon illumination of the p-Si interface, photogenerated electrons reduced the surface-confined viologen centers. These electrons were then "passed" on to the platinum particles catalyzing the reduction of water to $H_2$.

*Surface and Sub-Band-Gap States*

In addition to photodecomposition processes, real electrode behavior is often affected by the fact that there is a variety of localized energy states present on the semiconductor surface. If these states are energetically disposed to communicate with the semiconductor bulk (i.e., their energy is such that charge-transfer overlap with a band can occur), then interfacial charge transfer may proceed through one or more of these states. Orbitals that are spatially localized at the electrode surface (i.e., the molecular orbitals associated with a surface species) are known as *surface states*. In addition to arising from pure lattice species, surface states may be generated by the chemical reactions occurring between the semiconductor surface and the local environment. For example, clean surfaces of silicon have intrinsic surface states due to the fact that the Si atoms on the semiconductor surface are not four-coordinate, as expected for an $sp^3$-hybridized Si. Thus, surface silicon atoms have nonbonded valence electrons. In addition to the intrinsic surface states, Si readily reacts with oxygen to generate an interfacial layer of $SiO_x$. This material often contains molecular orbitals of appropriate energy to interact with the Si bands.

The existence of surface states in general can lead to a variety of nonidealities in the output parameters associated with semiconductor-electrolyte junctions. Figure 28.6 provides the current–potential response for a photoelectrochemical cell containing a cadmium ferrocyanide-modified n-CdS electrode in an aqueous ferri/ferrocyanide electrolyte. Although open-circuit and

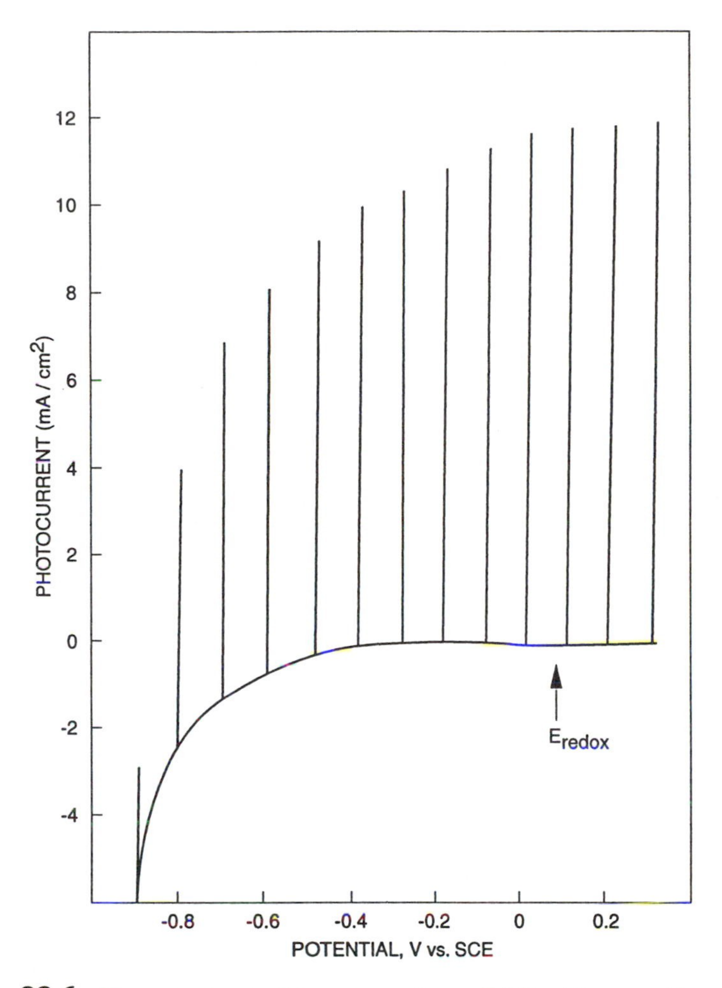


**Figure 28.6** The current–potential response of an n-CdS/ferri/ferrocyanide cell under chopped irradiation in (488 nm, 40 mW/cm$^2$). The ferri/ferrocyanide redox potential is marked in the diagram. Note that even when a dark cathodic current is present (solid line), a photocurrent is observed (spikes).

capacitance measurements indicate that the flatband potential is approximately −1.2 V vs. SCE, the dark current scan clearly shows a cathodic current at potentials significantly positive of this potential [20]. Thus, the n-CdS electrode is able to carry out a dark reduction at potentials where the space charge region is still in depletion (i.e., no accumulation layer exists). This can be accounted for by assuming a set of electrochemically active surface states exists at potentials just positive of (below) the conduction band edge. Such states could shuttle charge from the conduction band to the interface even though a depletion-type band bending exists. Note that the net effect of this situation is that under illumination, the onset potential (the point where cathodic and anodic currents are equal) is nowhere near the flatband potential, and thus the current-potential response cannot be used to estimate the flatband potential.

Since charge carriers can reside in surface states for prolonged time periods, such states often act as electron–hole recombination sites leading to increased cell inefficiency. This is often manifest as a poor fill factor or a low quantum yield for electron flow. In addition to mediating band-to-interface charge-transfer processes, if an extremely high density of surface states is present, they can in theory dominate the observed electrochemistry via a *Fermi level pinning* mechanism. This scenario involves the establishment of an equilibrium between the bulk lattice Fermi level and a Fermi level that represents the average energy of the surface state distribution. The existence of a surface state Fermi level implies a sufficient density of surface states to make a statistical model of these states reasonable. If this criterion is met, then the semiconductor will maintain an internal degree of band bending at the bulk–surface state interface which is independent of the electrolyte employed. A second equilibrium will be established between the surface state population and the electrochemically active solution states. This interaction will appear more or less metallic in nature if a sufficiently broad distribution of surface states is present. The net effect of this interaction will be a photopotential that is invariant with the redox potential of the electrolyte. In terms of a current–potential response, this will give rise to a photocurrent onset that is offset from the redox potential by a fixed amount (the difference between the bulk Fermi level and the surface state Fermi level). Thus, as the nature of the electroactive species is varied, the current–potential response varies. The possibility that this might occur was first discussed by Wrighton and Bard [24]. It has been pointed out that other processes can give rise to identical current–potential properties; thus, whether or not Fermi level pinning is prevalent remains a controversial issue.

### *Majority Carrier Processes*

Phenomenologically similar behavior can be obtained if the redox potential is sufficiently close to the conduction (or valence) band edge to allow for majority carriers to cross the band bending barrier either via a tunneling mechanism or by direct scaling of the barrier. Lewis was the first to demonstrate that the photovoltage of Si photoelectrodes is limited by the kinetics of photogenerated interfacial charge transfer, coupled with the flow of carriers over the semiconductor–electrolyte barrier [18]. This was only observed when the electrolyte couple employed was of an appropriate energy to provide good charge-transfer overlap between the redox species and the semiconductor band states.

Another mechanism by which charge carriers can overcome the interfacial barrier established by the charge-transfer equilibrium involves the photochemical population of conduction band states that lie above the top of the barrier. Conventional wisdom holds that once populated, rapid nonradiative transfer from such states should lead to the almost immediate transfer of excited electrons to states near the conduction band edge. Thus, thermally *hot electrons* should not exist for a sufficient period of time to allow for interfacial charge transfer. Nozik has proposed, however, that under conditions where the excited electrons are generated near the semiconductor surface, interfacial charge transfer may compete with nonradiative deactivation, allowing hot electrons to move across the semiconductor–electrolyte interface [25]. Recently, Koval has provided experimental evidence showing the production of hot electrons at an illuminated InP interface [26]. The production of hot electrons suggests that reducing potentials well beyond those predicted based on band edge positions can be accessed at the semiconductor interface.

A combination of majority and minority carrier processes has been observed to produce quantum yields in excess of one at both n- and p-type interfaces. In all cases where this has been noted, the redox species employed have been capable of multiple-electron processes. This type of behavior is often seen for the oxidation of carboxylic acids at n-type semiconductors (a two-electron process). It has also been noted for hydrazine oxidation (a four-electron process) and the reduction of hydrogen peroxide.

This phenomenon is known as *current doubling*, although it is not limited to two-electron processes [7]. Using an n-type system as an example, the minimum requirement for current doubling is the generation of a reactive intermediate upon reaction with a valence band hole that can be further oxidized by injection of an electron into the conduction band. For the oxidation of cyanide, the initially generated cyanogen free radical ($CN^- + \text{hole} \rightarrow CN^0$) generates a second redox potential ($CN^0/CN0^-$) that lies well above the n-CdS conduction band edge [113]. Thus, once formed, the cyanogen radical can transfer an electron to an empty conduction band orbital. Hence two charge carriers, a hole in the valence band and an electron in the conduction band, are produced upon absorption of a single photon. Injection of an electron into the conduction band competes with oxidation of $CN^0$ by a second photogenerated hole. This latter process has a limiting quantum yield of one, since the two-electron oxidation via holes (alone) requires two photons to be absorbed. As a result of these competitive charge-transfer pathways, increased illumination intensity (which produces a higher concentration of interfacial holes) gives rise to a decrease in the observed quantum yield.

### D. Polycrystalline Semiconductors

Although most solid-state junctions depend on single-crystal materials to avoid massive electron–hole recombination, photoelectrodes having reasonable efficiency have been produced using less ordered materials. This possibility is due in part to the fact that the photoelectrode is a front junction device. Also important is the role of the electrolyte in avoiding recombination processes. Once the redox-active species has been oxidized (or reduced), it is not subject to further chemistry at the semiconductor interface. As a result, while one observes a lowered quantum yield using electrodes composed of polycrystalline materials, the value is not so low as to be unattractive.

The use of polycrystalline materials is not limited to macroscopic electrodes. Suspended semiconductor particles can also be employed to carry out photoelectrochemical processes [27,28]. In this situation, each semiconducting particle acts like a miniature set of electrodes. If an n-type material is employed, oxidation takes place at illuminated portions of the particle, with reduction occurring in dark regions. As in the case of actual electrodes, kinetic overpotential can dramatically limit charge-transfer processes. Several authors have demonstrated that interfacial charge transfer can be enhanced by placing small islands of electrocatalysts on the particle surface [29–34]. Metal islands (typically Pt) have been utilized both as electron and hole catalysts, while metal oxides such as $RuO_2$ have proven to be a good oxidation catalyst. One outcome from powder photoelectrochemistry that is not observed in classical photoelectrochemical cells, due to the spatial separation of anode and cathode, is the relative ease with which a material that has been oxidized on an illuminated portion of the particle can be reduced at a dark portion. In cases where fast chemical reactions follow the initial charge-transfer process, this can lead to new product species. To date, this reactivity has been exploited to carry out a variety of sophisticated organic transformations [35].

In addition to providing new synthetic tools, it has been suggested that semiconducting powders may be of use in the natural decontamination of polluted waters. Solar irradiation of natural water supplies that have been treated with semiconducting powders could be used to oxidatively degrade pollutants. For example, the oxidation of cyanide to isocyanate, which decomposes into nitrogen and carbon dioxide, has been demonstrated. The oxidation of halides to atomic halogen and the oxidation of hydroxide to $OH^0$ has also been observed. These species can be employed as biocides to provide potable water.

## E. Dye-Sensitized Semiconducting Electrodes

One solution to the problem of semiconductor photodecomposition is to modify the spectral response of a stable wide-band-gap semiconductor so that solar energy can be efficiently utilized. This can be accomplished by adding to the electrolyte a dye that has absorption features that overlap the solar spectrum. The short excited-state lifetimes of molecular systems limit the distance an excited state can be expected to diffuse prior to nonradiative deactivation. Thus, unless a photon is absorbed when the dye molecule is near the semiconductor surface, or perhaps adsorbed on the semiconductor, productive charge transfer cannot be expected. As a result, a low concentration of dye must be employed to insure that the bulk of the absorption does not occur in regions of the electrochemical cell that are far from the electrode.

## F. Electrolytic Behavior of Adsorbed, Excited Photoreceptors

If a semiconductor electrode such as an n-ZnO crystal is illuminated in the presence of a small amount of dye (e.g., $5 \times 10^{-5}$ *M* rhodamine B), one can observe an enhanced photocurrent in the anodic region. If illumination is carried out by a tunable, monochromatic source, it is possible to obtain a photocurrent spectrum like that shown in Figure 28.7. In a typical experiment, after removing dissolved $O_2$ by nitrogen purging, one sets the potential of the semiconductor at a value in the region of limiting photocurrent (ca. +0.5 V vs. SCE) in the absence of dye. This current is simply due to the photoevolution of oxygen as described in Section II.B. If a sensitizing molecule is added to the cell, there is an additional photocurrent whenever illumination occurs at a wavelength absorbed by the sensitizer. In the long-wavelength region, the background photocurrent is often quite small; hence the photocurrent spectrum is essentially identical to the absorption spectrum of the sensitizer.

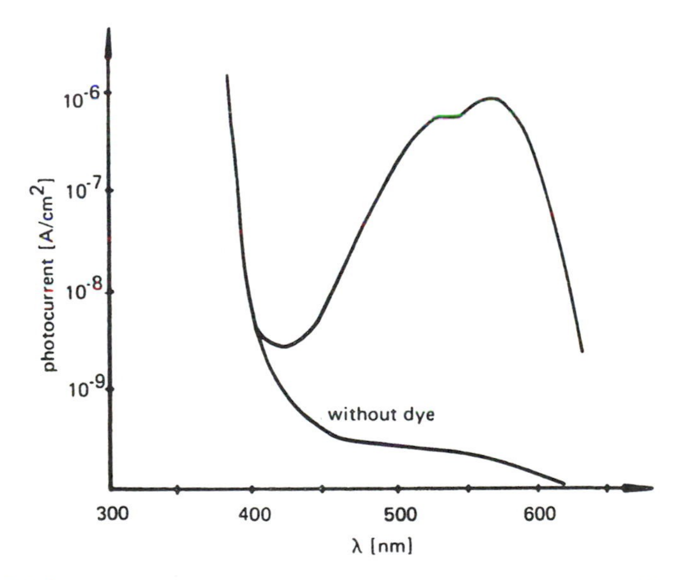


**Figure 28.7** Sensitized and unsensitized photocurrent spectrum at a ZnO electrode (doped with Cu). Electrolyte: 1 *M* KCl, pH 2; dye: $10^{-4}$ *M* rhodamine B. [From Ref. 49, reprinted with permission.]

This phenomenon arises from charge-transfer reactions between the semiconductor and excited dye molecules adsorbed on its surface. If the semiconductor band gap is large compared to the dye's excitation energy, electron transfer between the dye and the electrode may involve the highest normally filled level or the excited level of the dye, but usually not both. One of the dye levels will not be electroactive because it will match some energy in the electrode's band-gap region.

For our specific case, rhodamine B on n-ZnO at +0.5 V vs. SCE, there is no dye electroactivity without illumination, yet an anodic current oxidizes the dye when the light shines. The absence of the dark reaction implies that the highest occupied orbital of the dye is at an energy in the band gap or the valence band. The photocurrent then arises because the absorption raises a dye electron to an excited level that overlaps the conduction band. Thus the net effect is injection of electrons into the conduction band. Hole neutralization cannot explain the observations, because light of smaller energy than the band gap energy will create the effect. Moreover, hole neutralization is fully accounted for by the saturation level of $O_2$ generation, suggesting the energetic scheme shown in Figure 28.8.

It is interesting to compare this behavior with that expected at a metal electrode. If the Fermi level lies above the highest normally filled dye level, but below the lowest excited level, no dark redox process will occur. If the dye is

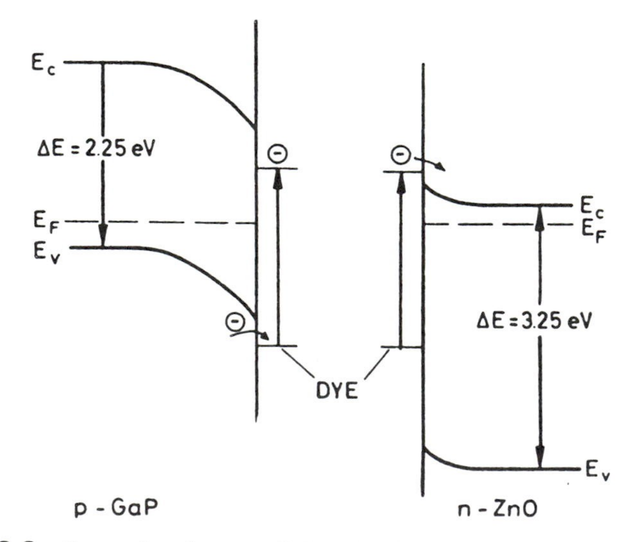


**Figure 28.8** Comparison between GaP and ZnO surfaces. [From Ref. 48, reprinted with permission.]

then excited, oxidation can occur by removal of an electron from the excited level; but simultaneously, electron transfer from the filled metal levels will reduce the vacancy in the highest normally filled level. The net result is quenching of the excited state by the electrode, but no external current is measured. These consequences follow from the absence of a band gap in the metal.

It has been found that the photosensitized electrochemical process is also influenced by other substances dissolved in solution. For example, the photocurrent from the rhodamine B/n-ZnO system increases when reducing agents such as hydroquinone are added to the system [36]. The hydroquinone serves to restore the adsorbed rhodamine B to its active reduced, ground-state form so that it can undergo repeated absorption acts leading to charge transfer. Several competing mechanisms are now being investigated for this *supersensitization* phenomenon. In contrast, oxygen often reduces (or *desensitizes*) the photocurrent. Typically this is due to quenching of the dye-excited state.

Photosensitized electrochemical reactions using ZnO single crystals with other dyes and solutes have been studied extensively by Gerischer and Tributsch [36,37], Hauffe and co-workers [38], and other groups [39–41]. Photosensitization effects with n-CdS [42,43] and n-$TiO_2$ [44] are reported to be similar to those with n-ZnO. This type of work has been extended to include covalently bound sensitizers on n-$SnO_2$ and n-$TiO_2$ [45–47].

For p-type semiconductors, such as GaAs, GaP, and $Cu_2O$, sensitized photoelectrochemical behavior is complementary to that seen with n-type material. Electrons are transferred from the valence band to the dye; hence holes are injected into the electrode, and a cathodic current flows. Arguments like those just presented lead to a picture like that presented in Figure 28.8. Results are displayed in Figure 28.9 for p-GaP with *N,N'*-diethylpseudocyanine as sensitizer. In this system, $O_2$ acts as a supersensitizer. Major studies of p-type electrodes have been carried out by Memming and Gerischer and their co-workers [42,48,49].

Quantum yields for dye-sensitized electrodes are typically quite low, if calculated based on number of photons *incident* on the electrode surface, which is the procedure utilized for reporting quantum yields of pure semiconductor interfaces. However, quantum yields based on the number of photons *absorbed* (the standard approach utilized with molecular systems) are estimated to be about 10%. This value is affected by the actual dye employed, as well as the redox reagents and electrode material [50]. The low quantum yield per incident photon is directly related to the limited number of dye molecules present on an electrode surface. This physical limitation has led some to speculate that dye sensitization will not produce a practically useful photoelectrochemical cell. However, Grätzel has recently announced that $TiO_2$-based cells utilizing a dye that is a multimetal analog of $[Ru(CN)_2(2,2'\text{-bipyridine})_2]$ (several ruthenium centers are bridged together via bridging cyanide ligands) yield solar conver-

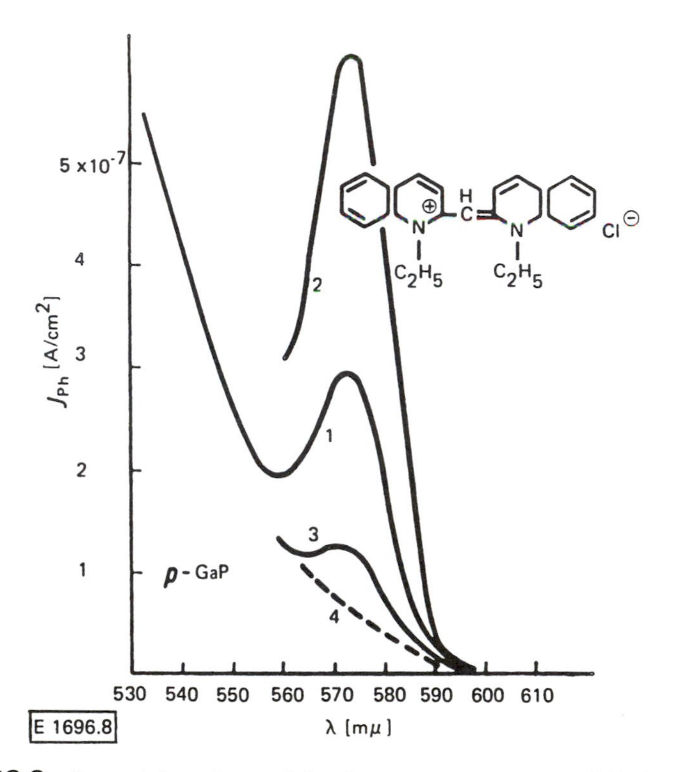


**Figure 28.9** Spectral dependence of the photocurrents at a p-type GaP electrode with sensitization by *N,N'*-diethylpseudoisocyanine. [Cathodic photocurrent, electrolyte: 1 *M* KCl + *N,N'*-diethylpseudoisocyanine chloride (1 mL of $10^{-2}$ *M* ethanol to 10 mL of water).] 1, $N_2$ flushing; 2, $O_2$ flushing; 3, addition of piperidine ($10^{-2}$ *M*); 4, trace of the background photocurrent at p-type GaP without dye. [From Ref. 42.]

sion efficiencies on the order of 10%. This level of efficiency is comparable to that obtained from direct semiconductor–electrolyte devices. This impressive efficiency is apparently obtained by utilizing an ultrahigh surface area form of $TiO_2$. The large surface area allows for the chemisorption of a large quantity of dye. In addition, the roughness of the electrode surface enhances absorption over reflection processes.

Arden and Fromherz [51–53] have reported elegant work involving monolayers of surfactants transferred to electrodes by the Langmuir–Blodgett method. Some results are shown in Figure 28.10. Two separate cationic dyes, thiacyanine (A) and oxacyanine (D), were synthesized with $C_{18}$ alkyl chains, so that they would form oriented monolayers on water–air interfaces. By standard techniques, these were transferred alternatively or in sequence to n-$In_2O_3$ electrodes, as

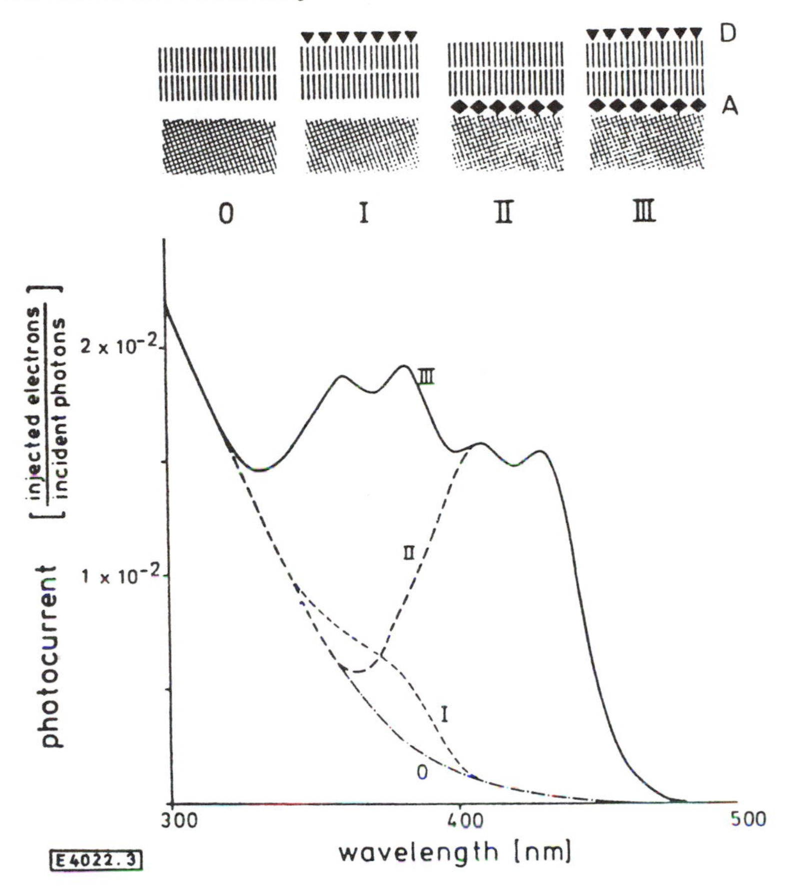


**Figure 28.10** Photocurrent per incident photon for various monolayer assemblies on $In_2O_3$. Configuration 0 is for a bilayer of surfactant molecules without dye deposited on the electrode (represented as the lower shaded area). Configuration I involves the donor dye in the outer layer and configuration II features the acceptor dye in the inner layer. In each of these cases, the remaining layer is made up of a surfactant without dye. Configuration III has both dyes together in the arrangement shown. The electrolyte was 5 m*M* borate buffer of pH 10 with 0.5 *M* allylthiourea as supersensitizer. [From Ref. 53.]

shown at the top of Figure 28.10. The particular orientations shown allow the polar head groups of A and D to interact, respectively, with the polar oxide surface and the aqueous solution, while hydrophobic paraffinic segments interact with each other in the interior of the bilayer. Much optical evidence can be cited for the existence and stability of this type of ordered arrangement [54].

The results in Figure 28.10 show that chromophore A sensitizes photocurrent (curve II), whereas chromophore D has little effect by itself (curve I). The apparent reason for the difference is the remoteness of chromophore A. In the tandem arrangement (III), however, one finds that light absorbed by D (at wavelengths below 380 nm) gives rise to a significant photocurrent. Arden and Fromherz interpret these results to imply that energy absorbed by D is transferred (by the Förster process) to A, which then undergoes charge transfer with the electrode. Without the A underlayer, D cannot sensitize photocurrent; without the D overlayer, A cannot produce significant effects below 380 nm. Luminescence from the D layer can also be observed, and by exploiting it, together with the electrochemical data, Fromherz and Arden have produced some very detailed pictures of interfacial charge-transfer dynamics.

## III. PHOTOEMISSION FROM METAL ELECTRODES

Cathodic currents are stimulated by the illumination of metal electrodes [55–59]. These currents are often strongly enhanced by the presence in solution of known electron scavengers such as $N_2O$ and $H_3O^+$. Research carried out by Barker [59–61], Pleskov [62–64], Delahay [65], and their co-workers indicates rather strongly that the impinging photons eject electrons from the electrode surface. They appear to travel some distance ($\sim 50$ Å) before becoming solvated [55,59,62,65]. If a scavenger is present, the solvated (usually aquated) electrons may react with it irreversibly. For example,

$$e_{aq} + N_2O + H^+ \rightarrow OH^\bullet + N_2 \tag{28.9}$$

The photoelectrons are then lost permanently from the electrode, and in this case the cathodic current is actually enhanced further by faradaic conversion of $OH^\bullet$ to $OH^-$ [55,59,60]. Without the scavenger, the photocurrent is small, because the photoelectrons return to the electrode, and the net charge transfer is almost zero.

Naturally, one must supply a minimum energy to remove an electron from the electrode; thus there is a *red-limit* wavelength above which photoejection is very improbable [33,34,55]. Since the Fermi level depends on electrode potential, this limit shifts to shorter wavelengths as the potential becomes more positive. Thus there is a threshold potential for photoemission and the emission becomes more probable at more negative potentials [59,62].

Various experimental approaches have been developed for studies of these phenomena [55,59–61,63,65]. Early techniques advanced by Barker's group featured irradiation of a dropping mercury electrod (DMA) by chopped light from low-, medium-, and high-pressure mercury lamps [59]. Pulse polarographic monitoring circuitry synchronized to the light chopper was used for measurement of the photocurrent. Note, however, that the time scale of a photoemission experiment is inherently very short. The diffusion time for return of the ejected electrons is on the order of 100 ns; hence there is a great potential for studies of very fast electrode processes. Because one cannot realize this potential with low-frequency chopping techniques, the emphasis has shifted toward flash illumination methods.

Barker later described some work that involved apparatus like that shown in Figure 28.11. Light was supplied to a continuously renewed mercury pool electrode by a Q-switched, frequency-doubled ruby laser with a pulse width of ~15 ns. The electrode was set initially at any desired potential by a simple polarizing circuit, the response of which was slow enough that the electrode's reaction to the flash could be monitored as a coulostatic transient, ΔE (measured with respect to the initial potential) versus time. The difference in charge with respect to the initial condition is straightforwardly related to ΔE,

$$\Delta q = C_d \, \Delta E \tag{28.10}$$

where $C_d$ is the differential capacitance of the interface. Since ΔE is only a few millivolts, $C_d$ is effectively constant for a given experiment. Useful information could be obtained for times as short as a few tens of nanoseconds after the flash.

The solution chemistry that can be studied by photoemission techniques is largely that instigated by solvated electrons. It therefore has much in common

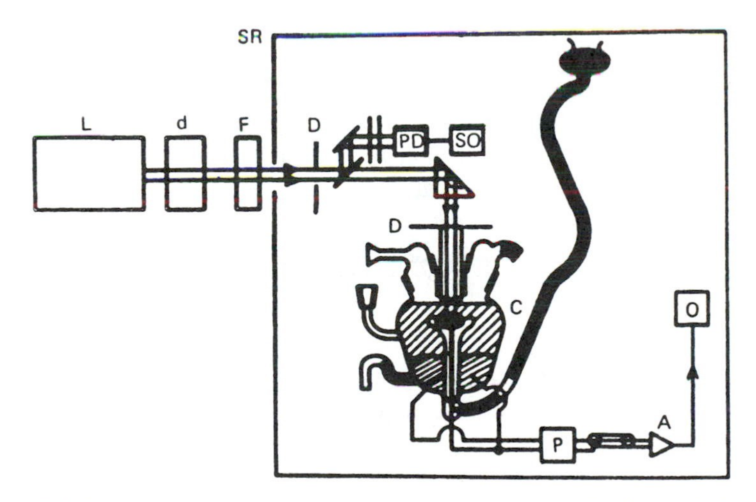


**Figure 28.11** Apparatus for photoemission measurements. L, Q-switched ruby laser; d, frequency doubler; F, $CuSO_4$ solution filter; SR, screened room; D, diaphragm; PD, photodiode; SO, storage oscilloscope; C, cell; P, polarizing circuit; A, wideband amplifier; O, oscilloscope. The mercury pool working electrode is renewed continuously from the reservoir at the upper right. [From Ref. 61.]

with the domain of pulse radiolysis, although the photoemission method advantageously avoids extensive radiolytic damage to the solution components. For electrochemistry, there is a very useful ability to generate, near the electrode, radicals and other intermediates of electrochemical importance. Purely electrochemical experiments often cannot provide the insight into electrode processes that one can obtain in this way.

An important example relates to hydrogen discharge on Hg [55,58,59,61,64]. It has been advocated that this process occurs in two steps, with $H^\bullet$ as an intermediate:

$$H_3O^+ + e \xrightarrow{\text{slow}} H^\bullet + H_2O \tag{28.11}$$

$$H^\bullet + e + H_3O^+ \longrightarrow H_2 + H_2O \quad \text{(low pH)} \tag{28.12}$$

or

$$H^\bullet + e + H_2O \longrightarrow H_2 + OH^- \quad \text{(high pH)} \tag{28.13}$$

However, electrochemical studies alone have not been able to verify the participation of these reactions, because one cannot unambiguously isolate the electrochemistry of $H^\bullet$ within the whole discharge process. The Pleskov and Barker groups have very cleverly untangled the web by relying on photoelectrons to create $H^\bullet$ homogeneously [60,61,64]:

$$H_3O^+ + e_{aq} \longrightarrow H^\bullet + H_2O \tag{28.14}$$

Thus they have been able to study hydrogen atom electrochemistry cleanly and in impressive detail. The atoms seem to be adsorbed strongly on the electrode, and from the adsorbed state they undergo very rapid anodic or cathodic electrochemistry, depending on potential.

## IV. ELECTROCHEMICAL MONITORING OF PHOTOLYTIC INTERMEDIATES

Photochemistry often involves species that are electrochemically interesting. Excited states, radicals, and metals with light-influenced oxidation states are very common intermediates or products; thus it is not surprising that numerous attempts have been made to study photochemical processes with electrochemical tools [66,67].

Experimental techniques are conveniently divided according to the way in which the irradiation is carried out. The simplest approach is to shine a fairly intense source continuously on the solution, so that the overall photochemical transformation occurs approximately at steady state. Almost any electrometric method might serve to monitor the intermediates or products. During the first half of this century, potentiometric measurements were often made in this way [66], but these *photopotential* studies have been almost wholly supplanted by faradaic approaches, which have superior sensitivity and selectivity. Berg and his co-workers first introduced the faradaic methods, with emphasis on polarography and amperometry at the DME, in the early 1960s [68,69]. Kuwana's review gives good coverage to their work through about 1966 and to the earlier work on photopotentials [66].

There are several drawbacks to experiments done with continuous irradiation. High light fluxes cannot be introduced without serious heating; thus the concentrations of intermediates are low. Moreover, all intermediates are present simultaneously, since the reaction proceeds steadily. This situation makes it difficult to discern the order in which sequential intermediates appear, and it may create interference problems from overlapping electrochemical responses.

The desire for temporal resolution of photolysis led to the development of flash methods. In these experiments [70] the solution is exposed to a short ($\sim 10$ $\mu s$ width) burst of light at high intensity (several hundred joules dissipated in the flash lamp). Absorption by the photoactive solute creates a high initial concentration of the primary intermediate. Its decay with time often leads to the rise and fall of other transient species that appear later in the reaction scheme. Because these time dependencies tell much about the photolysis mechanism, flash methods are immensely valuable to photochemistry and have become very common. Usually, the intermediates are followed by UV or visible absorption spectroscopy. Berg and Schweiss were first to implement electrochemical monitoring [71], but Perone and his co-workers have been particularly active since the middle 1960s in the development and application of the technique [67,72–76].

In the most straightforward approach, one observes the current transient arising from the electrolysis of a flash-generated species at an electrode held at constant potential. However, there are two serious complications:

1. Electrolysis occurring before the flash may deplete the electrode region of the photoactive substance or generate obtrusive products.
2. The current decay is not simply linked to a time-dependent bulk concentration of the electroactive substance. Superimposed on that time function is the normal Cottrell decay from the diffusion process.

Perone's group surmounted these difficulties through a technique called time-delayed potentiostatic analysis [72,73]. Figure 28.12 depicts its operation. Before the flash and after it for a delay $\tau$, the working electrode is disconnected from ground by an open FET switch. At time $\tau$, the switch closes, electrolysis begins, and for a brief period thereafter, one sees a current decay that is essentially a Cottrell relation undistorted by reactant disappearance. An observation proportional to the bulk concentration of the transient species can be obtained by sampling the current at a fixed interval following $\tau$. Repeating the

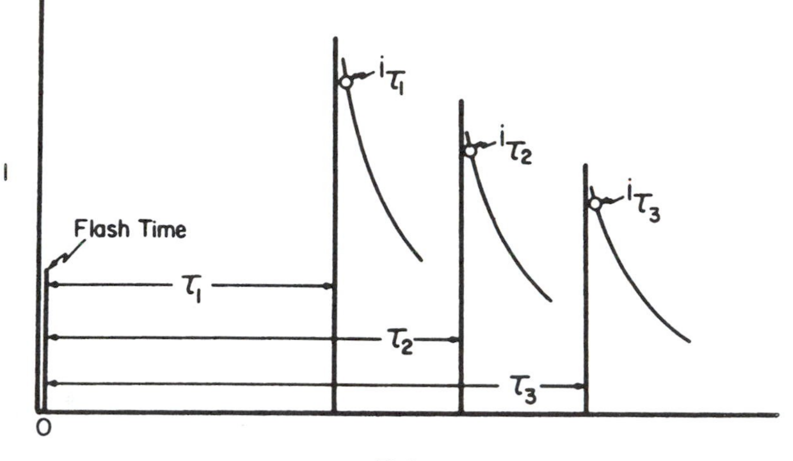


**Figure 28.12** Current–time curves observed with time-delayed potentiostatic analysis. $\tau_1$, $\tau_2$, and $\tau_3$ refer to three different delay times used for three different experiments. [From Ref. 67.]

experiment with different potentials or different $\tau$ values enables one to derive current–potential curves for a sequence of delay times; hence there is both qualitative and quantitative information about the transients. Observations as soon as 50 $\mu$s after the flash can be made in this way. The value of this qualitative information was demonstrated by a study of ferrioxalate photolysis [74,75]. Using simultaneous electrochemical and optical monitoring of flash transients [76], Perone and co-workers were able to make some strong inferences about the chemical nature of a key intermediate, and they found evidence for some additional mechanistic steps that could not be discerned solely from optical data.

The techniques described here for monitoring flash-photolyzed solutions are clearly applicable to other situations in which the solution's composition is modified suddenly. Henglein and co-workers have used them in some interesting electrochemical studies of free radicals generated by pulse radiolysis [77,78]. They have been able to resolve events on a $10^{-5}$ time scale.

Interest in continuous irradiation has been revived by the promising rotating electrode systems developed by two groups. Johnson and Resnick used a rotating ring-disk electrode design in which the disk was replaced by a quartz window, through which light was directed downward [79]. Photolytic products and intermediates are therefore swept past a Pt ring electrode, where they can be detected [79,80]. Albery et al. used a rotating disk electrode design featuring a transparent thin-film Pt disk through which the photolytic flux was directed

[81]. These workers have used their apparatus to carry out a detailed study of intermediates in the oxidation–reduction processes involving anthraquinones and hydroanthraquinones. The use of these rotated electrodes offers some advantages. The measurements are simpler than with flash methods, and for strongly absorbing solutions they feature a localized photolysis in a region near the electrode. Moreover, they hold out an intriguing promise for studying photolytic processes in substances that are produced electrochemically.

## V. ELECTROCHEMILUMINESCENCE

### A. Chemical Background

Research has uncovered a general tendency for very fast, energetic electron-transfer reactions to yield electronically excited products that may reveal themselves via luminescence [66,82–90]. A prototype of this chemistry is the reaction between the anion and cation radicals of 9,10-diphenylanthracene (DPA) in acetonitrile:

$$\mathrm{DPA}^{\overset{+}{\bullet}} + \mathrm{DPA}^{\overset{-}{\bullet}} \rightarrow {}^{1}\mathrm{DPA}^{*} + \mathrm{DPA} \tag{28.15}$$

The blue emission accompanying this reaction is easily identified with DPA's first excited singlet state $^{1}\mathrm{DPA}^{*}$, which clearly must be an ultimate reaction product.

Theories of electron transfer have rationalized redox excitation as a manifestation of the Franck–Condon principle [91–93]. The actual redox act in the DPA(+)/DPA(–) reaction occurs very quickly; hence the production of ground-state DPA molecules would require nearly instantaneous mechanical excitation of the molecular frames by the full reaction energy of 3.2 eV. Because the system has a limited ability to accept mechanical energy on a short time scale, very fast, energetic reactions may follow paths that limit the mechanical excitation of the products. Raising DPA to $^{1}\mathrm{DPA}^{*}$ requires about 3.0 eV; hence reaction 28.15 is nearly thermoneutral. Very little energy must be taken up in mechanical modes; thus we can see why that path competes significantly with the more exothermic production of ground-state molecules.

Various kinds of experimental evidence indicate that the DPA(+)/DPA(–) reaction does indeed yield its light by direct population of the emitting state. Other cases are not so straightforward. Consider the oxidation of $\mathrm{DPA}^{\overset{-}{\bullet}}$ by the cation radical of *N,N,N′,N′*-tetramethyl-*p*-phenylenediamine (TMPD). The free energy of that redox process, which also produces light from $^{1}\mathrm{DPA}^{*}$, is only 2.0 eV per transferred electron. Thus the energy emitted in a single photon greatly exceeds the energy available from one redox event. The emitting state must arise by a more complicated mechanism, and much evidence points to a path involving redox-excited triplets, $^3DPA^*$:

$$DPA^{\overset{-}{\bullet}} + TMPD^{\overset{+}{\bullet}} \rightarrow {}^3DPA^* + TMPD \tag{28.16}$$

$$^3DPA^* + {}^3DPA^* \rightarrow {}^1DPA^* + DPA \tag{28.17}$$

Reaction 28.17, called *triplet–triplet annihilation*, effectively localizes the energy from two electron transfers onto one DPA molecule.

One should not gain the impression from the foregoing that redox excitation is restricted to reactions between aromatic radical ions in aprotic solvents. Studies of such reactions have indeed dominated research because the optical and electrochemical properties of many aromatics are well known, but there are numerous cases of redox excitation outside this chemical domain. For example, singlet oxygen seems to arise from oxidations of superoxide ion in acetonitrile [94]. Similarly, luminescent tris(2,2′-bipyridyl)ruthenium(II) can arise in at least three ways: (1) from a kind of ion annihilation in $CH_3CN$ [95] or DMF,

$$Ru(bpy)_3^{+3} + Ru(bpy)_3^{+1} \rightarrow [Ru(bpy)_3^{+2}]^* + Ru(bpy)_3^{+2} \tag{28.18}$$

(2) from reduction of $Ru(bpy)_3^{+3}$ with radiolytically produced aquated electrons [96], and (3) from reduction of the same species in aqueous base containing hydrazine [97].

Experimental work has amply demonstrated the generality of redox excitation, and it has uncovered many details of the mechanisms by which chemiluminescence arises. Thus attention has been partly turned toward the quantitative aspects: the probabilities of photonic emission per redox event and the excitation yields for various energetically accessible product states. Understanding the factors governing the distribution of excitation yields is an intriguing fundamental issue, because it bears on the details of the intermolecular interaction giving rise to the redox process. Opportunities to focus sharply on the details of solution phase reactions are very rare; hence it is not surprising that redox excitation distributions are receiving serious theoretical and experimental study.

## B. Basic Experimental Considerations

Experimental work is often hampered by limited reactant stability. Any redox process that is sufficiently energetic to produce electronic excitation necessarily involves very potent redox agents. Only in a few cases is it feasible to isolate the reactants in bulk and react them by a flow method. More often they are electrolytically generated as needed from parent substances such as neutral aromatics. Electrochemical generation offers several advantages:

1. Electrode processes at controlled potential can be carried out selectively and with high current efficiency.
2. One has flexibility in the time domain, geometry, and sequencing of reactant generation.
3. It is possible to obtain accurate theoretical time profiles of the homogeneous redox rate; these are important for quantitative measurements and some mechanistic studies.

In the interests of improved electrochemical background limits and reactant stability, it is important to employ solvents that are as free as possible of nucleophiles and proton sources. Special attention always goes to the removal of water. The most important media are carefully purified acetonitrile, dimethylformamide, benzonitrile, and tetrahydrofuran. Popular supporting electrolytes are tetra-*n*-butylammonium perchlorate (TBAP) and fluoroborate ($TBABF_4$). Solutions are usually prepared by vacuum-line methods (Chap. 18) or in a dry box (Chap. 19) to exclude oxygen from the systems and to avoid contamination by water.

The most widely used technique for producing pairs of reactants involves the sequential generation of the oxidant and reductant at a single electrode [86,87,90]. Figure 28.13 displays the essentials of one such experiment in which the anion–cation annihilation is to be studied. The working electrode is a small Pt disk held initially at the rest potential in an unstirred solution of rubrene in benzonitrile. It is then carried through the potential program shown in Figure 28.13b. First the anions, then the cations are produced. Concentration profiles for the electroactive substances are depicted in Figure 28.14. During the second step, there is a zone at which anions and cations diffuse together and react (see Fig. 28.14c). Thus emission arises during the second step, but it decays as the initial reactant is consumed. Obviously, this experiment could be performed so as to produce a single transient as described, or one could continue indefinitely the alternate generation of reactants to yield a train of pulses.

An alternative to sequential generation is to use two separate electrodes to generate the reactants, then to move them together for reaction. An elegant implementation involves the rotating ring-disk electrode [87,98]. Consider, for example, an investigation of the rubrene reaction discussed earlier. One could produce the cation at the disk and the anion at the ring. The convective-diffusion pattern would then bring them together for reaction near the face of the ring. A steady reaction rate can be achieved by this method.

The steady-state method has advantages in its freedom from double-layer charging and in the simplicity of light and current measurements. The transient method is mechanically simpler and does not require a dual potentiostat (Chap. 6). Moreover, the rate of electrolysis is smaller, and hence solutions may suffer more slowly from the buildup of contaminants arising from side reactions.

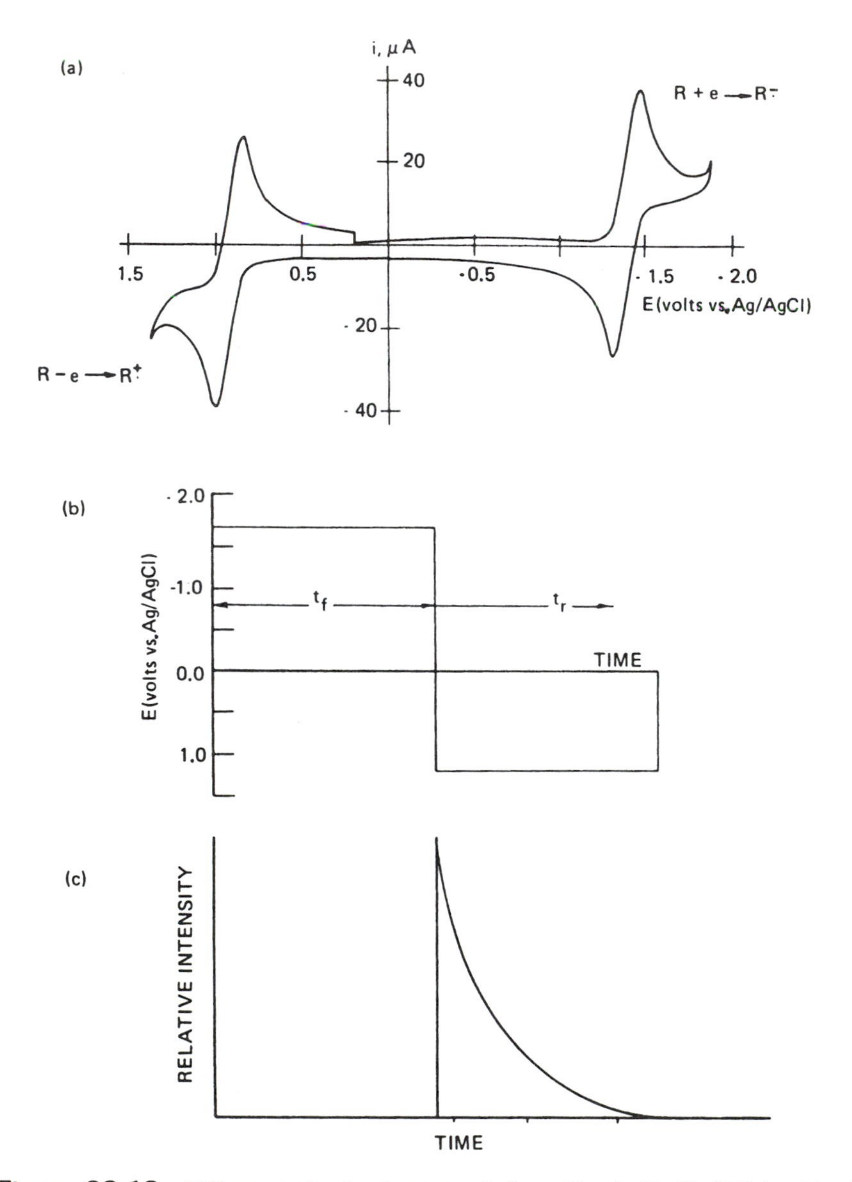


**Figure 28.13** ECL generation by the step technique. The Ag/AgCl, KCl (satd.) reference is –0.045 V vs. SCE. (a) Cyclic voltammetric curve (0.593 m*M* rubrene in benzonitrile with 0.1 *M* TBAP); (b) working electrode potential program; (c) emission intensity versus time.

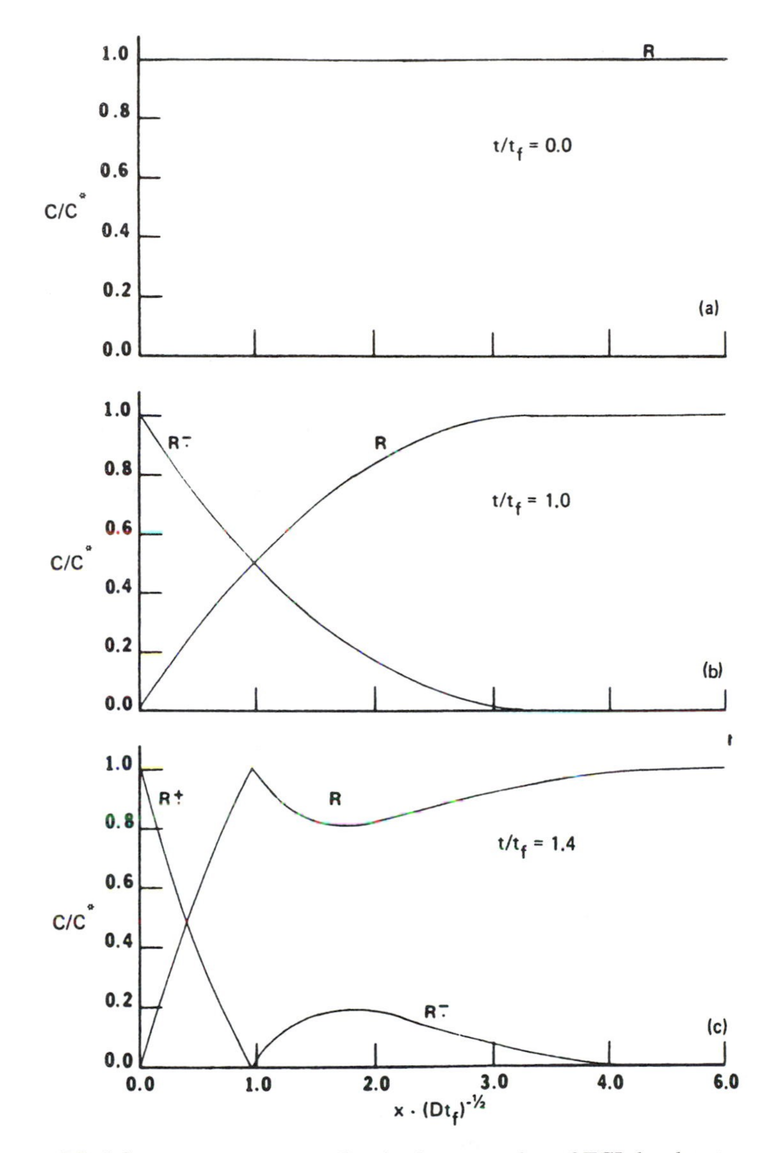


**Figure 28.14** Concentration profiles in the generation of ECL by the step method. The initial bulk concentrations of R are represented as C*. [From Ref. 87.]

Either method can be treated theoretically via digital simulation to obtain time dependencies of redox reaction rates, which are needed in some diagnostic and efficiency studies.

## C. Types of Experiments

Among the most important qualitative studies of any system are investigations of its emission spectra. One might wish to identify the emitting species or examine the spectral effects of certain additives. Spectral recording requires a light level that appears steady to the photometric instrument, which normally is a conventional fluorescence spectrometer. Steady-state generation of reactants straightforwardly satisfies this requirement. The continuous sequential method will also suffice if the period between light pulses is short compared to the photometric time constant.

One might use spectral recording in experiments in which one adds a triplet energy acceptor to the reacting system. Results from a study of the fluoranthene(–)/10-methylphenothiazine(+) reaction are shown in Figure 28.15 [86,99]. This process is an energetically deficient system that is thought to yield its blue fluoranthene-like emission via redox-produced fluoranthene (FA) triplets:

$$FA^{\bar{\bullet}} + 10\text{–}MP^{\overset{+}{\bullet}} \rightarrow {}^3FA^* + 10\text{–}MP \quad (28.19)$$

$${}^3FA^* + {}^3FA^* \rightarrow {}^1FA^* + FA \quad (28.20)$$

$${}^1FA^* \rightarrow FA + h\nu \quad (28.21)$$

Figure 28.15, curve a, displays the emission spectrum. Anthracene (An) can be added to this system without electrochemical interference, yet it alters the spectrum to the anthracene-like distribution of Figure 28.15, curve b. Apparently, the exothermic energy transfer

$$An + {}^3FA^* \rightarrow {}^3AN^* + FA \quad (28.22)$$

quenches ${}^3FA^*$ completely, and emission arises from ${}^1An^*$ produced by annihilation of anthracene triplets.

A quantitative variant of this interception approach is useful for evaluating the yields of triplet states from certain redox processes [86,87]. Consider, for example, the FA(–)/10-MP(+) reaction in the presence of trans-stilbene. No emission is seen because the following quenching process occurs:

$${}^3Fl^* + \textit{trans}\text{-stilbene} \rightarrow FL + {}^3(\text{stilbene})^* \quad (28.23)$$

$${}^3(\text{stilbene})^* \rightarrow \begin{cases} \textit{cis}\text{-stilbene} \\ \textit{trans}\text{-stilbene} \end{cases} \quad (28.24)$$

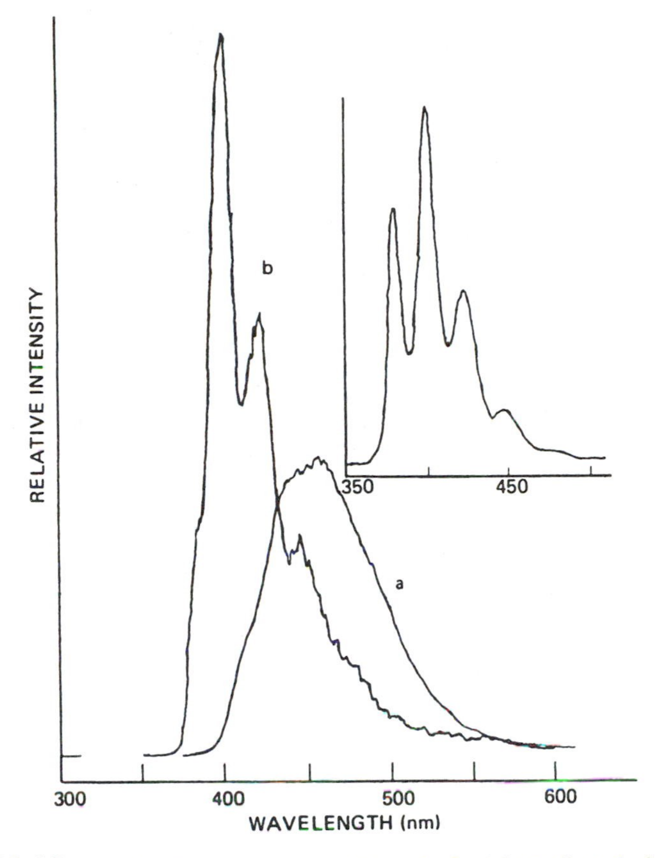


**Figure 28.15** a, Chemiluminescence spectrum obtained from electrolysis of a DMF solution containing 1 m*M* fluoranthene and 1 m*M* 10-MP. Alternating steps at −1.75 V and +0.88 V vs. SCE were used. b, Chemiluminescence spectrum under the same conditions, but with 1 m*M* anthracene added. Inset shows anthracene fluorescence spectrum for a $10^{-5}$ *M* DMF solution. Reabsorption reduces the 0,0 intensity in b. [From Ref. 99, adapted with permission.]

Coulometric generation of the reactants in bulk provides a known number of redox events plus a quantity of *cis*-stilbene that can be determined chromatographically. Since the cis–trans isomerization yields are known from photochemical studies, one can calculate the efficiency with which $^3Fl^*$ is formed. Triplet yields ranging from 2% to 80% have been recorded for various systems in this way [86,87].

Another interesting quantitative question is the probability of emission per redox event, $\phi_{ECL}$. Evaluating this parameter is very difficult because one requires *absolute* luminescence measurements. That is, one must know the total number of photons emitted at every wavelength in every direction. Absolute studies are complicated by geometric factors, variations of photometric response with wavelength, effects due to reabsorption of light before it emerges from solution, and uncertain reflectivity of the electrode. Many approaches to the problem have appeared in the literature [85–87,100–103]. In every such study, $\phi_{ECL}$ is obtained by comparing the photonic output with the faradaic charge used to generate the reactants. Steady-state and transient methods have both played important roles, and $\phi_{ECL}$ values ranging from 0.001% to about 20% have been reported.

The single-pulse transient experiment described earlier has also been studied extensively as a tool for mechanistic investigations. Feldberg was first to see the diagnostic value in the shape of the light decay curve [104,105], and he showed that it could be used to discriminate between systems producing singlets directly (S route) and those proceeding through triplet intermediates (T route). More recent theoretical and experimental work has explored the extraction of quantitative excitation yield information from the shapes of absolute decay curves [89,90,106–110]. Feldberg also noted that the curve shape for S-route emission could be used to evaluate the homogeneous electron-transfer rate constant if the time domain of the experiment could be made sufficiently short [104]. These rate parameters are very difficult to obtain because the reactions are so fast, yet they are of much current theoretical interest. Van Duyne's group followed Feldberg's lead by carrying out some difficult time-correlated single-photon-

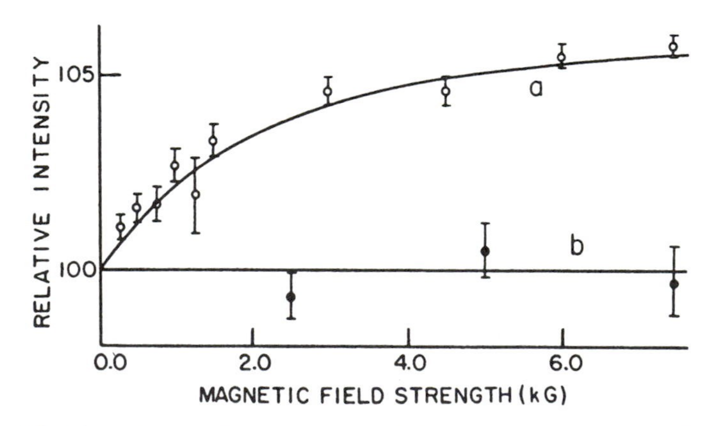


**Figure 28.16** Magnetic field effects on ECL intensity. a, From DPA(-)/TMPD(+); b, From DPA(+)/DPA(-). Light was generated at Pt by a train of alternating steps. [Adapted with permission from L.R. Faulkner and A.J. Bard, *J. Am. Chem. Soc. 91*:209 (1969). Copyright 1969 American Chemical Society.]

counting experiments involving 10-μs step widths [93]. They reported some rate data for the DPA(+)/DPA(-) reaction and compared the results with predictions from several theories.

Mechanistic questions have also been addressed via studies of magnetic field effects on electrogenerated chemiluminescence intensity [85-90,111]. Figure 28.16 contrasts the magnetic behavior for two systems. There is no significant field effect on DPA(+)/DPA(-) emission, but luminescence from DPA(-)/TMPD(+) is enhanced by the field. This enhancement apparently arises from field-dependent rate constants for triplet–triplet annihilation and for triplet quenching by radical ions. Both processes involve changes in spin angular momentum and, according to a theory of Merrifield [112], ought to have field-dependent encounter reaction probabilities. Extensive experimental work generally supports the view that a magnetic enhancement will generally appear whenever emission arises via the T route, and no effect will be seen when the S route applies [85–90].

In this area, as in semiconductor photoelectrochemistry, we have been able only to set forth major concepts and experimental tools. The chemical issues, which are often fascinating, are explored much more fully in several reviews [86–90].

## APPENDIX: STRUCTURES OF CITED MOLECULES

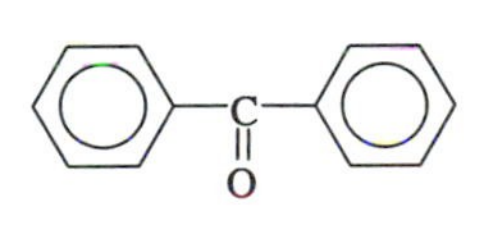

BP

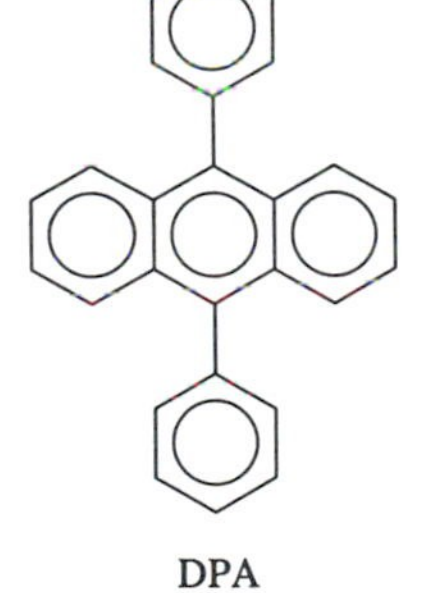
DPA

Fluoranthene

10-MP

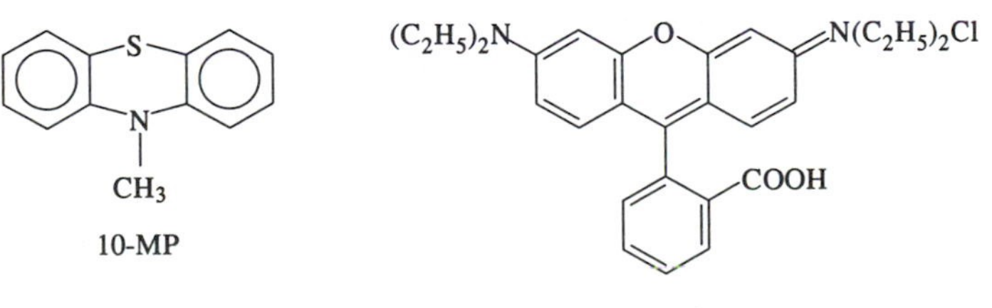

Rhodamine B

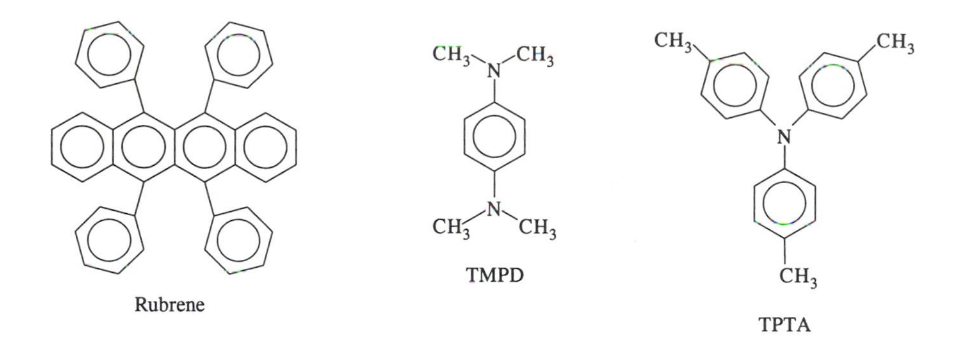


## REFERENCES


1. S. M. Sze, *Physics of Semiconductor Devices*, Wiley, New York, 1969.
2. C. Kittel, *Introduction to Solid-State Physics*, Wiley, New York, 1971.
3. V. A. Myamlin and Y. V. Pleskov, *Electrochemistry of Semiconductors*, Plenum Press, New York, 1967.
4. S. R. Morrison, *Electrochemistry at Semiconductor and Oxidized Metal Electrodes*, Plenum Press, New York, 1980.
5. H. Gerischer, in *Physical Chemistry: An Advanced Treatise*, Vol. 9A (H. Eyring, D. Henderson, W. Jost, eds.), Academic Press, New York, 1970, Chap. 5.
6. H. O. Finklea, *Semiconductor Electrodes: Studies in Physical and Theoretical Chemistry*, Vol. 55, Elsevier, Amsterdam, 1988.
7. L. M. Peters, *Chem. Rev. 90*:753–769 (1990).
8. A. J. Bard and M. S. Wrighton, *J. Electrochem. Soc. 124*:1706–1710 (1977).
9. H. Gerischer, *J. Electroanal. Chem. 82*:133–143 (1977).
10. R. Williams, *J. Chem. Phys. 32*:1505 (1960).
11. A. B. Ellis, S. W. Kaiser, and M. S. Wrighton, *J. Am. Chem. Soc. 98*:1635 (1976).
12. G. Hodes, J. Manassen, and D. Cahen, *Nature 261*:403 (1976).
13. B. Miller, and A. Heller, *Nature 262*:680 (1976).
14. H. C. Chang, A. Heller, B. Schwartz, S. Menezes, and B. Miller, *Science 196*:1097 (1977).
15. D. Cahen, G. Hodes, and J. Manassen, *J. Electrochem. Soc. 125*:1623 (1978).
16. A. Heller and B. Miller, *Adv. Chem. Ser. 184*:215 (1980).
17. H. D. Rubin, B. D. Humphrey, and A. B. Bocarsly, *Nature 308*:339 (1984).
18. C. M. Gronet, N. S. Lewis, G. Cogan, and J. Gibbons, *Proc. Natl. Acad. Sci. USA 80*:1152 (1983).
19. M. S. Wrighton, A. B. Bocarsly, J. M. Bolts, M. G. Bradley, A. B. Fischer, N. S. Lewis, M. C. Palazzotto, and E. G. Walton, *Adv. Chem. Ser. 184*:269 (1980).
20. H. D. Rubin, D. J. Arent, B. D. Humphrey, and A. B. Bocarsly, *J. Electrochem. Soc. 134*:93 (1987).
21. D. J. Arent, H. D. Rubin, Y. Chen, and A. B. Bocarsly, *J. Electrochem. Soc. 139*:2705–2712 (1992).

22. A. B. Bocarsly, E. G. Walton, and M. S. Wrighton, *J. Am. Chem. Soc. 102*:3390 (1980).
23. D. C. Bookbinder, J. A. Bruce, R. N. Dominey, N. S. Lewis, and M. S. Wrighton, *Proc. Natl. Acad. Sci. USA 77*:6280 (1980).
24. A. J. Bard, A. B. Bocarsly, F.-R. Fan, E. G. Walton, and M. S. Wrighton, *J. Am. Chem. Soc. 102*:3671 (1980).
25. A. J. Nozik, C. A. Parsons, D. J. Dunlavy, B. M. Keyes, and R. K. Ahrenkiel, *Solid State Commun. 75*:297 (1990).
26. C. A. Koval, and P. R. Segar, *J. Phys. Chem. 94*:2033–2039 (1990).
27. A. J. Bard, *Science 207*:139 (1980).
28. A. Heller, *Acc. Chem. Res. 23*:128 (1990).
29. S. N. Frank and A. J. Bard, *J. Phys. Chem. 81*:1484 (1977).
30. T. Fruend and W. P. Gomes, *Catal. Rev. 3*:1 (1969).
31. B. Kraeutler and A. J. Bard, *J. Am. Chem. Soc. 100*:2239 (1978).
32. B. Kraeutler and A. J. Bard, *J. Am. Chem. Soc. 100*:5985 (1978).
33. M. Grätzel, *Faraday Discuss. Chem. Soc. 70*:311 (1980).
34. M. Grätzel, *Acc. Chem. Res. 14*:376 (1981).
35. M. A. Fox, *Acc. Chem. Res. 16*:314 (1983).
36. H. Gerischer and H. Tributsch, *Ber. Bunsen-Ges. Phys. Chem. 72*:437 (1968).
37. H. Tributsch and H. Gerischer, *Ber. Bunsen-Ges. Phys. Chem. 73*:251 (1969); H. Tributsch, *Ber. Bunsen-Ges. Phys. Chem. 73*:582 (1969); H. Tributsch and H. Gerischer, *Ber. Bunsen-Ges. Phys. Chem. 73*:850 (1969); H. Gerischer, M. E. Michel-Beyerle, F. Rebentrost, and H. Tributsch, *Electrochim. Acta 13*:1509; H. Gerischer, *Photochem. Photobiol. 16*:243 (1972).
38. K. Hauffe, H. J. Danzmann, H. Pusch, J. Rauge, and H. Volz, *J. Electrochem. Soc. 117*:993 (1970).
39. H. Tributsch and M. Calvin, *Photochem. Photobiol. 14*:95 (1971).
40. A. Terenin and I. Akimov, *J. Phys. Chem. 69*:730 (1965).
41. W. P. Gomes and F. Cardon, *Ber. Bunsen-Ges. Phys. Chem. 75*:914 (1971).
42. H. Tributsch and H. Gerischer, *Ber. Bunsen-Ges. Phys. Chem. 73*:850 (1969).
43. A. Fujishima, T. Watanabe, O. Tatsuoki, and K. Honda, *Chem. Lett.* 13 (1975).
44. A. Fujishima, E. Hayashitani, and K. Honda, *J. Inst. Ind. Sci.* Univ. Tokyo *(Seisan Kenkyo) 23*:363 (1971).
45. T. Osa and M. Fujihira, *Nature 264*:349 (1976).
46. M. Fujihira, N. Ohishi, and T. Osa, *Nature 268*:226 (1977).
47. M. Fujihira, T. Osa, D. Hursh, and T. Kuwana, *J. Electroanal. Chem. 88*:285 (1978).
48. R. Memming and H. Tributsch, *J. Phys. Chem. 75*:562 (1971).
49. H. Gerischer, *Photochem. Photobiol. 16*:243 (1972).
50. H. Tributsch, *Photochem. Photobiol. 16*:261 (1972).
51. W. Arden and P. Fromherz, *Ber. Bunsen-Ges. Phys. Chem. 82*:868 (1978).
52. W. Arden and P. Fromherz, *J. Electrochem. Soc. 127*:370 (1980).
53. P. Fromherz and W. Arden, *Ber. Bunsen-Ges. Phys. Chem. 84*:1045 (1980).
54. H. Bücher, K. H. Drexhage, M. Fleck, H. Kuhn, D. Möbius, F.P. Schaefer, J. Sondermann, W. Sperling, P. Tillman, and J. Wiegand, *Mol. Cryst. 2*:199 (1967).

55. Yu. Ya. Gurevich, Yu. V. Pleskov, and Z. A. Rotenberg, *Photoelectrochemistry*, Plenum Press, New York, 1978.
56. Yu. V. Pleskov and Z. A. Rotenberg, *Adv. Electrochem. Eng. 11*:1 (1978).
57. M. Heyrovsky and R. G. W. Norrish, *Nature 200*:880 (1963); M. Heyrovsky, *Nature 206*:1356 (1965).
58. H. Berg and H. Schweiss, *Electrochim. Acta 9*:425 (1964).
59. G. C. Barker, A. W. Gardner, and D. C. Sammon, *J. Electrochem. Soc. 113*:1182 (1966).
60. G. C. Barker, *Ber. Bunsen-Ges. Phys. Chem. 75*:728 (1971).
61. G. C. Barker, D. McKeown, M. J. Williams, G. Bottura, and V. Concialini, *Faraday Discuss. Chem. Soc. 56*:41 (1974).
62. Yu. V. Pleskov and Z. A. Rotenberg, *J. Electroanal. Chem. 20*:1 (1969).
63. A. Brodsky and Yu. V. Pleskov, in *Progress in Surface Sciences*, Vol. 2, Part 1 (S. G. Davison, ed.), Pergamon Press, Oxford, 1972.
64. Yu.V. Pleskov, Z. A. Rotenberg, V. V. Eletsky, and V. I. Lakomov, *Faraday Discuss. Chem. Soc. 56*:52 (1974).
65. P. Delahay and V. S. Srinivasan, *J. Phys. Chem. 70*:420 (1966).
66. T. Kuwana, in *Electroanalytical Chemistry*, Vol. 1 (A. J. Bard, ed.), Marcel Dekker, New York, 1966, Chap. 3 and references therein.
67. S. P. Perone and H. D. Drew, in *Analytical Photochemistry and Photochemical Analysis: Solids, Solutions, and Polymers* (J. Fitzgerald, ed.), Marcel Dekker, New York, 1971, Chap. 7 and references therein.
68. H. Berg and H. Schweiss, *Naturwissenschaften 47*:513 (1960).
69. H. Berg, H. Schweiss, E. Stutter, and K. Weller, *J. Electroanal. Chem. 15*:415 (1967), and references therein.
70. G. Porter, in *Techniques of Organic Chemistry*, Vol. 3, Part II (A. Weissberger, ed.), Wiley-Interscience, New York, 1963, Chap. XIX.
71. H. Berg and H. Schweiss, *Nature 191*:1270 (1961).
72. S. P. Perone and J. R. Birk, *Anal. Chem. 38*:1589 (1966).
73. G. L. Kirschner and S. P. Perone, *Anal. Chem. 44*:443 (1972).
74. R. A. Jamieson and S. P. Perone, *J. Phys. Chem. 76*:830 (1972).
75. J. I. H. Patterson and S. P. Perone, *J. Phys. Chem. 77*:2437 (1973).
76. J. I. H. Patterson and S. P. Perone, *Anal. Chem. 44*:1978 (1972).
77. M. Grätzel and A. Henglein, *Ber. Bunsen-Ges. Phys. Chem. 77*:2 (1973); *77*:6 (1973); *77*:11 (1973); *77*:17 (1973).
78. A. Henglein, *Electroanal. Chem. 9*:163 (1976).
79. D. C. Johnson and E. W. Resnick, *Anal. Chem. 44*:637 (1972).
80. J. R. Lubbers, E. W. Resnick, P. R. Gaines, and D. C. Johnson, *Anal. Chem. 46*:865 (1974).
81. W. J. Albery, M. D. Archer, N. J. Field, and A. D. Turner, *Faraday Discuss. Chem. Soc. 56*:28 (1974).
82. A. J. Bard and L. R. Faulkner, *Electrochemical Methods*, Wiley, New York, 1980, Chap. 14.
83. E. A. Chandross, *Trans. N.Y. Acad. Sci., Ser. 2*:32, 571 (1969), and references therein.

84. D. M. Hercules, in *Physical Methods of Organic Chemistry*, Part II (A. Weissberger and B. Rossiter, eds.), Academic Press, New York, 1971, and references therein.
85. A. J. Bard, C. P. Keszthelyi, H. Tachikawa, and N. E. Tokel, in *Chemiluminescence and Bioluminescence* (D. M. Hercules, J. Lee, and M. Cormier, eds.), Plenum Press, New York, 1973, and references contained therein.
86. L. R. Faulkner, *Int. Rev. Sci., Phys. Chem., Ser. 29*:213 (1975).
87. L. R. Faulkner and A. J. Bard, *Electroanal. Chem. 10*:1 (1977).
88. F. Pragst, *Z. Chem. 18*:41 (1978).
89. L. R. Faulkner, *Methods Enzymol. 57*:494 (1978).
90. R. S. Glass and L. R. Faulkner, in *Chemical and Biological Generation of Excited States* (W. Adam and G. Cilento, eds.), Academic Press, New York, 1982, Chap. 6.
91. R. A. Marcus, *Annu. Rev. Phys. Chem. 15*:155 (1964).
92. R. A. Marcus, *J. Chem. Phys. 43*:2654 (1965); *52*:2083 (1970).
93. R. P. Van Duyne and S. F. Fischer, *J. Chem. Phys. 5*:183 (1974).
94. E. A. Mayeda and A. J. Bard, *J. Am. Chem. Soc. 95*:6223 (1973).
95. N. E. Tokel-Takvoryan, R. E. Hemingway, and A. J. Bard, *J. Am. Chem. Soc. 95*:6582 (1973).
96. J. E. Martin, E. J. Hart, A. W. Adamson, H. Gafney, and J. Halpern, *J. Am. Chem. Soc. 94*:9238 (1972).
97. D. M. Hercules and F. E. Lytle, *J. Am. Chem. Soc. 88*:4745 (1966).
98. J. T. Maloy, K. B. Prater, and A. J. Bard, *J. Am. Chem. Soc. 93*:5959 (1971).
99. D. J. Freed and L. R. Faulkner, *J. Am. Chem. Soc. 93*:2097 (1971).
100. C. P. Keszthelyi, N. E. Tokel-Takvoryan, and A. J. Bard, *Anal. Chem. 47*:249 (1975).
101. P. R. Michael and L. R. Faulkner, *Anal. Chem. 48*:1188 (1976).
102. P. R. Michael and L. R. Faulkner, *J. Am. Chem. Soc. 99*:7754 (1977).
103. W. L. Wallace and A. J. Bard, *J. Phys. Chem. 83*:1350 (1979).
104. S. W. Feldberg, *J. Am. Chem. Soc. 88*:390 (1966).
105. S. W. Feldberg, *J. Phys. Chem. 70*:3928 (1966).
106. L. R. Faulkner, *J. Electrochem. Soc. 124*:1724 (1977).
107. J. L. Morris, Jr. and L. R. Faulkner, *J. Electrochem. Soc. 125*:1079 (1978).
108. J. D. Luttmer and A. J. Bard, *J. Phys. Chem. 85*:1155 (1981).
109. R. S. Glass and L. R. Faulkner, *J. Phys. Chem. 85*:1160 (1981).
110. R. S. Glass and L. R. Faulkner, *J. Phys. Chem. 86*:1652 (1982).
111. L. R. Faulkner, H. Tachikawa, and A. J. Bard, *J. Am. Chem. Soc. 94*:691 (1972).
112. R. E. Merrifield, *J. Chem. Phys. 48*:4318 (1968).
113. G. Seshadri, J. K. M. Chun, and A. B. Bocarsly, *Nature, 352*:508 (1991).

# 29

# Principles and Techniques of Electrochemical–Electron Paramagnetic Resonance Experiments

**Ira B. Goldberg and Ted M. McKinney** *Rockwell Science Center, Thousand Oaks, California*

## I. MAGNETIC RESONANCE IN ELECTROCHEMICAL STUDIES

Earlier chapters described different electrochemical techniques useful in the elucidation of various reaction mechanisms. Often the conclusions drawn about a sequence of heterogeneous electron transfer and homogeneous chemical reactions are the result of ingenious exercises in deductive reasoning based on interpretation of the electrochemical response of the system. As the editor noted in his introductory chapter, some interpretations tend to be presented with a certainty that belies the shakiness of arguments based largely on circumstantial evidence.

Clearly, techniques that provide definitive identification of intermediate or product species can be a valuable adjunct in the study of complicated electrochemical reaction sequences. Almost every imaginable analytical method has been used, and spectroscopic techniques have proven to be particularly valuable; each particular method contributes a unique set of data for the experimentalist to interpret. Conversely, it should be recognized that electrochemistry has also aided spectroscopists by enabling them to prepare and study species that might otherwise be inaccessible.

This chapter provides a review of the principles of electron paramagnetic* resonance (EPR) as they apply to electrochemical studies. EPR is a spectroscopic

---

*For decades, there have been voluble and passionate arguments in favor of using each of the terms electron paramagnetic resonance (EPR) and electron spin resonance (ESR) to describe a single experimental technique. A group of scientists has banded together to form the International EPR

method that is sensitive to materials with unpaired electrons, such as organic and inorganic free radicals, radical anions and cations, and the complexes of various transition metal ions; many substances of biological significance fall into these categories.

As originally postulated by Michaelis, most electrode processes involve the transfer of *one* electron to or from an "electroactive" material. Consequently, either the parent compound or the electrode product contains at least one unpaired electron and it is said to be "paramagnetic." The electrochemical reduction or oxidation of neutral organic molecules represents a typical system:

$$R^{\circ} + e \rightarrow R^{-\bullet} \quad \text{(anion radical)} \tag{29.1a}$$

$$R^{\circ} - e \rightarrow R^{+\bullet} \quad \text{(cation radical)} \tag{29.1b}$$

Often the initial electrode product will undergo subsequent chemical or electron-transfer reactions, but the stability of the species depends greatly on the reactivity of the medium in that it is produced.

Nuclear magnetic resonance (NMR) is another spectroscopic technique that has many formal similarities to EPR. It is widely used for product analysis, as in studies leading to the development of electrosynthetic methods. The intrinsic sensitivity of NMR is several orders of magnitude lower than that of EPR, making simultaneous NMR–electrochemical experiments unattractive.

The nature of hyphenated techniques such as electrochemistry–EPR (EC-EPR) is that the studies of utilizing them can take many different directions. To be specific, one investigation might make use of EPR to verify the existence of a free radical intermediate, whereas some other study of the EPR parameters of a homologous series of radicals might require the use of electrochemistry to prepare a species inaccessible by conventional chemistry. A different study of the effect of substituents on the molecular orbital energies of a series of related compounds might rely on EC-EPR, while a different problem about the rate of homogeneous electron transfer might be solved by measuring the line width of the EPR absorptions of an electrochemically generated reactant.

In order to appreciate some of the variety of problems that can be aided with EC-EPR, it is necessary to understand the basic theory of EPR and the instrumentation required to perform an EPR experiment. It is also necessary to understand how an unpaired electron interacts with the atomic nuclei in a free radical, because this interaction produces the multiple lines characteristic of an EPR spectrum. The magnitude of the spacings of these lines can be related to the unpaired electron distribution in a free radical, so it is also helpful to have

---

(ESR) Society with parenthetical deference to the losers in the nomenclature battle. The Society recommends that authors continue to point out the equivalence of the two terms until EPR becomes universally accepted. The authors of this chapter remain evenly divided on the issue.

a rudimentary understanding of molecular orbital theory. Consequently, this chapter requires digression into several topics that were not addressed in previous chapters.

This chapter presents an elementary discussion of the theory, instrumentation, and practice of EPR–electrochemical studies. We recite the usual disclaimers about limitations of space to explain that the subject cannot be covered comprehensively here. The selected bibliography at the end of the chapter is broken down into broad categories to guide the interested reader to specific topics. The student who wishes a more thorough discussion of the general subject at an elementary level may find McKinney's review [1] helpful.

## II. MOLECULAR ORBITALS IN ELECTROCHEMISTRY AND EPR

Molecular orbital (MO) theory is widely used in the literature to describe both organic [2] and inorganic chemical systems [3]. The MOs of a given molecule can be calculated by a variety of methods at different levels of sophistication. Increased accuracy is gained at the expense of longer computation time, but the results all have the following properties in common:

1. Each MO is characterized by a particular energy level (an "eigenvalue" of the MO calculation) measured on a convenient energy scale. The zero of energy is fixed at some arbitrary point, analogous to the zero point of potential (reference electrode) in an electrochemical experiment.
2. Each MO represents a specific spatial distribution of probability for the electrons in that orbital; this is known as the "eigenfunction" for that orbital. This information can be used to evaluate the probability of finding a given electron in some particular volume of space (e.g., in the neighborhood of a certain atomic nucleus).

The electrochemical behavior of a molecule is directly related to the MO energy levels, while experimental EPR parameters correlate with the electron probability distribution.

We restrict our attention in this chapter to the simple but widely used Hückel MO (HMO) method for calculating orbitals for $\pi$-electron and aromatic molecules [2]. The HMO scheme assumes that a conjugated $\pi$-electron molecule consists of a network of $sp^2$-hybridized carbon atoms lying in a plane and each participating atom i has a 2p electron in an atomic orbital, $\phi_i$, perpendicular to this plane. Linear combinations of these atomic orbitals (LCAO) result in the molecular $\pi$ wavefunctions, $\psi_j$, each of which has a discrete energy, $E_j$. In terms of the parameters used in HMO computations,

$$E_j = \alpha + \lambda_j \beta^{\circ}_{CC} \tag{29.2}$$

where $\alpha$ is a reference point and can be ignored; then $\lambda_j$ becomes a measure of the energy of the jth MO in units of $\beta^{\circ}_{CC}$, which can be estimated from experimental data. For example, $\beta^{\circ}_{CC}$ has been assigned values between –2.04 and –3.13 eV for electrochemical oxidations and between –2.37 and –2.72 eV for electrochemical reductions [4]. This range of values reflects the crudeness of the HMO model itself, since electron correlations, electron repulsions, and solvation changes are neglected.

The energy levels correspond to bonding ($E_j < \alpha$), antibonding ($E_j > \alpha$), and in some molecules nonbonding ($E_j = \alpha$) orbitals. For example, the naphthalene molecule has only bonding and antibonding orbitals as shown in Figure 29.1. The net binding energy of the π-electron system can be evaluated by summing the energy of each electron in the occupied orbitals. The reader can

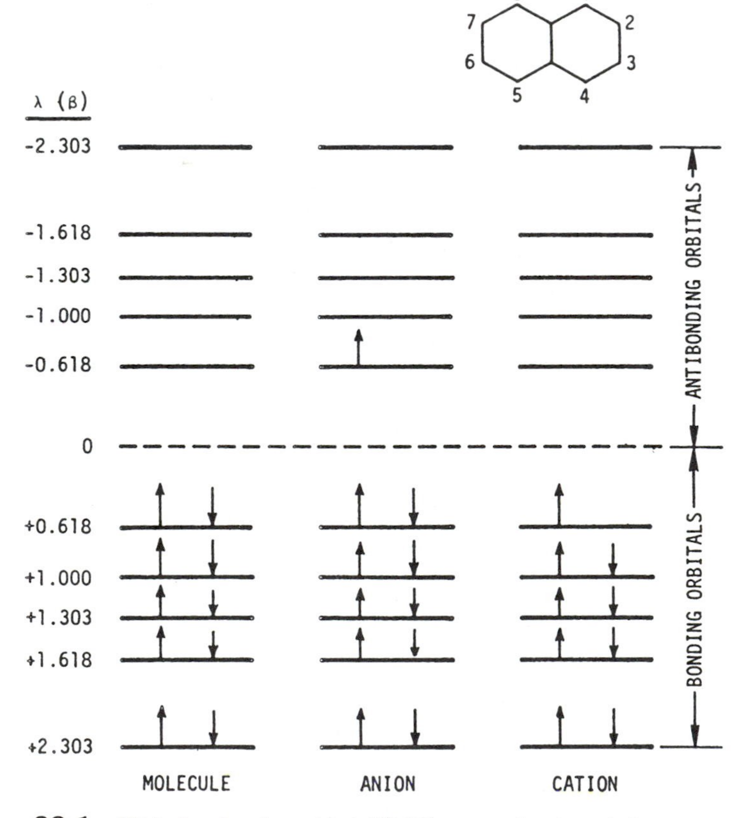


**Figure 29.1** Hückel molecular orbital (HMO) energy levels and electron occupation for naphthalene, its anion radical, and its cation radical. HMO calculations employ $\beta$ = –2.5 eV as the unit of energy so that $+\beta$ values correspond to negative or binding energy.

demonstrate that energy must be supplied to the naphthalene molecule to form either the anion or cation radical; that is, the change in free energy is positive for either reaction. The half-wave potential is closely related to the value $\lambda_j$ of the molecular orbital of the anion or cation. The relationship

$$E_{1/2} = A + B\lambda_j \tag{29.3}$$

has been well established [5,6]. The term A includes the difference between the zero points of the MO calculations and the reference electrode potential, and B is the empirical value of $\beta^{\circ}_{CC}$, expressed in electron-volts.

The spatial distribution of electrons in the jth MO is related to the numerical coefficients $c_i$ of the atomic orbitals $\phi_i$ whose linear combination yields the $\pi$ orbital, $\psi_j$:

$$\Psi_j = c_{j1}\phi_1 + c_{j2}\phi_2 + \cdots + c_{jn}\phi_n = \sum_{i=1}^{n} c_{ji}\phi_i \tag{29.4}$$

When $\psi_j$ is a normalized wavefunction, the atomic orbital coefficients fulfill the condition

$$\sum_{i=1} c_{ji}^2 = 1 \tag{29.5}$$

and when $\psi_k$ denotes the particular MO containing an unpaired electron,

$$\rho_i = c_{ki}^2 \tag{29.6}$$

where $\rho_i$ is the probability of the unpaired electron being associated with atom i. The quantity $\rho$ is known as the unpaired electron density or spin density (advanced treatments [6] make quantitative distinction between these two terms) and its importance in the interpretation of EPR data becomes apparent later.

The simplest description of transition metal ions is given by crystal field theory [3], but any mathematical discussion is beyond the scope of this chapter. Consequently, only some qualitative results are given here. For more detail the reader is directed to any of the references in the bibliography listed under Inorganic Ions.

For a transition metal ion in a crystal or in solution, ligands, solvents, or associated counterions are treated as point charges around the ion. Octahedral and tetrahedral crystal fields are shown in Figure 29.2. The octahedral crystal field can be described by six negative point charges, one on each face of a cube, that act on a positive ion at the center of the cube (Fig. 29.2a). The faces of the cube are parallel to the xy, yz, and xz planes through the center ion. The $d_{z^2}$ and $d_{x^2-y^2}$ orbitals have greatest electron density along the axes of the coordinate system centered on the metal ion, while the $d_{xy}$, $d_{yz}$, and $d_{xz}$ orbitals have the greatest electron density in, respectively, the xy, yz, and xz planes. Con-

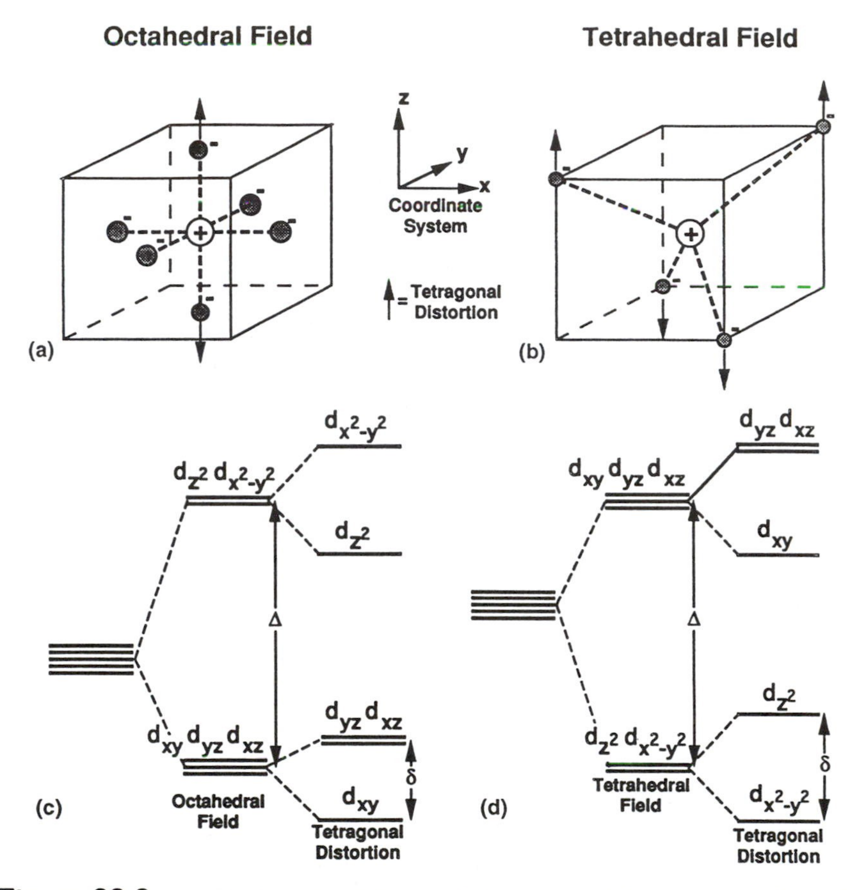


**Figure 29.2** (a) Octahedral and (b) tetrahedral crystal fields represented as point charges around a central ion. Arrows show the effect of a tetrahedral distortion to the crystal field. (c) d-Orbital energy level diagrams for octahedral crystal field and octahedral crystal field with tetragonal distortion, and (d) tetrahedral crystal field and tetrahedral crystal field with tetragonal distortion.

sequently, in the case of the octahedral crystal field, the $d_{xy}$, $d_{yz}$, and $d_{xz}$ orbitals are farther away from the negative point charges and are therefore lower in energy than the $d_{z^2}$ and $d_{x^2-y^2}$ orbitals. Tetragonal distortion further separates the $d_{xy}$ orbital from the $d_{xz}$ and $d_{yz}$ orbitals, and the $d_{z^2}$ orbital from the $d_{x^2-y^2}$ orbital. This is shown in Figure 29.2c. The opposite orbital splitting takes place in the tetrahedral crystal field where four point charges occupy the opposite corners of a cube as shown in Figure 29.2b. The energy levels for the tetrahedral fields are shown in Figure 29.2d.

In Figure 29.2, the splitting $\Delta$ is called the crystal field splitting, and $\delta$ is called the tetragonal field splitting. Additional distortion to the crystal field can split the remaining degenerate orbitals. The electronic configuration of the transition metal is commonly described by introducing the required number of electrons into the orbitals one at a time. For nondegenerate orbitals, the electrons fill the energy level diagram in a way similar to that shown in Figure 29.1—first "spin-up," then "spin-down." However, for an n-degenerate orbital, Hund's rule requires the first n electrons to be placed "spin-up," and then subsequent electrons are placed "spin-down." For example, $Fe^{3+}$, with five d-electrons, in the absence of a large crystal field has five unpaired electrons. Since each electron has a spin angular momentum quantum number of $S = ½$, the spin state of the $Fe^{3+}$ is $S = 5/2$ and is characterized as high-spin $Fe^{3+}$. However, the presence of a large octahedral crystal field splits the fivefold degeneracy into three lower energy levels and two higher energy levels as shown in Figure 29.2c. In this case there are three "spin-up" and two "spin-down" electrons occupying this threefold degenerate orbital. This creates an $S = ½$ spin state, known as low-spin $Fe^{3+}$.

## III. PRINCIPLES OF EPR

### A. Paramagnetic Resonance

Electrons and some nuclei have intrinsic angular momentum, which can be visualized as rotation of the particle, hence the trivial name "spin." This spin causes the charged particle to behave as a small magnet which is characterized by a magnetic moment $\mu$. Electrons in nonspherical orbitals possess additional orbital angular momentum, which is likewise manifested as an observable magnetic moment. Even molecules without unpaired electrons may have rotational angular momentum moments that contribute to the paramagnetic susceptibility of a sample; this is the reason for the decision to use the generic word "paramagnetic" in naming the EPR technique. In most of the examples described in this chapter, the dominant contribution to paramagnetism originates in electron spin angular momentum.

For electrons, the unit of measure of the magnetic moment is the Bohr magneton,

$$\mu_B = \frac{|e|\hbar}{2m_e c} = 9.274078 \times 10^{-24}\ \text{J/T}^* \tag{29.7a}$$

---

*Most of the literature cited in the bibliography uses cgs units, which have since been superseded by SI units. A conversion table with recommended units is given in Appendix II of this chapter.

(e is the electron charge, $\hbar$ is Planck's reduced constant, $m_e$ is the electron mass, and c the speed of light); for nuclei the unit is the nuclear magneton

$$\mu_N = \frac{|e|\hbar}{2m_pc} = 5.050824 \times 10^{-27} \text{ J/T} \tag{29.7b}$$

The magnetic moment of a free electron is related to its spin angular momentum **S** as

$$\boldsymbol{\mu}_e = -g_e\mu_B\mathbf{S} \tag{29.8a}$$

where the negative sign indicates opposite vector orientations of $\boldsymbol{\mu}_e$ and **S**, and $g_e = 2.00232$ is the free–electron g factor. Electrons on isolated atoms may also possess orbital angular momentum which augments the spin magnetic moment. This orbital contribution tends to be *quenched* by strong electrostatic fields such as those present in the covalent bonds of organic molecules or metal–ligand complexes. However, the quenching is seldom complete, and the net magnetic moment has some contribution from the residual orbital angular momentum. This can be accommodated by rewriting Equation 29.8a as

$$\mu_S = g\mu_BS \tag{29.8b}$$

and viewing the g factor as a parameter that is characteristic of the particular species.

All reported g values for organic radicals fall between 2.0007 (nitrosobenzene cation radical) and 2.0120 (tetraiodo-*p*-benzosemiquinone). Meticulous experimental technique is required to measure g values to high precision so that organic species are usually characterized in terms of other EPR parameters (see later discussion). A theoretical treatment has been developed by Stone [7] to explain the g values of organic radicals in terms of the results of HMO calculations, and simplified versions of this treatment can be found in the general references at the end of this chapter. Ions of the higher transition series, particularly the rare earth ions, exhibit g values that differ greatly from $g_e$. This fact can be very useful in identifying such species. Although these g values are well understood, the calculations are quite lengthy, and the reader is again referred to the general texts and review papers on metal ions listed at the end of the chapter. In general, the deviation of the g factor from $g_e$ depends on the crystal field energies, $\Delta$ and $\delta$, of Figure 29.2. Additional angular momentum that can add or subtract from the angular momentum of the unpaired electron arises from mixing contributions from excited states. Thus, the larger the deviation from $g_e$, the smaller the energy separations between the ground and excited states.

Many metal ions have more than one unpaired electron; similarly, certain organic species (diradicals and triplet and higher states) may have several unpaired spins. The effect of several electron spin dipoles in close proximity on a molecule is to broaden the spectral lines so drastically that the EPR spectrum usually cannot be observed in solution. Consequently, such species have little relevance to electrochemical–EPR studies, and we limit our attention to doublet (one unpaired electron) states. Exceptions to this are $Mn^{2+}$ and $Fe^{3+}$ with $S = 5/2$, and sometimes $Cr^{3+}$ with $S = 3/2$ and strong ligand coordination; however, in solution, these may be treated as if they had only one unpaired electron, although the EPR absorption intensity is stronger than expected for a one-electron species.

Many inorganic ions are not readily detected because they have orbitally degenerate ground states. Spin relaxation in these states is so rapid that the EPR spectrum cannot be observed at typical temperatures where electrochemical experiments are carried out. Temperatures of 4 K or lower are required to observe these ions. Table 29.1 lists transition metal ions that occur in nondegenerate ground states due to crystal field splittings and as such are amenable to observation.

The measurable component of electron spin has the magnitude of $\hbar/2$, which produces two allowed quantum spin states, $m_s = +½$ (designated by convention as $|\alpha>$ spin), and $m_s = -½$ (or $|\beta>$ spin). When a free electron is placed in a strong magnetic field, its magnetic moment $\boldsymbol{\mu}_s$ behaves like a tiny bar magnet. It will line up either parallel or antiparallel to the field **H**.* The energy of this interaction is

**Table 29.1** 3d Ions with Nondegenerate Ground States

| | | Crystal Field | | | | |
|---|---|---|---|---|---|---|
| State | Examples | None or small | Octahedral | Octahedral + tetragonal | Tetrahedral | Tetrahedral + tetragonal |
| $3d^1$ | $Ti^{3+}$, $VO^{2+}$ | — | — | $S = 1/2$ | — | $S = 1/2$ |
| $3d^2$ | $V^{3+}$ | — | — | $S = 0$ | $S = 1$ | $S = 0$ |
| $3d^3$ | $Cr^{3+}$ | — | $S = 3/2$ | — | — | $S = 1/2$ |
| $3d^4$ | $Cr^{2+}$, $Mn^{3+}$ | — | — | — | $S = 0$ | $S = 0$ |
| $3d^5$ | $Fe^{3+}$, $Mn^{2+}$ | $S = 5/2$ | — | — | — | $S = 1/2$ |
| $3d^6$ | $Fe^{2+}$ | — | $S = 0$ | $S = 0$ | — | $S = 0$ |
| $3d^7$ | $Co^{2+}$ | — | — | $S = 1/2$ | — | — |
| $3d^8$ | $Co^{3+}$, $Ni^{2+}$ | — | $S = 1$ | $S = 0$ | — | — |
| $3d^9$ | Cu | — | — | $S = 1/2$ | — | — |

*The symbol H (which represents magnetic field) rather than B (which represents magnetic flux density) has improperly been used in the EPR literature as the parameter that determines the electron spin energy. To add more confusion, although H has been called magnetic field, it is correctly given in units of magnetic flux density, gauss (cgsm) or tesla (SI) (see Appendix II). The straightforward way to correct this situation is to change the symbol H to B in most EPR literature, but since the abuse is so firmly entrenched, we will perpetuate this minor sin.

$$E = \boldsymbol{\mu}_S \cdot \mathbf{H} = \pm \tfrac{1}{2} g\mu_B H \qquad (29.9)$$

where the plus sign corresponds to the energy of $|\alpha>$ spin and the minus to $|\beta>$ spin. As shown in Figure 29.3a, the energy difference between the two levels depends linearly on the strength of the applied field,

$$\Delta E = g\mu_B H \qquad (29.10)$$

Transitions between the two electron-spin states can be induced by application of electromagnetic radiation of the correct frequency (i.e., $\nu = \Delta E/h$). In practice, it is simpler to hold the frequency constant while scanning the magnetic field to produce the condition for resonant absorption of energy; see Figure

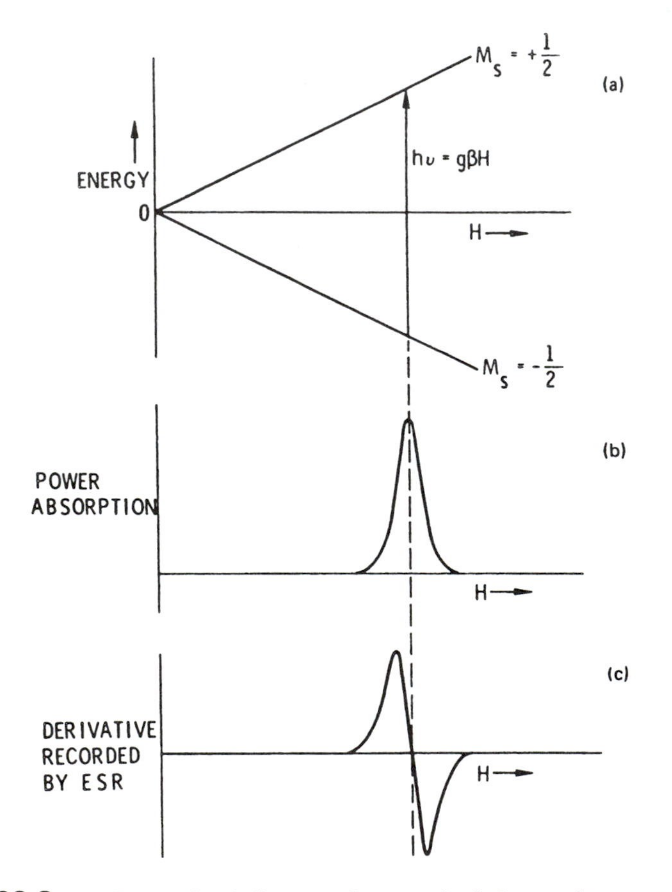


**Figure 29.3** (a) Energy-level diagram of an unpaired electron in a magnetic field. (b) Absorption at constant frequency and swept field. (c) Derivative signal recorded by EPR.

29.3b and c. The majority of EPR experiments are carried out with radiation at a frequency of about 9.5 GHz (X-band). For g values close to 2, this corresponds to a field strength of about 340 mT and the electron energy levels are separated by approximately $6.3 \times 10^{-24}$ J.

Typical samples for EPR spectroscopy contain about $10^{17}$ spins. The relative populations of the two spin states are given by the Boltzmann distribution,

$$\frac{N_\alpha}{N_\beta} = \exp\left(\frac{-\Delta E}{kT}\right) \tag{29.11}$$

where k is the Boltzmann constant and T the absolute temperature. At room temperature there are about 998 spins in the higher-energy level for every 1000 in the lower level. When the frequency of the electromagnetic radiation and the applied field satisfy the resonance condition

$$h\nu_0 = g\mu_B H_0 \tag{29.12}$$

the radiation excites both absorption and emission transitions with equal probabilities. The observation of a net absorption of energy requires that the Boltzmann distribution remain essentially undisturbed. Fortunately, there are pathways other than stimulated emission that permit spins in the upper state to return to the lower state. This phenomenon, known as spin relaxation, has received much attention but cannot be described in any detail here (see the references at the end of the chapter). For doublet-state species, the relaxation mechanisms are often relatively inefficient. When such a sample is subjected to excessive levels of microwave power, the two spin states tend to become equally populated and the net absorption of energy approaches zero. This phenomenon is known as power saturation.

Similar considerations are often mentioned in discussions of NMR spectroscopy because the two techniques are formally quite similar. Indeed, the reader might well inquire what the nuclei are doing while the unpaired electron spins are undergoing transitions. A simple calculation (with $\mu_N$ from Eq. 29.7b and $g_p = 5.585$ for the proton g value) reveals that the proton resonance frequency at 3400 G is about 14.5 MHz [i.e., in the much longer wavelength, radiofrequency (RF) region of the electromagnetic spectrum]. So although the external field does produce polarization of the proton spins, the microwave radiation used in EPR is completely ineffectual in causing nuclear transitions.

## B. Hyperfine Interactions

Nuclear spin does, in fact, play a significant role in EPR. It can interact with the unpaired electron spin to give rise to the hyperfine splittings (HFS), which greatly increase the amount of information that can be extracted from an EPR spectrum.

When a paramagnetic sample contains magnetic nuclei, the alignment of these nuclear magnetic moments relative to the external field produces small local fields that are capable of splitting the electron-spin energy levels into electron-nuclear hyperfine levels. The simplest example of this appears in the EPR spectrum of hydrogen *atoms*, H• (1s configuration). Protons have a nuclear spin of I = ½ (in units of $\hbar$) and a positive magnetic moment. This generates a magnetic field, $H_{local}$, which adds to or subtracts from the applied field at the electron. Since it is the total field strength that determines the spacing of the electron spin energy levels, the resonance condition, Equation 29.12, is achieved when $H_{total} = H_0$, or

$$\frac{h\nu_0}{g\mu_B} = H_0 = H_{external} \pm H_{local} \tag{29.13}$$

These arguments are summarized in Figure 29.4. The lower portion of the figure shows that the EPR spectrum of H• atoms consists of two hyperfine lines separated by 50.6 mT. This separation is a measure of the strength of the coupling between the electron and nuclear magnetic moments in this species (i.e., the value of $H_{local}$ experienced by the unpaired electrons). It is called the hyperfine splitting constant (HFSC), usually symbolized by $a_i$, where i refers to a set of nuclei. Actually, the word "constant" is slightly misleading: $a_H = 50.6$ mT is indeed a constant for H• atoms in the 1s configuration, but H• atoms in the 2s state display $a_H = 6.3$ mT. So the constants have different values for different species or for different electronic configurations of the same species.

In general, a nucleus of spin I will produce 2I + 1 lines in the EPR spectrum. Thus a single $^{14}N$ nucleus (I = 1) yields a three-line spectrum; $^{23}Na$ (I = 3/2) yields four lines, and so on. Many common nuclei (e.g., $^{12}C$ and $^{16}O$) have zero nuclear spin and do not interact with the unpaired electron.

Usually, an organic radical will have more than one magnetic nucleus. In such instances, the EPR spectrum can display rich hyperfine structure, which is useful both for identifying the species and for extracting information about the electronic structure of the radical. The origin of such hyperfine patterns is straightforward, but is most easily understood in terms of concrete examples.

Consider the planar methyl radical, $H_3C$•, which has three geometrically equivalent protons. Their spins combine to produce four distinct values for the local field:

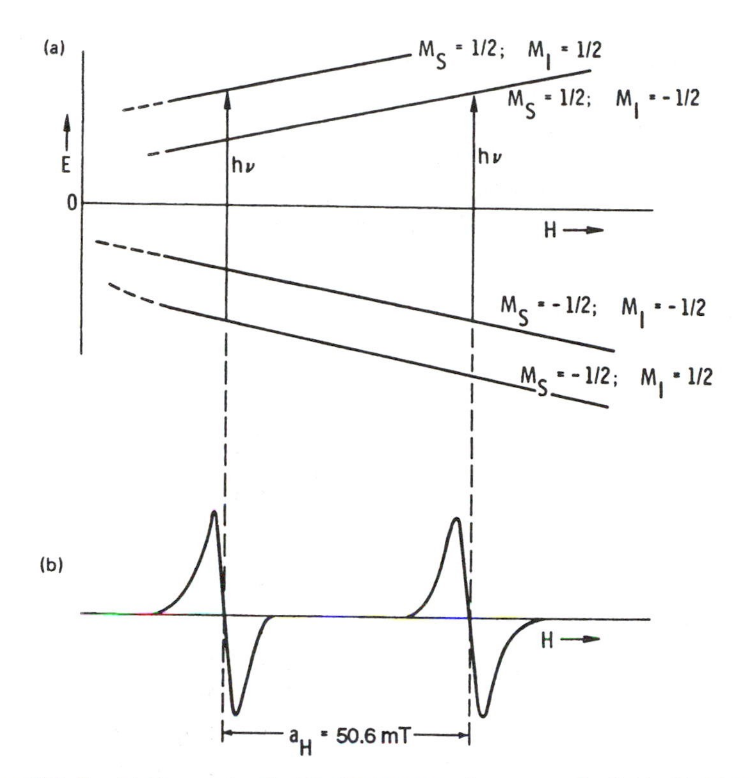


**Figure 29.4** (a) Energy level of hydrogen atom in a magnetic field. (b) EPR spectrum of hydrogen atom.

But the statistical weights for these combinations are in the ratio 1:3:3:1:

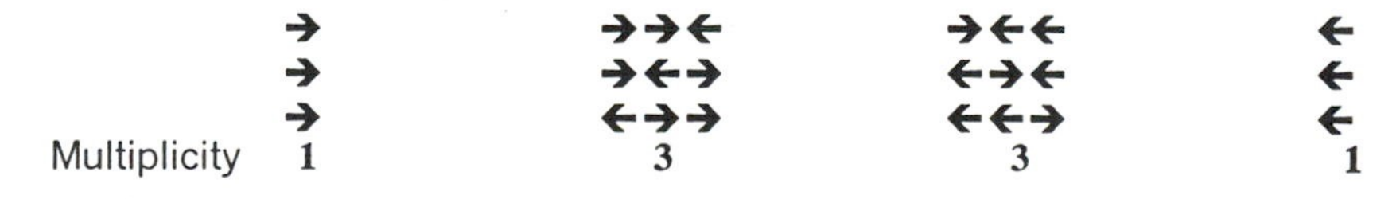

So the experimental spectrum will contain four lines with relative intensities in this ratio (see Fig. 29.5). The separation between adjacent lines is constant (to first order) and $a_H = 2.3$ mT.

In general, n equivalent nuclei of spin I will produce an EPR multiplet consisting of (2nI + 1) equally spaced lines. For spins of I = ½, the intensity ratios are given by the coefficients of the binomial expansion $(a + b)^n$; but for I > ½, the formula for intensities is much more complicated. For modest values of n it is easy to find the intensity ratios by sketching a branching diagram; thus

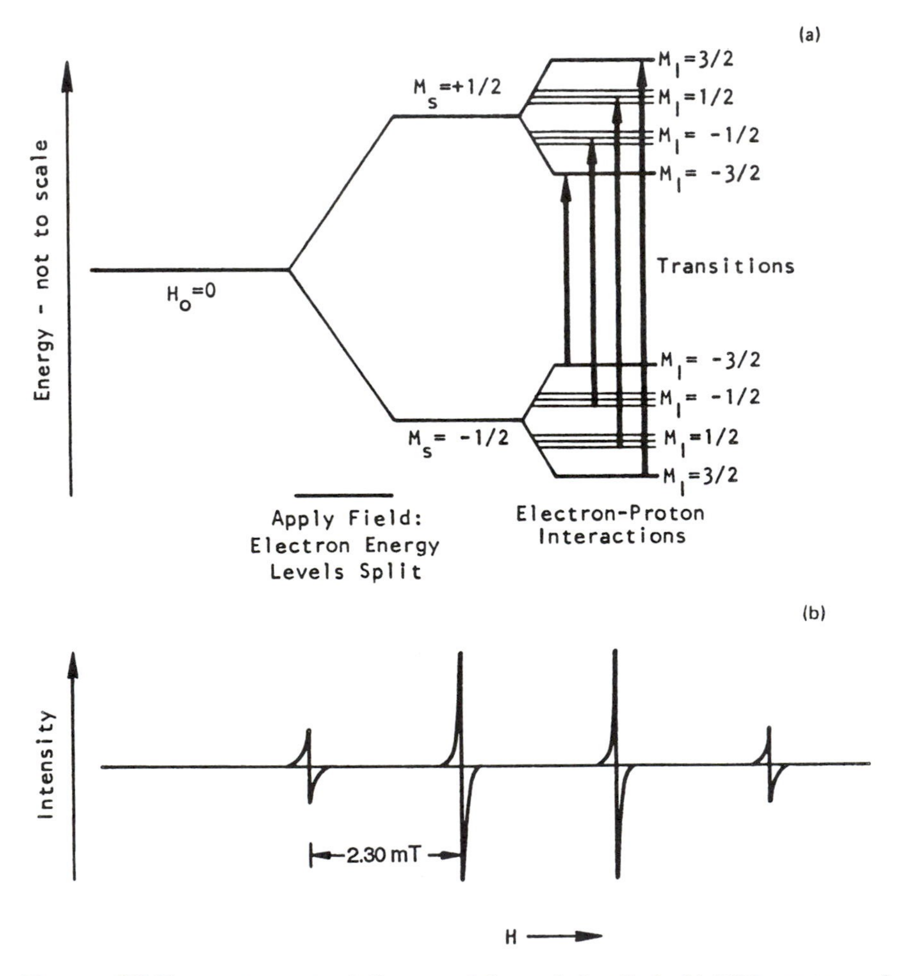


**Figure 29.5** (a) Energy-level diagram of the methyl radical. (b) EPR spectrum of the methyl radical.

for two equivalent $^{14}N$ nuclei, the first contributes a 1:1:1 triplet and each of these lines is subsequently split by the other nucleus into secondary 1:1:1 triplets of the same spacing. So the overall pattern is a quintet of 1:2:3:2:1 intensity.

Branching diagrams are often crucial to obtaining an assignment of a spectrum when a radical contains more than one set of equivalent nuclei. The spectrum of the naphthalene anion is shown in Figure 29.6 together with a diagram

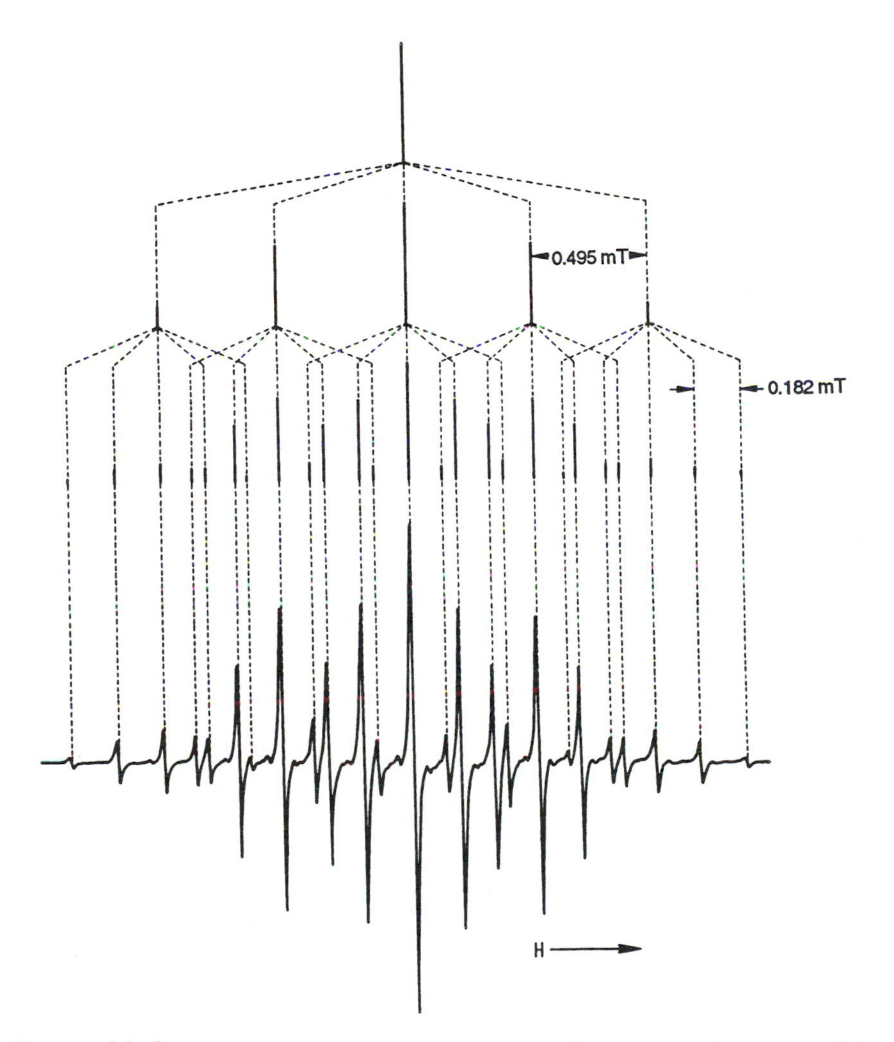


**Figure 29.6** Spectrum of the naphthalene radical anion and a reconstruction of the spectrum.

assigning each line position an intensity. This radical contains two sets of protons of four spins each, with $a_H = 0.495$ mT and $a_H = 0.182$ mT. The first set produces a quintet of intensity 1:4:6:4:1. Note that the intensity of each line of the secondary quintet is determined by the intensity of the line from which it branches.

When a radical contains several different sets of nuclei, the task of assigning the splitting constants can become arduous. The procedure is discussed in many textbooks on EPR.

The local field produced by a nuclear magnetic moment can interact with the electron spin by two different mechanisms. The first and most obvious is the familiar interaction of two magnetic dipoles separated by a distance vector **r**; the strength of interaction depends on the magnitude of $|\mathbf{r}|$. Accordingly, the dipolar interaction is *anisotropic* and its magnitude changes for different orientations of the radical relative to the direction of the external field. For fluid samples in a low-viscosity solvent, the radicals tumble freely so that the dipolar interaction averages to zero and need not be considered in electrochemical systems.

The other mechanism is called the Fermi contact interaction and it produces the *isotropic* splittings observed in solution-phase EPR spectra. Electrons in spherically symmetric atomic orbitals (s orbitals) have finite probability in the nucleus. (Mössbauer spectroscopy is another technique that depends on this fact.) Of course, the strength of interaction will depend on the particular s orbital involved. Orbitals of lower-than-spherical symmetry, such as p or d orbitals, have zero probability at the nucleus. But an unpaired electron in such an orbital can acquire a fractional quantity of s character through hybridization or by polarization of adjacent orbitals (configuration interaction). Some simple cases are described later.

*Proton Splittings*

**Spin Polarization.** The methyl radical $H_3C\cdot$ is a planar species with $C_{3v}$ symmetry. The unpaired electron resides in a carbon 2p orbital and the C–H bonds are $sp^2$ hybrids. The unpaired electron has zero probability in the plane of the hydrogen nuclei, yet the protons display an isotropic splitting of 2.3 mT, so we must rationalize this experimental observation.

The same phenomenon that leads to Hund's rule of maximum multiplicity in atoms (i.e., quantum-mechanical exchange stabilization) produces polarization of the electron spins in the C–H $\sigma$ bond. In a valence-bond treatment, the bond is comprised of one electron from a carbon $sp^2$ orbital and another from a hydrogen 1s orbital. Exchange forces act to polarize the $sp^2$ electron so that its spin is parallel to the unpaired spin in the carbon 2p orbital; this leaves the

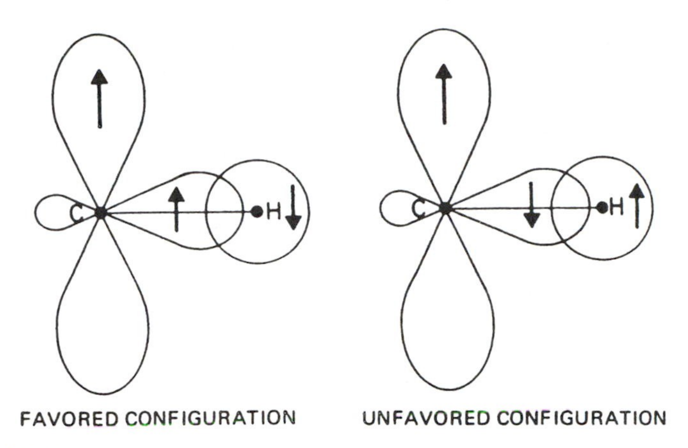


**Figure 29.7** Mechanism of hyperfine interactions by spin polarization in the CH fragment.

hydrogen 1s electron oriented in the opposite direction. Of course, the separation into atomic orbitals is far from complete, but this mechanism does place net spin (of opposite sign to the unpaired spin) in a spherical orbital centered on the adjacent atom (see Fig. 29.7).

This spin polarization mechanism was first advanced by McConnell and Chestnut [8], who proposed the equation

$$a_H = -Q\rho_C^{\pi} \tag{29.14}$$

to explain the hydrogen HFSC for aromatic hydrocarbons in general. The negative sign indicates that the spin density at the proton is oriented opposite to the unpaired spin in the carbon p orbital. (However, there is no simple experimental means for distinguishing positive and negative coupling constants.) Q is a semi-empirical parameter on the order of 2.4 mT and $\rho_C^{\pi}$ is the *spin density* in an orbital of $\pi$ symmetry on the carbon adjacent to the hydrogen atom. This concept is clarified by examination of the data in Table 29.2, where the spin densities can be evaluated by inspection. For radicals of lower symmetry, MO calculations are required.

It will be noted that Q does vary among the radicals in Table 29.2. Even more interesting is the fact that the molecules that form anion radicals upon reduction and cation radicals upon oxidation usually display larger proton splittings in the cationic species. Colpa and Bolton [9] proposed a more sophisticated equation based on the fact that negative charge in the p-type orbital will

**Table 29.2** Proton Hyperfine Splittings in High-Symmetry Radicals

| Radical | Splitting $a_H$ | Q (mT) | $\rho_C^\pi$ | $a_H$ (calc. Eq. 29.15) |
|---|---|---|---|---|
| $CH_3^{\cdot}$ | -2.304 | -2.304 | 1 | -2.7 |
| $C_5H_5^{\cdot}$ | -0.598 | -2.99 | 1/5 | -0.540 |
| $C_6H_6^{-\cdot}$ | -0.375 | -2.25 | 1/6 | -0.414 |
| $C_7H_6^{\cdot}$ | -0.391 | -2.74 | 1/7 | -0.385 |
| $C_8H_8^{-\cdot}$ | -0.321 | -2.57 | 1/8 | -0.317 |

cause the orbital to expand and thus decrease the exchange interaction. Their relationship is of the form

$$(a_H)_i = (Q - K\varepsilon_i)(\rho_C^\pi)_i \tag{29.15}$$

where Q and K have the empirical values of –2.70 and –1.29 mT, respectively, and $\varepsilon_i$ is the net excess charge in the p orbital on the ith carbon atom (to which $H_i$ is bonded) (e.g., for benzene anion radical $\varepsilon_i = -1/6$).

**Hyperconjugation.** Spin polarization is rapidly attenuated by interposition of tetrahedral carbon atoms between the atom carrying the unpaired electron and another atom possessing nuclear spin. Yet protons of a methyl group often display large splittings. For example, the ethyl radical $CH_3CH_2\cdot$ exhibits splittings from two protons, $a_H = 2.24$ mT, and three methyl protons of $a_H =$ 2.69 mT. A hyperconjugative mechanism, familiar from organic chemistry, is invoked to explain such observations:

$$H_3C - \dot{C}H_2 \rightleftarrows \overset{H\bullet}{\underset{H_2}{}}C = CH_2 \tag{29.16}$$

A more formal MO treatment of this process suggests that hyperconjugation can be viewed as a direct transfer of electron spin from the carbon p orbital to an s orbital centered on hydrogen (Fig. 29.8); the efficiency of transfer depends on the position of the so-called β-proton* relative to the p orbital of the adjacent trigonal carbon. An equation analogous to Equation 29.14 describes the coupling of β-protons,

$$a_H = (B_0 + B_2 \cos^2\theta)\rho_C^\pi \tag{29.17}$$

where $B_0$ (≅ –0.1 to +0.5 mT) is the contribution from residual spin polarization and $B_2$ (≅ +4.0 to +5.3 mT) is the direct-transfer term; the angle θ is defined in Figure 29.8b. The presence of the $\cos^2\theta$ term sometimes allows

*This use of β derives from the organic chemistry practice of using Greek letters to denote position, e.g., a β-lactone. It has nothing to do with |β> spin state.

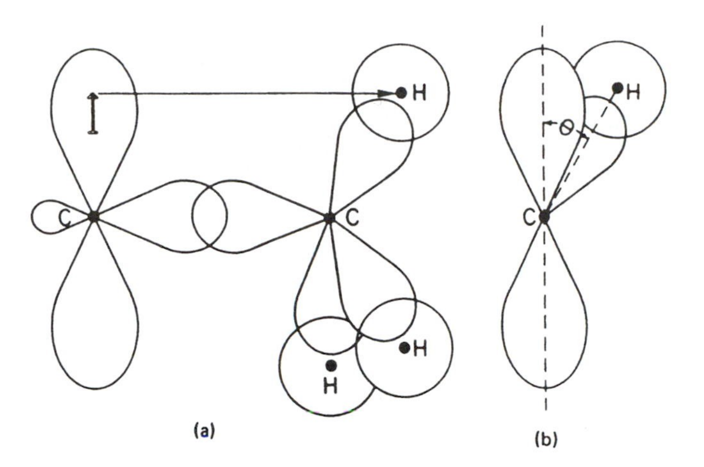


**Figure 29.8** Mechanism of hyperfine interaction by hyperconjugation.

extraction of steric information about rigid radicals or the preferred orientation of the β-protons of bulky substituents. For a freely rotating group, such as methyl, the time average value of $\cos^2 \theta$ is 0.5.

*Other Nuclei in π-Electron Systems*

Magnetic nuclei associated with orbitals in the π-electron system can also exhibit HFS. The interactions that describe the HFSC of such nuclei involve electron exchange between adjacent atoms as well as between the unpaired electron and the core electrons of the particular atom. These expressions are complicated and are derived on theoretical grounds in References 10 and 11, and empirical parameters for many heteroatoms are tabulated in References 1, 6, and 11.

*Transition Metal Nuclei*

In general, d-orbital electrons are highly anisotropic and have no spin density at the transition metal nucleus, so that the largest hyperfine interaction occurs between the electron magnetic dipole and the nuclear magnetic dipole. However, in solution, the anisotropic HFS averages to zero, or nearly zero. The isotropic HFS that are observed in solution occur in much the same way as the spin polarization of the C–H fragment shown in Figure 29.7. However, positive spin in the d-orbital creates positive spin in the s-orbital near the transition metal nucleus. Typical values are 7–8.2 mT for $Mn^{2+}$, 9–11 mT for $VO^{2+}$, and 4–9 mT for $Cu^{2+}$. Generally, the metal HFS decreases as the d-electron becomes more delocalized onto ligands associated with the transition metal ion. Incomplete averaging of the anisotropic HFS often causes lines of the different nuclear spin states to be of unequal intensities. In these cases the lines are sharpest toward the center of the spectrum and broaden toward the ends of the spectrum.

Some transition metal complexes exhibit HFS due to the ligand; this is called superhyperfine splittings (SHFS). Often SHFS are not completely resolved, but contribute to the line width. In such cases isotropic substitution of the ligand nuclei, such as deuterium for hydrogen, can allow the presence of SHFS to be confirmed.

## IV. EPR INSTRUMENTATION

### A. The Basic EPR Experiment

Figure 29.9 shows a comparison between the components of a rudimentary EPR spectrometer and the corresponding elements of a more familiar apparatus for visible spectrometry. In EPR, the source of excitation radiation is a microwave device called a klystron. The microwaves would disperse in free space and must therefore be conducted to the sample by waveguide or coaxial cable. The sample, contained in the sample tube, is held in a microwave cavity between the poles of a magnet. The detector is usually a diode that produces a dc output propor-

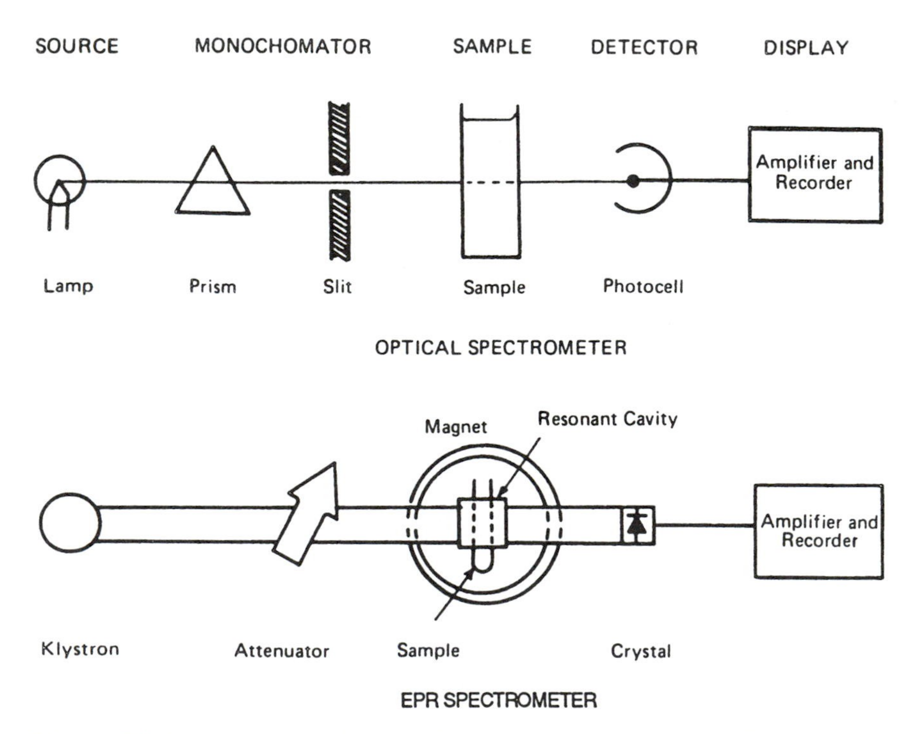


**Figure 29.9** Comparison of an optical spectrometer with a simple EPR spectrometer.

tional to the level of RF power incident on it. The dc from the detector is displayed on a recorder or oscilloscope. In principle, the EPR spectrometer could be operated in a mode analogous to visible spectrometers (i.e., by varying the frequency of the klystron while holding the magnetic field fixed). In practice, the klystron frequency is fixed because the microwave cavity acts as a high-Q tuned circuit and it is far easier to vary the magnetic field to attain the resonance condition. Although some instruments can operate in the mode outlined in Figure 29.9, it is not used widely because of its poor sensitivity. Various modifications involving modulation of the magnetic field and lock-in (phase-sensitive) or superheterodyne detection give much better results and are used in most research instruments.

## B. Components of the Basic Instrumentation

An understanding of EPR instrumentation requires a general knowledge of the operation of microwave components and magnet systems. A brief description of these is given next. Detailed discussion of the design of EPR spectrometers can be found in the references, particularly the books by Alger and Poole listed in Appendix I.

**Klystrons.** The most commonly used radiation source is a klystron; these tubes are available at discrete frequencies between 2.5 and 220 GHz. Many klystrons can be tuned over a range up to ±3% of the nominal frequency by a control that varies the physical dimensions of a resonant cavity inside the tube. Finer adjustment of the frequency is achieved by varying the voltage applied to the resonator and reflector electrodes. Thermal stability is obtained by immersion of the entire tube in an oil bath, or by water or air cooling. A feedback circuit provides automatic frequency control (AFC) to continuously correct the output frequency to the resonance frequency of the cavity. The power output of the klystrons used in EPR spectrometers is generally about 300–700 mW. The most widely used frequency for EPR spectrometers is 9.5 GHz, which is called X-band.

**Isolators.** Reflection of microwave power back into the klystron is prevented by an isolator. This consists of a strip of ferrite material that passes microwaves in one direction only and is necessary to stabilize the klystron frequency.

**Attenuators.** Klystrons operate at constant power, which can be attenuated by insertion of a piece of "lossy" material into the waveguide. The position determines the degree of attenuation. Another type of attenuation device relies on polarization of the microwave radiation.

**Waveguide.** The microwave radiation is usually conveyed to the sample and detector by a waveguide, which is a hollow, rectangular tube made of aluminum, copper, or brass, plated with silver or iridium.

**Cavities.** The sample is contained in a resonant cavity, which is a structure capable of supporting a standing-wave pattern. The cavity is analogous to a tuned circuit (e.g., a parallel RLC combination) used at lower frequencies and must be matched to the klystron frequency range. A measure of the quality of the cavity is its "Q" or "Q factor," which is defined as

$$Q = 2\pi \frac{\text{energy stored in cavity}}{\text{energy lost per cycle}} \tag{29.18}$$

and the sensitivity of the spectrometer is directly proportional to the value of Q. The standing wave is composed of both magnetic and electric fields at right angles to each other. Two commonly used cavities are the rectangular $TE_{102}$

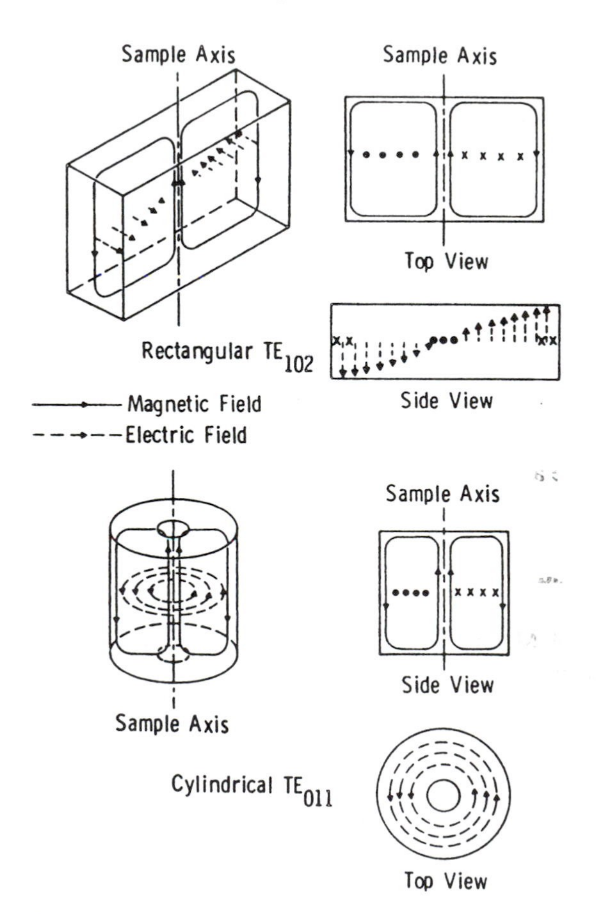


**Fi... 29.10** Magnetic and electric field distribution inside rectangular and cylindric... ...vities.

cavity and the cylindrical $TE_{011}$ cavity (Fig. 29.10). It is the magnetic field component that stimulates EPR transitions, so that the sample should be placed where the intensity of the magnetic field is maximum. The electric field also interacts with the sample through a process of dielectric loss. This arises from interaction of the microwave electric field vector with electric dipoles inside the cavity; in general, the greater the dielectric constant, the greater the loss. If the sample or solvent is "lossy" (i.e., absorbs microwave power), the Q of the cavity may be drastically decreased. This problem is diminished when the geometry of the sample holder and that of the cavity are matched so as to locate the sample in a region of high microwave magnetic field while minimizing its interaction with the electric field. Figure 29.10 shows that the electric field is smallest in a plane through the middle of the $TE_{102}$ cavity and along the cylindrical axis of the $TE_{011}$ cavity. For X-band spectrometers, tubing of 2–4 mm i.d. (sample volume about 0.15–0.5 mL) can be used in a rectangular cavity with low-loss solvents such as dimethylformamide, acetonitrile, or dimethyl sulfoxide. Aqueous solutions and biological tissue samples are so lossy that the only way to achieve a significant sample volume at X-band is to confine the sample to a thin sheet in the midplane of the rectangular cavity. Rectangular flat cells with a thickness of about 0.25 mm (sample volume of 0.05 mL) are suitable for aqueous samples in rectangular cavities. Clearly, the solvent selected for an electrochemical EPR study will determine the geometry of the sample holder and the cavity.

There may be special advantages in performing electrochemical–EPR experiments at frequencies lower than 9–9.5 GHz. Because cavities are larger, cells can occupy a greater volume. Many solvents exhibit smaller dielectric loss at 1–2 GHz than at 9 GHz. Also, the detector sensitivity increases at lower frequency, and since a greater sample size is possible, a greater signal-to-noise ratio may be obtained. However, very little work has been done in this frequency region.

**Couplers and Matching Screws.** Coupling microwaves between the components can be achieved by various methods. Frequently, an iris or an adjustable slot is used. Matching of waveguide elements (analogous to impedance matching in conventional circuits) is accomplished by using screws or stubs, which can be inserted into the waveguide or across the coupling iris or moved in lateral slots of the waveguide.

**Microwave Bridges.** In principle, EPR absorption could be observed as a small decrease in a large background signal. But a preferable arrangement utilizes a microwave bridge that functions like the familiar Wheatstone bridge. This permits much more efficient amplification of the EPR signal.

**Magnets.** An electromagnet capable of producing fields of at least 0.5 T is required for EPR at 9.5 GHz, which is the most common frequency for

EPR systems. For organic radicals in solution, the homogeneity of the field should be better than about 0.005 mT over the sample region. The homogeneity requirements are not as severe for transition metal ions, due to their broader spectral lines.

The EPR spectrum is recorded by slowly varying the magnetic field through the resonance condition. In most instruments, the magnetic field is regulated by using a feedback control circuit to sense deviations of the field strength from the anticipated value and to correct for these discrepancies. Accordingly, the field sweep involves the use of a slow ramp voltage, which is compared with the output of a field-strength sensor (e.g., a Hall effect probe or a rotating coil), which generates a signal proportional to the actual field strength.

**Modulation Coils.** It is easier to discriminate against noise in an ac (rather than dc) signal, so most EPR spectrometers employ magnetic field modulation to render the absorption as an ac waveform. This modulation is produced by supplying an adjustable alternating current to coils oriented parallel to the magnetic field coils. For low-frequency modulation (400 Hz or less) the coils can be mounted outside the cavity or perhaps on the magnet pole pieces. At higher frequencies (1 kHz or more) the modulation field cannot penetrate metal effectively, so either the cavity must be constructed of a nonmetallic material (e.g., quartz or plastic with a thin plating of silver to support the standing microwave pattern) or the coils must be embedded in the cavity walls.

**Detectors.** Most EPR spectrometers use a diode to rectify microwave energy to a signal that can be manipulated by conventional amplifier circuits. The magnitude of the diode output depends on the incident power, so the fluctuations resulting from field-modulated absorption by the sample exit from the diode as a pulsating current. Microwave rectification is inefficient at low power levels and noisy at high levels, so the diode performance is optimized by application of a constant-bias power level. The silicon crystal detectors (point contact diodes) that were originally used produce a fairly high noise level that is inversely proportional to the frequency of the detected signal (1/f noise); to minimize this effect the signal was typically detected with 100-kHz field modulation or by superheterodyne methods. However, 100-kHz modulation introduces absorption sidebands at $\pm 3.6\ \mu T$, and these distort the main absorption signal if the line width is less than about 6 $\mu$T (typical of organic radicals in solution).

More recently, Schottky-barrier diodes and backward diodes have been used as detectors. These do not require as much power to bias the diode to its optimum output and thus permit observation of EPR at lower incident power levels. They also have a much lower 1/f noise characteristic so that modulation frequencies between 6 and 25 kHz (equivalent to 200- to 900-$\mu$T sidebands) can yield the same sensitivity that 100 kHz provides with silicon diodes.

## C. Phase-Sensitive Detection

Although some spectrometers operate on the superheterodyne principle, the majority employ field modulation and phase-sensitive detection. Figure 29.11 shows how this method converts the absorption line to the first-derivative display, which is characteristic of EPR. The modulation field (represented by the sine waves enclosed in the vertical bands) is superimposed on the strong field being swept by the electromagnet; the net field thus produces a degree of microwave absorption that varies sinusoidally. The amplitude of this variation is proportional to the *slope* of the EPR absorption line at the nominal value of the sweep field (represented by the sine waves in the horizontal bands). The phase of the pulsating current produced by the detector is compared with the phase of the field modulation current; as the field sweeps through the absorption

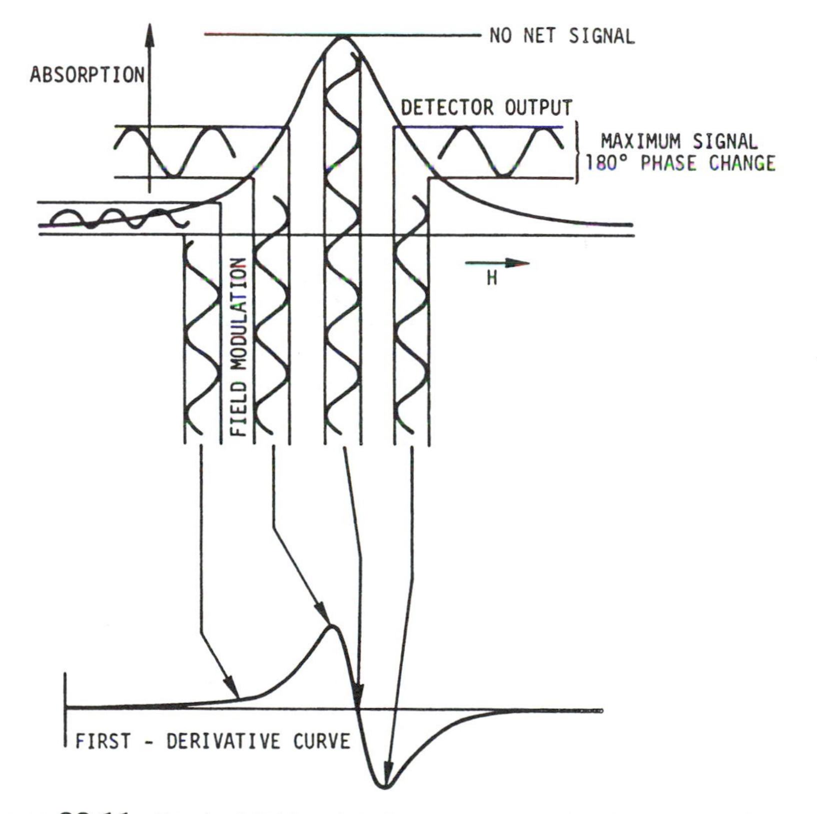


**Figure 29.11** Result of field modulation and phase-sensitive detection in EPR spectrometers.

maximum, a phase shift of 180° occurs. The output of the phase-sensitive detector is further amplified to produce the final presentation shown at the bottom of the figure.

One of the problems associated with this technique is that when the amplitude of the modulation field is too large, the detected line shape becomes distorted and decreases the resolution as well as the signal amplitude in extreme cases.

The components described above are assembled in a typical spectrometer as shown in Figure 29.12.

## D. Spectrometer Sensitivity

The sensitivity of a spectrometer is proportional to the magnetic susceptibility of the particular sample and to the values of various instrumental parameters.

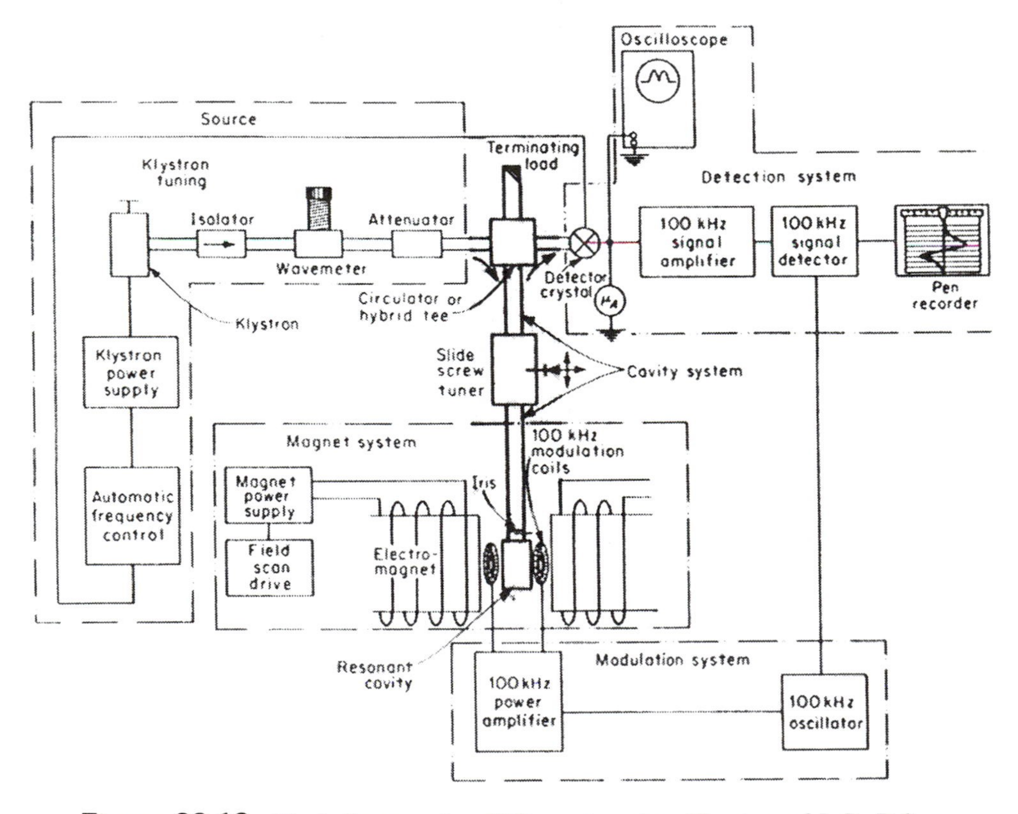


**Figure 29.12** Block diagram of an EPR spectrometer. [Courtesy of J. R. Bolton, University of Ontario, London, Ontario.]

Neglecting the susceptibility term, the minimum detectable number of spins, $N_{min}$,

$$N_{min} \propto \frac{\Delta\nu}{Q_0 \eta \nu_0} \left( \frac{\Delta f \, kT}{2P_0} \right)^{1/2} \tag{29.19}$$

where $\Delta\nu$ is the line width of the absorption, $\nu_0$ the frequency of the absorption, $Q_0$ the Q of the cavity off resonance, $\Delta f$ the bandwidth of the spectrometer response, k Boltzmann's constant, T the absolute temperature of the detector, $P_0$ the incident power, and $\eta$ the filling factor, which is the power density integrated over the cavity volume.

Several parameters in Equation 29.19 can be adjusted to optimize the sensitivity of the spectrometer. For example, the incident power can be increased provided that the absorption is not saturated, and the extent of RC filtering can be increased (i.e., the $\Delta f$ decreased). For small samples it is sometimes feasible to use a higher frequency (higher field) spectrometer and thus increase the population difference between the two energy levels (cf. Eq. 29.11). But it can be shown that if the same geometry is maintained, the minimum detectable *concentration* is theoretically proportional to $\nu_0^{-1/2}$, so for electrochemical experiments the sensitivity actually decreases at higher frequencies, all other factors being equal.

Finally, the filling factor and the cavity Q must be considered. For a given sample, the best signal-to-noise ratio is obtained when the presence of the sample reduces $Q_0$ to about two-thirds that of the Q of the empty cavity. The dielectric behavior of the sample puts some constraint on the filling factor, and one chooses the cavity geometry that provides the maximum $\eta$ with the minimum reduction in Q. As a general rule, rectangular geometry provides the best performance for lossy samples such as aqueous solutions.

## V. EXPERIMENTAL METHODS OF EPR-ELECTROCHEMICAL STUDIES

### A. General Description

It is convenient to subdivide EPR–electrochemical experiments into two broad categories. The first, called "external generation," involves electrolysis outside the EPR cavity, transfer of the radicals to the cavity, and observation of the EPR spectrum. This category includes both experiments that require a discrete transfer step and those in which the electrolyzed solution is pumped through the cavity for either continuous or stopped-flow measurements. The second classification is called "internal," "in situ," or "intra muros" generation. It involves placement of a working electrode directly inside the EPR cavity and permits recording the EPR signal with minimal time lapse. This method usually requires certain design compromises that degrade both the electrochemical and EPR performance of the cell.

Some experiments are designed primarily to obtain the spectra of radical species; electrochemistry provides a convenient means for producing the radicals. Other investigations are designed to elucidate the electrochemical behavior of a system, and EPR provides a simple method for detecting the presence, estimating the lifetimes, or determining the kinetics of the free-radical intermediates. The following sections describe many different kinds of electrochemical-EPR experiments. The details of a given experiment are often determined by the particular goals of the study.

**Oxygen Removal.** EPR and electrochemistry share a common concern about removal of oxygen from the experimental solutions, but for different reasons. In addition to the possibility of chemical reactions between oxygen and the free radicals, there is also the problem of magnetic interactions between the paramagnetic (ground-state triplet) $O_2$ molecules and the unpaired spins of interest. The resultant line broadening can cause loss of resolution in the EPR spectrum. Methods for deoxygenating EPR samples run the gamut from simple nitrogen sparging to sophisticated vacuum techniques.

*External Generation of Radicals*

The use of external generation requires long-term stability of the radicals because of the significant time lapse between generation and observation. Nonetheless, this method has distinct advantages in certain situations. For example, the solution can be exhaustively electrolyzed. This produces the maximum signal intensity from the sample and diminishes complications due to electron exchange reactions (see Sec. VI.F). Moreover, since the EPR tube is essentially independent of the electrochemical glassware, optimum design of both cells can be achieved. This means that the electrode materials and geometry can be chosen freely, with a view toward efficient electrolysis and negligible uncompensated cell resistance. Similarly, the only constraint on the EPR sample tube is that it provide the optimum balance between cavity Q and filling factor for the particular solvent system employed.

Accordingly, external generation is suitable for studies of stable free radicals when it is desirable to optimize signal and resolution, such as in relaxation studies and in the correlation of EPR parameters with MO calculations. Fraenkel and co-workers [12] developed the cell shown in Figure 29.13, used in studies of spin-density distribution in nitriles, semiquinones, and nitroaromatic anion radicals. Oxygen is pumped away from the sample, and supporting electrolyte and all subsequent operations are performed under high-vacuum conditions. The rather complicated manipulations are described in the original paper, but the

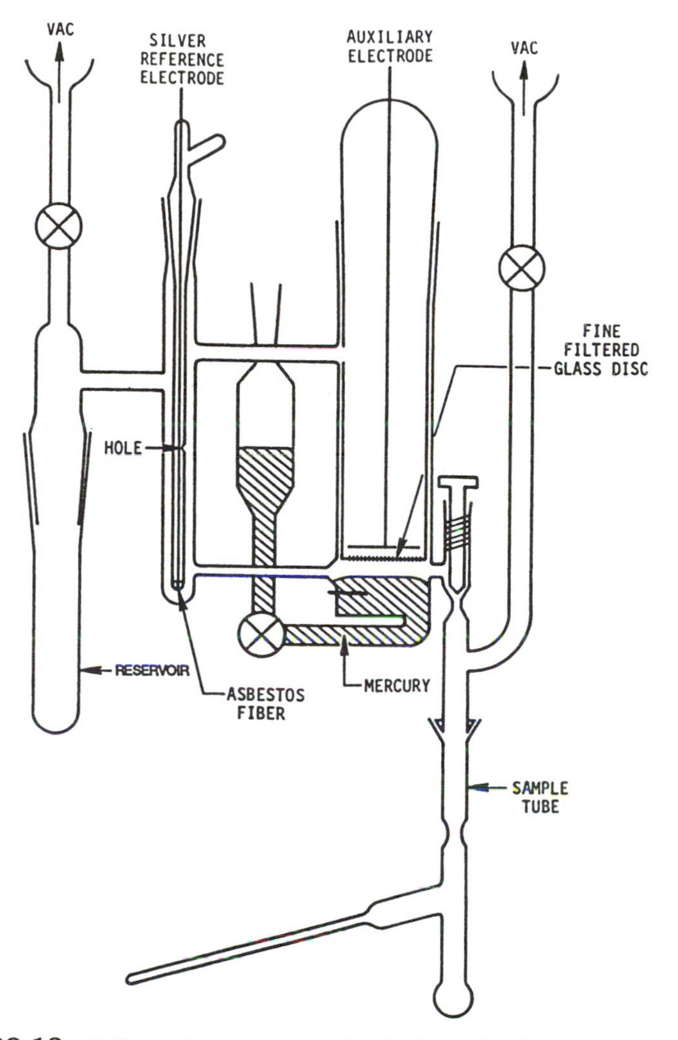


**Figure 29.13** Cell used for generation of radicals outside the spectrometer for EPR detection.

important points for this discussion are that the parallel orientation of the working and auxiliary electrodes results in uniform current density, the potential can be controlled accurately, and the fine frit prevents mixing of auxiliary electrode products with thc solution of free radicals. Upon completion of the generation step, the sample is transferred to the evacuated EPR tube, the solution is frozen, and the EPR cell is torch-sealed to prevent air contamination. Distillation of solvent between the reservoir bulb and the EPR tube permits investigation of concentration effects, and also permits an optimum signal to be obtained.

*Internal Generation*

Electrochemical generation directly inside the microwave cavity (in situ) permits observation of radicals that might otherwise be too unstable for observation. This method was originated by Maki and Geske [13], using a cell similar to that shown in Figure 29.14 to generate nitrobenzene anion in acetonitrile. These early experiments were designed primarily to detect the presence of radicals and to obtain EPR spectra. For similar systems in which electrochemical behavior is relatively uncomplicated, there need be little concern about the precise value of the working electrode potential as long as secondary reactions do not occur. Application of a voltage (corresponding to a point on the diffusion current plateau of the polarographic wave) between the reference and working electrodes produces a faradaic current that decays rapidly as the semi-infinite diffusion layer is established. Within minutes, the current drops to a few microamperes so that $E_{effective}$ (= $E_{applied}$ – iR) approaches the desired value. (The cell of Fig. 29.14 has a resistance of about 2 kΩ when filled with a 0.1 *M* solution of tetraalkylammonium perchlorate in acetonitrile.) The usual concentration of substrate is millimolar or less, so that the total quantity of charge transferred is only on the order of 1 C. Hence the massive, low-resistance reference electrode is not significantly polarized.

Adams and co-workers [14] and others [15] developed a flat cell suitable for aqueous solution studies. The Varian electrolysis apparatus shown in Figure 29.15 is modeled after their design and has been used widely because most EPR spectrometers come equipped with a $TE_{102}$ cavity. Even when using aqueous solutions, the cell resistance is high due to the thin cross section of electrolyte near the working electrode, and this problem is compounded with most nonaqueous solvents. With a mercury working electrode filling the lower half of the flat cuvette, the current density is nearly uniform and the current–time response is similar to that described above. (In three-electrode controlled-potential operation, the secondary reference electrode in Fig. 29.15 functions as the auxiliary electrode.)

The area of the mercury electrode surface in such a cell is only about 0.45 $mm^2$. Thus only a small amount of radical can be generated. To increase the surface area so that larger quantities can be generated, a foil or wire mesh electrode is used in place of the mercury. However, in this configuration, large potential gradients can promote convection, making the overall electrode behavior impossible to characterize.

Nonetheless, many significant EPR studies have been accomplished with flat cells. Most commercial versions are fabricated of quartz, since Pyrex is lossy and contains paramagnetic impurities that can be detected at the high modula-

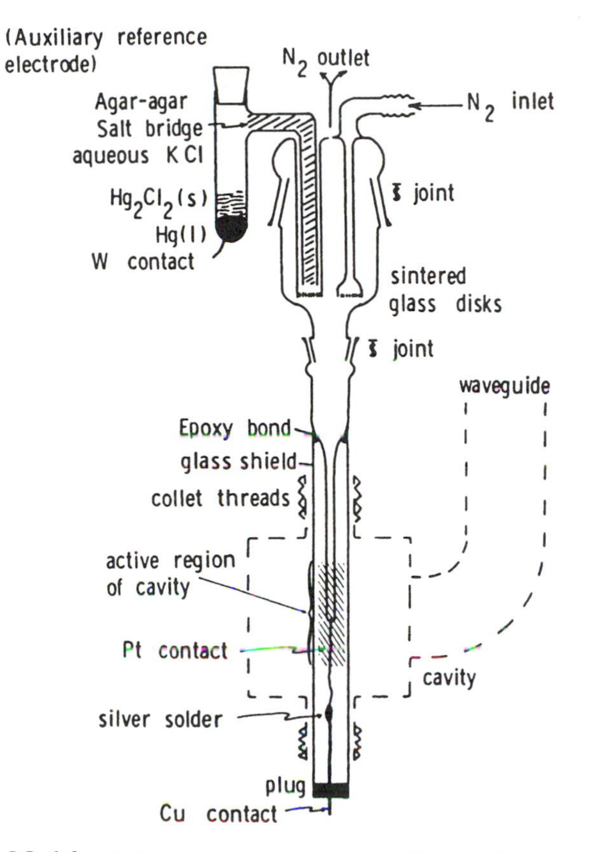


**Figure 29.14** Cell used for generation of radicals with spectrometer.

tion amplitudes employed in studies of transition metal ions. Since the majority of the loss is due to the solvent, and low modulation amplitudes are used to detect the narrower lines of organic radicals, less expensive Pyrex flat cells are frequently adequate. An even cheaper version, made of a flattened section of heat-shrinkable Teflon, has been reported [16] to be suitable for systems where the dimensions and placement of the cell are not critical. Of course, when irradiation studies are contemplated, quartz is the most suitable material. A cell for simultaneous irradiation, electrolysis, and EPR observation has been described [17].

The optimum orientation of the flat electrochemical cell is in the center of the rectangular cavity, so that the sample is in the region of the maximum magnetic field. In this configuration, the face of the flat cell is parallel to the end plate of the cavity. Usually the cell position along the length of the cavity

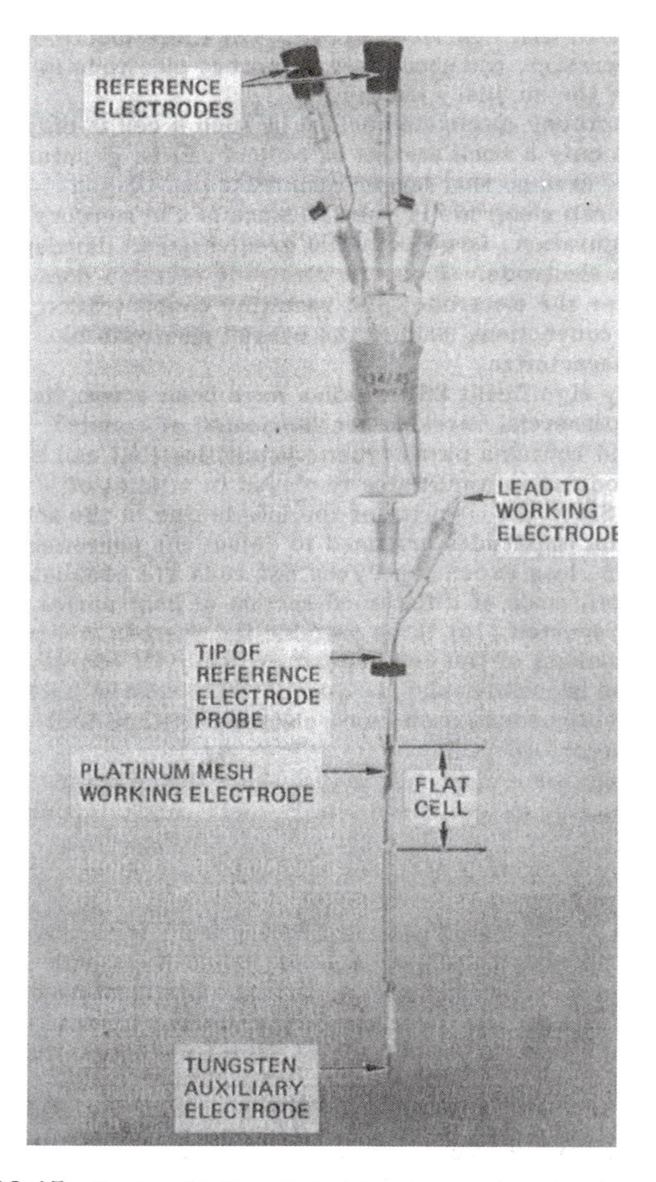


**Figure 29.15** Commercial flat cell for internal generation of radical ions. [Courtesy of Varian Associates, Palo Alto, CA.]

can be adjusted by screws in the mounting bracket. Experimentally, the cell is placed in the cavity and rotated and positioned to achieve the maximum Q-factor, which corresponds to the greatest "dip" in the microwave cavity resonance. As a cautionary note, when low-loss solvents are used with a flat cell, two discrete minima can be observed in the cavity resonance as the flat cell is rotated in the cavity. The first occurs as described earlier when the face of the flat cell is parallel to the cavity end plate. The second occurs when the cell is rotated 90° so that the cell is perpendicular to the cavity end plate. However, this produces lower EPR sensitivity than when the cell is in the optimum orientation. A mark placed on the cylindrical neck of the flat cell can help confirm the correct orientation.

There are many electrochemical systems in which the free radical is not sufficiently stable to maintain a steady-state concentration over the interval required for a complete scan of the EPR spectrum. A scan of at least 10 min is usually required to obtain even a moderately well resolved spectrum. Radical stability can often be enhanced by generating and observing the spectrum at low temperatures. A dewar jacketed tube running through the cavity holds the sample tube, which is bathed in a stream of chilled gas. The temperature of the sample can be held within close tolerances by a heating element that warms the gas to the desired temperature. (The same apparatus can be used for high-temperature studies when radical stability is not a problem.) Since the cavity dewar is cylindrical, a sample tube similar to that in Figure 29.14 (modified by removal of the supporting shield tube) has been developed for these studies. For example, we have used a 4-mm-o.d. Purex tube connected by a ground-glass joint to a reservoir similar to that shown in Figure 29.14. A platinum foil was sealed in the tube so that it is in the sensitive region of the cavity. The lead that penetrates the Pyrex seal is brought out of the cavity through the top of the dewar insert [18]. A more sophisticated system incorporating a helical electrode was used in order to obtain a large surface area with minimum loss due to the electrode and solution [19–21]. A cell designed specifically for electrolytic studies in liquid ammonia has also been described [22].

## B. Some Problems of Quantitative Measurements

Precise analytical measurements with EPR are difficult even though the technique is sensitive to low concentrations of paramagnetic species. The heterogeneous nature of electrochemical charge transfer makes the problem even more complicated, and only a few detailed investigations have been published. Nearly all of these have employed a flat cell in a $TE_{102}$ cavity, and the following discussions are based on this configuration.

*Sample Position*

Figure 29.16 shows the EPR signal amplitude as a function of sample position when a small speck of solid paramagnetic material is moved along the vertical axis of the cavity. Clearly, the sensitivity is greatest when the sample is at the geometric center. The change in response when the sample is moved from side to side in the plane of maximum $H_{rf}$ is much less than the up-and-down variation. The implication of Figure 29.16 is obvious: Even for a homogeneous solution of radicals in a flat cell, about 80% of the signal comes from those spins within 0.5 cm of the center of the cavity, and the determination of the absolute number of spins present is difficult. Moreover, when electrolytic generation is employed, the greatest concentration of radicals occurs at the edge of the elec-

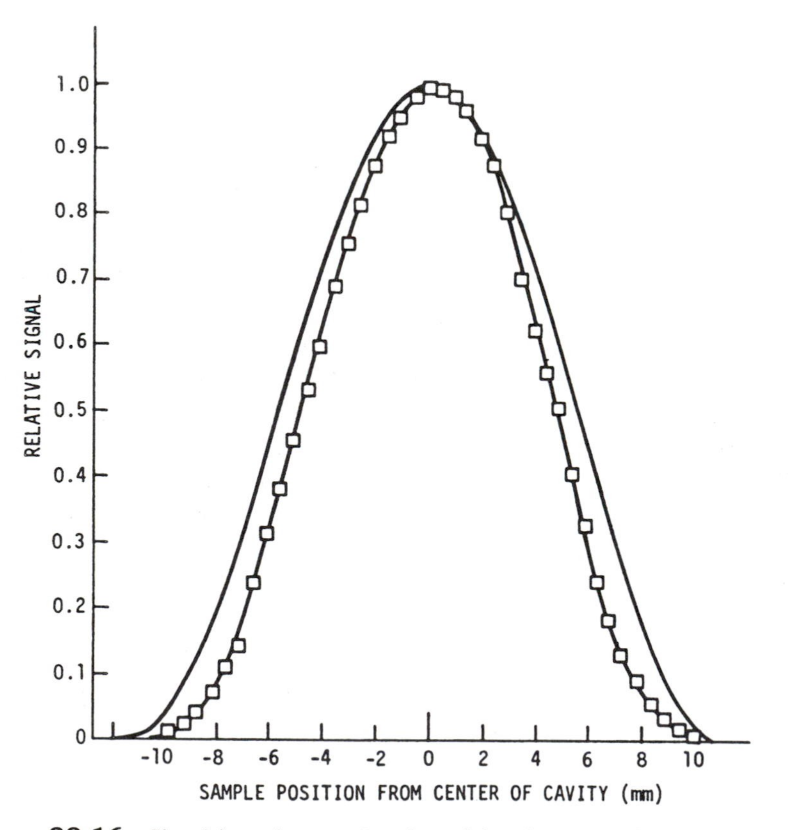


**Figure 29.16** Signal intensity as a function of the placement of a small paramagnetic sample along the axis of a $TE_{102}$ mode cavity. Squares are experimental points; solid line represents $\cos^2$ dependence of the RF field strength.

trode that is nearest to the auxiliary electrode, and it is important that this region of high concentration be near the center of the cavity. However, reproducible placement and generation of radicals in electrolytic cells is difficult.

*Line Widths*

The ultimate sensitivity of an EPR spectrometer is often quoted as

$$N_{min} = N_0(\Delta H) \tag{29.20}$$

(cf. Eq. 29.19), where the smallness of $N_0$ is a measure of the design quality and performance of the system and $\Delta H$ is the peak-to-peak width of a one-line spectrum.

The important functionality of Equation 29.20 is that the sensitivity is less for broad lines. We describe in a later section how electron exchange reactions cause lines to broaden when the electrochemical reaction is incomplete (see Sec. VI.F).

*Hyperfine Structure*

Equation 29.20 applies only to a one-line spectrum, and the sensitivity decreases when the radical sample exhibits HFS. As a simple case, reconsider the $H_3C\cdot$ spectrum, with its four lines; one of the outside lines represents only one-eighth of the total number of radicals in the sample because the 1:3:3:1 intensity ratios reflect the statistical weights of the various nuclear spin configurations. Thus the detection sensitivity falls off rapidly with progressive HFS.

## C. Dynamics of In Situ Generation

*Controlled-Potential Operation*

**Quiescent Solutions.** When the applied potential of the working electrode is set onto the plateau of the voltammetric curve, the concentration of electrochemical reactant is depleted at the electrode surface with concomitant increase of the product concentration. A diffusion profile (concentration gradient) is established, which extends progressively farther out into solution as the electrolysis continues, and the current decays as $t^{-1/2}$. For cells in which there is high electrolyte resistance, the largest proportion of the electrolysis current flows through that portion of the working electrode so that the thickness of the diffusion layer is not uniform [23]. Local thermal and density gradients evolve so that diffusion is augmented by convective transport which produces a steady-state current on the order of tens of microamperes. The EPR signal level becomes constant after several minutes if the radicals are sufficiently stable. Although the concentration profile is essentially indeterminate, the electrolysis is efficient because the working electrode potential is controlled. The current-carrying edge of the electrode should be positioned slightly lower than the center of the cavity.

**Flowing Solutions.** For unstable radicals, the rate of transport of parent material to the electrode is not sufficient to maintain a steady-state radical concentration in quiescent solutions. But the EPR signal can be maintained by forcing a controlled flow of fresh solution through the electrolysis cuvette, with the rate of flow determining the thickness of the diffusion layer. The nonuniform current distribution over the electrode surface makes quantitative measurements difficult. If the direction of solution flow is downward, the current-carrying edge of the electrode should be placed slightly above the cavity center so that the radicals will be swept into the sensitive region.

*Controlled-Current Operation*

**Quiescent Solutions.** Coulometry at constant current provides a simple method for measuring the quantity of electrogenerated species as long as the reaction proceeds with 100% current efficiency. However, this condition breaks down with depletion of the electroactive material in the diffusion layer (cf. chronopotentiometric transitions; see Fig. 4.3). For low values of the applied current, the thermal and density gradients supplement diffusion sufficiently to sustain electrolysis without the potential shifting to a different reaction. This mode of radical generation has been employed successfully in the study of stable species.

At higher values of the applied current, the effective potential across the double layer ultimately increases enough to produce solvent decomposition. But the ohmic potential drop through the solution along a solid electrode in a cell such as Figure 29.15 prevents this from occurring uniformly over the electrode surface. This cell was modeled in computer simulations [24]. One set of results is shown schematically in Figure 29.17 for the constant-current reduction of anthraquinone, which exhibits the following couples:

$$R + e \rightarrow R^{\bar{\bullet}} \qquad E_{1/2} = -0.83 \text{ V (vs. SCE)} \tag{29.21a}$$

$$R^{\bar{\bullet}} + e \rightarrow R^{2-} \qquad E_{1/2} = -1.4 \text{ V (vs. SCE)} \tag{29.21b}$$

$$\text{Solvent} + ne \rightarrow \text{products} \qquad E \text{ more negative than } -2.7 \text{ V (vs. SCE)} \tag{29.21c}$$

When the current pulse is initiated ($t = 0$), all the current goes to produce radical anion; after 25 s, the concentration of parent material is depleted at the end of the electrode closest to the auxiliary electrode ($X/L = 0$) and production of the dianion commences there while anion radical continues to be produced at the middle of the electrode. Further depletion causes the shifts in current profile shown at 50 and 75 s. At 100 s, the current in the portion of the working electrode closest to the auxiliary electrode again increases with the onset of solvent decomposition. Figure 29.18 compares the resulting time-dependent EPR sig-

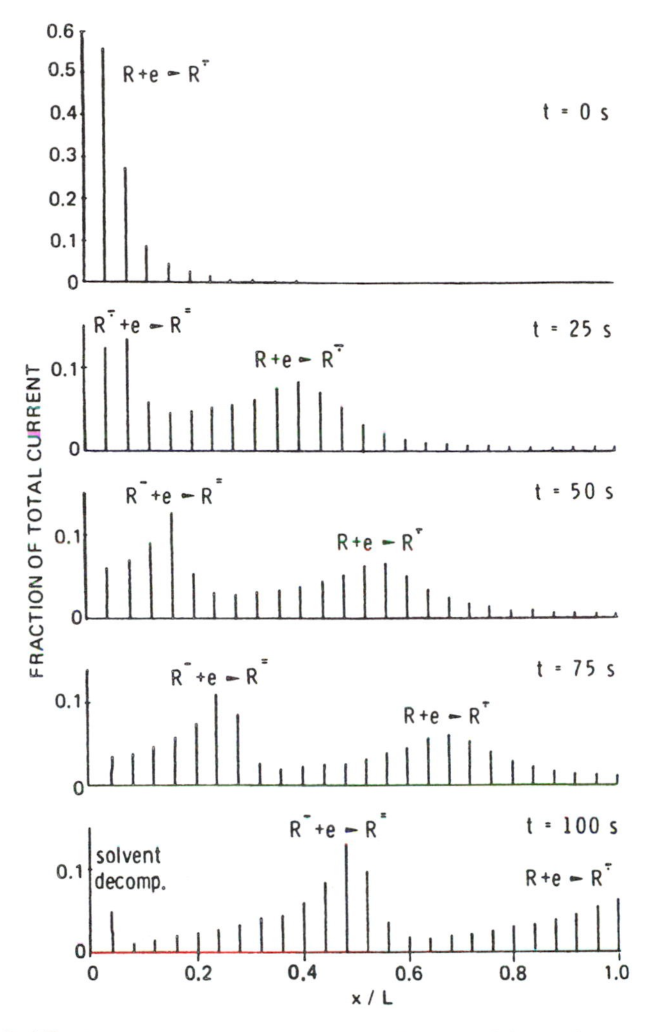


**Figure 29.17** Simulation results showing the fraction of the total current at each of 25 segments of the working electrode at different times during the constant-current electrolysis of 2.5 m*M* anthraquinone and in 0.1 *M* tetrabutylammonium iodide–dimethylformamide solution. Total current 100 $\mu$A; electrode dimensions 0.5 cm × 3 cm; flat cell dimensions 0.5 cm × 0.9 cm × 3 cm.

nal intensity with that of an ideal system for which the current efficiency of radical production is 100%. The real system shows a rapid rise in EPR signal when the current is interrupted. This is caused both by homogeneous solution reaction, $R + R^{2-} \rightarrow R^{\bar{\cdot}}$, as well as the heterogeneous reactions shown in the figure. The latter reactions produce an open-circuit current within the electrode in

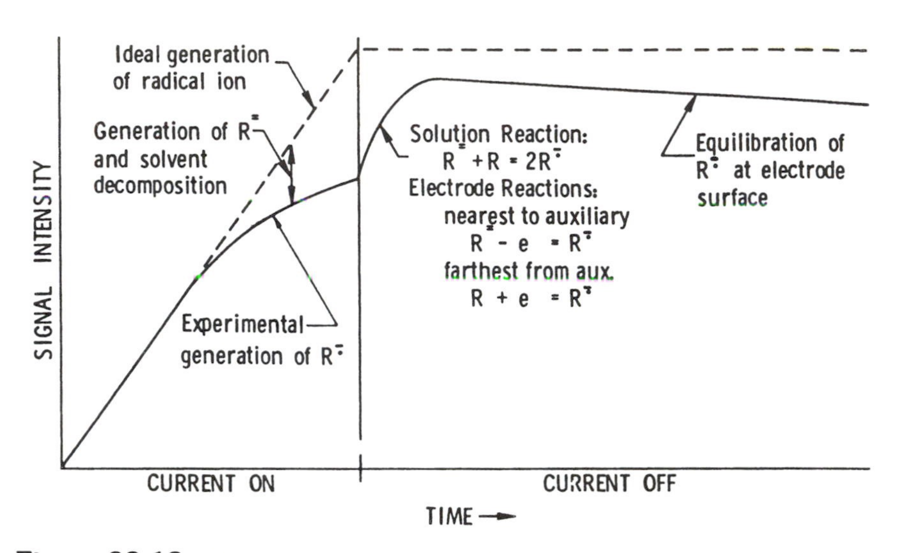


**Figure 29.18** Time dependence of the EPR signal during and following generation by constant current.

response to galvanic potentials arising from local concentration gradients along the electrode surface.

**Flowing Solutions.** When higher current densities are required, the mass transport can be supplemented by flowing solution through the cuvette. Dohrmann and Vetter [25] explored this arrangement to determine the minimum lifetime, reported as $\tau/\Delta H$ ($k = 1/\tau$ is the rate for first-order radical decay and $\Delta H$ is the peak-to-peak EPR line width), of electrogenerated radicals that could be detected. Their approximate calculations yielded values from $10^{-2}$ to $10^{-4}$ s/mT, depending on considerations of flow rate, current density, electrode area, electrolyte conductance, and substrate concentration, for a spectrum of one line.

*Simultaneous Electrochemistry–Electron Paramagnetic Resonance (SEEPR) Techniques*

Because of the difficulties described earlier, electroanalytical studies are usually performed separately from radical generation studies. But a flat cell has been designed [26] (Fig. 29.19) to permit simultaneous monitoring of the electrochemical and EPR response of a free-radical system (SEEPR). The auxiliary electrode extends along the edges of the working electrode, which diminishes the problems of iR drops and provides better uniformity of current density than is possible with conventional electrode placement. This cell is used primarily for short-term (on the order of seconds) electrochemical experiments, such as

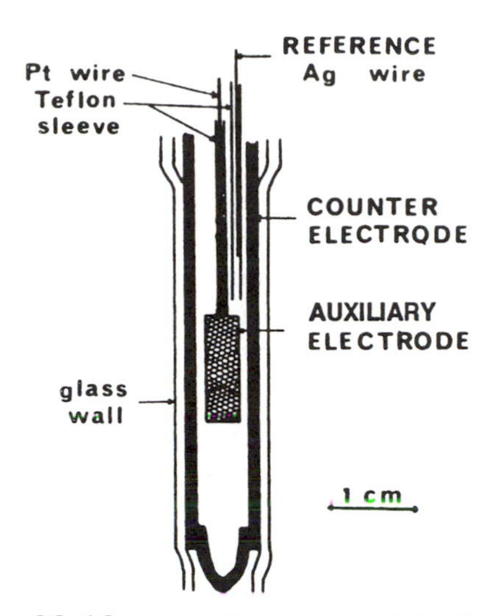


**Figure 29.19** Flat cell used to minimize iR losses.

cyclic voltammetry or current-pulse electrolysis, while recording the EPR signal intensity. When a current pulse is interrupted before the chronopotentiometric transition time, the decay of the EPR signal provides a measure of the radical instability, and the shape of the decay response can be related to kinetic parameters. Working curves can be calculated [27], using different assumptions about the mechanism and rate constant, and comparison with experimental observations permits extraction of kinetic data. This has been demonstrated for some dimerization reactions following the electron transfer step [28].

Coulometry is generally difficult with in situ EPR experiments because of convection and iR drop in the long, thin layer of solution. In order to determine the current efficiency of carbazole anion formation, an in situ measurement was carried out [29,30]. Carbazoles can either form soluble products or polymerize forming a conductive layer on the electrode. Results of the measurement showed that there was between 0.2 and 2% current efficiency for radical formation depending on the carbazole used. The system described, however, can only be used for coulometry over short periods of time because of the problems discussed earlier.

A cell design for a $TE_{011}$ mode cavity that minimizes cell resistance and uses a helical working electrode with the auxiliary electrode in the center was also designed [31]. If the helix is wound so that the spacings between the loops are of the same order as the thickness of the wire, the microwave field does not penetrate to the inside of the helix. As a result, the material generated at the auxiliary electrode is not detected by the EPR spectrometer. This principle was also employed in the design of the low-temperature cell mentioned previously [19].

An electrode configuration similar to that in Figure 29.19 has been designed for cylindrical samples [32] in a $TE_{012}$ cavity. It is available commercially, but the smaller working electrode area requires use of flowing solution and signal averaging techniques (Sec. V.E). A more ambitious approach to the iR drop–current density problem has been advanced by designing a cell containing a set of discrete electrodes, with each being controlled by a separate potential or current source [33]. All of these cells have yielded encouraging results for model systems, but the field of SEEPR measurements remains to be exploited.

## D. Electrochemical EPR Flow Cells: Design and Methodology

The need for flowing solutions with in situ generation was mentioned earlier. A simple cell fulfilling these requirements is shown in Figure 29.20a. Since the concentration of free radicals within the cavity can be evaluated by a potentiostatic reversal method [34], this cell is also useful for kinetic studies [35].

The cell shown schematically in Figure 29.20b permits external generation, followed by EPR detection. The solution can either be recirculated to the electrolysis cell or discarded after observation. Umemoto [36] used a similar apparatus to generate moderately stable radicals coulometrically, followed by stopped-flow measurements of the decay kinetics. Forno [37] used a more elaborate recirculating system with two electrochemical cells in series. The unstable product of the first electrolysis was pumped to the second electrolysis cell, where it was converted to a free radical and thence to the cavity for observation (Sec. VI.A).

The cell diagrammed in Figure 29.20c has been used extensively by Kastening to study the reaction mechanisms and kinetics when a stable electrochemically generated species is mixed with other reactants [38]. The calculations that were used to reduce the experimental data to kinetic parameters included the effects of flow rate, reaction mechanism, rate constant, and the dependence of the EPR spectrometer nonuniform sensitivity to the position of the radicals in the cavity. Unfortunately, the correction for the latter effect was introduced as the idealized cosine-squared function (solid line in Fig. 29.16) rather than the actual sensitivity variation that occurs with magnetic field modulation. It is important to be aware of this difficulty when performing experiments to determine accurate kinetic parameters.

External tubular electrodes have been used with flowing solutions in some recent electrochemical–EPR studies by Albery and co-workers [39–42]. The electrode was placed above the EPR cavity and the solution was allowed to flow

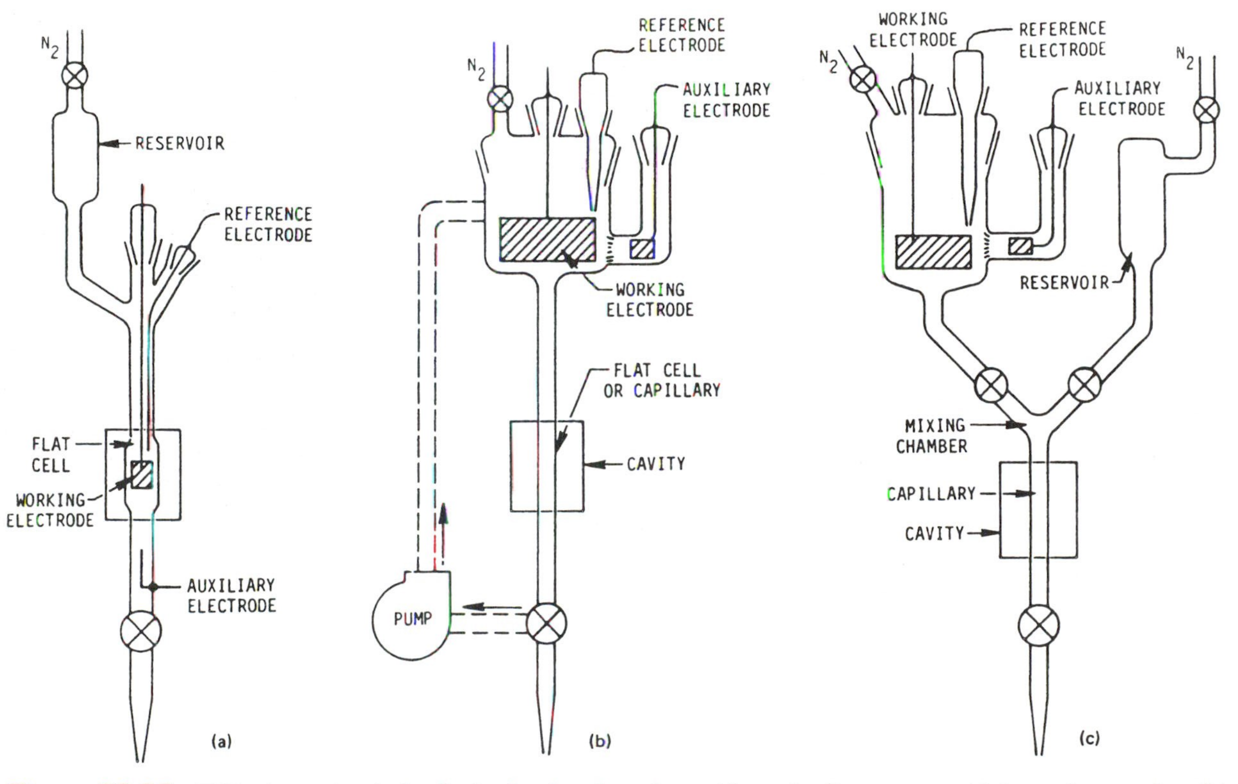


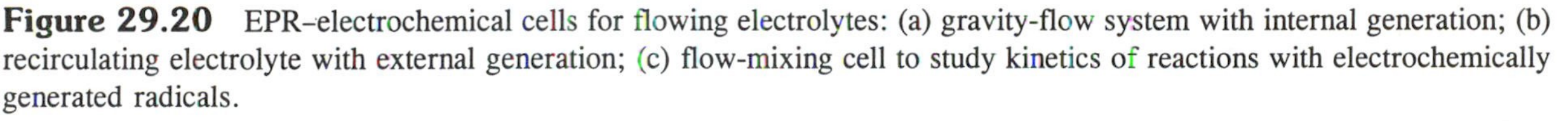

**Figure 29.20** EPR–electrochemical cells for flowing electrolytes: (a) gravity-flow system with internal generation; (b) recirculating electrolyte with external generation; (c) flow-mixing cell to study kinetics of reactions with electrochemically generated radicals.

past the electrode into the cavity. This electrode was fabricated by drilling a 1-mm-diameter hole through a 2-mm section of gold rod, and mounting between two lengths of 1-mm-i.d. capillary tubing. The experimental performance of this electrode was compared to the mathematical model for stable radicals [40], and for radicals that decay by first- [41] and second-order [42] processes. A qualitative review of the applications of this cell was published [43]. Many of these mathematical analyses require numerical solutions.

Interest in photoelectrochemistry has spurred the development of a photoelectrochemical EPR cell [44]. This cell is constructed in an analogous way to the flat cell. The semiconductor or metallic working electrode is mounted on one wall and the solution flows through a channel ($0.4 \times 6.0 \times 30$ mm) between the electrode and the outer cell wall. Electrical contact to the working electrode is made by a fine wire placed through a hole in the wall behind the electrode. The outside optically transparent wall is cemented to the channel containing the working electrode. The auxiliary electrode was a Pt mesh placed below the cell, and the reference electrode was placed above the cell. A theoretical model was developed to predict the EPR signal as a function of solution flow rate through the cell in the absence of homogeneous reactions. In the first case, a constant concentration of radical at the electrode was assumed, and in the second case, a uniform current density at the electrode was assumed. Performance of the cell was consistent with theoretical predictions, using the photoelectrochemical reduction of fluorescein in aqueous media, which has been well established [45].

## E. Data Acquisition and Signal Averaging

Data processing techniques are extremely useful in both pure EPR and electrochemical–EPR studies. Details of the EPR computer interface are unique to each system and to the goals of each experiment. Since the theory and methodology of these digital operations are similar to those described elsewhere in this book, the discussion will not be reiterated here. There are numerous examples of signal averaging for kinetic measurements and for spectral accumulation using rapid scans. Short-lived species may be studied by these techniques.

## F. Alternating-Current Generation

One of the earliest EPR–electrochemical attempts to utilize alternating current to generate radicals [46] was of questionable success, probably due to the low sensitivity of spectrometers of that period. Recently, under extreme conditions Denat and Gosse [47] observed that a signal could be obtained at potential modulation frequencies of 0.5–1000 Hz if high voltages (up to 1.3 kV) were used in high-resistance media. This method, when applied to practical systems, may allow enhancement of the signal-to-noise ratio, or permit determination of kinetics by modulation spectroscopy.

## VI. SELECTED APPLICATIONS OF EPR AND ELECTROCHEMICAL MEASUREMENTS

Attempts to reduce anthracene with an alkali metal in acetonitrile causes solvent decomposition, whereas controlled-potential electrolysis produces stable anion radicals. Thus the working electrode of a coulometric cell can be considered as a continuously adjustable "reagent," capable of producing a wide variety of radical species in diverse solvent systems. The versatility of electrochemical EPR methods is best illustrated by citing a few specific examples from the extensive literature. More complete compilations appear in the reviews listed in Appendix I, but the studies mentioned next provide some appreciation for the techniques.

### A. Identification of Radicals

The literature describes abundant examples in which the observed radical differs from the parent compound by only one electron. It should not be assumed that these studies represent a trivial application of electrochemical–EPR. Rather, the results have greatly enriched our understanding of electronic structure and molecular dynamics.

In more complicated cases, it may not be known whether the detected radicals are generated at the electrode or if the initial species reacts quickly to form other paramagnetic products. In some cases, the electrolysis initiates an intramolecular rearrangement so that the structure of the radical differs from that of the parent compound. Electrochemistry alone does not provide direct information on the structure of the product. But the HFS pattern of an EPR spectrum does.

One of the earliest applications of combined EPR and electrochemical measurements was the study of nitroalkane reductions [48]. Cyclic voltammetry revealed irreversible reduction waves, but some anodic peaks were observed on the reverse scan and were attributed to reaction intermediates. In situ generation produced an initial spectrum attributable to the nitroalkane anion radical, but after some time, the dialkylnitroxide spectrum was detected. This information, combined with analysis of the products formed during bulk electrolysis, suggested the following reaction sequence:

$$RNO_2 + e \longrightarrow RNO_2^{\bullet-} \qquad \text{(spectrum observed)}$$

$$RNO_2^{\bullet-} \xrightarrow{\text{slow}} R^{\bullet} + NO_2^- \qquad \text{(no spectrum, nitrite ion isolated)}$$

$$R^{\bullet} + RNO_2^{\bullet-} \longrightarrow R_2 + NO_2^- \qquad \text{(product isolated)}$$

$$2R^{\bullet} + RNO_2 \longrightarrow R_2 + NOR \qquad \text{(product isolated)}$$

$$R_2NO_2^- \xrightarrow{\text{HS}} R_2NO^{\bullet} + OH^- \qquad \text{(spectrum observed)}$$

Similar studies [49,50] on the cathodic reduction of halogenated nitrobenzenes revealed the following mechanism:

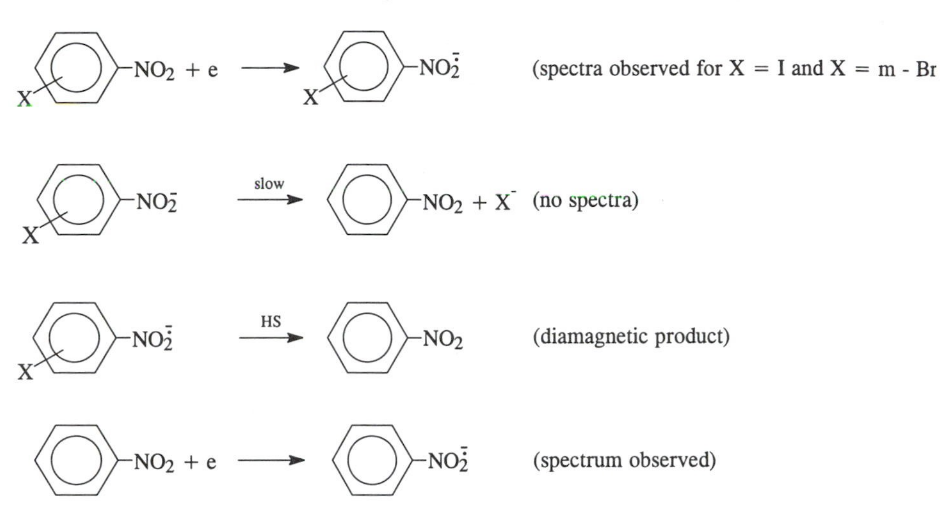


The rates of elimination followed the order I < *m*-Br < *o*-Br < *o*-Cl < *p*-Br < *p*-Cl.

The tetraphenylallene system represents a particularly elegant confluence of chemistry, electrochemistry, and EPR. The carbanion of 1,1,3,3-tetraphenylpropene is formed spontaneously in alkaline medium. Or it can be formed by the two-electron reduction of 2-ethoxy-1,1,3,3-tetraphenyl-prop-1-ene (at 2.2 V) or of tetraphenylallene (at –2.11 V). The tetraphenylallyl radical is then produced by anodic oxidation of the carbanion at –0.95 V. Its EPR spectrum was obtained using the two-stage flow cell described earlier [37].

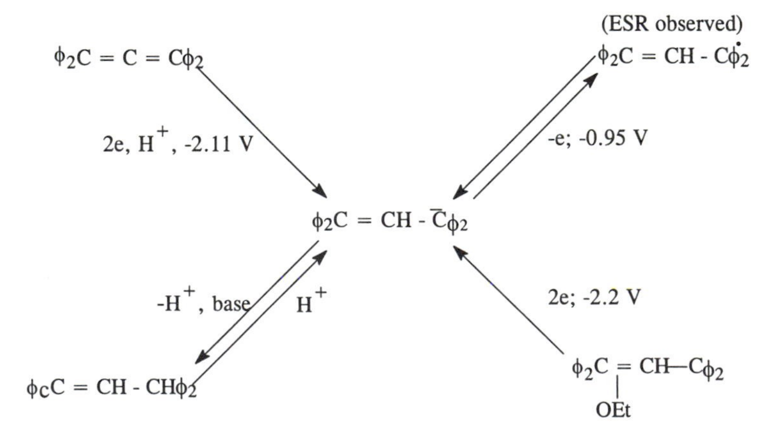

## B. Kinetic Measurements

Kinetic measurements by EPR are difficult, primarily because of the instrumental considerations mentioned in Sections V.B and V.C. Nonetheless, the specificity of EPR toward paramagnetic species has prompted workers to measure reaction rates of radicals generated by both in situ and external electrolysis, using both static and flowing solutions.

Umemoto [36] used external generation, followed by stopped-flow detection, to study the protonation of anthracene, anthraquinone, and benzophenone anions in dimethylformamide-water solutions using an apparatus similar to Figure 29.20c. The measured half-lives were about 1.5 min or more. In the case of anthracene, the decomposition rates agreed with those obtained from polarographic measurements and the following scheme was proposed.

$$R + e \rightleftarrows R^{\bar{\bullet}}$$

$$R^{\bar{\bullet}} + H_2O \longrightarrow RH^{\bullet} + OH^-$$

$$RH^{\bullet} + e \longrightarrow RH^- \quad \text{(at electrode)}$$

$$RH^{-} + R^{\bar{\bullet}} \longrightarrow RH^- + R \quad \text{(in solution)}$$

$$RH^- + H_2O \longrightarrow RH_2 + OH^-$$

However, the anions of anthraquinone and benzophenone appeared to undergo additional steps corresponding to disproportionation of the radicals followed by protonation of the dianion:

$$2R^{\bar{\bullet}} \rightleftarrows R^{2-} + R$$

$$R^{2-} + 2H_2O \rightarrow RH_2 + 2OH^-$$

and the EPR data did not agree with polarographic measurements. In the same vein, studies of the protonation rates of nitrobenzene anions, using a variety of flow and in situ techniques, were not consistent among themselves and are not discussed here.

The cell diagrammed in Figure 29.19 has been used to study the dimerization reactions of olefin anions [27]. The radicals were generated by constant-current electrolysis using a pulse of short duration (with respect to the radical half-life). The decay of the EPR signal was compared with digital simulations assuming various mechanisms and rate constants. In order to distinguish the correct mechanism, it was necessary to perform the individual experiments at different values of the applied current. Fresh solution was transferred to the electrochemical–EPR cuvette for each such measurement. Rate constants obtained in this fashion agreed with values obtained from cyclic voltammetry, double-potential-step, and rotating ring-disk electrode experiments. The dominant mechanism appeared to involve dimerization of two anion radicals followed by rapid protonation.

The kinetics of the reaction of nitrobenzene anion radicals at pH 9.3 was studied using a flow cell and varying the velocity and concentration of the solution [42]. The analysis showed that unsubstituted nitrobenzene anions may undergo a second-order reaction of the form

$$H_2O + 2\phi NO_2^{\bar{\cdot}} \xrightarrow[\text{slow}]{} \phi NO_2 + \phi NO + 2OH^-$$

$$2H_2O + \phi NO + 2\phi NO_2^{\bar{\cdot}} \xrightarrow[\text{fast}]{} 2\phi NO_2 + \phi NNOH + 2OH^-$$

where the rate of the slow reaction was 45 $M^{-1}$ $s^{-1}$, which is comparable to rates found by other EPR–electrochemical measurements [38,51].

An interesting study [52] of the protonation kinetics and equilibrium of radical cations and dications of three carotenoid derivatives involved cyclic voltammetry, rotating-disk electrolysis, and in situ controlled-potential electrochemical generation of the radical cations. Controlled-potential electrolysis in the EPR cavity was used to identify the electrode reactions in the cyclic voltammograms at which radical ions were generated. The concentrations of the radicals were determined from the EPR amplitudes, and the buildup and decay were used to estimate lifetimes of the species. To accomplish the correlation between the cyclic voltammetry and the formation of radical species, the relative current from cyclic voltammetry and the normalized EPR signal amplitude were plotted against potential. Electron transfer rates and the reaction mechanisms, EE or ECE, were determined from the electrochemical measurements. This study shows how nicely the various measurement techniques complement each other.

Cells such as those described in References 23, 24, and 29 are particularly suited to study of short-lived intermediates requiring in situ generation at accurately controlled potentials. When a conventional electrochemical cell was used to study the Kolbe synthesis oxidation of triphenylacetic acid [53], it was concluded that the initially formed radical was triphenylacetoxyl ($\phi_3$-CCOO•), based on the assignment of two *para*- and four *ortho*-proton splittings. A more careful study [54] using the cell described in Reference 23 showed that it is in fact the triphenylmethyl radical that is formed initially; the identity of the other species was not established, although it is clearly not the acetoxyl radical.

## C. Spin Trapping for Identification of Intermediates

Spin trapping is a method by which unstable radicals are allowed to react with a diamagnetic "trap" so that they form a secondary, more stable radical called the spin adduct [55,56]. Nitrones, such as phenyl-*t*-butylnitrone (Eq. 29.22), and nitroso compounds (Eq. 29.23) are generally used as traps. Nitrones offer greater stability than nitroso compounds; however, in the nitrone, the radical group is incorporated at a position farther from the nitroxyl group than the nitroso trap, so that distinctive spectral information is lost. Some ambiguity may result in the use of these traps. For example, the phenyl-*t*-butylnitrone was employed to study the electrolysis of water [57]. Benzyl-*t*-butyl nitroxide was formed during the reduction of water; however, this may be the result either of direct hydrogenation or the abstraction of a proton after formation of the radical anion [58]. Furthermore, during oxidation of water, the nitrone might cyclize [58], giving anomalously large proton splittings rather than a hydroxy adduct as originally supposed [57].

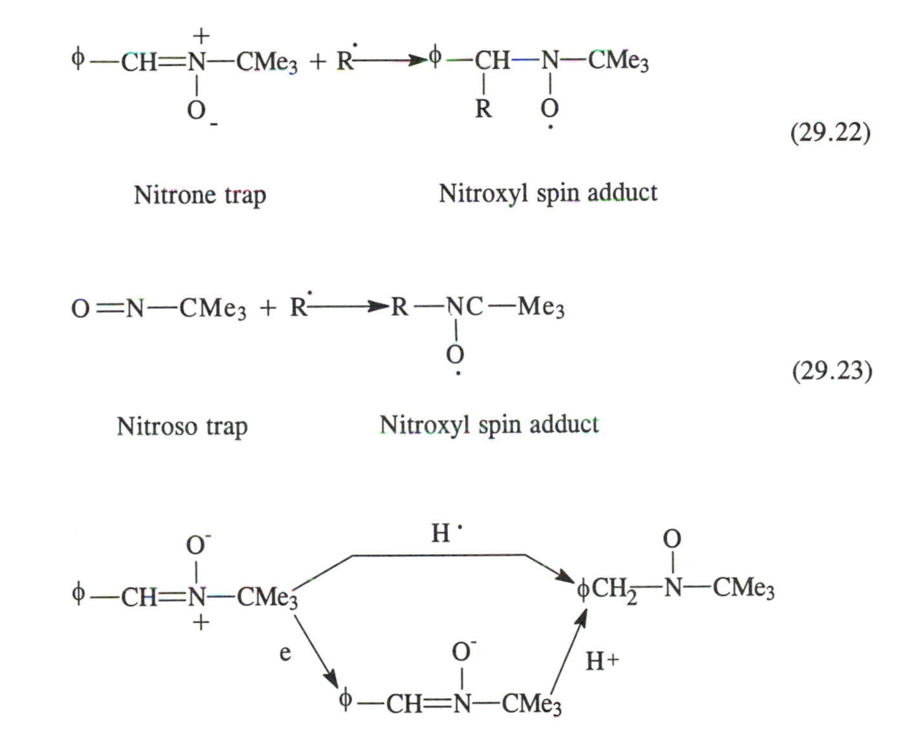


Electrochemical studies in nonaqueous media [59] indicate that phenyl-*t*-butyl nitrone is not electroactive between +1.5 and –2.4 V vs. SCE. Studies of the reduction of phenyldiazonium salts in the presence of the nitrone gave EPR spectra of the phenyl adduct. However, even here, a two-electron reduction may give the phenyl anion, which can add to the nitrone to give phenyl-*t*-butyl hydroxylamine, which can be reoxidized to the nitroxide. Chemical tests were used to show that the phenyl radical, rather than the anion, was produced.

Studies of the reactions of supporting electrolytes solely by electrochemical means are very difficult because of the extreme irreversibility of their oxidation and reduction. In one study [60], tetrabutylammonium tetrabutylborate oxidation ($E_{pa}$ +0.35 V) was studied using a phenyl-*t*-butylnitrone (Eq. 29.22) spin trap. In deoxygenated solutions, the *n*-butyl radical adduct was confirmed. When the solutions were not deoxygenated, mixtures of the *n*-butyl and *n*-butoxy radical adducts were found. These radicals can be distinguished by their different nitrogen and β-proton hyperfine splittings, which are 1.488 and 0.306 mT, and 1.380 and 0.227 mT, respectively, for the butyl and butoxy spin adducts. The results show that the alkyl radicals form on oxidation of the tetralkylborate anion, and that alkyl radical subsequently reacts with $O_2$.

While the spin-trap method has a great deal to offer in terms of mechanism diagnostics, the experimenter must be cautious to confirm the nature and origin of the spin adduct, and the electroinactivity of the spin trap or spin adduct in the potential range used in the experiments. In this regard, the electrochemical characteristics of a variety of nitrones and nitroso-*t*-butane were measured [61] in aqueous and acetonitrile solutions. The potential ranges of phenyl-*t*-butylnitrone and nitrosobutane were, respectively, +1.47 to –2.40 V vs. SCE and +1.82 to –1.77 V vs. SCE. However, nitrosobutane exhibited dimer formation in the anodic wave at +1.45 V in acetonitrile. In aqueous media, no oxidation waves were observed for these materials, but both exhibited reductions at about –1.8 V.

## D. The Use of EPR in Structural Studies

We mentioned earlier that the half-wave potential of an electrode reaction can be related to the energy of the HMO containing the unpaired electron (Eq. 29.3). In addition, we also noted that the EPR spectrum is dependent on the coefficients of this molecular orbital. $\beta^{\circ}_{CC}$ (Eq. 29.2) is usually called the resonance integral and it represents the energy of a π bond in the aromatic system. If we consider 9,10-diphenylanthracene as an example, the phenyl groups are not coplanar with the anthracene nucleus because of steric interactions between protons on the adjacent rings. This twist of angle θ diminishes the resonance energy of the bond between the phenyl and anthryl carbon atoms. In the HMO approximation, the resonance integral, $\beta_{rs}$, between those atoms becomes

$$\beta_{rs} = \beta^{\circ}_{CC} |\cos \theta| \tag{29.24}$$

The value of θ depends on the steric bulk of both groups. For example, phenyl is less bulky than 2,4,6-tri-*t*-butylphenyl.

Accordingly, HMO theory predicts that the half-wave potential and the HFSC of the anthryl protons should approach those of the unsubstituted anthracene ions when substitution causes θ to approach 90°. But for smaller values of θ, oxidation or reduction is facilitated and the unpaired electron can delocalize onto the substituent and produce additional hyperfine splittings.

The value of θ in a particular radical can be estimated by comparison of the experimental half-wave potential and HFSC with the results obtained from a series of HMO calculations using different assumed values of θ. A large number of phenyl-substituted aromatic compounds [62] and ethylenes [63] have been treated in this fashion. Similar evidence for the twisting of the nitro group in nitroaromatic anion radicals is summarized in Reference 1. Restricted rotation of alkyl substituents is also discussed in Reference 1, but this torsion does not significantly affect the electrochemical behavior.

## E. Inorganic Systems

Inorganic radicals and transition metal ions typically exhibit broad lines, and hence diminished sensitivity in the EPR method. Consequently, when dealing with small quantities of paramagnetic material, it is often more difficult to detect inorganic species. Several important studies have been reported, however. Kastening's study [64] of the reduction of $SO_2$ in dimethylformamide showed that an equilibrium was established between the $SO^{\bullet-}$ radical ion and the dimer $S_2O_4^{\bullet-}$,

$$SO_2 + SO_2^{\bullet-} \rightleftarrows S_2O_4^{\bullet-}$$

in which K ≈ 320 $M^{-1}$. The data show that $(SO_2)_2SO_2^{\bullet-}$ and higher adducts are not likely.

Gardiner and Casey [65] observed the radicals $O_3^{\bullet-}$, $O_2^{\bullet-}$, and $MO_2^{\bullet-}$, which are formed on various electrode surfaces during oxidation processes on different materials in KOH solutions. Studies of copper dissolution in pyrophosphate [66] resulted in polymeric copper pyrophosphate complexes.

The complex $[Cr(CO)_5F]^-$ provides a good example [67] of the use of EPR–electrochemical methods to understand the reactions of a transition metal complex. Rotating-disk voltammograms of [dibenzo-18-crown-6-K]$[Cr(CO)_5X]$, where X is a halogen, show oxidation steps at about +0.5 V and +1.1 V vs. SCE, depending on the solvent, electrode material, and the halide. The first oxidation step of the fluoride complex was examined by in situ electrochemical–EPR measurements. Two major EPR absorptions were observed corresponding to g = 1.982 and g = 1.969. Because fluorine, with I = ½, could give rise to SHFS as large as the 2.18 mT that separated the two EPR absorptions, it was necessary to rule out fluorine SHFS even though the amplitudes of the absorptions differed. As a result, the complex $Cr(CO)_6$ was also generated by in situ electrochemical oxidation. The radical formed by this process exhibited the same EPR spectrum as the g = 1.982 absorption. The two absorptions were therefore assigned to $[Cr(CO)_6]^+$ and $Cr(CO)_5F$. Based on the analysis, the oxidation of the fluoride complex was found to undergo the following reaction steps:

$$[Cr(CO)_5F]^- \rightarrow Cr(CO)_5F + e^-$$

$$2\ Cr(CO)_5F \rightarrow [Cr(CO)_5F]^- + [Cr(CO)_5F]^+$$

$$[Cr(CO)_5F]^+ \rightarrow Cr^{2+} + F^- + 5\ CO$$

$$Cr(CO)_5F + CO \rightarrow [Cr(CO)_6F]^+ + F^-$$

Based on Table 29.1, several points can be made about the species generated. First, both $Cr(CO)_6$ and $[Cr(CO)_5F]^-$ are neutral chromium species so that they can be considered $3d^6$ ions. In a strong octahedral field or a strong octahedral field with tetragonal distortion, the electrons would occupy the three lowest energy d-orbitals and are therefore diamagnetic ($S = 0$). $Cr^{2+}$ formed in the reaction would be $3d^4$. Whether in a strong or weak octahedral field, the $Cr^{2+}$ ions would have a degenerate ground state with a multiplet spin ($S > ½$) and would not be readily detected by EPR. However, the chromium in either $[Cr(CO)_6F]^+$ or $Cr(CO)_5F$ would have a $3d^5$ configuration. Since it is difficult to generate an octahedral field that is sufficiently strong to separate the energy levels in this state, the very strong EPR signal is most likely due to the $3d^5$ ($S = 5/2$) configuration.

## F. Homogeneous Electron-Transfer Reactions

The occurrence of homogeneous electron-transfer reactions between the radical ion and parent molecule,

$$R^{\overline{\bullet}} + R \rightleftarrows R + R^{\overline{\bullet}} \tag{29.25a}$$

or

$$R^{\overset{+}{\bullet}} + R \rightleftarrows R + R^{\overset{+}{\bullet}} \tag{29.25b}$$

has some important implications for electrochemical-EPR experiments. These reactions are usually characterized by large rate constants [$k_2 \approx 10^7$–$10^9\ M^{-1}\ s^{-1}$]. When the electron is transferred, the new radical that is formed will typically have a different configuration of nuclear spins, so that the EPR resonance condition shifts to a different hyperfine line. If the absolute rate is large, these shifts are manifested as abnormally broad EPR lines with concomitant loss of spectral resolution.

Consider a series of EPR spectra of a radical species in the presence of increasing amounts of the parent material. A dilute solution of the radical in the absence of parent shows a well-resolved spectrum. As parent compound is added, the individual hyperfine lines begin to broaden with onset of the electron transfer reaction. At higher concentrations of parent, the electron transfer rate increases. When the rate becomes very rapid, the average hyperfine field, $H_{local}$, begins to average toward zero and the individual hyperfine lines eventually merge into a single, broad EPR signal. Continued addition of parent causes the breadth of this line to decrease. In this fast-exchange limit the second-order rate constant can be evaluated [68,69]. The reader is directed to References 69 and 70 for details of this procedure.

The method of determination of the exchange rate is particularly suited to in situ electrogeneration methods, since one can add different, high concentrations of parent material to the cell, coulometrically generate a small amount of radical ion, and determine the exchange-narrowed line width. The occurrence of exchange reactions (Eq. 29.25) can also be a problem in routine electrochemical EPR studies, for unless the radical generation step is essentially complete, the spectral resolution will be less than ideal.

It is worth noting that Equations 29.25a and b do not represent chemical reactions in the usual sense. Indeed, there would be little reason for even contemplating them were it not for the EPR evidence. But closer scrutiny reveals an intriguing similarity with the transport mechanism (mobility) of hydrogen ion in water, and it might not be idle to speculate about the role of reaction 29.25 in biological processes. We can also speculate on those oxidation and reduction processes of importance in energy storage or in biological systems, which may someday be elucidated by the use of electrochemical–EPR methodology.

## G. Topics of Recent Interest

### *Conductive Polymers*

Conductive polymers have received a great deal of attention during the past decade because of their potential application in electronic devices. In general, polymers such as polyacetylene, polypyrroles, polythiophenes, and poly-*p*-phenylene sulfides are insulating or have poor conductivity until they are doped with electron acceptors or donors. Acceptors, such as $SbF_5$, remove an electron from the polymer structure and create a positive site or a "hole." Holes can then conduct electricity through the $\pi$-electron network of the polymer. Such conductive polymers are said to be p-type. Similarly, an electron donor such as an alkali metal or $Fe^{2+}$ gives to the polymer negative charge, which can also be conducted through the $\pi$-electron network. Such conductive polymers are said to be n-type. Doping can be accomplished electrochemically or by chemical reaction. Only electrochemical doping is considered for the purpose of this chapter.

Experiments have been done where the polymer film is grown on an electrode surface, doped, and then removed from the electrochemical cell and placed in the EPR spectrometer. Examples are given in References 71 and 72. In such experiments, the number of electron spins can be determined as a function of the applied potential or time. Other experiments have been carried out using in situ electrochemical doping. Examples of these experiments are given in References 73 and 74. By using in situ doping, the gradual change of the number of unpaired electrons can be monitored, and the transformation from an insulating to a highly conductive material can be monitored. An example of the changes in the EPR spectrum that are observed as polyacetylene is doped with gas phase $AsF_5$ is shown in Figure 29.21. At the start of doping, the polyacetylene exhibits a normal EPR line shape due to the defects or "dangling bonds" in the polyacetylene. However, as doping progresses, the line shape becomes asymmetric. The asymmetric line shape is known as dysonian, and it occurs in the unique case in which the film is much thicker than the skin depth, and the transit time of the electron through the skin depth is faster than the spin-spin relaxation time, commonly called $T_2$. The skin depth, $\delta$, is the penetration depth of the electric or magnetic field in a conductive material. For a nonmagnetic material, $\delta$ is given by

$$\delta = 503.3/\sigma^{1/2}\nu^{1/2} \text{ (meters)}$$

where $\sigma$ is the conductivity in mhos/m. $T_2$ governs the line width of the spectrum, and is proportional to the reciprocal of the line width after it is converted to frequency units,

$$T_2 = \frac{h}{\sqrt{3}\pi g \mu_B H_{pp}}$$

where $H_{pp}$ is the peak-to-peak line width of the EPR spectrum. Not every conductive polymer can be made sufficiently conductive that its EPR spectrum exhibits a dysonian line shape. In the transition between the normal and dysonian line shapes, the common assumptions that either the amplitude of the spectrum or the amplitude multiplied by the square of the line width is proportional to the spin concentration are no longer valid. Therefore, in most cases it is advantageous to avoid this by using a film that is thinner than the skin depth at the maximum conductivity.

*Buckminsterfullerenes*

Since the buckminsterfullerenes $C_{60}$ and $C_{70}$ were discovered in 1990, there has been extreme interest in these compounds and their derivatives. The first fulleride radical anion, $C_{60}^{-}$, was prepared by bulk electrolysis using tetraphenylphosphonium chloride supporting electrolyte in dichlorobenzene [76]. Since there are no protons, the spectrum exhibits only one line. Furthermore, because the spin density is so uniformly distributed in the molecule, HFS from naturally

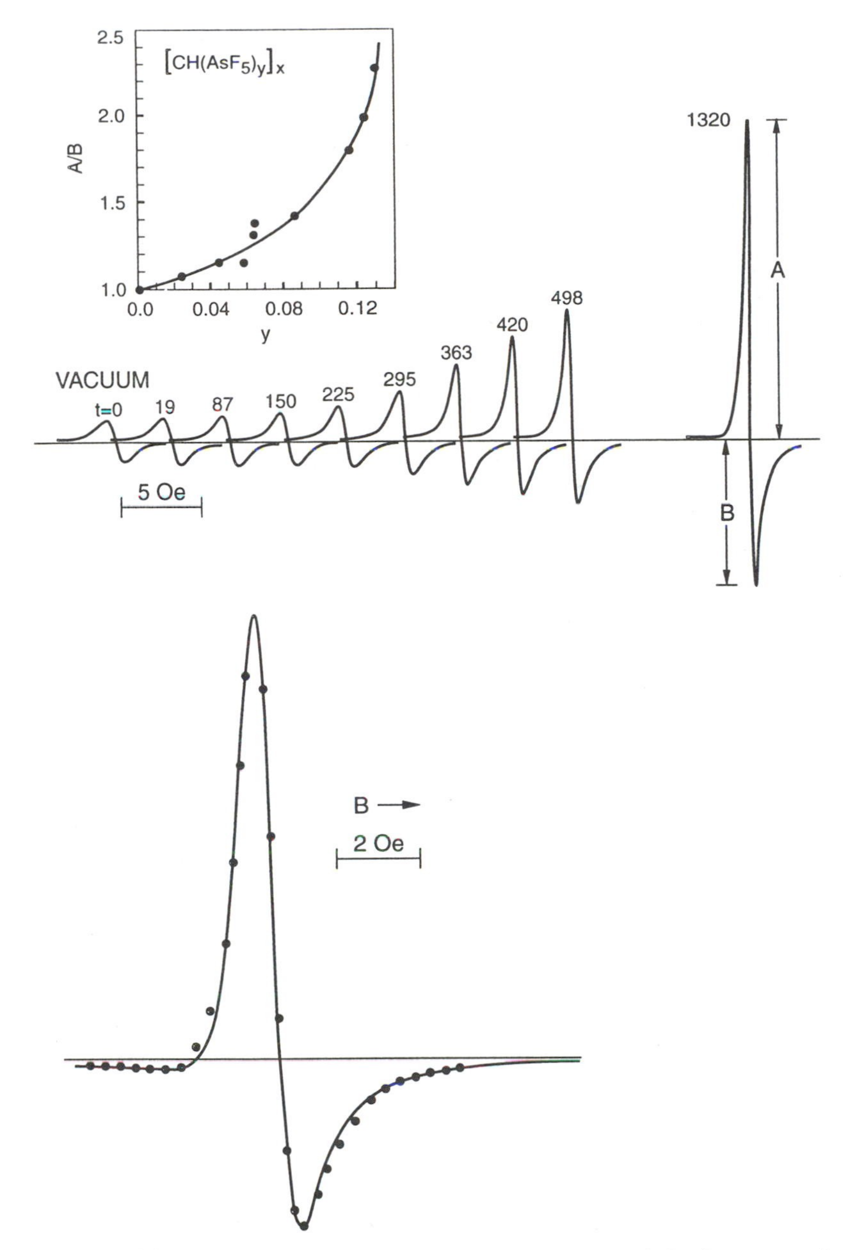


**Figure 29.21** The upper portion shows the EPR lines recorded prior to exposing polyacetylene film to gaseous $AsF_5$ and the transformation of a symmetric EPR line shape to a dysonian line for conductive material. The ratio of the low-field to high-field amplitudes is shown in the inset as a function of the composition of the doped film. The lower portion of the curve shows a single line fit to the theoretical dysonian line shape (circles).

abundant $C^{13}$ are too small to be observed. The ground state of this anion radical is in a degenerate state. However, the degeneracy of large molecules is easily lifted by small distortions (Jahn–Teller effect) so that the line width sharpens as the sample temperature is decreased. A more detailed study [77] confirmed these assertions. In some cases, up to six electrochemical reduction steps from the neutral $C_{60}$ or $C_{70}$ can be observed; however, not all of these species have been characterized by EPR spectroscopy [78]. Measurement of $C_{60}^{2-}$ generated by bulk electrolysis and measured by EPR as a frozen solution showed that the two electrons were unpaired so that the dianion was in a triplet (S = 1) state. This is possible because the two lowest energy states are degenerate, so the electrons could be unpaired. Since the molecule is large, the electrons were far apart so that electron–electron interaction was small. The spectrum tentatively assigned to the trianion, $C_{60}^{3-}$, showed no evidence of an S = 3/2 state, and was similar to EPR spectrum of $C_{60}^{-}$. This was possible because in the $\pi$-electron framework, if two energy levels are degenerate, then molecules containing one or three unpaired electrons in those degenerate states would each have similar degenerate ground states. This is shown in the diagram of the electron configurations among the degenerate orbitals.

| | | | | | | |
|---|---|---|---|---|---|---|
| Configuration I | ↑ | ___ | ↑ | ↑ | ↑↓ | ↑ |
| Configuration II | ___ | ↑ | ↑ | ↑ | ↑ | ↑↓ |
| | Monoanion (Degenerate) | | Dianion (Nondegenerate) | | Trianion (Degenerate) | |

A final noteworthy study [79] is that of $C_{60}H_2$. When $C_{60}H_2$ is reduced by bulk electrolysis, two superimposed spectra, each exhibiting a single line, are observed. These were assigned to $C_{60}H_2^{-}$ and $C_{60}^{-}$. Since most of the spectra recorded for fullerene ions exhibit only a single line, there is much that remains to be done in the future to understand better the various fullerene oxidation states.

## APPENDIX I: SELECTED BIBLIOGRAPHY

### Books

R. S. Alger, *Electron Paramagnetic Resonance: Techniques and Applications*, Wiley-Interscience, New York, 1967.

N. M. Atherton, *Electron Spin Resonance*, Hilger and Watts, London, 1975.

A. Carrington and A. D. McLachlan, *Introduction to Magnetic Resonance*, Harper & Row, New York, 1967.

F. Gerson, *High Resolution Electron Spin Resonance Spectroscopy*, Wiley, London, 1970.

C. P. Poole, *Electron Spin Resonance*, 2nd Ed., Wiley-Interscience, New York, 1983.

J. E. Wertz and J. R. Bolton, *Electron Spin Resonance: Elementary Theory and Practical Applications*, McGraw-Hill, New York, 1972.

J. Well, J. R. Bolton, and J. E. Wertz, *Electron Spin Resonance*, John Wiley & Sons, New York, in press.

## Reviews

### *Electrochemistry and EPR*

R. N. Adams, *J. Electroanal. Chem. 8*:252 (1964). "Application of Electron Paramagnetic Resonance Techniques in Electrochemistry."

I. B. Goldberg and A. J. Bard, in *Magnetic Resonance in Chemistry and Biology* (J. Herak, ed.), Marcel Dekker, New York, 1975. "The Application of Electron Spin Resonance Spectroscopy to Electrochemistry."

B. Kastening, *Chem.-Ing.-Tech. 42*:290 (1970). "Application of Electron Spin Resonance in Organic Electrochemistry."

B. Kastening, *Adv. Anal. Chem. Instrum./Electroanalytical Chemistry*, Vol. 20 (R. N. Wiley and H. W. Nurnberg, eds.), Wiley, New York, 1974, p. 421.

T. M. McKinney, *Electroanal. Chem. 10*:97 (1977). "Electron Spin Resonance and Electrochemistry."

### *Relaxation Processes in Solutions*

P. W. Atkins, *Adv. Mol. Relaxation Processes 2*:121 (1972). "Dynamics of Electron Spin Relaxation in Solution."

A. Hudson and G. R. Luckhurst, *Chem. Rev. 69*:191 (1969). "The Electron Resonance Lineshapes of Radicals in Solution."

W. B. Lewis and L. O. Morgan, *Transition Met. Chem. 33* (1968). "Paramagnetic Relaxation in Solution."

P. D. Sullivan and J. R. Bolton, *Adv. Magn. Reson. 4*:39 (1970). "The Alternating Line-Width Effect."

### *Organic Radicals in Solution*

A. Carrington, *Q. Rev. 17*:67 (1963). "Electron Spin Resonance Spectra of Aromatic Radicals and Radical-Ions."

D. H. Geske, *Prog. Phys. Org. Chem. 4*:125 (1967). "Conformation and Structure as Studied by Electron Spin Resonance Spectroscopy."

A. Hudson and K. D. J. Root, *Adv. Magn. Reson. 5*:1 (1971). "Halogen Hyperfine Interactions."

E. T. Kaiser and L. Levan, eds., *Radical Ions*, Wiley, New York, 1968. Chaps. 1–8, especially.

S. P. Solodovnikov and A. I. Prokof'ev, *Russ. Chem. Rev. 39*:591 (1970). "The Electronic Structure of Radical Anions."

### *Inorganic Ions*

B. A. Goodman and J. B. Raynor, *Adv. Inorg. Chem. Radiochem. 13*:235 (1970).

G. F. Kokoszka and G. Gordon, *Tech. Inorg. Chem. 7*:151 (1968). "Electron Paramagnetic Resonance."

E. Konig, in *Physical Methods in Advanced Inorganic Chemistry* (H. A. O. Hill and P. Day, eds.), Wiley, New York, 1968, Chap. 7. "Electron Paramagnetic Resonance."

H. A. Kuska and M. T. Rogers, in *Radical Ions* (E. T. Kaiser and L. Kevan, eds.), Wiley, New York, 1968. "Electron Spin Resonance of First Row Transition Metal Complex Ions."

B. R. McGarvey, *Transition Met. Chem. 3*:89 (1966). "Electron Spin Resonance of Transition-Metal Complexes."

T. F. Yen, ed., *Electron Spin Resonance of Metal Complexes*, Plenum Press, New York, 1969.

*Tabulations of EPR Data*

K. W. Bowers, *Adv. Magn. Reson. 1*:317 (1965). "Electron Spin Resonance of Radical Ions."

*Landolt-Bornstein Numerical Data and Functional Relationships in Science and Technology*, Springer-Verlag, New York, 1965–current. Has many volumes dealing with EPR data for organic and inorganic radicals and paramagnetic metal ions.

*General*

*Specialist Periodical Reports, Electron Spin Resonance*, Vols. 1–6, Chemical Society, London.

## APPENDIX II: UNITS

### Magnetic Field Units

SI units are the accepted international standard for reporting scientific data. However, the conversion from earlier systems is slow and far from complete. Historically, the cgsm system was used for most EPR work. Table A29.1 compares SI and cgsm units.

Much confusion arises from improper usage of the magnetic flux density, **B**, and the magnetic field strength, **H**, which are related by the equation

$$\mathbf{B} = \mu \mathbf{H}$$

where **B** and **H** are both vectors and $\mu$ is the permeability. The confusion in units arises from the fact that in the cgsm system the value of $\mu$ is 1 in vacuo (and very close to 1 in air also). The difference between actual value of the permeability and unity represents the contribution from the bulk magnetization of the sample. The conversion between tesla and gauss is $1\ \mathrm{T} = 10^4\ \mathrm{G}$. In fact, it is the value of **B** which is important in magnetic resonance since it is the parameter that influences the energy levels within the sample.

Strictly speaking, an EPR spectrometer system produces a field of known strength which should properly be reported in oersted or A/m. However, most EPR spectrometers refer to the magnetic field strength in units of gauss. Since the spectrometer has no way of knowing the permeability of the sample, this

**Table A29.1** Magnetic Units

| Quantity | Symbol | Name | |
|---|---|---|---|
| | | SI | cgsm |
| Magnetic flux | B | Tesla (T) | Gauss (G) |
| Magnetic field strength | H | Amperes/meter | Oersted (Oe) |
| Permeability in vacuo | $\mu_0$ | $1.25664 \times 10^{-6}$ Henries/meter | 1 |

usage is not strictly correct, particularly when this parameter is used to report hyperfine splitting values. But since the permeability of most EPR samples is close to unity (cgsm), the inaccuracy is generally insignificant. Frequency units are recommended for reporting fine and hyperfine structure. Some conversions are:

$$B\ (\mathrm{T}) = B\ (\mathrm{G})/10^4$$

$$H\ (\mathrm{A/m}) = 79.588\ H\ (\mathrm{Oe})$$

$$A\ (\mathrm{MHz}) = 2.80247(g/g_e)a\ (\mathrm{G})$$

$$g_e = 2.0023192\text{8}$$

$$A\ (\mathrm{MHz}) = (c \times 10^{-6})A\ (\mathrm{cm}^{-1})$$

$$A\ (\mathrm{MHz}) = 2.80247 \times 10^4(g/g_e)a\ (\mathrm{T})$$

where $a$ represents the hyperfine splitting with the magnetic units in parentheses. The conversion factors of $a$ to A also apply to other parameters derived from EPR.

Other useful constants are

$$c = 2.997925 \times 10^8\ \mathrm{m/s}$$

$$\mu_B = 9.27408 \times 10^{-24}\ \mathrm{J/T}$$

$$\mu_p = 1.410617 \times 10^{-26}\ \mathrm{J/T}$$

## REFERENCES


1. T. M. McKinney, *Electroanal. Chem. 10*:97 (1977).
2. A. Streitweiser, Jr., *Molecular Orbital Theory for Organic Chemists*, Wiley, New York, 1961.
3. T. M. Dunn, D. S. McClure, and R. G. Pearson, *Some Aspects of Crystal Field Theory*, Harper and Row, 1965.
4. See *ibid.*, pp. 173–188, and R. Zahrodnik and C. Parkanyi, *Talanta 12*:1289 (1965).

5. K. Mobius, *Z. Naturforsch. 20a*:1093, 1102 (1965).
6. J. R. Bolton, in *Radical Ions* (E. T. Kaiser and L. Levan, eds.), Wiley-Interscience, New York, 1968, pp. 1–32.
7. A. J. Stone, *Proc. R. Soc. A271*:424 (1963); *Mol. Phys. 6*:311, 509 (1963).
8. H. M. McConnell and D. B. Chestnut, *J. Chem. Phys. 28*:107 (1958).
9. J. R. Bolton, *J. Chem. Phys. 43*:309 (1965); J. P. Colpa and J. R. Bolton, *Mol. Phys. 6*:273 (1963).
10. M. Karplus and G. K. Fraenkel, *J. Chem. Phys. 35*:1312 (1961).
11. K. D. Sales, *Adv. Free Radical Chem. 3*:139 (1969).
12. P. H. Reiger, I. Bernal, W. H. Reinmuth, and G. K. Fraenkel, *J. Am. Chem. Soc. 85*:683 (1963); P. H. Reiger and G. K. Fraenkel, *J. Chem. Phys. 37*:2795 (1962); *37*:2811 (1963); *39*:609 (1963); J. R. Bolton and G. K. Fraenkel, *J. Chem. Phys. 40*:3307 (1964).
13. A. H. Maki and D. H. Geske, *J. Chem. Phys. 30*:1356 (1960); D. H. Geske and A. H. Maki, *J. Am. Chem. Soc. 82*:2671 (1960); A. H. Maki and D. H. Geske, *J. Am. Chem. Soc. 83*:1852 (1961); A. H. Maki and D. H. Geske, *J. Chem. Phys. 33*:825 (1960).
14. L. H. Piette, P. Ludwig, and R. N. Adams, *J. Am. Chem. Soc. 83*:2671 (1961); L. H. Piette, P. Ludwig, and R. N. Adams, *J. Am. Chem. 84*:4212 (1962).
15. T. Kubota, H. Miyazaki, and K. Nishikida, *Annu. Rep. Shionogi Res. Lab. 16*:1 (1966).
16. J. F. Ambrose, D. Dillard, A. K. Carpenter, and R. F. Nelson, *Anal. Chem. 42*:814 (1970).
17. C. M. Lang, J. Harbour, and A. V. Guzzo, *J. Phys. Chem. 75*:2861 (1971).
18. I. B. Goldberg and A. J. Bard, in *Magnetic Resonance in Chemistry and Biology* (J. N. Herak and K. . Adamic), Marcel Dekker, New York, 1975, Chap. 10.
19. H. Ohya-Nishiguchi, *Bull. Chem. Soc. Jpn. 52*:1064 (1979).
20. K. R. Fernando, A. J. McQuillan, B. M. Peake, and J. Wells, *J. Magn. Reson. 68*:551 (1986).
21. R. N. Bagghi, A. M. Bond, and R. Colton, *J. Electroanal. Chem. 199*:297 (1986).
22. D. H. Levy and R. J. Myers, *J. Chem. Phys. 41*:1062 (1962).
23. I. B. Goldberg and A. J. Bard, *J. Electroanal. Chem. 38*:313 (1971).
24. I. B. Goldberg, A. J. Bard, and S. W. Feldberg, *J. Phys. Chem. 76*:2550 (1972).
25. J. K. Dohrmann and K. J. Vetter, *J. Electroanal. Chem. 20*:23 (1969).
26. I. B. Goldberg and A. J. Bard, *J. Phys. Chem. 75*:3281 (1971).
27. I. B. Goldberg and A. J. Bard, *J. Phys. Chem. 78*:290 (1974).
28. I. B. Goldberg, D. Boyd, R. Hirasawa, and A. J. Bard, *J. Phys. Chem. 78*:295 (1974).
29. P. Kubacek, *Collect. Czech. Chem. Commun. 46*:40 (1981).
30. S. Kolar and P. Kubacek, *Chem. Listy 72*:1294 (1978).
31. R. D. Allendoerfer, G. A. Martinchek, and S. Bruckenstein, *Anal. Chem. 47*:890 (1975).
32. R. Hirasawa, T. Mukaibo, H. Hasegawa, Y. Kanda, and T. Maruyama, *Rev. Sci. Instrum. 39*:935 (1968); R. Hirasawa, T. Mukaibo, H. Hasegawa, N. Odan, and T. Maruyama, *J. Phys. Chem. 72*:2541 (1968).

33. B. Kastening, B. Gostisa-Mihelcic, and J. Divisek, *Disc. Faraday Soc. 56*:341 (1973); B. Kastening, J. Divisek, B. Gostisa-Mihelcic, and H.G. Muller, *Z. Phys. Chem. 87*:125 (1973).
34. R. Koopman and H. Gerischer, *Ber. Bunsen-Ges. Phys. Chem. 70*:118 (1966).
35. R. Koopman, *Ber. Bunsen-Ges. Phys. Chem. 70*:121 (1966); R. Koopman and H. Gerischer, *Ber. Bunsen-Ges. Phys. Chem. 70*:127 (1966).
36. K. Umemoto, *Bull. Chem. Soc. Jpn. 40*:1058 (1967).
37. A. E. J. Forno, *Chem. Ind. 1728* (1968).
38. B. Kastening, *Collect. Czech. Chem. Commun. 30*:4033 (1965); B. Kastening, *Z. Anal. Chem. 224*:296 (1967); B. Kastening and S. Vavricka, *Ber. Bunsen-Ges. Phys. Chem. 72*:27 (1968).
39. W. J. Albery, B. A. Coles, A. M. Couper, and K. M. Garnett, *J. Chem. Soc., Chem. Commun. 198* (1974).
40. W. J. Albery, B. A. Coles, and A. M. Couper, *J. Electroanal. Chem. 65*:901 (1975).
41. W. J. Albery, R. G. Compton, A. T. Chadwick, B. A. Coles, and J. A. Kenkait, *J. Chem. Soc. Faraday Soc. 1, 76*:1391 (1980).
42. W. J. Albery, A. T. Chadwick, B. A. Coles, and N.A. Hampson, *J. Electroanal. Chem. 75*:229 (1977).
43. R. G. Compton and A. R. Hillman, *Chem. Britain 22*:1088 (1986).
44. B. A. Coles and R. G. Compton, *J. Electroanal. Chem. 144*:87 (1983).
45. S. K. Yig and N. R. Bannerjee, *Electrochim. Acta 16*:157 (1971).
46. R. J. Collier and M. E. Peover, *Nature 220*:155 (1968).
47. A. Denat and B. Gosse, *C. R. Acad. Sci., Sec. C 290*:227 (1980).
48. A. K. Hoffman, W. G. Hodgson, D. L. Maricle, and W. H. Jura, *J. Am. Chem. Soc. 86*:631 (1964).
49. T. Kitagawa, T. P. Layloff, and R. N. Adams, *Anal. Chem. 35*:822 (1963).
50. T. Fujinaga, Y. Deguchi, and K. Umemoto, *Bull. Chem. Soc. Jpn. 37*:822 (1964).
51. D. Kolb, W. Wirths, and H. Gerischer, *Ber. Bunsen-Ges. Phys. Chem. 73*:148 (1969).
52. M. Khaled, A. Hadjipetrou, L. D. Kispert, and R. D. Allendorfer, *J. Phys. Chem. 95*:2438 (1991).
53. N. B. Kondrikov, V. V. Orlov, V. I. Ermakov, and M. Ya. Fioshin, *Sov. Electrochem. 8*:920 (1972).
54. R. D. Goodin, J. C. Gilbert, and A. J. Bard, *J. Electroanal. Chem. 59*:163 (1975).
55. M. J. Perkins, *Chem. Soc. Spec. Publ.* 24, 1970, p. 97.
56. E. G. Janzen, *Acc. Chem. Res. 4*:31 (1971).
57. P. H. Kasai and D. McLeod, *J. Phys. Chem. 82*:619 (1978).
58. E. G. Janzen, R. L. Dudley, E. G. Davis, and C. A. Evans, *J. Phys. Chem. 82*:2445 (1978).
59. A. J. Bard, J. C. Gilbert, and R. D. Goodin, *J. Am. Chem. Soc. 96*:620 (1974).
60. E. E. Bancroft, H. N. Blount, and E. G. Janzen, *J. Am. Chem. Soc. 101*:3692 (1979).
61. G. L. McIntire, H. N. Blount, H. J. Stronks, R. V. Shetty, and E. G. Janzen, *J. Phys. Chem. 84*:916 (1980).

62. L. O. Wheeler, K. S. V. Santhanam, and A. J. Bard, *J. Phys. Chem. 71*:2223 (1967); L. S. Marcoux, A. Lomax, and A. J. Bard, *J. Am. Chem. Soc. 92*:243 (1970) (see this paper for a complete list of phenyl-substituted hydrocarbons that have been studied); L. O. Wheeler and A. J. Bard, *J. Phys. Chem. 71*:4513 (1967); A. J. Bard, K. S. V. Santhanam, J. T. Maloy, J. Phelps, and L. O. Wheeler, *Discuss. Faraday Soc. 45*:167 (1968).
63. J. A. Valenzuela and A. J. Bard, *J. Phys. Chem. 72*:286 (1968); D. H. Geske and M. V. Merritt, *J. Am. Chem. Soc. 91*:6921 (1969).
64. B. Kastening and B. Gostisa-Mihelcic, *J. Electroanal. Chem. 100*:801 (1979).
65. C. L. Gardiner and E. J. Casey, *Can. J. Chem. 52*:930 (1974).
66. M. S. Shapnik, K. A. Zinkicheva, A. V. Ilyasov, and G. S. Vozdvizhenskii, *Sov. Electrochem. 8*:289 (1974).
67. R. N. Bagghi, A. M. Bond, R. Colton, D. L. Luscombe, and J. E. Moir, *J. Am. Chem. Soc. 108*:3352 (1986).
68. C. S. Johnson, Jr., in *Advances in Magnetic Resonance*, Vol. 1 (J. S. Waugh, ed.), Academic Press, New York, 1965, p.33.
69. R. Chang and C. S. Johnson, Jr., *J. Am. Chem. Soc. 88*:2338 (1966); C. S. Johnson, Jr. and J. B. Holz, *J. Chem. Phys. 50*:4420 (1969).
70. B. A. Kowert, L. Marcoux, and A. J. Bard, *J. Am. Chem. Soc. 94*:5538 (1972).
71. G. Tourillon, D. Courier, P. Garnier, and D. Vivien, *J. Phys. Chem. 88*:1049 (1984).
72. C. Arbizzani, M. Mastragostino, G. Dellepaine, P. Piaggio, and G. Zotti, *Synth. Methods. 55–57*:1354 (1993).
73. S. Lefrant, P. Bernier, R. B. Kaner, and A. G. MacDairmid, *Mol. Cryst. Liq. Cryst. 121*:233 (1985).
74. J. Chen and A. J. Heeger, *Synth. Methods 24*:311–327 (1988).
75. I. B. Goldberg, H. R. Crowe, P. R. Newman, A. J. Heeger, and A. G. MacDairmid, *J. Chem. Phys. 70*:1132 (1979).
76. P.-M. Allemand, G. Srdanov, A. Koch, K. Khemani, F. Wudl, Y. Rubin, F. Diederich, M. M. Alvarez, S. J. Anz, and R. L. Whetten, *J. Am. Chem. Soc. 113*:2780 (1991).
77. J. Stinchombe, A. Penicaud, P. Bhyrappa, P. D. W. Boyd, and C. A. Reed, *J. Am. Chem. Soc. 115*:5212 (1993).
78. D. Dubois, M. T. Jones, and K. M. Kadish, *J. Am. Chem. Soc. 115*:6446 (1992).
79. P. Boulas, F. D'Souza, C. C. Henderson, P. A. Cahill, M. T. Jones, and K. M. Kadish, *J. Phys. Chem. 97*:13435 (1993).

# Index